T0402436

THE VASCULOME

Companion Web Site:

https://www.elsevier.com/books-and-journals/book-companion/9780128225462

The Vasculome: From Many, One

Zorina S. Galis, Editor

Available Resources:

- Chapter Word cloud
- Book Word cloud
- Index

THE VASCULOME

FROM MANY, ONE

Edited by

ZORINA S. GALIS

Vascular Researcher, Bethesda, MD, United States

Academic Press is an imprint of Elsevier
125 London Wall, London EC2Y 5AS, United Kingdom
525 B Street, Suite 1650, San Diego, CA 92101, United States
50 Hampshire Street, 5th Floor, Cambridge, MA 02139, United States
The Boulevard, Langford Lane, Kidlington, Oxford OX5 1GB, United Kingdom

ISBN: 978-0-12-822546-2

For information on all Academic Press publications visit our website at https://www.elsevier.com/books-and-journals

Publisher: Stacy Masucci
Acquisitions Editor: Katie Chan
Editorial Project Manager: Kristi Anderson
Production Project Manager: Omer Mukthar
Cover Designer: Miles Hitchen

Contributions provided by Zorina S. Galis, Ph.D to the Work were written in a personal capacity and do not necessarily reflect the opinions or endorsement of the National Institutes of Health (NIH), Department of Human Health Services (HHS), or the Federal Government.
Cover image: Leonard Rubins for www.digitalartwork.org

Typeset by TNQ Technologies

Contents

PART II
INVESTIGATING THE VASCULOME: CONTEXT-DRIVEN METHODS, USES, AND LIMITATIONS

Section 1
Experimental and computational studies of the vasculome

12. The flow-dependent endotheliome: hemodynamic forces, genetic programs, and functional phenotypes

GUILLERMO GARCÍA-CARDEÑA, AND MICHAEL A. GIMBRONE, JR.

13. Intravital photoacoustic microscopy of microvascular function and oxygen metabolism

NAIDI SUN AND SONG HU

14. Systems biology modeling of endothelial cell and macrophage signaling in angiogenesis in human diseases

YU ZHANG, CHEN ZHAO, AND ALEKSANDER S. POPEL

15. Simulation of blood flow and oxygen transport in vascular networks

TIMOTHY W. SECOMB

16. Clinical investigations of vascular function

ARSHED A. QUYYUMI, ZAKARIA ALMUWAQQAT, AND SHABATUN J. ISLAM

Section 2
Realizing the promise of new high-content technologies

17. Angiodiversity—A tale retold by comparative transcriptomics

XIAOWU GU AND ONDINE CLEAVER

18. Using pattern recognition and discriminant analysis of functional perfusion data to create "angioprints" for normal and perturbed angiogenic microvascular networks

EVA K. LEE AND ZORINA S. GALIS

19. Artificial intelligence for the vasculome

LUIS EDUARDO JUAREZ-OROZCO, MING WAI YEUNG, JAN WALTER BENJAMINS, FATEMEH KAZEMZADEH, GONÇALO HORA DE CARVALHO, AND PIM VAN DER HARST

PART III

VASCULOME DYNAMICS: IN HEALTH AND IN SICKNESS

Section 1

Cooperating during development and organogenesis to create vasculome diversity

20. Vascular development and organogenesis: depots of diversity among conduits of connectivity define the vasculome

VICTORIA L. BAUTCH

21. Normal vascular identity (arteries, veins, and lymphatics) and malformations

LUIS GONZALEZ AND ALAN DARDIK

22. Sprouting angiogenesis in vascular and lymphatic development

ANNE EICHMANN AND JINYU LI

Section 2

Physiological and pathological remodeling of the vasculome

23. Enablers and drivers of vascular remodeling

JAY D. HUMPHREY

24. Extracellular matrix dynamics and contribution to vascular pathologies

AMANDA L. MOHABEER AND MICHELLE P. BENDECK

PART V

THE VASCULOME IN DIAGNOSIS, PREVENTION, AND TREATMENT OF OTHER DISEASES

PART VI

LOOKING FORWARD TOWARD "PRECISION HEALTH" FOR THE VASCULOME

Contributors

Bipul R. Acharya Department of Cell Biology, University of Virginia School of Medicine, Charlottesville, VA, United States; Cardiovascular Research Center, University of Virginia School of Medicine, Charlottesville, VA, United States; Wellcome Centre for Cell-Matrix Research, Faculty of Biology, Medicine and Health, University of Manchester, Manchester, United Kingdom

Dritan Agalliu Department of Neurology, Columbia University Irving Medical Center, New York, NY, United States; Department of Pathology and Cell Biology, Columbia University Irving Medical Center, New York, NY, United States

V.A. Alexandrescu Cardiovascular and Thoracic Surgery Department, CHU Sart-Tilman University Hospital, Liège, Belgium

Zakaria Almuwaqqat Emory Clinical Cardiovascular Research Institute, Division of Cardiology, Department of Medicine, Emory University School of Medicine, Atlanta, Georgia

Rheure Alves-Lopes Institute of Cardiovascular and Medical Sciences, BHF Glasgow Cardiovascular Research Centre, University of Glasgow, Glasgow, United Kingdom

Ken Arai Neuroprotection Research Laboratory, Department of Radiology and Neurology, Massachusetts General Hospital, Harvard Medical School, Boston, MA, United States

Wadih Arap Rutgers Cancer Institute of New Jersey, Newark, NJ, United States; Division of Hematology/Oncology, Department of Medicine, Rutgers New Jersey Medical School, Newark, NJ, United States

Victoria L. Bautch Department of Biology, University of North Carolina at Chapel Hill, Chapel Hill, NC, United States; Lineberger Comprehensive Cancer Center, University of North Carolina at Chapel Hill, Chapel Hill, NC, United States; Curriculum in Cell Biology and Physiology, University of North Carolina at Chapel Hill, Chapel Hill, NC, United States; McAllister Heart Institute, University of North Carolina at Chapel Hill, Chapel Hill, NC, United States

Lisa M. Becker Laboratory of Angiogenesis and Vascular Metabolism, Department of Oncology and Leuven Cancer Institute (LKI), KU Leuven, VIB Center for Cancer Biology, VIB, Leuven, Belgium

Michelle P. Bendeck Department of Laboratory Medicine and Pathobiology, University of Toronto, Toronto, ON, Canada; Translational Biology and Engineering Program, Ted Rogers Centre for Heart Research, University of Toronto, Toronto, ON, Canada

Jan Walter Benjamins Department of Cardiology, University of Groningen, University Medical Center Groningen, Groningen, the Netherlands

Saptarshi Biswas Department of Neurology, Columbia University Irving Medical Center, New York, NY, United States

E. Boesmans Cardiovascular and Thoracic Surgery Department, CHU Sart-Tilman University Hospital, Liège, Belgium

Livia L. Camargo Institute of Cardiovascular and Medical Sciences, BHF Glasgow Cardiovascular Research Centre, University of Glasgow, Glasgow, United Kingdom

Peter Carmeliet Laboratory of Angiogenesis and Vascular Metabolism, Department of Oncology and Leuven Cancer Institute (LKI), KU Leuven, VIB Center for Cancer Biology, VIB, Leuven, Belgium; Laboratory of Angiogenesis and Vascular Heterogeneity, Department of Biomedicine, Aarhus University, Aarhus, Denmark; State Key Laboratory of Ophthalmology, Zhongshan Ophthalmic Center, Sun Yat-Sen University, Guangzhou, Guangdong, P.R. China

Munir Chaudhuri Department of Internal Medicine, Emory University School of Medicine, Atlanta, GA, United States

Nicholas W. Chavkin Department of Cell Biology, University of Virginia School of Medicine, Charlottesville, VA, United States; Cardiovascular Research Center, University of Virginia School of Medicine, Charlottesville, VA, United States

Ondine Cleaver Department of Molecular Biology, University of Texas Southwestern Medical Center, Dallas, TX, United States

Clément Cochain Institute of Experimental Biomedicine, University Hospital Würzburg, Würzburg, Germany; Comprehensive Heart Failure Center, University Hospital Würzburg, Würzburg, Germany

Michael S. Conte Department of Surgery, Division of Vascular and Endovascular Surgery, University of California San Francisco, San Francisco, CA, United States

Azzurra Cottarelli Department of Neurology, Columbia University Irving Medical Center, New York, NY, United States; Department of Pathology and Cell Biology, Columbia University Irving Medical Center, New York, NY, United States

Christie L. Crandall Department of Mechanical Engineering and Materials Science, Washington University, St. Louis, MO, United States

Anne Cuypers Laboratory of Angiogenesis and Vascular Metabolism, Department of Oncology and Leuven Cancer Institute (LKI), KU Leuven, VIB Center for Cancer Biology, VIB, Leuven, Belgium

Andreas Daiber Department of Cardiology, Medical Center of the Johannes Gutenberg University, Mainz, Germany; German Center for Cardiovascular Research (DZHK), Mainz, Germany

Alan Dardik Department of Cellular and Molecular Biology, Yale School of Medicine, New Haven, CT, United States; Vascular Biology and Therapeutics Program, Yale School of Medicine, New Haven, CT, United States; Department of Surgery, Yale School of Medicine, New Haven, CT, United States; VA Connecticut Healthcare System, West Haven, CT, United States

Jui M. Dave Yale Cardiovascular Research Center, Section of Cardiovascular Medicine, Department of Medicine, Yale University, New Haven, CT, United States; Department of Genetics, Yale University, New Haven, CT, United States

J.O. Defraigne Cardiovascular and Thoracic Surgery Department, CHU Sart-Tilman University Hospital, Liège, Belgium

Wenjun Deng Neuroprotection Research Laboratory, Department of Radiology and Neurology, Massachusetts General Hospital, Harvard Medical School, Boston, MA, United States; Department of Neurology, Clinical Proteomics Research Center, Massachusetts General Hospital, Harvard Medical School, Boston, MA, United States

Robert J. DeStefano Department of Internal Medicine, Emory University School of Medicine, Atlanta, GA, United States

Devinder Dhindsa Emory Clinical Cardiovascular Research Institute, Division of Cardiology, Emory University School of Medicine, Atlanta, GA, United States

Danny J. Eapen Department of Medicine, Division of Cardiology, Emory University, Atlanta, GA, United States

Anne Eichmann Cardiovascular Research Center, Department of Internal Medicine, Yale University School of Medicine, New Haven, CT, United States; Department of Cellular and Molecular Physiology, Yale University School of Medicine, New Haven, CT, United States; Inserm U970, Paris Cardiovascular Research Center, Paris, France

Christian El Amm Department of Surgery, University of Oklahoma Health Sciences Center, Oklahoma City, OK, United States

Omotayo Eluwole Department of Physiology, University of Witwatersrand, Johannesburg, Gauteng, South Africa; Obafemi Awolowo University, Ile-Ife, Osun State, Nigeria

Christian Faaborg-Andersen Emory University School of Medicine, Atlanta, GA, United States

Steven A. Fisher Departments of Medicine (Cardiovascular Division) and Physiology, University of Maryland School of Medicine and Baltimore Veterans Administration Medical Center, Baltimore, MD, United States

Zorina S. Galis Vascular Researcher, Bethesda, MD, United States

Guillermo García-Cardeña Department of Pathology, Center for Excellence in Vascular Biology, Brigham and Women's Hospital and Harvard Medical School, Boston, MA, United States

Xin Geng Cardiovascular Biology Research Program, Oklahoma Medical Research Foundation, Oklahoma City, OK, United States

Michael A. Gimbrone, Jr. Department of Pathology, Center for Excellence in Vascular Biology, Brigham and Women's Hospital and Harvard Medical School, Boston, MA, United States

Luis Gonzalez Department of Cellular and Molecular Biology, Yale School of Medicine, New Haven, CT, United States; Vascular Biology and Therapeutics Program, Yale School of Medicine, New Haven, CT, United States

Daniel M. Greif Yale Cardiovascular Research Center, Section of Cardiovascular Medicine, Department of Medicine, Yale University, New Haven, CT, United States; Department of Genetics, Yale University, New Haven, CT, United States

Xiaowu Gu Department of Molecular Biology, University of Texas Southwestern Medical Center, Dallas, TX, United States

Shuzhen Guo Neuroprotection Research Laboratory, Department of Radiology and Neurology, Massachusetts General Hospital, Harvard Medical School, Boston, MA, United States

Tara L. Haas Faculty of Health, York University, Toronto, ON, Canada

Omar Hahad Department of Cardiology, Medical Center of the Johannes Gutenberg University, Mainz, Germany; German Center for Cardiovascular Research (DZHK), Mainz, Germany

Pim van der Harst Department of Cardiology, Heart and Lung Division, University Medical Centre Utrecht, Utrecht, the Netherlands; Department of Cardiology, University of Groningen, University Medical Center Groningen, Groningen, the Netherlands

Peter K. Henke Section of Vascular Surgery, Department of Surgery, University of Michigan, Ann Arbor, MI, United States

Karen K. Hirschi Department of Cell Biology, University of Virginia School of Medicine, Charlottesville, VA, United States; Cardiovascular Research Center, University of Virginia School of Medicine, Charlottesville, VA, United States; Departments of Medicine and Genetics, Yale University School of Medicine, Yale Cardiovascular Research Center, New Haven, CT, United States

C. Holemans Cardiovascular and Thoracic Surgery Department, CHU Sart-Tilman University Hospital, Liège, Belgium

Gonçalo Hora de Carvalho Department of Cardiology, University of Groningen, University Medical Center Groningen, Groningen, the Netherlands

Song Hu Department of Biomedical Engineering, Washington University in St. Louis, St. Louis, MO, United States

Jay D. Humphrey Department of Biomedical Engineering and Vascular Biology and Therapeutics Program, Yale University, New Haven, CT, United States

Shabatun J. Islam Emory Clinical Cardiovascular Research Institute, Division of Cardiology, Department of Medicine, Emory University School of Medicine, Atlanta, Georgia

Xinguo Jiang VA Palo Alto Health Care System, Palo Alto, CA, United States; Stanford University School of Medicine, Stanford, CA, United States

Luis Eduardo Juarez-Orozco Department of Cardiology, Heart and Lung Division, University Medical Centre Utrecht, Utrecht, the Netherlands; Department of Cardiology, University of Groningen, University Medical Center Groningen, Groningen, the Netherlands

Angelos D. Karagiannis Department of Internal Medicine, Emory University School of Medicine, Atlanta, GA, United States

Anita Kaw Division of Medical Genetics, Department of Internal Medicine, McGovern Medical School, University of Texas Health Science Center at Houston, Houston, TX, United States

Kaveeta Kaw Division of Medical Genetics, Department of Internal Medicine, McGovern Medical School, University of Texas Health Science Center at Houston, Houston, TX, United States

Fatemeh Kazemzadeh Department of Cardiology, University of Groningen, University Medical Center Groningen, Groningen, the Netherlands

A. Kerzmann Cardiovascular and Thoracic Surgery Department, CHU Sart-Tilman University Hospital, Liège, Belgium

Alexander S. Kim Department of Surgery, Division of Vascular and Endovascular Surgery, University of California San Francisco, San Francisco, CA, United States

Ageliki Laina Department of Clinical Therapeutics, National and Kapodistrian University of Athens, School of Medicine, Athens, Greece; Biosciences Institute, Vascular Biology and Medicine Theme, Faculty of Medical Sciences, Newcastle University, Newcastle upon Tyne, United Kingdom

Eva K. Lee Center for Operations Research in Medicine and Healthcare, Georgia Institute of Technology, Atlanta, GA, United States; Center for Bioinformatics and Computational Genomics, Georgia Institute of Technology, Atlanta, GA, United States

Jinyu Li Cardiovascular Research Center, Department of Internal Medicine, Yale University School of Medicine, New Haven, CT, United States; Department of Cellular and Molecular Physiology, Yale University School of Medicine, New Haven, CT, United States

Wenlu Li Neuroprotection Research Laboratory, Department of Radiology and Neurology, Massachusetts General Hospital, Harvard Medical School, Boston, MA, United States

Chien-Jung Lin Department of Cell Biology and Physiology, Washington University, St. Louis, MO, United States; Department of Internal Medicine (Cardiovascular Division), Washington University, St. Louis, MO, United States

Xiaolei Liu Center for Vascular and Developmental Biology, Feinberg Cardiovascular and Renal Research Institute, Feinberg School of Medicine, Northwestern University, Chicago, IL, United States

Eng H. Lo Neuroprotection Research Laboratory, Department of Radiology and Neurology, Massachusetts General Hospital, Harvard Medical School, Boston, MA, United States; Department of Neurology, Clinical Proteomics Research Center, Massachusetts General Hospital, Harvard Medical School, Boston, MA, United States

Josephine Lok Neuroprotection Research Laboratory, Department of Radiology and Neurology, Massachusetts General Hospital, Harvard Medical School, Boston, MA, United States; Pediatric Critical Care Medicine, Massachusetts General Hospital, Harvard Medical School, Boston, MA, United States

Mark W. Majesky Center for Developmental Biology & Regenerative Medicine, Seattle Children's Research Institute, University of Washington, Seattle, WA, United States; Departments of Pediatrics, University of Washington, Seattle, WA, United States; Laboratory Medicine & Pathology, University of Washington, Seattle, WA, United States

Ziad Mallat Department of Medicine, Division of Cardiovascular Medicine, University of Cambridge, Cambridge, United Kingdom; Université de Paris, PARCC, INSERM, Paris, France

Muzi J. Maseko Department of Physiology, University of Witwatersrand, Johannesburg, Gauteng, South Africa

Dianna M. Milewicz Division of Medical Genetics, Department of Internal Medicine, McGovern Medical School, University of Texas Health Science Center at Houston, Houston, TX, United States

Amanda L. Mohabeer Department of Laboratory Medicine and Pathobiology, University of Toronto, Toronto, ON, Canada; Translational Biology and Engineering Program, Ted Rogers Centre for Heart Research, University of Toronto, Toronto, ON, Canada

Augusto C. Montezano Institute of Cardiovascular and Medical Sciences, BHF Glasgow Cardiovascular Research Centre, University of Glasgow, Glasgow, United Kingdom

Giorgio Mottola Yale Cardiovascular Research Center, Section of Cardiovascular Medicine, Department of Medicine, Yale University, New Haven, CT, United States; Department of Genetics, Yale University, New Haven, CT, United States

Thomas Münzel Department of Cardiology, Medical Center of the Johannes Gutenberg University, Mainz, Germany; German Center for Cardiovascular Research (DZHK), Mainz, Germany

Daniel D. Myers Unit for Laboratory Animal Medicine, University of Michigan, Ann Arbor, MI, United States

Karla B. Neves Institute of Cardiovascular and Medical Sciences, BHF Glasgow Cardiovascular Research Centre, University of Glasgow, Glasgow, United Kingdom

Mark R. Nicolls VA Palo Alto Health Care System, Palo Alto, CA, United States; Stanford University School of Medicine, Stanford, CA, United States

MingMing Ning Neuroprotection Research Laboratory, Department of Radiology and Neurology, Massachusetts General Hospital, Harvard Medical School, Boston, MA, United States; Department of Neurology, Clinical Proteomics Research Center, Massachusetts General Hospital, Harvard Medical School, Boston, MA, United States

Andrea T. Obi Section of Vascular Surgery, Department of Surgery, University of Michigan, Ann Arbor, MI, United States

Guillermo Oliver Center for Vascular and Developmental Biology, Feinberg Cardiovascular and Renal Research Institute, Feinberg School of Medicine, Northwestern University, Chicago, IL, United States

Renata Pasqualini Rutgers Cancer Institute of New Jersey, Newark, NJ, United States; Division of Cancer Biology, Department of Radiation Oncology, Rutgers New Jersey Medical School, Newark, NJ, United States

Alessandra Pasut Laboratory of Angiogenesis and Vascular Metabolism, Department of Oncology and Leuven Cancer Institute (LKI), KU Leuven, VIB Center for Cancer Biology, VIB, Leuven, Belgium

Alexandra Pislaru Faculty of Science, York University, Toronto, ON, Canada

Aleksander S. Popel Department of Biomedical Engineering, School of Medicine, Johns Hopkins University, Baltimore, MD, United States

Raymundo A. Quintana Department of Medicine, Division of Cardiology, Emory University, Atlanta, GA, United States

Arshed A. Quyyumi Emory Clinical Cardiovascular Research Institute, Division of Cardiology, Department of Medicine, Emory University School of Medicine, Atlanta, Georgia

Francisco J. Rios Institute of Cardiovascular and Medical Sciences, BHF Glasgow Cardiovascular Research Centre, University of Glasgow, Glasgow, United Kingdom

Stanley G. Rockson Stanford University School of Medicine, Stanford, CA, United States

Martina Rudnicki Faculty of Health, York University, Toronto, ON, Canada

Junichi Saito Yale Cardiovascular Research Center, Section of Cardiovascular Medicine, Department of Medicine, Yale University, New Haven, CT, United States; Department of Genetics, Yale University, New Haven, CT, United States

Charles D. Searles, Jr. Department of Medicine, Division of Cardiology, Emory University, Atlanta, GA, United States

Timothy W. Secomb Department of Physiology and BIO5 Institute, University of Arizona, Tucson, AZ, United States

Cristina M. Sena Institute of Physiology, iCBR, Faculty of Medicine, University of Coimbra, Coimbra, Portugal

Richard L. Sidman Department of Neurology, Harvard Medical School, Boston, MA, United States

Federico Silva-Palacios Department of Internal Medicine, University of Oklahoma Health Sciences Center, Oklahoma City, OK, United States

Tracey L. Smith Rutgers Cancer Institute of New Jersey, Newark, NJ, United States; Division of Cancer Biology, Department of Radiation Oncology, Rutgers New Jersey Medical School, Newark, NJ, United States

Suman Sood Division of Hematology/Oncology, Department of Internal Medicine, University of Michigan, Ann Arbor, MI, United States

Laurence S. Sperling Emory Clinical Cardiovascular Research Institute, Division of Cardiology, Emory University School of Medicine, Atlanta, GA, United States; Katz Professor in Preventive Cardiology, Professor of Global Health, Hubert Department of Global Health, Rollins School of Public Health at Emory University, Atlanta, GA, United States

R. Sathish Srinivasan Cardiovascular Biology Research Program, Oklahoma Medical Research Foundation, Oklahoma City, OK, United States; Department of Cell Biology, University of Oklahoma Health Sciences Center, Oklahoma City, OK, United States

Kimon Stamatelopoulos Department of Clinical Therapeutics, National and Kapodistrian University of Athens, School of Medicine, Athens, Greece; Biosciences Institute, Vascular Biology and Medicine Theme, Faculty of Medical Sciences, Newcastle University, Newcastle upon Tyne, United Kingdom

Konstantinos Stellos Biosciences Institute, Vascular Biology and Medicine Theme, Faculty of Medical Sciences, Newcastle University, Newcastle upon Tyne, United Kingdom; Department of Cardiology, Newcastle Hospitals NHS Foundation Trust, Newcastle upon Tyne, United Kingdom

Naidi Sun Department of Biomedical Engineering, Washington University in St. Louis, St. Louis, MO, United States

Wen Tian VA Palo Alto Health Care System, Palo Alto, CA, United States; Stanford University School of Medicine, Stanford, CA, United States

Rhian M. Touyz Institute of Cardiovascular and Medical Sciences, BHF Glasgow Cardiovascular Research Centre, University of Glasgow, Glasgow, United Kingdom

Nikolaos I. Vlachogiannis Biosciences Institute, Vascular Biology and Medicine Theme, Faculty of Medical Sciences, Newcastle University, Newcastle upon Tyne, United Kingdom; Department of Cardiology, Newcastle Hospitals NHS Foundation Trust, Newcastle upon Tyne, United Kingdom

Jessica E. Wagenseil Department of Mechanical Engineering and Materials Science, Washington University, St. Louis, MO, United States

Thomas W. Wakefield Section of Vascular Surgery, Department of Surgery, University of Michigan, Ann Arbor, MI, United States

Charlotte R. Wayne Department of Neurology, Columbia University Irving Medical Center, New York, NY, United States

Changhong Xing Neuroprotection Research Laboratory, Department of Radiology and Neurology, Massachusetts General Hospital, Harvard Medical School, Boston, MA, United States; Department of Pathology, University of Texas Southwestern Medical Center, Dallas, TX, United States

Ming Wai Yeung Department of Cardiology, University of Groningen, University Medical Center Groningen, Groningen, the Netherlands

Yu Zhang Department of Biomedical Engineering, School of Medicine, Johns Hopkins University, Baltimore, MD, United States

Chen Zhao Department of Biomedical Engineering, School of Medicine, Johns Hopkins University, Baltimore, MD, United States

Preamble

"THE VASCULOME: FROM MANY, ONE" CONCEPT ZORINA S. GALIS

When a patient experiences a medical condition affecting a discrete organ, such as an eye or kidney, the choice of medical experts to be consulted appears clear, yet, frequently the underlying problem may be a local manifestation of a systemic vascular condition, such as hypertension, and that important connection frequently may go undiagnosed. This is especially true for the dysfunction of the microvasculature, which, in spite of its microscopic scale, is essential for every organ's survival and proper function, and makes up the majority of our vasculature, yet may indeed be *out of sight, out of mind*! Such everyday medical occurrences illustrate the important ramifications of local and systemic vascular health, respectively, malfunction and the continuous deepening of biomedical expertise needed to deal with the complexity of different parts of the body-wide vascular system, as it substantially participates in the function and malfunction of all the different organs. Furthermore, medical conditions clearly associated with malfunction of the vasculature continue to be the leading causes of mortality and disability in medium and high-income countries, and they are on the rise in the rest of the world. Our grasp of a unified vasculature, a main pillar of whole-person health, has been inadvertently fragmented by the evolution of highly specialized "knowledge silos." Vascular researchers and practitioners focusing on various areas of the same vascular system may spend their entire professional lives segregated into these knowledge silos, training and working in different academic and health care departments, using different scientific jargons, belonging to different professional societies, and attending different specialty conferences.

A life-long quest to understand the structure, function, and malfunction of the vasculature, in its many contexts, and to enable the design of effective clinical interventions has led the Editor of this book to the realization that further advances could well be accelerated by a collective reframing of future fundamental and clinical explorations of the vasculature so that we always consider any of its various parts as integral to its body-wide system. Since the vasculature is integrated not only in itself, but also within all the tissues across the entire body, one cannot fully grasp the developmental or functional underpinnings, including the malfunction, of any tissue or organ without considering its vasculature. All these considerations have led to our proposing the following concept of *the vasculome.*

The vasculome is an integrated, multiscale, multidimensional representation of the entire vasculature and its local and systemic interactions and functions within the body. The vasculome *concept* encompasses not only the entire physical blood and lymph vasculature, but also the body of knowledge obtained through different types of investigations about its local and unified nature and its essential importance to the human body's survival and function from the integrative perspective of "*E pluribus unum*: from many, one." The vasculature's systemic unity within the body is naturally achieved through the seamless integration of local structural and functional vascular variations. These variations are exquisitely adapted to support the survival and key functions across body's several size scales, from the single-cell level to the specialized multicellular organotypic functional tissue units (FTUs) that collectively support and define each organ's function, to whole tissues and organs—and ultimately to the entire body.

Answering fundamental and clinical questions about this complex integrated system usually requires breaking it down to manageable pieces to allow different types of investigations in various contexts. For instance, questions may require circumscribing investigations to one of the body areas or a defined size scale, at cell, tissue, or organ level, in specific research environments (e.g., clinical, experimental, in vivo, in situ, in vitro, or in silico), and using investigative tools applicable to those environments. The latest scientific and technological advancements have augmented the amount of multidimensional information we can amass about the vasculature. It is fair to say that our knowledge about the vasculome has many times circled back to the classical understanding of it, yet in an ever-expanding ascending spiral, knowledge is elevated at a new level when additional information is gained through the use of the latest sophisticated investigative methods. These new understandings compel us to find ways to rethink holistically about the essential contributions of the vasculome to human health and disease. They also support the vision that the vasculome could be personalized and predictive, which can help advance the clinical assessment and management of its local and systemic contributions in each one of us.

I am grateful to many leading vascular researchers who, in spite of the challenging times we have been weathering during the extended COVID pandemic, have answered the call to participate in the present Vasculome book project by providing their insights and examples to help flesh-in this concept, often challenging our existing assumptions about real or perceived boundaries of the vasculature. They remind our readers to approach knowledge about the vasculature guided by its nature-built integrative nature, for the vasculome is integrated within itself and within the body. Accordingly, this book highlights the fundamental bases of the "unity in diversity" of the vasculome, from its genesis in the coming together of various cell lineages during development, to its deceptively simple solution for architectural design: the efficient interplay of a few major types of building blocks that support vascular and other body-wide functions through highly specialized local functional variations. Finally, it hypothesizes the underpinnings of vascular dysfunction and its essential contributions to disease. To these ends, the book also presents some of the many investigative tools specific to the vasculome. The summaries appearing immediately below highlight main concepts presented in the various book sections; however, given the highly integrative nature of the vasculome, it is important to emphasize that these concepts could be organized—and should be connected by the reader—in many different ways. An interactive visualization tool has been built to show connections between the main topics of *The Vasculome* and is accessible from the book's web page.

PART I. THE ARCHITECTURAL DESIGN OF THE VASCULOME: BASIS FOR SYSTEM-WIDE FUNCTIONAL UNITY AND LOCAL DIVERSITY

The earliest descriptions of human dissections support the idea of a unified vasculature, the physical basis of the vasculome. A lot of our understating of the vasculature comes from the later establishment of many in vitro, in situ, and in vivo investigative environments and methods which have advanced our knowledge by studying the vasculature's components in isolation, as briefly reviewed by Bautch (2022).

As suggested above, the nature-built, seamless body-wide integration of the vasculome's structure and function is achieved through the efficient interplay of a few types of specialized building blocks, different cell types and extracellular matrix (ECM) components. These are assembled in various combinations and proportions to create the specialized structural units representing different types of vessels carrying blood or lymph.

Every blood and lymphatic vessel is built upon the shared foundation of an innermost, single-cell thick, contiguous layer formed through the assembly of endothelial cells (ECs). While physically contiguous, the endothelial layer is assembled from different EC types essential for the survival and specific function of not only each specialized vascular structure, but also of all other cells within the entire body (Acharya et al., 2022; Becker et al., 2022; Gu and Cleaver, 2022). For instance, organotypic ECs are the central components of the multicellular specialized functional tissue units (FTUs) of various organs, such as the kidney, brain, or muscle (Cottarelli et al., 2022).

In various segments of the vasculature, the endothelium is overlayed by different types of successive concentric vascular layers formed through the assembly of various combinations of other vascular cell types (Cochain and Mallat, 2022; Fisher, 2022; Majesky, 2022; Sena, 2022), embedded and held together by the mortar-like ECM (Wagenseil et al., 2022). This efficient structural design suggests that various vascular segments could be reengineered through the successive application of their typical cellular and ECM layers on top of their specialized endothelial layer. Furthermore, a 3D print of any organ perhaps might be produced, first by creating its tubular network of one-cell thick endothelial layer upon which all other vascular cell types and ECM layers could be applied to build the organ's vasculature first, then continuing to add consecutive layers of various types of neighboring cells, all dependent on vasculature for survival, to eventually building up an entire organ from the inside-out.

Section 1. The endothelium as the key unifying principle of the vasculome. Basis for endothelial systemic unity and for engineering of local specialization

The endothelium, which creates the interface between all tissues and circulating blood or lymph, is a master sensor, mediator, and integrator (sensitive and responsive) of local and systemic signals. The physical continuity of the endothelial layer, which covers the inside of all blood and lymph vessels and the heart, creates the structural and functional unity of the vasculome throughout the body. In all blood vessels, large and small, throughout the body, the endothelial layer performs the same main key functions. It mediates blood—tissue exchanges, such as those involving oxygen and essential nutrients. It enables body-wide communication and integration via sensing and mediating physical and biochemical signals, as well as the controlled transfer of molecules and cells between blood, lymph, and all other tissues. It maintains a nonthrombogenic surface, and it participates in control of blood flow and immunosurveillance. At the same time, specialized adaptations to the unique needs of each of the different tissues and organs, whose survival and proper function depend on endothelium-mediated exchanges, are achieved through local variations in EC structure and function. In this respect, the endothelial layer is structurally and functionally central to not only integrating all the parts of the vasculome, but also to the integration of the vasculome within each body tissue perfused by blood vessels and drained by lymph vessels. Yet, likely because of its thin, one-cell thick dimensions, the endothelium is too easily forgotten or remains elusive in most clinical investigations. The vasculome concept addresses this oversight head on by placing the endothelium front and center. Several chapters are dedicated to describing the various bases of the Vasculome's "unity in diversity" that can be directly related back to the endothelium.

New single-cell analyses are expanding and deepening the knowledge about the in situ structural and functional heterogeneity of the endothelium (Acharya et al., 2022), revealed over decades using different investigative techniques, principally electron microscopy, perhaps the original technology allowing single-cell level analysis of the EC's structure and function in its native environments (Galis, 2022).

Some aspects of EC diversity and identity, both within the various blood and lymphatic vessel types, and within various organs, are genetically determined. Some are acquired early during organogenesis (Bautch, 2022; Gu and Cleaver, 2022) and even into adulthood. Maintenance of ECs' identity depends on their environment, which also plays an important role in maintaining the identity of various types of vessels (Dardik and Gonzalez, 2022). For instance, the ability to integrate different combinations of mechanical forces—including shear, pulsatile, and hydrostatic forces—in the various active environments of different vascular segments and body locations, and to translate them into biochemical signals, plays an essential role in acquiring and maintaining EC diversity and healthy functioning. The close relation between those mechanical forces and the ability of ECs to acquire and maintain normal function is supported by research showing that certain forces have differing effects on different EC types. For instance, oscillatory shear stress (OSS) is developmentally needed for lymphatic ECs to form normal valves, yet OSS triggers pathogenic pathways in arterial ECs that need laminar shear for their normal function (García-Cardeña and Gimbrone, 2022).

The local variation of endothelial structure and function is exquisitely adapted to support tissues' specific metabolic needs and functions, best demonstrated by the structural and functional diversity of organotypic capillaries, the "simplest" vascular structures (Rudnicki et al., 2022). Examples of the diversity of EC permeability include (1) the tight continuous ECs limiting blood–tissue exchanges to controlled transcellular routes, the main basis for the tight blood–brain barrier (BBB) (Cottarelli et al., 2022, Gu and Cleaver, 2022, Xing et al., 2022); (2) the fenestrated ECs forming the selective glomerular filtration barrier (GFB) of the kidney (Gu and Cleaver, 2022); and (3) the discontinuous ECs of lymph nodes, which allow transfer of entire cells (Jiang and Rockson, 2022). These specialized ECs participate at the core of the various types of FTUs that define and support each organ's main function. Their "unity in diversity" supports the important notion that due to the contiguity of the endothelial layer, all identical FTUs within an organ, as well as all the different types of FTUs of all the other organs, are physically and likely functionally, connected to each other, and also integrated within the entire vasculome. Respectively, the organotypic FTUs directly contribute to the integration of the vasculome within the architecture, environment, and specific function of all organs, explaining why the health of the vasculome is central to the health of each organ, as well as health overall.

The endothelium recruits other types of tissue-specific cells, not only during the development of the vasculature itself (Eichmann and Li, 2022), but also during the organogenesis of various tissues (Bautch, 2022). ECs meet high energetic demands during organogenesis and angiogenesis by relying mainly on glycolysis. Oxygen availability and EC metabolic state are emerging as determinants of arterial–venous specification. Beyond driving and orchestrating vascular development, organogenesis, and tissue patterning, the endothelium continues to produce angiocrine factors that are essential for the survival, function, and regeneration of other cells throughout adulthood (Bautch, 2022; Gu and Cleaver, 2022). Developmental EC programs may be reactivated during this time, as an appropriate response needed for repair and regeneration, but also inappropriately, as in the pathogenesis of some genetic or other diseases that have endothelial dysfunction as a main driver. The capacity of ECs to transdifferentiate into other cell types is an area of active research and debate because of its relevance to clinically significant processes, including tissue regeneration and vascular pathologies.

The endothelium has to be able to dynamically adapt to fluctuating nutrients, to tissue metabolic demands, to hemodynamic forces, and other to stimuli from both its luminal and tissue sides. Understanding the variety of EC types, subsets, and their different metabolic states is important for understanding tissue and organ function and the ability to intervene therapeutically (Becker et al., 2022; Rudnicki et al., 2022). Endothelial cells also participate in immune surveillance, signaling the need for, and mediating, the inflammatory response. The EC inflammatory gene repertoire is organ-specific and related to organ function and resilience/susceptibility. "Activation" of the endothelium in response to some of these challenges can lead to dysfunction (Munzel et al., 2022). Not surprisingly, endothelial dysfunction, which can be measured clinically (Quyyumi et al., 2022), has been associated with many local and systemic clinical conditions.

Section 2. Vasculome's key building blocks—beyond the endothelium

Various types of blood or lymphatic vessels are created through the combinatorial assemblage of a few types of vascular cells: endothelial and vascular smooth muscle cells, pericytes, as well as and other cells residing in the vascular ECM, further contributing to the overall unity and local specialization of various vascular segments within the vasculome (Dave et al., 2022, Jiang and Rockson, 2022, Liu and Oliver, 2022, Rudnicki et al., 2022). The physical and functional interactions between all these different types of building blocks create the basis for additional functions of the vasculome, including vascular reactivity, resistance, and vessel wall remodeling. The specific structure of every individual blood or lymph vessel supports its function within the local environment and its integration within the body wide vasculome.

The vascular smooth muscle cells (VSMCs) are the most abundant vascular cell type that invests the walls of vasculome's blood and lymph vessels. Understanding of the VSMCs' great diversity has continuously evolved. Thanks to a variety of new investigative approaches, the VSMC origin and life trajectory has been connected to the EC-driven orchestration of vascular development and patterning (Bautch, 2022; Gu and Cleaver, 2022). VSMC precursors from different embryonic origins are dynamically recruited to the nascent vasculature in response to its evolving local needs and mechanical cues, contributing to VSMC heterogeneity, not only in various parts of the vasculature but also within the same vascular segment (Majesky, 2022). The fundamental understanding of how VSMCs acquire heterogeneity, the connection between VSMC positional identity and vascular disease, and the mechanisms that either help maintain or work to change the identity of VSMC in different vascular beds and in various contexts of the vasculome (Fisher, 2022) have created bases for guiding development of interventional strategies to address local manifestation of vascular disease within a systemic vasculature. Pericytes, the mural cells of the microvasculature, also play important roles, perhaps best exemplified in their participation in the neurovascular unit (NVU) (Cottarelli et al., 2022; Xing et al., 2022).

New technologies also have allowed a new appreciation of the contribution of other cell types found to be associated with vascular structures, yet not traditionally considered to be "vascular" cells. We now know that some immune cells are legitimate residents of the normal vasculome. Playing an important "surveillance" function, these contribute to its homeostasis, as well as participate in the normal or pathological tissue remodeling that occurs in response to vascular injury and the development and progression of vascular diseases (Cochain and Mallat, 2022). The macrophage, the main type of inflammatory cell traditionally investigated in relation to the diseased vasculature, is now known to come in different flavors. Some vascular resident macrophages are adapted to performing specific homeostatic functions in various normal vascular beds, while others are recruited postnatally from circulating monocytes in inflammatory conditions associated with the development and progression of various vascular conditions, including formation of atherosclerotic lesions. A small number of neutrophils, mast cells, dendritic cells, and T cells (CD4+ and CD8+) also can be found in the normal arterial wall, and various lymphoid cells reside in the intima and adventitia of arteries and in perivascular structures, such as the perivascular-associated adipose tissue (PVAT), as well as in the microvasculature. Normal PVAT that contains adipocytes, vascular and immune cells, fibroblasts, and multipotent mesenchymal cells contributes to the homeostasis and vasoactivity of large arteries (Sena, 2022). Under pathological conditions, secondary lymphoid structures can develop within the PVAT (Cochain and Mallat, 2022), contributing to vascular dysfunction. Again, the local diversity of vascular cellular constituents is the basis for the diversity of vascular functions. Finally, changes in the normal cellular composition and the overall vascular landscape are associated with disease processes and aging.

Another crucial aspect of the connection between the characteristic structure and the function of each type of vessel is the composition and organization of its vascular ECM, which provides the cells with a structural framework, mechanical properties, mechanosensing, and other signals contributing to the overall cellular environment needed to perform local functions and to participate in the systemic activities of the vasculome (Wagenseil et al., 2022). The "combinatorial" assembly of various ECM components contributes to the distinct structural and functional identity of different types of vascular segments in various locations. This assembly closely correlates with the specific cellular composition, tissue organization, geometry, and function, as well as the type and magnitude of mechanical forces vessel wall cells have to create and withstand, as highlighted by the dire consequences of inborn errors of

ECM-associated genes (Kaw et al., 2022). The ECM composition and assembly changes during development and aging, or in response to changes in the physical environment of blood vessels, which drive vascular tissue remodeling (Humphrey, 2022). These increase ECM turnover, triggering both synthesis of increased or different repertoire of ECM components, as well as increased production and action of ECM degrading enzymes. Such ECM changes are an essential component of normal adaptation and repair, but also contribute to the development and progression of various vascular diseases (Mohabeer and Bendeck, 2022). Far from being, as once thought, the amorphous "filler" between the cells, the ECM has emerged as one of the most fascinating and dauntingly multifaceted components of every tissue. Methods that would allow us to understand how this 3D complex assembled, yet dynamic, vascular ECM functions in its natural context are still lagging compared to those used to investigate cell function in situ.

Section 3. Putting it all together: vasculome integration across body scales and within tissues

The Vasculome concept calls for integration of vascular knowledge over the entire body and across its size scales, recapturing the nature-built vascular system design: seamlessly tapering from the large, stand-alone major blood vessels, which are among the longest and strongest anatomical structures in the body (macroscale), to the small vessels investing all organs (mesoscale), down to the single-cell scale of the smallest functional vascular structures that invest all tissues (microscale), the one EC-thick capillary tubes formed by one EC hugging itself.

The blood vasculature creates a body-wide closed circuit through the seamless hierarchical assembly of distinct structural units representing different types of blood vessels. Anatomical observations of the vascular system indicate once again its very efficient design: the road taken by oxygenated blood through the largest vessels to the smallest vessels only branches six times to reach any other cell within the body, and a similar, reversed branching pattern ensures blood's return to the heart from any location within the body. In addition, the lymphatic vasculature, mostly running in parallel to the blood vasculature, functions as an auxiliary backup system collecting fluid that has escaped into tissues and returning it via a unidirectional trip back into the blood vasculature (Jiang and Rockson, 2022).

This hierarchical design is most obvious in the building of the arterial system that tapers down from the aorta; the largest blood vessel, to large, elastic, and medium muscular arteries; then to smaller arterioles, to regulate blood flow and distribution to all organs and tissues. Arteries are built during development through the highly orchestrated integration of the greatest number and types of the various vasculome building blocks to create the anatomical structures needed to support pumping and conducting oxygenated blood, as well as to accomplish organ-specific functions (Dave et al., 2022). For instance, the hypoxic dilation of coronary arteries and sympathetic activation are needed to pump blood from the heart, while the hypoxic vasoconstriction of the pulmonary artery triggers blood oxygenation in the lungs. The serial design of the liver vasculature is needed for detoxification of blood received from the digestive organs, while in the kidney the afferent arterioles modulate hydrostatic pressure, and the efferent arterioles control the filtration rate. It is no surprise that arteries are the main contributors to vascular pathologies and systemic disease, such as atherosclerosis, aneurysm formation, peripheral arterial disease, or hypertension (Conte and Kim, 2022; Touyz et al., 2022). Mechanisms implicated in arterial pathology include reactivation of vascular developmental programs (Bautch, 2022), dysfunction of their various constituents, including the ECs (Becker et al., 2022; Munzel et al., 2022), VSMCs (Majesky, 2022), ECM (Kaw et al., 2022; Mohabeer and Bendeck, 2022), and vascular inflammation (Cochain and Mallat, 2022).

The theme of vascular diversity in the service of local tissue function is a major tenet of the vasculome, evident even at the smallest end of the vascular hierarchy and the simplest vascular structures, the capillaries, where most of the blood–tissue exchanges occur (Rudnicki et al., 2022). Organs usually contain different types of capillaries that support their survival, specific metabolic needs, and function (Gu and Cleaver, 2022). Much of the capillary specialization relies on the diversity of ECs, traditionally characterized in relation to their structural and functional features involved in transendothelial transport, i.e., continuous, fenestrated, or discontinuous endothelia, and a major basis for capillary specificity. New technologies have revealed that the variety of metabolic EC phenotypes contributes to normal EC identity in different tissues and to EC regenerative capacity, but also to EC transition to dysfunction and disease (Becker et al., 2022; Rudnicki et al., 2022). Such findings enhance the fundamental understanding of EC biology and suggest an important opportunity to use ECs to collect functional and spatial information about their local tissue milieu and neighboring cells (Galis, 2022).

Capillary specialization is essential for the function of various organ FTUs, which have organotypic ECs as their central elements. The intricate relationship between organ-specific ECs and other cell types forming their FTUs begins early in development and continues throughout the life of the FTUs. For instance, an intricate developmental "dance," spanning the embryonic and postnatal stage, was shown to be needed for the proper formation of the tight barriers between blood and tissues in the brain, spinal cord, and retina: essential in supporting the function of their NVUs and maintaining their immune privilege (Cottarelli et al., 2022). Any changes in the normal properties of their continuous-type ECs, e.g., changes in tight junction integrity, low levels of transcytosis, and low expression of adhesion molecules, central for the BBB/blood–retina barrier (BRB), and NVUs, are associated with organ dysfunction and many pathologies, including stroke, multiple sclerosis, or retinal disease. Thus, the ability to assess organotypic EC function provides a very useful and sensitive way to monitor the functional state of the various organ specific FTUs, and perhaps diagnose and manage any changes detectible before overt clinical symptoms (Galis, 2022). Other capillary characteristics—including at the single-cell level (e.g., type and density of accompanying mural cells), or at the tissue level (e.g., the typical spatial capillary density and branching within the tissues)—are also essential for their ability to support local function and tissue metabolic needs (Secomb, 2022). Each tissue's microvascular networks create 3D identifiable vascular patterns that are indicative of normal or diseased states (Galis, 2022; Lee and Galis, 2022).

Blood and lymph are collected from the tissue and returned to the heart via veins and lymphatic vessels, based on a similar hierarchical vascular organization, only this time moving back through small then large size vessels. The specific structure of veins and lymphatic vasculature is adapted to their particular function and mechanical environment, although it is constituted from similar types of building blocks as those on the arterial side. The veins are an integral part of the closed circuit of the blood vasculature, while the lymphatic vasculature is largely paralleling and complementing the blood vasculature, and nevertheless inextricably related to it, developmentally, functionally, and structurally. The lymphatic vasculature unidirectionally returns filtrated lymph to the vein side of blood circulation, helping maintain interstitial fluid balance between tissues and the blood. It also supports reverse cholesterol transport by removing cholesterol from peripheral tissues and providing circulatory conduits for immune cell trafficking and dietary lipid absorption (Jiang and Rockson, 2022). Lymph collection and unidirectional flow are supported by specialized vascular features, such as buttons, valves, lymphangions, and the hierarchical organization of the lymphatic vasculature. The lymphatic vasculature also supports the functional integration of the various components of the lymphatic system, including primary and secondary organs. For instance, the filtering of interstitial fluid collected from different tissues through numerous lymph nodes is essential for immune priming and induction of peripheral tolerance. Inborn defects of lymphatic vasculature or malfunction secondary to interventions are causes for known lymphatic pathologies (El Amm et al., 2022; Jiang and Rockson, 2022) and contribute to other diseases (Liu and Oliver, 2022).

PART II. INVESTIGATING THE VASCULOME: CONTEXT-DRIVEN METHODS, USES, AND LIMITATIONS

The vasculome is investigated using some of the methods commonly used for other body tissues and organs, yet its characteristics also warrant specialized approaches. For instance, the advantages and limitations of studying cells taken out of their in situ natural context and cultured in vitro are widely recognized. However, culturing vascular cells, especially ECs, warrants additional consideration, related to the difficulties of (1) harvesting viable and sufficient number of ECs from endothelial monolayers or from among the many other cell types of various tissue samples and (2) recreating in vitro the physically active environment necessary for maintaining vascular cell identity and function. A work-around has been to isolate human vascular cells from the more readily available vessels, e.g., ECs from discarded human umbilical veins or VSMCs of saphenous veins harvested for vascular grafts, and then to propagate and maintain these cells for several generations in the traditional static cell culture conditions.

Reports on the activities of specific types of vascular cells maintained and propagated under static conditions in vitro may not be representative of those of corresponding cell types of other vessels, e.g., of the ECs of other types of arteries, veins, or microvasculature, or even of their own their behaviors within their native mechanically active environments.

Section 1. Experimental and computational studies of the vasculome

The close spatial association between atherosclerotic lesions and areas of arteries experiencing complex blood flow patterns has led to early investigations of a potential causal effect of disturbed shear stress. Several in vitro systems have been built to directly investigate the effect of different shear stress patterns upon the structure and function of ECs (García-Cardeña and Gimbrone, 2022). The use of such in vitro systems has allowed researchers to demonstrate that steady laminar flow may be "atheroprotective" because it is associated with the expression of genes supporting antithrombotic, antiadhesive, antiinflammatory, and antioxidant activities, while the expression of these genes is reduced or absent in areas with disturbed flow, giving rise to the "Atheroprotective Gene Hypothesis." Through the application of various in vitro flow stimulator models, the Kruppel-like factor (KLF) 2 has emerged as a key atheroprotective transcription factor. Such experiments have also unraveled molecular pathways activated by blood flow, creating opportunities for new therapeutic approaches in atherosclerosis.

An advantage of in vivo experiments is that they allow for interrogation of the effects of systemic factors such as sex, race, age, and of systemic conditions such as obesity or diabetes on the structure and function of vasculature and its components in various body locations and contexts. These effects are still not systematically examined, although many have been found to be independently associated with cardiovascular disease. In vivo experiments also provide the most comprehensive way to study the effects of various therapeutic interventions. Most in vivo investigative methods are currently applicable only in experimental conditions. For instance, in vivo observations of blood flow, transendothelial transport, and regulatory processes are revealing, but particularly challenging in the dynamic metabolic and mechanical environments of the microvasculature. Label-free and minimally invasive in situ imaging of microvascular function and oxygen metabolism by intravital photoacoustic microscopy (Sun and Hu, 2022) has allowed multiparametric imaging of the cerebrovascular structures in the awake rat brain, including comprehensive and quantitative characterization of their structure (density, tortuosity), mechanical properties (e.g., wall shear stress, resistance, and reactivity), hemodynamics (e.g., blood perfusion, oxygenation, and flow speed), and the associated tissue oxygen extraction and metabolism. Despite any limitations, in vivo and in situ experimental methods provide important information that can be used in conjunction with theoretical modeling to predict parameters for vascular segments that cross boundaries and for which experimental information about flow and oxygen levels is incomplete (Secomb, 2022). Understanding these phenomena requires a network-oriented approach to analyzing vascular function.

Simulations of blood flow have to consider the hierarchical branching structure of the vasculome, which seamlessly connects the largest to the smallest vascular structures with inner diameters of the vascular space available to carry blood that vary over four orders of magnitude. Different dynamic regimens are at work in large vessels, as compared to those in smaller ones whose diameters approach the size of the red blood cells passing through them. Other geometrical parameters, such as vascular segment length, bifurcations, curvatures, etc., vary widely. All these variations result in complex blood flow patterns, creating variations in blood shear stress, which are sensed and acted upon by the inner EC lining of the vasculature. Pathological conditions involve various changes in vascular geometry, during which the physiological adaptive reaction of the vasculature may fail to restore normal blood flow, which brings other systems into play. Physics of oxygen transport simulation allows us to calculate values of parameters such as tissue perfusion and oxygen extraction, which can be used to predict the effects of various changes in blood flow or those due to variations in oxygen demand. These simulations are valuable for predicting key characteristics driving vascular function or dysfunction, especially for the natural in situ environments where experimental information is very difficult to obtain. They can also be used to inform experiments with isolated vascular cells or vessels. For instance, estimates of complex flow hemodynamic conditions have been used to develop in vitro systems allowing for testing the effects of different flow patterns upon cultured ECs (García-Cardeña and Gimbrone, 2022).

Combining simulation and in vitro experimentation has also been used to dissect intracellular pathways involved in interactions between ECs and other cell types, to enable predictions at the systemic level, as various ECs share many common "nodes": molecules that act as master regulators of multiple processes. For instance, models have been developed for cellular interactions between ECs and macrophages during angiogenesis, a multifactorial process driven by hypoxia (Eichmann and Li, 2022) that is essential in tissue development and repair, but also can fuel pathology, thus has important translational applications (Zhang et al., 2022).

Clinical investigations of endothelial and vascular function are essential to assessing overall health (Quyyumi et al., 2022).Various risk factors and genetics contribute to endothelial dysfunction, which has been shown to be an independent predictor of future development of hypertension, diabetes, or atherosclerosis, and a predictor of progression toward future adverse cardiovascular events in patients with cardiovascular diseases. Increased arterial stiffening was shown to be a precursor and predictor of the development of hypertension and increased damage of end-organs, including the heart, kidney, and the brain, thus indicating that finding treatments that reduce or prevent arterial stiffening should help prevent and control hypertension (Touyz et al., 2022). Arterial stiffening can be noninvasively assessed using pulse wave velocity (PWV). Current barriers to a widespread clinical use of noninvasive and invasive vascular function methods include both the need for a thorough understanding of their clinical applicability and practical clinical protocols that can help reduce operator-dependent variability.

Section 2. Realizing the promise of new high-content technologies

The use of new investigative technologies, including single-cell analysis, has greatly enhanced our understanding of the basis of cellular similarities and heterogeneity across the vasculome, including among ECs (Acharya et al., 2022; Becker et al., 2022; Gu and Cleaver, 2022), VSMCs (Majesky, 2022), and immune cells (Cochain and Mallat, 2022). New EC types have been identified, and many previously unsuspected transcriptomic differences have been found between ECs from different areas of the same organ, such as the kidney (Gu and Cleaver, 2022). Many examples of intraorgan heterogeneity, especially among ECs, have revealed the bases for angiodiversity, a reflection of the EC-tissue signaling reciprocity in the service of local function that begins early in embryogenesis and progresses in an increasingly complex manner during later development and even into adulthood (Gu and Cleaver, 2022). For instance, comparing the ECs of specialized central nervous system (CNS) vascular beds with or without a BBB (including the circumventricular organs and the choroid plexus) has allowed for the identification of EC genes associated with the BBB special properties (Gu and Cleaver, 2022). It is likely that such discovered differences can be exploited for translational applications that seek to overcome the challenge of delivering therapeutics to the immune-privileged NVUs of the brain (Cottarelli et al., 2022). Comparative transcriptomics is adding to our understanding of previously reported differences in the structure and function of ECs obtained by other investigative methods, revealing the molecular underpinnings of the overall diversity of vasculomes of various organs and their responses to various stresses, aging, or systemic conditions, such as hypertension or diabetes (Gu and Cleaver, 2022; Xing et al., 2022).

An ever-increasing challenge in biomedical research is managing many diverse types of high-content information collected in different contexts (in vitro, in situ, in vivo) through different investigative modalities (imaging, multiple -omics). A challenge that is specific to the vasculome is the need to organize, analyze, and be able to interpret data that reflects its diverse-yet-perfectly-integrated structural and functional nature, across all the size scales and within all the various tissues throughout the body. This deeper understanding of a spatial and functional vasculome (Galis, 2022) will be essential for our ability to diagnose, predict, treat, or engineer vascular and tissue/organ function. Computational approaches, including machine learning and artificial intelligence, widely used in other areas of biomedical exploration, are likely to be of help, especially if they can account for the integrative nature of the vasculome at the level of intracellular pathways and single cells (Gu and Cleaver, 2022; Zhang et al., 2022), vascular structures (Lee and Galis, 2022; Secomb, 2022), or the entire body (Eapen et al., 2022, Galis, 2022, Juárez-Orozco et al., 2022).

PART III. VASCULOME DYNAMICS: IN HEALTH AND IN SICKNESS

Section 1. Cooperating during development and organogenesis to create vasculome diversity

The development of the vasculature is essential for the development of all other tissues. Immediately functional blood vessels develop early in embryogenesis and continue to grow, diversify, and be shaped to provide sustenance, temporal, and spatial cues for organ patterning, and differentiation during organogenesis (Bautch, 2022). Vascular development is reactivated during repair, but also during pathological processes. Investigations of vascular development under normal conditions or in situations where genetic mutations lead to vascular malformations provide important clues about the vasculome's essential role in tissue and organ development, functions, and diseases.

ECs largely drive the development and patterning of the blood and lymphatic vasculature (Bautch, 2022; Eichmann and Li, 2022), with hypoxia as a main driver. The vascular endothelial growth factor (VEGF) pathway has emerged as the main mediator of vasculogenesis, de novo vessel formation from EC precursors or angioblasts, and angiogenesis, when new vessels sprout or split from existing ones. A division of labor among ECs is the basis for sprouting angiogenesis, involving changes in EC metabolism, shape, and polarization, which turn some cells into the leading "tip" ECs that break through the ECM in response to growth factors, and others into the accompanying "stalk" type ECs that proliferate and lay the path for new vessels (Eichmann and Li, 2022). Additional mechanisms are thought to contribute to formation of blood vessels in tumors.

EC-secreted angiocrine factors play an important role in organogenesis and the formation of organ-specific stem cell niches near some vessels. Likewise, the various organs produce factors that influence EC differentiation (Gu and Cleaver, 2022). An instructive example for the local tissue–vasculature cross-talk in the CNS is illustrated by the recruitment of ECs within the tissue via sprouting angiogenesis and the formation of the characteristic BBB that tightly controls the transfer between blood and CNS tissues via a continuous EC layer (Cottarelli et al., 2022). The limited access across the BBB protects the brain against inflammation and edema; on the other hand, it also makes delivery of medication across the BBB extraordinarily difficult (Xing et al., 2022). The specificity of blood–tissue exchange function in the kidney is achieved by its organotypic ECs that are "fenestrated" to facilitate the transfer of molecules from blood into urine (Gu and Cleaver, 2022). Among the vascular specializations that depend on systemic changes is the orderly process that leads to blood investing the lung vasculature and initiation of the blood oxygenation in the newborn.

The hierarchical pattern of the vasculature is tailored to the size, metabolic needs, and function of each particular organ. The specific vascular patterning and identity is established and maintained through the dynamic relationships between the characteristics of the local hemodynamic environment (blood flow, pressure, etc.) and the vascular tissue stiffness (Bautch, 2022). The original specification of vascular identity occurs during vasculogenesis with the early establishment of EC heterogeneity, through modulations of expression/suppression of specific sets of identity markers leading to the angioblast differentiation into arterial or venous ECs. Three distinct vascular identities, arterial, venous, and lymphatic, are established during the development of the vasculome (Dardik and Gonzalez, 2022). Divergent pathways triggered in a primordial vessel lead to the formation of an initial artery and an initial vein. The sprouting of additional new blood vessels from the two large arterial and venous precursors, known as angiogenesis, also follows specific spatial and temporal patterns during organogenesis, with veins continuing to sprout later in the process and with the separation between the vein/blood and the lymphatic vasculatures unleashing the development of the lymphatic system. Further differentiation of the venous ECs gives rise to the lymphatic identity specification, through the action of Prox1 transcription factor, thought to be necessary and sufficient to downregulate the blood EC-specific genes and upregulate the lymph EC-specific genes. The development of the lymphatic vasculature occurs through lymphangiogenesis and lymphvasculogenesis after the establishment of the blood vasculature (Eichmann and Li, 2022). Apparently, the main source for lymphatic ECs participating in lymphangiogenesis is the cardinal vein ECs. Lymphatic ECs demonstrate organ-specific heterogeneity like blood vessel ECs. Some organ-specific lymphatic ECs that participate in lymphvasculogenesis were found to originate from other nonvenous ECs and even from other cellular sources, such as smooth muscle cells and macrophages. Defects in lymphangiogenesis during development result in congenital hypoplastic and primary lymphedema (Jiang and Rockson, 2022). Lymphangiogenesis is stimulated in pathological conditions such as inflammation and tumorigenesis.

Achieving and maintaining vascular identity, i.e., the structure and function specifically associated with a certain type of vessel, depends on the proper spatiotemporal orchestration of specific sets of molecular determinants, from their expression during early development to the lifetime maintenance of vascular identities within the different characteristic hemodynamic environments. Aberrations of vascular identity early on in development have been associated with pathological conditions, such as the occurrence of vascular malformations or vascular tumors (Bautch, 2022). These conditions range from benign, or self-resolving with age, to life-threatening, malignant, and aggressive. However, later alterations in the normal expression pattern of molecular determinants due to changes in the hemodynamic environment, such as those produced by surgical interventions, play a key adaptive role, and can result in loss or changes in vascular identity (Humphrey, 2022). For instance, after vein grafts used as substitutes for blocked arteries are placed in a new, fast flow and pulsatile arterial hemodynamic environment, they undergo adaptive changes resulting in loss of their original vein identity.

Section 2. Physiological and pathological remodeling of the vasculome

The vasculome senses and responds to local and systemic metabolic and physiological demands and stimuli. Such responses vary from immediate and short-lived, to long-term changes in vasculome permeability, reactivity, or structure (Humphrey, 2022). Its inability to respond adequately is generally referred to as vascular dysfunction, which is a major contributor to many local and systemic diseases. A dynamic ECM participation, including synthesis, degradation, or reassembly of its components, is an essential component of the physiological or pathological remodeling of the vasculome (Mohabeer and Bendeck, 2022). The dynamic interplay between the vascular ECM and the VSMCs, the major source of vascular ECM, has been implicated in the pathological changes affecting various arteries that are essential components of atherosclerosis, diabetes, or hypertension and include arterial stiffening, calcification, and aneurysm formation.

Understanding the connection between systemic and local manifestations of vasculome function and dysfunction is key to advancing clinical management of conditions affecting the vasculature. The same risk factors or the same disease can differently affect various segments of the vasculome and hence have heterogenous clinical presentations. Hypertension is a good example of the systemic–local interplay within the vasculome: in response to a systemic increase in blood pressure, the compensatory vascular adaptation occurs mainly through changes in the thickness of the arterial walls. Yet, locally, arterial remodeling differs depending on the type of vessel; changes in the arterial wall's thickness help muscular arteries maintain their lumen size, while lumen is decreased in arterioles and increased in elastic arteries that tend to dilate while adapting their wall thickness. These "nonideal" local adaptations have benefits, e.g., thickening of arterioles protects the microvasculature of end-organs from high pulse pressure, yet it contributes to increasing peripheral resistance that adds to the pathophysiology of high blood pressure (Humphrey, 2022). Accordingly, the vascular phenotype observed in hypertension is both a cause and a consequence of high blood pressure (Touyz et al., 2022). The most common type of hypertensive remodeling in mild essential hypertension is the inward eutrophic remodeling of resistance arterioles, while hypertrophic remodeling is commonly found in patients with secondary hypertension. The variety of vascular responses in hypertension explains the challenges faced by therapeutic interventions. Various antihypertensive medications target different processes, with some antihypertensive agents reversing remodeling of small resistance arteries.

The anomalies and dysfunction of the lymphatic vasculature present a complex landscape, being currently classified mostly based on clinical phenotype (El Amm et al., 2022). Many of these conditions may be due to common genetic mutations (e.g., abnormal activation of PI3K or MAPK pathways). Some mutations have been shown to be inherited, and some to be somatic in nature, while many lymphatic anomalies are still of unknown etiology. Due to the interconnected nature of the lymphatic vasculature, the local damage, especially in the upper segments, impairs the function of the entire system. Improving our ability to identify the underlining causes, including mutations, creates opportunities for therapeutic interventions.

PART IV. THE VASCULOME AS PERPETRATOR AND VICTIM IN LOCAL AND SYSTEMIC DISEASES

At the level of the entire vasculome, endothelial dysfunction has been identified as a common denominator for the development of many cardiovascular and other conditions. The value of invasive and noninvasive clinical measurements of endothelial dysfunction (Quyyumi et al., 2022) as an early predictor of various cardiovascular events, including mortality, has been supported by many studies (Munzel et al., 2022). A main cause of endothelial and vascular dysfunction is oxidative stress, which can be measured clinically and may offer future opportunities to intervene, especially based on the emerging fundamental understanding of reactive oxygen species (ROS) sources and interactions, including their interference with vascular activities of nitric oxide (NO) or leading to epigenetic modifications. Vascular dysfunction, including dysfunction due to activation of inflammatory pathways-associated ROS, contributes to hypertension, one of the most common systemic conditions (Touyz et al., 2022). Its pathophysiological mechanisms include increased vascular resistance, determined primarily by decreased endothelium-dependent vasorelaxation, increased vascular hyperreactivity, and arterial remodeling, with the participation of all the arterial wall constituents. Changes in PVAT composition and secretory activities contribute to vascular dysfunction (Sena, 2022), while dedifferentiation of VSMCs (Majesky, 2022) into a supercontractile phenotype is a major contributor to hypertensive arterial remodeling (Fisher, 2022).

Several chapters of the book illustrate the value of being able to effectively contrast and compare the local variations of vascular structure, respectively, vascular function and dysfunction, in order to therapeutically target various parts of an otherwise integrated vasculome. Such comparisons made within the vasculome have included investigations of blood versus lymphatic vasculature, arterial versus venous circulations, or their specific segments. Other approaches have defined and contrasted vascular territories feeding various body parts, as described by the angiosome or organ vasculome concepts. All these fundamental and clinical explorations have connected local, regional, and systemic vascular dysfunction to cardiovascular and other disease risk and severity and have paved the way current and future interventions. Many promising advances have been made in understanding both the importance of vascular health for overall health and how this can be measured for diagnostic purposes (Eapen et al., 2022). The ultimate utility of the vasculome will be in improving the diagnosis, prevention, and treatment of many diseases, most of which have a vascular component.

For instance, investigation of aneurysmal disease has revealed significant differences in various arteries, or even in the same artery, challenging us to understand why and how. Specifically, the inherited risk of aneurysm formation differs within regions of the same artery, i.e., the genetically inherited risk of developing aneurysms in the thoracic aorta does not appear to increase the risk for their development in the abdominal areas of aorta, where formation of aneurysms is associated with specific risk factors related to lifestyle, hypertension, and male sex (Kaw et al., 2022). Further investigations of single gene effects have highlighted structures and signaling pathways critical for preserving the thoracic aorta integrity, including the intricate relationship between VSMC and ECM components, specifically the VSMC elastin-contractile unit and its essential role in supporting vascular mechanotransduction (Mohabeer and Bendeck, 2022). Detailed understanding of the molecular bases of genetically triggered aneurysm allows us to assess risk and prevent catastrophic dissection in these medical conditions that usually progress asymptomatically.

Lymphatic vasculature dysfunction has been traditionally known to lead to primary and secondary lymphedema due to perturbations in tissue fluid homeostasis (Jiang and Rockson, 2022). More recent investigations of various parts of the lymphatic vasculature have helped been identify important contributions to immune surveillance and dietary fat absorption in the gastrointestinal organs, and lymphatic function impairment has been associated with a variety of medical conditions, including obesity, cardiovascular disease, glaucoma, and neurological disorders (Liu and Oliver, 2022). Thus, therapeutic restoration of lymphatic vasculature's healthy function provides alternative strategies for treating these conditions.

Systemic atherosclerosis has distinct local presentations, such as impairment of blood flow toward the lower extremities, known as peripheral arterial disease (PAD). PAD itself has a broad spectrum of clinical presentations, ranging from asymptomatic to clinical symptoms of intermittent pain with impaired physical activity and ischemic pain at rest, to chronic limb-threatening ischemia (CLTI), and tissue or entire limb loss. Common patterns of PAD localization have been associated with various key vascular risk factors, including diabetes, smoking, hypertension, dyslipidemia, and obesity (Conte and Kim, 2022). Optimal medical management of PAD needs to consider the underlining systemic conditions and the local presentation, ranging from control of systemic illness to surgical interventions, to increase local blood flow through endoscopic or open surgical revascularization procedures.

Arterial and venous thrombotic disorders traditionally have been viewed as different entities, although they share many commonalities, including major mediators, signaling pathways, and clinical risk factors that can be further boiled down to endothelial dysfunction and inflammation. In the peripheral venous side of the circulation, thromboembolic phenomena are main culprits for disease. These occur in the specific structural and biomechanical environment of the veins, whose main function is to return blood to the heart. Basic and clinical observations have shown that venous thrombosis involves stasis, endothelial injury, hypercoagulability, and sterile inflammation, pointing to the need for development and testing of new specific biomarkers and therapeutic interventions (Obi et al. 2022).

Systemic conditions can affect differently not only different types of vessels, but also the entire vascular structures investing specific organs or limbs. For instance, hypertension, diabetes, or aging differently affect the vascular transcriptome, proteome, secretome, and metabolome of the brain, heart, lung, or kidney (Xing et al., 2022). From this point of view, each organ (and perhaps each limb) can be regarded as having its own "vasculome" interconnected with that of other organs and body parts within the body wide vasculome. The brain vasculome was previously defined as "the transcriptional profile of the brain microvascular network" (Xing et al., 2022), and has helped to shed light on the important, but frequently overlooked contribution of the vasculature to brain health, aging, and disease. Findings hold out the promise of our being able to reverse some of the brain vasculature damage due to aging or disease (e.g., through exercise). Yet, the ability to integrate the regional functional vascular responses to systemic conditions, or to therapeutic and life-style interventions, such as exercise, is still needed if we are to be able to effectively manage clinically the whole person. Understanding the circadian rhythm modulation of the vasculome function and its effects on the efficacy of any vascular interventions are also important areas due for further in-depth exploration.

Knowledge of the vasculome is essential for any surgical interventions. Our ability to precisely identify the specific vascular structures responsible for supplying blood to distinct body structures creates the opportunity for targeted interventions and can help determine how certain body organs or parts can be removed, spared, or restored. Perfusion through the lower limbs offers a great illustration of the clinical applicability of considering the fractal nature of the vasculome: the arterial trunk successively divides into smaller vessels to generate a total wider cross-functional area for outward blood flow toward the peripheral tissues. Six levels are described from the largest arteries (iliac and femoral arteries) to the capillary level, with the intermediate levels being the most likely useful for targeting treatment of CLTI. Collateral vessels, which are essential for maintaining flow when the main arterial structure is occluded, can be also affected by inherited interindividual variability and can be damaged in diabetes mellitus or renal insufficiency and, themselves, can cause CLTI pathologies. One lower limb revascularization approach takes advantage of the functional angiosome concept, which holds that the body can be divided in main vascular territorial "watersheds," or angiosomes. Each angiosome consists of arterial and venous components; includes collateral, anastomotic, and other vessels; and services distinct areas of the body, exemplifying the integrative nature of the vasculome (Alexandrescu et al., 2022). The revised version of the original angiosome concept, initially described as a means to topographically delineate skin areas based on their specific vascular watershed for use in plastic surgery applications, describes the functional angiosome, which includes the collateral, anastomotic, and other vessels that assist the main angiosomes. The angiosome concept offers an additional approach to assessing the body territories that are serviced by the various parts of the vasculome, with great applicability for clinical interventions. Accordingly, this concept has seen increased adoption in the clinical management of wound healing and limb salvage, through the targeted recanalization of vascular structures responsible for supplying ischemic areas, as well as in the planning of surgical interventions for internal organs, which need either to recanalize or spare specific vessels. The angiosome approach holds promise for the therapeutic manipulation of collateral blood flow. However, standardized definitions, functional diagnostic methods, and more systematic corroborative evidence from large multicenter studies are needed.

The precise targeting of specific organs, including their diseased tissues, for clinical purposes has been achieved by taking advantage of the discovery of endothelial "zip codes" displayed by different ECs, either due to their location or their functional status. Such molecular endothelial "zip codes"—associated with specific areas of normal vasculature or with specific disease conditions identified through combinatorial phage display—have allowed the development of peptide libraries used for several clinical applications targeting cancer of the prostate, breast, or lung, retinopathy, and white fat, and are already approved or undergoing the process for FDA approval (Smith et al., 2022). The ability to widely identify and characterize EC "zip codes" across the vasculome could be useful for therapeutic targeting of any EC and its neighboring cells and for producing maps of the entire human body at single-cell resolution (Galis, 2022).

The development of new therapeutic approaches targeting the vasculome will be based on our improved understanding of fundamental biological processes applied to its specific nature. For instance, the discovery that various RNAs provide an additional level of regulation for gene expression and function opened up a variety of options for RNA therapies (Laina et al., 2022). These can be used to control the expression of a specific molecule, such as therapies targeting specific mRNAs, or to affect entire biological processes by targeting various types of noncoding RNAs. The type of molecules developed for therapeutic effect includes antisense oligonucleotides (ASOs), small interfering RNAs (siRNAs), and aptamer-based therapies. Some of the first RNA-based therapies targeting various components of the lipid metabolism, traditionally associated with cardiovascular diseases (CVDs), have reached the clinical stage of development. New RNA-based interventions for cardiomyopathy are also being evaluated in patients. Ongoing technological developments in refining RNA-based interventions hold great promise for the ability to intervene in medical conditions for which there has been limited success using traditional therapeutics based on small molecule and biologicals, by overcoming some of the current barriers, such as the limited intracellular access of monoclonal antibodies, or large payloads necessary to achieve therapeutic effect.

PART V. LOOKING FORWARD TOWARD "PRECISION HEALTH" FOR THE VASCULOME

The vasculome is essential for the healthy functioning of every body tissue and organ. It is equally essential that we define and measure a healthy vasculome. Are currently available approaches mete to the task?

The American Heart Association has defined "Ideal cardiovascular health" (CVH) as the presence of optimal levels of all of the "Life Simple 7:" blood pressure, total cholesterol, fasting blood glucose, smoking, physical activity, body mass index, and diet (Eapen et al., 2022). Many of these are modifiable risk factors associated with atherosclerosis, a disease of medium and large arteries, responsible for a significant share of CVD. One line of reasoning is that maintenance of vascular health may be centered on optimizing the shared and highly correlated risk factors associated with atherosclerosis in the main affected vascular beds, i.e., the coronary, cerebrovascular, peripheral vascular, and the central aortic vascular beds (Eapen et al., 2022). These include both traditional and nontraditional risk factors. For example, carotid-intima thickness has been reported to be a predictor of CVD. However, lack of standardization has led to that factors not being included in routine risk assessment. Coronary artery calcium (CAC) has been found to predict CVD in a variety of cohorts, regardless of age, gender, and ethnicity. It is used in practice to assess CVD risk and to direct preventive and therapeutic interventions (e.g., statins, aspirin). Enhancing cardiovascular health increases resilience to disease in general and must become a goal at every level: individual, family, community, and general population. Regardless of individual differences, improvements in cardiovascular health are already possible through common life-style interventions, such as exercise and diet, known to directly improve endothelial function. Further health advances will come from researchers, health professionals, and the general populous actively working to lower these modifiable risk factors and from our leveling of the societal playing field in access to clean air, healthy food, and health care.

The vasculome is a true tree of life for our entire body and holds the key to our whole-person health. New technologies bring the promise of our being able in the near future to use the vasculome to precisely detect the first signs of dysfunction at the level of cells and functional units in various organs (Galis, 2022), creating the opportunity to diagnose, prevent, intercept, and potentially reverse dysfunction before the emergence of clinically overt consequences of disease. The definition of a healthy vasculome will likely be based on several key parameters that together create the picture of local (e.g., at the level of various organ FTUs or organs) and of systemic vascular function. These values can serve as sensitive harbingers of subtle functional changes, even before clinical symptoms manifest themselves. As with the currently measured clinical parameters, e.g., blood pressure, cholesterol, or blood sugar, the range of "healthy" (normal) values will likely depend on several characteristics of each vasculome's owner, including the age, sex, or ethnic background.

Advances in understanding the vasculome generate the bright promise of both our research and translational communities being able to gain a shared, more nuanced (personalized) and in-depth (subclinical) characterization of our vascular and overall health (Galis, 2022).

ACKNOWLEDGMENTS

The concept of the Vasculome is a corollary of over 30 years of work, in which this book's Editor has pursued explorations in vascular biology and medicine: training, researching, teaching, and developing therapeutic approaches and major scientific collaborative initiatives within the academia, industry, and government. I thank all those who have challenged me, and from whom I have learned: mentors, mentees, colleagues, scientific friends, reviewers of my manuscripts and funding applications, and all those who have inspired me to keep going: family, friends, and patients. My research career has been possible due to the blessing of my late parents, who encouraged me to take a chance and defect from Communist Romania during the rare opportunity to attend a conference in Canada as a Young Investigator Awardee, in hopes of better personal and professional opportunities, while knowing full well that they would be sacrificing their own careers. My quest for further scientific training and knowledge would have not been possible without the full cooperation of the best son anyone could wish for. If it were not for his brilliance and self-sufficiency, I would never have been able to undertake the many hours of research training and lab work required and raise him at the same time.

Special thanks to each of the authors of *The Vasculome's* chapters summarized above and appearing in full below, who were willing and able to join this collective project even through and in spite of the difficult times of the past 2 years. We are all and forever in your debt.

References

Acharya BR, Chavkin NW, Hirschi KK. Developmental of and environmental impact on, endothelial cell diversity. In: Galis ZS, ed. *The Vasculome. From Many, One*. Elsevier; 2022.

Alexandrescu VA, Kerzmann A, Boesmans E, Holemans C, Defraigne J-O. Angiosome concept for vascular interventions. In: Galis ZS, ed. *The Vasculome. From Many, One*. Elsevier; 2022.

Bautch VL. Vascular development and organogenesis: depots of diversity among conduits of connectivity define the vasculome. In: Galis ZS, ed. *The Vasculome. From Many, One*. Elsevier; 2022.

Becker L, Pasut A, Cuypers A, Carmeliet P. Endothelial cell heterogeneity in health and disease – new insights from single cell studies. In: Galis ZS, ed. *The Vasculome. From Many, One*. Elsevier; 2022.

Cochain C, Mallat Z. Resident vascular immune cells in health and atherosclerotic disease. In: Galis ZS, ed. *The Vasculome. From Many, One*. Elsevier; 2022.

Conte MS, Kim AS. Peripheral arterial disease (pathophysiology, presentation, prevention/management). In: Galis ZS, ed. *The Vasculome. From Many, One*. Elsevier; 2022.

Cottarelli A, Wayne C, Agalliu D, Biswas S. The neurovascular unit and blood-CNS barriers in health and disease. In: Galis ZS, ed. *The Vasculome. From Many, One*. Elsevier; 2022.

Dardik A, Gonzalez L. Normal vascular identity (arteries, veins, lymphatics) and malformations. In: Galis ZS, ed. *The Vasculome. From Many, One*. Elsevier; 2022.

Dave J, Saito J, Mottola G, Greif DM. Out to the tissues: the arterial side (arteries, arterioles - development, structure, functions, pathologies). In: Galis ZS, ed. *The Vasculome. From Many, One*. Elsevier; 2022.

Eapen DJ, Faaborg-Andersen C, DeStefano RJ, Karagiannis AD, Quintana RA, Dhindsa D, Chaudhuri M, Searles Jr CD, Sperling LS. Defining and optimizing vascular health. In: Galis ZS, ed. *The Vasculome. From Many, One*. Elsevier; 2022.

Eichmann A, Li J. Sprouting angiogenesis in vascular and lymphatic development. In: Galis ZS, ed. *The Vasculome. From Many, One*. Elsevier; 2022.

El Amm C, Silva-Palacios F, Geng X, Srinivasan RS. Lymphatic vascular anomalies and dysfunction. In: Galis ZS, ed. *The Vasculome. From Many, One*. Elsevier; 2022.

Fisher S. Smooth Muscle Cell contractile diversity. In: Galis ZS, ed. *The Vasculome. From Many, One*. Elsevier; 2022.
Galis ZS. The vasculome provides a body-wide cellular positioning system and functional barometer. The "vasculature as common coordinate frame (CCF)" concept. In: Galis ZS, ed. *The Vasculome. From Many, One*. Elsevier; 2022.
García-Cardeña G, Gimbrone Jr MA. The Flow-dependent Endotheliome: Hemodynamic Forces, Genetic Programs, and Functional Phenotypes. In: Galis ZS, ed. *The Vasculome. From Many, One*. Elsevier; 2022.
Gu X, Cleaver O. Angiodiversity – A tale retold by comparative transcriptomics. In: Galis ZS, ed. *The Vasculome. From Many, One*. Elsevier; 2022.
Humphrey JD. Enablers and drivers of vascular remodeling. In: Galis ZS, ed. *The Vasculome. From Many, One*. Elsevier; 2022.
Jiang X, Rockson SG. Lymphatic biology and medicine. In: Galis ZS, ed. *The Vasculome. From Many, One*. Elsevier; 2022.
Juárez-Orozco LE, Yeung MW, Benjamins JW, Kazemzadeh F, Carvalho G, van der Harst P. Artificial intelligence (AI) for the vasculome. In: Galis ZS, ed. *The Vasculome. From Many, One*. Elsevier; 2022.
Kaw K, Kaw A, Milewicz DM. Extracellular matrix genetics of thoracic and abdominal aortic diseases. In: Galis ZS, ed. *The Vasculome. From Many, One*. Elsevier; 2022.
Laina A, Vlachogiannis NI, Stamatelopoulos K, Stellos K. RNA therapies for cardiovascular disease. In: Galis ZS, ed. *The Vasculome. From Many, One*. Elsevier; 2022.
Lee EK, Galis ZS. Using pattern recognition and discriminant analysis of functional perfusion data to create "angioprints" for normal and perturbed angiogenic microvascular networks. In: Galis ZS, ed. *The Vasculome. From Many, One*. Elsevier; 2022.
Liu X, Oliver G. Functional roles of lymphatics in health and disease. In: Galis ZS, ed. *The Vasculome. From Many, One*. Elsevier; 2022.
Majesky MW. The remarkable diversity of vascular smooth muscle in development and disease: a paradigm for mesenchymal cell types. In: Galis ZS, ed. *The Vasculome. From Many, One*. Elsevier; 2022.
Mohabeer AL, Bendeck MP. Extracellular matrix dynamics and contribution to vascular pathologies. In: Galis ZS, ed. *The Vasculome. From Many, One*. Elsevier; 2022.
Munzel T, Hahad O, Daiber A. Endothelial dysfunction: basis for many local and systemic conditions. In: Galis ZS, ed. *The Vasculome. From Many, One*. Elsevier; 2022.
Obi AT, Myers Jr DD, Henke PK, Sood S, Wakefield TW. Venous diseases including thromboembolic phenomena. In: Galis ZS, ed. *The Vasculome. From Many, One*. Elsevier; 2022.
Quyyumi AA, Almuwaqqat Z, Islam SJ. Clinical Investigations of vascular function. In: Galis ZS, ed. *The Vasculome. From Many, One*. Elsevier; 2022.
Rudnicki M, Pislaru A, Haas TL. Capillary diversity: endothelial cell specializations to meet tissue metabolic needs. In: Galis ZS, ed. *The Vasculome. From Many, One*. Elsevier; 2022.
Secomb TW. Simulation of blood flow and oxygen transport in vascular networks. In: Galis ZS, ed. *The Vasculome. From Many, One*. Elsevier; 2022.
Sena C. Perivascular adipose tissue (physiology, pathology). In: Galis ZS, ed. *The Vasculome. From Many, One*. Elsevier; 2022.
Smith TL, Sidman RL, Arap W, Pasqualini R. Targeting vascular zip codes: from combinatorial selection to drug prototypes. In: Galis ZS, ed. *The Vasculome. From Many, One*. Elsevier; 2022.
Sun N, Hu S. Intravital photoacoustic microscopy of microvascular function and oxygen metabolism. In: Galis ZS, ed. *The Vasculome. From Many, One*. Elsevier; 2022.
Touyz RM, Rios FJ, Montezano AC, Neves K, Eluwole O, Maseko MJ, Alves-Lopes R, Camargo LL. The vascular phenotype in hypertension. In: Galis ZS, ed. *The Vasculome. From Many, One*. Elsevier; 2022.
Wagenseil J, Lin C-J, Crandall C. Major vascular extracellular matrix components, differential distribution supporting structure and functions of the Vasculome. In: Galis ZS, ed. *The Vasculome. From Many, One*. Elsevier; 2022.
Xing C, Guo S, Li W, Deng W, Ning M, Lok J, Arai K, Lo EH. The brain vasculome: an integrative model for CNS function and disease. In: Galis ZS, ed. *The Vasculome. From Many, One*. Elsevier; 2022.
Zhang Y, Zhao C, Popel AS. Systems biology modeling of endothelial cell and macrophage signaling in angiogenesis in human diseases. In: Galis ZS, ed. *The Vasculome. From Many, One*. Elsevier; 2022.

Acknowledgment

I wish to thank graduate students Lalith Dupathi, Satyajeet Sarfare, Ruchir Tatar, Sripad Joshi, and Churchil Moondra, for mining and organizing *The Vasculome's* abbreviations, key words, and chapters content, and for producing the book's interactive visual index in fulfilment of their requirements for the ENGR-E 583/483 Information Visualization class, of the Luddy School of Informatics, Computing, and Engineering at Indiana University. To our knowledge this represents pioneering work in the realm of electronic publishing.

List of abbreviations

Chapter	Abbreviations	Term
27	$[Ca^{2+}]_i$	Intracellular free concentration of Ca^{2+}
2	AA	Amino acid
8; 24; 29;	AAA	Abdominal aortic aneurysm
24	AAs	Aortic aneurysm
32	AAV	Adeno-associated virus
32	AAVP	Adeno-associated virus/phage
30; 33	ABI	Ankle-brachial index
33	AC	Angiosome concept
16	ACCT	Anglo-Cardiff Collaborative Trial
8; 26	ACE	Angiotensin-converting enzyme
8	ACE2	Angiotensin-converting enzyme 2
26	ACh	Acetylcholine
24	αSMA	Alpha (α) smooth muscle actin
34	ADAR2	Adenosine deaminase acting on RNA-2
32	ADC	Antibody–drug conjugate
13	ADM	Acellular dermal matrix
28	Adm	Adrenomedullin
16; 26	ADMA	Asymmetric dimethylarginine
27	ADRFs	Adipocyte-derived relaxing factors
8	AdvSca1	Adventitial stem cell antigen-1
2	AFE	Alveolar fibroelastosis
24; 30	AGEs	Advanced glycation end products
1	AGM	Aorta-gonad-mesonephros
28	AH	Aqueous humor
36	AHA	American Heart Association
19; 30; 37	AI	Artificial intelligence
1	AKT	A serine/threonine protein kinase
30	ALI	Acute limb ischemia
22	ALK1	Activin receptor-like kinase 1
32	AMD	Age-related macular degeneration
27	AMPK	Protein kinase, AMP-activated
11	ANG	Angiopoietin
14	Ang	Angiopoietin

(*Continued*)

Chapter	Abbreviations	Term
20	Ang1	Angiopoietin-1
4	Ang II	Angiotensin II
1	ANGPT1	Angiopoietin 1
34	ANGPTL3	Angiopoietin-like 3
9	Angptl4	Angiopoietin-like 4
28	Angptl7	Angiopoietin-like protein 7
32	ANX2	Annexin A2
3	AoA	Aortic arch
10	APCs	Astrocyte precursor cells
9	APLN	Apelin
36	ApoB	Apolipoprotein B
35	ApoD	Apolipoprotein D
35	ApoE	Apolipoprotein E
10	Aqp4	Aquaporin-4
8	ARBs	Angiotensin receptor blockers
2	ARDS	Acute respiratory distress syndrome
16; 36	ARIC	Atherosclerosis Risk in Communities
3	AsAo	Ascending aorta
7	ASC	Ascending aorta
16	ASCOT	Anglo-Scandinavian Cardiac Outcomes Trial
36	ASCVD	Atherosclerotic cardiovascular disease
34	ASGPR	Asialoglycoprotein receptor
34	ASO	Antisense oligonucleotide
2	ATAC	Assay for transposase-accessible chromatin
27	ATF2	Activating transcription factor 2
5	ATLOs	Artery tertiary lymphoid organs
1	ATP	Adenosine triphosphate
35	Atp1a2	ATPase, Na+/K+ transporting, alpha two polypeptide
20; 21; 22	AVM	Arterial—venous malformation
17	AVR	Ascending vasa recta
10; 17; 32; 35	BBB	Blood—brain barrier
27	BeAT	Beige adipose tissue
16; 26	BH4	Tetrahydrobiopterin
27	BHF	British Heart Foundation
26	BK	K+
17	BM	Basement membrane
37	BMI	Body mass index
1; 22; 35	BMP	Bone morphogenic protein
9	BMP6	Bone morphogenic protein 6

Chapter	Abbreviations	Term
11	BMP9	Bone morphogenetic protein 9
26	BNP	Brain natriuretic peptide
4	BP	Blood pressure
10; 17	BRB	Blood–retinal barrier
1	BRG1	Brahma-related gene 1
10	BSCB	Blood–spinal cord barrier
35	Bsg	Basigin
27	Ca2+	Calcium
27	[Ca2+]i	Intracellular free concentration of Ca2+
36	CAC	Coronary artery calcium
31	CACTUS	Anticoagulant therapy for symptomatic calf deep vein thrombosis
36	CAD	Coronary artery disease
16	CAFÉ	Conduit Artery Function Evaluation Study
28	Calcrl	Calcitonin receptor-like receptor
27	cAMP	Cyclic adenosine monophosphate
10	CAMs	Cell adhesion molecules
26	CANTOS	Canakinumab for the management of cardiovascular disease
36	CAPRIE	Clopidogrel versus aspirin in patients at risk of ischemic events
36	CAPS	Carotid Atherosclerosis Progression Study
36	CARDIA	Coronary Artery Risk Development in Young Adults
33	CAs	Cutaneous arteries
33	CATI	Chronic angiosomal threatening ischemia
17; 37	CCF	Common Coordinate Frame
25	CCLA	Central conducting lymphatic anomaly
20; 21	CCM	Cerebral cavernous malformation
5	cDC1	Type 1 cDC
5	cDCs	Conventional (or Classical) DCs
1	Cdh5	Cadherin 5
36	CETP	Cholesteryl ester transfer protein
4; 26; 27	cGMP	Cyclic guanosine monophosphate
1	CHD4	Chromodomain helicase DNA-binding protein 4
8	CHIP	Clonal hematopoiesis of indeterminate potential
36	CHS	Cardiovascular Health Study
36	CIMT	Carotid intima-media thickness
30	CKD	Chronic kidney disease
1	cKit	Tyrosine-protein kinase KIT
33	CLI	Critical limb ischemia

(*Continued*)

Chapter	Abbreviations	Term
27	CLRs	C-type lectin receptors
30; 33	CLTI	Critical limb-threatening ischemia
13	CMRO2	Cerebral metabolic rate of oxygen
18	CN	Crossing number
10; 17; 20; 35	CNS	Central nervous system
2	CNV	Choroidal neovascularization
16	CO	Carbon monoxide
24	COL3A1	Collagen III
16	COX	Cyclooxygenase
33	CPs	Cutaneous perforators
36	CR	Cardiac rehabilitation
17	cRECs	Cortex RECs
36	CRF	Cardiorespiratory fitness
16	CRIC	Chronic renal insufficiency cohort study
26; 30	CRP	C-reactive protein
7	CS	Chondroitin sulfate
10	CSD	Cortical spreading depressions
10; 28	CSF	Cerebrospinal fluid
19	CT	Computed tomography
30	CTA	Computed tomography angiography
33	CTO	Chronic total occlusions
2	CTP1A	Carnitine palmitoyl-transferase 1A
31	CTR	Nontreated control
17; 33	CV	Choke vessels
34	CV	Cardiovascular
16; 34; 36	CVD	Cardiovascular disease
17	CVOs	Circumventricular organs
1	Cx	Connexin
1	Cx37	Connexin 37
1	Cx43	Connexin 43
16	CYP4502c	Cytochrome P450 2C
5	CyTOF	Cytometry by time of flight
3; 17	DA	Ductus arteriosus
22	DAB2	Disabled homologue 2
27	DAG	Diacylglycerol
18	DAMIP	Discriminant analysis model based on mixed-integer programming
27	DAMPs	Damage-associated molecular patterns

Chapter	Abbreviations	Term
3	DAoA	Descending portion of the AoA
36	DASH	Dietary Approaches to Stop Hypertension
11	DCs	Dendritic cells
24	ddr1	Discoidin domain receptor tyrosine kinases
3	DFs	Dermal fibroblasts
17	DHA	Docosahexaenoic acid
21	DLL	Delta-like
1	DLL4	Delta-like ligand 4
28	Dll4	Delta-like ligand 4
24; 30	DM	Diabetes mellitus
1; 12	DNMT	DNA methyltransferase
1	dNTP	Deoxynucleotide triphosphate
10	DR	Diabetic retinopathy
7	DRGs	Differentially regulated genes
7	DS	Dermatan sulfate
30	DSA	Digital subtraction angiography
7	DSC	Descending aorta
4	DSCAM	Downs Syndrome Cell Adhesion Molecule gene
30	DUS	Duplex ultrasound
17	DVR	Descending vasa recta
31	DVT	Deep vein thrombosis
4	E	Exon
1	E2F1	E2F transcription factor 1
10	EAE	Experimental autoimmune encephalomyelitis
33	EAOD	End-artery occlusive disease
36	EBCT	Electron-beam computed tomography
1; 2;7; 8;9; 10; 11; 17; 21; 23; 30; 35; 37	EC	Endothelial cell
1	ECE1	Endothelin converting enzyme 1
7; 8;10; 16; 22; 23; 24; 29; 35	ECM	Extracellular matrix
16	EDHF	Endothelium-derived hyperpolarizing factor
24	EDPs	Elastin-derived peptides
26	EDRF	Endothelium-derived relaxing factor
4	EDRF/NO	Endothelial-derived relaxing factor/nitric oxide
7; 8	EEL	External elastic lamina
16	EETs	Eicosatrienoic acids
1	Efnb2	Ephrin B2

(*Continued*)

Chapter	Abbreviations	Term
29	EGF	Epidermal growth factor
1	EGLN3	Egl-9 family hypoxia inducible factor 3
1	EHT	Endothelial-to-hematopoietic transition
31	EIM	Electrolytic inferior vena cava model
2	EIT	Epithelial immune cell-like transition
8	ELN	Elastin
7	ElnEKO	Eliminating elastin expression from ECs
1	EMCN	Endomucin
16	EMPs	Endothelial microparticles
17	EMPs	Erythro-myeloid progenitors
3	EMT	Epithelial—mesenchymal transition
1	EndMT	Endothelial-to-mesenchymal transition
1; 6;16; 26; 27; 30;	eNOS	Endothelial nitric oxide synthase
16	EPCs	Endothelial progenitor cells
32	EphA5	Ephrin A5
8	EphB2	EphrinB2
1	Ephb4	Ephrin-type b receptor 4
8	EphB4	EphrinB4
4	ER	ER, estrogen receptor
34	ERC	European Research Council
10; 29	ERG	Electrocardiogram
11	ESAM	Endothelial cell selective adhesion molecule
27	ESH	European Society of Hypertension
30	ESRD	End stage renal disease
30	ESVS	European Society for Vascular Surgery
1	Et	Endothelin-1
35	ETS	E26 transformation-specific
35	EVs	Extracellular vesicles
33	EVTs	Endovascular techniques
1; 2; 9	FA	Fatty acid
17; 35	FACS	Fluorescence-activated cell sorting
2	FAO	Fatty acid oxidation
2	FASN	Fatty acid synthase
26	FATE	The Firefighters and Their Endothelium
24; 29	FBN1	Fibrillin-1
23	FBs	Fibroblasts
34	FCS	Amilial chylomicronemia syndrome
31; 32	FDA	Food and Drug Administration
14; 30	FGF	Fibroblast growth factor

Chapter	Abbreviations	Term
1	FGF2	Fibroblast growth factor 2
9	FGF9	Fibroblast growth factor 9
1	FLT4	Fms-related receptor tyrosine kinase 4
31	FMC	Fibrin monomer complex
16; 26; 36	FMD	Flow-mediated dilation
17; 35	FOX	Forkhead box
1	FOXC2	Forkhead box protein C2
1	FOXO1	Forkhead box protein O1
9	FoxO1	Forkhead box protein O1
30	FP	Femoropopliteal
37	FTU	Functional tissue unit
7	GAGs	Glycosaminoglycans
34	GalNAc	triantennary N-acetylgalactosamine
1	GATA2	GATA binding protein 2
17	GBM	Glomerular basement membrane
32	GCV	Ganciclovir
2	GEM	Genome-scale metabolic model
4	GEO	Gene expression omnibus
10; 35	GFAP	Glial fibrillary acidic protein
17; 37	GFB	Glomerular filtration barrier
26	GHO	Global Health Observatory
1	Gja	Gap junction protein alpha
6	GK	Goto-Kakizaki
25	GLA	Generalized lymphatic anomaly
35	Glul	Glutamate-ammonia ligase
1; 9	GLUT1	Glucose transporter 1
1	GLUT3	Glucose transporter 3
26	GMP	Guanosine monophosphate
31	GP1a	Glycoprotein 1b
4	GPCRs	G-protein-coupled receptors
32	GPR78	Glucose-regulated protein
13	GPU	Graphics processing unit
17	gRECs	Glomerular RECs
25	GSD	Gorham–Stout disease
29	GWAS	Genome wide association studies
2	H1N1	Influenza A virus subtype H1N1
16; 27	H2O2	Hydrogen peroxide
6; 27	H2S	Hydrogen sulfide

(Continued)

Chapter	Abbreviations	Term
7	HA	Hyaluronic acid
34	hATTR	Hereditary transthyretin-mediated
1	HDAC	Histone deacetylase
1	HDAC3	Histone deacetylase 3
28; 34	HDL	High-density lipoprotein
34	HeFH	Hypercholesterolemia
2; 20	HEV	High-endothelial venule
1	HEY2	Hairy/enhancer-of-split related with YRPW motif protein 2
34	HFpEF	Heart failure with preserved ejection fraction
28	HFSCs	Hair follicle stem cells
30	HgbA1c	Hemoglobin A1c
14; 30	HGF	Hepatocyte growth factor
37	HHS	Human Health Services
1; 20; 21; 22	HHT	Hereditary hemorrhagic telangiectasia
30	HICs	High-income countries
1; 11; 14;	HIF	Hypoxia-inducible factor
36	HIV	Human immunodeficiency virus
1	HK	Hexokinase
21	HLTS	Hypotrichosis-lymphedema-telangiectasia syndrome
31	HMGB	High-mobility group box
36	HNR	Heinz Nixdorf Recall
34	HoFH	Homozygous FH
1	HOTAIR	HOX antisense intergenic RNA
2	hPSCs	Human pluripotent stem cells
36	hsCRP	High sensitivity C-reactive protein
7	HSP	Heparan sulfate
1	HSPC	Hematopoietic stem and progenitor cells
14	HSS	Hypoxia serum starvation
32	HSVtk	Herpes simplex virus thymidine kinase
24; 29	HTAD	Hereditable thoracic aortic disease
37	HuBMAP	Human Biomolecular Atlas Program
2	HUVECs	Human umbilical vein endothelial cells
30	IC	Intermitten claudication
2; 10; 31	ICAM	Intercellular cell adhesion molecule
7; 8	IEL	Internal elastic lamina
9	IGF1R	Insulin-like growth factor 1 receptor
21	IH	Infantile hemangiomas
32	IHC	Immunohistochemical

Chapter	Abbreviations	Term
6; 31	IL	Interleukin
5	ILCs	Innate lymphoid cells
28	ILK	Integrin-linked kinase
24	ILT	Intraluminal thrombus
2	IMECs	Immunomodulatory ECs
10	iNOS	Inducible nitric oxide synthase
28	IOP	Intraocular pressure
26	IP3	Inositol trisphosphate
2	IPF	Idiopathic pulmonary fibrosis
9	IR	Insulin receptor
33	IR	Indirect revascularization
26	IRAG	Inositol trisphosphate (IP3)-receptor-associated G-kinase substrate
10	IRBP	Retinoid-binding protein
27	IRF	Interferon-regulatory factor
3	Isl1	Islet1
27	IsoLGs	Isolevuglandins
21; 25	ISSVA	International Society for the Study of Vascular Anomalies
31	IVC	Inferior vena cava
1	JAG1	Jagged1
36	JUPITER	Justification for the Use of Statins in Primary Prevention: An Intervention Trial Evaluating Rosuvastatin
16	K + IR	Potassium channels
26	KATP	ATP-dependent potassium channels
16	KCa	Potassium
25	KHE	Kaposiform hemangioendothelioma
25	KLA	Kaposiform lymphangiomatosis
1; 12; 27	KLF2	Kruppel-like factor 2
3	KLF4	Krüpple-like factor 4
7; 21	KS	Keratan sulfate
25	KTS	Klippel–Trenaunay syndrome
25	LAM	Lymphangioleiomyomatosis
10	Lama2	Laminin α2 chain gene
3	LCA	Left subclavian artery
34; 36	LDL	Low-density lipoprotein
29 34	LDLR	Low-density lipoprotein receptor
24; 29	LDS	Loeys–Dietz syndrome
2	LEC	Lymph EC
11; 17; 25	LECs	Lymphatic endothelial cells

(Continued)

Chapter	Abbreviations	Term
11; 25	LMC	Lymphatic muscle cell
30	LMIC	Low/middle-income country
25	LMs	Congenital lymphatic malformations
1	lncRNA	Long noncoding RNA
25	LNs	Lymph nodes
24	LOX	Lysyl oxidase
9; 34	LPL	Lipoprotein lipase
17	LPM	Lateral plate mesoderm
11	LTB4	Leukotriene B4
25	LVs	Lymphatic valves
25	LVVs	Lymphovenous valves
1; 25; 28	LYVE1	Lymphatic vessel endothelial hyaluronan receptor 1
4	LZ	Leucine zipper
4	MA	Mesenteric artery
34	mAbs	Monoclonal antibodies
30	MACE	Major adverse cardiac event
1	Mad	Mothers against decapentaplegic
30	MALE	Major adverse limb event
26	MAOs	Monoamine oxidases
14; 25	MAPK	Mitogen-activated protein kinase
36	MDCS	Malmo Diet and Cancer Study
36	MDCT	Multidetector computed tomography
17	ME	Median eminence
1	MEG3	Maternally expressed 3
24; 36	MESA	Multi-Ethnic Study of Atherosclerosis
24; 29;	MFS	Marfan syndrome
24	MGP	Matrix glia protein
9	MHC	Major histocompatibility complex
28; 34; 36	MI	Myocardial infarction
1	MiR	microRNA
37	ML	Machine learning
4; 27; 29	MLCK	Myosin light chain kinase
27	MLCP	MLC phosphatase
1; 8; 14; 18; 23; 24; 31	MMP	Matrix metalloprotease
5	MMP9	Matrix metalloproteinase-9
31	Mo/Mø	Monocyte/macrophage
5	MoDC	Monocyte-derived dendritic cell
4	MP	Myosin phosphatase
26	mPTP	Mitochondrial permeability transition pore

Chapter	Abbreviations	Term
21	MPV	Median porencephalic vein
19	MR	Magnetic resonance
30	MRA	Magnetic resonance angiography
17	mRECs	Medullary RECs
5	mregDCs	Immunoregulatory molecules
10; 16; 26	MRI	Magnetic resonance imaging
34	mRNAs	Messenger RNAs
24	MRTFA	Myocardin-related transcription factor A
31	MRV	Magnetic resonance venography
10	MS	Multiple sclerosis
3	MSC	Mesenchymal stem cell
35	MudPIT	Multidimentional Protein Identification Technology
30	MVD	Microvascular disease
4	Mypt	Myosin phosphatase targeting subunit 1
24	MYH11	Myosin heavy chain 11
4	Myl6	Myosin light chain
4	MYLK	Myosin kinase
24	MYLK	Myosin light chain kinase
3	Myocd	Myocardin
4	NCBI	National Center for Biotechnology Information
26	NCDs	Noncommunicable diseases
36	NCEP	National Cholesterol Education Program
34	ncRNAs	Noncoding RNAs
27	NCX	Sodium–calcium exchanger
13	ND	Neutral density
4	NE	Norepinephrine
31	NETs	Neutrophil extracellular traps
27	NFAT	Nuclear factor of activated T-cells
17	NGS	Next-generation sequencing
36	NH	Non-Hispanic
36	NHANES	Nutrition Examination Survey
37	NHLBI	National Heart Lung and Blood Institute
2; 22	NICD	Notch intracellular domain
21; 37	NIH	National Institutes of Health
5	NK	Natural Killer
5	NKT	Natural killer T
36	NLA	National Lipid Association

(*Continued*)

Chapter	Abbreviations	Term
27	NLRs	Leucine-rich repeat-containing protein receptors
26	NMD	Nitroglycerin
1; 6; 9; 16; 23; 27; 30; 36	NO	Nitric oxide
1	Nos3	Nitric oxide synthase 3
16	NOx	NADPH oxidase catalytic subunit (isoforms 1, 2, and 4)
26	NOX2	Phagocytic
9	NRG1	Neuregulin-1
1; 21; 22	NRP	Neuropilin
9; 10; 28	NRP1	Neuropilin-1
25	NRP2	Coreceptor neuropilin 2
2	NSCLC	Nonsmall cell lung cancer
2	NSIP	Nonspecific interstitial pneumonia
4	nt	Nucleotide
26	NTG	Nitroglycerin
10; 35	NVU	Neurovascular unit
16	O2	Oxygen
2	OE	Overexpression
10; 35	OGD	Oxygen and glucose deprivation
10	OIR	Oxygen-induced retinopathy
34	OLE	Open-label extension
26	ONOO−	Intermediate peroxynitrite
5	OSEs	Oxidation-specific epitopes
1	OSS	Oscillatory shear stress
2	OXPHOS	Oxidative phosphorylation
5	oxPL	Oxidized phospholipid
4	P2R	Purinergic receptors
14; 0; 36	PAD	Peripheral arterial disease
31	PAI	Plaminogen activator inhibitor
13	PAM	Photoacoustic microscopy
27	PAME	Methyl palmitate
27	PAMPS	Pathogen-associated molecular patterns
22	PAR3	Partitioning defective 3 homologue
16; 26	PAT	Peripheral arterial tonometry
2	PBECs	Peri-bronchial ECs
34	PCSK9	Proprotein convertase subtilisin/kexin type 9
26	PDE	Partial differential equation
22	Pdpn	Podoplanin
3	PE	Proepicardium

Chapter	Abbreviations	Term
26	PEG	Polyethylene glycol
28	PF4/CXCL4	Platelet factor 4
33	PFFRs	Peripheral fractional flow resistances
1	PFKFB3	6-Phosphofructo-2-kinase/fructose-2,6-bisphosphatase 3
35	PG	Perineural glia
11	PGD	Primary graft dysfunction
9	PGF	Placental growth factor
6; 16; 26	PGI2	Prostacyclin
26	PGI2S	Prostacyclin synthase
8	PH	Pulmonary hypertension
1	PI3K	Phosphoinositide-3 kinase
27	PKA	Protein kinase A
26; 27; 35	PKC	Protein kinase C
4; 27	PKG	Protein kinase G
2	PKM2	M2 isoform of pyruvate kinase
16; 27	PLC	Phospholipase C
26	PLM	Passive leg movement
10	pMCAO	Permanent middle cerebral artery occlusion
31	PMN	Polymorphonuclear leukocyte, neutrophil
31	PPV	Positive predictive value
4; 24	PRKG1	Protein kinase G 1
4	PRKGIa	cGMP-dependent protein kinase
25	PROS	PIK3CA-related overgrowth syndromes
1	PROX1	Prospero-related homeobox 1
27	PRRs	Pattern recognition receptors
11	PTAs	Peripheral tissue antigens
35	Ptgds	Prostaglandin d synthase
6; 8	PVAT	Perivascular adipose tissue
25	PWS	Parkes Weber syndrome
16; 24; 26	PWV	Pulse wave velocity
17	PXM	Paraxial mesoderm
14	QSP	Quantitative systems pharmacology
29	R149C/+	Resistance of Acta2
1	RA	Retinoic acid
36	RA	Rheumatoid arthritis
26	RAAS	Renin–Angiotensin–Aldosterone System
24; 30	RAGE	Receptor for advanced glycation end products
25	RASA1	RAS P21 Protein Activator 1

(Continued)

Chapter	Abbreviations	Term
15	RBCs	Red blood cells
11	RCT	Randomized controlled trial
17	RECs	Renal endothelial cells
2	rESCs	Resident endothelial stem cells
9	RGS4	'Regulator of G protein signaling 4'
34	RISC	RNA-induced silencing complex
29	RLC	Regulatory light chain
31	rNAP5	Recombinant nematode anticoagulant peptide
4	RNASeq	Single cell RNA sequencing
27	RNS	Reactive nitrogen species
27	ROC	Receptor-operated channels
27	ROCK	Rho-associated protein kinase
10	ROP	Retinopathy of prematurity
4; 14; 24; 26; 27	ROS	Reactive oxygen species
17	RPE	Retinal pigment epithelium
5	RTMs	Resident tissue macrophages
1	RUNX1	Rant-related transcription factor 1
8	S1P	Sphingosine 1-phosphate
33	SA	Angiosome-targeted arteries
2	SADC	Solid lung adenocarcinoma
13	SAM	Scanning acoustic microscopy
28	SC	Subcutaneous
3	Sca1/Ly6a	Stem cell antigen-1
32	scFv	Single chain fragmen
3	Sclerotome	Somitic paraxial mesoderm
2	scRNAseq	Single-cell RNA sequencing
4	scRNAseq	Single cell RNAseq
27	SERCA	Sarcoplasmic/endoplasmic reticulum Ca2+-ATPase
36	SES	Socioeconomic status
26; 27	sGC	Soluble guanylate cyclase
30	SGLT2i	Sodium-glucose transporter 2 inhibitor
3	SHF	Second heart field
1	SHH	Sonic hedgehog
8	Shh	*Sonic hedgehog*
27; 35	SHR	Spontaneously hypertensive rats
35	Silac	Stable isotope labeling with aminoacids in cell culture
34	siRNAs	Small interfering RNAs
5	Sirpα	Signal regulatory protein α
1	SIRT1	Silent mating type information regulation 2 homolog 1

Chapter	Abbreviations	Term
4	sk	Skeletal muscle
36	SLE	Systemic lupus erythematosus
4; 29	SM	Smooth muscle
1	SMAD	*C. elegans* gene Sma
24	SMAD3	Decapentaplegic drosophila homolog 3
4; 7; 8; 11; 23; 29	SMCs	Smooth muscle cell
4	SMcKO	Smooth muscle conditional knock out
29	SMDS	Smooth muscle dysfunction syndrome
4	SMMHC	Smooth muscle myosin heavy chain
4	SMTN	Smoothelin
4	SMTNL1	Smoothelin-like 1
13	SNR	Signal-to-noise ratio
13	sO2	Oxygen saturation of hemoglobin
27	SOC	Store-operated channels
16	SOD	Superoxide dismutase
1	SOX	SRY-box transcription factor
35	Sox	SRY-Box
32	SPARTA	Selection of phage-displayed accessible recombinant targeted antibodies
35	SPG	Subperineural glia
35	Spock2	Osteonectin
30	SPP	Segmental perfusion pressures
33	SPP	Skin perfusion pressure
4	SRA	Small resistance artery
3; 24	SRF	Serum response factor
36	SSBs	Sugar-sweetened beverages
12	SSREs	Shear stress responsive elements
29	STAD	Sporadic thoracic aortic disease
36	suPAR	Soluble urokinase plasminogen activator receptor
8	SVAS	Supravalvular aortic stenosis
19	SVM	Support vector machine
26	T2DM	Type 2 diabetes mellitus
29	TAAD	Thoracic aortic aneurysm and dissections
24	TAD	Thoracic aortic disease
24	TAZ	Transcriptional coactivator with PDZ-binding motif
30	TBI	Toe-brachial index
35	Tbx1	T-box transcription factor

(*Continued*)

Chapter	Abbreviations	Term
30	TC	Total cholesterol
1	TCA	Tricarboxylic acid
30	TcPO2	Transcutaneous oximetry
2	TECs	Tumor ECs
17	TECs	Transendothelial channels
3	TET2	Ten-eleven translocation-2
32	Tf	Transferrin
3; 24	TFs	Transcription factors
34	TGs	Triglycerides
8	TGF	Transforming growth factor
29	TGFB2	TGF-β2
10	Tgfbr2	TGF-β receptor 2
24	TGFβ	Transforming growth factor beta
22	TGF-β	Transforming growth factor β
9	TGFβ1	Transforming growth factor beta 1
24	TGFβR1	TGFβ-receptor type 1
24	TGFβR2	TGFβ-receptor type 2
10	Th1	T-helper 1
10	Th17	T-helper 17
12	TIMP3	Tissue inhibitor of metalloproteinase 3
24	TIMPs	Tissue inhibitors of metalloproteinases
29	TIPS	Training Interdisciplinary Pharmacology Scientists
10; 17	TJs	Tight junctions
27; 31	TLR	Toll-like receptor
28	TM	Trabecular meshwork
10	tMCAO	Transient middle cerebral artery occlusion
30	TP	Tibiopedal
4	Tpm1	Tropomyosin-alpha
5; 27	Treg	Regulatory T cell
5	TRM	Tissue resident memory
27	TRP	Transient receptor potential cation
22	TSAd	T cell-specific adapter
25	TSC	Tumor suppressors tuberous sclerosis
14	TSP1	Thrombospondin 1
34	TTR	Transthyretin
24	UHRF1	Ubiquitin like with PHD and ring finger domains 1
2	UIP	Usual interstitial pneumonia
26	VASP	Vasodilator-stimulated phosphoprotein

Chapter	Abbreviations	Term
10	vBM	Vascular basement membrane
10	VCAM	Vascular cell adhesion molecule-1
17	VCCF	Vascular Common Coordinate Framework
35	VEF	Vascular endothelial function
1; 8; 10; 17; 21; 22; 30; 32	VEGF	Vascular endothelial growth factor
28	VEGFR1	Vascular endothelial growth factor receptor-1
9; 10	VEGFR2	Vascular endothelial growth factor receptor-2
25	VEGFR3	Vascular endothelial growth factor receptor-3
19	VesSAP	Vessel Segmentation and Analysis Pipeline
33	VFR	Vascular flow reserve
21	VGAM	Vein of Galen aneurysmal malformations
5	VIP	Vasoactive peptides
10	VLA	Very Late Activation Antigen
34; 36	VLDL	Very-low-density lipoprotein
27	VOC	Voltage-operated
17	VPs	Vascular progenitors
4; 7	VSM	Vascular smooth muscle
3; 4; 5; 6; 10; 24; 27; 35	VSMC	Vascular smooth muscle cell
31	VT	Venous thrombosis
31	VTE	Venous thromboembolism
17	VVO	Vesiculo-vacuolar organelles
1; 31	vWF	von Willebrand factor
27; 32	WAT	White adipose tissue
8	WBS	Williams–Beuren syndrome
33	WDR	Wound-directed revascularization
30	WIfI	Wound, ischemia, foot infection
35	WKY	Wistar–Kyoto rats
1	Wnt	Wingless-related integration site
26	XDH	Xanthine dehydrogenase
26	XO	Xanthine oxidase
24	YAP	Yes associated protein

PART I

The architectural design of the vasculome: basis for system-wide functional unity and local diversity

SECTION 1

The endothelium: key unifying principle of the vasculome. Basis for systemic unity and for engineering of local specialization

CHAPTER

1

Development of, and environmental impact on, endothelial cell diversity

Bipul R. Acharya[1,2,3,a], *Nicholas W. Chavkin*[1,2,a] *and Karen K. Hirschi*[1,2,4]

[1]Department of Cell Biology, University of Virginia School of Medicine, Charlottesville, VA, United States
[2]Cardiovascular Research Center, University of Virginia School of Medicine, Charlottesville, VA, United States
[3]Wellcome Centre for Cell-Matrix Research, Faculty of Biology, Medicine and Health, University of Manchester, Manchester, United Kingdom [4]Departments of Medicine and Genetics, Yale University School of Medicine, Yale Cardiovascular Research Center, New Haven, CT, United States

Introduction

The de novo emergence of a closed-loop blood circulatory system composed of the contractile heart and arterial, venous, and capillary vessels, which transport nutrients and oxygen to growing organs and remove metabolic waste products, is critical for embryonic development. The developing lymphatic system collaborates with the cardiovascular system to maintain fluid and immune homeostasis via the transport of lymph fluid unidirectionally from tissue capillaries into the venous circulation through the subclavian vein.[1,2] The luminal surface of all blood and lymphatic vessels in these networks is lined by endothelial cells (ECs), which enable selective permeability of macromolecules and immune cells, and also confer adaptability of the vessels to cope with tissue-specific microenvironments. The specification of these different phenotypes during development establishes the overall heterogeneity of ECs that exhibit distinct morphology, gene expression, and cellular function.[3] Additionally, organ-specific EC phenotypes, especially in capillaries, initiate during organogenesis and are maintained postnatally to enable organ-specific vascular functions[4] (depicted in Fig. 1.1).

Initial development and maintenance of EC diversity and heterogeneity is promoted by extrinsic factors including signaling molecules, hemodynamic forces, oxygen availability, and nutrient sources. This collective physio-chemical microenvironment contributes to the establishment of specific EC phenotypes during organogenesis and in organ-specific vasculature. Disruptions in any of these regulatory processes can lead to impaired acquisition of EC identity, defects in vascular function, and disease progression.[5]

In this chapter, we briefly highlight mechanisms of EC specification (reviewed in depth in Ref. 6), review EC diversity in adult organs, and discuss the impact of hemodynamic forces, hypoxia, metabolic state, epigenetic regulators, and noncoding RNAs on the regulation of EC diversity during organogenesis and in adult tissues. Such insights can further our understanding of the broad range of vascular functions in the body and reveal therapeutic targets for specific EC subtypes in different vascular diseases.

Diversity of endothelial cell types

EC differentiation is an early and critical step during organogenesis. Blood endothelial precursor cells (angioblasts) are derived from mesodermal progenitors and form primitive vascular plexi in a process referred to as vasculogenesis. Plexi are then remodeled into vascular networks made up of arterial, venous, and capillary ECs via angiogenesis. During lymphatic development, lymphatic ECs can be derived from a subset of ECs in the cardinal vein,[7] ECs in other venous vessels,[8,9] and mesenchymal progenitors.[10,11]

[a] Authors contributed equally.

The Vasculome
https://doi.org/10.1016/B978-0-12-822546-2.00013-7

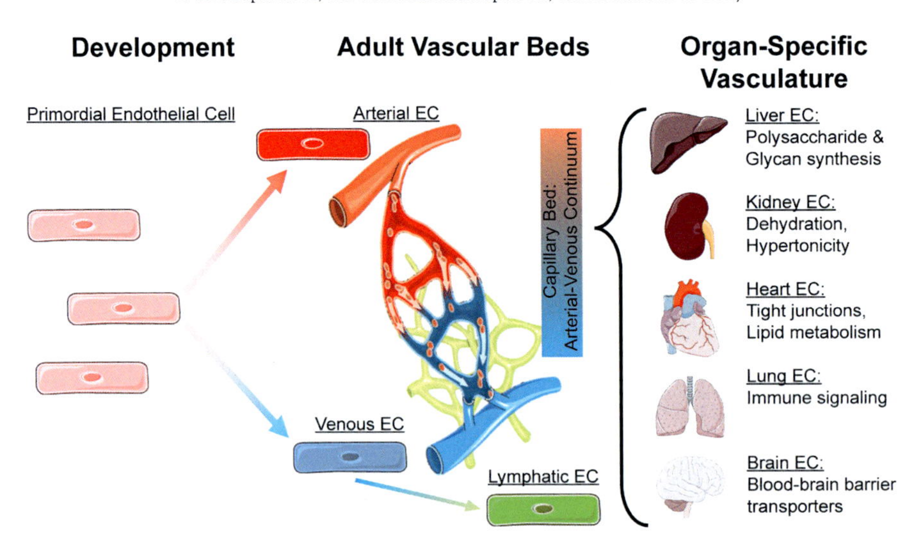

FIGURE 1.1 **Diversity of endothelial cell phenotypes.** During development, primordial ECs become specified toward blood and lymphatic phenotypes. Adult vascular beds contain arterial, venous, and lymphatic ECs, with capillary ECs exhibiting intermediate phenotypes among an arterial–venous continuum. Different organs contain specialized capillary ECs that enable organ vascular-specific functions.

These specialized ECs are structurally and functionally different. For example, elongated arterial ECs have tighter intracellular junctions than short and wide venous ECs that form valves, and both EC types have distinct inflammatory and angiocrine responses.[12] Capillary ECs have unique coagulation and permeability properties,[13,14] relative to arterial and venous ECs. Lymphatic ECs possess specialized discontinuous cell junctions for continuous fluid flow and immune cell influx.[15]

A small subset of ECs can also give rise to other cell types; for example, during development, ECs in the extraembryonic yolk sac and placenta, and the embryonic aorta–gonad–mesonephros (AGM) region, undergo endothelial-to-hematopoietic transition (EHT) and give rise to hematopoietic stem and progenitor cells (HSPCs) (reviewed in Ref. 16). Additionally, during cardiac development, ECs in the endocardium go through endothelial-to-mesenchymal transition (EndMT), resulting in a mesenchymal phenotype that contributes to structural and vascular mural cell types (pericytes, vascular smooth muscle cells) in the heart.[17,18] All of these EC specializations are promoted by a variety of extrinsic signals that we discuss throughout.

Arterial–venous specification

In the early stage of vascular plexus remodeling, hypoxic areas of developing organs, that require more oxygen and nutrients, exhibit increased vascular endothelial growth factor-A (VEGF-A), which promotes angiogenesis.[19,20] During angiogenesis, ECs transit through intermediate tip- and stalk-cell phenotypes.[21] Endothelial tip cells migrate toward VEGF-A,[22] while endothelial stalk cells support lumen formation behind the migrating front.[23,24] Tip- and stalk-cell identity and function are controlled through a balance of Notch signaling between the two cell types mediated by Notch receptor ligands jagged1 (JAG1) and delta-like ligand 4 (DLL4). JAG1 is proangiogenic and enhances tip- and stalk-cell phenotypes and DLL4 is antiangiogenic and promotes EC maturation.[25,26] As ECs are migrating into tissues and primitive blood vessels are forming, ECs become specified toward arterial and venous phenotypes that make up the circulatory network.[27]

Arterial specification is promoted by Notch signaling which induces the activity of transcription factors SOX9 and SOX17,[28] leading to expression of arterial-enriched genes (i.e., *Efnb2, Gja4, Gja5, Hey2, Nrp1*).[29–32] Notch signaling can be activated downstream of VEGF-A and SHH signaling.[33] Arterial shear flow forces also activate Notch signaling and this leads to upregulation of connexin (Cx)37 and downstream cell cycle inhibitor p27, to promote growth arrest, which enables the upregulation of arterial genes.[34] The transforming growth factor/bone morphogenic protein (TGF-β/BMP) signaling pathways cross-talk with Notch signaling and also play critical roles in arterial EC specification.[35] For example, Notch-induced activation of SMAD6 inhibits BMP2- and BMP6-induced angiogenic sprouting of venous ECs,[26] and BMP9 ligand induces the arterial-specific gene *Efnb2*.[36] Additionally, downstream of these pathways, tip-cells may undergo arterial specification and become part of the developing arterial vessels.[37,38]

Venous specification and the expression of venous EC-enriched genes (i.e., *Ephb4, Nrp2, Emcn*)[29] is promoted by the PI3K/AKT signaling pathway,[39] BMP signaling and downstream SMAD transcriptional regulators,[40] as well as transcription factor COUP-TFII, which inhibits Notch signaling and arterial identity.[41,42] A recent study investigating enhancer regions of the *Ephb4* gene show that its transcription is promoted by the activation of SMAD1/5.[40] However, the transcriptional regulation of venous EC-specific genes, overall, is not well understood.

Importantly, ECs in a mature artery–capillary–vein circulatory network are now known to exist in a continuum of phenotypes between arterial and venous identities.[41,43] Therefore, signaling pathways that promote arterial versus venous phenotypes must be intricately balanced throughout the networks. These may include TGF-β/BMP signaling, which is found to be disrupted in patients with Hereditary Hemorrhagic Telangiectasia that exhibit arterial–venous malformations.[44] Indeed, transgenic animal models have shown that the TGF-β and BMP signaling are required for proper arterial–venous specification, and disruption of these pathways leads to arterial–venous malformations.[45–50] Additionally, signaling that promotes arterial–venous specification must continue after mature blood vessels have formed, as ECs maintain plasticity in adult organs and loss of specification signaling leads to dedifferentiation.[51,52]

Hematopoietic and mesenchymal transition

During development, a small subset of blood ECs in the yolk sac, placenta, and AGM become specified as hemogenic ECs that undergo EHT and give rise to HSPC (reviewed in Ref. 16). These HSPCs then migrate to the fetal liver, and then to fetal bone marrow, where they remain postnatally and give rise to all blood cells throughout life.[53] Hemogenic EC specification is initiated via retinoic acid (RA) signaling in the yolk sac[54] and the AGM,[55] and requires downstream regulation of cKit, Notch, and p27.[54] Recent studies in human embryonic stem cells revealed that RA and EC cell cycle control are also necessary to enable the upregulation of hematopoietic genes and downregulation of EC genes during human EHT.[56] Blood shear forces also play a role in hemogenic specification, and definitive hematopoiesis occurs during the same time frame in embryogenesis as the onset of cardiac contraction.[57,58]

During cardiac development, a subset of ECs can undergo EndMT and give rise to mesenchymal cells that contribute to structural and mural cell types (pericytes, vascular smooth muscle cells) of the heart.[17] Additionally, the mitral valve, the tricuspid valve, and the ventral septum of the heart are formed by invading cells of endothelial origin that undergo EndMT and contribute to the valve interstitial cells.[59,60] Thus, ECs and their diverse subtypes are not only essential for blood and lymphatic vascular development, but also hematopoietic and cardiac development, as well.

Lymphatic specification

Lymphatic circulation is essential for transporting extracellular fluids, proteins, and cells from the interstitial space of organs to blood vessels for excretion or recycling. To form the lymphatic vasculature during organogenesis, a subset of ECs in the cardinal vein undergoes specification toward a lymphatic EC phenotype.[7] Lymphatic ECs can also be derived from other venous vessels[8,9] and mesenchymal progenitors.[10,11] The process of lymphatic specification is initiated via COUP-TFII- and SOX18-mediated upregulation of transcription factor PROX1,[61–64] which is a master regulator of lymphatic EC development.[65,66] PROX1 induces the expression of FLT4/VEGFR3, allowing lymphatic ECs to respond to VEGF-C that promotes their proliferation and migration.[67] Further lymphatic specification occurs with induction of LYVE-1, which is a lymphatic-specific hyaluronan receptor that plays a role in the inflammatory response, in part, by binding to immune cells to assist cell translocation through lymph vessels.[68–70] Like veins, lymphatic vessels form valves, which is achieved via an interplay between mechanotransduction, AKT, and EPHB4 signaling,[71–73] and requires expression of Cx37 and Cx43.[74] Collectively, the specification of blood and lymphatic ECs enables vascular systems to function coordinately to maintain tissue survival, fluid balance, and immune surveillance.

Endothelial cell diversity within adult organs

ECs exhibit organism-wide morphological and functional heterogeneity. Although every adult organ contains each EC phenotype (arterial, venous, capillary, lymphatic), different organs possess distinct phenotypes among those subtypes. For example, ECs of the brain and heart turnover at a slow rate and highly express tight junction proteins, which enables stable barrier protection.[75,76] In contrast, liver and kidney ECs have a higher proliferative rate,[76] and exhibit irregularly sized, convoluted, and intercalated junctions that allow passage and filtration of large particles.[4] Thus, these organ-specific characteristics enable diverse organ-specific EC functions.

Recent advanced technologies have further elucidated the EC diversity within and between organs. For example, recent studies using single-cell analyses have highlighted the broad diversity among vascular cells within organs, such as the brain, and also revealed that they exhibit a continuum of phenotypes from arterial to capillary to venous, as previously mentioned,[41,43] and reviewed in Ref. 77. A broader analysis of ECs in multiple mouse organs confirmed the existence of an arterial–capillary–venous phenotypic continuum within vascular networks in different organs.[78] Establishing and maintaining this continuum may be critical for maintaining different functions of distinct regions of the arterial and venous branches. Disruption of endothelial arterial–venous specification results in disorganization of the vascular network and leads to arterial–venous malformations.[45–50] These studies have highlighted the high degree of EC phenotypic complexity within vascular networks within all organs.

Of note, specialization of capillary ECs seems to be the most important for enabling organ-specific vascular functions. For example, a single-cell analysis of many different mouse organs found that brain capillary ECs express transporters to maintain the blood–brain barrier, while kidney capillary ECs express genes that enable responsive to dehydration and hypertonicity.[78] Additionally, the same study found that capillary ECs make up 50%–85% of all ECs, and exhibit the greatest variation between organs; ECs from arteries, and veins had less variation between organs.[78] This multiorgan study also found enrichment of gene sets in capillary ECs associated with organ-specific functions. For instance, the main genes expressed in brain ECs include membrane transporters, liver ECs include glycan and polysaccharide synthesis, lung ECs include immune signaling, and muscle ECs (heart and skeletal) include lipid metabolism. Together, these single-cell analysis studies suggest that capillary ECs exhibit the largest phenotypic variation among different organs, and these variations are consistent with maintaining distinct vascular functions in each organ.

Adult organs also contain lymphatic vessels for draining fluid and substances from interstitial spaces. Variations in organ-specific lymphatic vessel function suggest that a similar level of diversity exists for lymphatic ECs.[79] The discovery of nonvenous origin(s) of lymphatic ECs that constitute the skin, mesentery, and heart lymphatics has insinuated the possibility that distinct embryonic origins of organ-specific lymphatic ECs may contribute to their organ-specific functions.[10,80,81] However, relatively fewer studies of lymphatic ECs have not yet discerned their heterogeneity within and between organs; thus, a more in-depth characterization of lymphatic ECs is needed.

Impact of environmental factors on endothelial cell specification

Initial specification of some primordial ECs in large vessels occurs before cardiac beating begins.[82] Further specification and maturation into distinct subtypes continues after, and is sometimes dependent on, the onset of blood flow.[34,83] Thus, the specification of ECs is shaped by their circumambient physio-chemical environment; that is, circulating blood, status of oxygen availability, and metabolic state influence genetic and epigenetic control of EC growth and identity (depicted in Fig. 1.2). Effects of environmental factors continue to maintain different EC phenotypes, given that vessel-specific ECs retain the ability to permute into different phenotypes under changing molecular environments.[42,84]

Hemodynamic forces

Various magnitudes and patterns of flow forces contribute to the diversity of EC phenotypes during development, as well as maintenance in adult tissues. Soon after the onset of vasculogenesis, initial cardiac contractions propel blood flow in the growing vessels and generate a variety of mechanical forces.[85] These include fluid shear stress from the frictional force exerted by the blood flow in parallel to the lumen surface, circumferential cyclic strain from the pulsatile stretch that dilates blood vessels in response to increased flow, and the hydrostatic pressure of the flowing blood. ECs sense these forces on the luminal surface of vessels and transduce them into biochemical signals through a mechanotransduction process that involves transmembrane receptors, junctional proteins, ion channels, primary cilia, caveolae, and glycocalyx.[86–88] Patterns of forces (laminar, pulsatile, and oscillatory) impose diverse magnitudes of shear stress on ECs in the capillary plexus (1–5 dyne/cm^2), and arterial (10–70 dyne/cm^2), venous, and lymphatic (1–6 dyne/cm^2) vessels.[89–91] Pulsatile hemodynamic flow exerts mechanical force on ECs by stretching the vessel (~5%–8%, physiological), which is often abnormal in diseases (~20%, hypertension) and impairs EC functions.[92] These different patterns of forces differentially affect EC phenotype.

In remodeling vessels, the role of hemodynamic force-regulated EC specification and maintenance has been investigated through elegant studies tracking the molecular markers of distinct EC fates. The transcription factor KLF2 has emerged as a likely candidate for relaying shear stress into downstream mechanotransduction signaling that determines EC fate in response to flow.[93,94] More specifically, laminar flow-mediated shear stress upregulates VEGF-A signaling in ECs in developing embryonic vasculature via KLF2, which modulates VEGFR2 and downstream genes.[95–97] As previously mentioned, shear stress at arterial levels (12–18 dyne/cm^2) also activates a Notch/Cx37/p27 signaling axis that induces G1 cell cycle arrest of ECs, which enables the upregulation of arterial genes.[31]

Less is known about hemodynamic force-regulated venous EC specification. Low oscillatory shear stress (OSS) and reduced pulsatility may inhibit Notch

Environmental Factors Influencing Endothelial Specification

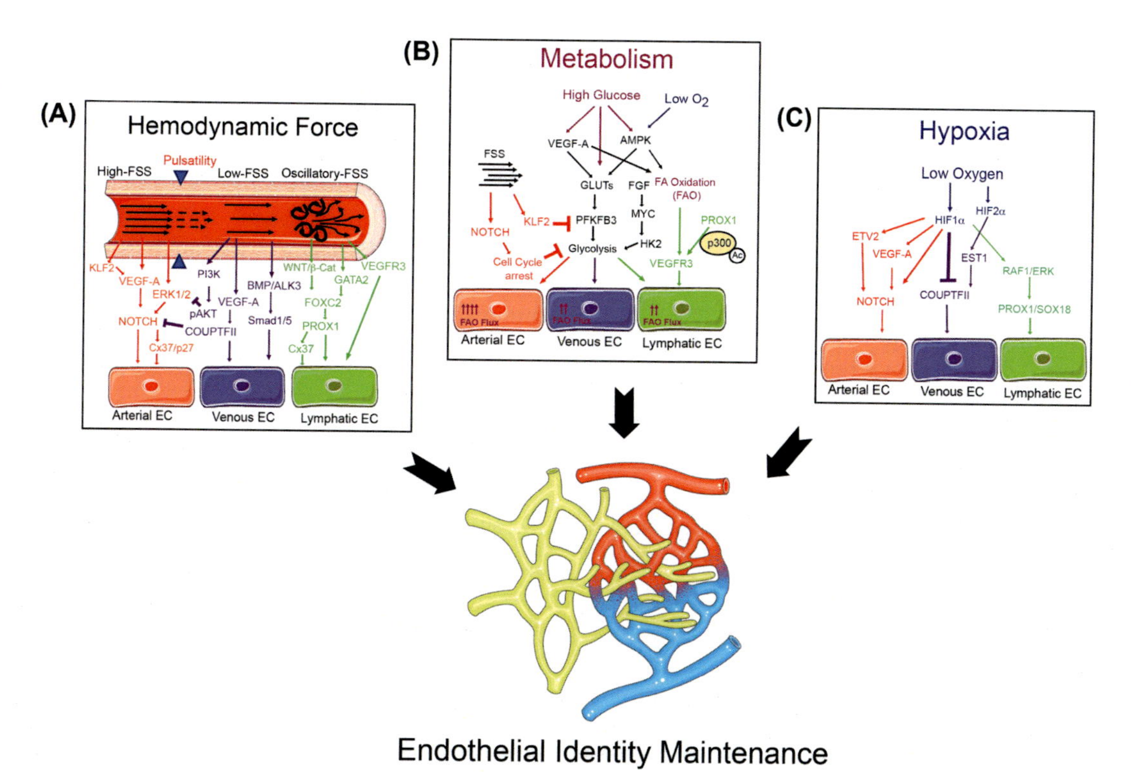

FIGURE 1.2 Environmental factors influencing EC specification and identity maintenance. (A) Hemodynamic force (high pulsatile/laminar) induces KLF2, VEGF-A, and Notch signaling for arterial specification. Low laminar flow induces VEGF-A-mediated COUPTFII expression and BMP/SMAD signaling for venous specification. Low oscillatory flow induces Wnt and VEGF-C-mediated PROX1 signaling for lymphatic specification. (B) High glucose induces VEGF-A-mediated glycolysis via GLUTs and PFKFB3 upregulation. VEGF-A and AMPK also induce fatty acid (FA) oxidation in developing ECs that upregulate VEGFR3/FGF/HK signaling axis for venous and lymphatic specification. Low oxygen and fluid shear stress also influence EC specification by regulating glycolysis and FA oxidation. (C) Hypoxia or low oxygen induces HIF1 expression that regulates downstream arterial-promoting Notch signaling, venous-promoting COUPTFII, and lymphatic-promoting PROX1.

signaling by either activating AKT downstream of PI3K signaling, or COUP-TFII downstream of VEGF-A signaling, resulting in further upregulation of COUP-TFII and subsequent venous specification.[39,42] As previously mentioned, recent studies identified BMP signaling through SMAD1/5 as a regulator of venous identity in embryonic ECs of mice and zebrafish, where signal perturbations result in aberrant venous formation, but have no effect on arterial identity.[40] Although shear stress can modulate these signaling pathways in ECs, whether hemodynamic forces induce these pathways to promote venous EC specification is still an open question.

OSS is a critical regulator of lymphatic EC patterning and lymphatic valve formation.[98] OSS activates Wnt/β-catenin signaling which, in turn, upregulates FOXC2 and Cx37. A parallel study showed that OSS activates GATA2, a transcription factor that promotes the FOXC2-mediated activation of the PROX1 transcriptional program that promotes lymphatic EC specification. In addition, interstitial fluid imposes hydrostatic pressure on ECs, and promotes their lymphatic specification via VEGFR3 and β1-integrin.[99,100]

Shear stress also modulates EHT; that is, in both mouse and zebrafish embryos, heartbeat initiation and ensuing blood flow upregulates RUNX1 expression, which is a critical transcriptional regulator of embryonic hematopoiesis.[101,102] Furthermore, in zebrafish, flow-induced KLF2/NO signaling, as well as mechanosensory primary cilia, promotes HSPC generation.[103,104] EndMT, which regulates cardiac valve formation in response to TGFβ signaling,[105] is also regulated by low OSS downstream of KLF2.[106] Although this area of research is emerging and requires further investigation, the highlighted studies illuminate that hemodynamic forces modulate endothelial specification to contribute to EC heterogeneity. Gaining further understanding of how ECs respond to hemodynamic mechanical stimuli, in concert with soluble signals, will help better define mechanisms underlying EC diversity.

Metabolic substrates

Vessel remodeling is a high energy-consuming process, and ECs must balance metabolic needs to meet elevated energetic and biosynthetic needs. Highly glycolytic primordial ECs rely on glycolysis for their adenosine triphosphate (ATP) production, since this process is faster than oxidative phosphorylation and keeps oxygen availability high for developing cells during organogenesis. Glycolysis supports ~85% of the energy requirement of blood vascular ECs and ~70% of lymphatic ECs.[107] The metabolic state of ECs is emerging as a determining factor for arterial–venous specialization. The best characterized metabolic pathways that impact vascular development are glycolysis, fatty acid (FA) oxidation, and glutamine metabolism.

The VEGF/VEGFR2 signaling axis stimulates EC glycolytic flux by increasing the expression of glucose transporter 1 (GLUT1), and by activating 6-phosphofructo-2-kinase/fructose-2,6-bisphosphatase 3 (PFKFB3)–driven glycolysis for the generation of ATP. Transcriptomic analysis of human umbilical arterial ECs showed increased expression of both GLUT1/GLUT3 along with arterial-specific genes.[108] Notch, an arterial-promoting signaling pathway, is able to arrest EC cell cycle progression by reducing PFKFB3 levels and their glycolytic flux, which may provide a window of opportunity for specification.[109] Significant metabolic heterogeneity exists among arterial and venous ECs in mature vessels, such as lower flux of FA oxidation in venous ECs compare to arterial ECs.[110,111]

In sprouting angiogenesis, metabolic pathways also promote EC specification into tip- and stalk cells. In tip cells, VEGFR2-induced glycolysis increases localized ATP generation that remodels lamellipodia and filopodia for tip-cell migration.[112] VEGFR3/FGF2 signaling axis also enhances EC glycolysis during angiogenesis by modulating glycolytic enzyme hexokinase 2.[113] Inhibition of glycolytic pathways impairs new vessel formation and tip-cell specification. FA metabolism controls stalk-cell proliferation by producing deoxynucleotide triphosphate (dNTP) biomass. Once FAs are imported into ECs, they are oxidized in the mitochondria and subsequently synthesized into dNTPs for DNA replication in proliferating stalk cells.[110] A gradient of VEGFA–Notch–FOXO1 signaling in ECs regulates the expression of FA binding proteins, which help ECs to import FAs from blood into the cytoplasm; deficiency of this protein impairs EC proliferation and sprouting.[114,115] Amino acid glutamine metabolism in the blood and ECs is also critical for tip- and stalk-cell dynamics during angiogenesis. Inhibition of glutamine metabolism causes EC malfunction and vessel sprouting defects.[116] In proliferating ECs, glutamine serves as an anaplerotic source of carbons to replenish the TCA cycle to support protein and nucleotide synthesis.[117]

Lymphatic EC specification from venous ECs is also metabolically regulated. During lymphangiogenesis, glycolysis supplies energy to the growing lymphatic vessels, while FA oxidation promotes nucleotide synthesis for EC proliferation.[107] Additionally, during lymphatic EC specification, PROX1 stimulates FA oxidation that promotes epigenetic changes to key lymphatic genes (i.e., *Flt4/Vegfr3*) in venous ECs to promote lymphatic specification.[118] Although this area needs further study, defining a detailed metabolic landscape of blood and lymphatic ECs will significantly enhance our understanding of vascular development.

Hypoxia

Vascularization in early embryonic tissues occurs in hypoxic conditions where $\leq$3% O_2 is available in the uterine environment.[119] Hypoxia-inducible factors (HIFs) are highly conserved and control the expression of genes needed for vasculogenesis and angiogenesis. For example, HIF-1α upregulates matrix metalloproteinases (MMPs) that degrade the extracellular matrix and facilitate tip-cell migration.[120] HIF-1α also induces the expression of VEGF-A, which promotes the survival and expansion of ECs.[121] High levels of VEGF-A are also known to promote arterial specification via SHH and Notch signaling.[33] Interestingly, even without VEGF-A, hypoxia can induce arterial EC specification by upregulating Notch signaling, DLL4 expression, and downstream HEY2 expression.[122] During lymphangiogenesis, VEGF-C/VEGFR3 signaling is also enhanced during hypoxia, and PROX1 expression is directly upregulated by HIF-1α/2α.[123–125] Thus, oxygen availability significantly contributes to blood and lymphatic EC development.

Epigenetic and noncoding RNA regulation of endothelial cell phenotype

Beyond gene regulation by transcription factors, epigenetic remodeling of chromatin can impact gene expression without alterations in DNA sequences. For example, DNA methylation and posttranslational modification of nucleosome–histones can limit chromatin accessibility to transcription factors and cofactors.[126] Epigenetic regulation during EC differentiation has been found to play an important role in vascular development. Promoters of EC-specific genes such as *Vwf*, *Nos3* (encoding eNOS), *Cdh5*, and *Vegfa* are epigenetically silenced by histone methylation, which contributes to the regulation of EC fate and function.[127] During endothelial differentiation, inhibition of DNA methyltransferases (DNMT1 and DNMT3A) promotes expression of arterial-specific genes, such as *Dll4, Efnb2, Hey2,* and *Notch1*, via E2F1-mediated transcriptional upregulation, thereby specifying an arterial fate.[128]

Histone acetylation also controls the expression of genes important for EC differentiation and specification, including *Vwf, Notch1,* and *Nos3*,[129] and histone deacetylase (HDAC) activity is critical for both embryonic and adult EC differentiation.[130] In mice, EC-specific knockout of HDAC7 leads to defective EC–EC and EC-mural cell contacts, resulting in vascular dilatation, rupture, and embryonic lethality.[131] SIRT1 deacetylase–mediated fine-tuning of Notch signaling in ECs is also required for normal angiogenic sprouting in mice and zebrafish.[132]

Venous EC specification is also subjected to epigenetic regulation; an ATP-dependent active histone modifier, Brahma-related gene 1, regulates COUP-TFII expression in developing veins.[133] Chromodomain helicase DNA-binding protein 4 and histone deacetylase 3 are essential for lymphatic EC integrity and lymphatic valve formation.[134,135] Histone acetyltransferase p300 was shown to promote lymphatic EC specification through the activation of lymphatic-specific genes. In mice, another H3K79 methyltransferase Dot1 epigenetically primes lymphatic gene expression in their EC progenitors during lymphatic development.[136]

Noncoding RNAs (miRNAs and lncRNAs) also induce posttranscriptional changes that modulate vascular development. miRNAs can target key molecular drivers of EC differentiation, specification, proliferation, and migration, including MMPs, HIF-1, cytokines, and VEGF-A.[137,138] MiR-214 and Mir-126 promote angiogenesis and EC differentiation by upregulating NRP1 and VEGFR2.[139,140] MiR-101, MiR-20a, and MiR-200b influence the expression of genes such as *Hif1, Vegfc, Vegfa, Angpt1,* and *Egln3*, and negatively regulate developmental EC specification and angiogenic programs.[141] MiR-221 regulates stalk-cell proliferation, while MiR-27b regulates venous and tip-cell specification.[142,143] In addition, miR-223 was recently found to control EC protein N-glycosylation that regulates EHT.[144] lncRNAs, such as H19, MEG3, and HOTAIR, regulate VEGF-A signaling in ECs and MANTIS regulates the expression of COUP-TFII, SOX18, and SMAD6, required for EC specification.[141,145] MiR-31 and MiR-181a regulate FOXC2, PROX1, and other lymphatic EC-specific drivers including VEGFR-3, and thus influence lymphatic lineage specification.[146,147] In sum, these studies suggest that epigenetic mechanisms play important roles in EC differentiation and specification during vascular development.

Epigenetic regulation of EC-specific genes is also governed by environmental factors. For example, laminar shear stress augments global histone methylation to promote mouse embryonic stem cell differentiation into ECs.[148] At the onset of blood flow, KLF2a induces MiR-126 expression in embryonic ECs, which subsequently activates VEGF-A signaling and promotes aortic arch sprouting angiogenesis.[149] Analogously, hypoxia induces a global pattern of decreased histone acetylation and increased HDAC activity.[150] More specifically, in hypoxic conditions, HDAC7 is transported back to the nucleus with HIF-1α where it upregulates the transcriptional activity of HIF-1α.[151] Hypoxia also induces eNOS expression in EC precursors by removing the repressive H3K27me3 and DNA methylation marks on eNOS promoters, consequently enhancing EC differentiation.[126]

Various metabolic states can also confer variations in epigenetic modification patterns of specific genes in ECs. For example, ECs exposed to hyperglycemia increase p300 expression, which upregulates the expression of endothelin-1 and VEGF-A by inducing histone acetylation of their promoters.[152] In human aortic ECs, histone methyltransferase SET7 significantly increases p65 subunit of NF-kappa B under acute hyperglycemic exposure; this might influence atherogenic fates of EC in diabetic aorta.[153] Future studies are needed to understand how lipid and amino acid metabolism modulates epigenetic regulators of EC differentiation and specification.

Summary and conclusions

ECs acquire diverse phenotypes during organogenesis, which is controlled by a broad range of signaling pathways and environmental factors. This EC heterogeneity is necessary for the formation of cardiovascular and lymphatic networks, their further specialization to enable organ-specific functions, as well as hematopoietic and cardiac development.

EC diversity also appears to be critical for vascular responses postnatally, as clonal expansion of subsets of ECs has been observed in response to vascular injury and disease. For example, a small number of vessel-resident EC progenitors clonally expand after vascular injury and promote vessel regeneration.[154] Clonal expansion of ECs also contributes to the formation of cerebral cavernous malformations.[155,156] In addition, subsets of ECs undergo EndMT and contribute to the progression of atherosclerosis and fibrotic diseases. Interestingly, clonal expansion of a subset of regenerative ECs seems to be restricted to vascular disease response, as normal vascular development does not show clonal properties and instead shows random distribution of lineage-marked ECs.[157]

Although our understanding of EC heterogeneity and its regulation are still limited, gaining a better understanding of the molecular mechanisms that govern EC diversity and fate transitions will reveal exciting new mechanisms of vascular development and regeneration, and may enable development of therapies to treat vascular pathologies where these processes have gone awry.

References

1. Liao JK. Linking endothelial dysfunction with endothelial cell activation. *J Clin Invest*. 2013;123(2):540–541.
2. Semo J, Nicenboim J, Yaniv K. Development of the lymphatic system: new questions and paradigms. *Development*. 2016;143(6): 924–935.
3. Dejana E, Hirschi KK, Simons M. The molecular basis of endothelial cell plasticity. *Nat Commun*. 2017;8:14361.
4. Marcu R, Choi YJ, Xue J, et al. Human organ-specific endothelial cell heterogeneity. *iScience*. 2018;4:20–35.
5. Jambusaria A, Hong Z, Zhang L, et al. Endothelial heterogeneity across distinct vascular beds during homeostasis and inflammation. *Elife*. 2020;9.
6. Qiu J, Hirschi KK. Endothelial cell development and its application to regenerative medicine. *Circ Res*. 2019;125(4):489–501.
7. Yang Y, García-Verdugo JM, Soriano-Navarro M, et al. Lymphatic endothelial progenitors bud from the cardinal vein and intersomitic vessels in mammalian embryos. *Blood*. 2012;120(11): 2340–2348.
8. Hägerling R, Pollmann C, Andreas M, et al. A novel multistep mechanism for initial lymphangiogenesis in mouse embryos based on ultramicroscopy. *EMBO J*. 2013;32(5):629–644.
9. Yang Y, Oliver G. Development of the mammalian lymphatic vasculature. *J Clin Invest*. 2014;124(3):888–897.
10. Martinez-Corral I, Ulvmar MH, Stanczuk L, et al. Nonvenous origin of dermal lymphatic vasculature. *Circ Res*. 2015;116(10): 1649–1654.
11. Ulvmar MH, Martinez-Corral I, Stanczuk L, Mäkinen T. Pdgfrb-Cre targets lymphatic endothelial cells of both venous and non-venous origins. *Genesis*. 2016;54(6):350–358.
12. Fish JE, Wythe JD. The molecular regulation of arteriovenous specification and maintenance. *Dev Dynam*. 2015;224:391–409.
13. Pober JS, Sessa WC. Evolving functions of endothelial cells in inflammation. *Nat Rev Immunol*. 2007;7(10):803–815.
14. van Hinsbergh VW. Endothelium–role in regulation of coagulation and inflammation. *Semin Immunopathol*. 2012;34(1): 93–106.
15. Murfee WL, Rappleye JW, Ceballos M, Schmid-Schönbein GW. Discontinuous expression of endothelial cell adhesion molecules along initial lymphatic vessels in mesentery: the primary valve structure. *Lymphatic Res Biol*. 2007;5(2):81–89.
16. Gritz E, Hirschi KK. Specification and function of hemogenic endothelium during embryogenesis. *Cell Mol Life Sci*. 2016;73(8): 1547–1567.
17. Chen Q, Zhang H, Liu Y, et al. Endothelial cells are progenitors of cardiac pericytes and vascular smooth muscle cells. *Nat Commun*. 2016;7:12422.
18. Kovacic JC, Dimmeler S, Harvey RP, et al. Endothelial to mesenchymal transition in cardiovascular disease: JACC state-of-the-art review. *J Am Coll Cardiol*. 2019;73(2):190–209.
19. Chung AS, Ferrara N. Developmental and pathological angiogenesis. *Annu Rev Cell Dev Biol*. 2011;27:563–584.
20. Rattner A, Williams J, Nathans J. Roles of HIFs and VEGF in angiogenesis in the retina and brain. *J Clin Invest*. 2019;129(9):3807–3820.
21. Chen W, Xia P, Wang H, et al. The endothelial tip-stalk cell selection and shuffling during angiogenesis. *J Cell Commun Signal*. 2019;13(3):291–301.
22. Gerhardt H, Golding M, Fruttiger M, et al. VEGF guides angiogenic sprouting utilizing endothelial tip cell filopodia. *J Cell Biol*. 2003;161(6):1163–1177.
23. Sainson RC, Aoto J, Nakatsu MN, et al. Cell-autonomous notch signaling regulates endothelial cell branching and proliferation during vascular tubulogenesis. *FASEB J*. 2005;19(8):1027–1029.
24. Kamei M, Saunders WB, Bayless KJ, Dye L, Davis GE, Weinstein BM. Endothelial tubes assemble from intracellular vacuoles in vivo. *Nature*. 2006;442(7101):453–456.
25. Benedito R, Roca C, Sorensen I, et al. The notch ligands Dll4 and Jagged1 have opposing effects on angiogenesis. *Cell*. 2009;137(6): 1124–1135.
26. Mouillesseaux KP, Wiley DS, Saunders LM, et al. Notch regulates BMP responsiveness and lateral branching in vessel networks via SMAD6. *Nat Commun*. 2016;7:13247.
27. Fang JS, Hirschi KK. Molecular regulation of arteriovenous endothelial cell specification. *F1000Res*. 2019:8.
28. Sacilotto N, Monteiro R, Fritzsche M, et al. Analysis of Dll4 regulation reveals a combinatorial role for Sox and Notch in arterial development. *Proc Natl Acad Sci USA*. 2013;110(29):11893–11898.
29. dela Paz NG, D'Amore PA. Arterial versus venous endothelial cells. *Cell Tissue Res*. 2009;335(1):5–16.
30. Wythe JD, Dang LT, Devine WP, et al. ETS factors regulate Vegf-dependent arterial specification. *Dev Cell*. 2013;26(1):45–58.
31. Robinson AS, Materna SC, Barnes RM, De Val S, Xu SM, Black BL. An arterial-specific enhancer of the human endothelin converting enzyme 1 (ECE1) gene is synergistically activated by Sox17, FoxC2, and Etv2. *Dev Biol*. 2014;395(2):379–389.
32. Chiang IK, Fritzsche M, Pichol-Thievend C, et al. SoxF factors induce Notch1 expression via direct transcriptional regulation during early arterial development. *Development*. 2017;144(14): 2629–2639.
33. Lawson ND, Vogel AM, Weinstein BM. Sonic hedgehog and vascular endothelial growth factor act upstream of the Notch pathway during arterial endothelial differentiation. *Dev Cell*. 2002;3(1):127–136.
34. Fang JS, Coon BG, Gillis N, et al. Shear-induced Notch-Cx37-p27 axis arrests endothelial cell cycle to enable arterial specification. *Nat Commun*. 2017;8(1):2149.
35. Larrivee B, Prahst C, Gordon E, et al. ALK1 signaling inhibits angiogenesis by cooperating with the Notch pathway. *Dev Cell*. 2012;22(3):489–500.
36. Kim JH, Peacock MR, George SC, Hughes CCW. BMP9 induces EphrinB2 expression in endothelial cells through an Alk1-BMPRII/ActRII-ID1/ID3-dependent pathway: implications for hereditary hemorrhagic telangiectasia type II. *Angiogenesis*. 2012; 15(3):497–509.
37. Xu C, Hasan SS, Schmidt I, et al. Arteries are formed by vein-derived endothelial tip cells. *Nat Commun*. 2014;5:5758.
38. Pitulescu ME, Schmidt I, Giaimo BD, et al. Dll4 and Notch signalling couples sprouting angiogenesis and artery formation. *Nat Cell Biol*. 2017;19(8):915–927.
39. Hong CC, Peterson QP, Hong JY, Peterson RT. Artery/vein specification is governed by opposing phosphatidylinositol-3 kinase and MAP kinase/ERK signaling. *Curr Biol*. 2006;16(13):1366–1372.
40. Neal A, Nornes S, Payne S, et al. Venous identity requires BMP signalling through ALK3. *Nat Commun*. 2019;10(1):453.
41. Su T, Stanley G, Sinha R, et al. Single-cell analysis of early progenitor cells that build coronary arteries. *Nature*. 2018;559(7714): 356–362.
42. You LR, Lin FJ, Lee CT, DeMayo FJ, Tsai MJ, Tsai SY. Suppression of Notch signalling by the COUP-TFII transcription factor regulates vein identity. *Nature*. 2005;435(7038):98–104.
43. Vanlandewijck M, He L, Mae MA, et al. A molecular atlas of cell types and zonation in the brain vasculature. *Nature*. 2018; 554(7693):475–480.

44. Fernández -LA, Sanz-Rodriguez F, Blanco FJ, Bernabéu C, Botella LM. Hereditary hemorrhagic telangiectasia, a vascular dysplasia affecting the TGF-beta signaling pathway. *Clin Med Res*. 2006;4(1):66–78.
45. Bourdeau A, Dumont DJ, Letarte M. A murine model of hereditary hemorrhagic telangiectasia. *J Clin Invest*. 1999;104(10): 1343–1351.
46. Srinivasan S, Hanes MA, Dickens T, et al. A mouse model for hereditary hemorrhagic telangiectasia (HHT) type 2. *Hum Mol Genet*. 2003;12(5):473–482.
47. Torsney E, Charlton R, Diamond AG, Burn J, Soames JV, Arthur HM. Mouse model for hereditary hemorrhagic telangiectasia has a generalized vascular abnormality. *Circulation*. 2003; 107(12):1653–1657.
48. Ruiz S, Zhao H, Chandakkar P, et al. A mouse model of hereditary hemorrhagic telangiectasia generated by transmammary-delivered immunoblocking of BMP9 and BMP10. *Sci Rep*. 2016;5.
49. Crist AM, Lee AR, Patel NR, Westhoff DE, Meadows SM. Vascular deficiency of Smad4 causes arteriovenous malformations: a mouse model of Hereditary Hemorrhagic Telangiectasia. *Angiogenesis*. 2018;21(2):363–380.
50. Ola R, Kunzel SH, Zhang F, et al. SMAD4 prevents flow induced arteriovenous malformations by inhibiting casein kinase 2. *Circulation*. 2018;138(21):2379–2394.
51. Murphy PA, Kim TN, Huang L, et al. Constitutively active Notch4 receptor elicits brain arteriovenous malformations through enlargement of capillary-like vessels. *Proc Natl Acad Sci USA*. 2014;111(50):18007–18012.
52. Baeyens N, Larrivée B, Ola R, et al. Defective fluid shear stress mechanotransduction mediates hereditary hemorrhagic telangiectasia. *J Cell Biol*. 2016;214(7):807–816.
53. Mazo IB, Massberg S, von Andrian UH. Hematopoietic stem and progenitor cell trafficking. *Trends Immunol*. 2011;32(10):493–503.
54. Marcelo KL, Sills TM, Coskun S, et al. Hemogenic endothelial cell specification requires c-Kit, Notch signaling, and p27-mediated cell-cycle control. *Dev Cell*. 2013;27(5):504–515.
55. Chanda B, Ditadi A, Iscove NN, Keller G. Retinoic acid signaling is essential for embryonic hematopoietic stem cell development. *Cell*. 2013;155(1):215–227.
56. Qiu J, Nordling S, Vasavada HH, Butcher EC, Hirschi KK. Retinoic acid promotes endothelial cell cycle early G1 state to enable human hemogenic endothelial cell specification. *Cell Rep*. 2020; 33(9):108465.
57. Palis J, Robertson S, Kennedy M, Wall C, Keller G. Development of erythroid and myeloid progenitors in the yolk sac and embryo proper of the mouse. *Development*. 1999;126(22):5073–5084.
58. Kaimakis P, Crisan M, Dzierzak E. The biochemistry of hematopoietic stem cell development. *Biochim Biophys Acta*. 2013; 1830(2):2395–2403.
59. Markwald RR, Fitzharris TP, Manasek FJ. Structural development of endocardial cushions. *Am J Anat*. 1977;148(1):85–119.
60. Kisanuki YY, Hammer RE, Miyazaki J, Williams SC, Richardson JA, Yanagisawa M. Tie2-Cre transgenic mice: a new model for endothelial cell-lineage analysis in vivo. *Dev Biol*. 2001;230(2):230–242.
61. François M, Caprini A, Hosking B, et al. Sox18 induces development of the lymphatic vasculature in mice. *Nature*. 2008; 456(7222):643–647.
62. Lee S, Kang J, Yoo J, et al. Prox1 physically and functionally interacts with COUP-TFII to specify lymphatic endothelial cell fate. *Blood*. 2009;113(8):1856–1859.
63. Aranguren XL, Beerens M, Vandevelde W, Dewerchin M, Carmeliet P, Luttun A. Transcription factor COUP-TFII is indispensable for venous and lymphatic development in zebrafish and *Xenopus laevis*. *Biochem Biophys Res Commun*. 2011;410(1): 121–126.
64. Aranguren XL, Beerens M, Coppiello G, et al. COUP-TFII orchestrates venous and lymphatic endothelial identity by homo- or hetero-dimerisation with PROX1. *J Cell Sci*. 2013;126(Pt 5): 1164–1175.
65. Hong YK, Detmar M. Prox1, master regulator of the lymphatic vasculature phenotype. *Cell Tissue Res*. 2003;314(1):85–92.
66. Johnson NC, Dillard ME, Baluk P, et al. Lymphatic endothelial cell identity is reversible and its maintenance requires Prox1 activity. *Genes Dev*. 2008;22(23):3282–3291.
67. Hong YK, Harvey N, Noh YH, et al. Prox1 is a master control gene in the program specifying lymphatic endothelial cell fate. *Dev Dynam*. 2002;225(3):351–357.
68. Banerji S, Ni J, Wang SX, et al. LYVE-1, a new homologue of the CD44 glycoprotein, is a lymph-specific receptor for hyaluronan. *J Cell Biol*. 1999;144(4):789–801.
69. Luong MX, Tam J, Lin Q, et al. Lack of lymphatic vessel phenotype in LYVE-1/CD44 double knockout mice. *J Cell Physiol*. 2009;219(2):430–437.
70. Johnson LA, Banerji S, Lawrance W, et al. Dendritic cells enter lymph vessels by hyaluronan-mediated docking to the endothelial receptor LYVE-1. *Nat Immunol*. 2017;18(7):762–770.
71. Zhou F, Chang Z, Zhang L, et al. Akt/Protein kinase B is required for lymphatic network formation, remodeling, and valve development. *Am J Pathol*. 2010;177(4):2124–2133.
72. Sabine A, Petrova TV. Interplay of mechanotransduction, FOXC2, connexins, and calcineurin signaling in lymphatic valve formation. *Adv Anat Embryol Cell Biol*. 2014;214:67–80.
73. Zhang G, Brady J, Liang WC, Wu Y, Henkemeyer M, Yan M. EphB4 forward signalling regulates lymphatic valve development. *Nat Commun*. 2015;6:6625.
74. Kanady JD, Dellinger MT, Munger SJ, Witte MH, Simon AM. Connexin37 and Connexin43 deficiencies in mice disrupt lymphatic valve development and result in lymphatic disorders including lymphedema and chylothorax. *Dev Biol*. 2011;354(2):253–266.
75. Wolburg H, Lippoldt A. Tight junctions of the blood-brain barrier: development, composition and regulation. *Vasc Pharmacol*. 2002; 38(6):323–337.
76. Kliche K, Jeggle P, Pavenstädt H, Oberleithner H. Role of cellular mechanics in the function and life span of vascular endothelium. *Pflügers Archiv*. 2011;462(2):209–217.
77. Chavkin NW, Hirschi KK. Single cell analysis in vascular biology. *Front. Cardiovasc. Med*. 2020;7.
78. Kalucka J, de Rooij L, Goveia J, et al. Single-cell transcriptome atlas of murine endothelial cells. *Cell*. 2020.
79. Petrova TV, Koh GY. Organ-specific lymphatic vasculature: from development to pathophysiology. *J Exp Med*. 2018;215(1):35–49.
80. Klotz L, Norman S, Vieira JM, et al. Cardiac lymphatics are heterogeneous in origin and respond to injury. *Nature*. 2015;522(7554): 62–67.
81. Stanczuk L, Martinez-Corral I, Ulvmar MH, et al. cKit lineage hemogenic endothelium-derived cells contribute to mesenteric lymphatic vessels. *Cell Rep*. 2015;10(10):1708–1721.
82. Chong DC, Koo Y, Xu K, Fu S, Cleaver O. Stepwise arteriovenous fate acquisition during mammalian vasculogenesis. *Dev Dynam*. 2011;240(9):2153–2165.
83. le Noble F, Moyon D, Pardanaud L, et al. Flow regulates arterial-venous differentiation in the chick embryo yolk sac. *Development*. 2004;131(2):361–375.
84. Yuan L, Chan GC, Beeler D, et al. A role of stochastic phenotype switching in generating mosaic endothelial cell heterogeneity. *Nat Commun*. 2016;7:10160.
85. Lucitti JL, Jones EA, Huang C, Chen J, Fraser SE, Dickinson ME. Vascular remodeling of the mouse yolk sac requires hemodynamic force. *Development*. 2007;134(18):3317–3326.
86. Butler PJ, Tsou TC, Li JY, Usami S, Chien S. Rate sensitivity of shear-induced changes in the lateral diffusion of endothelial cell

membrane lipids: a role for membrane perturbation in shear-induced MAPK activation. *FASEB J*. 2002;16(2):216–218.
87. Weinbaum S, Zhang X, Han Y, Vink H, Cowin SC. Mechanotransduction and flow across the endothelial glycocalyx. *Proc Natl Acad Sci USA*. 2003;100(13):7988–7995.
88. Hahn C, Schwartz MA. Mechanotransduction in vascular physiology and atherogenesis. *Nat Rev Mol Cell Biol*. 2009;10(1):53–62.
89. Chiu JJ, Usami S, Chien S. Vascular endothelial responses to altered shear stress: pathologic implications for atherosclerosis. *Ann Med*. 2009;41(1):19–28.
90. Chiu JJ, Chien S. Effects of disturbed flow on vascular endothelium: pathophysiological basis and clinical perspectives. *Physiol Rev*. 2011;91(1):327–387.
91. Sweet DT, Jiménez JM, Chang J, et al. Lymph flow regulates collecting lymphatic vessel maturation in vivo. *J Clin Invest*. 2015; 125(8):2995–3007.
92. Anwar MA, Shalhoub J, Lim CS, Gohel MS, Davies AH. The effect of pressure-induced mechanical stretch on vascular wall differential gene expression. *J Vasc Res*. 2012;49(6):463–478.
93. Dekker RJ, van Thienen JV, Rohlena J, et al. Endothelial KLF2 links local arterial shear stress levels to the expression of vascular tone-regulating genes. *Am J Pathol*. 2005;167(2):609–618.
94. Parmar KM, Larman HB, Dai G, et al. Integration of flow-dependent endothelial phenotypes by Kruppel-like factor 2. *J Clin Invest*. 2006;116(1):49–58.
95. Lee JS, Yu Q, Shin JT, et al. Klf2 is an essential regulator of vascular hemodynamic forces in vivo. *Dev Cell*. 2006;11(6):845–857.
96. Meadows SM, Salanga MC, Krieg PA. Kruppel-like factor 2 cooperates with the ETS family protein ERG to activate Flk1 expression during vascular development. *Development*. 2009;136(7):1115–1125.
97. Heckel E, Boselli F, Roth S, et al. Oscillatory flow modulates mechanosensitive klf2a expression through trpv4 and trpp2 during heart valve development. *Curr Biol*. 2015;25(10):1354–1361.
98. Dixon JB, Greiner ST, Gashev AA, Cote GL, Moore JE, Zawieja DC. Lymph flow, shear stress, and lymphocyte velocity in rat mesenteric prenodal lymphatics. *Microcirculation*. 2006; 13(7):597–610.
99. Sabine A, Agalarov Y, Maby-El Hajjami H, et al. Mechanotransduction, PROX1, and FOXC2 cooperate to control connexin37 and calcineurin during lymphatic-valve formation. *Dev Cell*. 2012;22(2):430–445.
100. Cha B, Geng X, Mahamud MR, et al. Mechanotransduction activates canonical Wnt/β-catenin signaling to promote lymphatic vascular patterning and the development of lymphatic and lymphovenous valves. *Genes Dev*. 2016;30(12):1454–1469.
101. Adamo L, Naveiras O, Wenzel PL, et al. Biomechanical forces promote embryonic haematopoiesis. *Nature*. 2009;459(7250): 1131–1135.
102. North TE, Goessling W, Peeters M, et al. Hematopoietic stem cell development is dependent on blood flow. *Cell*. 2009;137(4): 736–748.
103. Wang L, Zhang P, Wei Y, Gao Y, Patient R, Liu F. A blood flow-dependent klf2a-NO signaling cascade is required for stabilization of hematopoietic stem cell programming in zebrafish embryos. *Blood*. 2011;118(15):4102–4110.
104. Liu Z, Tu H, Kang Y, et al. Primary cilia regulate hematopoietic stem and progenitor cell specification through Notch signaling in zebrafish. *Nat Commun*. 2019;10(1):1839.
105. Camenisch TD, Molin DG, Person A, et al. Temporal and distinct TGFbeta ligand requirements during mouse and avian endocardial cushion morphogenesis. *Dev Biol*. 2002;248(1):170–181.
106. Vermot J, Forouhar AS, Liebling M, et al. Reversing blood flows act through klf2a to ensure normal valvulogenesis in the developing heart. *PLoS Biol*. 2009;7(11):e1000246.
107. De Bock K, Georgiadou M, Carmeliet P. Role of endothelial cell metabolism in vessel sprouting. *Cell Metabol*. 2013;18(5):634–647.
108. Aranguren XL, Agirre X, Beerens M, et al. Unraveling a novel transcription factor code determining the human arterial-specific endothelial cell signature. *Blood*. 2013;122(24):3982–3992.
109. Kalucka J, Missiaen R, Georgiadou M, et al. Metabolic control of the cell cycle. *Cell Cycle*. 2015;14(21):3379–3388.
110. Schoors S, Bruning U, Missiaen R, et al. Fatty acid carbon is essential for dNTP synthesis in endothelial cells. *Nature*. 2015;520(7546): 192–197.
111. Parra-Bonilla G, Alvarez DF, Al-Mehdi AB, Alexeyev M, Stevens T. Critical role for lactate dehydrogenase A in aerobic glycolysis that sustains pulmonary microvascular endothelial cell proliferation. *Am J Physiol Lung Cell Mol Physiol*. 2010;299(4): L513–L522.
112. Rohlenova K, Veys K, Miranda-Santos I, De Bock K, Carmeliet P. Endothelial cell metabolism in health and disease. *Trends Cell Biol*. 2018;28(3):224–236.
113. Yu P, Wilhelm K, Dubrac A, et al. FGF-dependent metabolic control of vascular development. *Nature*. 2017;545(7653):224–228.
114. Elmasri H, Ghelfi E, Yu CW, et al. Endothelial cell-fatty acid binding protein 4 promotes angiogenesis: role of stem cell factor/c-kit pathway. *Angiogenesis*. 2012;15(3):457–468.
115. Harjes U, Bridges E, McIntyre A, Fielding BA, Harris AL. Fatty acid-binding protein 4, a point of convergence for angiogenic and metabolic signaling pathways in endothelial cells. *J Biol Chem*. 2014;289(33):23168–23176.
116. Huang H, Vandekeere S, Kalucka J, et al. Role of glutamine and interlinked asparagine metabolism in vessel formation. *EMBO J*. 2017;36(16):2334–2352.
117. Spolarics Z, Lang CH, Bagby GJ, Spitzer JJ. Glutamine and fatty acid oxidation are the main sources of energy for Kupffer and endothelial cells. *Am J Physiol*. 1991;261(2 Pt 1): G185–G190.
118. Wong BW, Wang X, Zecchin A, et al. The role of fatty acid β-oxidation in lymphangiogenesis. *Nature*. 2017;542(7639):49–54.
119. Simon MC, Keith B. The role of oxygen availability in embryonic development and stem cell function. *Nat Rev Mol Cell Biol*. 2008; 9(4):285–296.
120. Ben-Yosef Y, Miller A, Shapiro S, Lahat N. Hypoxia of endothelial cells leads to MMP-2-dependent survival and death. *Am J Physiol Cell Physiol*. 2005;289(5):C1321–C1331.
121. Liu Y, Cox SR, Morita T, Kourembanas S. Hypoxia regulates vascular endothelial growth factor gene expression in endothelial cells. Identification of a 5′ enhancer. *Circ Res*. 1995; 77(3):638–643.
122. Lanner F, Lee KL, Ortega GC, et al. Hypoxia-induced arterial differentiation requires adrenomedullin and notch signaling. *Stem Cell Dev*. 2013;22(9):1360–1369.
123. Nilsson I, Shibuya M, Wennström S. Differential activation of vascular genes by hypoxia in primary endothelial cells. *Exp Cell Res*. 2004;299(2):476–485.
124. Zhou B, Si W, Su Z, Deng W, Tu X, Wang Q. Transcriptional activation of the Prox1 gene by HIF-1α and HIF-2α in response to hypoxia. *FEBS Lett*. 2013;587(6):724–731.
125. Morfoisse F, Kuchnio A, Frainay C, et al. Hypoxia induces VEGF-C expression in metastatic tumor cells via a HIF-1α-independent translation-mediated mechanism. *Cell Rep*. 2014;6(1):155–167.
126. Ohtani K, Vlachojannis GJ, Koyanagi M, et al. Epigenetic regulation of endothelial lineage committed genes in pro-angiogenic hematopoietic and endothelial progenitor cells. *Circ Res*. 2011; 109(11):1219–1229.
127. Shirodkar AV, St Bernard R, Gavryushova A, et al. A mechanistic role for DNA methylation in endothelial cell (EC)-enriched gene expression: relationship with DNA replication timing. *Blood*. 2013;121(17):3531–3540.
128. Zhang R, Wang N, Zhang LN, et al. Knockdown of DNMT1 and DNMT3a promotes the angiogenesis of human mesenchymal

stem cells leading to arterial specific differentiation. *Stem Cell.* 2016;34(5):1273–1283.
129. Nakato R, Wada Y, Nakaki R, et al. Comprehensive epigenome characterization reveals diverse transcriptional regulation across human vascular endothelial cells. *Epigenet Chromatin.* 2019;12(1):77.
130. Rössig L, Urbich C, Brühl T, et al. Histone deacetylase activity is essential for the expression of HoxA9 and for endothelial commitment of progenitor cells. *J Exp Med.* 2005;201(11):1825–1835.
131. Chang S, Young BD, Li S, Qi X, Richardson JA, Olson EN. Histone deacetylase 7 maintains vascular integrity by repressing matrix metalloproteinase 10. *Cell.* 2006;126(2):321–334.
132. Guarani V, Deflorian G, Franco CA, et al. Acetylation-dependent regulation of endothelial Notch signalling by the SIRT1 deacetylase. *Nature.* 2011;473(7346):234–238.
133. Davis RB, Curtis CD, Griffin CT. BRG1 promotes COUP-TFII expression and venous specification during embryonic vascular development. *Development.* 2013;140(6):1272–1281.
134. Crosswhite PL, Podsiadlowska JJ, Curtis CD, et al. CHD4-regulated plasmin activation impacts lymphovenous hemostasis and hepatic vascular integrity. *J Clin Invest.* 2016;126(6): 2254–2266.
135. Janardhan HP, Milstone ZJ, Shin M, Lawson ND, Keaney JF, Trivedi CM. Hdac3 regulates lymphovenous and lymphatic valve formation. *J Clin Invest.* 2017;127(11):4193–4206.
136. Yoo H, Lee YJ, Park C, et al. Epigenetic priming by Dot1l in lymphatic endothelial progenitors ensures normal lymphatic development and function. *Cell Death Dis.* 2020;11(1):14.
137. Hua Z, Lv Q, Ye W, et al. MiRNA-directed regulation of VEGF and other angiogenic factors under hypoxia. *PLoS One.* 2006;1:e116.
138. Chen HX, Xu XX, Tan BZ, Zhang Z, Zhou XD. MicroRNA-29b inhibits angiogenesis by targeting VEGFA through the MAPK/ERK and PI3K/Akt signaling pathways in endometrial carcinoma. *Cell Physiol Biochem.* 2017;41(3):933–946.
139. Wang S, Aurora AB, Johnson BA, et al. The endothelial-specific microRNA miR-126 governs vascular integrity and angiogenesis. *Dev Cell.* 2008;15(2):261–271.
140. Descamps B, Saif J, Benest AV, et al. BDNF (Brain-Derived neurotrophic factor) promotes embryonic stem cells differentiation to endothelial cells via a molecular pathway, including MicroRNA-214, EZH2 (Enhancer of Zeste Homolog 2), and eNOS (endothelial Nitric Oxide Synthase). *Arterioscler Thromb Vasc Biol.* 2018;38(9):2117–2125.
141. Hernández-Romero IA, Guerra-Calderas L, Salgado-Albarrán M, Maldonado-Huerta T, Soto-Reyes E. The regulatory roles of non-coding RNAs in angiogenesis and neovascularization from an epigenetic perspective. *Front Oncol.* 2019;9:1091.
142. Nicoli S, Knyphausen CP, Zhu LJ, Lakshmanan A, Lawson ND. miR-221 is required for endothelial tip cell behaviors during vascular development. *Dev Cell.* 2012;22(2):418–429.
143. Biyashev D, Veliceasa D, Topczewski J, et al. miR-27b controls venous specification and tip cell fate. *Blood.* 2012;119(11): 2679–2687.
144. Kasper DM, Hintzen J, Wu Y, et al. The N-glycome regulates the endothelial-to-hematopoietic transition. *Science.* 2020;370(6521): 1186–1191.
145. Leisegang MS, Fork C, Josipovic I, et al. Long noncoding RNA MANTIS facilitates endothelial angiogenic function. *Circulation.* 2017;136(1):65–79.
146. Pedrioli DM, Karpanen T, Dabouras V, et al. miR-31 functions as a negative regulator of lymphatic vascular lineage-specific differentiation in vitro and vascular development in vivo. *Mol Cell Biol.* 2010;30(14):3620–3634.
147. Kazenwadel J, Michael MZ, Harvey NL. Prox1 expression is negatively regulated by miR-181 in endothelial cells. *Blood.* 2010; 116(13):2395–2401.
148. Illi B, Scopece A, Nanni S, et al. Epigenetic histone modification and cardiovascular lineage programming in mouse embryonic stem cells exposed to laminar shear stress. *Circ Res.* 2005;96(5): 501–508.
149. Nicoli S, Standley C, Walker P, Hurlstone A, Fogarty KE, Lawson ND. MicroRNA-mediated integration of haemodynamics and Vegf signalling during angiogenesis. *Nature.* 2010;464(7292): 1196–1200.
150. Schleithoff C, Voelter-Mahlknecht S, Dahmke IN, Mahlknecht U. On the epigenetics of vascular regulation and disease. *Clin Epigenet.* 2012;4(1):7.
151. Kato H, Tamamizu-Kato S, Shibasaki F. Histone deacetylase 7 associates with hypoxia-inducible factor 1alpha and increases transcriptional activity. *J Biol Chem.* 2004;279(40): 41966–41974.
152. Chen Q, Miller LJ, Dong M. Role of N-linked glycosylation in biosynthesis, trafficking, and function of the human glucagon-like peptide 1 receptor. *Am J Physiol Endocrinol Metab.* 2010; 299(1):E62–E68.
153. El-Osta A, Brasacchio D, Yao D, et al. Transient high glucose causes persistent epigenetic changes and altered gene expression during subsequent normoglycemia. *J Exp Med.* 2008;205(10): 2409–2417.
154. Hirschi KK, Dejana E. Resident endothelial progenitors make themselves at home. *Cell Stem Cell.* 2018;23(2):153–155.
155. Detter MR, Snellings DA, Marchuk DA. Cerebral cavernous malformations develop through clonal expansion of mutant endothelial cells. *Circ Res.* 2018;123(10):1143–1151.
156. Malinverno M, Maderna C, Abu Taha A, et al. Endothelial cell clonal expansion in the development of cerebral cavernous malformations. *Nat Commun.* 2019;10(1):2761.
157. Manavski Y, Lucas T, Glaser SF, et al. Clonal expansion of endothelial cells contributes to ischemia-induced neovascularization. *Circ Res.* 2018;122(5):670–677.

CHAPTER

2

Endothelial cell heterogeneity and plasticity in health and disease—new insights from single-cell studies

Lisa M. Becker[1], *Alessandra Pasut*[1], *Anne Cuypers*[1] *and Peter Carmeliet*[1,2,3]

[1]Laboratory of Angiogenesis and Vascular Metabolism, Department of Oncology and Leuven Cancer Institute (LKI), KU Leuven, VIB Center for Cancer Biology, VIB, Leuven, Belgium [2]Laboratory of Angiogenesis and Vascular Heterogeneity, Department of Biomedicine, Aarhus University, Aarhus, Denmark [3]State Key Laboratory of Ophthalmology, Zhongshan Ophthalmic Center, Sun Yat-Sen University, Guangzhou, Guangdong, P.R. China

Introduction

The vascular endothelium is bestowed with many different functions including delivery and exchange of oxygen/nutrients, vasoregulation and hemostasis, trafficking of immune cells, and transport of proteins.[1–3] This functional heterogeneity is reflected at the cellular and molecular level. Recent single-cell studies confirmed that endothelial cells (ECs) are highly heterogeneous at the transcriptional level and show distinct gene signatures both across tissues and vascular beds. In addition, these studies also led to the identification of new EC subtypes in physiological and pathological states and have brought new insights on the role of ECs and their heterogeneity.[4–12] Focusing on recent data from single-cell RNA sequencing (scRNAseq) studies, here we highlight two main examples of EC plasticity and heterogeneity: (1) the specification of ECs into distinct subtypes (artery, vein, capillaries) and (2) the ability of ECs to transition from a quiescent to an angiogenic phenotype. Without the intention to give an all-encompassing overview, here we will describe the mechanisms that drive EC plasticity in health and disease with a focus on metabolic plasticity and heterogeneity.[13]

Endothelial cell heterogeneity and plasticity

Diversification of endothelial cells

During embryonic development, ECs specify into functionally distinct types, namely artery, vein, capillary, and lymphatic ECs.[2,14–17] The interplay between signaling pathways, transcription factors, and growth factors drives ECs' abilities to acquire and maintain these distinct phenotypes. Early lineage tracing experiments documented that Notch and VEGF maintain the arterial fate of ECs, while COUP-TF2 is required for the acquisition of the venous phenotype by inhibiting Notch signaling.[15,16,18–20]

A single-cell transcriptomics study in the brain described the gene expression signature of arterial, venous, and capillary ECs and showed that arterial ECs were enriched in transcription factors, while ECs in capillaries and veins displayed increased expression of transporter genes, suggesting that brain ECs follow a "cellular zonation" pattern.[10] One of the largest scRNAseq studies performed to date described 78 EC populations across 11 tissues[4] and provided several new insights.[4] First, ECs from several tissues clustered together at the transcriptome level[4] (Fig. 2.1a). For example, brain and testis ECs were transcriptionally more similar to one another than to ECs derived from other organs, possibly because ECs in these tissues share similar functions.[4] Second, capillary ECs displayed remarkable heterogeneity across tissues and overall shared fewer genes compared to arterial and vein subtypes, possibly reflecting organ-specific metabolic needs. Identified capillary clusters were capillary 1 and 2, capillary arterial and venous, as well as capillary Aqp7^{+}, which were only found in colon and small intestine (Fig. 2.1b).[4] Third, EC phenotypes including tissue-restricted phenotypes (choroid plexus or glomerular

The Vasculome
https://doi.org/10.1016/B978-0-12-822546-2.00017-4

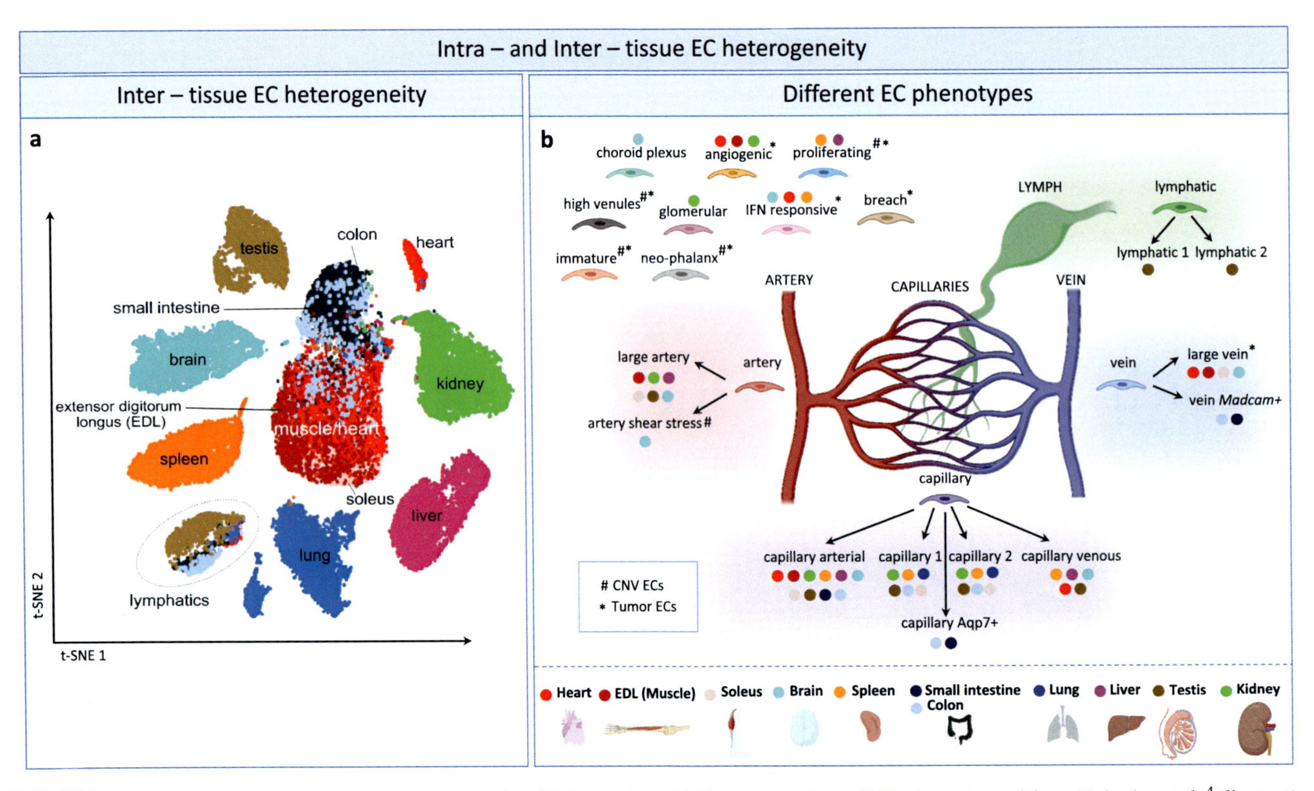

FIGURE 2.1 **Intra- and Inter-tissue EC heterogeneity.** (A) Inter-tissue EC heterogeneity: t-SNE plot, adapted from Kalucka et al.,[4] illustrating the inter-tissue EC heterogeneity, with ECs from some tissues clustering closer together (e.g., heart and muscle) than others. (B) Distinct EC subclusters identified by different single cell studies. Colored dots indicate in which organs the EC subset can be found. * indicates that the EC subset is also expressed in (lung) tumors, # indicates expression in choroidal neovascularization (CNV). Parts of the figure were generated with Biorender.

ECs) were described. Lymphatic ECs subclustered into two clusters only in testes. Artery ECs harbored two subsets: large artery ECs could be found in several organs, whereas artery shear stress ECs were only detected in the brain. The two vein subclusters, large vein and vein *Madcam+*, were only expressed in the brain, heart, muscle, and small intestine and colon, respectively (Fig. 2.1b). Among the unexpected EC phenotypes, proliferating ECs were detected in the liver and spleen.[4] Angiogenic ECs were detected in the kidney, heart, muscle, and testis and IFN-activated ECs in the brain, heart, muscle and spleen[4,6] (Figure 2.1b). Fourth, the tissue type, rather than the vascular tree, contributes to the observed heterogeneity.[4,9–11] The organ-specific diversification of ECs has also been captured by other single-cell and microarray analyses, including a reanalysis of ECs from the Tabula Muris consortium,[8] which showed a tissue-specific enrichment of specific genes in ECs in particular tissues. Also, the location within the tissue determines the heterogeneity of ECs. For example, in the kidney, medullary, cortical, and glomerular ECs are transcriptomically heterogenous and express a distinct set of genes depending on their position in the kidney.[9] This "tissue zonation" signature is also a feature of liver ECs.[11]

Single-cell transcriptomic studies in combination with lineage tracing strategies recently provided new insights into the development of the coronary artery.[12] Contrary to the general view that the coronary vasculature develops after onset of blood flow, reconstruction of single EC trajectories identified a population of venous ECs that could adopt a prearterial fate before the onset of blood flow and later acquire a fully differentiated arterial fate.[12] This finding may deserve further attention for translation into cardiovascular and regenerative medicine.[21]

Differentiation of endothelial cells into other cell types

Another example of EC plasticity is represented by the differentiation of ECs to other cell types. For instance, adult ECs can differentiate into hematopoietic stem cells via reprogramming with a combination of four transcription factors: FOSB, GFI1, RUNX1, and SPI1.[22] Of note, coculturing reprogrammed clones with human umbilical vein endothelial cells (HUVECs), overexpressing E4ORF1 (suggested to represent "vascular niche" cells), increased their self-renewing capacity, suggesting a cell–cell cross-talk. Indeed, HUVEC-produced angiocrine signals, such as BMP and CXCL12 signaling, promoted the self-renewing phenotype.[22] The combination of single-cell transcriptomic studies with reprogramming strategies provides an additional opportunity to investigate the reprogramming of ECs into other cell types. In one such study, ECs obtained from human pluripotent stem cells were reprogrammed into hematopoietic cells mainly via TAL1, RUNX1 and GATA2.[23] Analysis of the trajectories of single reprogrammed cells showed that cell cycle genes might play a role in regulating the reprogramming process.[23]

ECs can also differentiate into other mesodermal lineages including skeletal and cardiac muscle cells.[24–26] A scRNAseq study in zebrafish demonstrated that the transcription factor Etv2, known to regulate the EC-hematopoietic axis, prevents ECs from acquiring alternative fates.[24] In Etv2 knockout embryos, ECs upregulated myogenic makers at the expense of the vascular and myeloid fate. In other vertebrate species, ECs have the same developmental origin than cells in somites, are capable of differentiating into myogenic cells, and participate in muscle repair.[27,28] Overall, these findings underscore the plasticity of ECs and raise the question whether ECs can be used in cell therapeutic settings, etc.[29,30]

Angiogenic EC phenotypes

The formation of new vessels from preexisting ones is termed angiogenesis. This process can either occur via sprouting of vessels (sprouting angiogenesis, SA) or an existing vessel can split into two daughter vessels (intussusceptive angiogenesis (IA)).[1,17,31,32] IA allows new vessel growth via duplication (intussusceptive microvascular growth), and reshape the vascular tree by means of arborization (intussusceptive arborization) as well as pruning (intussusceptive pruning).[32] Angiogenesis occurs mainly during embryonic development and is almost absent in adult tissue homeostasis, with exceptions in skeletal muscles during exercise or in the endometrium during the estrus cycle.[33,34] Nevertheless, upon injury of adult tissues, SA and IA are activated to ensure growth of new vessels.

Retinal vascularization is a model for SA,[35–39] which has been widely used to decipher the sprouting angiogenic process. Angiogenic growth factors (for instance,-VEGF) stimulate a quiescent "phalanx" EC to acquire an invasive/migratory phenotype ("tip" phenotype), which then navigates the nascent sprout. The tip cell, which possesses filopodia to sense angiogenic signaling, then leads the nascent vessel sprout toward these signals. Proliferating "stalk" cells extent the sprout. Of note, tip and stalk cells do not represent genetically fixed states, but rather plastic EC phenotypes. During the angiogenic process, they dynamically differentiate into one another, ensuring the leading position in the sprouting process to the fittest EC.[36,38–40] Once vessel formation is complete, ECs differentiate back into quiescent phalanx ECs.

ScRNAseq offers opportunities to dive deeper into EC heterogeneity and to identify additional EC subsets. Analysis of ECs in cancer and choroidal neovascularization (CNV) provided insights into EC heterogeneity of SA and allowed identification of various EC clusters.[6,41] These EC subsets included known angiogenic ECs (proliferating ECs; tip ECs), along with mature phalanx ECs. Interestingly, typical stalk EC phenotypes were not found in either model, although EC phenotypes with stalk-like features were identified.[6,41] Novel EC subtypes comprised immature and neophalanx ECs (Fig. 2.1b).[41] Moreover, breach cells were identified as ECs (Fig. 2.1b) expressing tip cell markers, but simultaneously also genes associated with podosome rosette formation, presumably aiding tip cell migration.[6] The discovery of these novel phenotypes improves our understanding of SA. However, to date, single-cell studies investigating EC heterogeneity during IA remain elusive.

While genetic lineage tracing has been the traditional method to track the differentiation of cells, pseudotime analyses of single-cell transcriptomic studies enable in silico tracking of EC subsets during SA. Such an approach revealed postcapillary venule ECs as origin of vessel sprouting initiation. These ECs then differentiate into tip cells (with intermediate states in between these two phenotypes) and finally turn into phalanx-like ECs upon completion of sprouting.[41] Interestingly, postcapillary venules in tumors were enriched in a subset of cells with a resident endothelial stem cells (rESCs) gene signature.[6] It is tempting to speculate that rESCs possibly contribute to the formation of new vessels and vessel regeneration, as demonstrated previously.[42,43] In fact, anti-VEGF treatment caused expansion of venous ECs in tumors.[6] Since venous ECs contain the rESC population, this finding raises the question whether rESCs might contribute to a new resistance mechanism against anti-angiogenic therapy.

In the setting of embryonic development, angiogenesis is controlled by Notch and VEGF.[44,45] During SA, VEGF promotes expression of Dll4 in tip cells, which in turn activates Notch in neighboring stalk cells. Notch signaling prevents transitioning of stalk cells into a tip phenotype.[36,37] Notch and COUP-TF2 codetermine EC sprouting versus differentiation into arteries, whereby COUP-TF2 is responsible for upregulation of cell cycle genes and impairing arterial fate.[15,18,46] Notch, on the other hand, inhibits transcription of Myc and downstream metabolic and cell cycle genes, as demonstrated using fate mapping in mosaic Notch mice.[20] Thus, cell cycle as well as cellular metabolism might play a role in regulating EC identity during SA.[20]

EC metabolic heterogeneity

EC metabolism during angiogenesis

More than 10 years ago, the hypothesis was postulated that ECs need to adapt their metabolism to respond to angiogenic signals and to grow new blood vessels.[47] The critical role of EC metabolism for vessel formation was demonstrated by genetic studies, manipulating the expression of the glycolytic activator PFKFB3 without altering the expression of angiogenic signals.[48] As highlighted above, tip and stalk EC phenotypes are not genetically predetermined fates but rather interchangeable EC phenotypes; with stalk ECs being capable of taking over the tip cell position in a Notch-dependent process (in which high Notch signaling is associated with the stalk EC phenotype).[37,38] Notably, mosaic overexpression of PFKFB3 in an in vivo zebrafish SA model converted stalk into tip cells, even overruling enforced Notch signaling (via Notch intracellular domain overexpression)[48] (Fig. 2.2B). Conversely, gene silencing of PFKFB3 impaired tip cell competition in an in vitro mosaic EC spheroid SA model.[48–50]

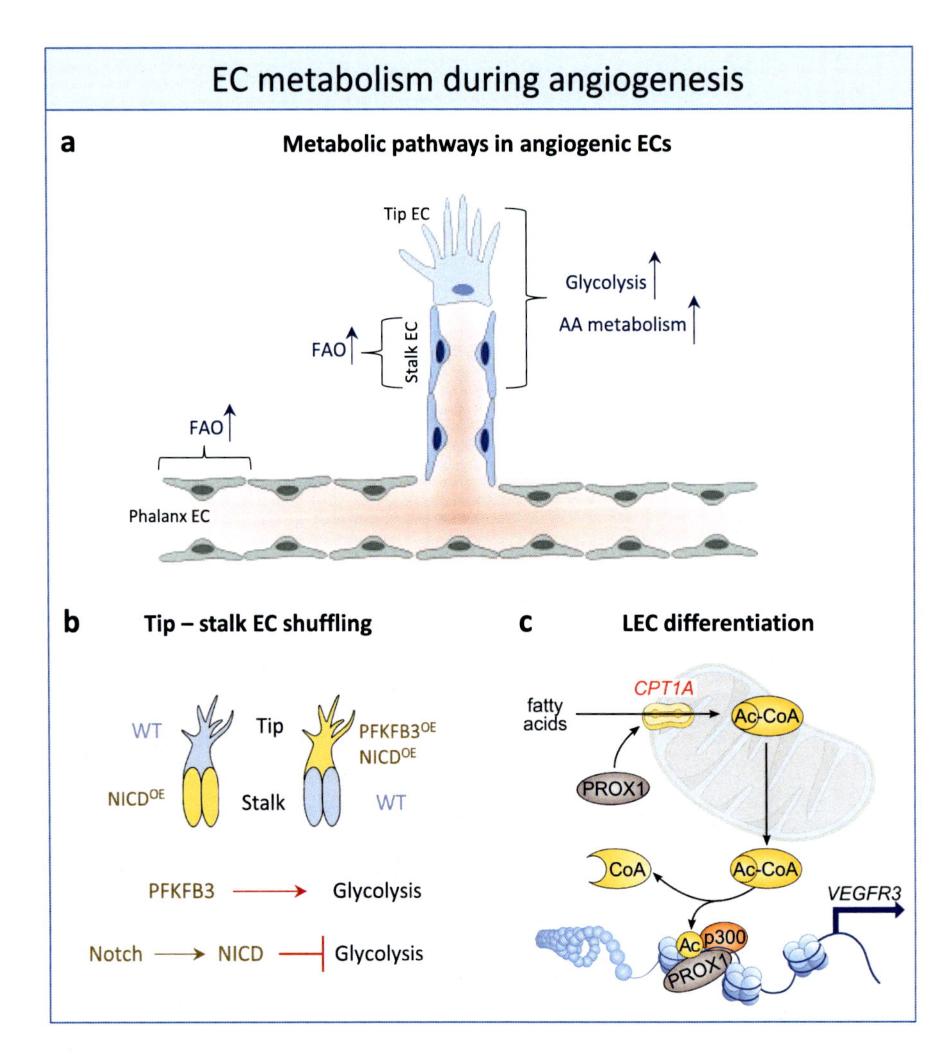

FIGURE 2.2 EC metabolism during angiogenesis. (A) Metabolic pathways in angiogenic ECs: Different metabolic pathways are upregulated in angiogenic ECs, including glycolysis (stalk and tip ECs), amino acid (glutamine/asparagine) metabolism (stalk and tip ECs), and fatty acid oxidation (stalk and phalanx ECs). (B) Tip—stalk EC shuffling: Stalk ECs differentiate into tip ECs upon upregulation of the glycolytic activator PFKFB3. Notch signaling (via overexpression (OE) of the signaling Notch intracellular domain (NICD)) maintains the stalk cell phenotype, and lowers glycolysis. Notch signaling can be overruled by PFKFB3 overexpression (OE), when both (NICD and PFKFB3) are overexpressed in the same cell. (C) Lymph EC (LEC) differentiation: Upregulation of FAO in vein ECs causes differentiation to LECs via increased Prox-1 and subsequent VEGFR3 upregulation. *AA*, amino acid; *FAO*, fatty acid oxidation; *NICD*, Notch intracellular domain; *OE*, overexpression; *WT*, wild type. *Schematics in panels B,C are adapted from Li X, Sun X, Carmeliet P. Hallmarks of endothelial cell metabolism in health and disease.* Cell Metabol. *2019;30: 414–433. https://doi.org/10.1016/j.cmet.2019.08.011.*

Apart from PFKFB3, several other metabolic enzymes control EC plasticity. For example, carnitine palmitoyl transferase 1A (CPT1A), a fatty acid oxidation (FAO) rate-controlling enzyme, imports fatty acids into mitochondria where they become oxidized into acetyl-CoA, which enters the TCA cycle. This process is essential for nucleotide biosynthesis in proliferating ECs[51] and for maintenance of redox homeostasis in the quiescent phenotype.[52] Notably, FAO increases histone acetylation of lymphangiogenic genes (upregulating Prox-1, the lymphatic EC master transcription factor), thereby stimulating differentiation of venous into lymphatic ECs during development (Fig. 2.2c).[53,54] In this process, VEGFR3 expression is induced upon Prox-1 upregulation.

Tip and stalk cell activities are also fueled by other metabolic pathways. Tip cells highly rely on glycolysis and amino acid (AA) metabolism. For example, glutamine catabolism increases the competitiveness of tip cells during vessel sprouting.[55] Stalk cells also rely on glycolysis and AA metabolism, but also use FAO for biomass production. In phalanx ECs, FAO contributes to maintain the quiescent phenotype (see above, Fig. 2.2a). Of note, ketone bodies are essential for EC proliferation in both lymphatic and blood vessels.[56,57] In the mouse aorta, when stem/progenitor cells differentiate to ECs, oxidative metabolism and glycolysis are increased, a mechanism potentially controlled by AKT/mTOR signaling.[58] Furthermore, FOXO1 suppresses Myc and cellular metabolism to regulate EC quiescence,[59] while KLF2, a proquiescence transcription factor, suppresses EC glycolysis.[60] Fatty acid synthase (FASN), a lipid synthesis enzyme, stimulates EC proliferation and angiogenesis via mTOR malonylation.[61] The M2 isoform of pyruvate kinase (PKM2) is another determining metabolic enzyme for EC proliferation to secure cell cycle progression (by NF-k/p53 signaling inhibition)[62,63] and EC quiescence to preserve vascular barrier function. Strikingly, PKM2 also indirectly modulates EC innate immune-modulatory functions epigenetically by adjusting S-adenosyl-methionine synthesis.[62] For a more in-depth characterization of regulation of EC metabolism, we refer to other reviews.[13,54,64]

Metabolic gene signatures of ECs

Single-cell studies uncovered an unforeseen metabolic transcriptome heterogeneity in different EC phenotypes. Indeed, when characterizing metabolic gene expression by ECs from various healthy mouse tissues, a specific metabolic gene signature was found for each EC subtype in these tissues.[4] A similar observation was made for ECs in disorders, in which new vessel formation occurs (CNV in the eye, lung tumors).[6,41] Remarkably, ECs from several tissues expressed highly similar metabolic transcriptome signatures, likely because these cells share similar biological functions (see above), indicating that the metabolic transcriptome signature contributes to EC identity.[4,41] Moreover, ECs from diverse tissues displayed metabolic transcriptome heterogeneity. For instance, brain ECs showed gene expression associated with glucose and AA transporters, while ECs from the spleen have a higher expression of cholesterol metabolism.[4,13] In turn, in lung ECs, metabolic gene expression suggests increased cAMP metabolism, whereas muscle and cardiac ECs upregulate genes related to metabolism and lipid uptake.[4,65] Additionally, ECs exhibit metabolic transcriptome heterogeneity within the vascular trees of specific tissues. For example, liver ECs express metabolic genes distinct for each vascular bed (capillaries, veins, arteries), while brain ECs express similar metabolic markers or markers specific to the vascular bed.[4] Brain ECs from particular vascular beds display heterogeneity of expression of transporter genes.[10,66,67] Also, renal ECs display zone-dependent metabolic heterogeneity on the transcriptome level. For instance, compared to cortical and glomerular ECs, medullary ECs showed enrichment in oxidative phosphorylation (OXPHOS) genes, a finding confirmed via metabolic flux analysis.[9] Furthermore, medullary ECs relied on oxidative metabolism to resist dehydration stress.[9]

In murine CNV and lung tumors, proliferating ECs exhibit an upregulated set of genes involved in nucleotide synthesis, one-carbon metabolism, glycolysis, OXPHOS, and TCA cycle.[41] Apart from the relevance of glycolysis for the functions of ECs,[47,48,50] OXPHOS is necessary for nucleotide synthesis during angiogenesis in the in vivo setting.[68,69] In human non-small cell lung cancer (NSCLC), various metabolic transcriptome adaptations were identified in subsets of tumor ECs (TECs). For instance, in capillary TECs, lipid metabolism–related gene expression is increased, whereas prostaglandin (metabolic) transcriptome signature is upregulated in venous TECs.[41] In addition, compared to TECs from early-stage ground glass nodules adenocarcinoma (GGN-ADC), TECs from late-stage solid lung adenocarcinoma (SADC) upregulate gene signatures of metabolic processes.[70] Comparison of three different cancers (colorectal, lung, ovarian) by single-cell analysis revealed increased expression of glycolytic- and OXPHOS-related genes in tip cells.[71] Lung TECs exhibit a downregulation of carbonic acid metabolic genes (a characteristic associated with alveolar ECs), but upregulate genes associated with glycolysis as well as OXPHOS.[71] Use of an integrated multiomics approach, including an EC-tailored genome-scale metabolic model (a mathematical model, predicting the essentiality of metabolic reactions), combined with functional validation, identified the metabolic genes *Aldh18a1* and *Sqle* (involved in, respectively, collagen and cholesterol synthesis) as novel angiogenic targets in CNV and lung tumors.[41]

EC heterogeneity in tissue repair

Vessel regeneration

EC dysfunction is a hallmark of many diseases, such as diabetes or atherosclerosis.[47] In addition, even in healthy individuals, the blood contains pathogens, and potentially toxic substances. Since ECs line blood vessels, they are continuously exposed to such hazards, which can cause endothelial damage, requiring repair/regeneration. Interestingly, despite the absence of angiogenesis in healthy (normal) quiescent tissues, a small population of angiogenic ECs was detected.[4] These cells might potentially be involved in endothelial regeneration following EC injury. In fact, ECs display several cycles of cell turnover throughout the entire adulthood[43,72] and thus are capable of self-renewal.

Several aspects of vascular repair still remain poorly understood, but single-cell studies are shedding light on the mechanisms of EC regeneration.[42,73] It is, for instance, debated whether stem-like ECs exist solely in large vessels or instead in all vascular beds.[42,73–78] It is also unclear whether other (non-endothelial) cell types (such as bone marrow–derived progenitors) contribute, or are even essential during vessel repair.[42,73–80] Vessel wall-resident ECs have been documented to contribute to vessel repair in studies, excluding substantial contributions of circulating stem cells or hematopoietic progenitors.[77,78,81,82] Alternatively, endothelial regeneration can rely on a rare population of $CD45^{neg}$ rESCs (characterized by expression of CD31, CD157, CD200) in large vessels of multiple organs.[73] $CD157^{+}$ ECs are able to activate the angiogenic program to repair damaged vessels, as observed in models of liver regeneration and hindlimb ischemia.[73] In another study investigating aortic injury, resident ECs (but not circulating progenitor cells) were responsible for endothelial regeneration.[42] Notably, these ECs displayed a distinct gene expression signature (dissimilar from sprouting ECs) with upregulation of genes involved in cell junctional complexes, and lack of migratory genes.[42] This divergent gene signature is reminiscent of that of cells that rapidly switch from a quiescent to a proliferative state. Moreover, endovascular progenitors were identified in the aorta by single-cell sequencing. Similar to other stem or progenitor cells, these endovascular progenitors show enriched mesenchymal gene expression and a high mitochondrial content,[74,83,84] and are able to transition to differentiated ECs, namely transit amplifying and mature ECs, as shown by pseudotime trajectory analysis.[74] Future studies will reveal whether vessel wall-resident progenitor ECs indeed contribute to vessel repair. Notably, stem-like ECs might also be present in the microvasculature.[75] A subset of capillary ECs with self-renewing capacity and other stem-like features differentiated into different EC phenotypes upon lung injury.[75,85] A subpopulation of capillary ECs harboring stem-like characteristics has also been detected in lymphoid tissues.[76] Future functional studies will characterize the contribution of such stem-like ECs to vessel repair.

Tissue repair

In 2019, the emergence of a new coronavirus strain, SARS-CoV-2, started a pandemic of global proportion. SARS-CoV-2 infection provokes COVID-19 disease, which in severe cases is characterized by critical acute respiratory distress syndrome (ARDS), which manifests in the lungs as well as widespread multiorgan dysfunction, progressing to fatal organ failure.[86] COVID-19 harbors multiple vascular ramifications, such as inflammation, thrombosis, and vascular leakage, which (at least in part) are likely caused by EC dysfunction.[86–91] Interestingly, a switch from SA to IA was detected in COVID-19 lungs.[88,89] Other lung afflictions, such as idiopathic pulmonary fibrosis (IPF)[92] or acute lung injury,[93] display similar vascular pathologies and neovascularization. We will discuss recent insights into the roles that EC subtypes play in such lung maladies.

First, "immunomodulatory ECs (IMECs)" are capable of interacting with different immune cells, thereby modulating the immune response.[4,66,94] In a murine model of systemic inflammation (by LPS treatment), ECs from the heart, lung, and brain showed enriched expression of genes associated with leukocyte trafficking.[66] Interestingly, despite the inter-tissue EC heterogeneity, LPS treatment of ECs isolated from different tissues induced transcriptomic signatures more similar to one another, as when compared to untreated ECs from different organs.[66] Notably, this study investigated ribosome-bound mRNAs, which are more likely to indicate functionally relevant transcriptional changes.[66] Single-cell studies identified gene expression signatures in ECs associated with immunoregulation, immune surveillance, scavenging, and interferon signaling.[4,8]

Interestingly, combinatorial single-cell RNA and Assay for Transposase-Accessible Chromatin (ATAC) sequencing of ECs from a partial carotid ligation model of atherosclerosis revealed that ECs are reprogrammed by disturbed flow, with enriched expression of immune-related genes (a phenotype coined "EndICLT").[95] This reprogramming seemed to underlie epigenetic regulation. Of note, carotid ECs did not express tissue-specific EC genes, reported in ECs from other organs,[4] highlighting further EC heterogeneity.[95]

Importantly, the reprogramming into the EndICLT phenotype did not rely on immune cells, as demonstrated in in vitro experiments with human aortic ECs exposed to disturbed flow (without immune cells).[95] Similarly, parallel bulk transcriptome and epigenome sequencing of ECs from 12 different organs revealed epigenetic reprogramming of ECs toward an immune-like phenotype, in response to inflammation (for instance, during viral infection).[96] ECs have previously been shown to engage in immune responses.[97,98] Importantly, these new insights reveal that ECs are extremely plastic cells capable of transitioning into immunomodulatory phenotypes; even though these phenotypes may well represent reversible transitory states, they nonetheless can be identified as a distinct statistically separable EC subset by single-cell analysis.[95] The transition of ECs into an immunomodulatory cell type is no unique observation. Epithelial cells in tumors have shown similar adaptations toward a phenotype with immune features.[99] The precise immunomodulatory functions of these IMECs in different contexts, such as normal tissue homeostasis, graft rejection reactions, and tumor immunity remain to be determined.[100,101]

Second, vascular repair can involve several different mechanisms of vascularization. For instance, proliferating ECs, a cluster enriched in metabolic and cell cycle genes, can contribute to vessel repair. In models of ischemia-induced neovascularization and choroidal SA, proliferative ECs clonally expand to ensure vessel repair.[102] IA represents another alternative mode of vessel growth during tissue repair. While SA is the predominant vascularization method detected in influenza (H1N1)-infected lungs[88] and usual interstitial pneumonia,[103] IA was found in rat models of colitis[104] and glomerulonephritis,[105] as well as in lungs from patients suffering alveolar fibroelastosis,[103] nonspecific interstitial pneumonia, and COVID-19 disease.[88]

Third, during vascular repair, EC subsets might migrate and spatially redistribute. Examples are COL15A1$^+$ peri-bronchial ECs (PBECs). In healthy lungs, PBECs can only be found in the bronchial vasculature adjacent to large airways, whereas in IPF, PBECs are enriched in areas of fibrosis and bronchiolization.[106] Possibly, neo-angiogenesis of the peri-bronchial vasculature is a consequence of the injury of peripheral pulmonary capillaries, which leads to increased pulmonary pressure and hypertrophy.

EC heterogeneity in cancer

Angiogenesis is one of the hallmarks of cancer. Using targeted approaches or bulk RNA sequencing, early studies demonstrated differences between TECs and ECs from healthy organs.[107–111] For instance, combined microarray analysis and single-cell sequencing unveiled distinct gene expression signatures in TECs from different human cancer types (lymphoma, colorectal, and breast cancer) compared to ECs from the respective healthy organs. Genes enriched in TECs included those associated with extracellular matrix metabolism, collagen deposition, diverse signaling pathways, as well as contraction of vascular smooth muscle cells.[112]

Single-cell analyses expanded our understanding of gene signatures in TECs. TECs from human NSCLC displayed distinct transcriptome signatures when compared to peritumoral lung ECs.[113] Expression levels of genes associated with Myc targets were upregulated in TECs, whereas those of antigen presentation genes (MHC class I/II) alongside genes encoding for chemokines and cytokines (CCL2, IL6, CCL18), and immune cell adhesion (ICAM1) were decreased in TECs. In line with these findings, TECs in SADC of the lung exhibited lower expression levels of antigen presentation genes compared to GGN-ADC, an earlier stage lung cancer.[70] These findings suggest that TECs might play a role in tumor immune evasion.

In-depth single-cell analysis comparing enriched TECs from NSCLC to their healthy (normal EC) counterparts provided additional insights.[6] Here, human NSCLC TECs and paired ECs from peritumoral tissues were investigated, in parallel with TECs from a murine lung cancer model (which were compared to healthy murine lung ECs). This analysis revealed a total of 33 distinct EC phenotypes in lung tumors and healthy lungs and provided key insights in TEC heterogeneity.[6] First, angiogenic EC phenotypes, such as tip or proliferating ECs (i.e., the angiogenic ECs), account for less than 10% of all TECs in human NSCLC, which potentially might (partly) explain the limited efficacy of anti-angiogenic therapy.[6] Interestingly, in tumors, traditional stalk cells were absent.[6] Second, in healthy lungs, a gene expression signature reminiscent of semi-professional antigen-presenting cells was detected in capillary ECs. In agreement with other single cell studies, these ECs were decreased in number in tumors.[6] Conversely, high-endothelial venules (HEVs), a subtype of ECs with immunoregulatory functions involved in immune cell recruitment, were expanded in tumors (Fig. 2.1B)[71]; thus, the interaction of the immune system with ECs in tumors seems to be complex and requires further elucidation. Third, in addition to traditional tip ECs, breach ECs were detected in tumors, likely assisting migrating tip cells (see above).[6] Fourth, the most sensitive TEC subsets to anti-angiogenic therapy (anti-VEGF therapy in tumor-bearing mice) were tip and breach TECs. However, proliferating TECs did not seem more sensitive to therapy, suggesting that TEC proliferation is stimulated by signals additional to or other than VEGF.[6] Venous TECs, on the other hand, expanded upon anti-VEGF therapy.

The observed expansion of venous TECs might contribute to anti-VEGF therapy resistance, considering that rESCs reside in the venous EC population (see above).[6]

ScRNAseq generates a tsunami of data and has enabled the identification of different EC subtypes based on distinct gene expression profiles. However, it remains challenging to distill relevant novel angiogenic targets from these tidal waves of data. Recently, an approach was put forward based on the hypothesis that congruent gene signatures between different datasets, across different disease models, species, and experimental conditions, would more likely represent biologically important transcriptome profiles and thereby identify relevant candidate target genes and EC subsets.[6] In fact, in such an integrated multi-disciplinary study, encompassing human and murine lung cancer tissues, a multi-step meta-analysis was performed on in-house generated scRNAseq, bulk transcriptomics, and proteomics, as well as publicly available bulk transcriptomics datasets of five different cancer types. This integrated analysis, combined with functional validation, identified the novel angiogenic targets PLOD1 and PLOD2.[6] Future integrated approaches will likely also focus on metabolic transcriptome signatures. Metabolic targets within the tumor vasculature hold promise for future therapy, as demonstrated by preclinical models (see above).

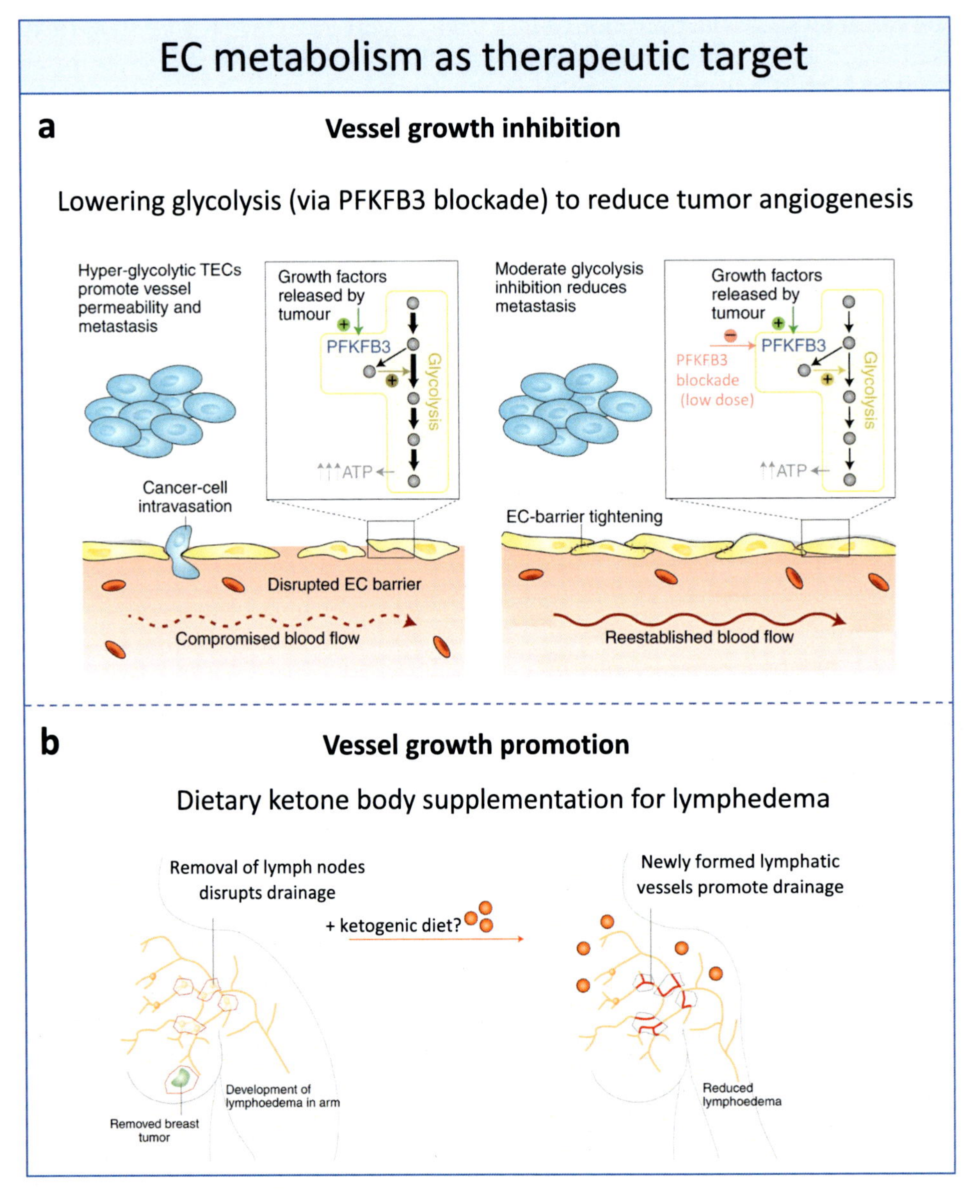

FIGURE 2.3 **EC metabolism as therapeutic target.** (A) Therapeutic targeting of EC metabolism might be used to inhibit vessel growth: lowering glycolysis by partial blockade of PFKFB3 induces tumor vessel normalization and reduces metastasis, partly by reestablishing the tumor EC (TEC) barrier. (B) A current clinical trial is evaluating whether a ketogenic diet reduces lymphedema by improving reconnection of disrupted lymph vessels through stimulation of LEC growth. *Both schematics are adapted from Falkenberg KD, Rohlenova K, Luo Y, Carmeliet P. The metabolic engine of endothelial cells.* Nat Metab. *2019;1:937–946. https://doi.org/10.1038/s42255-019-0117-9.*

Perspectives and therapeutic implications

Historically, ECs were classified according to their resident location within the vascular tree (e.g., capillaries, arteries, veins) and today the classification further includes their position within the vessel sprout (e.g., phalanx, stalk, tip ECs). Bulk omics studies uncovered that ECs displayed gene signatures that are tissue specific, while scRNAseq illustrated that ECs are even more heterogenous and demonstrate a greater plasticity than originally anticipated. For example, healthy ECs are flexible to adapt and thus change their transcriptome to differentiate temporarily into new EC phenotypes (e.g., immuno-modulatory ECs). Notably, single-cell studies identified an unforeseen metabolic transcriptome heterogeneity in ECs of distinct tissues but also within the vascular tree of a single tissue. Within the near future, these distinct metabolic and functional EC phenotypes might become potential targets for new therapies. A major challenge will be to decode how to target these EC phenotypes specifically (without affecting other non-EC cell types). Given the nonfixed, transient state of EC phenotypes, an interesting, unexplored strategy might be to "retune" these functional phenotypes, instead of the "pruning" strategy of current anti-angiogenic therapies.

As highlighted here, single-cell analyses generate long lists of multiple possible candidates. Therefore, another challenge is to determine and prioritize targets for clinical translation. A powerful approach to discover biologically relevant targets, presents the combination of single cell approaches with analysis of congruent gene expression across different species, conditions, tissues, etc. Furthermore, multi-omics approaches (combination of transcriptomic signatures with spatial and/or epigenetic information) might be instrumental as an additional novel strategy to further elucidate the mechanisms underlying this heterogeneity.[114]

Because of its critical role, EC metabolism might represent a possible new therapeutic angiogenic target. Because tumor vessels are structurally and functionally abnormal, they facilitate cancer cell intravasation through their leaky vessel wall and thereby metastasis.[111,115] TECs are even more hyperglycolytic than healthy ECs.[54,116] Treatment with a low dose of a PFKFB3-blocker lowered TEC glycolysis by 30%, nonetheless sufficient to normalize the structure and function of tumor vessels, to tighten the vascular barrier and to impair cancer cell intravasation[116] (Fig. 2.3A). On the other hand, metabolite delivery might also become a new treatment option for diseases, resulting from impaired vessel growth. Lymphedema (swelling of, for instance, arms) in breast cancer patients results from interruption of lymph drainage due to surgical resection of the cancer cell–infested lymph nodes. This condition is currently uncurable, yet represents a substantial invalidating challenge. Based on fundamental insights on the role of ketone bodies in lymphatic vessel growth[56,57] (see above), a clinical trial (NCT03991897) is currently evaluating the possible therapeutic effect of a ketogenic diet to reduce lymphedema by stimulating lymph vessel growth to reconnect the interrupted lymphatics and restore lymph drainage (Fig. 2.3B).

The efficacy of current anti-angiogenic therapy needs improvement. Focusing on EC metabolism, heterogeneity and alternative functions (such as its immunomodulatory functions) may offer new alternative opportunities to achieve this goal. Regardless of the challenge to develop new vascular medicines, the journey to unravel the scientific insights, underlying these processes, promises to be exciting and rewarding.

References

1. Aird WC. Phenotypic heterogeneity of the endothelium: I. Structure, function, and mechanisms. *Circ Res*. 2007;100:158–173. https://doi.org/10.1161/01.RES.0000255691.76142.4a.
2. Potente M, Makinen T. Vascular heterogeneity and specialization in development and disease. *Nat Rev Mol Cell Biol*. 2017;18: 477–494. https://doi.org/10.1038/nrm.2017.36.
3. Aird WC. Endothelial cell heterogeneity. *Cold Spring Harb Perspect Med*. 2012;2:a006429. https://doi.org/10.1101/cshperspect.a006429.
4. Kalucka J, de Rooij LPMH, Goveia J, et al. Single-cell transcriptome atlas of murine endothelial cells. *Cell*. 2020;180. https://doi.org/10.1016/j.cell.2020.01.015, 764–779. e720.
5. Jakab M, Augustin HG. Understanding angiodiversity: insights from single cell biology. *Development*. 2020;147. https://doi.org/10.1242/dev.146621.
6. Goveia J, Rohlenova K, Taverna F, et al. An integrated gene expression landscape profiling approach to identify lung tumor endothelial cell heterogeneity and angiogenic candidates. *Cancer Cell*. 2020;37:421. https://doi.org/10.1016/j.ccell.2020.03.002.
7. Chavkin NW, Hirschi KK. Single cell analysis in vascular biology. *Front Cardiovasc Med*. 2020;7:42. https://doi.org/10.3389/fcvm.2020.00042.
8. Paik DT, Tian L, Williams IM, et al. Single-cell RNA sequencing unveils unique transcriptomic signatures of organ-specific endothelial cells. *Circulation*. 2020;142:1848–1862. https://doi.org/10.1161/CIRCULATIONAHA.119.041433.
9. Dumas SJ, Meta E, Borri M, et al. Single-cell RNA sequencing reveals renal endothelium heterogeneity and metabolic adaptation to water deprivation. *J Am Soc Nephrol*. 2020;31:118–138. https://doi.org/10.1681/ASN.2019080832.
10. Vanlandewijck M, He L, Mae MA, et al. A molecular atlas of cell types and zonation in the brain vasculature. *Nature*. 2018;554: 475–480. https://doi.org/10.1038/nature25739.
11. Koch PS, Lee KH, Goerdt S, Augustin HG. Angiodiversity and organotypic functions of sinusoidal endothelial cells. *Angiogenesis*. 2021. https://doi.org/10.1007/s10456-021-09780-y.
12. Su T, Stanley G, Sinha R, et al. Single-cell analysis of early progenitor cells that build coronary arteries. *Nature*. 2018;559:356–362. https://doi.org/10.1038/s41586-018-0288-7.
13. Dumas SJ, Garcia-Caballero M, Carmeliet P. Metabolic signatures of distinct endothelial phenotypes. *Trends Endocrinol Metab*. 2020; 31:580–595. https://doi.org/10.1016/j.tem.2020.05.009.
14. Augustin HG, Koh GY. Organotypic vasculature: from descriptive heterogeneity to functional pathophysiology. *Science*. 2017;357. https://doi.org/10.1126/science.aal2379.

15. You LR, Lin FJ, Lee CT, et al. Suppression of Notch signalling by the COUP-TFII transcription factor regulates vein identity. *Nature.* 2005;435:98–104. https://doi.org/10.1038/nature03511.
16. Red-Horse K, Ueno H, Weissman IL, Krasnow MA. Coronary arteries form by developmental reprogramming of venous cells. *Nature.* 2010;464:549–553. https://doi.org/10.1038/nature08873.
17. Conway EM, Collen D, Carmeliet P. Molecular mechanisms of blood vessel growth. *Cardiovasc Res.* 2001;49:507–521. https://doi.org/10.1016/s0008-6363(00)00281-9.
18. Nicenboim J, Malkinson G, Lupo T, et al. Lymphatic vessels arise from specialized angioblasts within a venous niche. *Nature.* 2015; 522:56–61. https://doi.org/10.1038/nature14425.
19. Mack JJ, Iruela-Arispe ML. NOTCH regulation of the endothelial cell phenotype. *Curr Opin Hematol.* 2018;25:212–218. https://doi.org/10.1097/MOH.0000000000000425.
20. Luo W, Garcia-Gonzalez I, Fernandez-Chacon M, et al. Arterialization requires the timely suppression of cell growth. *Nature.* 2020. https://doi.org/10.1038/s41586-020-3018-x.
21. Rajendran P, Rengarajan T, Thangavel J, et al. The vascular endothelium and human diseases. *Int J Biol Sci.* 2013;9:1057–1069. https://doi.org/10.7150/ijbs.7502.
22. Lis R, Karrasch CC, Poulos MG, et al. Conversion of adult endothelium to immunocompetent haematopoietic stem cells. *Nature.* 2017;545:439–445. https://doi.org/10.1038/nature22326.
23. Canu G, Athanasiadis E, Grandy RA, et al. Analysis of endothelial-to-haematopoietic transition at the single cell level identifies cell cycle regulation as a driver of differentiation. *Genome Biol.* 2020;21:157. https://doi.org/10.1186/s13059-020-02058-4.
24. Chestnut B, Casie Chetty S, Koenig AL, Sumanas S. Single-cell transcriptomic analysis identifies the conversion of zebrafish Etv2-deficient vascular progenitors into skeletal muscle. *Nat Commun.* 2020;11:2796. https://doi.org/10.1038/s41467-020-16515-y.
25. De Val S, Chi NC, Meadows SM, et al. Combinatorial regulation of endothelial gene expression by ets and forkhead transcription factors. *Cell.* 2008;135:1053–1064. https://doi.org/10.1016/j.cell.2008.10.049.
26. Sumanas S, Choi K. ETS transcription factor ETV2/ER71/Etsrp in hematopoietic and vascular development. *Curr Top Dev Biol.* 2016; 118:77–111. https://doi.org/10.1016/bs.ctdb.2016.01.005.
27. Kardon G, Campbell JK, Tabin CJ. Local extrinsic signals determine muscle and endothelial cell fate and patterning in the vertebrate limb. *Dev Cell.* 2002;3:533–545. https://doi.org/10.1016/s1534-5807(02)00291-5.
28. Mayeuf-Louchart A, Lagha M, Danckaert A, et al. Notch regulation of myogenic versus endothelial fates of cells that migrate from the somite to the limb. *Proc Natl Acad Sci USA.* 2014;111: 8844–8849. https://doi.org/10.1073/pnas.1407606111.
29. Qiu J, Hirschi KK. Endothelial cell development and its application to regenerative medicine. *Circ Res.* 2019;125:489–501. https://doi.org/10.1161/CIRCRESAHA.119.311405.
30. Ng AHM, Khoshakhlagh P, Rojo Arias JE, et al. A comprehensive library of human transcription factors for cell fate engineering. *Nat Biotechnol.* 2020. https://doi.org/10.1038/s41587-020-0742-6.
31. Augustin HG. Tubes, branches, and pillars: the many ways of forming a new vasculature. *Circ Res.* 2001;89:645–647.
32. De Spiegelaere W, Casteleyn C, Van den Broeck W, et al. Intussusceptive angiogenesis: a biologically relevant form of angiogenesis. *J Vasc Res.* 2012;49:390–404. https://doi.org/10.1159/000338278.
33. Fan X, Bialecka M, Moustakas I, et al. Single-cell reconstruction of follicular remodeling in the human adult ovary. *Nat Commun.* 2019;10:3164. https://doi.org/10.1038/s41467-019-11036-9.
34. Wang W, Vilella F, Alama P, et al. Single-cell transcriptomic atlas of the human endometrium during the menstrual cycle. *Nat Med.* 2020;26:1644–1653. https://doi.org/10.1038/s41591-020-1040-z.
35. Potente M, Gerhardt H, Carmeliet P. Basic and therapeutic aspects of angiogenesis. *Cell.* 2011;146:873–887. https://doi.org/10.1016/j.cell.2011.08.039.
36. Hellstrom M, Phng LK, Hofmann JJ, et al. Dll4 signalling through Notch1 regulates formation of tip cells during angiogenesis. *Nature.* 2007;445:776–780. https://doi.org/10.1038/nature05571.
37. Hellstrom M, Phng LK, Gerhardt H. VEGF and Notch signaling: the yin and yang of angiogenic sprouting. *Cell Adhes Migrat.* 2007;1:133–136. https://doi.org/10.4161/cam.1.3.4978.
38. Jakobsson L, Franco CA, Bentley K, et al. Endothelial cells dynamically compete for the tip cell position during angiogenic sprouting. *Nat Cell Biol.* 2010;12:943–953. https://doi.org/10.1038/ncb2103.
39. Tammela T, Zarkada G, Nurmi H, et al. VEGFR-3 controls tip to stalk conversion at vessel fusion sites by reinforcing Notch signalling. *Nat Cell Biol.* 2011;13:1202–1213. https://doi.org/10.1038/ncb2331.
40. Boareto M, Jolly MK, Ben-Jacob E, Onuchic JN. Jagged mediates differences in normal and tumor angiogenesis by affecting tip-stalk fate decision. *Proc Natl Acad Sci USA.* 2015;112: E3836–E3844. https://doi.org/10.1073/pnas.1511814112.
41. Rohlenova K, Goveia J, Garcia-Caballero M, et al. Single-cell RNA sequencing maps endothelial metabolic plasticity in pathological angiogenesis. *Cell Metabol.* 2020;31:862–877. e814. https://doi.org/10.1016/j.cmet.2020.03.009.
42. McDonald AI, Shirali AS, Aragon R, et al. Endothelial regeneration of large vessels is a biphasic process driven by local cells with distinct proliferative capacities. *Cell Stem Cell.* 2018;23. https://doi.org/10.1016/j.stem.2018.07.011, 210–225. e216.
43. Wang Z, Cui M, Shah AM, et al. Cell-type-specific gene regulatory networks underlying murine neonatal heart regeneration at single-cell resolution. *Cell Rep.* 2020;33:108472. https://doi.org/10.1016/j.celrep.2020.108472.
44. Blanco R, Gerhardt H. VEGF and Notch in tip and stalk cell selection. *Cold Spring Harb Perspect Med.* 2013;3:a006569. https://doi.org/10.1101/cshperspect.a006569.
45. Chen W, Xia P, Wang H, et al. The endothelial tip-stalk cell selection and shuffling during angiogenesis. *J Cell Commun Signal.* 2019;13:291–301. https://doi.org/10.1007/s12079-019-00511-z.
46. Herbert SP, Huisken J, Kim TN, et al. Arterial-venous segregation by selective cell sprouting: an alternative mode of blood vessel formation. *Science.* 2009;326:294–298. https://doi.org/10.1126/science.1178577.
47. Li X, Sun X, Carmeliet P. Hallmarks of endothelial cell metabolism in health and disease. *Cell Metabol.* 2019;30:414–433. https://doi.org/10.1016/j.cmet.2019.08.011.
48. De Bock K, Georgiadou M, Schoors S, et al. Role of PFKFB3-driven glycolysis in vessel sprouting. *Cell.* 2013;154:651–663. https://doi.org/10.1016/j.cell.2013.06.037.
49. De Bock K, Georgiadou M, Carmeliet P. Role of endothelial cell metabolism in vessel sprouting. *Cell Metabol.* 2013;18:634–647. https://doi.org/10.1016/j.cmet.2013.08.001.
50. Schoors S, De Bock K, Cantelmo AR, et al. Partial and transient reduction of glycolysis by PFKFB3 blockade reduces pathological angiogenesis. *Cell Metabol.* 2014;19:37–48. https://doi.org/10.1016/j.cmet.2013.11.008.
51. Schoors S, Bruning U, Missiaen R, et al. Fatty acid carbon is essential for dNTP synthesis in endothelial cells. *Nature.* 2015;520: 192–197. https://doi.org/10.1038/nature14362.
52. Kalucka J, Bierhansl L, Vasconcelos Conchinha N, et al. Quiescent endothelial cells upregulate fatty acid beta-oxidation for vasculoprotection via redox homeostasis. *Cell Metabol.* 2018;28:881–894. e813. https://doi.org/10.1016/j.cmet.2018.07.016.
53. Wong BW, Wang X, Zecchin A, et al. The role of fatty acid beta-oxidation in lymphangiogenesis. *Nature.* 2017;542:49–54. https://doi.org/10.1038/nature21028.

54. Eelen G, de Zeeuw P, Treps L, et al. Endothelial cell metabolism. *Physiol Rev.* 2018;98:3–58. https://doi.org/10.1152/physrev.00001.2017.
55. Huang H, Vandekeere S, Kalucka J, et al. Role of glutamine and interlinked asparagine metabolism in vessel formation. *EMBO J.* 2017;36:2334–2352. https://doi.org/10.15252/embj.201695518.
56. Falkenberg KD, Rohlenova K, Luo Y, Carmeliet P. The metabolic engine of endothelial cells. *Nat Metab.* 2019;1:937–946. https://doi.org/10.1038/s42255-019-0117-9.
57. Garcia-Caballero M, Zecchin A, Souffreau J, et al. Role and therapeutic potential of dietary ketone bodies in lymph vessel growth. *Nat Metab.* 2019;1:666–675. https://doi.org/10.1038/s42255-019-0087-y.
58. Sun S, Chen S, Liu F, et al. Constitutive activation of mTORC1 in endothelial cells leads to the development and progression of lymphangiosarcoma through VEGF autocrine signaling. *Cancer Cell.* 2015;28:758–772. https://doi.org/10.1016/j.ccell.2015.10.004.
59. Wilhelm K, Happel K, Eelen G, et al. FOXO1 couples metabolic activity and growth state in the vascular endothelium. *Nature.* 2016; 529:216–220. https://doi.org/10.1038/nature16498.
60. Doddaballapur A, Michalik KM, Manavski Y, et al. Laminar shear stress inhibits endothelial cell metabolism via KLF2-mediated repression of PFKFB3. *Arterioscler Thromb Vasc Biol.* 2015;35: 137–145. https://doi.org/10.1161/ATVBAHA.114.304277.
61. Bruning U, Morales-Rodriguez F, Kalucka J, et al. Impairment of angiogenesis by fatty acid synthase inhibition involves mTOR malonylation. *Cell Metabol.* 2018;28:866–880 e815. https://doi.org/10.1016/j.cmet.2018.07.019.
62. Stone OA, El-Brolosy M, Wilhelm K, et al. Loss of pyruvate kinase M2 limits growth and triggers innate immune signaling in endothelial cells. *Nat Commun.* 2018;9:4077. https://doi.org/10.1038/s41467-018-06406-8.
63. Kim B, Jang C, Dharaneeswaran H, et al. Endothelial pyruvate kinase M2 maintains vascular integrity. *J Clin Invest.* 2018;128: 4543–4556. https://doi.org/10.1172/JCI120912.
64. Teuwen LA, Geldhof V, Carmeliet P. How glucose, glutamine and fatty acid metabolism shape blood and lymph vessel development. *Dev Biol.* 2019;447:90–102. https://doi.org/10.1016/j.ydbio.2017.12.001.
65. Son NH, Basu D, Samovski D, et al. Endothelial cell CD36 optimizes tissue fatty acid uptake. *J Clin Invest.* 2018;128:4329–4342. https://doi.org/10.1172/JCI99315.
66. Jambusaria A, Hong Z, Zhang L, et al. Endothelial heterogeneity across distinct vascular beds during homeostasis and inflammation. *Elife.* 2020;9. https://doi.org/10.7554/eLife.51413.
67. He L, Vanlandewijck M, Mae MA, et al. Single-cell RNA sequencing of mouse brain and lung vascular and vessel-associated cell types. *Sci Data.* 2018;5:180160. https://doi.org/10.1038/sdata.2018.160.
68. Schiffmann LM, Werthenbach JP, Heintges-Kleinhofer F, et al. Mitochondrial respiration controls neoangiogenesis during wound healing and tumour growth. *Nat Commun.* 2020;11:3653. https://doi.org/10.1038/s41467-020-17472-2.
69. Diebold LP, Gil HJ, Gao P, et al. Mitochondrial complex III is necessary for endothelial cell proliferation during angiogenesis. *Nat Metab.* 2019;1:158–171. https://doi.org/10.1038/s42255-018-0011-x.
70. Lu T, Yang X, Shi Y, et al. Single-cell transcriptome atlas of lung adenocarcinoma featured with ground glass nodules. *Cell Discov.* 2020;6:69. https://doi.org/10.1038/s41421-020-00200-x.
71. Qian J, Olbrecht S, Boeckx B, et al. A pan-cancer blueprint of the heterogeneous tumor microenvironment revealed by single-cell profiling. *Cell Res.* 2020;30:745–762. https://doi.org/10.1038/s41422-020-0355-0.
72. Bergmann O, Zdunek S, Felker A, et al. Dynamics of cell generation and turnover in the human heart. *Cell.* 2015;161:1566–1575. https://doi.org/10.1016/j.cell.2015.05.026.
73. Wakabayashi T, Naito H, Suehiro JI, et al. CD157 marks tissue-resident endothelial stem cells with homeostatic and regenerative properties. *Cell Stem Cell.* 2018;22:384–397 e386. https://doi.org/10.1016/j.stem.2018.01.010.
74. Lukowski SW, Patel J, Andersen SB, et al. Single-cell transcriptional profiling of aortic endothelium identifies a hierarchy from endovascular progenitors to differentiated cells. *Cell Rep.* 2019; 27:2748–2758 e2743. https://doi.org/10.1016/j.celrep.2019.04.102.
75. Gillich A, Zhang F, Farmer CG, et al. Capillary cell-type specialization in the alveolus. *Nature.* 2020. https://doi.org/10.1038/s41586-020-2822-7.
76. Brulois K, Rajaraman A, Szade A, et al. A molecular map of murine lymph node blood vascular endothelium at single cell resolution. *Nat Commun.* 2020;11:3798. https://doi.org/10.1038/s41467-020-17291-5.
77. Deng J, Ni Z, Gu W, et al. Single-cell gene profiling and lineage tracing analyses revealed novel mechanisms of endothelial repair by progenitors. *Cell Mol Life Sci.* 2020;77:5299–5320. https://doi.org/10.1007/s00018-020-03480-4.
78. Chen Q, Yang M, Wu H, et al. Genetic lineage tracing analysis of c-kit(+) stem/progenitor cells revealed a contribution to vascular injury-induced neointimal lesions. *J Mol Cell Cardiol.* 2018;121: 277–286. https://doi.org/10.1016/j.yjmcc.2018.07.252.
79. Caballero S, Hazra S, Bhatwadekar A, et al. Circulating mononuclear progenitor cells: differential roles for subpopulations in repair of retinal vascular injury. *Invest Ophthalmol Vis Sci.* 2013; 54:3000–3009. https://doi.org/10.1167/iovs.12-10280.
80. Hu Y, Davison F, Zhang Z, Xu Q. Endothelial replacement and angiogenesis in arteriosclerotic lesions of allografts are contributed by circulating progenitor cells. *Circulation.* 2003;108: 3122–3127. https://doi.org/10.1161/01.CIR.0000105722.96112.67.
81. Ingram DA, Mead LE, Moore DB, et al. Vessel wall-derived endothelial cells rapidly proliferate because they contain a complete hierarchy of endothelial progenitor cells. *Blood.* 2005;105:2783–2786. https://doi.org/10.1182/blood-2004-08-3057.
82. Naito H, Kidoya H, Sakimoto S, Wakabayashi T, Takakura N. Identification and characterization of a resident vascular stem/progenitor cell population in preexisting blood vessels. *EMBO J.* 2012;31:842–855. https://doi.org/10.1038/emboj.2011.465.
83. Khacho M, Slack RS. Mitochondrial activity in the regulation of stem cell self-renewal and differentiation. *Curr Opin Cell Biol.* 2017;49:1–8. https://doi.org/10.1016/j.ceb.2017.11.003.
84. Zhang H, Ryu D, Wu Y, et al. NAD(+) repletion improves mitochondrial and stem cell function and enhances life span in mice. *Science.* 2016;352:1436–1443. https://doi.org/10.1126/science.aaf2693.
85. Vila Ellis L, Cain MP, Hutchison V, et al. Epithelial Vegfa specifies a distinct endothelial population in the mouse lung. *Dev Cell.* 2020;52:617–630 e616. https://doi.org/10.1016/j.devcel.2020.01.009.
86. Wadman M, Couzin-Frankel J, Kaiser J, Matacic C. A rampage through the body. *Science.* 2020;368:356–360. https://doi.org/10.1126/science.368.6489.356.
87. Teuwen LA, Geldhof V, Pasut A, Carmeliet P. COVID-19: the vasculature unleashed. *Nat Rev Immunol.* 2020;20:389–391. https://doi.org/10.1038/s41577-020-0343-0.
88. Ackermann M, Verleden SE, Kuehnel M, et al. Pulmonary vascular endothelialitis, thrombosis, and angiogenesis in Covid-19. *N Engl J Med.* 2020;383:120–128. https://doi.org/10.1056/NEJMoa2015432.
89. Ackermann M, Mentzer SJ, Kolb M, Jonigk D. Inflammation and intussusceptive angiogenesis in COVID-19: everything in and

out of flow. *Eur Respir J*. 2020;56. https://doi.org/10.1183/13993003.03147-2020.
90. Price LC, McCabe C, Garfield B, Wort SJ. Thrombosis and COVID-19 pneumonia: the clot thickens!. *Eur Respir J*. 2020;56. https://doi.org/10.1183/13993003.01608-2020.
91. Rovas A, Osiaevi I, Buscher K, et al. Microvascular dysfunction in COVID-19: the MYSTIC study. *Angiogenesis*. 2020. https://doi.org/10.1007/s10456-020-09753-7.
92. Cosgrove GP, Brown KK, Schiemann WP, et al. Pigment epithelium-derived factor in idiopathic pulmonary fibrosis: a role in aberrant angiogenesis. *Am J Respir Crit Care Med*. 2004;170: 242–251. https://doi.org/10.1164/rccm.200308-1151OC.
93. Minamino T, Komuro I. Regeneration of the endothelium as a novel therapeutic strategy for acute lung injury. *J Clin Invest*. 2006;116:2316–2319. https://doi.org/10.1172/JCI29637.
94. Gola A, Dorrington MG, Speranza E, et al. Commensal-driven immune zonation of the liver promotes host defence. *Nature*. 2020. https://doi.org/10.1038/s41586-020-2977-2.
95. Andueza A, Kumar S, Kim J, et al. Endothelial reprogramming by disturbed flow revealed by single-cell RNA and Chromatin accessibility study. *Cell Rep*. 2020;33:108491. https://doi.org/10.1016/j.celrep.2020.108491.
96. Krausgruber T, Fortelny N, Fife-Gernedl V, et al. Structural cells are key regulators of organ-specific immune responses. *Nature*. 2020;583:296–302. https://doi.org/10.1038/s41586-020-2424-4.
97. Zhao Q, Del Pilar Molina-Portela M, Parveen A, et al. Heterogeneity and chimerism of endothelial cells revealed by single-cell transcriptome in orthotopic liver tumors. *Angiogenesis*. 2020;23: 581–597. https://doi.org/10.1007/s10456-020-09727-9.
98. Pober JS, Merola J, Liu R, Manes TD. Antigen presentation by vascular cells. *Front Immunol*. 2017;8:1907. https://doi.org/10.3389/fimmu.2017.01907.
99. Choi SY, Gout PW, Collins CC, Wang Y. Epithelial immune cell-like transition (EIT): a proposed transdifferentiation process underlying immune-suppressive activity of epithelial cancers. *Differentiation*. 2012;83:293–298. https://doi.org/10.1016/j.diff.2012.02.005.
100. Cai Q, Moore SA, Hendricks AR, Torrealba JR. Upregulation of endothelial HLA class II is a marker of antibody-mediated rejection in heart allograft biopsies. *Transplant Proc*. 2020;52: 1192–1197. https://doi.org/10.1016/j.transproceed.2020.01.049.
101. Al-Lamki RS, Bradley JR, Pober JS. Endothelial cells in allograft rejection. *Transplantation*. 2008;86:1340–1348. https://doi.org/10.1097/TP.0b013e3181891d8b.
102. Manavski Y, Lucas T, Glaser SF, et al. Clonal expansion of endothelial cells contributes to ischemia-induced neovascularization. *Circ Res*. 2018;122:670–677. https://doi.org/10.1161/CIRCRESAHA.117.312310.
103. Ackermann M, Stark H, Neubert L, et al. Morphomolecular motifs of pulmonary neoangiogenesis in interstitial lung diseases. *Eur Respir J*. 2020;55. https://doi.org/10.1183/13993003.00933-2019.
104. Konerding MA, Turhan A, Ravnic DJ, et al. Inflammation-induced intussusceptive angiogenesis in murine colitis. *Anat Rec (Hoboken)*. 2010;293:849–857. https://doi.org/10.1002/ar.21110.
105. Notoya M, Shinosaki T, Kobayashi T, Sakai T, Kurihara H. Intussusceptive capillary growth is required for glomerular repair in rat Thy-1.1 nephritis. *Kidney Int*. 2003;63:1365–1373. https://doi.org/10.1046/j.1523-1755.2003.00876.x.
106. Adams TS, Schupp JC, Poli S, et al. Single-cell RNA-seq reveals ectopic and aberrant lung-resident cell populations in idiopathic pulmonary fibrosis. *Sci Adv*. 2020;6:eaba1983. https://doi.org/10.1126/sciadv.aba1983.
107. St Croix B, Rago C, Velculescu V, et al. Genes expressed in human tumor endothelium. *Science*. 2000;289:1197–1202. https://doi.org/10.1126/science.289.5482.1197.
108. Viallard C, Larrivee B. Tumor angiogenesis and vascular normalization: alternative therapeutic targets. *Angiogenesis*. 2017;20: 409–426. https://doi.org/10.1007/s10456-017-9562-9.
109. Seaman S, Stevens J, Yang MY, et al. Genes that distinguish physiological and pathological angiogenesis. *Cancer Cell*. 2007;11: 539–554. https://doi.org/10.1016/j.ccr.2007.04.017.
110. Luo W, Hu Q, Wang D, et al. Isolation and genome-wide expression and methylation characterization of $CD31^+$ cells from normal and malignant human prostate tissue. *Oncotarget*. 2013;4: 1472–1483. https://doi.org/10.18632/oncotarget.1269.
111. Carmeliet P, Jain RK. Principles and mechanisms of vessel normalization for cancer and other angiogenic diseases. *Nat Rev Drug Discov*. 2011;10:417–427. https://doi.org/10.1038/nrd3455.
112. Sun Z, Wang CY, Lawson DA, et al. Single-cell RNA sequencing reveals gene expression signatures of breast cancer-associated endothelial cells. *Oncotarget*. 2018;9:10945–10961. https://doi.org/10.18632/oncotarget.23760.
113. Lambrechts D, Wauters E, Boeckx B, et al. Phenotype molding of stromal cells in the lung tumor microenvironment. *Nat Med*. 2018; 24:1277–1289. https://doi.org/10.1038/s41591-018-0096-5.
114. Cusanovich DA, Hill AJ, Aghamirzaie D, et al. A single-cell atlas of in vivo mammalian Chromatin accessibility. *Cell*. 2018;174: 1309–1324 e1318. https://doi.org/10.1016/j.cell.2018.06.052.
115. Jain RK. Normalization of tumor vasculature: an emerging concept in antiangiogenic therapy. *Science*. 2005;307:58–62. https://doi.org/10.1126/science.1104819.
116. Cantelmo AR, Conradi LC, Brajic A, et al. Inhibition of the glycolytic activator PFKFB3 in endothelium induces tumor vessel normalization, impairs metastasis, and improves chemotherapy. *Cancer Cell*. 2016;30:968–985. https://doi.org/10.1016/j.ccell.2016.10.006.

SECTION 2

Vasculome's key building blocks—beyond the endothelium

CHAPTER

3

The remarkable diversity of vascular smooth muscle in development and disease: a paradigm for mesenchymal cell types

Mark W. Majesky[1,2,3]

[1]Center for Developmental Biology & Regenerative Medicine, Seattle Children's Research Institute, University of Washington, Seattle, WA, United States [2]Departments of Pediatrics, University of Washington, Seattle, WA, United States [3]Laboratory Medicine & Pathology, University of Washington, Seattle, WA, United States

Introduction

In the mid-1960s, the long-held belief that the arterial media is composed of a single type of vascular smooth muscle cell (VSMC)[1,2,3] was being challenged.[4-6] This began a period in vascular biology when existing dogma were contested and key discoveries were made. These advances were accelerated by new technologies and produced a new vision of how blood and lymphatic vessels formed and functioned. For VSMCs, this period saw different models of SMC heterogeneity proposed, debated, rejected, or revised (reviewed in Ref. 7). Looking back, the great value of that debate was the series of experiments it inspired that led to our current understanding of the remarkable diversity of VSMC types within the vascular and lymphatic systems and the extreme plasticity of the SMC itself. Looking ahead, it is now possible to dissect the essential features of VSMC diversity at a whole genome level by applying rapid advances in DNA sequencing, single-cell technologies, and computational biology. The lessons learned suggest that new insights about VSMC diversity may also become a paradigm for a broad range of equally diverse but poorly understood mesenchymal cell types.

The vascular system is a developmental mosaic

The advanced technologies referred to above have proven most insightful when applied to questions of a type and scale that were not previously accessible. Yet, at the outset, it is important to keep in mind that biology rests on chemistry. For vascular development, the diffusion distance of oxygen in biological tissues is between 30 μm and 100 μm, and is often much less.[8,9] Since most cells in vertebrate embryos require a continuous supply of oxygen to grow, metabolize, divide, and function,[10] this diffusion distance requirement places physical–chemical constraints on morphogenesis since rapidly dividing cells with high metabolic rates must be positioned near a capillary vessel. As a result, the developing vascular system rapidly extends to all parts of the growing embryo and serves to pattern new organs as they form by organizing cells around nascent capillary networks.[11] As development proceeds, the heart becomes a more efficient pumping organ and intravascular pressures and blood flow rates increase to the point where a mural cell coating is required to provide mechanical strength and vasomotor control.[12,162] Since the timing of mural cell investment differs significantly depending on vessel caliber and distance from the heart, VSMC recruitment and differentiation are local events that collectively extend over many days or weeks of embryonic and even perinatal development. It is therefore not possible for a single source of progenitors to supply sufficient numbers of VSMCs to satisfy these dynamic spatial and temporal requirements. In fact, fate mapping studies now identify a surprising number of independent origins for SMCs in vascular and lymphatic development, a feature of embryonic and perinatal life that has important implications for vascular homeostasis and disease in adults.[14–17,18] Thus, the application of Cre-, Dre-, and CRISPR-site specific recombinases and gene editing tools to cell fate mapping in

The Vasculome
https://doi.org/10.1016/B978-0-12-822546-2.00030-7

vascular development has produced what could be called a "second anatomy" of the vascular system where the mosaic nature and clonal structure of blood vessels is clearly revealed.

Assembly of the vessel wall in development

Blood vessel walls assemble in a radial fashion that is highly responsive to biomechanical cues.[19–22] For conduit arteries, vessel diameters are proportional to blood flow rates[23] and the number of VSMC layers is that number required to maintain constant transmural wall strain.[24–26] Investment of developing blood vessels with VSMCs is accomplished by recruitment from surrounding mesenchymal progenitor cells. However, different progenitor cells are available for recruitment at different times and in different locations in vascular development leading to extensive VSMC mosaicism in vivo.[14,15,27] For example, the thoracic aorta develops with a heterogeneous distribution of origin-specific VSMC subtypes in a discontinuous gradient from the root to the ascending segments and reaching into the aortic arch[28–30] (Fig. 3.1). A numerically dominant population of second heart field–derived SMCs (SHF-SMC) in the aortic root and ascending aorta (AsAo) reflects a major contribution of SHF progenitors to elongation of the outflow tract in cardiac development.[31,32,33-35] As the AsAo transitions into the aortic arch, the distribution of SHF-SMCs abruptly terminates at the origin of the innominate artery, and cardiac neural crest–derived SMCs (CNC-SMC) become the dominant subtype, so that the arch itself is composed almost purely of CNC-SMCs.[29,36] Substituting trunk neural crest for premigratory cranial neural crest in mid-gestation embryos does not produce vascular SMCs in aortic arch arteries or elsewhere.[37] Thus, inductive cues that operate in the pharyngeal arch are either not perceived or their signal transduction pathways are blocked in non-CNC progenitors possibly as a result of incompatible anterior–posterior positional identity (discussed below). The descending aorta by contrast is composed of VSMCs derived from somitic paraxial mesoderm (sclerotome).[38,39,40] Renal arteries that emerge from the mid-descending aorta acquire VSMCs from Foxd1-positive cells of the metanephric mesenchyme.[41] VSMCs in mesenteric and gut blood vessels are derived by epithelial–mesenchymal transition (EMT) from serosal mesothelium.[42] VSMCs in the lung arise by a similar EMT from pleural mesothelium.[43] Coronary VSMCs are also mesothelium derived in this case arising from

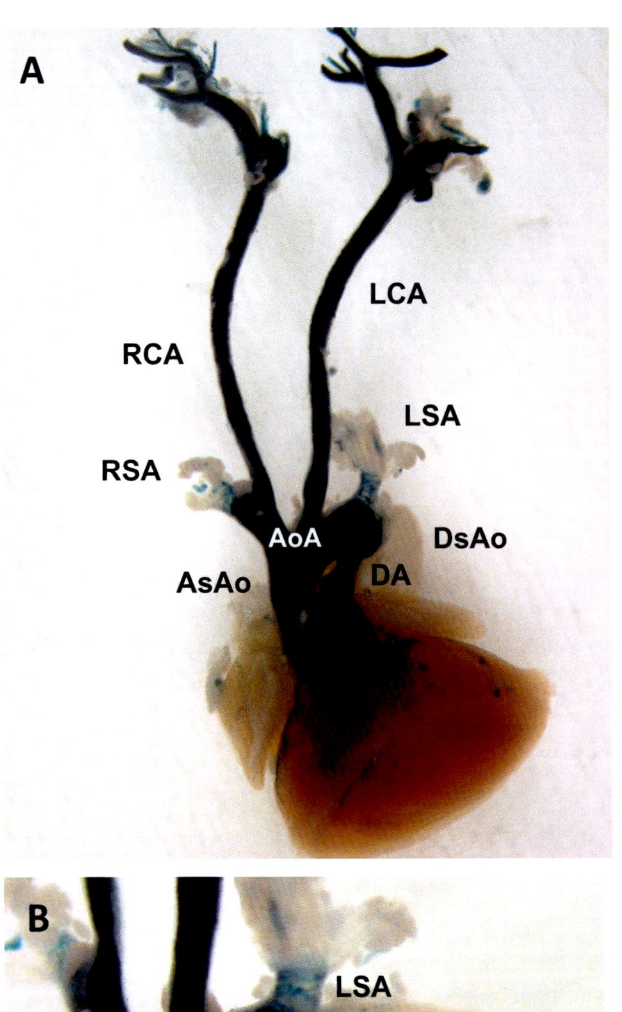

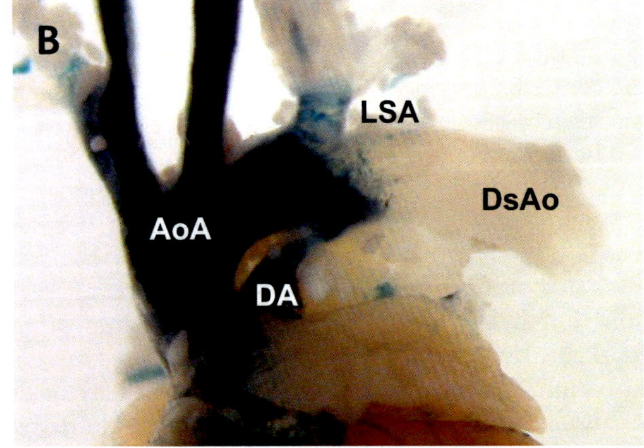

FIGURE 3.1 Vascular smooth muscle cells (VSMCs) are a developmental mosaic. Panel A: Wnt1cre; R26RlacZ whole mount stained major elastic arteries at 3 months of age. Panel B: Higher magnification view of the aortic arch region of the whole mount vessels in Panel A. Note that neural crest-derived VSMCs terminate abruptly at the origin of the left subclavian artery (LSA) on the rostral side and at the insertion site of the ductus arteriosus (DA) on the caudal side of the aortic arch (AoA). The nonlabeled aortic SMCs in the descending portion of the AoA (DsAo) have been shown to arise from paraxial mesoderm of the sclerotome.[38,39] *AsAo*, ascending aorta; *DsAo*, descending aorta; *LCA*, left carotid artery; *LSA*, left subclavian artery; *RCA*, right carotid artery; *RSA*, right subclavian artery.

the proepicardium (PE) after the PE first forms the epicardial covering of the heart.[44,45] These are just a few examples of a much larger number of independent VSMC origins in the vascular and lymphatic systems.[14,17] The ability to produce VSMCs from such a wide variety of progenitors via a process of convergent differentiation suggests that competence for VSMC differentiation must be a widespread trait of the vertebrate mesoderm subject to threshold regulation controlled by baseline repressor pathways.

Self-organization guides artery wall development

The assembly of pre-SMCs from any of the origins discussed above into an artery wall occurs in several steps. Endothelial cells initiate mural cell development via formation of adhesive contacts with smooth muscle progenitor cells mediated by N-cadherin.[46] As suggested above, this is a direct result of increasing physical forces of blood flow and shear stress acting on the endothelial surface.[47] This brings VSMC progenitors into contact with endothelial cells and allows for cell surface signaling molecules to be engaged on both cell types. One such signaling interaction is via jagged1 on endothelial cells and notch receptors on VSMC progenitors.[48] Jagged1-notch signaling thus activated promotes VSMC differentiation and initiates tunica media formation.[49–51] The next step promotes formation of multiple layers of VSMCs to form the bulk of the artery wall. Layer formation occurs by two coordinated processes. The first is the activation of Jagged1 expression by notch signaling in the group of VSMC progenitors in close contact with the first layer VSMCs.[49,52] This sequential relay of jagged1-Notch signaling is sufficient to produce multiple layers of SMCs in the developing artery wall.[49,52] Briot et al. reported that jagged1 signaling is also required during and after artery wall morphogenesis to continuously suppress the default chondrogenic fate of aortic SMCs derived from progenitors in the sclerotome.[40] This is a specific example of a more general role of Notch signaling to prevent cells from taking alternate paths in development.[53] The second is a limited and controlled invasion of VSMCs from the inner layers into the outer layers.[21] This process involves reorientation of circumferential SMCs and radial migration into the next layer. This step occurs early in wall formation and well before extensive deposition and cross-linking of elastic fibers between SMC layers that define the lamellae of postnatal artery walls.[24]

For blood vessels like the AsAo whose media is composed of SMCs from multiple embryonic origins, yet a third step is involved. As discussed above, the AsAo is composed of SMCs that differentiate from two independent embryonic origins, i.e., CNC and SHF. While developmental timing of SMC progenitor cell migration to the developing OFT can account for the bulk distribution of CNC-SMCs to the inner layers and SHF-SMCs to the outer layers, the lack of mixing of SMCs along the boundary requires a different explanation. For both SMC origins in the AsAo, SMC progenitor cells must migrate from their point of origin in the CNC or SHF to the aortic wall. While chemoattractants released by endothelial cells can provide general instructional cues for cell migration of SMC progenitors,[54] fine guidance to and within the vessel wall is most likely provided by chemorepulsive and contact-repulsive molecules similar to what is described for axon guidance.[55] Although yet to be tested, it is reasonable to propose that the lack of mixing is due to contact-repulsive molecules used by CNC-SMC and SHF-SMC progenitors during their migration to reach the aorta during wall formation.

Positional identity and VSMC diversity

In addition to lineage diversity, and at an even more fundamental level, cells in early embryos are patterned by their position with respect to the major axes of the embryo, dorsal–ventral, right–left, and anterior–posterior[56–60,61] (Fig. 3.2). Positional identity, conferred by expression of a *Hox code*,[57] is retained in the adult vasculature[62,63,64] and may contribute in important ways to site-specific origins of arterial disease, including aneurysms[65] and atherosclerosis.[66,67,68,69] Considering the presence of blood vessels throughout the body and the evidence for location-dependent vascular disease despite systemic risk factors, additional emphasis on mechanisms that establish and maintain positional identity in VSMCs is needed.

VSMC differentiation, phenotypic switching, and reprogramming

Vascular SMC differentiation is controlled by signal-responsive networks of positive and negative transcription factors, chromatin modifiers, and noncoding RNAs.[71–74] These networks ultimately control the interactions of serum response factor (SRF) with a CArG box motif [CC(AT$_6$)GG] present in most SMC-specific genes[75,76,77,78] (Fig. 3.3). The major positive cofactors are members of the myocardin gene family[79,80] and epigenetic regulators including Tet2.[81,82] The major repressors include Krüpple-like factor-4 (KLF4),[83–85]; Msx1[86]; and ternary complex factors including Elk1[87,88] and Foxo4.[84,85] The positive pathway is reinforced by micro-RNAs, particularly miR-143 and

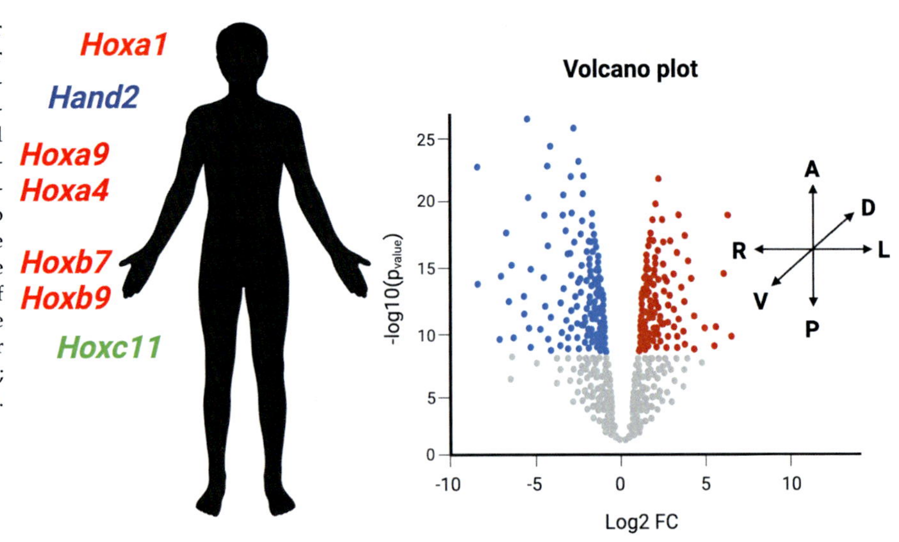

FIGURE 3.2 Positional identity of vascular smooth muscle cells and its relation to vascular disease. Vascular cells acquire positional identity (3D axes, far right) during early development via expression of a *Hox* code.[56,58,62] Panel A: Expression of these positional identify determinants (far left) is detected in adult transcriptional profiling studies of VSMCs in vivo (center right, typical volcano plot).[62,63,70] Some of these positional identity determinants have been linked to region-specific phenotypes of loss of function *Hox* mutants,[64] and the restricted distribution and severity of vascular disease (center left).[66,67] *A*, anterior; *D*, dorsal; *L*, left; *P*, posterior; *R*, right; *V*, ventral. *Figure created in Biorender.*

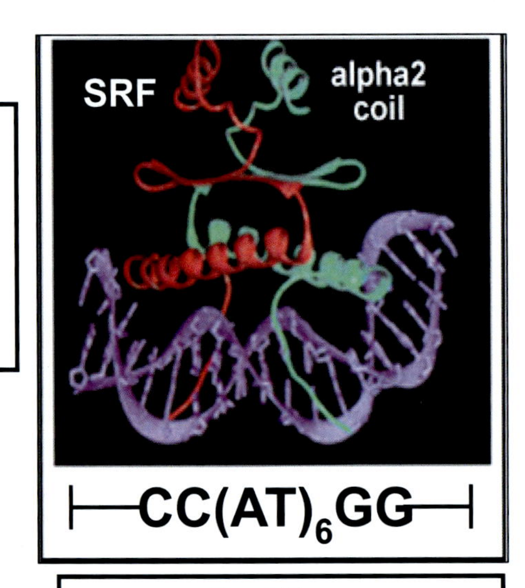

FIGURE 3.3 Transcriptional control of vascular smooth muscle cell differentiation. Serum response factor (SRF) binds to a cis-regulatory element [CC(AT)6GG] called a CArG box (center panel) found in many VSMC marker genes (lower panel, blue).[99] The alpha 2 coils of the central MADS box domain of SRF extend away from the DNA binding domain and serve as a platform for protein–protein interaction.[100] Among the hundreds of proteins reported to interact with SRF[99] are some of the activators (left, green)[101–103] and some of the repressors (right, red)[104–109] that are suggested to have a critical role in control of VSMC differentiation. Among the activators are a recently identified class of long noncoding RNAs *Myoslid*, *Carmn*, and *Miat* that facilitate protein–protein interactions on and around the CArG box.[90,91,93–95] Structure of SRF-MADS box (center panel) reproduced with permission.[100]

miR-145,[89] long noncoding RNAs[74,90–95] and epigenetic marks including H3K4me2.[96] For VSMC progenitors in the postnatal adventitia, the critical step to initiate SMC differentiation is to first downregulate the repressors of myocardin-SRF-CArG box interaction.[97] Similar evidence was reported for the transcriptional repressors Msx1[86] and the basic helix-loop-helix factor TCF21.[77,98]

Once the mature differentiated state is destabilized, additional factors in the local environment act to direct dedifferentiated VSMCs to adopt transcriptional landscapes characteristic of other mesenchymal cell types. This VSMC reprogramming process has been shown to produce osteogenic cells,[110,111] fibroblast-like cells,[112–114] artery wall-resident progenitor cells,[63,115,116] and cells that express macrophage marker genes.[117–119] Mechanisms that control VSMC differentiation and phenotypic switching have been recently reviewed.[73,120,121,122]

VSMC reprogramming to multiple mesenchymal cell types

What was unanticipated when VSMC cre-recombinase–mediated fate tracking experiments were first begun was the remarkable extent of in vivo reprogramming that arterial SMCs can undergo where all traces of VSMC marker gene expression are extinguished. Such studies now provide clear evidence of VSMC reprogramming to multiple mesenchymal cell types in vivo. Among the earliest studies pointing to VSMC reprogramming in arterial tissues were those documenting conversion to an osteogenic cell.[110,123] This conversion was later shown to be Runx2 dependent.[111,124] The rate and extent of aortic calcification was also shown to be VSMC embryonic origin dependent.[125] In addition, VSMCs have been reported to express macrophage-like marker genes when exposed to cholesterol-rich atherogenic environments in vitro[117,119,126] or in vivo.[115,118,127,128] Single-cell transcriptome analyses identified additional VSMC reprogramming fates including a fibroblast-like cell, called a fibromyocyte,[77,112] a chondrocyte-like cell,[129,130] and an adipocyte-like cell.[130] While these SMC reprogrammed states might not be perfect copies of the adopted cell types,[119] they nevertheless exhibit an extensive degree of cell type conversion while shedding nearly all of their previous VSMC identity.[120,131] This extreme plasticity may not be limited to VSMCs. Fibroblasts are an equally diverse and widely distributed cell type in the body with multiple embryonic origins from broadly competent mesoderm.[132,133] At sites of injury, fibroblasts can differentiate to myofibroblasts which are then capable of further transitions to dermal papilla cells in hair follicles[134] or dermal adipocytes.[135,133]

Role of Krüppel-like factor 4 in VSMC reprogramming

Many of these VSMC fate transitions have been shown to be dependent upon expression of the Krüppel-like zinc finger transcription factor KLF4.[136] KLF4 is well known for its role in reprogramming of adult fibroblasts to induced pluripotent stem cells.[137] In VSMCs, expression of KLF4 leads to downregulation of the mature differentiated phenotype and acquisition of a more permissive state of a type that may be required for additional reprogramming steps.[138] Downregulation of the VSMC differentiated phenotype by KLF4 occurs via multiple mechanisms including direct binding of KLF4 to SMC promoter–enhancer sequences, reduced binding of SRF to CArG box motifs, decreased expression of myocardin (Myocd) via KLF4 binding to cis-regulatory elements in the Myocd locus, and recruitment of HDACs that promote chromatin remodeling leading to transcriptional silencing (reviewed in Ref. 139,140). KLF4 expression in SMCs is controlled by complex gene regulatory networks that include miRs-143/145,[89] miR-506-3p,[92] long noncoding RNA *MIAT*,[95] and myocardin-related transcription factor-A.[141] In addition, evidence now suggests that VSMC expression of KLF4 activates a molecular fate-switch that opens a window for reprogramming to other mesenchymal cell fates described above.[130,136] KLF2 and KLF5 are part of a core KLF4 network in ES cells and may play similar cross-regulatory roles in VSMC reprogramming.[130,142] For example, Shankman et al. reported that VSMC-specific conditional deletion of KLF4 in atherosclerotic plaques in *Apoe(−/−)* mice produced marked reductions in VSMC-derived macrophage-like cells and mesenchymal stem cell (MSC)-like cells in the lesions.[115] Likewise, knockdown of KLF4 blocked osteogenic cell conversion of VSMCs exposed to high extracellular phosphate *in vitro*.[143] A pool of VSMC progenitor cells resides in a sonic hedgehog signaling niche in the adventitia.[97] These adventitial cells express KLF4 and stem cell antigen-1 (*Sca1/Ly6a*) and exhibit multiple cell fates in vitro[97,144] and *in vivo*.[116,145,146,147] Using *Myh11-creERT2;ROSA-YFP* mice, a fraction of these adventitial progenitor cells was shown to be derived from mature differentiated VSMCs *in vivo*.[116] This transition of mature VSMCs to adventitial MSC-like progenitor cells was dependent on expression of KLF4 in SMCs,[116] reminiscent of what was described during atherogenesis.[115] The presence of a small number of *Ly6a/Sca1*-positive SMCs in normal healthy aortic media was reported by Ref. 63. Similarly, a small number of MSC-like "disease-prone" cells in the normal aorta were reported by Ref. 130. It is possible that these medial *Sca1/Ly6a*-positive MSC-like cells appear and disappear stochastically in normal artery wall and are forerunners of the *Sca1/Ly6a*-positive, KLF4-positive SMC progenitor cells in the adventitia,[97,116,146] as well as those with a similar MSC marker profile appearing in atherosclerotic plaques[63,113,115] and aortic aneurysms.[130] Thus, VSMC plasticity not only encompasses contractile to synthetic phenotypic switching, but it also includes conversion to non-VSMC mesenchymal cell types via KLF4-dependent acquisition of an MSC-like intermediate phenotype in vivo.

VSMC single-cell transcriptomes and VSMC reprogramming

Evidence to support such a transition state was obtained in single-cell transcriptomes from atherosclerotic plaque VSMCs in mice. Four studies report characterization of an intermediate *Sca1/Ly6a*-positive cell that appears to be a precursor for VSMC-derived osteogenic cells, proinflammatory cells, and fibroblast-like cells in the plaque.[63,113,114,148] In VSMC-restricted KLF4 knockout mice, reductions in the number of *Lgals3*$^+$, *Ly6a/Sca1*$^+$ cells in the plaque correlated with reductions in *Runx2*$^+$, *Sox9*$^+$ osteogenic cells and increases in VSMC contractile marker-positive cells consistent with a required KLF4-dependent switch to an MSC-like intermediate state.[113,149] The previously described *Lgals3*$^{+}$ macrophage-like cell[115] has been reinterpreted to be an *Lgals3*$^{+}$ intermediate state, and renamed a "pioneer cell," reflective of its early appearance in plaque development.[113] Similar results were reported independently by Pan et al. using *Ldlr*-null mice.[114] In that study, an *Lgals3*$^{+}$, *Ly6a/Sca1*$^{+}$, *and Ly6c1*$^{+}$ transition state for VSMCs in the plaque was also identified and called "SEM" (stem, endothelial, macrophage) cells. An atherogenic-like milieu promotes endoplasmic reticulum stress, activates the unfolded protein response, and leads to increased expression of KLF4 in VSMCs in vitro suggesting a pathway by which a cholesterol-rich environment activates the KLF4-dependent molecular fate-switch in plaque VSMCs in vivo.[126] In addition, exposure of human VSMCs to hypoxia is reported to increase KLF4 expression in an HIF-1α-dependent manner.[150] Both hypoxia and elevated cholesterol content are common features of developing atherosclerotic plaques. Taken together, multiple single-cell studies now show VSMC reprogramming to other mesenchymal cell fates is readily detected in murine and human atherosclerosis.

VSMC single-cell transcriptomes and embryonic origins

Returning to the role of VSMC embryonic origins in position-specific vascular diseases, we can now ask whether single-cell transcriptomes shed any additional light on such a role in vivo. In one study of normal healthy arteries,[63] it is reported that VSMC scRNA profiles in adult mouse aorta reflected evidence of region-specific developmental history even after VSMC differentiation and maturation events have occurred in vivo. Adult aortic arch SMCs (cardiac neural crest derived) and descending thoracic aorta SMCs (paraxial mesoderm derived) produce transcriptomes with both embryonic origin- and position-dependent signatures.[63] As reviewed above, this analysis also showed that a low frequency (0.2%–1.6%) of aortic medial SMCs expresses the *Ly6a/Sca1* marker[63] commonly found in VSMC progenitor cells including those that normally reside in the adventitia.[97,116,146,151] The *Ly6a/Sca1*-positive VSMCs in healthy aortic media may represent a cell population that is poised to rapidly respond to inflammation or injury, clonally expand, and initiate repair responses *in vivo*.[63,148] Yet, for atherosclerosis at least, evidence to the contrary has been reported.[152] It will be interesting to determine if the frequency of *Ly6a/Sca1*-positive VSMCs in adult media differs among VSMCs of different embryonic origins in vivo. Given the central role of SMC TGFβ signaling in silencing KLF4 expression in vivo[130] and the VSMC origin-specific differences in TGFβ signaling[153,154,155], there is reason to investigate this possibility further. Taken together, the diversity of VSMC phenotypes in normal artery wall reflects origin-specific embryonic ancestry, position globally within the anterior–posterior, right–left, and dorsal–ventral axes of the body as well as locally within the artery wall, and conversion between contractile and MSC-like VSMC transcriptional states.

Vascular smooth muscle clonal expansions in vivo

A different form of cell–cell communication must be involved in the mono- or oligo-clonal expansion of VSMCs observed in neointima formation after arterial injury[156], aneurysm formation,[157] hypoxic pulmonary artery remodeling,[158,159] or during formation of atherosclerotic plaques.[127,129,158,160,161] The first report of VSMC clonality in vivo was by Benditt et al. in 1973. At the time, the authors interpreted clonality of the end stage atherosclerotic lesion as indicative of a single cell origin. This interpretation was challenged by those who favored a multicellular origin followed by selection of a dominant clone during plaque growth. The controversy lingered for decades until the advent of cre-recombinase–mediated multicolor lineage tracking technology. Using this technology, reports of VSMC clonal expansions multiplied quickly to include the varied examples cited above.

Yet the question, monoclonal origin or clonal selection, has proven difficult to resolve. One reason for the difficulty is the lack of a generally accepted mechanism for SMC clonal expansion in vivo.[162,163] Among the proposals are (1) lateral inhibition of surrounding cell proliferation by the expanding clone;[164] (2) restricted SMC migration into the intima by fenestrae of the internal elastic lamina;[30,165] (3) a preexisting small number of "primed SMCs" that preferentially respond to mitogenic stimuli;[63,159,166] and (4) the rapid appearance of a small number of SMC-derived proliferative MSC-like transitional cells in injured or diseased vessels.[113,114,115,129] The report that deletion of integrin β3 in myeloid cells from bone marrow resulted in polyclonal SMC expansions in hypoxic pulmonary arteries might argue against the rare SMC capable of clonal expansion.[158] However, the clonal expansion of notch3-expressing SMCs into the intima from a small pool of favorably positioned cells near the media-intima border suggests that preexisting SMC subsets will be critical to characterize and map within artery walls.[167] These objectives will be greatly facilitated by application of spatial transcriptomics[168–170] to models of SMC clonal expansion in vivo to assess how location of SMCs within the artery wall maps onto SMC transcriptomes to further evaluate the proposed mechanisms described above. Going forward, key questions for this field are (1) as one VSMC begins to clonally expand, is there an active mechanism to prevent its neighbors from also expanding and what is the molecular identity of that mechanism, (2) are "primed VSMCs" expressing the Ly6a/Sca1 antigen widely distributed within the media or are they clustered together, and (3) do Ly6a/Sca1-expressing VSMCs exist as a fixed number in the artery wall or are they constantly appearing and disappearing in a stochastic pattern in vivo?

SMC diversity: one genome, many epigenomes

Cell identity can be simply defined as the set of genes expressed by a given cell. Gene expression is a function of both the available combination of transcription factors (TFs) and TF cofactors (both activators and repressors) and the context for DNA binding and transcriptional activity enabled by chromatin modifiers and noncoding RNAs.[171,172] It follows then that cells can respond in markedly different ways to the same stimulus depending on the complement of expressed TFs and their nuclear context. The implications of the SMC lineage diversity model for SMC function and disease rest on two assumptions: (1) that SMCs of different embryonic origins have different epigenomes and (2) that origin-dependent epigenomes are at least partially retained through cell division and differentiation in the embryo and through maturation steps in adult vessels, as suggested by the data in Ref. 63.

Remodeling of chromatin through covalent histone modifications into transcriptionally active or -inactive nucleosomes and condensation of chromatin into heterochromatin are several ways that "footprints" of lineage history are left on the postnatal genome. Multiple examples from other cell types suggest this is likely also true for VSMCs. For example, Islet1 (Isl1) is an LIM-homeodomain pioneer transcription factor that is critical for heart development.[173] *Isl1* is transiently expressed in proliferating SHF progenitor cells during their specification to right ventricular cardiac myocytes, atrial myocardium, and the cardiac outflow tract.[174] Isl1 binds to cardiomyocyte structural genes well before these genes become transcriptionally active in differentiating myocytes. In fact, by the time these genes become maximally expressed in heart development, *Isl1* itself has switched off. Thus, cardiomyocyte identity is specified in cardiogenic fields in the early embryo where Isl1 target genes are epigenetically marked for expression later in the right ventricle. This demarcation is inherited in Isl1 progeny during heart development.[173] Another example is that of CD8$^+$ T cell fate during infection.[175] The prevailing view of T cell responses to infection held that a single lineage of naive T cells gives rise to different effector or memory T cells as a function of the type and amount of stimulation during infection. Smith et al. reported that multiple populations of naive CD8$^+$ T cells exist that are distinguished by their developmental origin, distinct chromatin landscapes, and different responses to challenge by microbial infections.[175]

VSMCs—a paradigm for mesenchymal cell types

Extreme plasticity and developmental diversity reported for VSMCs are also features shared by other mesenchymal cell types. Single-cell RNA sequencing provides a sensitive window into the complexity and heterogeneity of mesenchymal cell types in solid tissues. For example, Kuppe et al. performed a detailed study of the heterogeneity of kidney mesenchyme. Multiple cell states were identified for interstitial fibroblasts, myofibroblasts, and pericytes, with differing contributions to renal interstitial fibrosis.[176] Likewise, Tsukui et al. reported a collagen-producing cell atlas for bleomycin-induced pulmonary fibrosis that identifies 4 clusters of alveolar fibroblasts, 2 clusters of adventitial fibroblasts, and 1 cluster of peribronchial fibroblast among 12 clusters of collagen-producing cells in the fibrotic lung.[177] Of particular interest, dermal fibroblasts (DFs) have been shown to consist of multiple subtypes with diverse embryonic origins, gene expression and chromatin landscapes, anatomical positions, and differences in abilities for repair and regeneration.[132,178–181] DFs that arise from *engrailed1*-positive progenitors selectively produce fibrotic matrix in injured dorsal skin of mice, while *engrailed1*-negative DFs residing in the same environment contribute little to this fibrotic response.[132] The similarities between fibroblast heterogeneity and SMC diversity highlighted by these studies are striking.

Does VSMC reprogramming extend "the vasculome" beyond the vascular system?

Evidence suggests that VSMCs and pericytes, present in all organs, may give rise to one or more of the myofibroblast subsets in fibrotic kidneys, lungs, retina, and other organs.[146,182–185] Single-cell transcriptomes from atherosclerotic plaque VSMCs identify fibroblast-like, matrix-producing "fibromyocytes" as VSMC-derived mesenchymal cells in the plaque.[112–114] Likewise, mature VSMCs give rise to a subset of *Ly6A/Sca1*$^+$ adventitial MSC-like progenitor cells that, in turn, differentiate to fibrotic matrix-producing myofibroblasts after arterial injury.[116,146,186] These data raise the intriguing possibility that, through mural cell reprogramming, the *vasculome* may extend well beyond the vascular system per se. It is also possible that a similar capacity for reprogramming is just starting to be discovered for non-SMC mesenchymal cells *in vivo*.[135,133] Either way, the unexpected and extreme diversity of VSMC origins and fates revealed by advances in transgenic, genomic, and computational technologies heralds a new understanding of *the vasculome* and a greater appreciation of multiple ways that VSMCs contribute to pervasive vascular diseases.

Acknowledgments

Support provided by NIH grants 1R01HL-121877; 1R01HL-133723, the Loie Power Robinson Stem Cell and Regenerative Medicine Fund, the W.M. Keck Foundation-Medical Research Program, and the Seattle Children's Research Institute. This article is dedicated to the memory of Stephen M. Schwartz, MD, PhD.

References

1. Pease DC, Paule WJ. Electron microscopy of elastic arteries: the thoracic aorta of the rat. *J Ultrastruct Res*. 1960;3:469–483. https://doi.org/10.1016/s0022-5320(60)90023-x. pmid: 14431272.
2. Haust MD, More RH, Bencosme SA, Balis JU. Elastogenesis in human aorta: an electron microscopic study. *Exp Mol Pathol*. 1965;4: 508–524. https://doi.org/10.1016/0014-4800(65)90015-8. pmid: 5847769.
3. Wissler RW. The arterial medial cell, smooth muscle, or multifunctional mesenchyme? *Circulation*. 1967;36:1–4. https://doi.org/10.1161/01.cir.36.1.1. pmid: 6028742.
4. Takagi K, Kawase O. An electron microscopic study of elastogenesis in embryonic chick aorta. *J Electron Microsc (Tokyo)*. 1967;16: 330–339. pmid: 5589751.
5. Cooke PH, Smith SC. Smooth muscle cells: the source of foam cells in atherosclerotic white Carneau pigeons. *Exp Mol Pathol*. 1968;8: 171–189. https://doi.org/10.1016/0014-4800(68)90014-2. pmid: 5647247.
6. Moss NS, Benditt EP. Spontaneous and experimentally induced arterial lesions. I. An ultrastructural survey of the normal chicken aorta. *Lab Invest*. 1970;22:166–183. pmid: 5416695.
7. Campbell JH, Campbell GR. Smooth muscle phenotypic modulation: a personal experience. *Arterioscler Thromb Vasc Biol*. 2012;32: 1784–1789. https://doi.org/10.1161/ATVBAHA.111.243212. PMID: 22815344.
8. Krogh A. The supply of oxygen to the tissues and the regulation of the capillary circulation. *J Physiol*. 1919;52:457–474. https://doi.org/10.1113/jphysiol.1919.sp001844. PMID: 16993410.
9. Ketty SS. The theory and applications of the exchange of inert gas at the lungs and tissues. *Pharmacol Rev*. 1951;3:1–41. PMID: 14833874.
10. Hsia CCW, Schmitz A, Lambertz M, Perry SF, Maina JN. Evolution of air breathing: oxygen homeostasis and the transitions from water to land and sky. *Comp Physiol*. 2013;3:849–915. https://doi.org/10.1002/cphy.c120003. PMID: 23720333.
11. Lazarus A, Del-Moral PM, Ilovich O, Mishani E, Warburton D, Keshet E. A perfusion-independent role of blood vessels in determining branching stereotypy of lung airways. *Development*. 2011; 138:2359–2368. https://doi.org/10.1242/dev.060723. PMID: 21558382.
12. Risau W, Flamme I. Vasculogenesis. *Ann Rev Cell Dev Biol*. 1995;11: 73–91. https://doi.org/10.1146/annurev.cb.11.110195.000445. PMID: 8689573.
13. Hungerford JE, Little CD. Developmental biology of the vascular smooth muscle cell: building a multilayered vessel wall. *J Vasc Res*. 1999;36:2–27. https://doi.org/10.1159/000025622. pmid: 10050070.
14. Donadon M, Santoro MM. The origin and mechanisms of smooth muscle cell development in vertebrates. *Development*. 2021;148: dev197384. https://doi.org/10.1242/dev.197384. PMID: 33789914.
15. Majesky MW. Developmental basis of vascular smooth muscle diversity. *Arterioscler Thromb Vasc Biol*. 2007;27:1248–1258. https://doi.org/10.1161/ATVBAHA.107.141069. PMID: 17379839.
16. Shen M, Quertermous T, Fischbein MP, Wu JC. Generation of vascular smooth muscle cells from induced pluripotent stem cells: methods, applications, and considerations. *Circ Res*. 2021;128: 670–686. https://doi.org/10.1161/CIRCRESAHA.120.318049. PMID: 33818124.
17. Kenney HM, Bell RD, Masters EA, Xing L, Ritchlin CT, Schwarz EM. Lineage tracing reveals evidence of a popliteal lymphatic muscle progenitor cell that is distinct from skeletal and vascular muscle progenitors. *Sci Rep*. 2020;10:18088. https://doi.org/10.1038/s41598-020-75190-7. PMID: 33093635.
18. Sinha S, Iyer S, Granata A. Embryonic origins of human vascular smooth muscle cells: implications for in vitro modeling and clinical application. *Cell Mol Life Sci*. 2014;71:2271–2288. https://doi.org/10.1007/s00018-013-1554-3. pmid: 24442477.
19. Pexider T. Blood pressure in the third and fourth aortic arch and morphogenetic influence of laminar blood streams in development of the vascular system of the chick embryo. *Folia Morphol*. 1969;17:273–290. PMID: 5802137.
20. Shadwick RE. Mechanical design in arteries. *J Exp Biol*. 1999;202: 3305–3313. PMID: 10562513.
21. Greif DM, Kumar M, Lighthouse JK, et al. Radial construction of an arterial wall. *Dev Cell*. 2012;23:482–493. https://doi.org/10.1016/j.devcel.2012.07.009. PMID: 22975322.
22. Douguet D, Patel A, Xu A, Vanhoutte PM, Honoré E. Piezo ion channels in cardiovascular mechanobiology. *Trends Pharmacol Sci*. 2019;40:956–970. https://doi.org/10.1016/j.tips.2019.10.002. PMID: 31704174.
23. Langille BL. Arterial remodeling: relation to hemodynamics. *Can J Physiol Pharmacol*. 1996;74:834–841. PMID: 8946070.
24. Wolinsky H, Glagov S. A lamellar unit of aortic medial structure and function in mammals. *Circ Res*. 1967;20:99–111. https://doi.org/10.1161/01.res.20.1.99. PMID: 4959753.
25. Wagenseil JE, Mecham RP. Vascular extracellular matrix and arterial mechanics. *Physiol Rev*. 2009;89:957–989. https://doi.org/10.1152/physrev.00041.2008. PMID: 19584318.
26. Cook BL, Chao CJ, Alford PW. Architecture-dependent mechano-adaptation in single vascular smooth muscle cells. *J Biomech Eng*. 2021;143:101002. https://doi.org/10.1115/1.4051117. PMID: 33972987.
27. Majesky MW, Dong XR, Regan JN, Hoglund VJ. Vascular smooth muscle progenitor cells: building and repairing blood vessels. *Circ Res*. 2011;108:365–377. https://doi.org/10.1161/CIRCRESAHA.110.223800. pmid: 21293008.
28. Waldo K, Hutson M, Ward C, et al. Secondary heart field contributes myocardium and smooth muscle to the arterial pole of the developing heart. *Dev Biol*. 2005;281:78–90. https://doi.org/10.1016/j.ydbio.2005.02.012. PMID: 15848390.

29. Sawada H, Rateri DL, Moorleghen JJ, Majesky MW, Daugherty A. Smooth muscle cells derived from the second heart field and cardiac neural crest reside in spatially distinct domains in the media of the ascending aorta – brief report. *Arterioscler Thromb Vasc Biol.* 2017;37:1722–1726. https://doi.org/10.1161/ATVBAHA.117.309599. PMID: 28663257.
30. Lin CJ, Hunkins B, Roth R, Lin CY, Wagenseil JE, Mecham RP. Vascular smooth muscle cell subpopulations and neointimal formation in mouse models of elastin insufficiency. *Arterioscler Thromb Vasc Biol.* 2021;41:2980. https://doi.org/10.1161/ATVBAHA.120.315681, 2905 pmid: 34587758.
31. Dyer LA, Kirby ML. The role of secondary heart field in cardiac development. *Dev Biol.* 2009;336:137–144. https://doi.org/10.1016/j.ydbio.2009.10.009. PMID: 19835857.
32. Kelly RG. The second heart field. *Curr Top Dev Biol.* 2012;100: 33–65. https://doi.org/10.1016/B978-0-12-387786-4.00002-6. PMID: 22449840.
33. Dodou E, Verzi MP, Anderson JP, Xu SM, Black BL. Mef2c is a direct transcriptional target of ISL1 and GATA factors in the anterior heart field during mouse embryonic development. *Development.* 2004;131:3931–3942. https://doi.org/10.1242/dev.01256. pmid: 15253934.
34. Verzi MP, McCulley DJ, De Val S, Dodou E, Black BL. The right ventricle, outflow tract, and ventricular septum comprise a restricted expression domain with the secondary heart field. *Dev Biol.* 2005;287:134–145. https://doi.org/10.1016/y.jdbio.2005.08.041. pmid: 16188249.
35. Ma Q, Zhou B, Pu WT. Reassessment of Isl1 and Nkx2-5 cardiac fate maps using a Gata4-based reporter of Cre activity. *Dev Biol.* 2008;323:98–104. pmid: 18775691.
36. Jiang X, Rowitch DH, Soriano P, McMahon AP, Sucov HM. Fate of the mammalian cardiac neural crest. *Development.* 2000;127: 1607–1616. PMID: 10725237.
37. Kirby ML. Plasticity and predetermination of mesencephalic and trunk neural crest transplanted into the region of the cardiac neural crest. *Dev Biol.* 1989;134:402–412. https://doi.org/10.1016/0012-1606(89)90112-7. PMID: 2744240.
38. Pouget C, Pottin K, Jaffredo T. Sclerotomal origin of vascular smooth muscle cells and pericytes in the chick embryo. *Dev Biol.* 2008;315:437–447. https://doi.org/10.1016/j.ydbio.2007.12.045. PMID: 18255054.
39. Wasteson P, Johansson BR, Jukkola T, et al. Developmental origin of smooth muscle cells in the descending aorta in mice. *Development.* 2008;135:1823–1832. https://doi.org/10.1242/dev.020958. PMID: 18417617.
40. Briot A, Jaroszewicz A, Warren CA, et al. Repression of Sox9 by Jag1 is continuously required to suppress the default chondrogenic fate of vascular smooth muscle cells. *Dev Cell.* 2014;31: 707–721. https://doi.org/10.1016/j.devcel.2014.11.023. PMID: 25535917.
41. Sequeira-Lopez MLS, Lin EE, Li M, Hu Y, Sigmund CD, Gomez RA. The earliest metanephric arteriolar progenitors and their role in kidney vascular development. *Am J Physiol Regul Integr Comp Physiol.* 2015;308:R138–R149. https://doi.org/10.1152/ajpregu.00428.2014. PMID: 25427768.
42. Wilm B, Ipenberg A, Hastie ND, Burch JBE, Bader DM. The serosal mesothelium is a major source of smooth muscle cells of the gut vasculature. *Development.* 2005;132:5317–5328. https://doi.org/10.1242/dev.02141. PMID: 16284122.
43. Que J, Wilm B, Hasegawa H, Wang F, Bader D, Hogan BLM. Mesothelium contributes to vascular smooth muscle and mesenchyme during lung development. *Proc Natl Acad Sci USA.* 2008; 105:16626–16630. https://doi.org/10.1073/pnas.0808649105. PMID: 18922767.
44. Mikawa T, Gourdie RG. Pericardial mesoderm generates a population of coronary smooth muscle cells migrating into the heart along with ingrowth of the epicardial organ. *Dev Biol.* 1996;174: 221–232. https://doi.org/10.1006/dbio.1996.0068. PMID: 8631495.
45. Majesky MW. Development of coronary vessels. *Curr Top Dev Biol.* 2004;62:225–259. https://doi.org/10.1016/S0070-2153(04)62008-4. PMID: 15522744.
46. Paik J, Skoura A, Chae S, et al. Sphingosine-1-phosphate receptor regulation of N-cadherin mediates vascular stabilization. *Genes Dev.* 2004;18:2392–2403. https://doi.org/10.1101/gad.1227804. PMID: 15371328.
47. Baeyens N, Bandyopadhyay C, Coon BG, Yun S, Schwartz MA. Endothelial fluid shear stress sensing in vascular health and disease. *J Clin Invest.* 2016;126:821–828. https://doi.org/10.1172/JCI83083. PMID: 26928035.
48. High FA, Zhang M, Proweller A, et al. An essential role for Notch in neural crest during cardiovascular development and smooth muscle differentiation. *J Clin Invest.* 2007;117:353–363. https://doi.org/10.1172/JCI30070. PMID: 17273555.
49. Manderfield LJ, High FA, Engleka KA, et al. Notch activation of Jagged1 contributes to the assembly of the arterial wall. *Circulation.* 2012;125:314–323. https://doi.org/10.1161/CIRCULATIONAHA.111.047159. PMID: 22147907.
50. Scheppke L, Murphy EA, Zarpellon A, et al. Notch promotes vascular maturation by inducing integrin-mediated smooth muscle cell adhesion to the endothelial basement membrane. *Blood.* 2012;119:2149–2158. https://doi.org/10.1182/blood-2011-04-348706. PMID: 22134168.
51. Baeten JT, Lilly B. Notch signaling in vascular smooth muscle cells. *Adv Pharmacol.* 2017;78:351–382. https://doi.org/10.1016/bs.apha.2016.07.002. PMID: 28212801.
52. Hoglund VJ, Majesky MW. Patterning the artery wall by lateral induction of Notch signaling. *Circulation.* 2012;125:212–215. https://doi.org/10.1161/CIRCULATIONAHA.111.075937. pmid: 22147908.
53. Ho DM, Guruharsha KG, Artavanis-Tsakonas S. The Notch interactome: complexity in signaling circuitry. *Adv Exp Med Biol.* 2018; 1066:125–140. https://doi.org/10.1007/978-3-319-89512-3_7. PMID: 30030825.
54. Zerwes HG, Risau W. Polarized secretion of a platelet-derived growth factor-like chemotactic factor by endothelial cells in vitro. *J Cell Biol.* 1987;105:2037–2041. https://doi.org/10.1083/jcb.105.5.2037. PMID: 3680370.
55. Tessier-Lavigne M, Goodman CS. The molecular biology of axon guidance. *Science.* 1996;274:1123–1133. https://doi.org/10.1126/science.274.5290.1123. PMID: 8895455.
56. Chang HY. Anatomic demarcation of cells: genes to patterns. *Science.* 2009;326:1206–1207. https://doi.org/10.1126/science.1175686. PMID: 19965461.
57. Kessel M, Gruss P. Homeotic transformations of murine vertebrae and concomitant alteration of Hox codes induced by retinoic acid. *Cell.* 1991;67:89–104. https://doi.org/10.1016/0092-8674(91)90574-i. PMID: 1680565.
58. McGinnis W, Krumlauf R. Homeobox genes and axial patterning. *Cell.* 1992;68:283–302. https://doi.org/10.1016/0092-8674(92)90471-n. PMID: 1346368.
59. Nüsslein-Volhard C, Wieschaus E. Mutations affecting segment number and polarity in *Drosophila*. *Nature.* 1980;287:795–801. https://doi.org/10.1038/287795a0. PMID: 6776413.
60. Wolpert L. Positional information and the spatial pattern of cellular differentiation. *J Theor Biol.* 1969;25:1–47. https://doi.org/10.1016/s0022-5193(69)80016-0. PMID: 4390734.
61. Gerhart J., Kirschner M. Cells, Embryos, and Evolution. Malden, MA: Blackwell Science; 1997:238–295.
62. Pruett ND, Visconti RP, Jacobs DF, et al. Evidence for Hox-specified positional identities in adult vasculature. *BMC Dev Biol.* 2008;8:93. https://doi.org/10.1186/1471-213X-8-93. PMID: 18826643.

63. Dobnikar L, Taylor AL, Chappell J, et al. Disease-relevant transcriptional signatures identified in individual smooth muscle cells from healthy mouse vessels. *Nat Commun*. 2018;9:4567. https://doi.org/10.1038/s41467-018-06891-x. PMID: 30385745.
64. Pruett ND, Hajdu Z, Zhang J, et al. Changing topographic Hox expression in blood vessels results in regionally distinct vessel wall remodeling. *Biol Open*. 2012;1:430–435. https://doi.org/10.1242/bio.2012039. PMID: 23213434.
65. Lillvis JH, Erdman R, Schworer CM, et al. Regional expression of HOXA4 along the aorta and its potential role in human abdominal aortic aneurysms. *BMC Physiol*. 2011;11:9. https://doi.org/10.1186/1472-6793-11-9. PMID: 21627813.
66. Trigueros-Motos L, González-Granado JM, Cheung C, et al. Embryological-origin-dependent differences in homeobox expression in adult aorta: role in regional phenotypic variability and regulation of NF-κB activity. *Arterioscler Thromb Vasc Biol*. 2013;33:1248–1256. https://doi.org/10.1161/ATVBAHA.112.300539. PMID: 23448971.
67. Visconti RP, Alexander Awgulewitsch A. Topographic patterns of vascular disease: HOX proteins as determining factors? *World J Biol Chem*. 2015;6:65–70. https://doi.org/10.4331/wjbc.v6.i3.65. PMID: 26322165.
68. Haimovici H, Maier N. Fate of aortic homografts in canine atherosclerosis. 3. study of fresh abdominal and thoracic aortic implants into thoracic aorta: role of tissue susceptibility in atherogenesis. *Arch Surg*. 1964;89:961–969. https://doi.org/10.1001/archsurg.1964.01320060029006. PMID: 14208469.
69. Kayashima Y, Maeda-Smithies N. Atherosclerosis in different vascular locations unbiasedly approached with mouse genetics. *Genes*. 2020;11:E1427. https://doi.org/10.3390/genes11121427. PMID: 33260687.
70. Romay MC, Ma F, Hernandez GE, et al. Positional transcriptomics shed light on site-specific pathologies of the aorta. *Research Square preprint*. October 28, 2021. https://doi.org/10.21203/rs.3.rs-951818/v1.
71. Kawai-Kowase K, Owens GK. Multiple repressor pathways contribute to phenotypic switching of vascular smooth muscle cells. *Am J Physiol Cell Physiol*. 2007;292:C59–C69. https://doi.org/10.1152/ajpcell.00394.2006. PMID: 16956962.
72. Mack CP. Signaling mechanisms that regulate smooth muscle cell differentiation. *Arterioscler Thromb Vasc Biol*. 2011;31:1495–1505. https://doi.org/10.1161/ATVBAHA.110.221135. PMID: 21677292.
73. Shi N, Chen SY. Smooth muscle cell differentiation: model systems, regulatory mechanisms, and vascular diseases. *J Cell Physiol*. 2016;231:777–787. https://doi.org/10.1002/jcp.25208. PMID: 26425843.
74. Kok F, Baker AH. The function of long non-coding RNAs in vascular biology and disease. *Vasc Pharmacol*. 2019;114:23–30. https://doi.org/10.1016/j.vph.2018.06.004. PMID: 30919128.
75. Gustafson TA, Miwa T, Boxer LM, Kedes L. Interaction of nuclear proteins with muscle-specific regulatory sequences of the human cardiac alpha-actin promoter. *Mol Cell Biol*. 1988;8:4110–4119. https://doi.org/10.1128/mcb.8.10.4110-4119.1988. PMID: 3185542.
76. Mohun TJ, Chambers AE, Towers N, Taylor MV. Expression of genes encoding the transcription factor SRF during early development of *Xenopus laevis*: identification of a CArG box-binding activity as SRF. *EMBO J*. 1991;10:933–940. PMID: 2009862.
77. Nagao M, Lyu Q, Zhao Q, et al. Coronary disease-associated gene TCF21 inhibits smooth muscle cell differentiation by blocking the myocardin-serum response factor pathway. *Circ Res*. 2020;126:517–529. https://doi.org/10.1161/CIRCRESAHA.119.315968. PMID: 31815603.
78. Miano JM. Serum response factor: toggling between disparate programs of gene expression. *J Mol Cell Cardiol*. 2003;35:577–593. https://doi.org/10.1016/s0022-2828(03)00110-x. pmid: 12788374.
79. Wang DZ, Olson EN. Control of smooth muscle development by the myocardin family of transcriptional coactivators. *Curr Opin Genet Dev*. 2004;14:558–566. https://doi.org/10.1016/j.gde.2004.08.003. PMID: 15380248.
80. Miano JM. Myocardin in biology and disease. *J Biomed Res*. 2015;29:3–19. https://doi.org/10.7555/JBR.29.20140151. PMID: 25745471.
81. Liu R, Jin Y, Tang WH, et al. Ten-eleven translocation-2 (TET2) is a master regulator of smooth muscle plasticity. *Circulation*. 2013;128:2047–2057. https://doi.org/10.1161/CIRCULATIONAHA.113.002887. PMID: 24077167.
82. Ostriker AC, Xie Y, Chakraborty R, et al. TET2 protects against vascular smooth muscle cell apoptosis and intimal thickening in transplant vasculopathy. *Circulation*. 2021;144:455–470. https://doi.org/10.1161/CIRCULATIONAHA.120.050553. PMID: 34111946.
83. Adam PJ, Regan CP, Hautmann MB, Owens GK. Positive- and negative-acting Kruppel-like transcription factors bind a transforming growth factor beta control element required for expression of the smooth muscle cell differentiation marker SM22alpha in vivo. *J Biol Chem*. 2000;275:37798–37806. https://doi.org/10.1074/jbc.M006323200. PMID: 10954723.
84. Liu Y, Sinha S, McDonald OG, Shang Y, Hoofnagle MH, Owens GK. Kruppel-like factor 4 abrogates myocardin-induced activation of smooth muscle gene expression. *J Biol Chem*. 2005;280:9719–9727. https://doi.org/10.1074/jbc.M412862200. PMID: 15623517.
85. Liu ZP, Wang Z, Yanagisawa H, Olson EN. Phenotypic modulation of smooth muscle cells through interaction of Foxo4 and myocardin. *Dev Cell*. 2005;9:261–270. https://doi.org/10.1016/j.devcel.2005.05.017. PMID: 16054032.
86. Hayashi K, Nakamura S, Nishida W, Sobue K. Bone morphogenetic protein-induced MSX1 and MSX2 inhibit myocardin-dependent smooth muscle gene transcription. *Mol Cell Biol*. 2006;26:9456–9470. https://doi.org/10.1128/MCB.00759-06. PMID: 17030628.
87. Wang Z, Wang DZ, Hockemeyer D, McAnally J, Nordheim A, Olson EN. Myocardin and ternary complex factors compete for SRF to control smooth muscle gene expression. *Nature*. 2004;428:185–189. https://doi.org/10.1038/nature02382. PMID: 15014501.
88. Zhou J, Hi G, Herring BP. Smooth muscle-specific genes are differentially sensitive to inhibition by Elk1. *Mol Cell Biol*. 2005;25:9874–9885. https://doi.org/10.1128/MCB.25.22.9874-9885.2005. PMID: 16260603.
89. Cordes KR, Sheehy NT, White MP, et al. miR-145 and miR-143 regulate smooth muscle cell fate and plasticity. *Nature*. 2009;460:705–710. https://doi.org/10.1038/nature08195. PMID: 19578358.
90. Zhao J, Zhang W, Lin M, et al. MYOSLID is a novel serum response factor-dependent long noncoding RNA that amplifies the vascular smooth muscle differentiation program. *Arterioscler Thromb Vasc Biol*. 2016;36:2088–2099. https://doi.org/10.1161/ATVBAHA.116.307879. PMID: 27444199.
91. Ni H, Haemmig S, Deng Y, et al. A smooth muscle cell-enriched long noncoding RNA regulates cell plasticity and atherosclerosis by interacting with serum response factor. *Arterioscler Thromb Vasc Biol*. 2021;41:2399–2416. https://doi.org/10.1161/ATVBAHA.120.315911. PMID: 34289702.
92. Dong H, Jiang G, Zhang J, Kang Y. Mir-506-3p promotes the proliferation and migration of vascular smooth muscle cells via targeting KLF4. *Pathobiology*. 2021;88:277–288. https://doi.org/10.1159/000513506. PMID: 33882484.
93. Dong K, Shen J, He X, et al. CARMN is an evolutionarily conserved smooth muscle cell-specific LncRNA that maintains contractile phenotype by binding myocardin. *Circulation*. 2021;

144:1856–1875. https://doi.org/10.1161/CIRCULATIONAHA.121.055949. pmid: 34694145.

94. Vacante F, Rodor J, Lalwani MK, et al. CARMN loss regulates smooth muscle cells and accelerates atherosclerosis in mice. *Circ Res*. 2021;128:1258–1275. https://doi.org/10.1161/CIRCRESAHA.12.318688. PMID: 33622045.
95. Fasolo F, Jin H, Winski G, et al. Long non-coding RNA MIAT controls advanced atherosclerotic lesion formation and plaque destabilization. *Circulation*. 2021. https://doi.org/10.1161/CIRCULATIONAHA.120.152023. PMID: 34647815.
96. Liu M, Espinosa-Diez C, Mahan S, et al. H3K4 di-methylation governs smooth muscle lineage identity and promotes vascular homeostasis by restraining plasticity. *Dev Cell*. 2021;56:2765–2782. https://doi.org/10.1016/j.dev.cel.2021.09.001. PMID: 34582749.
97. Passman JN, Dong XR, Wu SP, et al. A sonic hedgehog signaling domain in the arterial adventitia supports resident Sca1$^+$ smooth muscle progenitor cells. *Proc Natl Acad Sci USA*. 2008;105:9349–9354. https://doi.org/10.1073/pnas.0711382105. PMID: 18591670.
98. Xie Y, Martin KA. TCF21: Flipping the phenotypic switch in SMC. *Circ Res*. 2020;126:530–532. https://doi.org/10.1061/CIRCRESAHA.120.316533. PMID: 32078454.
99. Miano JM, Long X, Fujiwara K. Serum response factor: master regulator of the actin cytoskeleton and contractile apparatus. *Am J Physiol Cell Physiol*. 2007;292:C70–C81. https://doi.org/10.1152/ajpcell.00386.2006. PMID: 16928770.
100. Pellegrini L, Tan S, Richmond TJ. Structure of serum response factor core bound to DNA. *Nature*. 1995;376:490–498. https://doi.org/10.1038/376490a0. PMID: 7637780.
101. Yin F, Herring BP. GATA-6 can act as a positive or negative regulator of smooth muscle-specific gene expression. *J Biol Chem*. 2005;280:4745–4752. https://doi.org/10.1074/jbc.M411585200. pmid: 15550397.
102. Hinson JS, Medlin MD, Taylor JM, Mack CP. Regulation of myocardin factor protein stability by the LIM-only protein FHL2. *Am J Physiol Heart Circ Physiol*. 2008;295:H1067–H1075. https://doi.org/10.1152/ajpheart.91421.2007. pmid: 18586895.
103. Shi N, Guo X, Chen SY. Olfactomedin 2, a novel regulator for transforming growth factor-b-induced smooth muscle differentiation of human embryonic stem cell-derived mesenchymal cells. *Mol Biol Cell*. 2014;25:4106–4114. https://doi.org/10.1091/mbc.E14-08-1255. pmid: 25298399.
104. Doi H, Iso T, Yamazaki M, et al. HERP1 inhibits myocardin-induced vascular smooth muscle cell differentiation by interfering with SRF binding to CArG box. *Arterioscler Thromb Vasc Biol*. 2005;25:2328–2334. https://doi.org/10.1161/01.ATV.0000185829.47163.32. pmid: 16151017.
105. Wu SP, Dong XR, Regan JN, Su C, Majesky MW. Tbx18 regulates development of the epicardium and coronary vessels. *Dev Biol*. 2013;383:307–320. https://doi.org/10.1016/j.ydbio.2013.08.019.
106. Zhou J, Hi G, Herring BP. Smooth muscle-specific genes are differentially sensitive to inhibition by Elk1. *Mol Cell Biol*. 2005;25:9874–9885. https://doi.org/10.1128/MCB.25.22.9874-9885.2005. pmid: 16260603.
107. Zheng JP, He X, Liu F, et al. YY1 directly interacts with myocardin to repress the triad myocardin/SRF/CArG box-mediated smooth muscle gene transcription during smooth muscle phenotypic modulation. *Sci Rep*. 2020;10:21781. https://doi.org/10.1038/s41598-020-78544-3. pmid: 33311559.
108. Yap C, Mieremet A, de Vries CJM, Micha D, de Waard V. Six shades of vascular smooth muscle cells illuminated by KLF4 (Krüppel-like factor 4). *Arterioscler Thromb Vasc. Biol*. 2021;41:2693–2707. https://doi.org/10.1161/ATVBAHA.121.316600. pmid: 34470477.
109. Zhou J, Hu G, Wang X. Repression of smooth muscle differentiation by a novel high mobility group box-containing protein, HMG2L1. *J Biol Chem*. 2010;285:23177–23185. https://doi.org/10.1074/jbc.M110.109868. pmid: 20511232.
110. Speer MY, Yang HY, Brabb T, et al. Smooth muscle cells give rise to osteochondrogenic precursors and chondrocytes in calcifying arteries. *Circ Res*. 2009;104:733–741. https://doi.org/10.1161/CIRCRESAHA.108.183053. PMID: 19197075.
111. Sun Y, Byon CH, Yuan K, et al. Smooth muscle cell-specific runx2 deficiency inhibits vascular calcification. *Circ Res*. 2012;111:543–552. https://doi.org/10.1161/CIRCRESAHA.112.267237. PMID: 22773442.
112. Wirka RC, Wagh D, Paik DT, et al. Atheroprotective roles of smooth muscle cell phenotypic modulation and the TCF21 disease gene as revealed by single-cell analysis. *Nat Med*. 2019;25:1280–1289. https://doi.org/10.1038/s41591-019-0512-5. PMID: 31359001.
113. Alencar GF, Owsiany KM, Santos K, et al. The stem cell pluripotency genes Klf4 and Oct4 regulate complex SMC phenotypic changes critical in late-stage atherosclerotic lesion pathogenesis. *Circulation*. 2020;142:2045–2059. https://doi.org/10.1161/CIRCULATIONAHA.120.046672. PMID: 32674599.
114. Pan H, Xue C, Auerbach BJ, et al. Single-cell genomics reveals a novel cell state during smooth muscle cell phenotypic switching and potential therapeutic targets for atherosclerosis in mouse and human. *Circulation*. 2020;142:2060–2075. https://doi.org/10.1161/CIRCULATIONAHA.120.048378. PMID: 32962412.
115. Shankman LS, Gomez D, Cherepanova OA, et al. KLF4-dependent phenotypic modulation of smooth muscle cells has a key role in atherosclerotic plaque pathogenesis. *Nat Med*. 2015;21:628–637. https://doi.org/10.1038/nm.3866. PMID: 25985364.
116. Majesky MW, Horita H, Ostriker A, et al. Differentiated smooth muscle cells generate a subpopulation of resident vascular progenitor cells in the adventitia regulated by Klf4. *Circ Res*. 2017;120:296–311. https://doi.org/10.1161/CIRCRESAHA.116.309322. PMID: 27834190.
117. Rong JX, Shapiro M, Trogan E, Fisher EA. Transdifferentiation of mouse aortic smooth muscle cells to a macrophage-like state after cholesterol loading. *Proc Natl Acad Sci USA*. 2003;100:13531–13536. https://doi.org/10.1073/pnas.1735526100. PMID: 14581613.
118. Allahverdian S, Chehroudi AC, McManus BM, Abraham T, Francis GA. Contribution of intimal smooth muscle cells to cholesterol accumulation and macrophage-like cells in human atherosclerosis. *Circulation*. 2014;129:1551–1559. https://doi.org/10.1161/CIRCULATIONAHA.113.005015. PMID: 24481950.
119. Vengrenyuk Y, Nishi H, Long X, et al. Cholesterol loading reprograms the microRNA-143/145-myocardin axis to convert aortic smooth muscle cells to a dysfunctional macrophage-like phenotype. *Arterioscler Thromb Vasc Biol*. 2015;35:535–546. https://doi.org/10.1161/ATVBAHA.114.304029. PMID: 25573853.
120. Liu M, Gomez D. Smooth muscle cell phenotypic plasticity. *Arterioscler Thromb Vasc Biol*. 2019;39:1715–1723. https://doi.org/10.1161/ATVBAHA.119.312131. PMID: 31340668.
121. Chakraborty R, Chatterjee P, Dave JM, et al. Targeting smooth muscle phenotypic switching in vascular disease. *JVS Vasc Sci*. 2021;2:79–94. https://doi.org/10.1016/j.jvssci.2021.04.001. PMID: 34617061.
122. Gomez D, Swiatlowska P, Owens GK. Epigenetic control of smooth muscle cell identity and lineage memory. *Arterioscler Thromb Vasc Biol*. 2015;35:2508–2516. https://doi.org/10.1161/ATVBAHA.115.305044. pmid: 26449751.

123. Luo G, Ducy P, McKee MD, et al. Spontaneous calcification of arteries and cartilage in mice lacking matrix GLA protein. *Nature*. 1997;386:78–81. https://doi.org/10.1038/386078a0. PMID: 9052783.
124. Lin ME, Chen T, Leaf EM, Speer MY, Giachelli CM. Runx2 expression in smooth muscle cells is required for arterial medial calcification in mice. *Am J Pathol*. 2015;185:1958–1969. https://doi.org/10.1016/j.ajpath.2015.03.020. PMID: 25987250.
125. Leroux-Berger M, Queguiner I, Maciel TT, Ho A, Relaix F, Kempf H. Pathologic calcification of adult vascular smooth muscle cells differs on their crest or mesodermal embryonic origin. *J Bone Miner Res*. 2011;26:1543–1553. https://doi.org/10.1002/jbmr.382. PMID: 21425330.
126. Chattopadhyay A, Kwartler CS, Kaw K, et al. Cholesterol-induced phenotypic modulation of smooth muscle cells to macrophage/fibroblast-like cells is driven by an unfolded protein response. *Arterioscler Thromb Vasc Biol*. 2021;41:302–316. https://doi.org/10.1161/ATVBAHA.120.315164. PMID: 33028096.
127. Feil S, Fehrenbacher B, Lukowski R, et al. Transdifferentiation of vascular smooth muscle cells to macrophage-like cells during atherogenesis. *Circ Res*. 2014;115:662–667. https://doi.org/10.1161/CIRCRESAHA.115.304634. PMID 25070003.
128. Albarrán-Juárez J, Kaur H, Grimm M, Offermanns S, Wettschureck N. Lineage tracing of cells involved in atherosclerosis. *Atherosclerosis*. 2016;251:445–453. https://doi.org/10.1016/j.atherosclerosis.2016.06.012. PMID: 27320174.
129. Jacobsen K, Lund MB, Shim J, et al. Diverse cellular architecture of atherosclerotic plaque derives from clonal expansion of a few medial SMCs. *JCI Insight*. 2017;2:e95890. https://doi.org/10.1172/jci.insight.95890. PMID: 28978793.
130. Chen PY, Qin L, Li Q, et al. Smooth muscle cell reprogramming in aortic aneurysms. *Cell Stem Cell*. 2020;26:542–557. https://doi.org/10.1016/j.stem.2020.02.013. PMID: 32243809.
131. Basatemur GL, Jørgensen HF, Clarke MCH, Bennett MR, Mallat Z. Vascular smooth muscle cells in atherosclerosis. *Nat Rev Cardiol*. 2019;16:727–744. https://doi.org/10.1038/s41569-019-0227-9. pmid: 31243391.
132. Rinkevich Y, Walmsley GG, Hu MS, et al. Identification and isolation of a dermal lineage with intrinsic fibrogenic potential. *Science*. 2015;348:aaa2151. https://doi.org/10.1126/science.aaa2151. PMID: 25883361.
133. Plikus MV, Wang X, Sinha S, et al. Fibroblasts: origins, definitions, and functions in health and disease. *Cell*. 2021;184:3852–3872. https://doi.org/10.1016/j.cell.2021.06.024. PMID: 34297930.
134. Lim CH, Sun Q, Ratti K, et al. Hedgehog stimulates hair follicle neogenesis by creating inductive dermis during murine skin wound healing. *Nat Commun*. 2018;9:4903. https://doi.org/10.1038/s41467-018-07142-9. PMID: 30464171.
135. Plikus MV, Guerrero-Juarez CF, Ito M, et al. Regeneration of fat cells from myofibroblasts during wound healing. *Science*. 2017;355:748–752. https://doi.org/10.1126/science.aai8792. PMID: 28059714.
136. Yap C, Mieremet A, de Vries CJM, Micha D, de Waard V. Six shades of vascular smooth muscle cells illuminated by KLF4 (Krüppel-like factor 4). *Arterioscler Thromb Vasc Biol*. 2021. https://doi.org/10.1161/ATVBAHA.121.316600. PMID: 34470477.
137. Takahashi K, Yamanaka S. Induction of pluripotent stem cells from mouse embryonic and adult fibroblast cultures by defined factors. *Cell*. 2006;126:663–676. https://doi.org/10.1016/j.cell.2006.07.024. PMID: 16904174.
138. Ye Z, Li W, Jiang Z, Wang E, Wang J. An intermediate state in trans-differentiation with proliferation, metabolic, and epigenetic switching. *iScience*. 2021;24:103057. https://doi.org/10.1016/j.isci.2021.103057. PMID: 34541470.
139. Gomez D, Owens GK. Smooth muscle cell phenotypic switching in atherosclerosis. *Cardiovasc Res*. 2012;95:156–164. https://doi.org/10.1093/cvr/cvs115. PMID: 22406749.
140. Bennett MR, Sinha S, Owens GK. Vascular smooth muscle cells in atherosclerosis. *Circ Res*. 2016;118:692–702. https://doi.org/10.1161/CIRCRESAHA.115.306361. PMID: 26892967.
141. Wang D, Rabhi N, Yet SF, Farmer SR, Layne MD. Aortic carboxypeptidase-like protein regulates vascular adventitial progenitor and fibroblast differentiation through myocardin related transcription factor A. *Sci Rep*. 2021;11:3948. https://doi.org/10.1038/s41598-021-82941-7. PMID: 33597582.
142. Jiang J, Chan YS, Loh YH, et al. A core Klf circuitry regulates self-renewal of embryonic stem cells. *Nat Cell Biol*. 2008;10:353–360. https://doi.org/10.1038/ncb1698. PMID: 18264089.
143. Yoshida T, Yamashita M, Hayashi M. Krüppel-like factor 4 contributes to high phosphate-induced phenotypic switching of vascular smooth muscle cells into osteogenic cells. *J Biol Chem*. 2012;287:25706–25714. https://doi.org/10.1074/jbc.M112.361360. PMID: 22679022.
144. Psaltis PJ, Harbuzariu A, Delacroix S, et al. Identification of a monocyte-predisposed hierarchy of hematopoietic progenitor cells in the adventitia of postnatal murine aorta. *Circulation*. 2012;125:592–603. https://doi.org/10.1161/CIRCULATIONAHA.111.059360. PMID: 22203692.
145. Toledo-Flores D, Williamson A, Schwarz N, et al. Vasculogenic properties of adventitial Sca1^{+}CD45^{+} progenitor cells in mice: a potential source of vasa vasorum in atherosclerosis. *Sci Rep*. 2019;9:7286. https://doi.org/10.1038/s41598-019-43765-8. PMID: 31086203.
146. Lu S, Jolly AJ, Strand KA, et al. Smooth muscle-derived progenitor cell myofibroblast differentiation through KLF4 downregulation promotes arterial remodeling and fibrosis. *JCI Insight*. 2020;5(23):139445. https://doi.org/10.1172/jci.insight.139445. PMID: 33119549.
147. Tang J, Wang H, Huang X, et al. Arterial Sca1+ vascular stem cells generate de novo smooth muscle for artery repair and regeneration. *Cell Stem Cell*. 2020;26:81–94. https://doi.org/10.1016/j.stem.2019.11.010. pmid: 31883835.
148. Worssam MD, Lambert J, Oc S, et al. Primed smooth muscle cells acting as first responder cells in disease. *bioRxiv*. 2020;10(19):345769. https://doi.org/10.1101/2020.10.19.345769.
149. Miano JM, Fisher EA, Majesky MW. Fate and state of vascular smooth muscle cells in atherosclerosis. *Circulation*. 2021;143:2110–2116. https://doi.org/10.1161/CIRCULATIONAHA.120.049922. PMID: 34029141.
150. Shan F, Huang Z, Xiong R, Huang QY, Li J. HIF1α-induced upregulation of KLF4 promotes migration of human vascular smooth muscle cells under hypoxia. *J Cell Physiol*. 2020;235:141–150. https://doi.org/10.1002/jcp.28953. PMID: 31270801.
151. Hu Y, Zhang Z, Torsney E, et al. Abundant progenitor cells in the adventitia contribute to atherosclerosis of vein grafts in ApoE-deficient mice. *J Clin Invest*. 2004;113:1258–1265.
152. Wang H, Zhao H, Zhu H, et al. Sca1^{+} cells minimally contribute to smooth muscle cells in atherosclerosis. *Circ Res*. 2021;128:133–135. https://doi.org/10.1161/CIRCRESAHA.120.317972. PMID: 33146591.
153. Topouzis S, Majesky MW. Smooth muscle lineage diversity in the chick embryo. Two types of aortic smooth muscle cell differ in growth and receptor-mediated transcriptional responses to transforming growth factor-beta. *Dev Biol*. 1996;178:430–445. https://doi.org/10.1006/dbio.1996.0229. PMID: 8830742.
154. Gadson Jr PF, Dalton ML, Patterson E, et al. Differential response of mesoderm- and neural crest-derived smooth muscle to TGF-beta1: regulation of c-myb and alpha1 (I) procollagen genes. *Exp Cell Res*. 1997;230:169–180. https://doi.org/10.1006/excr.1996.3398. pmid: 9024776.

155. MacFarlane EG, Parker SJ, Shin JY, et al. Lineage-specific events underlie aortic root aneurysm pathogenesis in Loeys-Dietz syndrome. *J Clin Invest*. 2019;129:659–675. https://doi.org/10.1172/JCI1123547. pmid: 30614814.
156. Chappell J, Harman JL, Narasimhan VM, et al. Extensive proliferation of a subset of differentiated, yet plastic, medial vascular smooth muscle cells contributes to neointimal formation in mouse injury and atherosclerosis models. *Circ Res*. 2016;119:1313–1323. https://doi.org/10.1161/CIRCRESAHA.116.309799. pmid: 27682618.
157. Clément M, Chappell J, Raffort J, et al. Vascular smooth muscle cell plasticity and autophagy in dissecting aortic aneurysms. *Arterioscler Thromb Vasc Biol*. 2019;39:1149–1159. https://doi.org/10.1161/ATVBAHA.118.311727. PMID: 30943775.
158. Misra A, Feng Z, Chandran RR, et al. Integrin beta3 regulates clonality and fate of smooth muscle-derived atherosclerotic plaque cells. *Nat Commun*. 2018;9:2073. https://doi.org/10.1038/s41467-018-04447-7. PMID: 29802249.
159. Sheikh AQ, Saddouk FZ, Ntokou A, Mazurek R, Greif DM. Cell autonomous and non-cell autonomous regulation of SMC progenitors in pulmonary hypertension. *Cell Rep*. 2018;23:1152–1165. https://doi.org/10.1016/j.celrep.2018.03.043. PMID: 29694892.
160. Benditt EP, Benditt JM. Evidence for a monoclonal origin of human atherosclerotic plaques. *Proc Natl Acad Sci USA*. 1973;70:1753–1756. https://doi.org/10.1073/pnas.70.6.1753. PMID: 4515934.
161. Wang Y, Nanda V, Direnzo D, et al. Clonally expanding smooth muscle cells promote atherosclerosis by excaping efferocytosis and activating the complement cascade. *Proc Natl Acad Sci USA*. 2020;117:15818–15826. https://doi.org/10.1073/pnas.2006348117. PMID: 32541024.
162. Schwartz S.M., Virmani R., Majesky M.W. An update on clonality: what smooth muscle cell type makes up the atherosclerotic plaque? F1000Res. 2018;7:F1000. Faculty Rev-1969. PMID: 30613386. doi:10.12688/f1000research.15994.1.
163. Espinosa-Diez C, Mandi V, Du M, Liu M, Gomez D. Smooth muscle cells in atherosclerosis: clones but not carbon copies. *JVS Vasc Sci*. 2021;2:136–148. https://doi.org/10.1016/j.jvssci.2021.02.002. pmid: 34617064.
164. Bray SJ. Notch signaling in context. *Nat Rev Mol Cell Biol*. 2016;17:722–735. https://doi.org/10.1038/nrm.2016.9.4. PMID: 27507209.
165. Lu YW, Lowery AM, Sun LY, et al. Endothelial myocyte enhancer factor 2c inhibits migration of smooth muscle cells through fenestrations in the internal elastic lamina. *Arterioscler Thromb Vasc Biol*. 2017;37:1380–1390. https://doi.org/10.1161/ATVBAHA.117.309180. pmid: 28473437.
166. Worssam MD, Jørgensen HF. Mechanisms of vascular smooth muscle cell investment and phenotypic diversification in vascular diseases. *Biochem Soc Trans*. 2021;49:2101–2111. https://doi.org/10.1042/BST20210138. pmid: 34495326.
167. Steffes LC, Froistad AA, Andurska A, et al. A notch3-marked subpopulation of vascular smooth muscle cells is the cell of origin for occlusive pulmonary vascular lesions. *Circulation*. 2020;142:1545–1561. https://doi.org/10.1161/CIRCULATIONAHA.120.045750. PMID: 32794408.
168. Ståhl PL, Salmén F, Vickovic S, et al. Visualization and analysis of gene expression in tissue sections by spatial transcriptomics. *Science*. 2016;353:78–82. https://doi.org/10.1126/science.aaf2403.
169. Rodriques SG, Stickels RR, Goeva A, et al. Slide-seq: a scalable technology for measuring genome-wide expression at high spatial resolution. *Science*. 2019;363:1463–1467. https://doi.org/10.1126/science.aaw1219. PMID: 30923225.
170. Srivatsan SR, Regier MC, Barkan E, et al. Embryo-scale, single cell spatial transcriptomics. *Science*. 2021;373:111–117. https://doi.org/10.1126/science.abb9536. PMID: 34210887.
171. Fisher AG. Cellular identity and lineage choice. *Nat Rev Immunol*. 2002;2:977–982. https://doi.org/10.1038/nri958. PMID: 12461570.
172. Anderson DM, Anderson KM, Nelson BR, et al. A myocardin-adjacent lncRNA balances SRF-dependent gene transcription in the heart. *Genes Dev*. 2021;35:835–840. https://doi.org/10.1101/gad.348304.121. PMID: 33985971.
173. Gao R, Liang X, Cheedipudi sm Cordero J, et al. Pioneering function of Isl1 in the epigenetic control of cardiomyocyte cell fate. *Cell Res*. 2019;29:486–501. https://doi.org/10.1038/s41422-019-0168-1. PMID: 31024170.
174. Cai CL, Liang X, Shi Y, et al. Isl1 identifies a cardiac progenitor population that proliferates prior to differentiation and contributes a majority of cells to the heart. *Dev Cell*. 2003;5:877–889. https://doi.org/10.1016/s1534-5807(03)00363-0. PMID: 14667410.
175. Smith NL, Patel RK, Reynaldi A, et al. Developmental origin governs $CD8^+$ T cell fate decisions during infection. *Cell*. 2018;174:117–130. https://doi.org/10.1016/j.cell.2018.05.029. PMID: 29909981.
176. Kuppe C, Ibrahim MM, Kranz J, et al. Decoding myofibroblast origins in human kidney fibrosis. *Nature*. 2020;589:281–286. https://doi.org/10.1038/s41586-020-2941-1. PMID: 33176333.
177. Tsukui T, Sun KH, Wetter JB, et al. Collagen-producing lung cell atlas identifies multiple subsets with distinct localization and relevance to fibrosis. *Nat Commun*. 2020;11:1920. https://doi.org/10.1038/s41467-020-15647-5. PMID: 32317643.
178. Driskell RR, Lichtenberger BM, Hoste E, et al. Distinct fibroblast lineages determine dermal architecture in skin development and repair. *Nature*. 2013;504:277–281. https://doi.org/10.1038/nature12783. PMID: 24336287.
179. Philippeos C, Telerman SB, Oules B, et al. Spatial and single-cell transcriptional profiling identifies functionally distinct human dermal fibroblast subpopulations. *J Invest Dermatol*. 2018;138:811–825. https://doi.org/10.1016/i.jid.2018.01.016. PMID: 29391249.
180. Guerrero-Juarez CF, Dedhia PH, Jin S, et al. Single-cell analysis reveals fibroblast heterogeneity and myeloid-derived adipocyte progenitors in murine skin wounds. *Nat Commun*. 2019;10:650. https://doi.org/10.1038/s41467-018-08247-x. PMID: 30737373.
181. Mascharak S, desJardins-Park HE, Davitt MF, et al. Preventing engrailed-1 activation in fibroblasts yields wound regeneration without scarring. *Science*. 2021;372:eaba2374. https://doi.org/10.1126/science.aba2374. PMID: 33888614.
182. Humphreys BD, Lin SL, Kobayashi A, et al. Fate tracing reveals the pericyte and not epithelial origin of myofibroblasts in kidney fibrosis. *Am J Pathol*. 2010;176:85–97. https://doi.org/10.2353/ajpath.2010.090517. PMID: 20008127.
183. Hung C, Linn G, Chow YH, et al. Role of lung pericytes and resident fibroblasts in the pathogenesis of pulmonary fibrosis. *Am J Respir Crit Care Med*. 2013;188:820–830. https://doi.org/10.1164/rccm.201212-22970C. PMID: 23924232.
184. Kramann R, Schneider RK, DiRocco DP, et al. Perivascular $GLi1^+$ progenitors are key contributors to injury-induced organ fibrosis. *Cell Stem Cell*. 2015;16:51–66. https://doi.org/10.1016/j.stem.2014.11.004. PMID: 25465115.
185. Ray HC, Corliss BA, Bruce AC, et al. $Myh11^+$ microvascular mural cells and derived mesenchymal cells promote retinal fibrosis. *Sci Rep*. 2020;10:15808. https://doi.org/10.1038/s41598-020-72875-x. PMID: 32978500.
186. Wu J, Montaniel KRC, Saleh MA, et al. *Hypertension*. 2016;67:461–468. https://doi.org/10.1161/HYPERTENSIONAHA.115.06123. PMID: 26693821.

CHAPTER

4

Smooth muscle diversity in the vascular system

Steven A. Fisher

Departments of Medicine (Cardiovascular Division) and Physiology, University of Maryland School of Medicine and Baltimore Veterans Administration Medical Center, Baltimore, MD, United States

Introduction

The mammalian vascular system is incredibly diverse, anatomically, functionally, and in its cell and molecular composition. This diversity is apparent when comparing the vascular beds of the different organ systems (lungs, brain, kidney, heart, splanchnic, etc.) and within a single vascular bed. This topic has been reviewed by myself and others over the past decade (see Refs. 1–3). Here I will briefly review this topic of smooth muscle cell (SMC) diversity in relation to vascular function followed by a more expansive review of the role of unique gene programs in wiring and rewiring of signaling pathways that control vessel tone and thus blood flow. This is followed by a review of some of the newer technologies that will advance the study of vascular smooth muscle phenotypic diversity. The emphasis here is on vascular function, i.e., how does variation in the SMC gene program determine vessel- and vascular bed-specific function.

Definition of the problem: diversity within the vascular system

Smooth muscle is dispersed throughout the vascular system, in large conduit arteries and veins (aorta, vena cava), and in small arteries, arterioles, and venules within each vascular bed. Each vascular bed, e.g., kidney, lung, splanchnic, muscle, and brain has unique structure, function, and vasomotor regulation (reviewed in Ref. 3). Because of the steep inverse relationship between vessel radius and resistance to blood flow (r^{-4}), it is the smaller vessels (<500 micron) where the major resistance to blood flow is located and regulated according to the needs of the tissue and more globally to maintain systemic blood pressure. The dispersion of SMC in small vessels buried within organs has historically confounded attempts to define organotypic vascular smooth muscle cell (VSMC) phenotype and their diversity. As will be reviewed below, this is being overcome using sophisticated techniques of cell labeling and single cell RNA sequencing (RNASeq).

A complete understanding of VSMC diversity and its impact on vascular function would require (1) defining the entire gene program mapped onto every cell of every vessel throughout the body (2) using what is already known or can be gleaned from databases to generate hypotheses as to how variable expression of gene products influences VSMC function in the regulation of blood flow and pressure and, (3) testing these hypotheses through gain/loss/change-of-function in vivo.

Gene programs

Extensive sequencing of organismal genomes/transcriptomes coupled to bioinformatic analyses estimates that ~22,000 mammalian genes encode several hundred thousand proteins, with estimates varying depending on the method of counting.[4] Diversity is in part generated by expansion of gene families, as for example, the numerous α- and β-subunits of the calcium channels; however, variable exon usage through alternative transcriptional start sites and alternative splicing of exons, and the highly regulated expression of these diverse gene products, is the major driver of diversity of cell types [5–8] (see Ref. 9 for an opposing view). In part, this is a function of simple math: while each new gene in a family has an additive effect (+1), variable combinations of alternative exons are multiplicative, as for example, the 18 distinct transcripts generated from 4 variable exons in the human tropomyosin-alpha (Tpm1) gene (reviewed in Ref. 10) and the record 37,000 unique potential transcripts from the fly Downs Syndrome Cell Adhesion Molecule

The Vasculome
https://doi.org/10.1016/B978-0-12-822546-2.00014-9

gene.[11] On average, each gene locus produces ~4 unique transcripts through alternative exon usage. Also, as will be discussed below, alternative splicing of exons provides a clever way to add or remove motifs so as to modify a subfunction of the protein while retaining the main functions intact. Likewise, while early studies focused on transcriptional control of gene expression as the primary means of control and generation of diversity, the subsequent decades of research discovered a plethora of regulatory mechanisms including splicing of exons, noncoding RNAs regulating RNA turnover and translation, and protein modifications regulating stability, e.g., ubiquitination, and function, that in sum regulate the expression and function of the end product/protein in a highly cell-specific manner.

Vascular smooth muscle functional diversity

There are many classification schemes for smooth muscle; the one most relevant to this review is based on contractile properties.[12,13] In tonic (slow) VSM present in large arteries and veins, force develops slowly, and a resting tone is maintained and modulated by signaling pathways. The resting tone is necessary for these VSMCs to withstand the high pressures (forces) of the mammalian vascular system. In the phasic (fast) VSM of the portal vein, gut, and bladder, the rate of force development is ~threefold faster with phasic contractions with a periodicity in the range of 0.1 Hz, at least an order of magnitude slower than in striated muscle. The purpose of the phasic contractions in the gastrointestinal/genito-urinary systems is obvious: to propel/expel food/urine. The purpose of the phasic contractions of the portal vein is less certain but presumably serves to propel blood flow in the unique dual parallel hepatic circulation (reviewed in Ref. 3). Resistance artery VSM has mixed properties, with some resting tone and superimposed phasic contractions termed vasomotion, the purpose of which is uncertain. [14,15] As noted above, because the VSMCs of these small arteries is where blood flow and pressure are predominately regulated, these have received increasing attention as of late for defining their gene programs and differences with those of the large arteries. Smooth muscle of the lymphatic vascular system also displays phasic contractile activity required to propel lymph; its smooth muscle is unique but beyond the scope of this review (see Ref. 16).

Vascular smooth muscle cell diversity: contractile properties

SMC myosin (Myh11) and smoothelin (SMTN) are gold standards as markers for the fully differentiated SMC phenotype.[17] Since their expression is specific to SMC, they serve as useful tools for the study of SMC subphenotypes (diversity), with the obvious limitation that the expression of a few genes does not a gene program make. Many SMC contractile genes are also expressed in nonmuscle cells, where they function in cell motility and cytokinesis. This can confound the analysis of the SMC gene program and phenotypic diversity when studied at the level of a given tissue composed of complex mixtures of cell types (discussed further below). Of note, unlike the case for striated muscle, there is no in vitro system in which the transition of cells from an undifferentiated to fully differentiated SMC state can be studied. A few in vitro systems for the study of SMC differentiation have been described,[18] with a recent one using human uterine derived immortalized cells claiming to be the first specific for the visceral SMC phenotype.[19]

Myh11

The distinct properties of myosin isozymes often play a dominant role in determining the contractile properties, i.e., fast versus slow contractile velocities, in muscle subphenotypes.[20,21] The fast isoform of Myh11 (SM-B) is generated by the inclusion of a 21-nucleotide alternative exon (E8) in the myosin head. Its expression correlates well with the several-fold higher ATPase activity and maximum velocity of shortening in phasic versus tonic smooth muscle.[22] The tonic smooth muscle of the large arteries and veins expresses almost exclusively the slow isoform of MYH11 (E8 skipped). The fast isoform of MYH11 (E8 included) was detected in small arteries of the heart,[23] lung,[24] muscular femoral artery,[25] and mesenteric arteries.[26] Whether the level of expression of this MYH11 isoform is significant both with regards to its level of expression and vascular function has not been determined.

The afferent versus efferent arterioles of the renal circulation serve as an exemplary case study of VSM phenotypic diversity. The afferent arteriole responds rapidly to changes in pressure and flow with a myogenic contractile response that autoregulates blood flow, as well as a tubulo-glomerular feedback mechanism that also autoregulates blood flow in the kidney.[27] The myogenic response is characteristic of resistance arteries and absent in the tonic smooth muscle of conduit vessels. It is also absent in the efferent arteriole, which maintains tone and regulates glomerular filtration through changes in tone, i.e., tonic smooth muscle phenotype. Consistent with these functional differences, the fast (phasic) isoform of MYH11 is expressed in the afferent arteriole, while only the slow (tonic) isoform is expressed in the efferent arteriole.[28,29] This correlates with the higher myosin ATPase activity and shortening velocity of agonist AngiotensinII (AngII) and Norepinephrine-induced contractions of rat and mouse afferent versus efferent arterioles. However, inactivation of the fast (SM-B) exon did not equalize contraction velocities in mouse afferent and efferent arterioles, suggesting other undefined molecular differences as contributory to these functional differences.[28] In contrast, using the same mouse model, it was reported that loss of the fast isoform of Myh11 reduced the shortening velocity of small mesenteric arteries, but oddly enough, also of the aorta, where its expression was not detected.[26] This could reflect a secondary phenomenon, an issue in all experiments involving genetic manipulation of the mouse.

Smoothelin

SMTN is also a late and specific marker of fully differentiated SMCs (reviewed in Ref. 30,31). SMTN is part of a multigene family of cytoskeletal proteins that includes SMTN-like 1 and 2. SMTN is a fascinating locus with alternative transcriptional start sites that are specific to vascular (tonic; Exon 1) and visceral (phasic; Exon 10) smooth muscle.[32,33] The unique regulatory programs in vascular versus visceral smooth muscle driving these differences in transcription are unknown. In addition, exon 20 (counting from the full length transcript), located within the C-terminal calponin homology domain, is alternatively spliced, with a switch from exon inclusion to exon skipping with developmental differentiation of vascular and visceral smooth muscle[31,34] and switch to exon inclusion with dedifferentiation of smooth muscle.[35] The functional significance of these isoforms in this cytoskeletal protein is unknown. Specific knock out of the long (A; vascular) isoform resulted in an interesting phenotype of reduced contractility of the saphenous and femoral arteries but increased mean arterial pressure by 20 mmHg.[36] Whether this could reflect a quirk of gene/isoform compensation or the differing expression and activities of SMTN A v B isoforms throughout the vascular system has not been fully investigated. The inactivation of the SMTNL1 gene also has interesting effects on VSM/SM function in pregnancy, thought to at least in part involve changes in Myosin Phosphatase (MP) expression and activity.[37–39] However, specific mechanisms have yet to be defined.

Broader smooth muscle gene program

The above are just two examples of the thousands of genes expressed and hundreds of thousands of differentially expressed gene products in any cell. While there are other worthy examples of differential expression within VSM, the functional significance is not as well established and space constraints will not allow their presentation here (see Ref. 1). More recently, investigators have turned to unbiased high throughput methodologies to identify differences between various smooth muscle tissues/cells or changes in gene programs in disease states. Typing "smooth muscle" into the NCBI Gene Expression Omnibus database retrieves 5582 data sets (3515 human, 1162 mouse). Many of these are not relevant here, as they compare different regions of the aorta, or are focused on dedifferentiation of smooth muscle and changes in VSM in atherosclerosis. A few data sets that the author is familiar with that are relevant include (1) our own, in which we compared gene expression in rat aorta versus mesenteric arteries by deep RNASeq,[40] (2) a study in which the effect of exercise and obesity on gene expression by RNASeq in rat skeletal muscle small arteries were examined,[41,42] and (3) a study by Llorian et al.[35] used exon tiling arrays to compare splice variants in mouse tonic aorta versus phasic bladder, and again when each are placed into culture to undergo proliferation/dedifferentiation. This last study was particularly revealing in its 2 × 2 design. It carefully identified splice variant isoform differences between aortic and bladder smooth muscle, and again between differentiated and dedifferentiated smooth muscle. The latter allows the determination of the extent to which differential splice variant isoform expression truly reflects a program of smooth muscle differentiation versus the more trivial explanation of contamination by nonmuscle cells.

Vascular smooth muscle cell diversity in vasomotor control and signaling networks

Introduction

Unlike striated muscle, the state of contraction of smooth muscle is primarily determined by the opposing activities of the myosin kinase (MYLK) and multisubunit MP (reviewed in Ref. 43). MYLK under the control of calcium/calmodulin phosphorylates the regulatory myosin light chain (Myl6) thereby activating Myosin ATPase activity and causing contraction. MP is a multisubunit enzyme which in smooth muscle is composed of catalytic (PP1cβ), regulatory (Mypt1 = PPP1R12a), small (M21 = PPP1r12b), and inhibitory (CPI17 = PPP1R14a) subunits.[44,45] Experiments in the 1990s identified MP as a critical end-target and effector of signaling pathways that vasoconstrict through inhibition of MP[46] and vasodilate through activation of MP.[47,48] In these experiments, smooth muscle tissues were permeabilized and force activated at fixed concentrations of calcium (= fixed activity of MLCK), such that any change in force induced by a given signal resulted from a change in

MP activity. Historically, there was also an appreciation that smooth muscle tissues differ in their response to these signaling pathways. Around the time of the discovery of endothelial-derived relaxing factor/nitric oxide (EDRF/NO) and its second messenger cGMP in mediating smooth muscle relaxation, experiments on isolated vascular and nonvascular smooth muscle tissues indicated resistance in some tissue preparations including dog portal vein,[49] chicken gizzard,[50] and nonvascular smooth muscle tissues[51] (see also Ref. 52). Subsequent studies supported the concept that classic EDRF/NO plays a lesser and endothelial-derived hyperpolarizing factors a greater role as artery size decreases in regional circulations,[53–55] including in humans.[56] However, a possible role for smooth muscle phenotypic diversity in determining responses to endogenous or pharmacological activators of this pathway was mostly overlooked, as the field focused on bioavailability of NO and acute or chronic desensitization of this pathway (see, for example, Refs. 57,58).

Splice variant isoforms of the MP regulatory subunit code for variable presence of a C-terminal leucine zipper motif

A critical discovery in signaling control of vasoreactivity was a molecular mechanism for activation of MP by the second messenger of NO signaling, cGMP. Surks, Mendelsohn, and coworkers showed that the cGMP-dependent protein kinase (PRKGIα) is targeted to MP through interactions mediated by the leucine zipper (LZ) dimerization motifs in the C-terminus of Mypt1 and N-terminus of PRKGIα[59] (Fig. 4.1). Yet, the specific mechanism by which PRKGIα activates MP is still uncertain,[60–63] but presumably involves phosphorylation of Ser-Thr residues of Mypt1 or other MP subunits. Subsequent to the Surks et al. discovery, experimentation in my laboratory supported a model in which variable expression of the Mypt1 LZ motif determines sensitivity to NO/cGMP-mediated vasorelaxation. A 31 nt Exon 24 (E24) 3′ splice variant of Mypt1 had been described that changes the reading frame and thereby codes for a Mypt1 variant that lacks the C-terminal LZ motif (LZ-)[64,65] (Fig. 4.1A). We subsequently showed that the expression of the Mypt1 E24/LZ ± isoforms, that serve critically important functions (discussed below), is evolutionarily conserved, developmentally regulated, tissue specific, and modulates in disease (reviewed in Refs. 3,44). Mypt1 mRNAs containing the E24 alternative exon coding for the LZ-isoform (E24+/LZ−) are exclusively or highly expressed in phasic smooth muscle of the rat and mouse portal vein, bladder, and intestine,[66–68] and are predominant in small resistance arteries (~80:20).[34,69–74] In contrast in

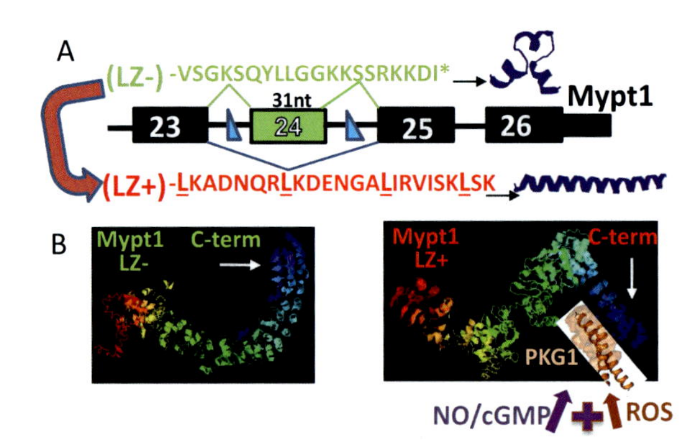

FIGURE 4.1 Alternative splicing of the Myosin phosphatase regulatory subunit (Mypt1) alternative exon 24 tunes sensitivity to NO/cGMP mediated vasodilation. (A) Isoforms of Mypt1 are generated by the alternative splicing of the 31 nt Exon 24 (24). Skipping of E24 generates an mRNA that codes for the C-terminal LZ motif (LZ+). Inclusion of E24 shifts the reading frame and codes for a distinct C-terminus lacking the LZ motif as well as generating a premature termination codon (LZ−). This LZ-variant of Mypt1 is more highly expressed in the smaller arteries and in phasic smooth muscle. Also shown are the ribbon diagrams as predicted by PEP-FOLD3 for the unique C-terminal sequence of the LZ+ and LZ− isoforms. The LZ+ isoform has the LZ motif within a well-ordered a-helix, while the LZ− C-terminus is disordered. The triangles indicate LoxP sequences that we inserted into the introns flanking E24 for the purpose of using Cre-Lox to delete the exon and shift expression from the LZ− to the LZ+ isoform of Mypt1. (B) Ribbon diagrams of the entire Mypt1 LZ− and LZ+ isoforms as predicted by I-TASSER and PEPFOLD-3. Protein Kinase GIa targets to MP via LZ-mediated heterodimerization with Mypt1. This is required for NO/cGMP, and presumably for ROS, mediated activation of MP, a critical signaling pathway in NO/cGMP/ROS mediated vasodilation in the control of blood flow and blood pressure. The Mypt LZ-isoform is resistant to NO/cGMP mediated activation.

the larger arteries and veins such as the aorta, vena cava, and carotid arteries, the Mypt1 E24-variant coding for LZ+ isoform predominates (but see below).

Leucine zipper motifs and role of variable expression in vasodilator responses

The LZ dimerization motif was originally discovered in what are now called the basic-LZ class of transcription factors (Fos, Jun, and others).[75] LZ motifs mediate dimerization via interdigitation of complementary α-helix/LZ structures. It was subsequently shown that variation in the sequence, structure, and deployment of the LZ in these families of transcription factors generates incredible diversity in transcriptional programming by mediating specific interactions with unique partners containing LZ motifs within an α-helical structure.[76–78] LZ dimerization motifs mediating various aspects of cell signaling and response have been identified in many proteins, with newer refinements for bioinformatic prediction of these motifs likely to add to these numbers.[79] LZ motifs have been identified to play critical roles in the function of a number of muscle proteins,[80–83] while for most of these their role in diversity and specificity in signaling in the control of muscle function has not been demonstrated.

Regarding the variable presence of the LZ motif in Mypt1, we demonstrated a correlation between its expression in vascular and nonvascular smooth muscle tissues with the ability of cGMP to activate MP and cause relaxation in permeabilized tissues under calcium clamp.[66–68] These physiological studies, supported by additional biochemical assays, supported the hypothesis that variable expression of the Mypt1 LZ determined sensitivity to NO/cGMP-mediated relaxation under calcium clamp. This is referred to as calcium desensitization of force production and represents one of the two arms of vasorelaxant signaling, the other being suppression of calcium flux.[84]

We next genetically engineered the mouse to directly test the role of Mypt1 E24 splice variants in determining VSM sensitivity to NO/cGMP in vivo. LoxP sites were inserted into the introns flanking Mypt1 E24, ~300 nt from the splice sites so as to avoid identified and conserved splicing regulatory sequences[66,85,86] (Fig. 4.1A). Crossing of these floxed mice with SMMHCCre[ER,87] mice enabled tamoxifen-inducible smooth muscle–specific deletion of E24 (SMcKO E24) with high efficiency. This shifts expression of Mypt1 isoforms from LZ− to LZ+.[72,73] The small mesenteric arteries of these mice were ~100- and ~10-fold more sensitive to the relaxing effect of an NO donor and cGMP analogue when studied by wire myography.[72,73] The SMcKO E24 mice also had ambulatory blood pressure that was ~20 mmHg lower as measured by telemetry. These experiments establish the role of the Mypt1 E24 splice variants in vasodilator responsiveness and regulation of BP and support a model in which toggling of Mypt1 splice variants tunes sensitivity to NO/cGMP-mediated vasorelaxation (Fig. 4.1). We and others have observed shifts in Mypt1 E24/LZ isoforms in a number of disease models including sepsis,[73] portal hypertension,[71] flow-induced small artery remodeling,[74,88] and heart failure. [69,89] In all but the latter, it is proposed that the switch to the E24-/LZ+ isoform represents a toggle switch by which VSM can call on vasodilator reserve to maximize NO responsiveness and blood flow under these disease conditions, though this has not been formally demonstrated. In recent experiments, we have also found that E24cKO suppresses the hypertensive response to chronic infusion of AngII, and that this effect is blocked by cotreatment with inhibition of NO synthase with oral L-nitroarginine methyl ester.[122] A shift to the E24+/LZ− isoform during normal arterial maturation is proposed to allow for the increased vascular resistance of the systemic circulation during postnatal maturation,[34,90] a necessary condition for the highly regulated high pressure/high resistance state of the mature vascular system. The skipping of Mypt1 E24 coding for the Mypt1 LZ+ isoform is the evolutionarily primordial and default pattern,[85] suggesting that splicing of Mypt1 E24 arose in parallel with the evolution of high pressure/high resistance circulations. Lastly, the expression of Mypt1 E24−/LZ + isoforms is also highly sensitive to the differentiated state of the smooth muscle,[35,67] so that the extent to which changes in Mypt1 E24/LZ ± isoforms represent modulation of contractile phenotype versus dedifferentiation of the smooth muscle requires further study.

Activation of this vasodilator pathway by reactive oxygen species

Subsequent to the discovery of cGMP as the critical second messenger for NO/EDRF-mediated vasorelaxation, it was shown that this pathway is also activated by reactive oxygen species (ROS)[73,91–94] (Fig. 4.1B). Molecular diversity again plays a role, as the reactive cysteine that transduces the ROS signal is present in the PRKGIα but not the PRGK1β isoform, with crystal structure showing that oxidation of the Cysteine 42 disulfide bond stabilizes PRKGIα.[95] PRKGIα and PRKGIβ differ only in their first exons coding for unique N-terminal LZ motifs. These isoforms are generated by alternative transcriptional start sites, sometimes mistakenly referred to as alternative splicing. While there has been some suggestion that these unique LZ motifs may target the isoforms to different substrates, this remains an open question.[96,97] There is additional diversity all through this vasodilator pathway, including the guanylate cyclases that generate cGMP, phosphodiesterases that catabolize it, PRKGI effectors, and downstream targets, as well as noncanonical pathways, that are heavily pharmacologically targeted but cannot feasibly be reviewed here.

Diversity in other signaling pathways controlling VSM tone

Recent experiments to address the question of diversity in VSM function and vasomotor control have used unbiased high throughput assays of gene expression in various vascular tissues. Kaur and coworkers analyzed gene expression by RT-PCR in single cells isolated from mouse aorta (ao) versus skeletal muscle (sk) small arteries.[98] There was significant variation between cells within a sample in the expression of the 367 nonolfactory G-protein–coupled receptors (GPCRs) assayed, as well as significant differences between ao and sk SMCs. A number of the peptide hormone receptors functioning in the regulation of blood flow (Ang, ET, NPY, purinergic), as well as some orphan receptors (Olfr78, Gpr126) were much more highly expressed in skSMC, consistent with ours and other prior studies of whole tissue RNA pools,[40,99] while other GPCRs (GPR133, P2yR6) predominated in aoSMC. GPCR expression was very similar in mesenteric arterial SM versus skeletal arterial SM, which differed considerably from ao and bladder SM (see also[100,101]; reviewed in Ref. 102). In the Reho study,[40] the purinergic P2Rx1 receptor was ~50-fold higher than P2Rx5 in rat small mesenteric arteries, while P2Rx1 was ~10-fold higher in the mesenteric arteries versus aorta. In the rat aorta, P2rx1 and P2Rx5 were expressed at similar levels, and P2Rx5 was ~fourfold higher in rat aorta versus mesenteric arteries. The functional significance of these differences in GPCR expression with respect to control of vascular function with few exceptions has not been tested.

Alternative splicing to rewire signaling networks

The model that we have proposed for alternative splicing of Mypt1 E24 as a toggle to tune reactivity to PRKGIα fits nicely with recent observations using higher throughput assays in other systems. These studies support the premise that regulated splicing of alternative exons remodels protein interactions to establish tissue-specific protein–protein interaction networks (see, for example, Ref. 103). These principles may be more universally applicable in examining how isoform expression changes signaling pathways in VSM. First, the inclusion of E24 into the Mypt1 converts the C-terminus from a highly structured α-helix coiled coil with an LZ motif to a relatively unstructured region as predicted by I-TASSER[104] and PEP-FOLD3[105] modeling programs (Fig. 4.1A and B). Intrinsically disordered proteins/disordered regions lack well-defined tertiary structures, and are often associated with cell signaling as their flexibility allows them to bind with a variety of partners. They often occur at the C- (or N-) terminus of proteins, which are often exposed and thus available for protein–protein interactions, and are preferred loci for alternative splicing.[106,107] The C-terminus of proteins has been recognized as being more frequently subject to isoform generation by alternative splicing to increase protein diversity.[108] The C-termini contain conserved motifs/minimotifs inferring function that for the vast majority is not known (see Ref. 109). Two studies using high throughput approaches concluded that tissue-specific alternative splicing of exons is a common mechanism to rewire protein interaction networks.[110,111] The bioinformatic study by Buljan and coworkers found that tissue-specific alternative exons more commonly map to disordered regions of proteins, i.e., have no predicted functional motif, and are more frequently subject to posttranslational modifications, mostly phosphorylation. The latter observation is consistent with a high throughput analysis of cell diversity by RNAseq in which the #1 Swiss-Prot keyword hit was phosphorylation.[6] The authors concluded that tissue-specific alternative splicing is often used to alter the potential for protein phosphorylation, i.e., to rewire signaling networks. Yang and coworkers used a high throughput experimental approach (Open Read Frame-Seq) to reach the conclusion that alternative splicing of exons markedly influences protein interaction networks.[111] While the experimental method is too complex to be succinctly summarized, they observed that (1) the majority of the ~1500 selected isoform pairs share less than 50% of their protein–protein interactions (2) interaction partners specific to the alternative isoforms tend to be highly tissue-specific and belong to distinct functional modules. They concluded that alternative splicing of exons leads to widespread expansion of protein interaction capabilities, and that alternatively spliced isoforms tend to behave like completely distinct genes in interactome networks rather than minor variants of each other, for which they coined the term functional alloforms. As applied to SM, it seems likely that diversity and specificity in signaling pathways have mostly yet to be discovered.

Newer methodologies for the study of smooth muscle diversity

As noted at the outset of this review, two substantial barriers to advancing our understanding of the role of VSM phenotypic diversity in vascular function are (1) the difficulty in isolating pure populations of VSMCs from specific vessels in the different vascular beds and (2) the difficulty in making predictions about how differences in gene expression may determine vascular function in the different regional circulations. Application of newer technologies promises to overcome these hurdles and significantly advance this field of research.

Single cell isolation and RNA sequencing

A study from "the old days" (1990s) found a correlation between expression of Myh11 isoforms and contractile kinetics in single SMCs isolated from the wall of the rodent aorta.[112] Fast forward to today and the combination of several recently developed tools allow for a significantly more sophisticated examination of this question. First, transgenic mice were developed that place fluorescent tags on specific cell populations which can then be purified by cell sorting.[113] These mice have a transgene that will ubiquitously express green or red (or other) but not both, fluorescent reporter proteins. The first of the reporters is flanked by a stop-flox cassette, such that in the presence of an SMC specific Cre, e.g., smMHCCre[87] the first reporter (e.g., green) and stop cassette is excised and the SMCs are specifically converted to expressing the second reporter (e.g., red). This and related models have been used by investigators to obtain purified population of SMCs from various tissues[98,114] and to track SMC fates in vivo.[115] However, this still leaves unresolved the diversity of cell types within a population of SMCs purified from a given tissue. Single cell RNAseq (scRNAseq) was developed to identify the unique subpopulations of cells within a larger population of cells. Methods have been devised to sequence RNA pools, converted into cDNAs, from individual cells in either plate or droplet format, consideration of the technical aspects of scRNAseq were summarized by Chavkin and Hirschi.[102] This is most useful for identifying unique clusters of cell subpopulations (gene programs) from within a complex population of cells. This was done by Dobnikar and coworkers to identify unique VSM subpopulations and transcriptional profiles in different regions of the thoracic mouse aorta.[100] It should be particularly revealing to define unique VSMC subphenotypes within regional circulations, for example, in renal afferent versus efferent arterioles, within the lungs and brain (see[123]), and other regional circulations.

Bioinformatics

While it probably is still a pipe dream that one could go to the computer and find out everything about anything, the amount of data in online databases is staggering ("integrative proteogenomics," see Refs. 116,117). This has given rise to its own discipline termed bioinformatics of which the author has limited expertise. Current issues include (1) having the expertise to navigate these databases, (2) databases arranging their data in a manner so that they are readily accessible to the scientific community, and (3) the limitation that functional significance of most protein isoforms is neither known nor predicted. Large databases such as NCBI, UCSC Genome browser, Ensembl and GTEx Portal have extensive data on genes, gene products, and expression. The Vastdb database is specific to alternative splice variant isoforms and provides relative expression levels (splicing index) for each exon.[118] tappAS is a web program dedicated to defining the functional significance of splice variants.[119] Thakur[120] and Sulakhe[117] and coworkers review different aspects of bioinformatics applied to the study of alternative splicing and its functional significance, while Glusman[116] reviews how to map mutations onto proteins to predict the effect on structure and function. Other useful databases for predicting effect of splice variants and sequence variation on protein structure and function include Uniprot and its links, Swiss-model Repository, I-TASSER, AS-ALPS, and PEP-FOLD3. Phosphositeplus is particularly useful for knowing phospho-residues determined by high and low-throughput assays, key to understanding kinase signaling pathways, and specificity of signaling. IIIdb is a database for isoform–isoform protein interactions and isoform network modules.

Conclusion

There is great functional diversity of smooth muscle within the vascular system that is a function of the diversity in the gene products that control vascular function. Here we have presented some guiding principles and a few examples to highlight this field of study. The ultimate goal is to fully map all of the diversity generated from the genome onto the vascular diversity and to understand its significance with respect to organotypic vascular function in normal and disease states. The rather limited studies done to date support the premise that exon variant isoforms, by altering protein–protein interaction domains, will play a significant role in rewiring signaling networks in a vessel-specific manner analogous to that of neural[103] and other cell types.[111] Hopefully, improved computational methods and machine learning will speed the rate of predicting the functional significance of exon variant isoforms. This coupled with new and more expedient systems for functional testing will enable full mapping of genomic complexity onto vascular diversity. This information will open up a whole new paradigm for treating vascular disease: use of powerful gene editing tools for gain-or-loss of exon splicing[121] to cause a shift in the expression of naturally occurring isoforms as a means to treat human disease.

Acknowledgment

S.A.F. has provisional patent #16/078,332 related to this work. This work was supported by NIH grant R01HL142971-01A1 and VA MERIT award BX004443. I would like to thank Daniel Kang participating in the UM Scholars program for bioinformatic research and help in the preparation of Fig. 4.1.

References

1. Fisher SA. Vascular smooth muscle phenotypic diversity and function. *Physiol Genom.* 2010;42A(3):169–187.
2. Majesky MW. Developmental basis of vascular smooth muscle diversity. *Arterioscler Thromb Vasc Biol.* 2007;27(6):1248–1258.

3. Reho JJ, Zheng X, Fisher SA. Smooth muscle contractile diversity in the control of regional circulations. *Am J Physiol Heart Circ Physiol*. 2014;306(2):H163–H172.
4. Yates AD, Achuthan P, Akanni W, et al. Ensembl 2020. *Nucleic Acids Res*. 2019;48(D1):D682–D688.
5. Johnson JM, Castle J, Garrett-Engele P, et al. Genome-Wide survey of human alternative pre-mRNA splicing with exon junction microarrays. *Science*. 2003;302(5653):2141–2144.
6. Merkin J, Russell C, Chen P, Burge CB. Evolutionary dynamics of gene and isoform regulation in mammalian tissues. *Science*. 2012; 338(6114):1593–1599.
7. Pal S, Gupta R, Kim H, et al. Alternative transcription exceeds alternative splicing in generating the transcriptome diversity of cerebellar development. *Genome Res*. 2011;21(8):1260–1272.
8. Wang ET, Sandberg R, Luo S, et al. Alternative isoform regulation in human tissue transcriptomes. *Nature*. 2008;456(7221):470–476.
9. Tress ML, Abascal F, Valencia A. Alternative splicing may not be the key to proteome complexity. *Trends Biochem Sci*. 2017;42(2):98–110.
10. Gooding C, Smith CW. Tropomyosin exons as models for alternative splicing. *Adv Exp Med Biol*. 2008;644:27–42.
11. Schmucker D, Chen B. Dscam and DSCAM: complex genes in simple animals, complex animals yet simple genes. *Genes Dev*. 2009;23(2):147–156.
12. Somlyo AP, Somlyo AV. Signal transduction and regulation in smooth muscle. *Nature*. 1994;372(17 November):231–236.
13. Somlyo AV, Somlyo AP. Electromechanical and pharmacomechanical coupling in vascular smooth muscle. *J Pharmacol Exp Therapeut*. 1968;159:129–145.
14. Aalkjaer C, Boedtkjer D, Matchkov V. Vasomotion - what is currently thought? *Acta Physiol (Oxf)*. 2011;202(3):253–269. https://doi.org/10.1111/j.1748-1716.2011.02320.x.
15. Haddock RE, Hill CE. Rhythmicity in arterial smooth muscle. *J Physiol Online*. 2005;566(3):645–656.
16. Chakraborty S, Davis MJ, Muthuchamy M. Emerging trends in the pathophysiology of lymphatic contractile function. *Semin Cell Dev Biol*. 2015;38:55–66.
17. Owens GK, Kumar MS, Wamhoff BR. Molecular regulation of vascular smooth muscle cell differentiation in development and disease. *Physiol Rev*. 2004;84(3):767–801.
18. Rothman A, Kulik TJ, Taubman MB, Berk BC, Smith CWJ, Nadal-Ginard B. Development and characterization of a cloned rat pulmonary arterial smooth muscle cell line that maintains differentiated properties through multiple subcultures. *Circulation*. 1992;86:1977–1986.
19. Vaes RDW, van den Berk L, Boonen B, van Dijk DPJ, Olde Damink SWM, Rensen SS. A novel human cell culture model to study visceral smooth muscle phenotypic modulation in health and disease. *Am J Physiol Cell Physiol*. 2018.
20. Babu GJ, Warshaw DM, Periasamy M. Smooth muscle myosin heavy chain isoforms and their role in muscle physiology. *Microsc Res Tech*. 2000;50(6):532–540.
21. Karagiannis P, Brozovich FV. The kinetic properties of smooth muscle: how a little extra weight makes myosin faster. *J Muscle Res Cell Motil*. 2003;24(2–3):157–163.
22. Kelley CA, Takahashi M, Yu JH, Adelstein RS. An insert of seven amino acids confers functional differences between smooth muscle myosins from the intestines and vasculature. *J Biol Chem*. 1993; 268(17):12848–12854.
23. Wetzel U, Lutsch G, Haase H, Ganten U, Morano I. Expression of smooth muscle myosin heavy chain B in cardiac vessels of normotensive and hypertensive rats. *Circ Res*. 1998;83(2):204–209.
24. Low RB, Mitchell J, Woodcock-Mitchell J, Rovner AS, White SL. Smooth-muscle myosin heavy-chain SM-B isoform expression in developing and adult rat lung. *Am J Respir Cell Mol Biol*. 1999; 20(4):651–657.
25. DiSanto ME, Cox RH, Wang Z, Chacko S. NH_2-terminal-inserted myosin II heavy chain is expressed in smooth muscle of small muscular arteries. *Am J Physiol*. 1997;272(5 Pt 1):C1532–C1542.
26. Babu GJ, Pyne GJ, Zhou Y, et al. Isoform switching from SM-B to SM-A myosin results in decreased contractility and altered expression of thin filament regulatory proteins. *AJP Cell Physiol*. 2004;287(3):C723–C729.
27. Navar LG, Arendshorst WI, Pallone TL, Inscho EW, Imig JD, Bell PD. *The Renal Microcirculation*. 2 ed. San Diego: Academic Press; 2008.
28. Patzak A, Petzhold D, Wronski T, et al. Constriction velocities of renal afferent and efferent arterioles of mice are not related to SMB expression. *Kidney Int*. 2005;68(6):2726–2734.
29. Shiraishi M, Wang X, Walsh MP, Kargacin G, Loutzenhiser K, Loutzenhiser R. Myosin heavy chain expression in renal afferent and efferent arterioles: relationship to contractile kinetics and function. *FASEB J*. 2003, 03–0096fje.
30. Murali M, MacDonald JA. Smoothelins and the control of muscle contractility. *Adv Pharmacol*. 2018;81:39–78.
31. van Eys GJ, Niessen PM, Rensen SS. Smoothelin in vascular smooth muscle cells. *Trends Cardiovasc Med*. 2007;17(1):26–30.
32. Kramer J, Aguirre-Arteta AM, Thiel C, et al. A novel isoform of the smooth muscle cell differentiation marker smoothelin. *J Mol Med (Berl)*. 1999;77(2):294–298.
33. Rensen SS, Thijssen VL, De Vries CJ, Doevendans PA, Detera-Wadleigh SD, Van Eys GJ. Expression of the smoothelin gene is mediated by alternative promoters. *Cardiovasc Res*. 2002;55(4): 850–863.
34. Zheng X, Reho JJ, Wirth B, Fisher SA. TRA2beta controls Mypt1 exon 24 splicing in the developmental maturation of mouse mesenteric artery smooth muscle. *Am J Physiol Cell Physiol*. 2015; 308(4):C289–C296.
35. Llorian M, Gooding C, Bellora N, et al. The alternative splicing program of differentiated smooth muscle cells involves concerted non-productive splicing of post-transcriptional regulators. *Nucleic Acids Res*. 2016;44(18):8933–8950.
36. Rensen SS, Niessen PM, van Deursen JM, et al. Smoothelin-B deficiency results in reduced arterial contractility, hypertension, and cardiac hypertrophy in mice. *Circulation*. 2008;118(8):828–836.
37. Borman MA, Freed TA, Haystead TA, Macdonald JA. The role of the calponin homology domain of smoothelin-like 1 (SMTNL1) in myosin phosphatase inhibition and smooth muscle contraction. *Mol Cell Biochem*. 2009;327(1–2):93–100. https://doi.org/10.1007/s11010-009-0047-z.
38. Lontay B, Bodoor K, Weitzel DH, et al. Smoothelin-like 1 protein regulates myosin phosphatase-targeting subunit 1 expression during sexual development and pregnancy. *J Biol Chem*. 2010; 285(38):29357–29366.
39. Turner SR, Chappellaz M, Popowich B, et al. Smoothelin-like 1 deletion enhances myogenic reactivity of mesenteric arteries with alterations in PKC and myosin phosphatase signaling. *Sci Rep*. 2019;9(1):481.
40. Reho JJ, Shetty A, Dippold RP, Mahurkar A, Fisher SA. Unique gene program of rat small resistance mesenteric arteries as revealed by deep RNA sequencing. *Phys Rep*. 2015;3(7). https://doi.org/10.14814/phy2.12450.
41. Jenkins NT, Padilla J, Thorne PK, et al. Transcriptome-wide RNA sequencing analysis of rat skeletal muscle feed arteries. I. Impact of obesity. *J Appl Physiol*. 2014;116(8):1017–1032.
42. Padilla J, Jenkins NT, Thorne PK, et al. Transcriptome-wide RNA sequencing analysis of rat skeletal muscle feed arteries. II. Impact of exercise training in obesity. *J Appl Physiol*. 2014;116(8):1033–1047.
43. Brozovich FV, Nicholson CJ, Degen CV, Gao YZ, Aggarwal M, Morgan KG. Mechanisms of vascular smooth muscle contraction and the basis for pharmacologic treatment of smooth muscle

disorders. *Pharmacol Rev.* 2016;68(2):476–532. https://doi.org/10.1124/pr.115.010652.

44. Dippold RP, Fisher SA. Myosin phosphatase isoforms as determinants of smooth muscle contractile function and calcium sensitivity of force production. *Microcirculation.* 2014b;21(3):239–248.
45. Hartshorne DJ, Ito M, Erdodi F. Role of protein phosphatase type 1 in contractile functions: myosin phosphatase. *J Biol Chem.* 2004; 279(36):37211–37214.
46. Morgan JP, Morgan KG. Stimulus-specific patterns of intracellular calcium levels in smooth muscle of ferret portal vein. *J Physiol.* 1984;351:155–167.
47. Lee MR, Li L, Kitazawa T. Cyclic GMP causes Ca^{2+} desensitization in vascular smooth muscle by activating the myosin light chain phosphatase. *J Biol Chem.* 1997;272:5063–5068.
48. Wu X, Somlyo AV, Somlyo AP. c-GMP-dependent stimulation reverses G-protein-coupled inhibition of smooth muscle myosin light chain phosphatase. *Biochem Biophys Res Commun.* 1996;220: 658–663.
49. Feletou M, Hoeffner U, Vanhoutte PM. Endothelium-dependent relaxing factors do not affect the smooth muscle of portal vein. *Blood Ves.* 1989;26:21–32.
50. Pfitzer G, Merkel L, Ruegg JC, Hofmann F. Cyclic GMP-dependent protein kinase relaxes skinned fibers from Guinea pig taenia coli but not from chicken gizzard. *Pflügers Archiv.* 1986;407(1):87–91.
51. Diamond J. Lack of correlation between cyclic GMP elevation and relaxation of nonvascular smooth muscle by nitroglycerin, nitroprusside, hydroxylamine and sodium azide. *J Pharmacol Exp Therapeut.* 1983;225(2):422–426.
52. Marsh N, Marsh A. A short history of nitroglycerine and nitric oxide in pharmacology and physiology. *Clin Exp Pharmacol Physiol.* 2000;27(4):313–319.
53. Nagao T, Illiano S, Vanhoutte PM. Heterogeneous distribution of endothelium-dependent relaxations resistant to NG-nitro-L-arginine in rats. *Am J Physiol.* 1992;263(4 Pt 2):H1090–H1094.
54. Sellke FW, Myers PR, Bates JN, Harrison DG. Influence of vessel size on the sensitivity of porcine coronary microvessels to nitroglycerin. *Am J Physiol.* 1990;258(2 Pt 2):H515–H520.
55. Shimokawa H, Yasutake H, Fujii K, et al. The importance of the hyperpolarizing mechanism increases as the vessel size decreases in endothelium-dependent relaxations in rat mesenteric circulation. *J Cardiovasc Pharmacol.* 1996;28(5):703–711.
56. Fok H, Jiang B, Clapp B, Chowienczyk P. Regulation of vascular tone and pulse wave velocity in human muscular conduit arteries: selective effects of nitric oxide donors to dilate muscular arteries relative to resistance vessels. *Hypertension.* 2012;60(5):1220–1225.
57. Gimbrone Jr MA, Garcia-Cardena G. Endothelial cell dysfunction and the pathobiology of atherosclerosis. *Circ Res.* 2016;118(4): 620–636.
58. Steinhorn BS, Loscalzo J, Michel T. Nitroglycerin and nitric oxide–A Rondo of themes in cardiovascular therapeutics. *N Engl J Med.* 2015;373(3):277–280. https://doi.org/10.1056/NEJMsr1503311.
59. Surks HK, Mochizuki N, Kasai Y, et al. Regulation of myosin phosphatase by a specific interaction with cGMP- dependent protein kinase Ialpha. *Science.* 1999;286(5444):1583–1587.
60. Grassie ME, Sutherland C, Ulke-Lemée A, et al. Cross-talk between Rho-associated kinase and cyclic nucleotide-dependent kinase signaling pathways in the regulation of smooth muscle myosin light chain phosphatase. *J Biol Chem.* 2012;287(43): 36356–36369.
61. Nakamura K, Koga Y, Sakai H, Homma K, Ikebe M. cGMP-dependent relaxation of smooth muscle is coupled with the change in the phosphorylation of myosin phosphatase. *Circ Res.* 2007;101(7):712–722.
62. Wooldridge AA, MacDonald JA, Erdodi F, et al. Smooth muscle phosphatase is regulated in vivo by exclusion of phosphorylation of Threonine 696 of MYPT1 by phosphorylation of serine 695 in response to cyclic nucleotides. *J Biol Chem.* 2004;279(33): 34496–34504.
63. Yuen SL, Ogut O, Brozovich FV. Differential phosphorylation of $LZ^{+}/LZ-$ MYPT1 isoforms regulates MLC phosphatase activity. *Arch Biochem Biophys.* 2014;562:37–42.
64. Chen YH, Chen MX, Alessi D, et al. Molecular cloning of cDNA encoding the 110 kDa and 21 kDa regulatory subunits of smooth muscle protein phosphatase 1. *FEBS (Fed Eur Biochem Soc) Lett.* 1994;356:51–55.
65. Johnson D, Cohen P, Chen YH, Chen MX, Cohen PTW. Identification of the regions on the M110 subunit of protein phosphatase 1M that interacts with the M21 subunit and with myosin. *Eur J Biochem.* 1997;244:931–939.
66. Fu K, Mende Y, Bhetwal BP, et al. Tra2beta protein is required for tissue-specific splicing of a smooth muscle myosin phosphatase targeting subunit alternative exon. *J Biol Chem.* 2012;287(20): 16575–16585.
67. Khatri JJ, Joyce KM, Brozovich FV, Fisher SA. Role of myosin phosphatase isoforms in cGMP-mediated smooth muscle relaxation. *J Biol Chem.* 2001;276(40):37250–37257.
68. Payne MC, Zhang HY, Prosdocimo T, et al. Myosin phosphatase isoform switching in vascular smooth muscle development. *J Mol Cell Cardiol.* 2006;40(2):274–282.
69. Karim SM, Rhee AY, Given AM, Faulx MD, Hoit BD, Brozovich FV. Vascular reactivity in heart failure: role of myosin light chain phosphatase. *Circ Res.* 2004;95(6):612–618.
70. Lu Y, Zhang H, Gokina N, et al. Uterine artery myosin phosphatase isoform switching and increased sensitivity to SNP in a rat L-NAME model of hypertension of pregnancy. *Am J Physiol Cell Physiol.* 2008;294(2):C564–C571.
71. Payne MC, Zhang HY, Shirasawa Y, et al. Dynamic changes in expression of myosin phosphatase in a model of portal hypertension. *Am J Physiol Heart Circ Physiol.* 2004;286(5): H1801–H1810.
72. Reho JJ, Kenchegowda D, Asico LD, Fisher SA. A splice variant of the myosin phosphatase regulatory subunit tunes arterial reactivity and suppresses response to salt loading. *Am J Physiol Heart Circ Physiol.* 2016;310(11):H1715–H1724.
73. Reho JJ, Zheng X, Asico LD, Fisher SA. Redox signaling and splicing dependent change in myosin phosphatase underlie early versus late changes in NO vasodilator reserve in a mouse LPS model of sepsis. *Am J Physiol Heart Circ Physiol.* 2015;308(9): H1039–H1050.
74. Zhang H, Fisher SA. Conditioning effect of blood flow on resistance artery smooth muscle myosin phosphatase. *Circ Res.* 2007; 100(5):730–737.
75. Landschulz W, Johnson P, McKnight S. The leucine zipper: a hypothetical structure common to a new class of DNA binding proteins. *Science.* 1988;240(4860):1759–1764.
76. Miller M. The importance of being flexible: the case of basic region leucine zipper transcriptional regulators. *Curr Protein Pept Sci.* 2009;10(3):244–269.
77. Murphy TL, Tussiwand R, Murphy KM. Specificity through cooperation: BATF-IRF interactions control immune-regulatory networks. *Nat Rev Immunol.* 2013;13(7):499–509.
78. Vinson C, Acharya A, Taparowsky EJ. Deciphering B-ZIP transcription factor interactions in vitro and in vivo. *Biochim Biophys Acta.* 2006;1759(1–2):4–12.
79. Osmankovic D, Doric S, Pojskic N, Lukic Bilela L. New approach to detect coiled coil and leucine zipper motifs in protein sequences. *J Comput Biol.* 2018;25(11):1278–1283.
80. Marks AR, Marx SO, Reiken S. Regulation of ryanodine receptors via macromolecular complexes: a novel role for leucine/isoleucine zippers. *Trends Cardiovasc Med.* 2002;12(4): 166–170.

81. Syme CA, Hamilton KL, Jones HM, et al. Trafficking of the Ca^{2+}-activated K^+ channel, hIK1, is dependent upon a C-terminal leucine zipper. *J Biol Chem*. 2003;278(10):8476–8486. https://doi.org/10.1074/jbc.M210072200.
82. Tian L, Coghill LS, MacDonald SH, Armstrong DL, Shipston MJ. Leucine zipper domain targets cAMP-dependent protein kinase to mammalian BK channels. *J Biol Chem*. 2003;278(10): 8669–8677.
83. Wemhöner K, Silbernagel N, Marzian S, et al. A leucine zipper motif essential for gating of hyperpolarization-activated channels. *J Biol Chem*. 2012;287(48):40150–40160.
84. Hofmann F, Feil R, Kleppisch T, Schlossmann J. Function of cGMP-dependent protein kinases as revealed by gene deletion. *Physiol Rev*. 2006;86(1):1–23.
85. Dippold RP, Fisher SA. A bioinformatic and computational study of myosin phosphatase subunit diversity. *Am J Physiol Regul Integr Comp Physiol*. 2014a;307(3):R256–R270.
86. Shukla S, Fisher SA. Tra2beta as a novel mediator of vascular smooth muscle diversification. *Circ Res*. 2008;103(5):485–492.
87. Wirth A, Benyo Z, Lukasova M, et al. G12-G13-LARG-mediated signaling in vascular smooth muscle is required for salt-induced hypertension. *Nat Med*. 2008;14(1):64–68.
88. Zhang H, Pakeerappa P, Lee HJ, Fisher SA. Induction of PDE5 and de-sensitization to endogenous NO signaling in a systemic resistance artery under altered blood flow. *J Mol Cell Cardiol*. 2009; 47(1):57–65.
89. Chen FC, Ogut O, Rhee AY, Hoit BD, Brozovich FV. Captopril prevents myosin light chain phosphatase isoform switching to preserve normal cGMP-mediated vasodilatation. *J Mol Cell Cardiol*. 2006;41(3):488–495.
90. Reho JJ, Fisher SA. The stress of maternal separation causes misprogramming in the postnatal maturation of rat resistance arteries. *Am J Physiol Heart Circ Physiol*. 2015;309(9): H1468–H1478.
91. Burgoyne JR, Madhani M, Cuello F, et al. Cysteine redox sensor in PKGIa enables oxidant-induced activation. *Science*. 2007; 317(5843):1393–1397.
92. Burgoyne JR, Oka S, Ale-Agha N, Eaton P. Hydrogen peroxide sensing and signaling by protein kinases in the cardiovascular system. *Antioxidants Redox Signal*. 2013;18(9):1042–1052. https://doi.org/10.1089/ars.2012.4817.
93. Prysyazhna O, Rudyk O, Eaton P. Single atom substitution in mouse protein kinase G eliminates oxidant sensing to cause hypertension. *Nat Med*. 2012;18(2):286–290.
94. Rudyk O, Prysyazhna O, Burgoyne JR, Eaton P. Nitroglycerin fails to lower blood pressure in redox-dead Cys42Ser PKG1alpha knock-in mouse. *Circulation*. 2012;126(3):287–295.
95. Qin L, Reger AS, Guo E, et al. Structures of cGMP-dependent protein kinase (PKG) iα leucine zippers reveal an interchain disulfide bond important for dimer stability. *Biochemistry*. 2015;54(29): 4419–4422.
96. Schlossmann J, Ammendola A, Ashman K, et al. Regulation of intracellular calcium by a signalling complex of IRAG, IP3 receptor and cGMP kinase Ibeta. *Nature*. 2000;404(6774):197–201.
97. Weber S, Bernhard D, Lukowski R, et al. Rescue of cGMP kinase I knockout mice by smooth muscle specific expression of either isozyme. *Circ Res*. 2007;101(11):1096–1103.
98. Kaur H, Carvalho J, Looso M, et al. Single-cell profiling reveals heterogeneity and functional patterning of GPCR expression in the vascular system. *Nat Commun*. 2017;8:15700.
99. Pluznick JL, Protzko RJ, Gevorgyan H, et al. Olfactory receptor responding to gut microbiota-derived signals plays a role in renin secretion and blood pressure regulation. *Proc Natl Acad Sci USA*. 2013;110(11):4410–4415.
100. Dobnikar L, Taylor AL, Chappell J, et al. Disease-relevant transcriptional signatures identified in individual smooth muscle cells from healthy mouse vessels. *Nat Commun*. 2018;9(1). https://doi.org/10.1038/s41467-018-06891-x, 4567.
101. Hwang B, Lee JH, Bang D. Single-cell RNA sequencing technologies and bioinformatics pipelines. *Exp Mol Med*. 2018;50(8):96. https://doi.org/10.1038/s12276-018-0071-8.
102. Chavkin NW, Hirschi KK. Single cell analysis in vascular biology. *Front Cardiovasc Med*. 2020;7. https://doi.org/10.3389/fcvm.2020.00042, 42.
103. Ellis JD, Barrios-Rodiles M, Colak R, et al. Tissue-specific alternative splicing remodels protein-protein interaction networks. *Mol Cell*. 2012;46(6):884–892.
104. Yang J, Zhang Y. I-TASSER server: new development for protein structure and function predictions. *Nucleic Acids Res*. 2015; 43(W1):W174–W181.
105. Lamiable A, Thévenet P, Rey J, Vavrusa M, Derreumaux P, Tufféry P. PEP-FOLD3: faster de novo structure prediction for linear peptides in solution and in complex. *Nucleic Acids Res*. 2016;44(W1):W449–W454.
106. Babu MM. The contribution of intrinsically disordered regions to protein function, cellular complexity, and human disease. *Biochem Soc Trans*. 2016;44(5):1185–1200. https://doi.org/10.1042/bst20160172.
107. Zhou J, Zhao S, Dunker AK. Intrinsically disordered proteins link alternative splicing and post-translational modifications to complex cell signaling and regulation. *J Mol Biol*. 2018;430(16): 2342–2359.
108. Sharma S, Schiller MR. The carboxy-terminus, a key regulator of protein function. *Crit Rev Biochem Mol Biol*. 2019;54(2):85–102. https://doi.org/10.1080/10409238.2019.1586828.
109. Minde DP, Dunker AK, Lilley KS. Time, space, and disorder in the expanding proteome universe. *Proteomics*. 2017;17(7).
110. Buljan M, Chalancon G, Eustermann S, et al. Tissue-specific splicing of disordered segments that embed binding motifs rewires protein interaction networks. *Mol Cell*. 2012;46(6):871–883.
111. Yang X, Coulombe-Huntington J, Kang S, et al. Widespread expansion of protein interaction capabilities by alternative splicing. *Cell*. 2016;164(4):805–817.
112. Meer DP, Eddinger TJ. Expression of smooth muscle myosin heavy chains and unloaded shortening in single smooth muscle cells. *Am J Physiol*. 1997;273(4 Pt 1):C1259–C1266.
113. Muzumdar MD, Tasic B, Miyamichi K, Li L, Luo L. A global double-fluorescent Cre reporter mouse. *Genesis*. 2007;45(9): 593–605.
114. Paez-Cortez J, Krishnan R, Arno A, et al. A new approach for the study of lung smooth muscle phenotypes and its application in a murine model of allergic airway inflammation. *PLoS One*. 2013; 8(9):e74469.
115. Sheikh AQ, Misra A, Rosas IO, Adams RH, Greif DM. Smooth muscle cell progenitors are primed to muscularize in pulmonary hypertension. *Sci Transl Med*. 2015;7(308).
116. Glusman G, Rose PW, Prlić A, et al. Mapping genetic variations to three-dimensional protein structures to enhance variant interpretation: a proposed framework. *Genome Med*. 2017;9(1):113.
117. Sulakhe D, D'Souza M, Wang S, et al. Exploring the functional impact of alternative splicing on human protein isoforms using available annotation sources. *Briefings Bioinf*. 2019;20(5): 1754–1768.
118. Tapial J, Ha KCH, Sterne-Weiler T, et al. An atlas of alternative splicing profiles and functional associations reveals new regulatory programs and genes that simultaneously express multiple major isoforms. *Genome Res*. 2017;27(10):1759–1768.
119. de la Fuente L, Arzalluz-Luque Á, Tardáguila M, et al. tappAS: a comprehensive computational framework for the analysis of the functional impact of differential splicing. *Genome Biol*. 2020;21(1):119. https://doi.org/10.1186/s13059-020-02028-w.

120. Thakur PK, Rawal HC, Obuca M, Kaushik S. Bioinformatics approaches for studying alternative splicing. In: Ranganathan S, Gribskov M, Nakai K, Schönbach C, eds. *Encyclopedia of Bioinformatics and Computational Biology.* Oxford: Academic Press; 2019: 221–234.
121. Yuan J, Ma Y, Huang T, et al. Genetic modulation of RNA splicing with a CRISPR-guided cytidine deaminase. *Mol Cell.* 2018;72(2), 380–394. e387.
122. Htet Myo, et al. Editing of the myosin phosphatase regulatory subunit suppresses angiotensin II induced hypertension via sensitization to nitric oxide mediated vasodilation. *Pflugers Arch.* 2021;473:611–622.
123. He Liqun, et al. Single-cell RNA sequencing of mouse brain and lung vascular and vessel-associated cell types. *Sci. Data.* 2018;5: 180160.

CHAPTER

5

Resident vascular immune cells in health and atherosclerotic disease

Clément Cochain[1,2] *and Ziad Mallat*[3,4]

[1]Institute of Experimental Biomedicine, University Hospital Würzburg, Würzburg, Germany [2]Comprehensive Heart Failure Center, University Hospital Würzburg, Würzburg, Germany [3]Department of Medicine, Division of Cardiovascular Medicine, University of Cambridge, Cambridge, United Kingdom [4]Université de Paris, PARCC, INSERM, Paris, France

Vascular diseases are a leading cause of death worldwide, in particular, atherosclerosis and its related complications such as myocardial infarction and stroke.[1] It is established that inflammation of the vascular wall is a critical driving force not only of atherosclerosis but also of other major vascular diseases such as aortic aneurysm.[2] In addition, evidence has emerged over the past years that resident vascular immune populations, in particular, resident tissue macrophages (RTMs), play crucial homeostatic roles in the maintenance of normal vascular function. Our understanding of the vascular immune landscape, and how it contributes to vessel homeostasis and disease development, has been closely intertwined with technological advances increasing the resolution at which we can investigate the immune system. Immunohistochemical analysis of human atherosclerotic lesions with monoclonal antibodies against immune cell populations was instrumental in pinpointing inflammation as an essential driver of atherosclerosis.[3] In the 2000s, multiparameter flow cytometry analysis identified all major immune cell lineages in experimental mouse atherosclerosis.[4] Cytometry by time of flight (CyTOF) now allows measuring the expression of dozens of cell surface markers on immune cells.[5,6] More recently, single-cell RNA-sequencing (scRNA-seq) has been used to evaluate the transcriptome of individual cells in mouse and human vessels and revealed an ever-increasing complexity and heterogeneity in the vascular immune landscape.[7,8]

In this chapter, we aim to provide a brief overview of the current state of knowledge about the immune cell landscape in normal arteries, the origin and proposed functions of vascular resident immune cell subsets, and how the vascular landscape shifts in disease conditions with a focus on atherosclerosis and aortic aneurysm. Mice have been the most commonly employed model to decipher the role of the immune system in vascular health and disease, thus we will mostly discuss findings from mouse studies, but will provide comparative insights into human biology wherever possible.

Macrophages

Macrophages populate arteries in homeostatic conditions, and have a critical role in virtually all vascular pathologies.[9] A long standing paradigm placed monocytes as the precursors of all macrophages in tissues.[10] However, fate-mapping studies in mice revealed that RTMs could be seeded during embryonic development and did not depend on postnatal hematopoietic progenitors, which was first shown for brain RTM, namely microglia.[11] Following this seminal discovery, many fate-mapping studies in the 2010s established the ontogeny of RTM populations across tissues and revealed that RTM can originate from yolk sac–derived erythromyeloid progenitors and fetal hematopoietic progenitors, seed organs before birth, and self-renew mostly independently of postnatal monocyte input.[12] The precise developmental origin of RTM and their renewal rate by postnatal monocytes varies widely across organs.[12] Whether the function of RTM is imprinted by their developmental origin, their tissular niche of residence, or a combination of both, is currently under intense scrutiny.[13,14] While the bulk of our knowledge on RTM ontogeny is based on animal studies, recent single-cell analysis of macrophages in human embryos suggest similar prenatal seeding of RTM in human tissues.[15]

The Vasculome
https://doi.org/10.1016/B978-0-12-822546-2.00032-0

The aorta contains specific populations of RTM with specific ontogeny, tissue localization, and function. In pathological conditions, there is a shift in the vascular macrophage landscape driven by influx of circulating monocytes that adopt heterogeneous phenotypes. Below, we provide an overview of the main characteristic and functions of arterial resident macrophages, and how the macrophage landscape shifts in disease conditions.

Adventitial resident tissue macrophages of the aorta

While it had been long known that macrophages populate the adventitia of healthy arteries, a study by Ensan et al.[16] was the first to perform a thorough characterization of the specific signature and developmental origin of adventitial RTM. Using transcriptomic analysis, they showed that arterial RTMs have a gene expression signature distinct from RTM in other tissues, and express surface markers such as LYVE1, TIMD4, or MHCII. Additional cell surface and transcript markers defining the adventitial RTM signature have been further defined in scRNA-seq[17,18] and CyTOF[5] studies, indicating that they may comprise different subsets including CD209b^{+} RTM with unknown functions[5] (Fig. 5.1A). Adventitial RTM originate from a mixture of yolk sac–derived erythromyeloid progenitors, fetal liver progenitors, and postnatal monocytes seeding the

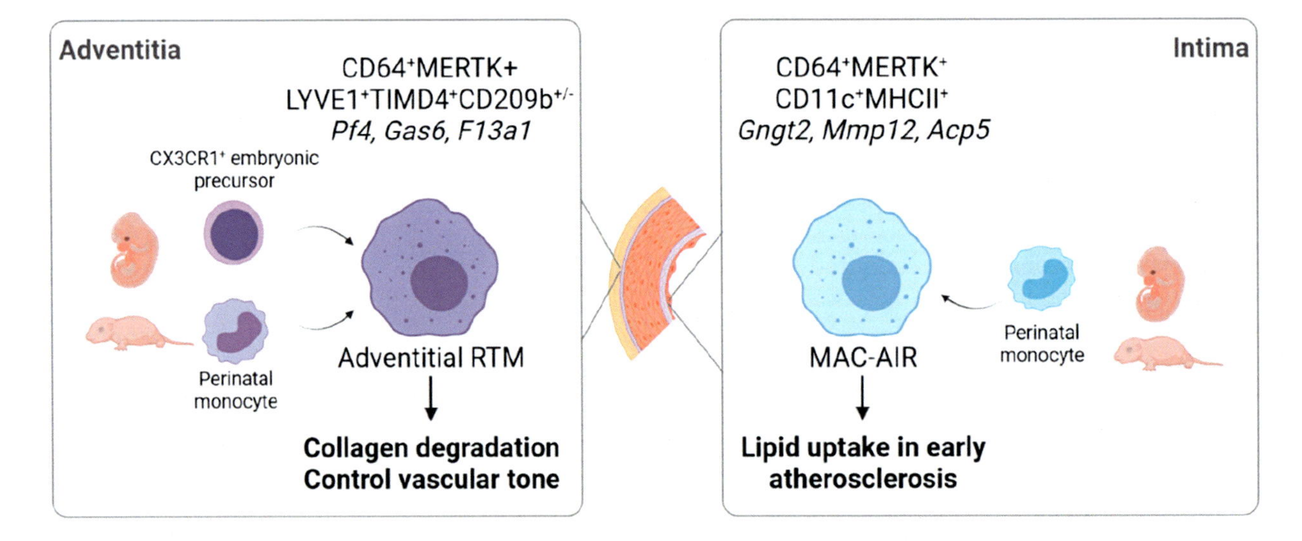

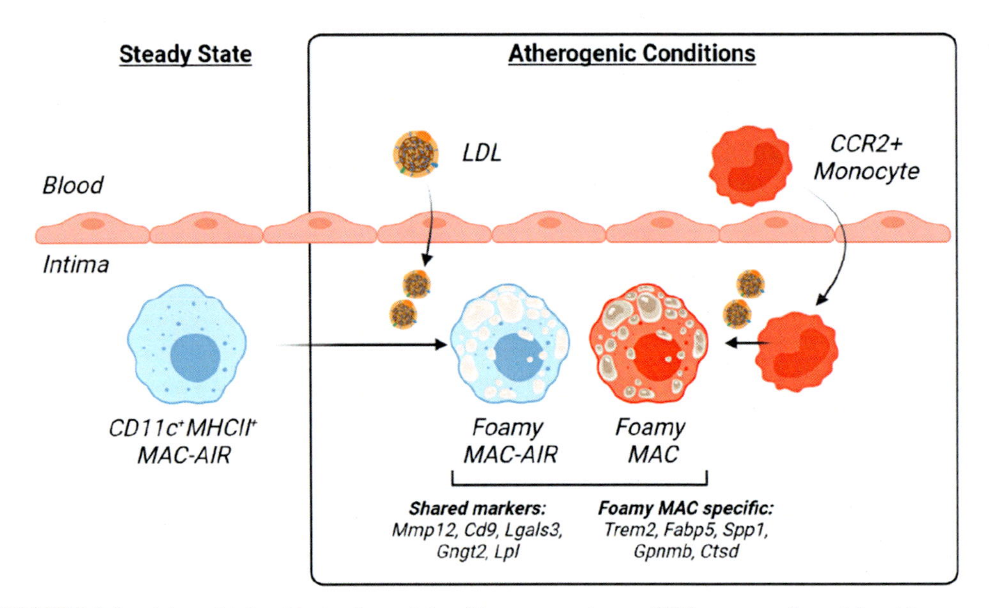

FIGURE 5.1 Adventitial and intimal arterial resident macrophages. (A) Summary of arterial resident macrophage subsets, their markers, and functions; (B) Commonalities and distinction between MAC-AIR and atherosclerosis-associated foamy/TREM2hi macrophages in the steady state and under hypercholesterolemic conditions.

aorta in the first weeks after birth.[16,19] Adventitial RTMs self-maintain independently of monocyte input in adult mice, both in homeostatic conditions and following their partial depletion in response to bacterial infection, and their survival is dependent on CX3CL1-CX3CR1 interactions.[16] LYVE1$^+$ adventitial RTM are conserved across species and have notably been observed in human arteries.[20,21] Via LYVE1, adventitial RTM interact with hyaluronan at the surface of vascular smooth muscle cells (VSMCs) and produce matrix metalloproteinase-9 (MMP9) to degrade VSMC-expressed type 1 collagen.[20] In mice, genetic or pharmacological depletion of LYVE1$^+$ adventitial RTM led to increased aortic collagen deposition and arterial stiffness, indicating a crucial role of adventitial RTM in maintaining large vessel homeostasis via regulation of the extracellular matrix.[20] The accumulation of CXCR3$^+$ perivascular adventitial macrophages highly expressing metalloproteinases and Factor XIII (encoded by *F13a1*, a marker transcript of arterial resident macrophages[17]) has also been proposed to mediate flow-dependent inward vascular remodeling.[22] The correspondence between CXCR3$^+$ and LYVE1$^+$ perivascular macrophages has not been formally addressed so far, and future integration of single-cell RNA-seq transcriptome from different studies including recent data from mouse carotid arteries under disturbed flow[23] could help address this issue.

Intimal resident macrophages: MAC-AIR

Recently, a subset of aortic intimal resident macrophages (MAC-AIR) was identified via scRNA-seq of healthy mouse arteries.[18] Imaging studies indicated that MAC-AIR are abundant in atherosclerosis-prone regions of the mouse aorta, and fate-mapping analyses indicated that they derived from CCR2$^+$ monocytes locally differentiating in CCR2$^-$ MAC-AIR in the aortic intima in the perinatal period. Transcriptomic profiling of these cells revealed a specific signature and expression of markers usually characteristic of dendritic cells, in particular, CD11c and MHCII. Interestingly, earlier studies had also observed tissue-resident CD11c$^+$MHCII$^+$ cells in the cardiac valves and intima of the mouse aorta,[24,25] but had identified these as dendritic cells based on the CD11c$^+$MHCII$^+$ marker combination then thought to be DC-specific. CD11c$^+$MHCII$^+$ MAC-AIR take up lipid upon the onset of atherogenic conditions, and their depletion reduces the accumulation of intimal foam cells in early atherosclerosis.[25,18] Whether this phenomenon is intrinsically pathogenic or is rather a protective mechanism to clear up intimal infiltration of LDL during, e.g., transient hyperlipidemia remains to be determined. While MAC-AIR show some degree of proliferative ability, they are rapidly outnumbered by foam cells differentiating from infiltrating monocytes in atherogenic conditions.[18] Interestingly, MAC-AIR and atherosclerosis-associated Foamy/*Trem2*^{hi} macrophages (see also next paragraph) share some specific transcriptomic markers that might be indicative of residence in the vascular intimal niche, but Foamy/*Trem2*^{hi} macrophages express a specific set of disease-associated genes[18,21] (Fig. 5.1B).

Macrophages in the diseased vessel wall

In pathological settings, vascular macrophages drastically increase their number and diversity. This phenomenon is mostly driven by infiltration of blood-borne inflammatory monocytes (in mice, Ly6C^{hi}CCR2$^+$ monocytes) that are mobilized from their reservoirs to the bloodstream and infiltrate inflamed vessels where they differentiate into macrophages.[26] Ly6C^{hi} monocytes are mobilized from the bone marrow in response to, e.g., CCL2/CCR2[27] or sympathetic nervous system activation,[28] and from a splenic reservoir from which they are mobilized in an angiotensin II-dependent manner both in AAA[29] and atherosclerosis.[30] Recruitment of monocytes to the vessel wall is mediated by various chemokine/chemokine receptor interactions, adhesion molecules, and integrins.[31] In advanced atherosclerotic disease, accumulation of vascular macrophages also occurs via local proliferation,[32] and while it had been proposed that VSMCs could locally transdifferentiate into macrophage-like cells upon lipid loading in plaques,[33] recent analyses combining fate-mapping of VSMCs with scRNA-seq indicated that phenotypically modulated VSMCs are clearly distinct from the macrophage lineage.[34]

Macrophage heterogeneity in atherosclerosis has long been investigated using the simplistic M1/M2 macrophage polarization model.[35,36] However, lesion macrophages are exposed to a plethora of stimuli controlling their gene expression profile and function, including, e.g., cholesterol crystals activating the NLRP3-inflammasome and IL1β production,[37,38] cell/cell interactions, cytokines,[35] and lipids including modified lipids such as oxidized phospholipids (oxPLs) recently shown to drive an hypermetabolic inflammatory state in macrophages.[39] A critical function of macrophages in atherosclerotic lesions is to scavenge lipids infiltrated in the intima, which are stored intracellularly in cholesteryl ester–containing lipid droplets, giving macrophages their characteristic foam cell appearance.[35] Macrophage lipid-loading has been shown to be intrinsically antiinflammatory, in particular, via activation of the liver-x-receptor pathway.[40] In lesions, macrophages also perform efferocytosis, i.e., the noninflammatory removal of dead cells, a process depending on specific proteins such as MerTK or MFG-E8.[41] Reduced efferocytic capacity of macrophages in advanced atherosclerotic plaques is thought to drive accumulation of secondary necrotic cells and formation of the plaque necrotic core.[42] Lesion macrophages also secrete pro- and antiinflammatory cytokines modulating the local immune response, and secrete proteases driving destabilization of plaques and their rupture.[35]

Recent studies employing scRNA-seq of mouse and human atherosclerotic lesions have increased our understanding of macrophage heterogeneity in atherosclerosis and revealed macrophage subsets with discrete gene expression patterns (Fig. 5.2). The major subsets of arterial macrophages in mouse atherosclerosis comprise resident macrophages (see above), proinflammatory macrophages expressing, e.g., *Il1b*, a discrete proinflammatory macrophage population with a type 1 interferon signature response, and $Trem2^{hi}$/Foamy macrophages with a gene expression signature indicating noninflammatory and lipid-handling capabilities.[7] Importantly, integration of mouse and human scRNA-seq data indicates that these major subsets are conserved in human carotid and coronary artery lesions.[8,21] In mice, an additional $Cd226^{+}Ccr2^{+}MHCII^{+}$ subset resembling small peritoneal macrophages and termed "cavity macrophages" has been described, but its role and specific localization within arteries is unknown.[7]

In experimental mouse AAA, single-cell analysis has evidenced the presence of diverse macrophage subsets both of tissue resident and blood-borne monocyte origin.[43] Single-cell analysis of human AAA tissue has also revealed diverse populations of monocytes and macrophages presenting distinct states resembling M1-like or M2-like polarized macrophages.[44] How the macrophage populations in AAA relate to those described in atherosclerosis has yet to be formally evaluated.

Granulocytes

Neutrophils: Although usually considered short-lived first responders to infection or tissue injury, recent evidence indicate that neutrophils populate organs in homeostatic conditions, where they adopt tissue-specific phenotypes and perform noncanonical functions dictated by the tissue microenvironment.[45] While small numbers of neutrophils are detected in healthy mouse aortas by flow cytometry or scRNA-seq,[46] their exact phenotype and function in healthy vessels is unknown. In pathological conditions, including atherosclerosis[47] and AAA,[48] neutrophils have been assigned a pathogenic role. They induce chronic vascular inflammation by triggering VSMC death via externalized histone H4,[49] and license macrophages for proinflammatory functions and inflammasome activation.[50] Recently, scRNA-seq analysis and flow cytometry revealed that mouse atherosclerotic arteries contain two subsets of neutrophils with distinct transcriptomic signatures, including neutrophils expressing SiglecF, a marker usually associated with mouse eosinophils.[46] The functions of these neutrophil populations in disease development, and whether human counterparts of these mouse subsets exist, are still unknown.

Eosinophils: Eosinophils have not been described in healthy arteries and as of yet not been observed in scRNA-seq analyses of vascular tissue, which might be caused by technical difficulties in capturing these cells with a low RNA content. Recent studies using a similar mouse model of eosinophil deficiency (the $\Delta dblGATA1^{-/-}$ mice) showed contrasting effect in experimental atherosclerosis and AAA: while eosinophils appeared as proatherogenic,[51] they were protective in an AAA model via IL-4-mediated macrophage polarization.[52] Eosinophils were not observed in murine atherosclerotic plaques, suggesting a rather systemic role,[51] while they accumulated in murine or human AAA samples, indicating a potential local effect,[52] which might underlie their context-dependent functions and pathogenicity. Of note, eosinophils were defined as $SiglecF^{+}$ in these studies, which might have led to identification of both $Ly6G^{-}SiglecF^{+}$ bona fide eosinophils and $Ly6G^{+}SiglecF^{+}$ neutrophils.

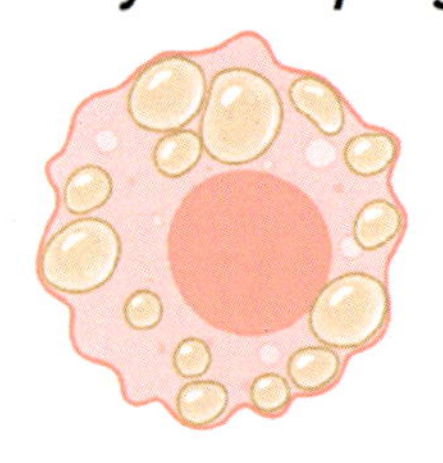

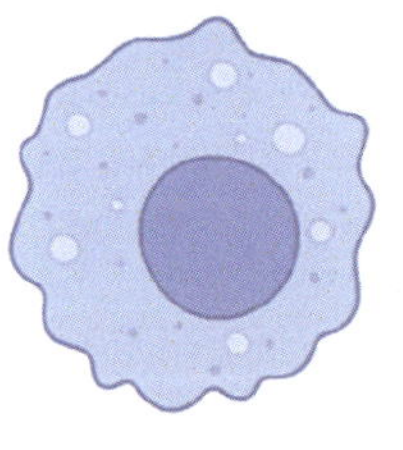

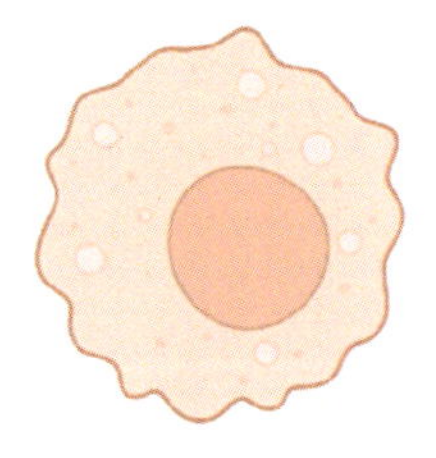

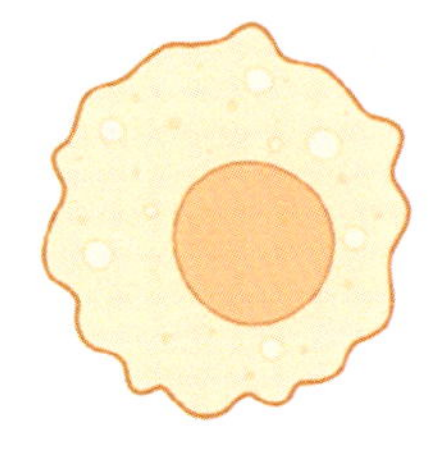

FIGURE 5.2 **Macrophage subsets detected by scRNA-seq in atherosclerosis and their marker transcripts.**

Mast cells: Mast cells reside in normal vessels in mice and humans.[53] Their role in vascular diseases is thought to be mostly pathogenic, with experimental evidences in atherosclerosis[54] and AAA.[55] Proteases delivered from mast cells granules have notably been involved in AAA formation.[56] Evidence in humans indicate that mast cells accumulate in the intima and adventitia during atheroprogression,[57] and that mast cells accumulation and activity are correlated to aortic aneurysm growth rate.[58]

Basophils: The role of basophils in the healthy and diseased vascular wall has not yet been thoroughly investigated. A recent study proposed that basophil-derived cytokines polarized macrophages toward protective phenotypes in myocardial repair after cardiac ischemia.[59] Whether such a mechanism may also be relevant to vascular pathologies is unknown.

Dendritic cells

Dendritic cells are categorized based on their ontogeny, dependence on specific transcription factors, and functional features.[60] Conventional (or classical) DCs (cDCs) originate from hematopoietic progenitors of restricted cDC potential and comprise two main populations: type 1 cDC (cDC1) depend on the transcription factor IRF8 and their main function is to cross-present antigens to $CD8^+$ T cells, while cDC2 depend on IRF4 and drive $CD4^+$ T cell responses.[60] Plasmacytoid dendritic cells are specialized in type I interferon production in response to viral infection,[60] but have been recently demonstrated to actually be of lymphoid rather than dendritic cell origin.[61] In addition, monocyte-derived dendritic cells (MoDCs) are thought to arise in inflammatory conditions and acquire DC functions including antigen presentation, but their functional overlap with bona fide cDCs remains unclear.[60]

Vascular dendritic cells have long been assumed to be important in the physiopathology of vascular diseases,[62] but many studies may in fact be highly confounded by misidentification of macrophages as dendritic cells due to the expression of common markers.[63] In particular, CD11c (encoded by *Itgax*) has often been used as a surface marker to define mouse DCs, or as a promoter driving Cre or DTR expression to target or deplete DCs.[64] However it is now known that CD11c is widely expressed by specific subsets of vascular macrophages. More precise combinations of markers now allow a better identification of cDCs versus monocyte-derived cells and macrophages, as cDCs are $CD26^+CD88^-$ while monocyte-derived cells and macrophages are $CD26^-CD88^+$.[65,66] Additionally, cDC1 express XCR1, CLEC9A (DNGR1), and either CD103 or CD8α depending on their tissue of residence, while cDC2 express CD11b and CD172a (Sirpα)[60] (Fig. 5.3).

Dendritic cell populations in arteries

Single-cell RNA-seq studies have identified several populations of vascular cDC in healthy and diseased vessels[7,21,17] (Fig. 5.3). A population of cDC1 expressing *Irf8*, *Xcr1*, and *Clec9a* was identified in arteries, alongside *Itgax*$^+$*CD209a*$^+$ DCs, initially annotated as MoDC/DC[17] but later identified as cDC2 based on expression of CD26 (*Dpp4*) and lack of CD88 (*C5ar1*).[21] CyTOF analysis of *Apoe*$^{-/-}$ aortas also identified cDC1 and cDC2 populations.[5] A third discrete population of DCs expressing markers of maturation (*Ccr7*, *Fscn1*) and a set of transcripts characteristic of a recently described "mature

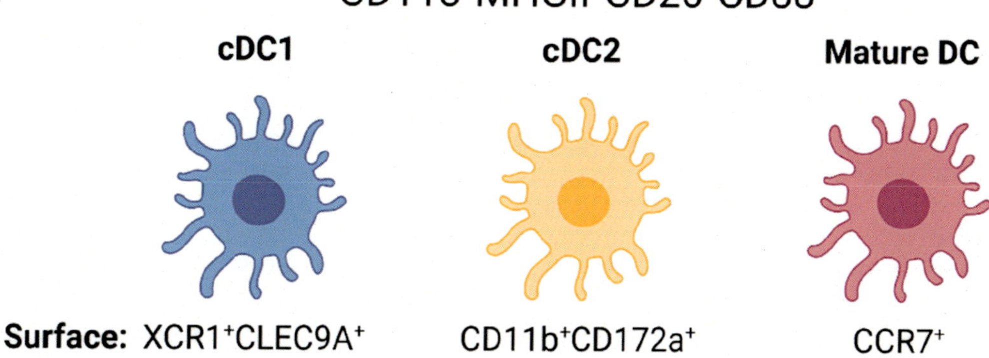

FIGURE 5.3 **Dendritic cell subsets observed in the mouse vasculature and their proposed markers (cell surface and transcripts).** Alternative names found in the literature include: CLEC9A = DNGR1, CD172a = SIRPα, Cd209a = DC-SIGN.

DCs enriched in immunoregulatory molecules" (mregDCs)[67] signature (e.g., *Ccl17, Il12b, Cd274*) was also observed.[7,17,21] While the contribution of cDC1 and *Ccr7*$^+$ mature DCs to the aortic immune landscape in health versus disease has not yet been formally investigated in scRNA-seq analysis due to the relative paucity of these cells, the level of *Cd209a*$^+$cDC2 was shown to drastically increase in the aortas of atherosclerotic high-fat diet fed *Ldlr*$^{-/-}$ mice compared to steady-state conditions.[17] Previous studies using flow cytometry also showed increases in aortic cDC1 levels in atherogenic conditions.[68] Importantly, human atherosclerotic vessels also contain cDC1 and cDC2 populations, and may contain mature DCs with a mregDC signature, although sampling of larger number of cells would be necessary to confirm these observations.[21]

Function of dendritic cells in arteries

While DCs functionalities in steering T cell responses are spatially restricted to lymphoid organs, the presence of DCs in arteries raises the possibility that they could locally drive protective or pathogenic T cells responses directly in the vasculature, and many studies have attempted to address this hypothesis. However, the local role of cDCs in vascular health and disease remains enigmatic. Most studies from the past three decades require reinterpretation in light of recent developments in the field. For instance, "intimal resident DCs" responsible for lipid uptake in early atherosclerosis[24,25] have now been identified as MAC-AIR[18] (see above), and earlier cell depletion studies in CD11c-DTR mice[64] might have targeted foamy macrophages and MAC-AIR. Other studies have attempted to specifically target cDC1 in experimental atherosclerosis by invalidating the *Flt3* or *Batf3* pathways required for their development. Flt3-dependent cDC1 were shown to reside in normal vessels, increase in number in atherogenic conditions, and proposed to induce atheroprotective Treg responses.[68] In contrast, mice deficient in *Batf3* also lacking cDC1 did not show any differences in atherosclerotic lesion formation.[69] These studies might however be confounded by the impact of the *Flt3* and *Batf3* pathways on other immune cell types such as, e.g., B cells[70] or CD8$^+$ T cells,[71] respectively. Recent studies using more precise genetic targeting strategies to target cDC1 have suggested a proatherogenic role of these cells,[72,73] but it is difficult to pinpoint effects related to systemic activity of dendritic cells (e.g., T cell priming in lymphoid organs) versus their local effect in the vascular wall. Studies using imaging of antigen-presenting cell interaction with T cells in the mouse aorta,[74] or based on the injection of ovalbumin in OT-II transgenic mice,[75] have suggested local priming of T cell responses directly in the aortic wall by antigen-presenting cells. Additionally, a CCL17$^+$ subset populates atherosclerotic vessels and is proatherogenic via restraining Treg homeostasis.[76] Single-cell analysis indicates that *Ccl17* expression is restricted to mature *Ccr7*$^+$ DCs in the mouse aorta, suggesting that mature immune regulatory DCs may have complex pleiotropic and opposing role in controlling atherosclerosis-associated T cell responses, on one hand promoting protective anti-inflammatory response via immune regulatory pathway, and on the other hand inhibiting protective Treg response via, e.g., *Ccl17*. Altogether, the role of Ccr7$^+$ mature DCs with an mREG signature in atherosclerosis still needs to be deciphered also taking into account specific contexts (e.g., early vs. late plaques). In AAA, the role of cDC subsets has not been precisely explored. Depletion of CD11c$^+$ cells was shown to attenuate the development of experimental AAA in mice, but here again this strategy might have targeted both cDCs and some macrophage subsets.[77]

CD4$^+$ and CD8$^+$ T cells

CD3$^+$ T cells reside within the adventitia of normal vessels in mice[4] and can also be observed in nondiseased human arteries, at sites of intimal thickening, already in children.[53] The development of atherosclerosis is associated with increased T cell accumulation both in the intima and the adventitia, in association with the presence of abundant cells expressing HLA-DR molecules, suggesting activation by antigen-presenting cells.[3,78] Indeed, early work by Hansson and colleagues showed that plaque-derived T cells recognize and are activated by oxidatively modified LDL in an HLA-DR-dependent manner.[79] Since then, numerous studies have evaluated the association between (antigen-specific) T cell subsets and disease progression.[80,81] These studies revealed that the development of vascular diseases is associated with a strong expansion and diversification of the T cell compartment in the vasculature, and various T cells subsets have been ascribed protective or pathogenic roles in aortic aneurysm formation and atherosclerosis (reviewed in details in Ref. 82 and Ref. 80, respectively). More recently, the heterogeneity of the T cell compartment in human carotid lesions was characterized by high-dimensional methods, i.e., CyTOF and single-cell RNA-seq,[83] revealing that human lesions contain many discrete T cell subsets that are, in comparison to circulating T cells, characterized by surface marker profiles suggestive of chronic activation, differentiation, and exhaustion.[83] Below we provide a brief overview of the main T cell subsets involved in vascular diseases.

Tissue-resident memory (TRM) cells: CD4$^+$ and CD8$^+$ TRM have been described in many tissues in mice and humans, where they provide protection against infection but have also been implicated in proinflammatory autoimmune responses.[84] Their presence and physiopathological roles in the vasculature have not been studied in detail. A study evidenced a minute population of S1PR1$^-$CD103$^+$CD44$^+$CD69$^+$ CD4$^+$ putative TRM in the aorta of Apoe$^{-/-}$ mice, but not in control animals, suggesting that CD4$^+$ TRM may arise under atherogenic conditions.[75]

CD4+ T helper cells

CD4+ T helper cells comprise subsets with functionally specialized phenotypes characterized by expression of specific cytokines and driven by master-regulator transcription factors. Th1 T cells are driven by T-bet, produce IFNγ, and have been uniformly demonstrated as being pathogenic in atherosclerosis[80] and AAA.[82] Th2 T cells, driven by GATA3 and producing type 2 cytokines (IL-4, IL-5, IL-13) are mostly thought to be protective in the vasculature,[82,80] although some site-specific proatherogenic effects of IL-4 have been reported.[85] The role of Th17 cells (RORγ dependent and producing IL-17) in the vasculature is controversial[86] as some studies have proposed atheroprotective roles, including the promotion of a stable lesion phenotype,[87] while other reports proposed a pathogenic role linked to promotion of myeloid cell recruitment.[88] Similarly, in AAA, protective[89] or pathogenic[90] roles of IL-17 have been proposed. The exact context-dependent roles of IL-17 in vascular pathophysiology and the mechanisms underlying drastically conflicting reports in experimental models are unknown. Other discrete Th cell subsets (e.g., Th9) may also be involved in vascular pathologies.[80] Antiinflammatory Foxp3$^+$ regulatory T cells (Tregs) have uniformly been demonstrated to be protective in vascular pathologies.[82,80]

Recent evidence using single-cell profiling of gut T cells indicates that the rigid nomenclature grouping specialized T helper cell subsets into Th archetypes (e.g., Th1, Th2, Th17, etc.) might not reflect the true transcriptional diversity of T cells, which were rather shown to adopt states across a polarized continuum in response to specific stimuli (in that case, microbes) and independently of classical master-regulator transcription factors (e.g., Tbet for Th1, GATA3 for Th2, and RORγ for Th17).[91,92] Whether this more nuanced view of T cell polarization and acquisition of stimulus-dependent states is relevant to other diseases, and to vascular diseases in particular, remains to be determined. In inflammatory conditions, Foxp3+ Tregs also show some degree of plasticity,[93] and Treg conversion to a dysfunctional and pathogenic Th1/Treg subset has been observed in experimental atherosclerosis.[94] A recent study dissecting $ApoB_{100}$-specific T cells at the single-cell level provided novel evidence that $ApoB_{100}$-specific T cells initially present a protective Treg phenotype, but then progressively convert to pathogenic Th1/Th17 cells with atherosclerosis progression.[95]

CD8$^+$ T cells

CD8+ cytotoxic T cells are present in human and murine atherosclerotic vessels where they secrete proinflammatory cytokines (TNF, IFNγ) and express cytotoxic enzymes (Granzyme-B, Perforin).[96,97] In single-cell RNA-seq analysis of murine aortas, a drastic increase in CD8$^+$ T cell proportions among total leukocytes is observed in atherogenic conditions.[17] Cell depletion studies in mice have evidenced a proatherogenic role of CD8$^+$ T cells, mediated by local proinflammatory and cytotoxic activities, but also by their role in systemic production and trafficking of proinflammatory Ly6Chi monocytes.[96,97] However, other studies have proposed protective roles of CD8+ T cells,[98] in particular of CD8+CD25+ T cells with an immune regulatory -phenotype.[99] Similar to CD4$^+$ responses, atherosclerosis-associated CD8$^+$ T cell responses are thought to be antigen-specific with evidence for the presence of ApoB100-specific CD8$^+$ T cells.[100]

Other T cell subsets

γδT cells

γδT cells represent a specific lineage of T cells with a distinct TCR composed of a γ and a δ chain, in contrast to "conventional" T cells expressing an αβTCR. They originate in the thymus, which produces distinct waves of functionally preprogrammed γδT cells poised to produce specific cytokines, in particular, IL-17A or IFNγ.[101] Specific subsets of γδT cells seed organs during embryonic development and in the perinatal period and establish residence in tissues where they have been proposed to perform various functions well beyond immune surveillance, e.g., promoting tissue repair in the skin or affecting adipose tissue function in obesity.[101] γδT cells are found at low numbers in the normal mouse aorta[102,103] and in aortic valves,[104] and have been detected by scRNA-seq in normal and atherosclerotic murine aorta as *Cxcr6*$^+$*Il17*$^+$ cells.[7] γδT cells have been observed in human atherosclerotic lesions,[105] but their role in vascular disease is unclear, as studies using *Tcrd*$^{-/-}$ mice lacking γδT cells crossed to atherosclerosis-prone *Apoe*$^{-/-}$ mice showed either a site-specific decrease in lesion formation[102] or no effect.[103]

Natural killer T cells

Natural killer T (NKT) cells express surface markers of NK cells (mouse: NK1.1, human: CD161) together with CD3 and T cell receptors.[106] NKT cells can respond to lipid antigens, and studies using mice genetically deficient in various subsets of NKT cells suggested roles in the control of systemic lipid metabolism.[106] In ApoE−/− mice, administration of the synthetic glycolipid αGalCer, which activates invariant NKT cells, led to increased lesion formation, suggesting a pathogenic role of this cell type.[106,107]

B cells

B cells can modulate cardiovascular disease processes in many ways.[108,109,110] Effector functions associated with antibody production do not require local B cells, because specific antibodies produced in the bone marrow or spleen can reach plaques via the circulation. Nevertheless, local secretion may also be relevant.[111] While B cells are largely absent in the intima, they are found in the adventitia of normal arteries to which they are recruited via an L-selectin-dependent mechanism.[4] B cells are also important components of perivascular FALCs (Fat-associated lymphoid clusters), which also contain myeloid (macrophages, DCs) innate lymphoid cells (ILCs) and T cells, organized around different types of stromal cells, and could be involved in immune surveillance of the vascular wall. Most B cells found in the adventitia and perivascular FALCs of normal arteries are innate B1a and B1b cells with potential homeostatic functions exerted through the production of natural IgM antibodies, e.g., IgM against oxidation-specific epitopes (OSEs). Vascular accumulation of B1 cells and their production of anti-OSE IgM antibodies is further modulated during the development of atherosclerosis in part through the transcriptional regulator Id3 via CCR6 expression.[112,113] However, the relative proportion of vascular B1 cells decreases with disease progression, and atherosclerotic arteries show increased accumulation of B2-like cells.[114] This is supported by data from CyTOF and scRNA-seq analyses of aortic leukocytes, which identified four different B cell clusters, including B1-like and B2-like clusters.[6,7] In advanced disease, B cells in the adventitia become organized in artery tertiary lymphoid organs (ATLOs), which contain different subsets of B cells that participate in germinal center responses, along with a variety of dendritic cells, CD4+ and CD8+ T cells[115,116,117] (Fig. 5.4). Deficient ATLO formation in $ApoE^{-/-}$ mice lacking lymphotoxin β receptors in VSMCs was associated with increased lesion formation, suggesting that ATLO formation and the associated local modulation of adaptive immune response are protective.[117]

Resident B1-like cells are also found in the pericardial adipose tissue as part of pericardial FALCs,[118,119] and both B1-like and B2-like cells have been described in the myocardium.[120,121,122] Resident B1-like cells appear to support the expression of MHC-II on resident cardiac macrophages.[123] More B cells accumulate in the myocardium after MI. However, their roles relative to systemic B cells remain unclear. Both antibody- and cytokine-mediated effects of B cells could be linked to systemic rather than local resident B cells.[124,119,125]

Other lymphoid cells

ILCs: ILCs arise from the common lymphoid progenitor but do not express recombined antigen receptors and lack most lineage defining surface antigens. Their identification relies on coexpression of defined markers (e.g., CD127, ICOS, KLRG1, and CD25) in lineage-negative cells. Bridging the gap between innate and adaptive immunity, ILC are assigned to one of 3 groups, ILC1, ILC2, and ILC3,[126] that mirror the Th1, Th2, and Th17 paradigm and share effector cytokines and prominent transcription factors. ILC2 most closely resemble Th2-committed T cells and thus, under the control of the transcription factor GATA3, they secrete large amounts of type 2 cytokines such as IL5 and IL13 upon stimulation.[127,128] ILC2 integrate multiple signals through the expression of an array of activatory and inhibitory receptors. ILC2 are activated by alarmins (IL25, IL33, TSLP), cytokines (IL4, IL7), lipid mediators (prostaglandins, lipoxins), and vasoactive intestinal peptides (VIPs), which induce the secretion of type 2 cytokines (IL5, IL13)[127,129,130] and molecules that impact tissue repair and metabolic processes (amphiregulin, methionine-enkephalin).[131,132] ILC2 are further supported by lymphocyte-derived IL-2,[133] and modulated by their interaction with other immune cells through the expression of selective surface receptors (ICOS, KLRG1). They are inhibited by signals generated during chronic inflammation (IL1β, IL12, IFNα, IFNγ).[127,128]

Although ILC2s are present in secondary lymphoid tissues, they are 10-fold more prevalent in barrier tissues. ILC2 show close association with the small vasculature and adipose tissue,[134] and are also found in the adventitia and peri-vascular adipose tissue of medium-sized and large arteries[135] (Fig. 5.4). Within the perivascular adipose tissue, ILC2 may be involved in the formation of FALCs.

ILC2 play critical roles in several processes with high relevance to CVD. Through their production of type 2 cytokines, mainly IL-13, ILC2s regulate macrophage behavior, promoting alternative macrophage activation

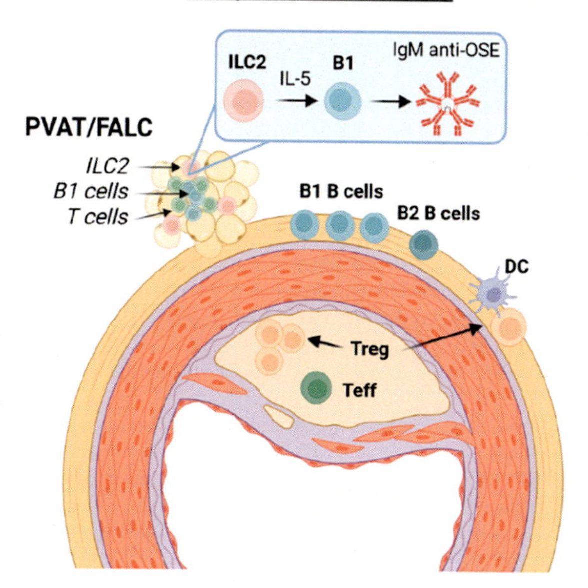

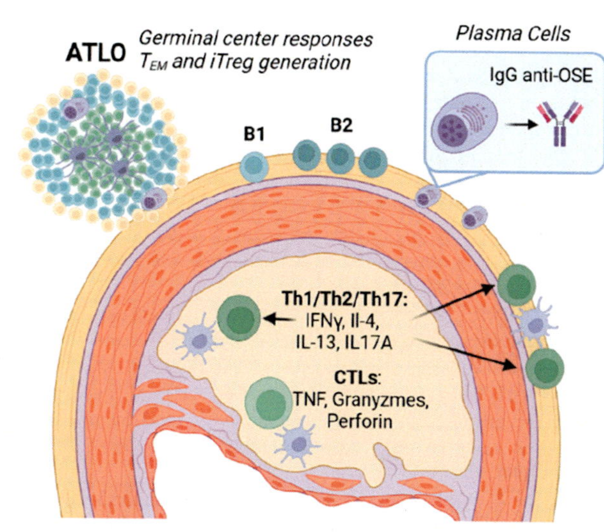

FIGURE 5.4 **Lymphoid cells in arteries and perivascular structures in health and disease.** In the steady state, various lymphoid cells reside in the intima and adventitia of arteries and in perivascular structures, such as perivascular-associated adipose tissue (PVAT), and fat-associated lymphoid clusters (FALCs) within PVAT. At the initial stages of atherosclerosis, developing T cell responses are thought to be mostly protective and dominated by regulatory T cells (Treg) over effector T cells (Teff). B1 cells dominate over B2 cells, and interactions between ILC2 and B1 cells in PVAT/FALC may drive production of immunoglobulin M (IgM) recognizing oxidation-specific epitopes (OSEs). During the progression of atherosclerosis, pathogenic immune responses start to dominate with, e.g., Th1 CD4+ T cells producing IFNγ, and cytotoxic T lymphocytes (CTLs) producing proinflammatory cytokines (TNF) and cytotoxic enzymes. B cell responses shift to being dominated by B2 and plasma cells producing IgG antibodies against OSE and other modified self-antigens or neo-self. At late stages of atherosclerosis, arterial tertiary lymphoid organs can develop, where germinal center responses and the generation of effector memory (T_{EM}) and induced regulatory T cells (iTreg) occur. Intriguingly, ATLOs were shown to limit atherosclerosis progression, providing a late counter-regulatory mechanism. At all stages, dendritic cells (DCs) contribute to recall T cell responses in the vascular wall.

during the development of atherosclerosis, thereby limiting the development of "unstable" plaques.[135] Whether part of the atheroprotective role of ILC2 could be mediated through their activation of B1 cells (e.g., within FALCs) is currently unknown and merits further consideration. ILC2s also accumulate in pericardial FALCs and mediastinum, and a recent work has shown an important role for these ILC2s in the response to MI.[136] ILC2s proliferate in the pericardial adipose tissue in the very early immune response after MI in mice, and their activation status is under the control of IL-33 and IL-2. ILC2 deficiency results in larger scar formation and reduced heart function after MI, and supplementation with IL-2 expands ILC2s and improves the recovery of heart function. Thus, ILC2s are resident innate immune cells of cardiovascular tissues and appear to be beneficial in multiple aspects of cardiovascular diseases. Targeting ILC2s, e.g., using low-dose IL-2, could constitute a new therapeutic strategy to improve cardiac remodeling and the acceleration of atherosclerosis in post-MI setting.

Natural killer (NK) cells: NK cells are cytotoxic, proinflammatory ILCs (a subset of ILC1) present in low numbers in diseased atherosclerotic vessels.[137] While some studies had proposed pathogenic activities of these cells in atherosclerosis,[137] a study using refined models of NK cell deficiency or hyperactivation showed no effect on lesion formation.[138]

Conclusion

Recent studies, notably employing high dimensional single-cell technologies, have demonstrated an ever-increasing complexity of vascular immune cell types in health and disease. Advances in "single-cell multiomics" technologies, including spatial single-cell omics, are developing at a fast-pace[139,140,141] along with advances in imaging methodologies (e.g., light sheet fluorescence microscopy[142]) and the development of new genetic models to better trace and target specific immune cell populations.[143,144] The combination of these methodologies is expected to illuminate our understanding of the role of immune cells in vascular homeostasis and vascular response to injury.

References

1. Roth GA, Mensah GA, Johnson CO, et al. Global burden of cardiovascular diseases and risk factors, 1990–2019: update from the GBD 2019 study. *J Am Coll Cardiol*. 2020;76:2982–3021.
2. Raffort J, Lareyre F, Clement M, Hassen-Khodja R, Chinetti G, Mallat Z. Monocytes and macrophages in abdominal aortic aneurysm. *Nat Rev Cardiol*. 2017;14:457–471.
3. Jonasson L, Holm J, Skalli O, Bondjers G, Hansson GK. Regional accumulations of T cells, macrophages, and smooth muscle cells in the human atherosclerotic plaque. *Arteriosclerosis*. 1986;6: 131–138.
4. Galkina E, Kadl A, Sanders J, Varughese D, Sarembock IJ, Ley K. Lymphocyte recruitment into the aortic wall before and during development of atherosclerosis is partially L-selectin dependent. *J Exp Med*. 2006;203:1273–1282.
5. Cole JE, Park I, Ahern DJ, et al. Immune cell census in murine atherosclerosis: cytometry by time of flight illuminates vascular myeloid cell diversity. *Cardiovasc Res*. 2018;114:1360–1371.
6. Winkels H, Ehinger E, Vassallo M, et al. Atlas of the immune cell repertoire in mouse atherosclerosis defined by single-cell RNA-sequencing and mass cytometry. *Circ Res*. 2018;122: 1675–1688.
7. Zernecke A, Winkels H, Cochain C, et al. Meta-analysis of leukocyte diversity in atherosclerotic mouse aortas. *Circ Res*. 2020;127: 402–426.
8. Vallejo J, Cochain C, Zernecke A, Ley K. Heterogeneity of immune cells in human atherosclerosis revealed by scRNA-Seq. *Cardiovasc Res*. 2021;117:2537–2543.
9. Mallat Z. Macrophages. *Arterioscler Thromb Vasc Biol*. 2017;37: e92–e98.
10. Randolph GJ. Tissue macrophages break dogma. *Nat Rev Immunol*. 2021;21:625.
11. Ginhoux F, Greter M, Leboeuf M, et al. Fate mapping analysis reveals that adult microglia derive from primitive macrophages. *Science*. 2010;330:841–845.
12. Ginhoux F, Guilliams M. Tissue-resident macrophage ontogeny and homeostasis. *Immunity*. 2016;44:439–449.
13. Guilliams M, Thierry GR, Bonnardel J, Bajenoff M. Establishment and maintenance of the macrophage niche. *Immunity*. 2020;52: 434–451.
14. Guilliams M, Svedberg FR. Does tissue imprinting restrict macrophage plasticity? *Nat Immunol*. 2021;22:118–127.
15. Bian Z, Gong Y, Huang T, et al. Deciphering human macrophage development at single-cell resolution. *Nature*. 2020;582:571–576.
16. Ensan S, Li A, Besla R, et al. Self-renewing resident arterial macrophages arise from embryonic CX3CR1(+) precursors and circulating monocytes immediately after birth. *Nat Immunol*. 2016;17: 159–168.
17. Cochain C, Vafadarnejad E, Arampatzi P, et al. Single-cell RNA-seq reveals the transcriptional landscape and heterogeneity of aortic macrophages in murine atherosclerosis. *Circ Res*. 2018;122: 1661–1674.
18. Williams JW, Zaitsev K, Kim KW, et al. Limited proliferation capacity of aortic intima resident macrophages requires monocyte recruitment for atherosclerotic plaque progression. *Nat Immunol*. 2020;21:1194–1204.
19. Weinberger T, Esfandyari D, Messerer D, et al. Ontogeny of arterial macrophages defines their functions in homeostasis and inflammation. *Nat Commun*. 2020;11:4549.
20. Lim HY, Lim SY, Tan CK, et al. Hyaluronan receptor LYVE-1-expressing macrophages maintain arterial tone through hyaluronan-mediated regulation of smooth muscle cell collagen. *Immunity*. 2018;49:326–341 e7.
21. Zernecke A, Erhard F, Weinberger T, et al. Integrated scRNA-seq analysis identifies conserved transcriptomic features of mononuclear phagocytes in mouse and human atherosclerosis. *bioRxiv*. 2020. https://doi.org/10.1101/2020.12.09.417535.
22. Zhou J, Tang PC, Qin L, et al. CXCR3-dependent accumulation and activation of perivascular macrophages is necessary for homeostatic arterial remodeling to hemodynamic stresses. *J Exp Med*. 2010;207:1951–1966.
23. Li F, Yan K, Wu L, et al. Single-cell RNA-seq reveals cellular heterogeneity of mouse carotid artery under disturbed flow. *Cell Death Dis*. 2021;7:180.
24. Choi JH, Do Y, Cheong C, et al. Identification of antigen-presenting dendritic cells in mouse aorta and cardiac valves. *J Exp Med*. 2009;206:497–505.
25. Paulson KE, Zhu SN, Chen M, Nurmohamed S, Jongstra-Bilen J, Cybulsky MI. Resident intimal dendritic cells accumulate lipid and contribute to the initiation of atherosclerosis. *Circ Res*. 2010; 106:383–390.
26. Poller WC, Nahrendorf M, Swirski FK. Hematopoiesis and cardiovascular disease. *Circ Res*. 2020;126:1061–1085.
27. Combadiere C, Potteaux S, Rodero M, et al. Combined inhibition of CCL2, CX3CR1, and CCR5 abrogates Ly6C(hi) and Ly6C(lo) monocytosis and almost abolishes atherosclerosis in hypercholesterolemic mice. *Circulation*. 2008;117:1649–1657.
28. Dutta P, Courties G, Wei Y, et al. Myocardial infarction accelerates atherosclerosis. *Nature*. 2012;487:325–329.
29. Mellak S, Ait-Oufella H, Esposito B, et al. Angiotensin II mobilizes spleen monocytes to promote the development of abdominal aortic aneurysm in Apoe−/− mice. *Arterioscler Thromb Vasc Biol*. 2015;35:378–388.
30. Leuschner F, Panizzi P, Chico-Calero I, et al. Angiotensin-converting enzyme inhibition prevents the release of monocytes from their splenic reservoir in mice with myocardial infarction. *Circ Res*. 2010;107:1364–1373.
31. Mestas J, Ley K. Monocyte-endothelial cell interactions in the development of atherosclerosis. *Trends Cardiovasc Med*. 2008;18:228–232.
32. Robbins CS, Hilgendorf I, Weber GF, et al. Local proliferation dominates lesional macrophage accumulation in atherosclerosis. *Nat Med*. 2013;19:1166–1172.
33. Feil S, Fehrenbacher B, Lukowski R, et al. Transdifferentiation of vascular smooth muscle cells to macrophage-like cells during atherogenesis. *Circ Res*. 2014;115:662–667.
34. Wirka RC, Wagh D, Paik DT, et al. Atheroprotective roles of smooth muscle cell phenotypic modulation and the TCF21 disease gene as revealed by single-cell analysis. *Nat Med*. 2019;25: 1280–1289.
35. Cochain C, Zernecke A. Macrophages in vascular inflammation and atherosclerosis. *Pflugers Arch*. 2017;469:485–499.
36. Nahrendorf M, Swirski FK. Abandoning M1/M2 for a network model of macrophage function. *Circ Res*. 2016;119:414–417.
37. Duewell P, Kono H, Rayner KJ, et al. NLRP3 inflammasomes are required for atherogenesis and activated by cholesterol crystals. *Nature*. 2010;464:1357–1361.
38. Rajamaki K, Lappalainen J, Oorni K, et al. Cholesterol crystals activate the NLRP3 inflammasome in human macrophages: a novel link between cholesterol metabolism and inflammation. *PLoS One*. 2010;5:e11765.
39. Di Gioia M, Spreafico R, Springstead JR, et al. Endogenous oxidized phospholipids reprogram cellular metabolism and boost hyperinflammation. *Nat Immunol*. 2020;21:42–53.
40. Spann NJ, Garmire LX, McDonald JG, et al. Regulated accumulation of desmosterol integrates macrophage lipid metabolism and inflammatory responses. *Cell*. 2012;151:138–152.
41. Doran AC, Yurdagul Jr A, Tabas I. Efferocytosis in health and disease. *Nat Rev Immunol*. 2020;20:254–267.
42. Cai B, Thorp EB, Doran AC, et al. MerTK receptor cleavage promotes plaque necrosis and defective resolution in atherosclerosis. *J Clin Invest*. 2017;127:564–568.

43. Zhao G, Lu H, Chang Z, et al. Single-cell RNA sequencing reveals the cellular heterogeneity of aneurysmal infrarenal abdominal aorta. *Cardiovasc Res*. 2021;117:1402–1416.
44. Li Y, Ren P, Dawson A, et al. Single-cell transcriptome analysis reveals dynamic cell populations and differential gene expression patterns in control and aneurysmal human aortic tissue. *Circulation*. 2020;142:1374–1388.
45. Ballesteros I, Rubio-Ponce A, Genua M, et al. Co-option of neutrophil fates by tissue environments. *Cell*. 2020;183:1282–1297 e18.
46. Vafadarnejad E, Rizzo G, Krampert L, et al. Dynamics of cardiac neutrophil diversity in murine myocardial infarction. *Circ Res*. 2020;127:e232–e249.
47. Zernecke A, Bot I, Djalali-Talab Y, et al. Protective role of CXC receptor 4/CXC ligand 12 unveils the importance of neutrophils in atherosclerosis. *Circ Res*. 2008;102:209–217.
48. Eliason JL, Hannawa KK, Ailawadi G, et al. Neutrophil depletion inhibits experimental abdominal aortic aneurysm formation. *Circulation*. 2005;112:232–240.
49. Silvestre-Roig C, Braster Q, Wichapong K, et al. Externalized histone H4 orchestrates chronic inflammation by inducing lytic cell death. *Nature*. 2019;569:236–240.
50. Warnatsch A, Ioannou M, Wang Q, Papayannopoulos V. Inflammation. Neutrophil extracellular traps license macrophages for cytokine production in atherosclerosis. *Science*. 2015;349:316–320.
51. Marx C, Novotny J, Salbeck D, et al. Eosinophil-platelet interactions promote atherosclerosis and stabilize thrombosis with eosinophil extracellular traps. *Blood*. 2019;134:1859–1872.
52. Liu CL, Liu X, Zhang Y, et al. Eosinophils protect mice from angiotensin-II perfusion-induced abdominal aortic aneurysm. *Circ Res*. 2021;128:188–202.
53. Wick G, Romen M, Amberger A, et al. Atherosclerosis, autoimmunity, and vascular-associated lymphoid tissue. *FASEB J*. 1997;11: 1199–1207.
54. Kovanen PT, Bot I. Mast cells in atherosclerotic cardiovascular disease - activators and actions. *Eur J Pharmacol*. 2017;816:37–46.
55. Swedenborg J, Mayranpaa MI, Kovanen PT. Mast cells: important players in the orchestrated pathogenesis of abdominal aortic aneurysms. *Arterioscler Thromb Vasc Biol*. 2011;31:734–740.
56. Wang Y, Shi GP. Mast cell chymase and tryptase in abdominal aortic aneurysm formation. *Trends Cardiovasc Med*. 2012;22: 150–155.
57. Bot I, Shi GP, Kovanen PT. Mast cells as effectors in atherosclerosis. *Arterioscler Thromb Vasc Biol*. 2015;35:265–271.
58. Shi GP, Lindholt JS. Mast cells in abdominal aortic aneurysms. *Curr Vasc Pharmacol*. 2013;11:314–326.
59. Sicklinger F, Meyer IS, Li X, et al. Basophils balance healing after myocardial infarction via IL-4/IL-13. *J Clin Invest*. 2021;131.
60. Cabeza-Cabrerizo M, Cardoso A, Minutti CM, Pereira da Costa M, Reis ESC. Dendritic cells revisited. *Annu Rev Immunol*. 2021;39:131–166.
61. Dress RJ, Dutertre CA, Giladi A, et al. Plasmacytoid dendritic cells develop from Ly6D(+) lymphoid progenitors distinct from the myeloid lineage. *Nat Immunol*. 2019;20:852–864.
62. Zernecke A. Dendritic cells in atherosclerosis: evidence in mice and humans. *Arterioscler Thromb Vasc Biol*. 2015;35:763–770.
63. Ericson JA, Duffau P, Yasuda K, et al. Gene expression during the generation and activation of mouse neutrophils: implication of novel functional and regulatory pathways. *PLoS One*. 2014;9: e108553.
64. Gautier EL, Huby T, Saint-Charles F, et al. Conventional dendritic cells at the crossroads between immunity and cholesterol homeostasis in atherosclerosis. *Circulation*. 2009;119:2367–2375.
65. Leach SM, Gibbings SL, Tewari AD, et al. Human and mouse transcriptome profiling identifies cross-species homology in pulmonary and lymph node mononuclear phagocytes. *Cell Rep*. 2020; 33:108337.
66. Bosteels C, Neyt K, Vanheerswynghels M, et al. Inflammatory type 2 cDCs acquire features of cDC1s and macrophages to orchestrate immunity to respiratory virus infection. *Immunity*. 2020;52:1039–10356.e9.
67. Maier B, Leader AM, Chen ST, et al. A conserved dendritic-cell regulatory program limits antitumour immunity. *Nature*. 2020; 580:257–262.
68. Choi JH, Cheong C, Dandamudi DB, et al. Flt3 signaling-dependent dendritic cells protect against atherosclerosis. *Immunity*. 2011;35:819–831.
69. Gil-Pulido J, Cochain C, Lippert MA, et al. Deletion of Batf3-dependent antigen-presenting cells does not affect atherosclerotic lesion formation in mice. *PLoS One*. 2017;12:e0181947.
70. Svensson MN, Andersson KM, Wasen C, et al. Murine germinal center B cells require functional Fms-like tyrosine kinase 3 signaling for IgG1 class-switch recombination. *Proc Natl Acad Sci U S A*. 2015;112:E6644–E6653.
71. Ataide MA, Komander K, Knopper K, et al. BATF3 programs CD8(+) T cell memory. *Nat Immunol*. 2020;21:1397–1407.
72. Haddad Y, Lahoute C, Clement M, et al. The dendritic cell receptor DNGR-1 promotes the development of atherosclerosis in mice. *Circ Res*. 2017;121:234–243.
73. Clement M, Haddad Y, Raffort J, et al. Deletion of IRF8 (interferon regulatory factor 8)-dependent dendritic cells abrogates proatherogenic adaptive immunity. *Circ Res*. 2018;122:813–820.
74. Koltsova EK, Garcia Z, Chodaczek G, et al. Dynamic T cell-APC interactions sustain chronic inflammation in atherosclerosis. *J Clin Invest*. 2012;122:3114–3126.
75. MacRitchie N, Grassia G, Noonan J, et al. The aorta can act as a site of naive CD4+ T-cell priming. *Cardiovasc Res*. 2020;116:306–316.
76. Weber C, Meiler S, Doring Y, et al. CCL17-expressing dendritic cells drive atherosclerosis by restraining regulatory T cell homeostasis in mice. *J Clin Invest*. 2011;121:2898–2910.
77. Krishna SM, Moran CS, Jose RJ, Lazzaroni S, Huynh P, Golledge J. Depletion of CD11c+ dendritic cells in apolipoprotein E-deficient mice limits angiotensin II-induced abdominal aortic aneurysm formation and growth. *Clin Sci (Lond)*. 2019;133:2203–2215.
78. Jonasson L, Holm J, Skalli O, Gabbiani G, Hansson GK. Expression of class II transplantation antigen on vascular smooth muscle cells in human atherosclerosis. *J Clin Invest*. 1985;76:125–131.
79. Stemme S, Faber B, Holm J, Wiklund O, Witztum JL, Hansson GK. T lymphocytes from human atherosclerotic plaques recognize oxidized low density lipoprotein. *Proc Natl Acad Sci U S A*. 1995; 92:3893–3897.
80. Saigusa R, Winkels H, Ley K. T cell subsets and functions in atherosclerosis. *Nat Rev Cardiol*. 2020;17:387–401.
81. Ammirati E, Moroni F, Magnoni M, Camici PG. The role of T and B cells in human atherosclerosis and atherothrombosis. *Clin Exp Immunol*. 2015;179:173–187.
82. Yuan Z, Lu Y, Wei J, Wu J, Yang J, Cai Z. Abdominal aortic aneurysm: roles of inflammatory cells. *Front Immunol*. 2020;11:609161.
83. Fernandez DM, Rahman AH, Fernandez NF, et al. Single-cell immune landscape of human atherosclerotic plaques. *Nat Med*. 2019;25:1576–1588.
84. Sasson SC, Gordon CL, Christo SN, Klenerman P, Mackay LK. Local heroes or villains: tissue-resident memory T cells in human health and disease. *Cell Mol Immunol*. 2020;17:113–122.
85. King VL, Szilvassy SJ, Daugherty A. Interleukin-4 deficiency decreases atherosclerotic lesion formation in a site-specific manner in female LDL receptor−/− mice. *Arterioscler Thromb Vasc Biol*. 2002;22:456–461.
86. Taleb S, Tedgui A, Mallat Z. IL-17 and Th17 cells in atherosclerosis: subtle and contextual roles. *Arterioscler Thromb Vasc Biol*. 2015;35:258–264.
87. Gistera A, Robertson AK, Andersson J, et al. Transforming growth factor-beta signaling in T cells promotes stabilization of

atherosclerotic plaques through an interleukin-17-dependent pathway. *Sci Transl Med*. 2013;5:196ra00.
88. Butcher MJ, Gjurich BN, Phillips T, Galkina EV. The IL-17A/IL-17RA axis plays a proatherogenic role via the regulation of aortic myeloid cell recruitment. *Circ Res*. 2012;110:675–687.
89. Romain M, Taleb S, Dalloz M, et al. Overexpression of SOCS3 in T lymphocytes leads to impaired interleukin-17 production and severe aortic aneurysm formation in mice–brief report. *Arterioscler Thromb Vasc Biol*. 2013;33:581–584.
90. Sharma AK, Lu G, Jester A, et al. Experimental abdominal aortic aneurysm formation is mediated by IL-17 and attenuated by mesenchymal stem cell treatment. *Circulation*. 2012;126: S38–S45.
91. Kiner E, Willie E, Vijaykumar B, et al. Gut CD4(+) T cell phenotypes are a continuum molded by microbes, not by TH archetypes. *Nat Immunol*. 2021;22:216–228.
92. van Beek JJP, Rescigno M, Lugli E. A fresh look at the T helper subset dogma. *Nat Immunol*. 2021;22:104–105.
93. Sakaguchi S, Vignali DA, Rudensky AY, Niec RE, Waldmann H. The plasticity and stability of regulatory T cells. *Nat Rev Immunol*. 2013;13:461–467.
94. Butcher MJ, Filipowicz AR, Waseem TC, et al. Atherosclerosis-driven Treg plasticity results in formation of a dysfunctional subset of plastic IFNgamma+ Th1/tregs. *Circ Res*. 2016;119: 1190–1203.
95. Wolf D, Gerhardt T, Winkels H, et al. Pathogenic autoimmunity in atherosclerosis evolves from initially protective apolipoprotein B100-reactive CD4(+) T-regulatory cells. *Circulation*. 2020;142: 1279–1293.
96. Cochain C, Koch M, Chaudhari SM, et al. CD8+ T cells regulate monopoiesis and circulating Ly6C-high monocyte levels in atherosclerosis in mice. *Circ Res*. 2015;117:244–253.
97. Kyaw T, Winship A, Tay C, et al. Cytotoxic and proinflammatory CD8+ T lymphocytes promote development of vulnerable atherosclerotic plaques in apoE-deficient mice. *Circulation*. 2013;127: 1028–1039.
98. Schafer S, Zernecke A. CD8(+) T cells in atherosclerosis. *Cells*. 2020;10.
99. Zhou J, Dimayuga PC, Zhao X, et al. CD8(+)CD25(+) T cells reduce atherosclerosis in apoE(−/−) mice. *Biochem Biophys Res Commun*. 2014;443:864–870.
100. Dimayuga PC, Zhao X, Yano J, et al. Identification of apoB-100 peptide-specific CD8+ T cells in atherosclerosis. *J Am Heart Assoc*. 2017;6.
101. Ribot JC, Lopes N, Silva-Santos B. Gammadelta T cells in tissue physiology and surveillance. *Nat Rev Immunol*. 2021;21: 221–232.
102. Vu DM, Tai A, Tatro JB, Karas RH, Huber BT, Beasley D. gammadeltaT cells are prevalent in the proximal aorta and drive nascent atherosclerotic lesion progression and neutrophilia in hypercholesterolemic mice. *PLoS One*. 2014;9:e109416.
103. Cheng HY, Wu R, Hedrick CC. Gammadelta (gammadelta) T lymphocytes do not impact the development of early atherosclerosis. *Atherosclerosis*. 2014;234:265–269.
104. Reinhardt A, Yevsa T, Worbs T, et al. Interleukin-23-dependent gamma/delta T cells produce interleukin-17 and accumulate in the enthesis, aortic valve, and ciliary body in mice. *Arthritis Rheumatol*. 2016;68:2476–2486.
105. Kleindienst R, Xu Q, Willeit J, Waldenberger FR, Weimann S, Wick G. Immunology of atherosclerosis. Demonstration of heat shock protein 60 expression and T lymphocytes bearing alpha/beta or gamma/delta receptor in human atherosclerotic lesions. *Am J Pathol*. 1993;142:1927–1937.
106. Getz GS, Reardon CA. Natural killer T cells in atherosclerosis. *Nat Rev Cardiol*. 2017;14:304–314.
107. Nakai Y, Iwabuchi K, Fujii S, et al. Natural killer T cells accelerate atherogenesis in mice. *Blood*. 2004;104:2051–2059.
108. Sage AP, Tsiantoulas D, Binder CJ, Mallat Z. The role of B cells in atherosclerosis. *Nat Rev Cardiol*. 2019;16:180–196.
109. Porsch F, Mallat Z, Binder CJ. Humoral immunity in atherosclerosis and myocardial infarction: from B cells to antibodies. *Cardiovasc Res*. 2021;117:2544–2562.
110. Dingwell LS, Shikatani EA, Besla R, et al. B-cell deficiency lowers blood pressure in mice. *Hypertension*. 2019;73:561–570.
111. Pattarabanjird T, Li C, McNamara C. B cells in atherosclerosis: mechanisms and potential clinical applications. *JACC Basic Transl Sci*. 2021;6:546–563.
112. Doran AC, Lipinski MJ, Oldham SN, et al. B-cell aortic homing and atheroprotection depend on Id3. *Circ Res*. 2012; 110:e1–12.
113. Srikakulapu P, Upadhye A, Rosenfeld SM, et al. Perivascular adipose tissue harbors atheroprotective IgM-producing B cells. *Front Physiol*. 2017;8:719.
114. Hamze M, Desmetz C, Berthe ML, et al. Characterization of resident B cells of vascular walls in human atherosclerotic patients. *J Immunol*. 2013;191:3006–3016.
115. Mohanta SK, Yin C, Peng L, et al. Artery tertiary lymphoid organs contribute to innate and adaptive immune responses in advanced mouse atherosclerosis. *Circ Res*. 2014;114:1772–1787.
116. Srikakulapu P, Hu D, Yin C, et al. Artery tertiary lymphoid organs control multilayered territorialized atherosclerosis B-cell responses in aged ApoE−/− mice. *Arterioscler Thromb Vasc Biol*. 2016;36:1174–1185.
117. Hu D, Mohanta SK, Yin C, et al. Artery tertiary lymphoid organs control aorta immunity and protect against atherosclerosis via vascular smooth muscle cell lymphotoxin beta receptors. *Immunity*. 2015;42:1100–1115.
118. Horckmans M, Bianchini M, Santovito D, et al. Pericardial adipose tissue regulates granulopoiesis, fibrosis, and cardiac function after myocardial infarction. *Circulation*. 2018;137:948–960.
119. Wu L, Dalal R, Cao CD, et al. IL-10-producing B cells are enriched in murine pericardial adipose tissues and ameliorate the outcome of acute myocardial infarction. *Proc Natl Acad Sci U S A*. 2019;116: 21673–21684.
120. Adamo L, Rocha-Resende C, Lin CY, et al. Myocardial B cells are a subset of circulating lymphocytes with delayed transit through the heart. *JCI Insight*. 2020;5.
121. Rocha-Resende C, Yang W, Li W, Kreisel D, Adamo L, Mann DL. Developmental changes in myocardial B cells mirror changes in B cells associated with different organs. *JCI Insight*. 2020;5.
122. Adamo L, Rocha-Resende C, Mann DL. The emerging role of B lymphocytes in cardiovascular disease. *Annu Rev Immunol*. 2020; 38:99–121.
123. Rocha-Resende C, Pani F, Adamo L. B cells modulate the expression of MHC-II on cardiac CCR2(-) macrophages. *J Mol Cell Cardiol*. 2021;157:98–103.
124. Zouggari Y, Ait-Oufella H, Bonnin P, et al. B lymphocytes trigger monocyte mobilization and impair heart function after acute myocardial infarction. *Nat Med*. 2013;19:1273–1280.
125. Haas MS, Alicot EM, Schuerpf F, et al. Blockade of self-reactive IgM significantly reduces injury in a murine model of acute myocardial infarction. *Cardiovasc Res*. 2010;87:618–627.
126. Spits H, Artis D, Colonna M, et al. Innate lymphoid cells—a proposal for uniform nomenclature. *Nat Rev Immunol*. 2013;13: 145–149.
127. Artis D, Spits H. The biology of innate lymphoid cells. *Nature*. 2015;517:293–301.
128. McKenzie ANJ, Spits H, Eberl G. Innate lymphoid cells in inflammation and immunity. *Immunity*. 2014;41:366–374.

129. Eberl G, Colonna M, Di Santo JP, McKenzie AN. Innate lymphoid cells. Innate lymphoid cells: a new paradigm in immunology. *Science*. 2015;348:aaa6566.
130. Klose CS, Artis D. Innate lymphoid cells as regulators of immunity, inflammation and tissue homeostasis. *Nat Immunol*. 2016; 17:765–774.
131. Brestoff JR, Kim BS, Saenz SA, et al. Group 2 innate lymphoid cells promote beiging of white adipose tissue and limit obesity. *Nature*. 2015;519:242–246.
132. Rak GD, Osborne LC, Siracusa MC, et al. IL-33-Dependent group 2 innate lymphoid cells promote cutaneous wound healing. *J Invest Dermatol*. 2016;136:487–496.
133. Roediger B, Kyle R, Tay SS, et al. IL-2 is a critical regulator of group 2 innate lymphoid cell function during pulmonary inflammation. *J Allergy Clin Immunol*. 2015;136:1653–16563 e7.
134. Moro K, Yamada T, Tanabe M, et al. Innate production of T(H)2 cytokines by adipose tissue-associated c-Kit(+)Sca-1(+) lymphoid cells. *Nature*. 2010;463:540–544.
135. Newland SA, Mohanta S, Clement M, et al. Type-2 innate lymphoid cells control the development of atherosclerosis in mice. *Nat Commun*. 2017;8:15781.
136. Yu X, Newland SA, Zhao TX, et al. Innate lymphoid cells promote recovery of ventricular function after myocardial infarction. *J Am Coll Cardiol*. 2021;78:1127–1142.
137. Winkels H, Ley K. Natural killer cells at ease: atherosclerosis is not affected by genetic depletion or hyperactivation of natural killer cells. *Circ Res*. 2018;122:6–7.
138. Nour-Eldine W, Joffre J, Zibara K, et al. Genetic depletion or hyperresponsiveness of natural killer cells do not affect atherosclerosis development. *Circ Res*. 2018;122:47–57.
139. Method of the year 2019: single-cell multimodal omics. *Nat Methods*. 2020;17:1.
140. Marx V. Method of the year: spatially resolved transcriptomics. *Nat Methods*. 2021;18:9–14.
141. Fernandez DM, Giannarelli C. Immune cell profiling in atherosclerosis: role in research and precision medicine. *Nat Rev Cardiol*. 2021;19:43–58.
142. Buglak NE, Lucitti J, Ariel P, Maiocchi S, Miller FJ, Bahnson ESM. Light sheet fluorescence microscopy as a new method for unbiased three-dimensional analysis of vascular injury. *Cardiovasc Res*. 2021;117:520–532.
143. Liu Z, Gu Y, Chakarov S, et al. Fate mapping via Ms4a3-expression history traces monocyte-derived cells. *Cell*. 2019;178: 1509–1525 e19.
144. Andrews LP, Vignali KM, Szymczak-Workman AL, et al. A Cre-driven allele-conditioning line to interrogate CD4(+) conventional T cells. *Immunity*. 2021;54:2209–2217.e6.

CHAPTER

6

Perivascular adipose tissue: friend or foe?

Cristina M. Sena

Institute of Physiology, iCBR, Faculty of Medicine, University of Coimbra, Coimbra, Portugal

Introduction

The state of the Vasculome is likely to be influenced by many factors including hemodynamics, metabolism, or redox state contributed by the different cells within or in the vicinity of the vessel wall, such as those that form the perivascular adipose tissue (PVAT). This represents a unique ectopic fat depot due to its strategic location, secretome, and function that confers structural support and exerts both protective and deleterious effects on blood vessels depending on the conditions.[1–4] Phenotypically, this adipose tissue can have brown, beige, or white characteristics depending on its location along the vascular system[5–7] and the environmental/metabolic context. PVAT also exhibits different vascularization, innervation, and adipokine profiles in different vessel types, location, or various species. Abdominal PVAT bears more similarity to white adipose tissue, while thoracic PVAT resembles brown adipose tissue.[8,9] In addition, PVAT has multiple cell types including adipocytes and preadipocytes, multipotent mesenchymal stem cells, fibroblasts, mast cells, macrophages, T cells, eosinophils, lymphocytes, and a network of capillaries and nerve cells.[10–13]

PVAT vascular functions

The cross-talk between PVAT and blood vessels is critical to maintaining the normal vascular function and homeostasis and balanced release of several paracrine factors, including anti- and proinflammatory markers, growth factors, and vasoactive molecules.[14]

Healthy PVAT exerts anticontractile effects with protective beneficial effects on blood vessels. Several beneficial functions of PVAT are summarized in Table 6.1.

PVAT contributes to the regulation of vascular tone through the secretion of relaxing factors and contractile factors. PVAT is able to release relaxing factors such as adiponectin, angiotensin 1-7, prostacyclin, hydrogen sulfide (H_2S), nitric oxide (NO), calcitonin gene–related peptide, and palmitic acid methyl esters.[15–17] Indeed, PVAT adipocytes express endothelial nitric oxide synthase and enzymes (cystathionine β-synthase and cystathionine γ-lyase) that catalyze the production of NO and H_2S, respectively.[18,19] In healthy individuals, PVAT has an anticontractile phenotype due to endothelium-dependent (associated with NO production) and endothelium-independent mechanisms (through H_2O_2 production). Other adipokines such as apelin, irisin, omentin-1, leptin, and visfatin have been previously identified as probable PVAT-derived relaxing factors due to their ability to promote endothelial-dependent vasorelaxation in blood vessels.[20,21] PVAT-derived contractile factors include chemerin, PGF2α, calpastatin and $O_2^{\bullet -}$ (originated from PVAT adipocytes), angiotensin II, serotonin, dopamine, and noradrenaline.[22–27] Thus, PVAT acts in coordination with neurohumoral systems to control vascular tone by targeting vascular smooth muscle cell (VSMC) receptors and/or channels.[28]

Endothelial cells are also influenced by the PVAT secretome through an outside-to-inside signaling although the mechanisms are not completely clarified. It was recently suggested that augmented Notch signaling in PVAT can change its phenotype by eliciting adipogenesis and lipid build-up and may contribute to a favourable milieu for vascular disease.[29] The endothelium, which is strategically positioned between circulating blood and the tissues, plays a vital role in health and disease.[30,31] The gene expression patterns in the vascular endothelium of the different vascular beds and different arteries (elastic vs. muscular) are region specific and change during various disease pathophysiologies.[32,33] The vascular network within each organ is unique and contributes to maintaining organ homeostasis under physiologic conditions. Moreover, an unhealthy endothelium contributes to organ

The Vasculome
https://doi.org/10.1016/B978-0-12-822546-2.00018-6

TABLE 6.1 Perivascular adipose tissue (PVAT) functions in health.

	Healthy PVAT
Functions	Protective vascular function • Paracrine vasodilator effect on vascular function • Protection against hypertension, inflammation and oxidative stress, triglyceride/fatty acids clearance • Protective effects against atherosclerosis, macrophage infiltration, and hypoxia • Protection against wave torsion and hypothermia • Regulation of muscle perfusion and insulin sensitivity • Beneficial effects on coronary artery bypass graft patency surgery • Anti-inflammatory effects

dysfunction during disease.[32,34–36] Elucidating the molecular pathways by which PVAT controls EC function in different vascular beds and conditions will allow future efforts to reverse endothelial dysfunction and to prevent vascular and other diseases.

Perivascular adipose tissue implications in pathology

Under pathological conditions such as obesity, hypertension, atherosclerosis, or type 2 diabetes, PVAT changes its phenotype, and becomes dysfunctional (Fig. 6.1), increasing its production of contractile factors and proinflammatory cytokines/chemokines (Table 6.2; Fig. 6.1).[21,37–42]

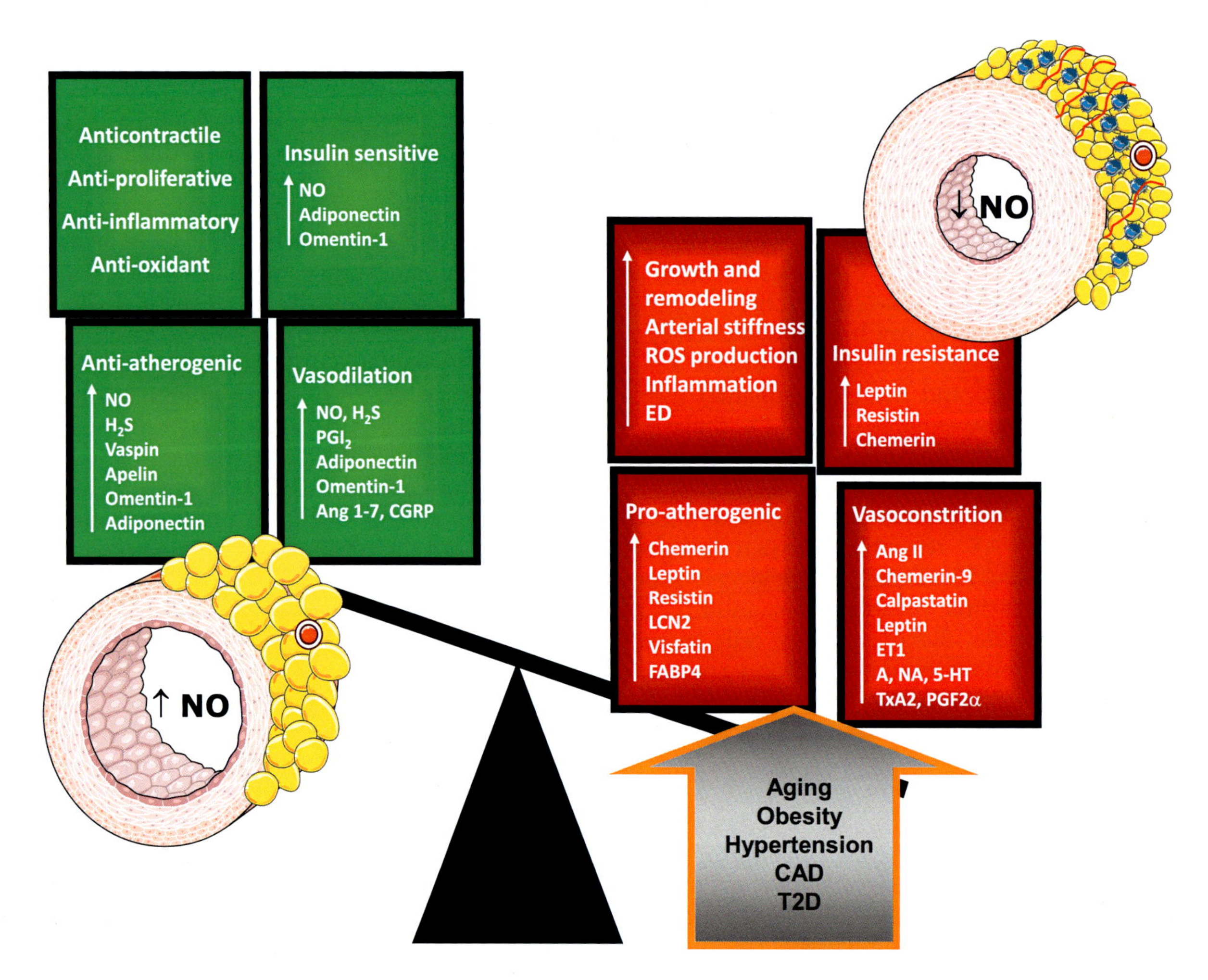

FIGURE 6.1 Perivascular adipose tissue balance can be thrown off by several vascular risk factors. The loss of anticontractile, antiinflammatory, antiatherogenic, and antiproliferative PVAT profile leads to growth, remodeling and VSMC proliferation, increased arterial stiffness, vascular insulin resistance; increased ROS and endothelin-1 production, endothelial dysfunction, vasoconstriction, increased inflammation (TNF-α, IL-6, TxA2, and CCL2), and proatherogenic effects (increment in ROS, endothelial dysfunction, and increased release of TNF-α, CCL2, TGF-β, IL-1β, and IL-6). *A*, adrenaline; *Ang 1-7*, angiotensin 1–7; *Ang II*, angiotensin II; *CAD*, coronary artery disease; *CCL2*, C–C motif chemokine ligand 2, *CGRP*, calcitonin gene–related peptide; *ED*, endothelial dysfunction; *ET1*, endothelin 1; *FABP4*, fatty acid binding protein 4; *H_2S*, hydrogen sulfide; *5-HT*, serotonin; *IL*, interleukin; *LCN2*, lipocalin-2; *NA*, noradrenaline; *NO*, nitric oxide; *PGF2α*, prostaglandin F2α; *PGI_2*, prostacyclin; *ROS*, reactive oxygen species; *PVAT*, perivascular adipose tissue; *TGF-β*, transforming growth factor beta; *TNF-α*, tumor necrosis alpha; *TxA2*, thromboxane A2; *T2D*, type 2 diabetes; *VSMC*, vascular smooth muscle cells.

TABLE 6.2 PVAT-derived factors in healthy and pathologic conditions.

	Healthy PVAT	Dysfunctional PVAT
PVAT-derived factors		
Anticontractile	NO, prostacyclin, adiponectin, leptin, Ang 1–7, H_2S, CGRP, PAME	
Contractile		Ang II, noradrenaline, 5-HT, $O_2^{\bullet -}$, chemerin, calpastatin, prostanoids (TxA2, PGF2α), leptin
Inflammatory mediators	Antiinflammatory: Adiponectin, NO, Ang 1–7, IL10, ADRF	Proinflammatory: Adipokines: Leptin, visfatin, resistin Cytokines: TNF-α, IL1β, IL6, IL17 Chemokines: CCL2, RANTES, CxCL3, CxCL10 Proinflammatory leukocytes accumulation Adhesion molecules: ICAM1, VCAM1 ROS: NADPH oxidases, $O_2^{\bullet -}$ IFN-γ, PAI-1

ADRF, adipocyte derived relaxing factor; *Ang 1-7*, angiotensin 1-7; *Ang II*, angiotensin II; *CCL2*, C–C motif chemokine ligand 2; *CGRP*, calcitonin gene-related peptide; *CxCL3*, C-X-C motif chemokine 3; *CxCL10*, C-X-C motif chemokine 10; *5-HT*, serotonin; *H_2S*, hydrogen sulfide; *ICAM1*, intercellular adhesion molecule 1; *IFN-γ*, interferon γ; *IL*, interleukin (IL); *NO*, Nitric oxide; *PAI-1*, plasminogen activator inhibitor-1; *PAME*, palmitic acid methyl esters; *PGF2α*, prostaglandin F2α; *PVAT*, perivascular adipose tissue; *RANTES*, regulated on activation normal T cell expressed and secreted; *ROS*, reactive oxygen species; *TNF-α*, tumor necrosis factor-α; *TxA2*, thromboxane A2; *VCAM1*, vascular cellular adhesion molecule 1.

The initially protective PVAT phenotype changes in the presence of pathological conditions, such as those associated with obesity, when PVAT promotes deleterious vascular effects that include the following: (1) loss or decrement of anticontractile PVAT profile; (2) amplified atherogenesis, oxidative stress, inflammation, vascular remodeling, hypertension, and calcification; (3) increased vascular insulin resistance, inflammation, and hypoxia; (4) participation in the development and destabilization of atherosclerotic plaques; and (5) influencing age-related arterial stiffness and increasing cardiovascular risk in menopause.[2,4,20,21,43,44]

Several PVAT-produced molecules involved in the regulation of vascular function and inflammation are summarized in Table 6.2. Indeed, PVAT exerts an impact on various cells of the vascular wall.[45] PVAT abuts the adventitia which allows paracrine diffusion of molecules and microvesicles through outside-to-inside signaling between PVAT and vascular cells. PVAT modulates VSMC tone, remodeling, proliferation, and migration.[28,46] The proinflammatory PVAT repertoire in aged Wistar rats was partially improved by supplementation with exogenous adiponectin[39], highlighting the importance of this adipokine in metabolic disorders. In addition, we previously showed that thoracic aortas from diabetic Goto-Kakizaki (GK) rats mounted in a myograph with PVAT lost their anticontractile properties.[42] This phenotype worsened endothelial dysfunction through oxidative stress and inflammation in a nonobese model of type 2 diabetes. PVAT of diabetic GK rats displayed higher oxidative stress, malonaldehyde levels and aldose reductase activity, macrophage infiltration, and reduced antioxidant defense enzymes.[41]

Conclusion

PVAT surrounds blood vessels and is an active endocrine/paracrine organ, directly influencing vascular tone and inflammation through the release of anti- and proinflammatory and vasoactive molecules. The normal anticontractile phenotype of PVAT is determined by the browning and inflammation status. Reversing the white characteristics of PVAT to a brown-like or beige phenotype might be a key approach to maintain a healthy vasculome. Further research is need on the signaling pathways originated from PVAT and its cross-talk with the underlying cells, including endothelial cells, VSMCs, immune cells, and fibroblasts under physiological and pathological conditions.

References

1. Nosalski R, Guzik TJ. Perivascular adipose tissue inflammation in vascular disease. *Br J Pharmacol*. 2017;174(20):3496–3513. https://doi.org/10.1111/bph.13705.
2. Kim HW, Belin de Chantemèle EJ, Weintraub NL. Perivascular adipocytes in vascular disease. *Arterioscler Thromb Vasc Biol*. 2019;39(11): 2220–2227. https://doi.org/10.1161/ATVBAHA.119.312304.
3. Liang X, Qi Y, Dai F, Gu J, Yao W. PVAT: an important guardian of the cardiovascular system. *Histol Histopathol*. 2020;35(8):779–787. https://doi.org/10.14670/HH-18-211.
4. Fernández-Alfonso MS, Somoza B, Tsvetkov D, Kuczmanski A, Dashwood M, Gil-Ortega M. Role of perivascular adipose tissue in health and disease. *Compr Physiol*. 2017;8(1):23–59. https://doi.org/10.1002/cphy.c170004.
5. Brown NK, Zhou Z, Zhang J, et al. Perivascular adipose tissue in vascular function and disease: a review of current research and animal models. *Arterioscler Thromb Vasc Biol*. 2014;34(8):1621–1630. https://doi.org/10.1161/ATVBAHA.114.303029.

6. Hildebrand S, Stümer J, Pfeifer A. PVAT and its relation to Brown, beige, and white adipose tissue in development and function. *Front Physiol*. 2018;9:70. https://doi.org/10.3389/fphys.2018.00070.
7. Samuel OO. Review on multifaceted involvement of perivascular adipose tissue in vascular pathology. *Cardiovasc Pathol*. 2020;49: 107259. https://doi.org/10.1016/j.carpath.2020.107259.
8. Fitzgibbons TP, Kogan S, Aouadi M, Hendricks GM, Straubhaar J, Czech MP. Similarity of mouse perivascular and brown adipose tissues and their resistance to diet-induced inflammation. *Am J Physiol Heart Circ Physiol*. 2011;301(4):H1425–H1437. https://doi.org/10.1152/ajpheart.00376.2011.
9. Padilla J, Jenkins NT, Vieira-Potter VJ, Laughlin MH. Divergent phenotype of rat thoracic and abdominal perivascular adipose tissues. *Am J Physiol Regul Integr Comp Physiol*. 2013;304(7): R543–R552. https://doi.org/10.1152/ajpregu.00567.2012.
10. Ruan CC, Zhu DL, Chen QZ, et al. Perivascular adipose tissue-derived complement 3 is required for adventitial fibroblast functions and adventitial remodeling in deoxycorticosterone acetate-salt hypertensive rats. *Arterioscler Thromb Vasc Biol*. 2010; 30(12):2568–2574. https://doi.org/10.1161/ATVBAHA.110.215525.
11. Campbell KA, Lipinski MJ, Doran AC, Skaflen MD, Fuster V, McNamara CA. Lymphocytes and the adventitial immune response in atherosclerosis. *Circ Res*. 2012;110(6):889–900. https://doi.org/10.1161/CIRCRESAHA.111.263186.
12. Withers SB, Forman R, Meza-Perez S, et al. Eosinophils are key regulators of perivascular adipose tissue and vascular functionality. *Sci Rep*. 2017;7:44571. https://doi.org/10.1038/srep44571.
13. Silva KR, Côrtes I, Liechocki S, et al. Characterization of stromal vascular fraction and adipose stem cells from subcutaneous, preperitoneal and visceral morbidly obese human adipose tissue depots. *PLoS One*. 2017;12(3). https://doi.org/10.1371/journal.pone.0174115. e0174115.
14. Ramirez JG, O'Malley EJ, Ho WSV. Pro-contractile effects of perivascular fat in health and disease. *Br J Pharmacol*. 2017;174(20): 3482–3495. https://doi.org/10.1111/bph.13767.
15. Fésüs G, Dubrovska G, Gorzelniak K, et al. Adiponectin is a novel humoral vasodilator. *Cardiovasc Res*. 2007;75(4):719–727. https://doi.org/10.1016/j.cardiores.2007.05.025.
16. Lee RM, Lu C, Su LY, Gao YJ. Endothelium-dependent relaxation factor released by perivascular adipose tissue. *J Hypertens*. 2009; 27(4):782–790. https://doi.org/10.1097/HJH.0b013e328324ed86.
17. Lee YC, Chang HH, Chiang CL, et al. Role of perivascular adipose tissue-derived methyl palmitate in vascular tone regulation and pathogenesis of hypertension. *Circulation*. 2011;124(10):1160–1171. https://doi.org/10.1161/CIRCULATIONAHA.111.027375.
18. Gao YJ. Dual modulation of vascular function by perivascular adipose tissue and its potential correlation with adiposity/lipoatrophy-related vascular dysfunction. *Curr Pharm Des*. 2007; 13(21):2185–2192. https://doi.org/10.2174/138161207781039634.
19. Wójcicka G, Jamroz-Wiśniewska A, Atanasova P, Chaldakov GN, Chylińska-Kula B, Bełtowski J. Differential effects of statins on endogenous H2S formation in perivascular adipose tissue. *Pharmacol Res*. 2011;63(1):68–76. https://doi.org/10.1016/j.phrs.2010.10.011.
20. Gollasch M. Adipose-vascular coupling and potential therapeutics. *Annu Rev Pharmacol Toxicol*. 2017;57:417–436. https://doi.org/10.1146/annurev-pharmtox-010716-104542.
21. Qi XY, Qu SL, Xiong WH, Rom O, Chang L, Jiang ZS. Perivascular adipose tissue (PVAT) in atherosclerosis: a double-edged sword. *Cardiovasc Diabetol*. 2018;17(1):134. https://doi.org/10.1186/s12933-018-0777-x.
22. Chang L, Xiong W, Zhao X, et al. Bmal1 in perivascular adipose tissue regulates resting-phase blood pressure through transcriptional regulation of angiotensinogen. *Circulation*. 2018;138(1):67–79. https://doi.org/10.1161/CIRCULATIONAHA.117.029972.
23. Owen MK, Witzmann FA, McKenney ML, et al. Perivascular adipose tissue potentiates contraction of coronary vascular smooth muscle: influence of obesity. *Circulation*. 2013;128(1):9–18. https://doi.org/10.1161/CIRCULATIONAHA.112.001238.
24. Kumar RK, Darios ES, Burnett R, Thompson JM, Watts SW. Fenfluramine-induced PVAT-dependent contraction depends on norepinephrine and not serotonin. *Pharmacol Res*. 2019;140:43–49. https://doi.org/10.1016/j.phrs.2018.08.024.
25. Watts SW, Dorrance AM, Penfold ME, et al. Chemerin connects fat to arterial contraction. *Arterioscler Thromb Vasc Biol*. 2013;33(6): 1320–1328. https://doi.org/10.1161/ATVBAHA.113.301476.
26. Ayala-Lopez N, Martini M, Jackson WF, et al. Perivascular adipose tissue contains functional catecholamines. *Pharmacol Res Perspect*. 2014;2(3):e00041. https://doi.org/10.1002/prp2.41.
27. Darios ES, Winner BM, Charvat T, Krasinksi A, Punna S, Watts SW. The adipokine chemerin amplifies electrical field-stimulated contraction in the isolated rat superior mesenteric artery. *Am J Physiol Heart Circ Physiol*. 2016;311(2):H498–H507. https://doi.org/10.1152/ajpheart.00998.2015.
28. Chang L, Garcia-Barrio MT, Chen YE. Perivascular adipose tissue regulates vascular function by targeting vascular smooth muscle cells. *Arterioscler Thromb Vasc Biol*. 2020;40(5):1094–1109. https://doi.org/10.1161/ATVBAHA.120.312464.
29. Boucher JM, Ryzhova L, Harrington A, et al. Pathological conversion of mouse perivascular adipose tissue by Notch activation. *Arterioscler Thromb Vasc Biol*. 2020;40(9):2227–2243. https://doi.org/10.1161/ATVBAHA.120.314731.
30. Sena CM, Pereira AM, Seiça R. Endothelial dysfunction - a major mediator of diabetic vascular disease. *Biochim Biophys Acta*. 2013; 1832(12):2216–2231. https://doi.org/10.1016/j.bbadis.2013.08.006.
31. Godo S, Shimokawa H. Endothelial functions. *Arterioscler Thromb Vasc Biol*. 2017;37(9):e108–e114. https://doi.org/10.1161/ATVBAHA.117.309813.
32. Bosetti F, Galis ZS, Bynoe MS, et al. "Small blood vessels: big health problems?": scientific recommendations of the National Institutes of Health Workshop. *J Am Heart Assoc*. 2016;5(11). https://doi.org/10.1161/JAHA.116.004389. e004389.
33. Acosta SA, Lee JY, Nguyen H, Kaneko Y, Borlongan CV. Endothelial progenitor cells modulate inflammation-associated stroke Vasculome. *Stem Cell Rev Rep*. 2019;15(2):256–275. https://doi.org/10.1007/s12015-019-9873-x.
34. Gorelick PB, Scuteri A, Black SE, et al. Vascular contributions to cognitive impairment and dementia: a statement for healthcare professionals from the american heart association/american stroke association. *Stroke*. 2011;42(9):2672–2713. https://doi.org/10.1161/STR.0b013e3182299496.
35. Hu X, De Silva TM, Chen J, Faraci FM. Cerebral vascular disease and neurovascular injury in ischemic stroke. *Circ Res*. 2017;120(3): 449–471. https://doi.org/10.1161/CIRCRESAHA.116.308427.
36. Koizumi K, Wang G, Park L. Endothelial dysfunction and amyloid-β-induced neurovascular alterations. *Cell Mol Neurobiol*. 2016;36(2): 155–165. https://doi.org/10.1007/s10571-015-0256-9.
37. Marchesi C, Ebrahimian T, Angulo O, Paradis P, Schiffrin EL. Endothelial nitric oxide synthase uncoupling and perivascular adipose oxidative stress and inflammation contribute to vascular dysfunction in a rodent model of metabolic syndrome. *Hypertension*. 2009;54(6):1384–1392. https://doi.org/10.1161/HYPERTENSIONAHA.109.138305.
38. Antonopoulos AS, Sanna F, Sabharwal N, et al. Detecting human coronary inflammation by imaging perivascular fat. *Sci Transl Med*. 2017;9(398). https://doi.org/10.1126/scitranslmed.aal2658. eaal2658.
39. Sena CM, Pereira A, Fernandes R, Letra L, Seiça RM. Adiponectin improves endothelial function in mesenteric arteries of rats fed a

high-fat diet: role of perivascular adipose tissue. *Br J Pharmacol.* 2017;174(20):3514–3526. https://doi.org/10.1111/bph.13756.

40. Kim HW, Shi H, Winkler MA, Lee R, Weintraub NL. Perivascular adipose tissue and vascular perturbation/atherosclerosis. *Arterioscler Thromb Vasc Biol.* 2020;40(11):2569–2576. https://doi.org/10.1161/ATVBAHA.120.312470.
41. Azul L, Leandro A, Boroumand P, Klip A, Seiça R, Sena CM. Increased inflammation, oxidative stress and a reduction in antioxidant defense enzymes in perivascular adipose tissue contribute to vascular dysfunction in type 2 diabetes. *Free Radic Biol Med.* 2020; 146:264–274. https://doi.org/10.1016/j.freeradbiomed.2019.11.002.
42. Leandro A, Queiroz M, Azul L, Seiça R, Sena CM. Omentin: a novel therapeutic approach for the treatment of endothelial dysfunction in type 2 diabetes. *Free Radic Biol Med.* 2020;S0891–S5849(20): 31295–31298. https://doi.org/10.1016/j.freeradbiomed.2020.10.021.
43. Lastra G, Manrique C. Perivascular adipose tissue, inflammation and insulin resistance: link to vascular dysfunction and cardiovascular disease. *Horm Mol Biol Clin Investig.* 2015;22(1):19–26. https://doi.org/10.1515/hmbci-2015-0010.
44. Ahmadieh S, Kim HW, Weintraub NL. Potential role of perivascular adipose tissue in modulating atherosclerosis. *Clin Sci (Lond).* 2020;134(1):3–13. https://doi.org/10.1042/CS20190577.
45. Queiroz M, Sena CM. Perivascular adipose tissue in age-related vascular disease. *Ageing Res Rev.* 2020;59:101040. https://doi.org/10.1016/j.arr.2020.101040.
46. Szasz T, Webb RC. Perivascular adipose tissue: more than just structural support. *Clin Sci (Lond).* 2012;122(1):1–12. https://doi.org/10.1042/CS20110151.

CHAPTER

7

Major vascular ECM components, differential distribution supporting structure, and functions of the vasculome

Christie L. Crandall[1], *Chien-Jung Lin*[2,3] *and Jessica E. Wagenseil*[1]

[1]Department of Mechanical Engineering and Materials Science, Washington University, St. Louis, MO, United States [2]Department of Cell Biology and Physiology, Washington University, St. Louis, MO, United States [3]Department of Internal Medicine (Cardiovascular Division), Washington University, St. Louis, MO, United States

Structure and function of the vasculome

ECM organization in different vascular beds

Blood vessels consist of an endothelial cell (EC) tube, surrounded by a basement membrane, differing amounts of mural cells (primarily smooth muscle cells, SMCs, in larger vessels), and ECM proteins. The relative amounts of SMCs and major ECM proteins, including elastin and collagen, depend on the vessel size, physiologic function, and hemodynamic environment (Fig. 7.1).[1] The EC tube is the first structure to form during vasculogenesis and angiogenesis.[2] The apical/luminal surface of the ECs has a proteoglycan and glycoprotein-rich glycocalyx that extends into the vessel lumen in contact with flowing blood,[3] while the basolateral/abluminal surface rests on a basement membrane composed of collagens, proteoglycans, and glycoproteins.[4]

Capillaries have only ECs, sparse mural cells (pericytes), and a basement membrane. In arteries and veins, the first ECM layer after the basement membrane is the internal elastic lamina (IEL). The IEL is a fenestrated, woven ECM sheet composed mostly of elastic fibers (elastin and additional proteins are necessary for assembly and cross-linking). The IEL provides a physical and mechanical substrate for cells within the vessel wall and modulates permeability[5] and physical and biochemical communication between ECs and SMCs.[6] There is evidence that IEL structure depends on vessel location and function.[7–9] In small arteries and in veins, the IEL is the only layer of elastic fibers, while in larger arteries, there is an external elastic lamina and additional elastic laminae layered with SMCs and collagen fibers throughout the medial layer of the wall (Fig. 7.2). The number of elastic laminae scales with physiologic blood pressure and vessel size in different animal species, and throughout the vascular tree, providing a constant tension/lamina ratio.[10] In addition to elastic laminae and collagen fibers, SMCs are also surrounded by a pericellular basement membrane.[11]

Outside the media, there is an adventitia composed of collagens, proteoglycans, fibroblasts, immune cells, and stem-like cells.[12] The adventitial layer in the aorta thickens with distance from the heart.[13] In large arteries, the adventitia has a vasa vasorum (the vasculature of the vessel) with capillaries that provide perfusion when the vascular wall is too thick for nutrients to be provided by simple diffusion. Generally, arteries with more than 29 layers of elastic laminae have a vasa vasorum.[14] The adventitia in large vessels is surrounded by perivascular adipose tissue that may play a critical role in regulating biologic processes.[15] The adventitia is connected to interstitial tissue to provide additional support and tethering of the vessels.[16]

Primary functions of different vascular beds

Although composed of similar building blocks, the entirety of the vasculome presents incredible variety in function. In development, mechanical signals drive differential ECM formation in arteries compared to veins

The Vasculome
https://doi.org/10.1016/B978-0-12-822546-2.00010-1

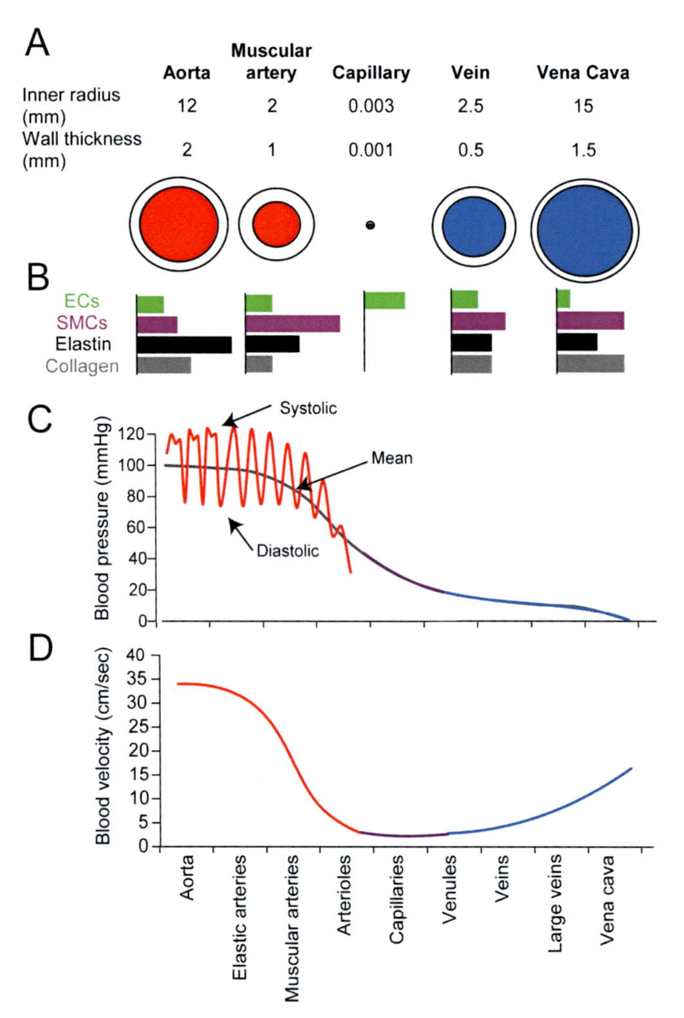

FIGURE 7.1 Structure–function relationships in the vasculature. Average inner radius and wall thickness of human blood vessels (A). Relative amounts of endothelial cells (ECs), smooth muscle cells (SMCs), elastic fibers (elastin), and collagen fibers (collagen) in the wall for each vessel (B). Blood pressure (C) and flow velocity (D) throughout the vascular tree. Systolic and diastolic pressures are shown in the aorta and elastic arteries, while mean pressure is shown throughout. *(B) Redrawn from Burton AC. Relation of structure to function of the tissues of the wall of blood vessels.* Physiol Rev. *1954;34: 619–642.*[101]

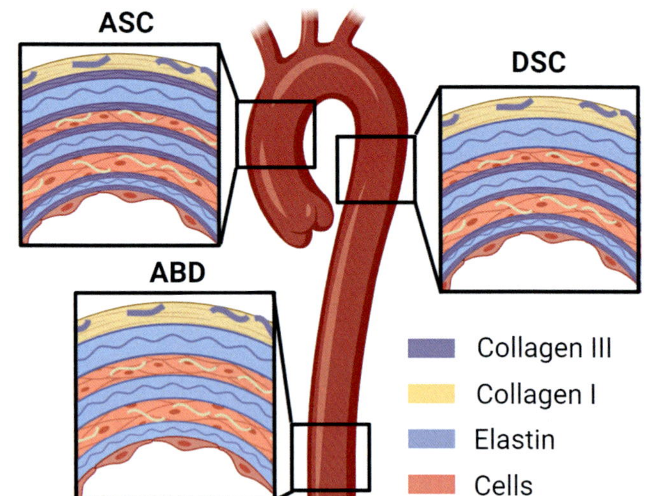

FIGURE 7.2 Schematic of differential radial collagen distribution in different parts of the aorta. In all parts of the aorta, there is a single layer of endothelial cells (ECs) at the lumen, concentric layers of smooth muscle cells (SMCs), and elastic laminae (elastin) in the media, and collagen I and III in the adventitia. In the ascending aorta (ASC), collagen III also surrounds the elastic laminae. In the descending aorta (DSC), there is collagen III around the elastic laminae, but primarily in the inner media. In the abdominal aorta (ABD) and in smaller muscular arteries, there is no collagen III in the medial layer.[30] *Created with BioRender.com.*

and in large arteries compared to small arteries. This is highly correlated with the geometry of and hemodynamic forces on the vessel wall (Fig. 7.1). Arteries are categorized as elastic or muscular according to their size, ECM composition, and function. Examples of elastic arteries are the aorta, pulmonary artery, and common carotids (Fig. 7.3A and B). These are larger in diameter, compared to muscular arteries, and contain concentric elastic laminae in the medial layer. Mechanical recoil of the structured medial layer allows elastic arteries to transform pulsatile flow exiting the heart into constant flow toward, and into the capillaries.[17] The function of capillaries is to exchange nutrients, gases and waste. This is made possible by diffusion through the single EC layer and its accompanying basement membrane. Differently, in muscular arteries, the medial layer is structured as a singular vascular smooth muscle (VSM) layer to control blood flow in response to physiologic demand. Examples of muscular arteries include coronary, cerebral, renal, and mesenteric arteries (Fig. 7.3C and D). Small arteries regulate local blood flow through SMC contraction and relaxation. Veins transport blood back toward the heart and, compared to arteries, have an overall thinner wall with a smaller medial layer and thicker adventitial layer with greater collagen content. This is a direct result from the lower blood pressures experienced in the veins.

The compositional makeup of the ECM in different portions of the vasculome allows for a tissue-level mechanical response to blood pressure, blood flow, and deformation of the vessel wall. The interplay between the mechanical response and ECM composition is critical in development, homeostasis, and disease. Elastic arteries, especially large arteries near the contracting heart, have both high stiffness and strength. The elastic laminae present in the medial layer allow the vessel to recoil with little energy loss and bundled collagen fibers become circumferentially aligned at high pressure to prevent rupture.[17,18] In contrast, being exposed to lower pressures, veins and capillaries have a thinner wall and lower stiffness. The larger diameter and higher extensibility of veins compared to capillaries allow the veins to act as a reservoir for about two-thirds

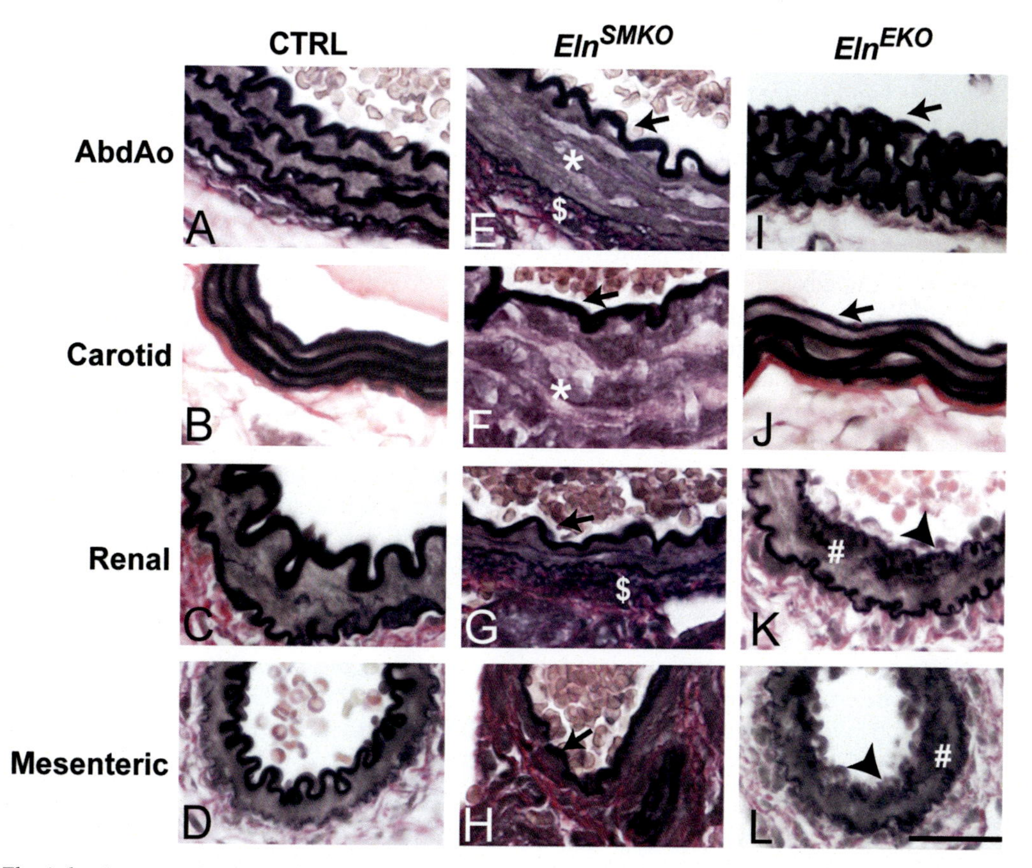

FIGURE 7.3 Elastic laminae organization and cellular contribution in different mouse arteries. Multiple layers of elastic laminae (black) are visible alternating with SMCs (purple). The number of layers typically increases with arterial size. Eliminating elastin expression from SMCs (Eln^{SMKO}) leaves the internal elastic lamina (IEL) near the ECs at the intimal surface intact (*arrows*) but eliminates or reduces elastic laminae in the medial layer of the wall (*). There are small patches of elastin near the adventitia in some Eln^{SMKO} arteries ($). Eliminating elastin expression from ECs (Eln^{EKO}) leaves the IEL intact in the larger arteries (*arrows*) but produces a fragmented IEL in the smaller arteries (*arrowheads*). There are also some scattered elastic fibers near the IEL (#). The results indicate that cellular contribution to elastic laminae formation depends on location in the arterial wall and in the arterial tree.[39] Scale = 25 μm.

of the cardiovascular system blood volume. The low pressure in the veins allows flow back to the heart to be relatively constant and the valves act to limit backflow.

Major vascular ECM proteins

In addition to the broadly observed differences in elastin and collagen amounts and cell types in different vascular beds (Fig. 7.1), there are more subtle differences in collagen types, organization of elastic fibers, and distribution of proteoglycans and glycoproteins with axial position in the vascular tree and radial position across the vessel wall that play a critical role in physiologic and pathophysiologic function within the Vasculome.

Collagens

There have been 28 different types of collagen identified to date. The common structural feature is a triple helix made possible by the characteristic repeating motif of glycine-X-Y, where X is typically proline, and Y is frequently 4-hydroxyproline.[19] The unique properties of each type of collagen are mainly due to sequences that interrupt the triple helix allowing them to form fibrils, sheets, or cross-link to other ECM proteins.[20] The proportion of collagen types and organization varies depending on axial and radial location within the vasculature.

In the basement membrane, the major collagen is network-forming collagen IV, with small amounts of multiplexin collagen XVIII.[4] Collagen IV is essential for basement membrane stability,[21] while collagen XVIII plays a role in maintaining basement membrane integrity.[22] Network-forming collagen VIII and beaded filament collagen VI are also associated with the basement membrane.[4] Collagen VIII may modulate EC[23] and SMC[24] phenotype. Collagen VI interacts with collagen IV,[25] collagen I,[26] and fibronectin,[27] providing an important bridge between the basement membrane and other ECM components in the vascular wall.

In the arterial media and adventitia, collagens I and III dominate, with small amounts of collagens IV, V, and VI present.[28] Collagens I and III are fibril-forming collagens that provide strength to the vessel wall. Collagen V regulates fibrillogenesis of collagens I and III.[29] The relative amounts and distribution of collagens I and III depend on vascular location and developmental stage. Howard and Macarak[30] found that small muscular arteries and the abdominal aorta of fetal calves have intense collagen I staining in media and weak collagen I and intense collagen III staining in the adventitia. The ascending aorta, however, has collagen III staining in the media, in addition to collagen I. Collagen III is in close apposition to elastic fibers and collagen I is near SMCs in the ascending aorta. The collagen distribution is similar in the adventitia of the ascending aorta, abdominal aorta, and the smaller arteries. The descending aorta also shows collagen III in the media, with more intense staining near the inner part of the wall (Fig. 7.2). The addition of collagen III to the medial layer may be related to the unique hemodynamic environment of the proximal aorta compared to the distal aorta and muscular arteries (Fig. 7.1).

The detection of aortic collagen III localization is antibody dependent, indicating that different epitopes may be accessible in different vascular locations.[30] Bartholomew and Anderson[31] found a similar localization of collagen III near elastic fibers and collagen I near SMCs in the media of bovine thoracic aorta (age and exact location not specified). They also found a decreasing gradient of collagen III intensity from the inner to outer media, consistent with Howard and Macarek.[30] In the adventitia, they found strong collagen I and weak collagen III staining, opposite to the results of Howard and Macarek.[30] Collagen I intensity in the media was increased by pretreatment with chondroitinase ABC, while collagen III intensity in the adventitia was increased by pretreatment with hyaluronidase, indicating that there are location-specific interactions with glycosaminoglycans (GAGs) that alter epitope availability.[31]

Elastin and elastic fibers

Elastin is secreted as soluble tropoelastin that coacervates on the cell surface, is transferred to extracellular microfibrils, and then is cross-linked by lysyl oxidase family members to become fully functional elastic fibers that provide reversible elasticity to the vessel wall. Up to 30 different proteins are associated with elastin, fibrillin microfibrils, and cross-linked elastic fibers.[32] Dense sheets of elastic fibers make up the IEL at the intima-medial border in arteries and veins and the additional laminae in the medial layer of large arteries. Differences in elastin amount, organization, and proportion of associated proteins vary with axial and radial vascular location.

The number of medial elastic laminae decreases with axial distance from the heart[33] and there is a corresponding decrease in elastin amounts.[34–36] As a result, the elastin/collagen ratio decreases with axial distance from the heart, altering the mechanical behavior of the vessel wall[13] as the hemodynamics change (Fig. 7.1). Radially, there is abundant elastin in the IEL and medial laminae, with small amounts of fine elastic fibers in the adventitia (Fig. 7.3A–D).[37,38] Cellular contribution to elastic laminae formation varies by radial and axial location. In mice, elastin expression in SMCs is required to form the medial elastic laminae in large, elastic arteries, but non-SMCs still form elastic fibers at the medial-adventitial border (Fig. 7.3E–H).[39] Elastin expression from ECs is required to form the complete IEL in small, muscular arteries, but is not required in most large, elastic arteries (Fig. 7.3I–L).

In mice, the lack of various fibulin family members associated with elastic fiber assembly has differing effects depending on radial and axial vessel location. Fibulin-2 knockout (*Fbln2*$^{-/-}$) mice display no vascular abnormalities,[40] while fibulin-5 knockout (*Fbln5*$^{-/-}$) mice have disrupted elastic laminae in the aortic media.[41] The double knockout mouse (*Fbln2*$^{-/-}$ *Fbln5*$^{-/-}$) additionally has a disrupted IEL in the thoracic aorta that is not observed in either single knockout mouse alone, indicating a role for fibulin-2 in IEL formation that is only evident when fibulin-5 is not expressed.[42] These results suggest that not all elastic laminae are identical, with unique cellular and protein contributions to the IEL in particular.

A genetic mutation in fibulin-4 associated with autosomal recessive cutis laxa type 1B causes disrupted elastic laminae in the large, elastic arteries, but the elastic laminae in the small, muscular arteries are unaffected in genetically engineered fibulin-4 mutant (*Fbln4*$^{E57K/E57K}$) mice.[43] The absence of fibulin-4 expression (*Fbln4*$^{-/-}$) in newborn mice causes disrupted elastic laminae in the ascending and descending aorta, but only the ascending aorta dilates to form an aneurysm[44] (Fig. 7.4A and B). Gene expression analyses show the highest number of differentially regulated genes in *Fbln4*$^{-/-}$ compared to *Fbln4*$^{+/+}$ ascending aorta, but the number is only about double the number of differentially regulated genes between ascending and descending aorta in control (*Fbln4*$^{+/+}$) mice (Fig. 7.4C), highlighting the importance of axial location in vascular phenotype.

Proteoglycans and glycosaminoglycans

Proteoglycans are macromolecules with a common domain structure consisting of a core protein with one

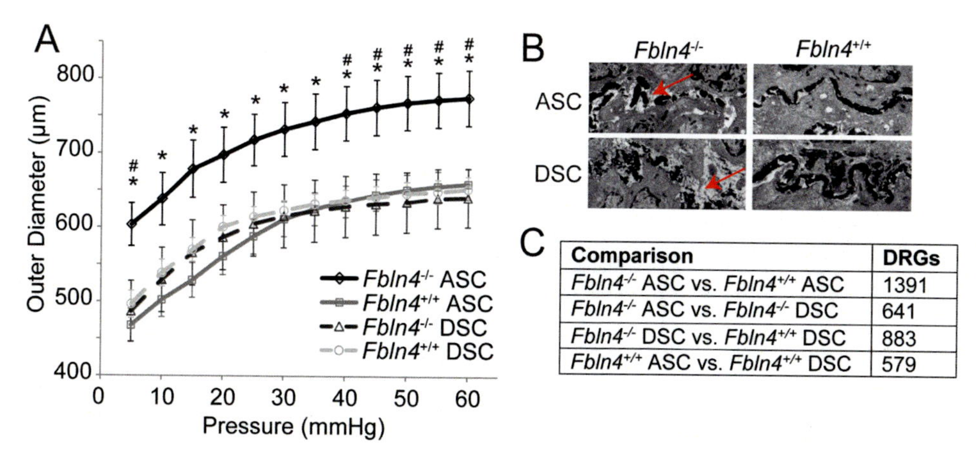

Comparison	DRGs
Fbln4-/- ASC vs. *Fbln4+/+* ASC	1391
Fbln4-/- ASC vs. *Fbln4-/-* DSC	641
Fbln4-/- DSC vs. *Fbln4+/+* DSC	883
Fbln4+/+ ASC vs. *Fbln4+/+* DSC	579

FIGURE 7.4 The absence of fibulin-4 has different effects on the ascending (ASC) and descending (DSC) aorta in newborn mice. *Fbln4*$^{-/-}$ mice develop aneurysms in the ASC only, as shown by the consistently larger diameter of *Fbln4*$^{-/-}$ ASC compared to *Fbln4*$^{+/+}$ ASC during pressure inflation experiments (A). *Fbln4*$^{-/-}$ and *Fbln4*$^{+/+}$ DSC pressure–diameter behaviors are indistinguishable from each other. However, both *Fbln4*$^{-/-}$ ASC and DSC have reduced amounts of elastin between SMC layers (*arrows*) as shown by electron microscopy (B). Differentially regulated genes (DRGs) were compared between each group (C). The aneurysmal *Fbln4*$^{-/-}$ ASC has the most DRGs compared to control *Fbln4*$^{+/+}$ ASC, but there are also a high number of DRGs between *Fbln4*$^{-/-}$ ASC and DSC and between control *Fbln4*$^{+/+}$ ASC and DSC, indicating large variations in gene expression and response to the absence of fibulin-4 due to vascular location alone. *Adapted from Kim J, Procknow JD, Yanagisawa H, Wagenseil JE. Differences in genetic signaling, and not mechanical properties of the wall, are linked to ascending aortic aneurysms in fibulin-4 knockout mice.* Am J Physiol Heart Circ Physiol *July 01, 2015;309(1):H103-H113. https://doi.org/10.1152/ajpheart.00178.2015.*

or more attached GAGs: chondroitin sulfate (CS), dermatan sulfate (DS), heparan sulfate (HS), or keratan sulfate.[45] GAGs only make up 2%–5% of the arterial dry weight,[46] but are important for vessel mechanical behavior.[47–49] Hyaluronic acid (HA) (also called hyaluronan) is a unique GAG that is not sulfated and is not attached to a core protein to form a proteoglycan. The proportions of different GAGs in the intima-media of the thoracic aorta vary by species, with human aorta having 41% CS, 31% HS, 23% HA, and 5% DS.[50] In contrast, the most abundant GAG in human veins is DS.[51]

Versican is a CS proteoglycan that has four splice variants providing differently sized molecules. Versican is present in the intima and media of the human fetal aorta, with little to no signal in the adventitia.[52] In contrast, all three layers of adult human aorta stain positively for versican.[53] The entire wall of adult human veins also stains positively for versican, but staining is restricted to the adventitia in small muscular arteries. Versican colocalizes with elastin[53] and interacts with elastic fiber proteins such as fibulins and fibrillin microfibrils.[54] Aggrecan is another CS proteoglycan that is typically present in small amounts in the human aorta. Aggrecan and versican are both HA-binding aggregating proteoglycans and are highly upregulated in human and mouse thoracic aortic aneurysms.[55] Differential aggrecan expression and breakdown may play a role in closure of the umbilical artery after birth, while the umbilical vein remains patent.[56]

Biglycan and decorin are small leucine-rich CS proteoglycans. Biglycan localizes with elastin and around the pericellular matrix, while decorin is enriched in the adventitial layer and localizes with collagen in the arterial wall.[45] Cultured bovine aortic SMCs express biglycan and decorin, while ECs express only biglycan.[57] ECs switch to decorin synthesis when stimulated to sprout or form tubes in vitro, indicating a pivotal role for decorin in angiogenesis.[58] There is differential availability/processing of the decorin precursor in arterial compared to venous SMCs.[59] Both biglycan and decorin modulate collagen fibrillogenesis.[60] Biglycan also regulates SMC proliferation and migration.[61] Differential distribution of biglycan and decorin depending on cell type, cell activity, and vascular location indicates myriad pathways through which vascular structure and function can be modulated through ECM organization.

HSPGs, including perlecan, agrin, and collagen XVIII (which has collagen and proteoglycan structural features), are abundant in the basement membrane.[4] Perlecan cross-links the collagen IV and laminin network.[4] Perlecan regulates cell adhesion, proliferation, and growth factor signaling and has different biochemical structure and bioactivity when secreted by human coronary SMCs compared to ECs.[62] Agrin is present in human aortic endothelium[63] and strengthens laminin anchorage in the basement membrane.[64] C-terminal fragments of collagen XVIII (endostatin) and perlecan (endorepellin) have antiangiogenic activity.[65]

HSPGs are the most common proteoglycans in the endothelial glycocalyx (50%–90%), with CS and DS being the next most common, and HA also playing a significant role.[66] The glycocalyx contains the HSPGs syndecan, glypican, and perlecan, the CSPGs versican, decorin, and biglycan, and the KSPG osteoglycan (also known as mimecan). Syndecans are type I transmembrane proteins that are a major component of the glycocalyx[67] and are expressed by all vascular cells.[68] Syndecans are involved in cell–cell and cell–matrix interactions that often work in concert with other cell surface receptors to modulate migration, proliferation, and differentiation. Glypicans are also transmembrane proteins and, like syndecans, interact with a multitude of ECM proteins, chemokines, and their cellular receptors.[69] Osteoglycan is present in the glycocalyx and is also expressed by aortic SMCs, inhibiting SMC proliferation.[70]

KSPGs in the vascular wall include lumican and fibromodulin.[45] Lumican and fibromodulin bind collagen I in vitro[71] and regulate TGF-beta signaling.[72] Lumican localizes to the outer media and adventitia of human coronary artery.[73] Fibromodulin binds directly to C1q and activates the classical complement pathway.[74]

Other glycoproteins

Laminins are a family of high molecular weight glycoproteins found in the basement membrane.[4] The laminin network is assembled first, then triggers deposition of nidogen and perlecan, which participate in cross-linking of the collagen IV network to form the core basement membrane structure. Laminins are classified according to their composition of heterotrimer chains and their vascular specificity varies by vessel type, location, and developmental stage.[75] Vascular ECs express laminin isoforms 411 and 511. In mice, laminin 411 is found in the endothelial basement membrane of different vessel types throughout development, while laminin 511 in embryonic and postnatal tissue is found only in the basement membrane of large vessels and is not present in capillaries until 3–4 weeks of age.[76] In VSM, laminins 411, 511, and in some cases, laminin 211 appear during development, with laminins 421, 521, and possibly 221 appearing at maturity.[75]

Fibronectin is incorporated between ECs and perivascular cells and is essential for blood vessel morphogenesis.[77] In the absence of fibronectin expression, the specification and localization of vascular ECs appears normal, but their interaction with surrounding mesenchyme is disturbed leading to severe vascular malformations and embryonic lethality.[78] Many vascular ECM proteins depend on fibronectin for assembly either through direct association with fibronectin fibrils or by using fibronectin as a provisional scaffold.[79] Fibronectin is required for fibrillin microfibril[80,81] and fibrillar collagen-1 deposition.[82] Fibronectin exists in two forms: plasma fibronectin and cellular fibronectin. In mice, the postnatal loss of cellular fibronectin expression affects collagen deposition but does not affect lifespan.[83] Surprisingly, plasma fibronectin was transported into the aortic wall in cellular fibronectin knockout mice, compensating for the loss of fibronectin in the wall. When both cellular and plasma fibronectin were eliminated postnatally, the mice died within 2 weeks with severe hemorrhage in the vascular cavity. These results demonstrate how circulating proteins can be incorporated into the vascular ECM.

The ECM and its interactions with cells

Substrate for cells and modulator of cell phenotype

Cells within the vasculome rely on the ECM as a substrate during development and while maintaining homeostasis in the adult. ECs control vascular tone, permeability, and modulate immune responses. SMCs provide contraction/relaxation capabilities to modulate vessel diameter in response to physiologic demand and hemodynamic changes. Both EC and SMC functions require strong, yet nuanced adherence to the ECM. During development, the endothelial basement membrane assembles two independent scaffolds of laminins and type IV collagen. Laminin 411 and 511 are present to anchor the ECs to the basement membrane. Development of the basement membrane is possible without collagen IV, yet full vessel formation is dependent on collagen IV to maintain wall stability.[21] SMCs similarly depend on collagen VI as an anchoring meshwork[25] and contract by pulling on the surrounding ECM substrate. Cultured SMCs lose their contractility unless supplemented with serum-derived ECM factors.[84]

Cell phenotype can be influenced by ECM components, as mentioned previously, and collagen VIII is one example.[23,24] The basement membrane proteins laminin and collagen IV promote a differentiated and contractile SMC phenotype. In vascular injury, fibronectin is exposed at the basement membrane, which promotes a proliferative and secretory SMC phenotype.[85] Reduced SMC contractility is associated with fragmented elastic fibers due to the absence of fibulin-4[86] and with alterations in radial aggrecan distribution.[56] This "dedifferentiation" of SMCs is the opposite of the developmental differentiation process and likely plays a role in vascular disease.[87]

Control transport and availability of signaling molecules

Transport of soluble extracellular molecules including proteins, small peptides, steroids, or dissolved gases is dependent on properties of the ECM. A main mechanism of this transport is bound water pathways, which result from the large water uptake ability of HA and proteoglycans. These ECM reservoirs can also serve as a local repository for growth factors and signaling molecules utilized as a mechanism for angiogenesis[88] and vessel remodeling.[89] Changes in permeability, specifically in the EC layer and the supporting ECM, allow for modulation of function, such as blood–tissue exchanges, or disease response. ECs utilize the ECM to regulate permeability by controlling adhesion. In large arteries, fenestrations in the elastic laminae control transport of water, nutrients, and electrolytes.[5,90] This is especially important in the medial layer of a large vessel, as these cells are relatively isolated from soluble factors compared to intimal and adventitial cells.

Dynamics of cell adhesion can be controlled through integrins and cell signaling. Outside-in signals, or ligand binding signals transduced from the ECM to cell cytoplasm, are equally important to cell adhesion as inside-out signaling.[91] Through integrin and ligand interactions, cell migration is made possible along the ECM scaffold and utilized in angiogenesis by ECs. Adhesion between cells and the ECM can be linked to cytoskeletal tension within a cell and can direct cell migration or alter cell shape.[92] For both ECs and SMCs, changes in integrin and ligand interactions modulate the immune response to disease or injury when the ECM content is altered.

Cleavage of ECM proteins by enzymes or proteases can also produce signaling molecules through the ECM or bloodstream. Collagen and elastin peptides liberated upon proteolytic degradation can affect cellular proliferation, migration, and apoptosis.[93,94] Vascular permeability and inflammation can be controlled by cleaved proteoglycans, such as from endothelial glycocalyx shedding.[95,96]

Cell-to-cell communication occurs by transport of signaling molecules, through gap junctions, or mechanotransduction pathways. Signaling pathways are constructed within and between layers within the vessel walls; for example, myoendothelial gap junctions allow direct communication of ECs to SMCs.[6] Cells can additionally use the ECM protein scaffold to send mechanical signals to neighboring cells.[97] Fluid shear stress[98] and transmural pressure or deformation[99] can influence cell behavior and dictate production and organization of ECM proteins. In turn, ECM rigidity can regulate both integrin signaling and cell response.[100]

Conclusions

The relative amounts of ECM components and cell types in the vessel wall vary with location within the vasculome and are tailored toward the specific structure and function of each vessel. There is exquisite control of ECM content with axial and radial position in the vasculome so that individual types and organization of collagen, elastin, proteoglycans, and glycoproteins vary according to the required vascular structure and function at multiple length scales, from interactions with individual cells to coordinated control of vessel lumen size. Each cell within the vessel wall is influenced by its local ECM environment and the resulting differences in physical support and biological signaling properties.

References

1. Jakab M, Augustin HG. Understanding angiodiversity: insights from single cell biology. *Development*. August 12, 2020;(15):147. https://doi.org/10.1242/dev.146621.
2. Drake CJ. Embryonic and adult vasculogenesis. *Birth Defects Res C Embryo Today*. February 2003;69(1):73–82. https://doi.org/10.1002/bdrc.10003.
3. Cosgun ZC, Fels B, Kusche-Vihrog K. Nanomechanics of the endothelial glycocalyx: from structure to function. *Am J Pathol*. April 2020;190(4):732–741. https://doi.org/10.1016/j.ajpath.2019.07.021.
4. Marchand M, Monnot C, Muller L, Germain S. Extracellular matrix scaffolding in angiogenesis and capillary homeostasis. *Semin Cell Dev Biol*. May 2019;89:147–156. https://doi.org/10.1016/j.semcdb.2018.08.007.
5. Penn MS, Saidel GM, Chisolm GM. Relative significance of endothelium and internal elastic lamina in regulating the entry of macromolecules into arteries in vivo. *Circ Res*. January 1994;74(1):74–82.
6. Heberlein KR, Straub AC, Isakson BE. The myoendothelial junction: breaking through the matrix? *Microcirculation*. May 2009;16(4):307–322. https://doi.org/10.1080/10739680902744404.
7. Song SH, Roach MR. Comparison of fenestrations in internal elastic laminae of canine thoracic and abdominal aortas. *Blood Vessels*. 1984;21(2):90–97.
8. Megens RT, Reitsma S, Schiffers PH, et al. Two-photon microscopy of vital murine elastic and muscular arteries. Combined structural and functional imaging with subcellular resolution. *J Vasc Res*. 2007;44(2):87–98. https://doi.org/10.1159/000098259.
9. Sandow SL, Gzik DJ, Lee RM. Arterial internal elastic lamina holes: relationship to function? *J Anat*. February 2009;214(2):258–266. https://doi.org/10.1111/j.1469-7580.2008.01020.x.
10. Wolinsky H, Glagov S. A lamellar unit of aortic medial structure and function in mammals. *Circ Res*. January 1967;20(1):99–111.
11. Aguilera CM, George SJ, Johnson JL, Newby AC. Relationship between type IV collagen degradation, metalloproteinase activity and smooth muscle cell migration and proliferation in cultured human saphenous vein. *Cardiovasc Res*. June 1, 2003;58(3):679–688. https://doi.org/10.1016/s0008-6363(03)00256-6.
12. Stenmark KR, Yeager ME, El Kasmi KC, et al. The adventitia: essential regulator of vascular wall structure and function. *Annu Rev Physiol*. 2013;75:23–47. https://doi.org/10.1146/annurev-physiol-030212-183802.
13. Sokolis DP. Passive mechanical properties and structure of the aorta: segmental analysis. *Acta Physiol*. August 2007;190(4):277–289. https://doi.org/10.1111/j.1748-1716.2006.01661.x.
14. Wolinsky H, Glagov S. Nature of species differences in the medial distribution of aortic vasa vasorum in mammals. *Circ Res*. April 1967;20(4):409–421. https://doi.org/10.1161/01.res.20.4.409.

15. Tinajero MG, Gotlieb AI. Recent developments in vascular adventitial pathobiology: the dynamic adventitia as a complex regulator of vascular disease. *Am J Pathol*. March 2020;190(3):520–534. https://doi.org/10.1016/j.ajpath.2019.10.021.
16. Moireau P, Xiao N, Astorino M, et al. External tissue support and fluid-structure simulation in blood flows. *Biomech Model Mechanobiol*. January 2012;11(1–2):1–18. https://doi.org/10.1007/s10237-011-0289-z.
17. Dobrin PB. Chapter 3: physiology and pathophysiology of blood vessels. In: Sidawy AN SB, DePalma RG, eds. *The Basic Science of Vascular Disease*. Futura Publishing; 1997:69–105.
18. Wolinsky H, Glagov S. Structural basis for the static mechanical properties of the aortic media. *Circ Res*. May 1964;14:400–413.
19. Ricard-Blum S. The collagen family. *Cold Spring Harb Perspect Biol*. January 1, 2011;3(1):a004978. https://doi.org/10.1101/cshperspect.a004978.
20. Bella J, Liu J, Kramer R, Brodsky B, Berman HM. Conformational effects of Gly-X-Gly interruptions in the collagen triple helix. *J Mol Biol*. September 15, 2006;362(2):298–311. https://doi.org/10.1016/j.jmb.2006.07.014.
21. Poschl E, Schlotzer-Schrehardt U, Brachvogel B, Saito K, Ninomiya Y, Mayer U. Collagen IV is essential for basement membrane stability but dispensable for initiation of its assembly during early development. *Development*. April 2004;131(7): 1619–1628. https://doi.org/10.1242/dev.01037.
22. Utriainen A, Sormunen R, Kettunen M, et al. Structurally altered basement membranes and hydrocephalus in a type XVIII collagen deficient mouse line. *Hum Mol Genet*. September 15, 2004;13(18): 2089–2099. https://doi.org/10.1093/hmg/ddh213.
23. Sage H, Iruela-Arispe ML. Type VIII collagen in murine development. Association with capillary formation in vitro. *Ann N Y Acad Sci*. 1990;580:17–31. https://doi.org/10.1111/j.1749-6632.1990.tb17914.x.
24. Hou G, Mulholland D, Gronska MA, Bendeck MP. Type VIII collagen stimulates smooth muscle cell migration and matrix metalloproteinase synthesis after arterial injury. *Am J Pathol*. February 2000;156(2):467–476. https://doi.org/10.1016/S0002-9440(10)64751-7.
25. Kuo HJ, Maslen CL, Keene DR, Glanville RW. Type VI collagen anchors endothelial basement membranes by interacting with type IV collagen. *J Biol Chem*. October 17, 1997;272(42): 26522–26529. https://doi.org/10.1074/jbc.272.42.26522.
26. Bonaldo P, Russo V, Bucciotti F, Doliana R, Colombatti A. Structural and functional features of the alpha 3 chain indicate a bridging role for chicken collagen VI in connective tissues. *Biochemistry*. February 6, 1990;29(5):1245–1254. https://doi.org/10.1021/bi00457a021.
27. Groulx JF, Gagne D, Benoit YD, Martel D, Basora N, Beaulieu JF. Collagen VI is a basement membrane component that regulates epithelial cell-fibronectin interactions. *Matrix Biol*. April 2011; 30(3):195–206. https://doi.org/10.1016/j.matbio.2011.03.002.
28. Kelleher CM, McLean SE, Mecham RP. Vascular extracellular matrix and aortic development. *Curr Top Dev Biol*. 2004;62:153–188.
29. Mak KM, Png CY, Lee DJ. Type V collagen in health, disease, and fibrosis. *Anat Rec*. May 2016;299(5):613–629. https://doi.org/10.1002/ar.23330.
30. Howard PS, Macarak EJ. Localization of collagen types in regional segments of the fetal bovine aorta. *Lab Invest*. November 1989; 61(5):548–555.
31. Bartholomew JS, Anderson JC. Investigation of relationships between collagens, elastin and proteoglycans in bovine thoracic aorta by immunofluorescence techniques. *Histochem J*. December 1983;15(12):1177–1190.
32. Kielty CM. Elastic fibres in health and disease. *Expet Rev Mol Med*. August 8, 2006;8(19):1–23. https://doi.org/10.1017/S146239940600007X.
33. Wolinsky H. Comparison of medial growth of human thoracic and abdominal aortas. *Circ Res*. October 1970;27(4):531–538.
34. Apter JT. Correlation of visco-elastic properties of large arteries with micorscopic structure. *Circ Res*. 1966;XIX.
35. Harkness ML, Harkness RD, McDonald DA. The collagen and elastin content of the arterial wall in the dog. *Proc R Soc Lond B Biol Sci*. June 25, 1957;146(925):541–551. https://doi.org/10.1098/rspb.1957.0029.
36. Looker T, Berry CL. The growth and development of the rat aorta. II. Changes in nucleic acid and scleroprotein content. *J Anat*. October 1972;113(Pt 1):17–34.
37. Chen H, Slipchenko MN, Liu Y, et al. Biaxial deformation of collagen and elastin fibers in coronary adventitia. *J Appl Physiol*. December 2013;115(11):1683–1693. https://doi.org/10.1152/japplphysiol.00601.2013.
38. White JV, Haas K, Phillips S, Comerota AJ. Adventitial elastolysis is a primary event in aneurysm formation. *J Vasc Surg*. February 1993;17(2):371–380. https://doi.org/10.1067/mva.1993.43023. Discussion 380-1.
39. Lin CJ, Staiculescu MC, Hawes JZ, et al. Heterogeneous cellular contributions to elastic laminae formation in arterial wall development. *Circ Res*. November 8, 2019;125(11):1006–1018. https://doi.org/10.1161/CIRCRESAHA.119.315348.
40. Sicot FX, Tsuda T, Markova D, et al. Fibulin-2 is dispensable for mouse development and elastic fiber formation. *Mol Cell Biol*. February 2008;28(3):1061–1067. https://doi.org/10.1128/MCB.01876-07.
41. Yanagisawa H, Davis EC, Starcher BC, et al. Fibulin-5 is an elastin-binding protein essential for elastic fibre development in vivo. *Nature*. January 10, 2002;415(6868):168–171.
42. Chapman SL, Sicot FX, Davis EC, et al. Fibulin-2 and fibulin-5 cooperatively function to form the internal elastic lamina and protect from vascular injury. *Arterioscler Thromb Vasc Biol*. January 2010; 30(1):68–74. https://doi.org/10.1161/ATVBAHA.109.196725.
43. Halabi CM, Broekelmann TJ, Lin M, Lee VS, Chu ML, Mecham RP. Fibulin-4 is essential for maintaining arterial wall integrity in conduit but not muscular arteries. *Sci Adv*. May 2017;3(5): e1602532. https://doi.org/10.1126/sciadv.1602532.
44. Kim J, Procknow JD, Yanagisawa H, Wagenseil JE. Differences in genetic signaling, and not mechanical properties of the wall, are linked to ascending aortic aneurysms in fibulin-4 knockout mice. *Am J Physiol Heart Circ Physiol*. July 01, 2015;309(1): H103–H113. https://doi.org/10.1152/ajpheart.00178.2015.
45. Wight TN. A role for proteoglycans in vascular disease. *Matrix Biol*. October 2018;71–72:396–420. https://doi.org/10.1016/j.matbio.2018.02.019.
46. Wight TN. Vessel proteoglycans and thrombogenesis. *Prog Hemost Thromb*. 1980;5:1–39.
47. Mattson JM, Turcotte R, Zhang Y. Glycosaminoglycans contribute to extracellular matrix fiber recruitment and arterial wall mechanics. *Biomech Model Mechanobiol*. February 2017;16(1): 213–225. https://doi.org/10.1007/s10237-016-0811-4.
48. Azeloglu EU, Albro MB, Thimmappa VA, Ateshian GA, Costa KD. Heterogeneous transmural proteoglycan distribution provides a mechanism for regulating residual stresses in the aorta. *Am J Physiol Heart Circ Physiol*. March 2008;294(3): H1197–H1205. https://doi.org/10.1152/ajpheart.01027.2007.
49. Roccabianca S, Bellini C, Humphrey JD. Computational modelling suggests good, bad and ugly roles of glycosaminoglycans in arterial wall mechanics and mechanobiology. *J R Soc Interface*. August 6, 2014;11(97):20140397. https://doi.org/10.1098/rsif.2014.0397.
50. Aikawa J, Munakata H, Isemura M, Ototani N, Yosizawa Z. Comparison of glycosaminoglycans from thoracic aortas of several mammals. *Tohoku J Exp Med*. May 1984;143(1):107–112. https://doi.org/10.1620/tjem.143.107.

51. Leta GC, Mourao PA, Tovar AM. Human venous and arterial glycosaminoglycans have similar affinity for plasma low-density lipoproteins. *Biochim Biophys Acta*. April 24, 2002;1586(3): 243–253. https://doi.org/10.1016/s0925-4439(01)00102-8.
52. Yao LY, Moody C, Schonherr E, Wight TN, Sandell LJ. Identification of the proteoglycan versican in aorta and smooth muscle cells by DNA sequence analysis, in situ hybridization and immunohistochemistry. *Matrix Biol*. April 1994;14(3):213–225. https://doi.org/10.1016/0945-053x(94)90185-6.
53. Bode-Lesniewska B, Dours-Zimmermann MT, Odermatt BF, Briner J, Heitz PU, Zimmermann DR. Distribution of the large aggregating proteoglycan versican in adult human tissues. *J Histochem Cytochem*. April 1996;44(4):303–312. https://doi.org/10.1177/44.4.8601689.
54. Isogai Z, Aspberg A, Keene DR, Ono RN, Reinhardt DP, Sakai LY. Versican interacts with fibrillin-1 and links extracellular microfibrils to other connective tissue networks. *J Biol Chem*. February 8, 2002; 277(6):4565–4572. https://doi.org/10.1074/jbc.M110583200.
55. Cikach FS, Koch CD, Mead TJ, et al. Massive aggrecan and versican accumulation in thoracic aortic aneurysm and dissection. *JCI Insight*. March 8, 2018;3(5). https://doi.org/10.1172/jci.insight.97167.
56. Nandadasa S, Szafron JM, Pathak V, et al. Vascular dimorphism ensured by regulated proteoglycan dynamics favors rapid umbilical artery closure at birth. *eLife*. Sep 10, 2020;9:e60683. https://doi.org/10.7554/eLife.60683 (accepted).
57. Jarvelainen HT, Kinsella MG, Wight TN, Sandell LJ. Differential expression of small chondroitin/dermatan sulfate proteoglycans, PG-I/biglycan and PG-II/decorin, by vascular smooth muscle and endothelial cells in culture. *J Biol Chem*. December 5, 1991; 266(34):23274–23281.
58. Jarvelainen HT, Iruela-Arispe ML, Kinsella MG, Sandell LJ, Sage EH, Wight TN. Expression of decorin by sprouting bovine aortic endothelial cells exhibiting angiogenesis in vitro. *Exp Cell Res*. December 1992;203(2):395–401. https://doi.org/10.1016/0014-4827(92)90013-x.
59. Franch R, Chiavegato A, Maraschin M, et al. Differential availability/processing of decorin precursor in arterial and venous smooth muscle cells. *J Anat*. September 2006;209(3):271–287. https://doi.org/10.1111/j.1469-7580.2006.00614.x.
60. Hocking AM, Shinomura T, McQuillan DJ. Leucine-rich repeat glycoproteins of the extracellular matrix. *Matrix Biol*. April 1998; 17(1):1–19. https://doi.org/10.1016/s0945-053x(98)90121-4.
61. Shimizu-Hirota R, Sasamura H, Kuroda M, Kobayashi E, Hayashi M, Saruta T. Extracellular matrix glycoprotein biglycan enhances vascular smooth muscle cell proliferation and migration. *Circ Res*. April 30, 2004;94(8):1067–1074.
62. Lord MS, Chuang CY, Melrose J, Davies MJ, Iozzo RV, Whitelock JM. The role of vascular-derived perlecan in modulating cell adhesion, proliferation and growth factor signaling. *Matrix Biol*. April 2014;35:112–122. https://doi.org/10.1016/j.matbio.2014.01.016.
63. Didangelos A, Yin X, Mandal K, Baumert M, Jahangiri M, Mayr M. Proteomics characterization of extracellular space components in the human aorta. *Mol Cell Proteomics*. September 2010;9(9):2048–2062. https://doi.org/10.1074/mcp.M110.001693.
64. Moll J, Barzaghi P, Lin S, et al. An agrin minigene rescues dystrophic symptoms in a mouse model for congenital muscular dystrophy. *Nature*. September 20, 2001;413(6853):302–307. https://doi.org/10.1038/35095054.
65. Iozzo RV. Basement membrane proteoglycans: from cellar to ceiling. *Nat Rev Mol Cell Biol*. August 2005;6(8):646–656. https://doi.org/10.1038/nrm1702.
66. Reitsma S, Slaaf DW, Vink H, van Zandvoort MA, oude Egbrink MG. The endothelial glycocalyx: composition, functions, and visualization. *Pflueg Arch Eur J Physiol*. June 2007;454(3): 345–359. https://doi.org/10.1007/s00424-007-0212-8.
67. Zeng Y. Endothelial glycocalyx as a critical signalling platform integrating the extracellular haemodynamic forces and chemical signalling. *J Cell Mol Med*. August 2017;21(8):1457–1462. https://doi.org/10.1111/jcmm.13081.
68. Alexopoulou AN, Multhaupt HA, Couchman JR. Syndecans in wound healing, inflammation and vascular biology. *Int J Biochem Cell Biol*. 2007;39(3):505–528. https://doi.org/10.1016/j.biocel.2006.10.014.
69. Iozzo RV, Sanderson RD. Proteoglycans in cancer biology, tumour microenvironment and angiogenesis. *J Cell Mol Med*. May 2011;15(5):1013–1031. https://doi.org/10.1111/j.1582-4934.2010.01236.x.
70. Gu XS, Lei JP, Shi JB, et al. Mimecan is involved in aortic hypertrophy induced by sinoaortic denervation in rats. *Mol Cell Biochem*. June 2011;352(1–2):309–316. https://doi.org/10.1007/s11010-011-0767-8.
71. Kalamajski S, Oldberg A. Homologous sequence in lumican and fibromodulin leucine-rich repeat 5-7 competes for collagen binding. *J Biol Chem*. January 2, 2009;284(1):534–539. https://doi.org/10.1074/jbc.M805721200.
72. Honardoust D, Eslami A, Larjava H, Hakkinen L. Localization of small leucine-rich proteoglycans and transforming growth factor-beta in human oral mucosal wound healing. *Wound Repair Regen*. November-December 2008;16(6):814–823. https://doi.org/10.1111/j.1524-475X.2008.00435.x.
73. Onda M, Ishiwata T, Kawahara K, Wang R, Naito Z, Sugisaki Y. Expression of lumican in thickened intima and smooth muscle cells in human coronary atherosclerosis. *Exp Mol Pathol*. April 2002;72(2):142–149. https://doi.org/10.1006/exmp.2002.2425.
74. Sjoberg A, Onnerfjord P, Morgelin M, Heinegard D, Blom AM. The extracellular matrix and inflammation: fibromodulin activates the classical pathway of complement by directly binding C1q. *J Biol Chem*. September 16, 2005;280(37):32301–32308. https://doi.org/10.1074/jbc.M504828200.
75. Yousif LF, Di Russo J, Sorokin L. Laminin isoforms in endothelial and perivascular basement membranes. *Cell Adh Migr*. January-February 2013;7(1):101–110. https://doi.org/10.4161/cam.22680.
76. Sorokin LM, Pausch F, Frieser M, Kroger S, Ohage E, Deutzmann R. Developmental regulation of the laminin alpha5 chain suggests a role in epithelial and endothelial cell maturation. *Dev Biol*. September 15, 1997;189(2):285–300. https://doi.org/10.1006/dbio.1997.8668.
77. Astrof S, Hynes RO. Fibronectins in vascular morphogenesis. *Angiogenesis*. 2009;12(2):165–175. https://doi.org/10.1007/s10456-009-9136-6.
78. George EL, Baldwin HS, Hynes RO. Fibronectins are essential for heart and blood vessel morphogenesis but are dispensable for initial specification of precursor cells. *Blood*. October 15, 1997; 90(8):3073–3081.
79. Singh P, Carraher C, Schwarzbauer JE. Assembly of fibronectin extracellular matrix. *Annu Rev Cell Dev Biol*. 2010;26:397–419. https://doi.org/10.1146/annurev-cellbio-100109-104020.
80. Kinsey R, Williamson MR, Chaudhry S, et al. Fibrillin-1 microfibril deposition is dependent on fibronectin assembly. *J Cell Sci*. August 15, 2008;121(Pt 16):2696–2704. https://doi.org/10.1242/jcs.029819.
81. Sabatier L, Chen D, Fagotto-Kaufmann C, et al. Fibrillin assembly requires fibronectin. *Mol Biol Cell*. February 2009;20(3):846–858. https://doi.org/10.1091/mbc.E08-08-0830.

82. Sottile J, Hocking DC. Fibronectin polymerization regulates the composition and stability of extracellular matrix fibrils and cell-matrix adhesions. *Mol Biol Cell*. October 2002;13(10):3546–3559. https://doi.org/10.1091/mbc.e02-01-0048.
83. Kumra H, Sabatier L, Hassan A, et al. Roles of fibronectin isoforms in neonatal vascular development and matrix integrity. *PLoS Biol*. July 2018;16(7):e2004812. https://doi.org/10.1371/journal.pbio.2004812.
84. Birukov KG, Frid MG, Rogers JD, et al. Synthesis and expression of smooth muscle phenotype markers in primary culture of rabbit aortic smooth muscle cells: influence of seeding density and media and relation to cell contractility. *Exp Cell Res*. January 1993;204(1):46–53. https://doi.org/10.1006/excr.1993.1007.
85. Thyberg J, Blomgren K, Roy J, Tran PK, Hedin U. Phenotypic modulation of smooth muscle cells after arterial injury is associated with changes in the distribution of laminin and fibronectin. *J Histochem Cytochem*. June 1997;45(6):837–846. https://doi.org/10.1177/002215549704500608.
86. Huang J, Davis EC, Chapman SL, et al. Fibulin-4 deficiency results in ascending aortic aneurysms: a potential link between abnormal smooth muscle cell phenotype and aneurysm progression. *Circ Res*. February 19, 2010;106(3):583–592. https://doi.org/10.1161/CIRCRESAHA.109.207852.
87. Raines EW. The extracellular matrix can regulate vascular cell migration, proliferation, and survival: relationships to vascular disease. *Int J Exp Pathol*. June 2000;81(3):173–182. https://doi.org/10.1046/j.1365-2613.2000.00155.x.
88. Ruhrberg C, Gerhardt H, Golding M, et al. Spatially restricted patterning cues provided by heparin-binding VEGF-A control blood vessel branching morphogenesis. *Genes Dev*. October 15, 2002;16(20):2684–2698. https://doi.org/10.1101/gad.242002.
89. Vlodavsky I, Folkman J, Sullivan R, et al. Endothelial cell-derived basic fibroblast growth factor: synthesis and deposition into subendothelial extracellular matrix. *Proc Natl Acad Sci U S A*. April 1987;84(8):2292–2296. https://doi.org/10.1073/pnas.84.8.2292.
90. Tada S, Tarbell JM. Internal elastic lamina affects the distribution of macromolecules in the arterial wall: a computational study. *Am J Physiol Heart Circ Physiol*. August 2004;287(2):H905–H913. https://doi.org/10.1152/ajpheart.00647.2003.
91. Luo BH, Carman CV, Springer TA. Structural basis of integrin regulation and signaling. *Annu Rev Immunol*. 2007;25:619–647. https://doi.org/10.1146/annurev.immunol.25.022106.141618.
92. Parker KK, Brock AL, Brangwynne C, et al. Directional control of lamellipodia extension by constraining cell shape and orienting cell tractional forces. *FASEB J*. August 2002;16(10):1195–1204. https://doi.org/10.1096/fj.02-0038com.
93. Ortega N, Werb Z. New functional roles for non-collagenous domains of basement membrane collagens. *J Cell Sci*. November 15, 2002;115(Pt 22):4201–4214. https://doi.org/10.1242/jcs.00106.
94. Heinz A. Elastases and elastokines: elastin degradation and its significance in health and disease. *Crit Rev Biochem Mol Biol*. June 2020;55(3):252–273. https://doi.org/10.1080/10409238.2020.1768208.
95. Rehm M, Bruegger D, Christ F, et al. Shedding of the endothelial glycocalyx in patients undergoing major vascular surgery with global and regional ischemia. *Circulation*. October 23, 2007;116(17):1896–1906. https://doi.org/10.1161/CIRCULATIONAHA.106.684852.
96. Yuan SY, Rigor RR. Regulation of Endothelial Barrier Function. *Colloquium Series on Integrated Systems Physiology: From Molecule to Function to Disease*. 2011;3(1):1–146. https://doi.org/10.4199/C00025ED1V01Y201101ISP013.
97. Martino F, Perestrelo AR, Vinarsky V, Pagliari S, Forte G. Cellular mechanotransduction: from tension to function. *Front Physiol*. 2018;9:824. https://doi.org/10.3389/fphys.2018.00824.
98. Shyy JY, Chien S. Role of integrins in endothelial mechanosensing of shear stress. *Circ Res*. November 1, 2002;91(9):769–775. https://doi.org/10.1161/01.res.0000038487.19924.18.
99. O'Callaghan CJ, Williams B. Mechanical strain-induced extracellular matrix production by human vascular smooth muscle cells: role of TGF-beta(1). *Hypertension*. September 2000;36(3):319–324. https://doi.org/10.1161/01.hyp.36.3.319.
100. Choquet D, Felsenfeld DP, Sheetz MP. Extracellular matrix rigidity causes strengthening of integrin-cytoskeleton linkages. *Cell*. January 10, 1997;88(1):39–48. https://doi.org/10.1016/s0092-8674(00)81856-5.
101. Burton AC. Relation of structure to function of the tissues of the wall of blood vessels. *Physiol Rev*. 1954;34:619–642.

SECTION 3

Putting it all together: vasculome integration across body scales and within tissues

CHAPTER

8

Out to the tissues: the arterial side (arteries, arterioles—development, structure, functions, and pathologies)

Jui M. Dave[1,2], *Junichi Saito*[1,2], *Giorgio Mottola*[1,2] *and Daniel M. Greif*[1,2]

[1]Yale Cardiovascular Research Center, Section of Cardiovascular Medicine, Department of Medicine, Yale University, New Haven, CT, United States [2]Department of Genetics, Yale University, New Haven, CT, United States

Introduction

The cardiovascular system, comprised of the heart and blood vessels, is the first organ system to function during development. The heart pumps oxygenated and nutrient-rich blood through a hierarchical network of arteries. The arterial network subsequently branches into microvessels known as capillaries that facilitate the exchange of nutrients and waste products as well as oxygen and carbon dioxide in every organ and tissue, and thus, supports the body's metabolic activity.

The anatomy of the arterial system resembles a ramified tree of vessels. The aorta is the largest artery of the body originating from the left ventricle of the heart and giving rise to major arteries that branch successively into smaller arterioles and then microscopic capillaries. Vessel caliber decreases and resistance to flow increases as blood flows from the heart to the arterioles. The high resistance and low flow rate in individual capillaries facilitate optimal transport across the capillary wall at the level of the tissue. Hence, the hierarchical arterial system is central to the regulation of arterial resistance and blood pressure and blood flow distribution and rate. Precise regulation of cardiovascular development and function is essential for physiological homeostasis, and dysregulation of molecular and cellular mechanisms is integral to cardiovascular pathologies. In this chapter, we discuss the organization, classification, and heterogeneity of the arterial system, structure of a typical arterial wall, and pathogenesis of select arterial diseases.

The arterial system

The aorta ascends from the left ventricle at the aortic root to the aortic arch and then descends along the spinal column. The descending aorta is subclassified into the thoracic and abdominal regions. Major arteries branch out from the aortic arch and descending aorta. Along the arterial tree, arteries are sequentially classified as elastic (large diameter, conduit), muscular (medium diameter, distributing), and arterioles (small diameter, branching to capillaries) (Fig. 8.1).

Elastic arteries (e.g., aorta and pulmonary artery) are highly extensible and resilient to variations in blood flow due to the pumping heart.[1] Muscular arteries receive blood from the elastic arteries, and through contraction and relaxation of their walls, control the distribution of blood to different regions of the body depending on metabolic demands.[1] Arterioles have relatively thin walls and small lumens and taken together are the primary regulators of vascular resistance and blood flow distribution to capillaries in specific tissue subcompartments.[1]

The arterial wall: development, structure, and function

A typical arterial wall is formed of three primary layers: the innermost *tunica intima*, the middle *tunica media*, and the outermost *tunica adventitia*.[2,3]

The Vasculome
https://doi.org/10.1016/B978-0-12-822546-2.00015-0

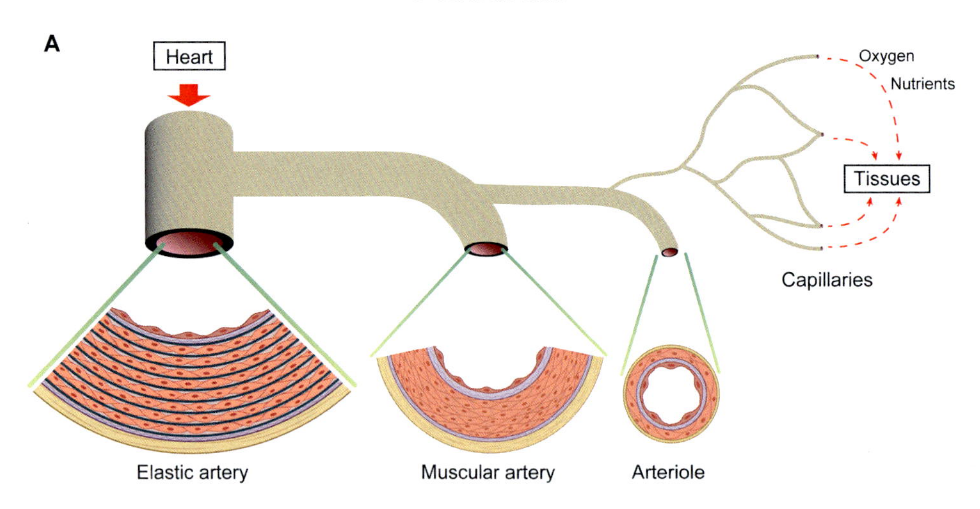

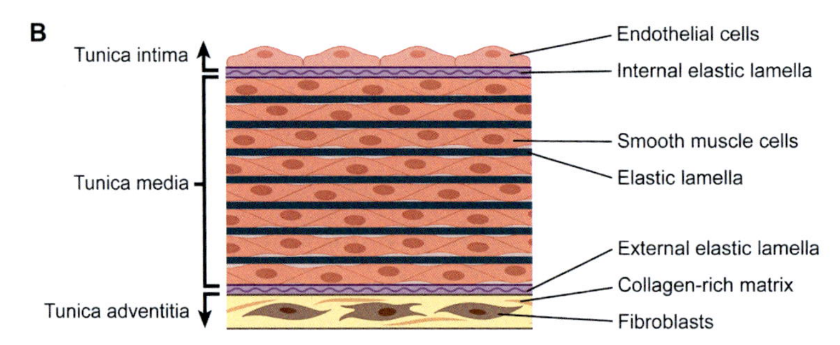

FIGURE 8.1 Schematic of the arterial system. (A) The vasculature branches along with the arterial tree. Oxygenated and nutrient-rich blood is pumped by the heart and transferred to the tissues through a series of arteries, arterioles, and capillaries. (B) Transverse section of an elastic artery highlighting major constituents of the arterial wall. *Graphical objects were created with BioRender (https://biorender.com/).*

Tunica intima

The *tunica intima* consists of a continuous monolayer of endothelial cells (ECs), adjacent basement membrane, intimal cells, subendothelial connective tissue, and internal elastic lamina (IEL). ECs line the entire lumen of the vasculature and are intimately involved in barrier function, coagulation, blood flow regulation, and inflammation.[4] ECs are the most widely investigated vascular cell type and studies utilizing specific cell markers (e.g., vascular endothelial-cadherin, Tie2, vascular endothelial growth factor receptor-2) in modern developmental and genetic approaches have greatly improved our understanding of EC origins and function. The basement membrane is a specialized sheet-like dense structure of extracellular matrix (ECM) which lines the abluminal side of ECs and provides structural support. Basement membrane is enriched in laminins, collagens, proteoglycans, and glycoproteins and is implicated in regulating blood vessel development and homeostasis.[5,6] A subendothelial layer of loose connective tissue rich in proteoglycans lies between the basement membrane and IEL of a human artery. Although little is known about its physiological role, this sublayer is more prominent in larger vessels and contains a heterogenous population of subendothelial intimal cells, including pericyte-like stellate cells, intimal fibroblasts, elongated smooth muscle cells (SMCs), and immune cells.[7,8] The origins and gene expression profiles of subendothelial intimal cells are not well described; however, these cells are suggested to be key players in arterial pathologies, such as atherosclerosis and stenosis.[9–11] The IEL is a fenestrated layer of elastic fibers which demarcates the outermost part of the *tunica intima* and forms a critical physical barrier between the intima and media. A recent study suggests that ECs are an important source of the IEL in muscular and resistance arteries.[12]

In the developing embryo, primitive ECs largely arise from lateral plate mesoderm by vasculogenesis and coalesce into early vascular tubes.[13,14] Thereafter, multiple nascent blood vessels arise from the preexisting early tubes through a dynamic multi-step process known as angiogenesis which entails EC proliferation, migration, invasion, lumenization, and stabilization.[15,16] Early ECs are further segregated into arterial and venous fates by NOTCH pathway mediated differential expression of *ephrinB2* and *ephrinB4 (EphB4)*, respectively.[17–19] Activation of NOTCH signaling pathway regulates early arterial fate by inducing expression of *sonic hedgehog* and *vascular endothelial growth factor (Vegf)* while downregulation of NOTCH signaling establishes venous identity in ECs by inducing expression of *EphB4* and *Flt4*.[20,21] The transcription factor, chicken ovalbumin upstream promoter-transcription factor II, positively regulates venous identity by deactivating NOTCH signaling.[22,23] Furthermore, during active angiogenesis, EC sprouts consists of the leading "tip" cells at the growing end of the vessel followed by "stalk" cells which form the vessel body.[24] Tip cells display long filopodia and sense angiogenic directional cues in the surrounding ECM through surface receptors that activate downstream signaling to facilitate migration in a particular direction. Stalk cells exhibit high proliferation to produce new ECs that are connected by tight junctions and form lumens.[25,26] This process is intricately coordinated by VEGF and NOTCH signaling pathways and gives rise to nascent vascular beds which require additional reinforcement to become mature vessels.[24,27,28]

The final step in vessel stabilisation occurs with recruitment of mural cells (SMCs in arteries and arterioles or pericytes in capillaries) which coat the neovessels and induce vessel quiescence by deactivating angiogenesis.[29] Major signaling pathways known to regulate EC-mural cell interactions include pathways involving: (1) platelet-derived growth factor B (PDGF-B), (2) NOTCH, (3) Angiopoietin-Tie2, (4) VEGF, (5) transforming growth factor (TGF)-β and (6) sphingosine 1-phosphate.[29–32,115] The uncoated vasculature is further refined by pruning and regression.[14] Once the mature vascular system is established, ECs remain mostly quiescent and are activated only during select physiological processes, such as wound healing and pregnancy, or pathological conditions.[16] Given the critical role of ECs in maintaining vascular homeostasis, it is not surprising that aberrant EC activation is a key feature of most cardiovascular diseases, including atherosclerosis, stroke, pulmonary hypertension (PH), and diabetic vasculopathy.[4]

Tunica media

The arterial *tunica media* primarily consists of multiple layers of circumferentially oriented SMCs and ECM, including collagen, elastin (ELN), and proteoglycans. The ECM provides structural support as well as elasticity to the blood vessel.[3] The developmental origins of vascular SMCs are diverse in different tissues and even in different regions of the same vessel (e.g., aorta).[33,34] For example, cardiac neural crest cells of the ectoderm give rise to SMCs of the aortic root and ascending aorta with extension into the arch just distal to the subclavian artery branch, and second heart field cells also contribute to root and ascending aorta SMCs with termination just proximal to the branch of the innominate artery.[34] SMCs of the inner medial layers of the ascending aorta are cardiac neural crest derived, and outer layer SMCs are second heart field derived.[34] Interestingly, aortic dissection often initiates at cell ontogeny boundaries.[35]

SMCs play diverse roles depending on their phenotype which have been broadly classified as contractile or synthetic.[36,37] Contractile SMCs have a long spindle shape with a single, elongated nucleus, while synthetic SMCs are less elongated and have a "cobblestone" morphology. The main function of contractile SMCs is to regulate vessel diameter and thus, blood pressure and tissue perfusion.[38] Activation of SMC contractility is primarily controlled by autonomic nerves that act on specific receptors expressed on the SMC surface. SMCs are connected to each other by tight and gap junctions, facilitating intercellular transport of signaling molecules and depolarizing currents and thereby coordinated control of vessel diameter. In addition, migration and proliferation, proinflammatory, and secretory responses of synthetic SMCs are important during vessel remodeling, injury, and disease.[37] Synthetic SMCs also secrete ECM and matrix metalloproteinases (MMPs) which facilitate ECM remodeling.

A highly organized ECM is critical for normal vascular development and function. The ECM shelters wall cells of large elastic arteries (e.g., aorta) from high wall stresses and is critical for the integrity of vessel structure.[39,40] SMCs and fibroblasts secrete ECM components and organize the ECM network in response to mechanical stimuli.[39,40] Transcriptomic analysis of the developing murine aorta shows that expression of most ECM components begins at embryonic day (E) 14, steadily increases until the early postnatal stage, and then gradual decreases throughout life.[41] Although the aortic ECM is comprised of many proteins, its mechanical properties primarily depend on two constituents: collagen and elastic fibers. Collagen fibers provide mechanical support with its material stiffness and strength and present integrin binding sites that facilitate cell adhesion.[42] Elastic fibers have a spring coil-like structure, dramatically extending under pressure and providing extensibility and resilience to the vascular wall.

The *tunica media* is separated from the intima and the adventitia by the IEL and external elastic laminae (EEL), respectively. Recent studies using *Eln* floxed mice show that IEL of most elastic arteries is produced by both SMCs and ECs, while IEL of muscular arteries is formed mainly by ECs.[12] SMC-specific deletion of *Eln* results in disrupted IEL in the ascending aorta and neointimal formation, suggesting that IEL plays a role as a barrier between the media and intima in large elastic arteries.[12] Although the role of EEL is not defined, this study demonstrates that SMCs are the major source of EEL in muscular arteries.[12]

Tunica adventitia

The adventitia, outermost layer of the vessels, is an assembly of loose ECM, vessels (*vasa vasorum*), autonomic nerves, and various cell types including fibroblasts, inflammatory cells, adipocytes, and progenitor cells.[43] The adventitia anchors the vessel to the surrounding tissue and provides mechanical strength to prevent vessel overexpansion. In contrast to the intimal–medial and medial–adventitial borders that are demarcated by elastic lamina, no distinctive anatomical structure marks the border between the adventitia and the surrounding connective tissues. Instead, the adventitia continues into a looser connective tissue rich in adipocytes. Fibroblasts in the adventitia play important roles during vascular remodeling via activation of NADPH oxidase.[44,45] During physiological wound healing and pathologies, such as atherosclerosis, fibroblasts are suggested to be an important source of differentiated myofibroblasts. Myofibroblast differentiation is characterized by TGFβ1-mediated de novo expression of SMA (α-smooth muscle actin).[46] Unlike rapid contraction of SMCs, myofibroblasts exhibit a long-lasting isometric tension resulting in a slow and irreversible retraction.[47]

Progenitor cells in the adventitia are heterogeneous in their potential for cell differentiation. Adventitial stem cell antigen-1 (AdvSca1)$^+$ cells can differentiate into macrophage-like cells or mural cells.[48–50] Indeed, it has recently been suggested that AdvSca1$^+$ cells are a back-up SMC source that are activated to contribute when preexisting SMCs are insufficient, such as following loss and/or impaired proliferation of local SMCs.[48] During such situations, AdvSca1$^+$ cells from the adventitia may act as facultative stem cells, migrating into the media and differentiating into SMCs for vascular repair and regeneration. In addition, the adventitia is contiguous—without a separating fascia layer—with the perivascular adipose tissue (PVAT). The PVAT supports the mechanical structure of underlying blood vessels and secretes a multitude of factors that regulate vascular homeostasis and the inflammatory state.[51–53] Changes in the profile of these mediators produced by the PVAT are observed in the context of obesity and atherosclerosis and implicated in modulating disease.[51–53]

Tissue-specific heterogeneity in the arterial system

Lung: The pulmonary arterial system supplies a high-density capillary network, which facilitates the uptake of oxygen and release of carbon dioxide by deoxygenated blood. Pulmonary arteries have a specialized mechanism regulating their tone known as hypoxic pulmonary vasoconstriction.[54,55] While systemic arteries typically dilate to hypoxia, pulmonary arteries constrict in response to alveolar hypoxia, diverting blood away from poorly ventilated regions of the lungs, and matching perfusion with ventilation. Hypoxia induces accumulation of potassium ions in pulmonary aortic SMCs through reduced redox balance in the mitochondria and closed Kv1.5 channels. This increase in potassium ions elevates the resting membrane potential and prompts calcium ion influx, resulting in SMC contraction.

Heart: The heart receives a rich blood supply from the coronary arteries. Systolic cardiac contraction strongly compresses the intramyocardial vessels so that ~85% of entire coronary blood flow occurs during diastole.[56] During increased oxygen demand, the coronary arteries dilate dramatically to allow for a pronounced increase of blood flow. This coronary dilation is caused by reduction of endothelin release from the coronary ECs and a sympathetically mediated β-receptor activation of coronary artery SMCs.[56,57]

Liver: The vasculature of the liver, spleen, pancreas, and abdominal gastrointestinal tract comprises the splanchnic circulation which receives ~25% of the cardiac output.[58] The afferent hepatic vasculature is unique in that it has both arterial and venous sources. The hepatic artery provides one-third of the blood supply to the liver with the remainder delivered by the portal vein which carries venous blood draining the stomach, small intestine, right half of the colon, spleen, and pancreas. Thus, the circulation of the liver is predominantly in series, not in parallel, with that of the digestive organs. This arrangement of the vasculature promotes hepatic uptake of nutrients and detoxification of foreign substances that have been absorbed during digestion.

Kidney: The renal arteries branch progressively to form afferent arterioles, leading to glomerular capillaries, where large amounts of fluid and solutes are filtered to initiate urine formation.[59] The distal ends of the capillaries of each glomerulus coalesce to form the efferent arteriole, which leads to a second capillary network, the peritubular capillaries surrounding the renal tubules. By adjusting the resistance of the afferent and efferent arterioles, kidneys regulate the hydrostatic pressure in both the glomerular and the peritubular capillary beds.[60] In addition, when arterial pressure decreases, renin is released from the juxtaglomerular cells, which are specialized arteriole SMCs. Renin catalyzes the conversion of angiotensinogen to angiotensin I, which is further processed to angiotensin II by angiotensin converting enzyme (ACE).

Arterial pathologies

Supravalvular aortic stenosis and Williams–Beuren syndrome

Supravalvular aortic stenosis (SVAS) is an arterial obstructive disorder affecting ~1:20,000 newborns and is caused by loss-of-function mutations in a single allele of the human *ELN* gene.[61,62] This mutant *ELN* results in defective elastic lamellae and arterial SMC hyperproliferation culminating in increased medial thickness, arterial resistance, and left ventricular pressure and thickness.[63] Similarly, $Eln^{(+/-)}$ mice accumulate more SMC layers and are hypertensive during their normal lifespan.[64] In contrast, late stage embryonic or early postnatal $Eln^{(-/-)}$ mice display hyperproliferative, excess, and disorganized medial SMCs that cause arterial lumen stenoses.[65] SVAS occurs independently or as an integral part of Williams-Beuren syndrome (WBS), the latter caused by heterozygous deletion of 26–28 genes (including *ELN*) on human chromosome 7.[66,67] WBS occurs in ~1:10,000 births and patients suffer from low-birth weight, developmental delay, and cardiovascular, kidney, facial, and dental abnormalities.[68] Patients with nonsyndromic SVAS or with WBS are at a high risk of sudden death, and surgery is the only treatment option for the obstructive arterial disease.

Unfortunately, insights into mechanisms linking defective elastic lamellae and excessive SMC accumulation are limited, which is a major roadblock for developing efficacious pharmacological therapies. We previously reported that integrin β3 is upregulated in the aortic media of ELN mutant mice and SVAS and WBS patients.[69] Our studies demonstrate that genetic or pharmacological inhibition of integrin β3 attenuates SMC hyperproliferation and aortic stenosis in ELN mutant mice. Further understanding how ELN deficiency causes increased integrin β3 levels and downstream signaling in aortic SMCs is critical for advancing therapies for SVAS and potentially other vasculoproliferative diseases associated with ELN abnormalities.

Atherosclerosis and restenosis

Atherosclerosis is a chronic inflammatory process of the arterial wall. Its primary clinical manifestations include ischemic heart disease and stroke, which are leading causes of death globally. Atherosclerosis is characterized by the formation of intimal lesions called plaques or atheromas. Structurally, an atheroma consists of a soft necrotic core of lipid-laden foam cells, cholesterol crystals, and cellular debris. The necrotic core sits deep to a fibrous cap which faces the vessel lumen and is comprised of SMCs, ECM, and inflammatory cells. An atheroma starts to form in response to EC injury as early as in childhood as an asymptomatic, almost flat lesion (known as a fatty streak) which can progressively thicken, generally over a period of many years. Sustained EC injury and inflammation promote progressive accumulation of lipids—mostly low-density lipoproteins or its oxidized form—within the lesion and the formation of an advanced plaque (Fig. 8.2).

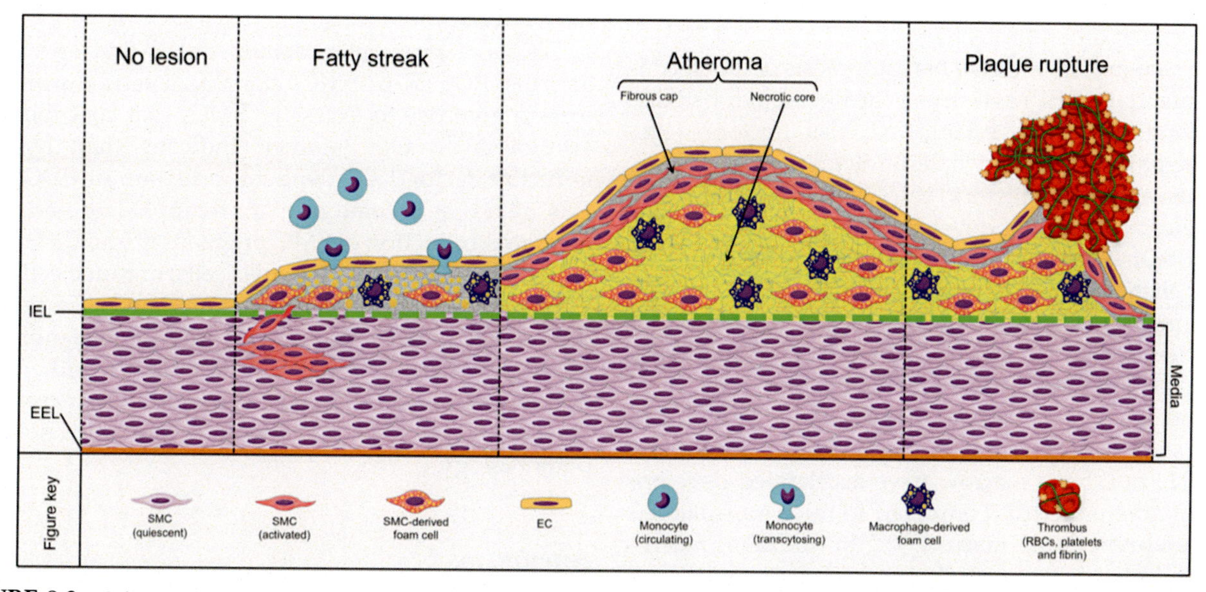

FIGURE 8.2 **Schematic representing the development of an atheroma and plaque rupture.** A fatty streak forms as macrophage and SMC-derived foam cells start to accumulate in the tunica intima. An atheroma successively develops and is composed of a fibrous cap and a necrotic core. An unstable atheroma is prone to rupture which results in thrombus formation and arterial stenosis or obstruction. *EC*, endothelial cell; *EEL*, external elastic lamina; *IEL*, internal elastic lamina; *SMC*, smooth muscle cell.

The key pathological event in the natural history of atherogenesis is rupture of the fibrous cap of the plaque which exposes prothrombotic material to the circulation. The ensuing thrombosis results in sudden partial or complete occlusion of the artery and acute ischemia of downstream tissues.[70] Plaques that are prone to rupture (i.e., unstable) are associated with a thin fibrous cap containing SMA$^+$ cells, a large number of galectin-3$^+$ cells, and a large necrotic core containing foam cells.[71,72] Until the last decade, SMCs and macrophages had been traditionally identified by cellular markers, such as SMA and galectin-3, respectively. This methodology led to the postulate that SMCs have an atheroprotective role, whereas macrophages were primarily responsible for plaque progression and rupture. Recently, this overly simplistic view has been disputed by lineage tracing studies which reveal that most cells within an advanced plaque are transdifferentiated SMCs, many of which transition through a galectin-3$^+$ state to take on diverse phenotypes.[73–76] Furthermore, studies utilizing multicolor recombination markers have facilitated clonal analysis of the SMC-derived plaque cells.[77] Several studies have demonstrated that a large portion of the cellularity of atheromas in mice derive from clonal expansion of rare SMCs.[78–80] By imaging murine plaques at different stages, our lab demonstrated that a SMC progenitor is able to migrate into the intima and give rise to SMCs of the fibrous cap by clonal expansion. Subsequently, SMC-derived cells migrate into the necrotic core and transdifferentiate.[80] Macrophages also have an ambivalent role in the development of atherosclerosis with several phenotypic subsets recently identified in murine models thought to be either proatherogenic or atheroprotective.[81]

The recognition of clonal hematopoiesis of indeterminate potential (CHIP) as a strong cardiovascular risk factor alongside more established risk factors (e.g., smoking, diabetes, hypercholesterolemia, hypertension, male sex, and family history) emphasizes the critical role of myeloid cells in the development of atherosclerosis. CHIP occurs when somatic mutations of genes involved in epigenetic regulation (e.g., *TET2, DNMT3A, ASXL1*) are acquired by hematopoietic cells and generally confer a proliferative advantage without causing frank malignancy. It is estimated that more than 10% of individuals over the age of 70 carry mutations seen in CHIP.[82,83] In recent studies, hypercholesterolemic mice transplanted with *Tet2* null bone marrow have accelerated development of atherosclerosis compared to mice transplanted with wild-type bone marrow.[84,85] In addition, macrophages isolated from *Tet2*$^{(-/-)}$ mice have increased expression of proinflammatory cytokines in culture.[84,85]

Numerous revascularization approaches are available to treat atherosclerotic plaques, but their long-term efficacy is diminished by intimal hyperplasia and subsequent restenosis. Intimal hyperplasia commonly occurs after balloon angioplasty, intraarterial stent placement or arterialization of a vein graft, and represents a stereotyped response to vascular injury, consisting of a complex interplay between inflammatory cells, platelets, ECs, and SMCs.[70,86–89] After vascular injury, SMCs from the *tunica media* migrate to the *tunica intima* where they proliferate and produce ECM, thereby thickening the vascular wall. Intimal hyperplasia frequently manifests clinically as the previously treated vessel reoccludes, often requiring a repeat intervention.[86,87,89,90] The development of drug eluting stents over the past 20 years has reduced the incidence of restenosis after endovascular intervention by coating stents with drugs that are antiinflammatory and antiproliferative (e.g., paclitaxel, sirolimus, and everolimus) at the site of vascular injury.[86,89]

Pulmonary hypertension

PH has a global prevalence of ~1% and causes ~20,000 deaths per year in the United States of America.[91,92] With limited treatment options, there is a dire need for therapies that can reverse or prevent progression of this disease. PH is defined by a mean pulmonary arterial pressure greater than 20 mmHg, and the World Health Organization classifies PH into five groups based on etiology.[93] A key pathological feature of PH is hypermuscularization of the lung vasculature, such as coating of normally unmuscularized distal pulmonary arterioles with excessive and aberrant SMCs, which reduces vessel compliance and increases pulmonary arterial pressure.[94–96]

We have previously demonstrated that specialized arteriole PDGFR-β$^+$ SMCs clonally expand during hypoxia to give rise to excessive SMCs that coat the distal arterioles.[97,98] Our studies indicate that hypoxia-inducible factor 1α-dependent induction of PDGF-B in ECs plays a pivotal role in Kruppel-like factor 4–mediated induction of SMC progenitors and distal muscularization.[99] Beyond vascular cells, immune cells such as monocytes/macrophages accumulate in the perivascular space during hypoxia and contribute significantly to the pathogenesis of PH.[100,101,116] Taken together, these findings emphasize the importance of cellular cross-talk between ECs, hematopoietic cells, and SMCs in PH pathobiology.

Aneurysms

An aneurysm is a ballooning of the arterial wall caused by localized mural weakness. Most aneurysms develop slowly without symptoms; however, large aneurysms increase the risk of arterial rupture, which is a catastrophic event resulting in uncontrolled bleeding and often death. Although aneurysms can occur throughout the body, the aorta, and especially the infrarenal portion of the aorta, is frequently affected. Abdominal aortic aneurysm (AAA) shares multiple risk factors, including older age, male sex, smoking, hypertension, and dyslipidemia, with atherosclerosis.[102] Moreover, major sequelae of atherosclerosis, ischemic heart disease and peripheral artery disease, are also risk factors for AAA, and thus, it is suggested that atherosclerosis may itself contribute to the pathogenesis of AAA. In addition, inflammation plays a key role in the development and progression of AAA.[102] Chronic aortic inflammation induces the production of proteolytic enzymes (e.g., MMPs), oxygen-derived free radicals and cytokines which in turn lead to SMC dysfunction and destruction of the aortic media.[102,103]

Beyond the aorta, aneurysms often involve intracranial arteries. These vessels are prone to aneurysm formation partly because they lack a connective tissue-rich adventitia and are instead surrounded by cerebrospinal fluid.[104] Intracranial aneurysms are often saccular bulges located at arterial bifurcations and are the underlying etiology of most cases of subarachnoid hemorrhage (SAH). Altered hemodynamic flow proximal to and within saccular aneurysms are important in the development and progression of these lesions. On the cellular level, key pathobiological features include EC dysfunction, SMC phenotypic modulation, recruitment of macrophages, and T-cells and result in the thinning of the tunica media, fragmentation of elastic lamellae, and degradation of vessel wall integrity.[104,105] A number of risk factors for rupture of an intracranial aneurysm have been identified, including, but not limited to, age >70 years, hypertension, smoking, and previous SAH from a distinct aneurysm.[104,106] Given the relative abundance of recently generated collagen type I in intracranial aneurysms of patients with a history of hypertension or smoking, it is suspected that there is ongoing ECM remodeling during aneurysm pathobiology.[107]

Coronavirus disease 2019

The global pandemic coronavirus disease 2019 (COVID-19) is caused by severe acute respiratory syndrome coronavirus 2 (SARS-CoV-2), and hypoxia due to acute respiratory distress syndrome is the primary complication of severe disease. Preexisting cardiovascular pathologies substantially worsen prognosis of COVID-19, increasing rates of thromboembolism, septic shock, and death.[108] SARS-CoV-2 enters cells via the transmembrane angiotensin-converting enzyme 2 (ACE2) which is expressed on lung epithelial cells, and in the vasculature, on pericytes as well as on select brain SMCs.[109,110] ACE2 is a negative regulator of the renin-angiotensin system. ACE catalyzes the conversion of angiotensin I into the potent vasoconstrictor angiotensin II, whereas ACE2 hydrolyzes angiotensin II into angiotensin-(1−7), which is a vasodilator.[111] Although ACE inhibitors and angiotensin receptor blockers (ARBs) have been demonstrated to enhance the expression of ACE2 in some animal experiments,[112] large population-based studies show no evidence that ACE inhibitors or ARBs alter the susceptibility to or severity of COVID-19.[113] Ongoing clinical trials for COVID-19 patients are evaluating diverse therapeutic modalities, including recombinant human ACE2.[114]

Conclusion

The arterial system is a hierarchal assembly of blood vessels designed primarily to support metabolism by delivering oxygenated and nutrient-rich blood from the heart to the capillary beds of the body's tissues. The structure and diameter of the arteries and/or arterioles vary greatly in organ- and tissue-specific manner to maintain a pressure gradient and serve specialized functions such as mechanical support and resilience (e.g., elastic arteries). A healthy vessel wall is critical and requires integration of diverse cell types, ECM components, and molecular signals to coordinate proper wall assembly and function. Indeed, major cardiovascular diseases display severe defects in the normal composition and architecture of the arterial wall. For example, SVAS, atherosclerosis, and PH are plagued with accumulation of aberrant and excessive SMCs. With modern developmental, genetic, and -omic approaches, our knowledge of regional-specific and spatiotemporal changes in vascular cells during vessel morphogenesis and pathogenesis has accelerated over the past decade. Yet, several important unanswered questions remain. For instance, what are the range of sources and underlying molecular mechanisms of aberrant arterial cells in arteriopathies? How do immune cells both negatively and positively impact the function of the arterial wall? What is the role of the adventitia in arterial disease? What are the cellular sources of ECM components in specific arteries, what are the roles of these components in disease, and how can they be manipulated? Cardiovascular disease continues to impose a massive burden on human health and only through continued rigorous and coordinated investigations of the arterial wall in development and disease will the biomedical community meet the dire need for novel and efficacious vascular therapies.

Acknowledgments

We thank Greif laboratory members for input. Funding was provided by the NIH (R35HL150766, R01HL125815, R01HL133016, R01HL142674, R21NS088854 and R21NS123469 to D.M.G., and G.M. was supported by a T32 Postdoctoral Training Grant [T32HL00795020]) and the American Heart Association (Established Investigator Award, 19EIA34660321 to D.M.G., and J.M.D. was supported by a Postdoctoral Fellowship [17POST33400071] and Career Development Award [856332]).

References

1. Aaronson PI, Ward JPT, Wiener CM. *The Cardiovascular System at a Glance*. 2nd ed. Malden, Mass: Blackwell; 2004.
2. Mazurek R, Dave JM, Chandran RR, Misra A, Sheikh AQ, Greif DM. Vascular cells in blood vessel wall development and disease. *Adv Pharmacol*. 2017;78:323–350.
3. Seidelmann SB, Lighthouse JK, Greif DM. Development and pathologies of the arterial wall. *Cell Mol Life Sci*. 2014;71(11): 1977–1999.
4. Cines DB, Pollak ES, Buck CA, et al. Endothelial cells in physiology and in the pathophysiology of vascular disorders. *Blood*. 1998;91(10):3527–3561.
5. Jayadev R, Sherwood DR. Basement membranes. *Curr Biol*. 2017; 27(6):R207–R211.
6. Davis GE, Senger DR. Endothelial extracellular matrix: biosynthesis, remodeling, and functions during vascular morphogenesis and neovessel stabilization. *Circ Res*. 2005; 97(11):1093–1107.
7. Orekhov AN, Bobryshev YV, Chistiakov DA. The complexity of cell composition of the intima of large arteries: focus on pericyte-like cells. *Cardiovasc Res*. 2014;103(4):438–451.
8. Stary HC, Blankenhorn DH, Chandler AB, et al. A definition of the intima of human arteries and of its atherosclerosis-prone regions. A report from the Committee on Vascular Lesions of the Council on Arteriosclerosis, American Heart Association. *Arterioscler Thromb*. 1992;12(1):120–134.
9. Schwartz SM, deBlois D, O'Brien ER. The intima. Soil for atherosclerosis and restenosis. *Circ Res*. 1995;77(3):445–465.
10. Orekhov AN, Andreeva ER, Andrianova IV, Bobryshev YV. Peculiarities of cell composition and cell proliferation in different type atherosclerotic lesions in carotid and coronary arteries. *Atherosclerosis*. 2010;212(2):436–443.
11. Orekhov AN, Andreeva ER, Mikhailova IA, Gordon D. Cell proliferation in normal and atherosclerotic human aorta: proliferative splash in lipid-rich lesions. *Atherosclerosis*. 1998;139(1):41–48.
12. Lin CJ, Staiculescu MC, Hawes JZ, et al. Heterogeneous cellular contributions to elastic laminae formation in arterial wall development. *Circ Res*. 2019;125(11):1006–1018.
13. Pouget C, Gautier R, Teillet MA, Jaffredo T. Somite-derived cells replace ventral aortic hemangioblasts and provide aortic smooth muscle cells of the trunk. *Development*. 2006;133(6):1013–1022.
14. Risau W, Flamme I. Vasculogenesis. *Annu Rev Cell Dev Biol*. 1995; 11:73–91.
15. Benjamin LE, Hemo I, Keshet E. A plasticity window for blood vessel remodelling is defined by pericyte coverage of the preformed endothelial network and is regulated by PDGF-B and VEGF. *Development*. 1998;125(9):1591–1598.
16. Duran CL, Howell DW, Dave JM, et al. Molecular regulation of sprouting angiogenesis. *Comp Physiol*. 2017;8(1):153–235.
17. Adams RH, Wilkinson GA, Weiss C, et al. Roles of ephrinB ligands and EphB receptors in cardiovascular development: demarcation of arterial/venous domains, vascular morphogenesis, and sprouting angiogenesis. *Genes Dev*. 1999;13(3):295–306.
18. Gerety SS, Wang HU, Chen ZF, Anderson DJ. Symmetrical mutant phenotypes of the receptor EphB4 and its specific transmembrane ligand ephrin-B2 in cardiovascular development. *Mol Cell*. 1999; 4(3):403–414.
19. Wang HU, Chen ZF, Anderson DJ. Molecular distinction and angiogenic interaction between embryonic arteries and veins revealed by ephrin-B2 and its receptor Eph-B4. *Cell*. 1998;93(5): 741–753.
20. Lawson ND, Scheer N, Pham VN, et al. Notch signaling is required for arterial-venous differentiation during embryonic vascular development. *Development*. 2001;128(19):3675–3683.
21. Lawson ND, Vogel AM, Weinstein BM. Sonic hedgehog and vascular endothelial growth factor act upstream of the Notch pathway during arterial endothelial differentiation. *Dev Cell*. 2002;3(1):127–136.
22. You LR, Lin FJ, Lee CT, DeMayo FJ, Tsai MJ, Tsai SY. Suppression of Notch signalling by the COUP-TFII transcription factor regulates vein identity. *Nature*. 2005;435(7038):98–104.
23. Vanlandewijck M, He L, Mae MA, et al. A molecular atlas of cell types and zonation in the brain vasculature. *Nature*. 2018; 554(7693):475–480.
24. Gerhardt H, Golding M, Fruttiger M, et al. VEGF guides angiogenic sprouting utilizing endothelial tip cell filopodia. *J Cell Biol*. 2003;161(6):1163–1177.
25. Dejana E, Tournier-Lasserve E, Weinstein BM. The control of vascular integrity by endothelial cell junctions: molecular basis and pathological implications. *Dev Cell*. 2009;16(2):209–221.
26. Iruela-Arispe ML, Davis GE. Cellular and molecular mechanisms of vascular lumen formation. *Dev Cell*. 2009;16(2):222–231.
27. Phng LK, Gerhardt H. Angiogenesis: a team effort coordinated by notch. *Dev Cell*. 2009;16(2):196–208.
28. Blanco R, Gerhardt H. VEGF and Notch in tip and stalk cell selection. *Cold Spring Harb Perspect Med*. 2013;3(1):a006569.
29. Gaengel K, Genove G, Armulik A, Betsholtz C. Endothelial-mural cell signaling in vascular development and angiogenesis. *Arterioscler Thromb Vasc Biol*. 2009;29(5):630–638.
30. Gerhardt H, Betsholtz C. Endothelial-pericyte interactions in angiogenesis. *Cell Tissue Res*. 2003;314(1):15–23.
31. Armulik A, Abramsson A, Betsholtz C. Endothelial/pericyte interactions. *Circ Res*. 2005;97(6):512–523.
32. Eilken HM, Dieguez-Hurtado R, Schmidt I, et al. Pericytes regulate VEGF-induced endothelial sprouting through VEGFR1. *Nat Commun*. 2017;8(1):1574.
33. Majesky MW. Developmental basis of vascular smooth muscle diversity. *Arterioscler Thromb Vasc Biol*. 2007;27(6):1248–1258.
34. Sawada H, Rateri DL, Moorleghen JJ, Majesky MW, Daugherty A. Smooth muscle cells derived from second heart field and cardiac neural crest reside in spatially distinct domains in the media of the ascending aorta-brief report. *Arterioscler Thromb Vasc Biol*. 2017;37(9):1722–1726.
35. Cheung C, Bernardo AS, Trotter MW, Pedersen RA, Sinha S. Generation of human vascular smooth muscle subtypes provides insight into embryological origin-dependent disease susceptibility. *Nat Biotechnol*. 2012;30(2):165–173.
36. Owens GK. Regulation of differentiation of vascular smooth-muscle cells. *Physiol Rev*. 1995;75(3):487–517.
37. Owens GK, Kumar MS, Wamhoff BR. Molecular regulation of vascular smooth muscle cell differentiation in development and disease. *Physiol Rev*. 2004;84(3):767–801.
38. Horowitz A, Menice CB, Laporte R, Morgan KG. Mechanisms of smooth muscle contraction. *Physiol Rev*. 1996;76(4):967–1003.
39. Wagenseil JE, Mecham RP. Vascular extracellular matrix and arterial mechanics. *Physiol Rev*. 2009;89(3):957–989.
40. Humphrey JD, Milewicz DM, Tellides G, Schwartz MA. Cell biology. Dysfunctional mechanosensing in aneurysms. *Science*. 2014;344(6183):477–479.

41. Kelleher CM, McLean SE, Mecham RP. Vascular extracellular matrix and aortic development. *Curr Top Dev Biol*. 2004;62:153–188.
42. Heino J. The collagen receptor integrins have distinct ligand recognition and signaling functions. *Matrix Biol*. 2000;19(4): 319–323.
43. Majesky MW, Dong XR, Hoglund V, Daum G, Mahoney Jr WM. The adventitia: a progenitor cell niche for the vessel wall. *Cells Tissues Organs*. 2012;195(1–2):73–81.
44. Haurani MJ, Pagano PJ. Adventitial fibroblast reactive oxygen species as autacrine and paracrine mediators of remodeling: bellwether for vascular disease? *Cardiovasc Res*. 2007;75(4):679–689.
45. Rey FE, Pagano PJ. The reactive adventitia: fibroblast oxidase in vascular function. *Arterioscler Thromb Vasc Biol*. 2002;22(12): 1962–1971.
46. Hinz B, Gabbiani G. Fibrosis: recent advances in myofibroblast biology and new therapeutic perspectives. *F1000 Biol Rep*. 2010; 2:78.
47. Follonier Castella L, Gabbiani G, McCulloch CA, Hinz B. Regulation of myofibroblast activities: calcium pulls some strings behind the scene. *Exp Cell Res*. 2010;316(15):2390–2401.
48. Tang J, Wang H, Huang X, et al. Arterial Sca1(+) vascular stem cells generate de novo smooth muscle for artery repair and regeneration. *Cell Stem Cell*. 2020;26(1):81–96.
49. Majesky MW. Adventitia and perivascular cells. *Arterioscler Thromb Vasc Biol*. 2015;35(8):e31–35.
50. Majesky MW, Dong XR, Hoglund V, Mahoney Jr WM, Daum G. The adventitia: a dynamic interface containing resident progenitor cells. *Arterioscler Thromb Vasc Biol*. 2011;31(7):1530–1539.
51. Ahmadieh S, Kim HW, Weintraub NL. Potential role of perivascular adipose tissue in modulating atherosclerosis. *Clin Sci (Lond)*. 2020;134(1):3–13.
52. Chatterjee TK, Stoll LL, Denning GM, et al. Proinflammatory phenotype of perivascular adipocytes: influence of high-fat feeding. *Circ Res*. 2009;104(4):541–549.
53. Brown NK, Zhou Z, Zhang J, et al. Perivascular adipose tissue in vascular function and disease: a review of current research and animal models. *Arterioscler Thromb Vasc Biol*. 2014;34(8): 1621–1630.
54. Sylvester JT, Shimoda LA, Aaronson PI, Ward JP. Hypoxic pulmonary vasoconstriction. *Physiol Rev*. 2012;92(1):367–520.
55. Dunham-Snary KJ, Wu D, Sykes EA, et al. Hypoxic pulmonary vasoconstriction: from molecular mechanisms to medicine. *Chest*. 2017;151(1):181–192.
56. Duncker DJ, Bache RJ. Regulation of coronary blood flow during exercise. *Physiol Rev*. 2008;88(3):1009–1086.
57. Beyer AM, Gutterman DD. Regulation of the human coronary microcirculation. *J Mol Cell Cardiol*. 2012;52(4):814–821.
58. Brauer RW. Liver circulation and function. *Physiol Rev*. 1963;43: 115–213.
59. Wright FS, Briggs JP. Feedback control of glomerular blood flow, pressure, and filtration rate. *Physiol Rev*. 1979;59(4):958–1006.
60. Carlstrom M, Wilcox CS, Arendshorst WJ. Renal autoregulation in health and disease. *Physiol Rev*. 2015;95(2):405–511.
61. Metcalfe K, Rucka AK, Smoot L, et al. Elastin: mutational spectrum in supravalvular aortic stenosis. *Eur J Hum Genet*. 2000; 8(12):955–963.
62. Li DY, Toland AE, Boak BB, et al. Elastin point mutations cause an obstructive vascular disease, supravalvular aortic stenosis. *Hum Mol Genet*. 1997;6(7):1021–1028.
63. Merla G, Brunetti-Pierri N, Piccolo P, Micale L, Loviglio MN. Supravalvular aortic stenosis: elastin arteriopathy. *Circ Cardiovasc Genet*. 2012;5(6):692–696.
64. Li DY, Faury G, Taylor DG, et al. Novel arterial pathology in mice and humans hemizygous for elastin. *J Clin Invest*. 1998;102(10): 1783–1787.
65. Li DY, Brooke B, Davis EC, et al. Elastin is an essential determinant of arterial morphogenesis. *Nature*. 1998;393(6682):276–280.
66. Pober BR. Williams-Beuren syndrome. *N Engl J Med*. 2010;362(3): 239–252.
67. Ewart AK, Jin W, Atkinson D, Morris CA, Keating MT. Supravalvular aortic stenosis associated with a deletion disrupting the elastin gene. *J Clin Invest*. 1994;93(3):1071–1077.
68. Pober BR, Johnson M, Urban Z. Mechanisms and treatment of cardiovascular disease in Williams-Beuren syndrome. *J Clin Invest*. 2008;118(5):1606–1615.
69. Misra A, Sheikh AQ, Kumar A, et al. Integrin beta3 inhibition is a therapeutic strategy for supravalvular aortic stenosis. *J Exp Med*. 2016;213(3):451–463.
70. Mitchell RN, Halushka MK. Blood vessels. In: *Robbins and Cotran Pathologic Basis of Disease*. 10th ed. Elsevier; 2021:485–525.
71. Bennett MR, Sinha S, Owens GK. Vascular smooth muscle cells in atherosclerosis. *Circ Res*. 2016;118:692–702.
72. Falk E. Pathogenesis of atherosclerosis. *J Am Coll Cardiol*. 2006; 47(8 Suppl):C7–C12.
73. Nguyen AT, Gomez D, Bell RD, et al. Smooth muscle cell plasticity: fact or fiction? *Circ Res*. 2013;112:17–22.
74. Shankman LS, Gomez D, Cherepanova OA, et al. KLF4-dependent phenotypic modulation of smooth muscle cells has a key role in atherosclerotic plaque pathogenesis. *Nat Med*. 2015; 21:628–637.
75. Alencar G, Owsiany K, K S, et al. The stem cell pluripotency genes Klf4 and Oct4 regulate complex SMC phenotypic changes critical in late-stage atherosclerotic lesion pathogenesis. *Circulation*. 2020; 142(21):2045–2059.
76. Wirka RC, Wagh D, Paik DT, et al. Atheroprotective roles of smooth muscle cell phenotypic modulation and the TCF21 disease gene as revealed by single-cell analysis. *Nat Med*. 2019;25: 1280–1289.
77. Basatemur GL, Jorgensen HF, Clarke MCH, Bennett MR, Mallat Z. Vascular smooth muscle cells in atherosclerosis. *Nat Rev Cardiol*. 2019;16(12):727–744.
78. Chappell J, Harman JL, Narasimhan VM, et al. Extensive proliferation of a subset of differentiated, yet plastic, medial vascular smooth muscle cells contributes to neointimal formation in mouse injury and atherosclerosis models. *Circ Res*. 2016;119(12): 1313–1323.
79. Jacobsen K, Lund MB, Shim J, et al. Diverse cellular architecture of atherosclerotic plaque derives from clonal expansion of a few medial SMCs. *JCI Insight*. 2017;2(19).
80. Misra A, Feng Z, Chandran RR, et al. Integrin beta3 regulates clonality and fate of smooth muscle-derived atherosclerotic plaque cells. *Nat Commun*. 2018;9(1):2073.
81. Barrett Tessa. Macrophages in atherosclerosis regression. *Arterioscler Thromb Vasc Biol*. 2020;40:20–33.
82. Libby P, Ebert BL. CHIP (clonal hematopoiesis of indeterminate potential): potent and newly recognized contributor to cardiovascular risk. *Circulation*. 2018;138:666–668.
83. Jaiswal S, Ebert BL. Clonal hematopoiesis in human aging and disease. *Science*. 2019;366(6465).
84. Fuster JJ, MacLauchlan S, Zuriaga MA, et al. Clonal hematopoiesis associated with TET2 deficiency accelerates atherosclerosis development in mice. *Science*. 2017;355:842–847.
85. Jaiswal S, Natarajan P, Silver AJ, et al. Clonal hematopoiesis and risk of atherosclerotic cardiovascular disease. *N Engl J Med*. 2017;377:111–121.
86. Grech ED. Percutaneous coronary intervention. I: history and development. *BMJ*. 2003;326:1080–1082.
87. Dick P, Wallner H, Sabeti S, et al. Balloon angioplasty versus stenting with nitinol stents in intermediate length superficial femoral artery lesions. *Cathet Cardiovasc Interv*. 2009;74:1090–1095.

88. Wu B, Mottola G, Schaller M, Upchurch Jr GR, Conte MS. Resolution of vascular injury: specialized lipid mediators and their evolving therapeutic implications. *Mol Aspect Med*. 2017;58:72–82.
89. Fernandez-Ruiz I. Interventional cardiology: drug-eluting or bare-metal stents? *Nat Rev Cardiol*. 2016;13(11):631.
90. Buccheri D, Piraino D, Andolina G, Cortese B. Understanding and managing in-stent restenosis: a review of clinical data, from pathogenesis to treatment. *J Thorac Dis*. 2016;8(10):E1150–E1162.
91. Hoeper MM, Humbert M, Souza R, et al. A global view of pulmonary hypertension. *Lancet Respir Med*. 2016;4(4):306–322.
92. George MG, Schieb LJ, Ayala C, Talwalkar A, Levant S. Pulmonary hypertension surveillance: United States, 2001 to 2010. *Chest*. 2014;146(2):476–495.
93. Simonneau G, Montani D, Celermajer DS, et al. Haemodynamic definitions and updated clinical classification of pulmonary hypertension. *Eur Respir J*. 2019;53(1).
94. Mahapatra S, Nishimura RA, Sorajja P, Cha S, McGoon MD. Relationship of pulmonary arterial capacitance and mortality in idiopathic pulmonary arterial hypertension. *J Am Coll Cardiol*. 2006;47(4):799–803.
95. Farber HW, Loscalzo J. Pulmonary arterial hypertension. *N Engl J Med*. 2004;351(16):1655–1665.
96. Rabinovitch M. Molecular pathogenesis of pulmonary arterial hypertension. *J Clin Invest*. 2008;118(7):2372–2379.
97. Sheikh AQ, Lighthouse JK, Greif DM. Recapitulation of developing artery muscularization in pulmonary hypertension. *Cell Rep*. 2014;6(5):809–817.
98. Sheikh AQ, Misra A, Rosas IO, Adams RH, Greif DM. Smooth muscle cell progenitors are primed to muscularize in pulmonary hypertension. *Sci Transl Med*. 2015;7(308):308ra159.
99. Sheikh AQ, Saddouk FZ, Ntokou A, Mazurek R, Greif DM. Cell autonomous and non-cell autonomous regulation of SMC progenitors in pulmonary hypertension. *Cell Rep*. 2018;23(4):1152–1165.
100. Florentin J, Dutta P. Origin and production of inflammatory perivascular macrophages in pulmonary hypertension. *Cytokine*. 2017;100:11–15.
101. Yu YA, Malakhau Y, Yu CA, et al. Nonclassical monocytes sense hypoxia, regulate pulmonary vascular remodeling, and promote pulmonary hypertension. *J Immunol*. 2020;204(6):1474–1485.
102. Golledge J. Abdominal aortic aneurysm: update on pathogenesis and medical treatments. *Nat Rev Cardiol*. 2019;16(4):225–242.
103. Michel JB, Martin-Ventura JL, Egido J, et al. Novel aspects of the pathogenesis of aneurysms of the abdominal aorta in humans. *Cardiovasc Res*. 2011;90(1):18–27.
104. Etminan N, Rinkel GJ. Unruptured intracranial aneurysms: development, rupture and preventive management. *Nat Rev Neurol*. 2016;12(12):699–713.
105. Austin G, Fisher S, Dickson D, Anderson D, Richardson S. The significance of the extracellular matrix in intracranial aneurysms. *Ann Clin Lab Sci*. 1993;23(2):97–105.
106. Greving JP, Wermer MJ, Brown Jr RD, et al. Development of the PHASES score for prediction of risk of rupture of intracranial aneurysms: a pooled analysis of six prospective cohort studies. *Lancet Neurol*. 2014;13(1):59–66.
107. Etminan N, Dreier R, Buchholz BA, et al. Age of collagen in intracranial saccular aneurysms. *Stroke*. 2014;45(6):1757–1763.
108. Inciardi RM, Adamo M, Lupi L, et al. Characteristics and outcomes of patients hospitalized for COVID-19 and cardiac disease in Northern Italy. *Eur Heart J*. 2020;41(19):1821–1829.
109. Chen L, Li X, Chen M, Feng Y, Xiong C. The ACE2 expression in human heart indicates new potential mechanism of heart injury among patients infected with SARS-CoV-2. *Cardiovasc Res*. 2020;116(6):1097–1100.
110. He L, Mäe MA, Sun Y, et al. Pericyte-specific vascular expression of SARS-CoV-2 receptor ACE2 – implications for microvascular inflammation and hypercoagulopathy in COVID-19 patients. *bioRxiv*. 2020. https://doi.org/10.1101/2020.05.11.088500.
111. Chamsi-Pasha MA, Shao Z, Tang WH. Angiotensin-converting enzyme 2 as a therapeutic target for heart failure. *Curr Heart Fail Rep*. 2014;11(1):58–63.
112. Ferrario CM, Jessup J, Chappell MC, et al. Effect of angiotensin-converting enzyme inhibition and angiotensin II receptor blockers on cardiac angiotensin-converting enzyme 2. *Circulation*. 2005;111(20):2605–2610.
113. Mancia G, Rea F, Ludergnani M, Apolone G, Corrao G. Renin-angiotensin-aldosterone system blockers and the risk of Covid-19. *N Engl J Med*. 2020;382(25):2431–2440.
114. Guzik TJ, Mohiddin SA, Dimarco A, et al. COVID-19 and the cardiovascular system: implications for risk assessment, diagnosis, and treatment options. *Cardiovasc Res*. 2020;116(10):1666–1687.
115. Dave JM, Mirabella T, Greif DM. ALK5/TIMP3 axis in pericytes contributes to endothelial morphogenesis in the developing brain. *Developmental Cell*. 2018;44:665.
116. Ntokou A, Dave JM, Kauffman AC, et al. Macrophage-derived PDGF-B induces muscularization in murine and human pulmonary hypertension. *JCI Insight*. 2021;6(6):e139067.

CHAPTER

9

Capillary diversity: endothelial cell specializations to meet tissue metabolic needs

Martina Rudnicki[1], *Alexandra Pislaru*[2] *and Tara L. Haas*[1]

[1]Faculty of Health, York University, Toronto, ON, Canada [2]Faculty of Science, York University, Toronto, ON, Canada

Introduction

Every tissue in the body is invested with a comprehensive network of capillaries on which they rely to provide nutrients and remove waste products. Capillaries are composed of a single layer of endothelial cells (ECs), surrounded by a meshwork of matrix proteins (the basement membrane) and typically are partially wrapped with mural cells known as pericytes. Three categories of capillaries have been described. Discontinuous capillaries exist in the liver, fenestrated capillaries are found in kidney, intestines, and endocrine glands, whereas all other tissues in the body contain continuous capillaries (as reviewed by Aird).[1]

Capillaries are the gateway through which all circulating energy substrates and hormones must pass to support the metabolic functions of the surrounding "end-user" parenchymal cells. The junctions between adjacent ECs in continuous capillaries restrict the movement of large molecules between plasma and the tissue interstitium. Consequently, many substrates must be transferred through the ECs by transcytosis and facilitative transport processes. The regulatory mechanisms that enable ECs to provide tissue-specific delivery of materials are gaining recognition as crucial elements in establishing tissue function and whole-body metabolic homeostasis. Furthermore, nutrient availability and tissue metabolic activity are dynamic variables; thus, structural and functional plasticity are necessary attributes of capillaries. Signals from parenchymal cells provide cues to remodel the capillary network to meet the functional needs of the tissue and angiocrine factors released from ECs influence parenchymal cell survival, metabolic activity, and regenerative capacity.

Analyses of tissue-specific EC transcriptomes have led to a portrait of the EC as having core elements that are common to EC across organs together with a distinct tissue-specific EC signature. This chapter will highlight some of the shared and specialized molecular signatures and roles of ECs within the heart, skeletal muscle, and adipose tissue. We also will discuss how capillary ECs contribute to the pivotal roles of these tissues in maintaining whole-body metabolic homeostasis.

Attributes of capillary endothelial cells

ECs express a cluster of canonical markers (i.e., Pecam1, Cadherin5, Tie1) that are commonly used to distinguish them from stromal and parenchymal cells. Analysis of murine ECs from multiple organs identified more than 20 transcription factors that are commonly expressed in EC populations from most tissues, leading to the speculation that these could represent "vascular EC identity factors" regulating canonical EC genes.[2] Moreover, single-cell studies of human and murine tissues have revealed the existence of subpopulations of ECs with diverse molecular features within an individual tissue.[3–11] The presence of ECs from arterioles, capillaries venules, and lymph capillaries within each tissue provides one reason for this heterogeneity, because ECs from each distinct vessel type display some characteristic molecular signatures.[12] Nonetheless, multiple subpopulations of capillary ECs with distinct transcript profiles can be detected within a single tissue.[6]

The functional implications of specialized subgroups of capillary ECs remain to be explored. Interestingly, there is growing functional evidence that not all capillary ECs possess the same angiogenic potential. In fact, lineage tracing showed that proliferation is restricted to a subpopulation of ECs in both skeletal muscle and the heart.[13,14] Single-cell sequencing of cardiac ECs

The Vasculome
https://doi.org/10.1016/B978-0-12-822546-2.00001-0

identified this proliferating subset to exhibit a progenitor-like transcript signature with enriched expression of plasmalemma vesicle-associated protein, a regulator of EC proliferation.[14] A recent analysis of cells within the human heart also detected a population of ECs enriched with proangiogenic markers.[7] Thus, it is feasible that a limited number of ECs that possess proliferative potential establish the angiogenic capacity of a given tissue, but this will require further investigation.

Another level of molecular heterogeneity is evident when comparing EC molecular fingerprints across tissues with capillary ECs displaying organ-specific profiles.[2, 4, 5, 6, 8] Analyses of canonical transcription factor binding elements suggest that EC from different organs relies on distinct sets of transcription factors to establish these tissue-specific expression profiles.[2] These analyses highlight the distinct roles of ECs in supporting the diverse needs of different organs. Factors within the local environment play a substantial role in establishing these gene expression profiles as they are not preserved when ECs are cultured.[3,15]

EC molecular fingerprints align most closely between organs that share a similar developmental origin of the EC (i.e., ectoderm, endoderm, or mesoderm)[4] or that have similar metabolic demands. The ECs of adipose, cardiac, and skeletal muscle originate from the mesoderm. These are tissues that display a high capacity to uptake fatty acid (FA) (for energy production or storage). ECs from skeletal muscle and heart in particular share a highly conserved transcriptome.[2,4] An overlapping signature also is present in adipose tissue, with the ECs from these tissues exhibiting an enriched expression of genes belonging to ontologies of angiogenesis and low-density lipoprotein particle-mediated signaling[4] when compared to EC from other tissues of dissimilar origin and function (i.e., brain, liver, pancreas).[8] Despite these transcriptome similarities, ECs from these tissues also express nonoverlapping gene sets. Below, we will discuss key functions of capillary ECs in maintaining tissue homeostasis, highlighting emerging knowledge of the molecular commonalities and distinctions of the cardiac, muscle, and adipose ECs.

Maintenance and expansion of capillary networks

Capillary density is an indisputable determinant of oxygen and nutrient delivery by establishing diffusion distances and the surface area available for exchange.[16] During development and maturation, capillary number increases to keep pace with tissue growth, thus maintaining adequate delivery of nutrients and removal of metabolic by-products. This occurs by the process of angiogenesis, the formation of new capillaries from preexisting ones. However, capillary networks in adult tissues are largely stable and angiogenesis is limited to occasions of increased tissue mass or metabolic need.

ECs within mature capillaries have a low proliferative rate and thus are considered to be quiescent.[1] This state is controlled by an active process involving the integration of extracellular cues with cell-intrinsic regulatory mechanisms. The pathways coordinated by Notch and the transcriptional factor Forkhead Box O1 (FoxO1) are pivotal contributors to the maintenance of EC quiescence by suppressing EC glucose metabolism and cell cycle progression.[17,18] While proliferating ECs are highly glycolytic, quiescent ECs rely on FA oxidation to sustain the tricarboxylic acid cycle and secure redox homeostasis through NADPH generation.[19] This protects them against the high oxygen levels and fluctuating nutrient levels in the plasma to which ECs are directly and continuously exposed. FA oxidation also serves to maintain EC identity, as inhibition of this metabolic pathway promotes excessive endothelial-to-mesenchymal transition.[20]

Sprouting angiogenesis, the most extensively-investigated form of microvascular remodeling, requires that ECs acquire "tip" and "stalk" cell phenotypes to support the processes of migration and proliferation needed for the generation of a new capillary.[21,22] These energy-intensive events are supported by rapid upregulation of glycolysis pathway components in the ECs.[23,24] This switch in cellular metabolism, which is initiated following EC activation by angiogenic factors, is necessary for successful capillary growth.[25] Although many molecules have been identified to induce or to support the angiogenic process, vascular endothelial growth factor-A (VEGF-A) is paramount as an effector for the induction of angiogenesis in all tissues. VEGF-A binding to the vascular endothelial growth factor receptor-2 (VEGFR2) activates numerous EC signaling pathways that coordinate the upregulation of glycolysis, proliferation, sprouting, and migration.[26] Activation of these pathways also can be achieved indirectly by upregulation of vascular endothelial growth factor-B (VEGF-B) or placental growth factor (PGF), which bind to VEGFR1, thus freeing the available VEGF-A to bind VEGFR2.[27] These angiogenic factors are induced in response to tissue specific biochemical signals or mechanical forces within the local environment (Fig. 9.1).

Capillary growth in healthy adult skeletal muscle occurs in response to a sustained increase in muscle activity (i.e., exercise training) or oxidative status, thus coupling capillarization with muscle metabolic activity and tissue oxidative capacity.[28–30] Although numerous proangiogenic factors can be induced in exercising muscle, skeletal myocyte production of VEGF-A is dominant in promoting muscle angiogenesis as its deletion in myocytes prevents exercise-induced angiogenesis.[31]

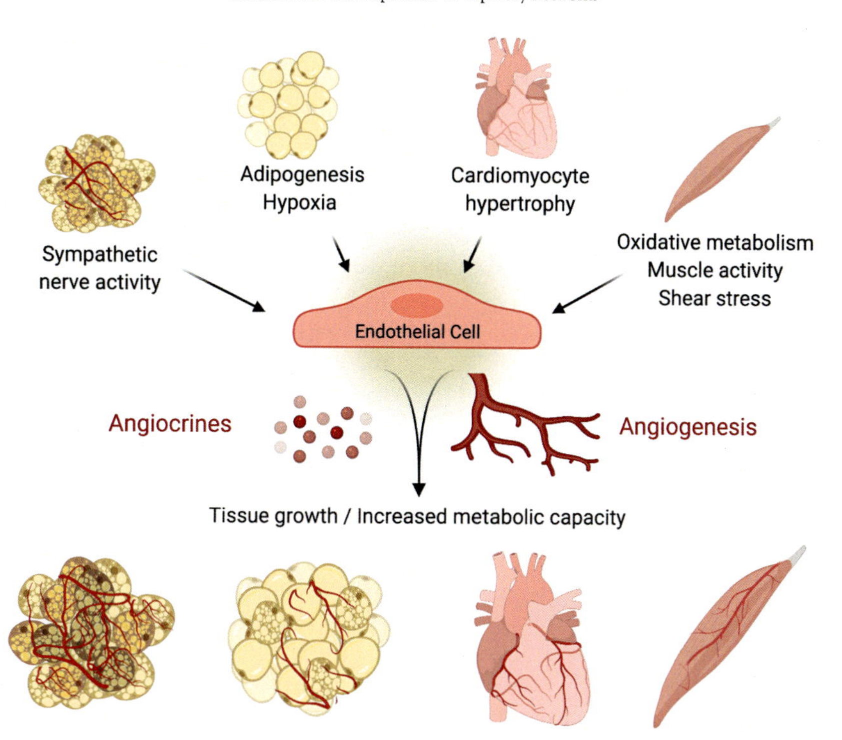

FIGURE 9.1 Endothelial cell roles in tissue homeostasis. Tissue-specific processes trigger an endothelial response leading to vascular remodeling by angiogenesis and/or the release of angiocrine factors, which ultimately supports tissue growth and enhances tissue metabolic capacity.

Skeletal muscle VEGF-A production is increased in response to tissue hypoxia, increased metabolites, and mechanical force associated with contraction, driven by the transcription factors HIF1α and PGC1α.[32–35] Notably, paracrine signals originating from ECs also instruct skeletal muscle to produce VEGF-A. This is evident during shear stress–induced angiogenesis, which is dependent on EC release of nitric oxide (NO).[36] This process also is reliant on VEGF-A derived from the skeletal myocytes.[37,38] Another paracrine factor, leptin, stimulates both skeletal myocyte FA oxidation and VEGF-A production,[39,40] thus coupling angiogenesis to a shift in cellular metabolism. Interestingly, there is evidence that nutrient-sensing pericytes contribute to this feedback loop by secretion of leptin.[40]

Cardiac myocytes in the adult heart are continuously active, highly oxidative and well invested with a dense capillary network.[28] Physiological hypertrophy of cardiomyocytes rather than increased oxidative capacity is the key stimulus for angiogenesis.[41–43] Generally, stimuli that initiate cardiomyocyte hypertrophy also increase the production of proangiogenic factors.[43] Intriguingly, volume overload–induced hypertrophy (which increases diastolic wall stress) is a more efficient stimulus for capillary growth compared to pressure overload–induced hypertrophy (associated with elevated systolic wall stress),[43–45] but the reasons for this are unclear. Growth factor signaling and mechano-transduction pathways initiated by stretch of the myocardial wall are thought to be activated within cardiomyocytes when the heart experiences volume overload.[46] These signals acutely activate Akt, which increases stabilization of HIF1α and leads to increased expression of proangiogenic factors VEGF-A, VEGF-B, and PGF.[47–50] VEGF-A is a critical factor for cardiac EC survival and angiogenesis, and its inhibition causes capillary rarefaction.[47] Interestingly, ECs from the heart are less responsive to angiogenic stimuli than those from skeletal muscle even when cultured,[51] but the underlying mechanism is not clear.

Angiogenic stimuli differ substantially between brown and white adipose tissues. Angiogenesis in brown adipose is regulated by sympathetic nerve activity, which is a potent inducer of VEGF-A production in brown adipocytes.[52–54] In this way, sympathetic nerve signaling coordinates angiogenesis and adipocyte metabolic activity.[55] Similarly, beiging of white adipose, which can occur in response to activation of β-adrenergic receptors, is coupled with angiogenesis, with both processes relying on VEGF-A signaling via VEGFR2.[53] In contrast to brown adipose, hypoxia has long been proposed as a key stimulus underlying capillary growth during expansion of white adipose. Indeed, adipocyte production of VEGF-A increases under hypoxic conditions. However, hypoxia is not prevalent in human adipose tissue;[56] thus, its relevance as an angiogenic stimulus in humans has been questioned. Notably, angiogenesis in white adipose tissue may be more closely coordinated with adipogenesis than adipocyte hypertrophy.[57,58] This is exemplified when comparing subcutaneous and visceral adipose depots because subcutaneous has a greater capillary density and the mesenchymal stromal cells in this depot exhibit a higher adipogenic differentiation potential and secrete more proangiogenic factors compared to visceral adipose.[59]

Endothelial control of nutrient delivery

In conjunction with capillary density, the delivery of metabolic substrates such as glucose and FAs is dynamically influenced by capillary EC functional capacity to move these substances from plasma to the tissue interstitium (Fig. 9.2).

Glucose uptake

ECs in the heart, skeletal muscle, and adipose tissues share glucose and insulin delivery characteristics that are similar to ECs of other organs; thus, we will discuss these functions generally. All ECs rely on Glucose transporter 1 (GLUT1), also known as "solute carrier family 2, facilitated glucose transporter member 1," to manage glucose uptake.[60] This transporter functions independently of insulin signaling and constitutively moves glucose in proportion to the concentration gradient. Of note, GLUT1 not only transports glucose into ECs, it also allows glucose to exit on the abluminal surface.[61] Indeed, modifying the level of EC GLUT1 concordantly alters glucose uptake by the heart and brain, reinforcing the relevance of ECs in supporting glucose handling by the tissues.[61] GLUT1 expression is increased by angiogenic mediators, thus supporting the increase in glucose metabolism that is required to allow EC proliferation and migration.[62] Conversely, Notch signaling uncouples glucose transcytosis from endothelial glycolysis in quiescent EC by suppressing expression of the glycolytic regulatory enzyme PFKFB3 (6-phosphofructo-2-kinase/fructose-2,6-biphosphatase 3).[63] This enables quiescent ECs to maintain their functional role in glucose transfer without concomitantly increasing glycolytic flux or enacting proliferation. GLUT1 function can be also critically affected by the membrane cholesterol content in ECs, which is affected by tissue release of VEGF-B.[60]

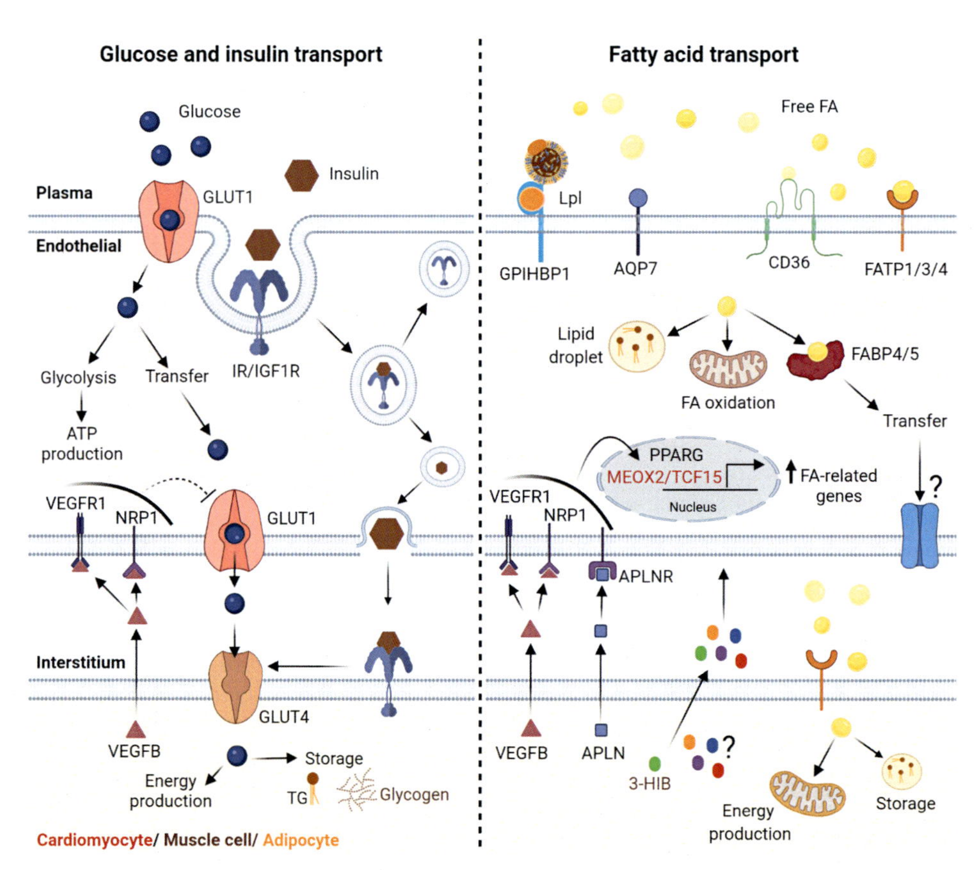

FIGURE 9.2 Nutrient delivery and its regulation. Endothelial cells govern glucose and insulin (**left side**) as well as fatty acid (**right side**) delivery to surrounding tissues. The transendothelial transport of nutrients depends on a bidirectional cross-talk between the endothelial and parenchymal cells.

Once glucose passes the EC barrier, uptake of glucose from the interstitium into skeletal myocytes, cardiomyocytes, and adipocytes is largely dependent on GLUT4, which is an insulin-dependent transporter.[64] Thus, tissue glucose uptake also depends on insulin delivery from plasma to the interstitium. Under physiological conditions, insulin transits through ECs via transcytosis, which is mediated by caveolae-localized insulin (IR) and Insulin Growth Factor 1 receptors.[65] EC-specific knockout of insulin receptors delays insulin-dependent signaling in skeletal muscle and adipose tissue.[66] Moreover, the knockout of Caveolin1, one of proteins required for caveolae formation and highly expressed by most ECs, compromises endothelial uptake of insulin and lowers the insulin sensitivity of the parenchymal target cells, e.g., skeletal myocytes.[67]

Fatty acid uptake

The process of FA and glycerol uptake from plasma starts with the triglyceride breakdown, which is performed by lipoprotein lipase (LPL). This enzyme localizes at the EC surface, tethered with a GPI-linked protein (GPIHBP1). Transporters such as FATP1/3/4, CD36, and aquaporin 7 (AQP7, a glycerol transporter) then mediate the entrance of FA and glycerol into the EC. Inside the EC, FA binds to intracellular lipid chaperones, including FA binding proteins FABP4 and FABP5. Internalized lipids can be stored in lipid droplets, consumed by the EC or exported to the interstitium.[68,69] Transcriptome analyses of ECs from heart, muscle, and adipose (both brown and white) display enrichment for many genes encoding proteins involved in lipid binding, transport and metabolism such as Cd36, Fatp1, Fatp3, Fatp4, Fabp4, Fabp5, Gpihbp1, Aqp7, and angiopoietin-like 4 (Angptl4).[4,8,15,70] This implies that ECs support an efficient supply system of FAs for these tissues. In fact, EC-specific deletion of both *Fabp4* and *Fabp5* significantly lowered FA oxidative metabolism capacity in skeletal and cardiac muscles.[71,72] Of note, the chaperone protein FABP4, also named as adipocyte protein 2, was initially thought to be an adipocyte-specific gene. However, it is now apparent that a large percentage of EC throughout the vascular tree express very high levels of this gene.

Genes encoding the transcription factors Meox2, Tcf15, and PPARγ, which are central regulators of cellular metabolism of FAs, are expressed at high levels in the EC from the heart, muscle, and adipose tissues, suggesting that they may establish the molecular signature related to the enrichment of FA handling machinery in EC.[6,15,73] Moreover, changes in their expression and/or transcriptional activity alter the expression of lipid binding/transporters and lipid uptake in microvascular ECs. For example, endothelial-specific deletion of *PPARγ* in mice decreases expression of *Cd36, Fabp4, Fatp3, Gpihbp1*, and impairs lipid uptake from plasma.[74] Meox2, and Tcf15 also are necessary to maintain the lipid fingerprint in heart EC, regulating the expression of FA transporters and binding proteins.[15] In the heart and skeletal muscle EC, Notch activity is an additional positive regulator of the expression of endothelial *Lpl, Angptl4, CD36, and Fabp4*.[75] Accordingly, the inhibition of EC Notch signaling reduces transendothelial lipid transport and leads to compensatory reliance on glucose, which causes heart hypertrophy and compromises heart function.[75] The impaired function of skeletal muscle and heart that is evident subsequent to EC-specific disruption of FA uptake and transport thus reinforces the importance of these EC functions in dictating tissue homeostasis.

EC FA uptake can be modulated by tissue paracrine signals, including VEGF-B, apelin (APLN) or 3-hydroxyisobutyrate, a catabolic intermediate of the branched chain amino acid valine.[76] The mechanisms of VEGF-B are best established. VEGF-B acting via VEGFR1/Neuropilin-1 stimulates the transcription of FA transport proteins, FATP3, and FATP4.[77] Accordingly, knockout of *Vegfb* in mice lowers lipid uptake in the muscle, heart, and brown adipose tissue.[77] VEGF-B synthesis is tightly associated with upregulation of mitochondrial genes, under the control of PGC1α,[77,78] which implies that skeletal and cardiac myocytes utilize VEGF-B to signal for increased FA uptake in accordance with their mitochondrial capacity to oxidize FAs. It is notable that VEGF-B signaling impedes GLUT1 function by lowering EC cholesterol content, which in turn decreases glucose EC transcytosis.[60] Thus, in these tissues VEGF-B appears to enforce the inverse relationship between FA oxidation and glucose consumption (as stipulated by the Randle cycle theorem).

Thus, the supply of nutrients to the tissue by ECs results from a complex bidirectional cross-talk of ECs with parenchymal cells that dynamically adjusts according to nutrient availability and tissue needs. While analysis of the molecular signatures of cardiac, skeletal muscle, and adipose ECs indicates they are specialized in providing FAs, these ECs also share homotypic features in delivering glucose and insulin, overall enabling tissue metabolic flexibility.

Release of angiocrine factors to support tissue physiology

Another important function of EC is the secretion of angiocrine factors that modify functions of cells within the microenvironment[79] (as indicated in Fig. 9.1). Some of these factors display tissue-specific expression profiles, while others are produced by most ECs but exert distinct effects dependent on the local microenvironment.[2]

Angiocrine signals have been studied extensively in the heart. VEGFR2 signaling induces ECs to release multiple factors that regulate cardiomyocyte contractility, survival, and hypertrophy.[80] For instance, activation of VEGFR2 increases EC production of NO, which in turn promotes cardiomyocyte hypertrophy by provoking degradation of "Regulator of G protein signaling 4", a repressor of Akt signaling.[81] Of note, NO production also is stimulated by VEGF-B, PGF, and macrophage-derived peptide PR39.[82,83] Neuregulin-1 (NRG1) and APLN are additional angiocrine molecules released by cardiac ECs that might exert protective effects in hypoxic cardiac muscle. Hypoxia induces NRG1 production by endocardial and coronary microvascular EC[84] and enhances APLN levels.[85] EC-selective deletion of *Nrg1* significantly decreases cardiac tolerance to ischemic insult. In turn, EC secretion of APLN promotes glucose uptake in cardiomyocytes by enhancing the production of GLUT4,[86] which may be beneficial for the heart when oxygen supply is insufficient.

A role for EC-derived APLN also is implicated in skeletal muscle based on reports that APLN can increase mitochondrial content in skeletal myocytes[87] and promote skeletal myocyte hypertrophy.[88] Other than APLN, putative angiocrine factors enriched in skeletal muscle include Angiopoietin 2, bone morphogenic protein 6 (BMP6), and fibroblast growth factor 9.[2] These factors may modulate vascular structural integrity and exert pleiotropic influences on surrounding cells. It is notable that BMP6, expressed by ECs of skeletal the muscle, heart, brown, and white adipose tissue,[89] can influence glucose sensitivity and oxidative metabolism in parenchymal cells.[90]

ECs from subcutaneous and visceral white adipose tissue of healthy mice secrete extracellular matrix proteins (collagens, thrombospondins, proteoglycans) and proteases that can shape the adipose environment and exert direct functional effects on neighboring cells.[91] Platelet-derived growth factor C (PDGF-C) secretion by EC may mediate the effects of VEGF-A in promoting the beiging of white adipocytes.[92] Overall, little is known about the nature of angiocrines from AT. Interestingly, EC-derived small extracellular vesicles provide an additional means of metabolic communication with adipocytes. Under fasting conditions, glucagon stimulates EC release of vesicles laden with proteins, lipids, and signaling components that may directly regulate adipocyte metabolism.[93] Adipose tissue ECs also accumulate small lipid droplets that are then released and taken up by preadipocytes/adipocytes.[68] These EC-secreted lipids act as PPARγ ligands within adipocytes, thus regulating adipocyte metabolic functions.[68]

EC-secreted factors also can dictate quiescence, self-renewal, and the fate of local stem cells in adipose tissue and muscle (Fig. 9.3). This is facilitated by the strategic

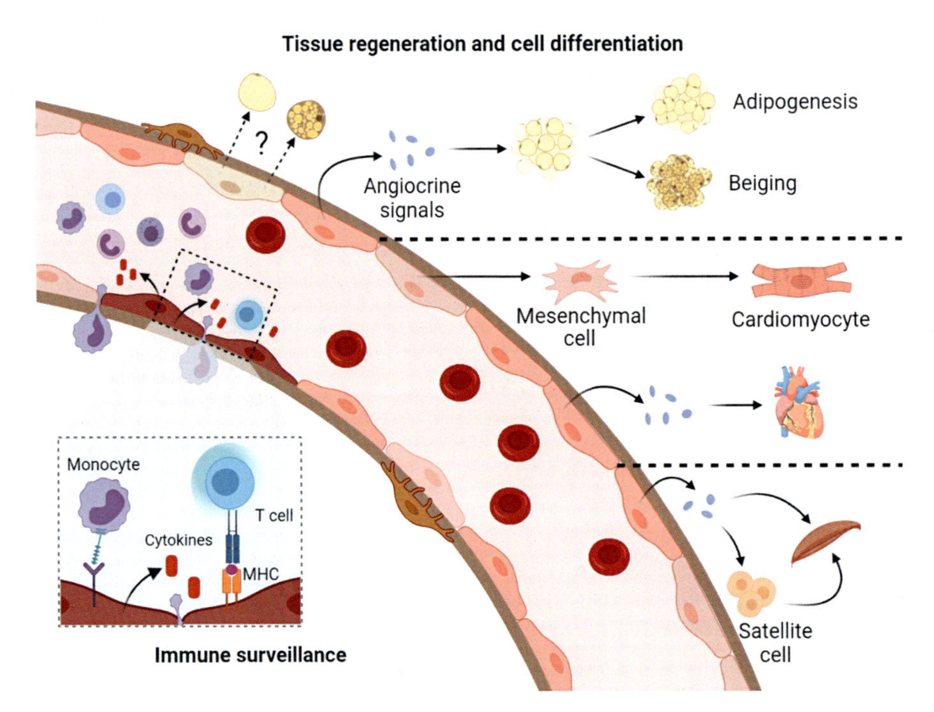

FIGURE 9.3 **Endothelial cells play pivotal roles in immunity and tissue regeneration.** Endothelial cells modulate inflammatory processes by controlling immune cells recruiting and trafficking, and antigen presentation **(left side)**. Endothelial cells release angiocrine signals that alter tissue processes (adipogenesis) or the stem-cell niche, as in skeletal muscle where these factors can alter the proliferation of satellite cells. Some endothelial cells also may have the capacity to differentiate into other tissue-resident cell types (**right side**).

positioning of tissue-resident stem and progenitor cells in proximity to capillary ECs. In white adipose tissue, EC-secreted factors regulate preadipocyte proliferation.[94] During development, BMP6 (produced by EC) controls both the differentiation of brown adipocytes from $Myf5^+$ precursors[90] and the proliferation of skeletal muscle satellite cells.[95] In the adult, EC production of transforming growth factor β1 and the Notch ligand Dll4 reinforce the quiescent state of adjacent skeletal muscle satellite cells.[96] EC angiocrine factors also are critical orchestrators of the regenerative process in muscle following injury by promoting the survival and proliferation of satellite cells and other muscle progenitor cell populations.[95,97,98] Furthermore, angiocrine signals can modulate the phenotype of immune resident cells, such as macrophages, during tissue remodeling. Indeed, ECs may recruit circulating immune cells to active muscle[99] and vascular endothelial Dll1 guides macrophage maturation in ischemic tissue by suppressing the induction of an inflammatory phenotype.[100] In addition, the metabolic activity of ECs also may indirectly affect cells within the local environment. For example, lactate released from highly glycolytic ECs was shown to direct macrophage polarization and muscle regeneration after ischemia.[101]

Endothelial cell role in immune surveillance

Inflammation is an essential physiological process that permits survival during infection and underlies tissue repair and regeneration. ECs play a pivotal role in this process as they maintain a noninflamed environment by controlling vascular permeability, inhibiting blood coagulation, and recruiting circulating cells to the peripheral tissue to support an inflammatory response.[102] Although it has been assumed that ECs do not express inflammatory-related genes until activated by cytokines, analyses of human and mouse samples have found that the heart, muscle, and adipose tissue possess EC subpopulations enriched for immune-related transcripts.[7–10,103] It is notable that the inflammatory-related signature of the cardiac and muscle ECs was observed in samples from healthy individuals and an increase in this immune-like EC population was detected in heart failure samples.[7,11] In healthy mice, ECs from both the muscle and heart display enrichment of interferon-induced genes, suggesting similarity in an inflammatory signature.[6] Cardiac ECs also displayed an overrepresentation of genes related to complement and coagulation systems when compared to ECs from skeletal muscle.[73] This is interesting as the heart is especially vulnerable to thrombotic events and can imply that the tissue-specific signature of EC can directly contribute to pathological states. Human white adipose tissue exhibits differential enrichment of inflammatory mediators in ECs from visceral when compared to subcutaneous adipose.[104] This coincides with the known prevalence of inflammatory cell infiltrates in visceral adipose from obese individuals.

More recently, subpopulations of ECs identified within several human tissues, including visceral adipose tissue (omentum), heart and muscle, expressed high levels of genes encoding major histocompatibility complex class II proteins.[105] These proteins, responsible for antigen presentation to trigger an adaptive immune response, are typically expressed on dendritic cells, macrophages, and B cells. Altogether, these studies reveal that some ECs may be specialized in participating in tissue immune surveillance, perhaps positioning the EC as a player in both innate and acquired immunity (Fig. 9.3).

Indicators of tissue-specific endothelial cell plasticity?

Several lines of evidence indicate that some ECs exhibit plasticity that may enable these cells to differentiate to other cell types to support tissue growth and regeneration (Fig. 9.3). For example, a subset of murine adipose ECs that express the transcription factor *Zfp423* (a multizinc finger transcriptional regulator that is a determinant of preadipocyte commitment) has been suggested to give rise to both white and brown adipocytes following activation of PPARγ.[106,107] *In vivo* lineage tracing of cells labeled initially by expression of the EC marker VE-cadherin was subsequently identified as preadipocytes and adipocytes.[107] However, the extent of this EC transdifferentiation potential in humans remains under debate as studies have provided conflicting results.[68,108]

A subpopulation of ECs from healthy hearts (of human and mouse) express canonical cardiomyocyte genes involved in processes such as cardiac muscle tissue development, myofibril assembly, and cardiac contraction,[7,103] Furthermore, cardiac ECs were found to have open chromatin at the expressed cardiomyocyte-specific genes, which appeared to be actively maintained by the surrounding cardiomyocytes.[109] This may reflect shared developmental origins of cardiomyocytes and cardiac ECs, but it also reinforces the experimental evidence that some ECs give rise to cardiomyocytes after injury through the process of endothelial-to-mesenchymal transition.[110] In contrast to the heart, ECs isolated from skeletal muscle do not express muscle-specific transcripts and there is no evidence to suggest that mature ECs can give rise to muscle progenitor cells in the adult, despite sharing a common somatic precursor.[111,112]

Consequences of compromised endothelial cell performance

Dysfunction of ECs contributes to the pathogenesis of many diseases. Disruption of key EC signal pathways associated with angiogenic responsiveness and the regulation of nutrient transport and angiocrine secretions occurs in aging, inflammation, and chronic nutrient excess. These conditions frequently correspond with reduced activation of Akt, which exerts broad-reaching downstream effects on EC functions. This dysregulation can occur secondary to impaired insulin sensitivity, inflammatory cytokines, or overactivation of intracellular suppressors of Akt.[113] Ultimately, decreased Akt activity permits an elevation in FoxO1, a powerful suppressor of EC metabolism and cell cycle that inhibits angiogenesis.[17,114–116] Concurrently, impaired Akt-dependent phosphorylation of endothelial NO synthase lowers the production of NO, which shifts EC contribution to inflammation, coagulation, and angiogenesis and also decreases angiocrine signals dependent on NO including vasodilation.[114,117,118]

Dysregulation of angiogenesis can lead to a mismatch between capillary network supply and tissue metabolic needs, which detrimentally affects tissue functions. For example, inhibition of angiogenesis during physiological cardiac hypertrophy culminates with cardiac contractile dysfunction.[47] In skeletal muscle, inadequate vascularization is linked closely with peripheral insulin resistance and impaired glucose uptake. Impaired capillary growth also is an early event in the progression of white adipose tissue dysfunction,[119] disrupting adipocyte functions and enhancing local inflammation. Adverse shifts in the balance of angiogenic regulators within the microenvironment can be one of the causes for failure of angiogenesis, as is known to occur in white adipose with long-term nutrient excess.[59,119–122]

Moreover, deficiencies in the uptake of lipids by the ECs lead to lower storage of FA by the adipose tissue, resulting in high circulating triglycerides, which accumulate in liver and peripheral tissues, contributing to the development of peripheral insulin resistance and type 2 diabetes mellitus.[69] Mutations in the gene encoding *GPIHBP1* that prevent its binding of LPL result in hypertriglyceridemia due to impaired lipolysis and EC uptake of FAs.[123] Overactivation of EC Notch signaling occurs with long-term high fat or high fat/high sugar diet and it downregulates transendothelial transport of insulin, significantly impairing skeletal muscle glucose uptake and contributing to systemic insulin resistance.[124]

EC senescence, which is accelerated by chronic inflammation and obesity,[104] is another mechanism through which EC functions are altered. Senescent ECs release an altered secretome that disrupts the functions of parenchymal cells.[125] In adipose, EC senescence–associated secretory profile provokes oxidative stress and premature senescence of adipocytes.[126] Remarkably, this EC senescence secretome exerts profound whole body effects, as it impaired insulin signaling in otherwise healthy animals.[126]

Conclusions

ECs are now recognized as leading mediators in the regulation of tissue-level and whole-body physiology. The regulation of capillary EC phenotype and angiogenesis is deeply intertwined; thus, it is difficult to unravel the independent influences of capillary nutrient transport, angiocrine signals, and angiogenesis. Together, these functions harmonize with the paracrine signals emanating from the local cells to fine-tune EC phenotype in accordance with fluctuating tissue needs. Impairments in these EC processes amplify the pathologies associated with systemic metabolic disturbances. Within this context, marked progress has been achieved in characterizing tissue-specific attributes of EC. The implications of these diverse and specialized subpopulations of ECs within each organ may be profound, but their functional significance needs further elucidation. As our understanding of EC biology grows, this knowledge holds promise to identify new biomarkers and molecular underpinnings of EC functions and/or disease.

Acknowledgments

The authors thank Drs. Emmanuel Nwadozi, Thomas Gustafsson, and Anthony Scimè for their review and discussion of the chapter. Schematic figures were created using BioRender.com. TLH is funded by the Natural Science and Engineering Council of Canada.

References

1. Aird WC. Phenotypic heterogeneity of the endothelium: I. Structure, function, and mechanisms. *Circ Res*. 2007;100(2):158–173. https://doi.org/10.1161/01.RES.0000255691.76142.4a.
2. Nolan DJ, Ginsberg M, Israely E, et al. Molecular signatures of tissue-specific microvascular endothelial cell heterogeneity in organ maintenance and regeneration. *Dev Cell*. 2013;26(2). https://doi.org/10.1016/j.devcel.2013.06.017.
3. Marcu R, Choi YJ, Xue J, et al. Human organ-specific endothelial cell heterogeneity. *iScience*. 2018;4:20–35. https://doi.org/10.1016/j.isci.2018.05.003.
4. Feng W, Chen L, Nguyen PK, Wu SM, Li G. Single cell analysis of endothelial cells identified organ-specific molecular signatures and heart-specific cell populations and molecular features. *Front Cardiovasc Med*. 2019;6. https://doi.org/10.3389/fcvm.2019.00165.
5. He S, Wang L-H, Liu Y, et al. Single-cell transcriptome profiling of an adult human cell atlas of 15 major organs. *Genome Biol*. 2020;21(1):294. https://doi.org/10.1186/s13059-020-02210-0.

6. Kalucka J, de Rooij LPMH, Goveia J, et al. Single-cell transcriptome atlas of murine endothelial cells. *Cell*. 2020;180(4). https://doi.org/10.1016/j.cell.2020.01.015, 764–779. e20.
7. Litviňuková M, Talavera-López C, Maatz H, et al. Cells of the adult human heart. *Nature*. 2020;588:466–472. https://doi.org/10.1038/s41586-020-2797-4.
8. Paik DT, Tian L, Williams IM, et al. Single-cell RNA sequencing unveils unique transcriptomic signatures of organ-specific endothelial cells. *Circulation*. 2020;142(19):1848–1862. https://doi.org/10.1161/CIRCULATIONAHA.119.041433.
9. Rubenstein AB, Smith GR, Raue U, et al. Single-cell transcriptional profiles in human skeletal muscle. *Sci Rep*. 2020;10(1):229. https://doi.org/10.1038/s41598-019-57110-6.
10. Vijay J, Gauthier M-F, Biswell RL, et al. Single-cell analysis of human adipose tissue identifies depot and disease specific cell types. *Nat Metab*. 2020;2(1):97–109. https://doi.org/10.1038/s42255-019-0152-6.
11. Wang L, Yu P, Zhou B, et al. Single-cell reconstruction of the adult human heart during heart failure and recovery reveals the cellular landscape underlying cardiac function. *Nat Cell Biol*. 2020;22(1):108–119. https://doi.org/10.1038/s41556-019-0446-7.
12. Adams RH. Molecular control of arterial-venous blood vessel identity. *J Anat*. 2003;202(1):105–112. https://doi.org/10.1046/j.1469-7580.2003.00137.x.
13. Manavski Y, Lucas T, Glaser SF, et al. Clonal expansion of endothelial cells contributes to ischemia-induced neovascularization. *Circ Res*. 2018;122(5):670–677. https://doi.org/10.1161/CIRCRESAHA.117.312310.
14. Li Z, Solomonidis EG, Meloni M, et al. Single-cell transcriptome analyses reveal novel targets modulating cardiac neovascularization by resident endothelial cells following myocardial infarction. *Eur Heart J*. 2019;40(30):2507–2520. https://doi.org/10.1093/eurheartj/ehz305.
15. Coppiello G, Collantes M, Sirerol-Piquer MS, et al. Meox2/Tcf15 heterodimers program the heart capillary endothelium for cardiac fatty acid uptake. *Circulation*. 2015;131(9):815–826. https://doi.org/10.1161/CIRCULATIONAHA.114.013721.
16. Poole DC, Pittman RN, Musch TI, Østergaard L. August Krogh's theory of muscle microvascular control and oxygen delivery: a paradigm shift based on new data. *J Physiol*. 2020;598(20):4473–4507. https://doi.org/10.1113/JP279223.
17. Wilhelm K, Happel K, Eelen G, et al. FOXO1 couples metabolic activity and growth state in the vascular endothelium. *Nature*. 2016;529(7585):216–220. https://doi.org/10.1038/nature16498.
18. Mack JJ, Mosqueiro TS, Archer BJ, et al. NOTCH1 is a mechanosensor in adult arteries. *Nat Commun*. 2017;8(1):1620. https://doi.org/10.1038/s41467-017-01741-8.
19. Kalucka J, Bierhansl L, Conchinha NV, et al. Quiescent endothelial cells upregulate fatty acid β-oxidation for vasculoprotection via redox homeostasis. *Cell Metabol*. 2018;28(6). https://doi.org/10.1016/j.cmet.2018.07.016, 881–894. e13.
20. Xiong J, Kawagishi H, Yan Y, et al. A metabolic basis for endothelial-to-mesenchymal transition. *Mol Cell*. 2018;69(4). https://doi.org/10.1016/j.molcel.2018.01.010, 689–698. e7.
21. Gerhardt H, Golding M, Fruttiger M, et al. VEGF guides angiogenic sprouting utilizing endothelial tip cell filopodia. *J Cell Biol*. 2003;161(6):1163–1177. https://doi.org/10.1083/jcb.200302047.
22. Gerhardt H, Betsholtz C. How do endothelial cells orientate? *Experientia Suppl*. 2005;(94):3–15. https://doi.org/10.1007/3-7643-7311-3_1.
23. Potente M, Carmeliet P. The link between angiogenesis and endothelial metabolism. *Annu Rev Physiol*. 2017;79:43–66. https://doi.org/10.1146/annurev-physiol-021115-105134.
24. Dumas SJ, García-Caballero M, Carmeliet P. Metabolic signatures of distinct endothelial phenotypes. *Trends Endocrinol Metabol*. 2020;31(8):580–595. https://doi.org/10.1016/j.tem.2020.05.009.
25. Li X, Kumar A, Carmeliet P. Metabolic pathways fueling the endothelial cell drive. *Annu Rev Physiol*. 2019;81:483–503. https://doi.org/10.1146/annurev-physiol-020518-114731.
26. Simons M, Gordon E, Claesson-Welsh L. Mechanisms and regulation of endothelial VEGF receptor signalling. *Nat Rev Mol Cell Biol*. 2016;17(10):611–625. https://doi.org/10.1038/nrm.2016.87.
27. Bry M, Kivelä R, Leppänen V-M, Alitalo K. Vascular endothelial growth factor-B in physiology and disease. *Physiol Rev*. 2014;94(3):779–794. https://doi.org/10.1152/physrev.00028.2013.
28. Hudlicka O, Brown M, Egginton S. Angiogenesis in skeletal and cardiac muscle. *Physiol Rev*. 1992;72(2):369–417. https://doi.org/10.1152/physrev.1992.72.2.369.
29. Egginton S. Physiological factors influencing capillary growth. *Acta Physiol (Oxf)*. 2011;202(3):225–239. https://doi.org/10.1111/j.1748-1716.2010.02194.x.
30. Gustafsson T. Vascular remodelling in human skeletal muscle. *Biochem Soc Trans*. 2011;39(6):1628–1632. https://doi.org/10.1042/BST20110720.
31. Olfert IM, Howlett RA, Wagner PD, Breen EC. Myocyte vascular endothelial growth factor is required for exercise-induced skeletal muscle angiogenesis. *Am J Physiol Regul Integr Comp Physiol*. 2010;299(4):R1059–R1067. https://doi.org/10.1152/ajpregu.00347.2010.
32. Gustafsson T, Puntschart A, Kaijser L, Jansson E, Sundberg CJ. Exercise-induced expression of angiogenesis-related transcription and growth factors in human skeletal muscle. *Am J Physiol*. 1999;276(2):H679–H685. https://doi.org/10.1152/ajpheart.1999.276.2.H679.
33. Gustafsson T, Rundqvist H, Norrbom J, Rullman E, Jansson E, Sundberg CJ. The influence of physical training on the angiopoietin and VEGF-A systems in human skeletal muscle. *J Appl Physiol*. 2007;103(3):1012–1020. https://doi.org/10.1152/japplphysiol.01103.2006.
34. Ameln H, Gustafsson T, Sundberg CJ, et al. Physiological activation of hypoxia inducible factor-1 in human skeletal muscle. *FASEB J*. 2005;19(8):1009–1011. https://doi.org/10.1096/fj.04-2304fje.
35. Arany Z, Foo S-Y, Ma Y, et al. HIF-independent regulation of VEGF and angiogenesis by the transcriptional coactivator PGC-1alpha. *Nature*. 2008;451(7181):1008–1012. https://doi.org/10.1038/nature06613.
36. Williams JL, Cartland D, Hussain A, Egginton S. A differential role for nitric oxide in two forms of physiological angiogenesis in mouse. *J Physiol*. 2006;570(Pt 3):445–454. https://doi.org/10.1113/jphysiol.2005.095596.
37. Rivilis I, Milkiewicz M, Boyd P, et al. Differential involvement of MMP-2 and VEGF during muscle stretch- versus shear stress-induced angiogenesis. *Am J Physiol Heart Circ Physiol*. 2002;283(4):H1430–H1438. https://doi.org/10.1152/ajpheart.00082.2002.
38. Uchida C, Nwadozi E, Hasanee A, Olenich S, Olfert IM, Haas TL. Muscle-derived vascular endothelial growth factor regulates microvascular remodelling in response to increased shear stress in mice. *Acta Physiol (Oxf)*. 2015;214(3):349–360. https://doi.org/10.1111/apha.12463.
39. Muoio DM, Dohm GL, Fiedorek FT, Tapscott EB, Coleman RA, Dohn GL. Leptin directly alters lipid partitioning in skeletal muscle. *Diabetes*. 1997;46(8):1360–1363. https://doi.org/10.2337/diab.46.8.1360.
40. Nwadozi E, Ng A, Strömberg A, et al. Leptin is a physiological regulator of skeletal muscle angiogenesis and is locally produced by PDGFRα and PDGFRβ expressing perivascular cells. *Angiogenesis*. 2019;22(1):103–115. https://doi.org/10.1007/s10456-018-9641-6.
41. Riley PR, Smart N. Vascularizing the heart. *Cardiovasc Res*. 2011;91(2):260–268. https://doi.org/10.1093/cvr/cvr035.
42. Anversa P, Levicky V, Beghi C, McDonald SL, Kikkawa Y. Morphometry of exercise-induced right ventricular hypertrophy

in the rat. *Circ Res*. 1983;52(1):57–64. https://doi.org/10.1161/01.res.52.1.57.
43. Oka T, Akazawa H, Naito AT, Komuro I. Angiogenesis and cardiac hypertrophy: maintenance of cardiac function and causative roles in heart failure. *Circ Res*. 2014;114(3):565–571. https://doi.org/10.1161/CIRCRESAHA.114.300507.
44. Wright AJ, Hudlicka O. Capillary growth and changes in heart performance induced by chronic bradycardial pacing in the rabbit. *Circ Res*. 1981;49(2):469–478. https://doi.org/10.1161/01.res.49.2.469.
45. Tomanek RJ, Torry RJ. Growth of the coronary vasculature in hypertrophy: mechanisms and model dependence. *Cell Mol Biol Res*. 1994;40(2):129–136.
46. Gogiraju R, Bochenek ML, Schäfer K. Angiogenic endothelial cell signaling in cardiac hypertrophy and heart failure. *Front Cardiovasc Med*. 2019;6. https://doi.org/10.3389/fcvm.2019.00020.
47. Shiojima I, Sato K, Izumiya Y, et al. Disruption of coordinated cardiac hypertrophy and angiogenesis contributes to the transition to heart failure. *J Clin Invest*. 2005;115(8):2108–2118. https://doi.org/10.1172/JCI24682.
48. Sano M, Minamino T, Toko H, et al. p53-induced inhibition of Hif-1 causes cardiac dysfunction during pressure overload. *Nature*. 2007;446(7134):444–448. https://doi.org/10.1038/nature05602.
49. Kivelä R, Hemanthakumar KA, Vaparanta K, et al. Endothelial cells regulate physiological cardiomyocyte growth via VEGFR2-mediated paracrine signaling. *Circulation*. 2019;139(22):2570–2584. https://doi.org/10.1161/CIRCULATIONAHA.118.036099.
50. Räsänen M, Sultan I, Paech J, et al. VEGF-B promotes endocardium-derived coronary vessel development and cardiac regeneration. *Circulation*. 2021;143:65–77. https://doi.org/10.1161/CIRCULATIONAHA.120.050635.
51. Kocijan T, Rehman M, Colliva A, et al. Genetic lineage tracing reveals poor angiogenic potential of cardiac endothelial cells. *Cardiovasc Res*. 2020;30. https://doi.org/10.1093/cvr/cvaa012. Published online January.
52. Xue Y, Petrovic N, Cao R, et al. Hypoxia-independent angiogenesis in adipose tissues during cold acclimation. *Cell Metabol*. 2009;9(1):99–109. https://doi.org/10.1016/j.cmet.2008.11.009.
53. Cao Y, Wang H, Wang Q, Han X, Zeng W. Three-dimensional volume fluorescence-imaging of vascular plasticity in adipose tissues. *Mol Metab*. 2018;14:71–81. https://doi.org/10.1016/j.molmet.2018.06.004.
54. Fredriksson JM, Lindquist JM, Bronnikov GE, Nedergaard J. Norepinephrine induces vascular endothelial growth factor gene expression in brown adipocytes through a beta -adrenoreceptor/cAMP/protein kinase A pathway involving Src but independently of Erk1/2. *J Biol Chem*. 2000;275(18):13802–13811. https://doi.org/10.1074/jbc.275.18.13802.
55. Cannon B, Nedergaard J. Brown adipose tissue: function and physiological significance. *Physiol Rev*. 2004;84(1):277–359. https://doi.org/10.1152/physrev.00015.2003.
56. Goossens GH, Bizzarri A, Venteclef N, et al. Increased adipose tissue oxygen tension in obese compared with lean men is accompanied by insulin resistance, impaired adipose tissue capillarization, and inflammation. *Circulation*. 2011;124(1):67–76. https://doi.org/10.1161/CIRCULATIONAHA.111.027813.
57. Cao Y. Angiogenesis modulates adipogenesis and obesity. *J Clin Invest*. 2007;117(9):2362–2368. https://doi.org/10.1172/JCI32239.
58. Nishimura S, Manabe I, Nagasaki M, et al. Adipogenesis in obesity requires close interplay between differentiating adipocytes, stromal cells, and blood vessels. *Diabetes*. 2007;56(6):1517–1526. https://doi.org/10.2337/db06-1749.
59. Ayaz-Guner S, Alessio N, Acar MB, et al. A comparative study on normal and obese mice indicates that the secretome of mesenchymal stromal cells is influenced by tissue environment and physiopathological conditions. *Cell Commun Signal*. 2020;18(1):118. https://doi.org/10.1186/s12964-020-00614-w.
60. Moessinger C, Nilsson I, Muhl L, et al. VEGF-B signaling impairs endothelial glucose transcytosis by decreasing membrane cholesterol content. *EMBO Rep*. 2020;21(7). https://doi.org/10.15252/embr.201949343.
61. Huang Y, Lei L, Liu D, et al. Normal glucose uptake in the brain and heart requires an endothelial cell-specific HIF-1α-dependent function. *Proc Natl Acad Sci USA*. 2012;109(43):17478–17483. https://doi.org/10.1073/pnas.1209281109.
62. De Bock K, Georgiadou M, Schoors S, et al. Role of PFKFB3-driven glycolysis in vessel sprouting. *Cell*. 2013;154(3):651–663. https://doi.org/10.1016/j.cell.2013.06.037.
63. Veys K, Fan Z, Ghobrial M, et al. Role of the GLUT1 glucose transporter in postnatal CNS angiogenesis and blood-brain barrier integrity. *Circ Res*. 2020;127(4):466–482. https://doi.org/10.1161/CIRCRESAHA.119.316463.
64. Richter EA, Hargreaves M. Exercise, GLUT4, and skeletal muscle glucose uptake. *Physiol Rev*. 2013;93(3):993–1017. https://doi.org/10.1152/physrev.00038.2012.
65. Yazdani S, Jaldin-Fincati JR, Pereira RVS, Klip A. Endothelial cell barriers: transport of molecules between blood and tissues. *Traffic*. 2019;20(6):390–403. https://doi.org/10.1111/tra.12645.
66. Konishi M, Sakaguchi M, Lockhart SM, et al. Endothelial insulin receptors differentially control insulin signaling kinetics in peripheral tissues and brain of mice. *Proc Natl Acad Sci USA*. 2017;114(40):E8478–E8487. https://doi.org/10.1073/pnas.1710625114.
67. Wang H, Wang AX, Barrett EJ. Caveolin-1 is required for vascular endothelial insulin uptake. *Am J Physiol Endocrinol Metab*. 2011;300(1):E134–E144. https://doi.org/10.1152/ajpendo.00498.2010.
68. Gogg S, Nerstedt A, Boren J, Smith U. Human adipose tissue microvascular endothelial cells secrete PPARγ ligands and regulate adipose tissue lipid uptake. *JCI Insight*. 2019;4(5). https://doi.org/10.1172/jci.insight.125914.
69. Hasan SS, Fischer A. The endothelium: an active regulator of lipid and glucose homeostasis. *Trends Cell Biol*. 2021;31(1):37–49. https://doi.org/10.1016/j.tcb.2020.10.003.
70. Briot A, Decaunes P, Volat F, et al. Senescence alters PPARγ (peroxisome proliferator-activated receptor gamma)-dependent fatty acid handling in human adipose tissue microvascular endothelial cells and favors inflammation. *Arterioscler Thromb Vasc Biol*. 2018;38(5):1134–1146. https://doi.org/10.1161/ATVBAHA.118.310797.
71. Iso T, Maeda K, Hanaoka H, et al. Capillary endothelial fatty acid binding proteins 4 and 5 play a critical role in fatty acid uptake in heart and skeletal muscle. *Arterioscler Thromb Vasc Biol*. 2013;33(11):2549–2557. https://doi.org/10.1161/ATVBAHA.113.301588.
72. Iso T, Haruyama H, Sunaga H, et al. Exercise endurance capacity is markedly reduced due to impaired energy homeostasis during prolonged fasting in FABP4/5 deficient mice. *BMC Physiol*. 2019;19(1):1. https://doi.org/10.1186/s12899-019-0038-6.
73. Lother A, Bergemann S, Deng L, Moser M, Bode C, Hein L. Cardiac endothelial cell transcriptome. *Arterioscler Thromb Vasc Biol*. 2018;38(3):566–574. https://doi.org/10.1161/ATVBAHA.117.310549.
74. Kanda T, Brown JD, Orasanu G, et al. PPARgamma in the endothelium regulates metabolic responses to high-fat diet in mice. *J Clin Invest*. 2009;119(1):110–124. https://doi.org/10.1172/JCI36233.
75. Jabs M, Rose AJ, Lehmann LH, et al. Inhibition of endothelial notch signaling impairs fatty acid transport and leads to metabolic and vascular remodeling of the adult heart. *Circulation*. 2018;137(24):2592–2608. https://doi.org/10.1161/CIRCULATIONAHA.117.029733.
76. Hwangbo C, Wu J, Papangeli I, et al. Endothelial APLNR regulates tissue fatty acid uptake and is essential for apelin's glucose-lowering effects. *Sci Transl Med*. 2017;9(407). https://doi.org/10.1126/scitranslmed.aad4000.

77. Hagberg CE, Falkevall A, Wang X, et al. Vascular endothelial growth factor B controls endothelial fatty acid uptake. *Nature*. 2010;464(7290):917–921. https://doi.org/10.1038/nature08945.
78. Mehlem A, Palombo I, Wang X, Hagberg CE, Eriksson U, Falkevall A. PGC-1α coordinates mitochondrial respiratory capacity and muscular fatty acid uptake via regulation of VEGF-B. *Diabetes*. 2016;65(4):861–873. https://doi.org/10.2337/db15-1231.
79. Rafii S, Butler JM, Ding B-S. Angiocrine functions of organ-specific endothelial cells. *Nature*. 2016;529(7586):316–325. https://doi.org/10.1038/nature17040.
80. Hemanthakumar KA, Kivelä R. Angiogenesis and angiocrines regulating heart growth. *Vasc Biol*. 2020;2(1):R93–R104. https://doi.org/10.1530/VB-20-0006.
81. Jaba IM, Zhuang ZW, Li N, et al. NO triggers RGS4 degradation to coordinate angiogenesis and cardiomyocyte growth. *J Clin Invest*. 2013;123(4):1718–1731. https://doi.org/10.1172/JCI65112.
82. Tirziu D, Chorianopoulos E, Moodie KL, et al. Myocardial hypertrophy in the absence of external stimuli is induced by angiogenesis in mice. *J Clin Invest*. 2007;117(11):3188–3197. https://doi.org/10.1172/JCI32024.
83. Silvestre J-S, Tamarat R, Ebrahimian TG, et al. Vascular endothelial growth factor-b promotes in vivo angiogenesis. *Circ Res*. 2003;93(2):114–123. https://doi.org/10.1161/01.RES.0000081594.21764.44.
84. Hedhli N, Huang Q, Kalinowski A, et al. Endothelium-derived neuregulin protects the heart against ischemic injury. *Circulation*. 2011;123(20):2254–2262. https://doi.org/10.1161/CIRCULATIONAHA.110.991125.
85. Eyries M, Siegfried G, Ciumas M, et al. Hypoxia-induced apelin expression regulates endothelial cell proliferation and regenerative angiogenesis. *Circ Res*. 2008;103(4):432–440. https://doi.org/10.1161/CIRCRESAHA.108.179333.
86. Zeng H, He X, Chen J. Endothelial sirtuin 3 dictates glucose transport to cardiomyocyte and sensitizes pressure overload-induced heart failure. *J Am Heart Assoc*. 2020;9(11). https://doi.org/10.1161/JAHA.120.015895.
87. Frier BC, Williams DB, Wright DC. The effects of apelin treatment on skeletal muscle mitochondrial content. *Am J Physiol Regul Integr Comp Physiol*. 2009;297(6):R1761–R1768. https://doi.org/10.1152/ajpregu.00422.2009.
88. Vinel C, Lukjanenko L, Batut A, et al. The exerkine apelin reverses age-associated sarcopenia. *Nat Med*. 2018;24(9):1360–1371. https://doi.org/10.1038/s41591-018-0131-6.
89. Tabula Muris Consortium. Single-cell transcriptomics of 20 mouse organs creates a Tabula Muris. *Nature*. 2018;562(7727):367–372. https://doi.org/10.1038/s41586-018-0590-4.
90. Sharma A, Huard C, Vernochet C, et al. Brown fat determination and development from muscle precursor cells by novel action of bone morphogenetic protein 6. *PLoS One*. 2014;9(3):e92608. https://doi.org/10.1371/journal.pone.0092608.
91. Hocking SL, Wu LE, Guilhaus M, Chisholm DJ, James DE. Intrinsic depot-specific differences in the secretome of adipose tissue, preadipocytes, and adipose tissue-derived microvascular endothelial cells. *Diabetes*. 2010;59(12):3008–3016. https://doi.org/10.2337/db10-0483.
92. Seki T, Hosaka K, Lim S, et al. Endothelial PDGF-CC regulates angiogenesis-dependent thermogenesis in beige fat. *Nat Commun*. 2016;7:12152. https://doi.org/10.1038/ncomms12152.
93. Crewe C, Joffin N, Rutkowski JM, et al. An endothelial-to-adipocyte extracellular vesicle axis governed by metabolic state. *Cell*. 2018;175(3). https://doi.org/10.1016/j.cell.2018.09.005, 695–708. e13.
94. Lau DC, Shillabeer G, Wong KL, Tough SC, Russell JC. Influence of paracrine factors on preadipocyte replication and differentiation. *Int J Obes*. 1990;14(Suppl 3):193–201.
95. Stantzou A, Schirwis E, Swist S, et al. BMP signaling regulates satellite cell-dependent postnatal muscle growth. *Development*. 2017;144(15):2737–2747. https://doi.org/10.1242/dev.144089.
96. Verma M, Asakura Y, Murakonda BSR, et al. Muscle satellite cell cross-talk with a vascular niche maintains quiescence via VEGF and notch signaling. *Cell Stem Cell*. 2018;23(4). https://doi.org/10.1016/j.stem.2018.09.007, 530–543. e9.
97. Christov C, Chrétien F, Abou-Khalil R, et al. Muscle satellite cells and endothelial cells: close neighbors and privileged partners. *Mol Biol Cell*. 2007;18(4):1397–1409. https://doi.org/10.1091/mbc.E06-08-0693.
98. Criswell TL, Corona BT, Wang Z, et al. The role of endothelial cells in myofiber differentiation and the vascularization and innervation of bioengineered muscle tissue in vivo. *Biomaterials*. 2013;34(1):140–149. https://doi.org/10.1016/j.biomaterials.2012.09.045.
99. Strömberg A, Rullman E, Jansson E, Gustafsson T. Exercise-induced upregulation of endothelial adhesion molecules in human skeletal muscle and number of circulating cells with remodeling properties. *J Appl Physiol*. 2017;122(5):1145–1154. https://doi.org/10.1152/japplphysiol.00956.2016.
100. Krishnasamy K, Limbourg A, Kapanadze T, et al. Blood vessel control of macrophage maturation promotes arteriogenesis in ischemia. *Nat Commun*. 2017;8. https://doi.org/10.1038/s41467-017-00953-2.
101. Zhang J, Muri J, Fitzgerald G, et al. Endothelial lactate controls muscle regeneration from ischemia by inducing M2-like macrophage polarization. *Cell Metabol*. 2020;31(6). https://doi.org/10.1016/j.cmet.2020.05.004, 1136–1153. e7.
102. Pober JS, Sessa WC. Evolving functions of endothelial cells in inflammation. *Nat Rev Immunol*. 2007;7(10):803–815. https://doi.org/10.1038/nri2171.
103. Jambusaria A, Hong Z, Zhang L, et al. Endothelial heterogeneity across distinct vascular beds during homeostasis and inflammation. *Elife*. 2020;9. https://doi.org/10.7554/eLife.51413.
104. Villaret A, Galitzky J, Decaunes P, et al. Adipose tissue endothelial cells from obese human subjects: differences among depots in angiogenic, metabolic, and inflammatory gene expression and cellular senescence. *Diabetes*. 2010;59(11):2755–2763. https://doi.org/10.2337/db10-0398.
105. Han X, Zhou Z, Fei L, et al. Construction of a human cell landscape at single-cell level. *Nature*. 2020;581(7808):303–309. https://doi.org/10.1038/s41586-020-2157-4.
106. Gupta RK, Mepani RJ, Kleiner S, et al. Zfp423 expression identifies committed preadipocytes and localizes to adipose endothelial and perivascular cells. *Cell Metabol*. 2012;15(2):230–239. https://doi.org/10.1016/j.cmet.2012.01.010.
107. Tran K-V, Gealekman O, Frontini A, et al. The vascular endothelium of the adipose tissue gives rise to both white and brown fat cells. *Cell Metabol*. 2012;15(2):222–229. https://doi.org/10.1016/j.cmet.2012.01.008.
108. Min SY, Kady J, Nam M, et al. Human "brite/beige" adipocytes develop from capillary networks, and their implantation improves metabolic homeostasis in mice. *Nat Med*. 2016;22(3):312–318. https://doi.org/10.1038/nm.4031.
109. Yucel N, Axsom J, Yang Y, Li L, Rhoades JH, Arany Z. Cardiac endothelial cells maintain open chromatin and expression of cardiomyocyte myofibrillar genes. *Elife*. 2020;9. https://doi.org/10.7554/eLife.55730.
110. Fioret BA, Heimfeld JD, Paik DT, Hatzopoulos AK. Endothelial cells contribute to generation of adult ventricular myocytes during cardiac homeostasis. *Cell Rep*. 2014;8(1):229–241. https://doi.org/10.1016/j.celrep.2014.06.004.
111. Kardon G, Campbell JK, Tabin CJ. Local extrinsic signals determine muscle and endothelial cell fate and patterning in the vertebrate limb. *Dev Cell*. 2002;3(4):533–545. https://doi.org/10.1016/s1534-5807(02)00291-5.

112. Ambler CA, Nowicki JL, Burke AC, Bautch VL. Assembly of trunk and limb blood vessels involves extensive migration and vasculogenesis of somite-derived angioblasts. *Dev Biol*. 2001; 234(2):352–364. https://doi.org/10.1006/dbio.2001.0267.
113. Koide M, Ikeda K, Akakabe Y, et al. Apoptosis regulator through modulating IAP expression (ARIA) controls the PI3K/Akt pathway in endothelial and endothelial progenitor cells. *Proc Natl Acad Sci USA*. 2011;108(23):9472–9477. https://doi.org/10.1073/pnas.1101296108.
114. Karki S, Farb MG, Ngo DTM, et al. Forkhead box O-1 modulation improves endothelial insulin resistance in human obesity. *Arterioscler Thromb Vasc Biol*. 2015;35(6):1498–1506. https://doi.org/10.1161/ATVBAHA.114.305139.
115. Rudnicki M, Abdifarkosh G, Nwadozi E, et al. Endothelial-specific FoxO1 depletion prevents obesity-related disorders by increasing vascular metabolism and growth. *Elife*. 2018;7:e39780. https://doi.org/10.7554/eLife.39780.
116. Nwadozi E, Roudier E, Rullman E, et al. Endothelial FoxO proteins impair insulin sensitivity and restrain muscle angiogenesis in response to a high-fat diet. *FASEB J*. 2016;30(9):3039–3052. https://doi.org/10.1096/fj.201600245R.
117. Akakabe Y, Koide M, Kitamura Y, et al. Ecscr regulates insulin sensitivity and predisposition to obesity by modulating endothelial cell functions. *Nat Commun*. 2013;4:2389. https://doi.org/10.1038/ncomms3389.
118. Pellegrinelli V, Rouault C, Veyrie N, Clément K, Lacasa D. Endothelial cells from visceral adipose tissue disrupt adipocyte functions in a three-dimensional setting: partial rescue by angiopoietin-1. *Diabetes*. 2014;63(2):535–549. https://doi.org/10.2337/db13-0537.
119. Crewe C, An YA, Scherer PE. The ominous triad of adipose tissue dysfunction: inflammation, fibrosis, and impaired angiogenesis. *J Clin Invest*. 2017;127(1):74–82. https://doi.org/10.1172/JCI88883.
120. Nugroho DB, Ikeda K, Barinda AJ, et al. Neuregulin-4 is an angiogenic factor that is critically involved in the maintenance of adipose tissue vasculature. *Biochem Biophys Res Commun*. 2018; 503(1):378–384. https://doi.org/10.1016/j.bbrc.2018.06.043.
121. Jiang H, Ding X, Cao Y, Wang H, Zeng W. Dense intra-adipose sympathetic arborizations are essential for cold-induced beiging of mouse white adipose tissue. *Cell Metabol*. 2017;26(4). https://doi.org/10.1016/j.cmet.2017.08.016, 686–692. e3.
122. Ratushnyy A, Ezdakova M, Buravkova L. Secretome of senescent adipose-derived mesenchymal stem cells negatively regulates angiogenesis. *Int J Mol Sci*. 2020;21(5). https://doi.org/10.3390/ijms21051802.
123. Beigneux AP, Franssen R, Bensadoun A, et al. Chylomicronemia with a mutant GPIHBP1 (Q115P) that cannot bind lipoprotein lipase. *Arterioscler Thromb Vasc Biol*. 2009;29(6):956–962. https://doi.org/10.1161/ATVBAHA.109.186577.
124. Hasan SS, Jabs M, Taylor J, et al. Endothelial Notch signaling controls insulin transport in muscle. *EMBO Mol Med*. 2020;12(4): e09271. https://doi.org/10.15252/emmm.201809271.
125. Yin H, Pickering JG. Cellular senescence and vascular disease: novel routes to better understanding and therapy. *Can J Cardiol*. 2016;32(5):612–623. https://doi.org/10.1016/j.cjca.2016.02.051.
126. Barinda AJ, Ikeda K, Nugroho DB, et al. Endothelial progeria induces adipose tissue senescence and impairs insulin sensitivity through senescence associated secretory phenotype. *Nat Commun*. 2020;11(1):481. https://doi.org/10.1038/s41467-020-14387-w.

CHAPTER

10

The neurovascular unit and blood–CNS barriers in health and disease

Azzurra Cottarelli[1,2], *Charlotte R. Wayne*[1], *Dritan Agalliu*[1,2] *and Saptarshi Biswas*[1]

[1]Department of Neurology, Columbia University Irving Medical Center, New York, NY, United States [2]Department of Pathology and Cell Biology, Columbia University Irving Medical Center, New York, NY, United States

The neurovascular unit and the blood–CNS barrier in the healthy state

The central nervous system (CNS) is vascularized mainly by angiogenesis, a process in which new blood vessels sprout from existing vessels. During CNS development, blood vessels are in close contact with both neuronal and glial progenitors and mature cells which leads to the formation of a neurovascular unit (NVU) (Fig. 10.1A). Reciprocal interactions between vascular components such as endothelial cells (ECs), endothelial-derived basement membrane (BM) and mural cells [pericytes (PCs) and vascular smooth muscle cells (VSMCs)], and neuroglial components including neurons, microglia, astroglia (astrocytes and Müller glia), and glia-derived BM of the NVU regulate proper angiogenesis and barrier properties unique to the neurovasculature, namely the blood–brain barrier (BBB), the blood–spinal cord barrier (BSCB), and the blood–retinal barrier (BRB) in the brain, the spinal cord, and the eye, respectively (Fig. 10.1A).

The neurovascular barrier establishment is mediated by two primary cell biological mechanisms that regulate the movement of nutrients, antibodies, and immune cells from the bloodstream into the CNS and vice versa. First, CNS ECs form several types of junctions between them including adherens, gap, and tight junctions (TJs) (Fig. 10.1B). TJs, which prevent the movement of small molecules between ECs, establish a paracellular vascular barrier (Fig. 10.1B). The major TJ transmembrane proteins at the BBB/BSCB/BRB are Claudin-5, -12, and Occludin, which interact with similar proteins in adjacent cells to form these specialized junctions. Zonula Occludens-1, -2, and -3 proteins (ZO-1, -2, and -3) are cytoplasmic TJ-associated proteins that form anchors with the cytoskeleton (Fig. 10.1B; reviewed in detail in 1,2). CNS ECs also form adherens and gap junctions, which play important roles in cell adhesion, cell-cell communication, response to shear stress, and regulation of neurovascular coupling (Fig. 10.1B; reviewed in detail in Refs. 3–5). Second, CNS ECs exhibit negligible transcellular permeability due to a low rate of receptor-mediated or absorptive transcytosis, regulated by a low number of caveolae combined with the absence of fenestrae (Fig. 10.1B; reviewed in detail in Refs. 6,7). These properties limit the movement of larger proteins, antibodies, and immune cells across CNS ECs. To overcome these two physical barriers, CNS ECs express a large number of passive and active transporters that shuttle nutrients and ions across blood vessels into the CNS and regulate the efflux of toxins and drugs out of the CNS (Fig. 10.1B).[1] Moreover, mature CNS ECs in the healthy brain suppress expression of cell adhesion molecules (CAMs) that mediate interactions with immune cells. These essential barrier properties are shared among all CNS ECs located in the brain, spinal cord and the retina.[8]

CNS angiogenesis and BBB/BRB development

Brain

Brain angiogenesis has mostly been studied in the context of the cerebral cortex. Starting from embryonic day (E) 10.0–11.5, neural progenitor cells differentiate into radial glia and cortical progenitors that will give

The Vasculome
https://doi.org/10.1016/B978-0-12-822546-2.00023-X

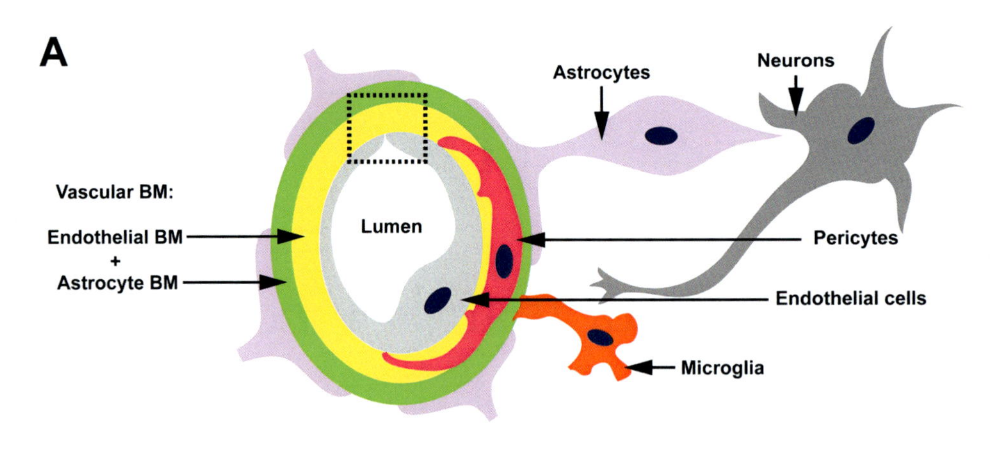

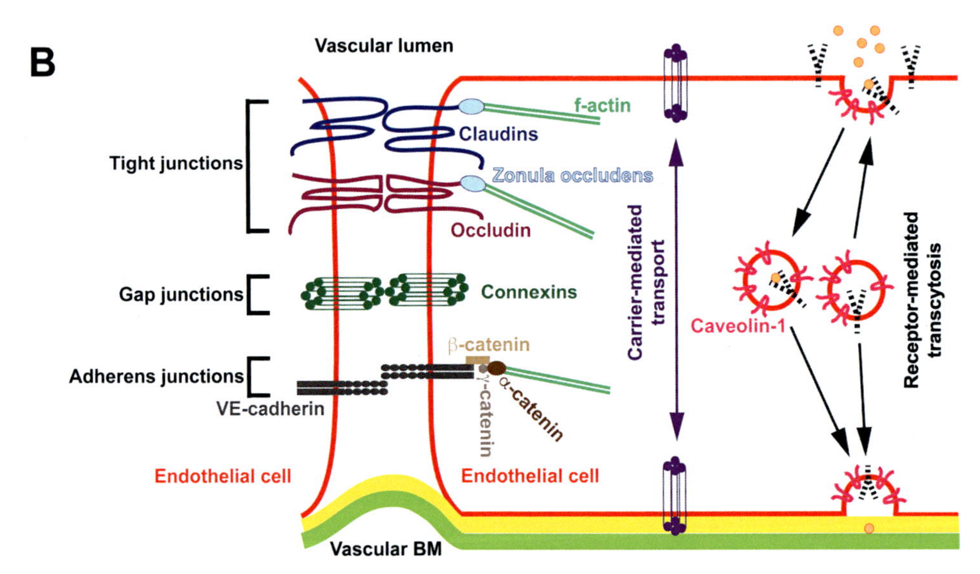

FIGURE 10.1 **Interactions among various cell types of the neurovascular unit are critical for the formation of neurovascular barriers.** (**A**) In the CNSs, vascular components including endothelial cells (light gray), endothelial cell–derived basement membrane (endothelial BM; yellow), and mural cells such as pericytes (magenta), together with astrocytes (light purple), astrocyte-derived BM (astrocyte BM; green), neurons (dark gray), and microglia (orange) interact anatomically to form the neurovascular unit (NVU). Boxed area is presented in greater detail in panel **B**. (**B**) A schematic of cell biological mechanisms that contribute to the neurovascular barrier properties of CNS endothelial cells. Tight junctions, formed by Claudins (dark blue), Occludin (maroon), and Zonula Occludens (sky blue), limit the movement of small molecules between endothelial cells (red outline), thereby forming a paracellular barrier. Tight junction proteins interact with the cytoskeleton (F-actin; green) via zonula occludens proteins (light blue). Adherens junctions, which are formed by VE-Cadherin (dark gray) as well as α- (dark brown), β- (medium brown), and γ-catenins (light brown), are critical for cell–cell interactions and shear stress sensing. Endothelial cells communicate with each other via Connexin-regulated gap junctions (dark green). Low rates of Caveolin-1 (pink)-dependent bulk transcytosis prevent trafficking of larger molecules and antibodies within CNS endothelial cells, establishing a transcellular barrier. Finally, CNS endothelial cells also contain various active and passive transporters (violet) to facilitate the movement of nutrients, but prevent drug transport, between the blood and the brain/spinal cord/retina. *Adapted from Biswas S, Cottarelli A, Agalliu D. Neuronal and glial regulation of CNS angiogenesis and barriergenesis.* Development. *2020;147.*

rise to both neurons (early) and oligodendrocytes and astrocytes (later) (Fig. 10.2A). The differentiation of radial glial cells into layer-specific neuronal subtypes begins at E11.5 and the inside-out migration of terminally differentiated neurons along radial glial processes in the cortex continues throughout the first postnatal week of development (Fig. 10.2A and A′) (reviewed in Refs. 9,10). The differentiation of oligodendrocytes from progenitor cells begins around E13.0–E14.5 in ventral regions of the brain; however, a second distinct wave of oligodendrogenesis has also been reported during postnatal development (Fig. 10.2A) (reviewed in Ref. 11). Around E18.5, radial glial cells start generating astrocyte precursors, which migrate to their final position and begin to acquire markers of fully mature astrocytes during postnatal development (Fig. 10.2A) [reviewed in Refs. 9,12]. The formation of synapses in the cortex starts around E16.5; however, synapses exhibit mature morphology and function only around the third postnatal week of development (reviewed in detail in Refs. 9,10).

Forebrain angiogenesis in the mouse was traditionally thought to begin at E10.0–E11.5, peak around birth and then gradually decline until postnatal day 25 (P25)

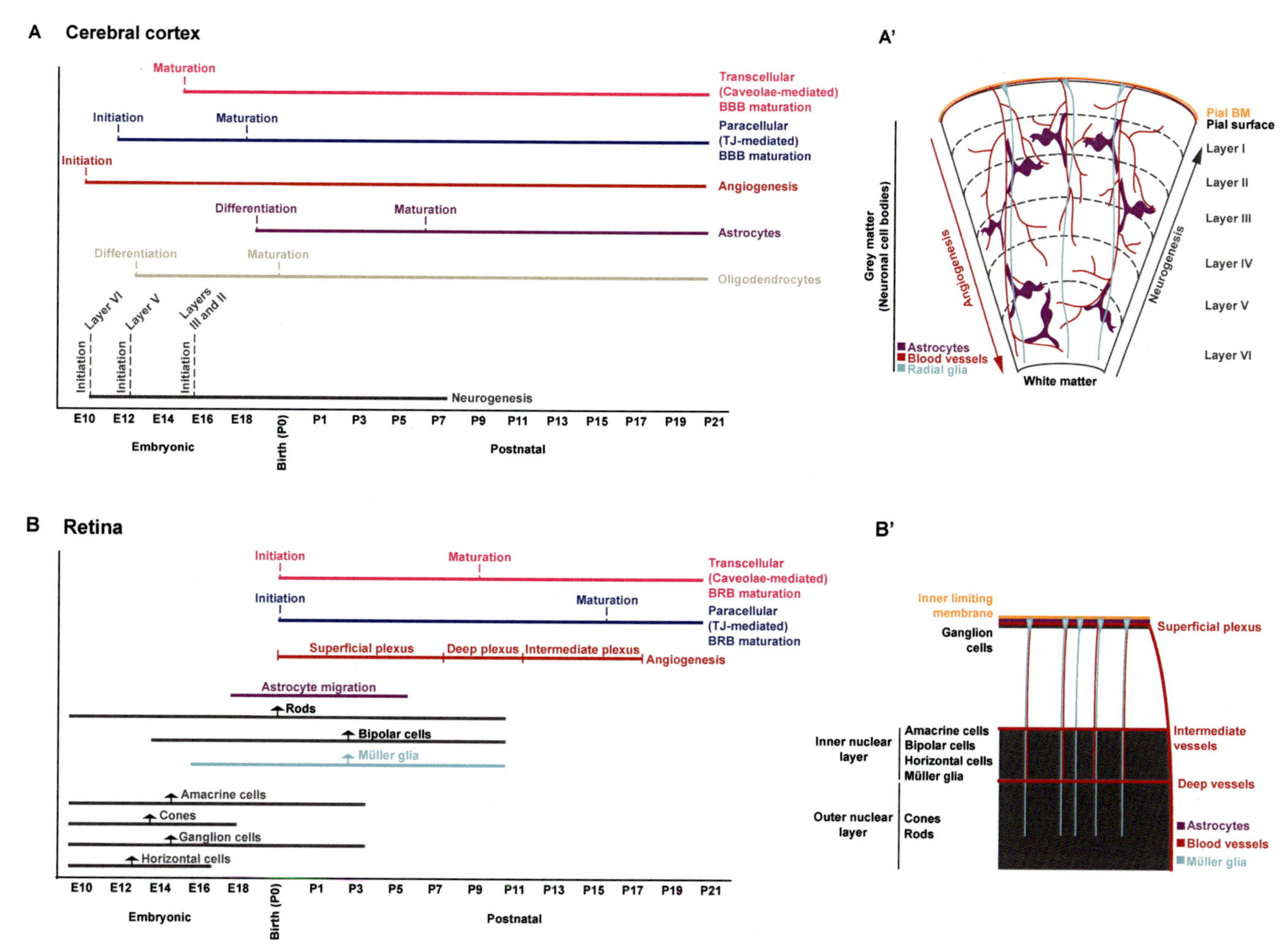

FIGURE 10.2 Developmental timelines for neurogenesis, gliogenesis, angiogenesis, and barriergenesis in the brain and the retina. (A) Developmental timeline for the development of neurons, glial cells, and blood vessels in the mouse cerebral cortex. Neurogenesis in the cerebral cortex begins at E11, whereas gliogenesis (oligodendrocyte and astrocyte formation) starts around E13 and E18.5, respectively. Neurogenesis is completed by P8–P10, whereas gliogenesis persists for prolonged periods during postnatal development corresponding to the expansion and maturation of astrocytes and ensheathment of blood vessels by astrocyte endfeet. Cortical angiogenesis also spans both the embryonic and postnatal stages of development and is completed by P25. The initiation of BBB maturation [establishment of both paracellular (pink) and transcellular (blue) barrier properties] starts soon after angiogenesis. However, by birth, both the paracellular and transcellular barrier properties of brain endothelial cells are mature. *Dotted lines* indicate the initiation and maturation of the endothelial barrier properties or the start of differentiation and appearance of mature astrocytes and oligodendrocytes. **(A′)** Neurogenesis in the cortex occurs in an inside-out fashion (*gray arrow*), whereas angiogenesis occurs in an outside-in fashion (*red arrow*). Mature astrocytes ensheath blood vessels to ensure the maintenance of the BBB. **(B)** Developmental timeline for the development of neurons, glial cells, and blood vessels in the mouse retina. Amacrine cells, cones, ganglion cells, and horizontal cells are born during the embryonic phase, whereas bipolar cells and Müller glia are born during the postnatal phase. Rod development occurs throughout both phases. Astrocytes migrate into the retina from E18.5 until P6. Growth of the superficial plexus (P1-8) follows an astrocyte template, over the inner limiting membrane. Sprouts from the superficial vessels, guided by Müller glia, then form the deep (P8–12) and intermediate plexuses (P12–21). The transcellular BRB matures by P10, whereas the paracellular barrier does not mature until P18. *Upward arrows* indicate peak of development for each process. *Dotted lines* indicate the initiation and maturation of BRB properties in retinal endothelial cells. **(B′)** Schematic diagram shows the relationship of distinct neuronal cell types with distinct vascular plexuses in the retina. Ganglion cell bodies reside in the ganglion cell layer, which is vascularized by the superficial plexus. Amacrine, bipolar, horizontal, and Müller cell bodies reside in the inner nuclear layer, which is vascularized by the intermediate and deep vascular plexuses. Photoreceptors occupy the outer nuclear layer, which makes contacts with the deep vascular plexus. *Adapted from Biswas S, Cottarelli A, Agalliu D. Neuronal and glial regulation of CNS angiogenesis and barriergenesis.* Development. *2020;147.*

(reviewed in Ref. 13). However, recent RNA sequencing studies suggest the existence of a second wave of postnatal brain angiogenesis.[14,15] These studies have shown that EC-specific mRNA transcripts encoding (a) angiogenic tip markers (e.g., *Egfl7*, Mcam, *Apln*, *Lamb1*); (b) signaling factors that regulate CNS angiogenesis such as components of the VEGF, Notch, and Wnt pathways; (c) cell proliferation markers; or (d) cell cycle markers are decreased between E15.5 and E18.5 in the forebrain as compared to E11.5–E13.5, but exhibit a rapid increase in their levels during the first week after birth (P5–P9), indicative of a potential second wave of angiogenesis[14,15] (Fig. 10.2A and A′). These molecular studies are supported by analyses of EC proliferation using in vivo two-photon imaging of EC sprouts in the postnatal murine cortical vasculature, which demonstrate increased EC proliferation and sprouting from P5–P10, followed by a gradual decline and quiescence by P25 in the cortex.[16] The first rapid embryonic phase of angiogenesis in the developing cortex (E10.5–E15.5) likely corresponds to the developmental period during which progenitor cells undergo extensive proliferation and differentiation into layer-specific neuronal subtypes that migrate from deep to superficial layers of the cortex as well as oligodendrogenesis (Fig. 10.2A and A′).[12,17] The decline in transcription of angiogenic markers in brain ECs, which takes place between E15.5 and E18.5,[14] occurs when astrocytes begin to differentiate.[18] Thus, initiation of astrocyte differentiation may promote the transition from angiogenesis to vascular maturation characterized by acquisition of BBB properties (Fig. 10.2A). In contrast, the second postnatal wave of angiogenesis (P5–P10) in the developing cortex may be driven by activity-dependent mechanisms (neuronal activity) to refine the CNS vasculature and match it to the metabolic demands of neurons.[19]

The development of the BBB begins immediately following the embryonic wave of angiogenesis in the forebrain. BBB specification starts with the induction of a subset of barrier-specific genes, including those encoding for TJ-associated proteins (Claudin-5, Occludin, ZO-1) and the glucose transporter (Glut-1/Slc2a1), at early embryonic stages (E10.0–E11.5) in response to inductive signals (e.g., Wnts) derived from both neuronal and glial progenitor cells.[20–22] Recent bulk RNA sequencing data have confirmed that the majority of BBB-associated transcripts start to be translated by E11.5–E12.5.[14] However, analysis of the translation of BBB-associated transcripts has revealed six important features about BBB maturation. First, distinct BBB transcripts are induced and translated at different developmental stages of CNS development (Fig. 10.2A). For example, transcripts and proteins for most TJ-associated proteins (e.g., Cadherin-5, Claudin-5, ZO-1, JAM-A, and Occludin) are highly expressed in early development, whereas transcripts for the majority of organic ion transporters (e.g., Slc28a8, Slco2b1, Slco1a4) are induced strongly in the postnatal stages (P5–P9).[14,23] Second, there is a large heterogeneity in BBB maturation even among vascular beds within the same CNS region.[14] Third, the decrease in angiogenesis-related genes between E14.5 and E16.5 corresponds to an increase in transcription of several key transcription factors (Foxf2, Foxl2, Foxq1, Lef1, and Sox17) that play critical roles in BBB maturation.[14,23,24] However, two additional key BBB maturation factors (Ppard and Zic3) only appear at E17.5–E18.5. Fourth, the expression of some of the BBB-inducing transcription factors in later gestation (E15.5–E18.5) coincides with maturation of the primitive BBB initiated earlier in development (E11.5–E13.5). For example, during late gestation, brain ECs suppress caveolae-mediated transport via induction of Mfsd2a and suppression of PLVAP (PV-1; Meca32) expression to ensure a decrease in caveolae number, loss of fenestrae, and maturation of the transcellular barrier (Fig. 10.2A).[25–27] Fifth, cerebral vessels located in the pia matter significantly increase their transendothelial electrical resistance, a functional readout of the mature paracellular barrier that is regulated by TJs, only prior to birth,[28] although these vessels express TJ proteins earlier in development. These findings suggest that the maturation of the paracellular barrier follows that of the transcellular barrier in the brain vasculature (Fig. 10.2A). Sixth, transcripts encoding a subset of BBB-associated transporters, including ABC (e.g., Abcb1a/Mdr1a, Abcc4), amino acid (e.g., Slc1a1, Slc1a3), and organic ion (e.g., Slco1a4, Slco2a1, Slcob1) transporters, are highly upregulated only during postnatal development[23] when astrocytes mature and ensheath brain capillaries. Thus, astrocytes may play a critical role in regulating the expression of transporters, despite the fact that they are not necessary for formation of the paracellular and transcellular barrier, since these properties develop even when astrocyte maturation in the cortex is delayed by elimination of FGF-2.[29] Thus, BBB development and maturation occur gradually, spanning both embryonic and postnatal phases of brain development (Fig. 10.2A).

Retina

Neuronal development in the murine retina occurs during two major phases, beginning at E10 and continuing until P11 (Fig. 10.2B).[30] Cone photoreceptors, ganglion, horizontal, and amacrine cells are born during the embryonic phase, whereas bipolar cells and Müller glia are born postnatally (Fig. 10.2B). Rod photoreceptors appear throughout both phases, peaking at birth.[31] In contrast to the brain, astrocytes are not born in the retina but they enter via the optic nerve beginning at E17.5–18.5 and migrate radially toward the periphery.[32]

Angiogenesis in the mouse retina occurs entirely during postnatal development. After birth, embryonic hyaloid vessels gradually regress as retinal vessels grow into the superficial, intermediate, and deep vasculature (Fig. 10.2B and B′). The superficial vasculature expands from the optic nerve head radially toward the retinal periphery over the ganglion cell layer (P0–P8), following a template established by astrocytes. Vascular sprouts from the superficial vessels then dive into the retina guided by Müller glial processes[33] and form the deep vascular plexus within the outer plexiform layer (P8–P12). Finally, the intermediate vascular plexus arises in the inner plexiform layer (P12 onward), completing vascularization by P21 (Fig. 10.2B and B′) [see Ref. 34 for a detailed review of retinal angiogenesis].

Structurally, there are two types of BRBs in the mammalian retina. The outer BRB is composed of the retinal pigment epithelium and its underlying BM, separating photoreceptors from choroidal circulation. The inner BRB separates neurons and glia from retinal circulation by means of a structural barrier formed by retinal ECs that possess the same cell biological mechanisms as those described above for brain ECs forming the BBB (Fig. 10.2B).[8] A recent study suggests that the inner BRB is established as early as P3–P5 in the more mature part of the superficial vascular plexus of the central retina, whereas immature vessels around the growing vascular front remain leaky. However, BRB maturation (establishment of both paracellular and transcellular barrier properties) is completed throughout the superficial and deep vascular beds by P10.[35] Moreover, it has been suggested that retinal ECs possess functional TJs (i.e., a mature paracellular barrier) at P1, and that establishment of the BRB along the expanding superficial vascular tree is achieved only by gradual suppression of transcellular permeability from P1 to P10,[35] a feature that seems unique to the retina and is not observed in the brain vasculature (see above). In contrast, using a small molecular weight tracer biocytin-tetramethylrhodamine (TMR) tracer (~890 Da), which crosses the vascular barrier via the paracellular pathway,[36–38] Mazzoni et al., demonstrated that the paracellular barrier of the superficial plexus is immature at P1 but gradually matures between P10 and P18 (Fig. 10.2B)[39] suggesting that the transcellular BRB matures before the paracellular BRB in the retina, similar to the developmental timeline for BBB maturation (Fig. 10.2A and B).

Vascular basement membrane and CNS vascular barrier maturation

Brain

Since the vascular basement membrane (vBM) plays a crucial role in NVU assembly and BBB maintenance, genetic and pathological conditions associated with changes in vBM composition lead to perturbation of BBB properties. Loss of the pro-Collagen type IV α 1 gene (*Col4a1*) in mice causes vBM disorganization and cerebral hemorrhage.[40] Similarly, deletion of the Laminin α2 chain gene (*Lama2*)[41] or astrocyte-specific deletion of the Laminin γ1 chain[42] causes BBB leakage, hemorrhagic stroke, deficits in pericyte differentiation and loss of astrocyte endfeet polarization around blood vessels, and is associated with disorganization of TJs.[43] Consistent with these phenotypes, neuronal- and glial-specific deletion of Dystroglycan, a major receptor for laminins, increases BBB permeability.[41] In addition, loss of Integrin αvβ8, an extracellular matrix (ECM) receptor responsible for Transforming Growth Factor-β1 (TGF-β1) activation, has been shown to be important for CNS angiogenesis and barriergenesis. Global deletion of Integrin αv shows increased CNS angiogenesis as well as intracerebral hemorrhage.[44] Similarly, global deletion of β8 also shows increased CNS angiogenesis and perivascular astrogliosis. However, endothelial TJ integrity remains unaffected in these mice.[45] This phenotype is similar to that of EC-specific TGF-β receptor 2 (*Tgfbr2*) knock-out mice.[46] On the other hand, Neuropilin-1, expressed by brain ECs, suppresses TGF-β1 activation by forming intercellular complex with Integrin β8.[47] In summary, these genetic loss-of-function studies of ECM proteins and their cognate receptors underlie the importance of the proper NVU assembly for neurovascular development and BBB function.

As mentioned above, the correct polarization of glial endfeet around the vasculature is critical for maintaining vascular barrier properties. In the brain, astrocytes are the only glial cells that express the water channel protein Aquaporin-4 (Aqp4) at their endfeet.[48] However, the role of Aqp4 in BBB function is actively under debate, since some studies have shown that glial-specific deletion of Aqp4 reduces brain water content without affecting BBB permeability.[49] Loss of pericytes also causes BBB impairment.[50,51] However, the phenotypes observed in pericyte-deficient mice are likely due to multiple factors, including changes in EC-specific gene expression, alterations in laminin deposition in the vBM, and abnormal glial endfeet polarization around CNS blood vessels.[50,51]

Retina

Within the retina, vBM proteins are also essential for NVU assembly and BRB function. Global deletion of Laminin β2 chain or β2 and γ3 chains together causes BRB leakage in the retina.[52] Similarly, EC-specific deletion of Integrin β1 perturbs the formation of adherens junctions, leading to aberrant polarization and lumen defects in retinal ECs[53] as well as vascular leakage.[54] Loss of pericytes also leads to BRB disruption.[55] Similar to what has been described in the brain, retinal astrocytes express Integrin αvβ8, and Nestin-Cre-mediated deletion of either αv or β8 shows retinal vascular abnormalities, including vascular tufts as well as intraretinal hemorrhage.[56] Additionally, Müller glia- and neuron-specific deletion of Integrin β8 also results in hypoactive

TGF-β1 signaling and leads to increased vascular growth in the superficial plexus as well as hemorrhage within the retina, which could be due to either a secondary effect of abnormal vascular growth or impaired BRB formation.[57]

Aqp4 is also implicated in BRB formation in the retina. However, in contrast to its expression in the brain, Aqp4 is expressed at the endfeet of two retinal glial cell types: astrocytes that surround the superficial vascular plexus and Müller glia that contact the deeper vascular plexus.[58] Interestingly, *Aqp4*$^{-/-}$ mice exhibit vascular leakage only in the deep retinal plexus,[58] suggesting that Aqp4 is critical in Müller glia, but not astrocytes, for BRB integrity. These contrasting results between the brain and the retina suggest a complex role for Aqp4 in maintaining the vascular barrier properties of these two CNS regions, which may depend on either cell-intrinsic properties of the vascular bed or its surrounding microenvironment.

The neurovascular unit and blood–CNS barriers in neurological and ocular diseases

Stroke

With 795,000 new cases every year, stroke is the fifth leading cause of mortality and the leading cause for long-term disability in the United States.[59] More than 85% of all strokes are ischemic, caused by the occlusion of one of the arteries supplying blood to the brain.[60] At the vascular level, stroke damage occurs in two distinct stages—ischemia and reperfusion, with the latter being responsible for a markedly biphasic pattern of increased BBB permeability.[61] Notably, the mechanisms underlying this biphasic permeability are somewhat controversial, as the timing and the modality of BBB impairment vary widely based on the duration of the ischemic phase, the degree of reperfusion, and the specific experimental model used to induce stroke.

Loosening of TJs between ECs has been traditionally regarded as the main mechanism for pathological BBB defects in stroke. In a rat model of transient middle cerebral artery occlusion (tMCAO), TJ degradation, decreased expression of TJ proteins (Claudin-5, Occludin, and ZO-1), leakage of Evan's blue (a marker of serum albumin extravasation), and hemorrhagic transformation were observed as early as 3 h after the onset of reperfusion.[62] However, the subsequent observation that the early (4–6 h postocclusion) leakage of FITC-Albumin into the CNS parenchyma occurs without any evidence of junctional impairment in ECs in a model of embolic stroke[63] prompted Knowland et al. to examine the dynamics of junctional changes after tMCAO, taking advantage of a transgenic mouse model displaying fluorescently labeled TJs (eGFP-Claudin-5).[36] *In vivo* two-photon imaging revealed that leakage of biocytin-TMR, a readout of increased paracellular permeability mediated by TJ dysfunction, does not occur until 24–48 h after reperfusion. Consistently, the increase in the number of junctional abnormalities in the ipsilateral cortex only peaks after 48 h of reperfusion, suggesting that junctional remodeling accounts for the second wave of reperfusion-induced increase in BBB permeability.[36] Conversely, increases in both the number of endothelial caveolae and the transcellular transport of Albumin-Alexa594 were detected as early as 6 h after reperfusion,[36,64] a time when there is no apparent junctional involvement, indicating that transcytosis is responsible for the first wave of reperfusion-induced increase in BBB permeability (refer to Fig. 10.3 for the timeline of BBB disruption after stroke). Interestingly, a similar pattern of Caveolin-1 (Cav-1) upregulation preceding junctional breakdown has also been observed in other studies of ischemic stroke and a model of cortical cold-injury,[65,66] suggesting that the sequential involvement of transcellular and paracellular pathways in BBB permeability may be a common mechanism in focal brain injuries (Fig. 10.3).

Cav-1, the main component of endothelial caveolae, plays a key role in the regulation of BBB permeability after stroke and seemingly has either beneficial or detrimental effects depending on the degree of reperfusion. In a model of permanent MCAO, genetic deletion of Cav-1 resulted in larger stroke volumes compared to wild-type animals.[67] Neoangiogenesis (i.e., the formation of pathological new vessels) has been shown to occur in the ischemic areas of the brain, mostly as a response to hypoxia-induced Vascular Endothelial Growth Factor (VEGF),[68] which is able to induce endothelial proliferation.[69] *Cav-1*$^{-/-}$ mice show a reduction in vascular density and EC proliferation in the ischemic brain, possibly as a result of disrupted VEGF Receptor 2 signaling.[67] Conversely, in a tMCAO model, Cav-1 expression and phosphorylation are increased, and *Cav1*$^{-/-}$ mice do not show increased transcytosis and BBB transcellular permeability in the early reperfusion phase.[36,65] Sadeghian et al. recently provided a mechanism for the activation of post-reperfusion transcellular pathway by showing that cortical spreading depressions, intense waves of neuroglial depolarization that occur spontaneously after stroke and other brain injuries, are able to trigger transcytosis in brain ECs, while they fail to do so in *Cav1*$^{-/-}$ mice.[66]

In vivo studies have shown that the late-onset increase in paracellular permeability is not only chronologically distinct, but also mechanistically independent of Cav-1-mediated transcytosis, as only the early wave of post-reperfusion BBB damage is rescued in *Cav1*$^{-/-}$ animals.[36] Proinflammatory cytokines, such as TNFα, can increase

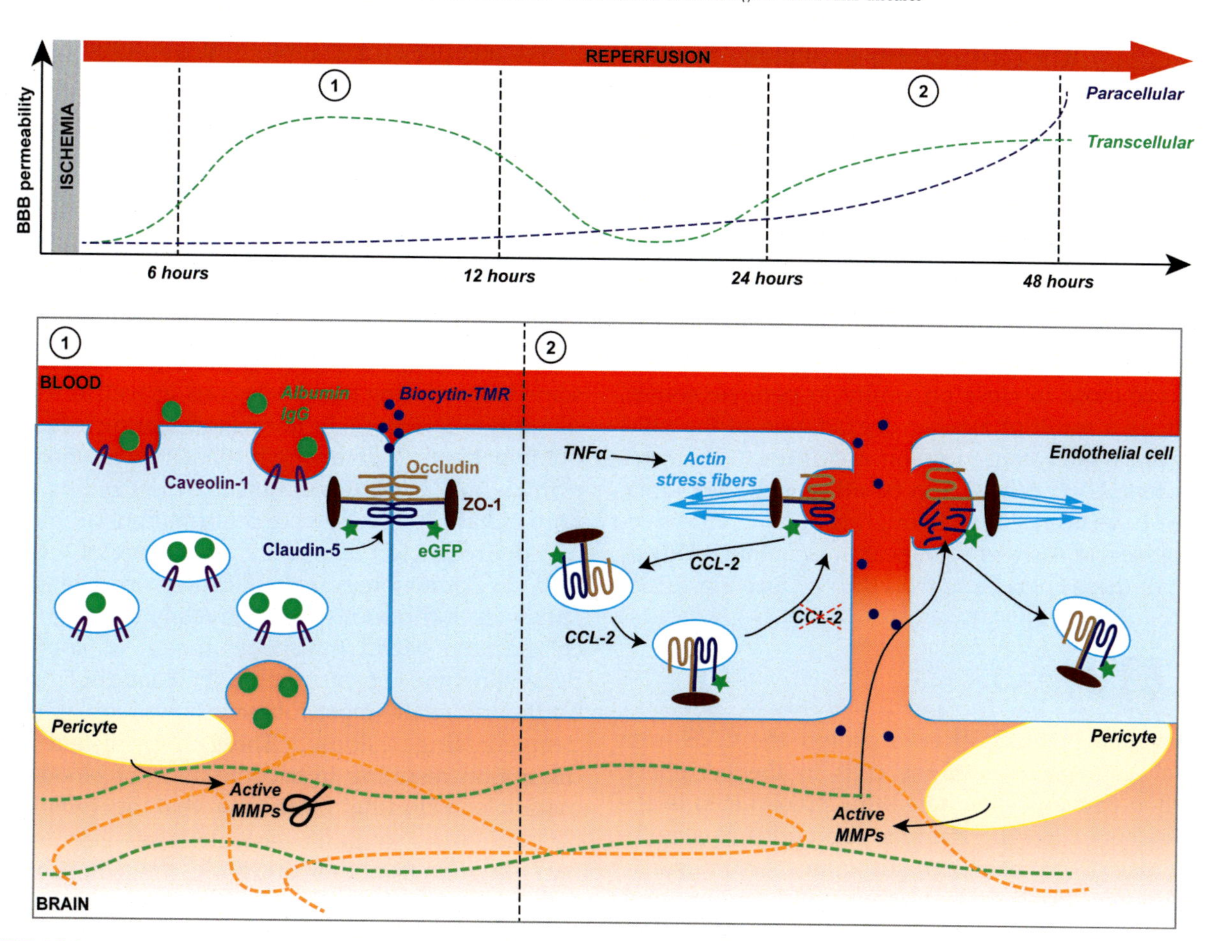

FIGURE 10.3 **Schematic diagram of mechanisms of BBB damage in ischemic stroke.** After ischemic stroke, the BBB undergoes two distinct waves of increased permeability. The first wave occurs between 6 and 12 h after the onset of reperfusion and is mostly due to an increase in transcellular permeability; the second wave occurs between 24 and 48 h after reperfusion and is mainly caused by increased paracellular permeability. Increased expression of Caveolin-1 in the early phase causes the passage of large molecular weight molecules, such as albumin and IgG, through receptor-mediated or bulk transcytosis. In the later phase, proinflammatory cytokines cause the internalization of TJ proteins via early endosomes to promote their degradation into the Pericytes contribute to BBB permeability by secreting metalloproteases (MMPs).

the formation of actin stress fibers,[70] which have been shown to be responsible for internalization of junctional proteins.[71] Cav-1 has been implicated in promoting internalization of Claudin-5 upon oxygen and glucose deprivation (OGD) in vitro[72] and shuttling junctional proteins toward the late endosome for degradation in response to the chemokine CCL2 (Fig. 10.3).[73] Overall, these in vitro studies suggest that junctional protein internalization may not necessarily culminate in their degradation, and a caveolar-independent mechanism may regulate junctional integrity at the BBB in conditions of ischemia/reperfusion (Fig. 10.3).

What are the molecular mechanisms underlying BBB breakdown after stroke? Several studies support a role for metalloproteases (MMPs) in this process. In vitro studies have demonstrated that matrix metalloproteinase-2 (MMP-2) secretion and activity are promoted by OGD; inhibition of MMP-2 is able to rescue OGD-dependent degradation of Occludin.[72] In vivo, during early reperfusion (2–3 h), MMP-2 expression and enzymatic activity are increased in the ipsilateral brain and colocalize with Glial Fibrillary Acidic Protein $(GFAP)^{+}$ reactive astrocytes in rats. Modest reductions in Claudin-5 and Occludin levels are also observed at this time point and are only partially rescued by the MMP inhibitor BB-1101,[74]. However, this may be due to a nonspecific and general effect of sudden hyperemia, since it is also observed in the contralateral brain.

Unlike MMP-2, MMP-9 expression and enzymatic activity were only increased in the ischemic brain 24 hours after reperfusion,[74] making it a more viable candidate for targeted regulation of changes in BBB permeability after stroke (Fig. 10.3). *Mmp9*$^{-/-}$ mice displayed a reduction in BBB impairment and a rescue in the degradation of ZO-1, but not of Occludin, after tMCAO.[75]

The mechanism through which metalloproteases can cause junctional dysfunction after stroke is still not clear, although it has been proposed that MMP-9 may promote degradation of Claudin-5 and the release of Claudin-5 fragments in the cerebrospinal fluid (Fig. 10.3) by activating the NF-κB pathway in a model of parasitic meningoencephalitis.[76]

In a photothrombosis cortical stroke model, pericytes were found to play a role in MMP-mediated BBB injury by promoting the secretion and activation of MMP-9. Inhibition of MMP-9, but not MMP-2, was able to significantly reduce leakage of large molecular weight Dextrans across the BBB.[77] Interestingly, the same study has also shown that the regions of BBB endothelium that were most susceptible to leakage after photothrombosis were those in close proximity to pericyte somas, possibly as a result of a faster degradation of the vBM due to the higher concentration of MMPs.[77] Pericytes are the first NVU cell type to be affected by ischemia, and they have been shown to migrate away from the cerebral vessels as early as 2 hours after ischemia, likely as a result of these changes in the vBM[78,79] (see Fig. 10.3).

Multiple sclerosis

A key role of the BBB is to protect the CNS from potential damage by immune cells in the periphery. Disruption of the BBB is therefore a pivotal step in the progression of CNS autoimmune conditions like multiple sclerosis (MS), neuromyelitis optica, neuropsychiatric systemic lupus erythematosus, and autoimmune encephalitis.[80,81] This section will focus on MS and its animal model experimental autoimmune encephalomyelitis (EAE).

MS is an autoimmune disease of the spinal cord and brain, in which the immune system targets myelin sheaths surrounding neurons. MS is the most common nontraumatic cause of disability in young adults, with upwards of 2.3 million people affected globally.[82] The pathogenic contribution of BBB permeability to MS initiation and evolution over time is a subject of intense interest.[1,83–87] Its animal model EAE is induced either by injection of myelin antigens combined with complete Freund's adjuvant, or through adoptive transfer of autoantigen-specific $CD4^+$ T-cells, resulting in extensive neuroinflammation and ascending paralysis.[88] The central role of the CNS vasculature to MS pathology was realized in the earliest descriptions of the disease, which noted the proximity of MS plaques to blood vessels in the brain and spinal cord,[89–91] and further studies revealed increased vascular permeability in MS patients.[92] Disruption of the BBB and BSCB is central to diagnosis and treatment of MS, as well as a leading focus of ongoing research.[93] Under disease conditions in both in MS and EAE, three crucial properties of CNS ECs (TJ integrity, low levels of transcytosis, and low expression of adhesion molecules), which together contribute to CNS immune privilege, are lost.

BBB disruption precedes symptom onset in both MS and EAE.[94,95] Magnetic resonance imaging (MRI), using gadolinium-based contrast agents, allows visualization of BBB permeability and is widely used to diagnose MS and monitor the progression of white matter lesions.[93,96] However, standard MRI does not detect all regions of BBB disruption, and enhancing lesions do not closely correlate with clinical exacerbations.[97] Postmortem studies reveal that TJ abnormalities and vascular permeability are not limited to actively demyelinating lesions,[98,99] reflecting persistent BBB deficits in some longstanding inactive lesions, as well as less severe disruption to vessels in normal-appearing white matter of MS patients.[100] In a longitudinal study in non-human primate EAE, BBB disruption as evidenced by gadolinium enhancement preceded formation of monocyte-rich demyelinated lesions by an average of 4 weeks.[101] BBB EC junctional destabilization precedes robust leukocyte infiltration and demyelination in a $CD8^+$ T cell -driven MOG transgenic model of spontaneous demyelination. In murine EAE, Claudin-5 and VE-Cadherin are the most robustly downregulated in pre-demyelinating lesions.[95] There is persistent BBB TJ disruption resulting in accumulation of blood-derived materials including fibrinogen that overwhelm local clearance capacity in the early days to weeks of disease onset in animal models of MS.[38,87] In acute EAE, strikingly, downregulated junctional proteins include Angulin-1,[102] Claudin-5,[95] VE-Cadherin,[95] and ZO-1.[103]

Junction remodeling also promotes an increase in transcellular EC permeability.[104,105] Caveolar transport is significantly increased in MS and EAE, allowing immune cells to enter the CNS; the number of intracellular vesicles within ECs correlates with disease severity.[106,107] Caveolins-1, -2, and -3 are upregulated in EAE,[108] and Cav-1 is required for T cell infiltration.[38,109] Cav-1 protein is increased in murine EAE,[108,109] and *Cav-1*$^{-/-}$ mice exhibit markedly attenuated neurological and histological signs of EAE.[38,109] Distinct T cell subtypes utilize distinct mechanisms of CNS entry. For example, T-helper 17 (Th17) cells enter early on in disease progression[110] through disrupted TJs, followed by the entry of T-helper 1 (Th1) cells, which selectively cross through caveolar transport[38] (Fig. 10.4).

The Wnt/β-catenin pathway, which drives BBB formation during development, is reactivated in CNS ECs in both MS and EAE, exerting multiple disease-ameliorating effects. In the absence of Wnt activity, Cav-1 is upregulated, promoting transcytosis, while higher Vascular Cell Adhesion Molecule (VCAM)-1 expression leads to increased $CD4^+$ T cell entry.[37] Higher expression of Intercellular Adhesion Molecule (ICAM)-1, which allows leukocytes to migrate against the direction of blood flow along the EC luminal surface, may

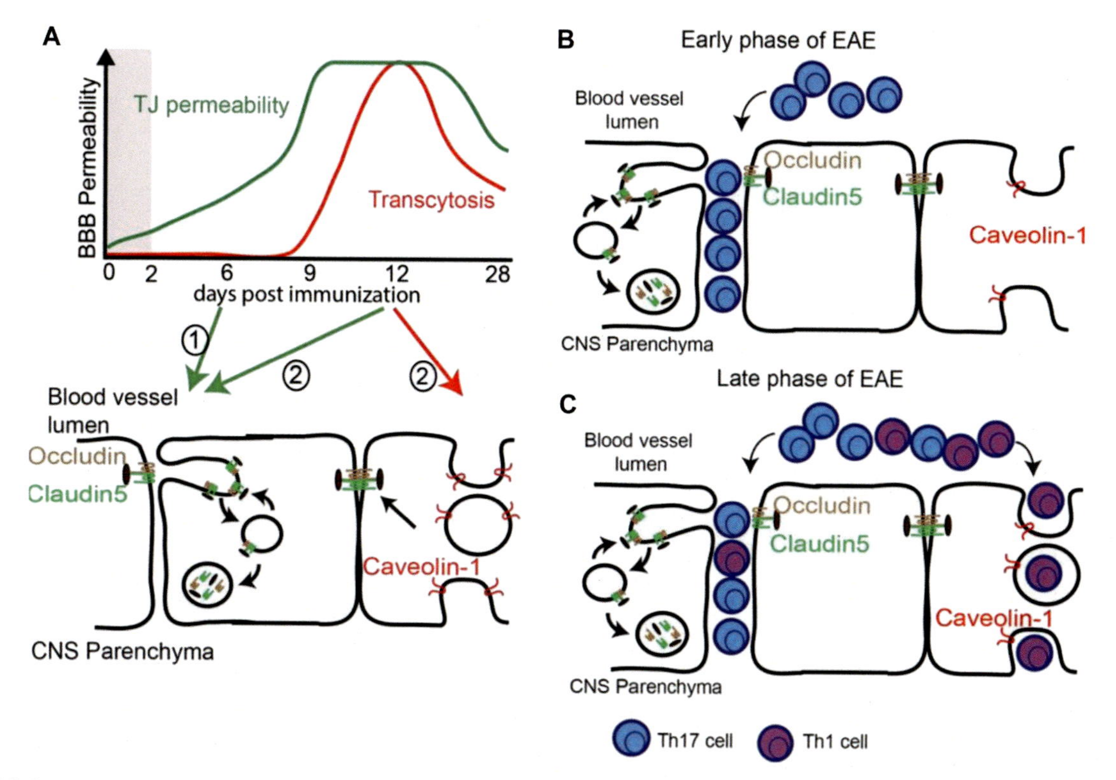

FIGURE 10.4 **Schematic presentation of BBB breakdown and lymphocyte trafficking across a damaged BBB in neuroinflammation.** (**A**) A working model of BBB disruption during EAE. Increased paracellular BBB permeability (*green line*) occurs prior to onset and persists throughout EAE. Transcellular BBB permeability (*red line*) transiently elevates only at acute disease. Paracellular permeability is due to rapid remodeling of TJ proteins, while transcellular permeability results from enhanced caveolar trafficking. (**B**) During the early phase of MS/EAE progression, there is an initial breakdown of endothelial cell TJs at the BBB that causes an increase in paracellular permeability. This allows the preferential entry of Th17 cells (blue) in the early phase of the disease, although some Th1 cells (purple) also enter through this route. (**C**) The increase in caveolar transport (transcellular permeability) through upregulation of Caveolin-1 is observed only during the late phase of neuroinflammation at the peak of the disease. Th1 cells (purple) preferentially use caveolae to cross the BBB due to enhanced transcellular permeability, whereas Th17 cells cross the BBB through disrupted TJs and as well as through a caveolae-dependent manner.

cause T cells to favor transcellular over paracellular diapedesis,[111] while Platelet Endothelial Cell Adhesion Molecule-1 (PECAM-1) expression favors paracellular over transcellular diapedesis.[112]

Upregulation of cell adhesion molecules: In contrast to those in the periphery, ECs of the CNS express very low levels of CAMs; however, under neuroinflammatory conditions CAM expression is upregulated.[113,114] Inflammatory cytokines like IL-1β, TNFα, and IFNγ drive endothelial expression of adhesion molecules, including P- and E-Selectins, VCAM-1, and ICAM-1.[115–118] Numerous CAMs mediate CNS entry of different specific immune cell types (reviewed in Ref.[88]). Very Late Activation Antigen (VLA)-4, comprised of α4/β1 Integrin subunits, mediates adhesion and migration of lymphocytes into the inflamed CNS through interaction with VCAM-1.[119,120] Antibody inhibition of α4-Integrin was shown to ameliorate EAE,[119,121] leading to the development of humanized anti-α4 Integrin monoclonal antibody Natalizumab as an approved treatment for MS.[122,123] Again, T cell subtypes preferentially rely on different mechanisms to enter the CNS: α4-Integrin is essential for infiltration by Th1 cells, but use of Lymphocyte function-associated antigen (LFA)-1 is sufficient for Th17 cell entry.[124,125]

Changes to the vBM: Immune cell infiltration correlates with laminin composition of the vBM in postcapillary venules. Stimulated encephalitogenic T-cells were found to bind primarily to Laminin α4β1γ1 in vitro, as well as to Laminin α5β1γ1, with no binding to either Laminin α1β1γ1 or Laminin α2β1γ1.[126] Correspondingly, in EAE brains, T-cell infiltrates were found to correlate with the presence of Laminin α4, and with the absence of Laminin α5.[126,127] Inflammation appears to shift vBM composition, with Integrin α6 immunoreactivity lost from the endothelium in advanced EAE, and *Lam5* mRNA upregulated by in vitro inflammatory cytokine treatment.[126] After breaching the vBM, T-cells rely on MMP-2 and MMP-9 to chew through the glia limitans/parenchymal BM and reach the neural parenchyma.[128]

Immune cells extravasate from leaky vessels through a process of capture, rolling, arrest, crawling, and diapedesis (reviewed in Ref.[129]). As immune cells push into cell junctions, transmembrane proteins, including Claudin-5 and VE-Cadherin, are forced aside, and a gap is created between cells.[104] This event is followed by rapid remodeling and resealing of junctions.[38,104] MS and EAE also evoke longer term TJ disruption. TJ adaptor proteins ZO-1 and Junctional Adhesion Molecule-A (JAM-A) are redistributed in MS and EAE.[103,130,131] In postmortem studies of MS patients, increased BBB permeability correlates with abnormalities in ZO-1 and Occludin localization[130,132] and loss of Claudin-5 staining in spinal venules is associated with increased leukocyte extravasation in EAE.[133] Conversely, overexpression of Claudin-1 in ECs reduces BBB permeability and clinical score in EAE.[134] Many of the changes to TJs are thought to be mediated by inflammatory cytokines upregulated in MS/EAE, including IL-1β, TNFα, IFNγ, IL-17, and CCL2 (reviewed in Ref. 135). In contrast to TJs, minimal changes are observed in adherens junctions in MS tissues.[131]

Retinal pathologies

Neurovascular diseases of the retina constitute the leading causes of acquired blindness. As of 2010, diabetic retinopathy (DR) affected around 100 million people globally,[136] making it the most frequent cause of adult acquired blindness in many countries.[137] About 40% of people with type 2 diabetes and about 86% of people with type 1 diabetes develop DR in the United States.[137] Retinopathy of prematurity (ROP) in preterm infants remains one of the major causes of child blindness worldwide,[138] although the prevalence rates vary significantly among countries even with similar neonatal care facilities.[139] Alterations in the retinal NVU as well as the BRB play significant roles in the progression of all these diseases. In this section, we will explore the main features of changes in the NVU that may present potential novel targets for treating these pathologies.

Diabetic retinopathy: One of the main pathologies of DR is retinal vBM thickening related to hyperglycemia-induced upregulation of Fibronectin and Collagen, along with impaired ECM degradation.[140] Another prominent pathology of DR is retinal pericyte loss and subsequent weakening of the capillaries that may lead to microaneurysms, which are among the first recognizable signs of DR.[141] Sometimes these microaneurysms accumulate circulating leukocytes, which also suggests a possible involvement of inflammation in their formation.[142] Retinal capillaries become progressively nonperfused in DR.[143] This leads to retinal hypoxia which is linked to VEGF upregulation among other factors. Upregulation of VEGF, in turn, leads to the pathological neovascularization and BRB disruption in the diabetic retina.[144]

Although vascular pathologies are the most explored aspect of DR, recent studies suggest that the primary effect of hypoxia may be on inner retinal neurons and glia, whereas vascular pathologies are secondary. Several retinal neuronal functional abnormalities such as abnormalities in contrast sensitivity, multifocal electroretinogram, and dark adaptation have been reported even before the onset of vascular pathologies.[145] One of the most affected cell types in DR are Müller glia that span the retinal thickness. The capillary-Müller glia interface becomes disrupted and Kir4.1 water channel expression in Müller glia endfeet becomes mislocalized in DR, contributing to BRB disruption.[146,147] Degeneration of retinal neurons in DR in postmortem human tissues has been reported as early as the 1960s.[144] Similarly, retinal ganglion cells undergo apoptosis in rat and mouse models of DR.[148] The body of evidence suggests that a number of factors including Nerve Growth Factor (NGF), Pigment Epithelium-Derived Factor (PEDF), Interphotoreceptor Retinoid-Binding Protein (IRBP), and Somatostatin have critical roles in neuronal apoptosis and subsequent vascular pathologies in DR.[144]

Immune response also plays a critical role in DR pathology. There is an upregulation in leukocyte adherence to ECs, termed leukostasis, likely due to upregulation of CAMs. This upregulation is observed within weeks of the onset of hyperglycemia in the rodent models of DR.[149] A similar phenomenon has also been reported in diabetic patients.[150] Consistently, genetic deletion of ICAM-1 and Integrin β2 and/or Inducible Nitric Oxide Synthase (iNOS) protected retinas from leukostasis and capillary death in animal models of DR.[149,151] Inflammation in the diabetic retina increases in a self-propagating manner. Chronic activation of microglia and expression of proinflammatory factors and cytokines (glutamate, iNOS, IL-1β, IL-6, TNFα, etc.) have been reported in the diabetic human retina and animal models of DR,[152] which may contribute to the apoptosis of retinal cells in DR and may be potential targets for therapeutic interventions.

Retinopathy of prematurity: The two-phase hypothesis of ROP postulates that the pathology develops first through delayed physiological vascular development (phase 1) and then by vasoproliferation into the vitreous (phase 2). In severe cases, this abnormal vascular growth in phase 2 generates a pull that results in retinal detachment. The most popular animal model of ROP is the oxygen-induced retinopathy (OIR) model, where the pathological progression also occurs in two phases—vaso-obliteration (phase 1) followed by pathological neovascularization (phase 2).[153] In both cases, phase 1 occurs due to the relative hyperoxia (due to the artificial high-oxygen environment) that hinders physiological vascular development, whereas phase 2 occurs due to the relative hypoxia (upon return to the room environment) that facilitates pathological neovascularization.[139]

Insulin-Like Growth Factor 1 (IGF-1), a growth factor critical for proper physiological vascular development, is absent in human infants born before completion of the third trimester, which possibly leads to the arrest of normal vascular growth during phase 1.[154] In rat models of OIR, JAK-STAT signaling pathway was also found to be activated during phase 1, which led to the downregulation of erythropoietin and subsequent delay in vascular development,[155] thus making this pathway a potential therapeutic target. Phase 2 involves relative hypoxia-induced increase in VEGF and Erythropoietin expression, which in turn leads to EC proliferation and pathological neovascularization.[156,157] These new vessels are poorly perfused and leaky, and lead to the formation of scar tissue, which as stated above, may cause retinal detachment.[139]

Pericytes also play an important role in the pathogenesis of ROP. One recent study demonstrated that abnormal, α-Smooth Muscle Actin (αSMA)-expressing pericytes cover pathological angiogenic sprouts and neovascular tufts in the OIR mouse model. Consistently, genetic depletion of pericytes decreased neovascular tuft formation.[158]

Neuronal components of the retina are also affected in ROP. ROP has effects on photoreceptor function that persist even after its active phase. Even mild ROP affects the development of the late-maturing central retina, residual effects of which are detectable years after ROP was active. Some deficits also occur in the peripheral retinal function and persist into adolescence and early adulthood.[159]

Finally, microglia have also been proposed to play a role in the pathology of ROP. One study reported that the depletion of microglial density during the vaso-obliteration phase of OIR may contribute to the decreased vascularity.[160] However, another study argued that it is not microglial density, but rather microglial activation state that is critical for OIR pathologies.[161]

A more detailed and clear understanding of the mechanisms underlying these alterations is critically important to develop effective interventions to treat these diseases.

References

1. Liebner S, Dijkhuizen RM, Reiss Y, Plate KH, Agalliu D, Constantin G. Functional morphology of the blood-brain barrier in health and disease. *Acta Neuropathol*. 2018;135:311–336.
2. Zhao Z, Nelson AR, Betsholtz C, Zlokovic BV. Establishment and dysfunction of the blood-brain barrier. *Cell*. 2015;163:1064–1078.
3. Iadecola C. The neurovascular unit coming of age: a journey through neurovascular coupling in health and disease. *Neuron*. 2017;96:17–42.
4. Wallez Y, Huber P. Endothelial adherens and tight junctions in vascular homeostasis, inflammation and angiogenesis. *Biochim Biophys Acta*. 2008;1778:794–809.
5. Baeyens N, Bandyopadhyay C, Coon BG, Yun S, Schwartz MA. Endothelial fluid shear stress sensing in vascular health and disease. *J Clin Invest*. 2016;126:821–828.
6. Langen UH, Ayloo S, Gu C. Development and cell biology of the blood-brain barrier. *Annu Rev Cell Dev Biol*. 2019;35:591–613.
7. Parton RG. Caveolae: structure, function, and relationship to disease. *Annu Rev Cell Dev Biol*. 2018;34:111–136.
8. Runkle EA, Antonetti DA. The blood-retinal barrier: structure and functional significance. *Methods Mol Biol*. 2011;686:133–148.
9. Farhy-Tselnicker I, Allen NJ. Astrocytes, neurons, synapses: a tripartite view on cortical circuit development. *Neural Dev*. 2018;13:7.
10. Thion MS, Garel S. On place and time: microglia in embryonic and perinatal brain development. *Curr Opin Neurobiol*. 2017;47:121–130.
11. Rowitch DH, Kriegstein AR. Developmental genetics of vertebrate glial-cell specification. *Nature*. 2010;468:214–222.
12. Martynoga B, Drechsel D, Guillemot F. Molecular control of neurogenesis: a view from the mammalian cerebral cortex. *Cold Spring Harb Perspect Biol*. 2012;4.
13. Engelhardt B, Liebner S. Novel insights into the development and maintenance of the blood-brain barrier. *Cell Tissue Res*. 2014;355:687–699.
14. Hupe M, Li MX, Kneitz S, et al. Gene expression profiles of brain endothelial cells during embryonic development at bulk and single-cell levels. *Sci Signal*. 2017;10.
15. Corada M, Orsenigo F, Bhat GP, et al. Fine-tuning of Sox17 and canonical Wnt coordinates the permeability properties of the blood-brain barrier. *Circ Res*. 2019;124:511–525.
16. Harb R, Whiteus C, Freitas C, Grutzendler J. In vivo imaging of cerebral microvascular plasticity from birth to death. *J Cerebr Blood Flow Metabol*. 2013;33:146–156.
17. Toma K, Hanashima C. Switching modes in corticogenesis: mechanisms of neuronal subtype transitions and integration in the cerebral cortex. *Front Neurosci*. 2015;9:274.
18. Bayraktar OA, Fuentealba LC, Alvarez-Buylla A, Rowitch DH. Astrocyte development and heterogeneity. *Cold Spring Harb Perspect Biol*. 2014;7:a020362.
19. Lacoste B, Comin CH, Ben-Zvi A, et al. Sensory-related neural activity regulates the structure of vascular networks in the cerebral cortex. *Neuron*. 2014;83:1117–1130.
20. Daneman R, Agalliu D, Zhou L, Kuhnert F, Kuo CJ, Barres BA. Wnt/beta-catenin signaling is required for CNS, but not non-CNS, angiogenesis. *Proc Natl Acad Sci USA*. 2009;106:641–646.
21. Stenman JM, Rajagopal J, Carroll TJ, Ishibashi M, McMahon J, McMahon AP. Canonical Wnt signaling regulates organ-specific assembly and differentiation of CNS vasculature. *Science*. 2008;322:1247–1250.
22. Liebner S, Corada M, Bangsow T, et al. Wnt/beta-catenin signaling controls development of the blood-brain barrier. *J Cell Biol*. 2008;183:409–417.
23. Martowicz A, Trusohamn M, Jensen N, et al. Endothelial beta-catenin signaling supports postnatal brain and retinal angiogenesis by promoting sprouting, tip cell formation, and VEGFR (vascular endothelial growth factor receptor) 2 expression. *Arterioscler Thromb Vasc Biol*. 2019;39:2273–2288.
24. Reyahi A, Nik AM, Ghiami M, et al. Foxf2 is required for brain pericyte differentiation and development and maintenance of the blood-brain barrier. *Dev Cell*. 2015;34:19–32.
25. Ben-Zvi A, Lacoste B, Kur E, et al. Mfsd2a is critical for the formation and function of the blood-brain barrier. *Nature*. 2014;509:507–511.

26. Hallmann R, Mayer DN, Berg EL, Broermann R, Butcher EC. Novel mouse endothelial cell surface marker is suppressed during differentiation of the blood brain barrier. *Dev Dynam*. 1995; 202:325−332.
27. Chow BW, Gu C. The molecular constituents of the blood-brain barrier. *Trends Neurosci*. 2015;38:598−608.
28. Butt AM, Jones HC, Abbott NJ. Electrical resistance across the blood-brain barrier in anaesthetized rats: a developmental study. *J Physiol*. 1990;429:47−62.
29. Saunders NR, Dziegielewska KM, Unsicker K, Ek CJ. Delayed astrocytic contact with cerebral blood vessels in FGF-2 deficient mice does not compromise permeability properties at the developing blood-brain barrier. *Dev Neurobiol*. 2016;76:1201−1212.
30. Rapaport DH, Wong LL, Wood ED, Yasumura D, LaVail MM. Timing and topography of cell genesis in the rat retina. *J Comp Neurol*. 2004;474:304−324.
31. Varshney S, Hunter DD, Brunken WJ. Extracellular matrix components regulate cellular polarity and tissue structure in the developing and mature retina. *J Ophthalmic Vis Res*. 2015;10:329−339.
32. Chan-Ling T, Chu Y, Baxter L, Weible Ii M, Hughes S. In vivo characterization of astrocyte precursor cells (APCs) and astrocytes in developing rat retinae: differentiation, proliferation, and apoptosis. *Glia*. 2009;57:39−53.
33. Stone J, Itin A, Alon T, et al. Development of retinal vasculature is mediated by hypoxia-induced vascular endothelial growth factor (VEGF) expression by neuroglia. *J Neurosci*. 1995;15: 4738−4747.
34. Stahl A, Connor KM, Sapieha P, et al. The mouse retina as an angiogenesis model. *Invest Ophthalmol Vis Sci*. 2010;51:2813−2826.
35. Chow BW, Gu C. Gradual suppression of transcytosis governs functional blood-retinal barrier formation. *Neuron*. 2017;93: 1325−1333 e1323.
36. Knowland D, Arac A, Sekiguchi KJ, et al. Stepwise recruitment of transcellular and paracellular pathways underlies blood-brain barrier breakdown in stroke. *Neuron*. 2014;82:603−617.
37. Lengfeld JE, Lutz SE, Smith JR, et al. Endothelial Wnt/beta-catenin signaling reduces immune cell infiltration in multiple sclerosis. *Proc Natl Acad Sci USA*. 2017;114:E1168−E1177.
38. Lutz SE, Smith JR, Kim DH, et al. Caveolin1 is required for Th1 cell infiltration, but not tight junction remodeling, at the blood-brain barrier in autoimmune neuroinflammation. *Cell Rep*. 2017; 21:2104−2117.
39. Mazzoni J, Smith JR, Shahriar S, Cutforth T, Ceja B, Agalliu D. The Wnt inhibitor Apcdd1 coordinates vascular remodeling and barrier maturation of retinal blood vessels. *Neuron*. 2017;96: 1055−1069 e1056.
40. Gould DB, Phalan FC, Breedveld GJ, et al. Mutations in Col4a1 cause perinatal cerebral hemorrhage and porencephaly. *Science*. 2005;308:1167−1171.
41. Menezes MJ, McClenahan FK, Leiton CV, Aranmolate A, Shan X, Colognato H. The extracellular matrix protein laminin alpha2 regulates the maturation and function of the blood-brain barrier. *J Neurosci*. 2014;34:15260−15280.
42. Chen ZL, Yao Y, Norris EH, et al. Ablation of astrocytic laminin impairs vascular smooth muscle cell function and leads to hemorrhagic stroke. *J Cell Biol*. 2013;202:381−395.
43. Yao Y, Chen ZL, Norris EH, Strickland S. Astrocytic laminin regulates pericyte differentiation and maintains blood brain barrier integrity. *Nat Commun*. 2014;5:3413.
44. Bader BL, Rayburn H, Crowley D, Hynes RO. Extensive vasculogenesis, angiogenesis, and organogenesis precede lethality in mice lacking all alpha v integrins. *Cell*. 1998;95:507−519.
45. Mobley AK, Tchaicha JH, Shin J, Hossain MG, McCarty JH. Beta8 integrin regulates neurogenesis and neurovascular homeostasis in the adult brain. *J Cell Sci*. 2009;122:1842−1851.
46. Nguyen HL, Lee YJ, Shin J, et al. TGF-beta signaling in endothelial cells, but not neuroepithelial cells, is essential for cerebral vascular development. *Lab Invest*. 2011;91:1554−1563.
47. Hirota S, Clements TP, Tang LK, et al. Neuropilin 1 balances beta8 integrin-activated TGFbeta signaling to control sprouting angiogenesis in the brain. *Development*. 2015;142:4363−4373.
48. Hubbard JA, Hsu MS, Seldin MM, Binder DK. Expression of the astrocyte water channel Aquaporin-4 in the mouse brain. *ASN Neuro*. 2015;7.
49. Haj-Yasein NN, Vindedal GF, Eilert-Olsen M, et al. Glial-conditional deletion of aquaporin-4 (Aqp4) reduces blood-brain water uptake and confers barrier function on perivascular astrocyte endfeet. *Proc Natl Acad Sci USA*. 2011;108:17815−17820.
50. Armulik A, Genove G, Mae M, et al. Pericytes regulate the blood-brain barrier. *Nature*. 2010;468:557−561.
51. Daneman R, Zhou L, Kebede AA, Barres BA. Pericytes are required for blood-brain barrier integrity during embryogenesis. *Nature*. 2010;468:562−566.
52. Gnanaguru G, Bachay G, Biswas S, Pinzon-Duarte G, Hunter DD, Brunken WJ. Laminins containing the beta2 and gamma3 chains regulate astrocyte migration and angiogenesis in the retina. *Development*. 2013;140:2050−2060.
53. Zovein AC, Luque A, Turlo KA, et al. Beta1 integrin establishes endothelial cell polarity and arteriolar lumen formation via a Par3-dependent mechanism. *Dev Cell*. 2010;18:39−51.
54. Yamamoto H, Ehling M, Kato K, et al. Integrin beta1 controls VE-cadherin localization and blood vessel stability. *Nat Commun*. 2015;6:6429.
55. Park DY, Lee J, Kim J, et al. Plastic roles of pericytes in the blood-retinal barrier. *Nat Commun*. 2017;8:15296.
56. Hirota S, Liu Q, Lee HS, Hossain MG, Lacy-Hulbert A, McCarty JH. The astrocyte-expressed integrin alphavbeta8 governs blood vessel sprouting in the developing retina. *Development*. 2011;138:5157−5166.
57. Arnold TD, Ferrero GM, Qiu H, et al. Defective retinal vascular endothelial cell development as a consequence of impaired integrin alphaVbeta8-mediated activation of transforming growth factor-beta. *J Neurosci*. 2012;32:1197−1206.
58. Nicchia GP, Pisani F, Simone L, et al. Glio-vascular modifications caused by Aquaporin-4 deletion in the mouse retina. *Exp Eye Res*. 2016;146:259−268.
59. Virani SS, Alonso A, Benjamin EJ, et al. Heart disease and stroke statistics-2020 update: a report from the American Heart Association. *Circulation*. 2020;141:e139−e596.
60. Dirnagl U, Iadecola C, Moskowitz MA. Pathobiology of ischaemic stroke: an integrated view. *Trends Neurosci*. 1999;22:391−397.
61. Sandoval KE, Witt KA. Blood-brain barrier tight junction permeability and ischemic stroke. *Neurobiol Dis*. 2008;32:200−219.
62. Jiao H, Wang Z, Liu Y, Wang P, Xue Y. Specific role of tight junction proteins claudin-5, occludin, and ZO-1 of the blood-brain barrier in a focal cerebral ischemic insult. *J Mol Neurosci*. 2011;44:130−139.
63. Krueger M, Hartig W, Reichenbach A, Bechmann I, Michalski D. Blood-brain barrier breakdown after embolic stroke in rats occurs without ultrastructural evidence for disrupting tight junctions. *PLoS One*. 2013;8. e56419.
64. Cipolla MJ, Crete R, Vitullo L, Rix RD. Transcellular transport as a mechanism of blood-brain barrier disruption during stroke. *Front Biosci*. 2004;9:777−785.
65. Nag S, Venugopalan R, Stewart DJ. Increased caveolin-1 expression precedes decreased expression of occludin and claudin-5 during blood-brain barrier breakdown. *Acta Neuropathol*. 2007; 114:459−469.
66. Sadeghian H, Lacoste B, Qin T, et al. Spreading depolarizations trigger caveolin-1-dependent endothelial transcytosis. *Ann Neurol*. 2018;84:409−423.

67. Jasmin JF, Malhotra S, Singh Dhallu M, Mercier I, Rosenbaum DM, Lisanti MP. Caveolin-1 deficiency increases cerebral ischemic injury. *Circ Res*. 2007;100:721–729.
68. Semenza GL. Regulation of tissue perfusion in mammals by hypoxia-inducible factor 1. *Exp Physiol*. 2007;92:988–991.
69. Manavski Y, Lucas T, Glaser SF, et al. Clonal expansion of endothelial cells contributes to ischemia-induced neovascularization. *Circ Res*. 2018;122:670–677.
70. Abdullah Z, Rakkar K, Bath PM, Bayraktutan U. Inhibition of TNF-alpha protects in vitro brain barrier from ischaemic damage. *Mol Cell Neurosci*. 2015;69:65–79.
71. Shi Y, Jiang X, Zhang L, et al. Endothelium-targeted overexpression of heat shock protein 27 ameliorates blood-brain barrier disruption after ischemic brain injury. *Proc Natl Acad Sci USA*. 2017;114:E1243–E1252.
72. Liu J, Jin X, Liu KJ, Liu W. Matrix metalloproteinase-2-mediated occludin degradation and caveolin-1-mediated claudin-5 redistribution contribute to blood-brain barrier damage in early ischemic stroke stage. *J Neurosci*. 2012;32:3044–3057.
73. Stamatovic SM, Keep RF, Wang MM, Jankovic I, Andjelkovic AV. Caveolae-mediated internalization of occludin and claudin-5 during CCL2-induced tight junction remodeling in brain endothelial cells. *J Biol Chem*. 2009;284:19053–19066.
74. Yang Y, Estrada EY, Thompson JF, Liu W, Rosenberg GA. Matrix metalloproteinase-mediated disruption of tight junction proteins in cerebral vessels is reversed by synthetic matrix metalloproteinase inhibitor in focal ischemia in rat. *J Cerebr Blood Flow Metabol*. 2007;27:697–709.
75. Asahi M, Wang X, Mori T, et al. Effects of matrix metalloproteinase-9 gene knock-out on the proteolysis of blood-brain barrier and white matter components after cerebral ischemia. *J Neurosci*. 2001;21:7724–7732.
76. Chiu PS, Lai SC. Matrix metalloproteinase-9 leads to claudin-5 degradation via the NF-kappaB pathway in BALB/c mice with eosinophilic meningoencephalitis caused by *Angiostrongylus cantonensis*. *PLoS One*. 2013;8. e53370.
77. Underly RG, Levy M, Hartmann DA, Grant RI, Watson AN, Shih AY. Pericytes as inducers of rapid, matrix metalloproteinase-9-dependent capillary damage during ischemia. *J Neurosci*. 2017;37:129–140.
78. Duz B, Oztas E, Erginay T, Erdogan E, Gonul E. The effect of moderate hypothermia in acute ischemic stroke on pericyte migration: an ultrastructural study. *Cryobiology*. 2007;55:279–284.
79. Gonul E, Duz B, Kahraman S, Kayali H, Kubar A, Timurkaynak E. Early pericyte response to brain hypoxia in cats: an ultrastructural study. *Microvasc Res*. 2002;64:116–119.
80. Platt MP, Agalliu D, Cutforth T. Hello from the other side: How autoantibodies circumvent the blood-brain barrier in autoimmune encephalitis. *Front Immunol*. 2017;8:442.
81. Shimizu F, Nishihara H, Kanda T. Blood-brain barrier dysfunction in immuno-mediated neurological diseases. *Immunol Med*. 2018; 41:120–128.
82. Thompson AJ, Baranzini SE, Geurts J, Hemmer B, Ciccarelli O. Multiple sclerosis. *Lancet*. 2018;391:1622–1636.
83. Engelhardt B, Carare RO, Bechmann I, Flugel A, Laman JD, Weller RO. Vascular, glial, and lymphatic immune gateways of the central nervous system. *Acta Neuropathol*. 2016;132:317–338.
84. Kamphuis WW, De Vries HE. Studying the blood-brain barrier will provide new insights into neurodegeneration - Yes. *Mult Scler*. 2018, 1352458518754367.
85. Petersen MA, Ryu JK, Akassoglou K. Fibrinogen in neurological diseases: mechanisms, imaging and therapeutics. *Nat Rev Neurosci*. 2018;19:283–301.
86. Pierson ER, Wagner CA, Goverman JM. The contribution of neutrophils to CNS autoimmunity. *Clin Immunol*. 2018;189:23–28.
87. Lee NJ, Ha SK, Sati P, et al. Spatiotemporal distribution of fibrinogen in marmoset and human inflammatory demyelination. *Brain*. 2018;141(6):1637–1649. https://doi.org/10.1093/brain/awy082.
88. Marchetti L, Engelhardt B. Immune cell trafficking across the blood-brain barrier in the absence and presence of neuroinflammation. *Vasc Biol*. 2020;2:H1–H18.
89. Charcot M. Histologie de la sclerose en plaque. *Gaz. Hosp*. 1868;41: 554–556.
90. Rindfleisch E. Histologisches Detail zu der grauen Degeneration von Gehirn und Rückenmark. (Zugleich ein Beitrag zu der Lehre von der Entstehung und Verwandlung der Zelle.). *Arch für Pathol Anat Physiol für Klin Med*. 1863;26:474–483.
91. Dawson JW. The Histology of disseminated sclerosis. *Edinb Med J*. 1916;17:377–410.
92. Broman T. Supravital analysis of disorders in the cerebral vascular permeability: two cases of multiple sclerosis. *Acta Psychiatr Scand*. 1947;22.
93. Milo R, Miller A. Revised diagnostic criteria of multiple sclerosis. *Autoimmun Rev*. 2014;13:518–524.
94. Kermode AG, Thompson AJ, Tofts P, et al. Breakdown of the blood-brain barrier precedes symptoms and other MRI signs of new lesions in multiple sclerosis. Pathogenetic and clinical implications. *Brain*. 1990;113(Pt 5):1477–1489.
95. Alvarez JI, Saint-Laurent O, Godschalk A, et al. Focal disturbances in the blood-brain barrier are associated with formation of neuroinflammatory lesions. *Neurobiol Dis*. 2015;74:14–24.
96. Ross JS, Delamarter R, Hueftle MG, et al. Gadolinium-DTPA-enhanced MR imaging of the postoperative lumbar spine: time course and mechanism of enhancement. *AJR Am J Roentgenol*. 1989;152:825–834.
97. Harris JO, Frank JA, Patronas N, McFarlin DE, McFarland HF. Serial gadolinium-enhanced magnetic resonance imaging scans in patients with early, relapsing-remitting multiple sclerosis: implications for clinical trials and natural history. *Ann Neurol*. 1991;29:548–555.
98. Barnes D, Munro PM, Youl BD, Prineas JW, McDonald WI. The longstanding MS lesion. A quantitative MRI and electron microscopic study. *Brain*. 1991;114(Pt 3):1271–1280.
99. Leech S, Kirk J, Plumb J, McQuaid S. Persistent endothelial abnormalities and blood-brain barrier leak in primary and secondary progressive multiple sclerosis. *Neuropathol Appl Neurobiol*. 2007; 33:86–98.
100. Kwon EE, Prineas JW. Blood-brain barrier abnormalities in longstanding multiple sclerosis lesions. An immunohistochemical study. *J Neuropathol Exp Neurol*. 1994;53:625–636.
101. Maggi P, Macri SM, Gaitan MI, et al. The formation of inflammatory demyelinated lesions in cerebral white matter. *Ann Neurol*. 2014;76:594–608.
102. Sohet F, Lin C, Munji RN, et al. LSR/angulin-1 is a tricellular tight junction protein involved in blood-brain barrier formation. *J Cell Biol*. 2015;208:703–711.
103. Bennett J, Basivireddy J, Kollar A, et al. Blood-brain barrier disruption and enhanced vascular permeability in the multiple sclerosis model EAE. *J Neuroimmunol*. 2010;229:180–191.
104. Winger RC, Koblinski JE, Kanda T, Ransohoff RM, Muller WA. Rapid remodeling of tight junctions during paracellular diapedesis in a human model of the blood-brain barrier. *J Immunol*. 2014;193:2427–2437.
105. Reijerkerk A, Kooij G, van der Pol SM, et al. Tissue-type plasminogen activator is a regulator of monocyte diapedesis through the brain endothelial barrier. *J Immunol*. 2008;181:3567–3574.
106. Claudio L, Brosnan CF. Effects of prazosin on the blood-brain barrier during experimental autoimmune encephalomyelitis. *Brain Res*. 1992;594:233–243.

107. Claudio L, Raine CS, Brosnan CF. Evidence of persistent blood-brain barrier abnormalities in chronic-progressive multiple sclerosis. *Acta Neuropathol*. 1995;90:228–238.
108. Shin T, Kim H, Jin JK, et al. Expression of caveolin-1, -2, and -3 in the spinal cords of Lewis rats with experimental autoimmune encephalomyelitis. *J Neuroimmunol*. 2005;165:11–20.
109. Wu H, Deng R, Chen X, et al. Caveolin-1 is critical for lymphocyte trafficking into central nervous system during experimental autoimmune encephalomyelitis. *J Neurosci*. 2016;36:5193–5199.
110. Murphy AC, Lalor SJ, Lynch MA, Mills KH. Infiltration of Th1 and Th17 cells and activation of microglia in the CNS during the course of experimental autoimmune encephalomyelitis. *Brain Behav Immun*. 2010;24:641–651.
111. Abadier M, Haghayegh Jahromi N, Cardoso Alves L, et al. Cell surface levels of endothelial ICAM-1 influence the transcellular or paracellular T-cell diapedesis across the blood-brain barrier. *Eur J Immunol*. 2015;45:1043–1058.
112. Wimmer I, Tietz S, Nishihara H, et al. PECAM-1 stabilizes blood-brain barrier integrity and favors paracellular T-cell diapedesis across the blood-brain barrier during neuroinflammation. *Front Immunol*. 2019;10:711.
113. Henninger DD, Panes J, Eppihimer M, et al. Cytokine-induced VCAM-1 and ICAM-1 expression in different organs of the mouse. *J Immunol*. 1997;158:1825–1832.
114. Daneman R, Zhou L, Agalliu D, Cahoy JD, Kaushal A, Barres BA. The mouse blood-brain barrier transcriptome: a new resource for understanding the development and function of brain endothelial cells. *PLoS One*. 2010;5. e13741.
115. Dustin ML, Rothlein R, Bhan AK, Dinarello CA, Springer TA. Induction by IL 1 and interferon-gamma: tissue distribution, biochemistry, and function of a natural adherence molecule (ICAM-1). *J Immunol*. 1986;137:245–254.
116. Pober JS, Gimbrone Jr MA, Lapierre LA, et al. Overlapping patterns of activation of human endothelial cells by interleukin 1, tumor necrosis factor, and immune interferon. *J Immunol*. 1986; 137:1893–1896.
117. Steffen BJ, Butcher EC, Engelhardt B. Evidence for involvement of ICAM-1 and VCAM-1 in lymphocyte interaction with endothelium in experimental autoimmune encephalomyelitis in the central nervous system in the SJL/J mouse. *Am J Pathol*. 1994;145: 189–201.
118. Greenwood J, Wang Y, Calder VL. Lymphocyte adhesion and transendothelial migration in the central nervous system: the role of LFA-1, ICAM-1, VLA-4 and VCAM-1. off. *Immunology*. 1995;86:408–415.
119. Elices MJ, Osborn L, Takada Y, et al. VCAM-1 on activated endothelium interacts with the leukocyte integrin VLA-4 at a site distinct from the VLA-4/fibronectin binding site. *Cell*. 1990;60: 577–584.
120. Vajkoczy P, Laschinger M, Engelhardt B. Alpha4-integrin-VCAM-1 binding mediates G protein-independent capture of encephalitogenic T cell blasts to CNS white matter microvessels. *J Clin Invest*. 2001;108:557–565.
121. Yednock TA, Cannon C, Fritz LC, Sanchez-Madrid F, Steinman L, Karin N. Prevention of experimental autoimmune encephalomyelitis by antibodies against alpha 4 beta 1 integrin. *Nature*. 1992; 356:63–66.
122. Tubridy N, Behan PO, Capildeo R, et al. The effect of anti-alpha4 integrin antibody on brain lesion activity in MS. The UK Antegren Study Group. *Neurology*. 1999;53:466–472.
123. Miller DH, Khan OA, Sheremata WA, et al. A controlled trial of natalizumab for relapsing multiple sclerosis. *N Engl J Med*. 2003; 348:15–23.
124. Glatigny S, Duhen R, Oukka M, Bettelli E. Cutting edge: loss of alpha4 integrin expression differentially affects the homing of Th1 and Th17 cells. *J Immunol*. 2011;187:6176–6179.
125. Rothhammer V, Heink S, Petermann F, et al. Th17 lymphocytes traffic to the central nervous system independently of alpha4 integrin expression during EAE. *J Exp Med*. 2011;208: 2465–2476.
126. Sixt M, Engelhardt B, Pausch F, Hallmann R, Wendler O, Sorokin LM. Endothelial cell laminin isoforms, laminins 8 and 10, play decisive roles in T cell recruitment across the blood-brain barrier in experimental autoimmune encephalomyelitis. *J Cell Biol*. 2001;153:933–946.
127. Wu C, Ivars F, Anderson P, et al. Endothelial basement membrane laminin alpha5 selectively inhibits T lymphocyte extravasation into the brain. *Nat Med*. 2009;15:519–527.
128. Agrawal S, Anderson P, Durbeej M, et al. Dystroglycan is selectively cleaved at the parenchymal basement membrane at sites of leukocyte extravasation in experimental autoimmune encephalomyelitis. *J Exp Med*. 2006;203:1007–1019.
129. Vestweber D. How leukocytes cross the vascular endothelium. *Nat Rev Immunol*. 2015;15:692–704.
130. Kirk J, Plumb J, Mirakhur M, McQuaid S. Tight junctional abnormality in multiple sclerosis white matter affects all calibres of vessel and is associated with blood-brain barrier leakage and active demyelination. *J Pathol*. 2003;201:319–327.
131. Padden M, Leech S, Craig B, Kirk J, Brankin B, McQuaid S. Differences in expression of junctional adhesion molecule-A and beta-catenin in multiple sclerosis brain tissue: increasing evidence for the role of tight junction pathology. *Acta Neuropathol*. 2007; 113:177–186.
132. Plumb J, McQuaid S, Mirakhur M, Kirk J. Abnormal endothelial tight junctions in active lesions and normal-appearing white matter in multiple sclerosis. *Brain Pathol*. 2002;12:154–169.
133. Paul D, Cowan AE, Ge S, Pachter JS. Novel 3D analysis of Claudin-5 reveals significant endothelial heterogeneity among CNS microvessels. *Microvasc Res*. 2013;86:1–10.
134. Pfeiffer F, Schafer J, Lyck R, et al. Claudin-1 induced sealing of blood-brain barrier tight junctions ameliorates chronic experimental autoimmune encephalomyelitis. *Acta Neuropathol*. 2011; 122:601–614.
135. Alvarez JI, Cayrol R, Prat A. Disruption of central nervous system barriers in multiple sclerosis. *Biochim Biophys Acta*. 2011;1812: 252–264.
136. Zheng Y, He M, Congdon N. The worldwide epidemic of diabetic retinopathy. *Indian J Ophthalmol*. 2012;60:428–431.
137. Cheung N, Mitchell P, Wong TY. Diabetic retinopathy. *Lancet*. 2010;376:124–136.
138. Hartnett ME. Retinopathy of prematurity: evolving treatment with anti-vascular endothelial growth factor. *Am J Ophthalmol*. 2020;218:208–213.
139. Hellstrom A, Smith LE, Dammann O. Retinopathy of prematurity. *Lancet*. 2013;382:1445–1457.
140. Roy S, Ha J, Trudeau K, Beglova E. Vascular basement membrane thickening in diabetic retinopathy. *Curr Eye Res*. 2010;35: 1045–1056.
141. Gardiner TA, Archer DB, Curtis TM, Stitt AW. Arteriolar involvement in the microvascular lesions of diabetic retinopathy: implications for pathogenesis. *Microcirculation*. 2007;14:25–38.
142. Stitt AW, Gardiner TA, Archer DB. Histological and ultrastructural investigation of retinal microaneurysm development in diabetic patients. *Br J Ophthalmol*. 1995;79:362–367.
143. Stitt AW, Curtis TM, Chen M, et al. The progress in understanding and treatment of diabetic retinopathy. *Prog Retin Eye Res*. 2016;51: 156–186.
144. Lechner J, O'Leary OE, Stitt AW. The pathology associated with diabetic retinopathy. *Vis Res*. 2017;139:7–14.
145. Heng LZ, Comyn O, Peto T, et al. Diabetic retinopathy: pathogenesis, clinical grading, management and future developments. *Diabet Med*. 2013;30:640–650.

146. Biedermann B, Bringmann A, Franze K, Faude F, Wiedemann P, Reichenbach A. GABA(A) receptors in Muller glial cells of the human retina. *Glia*. 2004;46:302–310.
147. Pannicke T, Iandiev I, Wurm A, et al. Diabetes alters osmotic swelling characteristics and membrane conductance of glial cells in rat retina. *Diabetes*. 2006;55:633–639.
148. Kusari J, Zhou S, Padillo E, Clarke KG, Gil DW. Effect of memantine on neuroretinal function and retinal vascular changes of streptozotocin-induced diabetic rats. *Invest Ophthalmol Vis Sci*. 2007;48:5152–5159.
149. Joussen AM, Poulaki V, Le ML, et al. A central role for inflammation in the pathogenesis of diabetic retinopathy. *FASEB J*. 2004;18: 1450–1452.
150. Chibber R, Ben-Mahmud BM, Chibber S, Kohner EM. Leukocytes in diabetic retinopathy. *Curr Diabetes Rev*. 2007;3:3–14.
151. Leal EC, Manivannan A, Hosoya K, et al. Inducible nitric oxide synthase isoform is a key mediator of leukostasis and blood-retinal barrier breakdown in diabetic retinopathy. *Invest Ophthalmol Vis Sci*. 2007;48:5257–5265.
152. Krady JK, Basu A, Allen CM, et al. Minocycline reduces proinflammatory cytokine expression, microglial activation, and caspase-3 activation in a rodent model of diabetic retinopathy. *Diabetes*. 2005;54:1559–1565.
153. Hartnett ME, Penn JS. Mechanisms and management of retinopathy of prematurity. *N Engl J Med*. 2012;367:2515–2526.
154. Hellstrom A, Engstrom E, Hard AL, et al. Postnatal serum insulin-like growth factor I deficiency is associated with retinopathy of prematurity and other complications of premature birth. *Pediatrics*. 2003;112:1016–1020.
155. Wang H, Byfield G, Jiang Y, Smith GW, McCloskey M, Hartnett ME. VEGF-mediated STAT3 activation inhibits retinal vascularization by down-regulating local erythropoietin expression. *Am J Pathol*. 2012;180:1243–1253.
156. Aiello LP, Pierce EA, Foley ED, et al. Suppression of retinal neovascularization in vivo by inhibition of vascular endothelial growth factor (VEGF) using soluble VEGF-receptor chimeric proteins. *Proc Natl Acad Sci USA*. 1995;92:10457–10461.
157. Watanabe D, Suzuma K, Matsui S, et al. Erythropoietin as a retinal angiogenic factor in proliferative diabetic retinopathy. *N Engl J Med*. 2005;353:782–792.
158. Dubrac A, Kunzel SE, Kunzel SH, et al. NCK-dependent pericyte migration promotes pathological neovascularization in ischemic retinopathy. *Nat Commun*. 2018;9:3463.
159. Fulton AB, Hansen RM, Moskowitz A, Akula JD. The neurovascular retina in retinopathy of prematurity. *Prog Retin Eye Res*. 2009; 28:452–482.
160. Checchin D, Sennlaub F, Levavasseur E, Leduc M, Chemtob S. Potential role of microglia in retinal blood vessel formation. *Invest Ophthalmol Vis Sci*. 2006;47:3595–3602.
161. Fischer F, Martin G, Agostini HT. Activation of retinal microglia rather than microglial cell density correlates with retinal neovascularization in the mouse model of oxygen-induced retinopathy. *J Neuroinflammation*. 2011;8:120.
162. Biswas S, Cottarelli A, Agalliu D. Neuronal and glial regulation of CNS angiogenesis and barriergenesis. *Development*. 2020:147.

CHAPTER

11

Lymphatic biology and medicine

Xinguo Jiang[1,2], *Wen Tian*[1,2], *Mark R. Nicolls*[1,2] and *Stanley G. Rockson*[2]
[1]VA Palo Alto Health Care System, Palo Alto, CA, United States [2]Stanford University School of Medicine, Stanford, CA, United States

Introduction

The lymphatic vasculature complements the blood circulatory system by returning capillary ultrafiltrate to the blood circulation.[1] Lymphatic vessels also provide conduits for immune cell trafficking and dietary lipid absorption, which influences energy metabolism.[1,2] An additional vital lymphatic function is the removal of peripheral tissue cholesterol through reverse cholesterol transport (RCT).[3] The prototypical lymphatic disorder is lymphedema, classified as either primary, due to hereditary or unknown factors, or secondary (acquired), caused by lymphatic injury originating from infection, cancer, cancer-related therapies, or physical trauma.[4,5] Recent research indicates that aberrant lymphatic remodeling, along with a decrease of lymphatic function, may play a pathogenic role in various chronic conditions, such as metabolic disorders, neoplasia, and cardiovascular, neurological, and ocular diseases.[1] Modulation of the lymphatic vasculature also influences the outcome in solid organ transplant rejection.[6,7]

To summarize recent advances of lymphatic biology in health and disease, we will first provide a brief overview of lymphatic anatomy, and the molecular and cellular mechanisms involved in lymphangiogenesis essential for pre- and postnatal lymphatic development and regeneration in adults. We will then discuss lymphedema pathophysiology and new insights into lymphatic biology in obesity, atherosclerosis, and solid organ transplant rejection. Updated lymphatic pathobiology in tumor, myocardial infarction, neurological disorders, glaucoma, Crohn's disease, and inflammation are reviewed elsewhere.[1,8–11]

Lymphatic vasculature

The lymphatic vasculature consists of a network of highly permeable, blind-ended capillaries, and larger collecting lymphatic vessels.[5,12] Lymphatic capillaries are comprised of a single layer of lymphatic endothelial cells (LECs) that directly tether to the elastic fibers of the extracellular matrix through anchoring filaments, lacking a continuous basement membrane or perivascular mural cell coverage.[1,5] Adherens junction protein VE-cadherin, and other junctional proteins such as Claudin, Zonula Occludens −1, Connexin, Occludin, and endothelial cell selective adhesion molecules form discontinuous button-like structures unique to lymphatic capillaries.[13] The buttons can open and close in response to the pulling of the anchoring filaments when interstitial pressure is high, and serve as *de facto* primary valves to control the uptake of interstitial fluid, macromolecules, and immune cells.[5,9] Lymphatic capillaries converge into precollectors. These vessels subsequently coalesce into collecting lymphatics, in which LECs are joined to one another by seamless zipper-like junctions and are ensheathed with continuous basement membrane, along with lymphatic muscle cell (LMC) coverage.[9,13] Intraluminal one-way lymphatic valves divide collecting lymphatics into contractile segments, designated as lymphangions, that form the structural basis for unidirectional lymph flow[8] (Fig. 11.1). The contractile function of each lymphangion is ensured by intrinsic forces within the vascular wall generated by phasic and tonic contractions of LMCs. The activity of lymphangions is further influenced by extrinsic forces induced by surrounding tissue movement (e.g., skeletal

The Vasculome
https://doi.org/10.1016/B978-0-12-822546-2.00009-5

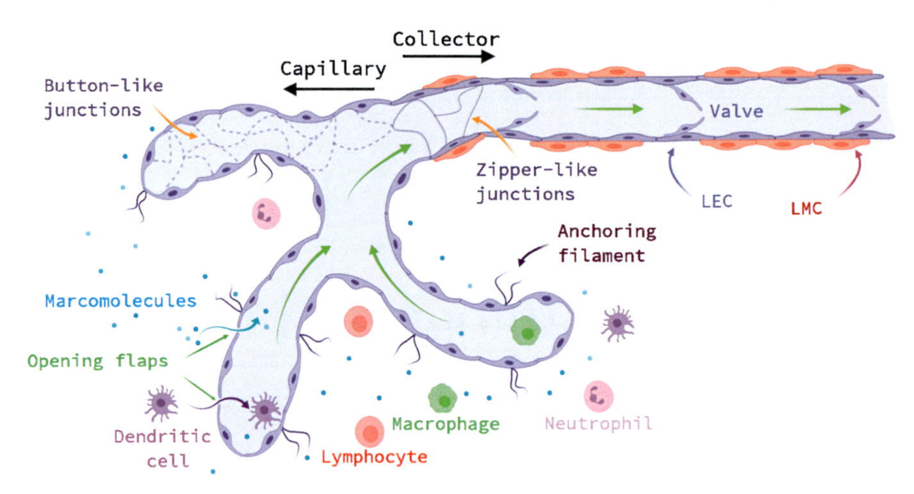

FIGURE 11.1 **Lymphatic vascular tree.** Lymphatic capillaries are comprised of a single layer of LECs (lymphatic endothelial cells) that tether directly to the extracellular matrix through anchoring filaments. Junctional proteins form button-like structures that interconnect capillary LECs. Those buttons enable the formation of overlapping LEC flaps, through which interstitial fluid, macromolecules, and immune cells enter the blind ends of the capillaries. Capillaries converge into collectors. LMCs (lymphatic muscle cells) cover the collectors sparing the regions adjacent to the intraluminal valves. Junctional proteins of the collecting lymphatics form uninterrupted structures, known as zipper-like junctions. Collecting lymphatics consist of lymphangions demarcated by intravascular valves; the segment between two valves is designated as one lymphangion. *All figures were created with BioRender.com.*

muscle contraction) and pressure gradients created by actions such as respiration, cardiac contraction, bowel evacuation, and arterial pulsation.[14] Collecting lymphatic vessels pass through chains of lymph nodes and converge into thoracic duct(s), through which the lymph is transported to the subclavian vein of the blood circulation.[5,8] Lymph nodes contain highly specialized anatomical compartments with morphologically unique lymphatic conduits; these are formed by LECs with distinct transcriptomics, providing unique microenvironments for immune priming and the induction of peripheral tolerance.[15–17]

Lymphangiogenesis

LEC specification

Lymphangiogenesis is fundamental for the generation of the lymphatic network during development and for lymphatic growth in adults during wound healing, inflammation, and tumorigenesis; it is the process of new lymphatic vessel formation from existing ones that involves coordinated LEC sprouting, migration, proliferation along with proper tube formation, and maturation.[18] Lineage tracing studies in developing mice embryos demonstrated that LECs originate from endothelial cells (ECs) of the cardinal vein at approximately E9.5, when SOX18 and COUP-TFII induce Prox-1 expression, which serves as the molecular switch for LEC specification.[19,20] A recent study demonstrates that LEC fate is imprinted during the transition through the paraxial mesoderm lineage,[21] providing new insights into earlier steps of LEC differentiation before the activation of the Sox18-COUP-TFII-Prox-1 axis. Transcriptomic analysis comparing LECs and BECs, isolated from mature mouse skin, demonstrated distinct gene expression profiles,[22] indicating lineage divergence of these two types of cells in adults. Interestingly, Prox-1 overexpression induces LEC-specific gene transcription in BECs,[23] and loss of Prox-1 expression promotes LEC de-differentiation into cells with BEC properties.[24] These experiments further support the notion that Prox-1 activity is essential for the induction and maintenance of LEC identity. Following LEC specification, primordial lymphatics form primary lymph sacs, which subsequently separate from blood circulation, requiring the expression of PDPN, Runx1, Clec2, Slp76, Syk, and Plcg2.[25] Although LECs originate primarily from venous endothelia,[19] other studies have shown that LECs may arise from cells of nonvenous sources,[26] including those from hemogenic endothelium,[27] and non-ECs.[28] The biological consequences of the existence of lymphatic structures comprised of LECs of heterogeneous origin, however, remain unknown.

Lymphatic sprouting

The best-characterized signaling involved in lymphatic sprouting is the VEGFC/VEGFR3 pathway.[29,30] VEGFR3 signaling is vital for human lymphatic development. VEGFR3 loss-of-function mutation causes congenital hypoplastic lymphedema, known as Milroy disease[31]; mutations in the VEGFC gene also cause primary lymphedema that has, essentially, the same clinical manifestations as those of Milroy lymphedema.[32] In mice, the lack of VEGFC leads to failure of primary lymph sac and lymphatic network formation, with embryonic lethality.[33,34] Additionally, VEGFC/VEGFR3 signaling mediates proper perinatal lymphangiogenesis and maintains the lacteals in adult mice[35]; reduced VEGFR3 signaling causes hypoplasia of Schlemm's canal (an eye lymphatic structure) and of the meningeal lymphatics.[36,37]

Furthermore, VEGFC-induced VEGFR3 activation promotes lymphangiogenesis in pathological conditions such as inflammation and tumor.[29] Available data support the concept that in lymphangiogenesis, as in angiogenesis, lymphatic growth is guided by leading tip cells that sense progrowth environmental cues formed by VEGFC gradients. In contrast, the highly proliferative lagging stalk cells fulfill vascular elongation.[29,38] Inhibition of NOTCH signaling by a soluble DLL4 (delta-like 4) leads to lymphatic vascular hypersprouting, which can be suppressed by a VEGFC/D trap,[39] indicating NOTCH activity suppresses the VEGFC/VEGFR3 pathway. These studies collectively support the notion that VEGFC/VEGFR3 and NOTCH pathways coordinate lymphangiogenesis in which the VEGFR3 signaling promotes tip cell phenotype, while NOTCH action enables stalk cell function.

Interestingly, studies also show that NOTCH signaling may enhance VEGFR3 activation by promoting the ephrin-B2-dependent VEGFR3 internalization,[40,41] implying context-dependent complex cross-talk between the NOTCH and VEGFC/VEGFR3 cascades. VEGFC is produced as a precursor protein, which requires the processing of C- and N-terminal propeptides to achieve full activity. Proprotein convertases furin, PC5, and PC7 mediate C-terminal cleavage,[42] whereas the metalloproteinase ADAMTS3, along with the scaffold protein CCBE1, is essential for the N-terminal processing.[43] Therefore, VEGFC production and activation coordinate the formation of the VEGFC gradient and guide lymphatic sprouting. Vascular smooth muscle cells (SMCs) and macrophages appear to be possible cellular sources for VEGFC production during development and inflammation, respectively.[29] NRP2 (neuropilin 2) plays a role in VEGFC/VEGFR3 signaling by acting as a coreceptor for VEGFR3,[30] suggesting that the axonal migration signal also plays a role in lymphangiogenesis. Taken together, these studies indicate that the VEGFC/VEGFR3 signaling, regulated at multiple levels, is fundamental for sprouting lymphangiogenesis.

Lymphatic maturation

Primitive lymphatic vessels undergo processes of maturation to form hierarchical networks that assure normal lymphatic function.[1] Lymphatic maturation is characterized by valve formation, mural cell recruitment, and basement membrane deposition of the collecting vessels. The angiopoietin (ANG)/TIE pathway plays a crucial role in this process.[30] Three ligands (ANG1, ANG2, ANG3/4) and two receptors (TIE1, TIE2) form the signaling system that regulates vascular maturation and quiescence.[44] Mice with ANG2 loss-of-function develop defective capillaries and abnormal collecting lymphatics with disorganized LMC coverage along with decreased numbers of intraluminal valves. These abnormal lymphatics, however, can be rescued by targeted ANG1 supplement,[45,46] indicating that the TIE2 signaling, activated by either ANG2 or ANG1, is required for lymphatic maturation. Further research is needed to determine the cell type–specific TIE2 activity involved in the maturation steps. TIE1 is a coreceptor for TIE2[47] and mediates proper lymphatic valve development and collecting lymphatic maturation,[48] supporting the concept that both TIE receptors play a role in lymphatic maturation. TIE1 expression may be essential for ANG2-induced TIE2 activation in LECs.[49–51] NOTCH signaling also regulates lymphatic valve development[52] and collecting lymphatic maturation.[40] In BEC, TIE2 activity regulates NOTCH signaling,[53] implying that the TIE2 and NOTCH pathways may cooperatively regulate lymphatic maturation. The transcription factor FOXC2 controls ANG2 expression[54] and mediates lymphatic valve formation and collecting lymphatic patterning,[55] further supporting the notion that the TIE2 signaling is a vital effector of the complex network regulating lymphatic maturation. Additionally, BMP9/ALK1 signaling appears to play a role in collector maturation.[56] Oscillatory fluid shear stress may also regulate valve formation by inducing the GATA2-FOXC2 cascade.[57–59] In summary, fluid pressure regulates lymphatic maturation by controlling the expression of master transcriptional factors such as GATA2 and FOXC2, which in turn activates the TIE2 and NOTCH pathways (Fig. 11.2).

Lymphedema

Lymphedema is a debilitating chronic state of lymphatic insufficiency, characterized by tissue swelling, inflammation, and dermal pathology, and no curative

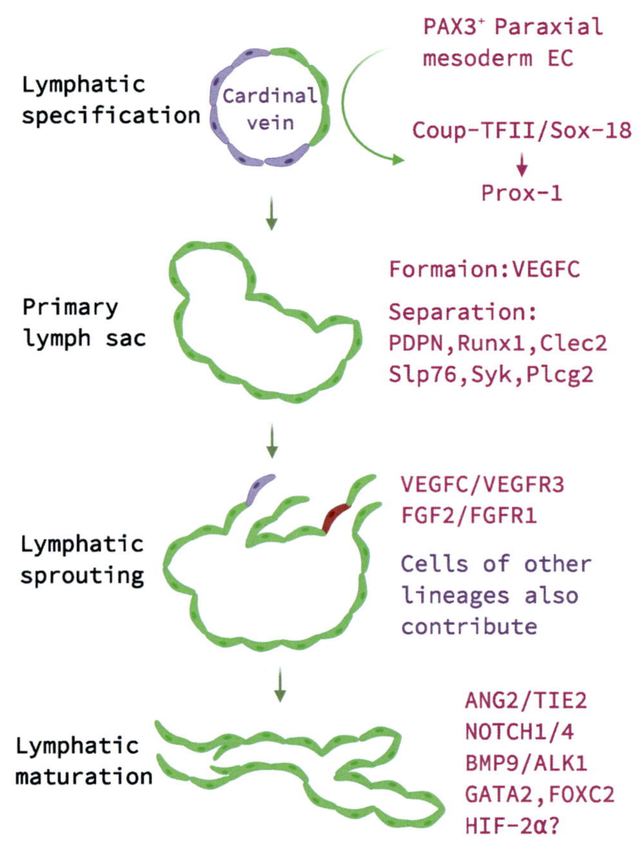

FIGURE 11.2 Lymphatic specification and sprouting. Expression of Coup-TFII and Sox-18 induce Prox-1 expression in a subset of cells in the cardinal vein. These cells derive from $PAX3^{+}$ mesoderm endothelial cells and subsequently acquire LEC fate. $Prox\text{-}1^{+}$ LECs proliferate toward the VEGFC gradient and migrate away from the vein to form the primary lymph sac. The separation of the lymph sac from the cardinal vein requires the expression of genes, including PDPN, Runx1, Clec2, Slp76, Syk, and Plcg2. VEGFC/VEGFR3 and FGF2/FGFR1 are key pathways promoting sprouting lymphangiogenesis. During this process, cells of other lineages, such as hemogenic endothelial or nonendothelial cells, can also adopt LEC fate and contribute to lymphangiogenesis. Subsequently, primitive lymphatic vessels undergo remodeling and maturation, which require TIE2, NOTCH, BMP9/ALK1 pathways and expression of transcription factors include GATA2, FOXC2, and HIF-2α. Abbreviations: *ALK1*, activin receptor-like kinase-1; *BMP*, bone morphogenetic protein; *FGF*, fibroblast growth factor; *FGFR*, FGF receptor; *FOXC2*, forkhead box C2; *GATA2*, GATA-binding factor 2; *HIF*, hypoxia-inducible factor; *PDPN*, podoplanin; *Plcg2*, phospholipase c gamma 2; *Slp76*, SH2 domain-containing leukocyte protein of 76 kDa; *Syk*, spleen tyrosine kinase.

pharmacological therapy.[1,5] Based on etiology, lymphedema can be classified as primary or secondary. Primary lymphedema occurs in either heritable or idiopathic fashion. Research in the past few decades has identified mutations of more than 30 genes that may independently cause primary lymphedema, including *VEGFC, VEGFR3, CCBE1, ADAMTS3, FAT4, PTPN14, KIF11, IKBKG, HGF, PTPN11, KRAS, EPHB4, ITGA9, SOX18, FOXC2, GATA2,* and *ANG2*,[1,4,12,60,61] with most of these genes notably participating in various steps of lymphatic network formation during development. Secondary lymphedema, by contrast, is the most abundant form of lymphedema, caused by lymphatic vessel damage as a result of cancer, cancer-related lymph node removal, radiation therapy, and parasitic infection.[1,5] In the United States, secondary lymphedema affects an estimated 6 million individuals, with most of them being cancer survivors.[62–64] For this reason, cancer-associated secondary lymphedema is the focus of the current discussion.

Risk factors

Extensive cancer treatment (e.g., lymph nodes removal along with regional radiation), obesity, and infection appear to increase the risk of developing lymphedema in breast cancer patients.[65,66] A recent meta-analysis concludes that patients with combined radiation therapy and axillary lymph node dissection are approximately three times more likely to develop lymphedema than those who experience lymph node removal alone.[67] A body mass index of >29 also increases lymphedema risk by an estimated threefold in breast cancer patients.[68] Additionally, individual genetic variations raise the likelihood of developing secondary lymphedema in cancer patients.[64] For instance, rare mutations in *HGF* or *MET* (participants in lymphangiogenesis) or *GJC2* (necessary for valve development and function) associate with increased risk of developing secondary lymphedema[69,70]; common variants of *IL-4, IL10, NFKB2, VEGFR2, VEGFR3,* and *RORC* also increase lymphedema risk in women treated for breast cancer.[71,72] Collectively, the severity of the lymphatic injury, patient metabolic status, and mutations or variations of genes associated with inflammation or lymphatic regeneration may predict the risk of developing lymphedema in cancer survivors (Fig. 11.3A).

Inflammation in lymphedema

The natural history of cancer-associated lymphedema is highly variable, with disease onset and progression ranging from latent in some patients to rapidly progressive in others.[73] Following an initial lymphatic insult, there is a chronic accumulation of protein- and lipid-rich interstitial fluid that promotes inflammation, adipose deposition, and tissue fibrosis.[5,74] Among these pathological responses, uncontrolled inflammation is probably the most critical factor in promoting lymphedema advancement.[5] High levels of proteins such as HSPs, HMGB1, and hyaluronan in the interstitial fluid may sensitize toll-like receptors and induce sterile inflammation,[75–77] which, if unresolved, sets the stage for the development of advanced skin pathology.

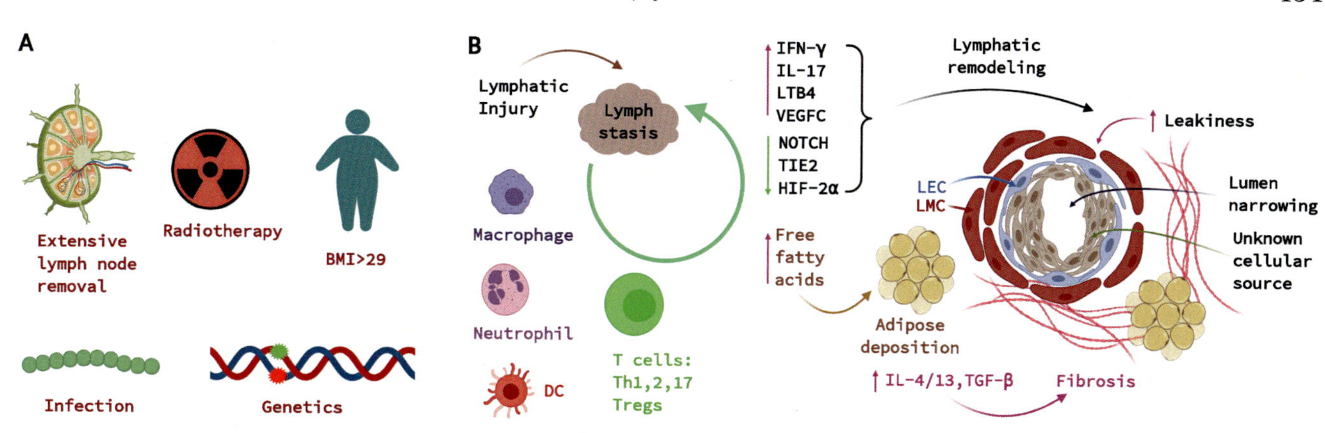

FIGURE 11.3 **Lymphedema pathobiology.** (A) Risk factors for lymphedema in cancer survivors include a greater degree of lymph node removal, radiotherapy, obesity, infection, and genetic predisposition. (B) Lymphatic injury leads to lymph stasis, which induces inflammation characterized by high infiltration of macrophages, neutrophils, dendritic cells, and T cells. Inflammatory cytokines IFN-γ (produced primarily by Th1 cells) and IL-17 (by Th17 cells) together with LTB_4 and VEGFC (produced by macrophage in abundance) promote pathological lymphatic remodeling characterized by leakiness increase, collector lumen narrowing with unknown cellular source comprising neointima; Th2 cytokines IL-4/13 and TGF-β promote tissue fibrotic remodeling. Free fatty acid-rich lymph enhances adipocyte differentiation and proliferation, which ultimately leads to excessive fat deposition in the diseased area. Abbreviations: *BMI*, body mass index; *HIF*, hypoxia-inducible factor; *IFN*, interferon; *IL*, interleukin; *LMC*, lymphatic muscle cell; *LTB_4*, leukotriene B_4; *TGF*, transforming growth factor; *Th*, T helper.

Inflammation of lymphedema tissues involves both innate and adaptive immune responses.[5]

In the setting of lymphatic injury, dendritic cells (DCs) that mature in the periphery are incapable of migrating efficiently to the draining lymph nodes.[78] Reduced DC trafficking not only promotes local inflammation by producing proinflammatory mediators, but also compromises protective antiinfection immunity.[79,80] Further studies are needed to determine whether DCs deliver autoantigens from the lymphedema tissue to induce autoantigen-responsive T cell maturation. Macrophages are master regulators of inflammation and fibrosis,[81] and unsurprisingly, lymphedema tissue exhibits a marked increase in macrophage infiltration.[78,82] Macrophages exhibit remarkable plasticity and can change their physiology in response to microenvironmental cues.[83] Accordingly, macrophages play complex pathophysiological roles in lymphedema. Macrophage-produced VEGFC is vital for lymphatic regeneration during the early stages of lymphatic insufficiency.[84,85] However, excessive VEGFC may induce nonproductive lymphangiogenesis and damage collecting lymphatic vessel function.[86,87] Macrophages may also influence the contraction of collectors by promoting iNOS (inducible nitric oxide synthase) production.[14,88] Macrophage-produced IL-6, by contrast, may suppress adipose deposition.[89,90] We found macrophages to be the primary source of the lipid mediator leukotriene B_4 (LTB_4) that promotes lymphedema progression by attracting additional proinflammatory immune cells and by suppressing the NOTCH signaling pathway that is required for lymphatic repair. Pharmacological antagonism of the LTB_4 pathway reverses preclinical lymphedema.[82] Consistently, ketoprofen, an NSAID that blocks the pathway involved in LTB_4 production, ameliorates both experimental and human clinical lymphedema.[91,92] Neutrophils, another critical cell type in innate immunity, may exacerbate lymphedema through elastase-mediated degradation of the ECM protein EMILIN1, effectively reducing the function of anchoring filaments.[93]

T cells, particularly $CD4^+$ T cells, including Treg, Th1, Th2, and Th17 subtypes, are essential mediators of lymphedema pathogenesis.[5,85] There is an increased infiltration of $CD4^+CD25^+$ Treg cells in preclinical and clinical lymphedema samples; Treg cell depletion aggravates, whereas adoptive transfer of Treg-enriched cells ameliorates, experimental lymphedema.[94] These experiments indicate that, in lymphedema, Treg activity is insufficient despite the fact that diseased tissues display increased numbers of these cells. Whether individual Treg cells are dysfunctional, or merely because effector cells overwhelm Treg cells, is not known. Th1 and Th17 cell promote excessive lymphangiogenesis by enhancing macrophage VEGFC production; suppressing Th1/Th17 functions alleviates lymphedema.[87] Th2 cells and type 2 cytokines, by contrast, are important pathogenic factors for tissue fibrosis and lymphatic dysfunction.[95] Thus, both innate and adaptive immunity coordinate to promote inflammation in lymphedema, an early intervention targeting the aberrant immune responses may represent a significant therapeutic opportunity for lymphedema patients.

Lymphatic remodeling in lymphedema

It has long been assumed that lymphatic injury and insufficient lymphatic vessel regeneration is the defining feature of lymphedema pathophysiology. The fact that VEGFC-mediated therapeutic lymphangiogenesis ameliorates preclinical mouse lymphedema supports this concept.[96,97] In clinical lymphedema, however, the lymphatics are characterized by excessive lymphangiogenesis, with newly generated vessels exhibiting profound functional defects. Indeed, more recent data demonstrated a paradoxical VEGFC increase in the edematous tissue and serum of patients with established lymphedema,[86,98] indicating that VEGFC may enhance pathological lymphatic remodeling and aggravate lymphedema progression.

In advanced lymphedema, lymphatic vessels (both capillaries and collecting lymphatics) in the affected limb display increased leakiness[80]; using sensitive imaging technology, such as ICG and MRI lymphangiography, studies have illustrated lymphatic hyperplasia with aberrant lymphatic channels and obstructive lesions across the lymphatic networks that, eventually, results in widespread interstitial fluid accumulation in the diseased area.[99] Mechanisms promoting pathological lymphatic remodeling during advanced stages of lymphedema are poorly understood. Available data suggest that this remodeling likely results from a combined effect of excessive lymphatic sprouting induced by VEGFC[86,87,98] and loss of LEC intrinsic signaling pathways that are required for lymphatic maturation and maintenance. We found that LEC NOTCH signaling is downregulated in lymphedema, and, furthermore, lymphedema induced in mice with LEC-specific NOTCH deficiency cannot be reversed by a therapy that reduces inflammation through blocking LTB_4 production. However, LTB_4 blockade ameliorates lymphedema in WT mice.[82] In addition, LEC TIE2 signaling is also reduced in lymphedema tissue, presumably resulting from elevated TNF-α expression.[49–51] Collectively, these studies indicate that there is an induced loss of signaling cascades that regulate lymphatic maturation and stability in lymphedema, and restoration of these pathways appears to be required for lymphatic normalization.

Although it is well recognized that hypoxia, through the functions of hypoxia-inducible factor (HIF)-1α and HIF-2α, plays a central role in governing angiogenesis,[100,101] the role of hypoxia in lymphangiogenesis and lymphatic remodeling is less well appreciated. Consistent with the concept that edema and inflammation reduce tissue oxygenation,[101–103] lymphedematous tissues are hypoxic and display increased HIF-1α expression.[51,104,105] Increased tissue or LEC expression of HIF-1α appears to promote lymphangiogenesis at the acute wound healing phase following lymphatic injury.[51,105] By contrast, LEC HIF-2α is required for proper embryonic lymphatic development, maintenance of adult lymphatic vessels, and lymphatic regeneration in lymphedema.[51] Collectively, hypoxia orchestrates both lymphangiogenesis and angiogenesis in health and disease. HIF-regulated pathways in blood vascular ECs may also play essential roles in modulating lymphatic function (Fig. 11.3B).

Pathological conditions with lymphatic defects

Obesity

Modern sedentary lifestyle has significantly increased the prevalence of obesity, and the projected obesity prevalence may reach nearly 50% of the population in the United States by 2030.[106] The chronic obese status predisposes individuals to develop metabolic syndrome and increases cardiovascular diseases, and therefore poses a significant health impact for the general population.[107,108] Available data demonstrate bidirectional cross-talk between the lymphatic vasculature and adipose tissue in homeostasis and in disease.[9,109] Zippering in the LEC junction of the lacteal, through a mechanism involving LEC VEGFA/VEGFR2 overactivation, decreases dietary lipid absorption and reduces body weight gain in mice fed with a high-fat diet.[110] Because the VEGFC/VEGFR3 pathway is critical for lacteal maintenance, mice with VEGFC deficiency are protected from diet-induced obesity and insulin resistance.[35] These studies indicate that the lymphatic vasculature can regulate body fat accumulation at the level of dietary lipid absorption. Other lines of evidence support the notion that defective lymphatics may promote adipose deposition and obesity. For instance, mice with Prox-1 heterozygosity (Prox-1$^{+/-}$), which have dilated and dysfunctional intestinal submucosal and mesenteric lymphatic vessels, develop spontaneous obesity that can be rescued upon restoration of lymphatic function.[111,112] A putative underlying mechanism is that leakage of lipid-rich lymph (free fatty acids) promotes adipocyte differentiation and proliferation.[112] Studies have also indicated that obesity promotes lymphatic structural and functional abnormalities, including decreased lymphatic capillary density, elevated lymphatic leakiness, and impaired pumping capacity of collecting lymphatics; however, these phenotypic changes are reversible with dietary modification and weight control.[113,114] Clinically, obesity is a risk factor for developing lymphedema in breast cancer patients following cancer-related therapy,[68] indirectly supporting the concept that obesity influences lymphatic function. Obesity may impair the lymphatic phenotype by creating a proinflammatory microenvironment (e.g., increased expression of leptin, TNF-α, and IL-6) that suppresses lymphatic regeneration and promotes pathological remodeling.[9,115] These studies indicate that modulating lymphatic function may represent a promising avenue to reduce the prevalence of obesity and its sequelae.

Atherosclerosis

Atherosclerosis is a chronic disorder characterized by inflammation and accumulation of lipids in the arterial wall.[116] It remains the leading cause of mortality in the United States and worldwide.[117,118] Lymphatic capillaries have been localized in the adventitial and periadventitial regions of the arterial wall,[118] and there is a growing interest in the role of the lymphatics in atherosclerosis research, presumably because the lymphatics play an essential role in RCT and immune modulation.[9,119] RCT refers to the mobilization of HDL cholesterol (HDL-C) from peripheral tissues through plasma, liver, and ultimately fecal excretion. In atherosclerosis, RCT begins with the efflux of free cholesterol from lipid-laden macrophages or SMCs (collectively known as foam cells) to cholesterol acceptors such as nascent or mature HDL to form HDL-C, which is subsequently removed through the adventitial/periadventitial lymphatic vessels.[3,120] Indeed, atherosclerosis tissues appear to have a decreased density of lymphatic vessels.[121] Restoration of lymphatic function in hypercholesterolemic mice improves cholesterol clearance,[122] and rescue of lymphatic function suppresses atherosclerosis progression.[123] Although lymphatic vasculature is essential for the uptake and transport of interstitial immune cells, lymphatic vessels around the atherosclerotic lesion do not appear to promote macrophage egress.[119] However, another report indicates that reducing the density of adventitial lymphatics leads to T cell accumulation in atherosclerotic lesions.[124] Collectively, functional adventitial lymphatic vasculature may promote cholesterol and inflammatory cell removal from plaques and ameliorate atherosclerosis.

Solid organ transplant rejection

Transplant immunobiology involves a concerted cascade of alloantigen recognition, T and B cell activation, and effector-mediated functions; alloimmune-induced rejection remains the main obstacle for the long-term survival of transplanted solid organs, and growing evidence suggests that the lymphatic vasculature may critically influence outcomes.[125] There is a perturbation of the donor lymphatic drainage following transplantation because current transplant surgery does not reconnect the donor and recipient lymphatics. On the other hand, when rejection is properly controlled, lymphatic flow reestablishes at approximately the second to the fourth postoperative week in solid organ transplants.[126,127] Lack of functional lymphatic drainage and the consequent tissue edema may promote primary graft dysfunction, the leading cause of short-term mortality of solid organ transplants.[128–131] In the lung, edema fluid contains high albumin and fibrinogen levels, which may inhibit the function of surfactants and promote alveolar collapse,[132] indicating that physiological lymphatic drainage is essential for fluid balance and normal functioning of the transplant. Indeed, augmenting lymphangiogenesis alleviates acute rejection and improves lung transplant function by enhancing the removal of low molecular weight hyaluronan.[133] This molecule promotes alloimmune responses as a danger signal and exacerbates tissue edema as a water-retaining hydrophilic molecule.[6] Consonant with this notion, kidney and heart transplants with a higher density of lymphatic vessels are less susceptible to severe forms of rejection.[134,135] It is worth mentioning that the effect of lymphangiogenesis in the cornea transplant is distinct from that of other solid organ transplants. That is, increased density of lymphatic vessels promotes rejection,[136] and suppressing lymphatic growth improves graft survival.[137] A plausible explanation for this distinct role of the lymphatics in cornea transplant is that normal cornea is devoid of blood and lymphatic vessels, and the growth of new lymphatics into cornea breaks the "immune privilege" of this organ.[138] Therefore, a generalizable concept is that normal lymphatic function is important for the health of transplanted organs that require lymphatic drainage under homeostatic conditions.

Transplant tolerance refers to the immune state in which the recipient's immune system ignores the transplanted allograft.[139] Clinically achievable tolerance following transplantation is known as operational tolerance, in which patients use minimal to no immunosuppression and have no sign of rejection.[140] Whereas mixed chimerism-based strategies induce the most robust allograft tolerance, their clinical application is limited by the logistic and medical challenges posed by required myeloablative induction and bone marrow transplantation. Thus, a more practical strategy improving transplant survival is to induce tolerance by harnessing the modulatory functions of immune components, including transplantation-relevant T and B cells.[139] While the lymphatic system is required for the development of alloimmunity, data from immune tolerance research suggest that the lymphatics may also contribute to transplant tolerance by inducing modulatory immune cell types. Treg cells play vital roles in transplant tolerance; they reside primarily in the draining lymph node,[141] consistent with a generalizable observation that Treg cells trafficked to the lymph node are potent immunosuppressors.[142–145] B cells are generally considered as precursors of plasma cells that generate alloantibodies and induce antibody-mediated rejection.[146] However, subsets of B cells that express high levels of IL-10, known as Breg cells, have an immunosuppressive function and promote transplant tolerance.[143] Interestingly, Breg cells also preferentially accumulate in lymph nodes.[147,148] These studies collectively indicate that the lymph node provides an important microenvironment for the induction of transplant tolerance. Prior investigations have also shown LECs of lymph nodes express peripheral tissue antigens (PTAs), and direct presentation of PTAs to $CD8^+$ T cells results in anergy or deletion due to enhanced PD-L1 engagement and a lack of costimulation; LECs may also induce $CD4^+$ T cell tolerance by presenting PTAs through MHC-II complex provided by DCs.[16] Whether lymph node LECs promotes the deletion of allospecific $CD4^+$ and $CD8^+$ T cells in transplantation awaits further investigation. To summarize, the lymphatic vascular system appears to act as a double-edged sword in transplant health, *i.e.*, the lymphatics are required for the alloimmune induction and contribute to tolerance induction. A strategy of careful modulation of the lymphatic vasculature may improve transplant health.

Concluding remarks

In summary, recent advances in lymphatic research have offered important insights into the fundamental mechanisms involved in lymphatic development and lymphangiogenesis. Investigations in lymphedema pathobiology have demonstrated that lymph stasis–induced inflammation plays a pivotal role in disease progression, and mitigating inflammation appears to be a promising therapy for this chronic, debilitating disorder. Restoring critical pathways that promote lymphatic function, such as the NOTCH and TIE2 signaling, may also offer a therapeutic advantage for lymphedema. Furthermore, novel insights into lymphatic pathophysiology in various chronic disorders provide new disease mechanisms and expand therapeutic opportunities. Lastly, the lymphatic vasculature appears to adopt hypoxic signaling to modulate its growth and remodeling. Future investigation into HIF-mediated pathways will benefit lymphatic research in health and disease.

References

1. Oliver G, Kipnis J, Randolph GJ, Harvey NL. The lymphatic vasculature in the 21(st) century: novel functional roles in homeostasis and disease. *Cell*. 2020;182(2):270–296.
2. Cifarelli V, Eichmann A. The intestinal lymphatic system: functions and metabolic implications. *Cell Mol Gastroenterol Hepatol*. 2019;7(3):503–513.
3. Martel C, Li W, Fulp B, et al. Lymphatic vasculature mediates macrophage reverse cholesterol transport in mice. *J Clin Invest*. 2013;123(4):1571–1579.
4. Brouillard P, Boon L, Vikkula M. Genetics of lymphatic anomalies. *J Clin Invest*. 2014;124(3):898–904.
5. Jiang X, Nicolls MR, Tian W, Rockson SG. Lymphatic dysfunction, leukotrienes, and lymphedema. *Annu Rev Physiol*. 2018;80: 49–70.
6. Cui Y, Liu K, Lamattina AM, Visner G, El-Chemaly S. Lymphatic vessels: the next frontier in lung transplant. *Ann Am Thorac Soc*. 2017;14(Supplement_3):S226–S232.
7. Edwards LA, Nowocin AK, Jafari NV, et al. Chronic rejection of cardiac allografts is associated with increased lymphatic flow and cellular trafficking. *Circulation*. 2018;137(5):488–503.
8. Brakenhielm E, Alitalo K. Cardiac lymphatics in health and disease. *Nat Rev Cardiol*. 2019;16(1):56–68.
9. Jiang X, Tian W, Nicolls MR, Rockson SG. The lymphatic system in obesity, insulin resistance, and cardiovascular diseases. *Front Physiol*. 2019;10:1402.
10. Jiang X. Lymphatic vasculature in tumor metastasis and immunobiology. *J Zhejiang Univ Sci B*. 2020;21(1):3–11.
11. Schwager S, Detmar M. Inflammation and lymphatic function. *Front Immunol*. 2019;10:308.
12. Aspelund A, Robciuc MR, Karaman S, Makinen T, Alitalo K. Lymphatic system in cardiovascular medicine. *Circ Res*. 2016; 118(3):515–530.
13. Baluk P, Fuxe J, Hashizume H, et al. Functionally specialized junctions between endothelial cells of lymphatic vessels. *J Exp Med*. 2007;204(10):2349–2362.
14. Scallan JP, Zawieja SD, Castorena-Gonzalez JA, Davis MJ. Lymphatic pumping: mechanics, mechanisms and malfunction. *J Physiol*. 2016;594(20):5749–5768.
15. Jalkanen S, Salmi M. Lymphatic endothelial cells of the lymph node. *Nat Rev Immunol*. 2020;20(9):566–578.
16. Randolph GJ, Ivanov S, Zinselmeyer BH, Scallan JP. The lymphatic system: integral roles in immunity. *Annu Rev Immunol*. 2016.
17. Xiang M, Grosso RA, Takeda A, et al. A single-cell transcriptional roadmap of the mouse and human lymph node lymphatic vasculature. *Front Cardiovasc Med*. 2020;7:52.
18. Tammela T, Alitalo K. Lymphangiogenesis: molecular mechanisms and future promise. *Cell*. 2010;140(4):460–476.
19. Srinivasan RS, Dillard ME, Lagutin OV, et al. Lineage tracing demonstrates the venous origin of the mammalian lymphatic vasculature. *Genes Dev*. 2007;21(19):2422–2432.
20. Koltowska K, Betterman KL, Harvey NL, Hogan BM. Getting out and about: the emergence and morphogenesis of the vertebrate lymphatic vasculature. *Development*. 2013;140(9):1857–1870.
21. Stone OA, Stainier DYR. Paraxial mesoderm is the major source of lymphatic endothelium. *Dev Cell*. 2019;50(2):247–255 e243.
22. Podgrabinska S, Braun P, Velasco P, Kloos B, Pepper MS, Skobe M. Molecular characterization of lymphatic endothelial cells. *Proc Natl Acad Sci USA*. 2002;99(25):16069–16074.
23. Petrova TV, Makinen T, Makela TP, et al. Lymphatic endothelial reprogramming of vascular endothelial cells by the Prox-1 homeobox transcription factor. *EMBO J*. 2002;21(17):4593–4599.
24. Johnson NC, Dillard ME, Baluk P, et al. Lymphatic endothelial cell identity is reversible and its maintenance requires Prox1 activity. *Genes Dev*. 2008;22(23):3282–3291.
25. Yang Y, Oliver G. Development of the mammalian lymphatic vasculature. *J Clin Invest*. 2014;124(3):888–897.
26. Ulvmar MH, Makinen T. Heterogeneity in the lymphatic vascular system and its origin. *Cardiovasc Res*. 2016;111(4):310–321.
27. Stanczuk L, Martinez-Corral I, Ulvmar MH, et al. cKit lineage hemogenic endothelium-derived cells contribute to mesenteric lymphatic vessels. *Cell Rep*. 2015;10(10):1708–1721.

28. Klotz L, Norman S, Vieira JM, et al. Cardiac lymphatics are heterogeneous in origin and respond to injury. *Nature*. 2015;522(7554):62–67.
29. Vaahtomeri K, Karaman S, Makinen T, Alitalo K. Lymphangiogenesis guidance by paracrine and pericellular factors. *Genes Dev*. 2017;31(16):1615–1634.
30. Zheng W, Aspelund A, Alitalo K. Lymphangiogenic factors, mechanisms, and applications. *J Clin Invest*. 2014;124(3):878–887.
31. Karkkainen MJ, Ferrell RE, Lawrence EC, et al. Missense mutations interfere with VEGFR-3 signalling in primary lymphoedema. *Nat Genet*. 2000;25(2):153–159.
32. Gordon K, Schulte D, Brice G, et al. Mutation in vascular endothelial growth factor-C, a ligand for vascular endothelial growth factor receptor-3, is associated with autosomal dominant milroy-like primary lymphedema. *Circ Res*. 2013;112(6):956–960.
33. Karkkainen MJ, Haiko P, Sainio K, et al. Vascular endothelial growth factor C is required for sprouting of the first lymphatic vessels from embryonic veins. *Nat Immunol*. 2004;5(1):74–80.
34. Hagerling R, Pollmann C, Andreas M, et al. A novel multistep mechanism for initial lymphangiogenesis in mouse embryos based on ultramicroscopy. *EMBO J*. 2013;32(5):629–644.
35. Nurmi H, Saharinen P, Zarkada G, Zheng W, Robciuc MR, Alitalo K. VEGF-C is required for intestinal lymphatic vessel maintenance and lipid absorption. *EMBO Mol Med*. 2015;7(11):1418–1425.
36. Aspelund A, Tammela T, Antila S, et al. The Schlemm's canal is a VEGF-C/VEGFR-3-responsive lymphatic-like vessel. *J Clin Invest*. 2014;124(9):3975–3986.
37. Aspelund A, Antila S, Proulx ST, et al. A dural lymphatic vascular system that drains brain interstitial fluid and macromolecules. *J Exp Med*. 2015;212(7):991–999.
38. Blanco R, Gerhardt H. VEGF and Notch in tip and stalk cell selection. *Cold Spring Harb Perspect Med*. 2013;3(1):a006569.
39. Zheng W, Tammela T, Yamamoto M, et al. Notch restricts lymphatic vessel sprouting induced by vascular endothelial growth factor. *Blood*. 2011;118(4):1154–1162.
40. Niessen K, Zhang G, Ridgway JB, et al. The Notch1-Dll4 signaling pathway regulates mouse postnatal lymphatic development. *Blood*. 2011;118(7):1989–1997.
41. Wang Y, Nakayama M, Pitulescu ME, et al. Ephrin-B2 controls VEGF-induced angiogenesis and lymphangiogenesis. *Nature*. 2010;465(7297):483–486.
42. Siegfried G, Basak A, Cromlish JA, et al. The secretory proprotein convertases furin, PC5, and PC7 activate VEGF-C to induce tumorigenesis. *J Clin Invest*. 2003;111(11):1723–1732.
43. Jeltsch M, Jha SK, Tvorogov D, et al. CCBE1 enhances lymphangiogenesis via A disintegrin and metalloprotease with thrombospondin motifs-3-mediated vascular endothelial growth factor-C activation. *Circulation*. 2014;129(19):1962–1971.
44. Eklund L, Kangas J, Saharinen P. Angiopoietin-Tie signalling in the cardiovascular and lymphatic systems. *Clin Sci*. 2017;131(1):87–103.
45. Gale NW, Thurston G, Hackett SF, et al. Angiopoietin-2 is required for postnatal angiogenesis and lymphatic patterning, and only the latter role is rescued by Angiopoietin-1. *Dev Cell*. 2002;3(3):411–423.
46. Dellinger M, Hunter R, Bernas M, et al. Defective remodeling and maturation of the lymphatic vasculature in Angiopoietin-2 deficient mice. *Dev Biol*. 2008;319(2):309–320.
47. Leppanen VM, Saharinen P, Alitalo K. Structural basis of Tie2 activation and Tie2/Tie1 heterodimerization. *Proc Natl Acad Sci USA*. 2017;114(17):4376–4381.
48. Qu X, Zhou B, Scott Baldwin H. Tie1 is required for lymphatic valve and collecting vessel development. *Dev Biol*. 2015;399(1):117–128.
49. Kim M, Allen B, Korhonen EA, et al. Opposing actions of angiopoietin-2 on Tie2 signaling and FOXO1 activation. *J Clin Invest*. 2016;126(9):3511–3525.
50. Korhonen EA, Lampinen A, Giri H, et al. Tie1 controls angiopoietin function in vascular remodeling and inflammation. *J Clin Invest*. 2016;126(9):3495–3510.
51. Jiang X, Tian W, Granucci EJ, et al. Decreased lymphatic HIF-2alpha accentuates lymphatic remodeling in lymphedema. *J Clin Invest*. 2020;130(10):5562–5575.
52. Murtomaki A, Uh MK, Kitajewski C, et al. Notch signaling functions in lymphatic valve formation. *Development*. 2014;141(12):2446–2451.
53. Zhang J, Fukuhara S, Sako K, et al. Angiopoietin-1/Tie2 signal augments basal Notch signal controlling vascular quiescence by inducing delta-like 4 expression through AKT-mediated activation of beta-catenin. *J Biol Chem*. 2011;286(10):8055–8066.
54. Xue Y, Cao R, Nilsson D, et al. FOXC2 controls Ang-2 expression and modulates angiogenesis, vascular patterning, remodeling, and functions in adipose tissue. *Proc Natl Acad Sci USA*. 2008;105(29):10167–10172.
55. Petrova TV, Karpanen T, Norrmen C, et al. Defective valves and abnormal mural cell recruitment underlie lymphatic vascular failure in lymphedema distichiasis. *Nat Med*. 2004;10(9):974–981.
56. Levet S, Ciais D, Merdzhanova G, et al. Bone morphogenetic protein 9 (BMP9) controls lymphatic vessel maturation and valve formation. *Blood*. 2013;122(4):598–607.
57. Sabine A, Bovay E, Demir CS, et al. FOXC2 and fluid shear stress stabilize postnatal lymphatic vasculature. *J Clin Invest*. 2015;125(10):3861–3877.
58. Kazenwadel J, Betterman KL, Chong CE, et al. GATA2 is required for lymphatic vessel valve development and maintenance. *J Clin Invest*. 2015;125(8):2979–2994.
59. Sweet DT, Jimenez JM, Chang J, et al. Lymph flow regulates collecting lymphatic vessel maturation in vivo. *J Clin Invest*. 2015;125(8):2995–3007.
60. Leppanen VM, Brouillard P, Korhonen EA, et al. Characterization of ANGPT2 mutations associated with primary lymphedema. *Sci Transl Med*. 2020;12(560).
61. Michelini S, Paolacci S, Manara E, et al. Genetic tests in lymphatic vascular malformations and lymphedema. *J Med Genet*. 2018;55(4):222–232.
62. Rockson SG, Rivera KK. Estimating the population burden of lymphedema. *Ann N Y Acad Sci*. 2008;1131:147–154.
63. Rockson SG. Lymphedema after breast cancer treatment. *N Engl J Med*. 2018;379(20):1937–1944.
64. Rockson SG, Keeley V, Kilbreath S, Szuba A, Towers A. Cancer-associated secondary lymphoedema. *Nat Rev Dis Primers*. 2019;5(1):22.
65. DiSipio T, Rye S, Newman B, Hayes S. Incidence of unilateral arm lymphoedema after breast cancer: a systematic review and meta-analysis. *Lancet Oncol*. 2013;14(6):500–515.
66. Helyer LK, Varnic M, Le LW, Leong W, McCready D. Obesity is a risk factor for developing postoperative lymphedema in breast cancer patients. *Breast J*. 2010;16(1):48–54.
67. Johnson AR, Kimball S, Epstein S, et al. Lymphedema incidence after axillary lymph node dissection: quantifying the impact of radiation and the lymphatic microsurgical preventive healing approach. *Ann Plast Surg*. 2019;82(4S Suppl 3):S234–S241.
68. Mehrara BJ, Greene AK. Lymphedema and obesity: is there a link? *Plast Reconstr Surg*. 2014;134(1):154e–160e.
69. Finegold DN, Schacht V, Kimak MA, et al. HGF and MET mutations in primary and secondary lymphedema. *Lymphatic Res Biol*. 2008;6(2):65–68.
70. Finegold DN, Baty CJ, Knickelbein KZ, et al. Connexin 47 mutations increase risk for secondary lymphedema following breast cancer treatment. *Clin Canc Res*. 2012;18(8):2382–2390.

71. Leung G, Baggott C, West C, et al. Cytokine candidate genes predict the development of secondary lymphedema following breast cancer surgery. *Lymphatic Res Biol*. 2014;12(1):10–22.
72. Newman B, Lose F, Kedda MA, et al. Possible genetic predisposition to lymphedema after breast cancer. *Lymphatic Res Biol*. 2012; 10(1):2–13.
73. Li CY, Kataru RP, Mehrara BJ. Histopathologic features of lymphedema: a molecular review. *Int J Mol Sci*. 2020;21(7).
74. Azhar SH, Lim HY, Tan BK, Angeli V. The unresolved pathophysiology of lymphedema. *Front Physiol*. 2020;11:137.
75. Zampell JC, Yan A, Avraham T, et al. Temporal and spatial patterns of endogenous danger signal expression after wound healing and in response to lymphedema. *Am J Physiol Cell Physiol*. 2011;300(5):C1107–C1121.
76. Liu NF, Zhang LR. Changes of tissue fluid hyaluronan (hyaluronic acid) in peripheral lymphedema. *Lymphology*. 1998;31(4): 173–179.
77. Chen GY, Nunez G. Sterile inflammation: sensing and reacting to damage. *Nat Rev Immunol*. 2010;10(12):826–837.
78. Rutkowski JM, Moya M, Johannes J, Goldman J, Swartz MA. Secondary lymphedema in the mouse tail: lymphatic hyperplasia, VEGF-C upregulation, and the protective role of MMP-9. *Microvasc Res*. 2006;72(3):161–171.
79. Randolph GJ, Angeli V, Swartz MA. Dendritic-cell trafficking to lymph nodes through lymphatic vessels. *Nat Rev Immunol*. 2005; 5(8):617–628.
80. Mortimer PS, Rockson SG. New developments in clinical aspects of lymphatic disease. *J Clin Invest*. 2014;124(3):915–921.
81. Wynn TA, Vannella KM. Macrophages in tissue repair, regeneration, and fibrosis. *Immunity*. 2016;44(3):450–462.
82. Tian W, Rockson SG, Jiang X, et al. Leukotriene B4 antagonism ameliorates experimental lymphedema. *Sci Transl Med*. 2017; 9(389).
83. Mosser DM, Edwards JP. Exploring the full spectrum of macrophage activation. *Nat Rev Immunol*. 2008;8(12):958–969.
84. Norrmen C, Tammela T, Petrova TV, Alitalo K. Biological basis of therapeutic lymphangiogenesis. *Circulation*. 2011;123(12): 1335–1351.
85. Ly CL, Kataru RP, Mehrara BJ. Inflammatory manifestations of lymphedema. *Int J Mol Sci*. 2017;18(1).
86. Gousopoulos E, Proulx ST, Bachmann SB, et al. An important role of VEGF-C in promoting lymphedema development. *J Invest Dermatol*. 2017;137(9):1995–2004.
87. Ogata F, Fujiu K, Matsumoto S, et al. Excess lymphangiogenesis cooperatively induced by macrophages and CD4(+) T cells drives the pathogenesis of lymphedema. *J Invest Dermatol*. 2016;136(3): 706–714.
88. Liao S, Cheng G, Conner DA, et al. Impaired lymphatic contraction associated with immunosuppression. *Proc Natl Acad Sci USA*. 2011;108(46):18784–18789.
89. Cuzzone DA, Weitman ES, Albano NJ, et al. IL-6 regulates adipose deposition and homeostasis in lymphedema. *Am J Physiol Heart Circ Physiol*. 2014;306(10):H1426–H1434.
90. Scheller J, Chalaris A, Schmidt-Arras D, Rose-John S. The pro- and anti-inflammatory properties of the cytokine interleukin-6. *Biochim Biophys Acta*. 2011;1813(5):878–888.
91. Rockson SG, Tian W, Jiang X, et al. Pilot studies demonstrate the potential benefits of antiinflammatory therapy in human lymphedema. *JCI Insight*. 2018;3(20).
92. Nakamura K, Radhakrishnan K, Wong YM, Rockson SG. Anti-inflammatory pharmacotherapy with ketoprofen ameliorates experimental lymphatic vascular insufficiency in mice. *PloS One*. 2009;4(12):e8380.
93. Pivetta E, Wassermann B, Del Bel Belluz L, et al. Local inhibition of elastase reduces EMILIN1 cleavage reactivating lymphatic vessel function in a mouse lymphoedema model. *Clin Sci*. 2016; 130(14):1221–1236.
94. Gousopoulos E, Proulx ST, Bachmann SB, et al. Regulatory T cell transfer ameliorates lymphedema and promotes lymphatic vessel function. *JCI Insight*. 2016;1(16):e89081.
95. Avraham T, Zampell JC, Yan A, et al. Th2 differentiation is necessary for soft tissue fibrosis and lymphatic dysfunction resulting from lymphedema. *FASEB J*. 2013;27(3):1114–1126.
96. Yoon YS, Murayama T, Gravereaux E, et al. VEGF-C gene therapy augments postnatal lymphangiogenesis and ameliorates secondary lymphedema. *J Clin Invest*. 2003;111(5):717–725.
97. Lahteenvuo M, Honkonen K, Tervala T, et al. Growth factor therapy and autologous lymph node transfer in lymphedema. *Circulation*. 2011;123(6):613–620.
98. Jensen MR, Simonsen L, Karlsmark T, Lanng C, Bulow J. Higher vascular endothelial growth factor-C concentration in plasma is associated with increased forearm capillary filtration capacity in breast cancer-related lymphedema. *Physiol Rep*. 2015;3(6).
99. Zaleska MT, Olszewski WL. Imaging lymphatics in human normal and lymphedema limbs-usefulness of various modalities for evaluation of lymph and edema fluid flow pathways and dynamics. *J Biophot*. 2018;11(8):e201700132.
100. Pugh CW, Ratcliffe PJ. Regulation of angiogenesis by hypoxia: role of the HIF system. *Nat Med*. 2003;9(6):677–684.
101. Lee P, Chandel NS, Simon MC. Cellular adaptation to hypoxia through hypoxia inducible factors and beyond. *Nat Rev Mol Cell Biol*. 2020;21(5):268–283.
102. Poeze M. Tissue-oxygenation assessment using near-infrared spectroscopy during severe sepsis: confounding effects of tissue edema on StO2 values. *Intensive Care Med*. 2006;32(5):788–789.
103. Palazon A, Goldrath AW, Nizet V, Johnson RS. HIF transcription factors, inflammation, and immunity. *Immunity*. 2014;41(4): 518–528.
104. Tabibiazar R, Cheung L, Han J, et al. Inflammatory manifestations of experimental lymphatic insufficiency. *PLoS Med*. 2006;3(7):e254.
105. Zampell JC, Yan A, Avraham T, Daluvoy S, Weitman ES, Mehrara BJ. HIF-1alpha coordinates lymphangiogenesis during wound healing and in response to inflammation. *FASEB J*. 2012; 26(3):1027–1039.
106. Ward ZJ, Bleich SN, Cradock AL, et al. Projected U.S. State-level prevalence of adult obesity and severe obesity. *N Engl J Med*. 2019;381(25):2440–2450.
107. Catrysse L, van Loo G. Inflammation and the metabolic syndrome: the tissue-specific functions of NF-kappaB. *Trends Cell Biol*. 2017;27(6):417–429.
108. O'Neill S, O'Driscoll L. Metabolic syndrome: a closer look at the growing epidemic and its associated pathologies. *Obes Rev*. 2015;16(1):1–12.
109. Escobedo N, Oliver G. The lymphatic vasculature: its role in adipose metabolism and obesity. *Cell Metabol*. 2017;26(4):598–609.
110. Zhang F, Zarkada G, Han J, et al. Lacteal junction zippering protects against diet-induced obesity. *Science*. 2018;361(6402): 599–603.
111. Harvey NL, Srinivasan RS, Dillard ME, et al. Lymphatic vascular defects promoted by Prox1 haploinsufficiency cause adult-onset obesity. *Nat Genet*. 2005;37(10):1072–1081.
112. Escobedo N, Proulx ST, Karaman S, et al. Restoration of lymphatic function rescues obesity in Prox1-haploinsufficient mice. *JCI Insight*. 2016;1(2).
113. Garcia Nores GD, Cuzzone DA, Albano NJ, et al. Obesity but not high-fat diet impairs lymphatic function. *Int J Obes*. 2016;40(10): 1582–1590.
114. Nitti MD, Hespe GE, Kataru RP, et al. Obesity-induced lymphatic dysfunction is reversible with weight loss. *J Physiol*. 2016;594(23): 7073–7087.

115. Ellulu MS, Patimah I, Khaza'ai H, Rahmat A, Abed Y. Obesity and inflammation: the linking mechanism and the complications. *Arch Med Sci*. 2017;13(4):851–863.
116. Back M, Yurdagul Jr A, Tabas I, Oorni K, Kovanen PT. Inflammation and its resolution in atherosclerosis: mediators and therapeutic opportunities. *Nat Rev Cardiol*. 2019;16(7):389–406.
117. Benjamin EJ, Muntner P, Alonso A, et al. Heart disease and stroke statistics-2019 update: a report from the American Heart Association. *Circulation*. 2019;139(10):e56–e528.
118. Kutkut I, Meens MJ, McKee TA, Bochaton-Piallat ML, Kwak BR. Lymphatic vessels: an emerging actor in atherosclerotic plaque development. *Eur J Clin Invest*. 2015;45(1):100–108.
119. Csanyi G, Singla B. Arterial lymphatics in atherosclerosis: old questions, new insights, and remaining challenges. *J Clin Med*. 2019;8(4).
120. Ouimet M, Barrett TJ, Fisher EA. HDL and reverse cholesterol transport. *Circ Res*. 2019;124(10):1505–1518.
121. Taher M, Nakao S, Zandi S, Melhorn MI, Hayes KC, Hafezi-Moghadam A. Phenotypic transformation of intimal and adventitial lymphatics in atherosclerosis: a regulatory role for soluble VEGF receptor 2. *FASEB J*. 2016;30(7):2490–2499.
122. Lim HY, Thiam CH, Yeo KP, et al. Lymphatic vessels are essential for the removal of cholesterol from peripheral tissues by SR-BI-mediated transport of HDL. *Cell Metabol*. 2013;17(5):671–684.
123. Milasan A, Smaani A, Martel C. Early rescue of lymphatic function limits atherosclerosis progression in Ldlr(-/-) mice. *Atherosclerosis*. 2019;283:106–119.
124. Rademakers T, van der Vorst EPC, Daissormont ITMN, et al. Adventitial lymphatic capillary expansion impacts on plaque T cell accumulation in atherosclerosis. *Sci Rep*. 2017;7(1):45263.
125. Wong BW. Lymphatic vessels in solid organ transplantation and immunobiology. *Am J Transplant*. 2020;20(8):1992–2000.
126. Ruggiero R, Muz J, Fietsam Jr R, et al. Reestablishment of lymphatic drainage after canine lung transplantation. *J Thorac Cardiovasc Surg*. 1993;106(1):167–171.
127. Brown K, Badar A, Sunassee K, et al. SPECT/CT lymphoscintigraphy of heterotopic cardiac grafts reveals novel sites of lymphatic drainage and T cell priming. *Am J Transplant*. 2011; 11(2):225–234.
128. Lee JC, Christie JD. Primary graft dysfunction. *Proc Am Thorac Soc*. 2009;6(1):39–46.
129. Belmaati E, Jensen C, Kofoed KF, Iversen M, Steffensen I, Nielsen MB. Primary graft dysfunction; possible evaluation by high resolution computed tomography, and suggestions for a scoring system. *Interact Cardiovasc Thorac Surg*. 2009;9(5): 859–867.
130. Khan SU, Salloum J, O'Donovan PB, et al. Acute pulmonary edema after lung transplantation: the pulmonary reimplantation response. *Chest*. 1999;116(1):187–194.
131. Singh SSA, Dalzell JR, Berry C, Al-Attar N. Primary graft dysfunction after heart transplantation: a thorn amongst the roses. *Heart Fail Rev*. 2019;24(5):805–820.
132. Nitta K, Kobayashi T. Impairment of surfactant activity and ventilation by proteins in lung edema fluid. *Respir Physiol*. 1994;95(1): 43–51.
133. Cui Y, Liu K, Monzon-Medina ME, et al. Therapeutic lymphangiogenesis ameliorates established acute lung allograft rejection. *J Clin Invest*. 2015;125(11):4255–4268.
134. Geissler HJ, Dashkevich A, Fischer UM, et al. First year changes of myocardial lymphatic endothelial markers in heart transplant recipients. *Eur J Cardio Thorac Surg*. 2006;29(5):767–771.
135. Stuht S, Gwinner W, Franz I, et al. Lymphatic neoangiogenesis in human renal allografts: results from sequential protocol biopsies. *Am J Transplant*. 2007;7(2):377–384.
136. Dietrich T, Bock F, Yuen D, et al. Cutting edge: lymphatic vessels, not blood vessels, primarily mediate immune rejections after transplantation. *J Immunol*. 2010;184(2):535–539.
137. Cursiefen C, Cao J, Chen L, et al. Inhibition of hemangiogenesis and lymphangiogenesis after normal-risk corneal transplantation by neutralizing VEGF promotes graft survival. *Invest Ophthalmol Vis Sci*. 2004;45(8):2666–2673.
138. Chauhan SK, Dohlman TH, Dana R. Corneal lymphatics: role in ocular inflammation as inducer and responder of adaptive immunity. *J Clin Cell Immunol*. 2014;5.
139. Rickert CG, Markmann JF. Current state of organ transplant tolerance. *Curr Opin Organ Transplant*. 2019;24(4):441–450.
140. Bishop GA, Ierino FL, Sharland AF, et al. Approaching the promise of operational tolerance in clinical transplantation. *Transplantation*. 2011;91(10):1065–1074.
141. Wei S, Kryczek I, Zou W. Regulatory T-cell compartmentalization and trafficking. *Blood*. 2006;108(2):426–431.
142. Huang MT, Lin BR, Liu WL, Lu CW, Chiang BL. Lymph node trafficking of regulatory T cells is prerequisite for immune suppression. *J Leukoc Biol*. 2016;99(4):561–568.
143. Geng S, Zhong Y, Zhou X, et al. Induced regulatory T cells superimpose their suppressive capacity with effector T cells in lymph nodes via antigen-specific S1p1-dependent egress blockage. *Front Immunol*. 2017;8:663.
144. Nakanishi Y, Ikebuchi R, Chtanova T, et al. Regulatory T cells with superior immunosuppressive capacity emigrate from the inflamed colon to draining lymph nodes. *Mucosal Immunol*. 2018;11(2):437–448.
145. Núñez NG, Tosello Boari J, Ramos RN, et al. Tumor invasion in draining lymph nodes is associated with Treg accumulation in breast cancer patients. *Nat Commun*. 2020;11(1):3272.
146. Peng B, Ming Y, Yang C. Regulatory B cells: the cutting edge of immune tolerance in kidney transplantation. *Cell Death Dis*. 2018; 9(2):109.
147. Ganti SN, Albershardt TC, Iritani BM, Ruddell A. Regulatory B cells preferentially accumulate in tumor-draining lymph nodes and promote tumor growth. *Sci Rep*. 2015;5:12255.
148. Natarajan P, Singh A, McNamara JT, et al. Regulatory B cells from hilar lymph nodes of tolerant mice in a murine model of allergic airway disease are CD5+, express TGF-β, and co-localize with CD4+Foxp3+ T cells. *Mucosal Immunol*. 2012;5(6):691–701.

PART II

Investigating the vasculome: context-driven methods, uses, and limitations

SECTION 1

Experimental and computational studies of the vasculome

CHAPTER

12

The flow-dependent endotheliome: hemodynamic forces, genetic programs, and functional phenotypes

Guillermo García-Cardeña and Michael A. Gimbrone, Jr.

Department of Pathology, Center for Excellence in Vascular Biology, Brigham and Women's Hospital and Harvard Medical School, Boston, MA, United States

Introduction

The vascular endothelium, comprising the interface between blood and the rest of the vessel wall, plays a fundamental role in the pathophysiology of the cardiovascular system.[1] Importantly, the unique functional plasticity of this single cell thick layer relies on the ability of endothelial cells to integrate and transduce both humoral and hemodynamic stimuli from their microenvironment.[2,3] For example, vascular endothelial cells can rapidly sense changes in blood flow and respond acutely, by secreting and metabolizing potent vasoactive substances, or chronically, by regulating transcriptional programs that lead to the modulation of their functional phenotype in both normal and pathological states.[4,5] Through the application of various in vitro model systems, multiple studies have documented that fluid shear stresses, comparable to those generated by the frictional force of blood flow on the endothelial lining in vivo, can directly influence cellular morphology, enzymatic activities, transcriptional and epigenetic events in cultured endothelial monolayers, resulting in adaptive and nonadaptive functional phenotypes.[4] Here we review efforts directed to defining the distinct hemodynamic environments present in different regions of the human vascular tree, and then replicating them in vitro, using various types of fluid mechanical devices—with the ultimate goal of gaining mechanistic insights into the "Flow-dependent Endotheliome."

Flow patterns in the human circulation

As the heart pumps blood throughout the branching vascular tree, the unique anatomical features of blood vessels and their relative location with respect to the heart result in distinct blood flow patterns at different sites in the circulatory system. As a result, a complex set of fluid mechanical forces (e.g., shear stress, pressure, and cyclic strain) is continuously imposed on the endothelial lining of these vessels.[3] To begin to understand the significance of these biomechanical stimuli on endothelium in health and disease, initially it was important to characterize the distinct flow patterns present in different regions of the circulatory system in vivo, and then to develop in vitro model systems to reproducibly expose cultured endothelial cells to these various hemodynamic stimuli, thus creating an experimental window on the Flow-dependent Endotheliome.

For large and medium size vessels, circulating blood cells are significantly smaller than the vessel diameter so blood flow can be modeled as a Newtonian fluid. Thus, meaningful measurements of blood flow velocity can be made, and the shear stresses that the endothelial cells sense are linearly related to the change in velocity profile from the center outward to the wall of the vessel. The most common approach for measuring flow velocity is Doppler ultrasound. Using this method, the changing frequencies of sound waves reflected from inside the vessel are used to determine the velocity flow profile of the blood at a given point.[6] Magnetic resonance

https://doi.org/10.1016/B978-0-12-822546-2.00034-4

imaging can also use to measure blood flow velocities.[7,8] Moreover, MRI and CT images of vessel structure can be combined with ultrasound measurements of flow velocities to more accurately compute the blood flow patterns as they change along the length of a given vessel. Using these quantitative imaging approaches, several distinct patterns of blood flow have been documented in different regions of the vascular tree.[9] Along the straight portions of medium and large size arteries (e.g., internal carotid artery, abdominal aorta), the direction of blood flow is parallel to the vessel wall. Typically, there is a pulse of fast flow acceleration (peak shear stresses from 10 to 70 dyn/cm^2)[10] to a steady-state velocity that then is maintained during the deceleration phase for the rest of the cardiac cycle. This pulsatile nature of blood flow, characterized by the rate at which blood accelerates to its peak speed, is a hallmark of arterial flow. In large and medium caliber straight veins (e.g., saphenous vein), blood flow is still predominantly one-dimensional, but it is much slower (peak shear stresses of approximately 1–2 dynes/cm^2). Blood travels at this speed in the forward direction for approximately two thirds of the cardiac cycle and in the reverse direction, at a similar speed, for the rest of the time. Notably, the acceleration and deceleration of the blood to reach its peak velocities in veins is much slower than in arteries.

Importantly, in the vicinity of branch points and curvatures in large vessels, these flow patterns no longer hold. In arteries, these geometries exhibit "disturbed flow"—a complex pattern characterized by three-dimensional whorls that often have a low time-average velocity in any given direction.[9] Notably, the aortic arch, carotid sinuses, and branch points of the aorta are pathophysiologically important regions that are exposed to these complex disturbed flow patterns.[11]

Although the flow patterns present in several vessel types have been defined in considerable detail, those in many vascular regions remain to be well characterized. For example, the coronary arteries are of great clinical interest due to the high prevalence of atherosclerosis in the vessels, but their smaller diameter and constant motion within the beating heart has made measuring flow patterns within the coronary vasculature especially challenging. Additionally, characterizing the shear stresses that endothelial cells experience in capillaries has also been challenging since whole blood can no longer be considered a Newtonian fluid owing to the relatively large size of its constituent cells compared to luminal diameter.

Modeling human hemodynamic forces in vitro

Important insights into the responses endothelial cells display when exposed to various types of shear stresses have come from in vitro experiments, where the effects of flow can be studied in a well-defined manner, in isolation from other hemodynamic factors, such as hydrostatic pressure and cyclic strain.[3] Thus, modeling flow patterns that better approximate those seen in disease-prone human arteries and then exposing cultured endothelial monolayers to these biomechanical forces using dynamic flow systems in vitro, has been an important, progressively evolving goal for investigators in the field.

In early in vitro flow experiments, cultured vascular endothelial cells were exposed to uniform laminar shear stresses similar in magnitude to physiological time-average values for various vessel types. These steady laminar flows initially were generated using simple parallel plate flow chambers.[12,13] Subsequently, more elaborate cone-plate apparatuses were developed.[14] In these devices, the rotation of the cone can be controlled to generate one-dimensional waveforms that more closely approximate the cyclic nature of physiological flows—sinusoidal waveforms or oscillating square waves. Cone-plate apparatuses were progressively refined to allow control by computerized motors to reproduce more complex physiological flow waveforms, captured from vessels in vivo by Doppler ultrasound and MRI.[15] Another approach to mimic physiological flows in a capillary geometry was to grow endothelial cells on the lining of small tubes and then pump fluid through them. Pumping fluid in a cyclic fashion exposes the endothelial layer to a pulsatile flow pattern. However, the precise definition of other fluid mechanical factors (e.g., cyclic stretch) present in these systems remains to be defined.[16]

Mimicking the spatially more complex two- and three-dimensional disturbed flow patterns seen in disease-prone sites, such as the atherosclerosis-susceptible carotid bulb, has been an important research focus. Certain dynamic flow cone-plate devices can replicate the waveforms present in these regions in vivo. In addition, models using a square step in the lower plate of a parallel plate chamber can be used to establish disturbed (i.e., spatial varying) flow patterns in the vicinity of the step.[17,18] Each of these models has been useful for probing the impact of distinct patterns of shear stress on the structure and function of vascular endothelial cells.

Unraveling flow-dependent endothelial gene expression and functional phenotypes

Early experiments employing a cone and plate flow device revealed that the application of unidirectional steady laminar shears stresses could induce dramatic time- and force-dependent cell shape changes in cultured endothelial cells.[13,14,19] These flow-induced shape changes were accompanied by reorganization of the actin cytoskeleton leading to cell alignment changes, thus mimicking the cellular morphology of the aortic endothelium in vivo.[20] Further studies also documented changes in the metabolic and synthetic activities of endothelial cells in response to shear stresses, including the production of vasoactive substances and extracellular matrix components, as well as the regulation of cell proliferation and cell motility. Importantly, physiological levels of laminar shear stresses were found to regulate the expression of several pathophysiologically relevant genes—including growth factors, such as PDGF-A, PDGF-B, VEGF, TGF-β; fibrinolytic factors, such as tPA; chemokines, such as MCP-1; and adhesion molecules, such as ICAM-1 and VCAM-1.[21–26] Moreover, detailed analysis of the promoter regions of certain of these genes revealed "shear stress responsive elements," thus providing an important molecular link between biomechanical stimulation and transcriptional regulation of endothelial gene expression.[27]

Based on the diversity of force dependencies and kinetic profiles for the regulation of various endothelial genes, it became evident that the molecular mechanisms linking mechano-activated signaling pathways at the cell surface to gene regulation at the transcriptional level were multiple and complex. Discrete, flow-activated ion channels and mechanically deformable cell surface receptors, as well as the "tensegrity" of the cellular cytoskeleton, all appeared to play a role in the sensing and transducing of flow-generated biomechanical forces into cellular responses in a given endothelial cell.[3,28,29]

Pathophysiologically interesting patterns of flow-dependent gene expression began to emerge from these in vitro experiments. For example, seminal studies by Topper et al. identified clusters of genes which were *differentially responsive* to the spatial and temporal properties of the applied shear stresses. Using differential mRNA display approaches, certain genes were found to show a coordinated and sustained upregulation in response to steady laminar flow. These included *eNOS, COX-2*, and *manganese-dependent SOD*, endothelial genes that encode enzymes with potent antithrombotic, antiadhesive, antiinflammatory, and antioxidant effects. In contrast, in response to highly disturbed (turbulent) shear stresses, these key genes showed reduced or absent expression.[30] These findings, coupled with the observation that certain regions of the arterial vasculature typically associated with steady laminar flow remain relatively resistant to the development of atherosclerotic lesions, led to the formulation of the "Atheroprotective Gene Hypothesis."[31,32] This hypothesis postulated that undisturbed laminar shear stresses upregulate the coordinated expression of a set of atheroprotective genes in endothelial cells, which then act locally in atherosclerosis-resistant areas to offset the effects of systemic risk factors in atherosclerotic cardiovascular disease. The critical testing of this hypothesis was aided by the following experimental approaches: (1) detailed characterization of the actual near-wall shear stress profiles in atheroprone and atheroprotected arterial geometries in vivo; (2) simulation of these complex flows, with spatial and temporal fidelity, over cultured human endothelial cell monolayers in vitro; (3) genome-wide comparative analyses of the resultant transcriptomes and their correlation with known pathogenic mechanisms in atherosclerosis; (4) validation of the expression of these genetic programs in endothelial cells in atherosclerosis-resistant and atherosclerosis-susceptible vascular geometries in vivo. The efforts of several groups generated a robust collection of experimental observations linking specific patterns of hemodynamic stimulation with the transcriptional regulation of functional phenotypes of endothelium in vitro and in vivo.[15,33–41] Collectively, these studies led to the recognition that a major driver of endothelial cell dysfunction in lesion-prone arterial geometries seemed to be the *absence of undisturbed laminar shear stress*, which normally serves as a positive regulator of a vasoprotective endothelial cell phenotype.[4]

These observations motivated in-depth bioinformatic analyses of the endothelial transcriptomes elicited by atheroprotective versus atheroprone flows. Attention was focused on a particular subset of genes, namely transcription factors, with the hope of identifying the factor(s) driving the atheroprotective endothelial cell phenotype. These efforts led to the identification of the zinc finger transcription factor, Kruppel-like factor 2 (KLF2) as the gene with the strongest differential upregulation by atheroprotective versus atheroprone waveform stimulation.[42] Further studies on the biological consequences of KLF2 expression in the vascular endothelium have defined KLF2 as a key integrator of the flow-mediated endothelial vasoprotective phenotype, and have shown that, both in vivo and in vitro, KLF2 expression is intimately linked to the type of shear stress acting on the endothelial surface.[42–44] In particular, atheroprotective shear stress waveforms, which are present in atherosclerosis-resistant regions in the human internal carotid artery, stimulate significant upregulation of KLF2 expression and suppression of proinflammatory and thrombogenic activity compared to shear stress waveforms derived from the atherosclerosis-prone human carotid sinus. Moreover, in vivo studies have documented a protective role for KLF2 expression during experimental atherosclerosis development, as well as a reduction in the proinflammatory state of monocytes and T cells in several pathological settings.[45]

Biomechanical control of KLF2 expression in vascular endothelium, however, remains to be fully characterized. Our current understanding of the flow-mediated signaling events that regulate KLF2 expression in endothelial cells is primarily restricted to the MEKK3/MEK5/ERK5/MEF2 pathway—in which MEKK3 activates MEK5, MEK5 activation leads to ERK5 phosphorylation and subsequent activation of the transcription factor MEF2, which binds and activates the KLF2 promoter.[42] Other modifiers of the MEK5/Erk5 pathway that have been identified in endothelial cells include AMPK, SIRT1, PKCζ, SENP2, and HDAC5.[46] Additional mechanisms of flow-induced KLF2 expression include PI3 kinase-dependent mRNA stabilization and chromatin remodeling.[47] The activation of the MEKK3/MEK5/ERK5/MEF2 cascade is both necessary and sufficient for flow-induced KLF2 transcriptional regulation, as determined by the manipulation of MEF2 and MEK5 activities, and regulation of ERK5 SUMOylation. Flow-mediated activation of the MEK5/ERK5/MEF2 axis has also been documented to be critical for inhibiting TNF-induced JNK phosphorylation and the downstream endothelial inflammatory response.[46] Besides flow, few natural regulators of KLF2 expression have been identified. Interestingly, the statins, a class of HMGCoA reductase inhibitors, which have several pleiotropic effects including the improvement of endothelial dysfunction, markedly upregulate the expression of KLF2 and its downstream transcriptional targets in vascular endothelial cells.[48,49] Cardiologists have speculated that this "off-target" pharmacologic effect may well explain the remarkable beneficial effects of this class of drugs in the prevention of atherosclerosis even in normo-cholesterolemic patients. In contrast, the proinflammatory cytokines tumor necrosis factor alpha (TNF-α) and interleukin-1 (IL-1) act as negative biological regulators of KLF2 in these cells.[50] Overall, activation of the MEKK3/MEK5/ERK5/MEF2 signaling cascade and the resulting increase in KLF2 transcription constitutes a critical molecular descriptor of endothelial vasoprotection and the endothelial response to atheroprotective flow. It is anticipated that additional effectors, acting either independently or along the MEKK3/MEK5/ERK5/MEF2 signaling axis, also modulate endothelial KLF2 expression. Therefore, unveiling these regulators and/or signaling pathways would expand understanding of the mechanisms controlling flow-mediated expression of KLF2 in vascular endothelial cells, and thus our ability to enhance endothelial functional integrity. Interestingly, the activation of the MEK5/Erk5 pathway per se leads to the endothelial cell vasoprotection in both KLF2-dependent and KLF2-independent manner, suggesting that additional targets of these kinases may be involved in mediating the beneficial effects of flow. One such target is the transcription factor KLF4, also upregulated by atheroprotective flow in cultured endothelial cells via MEK5/Erk5/MEF2 signaling.[51,52] KLF4 expression in cultured endothelial cells promotes endothelial vasoprotection, via the expression of several downstream transcriptional targets that are also activated by KLF2. Taken together, these observations have highlighted KLF2, and potentially KLF4, as attractive therapeutic targets for the treatment and/or prevention of cardiovascular disease.

Recent studies have documented that shear stresses regulate gene expression in cultured endothelial cells via two additional mechanisms: miRNAs and epigenetic modifications.[53–55] In cultured endothelial cells, a major action of miR10a is to downregulate the NF-kB proinflammatory pathway.[56] Atheroprotective flow also has been shown to upregulate the expression of miR19-a, miR-23b, and miR101 in cultured endothelial cells, leading to the suppression of endothelial cell proliferation.[57–59] In contrast, expression of miR92a, miR663, miR712, and miR34a is downregulated by atheroprotective flow and upregulated by atheroprone flow in cultured endothelial cells.[60–62] The atheroprotective flow-dependent suppression of miR92a expression results in the upregulation of KLF2 and KLF4 and certain of their downstream transcriptional targets in vitro and in vivo. Inhibition of miRNA-92a in endothelial cells modulates endothelial activation in response to shear stress and oxidized LDL and limits the development of atherosclerosis in LDL receptor-deficient mice, at least in part by increasing the expression of KLF2 and KLF4, thus suppressing endothelial activation. Inhibition of the expression of miR663 suppresses atheroprone flow-mediated endothelial activation.[63] miR712 downregulates the expression of tissue inhibitor of metalloproteinase 3 (TIMP-3) leading to the stimulation of endothelial inflammation, permeability, and atherosclerotic lesion formation in mice.[62] Similarly, the downregulation of miR34a contributes to the atheroprotective flow-mediated suppression of endothelial inflammation by downregulating NF-kB signaling.[64] Finally, atheroprotective flow also induces the secretion of the miR143-145 cluster via a KLF2-dependent pathway. These secreted miRNAs can act on neighboring vascular smooth muscle cells to regulate their turnover and phenotype.[65]

Recently, it has become apparent that hemodynamic forces also can modulate endothelial gene expression at the epigenetic level.[53,66] For example, two independent studies have shown that atheroprone flow significantly modulates endothelial DNA methylation patterns via alterations in DNA methyltransferase (DNMT) activity, in particular DNMT1. In addition, at the single gene level, disturbed flow was shown to increase methylation of the proximal promoter of *KLF4* in endothelial cells leading to the suppression of its expression by blocking a MEF2 binding site, and by regulating *DNMT3A*.[67] Considering the recent

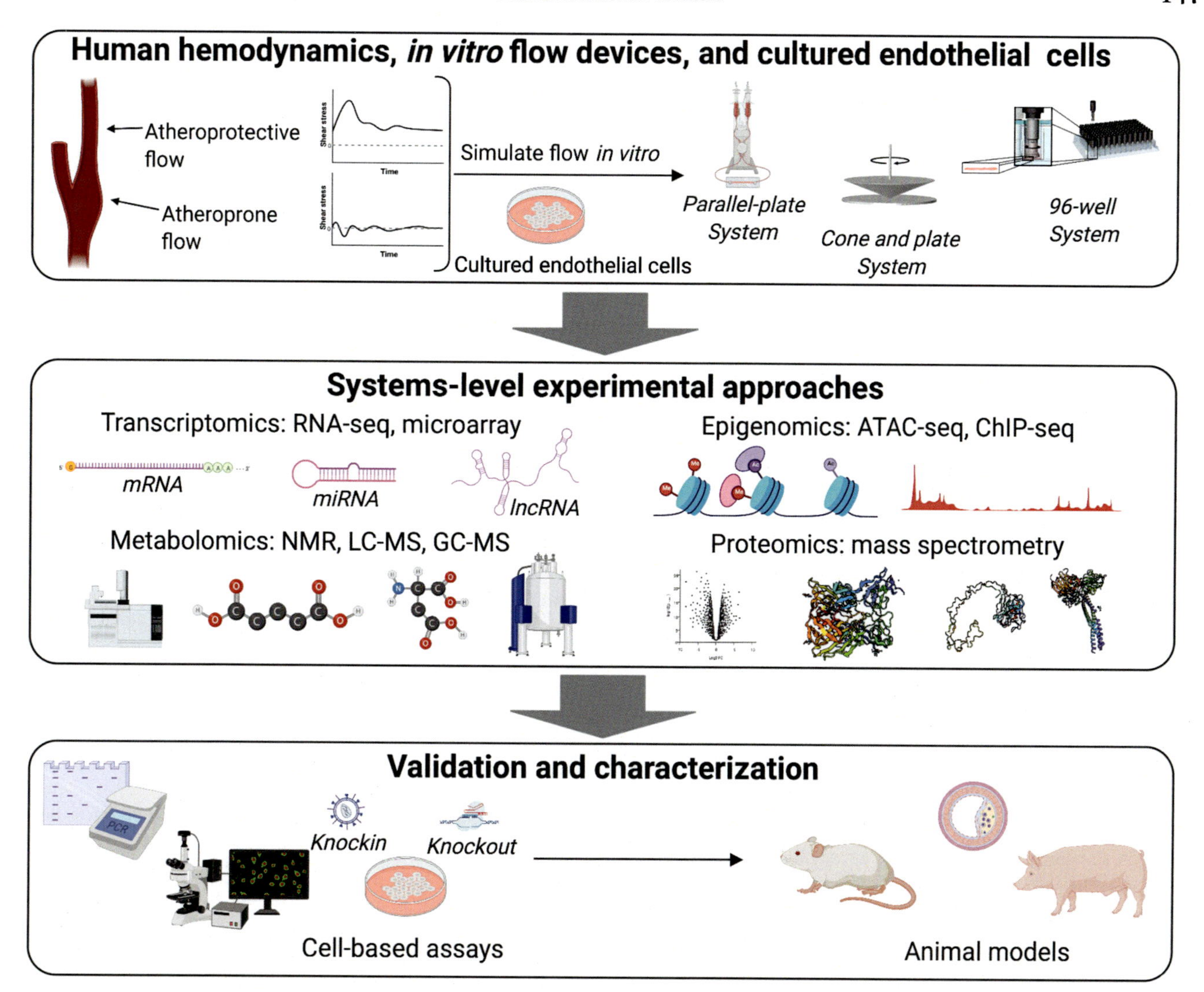

FIGURE 12.1 **Generalized experimental approach for defining and characterizing the flow-dependent endotheliome.** Cultured endothelial cells are exposed to pathophysiologically relevant patterns of shear stress using in vitro flow systems (top panel). Cells are analyzed to identify flow-dependent changes in mRNA/miRNA/lncRNA expression, epigenetic modifications, protein expression, and/or production of metabolites (middle panel). The functional relevance of differentially expressed molecules is assessed using gain-of-function and loss-of function approaches in cell-based assays and animal models of cardiovascular disease (bottom panel). Created with BioRender.com.

awareness of the global role of epigenetic changes in proinflammatory endothelial cell activation, this new area of flow-mediated gene regulation may offer additional opportunities for targeted pharmacological interventions in cardiovascular disease.

Conclusions and future directions

The ability to model pathophysiologically relevant hemodynamic environments in vitro coupled with the application of multiomics technologies continues to yield fundamental insights into the Flow-dependent Endotheliome (Fig. 12.1). The successful application of these new insights may lead to the development of pharmacomimetics of the natural, flow-mediated vasoprotective phenotype, and thus the creation of new approaches for the diagnosis and treatment of cardiovascular disease.

Acknowledgments

The authors wish to acknowledge past and present members of our laboratories in the Center for Excellence in Vascular Biology (Vascular Research Division, Department of Pathology) as well as our longtime collaborators Professors Forbes Dewey and Roger Kamm in the Fluid Mechanics Laboratory at the Massachusetts Institute of Technology. We thank Katherine Ho for assistance in the design of Fig. 12.1 and the support of research activities by grants from the National Institutes of Health (HL152367 and HL130624) to GG-C.

References

1. Gimbrone MA. Vascular endothelium: nature's blood-compatible container. *Ann N Y Acad Sci*. 1987;516(1). https://doi.org/10.1111/j.1749-6632.1987.tb33025.x.
2. Gimbrone MA, Nagel T, Topper JN. Biomechanical activation: an emerging paradigm in endothelial adhesion biology. *J Clin Invest*. 1997;99(8). https://doi.org/10.1172/JCI119346.
3. Davies PF. Flow-mediated endothelial mechanotransduction. *Physiol Rev*. 1995;75(3). https://doi.org/10.1152/physrev.1995.75.3.519.
4. Gimbrone MA, García-Cardeña G. Endothelial cell dysfunction and the pathobiology of atherosclerosis. *Circ Res*. 2016;118(4). https://doi.org/10.1161/CIRCRESAHA.115.306301.
5. Chiu JJ, Chien S. Effects of disturbed flow on vascular endothelium: pathophysiological basis and clinical perspectives. *Physiol Rev*. 2011;91(1). https://doi.org/10.1152/physrev.00047.2009.
6. Taylor KJW, Holland S, Doppler US. Part I. Basic principles, instrumentation, and pitfalls. *Radiology*. 1990;174(2). https://doi.org/10.1148/radiology.174.2.2404309.
7. Edelman RR. Basic principles of magnetic resonance angiography. *Cardiovasc Interv Radiol*. 1992;15(1). https://doi.org/10.1007/BF02733894.
8. Aguet J, Gill N, Tassos VP, Chavhan GB, Lam CZ. Contrast-enhanced body magnetic resonance angiography: how we do it. *Pediatr Radiol*. 2021. https://doi.org/10.1007/s00247-021-05020-z. Published online.
9. Younis HF, Kaazempur-Mofrad MR, Chung C, Chan RC, Kamm RD. Computational analysis of the effects of exercise on hemodynamics in the carotid bifurcation. *Ann Biomed Eng*. 2003;31(8). https://doi.org/10.1114/1.1590661.
10. Ali MH, Schumacker PT. Endothelial responses to mechanical stress: where is the mechanosensor?. In: *Critical Care Medicine*. Vol. 30. 2002. https://doi.org/10.1097/00003246-200205001-00005.
11. Hahn C, Schwartz MA. Mechanotransduction in vascular physiology and atherogenesis. *Nat Rev Mol Cell Biol*. 2009;10(1). https://doi.org/10.1038/nrm2596.
12. Koslow AR, Stromberg RR, Friedman LI, Lutz RJ, Hilbert SL, Schuster P. A flow system for the study of shear forces upon cultured endothelial cells. *J Biomech Eng*. 1986;108(4). https://doi.org/10.1115/1.3138625.
13. Levesque MJ, Nerem RM. The elongation and orientation of cultured endothelial cells in response to shear stress. *J Biomech Eng*. 1985;107(4). https://doi.org/10.1115/1.3138567.
14. Dewey CF, Bussolari SR, Gimbrone MA, Davies PF. The dynamic response of vascular endothelial cells to fluid shear stress. *J Biomech Eng*. 1981;103(3). https://doi.org/10.1115/1.3138276.
15. Blackman BR, García-Cardeña G, Gimbrone MA. A new in vitro model to evaluate differential responses of endothelial cells to simulated arterial shear stress waveforms. *J Biomech Eng*. 2002; 124(4). https://doi.org/10.1115/1.1486468.
16. Moore JE, Bürki E, Suciu A, et al. A device for subjecting vascular endothelial cells to both fluid shear stress and circumferential cyclic stretch. *Ann Biomed Eng*. 1994;22(4). https://doi.org/10.1007/BF02368248.
17. Hsu PP, Li S, Li YS, et al. Effects of flow patterns on endothelial cell migration into a zone of mechanical denudation. *Biochem Biophys Res Commun*. 2001;285(3). https://doi.org/10.1006/bbrc.2001.5221.
18. DePaola N, Gimbrone MA, Davies PF, Dewey CF. Vascular endothelium responds to fluid shear stress gradients. *Arterioscler Thromb*. 1992;12(11). https://doi.org/10.1161/01.atv.12.11.1254.
19. Remuzzi A, Forbes Dewey C, Davies PF, Gimbrone MA. Orientation of endothelial cells in shear fields in vitro. *Biorheology*. 1984; 21(4). https://doi.org/10.3233/BIR-1984-21419.
20. Wong AJ, Pollard TD, Herman IM. Actin filament stress fibers in vascular endothelial cells in vivo. *Science*. 1983;219(4586). https://doi.org/10.1126/science.6681677.
21. Resnick N, Collins T, Atkinson W, Bonthron DT, Dewey CF, Gimbrone MA. Platelet-derived growth factor B chain promoter contains a cis-acting fluid shear-stress-responsive element. *Proc Natl Acad Sci U S A*. 1993;90(10). https://doi.org/10.1073/pnas.90.10.4591.
22. Davis ME, Grumbach IM, Fukai T, Cutchins A, Harrison DG. Shear stress regulates endothelial nitric-oxide synthase promoter activity through nuclear factor κB binding. *J Biol Chem*. 2004;279(1). https://doi.org/10.1074/jbc.M307528200.
23. Korenaga E, Ando J, Kosaki K, Isshiki M, Takada Y, Kamiya A. Negative transcriptional regulation of the VCAM-1 gene by fluid shear stress in murine endothelial cells. *Am J Physiol Cell Physiol*. 1997;273(5 42–5). https://doi.org/10.1152/ajpcell.1997.273.5.c1506.
24. dela Paz NG, Walshe TE, Leach LL, Saint-Geniez M, D'Amore PA. Role of shear-stress-induced VEGF expression in endothelial cell survival. *J Cell Sci*. 2012;125(4). https://doi.org/10.1242/jcs.084301.
25. Shyy YJ, Hsieh HJ, Usami S, Chien S. Fluid shear stress induces a biphasic response of human monocyte chemotactic protein 1 gene expression in vascular endothelium. *Proc Natl Acad Sci U S A*. 1994;91(11). https://doi.org/10.1073/pnas.91.11.4678.
26. Diamond SL, Eskin SG, McIntire LV. Fluid flow stimulates tissue plasminogen activator secretion by cultured human endothelial cells. *Science*. 1989;243(4897). https://doi.org/10.1126/science.2467379.
27. Resnick N, Yahav H, Khachigian LM, et al. Endothelial gene regulation by laminar shear stress. *Adv Exp Med Biol*. 1997;430. https://doi.org/10.1007/978-1-4615-5959-7_13.
28. Wang N, Butler JP, Ingber DE. Mechanotransduction across the cell surface and through the cytoskeleton. *Science*. 1993;260(5111). https://doi.org/10.1126/science.7684161.
29. Ingber DE. Tensegrity: the architectural basis of cellular mechanotransduction. *Annu Rev Physiol*. 1997;59. https://doi.org/10.1146/annurev.physiol.59.1.575.
30. Topper JN, Cai J, Falb D, Gimbrone MA. Identification of vascular endothelial genes differentially responsive to fluid mechanical stimuli: cyclooxygenase-2, manganese superoxide dismutase, and endothelial cell nitric oxide synthase are selectively up-regulated by steady laminar shear stress. *Proc Natl Acad Sci U S A*. 1996; 93(19). https://doi.org/10.1073/pnas.93.19.10417.
31. Topper JN, Gimbrone MA. Blood flow and vascular gene expression: fluid shear stress as a modulator of endothelial phenotype. *Mol Med Today*. 1999;5(1). https://doi.org/10.1016/S1357-4310(98)01372-0.
32. Malek AM, Alper SL, Izumo S. Hemodynamic shear stress and its role in atherosclerosis. *J Am Med Assoc*. 1999;282(21). https://doi.org/10.1001/jama.282.21.2035.
33. Gimbrone MA, García-Cardeña G. Vascular endothelium, hemodynamics, and the pathobiology of atherosclerosis. *Cardiovasc Pathol*. 2013;22(1). https://doi.org/10.1016/j.carpath.2012.06.006.
34. García-Cardeña G, Comander J, Anderson KR, Blackman BR, Gimbrone MA. Biomechanical activation of vascular endothelium as a determinant of its functional phenotype. *Proc Natl Acad Sci U S A*. 2001;98(8). https://doi.org/10.1073/pnas.071052598.
35. McCormick SM, Eskin SG, McIntire Lv, et al. DNA microarray reveals changes in gene expression of shear stressed human umbilical vein endothelial cells. *Proc Natl Acad Sci U S A*. 2001;98(16). https://doi.org/10.1073/pnas.171259298.
36. Chen BP, Li YS, Zhao Y, et al. DNA microarray analysis of gene expression in endothelial cells in response to 24-h shear stress. *Physiol Genom*. 2001;7(1). https://doi.org/10.1152/physiolgenomics.2001.7.1.55.

37. Dai G, Kaazempur-Mofrad MR, Natarajan S, et al. Distinct endothelial phenotypes evoked by arterial waveforms derived from atherosclerosis-susceptible and -resistant regions of human vasculature. *Proc Natl Acad Sci U S A*. 2004;101(41). https://doi.org/10.1073/pnas.0406073101.
38. Passerini AG, Polacek DC, Shi C, et al. Coexisting proinflammatory and antioxidative endothelial transcription profiles in a disturbed flow region of the adult porcine aorta. *Proc Natl Acad Sci U S A*. 2004;101(8). https://doi.org/10.1073/pnas.0305938101.
39. Feng S, Bowden N, Fragiadaki M, et al. Mechanical activation of hypoxia-inducible factor 1a drives endothelial dysfunction at atheroprone sites. *Arterioscler Thromb Vasc Biol*. 2017;37(11). https://doi.org/10.1161/ATVBAHA.117.309249.
40. Wu D, Huang RT, Hamanaka RB, et al. HIF-1α is required for disturbed flow-induced metabolic reprogramming in human and porcine vascular endothelium. *eLife*. 2017;6. https://doi.org/10.7554/eLife.25217.
41. Ajami NE, Gupta S, Maurya MR, et al. Systems biology analysis of longitudinal functional response of endothelial cells to shear stress. *Proc Natl Acad Sci U S A*. 2017;114(41). https://doi.org/10.1073/pnas.1707517114.
42. Parmar KM, Larman HB, Dai G, et al. Integration of flow-dependent endothelial phenotypes by Kruppel-like factor 2. *J Clin Invest*. 2006;116(1). https://doi.org/10.1172/JCI24787.
43. Dekker RJ, van Thienen Jv, Elderkamp YW, et al. In vivo endothelial expression of lung krüppel-like factor (LKLF/KLF2) correlates with local biomechanical stresses. *Cardiovasc Pathol*. 2004;13(3). https://doi.org/10.1016/j.carpath.2004.03.570.
44. Dekker RJ, van Thienen Jv, Rohlena J, et al. Endothelial KLF2 links local arterial shear stress levels to the expression of vascular tone-regulating genes. *Am J Pathol*. 2005;167(2). https://doi.org/10.1016/S0002-9440(10)63002-7.
45. Atkins GB, Wang Y, Mahabeleshwar GH, et al. Hemizygous deficiency of krüppel-like factor 2 augments experimental atherosclerosis. *Circ Res*. 2008;103(7). https://doi.org/10.1161/CIRCRESAHA.108.184663.
46. Abe JI, Berk BC. Novel mechanisms of endothelial mechanotransduction. *Arterioscler Thromb Vasc Biol*. 2014;34(11). https://doi.org/10.1161/ATVBAHA.114.303428.
47. Huddleson JP, Ahmad N, Srinivasan S, Lingrel JB. Induction of KLF2 by fluid shear stress requires a novel promoter element activated by a phosphatidylinositol 3-kinase-dependent chromatin-remodeling pathway. *J Biol Chem*. 2005;280(24). https://doi.org/10.1074/jbc.M413839200.
48. Parmar KM, Nambudiri V, Dai G, Larman HB, Gimbrone MA, García-Cardeña G. Statins exert endothelial atheroprotective effects via the KLF2 transcription factor. *J Biol Chem*. 2005;280(29). https://doi.org/10.1074/jbc.C500144200.
49. Sen-Banerjee S, Mir S, Lin Z, et al. Kruppel-like factor 2 as a novel mediator of statin effects in endothelial cells. *Circulation*. 2005;112(5). https://doi.org/10.1161/CIRCULATIONAHA.104.525774.
50. SenBanerjee S, Lin Z, Atkins GB, et al. KLF2 is a novel transcriptional regulator of endothelial proinflammatory activation. *J Exp Med*. 2004;199(10). https://doi.org/10.1084/jem.20031132.
51. Villarreal G, Zhang Y, Larman HB, Gracia-Sancho J, Koo A, García-Cardeña G. Defining the regulation of KLF4 expression and its downstream transcriptional targets in vascular endothelial cells. *Biochem Biophys Res Commun*. 2010;391(1). https://doi.org/10.1016/j.bbrc.2009.12.002.
52. Ohnesorge N, Viemann D, Schmidt N, et al. Erk5 activation elicits a vasoprotective endothelial phenotype via induction of Krüppel-like factor 4 (KLF4). *J Biol Chem*. 2010;285(34). https://doi.org/10.1074/jbc.M110.103127.
53. Dunn J, Simmons R, Thabet S, Jo H. The role of epigenetics in the endothelial cell shear stress response and atherosclerosis. *Int J Biochem Cell Biol*. 2015;67. https://doi.org/10.1016/j.biocel.2015.05.001.
54. Man HSJ, Yan MS, Lee JJY, Marsden PA. Epigenetic determinants of cardiovascular gene expression: vascular endothelium. *Epigenomics*. 2016;8(7). https://doi.org/10.2217/epi-2016-0012.
55. Marin T, Gongol B, Chen Z, et al. Mechanosensitive microRNAs - role in endothelial responses to shear stress and redox state. *Free Radic Biol Med*. 2013;64. https://doi.org/10.1016/j.freeradbiomed.2013.05.034.
56. Fang Y, Shi C, Manduchi E, Civelek M, Davies PF. MicroRNA-10a regulation of proinflammatory phenotype in athero-susceptible endothelium in vivo and in vitro. *Proc Natl Acad Sci U S A*. 2010;107(30). https://doi.org/10.1073/pnas.1002120107.
57. Qin X, Wang X, Wang Y, et al. MicroRNA-19a mediates the suppressive effect of laminar flow on cyclin D1 expression in human umbilical vein endothelial cells. *Proc Natl Acad Sci U S A*. 2010;107(7). https://doi.org/10.1073/pnas.0914882107.
58. Wang KC, Nguyen P, Weiss A, et al. MicroRNA-23b regulates cyclin-dependent kinase-activating kinase complex through cyclin H repression to modulate endothelial transcription and growth under flow. *Arterioscler Thromb Vasc Biol*. 2014;34(7). https://doi.org/10.1161/ATVBAHA.114.303473.
59. Chen K, Fan W, Wang X, Ke X, Wu G, Hu C. MicroRNA-101 mediates the suppressive effect of laminar shear stress on mTOR expression in vascular endothelial cells. *Biochem Biophys Res Commun*. 2012;427(1). https://doi.org/10.1016/j.bbrc.2012.09.026.
60. Wu W, Xiao H, Laguna-Fernandez A, et al. Flow-dependent regulation of krüppel-like factor 2 is mediated by MicroRNA-92a. *Circulation*. 2011;124(5). https://doi.org/10.1161/CIRCULATIONAHA.110.005108.
61. Ni CW, Qiu H, Jo H. MicroRNA-663 upregulated by oscillatory shear stress plays a role in inflammatory response of endothelial cells. *Am J Physiol Heart Circ Physiol*. 2011;300(5). https://doi.org/10.1152/ajpheart.00829.2010.
62. Son DJ, Kumar S, Takabe W, et al. The atypical mechanosensitive microRNA-712 derived from pre-ribosomal RNA induces endothelial inflammation and atherosclerosis. *Nat Commun*. 2013;4. https://doi.org/10.1038/ncomms4000.
63. Loyer X, Potteaux S, Vion AC, et al. Inhibition of microRNA-92a prevents endothelial dysfunction and atherosclerosis in mice. *Circ Res*. 2014;114(3). https://doi.org/10.1161/CIRCRESAHA.114.302213.
64. Fan W, Fang R, Wu X, et al. Shear-sensitive microRNA-34a modulates flow-dependent regulation of endothelial inflammation. *J Cell Sci*. 2015;128(1). https://doi.org/10.1242/jcs.154252.
65. Hergenreider E, Heydt S, Tréguer K, et al. Atheroprotective communication between endothelial cells and smooth muscle cells through miRNAs. *Nat Cell Biol*. 2012;14(3). https://doi.org/10.1038/ncb2441.
66. Rashad S, Han X, Saqr K, et al. Epigenetic response of endothelial cells to different wall shear stress magnitudes: a report of new mechano-miRNAs. *J Cell Physiol*. 2020;235(11). https://doi.org/10.1002/jcp.29436.
67. Jiang YZ, Jiménez JM, Ou K, McCormick ME, Zhang LD, Davies PF. Hemodynamic disturbed flow induces differential DNA methylation of endothelial Kruppel-like factor 4 promoter in vitro and in vivo. *Circ Res*. 2014;115(1). https://doi.org/10.1161/CIRCRESAHA.115.303883.

CHAPTER

13

Intravital photoacoustic microscopy of microvascular function and oxygen metabolism

Naidi Sun and Song Hu

Department of Biomedical Engineering, Washington University in St. Louis, St. Louis, MO, United States

Introduction

Impairment of vascular function and oxygen metabolism occurs in a wide spectrum of human diseases, including coronary artery disease,[1] heart failure,[2] critical limb ischemia,[3] chronic diabetic ulcers with impaired wound healing[4]; ischemic stroke,[5] cancer,[6] Alzheimer's disease,[7] and other metabolism-related disorders.[8–10] Over the past decades, vascular-metabolic dysfunction has been extensively studied.[11–20] Capitalizing on the light absorption of endogenous hemoglobin, photoacoustic microscopy (PAM) has demonstrated unique ability for in vivo characterization of the microvasculature in experimental settings.[21–23] In PAM, spatially focused laser pulses are delivered into biological tissue, where part of the delivered photon energy is absorbed and converted into heat, causing transient thermoelastic expansion and acoustic emission. The generated ultrasonic waves are detected by acoustic sensors and analyzed to produce images. By providing label-free and minimally noninvasive imaging, PAM has enabled comprehensive quantification of microvascular function and oxygen metabolism in vivo, and has been employed for intravital imaging of different tissues and organs in animal models.

In this chapter, we introduce five intravital PAM approaches, with targets spanning from the superficial microvasculature to the vasculature of internal organs. We present detailed methodologies, review current and potential biomedical applications, and discuss the frontiers of intravital PAM of microvascular function and oxygen metabolism.

Multiparametric PAM: the working principles

Fig. 13.1A shows a typical multiparametric PAM system,[23] where 2 ns-pulsed lasers (wavelengths: 532 and 559 nm) provide dual-wavelength excitation. After a beam splitter combines the two orthogonally polarized outputs, the combined beam is attenuated by a neutral density (ND) filter. An iris, a condenser lens, and a pinhole enhance the fiber optic coupling efficiency. To compensate for fluctuations in laser intensity, ~5% of the laser energy is diverted by a beam sampler, measured by a high-speed photodiode, and used for fluence compensation. The spatially filtered beam is then coupled into a single-mode optical fiber through a microscope objective and delivered to the scanning head (Fig. 13.1B), where two identical doublets map the fiber output into the tissue to be imaged. A two-dimensional galvanometer scanner inserted between the doublets provides automatic confocal alignment of the optical and acoustic foci. A ring-shaped ultrasonic transducer is used for ultrasonic detection, and a correction lens placed before the transducer compensates for optical aberration at the air–water interface.

The system performs raster scanning, during which the two lasers are alternately triggered at 50-μs intervals to produce dual-wavelength A-line pairs (Fig. 13.1C). By analyzing the Brownian motion–induced amplitude fluctuation of erythrocyte signals in the A-line signals acquired at 532 nm, where the optical absorption coefficients of oxy-hemoglobin and deoxy-hemoglobin (HbO_2 and HbR, respectively) are nearly equal, the total concentration of hemoglobin (C_{Hb}) can be quantified in absolute values (Fig. 13.1D).[24] By adding a second measurement at 559 nm, PAM can differentiate HbO_2 and HbR based on their distinct absorption spectra, from which the oxygen saturation of hemoglobin (sO_2) can also be quantified in absolute values (Fig. 13.1E).[23] By quantifying the blood flow–induced decorrelation of the successively acquired A-line signals, the speed of blood

The Vasculome
https://doi.org/10.1016/B978-0-12-822546-2.00002-2

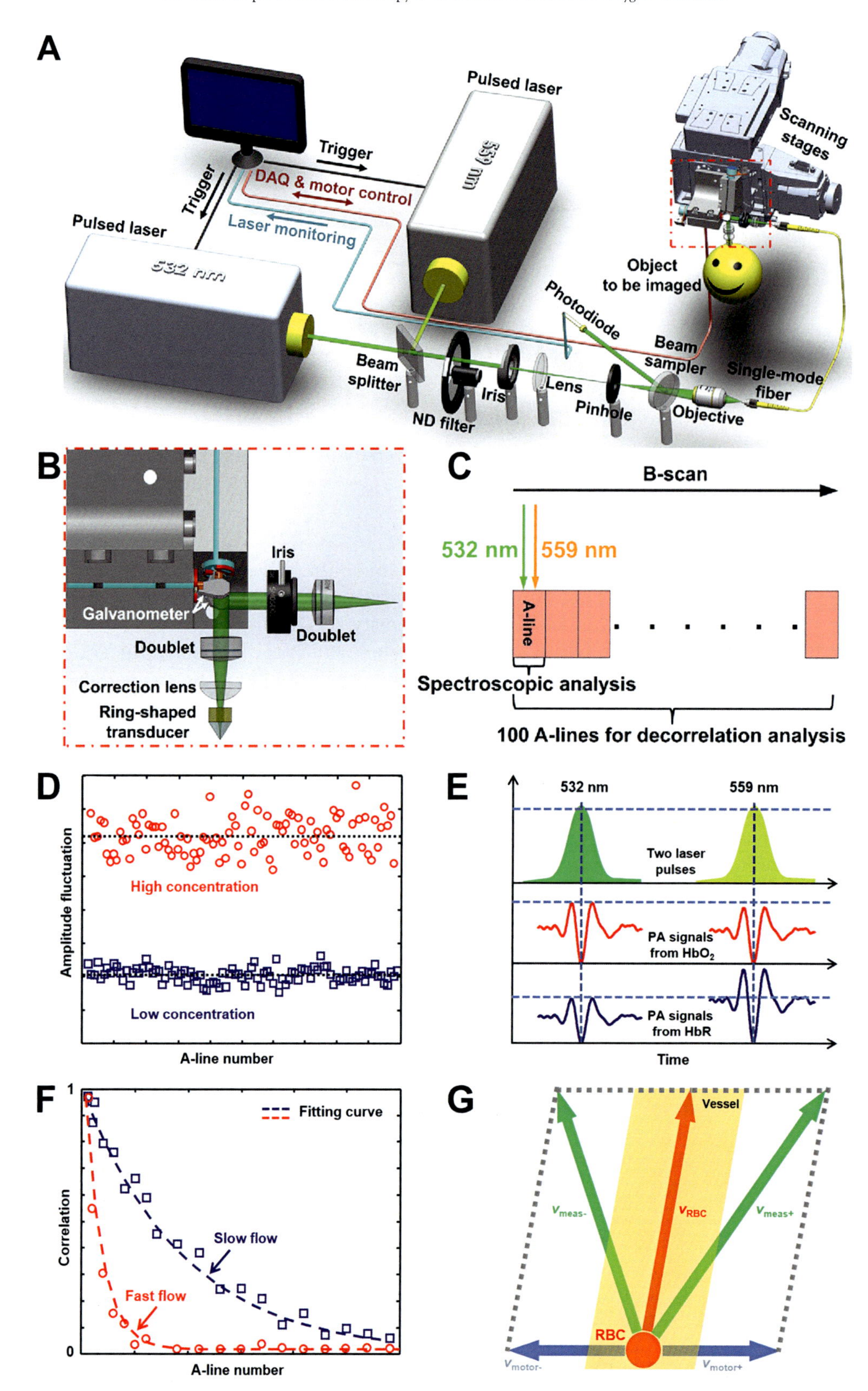

FIGURE 13.1 **Schematic and working principles of multiparametric PAM.** (A) System schematic. *DAQ*, data acquisition; *ND*, neutral density. (B) Blow-up of the scan head boxed in (A). (C) Scanning mechanism for simultaneous imaging of C_{Hb}, sO_2, and blood flow. (D–F) Statistical, spectroscopic, and correlation analyses for simultaneous PAM of C_{Hb}, sO_2, and blood flow speed, respectively. (G) Bidirectional analysis of the blood flow direction.

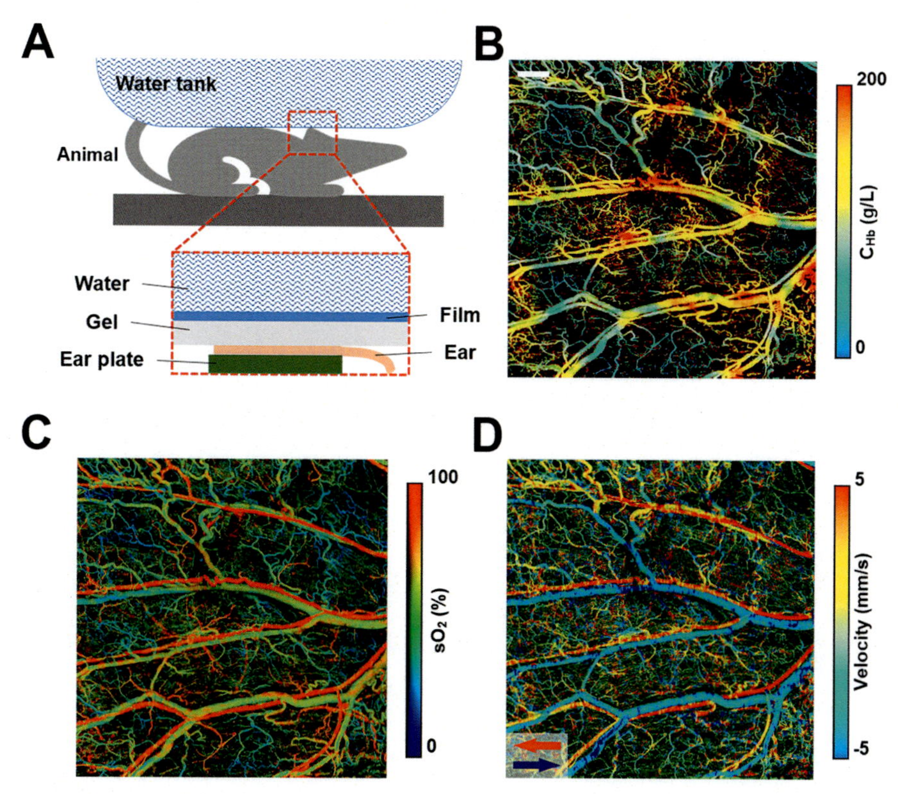

FIGURE 13.2 **Multiparametric PAM of the mouse ear.** (A) Experimental settings. (B–D) Multiparametric images of (B) total hemoglobin concentration (C_{Hb}), (C) oxygen saturation of hemoglobin (sO_2), and (D) blood flow velocity (speed plus direction). The *red and blue arrows* in (D) indicate the flow directions. Scale bars: 500 μm.

flow can be calculated (Fig. 13.1F).[23] Further, the direction of blood flow can be determined by identifying the difference between the flow speeds measured in adjacent forward and backward B-scans (Fig. 13.1G).[25]

Intravital PAM of the skin microvasculature

The mouse ear is often chosen for studying the skin microvasculature, predominantly because its vasculature is readily accessible and can be serially imaged over a large field of view without surgical intervention (e.g., installation of a window chamber). Fig. 13.2A shows multiparametric PAM images of the mouse ear captured in vivo by using an ear support plate and the settings shown.[26] The mouse ear is placed on a homemade plastic support plate and in gentle contact with the water tank, which is filled with temperature-maintained (37°C) deionized water. For effective acoustic coupling, thin layers of transparent ultrasound gel are applied between the ear, the plate, and the plastic membrane at the bottom of the water tank. Throughout the PAM experiment, the mouse is maintained under general anesthesia with 1.0%–1.5% isoflurane, and the body temperature is kept at 37°C using a temperature-controlled heating pad. The acquisition time is determined by the motor speed and step size in the y-direction (i.e., perpendicular to the direction of motor movement). For a typical setting, where the motor speed is 0.9 mm/s and the step size is 10 μm, the total acquisition time is ~70 min for an area of $6 \times 6\ mm^2$.

Using this technique, we have simultaneously imaged the microvascular C_{Hb}, sO_2, and blood flow velocity in a mouse ear (Fig. 13.2B–D). The multiparametric images clearly resolve the microvascular networks, showing dramatic differences in sO_2 and flow direction between arteries/arterioles and veins/venules. These results demonstrate the broad applicability of multiparametric PAM in studying microvascular function and oxygen metabolism in vivo. Additionally, other parameters, such as vessel tortuosity, volumetric flow, oxygen supply, resistance, and wall shear stress, can also be calculated by applying proper analyses.[26]

Capable of noninvasive imaging of the mouse ear in a label-free manner, PAM has shown considerable promises in a variety of microvascular research, including ligation-based hemodynamic redistribution and arteriogenesis (Fig. 13.3A),[26] the effects of therapeutics on the microvasculature (Fig. 13.3B),[27] and early cancer detection (Fig. 13.3C).[28]

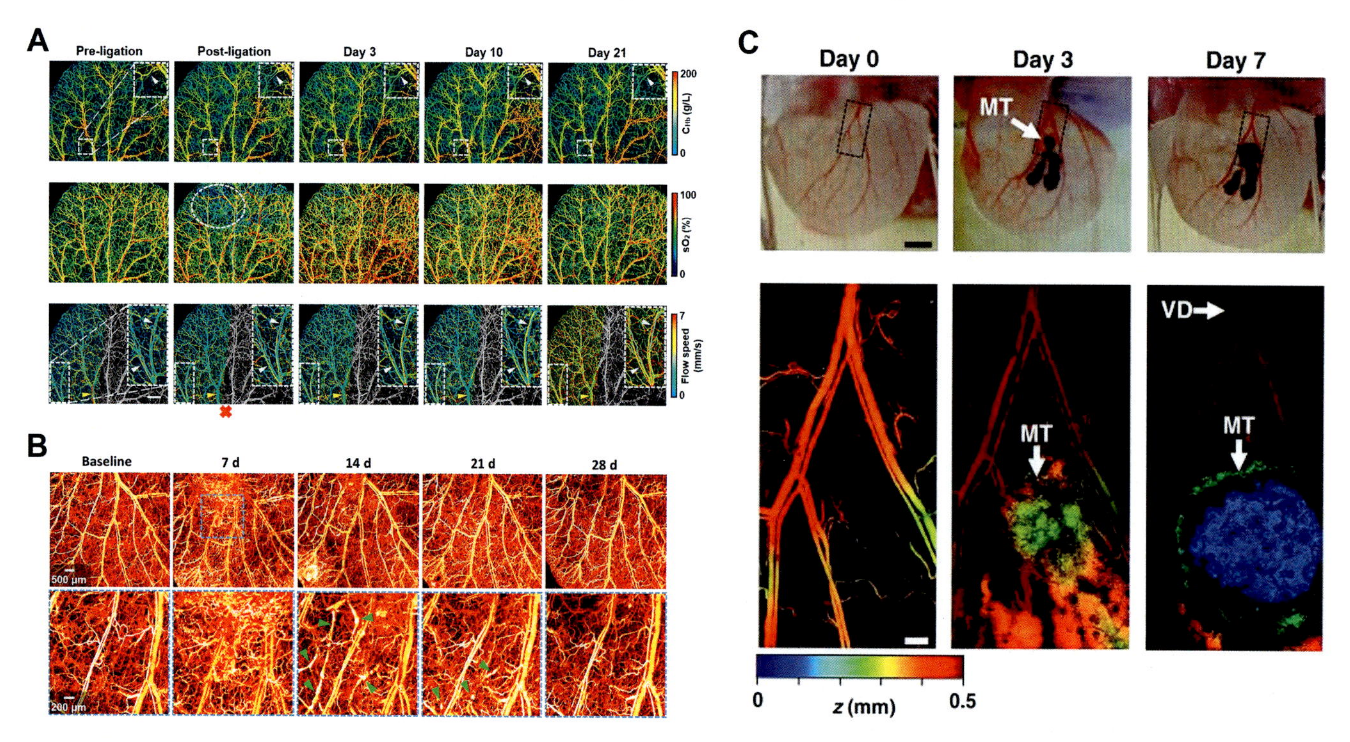

FIGURE 13.3 **Biomedical applications of mouse ear PAM imaging.** (A) Longitudinal monitoring and single-vessel analysis of hemodynamic redistribution and collateral remodeling across a microvascular network following arterial ligation (indicated by the *red cross*). Top, middle, and bottom rows are PAM images of C_{Hb}, sO_2, and blood flow speed before, right after, and up to 21 days postligation, respectively. The insets are enlargements of the white-boxed regions in the lower left corners of the two panels. Scale bar: 1 mm. (B) Intradermal injection of 100 μg AZD8601 (a modified mRNA, encoding vascular endothelial growth factor A) induces angiogenesis and neovascularization. Top row: microvascular structure of the whole field of view. Bottom row: Enlargement of the blue boxed region, showing a detailed view of the capillary angiogenesis on day 7 and neovessel formation on day 14 (*green arrows*). (C) PAM detection of early-stage melanoma. Top row: white-light photographs of a representative mouse ear before (day 0), and 3, and 7 days after the xenotransplantation of B16 melanoma tumor cells. Scale bar: 1 mm. Bottom row: PAM images of the tumor region (*dashed boxes* in top row) at 584 nm. The z dimension is coded by colors: blue (superficial) to red (deep). Scale bar: 125 μm.

Intravital PAM of the dorsal vasculature

Intravital PAM imaging of the mouse dorsal vasculature has been achieved by using a dorsal skin fold window chamber (Fig. 13.4A),[29] which allows high-resolution visualization of the vascularized adipofascial layer and subcutaneous dorsal microvasculature (Fig. 13.4B).[30] During imaging, the mouse was anesthetized, and the window chamber was fixed in place by a 3-D printed plate (not shown). With 1-mm penetration into the living tissue, PAM can visualize both the subdermal vasculature, providing quantitative insights into the C_{Hb}, sO_2, and blood flow.

An application of dorsal PAM is the investigation of acellular dermal matrix (ADM) biointegration.[30–32] The ADM provides a regenerative biomaterial scaffold to support host cell integration and collagen remodeling,[33] and PAM is well suited to help understand the interaction between the ADM and the host by assessing ADM-aided revascularization. Serial gross photographic images and PAM images of a window chamber with and without ADM implantation are shown in Fig. 13.4C.[30] The microvasculature in the ADM/vascular plexus border zone (outside the dashed circle) demonstrates a proliferation of small, highly branching vessels at early time points (day 10), followed by continued new vessel formation and vascular ingrowth into the center of the ADM over time (within the dashed circle at day 21). Besides, the assessment of biomaterial, intravital PAM imaging of subcutaneous dorsal vasculature can also be used in characterizing tumor angiogenesis.[34]

Intravital PAM of the cerebral microvasculature

The brain accounts for more than 20% of the oxygen consumption in the resting state.[35] Thus, the cerebral vascular metabolism plays a key role in brain disorders.[5,7] Combining light and ultrasound to image the cerebral microvasculature, PAM breaks the technical limitation of high-resolution functional imaging through the intact skull.[36] Because anesthesia is known to have profound effects on neuronal activities and cerebral hemodynamics, a novel application of PAM (Fig. 13.5A)

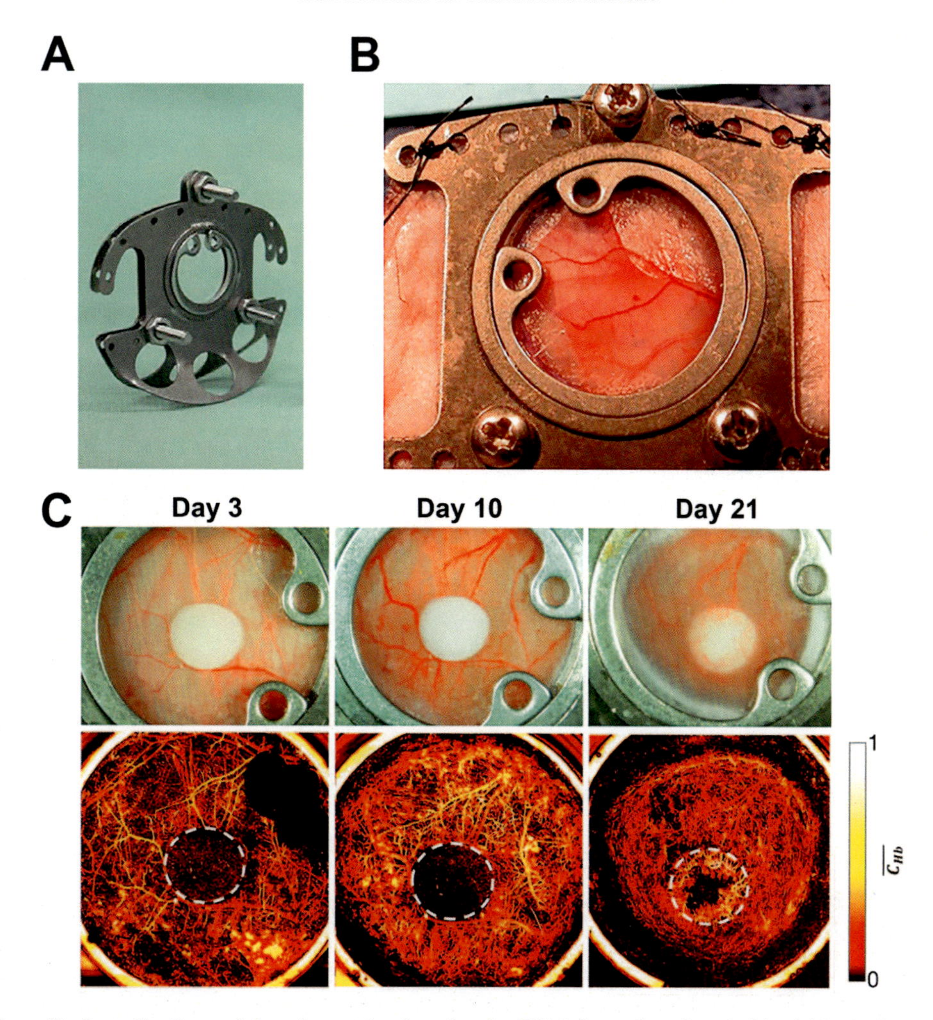

FIGURE 13.4 **Biomedical applications of dorsal vascular imaging by PAM through a dorsal skin fold window chamber.** (A) Window chamber, consisting of two symmetrical titanium frames for holding the dorsal skinfold of mice (chamber weight: ~2 g). (B) Close-up views showing the orientation of the dorsal skin-fold window chamber with native vascular plexus. (C) Serial gross photography (above) and in vivo PAM assessment (below) of the window chamber field of view, showing the microvasculature of the native vascular plexus at days 3, 10, and 21 after implantation of an acellular dermal matrix (ADM). *Dashed circle*: ADM position.

has been developed to comprehensively and quantitatively characterize cerebral hemodynamics and oxygen metabolism at the microscopic level in awake mice.[22] An angle- and height-adjustable head-restraint apparatus enables high-resolution PAM of the awake brain with minimal motion artifacts, and an air-floated spherical treadmill mitigates restraint-induced stress. Before PAM imaging and after hair removal under anesthesia, a surgical incision is made in the scalp to expose the skull. Dental cement is then applied to cover the exposed surface, except for the region of interest, and a nut is attached to the other side of the skull with dental cement. Once the cement has solidified, the skull over the region of interest is thinned with a surgical hand drill. After these preparations, multiparametric PAM images of the awake mouse brain can be acquired, as shown in Fig. 13.5B.[26]

For cortex-wide imaging, the depth of focus of the optical beam is often insufficient to accommodate the depth variation of the mouse brain surface, resulting in low spatial resolution and poor signal-to-noise ratio (SNR).[37,38] To overcome this limitation, the uneven surface contour can be extracted either through a prescan[36] or in real time.[39] In ultrasound-aided prescanning, a contour map of the mouse skull is first ultrasonically extracted (Fig. 13.6A). Then, to maintain high spatial resolution and sensitivity, PAM dynamically focuses on the underlying cortical vasculature when scanning across the uneven brain surface (Fig. 13.6B).[36] Fig. 13.6C is a schematic diagram of the real-time contour scanning.[39] Data acquisition starts with a conventional B-scan (B-scan #1) along the x-axis, without moving the vertical stage. Thus, the trace of the x−z stage appears as a horizontal line, which is denoted as trace 0. Immediately

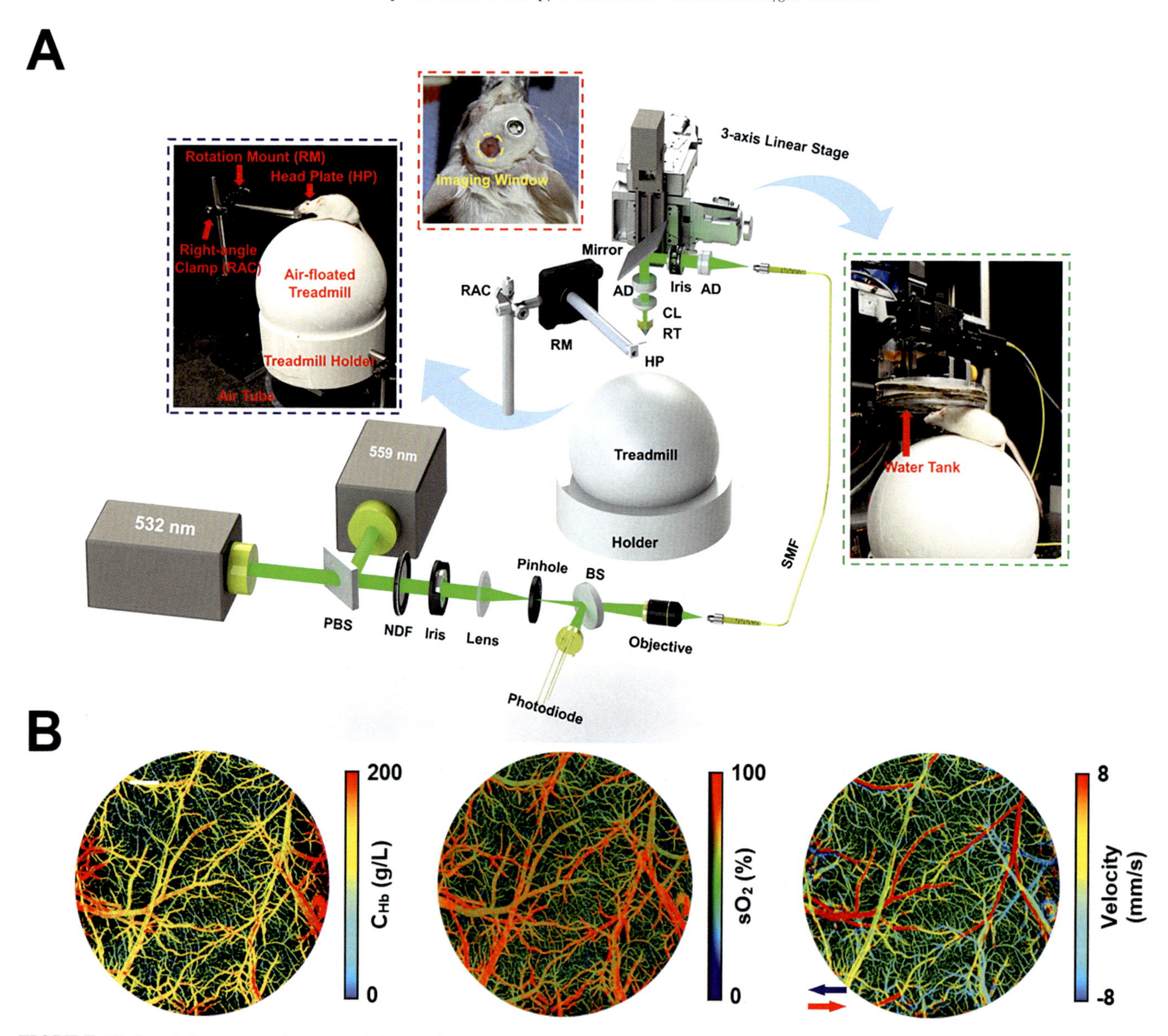

FIGURE 13.5 Multiparametric PAM of the awake mouse brain. (A) System schematic. *AD*, achromatic doublet; *BS*, beam sampler; *CL*, correction lens; *HP*, head plate; *NDF*, neutral density filter; *PBS*, polarizing beam splitter; *RAC*, right-angle clamp; *RM*, rotation mount; *RT*, ring-shaped ultrasonic transducer; *SMF*, single-mode fiber. Red-boxed inset: photograph of a mouse brain with a thinned-skull window and a nut attached using dental cement. Blue-boxed inset: photograph of the angle- and height-adjustable head-restraint apparatus. Green-boxed inset: mouse placement during imaging. (B) PAM images of C_{Hb}, sO_2, and blood flow speed acquired in the awake mouse brain. Scale bar: 500 μm.

after B-scan #1, the acquired data (data #1) are processed to extract the surface contour, which is denoted as trace 1. Then, the y stage makes a 10-μm step forward and starts B-scan #2 toward the negative x-axis, during which the z stage moves along trace 1 in reverse. After the completion of B-scan #2, its surface contour (trace 2) is extracted by combining trace 1 and data #2. The same procedure is repeated until the image acquisition is complete, and intravital imaging of the cortex-wide mouse brain is ultimately accomplished, as shown in Fig. 13.6D and E.[39]

Typical biomedical applications involving intravital imaging of the rodent brain include studying the influence of obesity on cerebrovascular hemodynamics and metabolism (Fig. 13.7A),[40] evaluating neuroprotective effects against cerebral hypoxia/ischemia (Fig. 13.7B),[41] and assessing blast exposure-induced cerebrovascular dysfunction (Fig. 13.7C).[42] For these applications, the PAM platform presented here enables comprehensive and quantitative characterization of the vascular structure (e.g., its density, tortuosity, and permeability), mechanical properties (e.g., wall shear

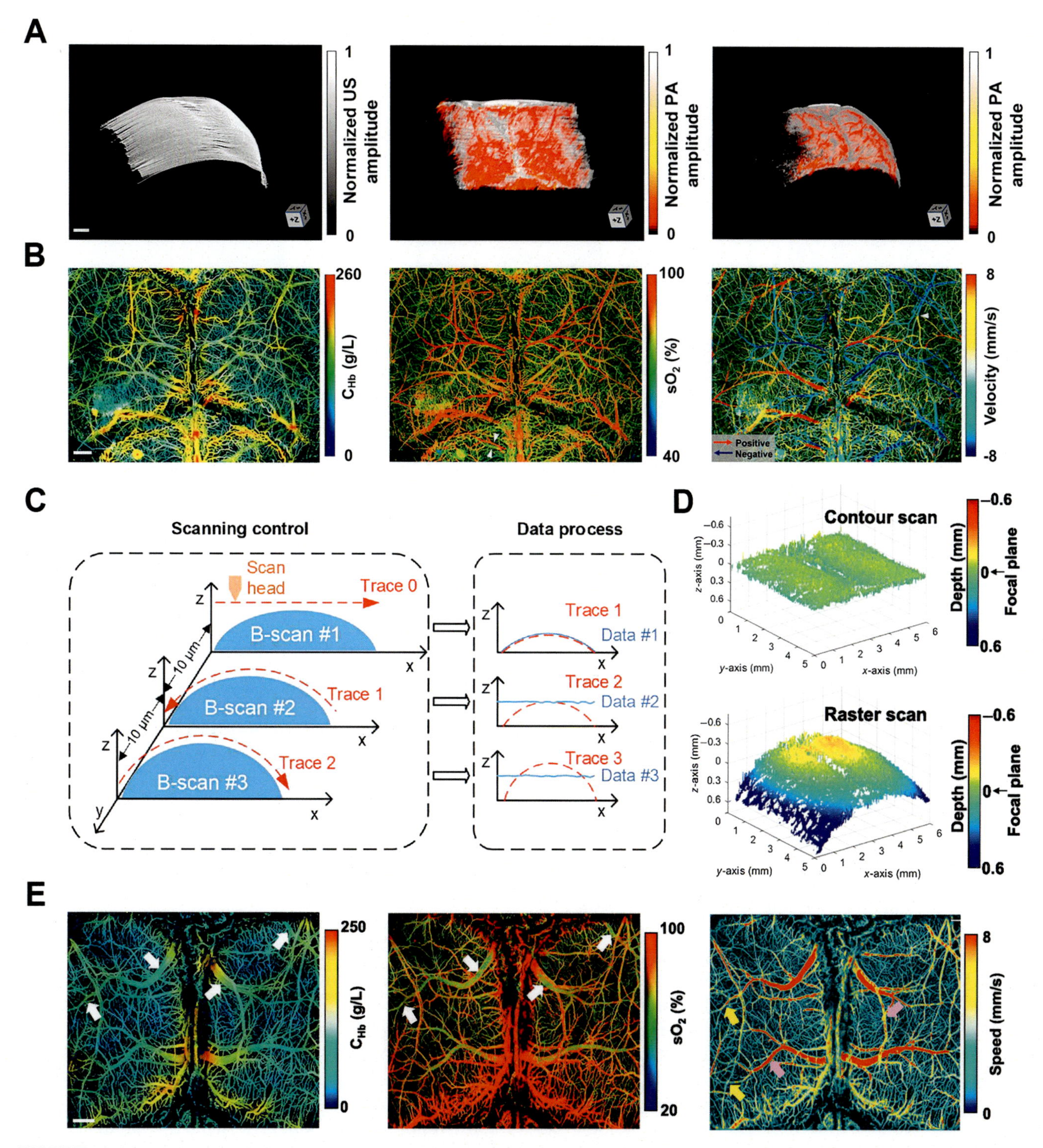

FIGURE 13.6 Cortex-wide, contour-scan, multiparametric PAM imaging of the mouse brain. (A) Left: 3D surface contour of the mouse skull, extracted by integrated scanning acoustic microscopy (SAM). Middle and right: 3D rendering of the mouse brain simultaneously imaged by SAM (gray) and PAM (hot), with and without the contour scan, respectively. Scale bar: 1 mm. (B) Ultrasound-aided multiparametric PAM of C_{Hb}, sO_2, and cerebral blood flow in a mouse brain through the intact skull. The *red and blue arrows* indicate the directions of the blood flow along the B-scan axis. Scale bar: 0.5 mm. (C) Real-time contour-scan scheme. (D) Surface contours of the entire mouse cortex, extracted from the PAM images acquired with and without the contour scan, respectively. (E) Multiparametric PAM images acquired by real-time contour scan. Scale bar: 0.5 mm.

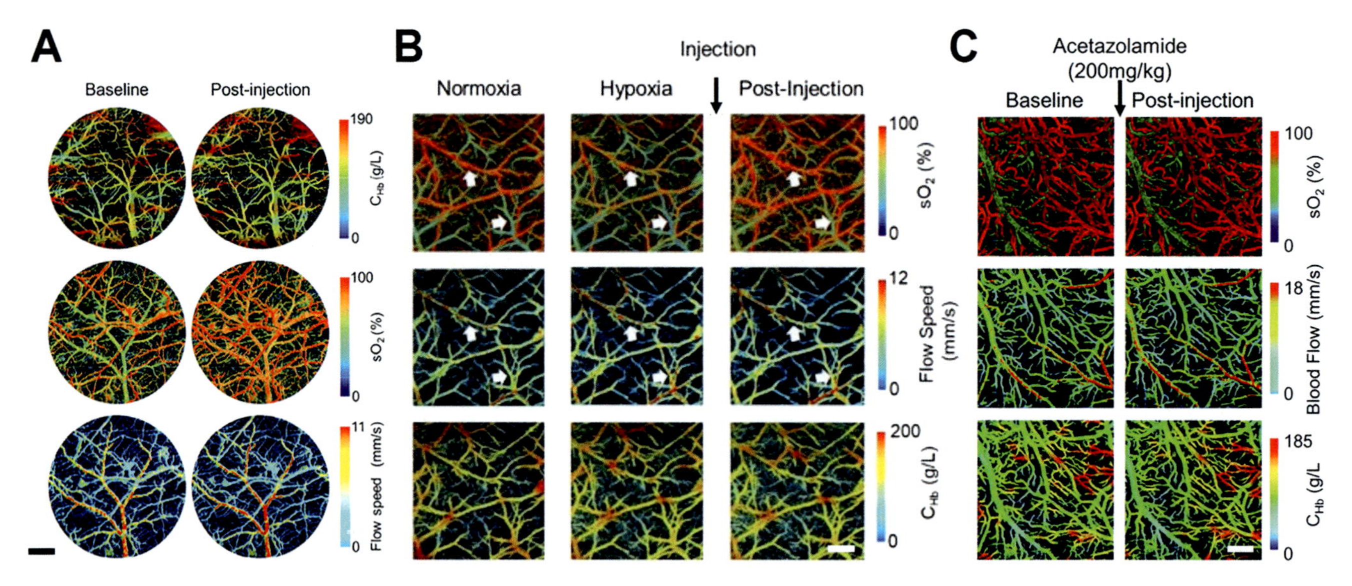

FIGURE 13.7 **Example applications of PAM imaging of the rodent brain.** (A) Obesity-induced change in cerebrovascular reactivity before and after intravenous injection of acetazolamide. Scale bar: 500 μm. (B) S1P-induced hemodynamic responses in the awake hypoxic mouse brain to the injection of a sphingosine kinase 2 inhibitor. The *white arrows* highlight the changes in sO_2 and blood flow speed, respectively. Scale bar: 500 μm. (C) Blast-induced changes in the cerebral vasculature in a rat subject to blast exposure (pressure: 196.5 kPa) to the injection of acetazolamide. Scale bar: 500 μm.

stress, resistance, and reactivity), hemodynamics (e.g., blood perfusion, oxygenation, and flow speed), and the associated tissue oxygen extraction and metabolism in the awake rodent brain.

Intravital PAM of the microvasculature in other tissues and organs

Beyond imaging the tissue/organ sites discussed above, PAM is also quite useful for experimental intravital imaging of muscles and the cornea in mice.[26,43] Before the muscle is imaged, the hair within the region of interest is removed and a surgical incision is made. Then, the subcutaneous fat is removed to expose the muscle under the skin. Using hindlimb imaging as an example, to stabilize the legs against respiration-induced movement, a clamp is also applied. With these procedures and implements, PAM images of mouse hindlimb are then acquired (Fig. 13.8A),[26] clearly showing the vasculature and muscle fibers. This technology enables the early detection of muscle ischemia injury[44] and studies of hypoxia in the rat femoral artery and skeletal muscle microcirculation,[45] among other subjects.

For cornea imaging, the vessel bed is adjusted perpendicular to the optical axis, guided by preimaging, since the temporal decorrelation method for blood flow quantification is primarily sensitive to transverse blood flow.[23] Thus, PAM enables visualization of individual capillaries throughout both the corneal and iris vasculatures, and the acquired PAM images of the mouse cornea include the hemoglobin content, sO_2, and blood velocity (Fig. 13.8B). The iris (green) and corneal (red) vessels can be separated to distinguish between the two vascular beds (Fig. 13.8C, left) by analyzing individual cross-sectional images (Fig. 13.8C, right). With this procedure, aspects of the vascular metabolism of the cornea can be studied, such as the retinal metabolic rate of oxygen[46] and corneal angiogenesis.[43]

Frontiers in intravital PAM of the microvasculature

Although PAM enables comprehensive quantification of the microvascular function and oxygen metabolism in vivo, further technical advances are required. First, the time-consuming scanning and offline data processing prevents conventional PAM from displaying quantitative images in real time, hindering the study of highly dynamic and time-sensitive biological phenomena. To address this challenge, by using laser scanning rather than conventional mechanical scanning, Wang et al. have developed 300-kHz A-line rate multiparametric PAM,[47] demonstrating high-resolution imaging at a speed 20 times faster than the previous generation. By

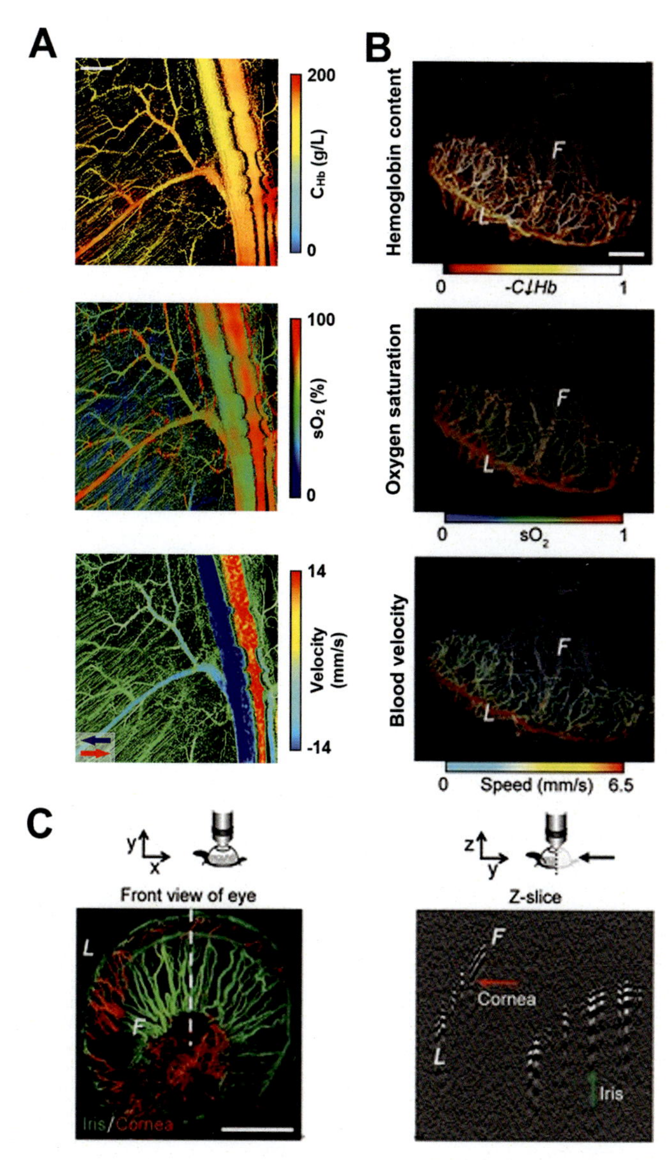

FIGURE 13.8 Intravital PAM imaging of the mouse muscle and cornea. (A) PAM of a mouse hindlimb, showing microvascular C_{Hb}, sO_2, and blood flow velocity. *Red and blue arrow* pairs: flow directions. Scale bars: 500 μm (B) PAM of a stimulated (i.e., chemically burned) mouse cornea, showing relative hemoglobin content, sO_2, and blood velocity in the neovascular network. The iris vasculature, pseudocolored in gray based on vessel depth, is visible in the background of each image and readily separable from the corneal neovessels. Scale bar: 1 mm. (C) PAM distinguishes between the iris (green) and corneal (red) vasculatures, due to their different depths. *Green arrow*: iris vessels; *Red arrow*: corneal neovessels in the z-slice.

using a MEMS mirror, Li et al. have further pushed the volumetric imaging rate of PAM to 2 Hz over a $2.5 \times 2.0 \times 0.5\ mm^3$ volume.[48] By using a cylindrically focused ultrasonic transducer, Zhong et al. have developed a novel PAM system that is 112 times faster than previous systems,[49] capturing functional images over an area of $4.5 \times 1\ mm^2$ at a frame rate of 1 Hz. On the data processing side, Xu et al. have reported a parallel computation design based on a graphics processing unit for high-speed quantification of blood flow in PAM,[50] achieving an 80-fold speed increase over previous algorithms and enabling real-time display of functional images over a large field of view.

Second, although the current PAM imaging platform enables in vivo imaging of a wide range of tissues and organs, further advancements are needed to meet the demands of biomedical studies and clinical translation. Toward this goal, Zhou et al. have developed a handheld PAM system to detect melanoma[51]; Oraevsky et al. have developed a linear array–based handheld photoacoustic system for functional imaging of the breast[52]; and Zemp et al. have developed a real-time PAM system to image the beating heart at a frame rate of ~50 Hz.[53]

A third limitation lies in the contact detection of ultrasonic waves. The current coupling media (e.g., water and ultrasound gel) require additional PAM system components (i.e., a water tank) that limit its application on freely moving subjects or in the clinic. To break this bottleneck, Hajireza et al. have developed a noncontact PAM system that captures in vivo images with a high SNR.[54] Further, Beard et al. have developed a Fabry-Perot sensor-based photoacoustic imaging method that also dispenses with the coupling medium.[55]

Despite the challenges, PAM has become a promising tool in biomedical research for intravital imaging of microvascular function and oxygen metabolism. Its many advantages include high spatial resolution, endogenous image contrast, and considerable tissue penetration. With further development of the instruments and complementary analytical tools, PAM will no doubt become a more versatile and attractive imaging modality.

Conclusion

In this chapter, we highlighted recent experimental studies that use PAM for intravital imaging of microvascular function and oxygen metabolism. PAM's biomedical applications have greatly expanded over the past decades, driven by both technical developments and a growing awareness of potential preclinical or clinical applications. Although challenges remain, PAM is expected to broadly impact both basic research and clinical practice by providing label-free, comprehensive, and quantitative functional-metabolic images.

Acknowledgments

This work is supported in part by the National Institutes of Health (NS099261) and American Heart Association (15SDG25960005) to S.H.

References

1. Nabel EG, Braunwald E. A tale of coronary artery disease and myocardial infarction. *N Engl J Med*. 2012;366(1):54–63. https://doi.org/10.1056/NEJMra1112570.
2. Landmesser U, Drexler H. Chronic heart failure: an overview of conventional treatment versus novel approaches. *Nat Clin Pract Cardiovasc Med*. 2005;2(12):628–638. https://doi.org/10.1038/ncpcardio0371.
3. Varu VN, Hogg ME, Kibbe MR. Critical limb ischemia. *J Vasc Surg*. 2010;51(1):230–241. https://doi.org/10.1016/j.jvs.2009.08.073.
4. Brem H, Tomic-Canic M. Cellular and molecular basis of wound healing in diabetes. *J Clin Invest*. 2007;117(5):1219–1222. https://doi.org/10.1172/JCI32169.
5. Girouard H, Iadecola C. Neurovascular coupling in the normal brain and in hypertension, stroke, and Alzheimer disease. *J Appl Physiol*. 2006;100(1):328–335. https://doi.org/10.1152/japplphysiol.00966.2005.
6. Kallinowski F, Schlenger KH, Runkel S, et al. Blood flow, metabolism, cellular microenvironment, and growth rate of human tumor xenografts. *Cancer Res*. 1989;49(14):3759–3764.
7. Iadecola C. Neurovascular regulation in the normal brain and in Alzheimer's disease. *Nat Rev Neurosci*. 2004;5(5):347–360. https://doi.org/10.1038/nrn1387.
8. Tadros T, Traber DL, Herndon DN. Hepatic blood flow and oxygen consumption after burn and sepsis. *J Trauma & Acute Care Surg*. 2000;49(1):101–108.
9. Cheng K, Ho K, Stokes R, et al. Hypoxia-inducible factor-1α regulates β cell function in mouse and human islets. *J Clin Invest*. 2010; 120(6):2171–2183. https://doi.org/10.1172/JCI35846.
10. Maquet P. Sleep function(s) and cerebral metabolism. *Behav Brain Res*. 1995;69(1):75–83. https://doi.org/10.1016/0166-4328(95)00017-N.
11. Frackowiak RSJ, Lenzi G-L, Jones T, Heather JD. Quantitative measurement of regional cerebral blood flow and oxygen metabolism in man using 15O and positron emission tomography: theory, procedure, and normal values. *J Comput Assist Tomogr*. 1980;4(6): 727–736.
12. Ibaraki M, Shimosegawa E, Miura S, et al. PET measurements of CBF, OEF, and CMRO2 without arterial sampling in hyperacute ischemic stroke: method and error analysis. *Ann Nucl Med*. 2004; 18(1):35. https://doi.org/10.1007/BF02985612.
13. Xu F, Ge Y, Lu H. Noninvasive quantification of whole-brain cerebral metabolic rate of oxygen (CMRO2) by MRI. *Magn Reson Med*. 2009;62(1):141–148. https://doi.org/10.1002/mrm.21994.
14. Zhou C, Choe R, Shah NS, et al. Diffuse optical monitoring of blood flow and oxygenation in human breast cancer during early stages of neoadjuvant chemotherapy. *JBO*. 2007;12(5):051903. https://doi.org/10.1117/1.2798595.
15. Gu Z, Taschereau R, Vu NT, et al. NEMA NU-4 performance evaluation of PETbox4, a high sensitivity dedicated PET preclinical tomograph. *Phys Med Biol*. 2013;58(11):3791–3814. https://doi.org/10.1088/0031-9155/58/11/3791.
16. Jones PB, Shin HK, Boas DA, et al. Simultaneous multispectral reflectance imaging and laser speckle flowmetry of cerebral blood flow and oxygen metabolism in focal cerebral ischemia. *J Biomed Opt*. 2008;13(4):044007. https://doi.org/10.1117/1.2950312.
17. Guevara E, Sadekova N, Girouard H, Lesage F. Optical imaging of resting-state functional connectivity in a novel arterial stiffness model. *Biomed Opt Expr*. 2013;4(11):2332–2346. https://doi.org/10.1364/BOE.4.002332.
18. Dizeux A, Gesnik M, Ahnine H, et al. Functional ultrasound imaging of the brain reveals propagation of task-related brain activity in behaving primates. *Nat Commun*. 2019;10(1):1400. https://doi.org/10.1038/s41467-019-09349-w.
19. Hall AM, Rhodes GJ, Sandoval RM, Corridon PR, Molitoris BA. In vivo multiphoton imaging of mitochondrial structure and function during acute kidney injury. *Kidney Int*. 2013;83(1):72–83. https://doi.org/10.1038/ki.2012.328.
20. Dunn KW, Sandoval RM, Kelly KJ, et al. Functional studies of the kidney of living animals using multicolor two-photon microscopy. *Am J Physiol Cell Physiol*. 2002;283(3):C905–C916. https://doi.org/10.1152/ajpcell.00159.2002.
21. Wang LV, Hu S. Photoacoustic tomography: in vivo imaging from organelles to organs. *Science*. 2012;335(6075):1458–1462. https://doi.org/10.1126/science.1216210.
22. Cao R, Li J, Ning B, et al. Functional and oxygen-metabolic photoacoustic microscopy of the awake mouse brain. *Neuroimage*. 2017; 150:77–87. https://doi.org/10.1016/j.neuroimage.2017.01.049.
23. Ning B, Kennedy MJ, Dixon AJ, et al. Simultaneous photoacoustic microscopy of microvascular anatomy, oxygen saturation, and blood flow. *Opt Lett OL*. 2015;40(6):910–913. https://doi.org/10.1364/OL.40.000910.
24. Zhou Y, Yao J, Maslov KI, Wang LV. Calibration-free absolute quantification of particle concentration by statistical analyses of photoacoustic signals in vivo. *J Biomed Opt*. 2014;19(3). https://doi.org/10.1117/1.JBO.19.3.037001.
25. Yao J, Maslov KI, Shi Y, Taber LA, Wang LV. In vivo photoacoustic imaging of transverse blood flow by using Doppler broadening of bandwidth. *Opt Lett OL*. 2010;35(9):1419–1421. https://doi.org/10.1364/OL.35.001419.
26. Sun N, Ning B, Bruce AC, et al. In vivo imaging of hemodynamic redistribution and arteriogenesis across microvascular network. *Microcirculation*. 2020;27(3):e12598. https://doi.org/10.1111/micc.12598.
27. Sun N, Ning B, Hansson KM, et al. Modified VEGF-A mRNA induces sustained multifaceted microvascular response and accelerates diabetic wound healing. *Sci Rep*. 2018;8(1):17509. https://doi.org/10.1038/s41598-018-35570-6.
28. Yao J, Maslov KI, Zhang Y, Xia Y, Wang LV. Label-free oxygen-metabolic photoacoustic microscopy in vivo. *J Biomed Opt*. 2011; 16(7). https://doi.org/10.1117/1.3594786.
29. Laschke MW, Vollmar B, Menger MD. The dorsal skinfold chamber: window into the dynamic interaction of biomaterials with their surrounding host tissue. *Eur Cell Mater*. 2011;22:147–164. https://doi.org/10.22203/ecm.v022a12. Discussion 164-167.
30. DeGeorge BR, Ning B, Salopek LS, et al. Advanced imaging techniques for investigation of acellular dermal matrix biointegration. *Plast Reconstr Surg*. 2017;139(2):395–405. https://doi.org/10.1097/PRS.0000000000002992.
31. Cottler PS, Sun N, Thuman JM, et al. The biointegration of a porcine acellular dermal matrix in a novel radiated breast reconstruction model. *Ann Plast Surg*. 2020;84(6S Suppl 5):S417–S423. https://doi.org/10.1097/SAP.0000000000002277.
32. Cottler PS, Olenczak JB, Ning B, et al. Fenestration improves acellular dermal matrix biointegration: an investigation of revascularization with photoacoustic microscopy. *Plast Reconstr Surg*. 2019; 143(4):971–981. https://doi.org/10.1097/PRS.0000000000005410.
33. Janis JE, Nahabedian MY. Acellular dermal matrices in surgery. *Plast Reconstr Surg*. 2012;130(5S-2):7S. https://doi.org/10.1097/PRS.0b013e31825f2d20.
34. Lin R, Chen J, Wang H, Yan M, Zheng W, Song L. Longitudinal label-free optical-resolution photoacoustic microscopy of tumor angiogenesis in vivo. *Quant Imag Med Surg*. 2015;5(1):23–29. https://doi.org/10.3978/j.issn.2223-4292.2014.11.08.

35. Raichle ME, Gusnard DA. Appraising the brain's energy budget. *Proc Natl Acad Sci USA*. 2002;99(16):10237–10239. https://doi.org/10.1073/pnas.172399499.
36. Ning B, Sun N, Cao R, et al. Ultrasound-aided multi-parametric photoacoustic microscopy of the mouse brain. *Sci Rep*. 2015;5(1): 18775. https://doi.org/10.1038/srep18775.
37. Yeh C, Soetikno B, Hu S, Maslov KI, Wang LV. Microvascular quantification based on contour-scanning photoacoustic microscopy. *J Biomed Opt*. 2014;19(9):96011. https://doi.org/10.1117/1.JBO.19.9.096011.
38. Yang J, Gong L, Shen Y, Wang LV. Synthetic bessel light needle for extended depth-of-field microscopy. *Appl Phys Lett*. 2018;113(18): 181104. https://doi.org/10.1063/1.5058163.
39. Xu Z, Sun N, Cao R, Li Z, Liu Q, Hu S. Cortex-wide multiparametric photoacoustic microscopy based on real-time contour scanning. *NPh*. 2019;6(3):035012. https://doi.org/10.1117/1.NPh.6.3.035012.
40. Cao R, Li J, Zhang C, Zuo Z, Hu S. Photoacoustic microscopy of obesity-induced cerebrovascular alterations. *Neuroimage*. 2019; 188:369–379. https://doi.org/10.1016/j.neuroimage.2018.12.027.
41. Cao R, Li J, Kharel Y, et al. Photoacoustic microscopy reveals the hemodynamic basis of sphingosine 1-phosphate-induced neuroprotection against ischemic stroke. *Theranostics*. 2018;8(22): 6111–6120. https://doi.org/10.7150/thno.29435.
42. Cao R, Zhang C, Mitkin VV, et al. Comprehensive characterization of cerebrovascular dysfunction in blast traumatic brain injury using photoacoustic microscopy. *J Neurotrauma*. 2019;36(10): 1526–1534. https://doi.org/10.1089/neu.2018.6062.
43. Kelly-Goss MR, Ning B, Bruce AC, et al. Dynamic, heterogeneous endothelial Tie2 expression and capillary blood flow during microvascular remodeling. *Sci Rep*. 2017;7(1):9049. https://doi.org/10.1038/s41598-017-08982-z.
44. Chen L, Ma H, Liu H, et al. Quantitative photoacoustic imaging for early detection of muscle ischemia injury. *Am J Transl Res*. 2017;9(5): 2255–2265.
45. Smith LM, Varagic J, Yamaleyeva LM. Photoacoustic imaging for the detection of hypoxia in the rat femoral artery and skeletal muscle microcirculation. *Shock*. 2016;46(5):527–530. https://doi.org/10.1097/SHK.0000000000000644.
46. Song W, Wei Q, Liu W, et al. A combined method to quantify the retinal metabolic rate of oxygen using photoacoustic ophthalmoscopy and optical coherence tomography. *Sci Rep*. 2014;4(1):6525. https://doi.org/10.1038/srep06525.
47. Wang T, Sun N, Cao R, et al. Multiparametric photoacoustic microscopy of the mouse brain with 300-kHz A-line rate. *Neurophotonics*. 2016;3(4). https://doi.org/10.1117/1.NPh.3.4.045006.
48. Lin L, Zhang P, Xu S, et al. Handheld optical-resolution photoacoustic microscopy. *J Biomed Opt*. 2017;22(4):41002. https://doi.org/10.1117/1.JBO.22.4.041002.
49. Zhong F, Zhong F, Bao Y, et al. High-speed wide-field multiparametric photoacoustic microscopy. *Opt Lett OL*. 2020;45(10): 2756–2759. https://doi.org/10.1364/OL.391824.
50. Xu Z, Wang Y, Sun N, Li Z, Hu S, Liu Q. Parallel computing for quantitative blood flow imaging in photoacoustic microscopy. *Sensors*. 2019;19(18):4000. https://doi.org/10.3390/s19184000.
51. Zhou Y, Xing W, Maslov KI, Cornelius LA, Wang LV. Handheld photoacoustic microscopy to detect melanoma depth in vivo. *Opt Lett*. 2014;39(16):4731–4734.
52. Oraevsky AA, Clingman B, Zalev J, Stavros AT, Yang WT, Parikh JR. Clinical optoacoustic imaging combined with ultrasound for coregistered functional and anatomical mapping of breast tumors. *Photoacoustics*. 2018;12:30–45. https://doi.org/10.1016/j.pacs.2018.08.003.
53. Zemp RJ, Song L, Bitton R, Shung KK, Wang LV. Realtime photoacoustic microscopy of murine cardiovascular dynamics. *Opt Expr OE*. 2008;16(22):18551–18556. https://doi.org/10.1364/OE.16.018551.
54. Hajireza P, Shi W, Bell K, Paproski RJ, Zemp RJ. Non-interferometric photoacoustic remote sensing microscopy. *Light Sci Appl*. 2017;6(6). https://doi.org/10.1038/lsa.2016.278. e16278-e16278.
55. Beard PC, Zhang EZY, Cox BT. Transparent Fabry-Perot polymer film ultrasound array for backward-mode photoacoustic imaging. In: *Photons Plus Ultrasound: Imaging and Sensing*. vol. 5320. International Society for Optics and Photonics; 2004:230–237. https://doi.org/10.1117/12.531332.

CHAPTER

14

Systems biology modeling of endothelial cell and macrophage signaling in angiogenesis in human diseases

Yu Zhang, Chen Zhao and Aleksander S. Popel

Department of Biomedical Engineering, School of Medicine, Johns Hopkins University, Baltimore, MD, United States

Introduction

Angiogenesis, the growth of new blood vessels from preexisting vessels, is an important biological process involved in development, growth, reproduction, and wound healing.[1–6] Angiogenesis is a highly regulated process and is tightly controlled by many cellular and molecular mediators.[7] Tissue experiencing a lack of oxygenation initiates hypoxia response via the hypoxia-inducible factor (HIF) pathway and induce angiogenesis.[8,9] Endothelial cells drive and regulate angiogenesis through various processes including tip-stalk cell selection, migration, proliferation, and their downstream signaling regulating vascular permeability and stability.[10–14] Macrophages are also actively involved during angiogenesis, modulating the immune and inflammatory homeostasis.[15]

Dysregulation of angiogenesis has been widely observed and implicated in the pathogenesis of various human diseases. In cancer, excessive and functionally aberrant (e.g., unorganized, leaky) tumor-associated angiogenesis is critically involved in tumor growth and metastasis.[3] In the eye, pathological neovascularization is a major contributing factor to vision loss in many ocular diseases affecting millions worldwide.[16] In ischemic diseases such as peripheral arterial disease (PAD), myocardial infarction, and cerebral ischemia, deficiency in the formation of stable, nonleaky blood vessels postinjury is a key reason behind persistent tissue damage and lack of natural tissue regeneration.[17–19] Given that angiogenesis is innately a multifactorial, multipathway, and multiscale process, systems biology modeling that mechanistically integrates knowledge across scales and studies can be a particularly useful approach to help understand the complex dynamic interaction and emergent principles that control angiogenesis in health and disease.[4] In this book chapter, we will review and discuss examples of systems biology modeling efforts applied to the integrative and translational investigation of angiogenesis-related intracellular pathways and cell-level dynamics, including computational models developed to study the regulation and signal transduction of quintessential angiogenesis modulators (e.g., HIFs, VEGFs, ANGs, fibroblast growth factors [FGFs]), as well as multipathway mechanistic models of major cell types (e.g., endothelial cells, macrophages) that dynamically drive angiogenesis in physiological and pathological tissue microenvironments.

Computational models of HIF stabilization and HIF-mediated cellular pathways in angiogenesis

One of the most significant drivers of angiogenesis in human diseases is hypoxia, and hypoxia-inducible factors (HIFs) are considered the master regulators in the hypoxic cellular response.[9,20] Functional HIFs are heterodimers composed of two subunits (α and β), while the α subunits (e.g., HIF1α, HIF2α) can undergo extensive oxygen-dependent degradation in normoxia. The biological processes responsible for HIFα degradation and stabilization under different oxygen tensions include numerous biochemical reaction steps with many molecular species interacting in a highly dynamic manner, and that therapeutic manipulation of HIF activity by

The Vasculome
https://doi.org/10.1016/B978-0-12-822546-2.00021-6

targeting these steps/species has been under extensive translational investigation in cancer and ischemic diseases.[21] Below we discuss several computational models that were specifically formulated to investigate different mechanistic aspects in HIF stabilization and HIF-mediated cellular pathways (model information is summarized in Table 14.1).

Regulation of HIF1α by hydroxylases

Interaction between HIFα and intracellular hydroxylases (e.g., PHDs, FIH) is a critical step that directs the subsequent recognition and targeted proteasomal degradation of HIFα by ubiquitin ligase complexes (e.g., VHL). Among a series of models proposed to study HIF stabilization (reviewed in[22]), Qutub et al. developed a mechanistic model that first included data-driven kinetic details of PHD2 association with cellular cofactors such as ascorbate, iron, and 2-oxoglutarate, in addition to oxygen.[33] The model predicted that the pattern of HIF1α response under hypoxia could vary considerably depending on the cofactor composition in the molecular environment, which suggested that the effectiveness of strategies targeting HIF1α hydroxylation may also vary across cell lines. Nguyen et al. further included mechanisms of FIH-mediated HIF1α asparagine hydroxylation and negative intracellular feedback through HIF1-induced PHD synthesis to simulate HIF1α transcriptional activities under physiological and pharmacological perturbations.[22] Their model analysis demonstrated an unexpected observation regarding the differential regulation of HIF1α stabilization and HIF1α transcriptional activity when PHD and FIH are both functionally inhibited, and this seemingly counterintuitive prediction was further confirmed in experiments and was mechanistically explained by a model-based regulatory axis.

HIF1α–ROS interaction

Cellular metabolic products such as reactive oxygen species (ROS) also participate in the regulation of HIF. Studies have shown that ROS is involved in the regulation of PHDs and also FIH through several distinct mechanisms (reviewed in[34]). A kinetics-based computational model that incorporated two possible mechanisms of ROS/HIF1α interaction was developed by Qutub et al. to quantitatively explain some seemingly contradictory data reported in the literature regarding the effect of ROS on HIF1α.[25] The authors further simulated the model under different concentration combinations of five molecular targets with functional ties to ROS and provided a theoretical basis to help understand the differential time-course HIF1α response in cancer versus ischemic conditions.

HIF1α–p53 interaction

The tumor suppressor protein p53, a well-known transcription factor with wide influences in a multitude of cellular processes, is also stabilized under hypoxia and can engage in HIF regulation in highly interactive manners.[35] Two computational models have been developed by researchers from the same group to mechanistically characterize the regulatory cross-talk between p53 and HIF1α pathways.[26,36] Their model stability analyses revealed a potential bifurcation between predominant activation of HIF1α versus p53 based on the severity of hypoxia that can be critical in the dynamic determination of cell fate and proangiogenic activation.

HIF-mediated regulation of pro- and antiangiogenic factors

Hypoxia, through the activation of HIFs, is known to initiate the programmed release of a number of pro- and antiangiogenic factors to shape angiogenesis.[20] Our group has recently formulated two computational models to investigate the coordinated cellular synthesis of VEGF and TSP1, respectively, under the control of hypoxia, cytokine signals, and posttranscriptional regulators (e.g., microRNAs).[23,24] Driven by extensive calibration against experimental data and analyses with respect to clinical pathophysiology, the models have proposed several novel mechanistic explanations for the disrupted balance between pro- and antiangiogenic factors in human diseases (e.g., insufficient induction of VEGF in PAD due to let-7 dysregulation, abnormal suppression of TSP1 production in cancers due to Myc overexpression), in addition to the quantitative model-based evaluation of various pathway-associated therapeutic interventions under pathological conditions.

Computational models of growth factor-mediated signaling pathways regulating angiogenesis

Angiogenesis is tightly regulated by many intracellular growth factor–mediated signal transduction pathways controlling different aspects of the vasculature, including vascular permeability, stability, proliferation, and quiescence.[1,37] The VEGF signaling pathway is a key regulator of vascular permeability, endothelial survival, and migration[38]; the Angiopoietin (Ang)–Tie signaling pathway is a major endothelial regulator of vascular stability and quiescence[39]; and fibroblast growth factor (FGF) is a key molecular promotor of

TABLE 14.1 Summary of selected computational models of major intracellular signaling pathways in angiogenesis, including HIF stabilization and HIF-mediated cellular pathways and growth factor-mediated pathways.

Pathway/axis modeled	Cell type modeled	Modeling method	Summary of model objectives/findings	References
HIF stabilization and HIF-mediated cellular pathways				
Hypoxia–HIF1α	General	Deterministic ODEs	Model showed that variations in intracellular molecular environment can lead to differential patterns of HIF1α expression	33
Hypoxia–HIF1α	General/validated experimentally in HEK293	Deterministic ODEs	Model showed that differential functions of PHD and FIH can result in nonlinear relationship between HIF1α protein stability and its transcriptional activity	22
HIF–microRNA–VEGF	Endothelial cells	Deterministic ODEs	Simulated cellular production of VEGF under the control of hypoxia and microRNAs; proposed a mechanistic explanation for insufficient VEGF in ischemic diseases	23
HIF–TGFβ–microRNA–TSP1	Endothelial cells, fibroblast	Deterministic ODEs	Simulated cellular TSP1 production controlled by hypoxia, TGFβ signals, and microRNAs; assessed strategies to therapeutically modulate TSP1 production under pathological conditions	24
HIF1α–ROS	General	Deterministic ODEs	Mechanistically explained how ROS affects HIF system through opposite mechanisms and the resulting differential response to ROS in tumor and ischemic conditions	25
HIF1α–p53	General	Deterministic ODEs	Simulated reciprocal regulation between HIF1a and p53 under different degrees of hypoxia	26
Growth factor–mediated intracellular signaling pathways				
VEGF–Akt–eNOS VEGF–Erk	Endothelial cells	Deterministic ODEs	Predicted TSP1/CD47 axis as potential target for rescuing VEGF downstream signaling through AKT and ERK	27,28
VEGFR–integrin	Endothelial cells	Deterministic ODEs	Simulated the integrin-VEGFR2 interaction and proposed potential mechanisms of integrin-modulating peptides	50
VEGFR–Neuropilin, VEGF isoforms	Endothelial cells	Deterministic ODEs	Characterized the role of neuropilin-1 and VEGF isoforms in VEGF signaling with applications in proangiogenic and antiangiogenic therapy	51,52,53
Ang–Tie	Endothelial cells	Deterministic ODEs	Simulated intracellular signaling of ANG-Tie pathway; examined VE-PTP's role in regulating Ang2 signaling; predicted synergisms for therapeutics targeting the pathway	64,65,66
FGF2	General	Deterministic PDEs	Simulated receptor binding and complex formation of FGF2	67
FGF–FGFR	General	Deterministic ODEs	Modeled FGF's surface binding, regulation, and downstream signaling to provide mechanistic explanation to its biphasic signaling response	68

Continued

TABLE 14.1 Summary of selected computational models of major intracellular signaling pathways in angiogenesis, including HIF stabilization and HIF-mediated cellular pathways and growth factor-mediated pathways.—cont'd

Pathway/axis modeled	Cell type modeled	Modeling method	Summary of model objectives/findings	References
FGF/VEGF	Endothelial cells	Deterministic ODEs	Characterized the intracellular signaling response by proangiogenic factors VEGF and FGF	29
FGF-2	Myocardium	Deterministic ODEs	Characterized the distribution and retention of FGF-2 following intracoronary administration	30
HGF	General	Deterministic ODEs	Performed in silico testing of various monotherapies and combination therapies targeting the HGF signaling axis	31
DLL–Notch	Endothelial cells	Deterministic ODEs	Simulated endothelial cell responses to different factors in the VEGF–DLL–Notch signaling axis during tip/stalk cell patterning	32

cell proliferation and migration during angiogenesis.[40] Here, we discuss several computational models focusing on selected individual growth factor–mediated signaling pathways (also summarized in Table 14.1).

VEGF

The VEGF signaling pathway consists of the VEGF family of cytokines, VEGF-A, VEGF-B, VEGF-D, VEGF-D, and PlGF,[41,42] and VEGF receptors VEGFR1/2/3.[43] VEGF signaling is regulated by the expression of ligands and receptors,[43] VEGF splicing,[44] coreceptor Neuropilins,[45] integrin signaling,[46] phosphatases,[47] receptor localization,[48] and expression of soluble forms of receptors.[49] Among the many computational models of different aspects of VEGF signaling that have greatly advanced our quantitative understanding of this essential pathway, Bazzazi et al. used a series of intracellular signaling pathway model to study VEGF's downstream signaling through ERK,[27] AKT,[28] and its regulation by TSP-1[27] and integrin[50]; Mac Gabhann et al.'s models explored various aspects of the VEGF signaling pathway, including the coreceptor of VEGFR, neuropilin, and VEGF isoforms[51–53]; and other models have also investigated VEGF's spatial distribution,[54] and the proteolysis of VEGF by matrix metalloproteinases (MMPs).[55,56]

ANG

The ANG–Tie signaling pathway consists of four ligands Ang1–4 and two receptors Tie2 and Tie1.[57] Ang1, the natural agonist of Tie2, is associated with vascular stability and quiescence, whereas Ang2 is considered as a context-dependent agonist of Tie2 and associated with vascular leakage and inflammation.[14,57,58] Ang3–4 are mouse and human interspecies of the ligand with unknown biological function.[59] Tie2's activation and downstream signaling are tightly regulated by many molecular regulatory mechanisms, including the Tie1 binding,[60] dephosphorylation by VE-PTP,[61] receptor extracellular domain shedding,[62,63] and junctional localization.[14] Computational modeling of the Ang–Tie signaling pathway allows for the quantitative investigations on the complex reaction network formed by the ligand–receptor interactions, the receptor dynamic and junctional localization, molecular regulatory mechanisms including VE-PTP and Tie1, and the downstream signaling at the systems level, and performing in silico analysis of molecular therapies that target this signaling pathway.[64–66]

FGF

FGFs and their receptors FGFRs are also key regulators in angiogenesis. FGF axis and its downstream signaling regulate cell proliferation, migration, and survival during angiogenesis.[40] Patel et al. used a partial differential equation-based computational model to simulate FGF's receptor binding and complex formation under fluid flow conditions.[67] Kanodia et al. used an ODE-based computational model to investigate FGF's tertiary surface interactions with its downstream signaling cascade to provide a mechanistic explanation of the biphasic response in ERK activation and Ras-Raf downstream signaling observed in in vitro experiments.[68] Song et al. investigated the intracellular signaling response of endothelial cells in response to proangiogenic factors such as VEGF and FGF using an ODE-based model.[29] Filion et al. used a compartmental ODE-based model to characterize the deposition and retention of FGF-2 after bolus and repetitive intracoronary delivery of FGF-2.[30]

In addition to the signaling pathways discussed above, some representative computational models for other major pathways in angiogenesis, including the hepatocyte growth factor signaling pathway[31] and the DLL–Notch signaling pathway[32] are summarized in Table 14.1.

Computational multipathway models of endothelial cells and macrophages in angiogenesis

Endothelial cells

Endothelial cells are actively involved in the induction and regulation of angiogenesis through a variety of processes, including tip-stalk cell selection and shuffling,[10] proliferation, and migration,[12,69] modulation of vascular integrity by controlling endothelial junctional integrity and formation,[70,71] and regulation of vascular permeability, stability, and quiescence via downstream signaling pathways.[38,72,73]

Perfahl et al.[74] and Walpole et al.[75] used agent-based models to characterize the response of endothelial cells to mechanical stimuli and pericyte coverage during vasculogenesis and angiogenesis. In addition, the various signaling pathways in endothelial cells often have interpathway cross-talk, shared downstream molecular regulatory mechanisms, and interpathway signal competition and redundancy that collectively form a highly complex reaction network.[72] Isolated models that study individual pathways sometimes cannot capture and resolve the network level response. Integrative models that combine multiple endothelial pathways allow for hypothesis testing, simulations, and predictions to be made at the systems level. Endothelial cell receptors such as VEGF receptors, integrin, and cadherin receptors share many downstream signaling nodes and cross-talk mechanisms in angiogenesis. Song et al. used ODE-based network models to quantitatively characterize the interactions between VEGF and FGF and their downstream activation of mitogen-activated protein kinase (MAPK) pathway.[29,76] Bauer et al. developed a Boolean network model including these pathways for in silico investigation of the receptor dynamics, activation, and their effects in the determination of cellular phenotypes, suggesting various molecular targets that are effective at a systems level for antiangiogenic therapy development.[77]

Endothelial regulation of vascular permeability and growth by the VEGF pathway and the ANG–Tie signaling pathway can also form a complex reaction network with various mechanisms of cross-talk and shared downstream signaling. ANG–Tie axis also regulates proinflammatory signaling downstream of TNF[78] (Fig. 14.1). VEGF affects Tie2's signaling by inducing

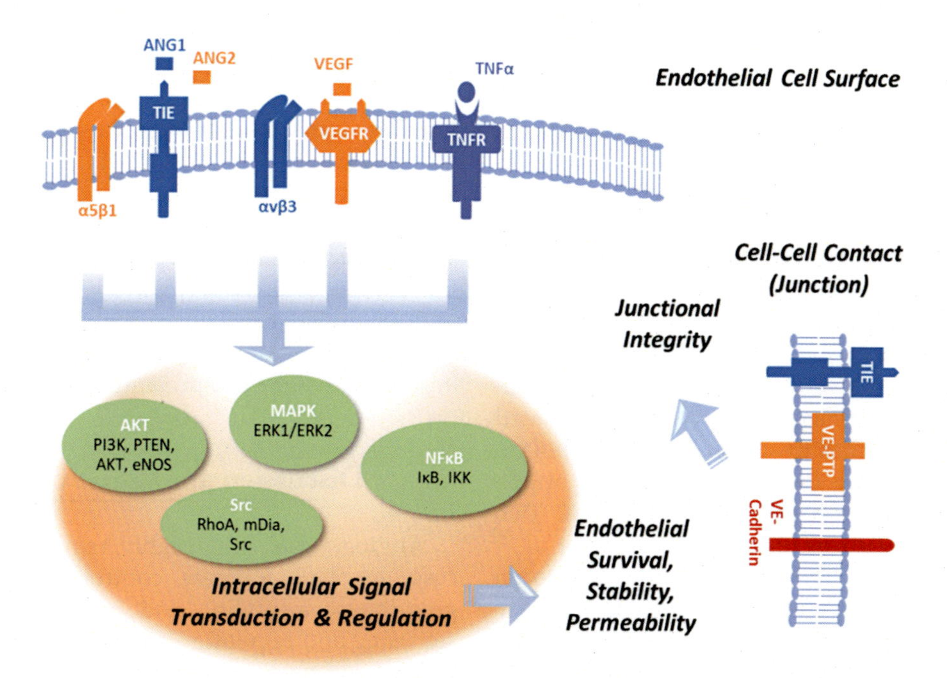

FIGURE 14.1 Overview of the multipathway model of endothelial cell signaling. Endothelial cell signaling regulating vascular growth, stability, permeability, and inflammation consists of multiple growth factor–driven pathways, their downstream signaling, integrin binding, and junctional localization that collectively control the phenotypes of the cell.

the extracellular domain shedding of Tie receptors and inducing the release of ligand Ang2.[62,79,80] ANG-induced activation of Tie2 can reverse VEGF-induced vascular permeability by sequestering Src kinase.[81,82] ANG and Tie are both regulated by integrins[46,83] and share downstream signaling nodes, including PI3K/AKT.[84] Various computational models have been developed for both the VEGF and the Ang signaling pathways (reviewed in the previous section). Integrative models that study these signaling pathways at the same time (Fig. 14.1) will allow for quantitative characterization of potential molecular strategies that can promote the desired states of vascular growth and stability in the treatment of a broad range of angiogenesis-related diseases.

Macrophages

Macrophages, an essential component in immune homeostasis, are versatile cells that are known to participate critically in various biological processes, including angiogenesis.[15] These cells can be flexibly educated by a wide range of signals in the microenvironment to dynamically shape their functional phenotypes, a process commonly referred to as polarization. The full spectrum of macrophage polarization can be described as a multidimensional continuum of highly diverse functional phenotypes with the classical M1-M2 notion describing the two extremes.[85] Within the spectrum, M2-like macrophages have been generally associated with significant antiinflammatory, proangiogenic and protumorigenic potential as revealed by numerous experimental studies in vitro and in vivo (reviewed in[15,86]). For integrative characterization of the underlying dynamics governing macrophage polarization in physiology (e.g., angiogenesis) and thereby exploiting its therapeutic value in human diseases, systems-level computational modeling is an ideal approach given the large number of signaling pathways, cross-talk and feedback mechanisms, phenotype readouts, and functional effector molecules involved.

Following this idea, several systems-level multipathway models of macrophages have been formulated using different approaches to study the general polarization process (including the regulation of the angiogenic potential of macrophages) at the cell level. Two recent models by Palma et al.[87] and Ramirez et al.,[88] respectively, employed a semimechanistic approach based on Boolean regulatory networks to identify macrophage activation patterns under simplified multipathway M1-M2 networks formed by protein–protein interactions. Another two models by Rex et al.[89] and Liu et al.,[90] respectively, used logic-based modeling combined with deterministic ODEs to simulate macrophage M1-M2 marker expression signatures in response to a range of exogenous stimuli. These models enabled the evaluation of the angiogenic potential of macrophages under different cytokine stimulation scenarios by focusing on the expression and activation profiles of macrophage-secreted cytokines with known proangiogenic properties (e.g., VEGF, FGF, MMPs) and also transcription factors broadly related to angiogenesis (e.g., NFκB, AP-1, STAT3).[91] In addition, our group, in two consecutive studies, has developed a large-scale multipathway computational model driven by experimental data and formulated using a mass-action kinetics-based system of ODEs to mechanistically characterize the complex biology and regulation in macrophage polarization and its impact on angiogenesis in health and disease (Fig. 14.2).[92,93] Calibrated and validated against more than 200 sets of quantitative experimental data (containing over 800 datapoints) compiled from literature and focused in-house experiments, our model is the first large-scale system that can predictively simulate, from quantitative, temporal, dose-dependent and single-cell perspectives, both the dynamic expression of an array of macrophage phenotype markers (e.g., iNOS, ARG1, TNFα, IFNγ, interleukins, CXCLs, VEGF) as well as the emergent activities of essential signaling hubs (e.g., MAPKs, PI3K/AKT), intermediate feedback regulators (e.g., SOCS, microRNAs), and key transcription factors (e.g., STATs, IRFs, NFκB, AP-1), together driven by a highly complex multipathway network spanning the M1-M2 spectrum. We created model-based macrophage polarization maps that offer systems-level insights on the dynamic phenotype and function of macrophages under a wide spectrum of polarizing conditions. In addition, we simulated the model under hypoxia serum starvation (HSS), an in vitro condition mimicking PAD, and showed mixed M1 and M2 marker responses which were supported by experimental findings at the level of individual macrophage phenotype markers under HSS.[93,94] In the model sensitivity analysis, we further proposed and examined in silico several targeted strategies that can potentially repolarize macrophages toward more antiinflammatory and proangiogenic phenotypes in HSS.

Discussion and future perspectives

Angiogenesis, as a highly regulated biological process, involves intricate control of vascular stability, growth, permeability, and quiescence. The complex signaling and regulatory networks formed by the multitude of cell types and driving pathways involved have posed a significant challenge in the mechanistic multiscale understanding of angiogenesis in health and disease. Systems biology modeling based on molecular

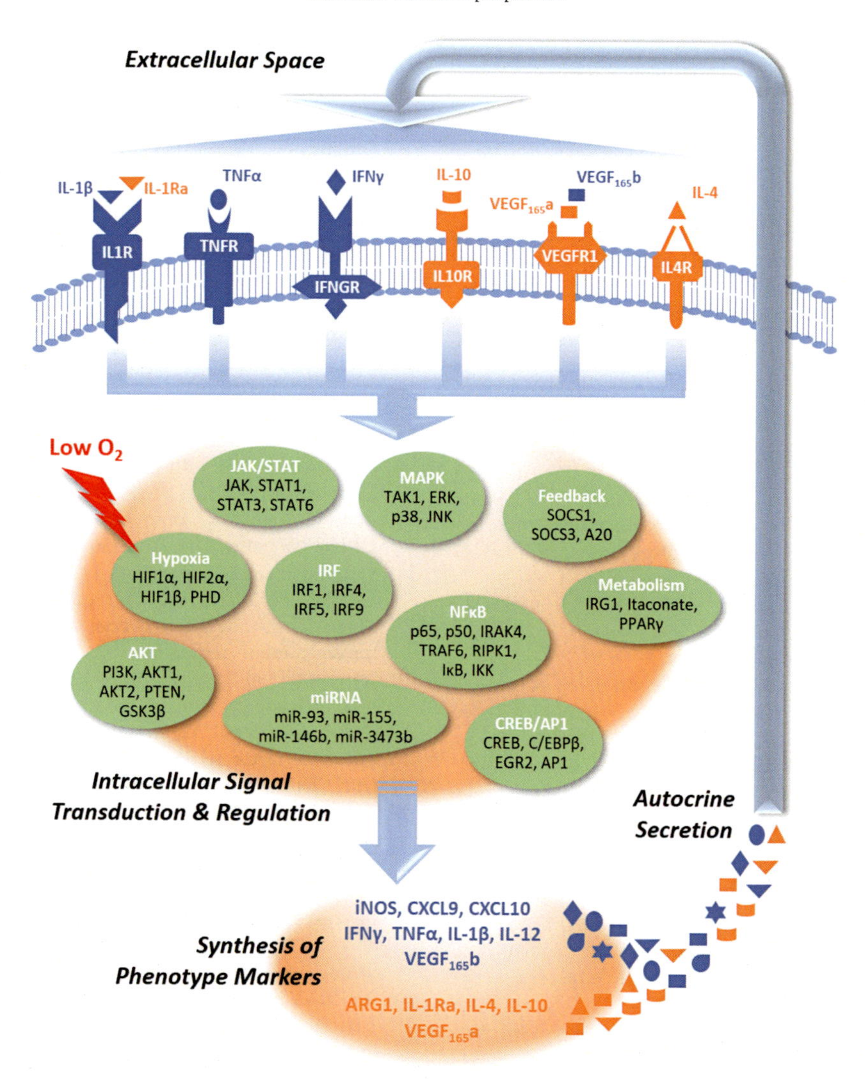

FIGURE 14.2 **Overview of a multipathway model of macrophage polarization.** The effects of different polarizing signals (e.g., cytokines, oxygen tension) are dynamically integrated in macrophages through a complex intracellular network of signal transduction and regulation that then leads to the synthesis and secretion of multiple markers associated with M1 and M2 phenotypes; some of these markers can also initiate continuous autocrine/paracrine signaling in macrophages to further regulate polarization and cell function. *Modified from Zhao C., Medeiros T.X., Sove R.J., Annex B.H., Popel A.S. A data-driven computational model enables integrative and mechanistic characterization of dynamic macrophage polarization. iScience. January 29, 2021;24(2):102112. https://doi.org/10.1016/j.isci.2021.102112.*

and cellular mechanisms is a powerful tool for the quantitative investigation of pathways and cells in angiogenesis. Models at the pathway scale that focus on a specific signaling pathway can provide mechanistic insights into its dynamic activation and nonlinear regulation. Models at the cellular level that integrate multiple pathways can enable quantitative analysis of the emergent properties within complex biochemical networks. Models at the cellular level can be further extended to study cell–cell communications and physiological processes involving multiple cell types; signaling pathway models can be further informed by bioinformatics data such as gene and protein expression data to examine context-specific behaviors.[95] Signaling pathway models and cell-level models can also be used to inform pharmacokinetics to develop mechanistically accurate quantitative systems pharmacokinetics models.[96] In addition to mechanistic modeling, systems biology modeling of angiogenesis can also be informed by bioinformatics. These data-driven models allow for the identification and characterization of the roles of angiogenesis-related genes and proteins in health and diseases at a systems level.[97–99]

In this book chapter, we briefly reviewed selected examples of mechanistic systems biology models of endothelial cell and macrophage signaling, at the pathway and cellular levels, in angiogenesis with implications in human diseases. Such models have enabled quantitative in silico investigation of the mechanistic and therapeutic aspects of signaling pathways and cellular dynamics in angiogenesis, as well as efficient generation of model-driven hypothesis to motivate further experimental studies.

References

1. Potente M, Gerhardt H, Carmeliet P. Basic and therapeutic aspects of angiogenesis. *Cell*. September 16, 2011;146(6):873–887. https://doi.org/10.1016/j.cell.2011.08.039.
2. Otrock ZK, Mahfouz RA, Makarem JA, Shamseddine AI. Understanding the biology of angiogenesis: review of the most important molecular mechanisms. *Blood Cells Mol Dis*. September-October 2007;39(2):212–220. https://doi.org/10.1016/j.bcmd.2007.04.001.
3. Carmeliet P, Jain RK. Principles and mechanisms of vessel normalization for cancer and other angiogenic diseases. *Nat Rev Drug Discov*. June 2011;10(6):417–427. https://doi.org/10.1038/nrd3455.
4. Logsdon EA, Finley SD, Popel AS, Mac Gabhann F. A systems biology view of blood vessel growth and remodelling. *J Cell Mol Med*. August 2014;18(8):1491–1508. https://doi.org/10.1111/jcmm.12164.
5. Fang Y, Kaszuba T, Imoukhuede PI. Systems biology will direct vascular-targeted therapy for obesity. *Front Physiol*. 2020;11:831. https://doi.org/10.3389/fphys.2020.00831.
6. Flegg JA, Menon SN, Byrne HM, McElwain DLS. A current perspective on wound healing and tumour-induced angiogenesis. *Bull Math Biol*. January 22, 2020;82(2):23. https://doi.org/10.1007/s11538-020-00696-0.
7. De Palma M, Biziato D, Petrova TV. Microenvironmental regulation of tumour angiogenesis. *Nat Rev Canc*. August 2017;17(8):457–474. https://doi.org/10.1038/nrc.2017.51.
8. de Heer EC, Jalving M, Harris AL. HIFs, angiogenesis, and metabolism: elusive enemies in breast cancer. *J Clin Invest*. October 1, 2020;130(10):5074–5087. https://doi.org/10.1172/JCI137552.
9. Semenza GL. Hypoxia-inducible factors in physiology and medicine. *Cell*. February 3, 2012;148(3):399–408. https://doi.org/10.1016/j.cell.2012.01.021.
10. Chen W, Xia P, Wang H, et al. The endothelial tip-stalk cell selection and shuffling during angiogenesis. *J Cell Commun Signal*. September 2019;13(3):291–301. https://doi.org/10.1007/s12079-019-00511-z.
11. Szymborska A, Gerhardt H. Hold me, but not too tight-endothelial cell-cell junctions in angiogenesis. *Cold Spring Harb Perspect Biol*. August 1, 2018;10(8). https://doi.org/10.1101/cshperspect.a029223.
12. Lamalice L, Le Boeuf F, Huot J. Endothelial cell migration during angiogenesis. *Circ Res*. March 30, 2007;100(6):782–794. https://doi.org/10.1161/01.RES.0000259593.07661.1e.
13. Gaengel K, Genove G, Armulik A, Betsholtz C. Endothelial-mural cell signaling in vascular development and angiogenesis. *Arterioscler Thromb Vasc Biol*. May 2009;29(5):630–638. https://doi.org/10.1161/ATVBAHA.107.161521.
14. Saharinen P, Eklund L, Miettinen J, et al. Angiopoietins assemble distinct Tie2 signalling complexes in endothelial cell-cell and cell-matrix contacts. *Nat Cell Biol*. May 2008;10(5):527–537. https://doi.org/10.1038/ncb1715.
15. Corliss BA, Azimi MS, Munson JM, Peirce SM, Murfee WL. Macrophages: an inflammatory link between angiogenesis and lymphangiogenesis. *Microcirculation*. February 2016;23(2):95–121. https://doi.org/10.1111/micc.12259.
16. Campochiaro PA. Ocular neovascularization. *J Mol Med (Berl)*. March 2013;91(3):311–321. https://doi.org/10.1007/s00109-013-0993-5.
17. Yoo SY, Kwon SM. Angiogenesis and its therapeutic opportunities. *Mediat Inflamm*. 2013;2013:127170. https://doi.org/10.1155/2013/127170.
18. Iyer SR, Annex BH. Therapeutic angiogenesis for peripheral artery disease: lessons learned in translational science. *JACC Basic Transl Sci*. October 2017;2(5):503–512. https://doi.org/10.1016/j.jacbts.2017.07.012.
19. Dragneva G, Korpisalo P, Yla-Herttuala S. Promoting blood vessel growth in ischemic diseases: challenges in translating preclinical potential into clinical success. *Dis Model Mech*. March 2013;6(2):312–322. https://doi.org/10.1242/dmm.010413.
20. Krock BL, Skuli N, Simon MC. Hypoxia-induced angiogenesis: good and evil. *Genes Cancer*. December 2011;2(12):1117–1133. https://doi.org/10.1177/1947601911423654.
21. Majmundar AJ, Wong WJ, Simon MC. Hypoxia-inducible factors and the response to hypoxic stress. *Mol Cell*. October 22, 2010;40(2):294–309. https://doi.org/10.1016/j.molcel.2010.09.022.
22. Nguyen LK, Cavadas MA, Scholz CC, et al. A dynamic model of the hypoxia-inducible factor 1alpha (HIF-1alpha) network. *J Cell Sci*. March 15, 2013;126(Pt 6):1454–1463. https://doi.org/10.1242/jcs.119974.
23. Zhao C, Popel AS. Computational model of microRNA control of HIF-VEGF pathway: insights into the pathophysiology of ischemic vascular disease and cancer. *PLoS Comput Biol*. November 2015;11(11):e1004612. https://doi.org/10.1371/journal.pcbi.1004612.
24. Zhao C, Isenberg JS, Popel AS. Transcriptional and post-transcriptional regulation of thrombospondin-1 expression: a computational model. *PLoS Comput Biol*. January 2017;13(1):e1005272. https://doi.org/10.1371/journal.pcbi.1005272.
25. Qutub AA, Popel AS. Reactive oxygen species regulate hypoxia-inducible factor 1alpha differentially in cancer and ischemia. *Mol Cell Biol*. August 2008;28(16):5106–5119. https://doi.org/10.1128/MCB.00060-08.
26. Wang P, Guan D, Zhang XP, Liu F, Wang W. Modeling the regulation of p53 activation by HIF-1 upon hypoxia. *FEBS Lett*. September 2019;593(18):2596–2611. https://doi.org/10.1002/1873-3468.13525.
27. Bazzazi H, Isenberg JS, Popel AS. Inhibition of VEGFR2 activation and its downstream signaling to ERK1/2 and calcium by thrombospondin-1 (TSP1): in silico investigation. *Front Physiol*. 2017;8:48. https://doi.org/10.3389/fphys.2017.00048.
28. Bazzazi H, Zhang Y, Jafarnejad M, Isenberg JS, Annex BH, Popel AS. Computer simulation of TSP1 inhibition of VEGF-Akt-eNOS: an angiogenesis triple threat. *Front Physiol*. 2018;9:644. https://doi.org/10.3389/fphys.2018.00644.
29. Song M, Finley SD. ERK and Akt exhibit distinct signaling responses following stimulation by pro-angiogenic factors. *Cell Commun Signal*. July 17, 2020;18(1):114. https://doi.org/10.1186/s12964-020-00595-w.
30. Filion RJ, Popel AS. Intracoronary administration of FGF-2: a computational model of myocardial deposition and retention. *Am J Physiol Heart Circ Physiol*. January 2005;288(1):H263–H279. https://doi.org/10.1152/ajpheart.00205.2004.
31. Jafarnejad M, Sové RJ, Danilova L, et al. Mechanistically detailed systems biology modeling of the HGF/Met pathway in hepatocellular carcinoma. *Npj Syst Biol & Appl*. August 16, 2019;5(1):29. https://doi.org/10.1038/s41540-019-0107-2.
32. Venkatraman L, Regan ER, Bentley K. Time to decide? Dynamical analysis predicts partial tip/stalk patterning states arise during angiogenesis. *PloS One*. 2016;11(11):e0166489. https://doi.org/10.1371/journal.pone.0166489.
33. Qutub AA, Popel AS. A computational model of intracellular oxygen sensing by hypoxia-inducible factor HIF1 alpha. *J Cell Sci*. August 15, 2006;119(Pt 16):3467–3480. https://doi.org/10.1242/jcs.03087.
34. Wong BW, Kuchnio A, Bruning U, Carmeliet P. Emerging novel functions of the oxygen-sensing prolyl hydroxylase domain enzymes. *Trends Biochem Sci*. January 2013;38(1):3–11. https://doi.org/10.1016/j.tibs.2012.10.004.

35. Obacz J, Pastorekova S, Vojtesek B, Hrstka R. Cross-talk between HIF and p53 as mediators of molecular responses to physiological and genotoxic stresses. *Mol Canc*. August 14, 2013;12(1):93. https://doi.org/10.1186/1476-4598-12-93.
36. Zhou CH, Zhang XP, Liu F, Wang W. Modeling the interplay between the HIF-1 and p53 pathways in hypoxia. *Sci Rep*. September 8, 2015;5:13834. https://doi.org/10.1038/srep13834.
37. Adams RH, Alitalo K. Molecular regulation of angiogenesis and lymphangiogenesis. *Nat Rev Mol Cell Biol*. June 2007;8(6): 464–478. https://doi.org/10.1038/nrm2183.
38. Koch S, Claesson-Welsh L. Signal transduction by vascular endothelial growth factor receptors. *Cold Spring Harb Perspect Med*. July 2012;2(7):a006502. https://doi.org/10.1101/cshperspect.a006502.
39. Augustin HG, Koh GY, Thurston G, Alitalo K. Control of vascular morphogenesis and homeostasis through the angiopoietin-Tie system. *Nat Rev Mol Cell Biol*. March 2009;10(3):165–177. https://doi.org/10.1038/nrm2639.
40. Bikfalvi A, Klein S, Pintucci G, Rifkin DB. Biological roles of fibroblast growth factor-2. *Endocr Rev*. February 1997;18(1):26–45. https://doi.org/10.1210/edrv.18.1.0292.
41. Tammela T, Enholm B, Alitalo K, Paavonen K. The biology of vascular endothelial growth factors. *Cardiovasc Res*. February 15, 2005;65(3):550–563. https://doi.org/10.1016/j.cardiores.2004.12.002.
42. Mac Gabhann F, Popel AS. Systems biology of vascular endothelial growth factors. *Microcirculation*. November 2008;15(8):715–738. https://doi.org/10.1080/10739680802095964.
43. Simons M, Gordon E, Claesson-Welsh L. Mechanisms and regulation of endothelial VEGF receptor signalling. *Nat Rev Mol Cell Biol*. October 2016;17(10):611–625. https://doi.org/10.1038/nrm.2016.87.
44. Nowak DG, Woolard J, Amin EM, et al. Expression of pro- and anti-angiogenic isoforms of VEGF is differentially regulated by splicing and growth factors. *J Cell Sci*. October 15, 2008;121(Pt 20):3487–3495. https://doi.org/10.1242/jcs.016410.
45. Staton CA, Kumar I, Reed MW, Brown NJ. Neuropilins in physiological and pathological angiogenesis. *J Pathol*. July 2007;212(3): 237–248. https://doi.org/10.1002/path.2182.
46. Mahabeleshwar GH, Feng W, Reddy K, Plow EF, Byzova TV. Mechanisms of integrin-vascular endothelial growth factor receptor cross-activation in angiogenesis. *Circ Res*. September 14, 2007; 101(6):570–580. https://doi.org/10.1161/CIRCRESAHA.107.155655.
47. Corti F, Simons M. Modulation of VEGF receptor 2 signaling by protein phosphatases. *Pharmacol Res*. January 2017;115:107–123. https://doi.org/10.1016/j.phrs.2016.11.022.
48. Domingues I, Rino J, Demmers JA, de Lanerolle P, Santos SC. VEGFR2 translocates to the nucleus to regulate its own transcription. *PloS One*. 2011;6(9):e25668. https://doi.org/10.1371/journal.pone.0025668.
49. Wu FT, Stefanini MO, Mac Gabhann F, Kontos CD, Annex BH, Popel AS. A systems biology perspective on sVEGFR1: its biological function, pathogenic role and therapeutic use. *J Cell Mol Med*. March 2010;14(3):528–552. https://doi.org/10.1111/j.1582-4934.2009.00941.x.
50. Bazzazi H, Zhang Y, Jafarnejad M, Popel AS. Computational modeling of synergistic interaction between alphaVbeta3 integrin and VEGFR2 in endothelial cells: implications for the mechanism of action of angiogenesis-modulating integrin-binding peptides. *J Theor Biol*. October 14, 2018;455:212–221. https://doi.org/10.1016/j.jtbi.2018.06.029.
51. Mac Gabhann F, Popel AS. Targeting neuropilin-1 to inhibit VEGF signaling in cancer: comparison of therapeutic approaches. *PLoS Comput Biol*. December 29, 2006;2(12):e180. https://doi.org/10.1371/journal.pcbi.0020180.
52. Clegg LE, Ganta VC, Annex BH, Mac Gabhann F. Systems pharmacology of VEGF165b in peripheral artery disease. *CPT Pharmacometr Syst Pharmacol*. December 2017;6(12):833–844. https://doi.org/10.1002/psp4.12261.
53. Clegg LE, Mac Gabhann F. A computational analysis of in vivo VEGFR activation by multiple co-expressed ligands. *PLoS Comput Biol*. March 2017;13(3):e1005445. https://doi.org/10.1371/journal.pcbi.1005445.
54. Mac Gabhann F, Ji JW, Popel AS. Computational model of vascular endothelial growth factor spatial distribution in muscle and pro-angiogenic cell therapy. *PLoS Comput Biol*. September 22, 2006; 2(9):e127. https://doi.org/10.1371/journal.pcbi.0020127.
55. Vempati P, Mac Gabhann F, Popel AS. Quantifying the proteolytic release of extracellular matrix-sequestered VEGF with a computational model. *PloS One*. July 29, 2010;5(7):e11860. https://doi.org/10.1371/journal.pone.0011860.
56. Vempati P, Popel AS, Mac Gabhann F. Extracellular regulation of VEGF: isoforms, proteolysis, and vascular patterning. *Cytokine Growth Factor Rev*. February 2014;25(1):1–19. https://doi.org/10.1016/j.cytogfr.2013.11.002.
57. Eklund L, Saharinen P. Angiopoietin signaling in the vasculature. *Exp Cell Res*. May 15, 2013;319(9):1271–1280. https://doi.org/10.1016/j.yexcr.2013.03.011.
58. Fukuhara S, Sako K, Noda K, Zhang J, Minami M, Mochizuki N. Angiopoietin-1/Tie2 receptor signaling in vascular quiescence and angiogenesis. *Histol Histopathol*. March 2010;25(3):387–396. https://doi.org/10.14670/HH-25.387.
59. Lee HJ, Cho CH, Hwang SJ, et al. Biological characterization of angiopoietin-3 and angiopoietin-4. *FASEB J*. August 2004;18(11): 1200–1208. https://doi.org/10.1096/fj.03-1466com.
60. Mueller SB, Kontos CD. Tie1: an orphan receptor provides context for angiopoietin-2/Tie2 signaling. *J Clin Invest*. September 1, 2016; 126(9):3188–3191. https://doi.org/10.1172/JCI89963.
61. Fachinger G, Deutsch U, Risau W. Functional interaction of vascular endothelial-protein-tyrosine phosphatase with the angiopoietin receptor Tie-2. *Oncogene*. October 21, 1999;18(43): 5948–5953. https://doi.org/10.1038/sj.onc.1202992.
62. Findley CM, Cudmore MJ, Ahmed A, Kontos CD. VEGF induces Tie2 shedding via a phosphoinositide 3-kinase/Akt dependent pathway to modulate Tie2 signaling. *Arterioscler Thromb Vasc Biol*. December 2007;27(12):2619–2626. https://doi.org/10.1161/ATVBAHA.107.150482.
63. Marron MB, Singh H, Tahir TA, et al. Regulated proteolytic processing of Tie1 modulates ligand responsiveness of the receptor-tyrosine kinase Tie2. *J Biol Chem*. October 19, 2007; 282(42):30509–30517. https://doi.org/10.1074/jbc.M702535200.
64. Zhang Y, Kontos CD, Annex BH, Popel AS. Angiopoietin-tie signaling pathway in endothelial cells: a computational model. *iScience*. October 25, 2019;20:497–511. https://doi.org/10.1016/j.isci.2019.10.006.
65. Alawo DOA. *Computational Modelling of the Angiopoietin and Tie Interactions*. University of Leicester; 2017. http://hdl.handle.net/2381/39222.
66. Zhang Y, Kontos CD, Annex BH, Popel AS. A systems biology model of junctional localization and downstream signaling of the Ang–Tie signaling pathway. *npj Syst. Biol Appl*. August 20, 2021; 7(1):34. https://doi.org/10.1038/s41540-021-00194-6.
67. Patel NS, Reisig KV, Clyne AM. A computational model of fibroblast growth factor-2 binding to endothelial cells under fluid flow. *Ann Biomed Eng*. January 2013;41(1):154–171. https://doi.org/10.1007/s10439-012-0622-4.
68. Kanodia J, Chai D, Vollmer J, et al. Deciphering the mechanism behind Fibroblast Growth Factor (FGF) induced biphasic signal-response profiles. *Cell Commun Signal*. May 15, 2014;12:34. https://doi.org/10.1186/1478-811X-12-34.

69. Norton KA, Popel AS. Effects of endothelial cell proliferation and migration rates in a computational model of sprouting angiogenesis. *Sci Rep*. November 14, 2016;6:36992. https://doi.org/10.1038/srep36992.
70. Dejana E, Orsenigo F, Lampugnani MG. The role of adherens junctions and VE-cadherin in the control of vascular permeability. *J Cell Sci*. July 1, 2008;121(Pt 13):2115–2122. https://doi.org/10.1242/jcs.017897.
71. Dejana E, Tournier-Lasserve E, Weinstein BM. The control of vascular integrity by endothelial cell junctions: molecular basis and pathological implications. *Dev Cell*. February 2009;16(2):209–221. https://doi.org/10.1016/j.devcel.2009.01.004.
72. Hofer E, Schweighofer B. Signal transduction induced in endothelial cells by growth factor receptors involved in angiogenesis. *Thromb Haemostasis*. March 2007;97(3):355–363.
73. Claesson-Welsh L, Dejana E, McDonald DM. Permeability of the endothelial barrier: identifying and reconciling controversies. *Trends Mol Med*. December 10, 2020. https://doi.org/10.1016/j.molmed.2020.11.006.
74. Perfahl H, Hughes BD, Alarcon T, et al. 3D hybrid modelling of vascular network formation. *J Theor Biol*. February 7, 2017;414:254–268. https://doi.org/10.1016/j.jtbi.2016.11.013.
75. Walpole J, Mac Gabhann F, Peirce SM, Chappell JC. Agent-based computational model of retinal angiogenesis simulates microvascular network morphology as a function of pericyte coverage. *Microcirculation*. November 2017;24(8). https://doi.org/10.1111/micc.12393.
76. Song M, Finley SD. Mechanistic insight into activation of MAPK signaling by pro-angiogenic factors. *BMC Syst Biol*. December 27, 2018;12(1):145. https://doi.org/10.1186/s12918-018-0668-5.
77. Bauer AL, Jackson TL, Jiang Y, Rohlf T. Receptor cross-talk in angiogenesis: mapping environmental cues to cell phenotype using a stochastic, Boolean signaling network model. *J Theor Biol*. June 7, 2010;264(3):838–846. https://doi.org/10.1016/j.jtbi.2010.03.025.
78. Hughes DP, Marron MB, Brindle NP. The antiinflammatory endothelial tyrosine kinase Tie2 interacts with a novel nuclear factor-kappaB inhibitor ABIN-2. *Circ Res*. April 4, 2003;92(6):630–636. https://doi.org/10.1161/01.RES.0000063422.38690.DC.
79. Fiedler U, Scharpfenecker M, Koidl S, et al. The Tie-2 ligand angiopoietin-2 is stored in and rapidly released upon stimulation from endothelial cell Weibel-Palade bodies. *Blood*. June 1, 2004;103(11):4150–4156. https://doi.org/10.1182/blood-2003-10-3685.
80. Matsushita K, Yamakuchi M, Morrell CN, et al. Vascular endothelial growth factor regulation of Weibel-Palade-body exocytosis. *Blood*. January 1, 2005;105(1):207–214. https://doi.org/10.1182/blood-2004-04-1519.
81. Gavard J, Patel V, Gutkind JS. Angiopoietin-1 prevents VEGF-induced endothelial permeability by sequestering Src through mDia. *Dev Cell*. January 2008;14(1):25–36. https://doi.org/10.1016/j.devcel.2007.10.019.
82. Cascone I, Audero E, Giraudo E, et al. Tie-2-dependent activation of RhoA and Rac1 participates in endothelial cell motility triggered by angiopoietin-1. *Blood*. October 1, 2003;102(7):2482–2490. https://doi.org/10.1182/blood-2003-03-0670.
83. Korhonen EA, Lampinen A, Giri H, et al. Tie1 controls angiopoietin function in vascular remodeling and inflammation. *J Clin Invest*. September 1, 2016;126(9):3495–3510. https://doi.org/10.1172/JCI84923.
84. DeBusk LM, Hallahan DE, Lin PC. Akt is a major angiogenic mediator downstream of the Ang1/Tie2 signaling pathway. *Exp Cell Res*. August 1, 2004;298(1):167–177. https://doi.org/10.1016/j.yexcr.2004.04.013.
85. Sica A, Mantovani A. Macrophage plasticity and polarization: in vivo veritas. *J Clin Invest*. March 2012;122(3):787–795. https://doi.org/10.1172/JCI59643.
86. Chen P, Bonaldo P. Role of macrophage polarization in tumor angiogenesis and vessel normalization: implications for new anticancer therapies. *Int Rev Cell Mol Biol*. 2013;301:1–35. https://doi.org/10.1016/B978-0-12-407704-1.00001-4.
87. Palma A, Jarrah AS, Tieri P, Cesareni G, Castiglione F. Gene regulatory network modeling of macrophage differentiation corroborates the continuum hypothesis of polarization states. *Front Physiol*. 2018;9:1659. https://doi.org/10.3389/fphys.2018.01659.
88. Ramirez R, Herrera AM, Ramirez J, et al. Deriving a Boolean dynamics to reveal macrophage activation with in vitro temporal cytokine expression profiles. *BMC Bioinf*. December 18, 2019;20(1):725. https://doi.org/10.1186/s12859-019-3304-5.
89. Rex J, Albrecht U, Ehlting C, et al. Model-based characterization of inflammatory gene expression patterns of activated macrophages. *PLoS Comput Biol*. July 2016;12(7):e1005018. https://doi.org/10.1371/journal.pcbi.1005018.
90. Liu X, Zhang J, Zeigler AC, Nelson AR, Lindsey ML, Saucerman JJ. Network analysis reveals a distinct axis of macrophage activation in response to conflicting inflammatory cues. *bioRxiv*. 2019:844464. https://doi.org/10.1101/844464.
91. Hamik A, Wang B, Jain MK. Transcriptional regulators of angiogenesis. *Arterioscler Thromb Vasc Biol*. September 2006;26(9):1936–1947. https://doi.org/10.1161/01.ATV.0000232542.42968.e3.
92. Zhao C, Mirando AC, Sove RJ, Medeiros TX, Annex BH, Popel AS. A mechanistic integrative computational model of macrophage polarization: implications in human pathophysiology. *PLoS Comput Biol*. November 2019;15(11):e1007468. https://doi.org/10.1371/journal.pcbi.1007468.
93. Zhao C, Medeiros TX, Sove RJ, Annex BH, Popel AS. A data-driven computational model enables integrative and mechanistic characterization of dynamic macrophage polarization. *iScience*. January 29, 2021;24(2):102112. https://doi.org/10.1016/j.isci.2021.102112.
94. Ganta VC, Choi M, Farber CR, Annex BH. Antiangiogenic VEGF165b regulates macrophage polarization via S100A8/S100A9 in peripheral artery disease. *Circulation*. January 8, 2019;139(2):226–242. https://doi.org/10.1161/CIRCULATIONAHA.118.034165.
95. Papin J, Subramaniam S. Bioinformatics and cellular signaling. *Curr Opin Biotechnol*. February 2004;15(1):78–81. https://doi.org/10.1016/j.copbio.2004.01.003.
96. Chelliah V, Lazarou G, Bhatnagar S, et al. Quantitative systems pharmacology approaches for immuno-oncology: adding virtual patients to the development paradigm. *Clin Pharmacol Ther*. July 19, 2020. https://doi.org/10.1002/cpt.1987.
97. Chu LH, Annex BH, Popel AS. Computational drug repositioning for peripheral arterial disease: prediction of anti-inflammatory and pro-angiogenic therapeutics. *Front Pharmacol*. 2015;6:179. https://doi.org/10.3389/fphar.2015.00179.
98. Chu LH, Rivera CG, Popel AS, Bader JS. Constructing the angiome: a global angiogenesis protein interaction network. *Physiol Genom*. October 2, 2012;44(19):915–924. https://doi.org/10.1152/physiolgenomics.00181.2011.
99. Chu LH, Vijay CG, Annex BH, Bader JS, Popel AS. PADPIN: protein-protein interaction networks of angiogenesis, arteriogenesis, and inflammation in peripheral arterial disease. *Physiol Genom*. August 2015;47(8):331–343. https://doi.org/10.1152/physiolgenomics.00125.2014.

CHAPTER

15

Simulation of blood flow and oxygen transport in vascular networks

Timothy W. Secomb

Department of Physiology and BIO5 Institute, University of Arizona, Tucson, AZ, United States

Introduction

The primary function of the vascular system is to transport materials throughout the body. This is accomplished by the flow of blood, driven by the pulsatile contraction of the heart. In single-cell organisms, materials can be distributed effectively by diffusion, i.e., the random movement of molecules driven by thermal energy, together with active transport processes of cell membranes. However, these processes are ineffective when materials must be transported over larger distances. Such transport is achieved by convection, in which substances are carried by a moving fluid, such as blood in the circulatory system.

Blood contains many substances and suspended elements, including red and white blood cells, platelets, metabolic substrates and waste products, and regulatory molecules. The transport of molecular oxygen is of particular significance and interest, for several reasons.[1] Firstly, the most immediate and severe consequence of a malfunction of the vascular system is lack of oxygen. Secondly, oxygen has a relatively low solubility in water, being a nonpolar molecule, and a specialized oxygen-carrying mechanism (red blood cells [RBCs]) is needed to achieve adequate rates of convective transport. Thirdly, the distance that blood can diffuse into oxygen-consuming tissue is very short, typically tens of μm.[2] This limitation places stringent demands on the vascular system, which has to bring oxygen by convection within a very short distance of all oxygen-consuming cells.

The hierarchical network structure of the circulatory system is crucial to its ability to provide adequate transport of oxygen and other materials. Convective transport over distances larger than about 1 mm is achieved most efficiently by vessels with relatively large diameters, since the power required to overcome viscous resistance to a given flow rate decreases with increasing diameter. On the other hand, a dense network of microscopic vessels is needed to ensure short distances for oxygen diffusion.[3] In most tissues, hierarchically branching arterial and venous trees with a wide range of vessel diameters are intermingled with a dense capillary network that extends throughout the tissue space.

Vascular networks are dynamic systems. The vascular system modulates blood flow acutely in response to changing local metabolic demands when regions of a tissue become more or less active. Regulation of blood flow is achieved mainly by changes in the contraction of vascular smooth muscle in the walls of small arteries and arterioles.[4] Over longer time scales, all vessels undergo structural remodeling of diameter and wall thickness.[5] These processes occur during growth and development, during normal physiological processes such as responses to exercise training, and in response to various disease states. The regulation of blood flow at a specific site requires not only control of local vessel diameters but also control of vessel diameters upstream and downstream along a flow pathway through the tissue.[6,7] The multiple mechanisms that contribute to the coordination of changes in vascular diameters along flow pathways are illustrated in Fig. 15.1. To understand the functioning of the vasculature, including the way in which these mechanisms are integrated to achieve short and long-term control of local blood flow, the structural characteristics of the vascular network must be considered.

Advances in imaging techniques have provided the ability to observe the structure of vascular networks in detail over a wide range of scales, from the major arteries to the capillaries. The ability to measure the distributions of blood flow velocities and substrate concentrations has also advanced, although such

The Vasculome
https://doi.org/10.1016/B978-0-12-822546-2.00007-1

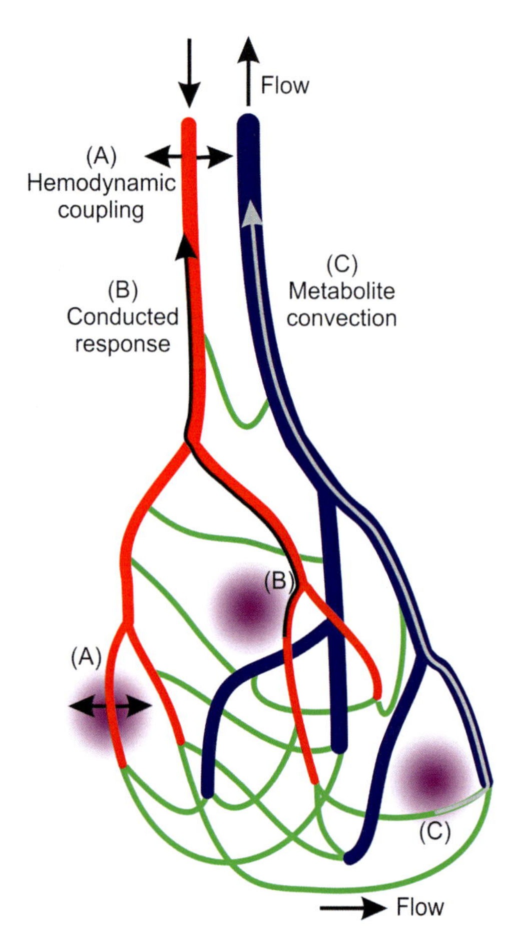

FIGURE 15.1 Interactions among segments in a vascular network, leading to acute or long-term diameter changes. Blood vessel types: red, arterioles; green, capillaries; blue, venules. Purple disks represent tissue regions generating metabolic signals, for instance due to hypoxia. (A) Information transfer by wall shear stress. Dilation of a terminal arteriole causes increased flow, increasing wall shear stress in an upstream arteriole, which may trigger vasodilation or structural increase of diameter. (B) Information transfer by upstream conducted responses. A metabolic signal at a peripheral location triggers a conducted response that travels upstream along vessel walls (long *black arrow*), which may elicit acute vasodilation and/or long-term structural increase of diameter. (C) Information transfer by metabolite convection. A metabolic signal at a peripheral location causes release of a metabolic signaling substance into the blood, which is carried downstream (long *gray arrow*), and may elicit long-term structural increase of diameter. (Venules have little or no smooth muscle, so acute vasodilation is not expected.)

measurements remain challenging, particularly in the smallest vessels. Exchange of materials between blood and tissues mainly occurs in the microcirculation, which includes the arterioles, capillaries, and venules. Flow and the mass transport processes occurring in the microcirculation are difficult to observe directly on microscopic scales. Theoretical approaches provide an alternative approach for studying the microcirculation, and have played an important role in such investigations for more than 100 years.[8]

The goal of this chapter is to provide a brief survey of theoretical models for simulation and analysis of transport processes in the vascular system, with emphasis on blood flow in microvascular networks and oxygen exchange between blood and tissue.

Simulation of blood flow

Characteristics of blood flow in the vasculature

The vascular system consists of an extensively branched network. Diameters range over four orders of magnitude in the human, from a few cm in the major arteries and veins to a few μm in the smallest capillaries. Segment lengths and flow velocities similarly show wide ranges. The dominant physical factors involved in blood flow depend strongly on vessel diameter. These factors are conveniently discussed in terms of the Reynolds number, defined as $\mathrm{Re} = \rho UD/\mu$ for tube flow, where ρ is fluid density, U is fluid velocity, D is tube diameter, and μ is fluid viscosity. When Re is much larger than 1, inertial effects are dominant, whereas when Re is much smaller than 1, effects of viscosity dominate those of inertia. Values of Re range from 10^3 or more in the largest arteries to less than 10^{-3} in the microcirculation.

In the arteries, because inertial effects are dominant, the distribution of flow velocity is sensitively dependent on vascular geometry.[9] Simulation of blood flow in the entrances of blood vessels, in curved or stenosed vessels, and in vessel bifurcations generally requires numerical methods. Computational fluid mechanics techniques for such problems are well developed.[10] Complex flow patterns can cause substantial variations in wall shear stress, i.e., the tangential force on vessel walls resulting from the motion of blood. Endothelial cells lining vessel walls exhibit biological responses to the magnitude and variation of wall shear stress. Such responses may include the development of atherosclerotic lesions.[11] In this inertia-dominated regime, the pressure drop along an artery depends on the blood flow rate, the acceleration of the flow, branching patterns, and variations in vessel diameter. Similar phenomena occur in the veins, with the difference that levels of velocity and wall shear stress are lower because veins have larger diameters than arteries that carry similar flow rates.

Flow at low Reynolds number presents a different set of fluid dynamic phenomena. The dominant effects of fluid viscosity result in smoothly varying velocity fields. Effects of acceleration are negligible, and the pressure drop Δp along a vessel segment is proportional to the volume flow rate Q. For a cylindrical tube with diameter D, this dependence is described by Poiseuille's law:

$$Q = \frac{\pi}{128}\frac{\Delta p \cdot D^4}{L\mu} \tag{15.1}$$

where μ is the viscosity. A homogeneous fluid such as water has a constant viscosity, dependent only on temperature. However, the fact that blood is a concentrated suspension of RBCs leads to more complex behaviors when vessels with diameters below about 0.3 mm are considered.[12] With decreasing vessel diameter, the finite size of RBCs relative to vessel diameter becomes increasingly important. RBCs tend to be distributed nonuniformly within the vessel cross-section, with cell-free or depleted layers near vessel walls. Such behavior causes a reduction in resistance to blood flow, termed the Fåhraeus–Lindqvist effect.[13] Blood flow in vivo is also affected by the presence of a layer of macromolecules lining the interior of vessel walls, termed the glycocalyx or endothelial surface layer. This layer reduces the effective diameter available for blood flow, and increases flow resistance.[14]

The effects of these phenomena on blood flow in microvessels can conveniently be represented by defining the apparent viscosity μ_{app} from Eq. (15.1):

$$\mu_{app} = \frac{\pi}{128} \frac{\Delta p \cdot D^4}{LQ} \tag{15.2}$$

Empirical equations describing the variation of μ_{app} with hematocrit (volume fraction of RBCs) and vessel diameter have been developed.[14,15] The conductance J of a vessel segment of length L and diameter D can then be defined as follows:

$$J = \frac{Q}{\Delta p} = \frac{\pi}{128} \frac{D^4}{L\mu_{app}} \tag{15.3}$$

A further consequence of blood's suspension characteristics is hematocrit partition in bifurcations, an effect also known as phase separation. At diverging bifurcations of microvessels, RBCs tend to preferentially enter the branch with higher flow, as a consequence of their nonuniform distribution in the parent vessel cross-section. This effect, which is compounded through multiple bifurcations, results in heterogeneous hematocrit levels among microvessels. Empirical equations have been developed to describe this effect.[15,16]

Calculation of flow rates in microvascular networks

The results described above provide a basis for predicting blood flow rates in experimentally observed microvascular networks, for which data on the diameters, lengths, and connectivity of all segments have been obtained.[17] Network topology is described by specifying the start and end node of each segment. Fluid exchange between vessels and surrounding tissue is neglected. If the hematocrit is known, then the conductance of each segment can be computed using Eq. (15.3). We consider first the situation where pressures are specified at all boundary nodes where flow enters or leaves the network. By conservation of mass, the net flow into each internal node must be zero. For example, if node 0 is connected to three other nodes 1, 2, and 3 by segments 1, 2, and 3, this condition gives

$$\begin{aligned} Q_1 + Q_2 + Q_3 &= 0 \\ &= J_1(p_1 - p_0) + J_2(p_2 - p_0) + J_3(p_3 - p_0) \end{aligned} \tag{15.4}$$

where Q_1, Q_2, and Q_3 are the flow rates into node 0, J_1, J_2, and J_3 are the segment conductances, and p_0, p_1, p_2, and p_3 are the nodal pressures. This holds at each internal node, yielding a set of linear equations for the nodal pressures. For large networks, this system is sparse and may be solved efficiently by iterative methods. If, instead of pressures, flows are specified at some boundary nodes, the same procedure can be applied with slight modifications.

When effects of hematocrit partition in diverging bifurcations are taken into account, the hematocrit varies among segments and depends on the distribution of flow rates, and this affects segment conductances. The method can be extended to include this phenomenon. In the initial step, a uniform hematocrit is assumed throughout the network, and the flows are calculated as described above. To predict updated values of segment hematocrits, a partial ordering of the nodes is performed, such that any two connected nodes appear in order of decreasing pressure. The empirical equations for phase separation[15] are then applied using this order at each bifurcation with diverging flows. The conductances are recomputed using the updated hematocrits.[15] This procedure is repeated until flows converge. A computer implementation of this algorithm is available at https://github.com/Secomb.

Flow estimation with incomplete boundary conditions

In most tissues, the capillaries form a continuous mesh throughout the tissue volume. Observed network structures in a given region show a large number of segments crossing the region boundary. Experimental determination of pressures or flow rates in all such segments is generally not feasible. To enable estimation of flow rates in such cases, a method was developed[18] in which a priori information about the typical distributions of wall shear stress and intravascular pressure[19,20] is used to set target values for these variables. The sum of squared deviations from these target values is minimized, subject to the constraint of conservation of flows at internal nodes. This approach has been found to yield results approximately consistent with available experimental data obtained from mesentery[18] and brain cortex.[21] A computer implementation is available at https://github.com/Secomb.

Simulation of oxygen transport

Physics of oxygen transport

The basic processes of oxygen transport to tissue have been extensively studied.[1,21–24] Availability of oxygen in blood and tissue is conveniently discussed in terms of the partial pressure of oxygen (P_{O2}), defined as the pressure of oxygen in the gaseous phase that would be in equilibrium with the blood or tissue. The concentration C of oxygen in tissue is given by Henry's law, $C = \alpha P$, where α is the solubility and P is the P_{O2}. Gradients of P in the tissue give rise to diffusive fluxes of oxygen according to Fick's first law of diffusion. Application of the principle of conservation of mass at steady state leads to the following:

$$D\alpha \cdot \nabla^2 P = M(P) \quad (15.5)$$

where D is oxygen diffusivity and ∇^2 is the Laplacian operator. The dependence of oxygen consumption rate $M(P)$ on P_{O2} is represented by a Michaelis–Menten relationship

$$M(P) = M_0 P/(P_0 + P) \quad (15.6)$$

where M_0 is oxygen demand and P_0 is P_{O2} at half-maximal consumption.

Tissue structures are heterogeneous at the microscale, consisting of aqueous and lipid fractions. The solubility and diffusivity of oxygen in lipids are comparable to those quantities in aqueous regions, so that tissues can be treated as homogeneous media with respect to oxygen transport, with effective values of diffusivity and solubility that reflect the contributions of the various tissue components. Vessel walls, including endothelial cells, can be treated as part of this continuum with regard to oxygen transport. This is in contrast to the transport of ions and polar molecules, where cell membranes impede diffusive transport, and the barrier function of the vessel wall must be taken into account.

The convective transport of oxygen is greatly enhanced by the presence of RBCs containing a high concentration of hemoglobin, a molecule with four binding sites for oxygen molecules. The fraction of occupied binding sites, the saturation S, can be represented as a function of blood P_{O2}, P_b, by the Hill equation:

$$S(P_b) = P_b^n / \left(P_b^n + P_{50}^n\right) \quad (15.7)$$

where P_{50} is the P_{O2} at 50% saturation and n is a constant. The rate of convective oxygen transport along a vessel is then given by

$$f(P_b) = Q\left[H_D C_0 S(Pb) + \alpha_{eff} P_b\right] \quad (15.8)$$

where Q is the flow rate, C_0 is the concentration of hemoglobin-bound oxygen in a fully saturated RBC and α_{eff} is the effective solubility of oxygen in blood. The discharge hematocrit H_D gives the volume flux of RBCs as a fraction of the total volume flux.[25] Conservation of mass implies that

$$df(P_b)/ds = -q_v(s) \quad (15.9)$$

where s is distance along the vessel and q_v is the rate of diffusive oxygen efflux per unit vessel length.

The diffusive efflux of oxygen from blood vessels requires radial gradients of P_{O2} within the vessels, such that the P_{O2} at the blood–tissue boundary (P_v) is lower than the mean P_{O2} within the RBCs (P_b). This can be represented approximately by the following:[26]

$$P_v(s) = P_b(s) - Kq_v(s) \quad (15.10)$$

where K represents intravascular resistance to radial oxygen transport and depends on vessel diameter. Hematocrit and flow velocity also influence this effect.[27]

Solution methods

Eq. (15.5) can be solved by various analytical and numerical methods, depending on the form of $M(P)$ and the geometry of the system being considered. For a single vessel, with constant M and other simplifying assumptions, the classic Krogh cylinder model is obtained.[8,24] For more complex geometries, finite-difference methods have been used.[28,29] When networks with hundreds or thousands of segments are considered, such methods become unwieldy, because resolution of steep gradients in P_{O2} around vessels requires fine spatial meshes. To circumvent this problem, methods using properties of the Laplacian operator have been developed.[30–32] Solutions to Eq. (15.5) can be represented as superposition of solutions to the same equation when the right-hand side is replaced by a delta function, which represents a point source or sink of oxygen. This "Green's function" method[22] is outlined here.

Green's function method

The network is represented as a set of segments connecting nodes with specified spatial coordinates. The

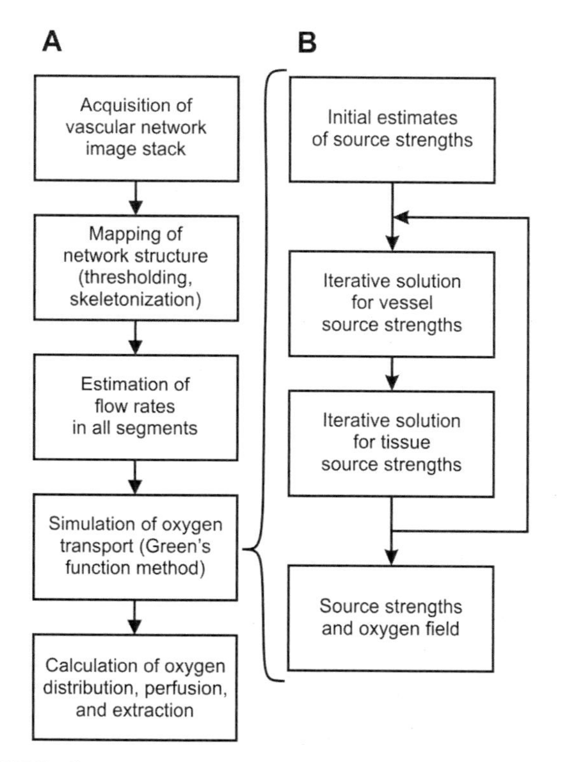

FIGURE 15.2 Overview of methods for simulating blood flow and oxygen transport in microvascular networks. (A) Typical steps in network mapping and simulation. (B) Steps in Green's function method. The procedures for updating vessel and tissue source strengths are applied in an alternating sequence until convergence is achieved within a specified tolerance. See text for details.

flow and hematocrit in each segment and the P_{O2} at each vessel entering the domain must also be specified. The network is subdivided into a set of short segments, each representing a source of oxygen. The tissue space is represented as an array of oxygen sinks at points arranged in a rectangular grid. As a consequence of the properties of the Laplacian operator, a relatively coarse grid of tissue points can be used while maintaining adequate resolution of oxygen gradients.[22] The P_{O2} at each vessel or tissue point is represented as a linear combination of the source and sink strengths, using the Green's function for a point source in an infinite domain. From Eq. (15.9), the convective oxygen flux in each segment is represented as a linear function of the vessel source strengths, and the corresponding vessel P_{O2} is given by inverting Eq. (15.8).

A two-level iterative procedure is then used to compute the unknown source and sink strengths (Fig. 15.2). Based on initial estimates of the source and sink strengths, the tissue P_{O2} values adjacent to each segment are computed. A linearized form of the governing equations is solved for updated estimates of the vessel source strengths, so as to achieve improved matching between these P_{O2} values and those obtained from Eq. (15.10). This "inner" iteration is repeated a specified number of times. Next, the P_{O2} values at each tissue point are computed by the Green's function method, and the sink strengths are recomputed using Eq. (15.6). This second "inner" iteration is also repeated a specified number of times. The whole process is then repeated until convergence of all unknowns is achieved to a specified tolerance. This typically takes between 20 and 100 "outer" iterations. Finally, the tissue oxygen field is computed to yield statistics of tissue and vessel P_{O2}, and graphical representations such as contour plots.

Estimation of oxygen boundary conditions

Mapped vascular network structures typically have multiple inflowing vessels on their boundaries, apart from the main feeding vessels. The Green's function method as described above requires specification of oxygen levels in these inflowing vessels. Utilization of fixed estimates for these parameters leads to inconsistent behavior when effects of varying blood flow or oxygen consumption rates are simulated, because such variations would affect P_{O2} levels in inflowing vessels. A method to overcome this challenge has recently been developed.[21] In this method, all segments crossing the region boundary are classified as arterioles, capillaries, and venules. The flow-weighted mean oxygen concentration in outflowing capillaries is used to set the oxygen level in inflowing capillaries, and oxygen levels in inflowing venules are specified similarly based on levels in outflowing venules. This approach has the additional advantage of allowing estimation of tissue perfusion and oxygen extraction without confounding effects due to dependence on domain size.[33] A computer implementation is available at https://github.com/Secomb.

An example of the results obtained by the application of the methods described above is shown in Fig. 15.3. In this network observed in the mouse cerebral cortex,[21,34] the pial vessels run across the upper surface of the domain, with penetrating arterioles and venules diving through the thickness of the cortex and feeding a dense network of capillaries. Numerous vessels cross the lateral and lower boundaries of the domain. This simulation predicts values of parameters such as tissue perfusion and oxygen extraction that agree well with experimental findings and allows exploration of the effects of varying blood flow and oxygen demand on the distribution of tissue oxygen levels.[21]

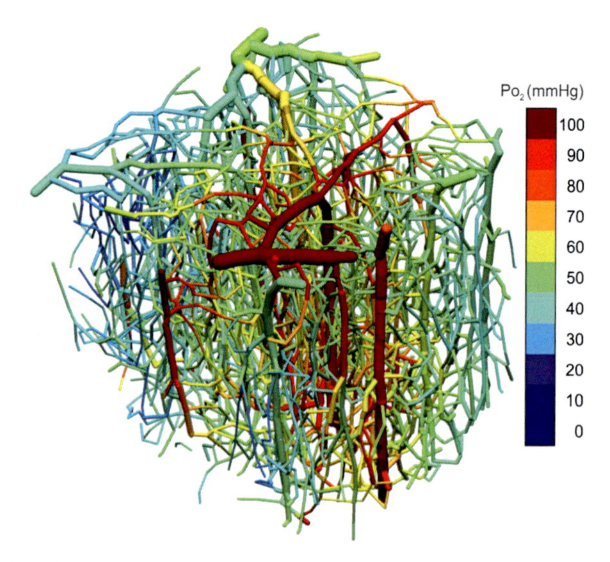

FIGURE 15.3 Computer-generated image of a three-dimensional microvascular network in a region approximately 600 × 600 × 660 μm of mouse cerebral cortex.[34] Image stack of network was obtained by two-photon microscopy of fluorescently labeled vessels. The derived network contains 4126 nodes and 4908 segments, including 208 terminal segments. Vessels are color coded according to predicted vessel P_{O2}, obtained using the methods outlined in the text.

Discussion

As implied by the "Vasculome" concept, the proper functioning of the vasculature depends on the integrated effects of numerous mechanisms acting on a range of physical scales. Because the vasculature consists of an extensive network of interconnected vessels, the delivery of blood flow to any point in the tissue is dependent on all the vessels forming a flow pathway from the heart to and from that point. The inverse fourth-power dependence of flow resistance on diameter according to Poiseuille's law implies the need for tight control of vessel diameters in segments that may be remote from the site being supplied. Understanding these phenomena requires a network-oriented approach to analyzing vascular function.

Direct experimental observation of transport and regulatory processes in the circulatory system is challenging, particularly in the microcirculation. Theoretical models can be used to predict aspects of vascular function that are not directly accessible experimentally and to provide quantitative tests of hypothesized biological mechanisms. Such models are most useful when integrated with available experimental information about the system. The theoretical approaches described in this chapter are intended to contribute to progress in understanding the normal function of the vasculature based on data obtained from advanced imaging methods, as well as providing a basis for understanding the pathological consequences of vascular dysfunction.[35]

Acknowledgments

This work was supported by NIH Grants R01 HL034555 and U01 HL136662.

References

1. Popel AS. Theory of oxygen transport to tissue. *Crit Rev Biomed Eng*. 1989;17(3):257–321.
2. Secomb TW, Hsu R, Dewhirst MW, Klitzman B, Gross JF. Analysis of oxygen transport to tumor tissue by microvascular networks. *Int J Radiat Oncol Biol Phys*. 1993;25(3):481–489.
3. Secomb TW, Alberding JP, Hsu R, Dewhirst MW, Pries AR. Angiogenesis: an adaptive dynamic biological patterning problem. *PLoS Comput Biol*. 2013;9(3):e1002983.
4. Duling BR, Hogan RD, Langille BL, et al. Vasomotor control: functional hyperemia and beyond. *Fed Proc*. 1987;46(2):251–263.
5. Pries AR, Reglin B, Secomb TW. Remodeling of blood vessels: responses of diameter and wall thickness to hemodynamic and metabolic stimuli. *Hypertension*. 2005;46(4):725–731.
6. Segal SS, Damon DN, Duling BR. Propagation of vasomotor responses coordinates arteriolar resistances. *Am J Physiol*. 1989; 256(3 Pt 2):H832–H837.
7. Secomb TW, Pries AR. Information transfer in microvascular networks. *Microcirculation*. 2002;9(5):377–387.
8. Krogh A. The number and the distribution of capillaries in muscle with the calculation of the oxygen pressure necessary for supplying the tissue. *J Physiol (Lond)*. 1919;52:409–515.
9. Ku DN. Blood flow in arteries. *Annu Rev Fluid Mech*. 1997;29: 399–434.
10. Taylor CA, Draney MT. Experimental and computational methods in cardiovascular fluid mechanics. *Annu Rev Fluid Mech*. 2004;36: 197–231.
11. Caro CG. Discovery of the role of wall shear in atherosclerosis. *Arterioscler Thromb Vasc Biol*. 2009;29(2):158–161.
12. Secomb TW. Blood flow in the microcirculation. *Annu Rev Fluid Mech*. 2017;49:443–461.
13. Pries AR, Neuhaus D, Gaehtgens P. Blood viscosity in tube flow: dependence on diameter and hematocrit. *Am J Physiol*. 1992;263: H1770–H1778.
14. Pries AR, Secomb TW, Gessner T, Sperandio MB, Gross JF, Gaehtgens P. Resistance to blood flow in microvessels in vivo. *Circ Res*. 1994;75(5):904–915.
15. Pries AR, Secomb TW. Microvascular blood viscosity in vivo and the endothelial surface layer. *Am J Physiol Heart Circ Physiol*. 2005;289(6):H2657–H2664.
16. Pries AR, Ley K, Claassen M, Gaehtgens P. Red cell distribution at microvascular bifurcations. *Microvasc Res*. 1989;38(1):81–101.
17. Pries AR, Secomb TW, Gaehtgens P, Gross JF. Blood flow in microvascular networks. Experiments and simulation. *Circ Res*. 1990; 67(4):826–834.
18. Fry BC, Lee J, Smith NP, Secomb TW. Estimation of blood flow rates in large microvascular networks. *Microcirculation*. 2012;19(6):530–538.
19. Pries AR, Secomb TW, Gaehtgens P. Structure and hemodynamics of microvascular networks: heterogeneity and correlations. *Am J Physiol*. 1995;269(5 Pt 2):H1713–H1722.
20. Pries AR, Secomb TW, Gaehtgens P. Design principles of vascular beds. *Circ Res*. 1995;77(5):1017–1023.
21. Celaya-Alcala JT, Lee GV, Smith AF, et al. Simulation of oxygen transport and estimation of tissue perfusion in extensive microvascular networks: application to cerebral cortex. *J Cerebr Blood Flow Metabol*. 2021;41(3):656–669.
22. Secomb TW, Hsu R, Park EY, Dewhirst MW. Green's function methods for analysis of oxygen delivery to tissue by microvascular networks. *Ann Biomed Eng*. 2004;32(11):1519–1529.

23. Goldman D. Theoretical models of microvascular oxygen transport to tissue. *Microcirculation*. 2008;15(8):795–811.
24. Middleman S. *Transport Phenomena in the Cardiovascular System*. New York: John Wiley & Sons; 1972.
25. Albrecht KH, Gaehtgens P, Pries A, Heuser M. The Fahraeus effect in narrow capillaries (i.d. 3.3 to 11.0 micron). *Microvasc Res*. 1979; 18(1):33–47.
26. Hellums JD, Nair PK, Huang NS, Ohshima N. Simulation of intraluminal gas transport processes in the microcirculation. *Ann Biomed Eng*. 1996;24(1):1–24.
27. Lucker A, Secomb TW, Weber B, Jenny P. The relative influence of hematocrit and red blood cell velocity on oxygen transport from capillaries to tissue. *Microcirculation*. 2017;24(3).
28. Beard DA, Bassingthwaighte JB. Modeling advection and diffusion of oxygen in complex vascular networks. *Ann Biomed Eng*. 2001; 29(4):298–310.
29. Goldman D, Popel AS. A computational study of the effect of capillary network anastomoses and tortuosity on oxygen transport. *J Theor Biol*. 2000;206(2):181–194.
30. Hsu R, Secomb TW. A Green's function method for analysis of oxygen delivery to tissue by microvascular networks. *Math Biosci*. 1989;96(1):61–78.
31. Groebe K. A versatile model of steady state O_2 supply to tissue. Application to skeletal muscle. *Biophys J*. 1990;57(3):485–498.
32. Hoofd L. Updating the Krogh model - assumptions and extensions. In: Egginton S, Ross HF, eds. *Oxygen Transport in Biological Systems: Modelling of Pathways from Environment to Cell*. Cambridge University Press; 1992:197–229.
33. Guibert R, Fonta C, Esteve F, Plouraboue F. On the normalization of cerebral blood flow. *J Cerebr Blood Flow Metabol*. 2013;33(5): 669–672.
34. Gagnon L, Sakadzic S, Lesage F, et al. Quantifying the microvascular origin of BOLD-fMRI from first principles with two-photon microscopy and an oxygen-sensitive nanoprobe. *J Neurosci*. 2015; 35(8):3663–3675.
35. Roy TK, Secomb TW. Effects of impaired microvascular flow regulation on metabolism-perfusion matching and organ function. *Microcirculation*. 2021;28, e12673.

CHAPTER

16

Clinical investigations of vascular function

Arshed A. Quyyumi, Zakaria Almuwaqqat and Shabatun J. Islam
Emory Clinical Cardiovascular Research Institute, Division of Cardiology, Department of Medicine, Emory University School of Medicine, Atlanta, Georgia

Vascular endothelial function

Endothelial function

The inner lining of the vascular wall called the endothelium is composed of a single layer of cells that constitute one of the largest organs in the human body. The endothelium is a physiologically dynamic organ that plays key roles in regulating the blood and lymph circulation and coordinating a response by a variety of endogenous agonists synthesized by the normal endothelium to various hemodynamic and bioactive signals such as shear stress and temperature changes. The healthy endothelium releases a variety of vasoactive substances including nitric oxide (NO), endothelin, prostacyclin, endothelium-derived hyperpolarizing factors (EDHFs) and carbon monoxide (CO) that regulate vasomotor tone via their paracrine effects on the vascular smooth muscle. In addition, by tonic release of some of these factors, the healthy endothelium promotes normal blood flow, inhibits clotting and platelet aggregation, and reduces adhesion of inflammatory cells to the vessel wall. When the endothelium is damaged or denuded, it is repaired and regenerated by local and circulating endothelial progenitor cells (EPCs) that originate in the bone marrow.

A loss of NO bioavailability, primarily due to increased oxidative stress, is the hallmark of dysfunctional endothelium. Due to injurious exposures to cardiometabolic risk factors and possible contribution by genetic factors, the protective characteristics of the healthy endothelium may be eliminated to varying extent. The resulting endothelial dysfunction is characterized by an imbalance between endogenous agents with vasoconstricting, prothrombotic, and proliferative characteristics and agents with vasodilating, antimitogenic, and antithrombogenic properties.[13]

Although endothelial dysfunction correlates with the cumulative risk factor burden overall, its magnitude varies in individuals with relatively similar risk factor profiles. This variability may be related to the severity and duration of risk factor exposure, the robustness of individual regenerative capacity, as well as other as yet unidentified risk factors and inherent genetic differences.[14–18]

Mediators of endothelial function

Nitric oxide (NO)

NO is a key endothelium-derived vasodilator that is endogenously synthesized as soluble gas by the constitutive action of endothelial NO synthase (eNOS) that utilizes L-arginine as a substrate Fig. 16.1.[19] eNOS normally synthesizes NO in a process that involves L-arginine oxidation, transfer of an electron to the ferrous–dioxygen complex in the oxygenase domain of tetrahydrobiopterin, leading to the formation of an iron–oxy species and conversion of tetrahydrobiopterin to 7,8-dihydrobiopterin.[20] Free radicals such as superoxide anions avidly interact with free NO to form peroxynitrite that can interact with other compounds to generate reactive oxygen species. These reactive oxygen species precipitate endothelial dysfunction as they oxidize NO to nitrite and nitrate, also uncouple eNOS leading to further production of reactive oxygen species.

Agonists such as acetylcholine, bradykinin, catecholamines, ischemia, temperature changes, and shear stress stimulate NO synthesis in the normal endothelium leading to vasodilation. NO diffuses into the vascular smooth muscle layer, activates intracellular guanylate cyclase, and leads to smooth muscle relation by a cyclic GMP-dependent mechanism, a pathway that is shared

The Vasculome
https://doi.org/10.1016/B978-0-12-822546-2.00019-8

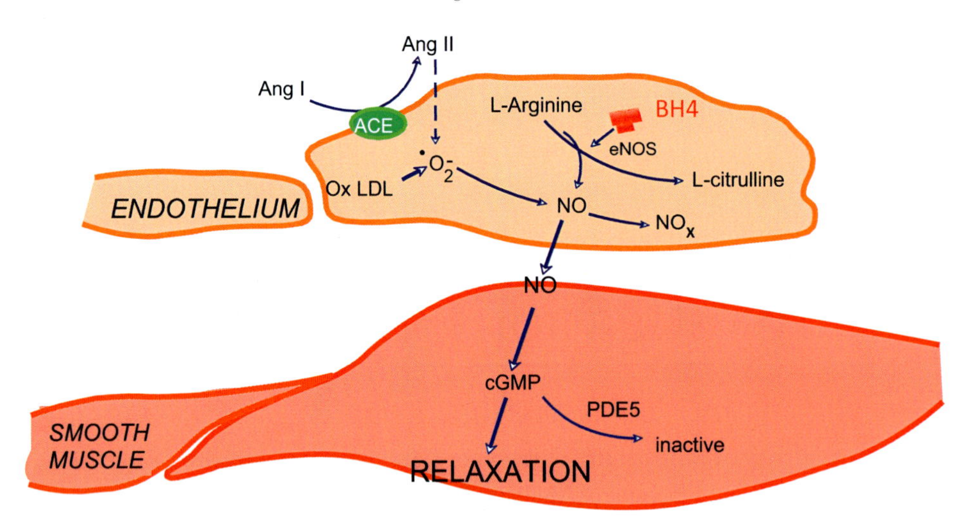

FIGURE 16.1 **Nitric oxide–stimulated smooth muscle relaxation.** NO is a soluble gas synthesized by endothelial NO synthase (eNOS) from L-arginine, a process which requires the cofactor tetrahydrobiopterin (BH4) for catalytic activity. Endothelial NOS is normally present as a dimer that acts in a "coupled" fashion to generate NO. Lack of tetrahydrobiopterin leads to the generation of superoxide anions (O_2–) and peroxynitrite (NO_x), resulting in a state of increased oxidative stress. This phenomenon is referred to as eNOS "uncoupling." Oxidative stress is also stimulated by angiotensin II. *ACE*, angiotensin-converting enzyme; *Ang I*, angiotensin I; *Ang II*, angiotensin II; *BH4*, tetrahydrobiopterin; *cGMP*, cyclic guanosine monophosphate; *eNOS*, endothelial nitric oxide synthase; *NO*, nitric oxide; NO_x, peroxynitrite; O_2–, superoxide anions; *oxLDL*, oxidized low density lipoprotein; *PDE5*, phosphodiesterase type 5. *Reprint permission Quyyumi A.A., Dakak N., Mulcahy D., et al. Nitric oxide activity in the atherosclerotic human coronary circulation. J Am Coll Cardiol. 1997;29(2):308–317.*

by exogenously administered nitrates. Thus, impairment of NO bioavailability can contribute to vasospasm, hypertension, and ischemia, as well as long-term remodeling in the vascular beds.

In addition to its immediate effects on the vascular smooth muscle cells (VSMCs), reduced NO bioavailability over the long term can trigger nuclear factor kappa B transcription factor (NF-κB)–dependent activation of adhesion molecules that incite vascular inflammation and atherosclerosis. Additionally, lack of NO triggers platelets and coagulation activation, increasing the propensity to a prothrombotic state.[21] Since NO assists with EPC release from the bone marrow, lack of NO could also lead to reduction in the number and functionality of EPCs.[22–24] As a result, decreased NO availability is considered as a key barometer of endothelial dysfunction.[25–29]

Prostaglandins, endothelium-derived hyperpolarizing factor, and other endothelium-derived vasoactive agents

The three main endothelial vasodilators, prostacyclin, NO, and EDHF act synergistically. For example, prostaglandins interact with the NO pathway and NO modulates cyclooxygenase activity and expression. Activation of the endothelial cyclooxygenase pathway leads to the generation of metabolites derived from arachidonic acid, including prostacyclin, a prostaglandin with vasodilatory and antiplatelet aggregating properties, and also thromboxane A2 with vasoconstrictor and proaggregatory properties (Fig. 16.2).[30] Prostacyclin induces vascular smooth muscle relaxation by calcium desensitization, independently of NO,[31] and contributes to fibrinolysis and inhibition of platelet aggregation and adhesion.[32]

An NO- and prostaglandin-independent vasodilator mechanism that primarily utilizes increasing potassium conductance, particularly in microcirculation, is mediated by EDHFs.[33] Endothelium-dependent hyperpolarization leads to depolarization of the VSMCs and its relaxation as a result of increased potassium conductance.[34–36] In conduit arteries, NO is the main mediator for endothelium-derived vasodilation, whereas in the microcirculation, EDHF appears to predominate.[37] EDHF may compensate for deficiencies in NO production in hypertension[38–40] and it is unclear whether EDHF plays a causal or compensatory role in endothelial dysfunction.

Other endothelial-derived mediators include endothelin-1 and carbon monoxide. Endothelin-1 is a known potent endothelium-derived endogenous vasoconstrictor agent that is produced and released primarily by endothelial cells.[41] Endothelin-1, the predominant isoform produced and released primarily by endothelial cells, manifests a profound vasoconstrictor effect, stimulates proliferation and hypertrophy of VSMCs, and promotes cell adhesion and thrombosis.[41,42] Endothelin-1 is released in response to physical and chemical stimuli such as stretch, shear stress and pH, and is upregulated with exercise, hypoxia, elevated levels of oxidized low density lipoprotein and glucose, estrogen deficiency, obesity, cocaine use, aging, and thrombin. Endothelin-1 causes vasoconstriction via stimulation of ET-A receptors and vasodilation as a result of activation of ET-B receptors on VSMCs.

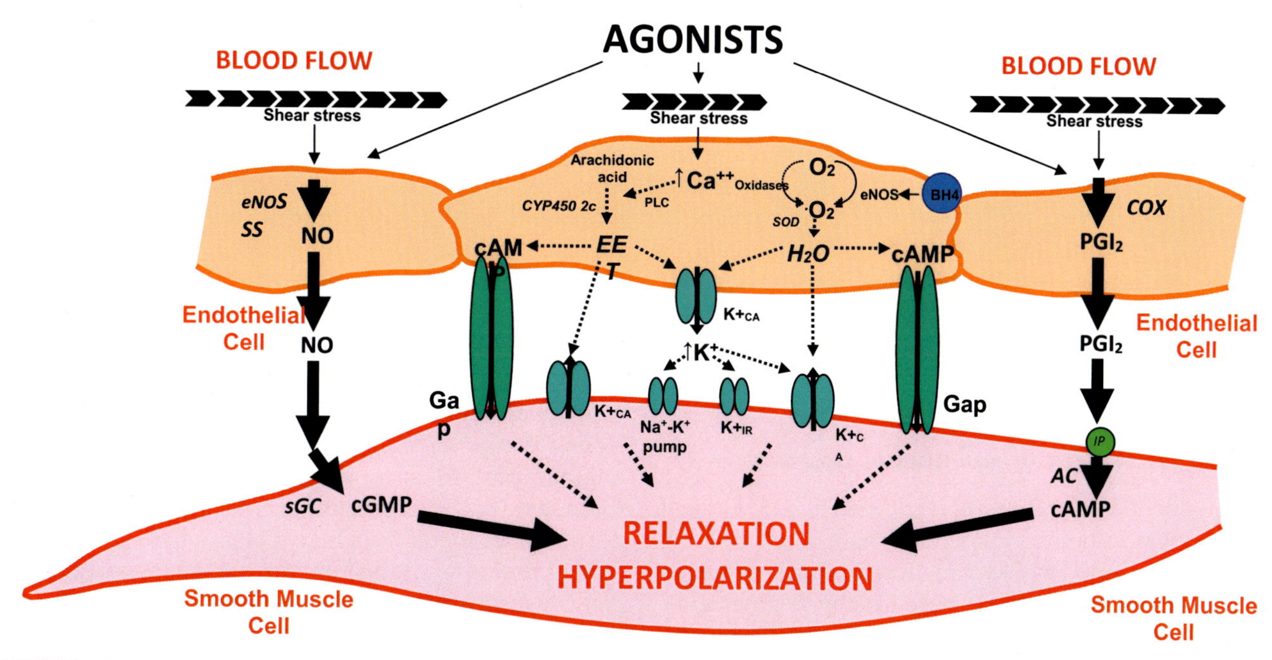

FIGURE 16.2 **Mechanisms underlying endothelial cell–mediated relaxation.** Agonists (bradykinin/acetylcholine/substance P) or shear stress increase the activity of endothelial NO synthase (eNOS) and cyclooxygenase (COX), providing nitric oxide (NO) and prostacyclin (PGI_2)-mediated dilation. There are multiple potential EDHF pathways. Increases in intracellular calcium activate phospholipase A2 (PLC) to produce arachidonic acid. Its metabolism by cytochrome P450 2C (CYP4502c) generates eicosatrienoic acids (EETs) that can stimulate calcium-dependent potassium (K_{Ca}) channels in endothelial and smooth muscle cells. EETs may also directly activate gap junctions (Gap). The increase in potassium in the interstitium may activate K_{Ca} channels, inwardly rectifying potassium channels (K^+_{IR}), or the Na^+-K^+ pump on smooth muscle cells and cause hyperpolarization. The action of eNOS (with cofactor tetrahydrobiopterin [BH4]) and oxidases on oxygen (O_2) produce the reactive oxygen species superoxide (O_2-). Hydrogen peroxide (H_2O_2) generated by dismutation of superoxide anions by superoxide dismutase (SOD) can also cause hyperpolarization by activating endothelial and smooth muscle K_{Ca} channels or by gap junctions. *AC*, Adenylyl cyclase; *cAMP*, cyclic Adenosine monophosphate; *cGMP*, cyclic guanosine monophosphate; *IP*, prostacyclin receptor; *sGC*, soluble guanylyl cyclase. *Reprint permission Quyyumi A.A., Dakak N., Mulcahy D., et al. Nitric oxide activity in the atherosclerotic human coronary circulation. J Am Coll Cardiol. 1997; 29(2):308–317.*

Carbon monoxide is a soluble gas with vasodilating properties in certain vascular beds and is synthesized endogenously by the breakdown of heme by the enzyme heme oxygenase and heme oxygenase genetic polymorphisms are linked to atherosclerosis.[43–46]

Other biomarkers of endothelial function: oxidative stress, ADMA, and endothelial microparticles

Abnormalities of endothelial function may result in changes to circulating levels of certain products of endothelial cells. These include measures of NO biology, measures of oxidative stress including aminothiols, inflammatory cytokines, adhesion molecules, regulators of thrombosis, as well as markers of endothelial damage and repair.

Direct measures of NO are fraught with difficulty and inaccuracy because of its short half-life and although levels of nitrites and nitrosylated proteins reflect NO generation, results may be misleading due to production from other NO sources including dietary NO. The balance of endogenous pro- and antioxidants appears to determine endothelial function by modulating NO bioactivity[47] and oxidative stress is implicated in the pathophysiology of CVD.[48,49] Proteins are susceptible to oxidation through alterations of reactive aminothiol residues that serve to alter their cellular signaling activity, thereby coupling redox modifications of aminothiols to functional activity.[50] Plasma aminothiols can be quantified to assess the oxidant burden in vivo.[51] Of these, cysteine constitutes the major extracellular pool that reacts readily with oxidants to form its oxidized disulphide, cystine. Intracellularly, glutathione is a major antioxidant that helps eliminate peroxides and maintain cellular redox, and its oxidized form is glutathione disulfide.[49] We have shown that increased OS, measured as higher cystine and lower glutathione levels, is cross-sectionally associated with cellular dysfunction, aging, risk factors for CVD including obesity, and subclinical CVD including endothelial dysfunction, microvascular dysfunction, arterial stiffness, wall thickness, and pulmonary hypertension.[2,52–60]

Asymmetric dimethylarginine (ADMA) is an endogenous metabolite that competes with L-arginine as a substrate for NO production.[61] Intraarterial infusion of ADMA impairs NO activity and impairs blood flow.[62,63] Moreover, ADMA levels are increased in renal failure,[64] heart failure,[65–67] pulmonary hypertension,[68] and, among pregnant women with preeclampsia, conditions associated with reduced NO bioavailability and endothelial dysfunction.[69–71] Other biomarkers of endothelial dysfunction include circulating levels of adhesion molecules such as E-selectin, vascular cell adhesion molecule-1, intercellular adhesion molecule-1, and P-selectin that are elevated in patients with endothelial cell activation.[28,72]

Clinical assessment of endothelial function

Assessment of blood vessel responses to biochemical or physical endothelium-dependent agonists is utilized for estimating endothelial function in vivo. Changes in vessel diameter can be assessed using angiography or ultrasound imaging, and blood flow can be measured using intracoronary Doppler assessment of coronary blood flow.[32,73] In subjects with normal endothelial function, acetylcholine, the most widely used endothelium-dependent vasodilating agent that stimulates eNOS to produce NO and also stimulates EDHF and prostacyclin production, induces dilation of both the conductance vessels and the microcirculation, thus increasing blood flow. However, in subjects with endothelial dysfunction, these vasodilator responses are attenuated. Endothelium-independent vascular smooth muscle function can be assessed using sodium nitroprusside or other donors of NO that do not require a normal endothelium to induce smooth muscle vasodilation. Techniques commonly used assess endothelial function clinically invasively and non-invasively are summarized in Table 16.1.

Coronary endothelial function

Ganz and colleagues were the first to assess of the functional role of endothelium using intracoronary infusions of acetylcholine[74] using quantitative coronary angiography to accurately assess changes in vessel diameter. With acetylcholine, epicardial vessels with normal endothelial function dilate, whereas those with dysfunctional endothelium do not dilate or paradoxically vasoconstrict.[74] Other endothelium-dependent agonists utilized include bradykinin, substance P, and salbutamol. Because these agonists are stimulating NO production in addition to releasing other endothelium-dependent vasodilators, specific bioavailability of NO can only be assessed by evaluating the responses to L-N^G monomethyl arginine (L-NMMA), an L-arginine analog, or other related antagonists of eNOS. Physical maneuvers that stimulate increases in

TABLE 16.1 Commonly used methods to assess endothelial function are invasively and noninvasively methods.

Method	Vascular bed	Advantages	Disadvantage	Provocation stimulus
Invasive methods for assessment of endothelial function				
Coronary epicardial vasoreactivity (QCA)	Epicardial conductance arteries	May help establish correlation with symptoms	Requires coronary angiography	Acetylcholine Exercise Pacing
Coronary microvascular function—Doppler wires	Coronary microvasculature	Assess coronary microvascular disease	Requires coronary angiography	Acetylcholine Adenosine
Venous occlusion plethysmography	Forearm Microvasculature	Easy access Vasoactive substances infused to generate a dose–response relationship Contralateral arm as a control	Requires cannulation of the brachial artery	Acetylcholine and other vasoactive substances
Commonly used methods to assess endothelial function noninvasively				
Flow-Mediated Dilation (FMD)	Brachial Artery Conduit artery	Easy access Correlation with coronary vascular function Many outcome studies Inexpensive	Challenging to perform Needs standardization	Reactive Hyperemia
EndoPAT	Finger Microvasculature	Easy to access and perform Automated Correlation with invasive microvascular vascular function	Expense of disposable finger probes PAT signal influenced by variable nonendothelial factors	Acetylcholine Adenosine

Reprint permission from Hasdai D., Lerman A. The assessment of endothelial function in the cardiac catheterization laboratory in patients with risk factors for atherosclerotic coronary artery disease. Herz. 1999;24(7):544–547.

coronary blood flow, such as pacing, exercise, exposure to cold, and hyperemic challenges can also be utilized to investigate endothelial integrity, as these stimuli lead to epicardial coronary vasodilation due to shear-mediated NO release in those with intact endothelial function.[75–79]

Coronary microvascular endothelial function can be assessed by measuring coronary blood flow responses to these endothelium-dependent agonists, where impaired vasodilation or lower coronary blood flow changes are pathognomonic of endothelial dysfunction.[80] Coronary flow reserve is assessed using adenosine that maximally dilates the coronary microcirculation, largely by non-NO mechanisms, and should be distinguished from reactivity measured in response to specific endothelium-dependent agonists.[81] As with the epicardial assessments, microvascular endothelium-independent responses can be assessed using NO donors such as sodium nitroprusside.

Peripheral endothelial function assessments

Employing forearm blood flow measurements using plethysmography and by infusing the aforementioned endothelium-dependent and -independent agonists into the cannulated brachial artery, peripheral endothelial studies can be performed.[82] Dose-dependent vasodilation can be measured and repeated after interventions. The small doses of agonists employed preclude any significant systemic effects. Similarly, intradermal administration of agonists and antagonists with measurements of skin blood flow changes using photoplethysmography or spectroscopy can also be a less invasive technique for studying peripheral endothelial function.

Flow-mediated dilation

The vasodilatory response to flow-induced shear stress of conductance arteries, measured as flow-mediated dilation (FMD), is a direct estimate of endothelial NO release. FMD is measured as the change in the diameter of the brachial or radial artery using high-resolution ultrasound where increased shear stress induces NO release in those with normal endothelial function, resulting in vasodilation. After a 5-minute episode of forearm ischemia, the brachial arterial NO-dependent vasodilation, measured as FMD, is a moderately reliable, yet highly operator-dependent measure of endothelial function.[28,83,84] A low FMD indicates endothelial dysfunction, and is often observed in atherosclerosis and those exposed to cardiovascular risk factors.[25,27,29] The FMD response is often compared with the endothelium-independent vasodilator response that assesses arterial smooth muscle function directly using NO donors such as nitroglycerin.

Alternative tests have been considered including systolic pulse contour analysis by radial artery tonometry.[85] Since NO effects can be triggered by β2 agonist, salbutamol, changes in augmentation index in response to salbutamol can be measured from the peripheral arterial waveform and used as a correlate for blood flow responses to acetylcholine.[5,85–88] Even though FMD has become the most widely used investigational technique for assessment of endothelial function, its application requires extensive training and standardization Table S1 **(information source)**.[89] Study preparation, site selection, image acquisition and the echo probe position, cuff occlusion time, and software selection are all essential elements to ensure study reproducibility.

Peripheral arterial tonometry

Peripheral arterial tonometry (PAT) is a known reproducible and relatively simple technique for assessment of digital microvascular function that is partly dependent on endothelial function by using finger plethysmography (EndoPAT, Itamar Medical).[90] Finger-tip hyperemic response ratio to the basal flow correlates with coronary endothelial function,[91] and with risk factors.[28] As with FMD, a pressure cuff is placed on the forearm and after obtaining baseline blood volume change measurements, the blood pressure cuff is inflated above systolic pressure and deflated after 5 min to induce reactive hyperemia on one arm. Augmentation of the pulse amplitude after reactive hyperemia reflects changes in digital microvascular dilation and the ratio of maximal hyperemic signal divided by baseline signal, adjusted for the changes in the control arm, is calculated as the reactive hyperemia index or RHI.[92] A lower RHI is associated with endothelial dysfunction and correlates with coronary endothelial dysfunction, and importantly is a predictor of future cardiovascular events.[92,93] However, RHI correlates modestly with FMD that is a measure of conductance vessel NO bioavailability.[94,95]

Clinical implications of endothelial dysfunction

Endothelial dysfunction is often precipitated by exposure of the vascular wall to risk factors for atherosclerosis and precedes development of any structural abnormalities in the vessel wall or signs and symptoms of end organ damage.[96] These risk factors include sedentary lifestyle, obesity, hypercholesterolemia, hypertension, diabetes mellitus, insulin resistance, tobacco smoking, aging, and others.[3,83,97,98] A greater burden of exposure to cardiovascular risk factors is proportionately associated with worse endothelial function, although there is considerable heterogeneity among individuals.[96] Other unaccounted risk factors and genetic influences, as well as the variable duration of exposure to individual risk factors and differences in regenerative capacity presumably account for some of this observed variability.[14–18]

Epicardial coronary segments with loss of NO activity and increase in endothelin-1 activity and are more likely to be sites of development of atherosclerotic plaque have more plaque burden, and the plaque is more likely to display features of vulnerability such as an increase in the necrotic lipid core.[99–101] Similarly, endothelial dysfunction affecting the microcirculation can mechanistically be linked to myocardial ischemia even in presence of nonobstructive atherosclerosis.[102]

Endothelial dysfunction is a predictor of future development of hypertension and diabetes and predicts development and progression of atherosclerosis and adverse cardiovascular events, independent of the risk factor burden.[103,104] Coronary and peripheral endothelial dysfunction is a predictor of adverse cardiovascular events in patients with cardiovascular disease such as coronary artery disease, microvascular angina, heart failure, and hypertension.[104–115] Despite overwhelming data suggesting the independent predictive value of endothelial function as a predictor for future cardiovascular events, it is not recommended for risk assessment at the present time, largely because the challenges involved in its measurement.

Therapy for endothelial dysfunction

Multiple studies have shown that cardioprotective medications such as statins and angiotensin antagonists and lifestyle changes including weight loss and exercise improve endothelial function and reduce cardiovascular events[24,116,117] Statins in addition to reducing oxidized LDL and lipid peroxidation also improve EPC counts and endothelial function, and the latter are considered to be important pleiotropic effects of statins.[118,119] Similarly, treatment of hypertension using conventional antihypertensive medications can improve endothelial dysfunction.[120,121] This improvement appears to be similar with various classes of antihypertensive medications including calcium channel blockers, certain β-receptor antagonists, and angiotensin inhibitors.[122–124] Angiotensin-converting enzyme inhibitors also increase EPC mobilization.[125] Beta blockers also improve NO production, reduce endothelin-1 levels, and improve FMD.[126] Treatment of diabetes using metformin improves endothelial function[127] and novel antiinflammatory agents such as infliximab also improve endothelial function and reduce serum nitrite levels.[128] Cell therapy using mobilization of EPCs using Granulocyte-Macrophage Colony Stimulating Factor appears to promote healthier endothelium.[129]

Persistence or worsening of endothelial dysfunction in the normal population has been associated with development of hypertension and diabetes during follow-up in some studies.[130–132] Similarly, among individuals with stable coronary artery disease, the persistence of low FMD despite optimized individualized therapy was predictive of higher risk of future events.[130]

Arterial stiffness

Mechanisms for arterial stiffness

There are three main layers—tunica intima, tunica media, and tunica adventitia—of large arteries and different processes in these layers can affect arterial stiffness. The main determinant, however, is the medial layer, which contains the extracellular matrix composed of elastin (contributing to elasticity) and collagen, VSMCs, cell-matrix connections, and matrix constituents, Fig. 16.3. With progressive distance from the heart, arteries have decreasing elastin-to-collagen ratio thereby increasing stiffness as one moves toward the periphery. Peripheral arteries have higher VSMC. Thus, hypertrophy and tone of smooth muscle is an important determinant of arterial stiffness for smaller arteries, such as the muscular type conduit and resistance arteries.

Multiple pathological processes impact arterial stiffness including elastin degradation, collagen deposition, hypertrophy of VSMCs, inflammation and endothelial dysfunction, and availability of nitric oxide.[133,134] Excess oxidant burden can alter DNA transcription leading to changes in numerous biochemical pathways involved in arterial remodeling. As such, oxidative stress has also been associated with increasing arterial stiffness in healthy subjects without any overt risk factors.[54] The pathophysiological mechanisms of arterial stiffness are likely secondary to a complex interplay of all these factors. Further research is ongoing to understand these processes and develop therapeutic approaches.[134]

Methods for evaluation of arterial elastic properties

Carotid-femoral **pulse wave velocity** (PWV) is a direct measure of arterial stiffness that is the current reference standard and can be measured noninvasively using reproducible techniques. It is calculated as the distance between two sites divided by the travel time of the pulse (transit time) from the proximal to the distal site.[134–137] Carotid-femoral PWV is the method most frequently employed in clinical studies and has been shown to correlate with clinical outcomes.

There are multiple ways to calculate PWV noninvasively.

Applanation tonometry (SphygmoCor, AtCor Medical) is a popular method for measuring carotid-femoral PWV, which involves sequential (nonsimultaneous)

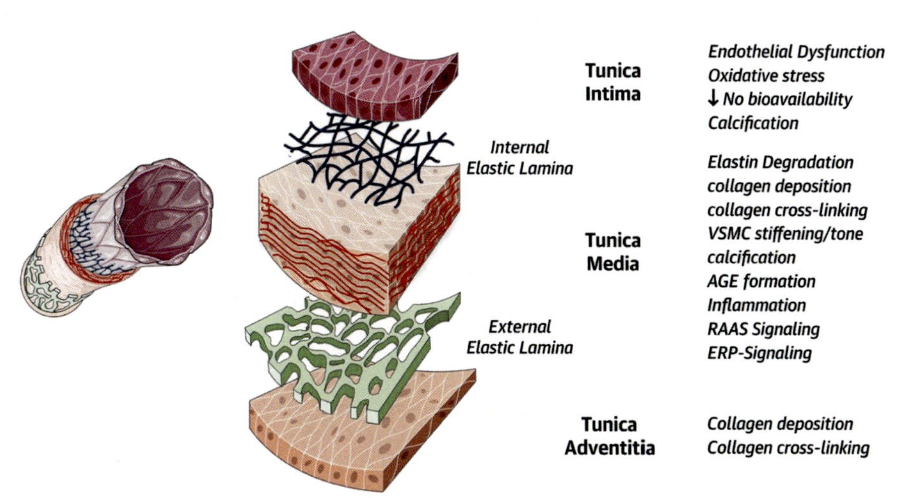

FIGURE 16.3 General Mechanisms of Arterial Stiffness. Various mechanisms can increase the stiffness of the intima, media, and adventitia. The media is, however, the most important layer in large arteries. *AGE*, advanced glycation end-products; *ERP*, elastin-related peptides; *NO*, nitric oxide; *RAAS*, renin-angiotensin-aldosterone system. *Reproduced from Chirinos JA, Segers P, Hughes T, Townsend R. Large-artery stiffness in health and disease: JACC state-of-the-art review.* J Am Coll Cardiol. *2019;74(9):1237–1263.*

detection and subsequent recording of pressure waveforms from the two arterial sites using sensitive tonometers (Millar instruments). Simultaneously, an electrocardiogram is recorded and the pulse transit time is calculated as the difference of the distal (femoral) waveform from the R wave minus the difference of the proximal (carotid) waveform from the R wave (right panel in Fig. 16.4). While abrupt changes in heart rate or the contractility can theoretically affect the accuracy due to the nonsimultaneous recording of the waveforms, this issue is mute since both arterial sites are examined only a few minutes apart.[4] The distance between two arterial sites used to calculate carotid femoral PWV can be the distance between the carotid and femoral sites, or this distance minus the sternal-carotid distance. There is no consensus

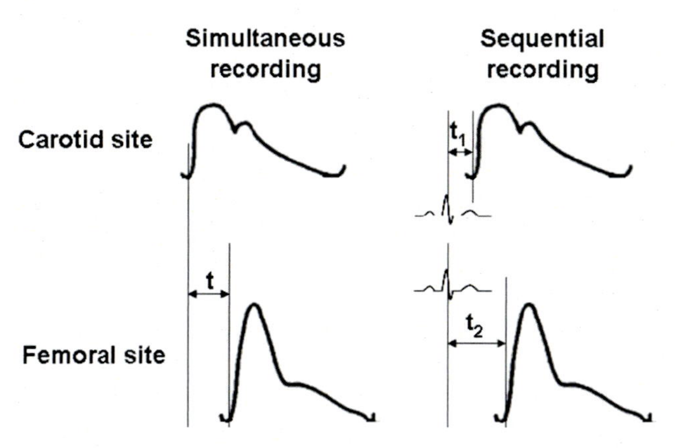

FIGURE 16.4 Measurement of pulse transit time t for calculation of carotid-femoral pulse wave velocity. For simultaneous recordings of the pressure waveform, t is measured directly with the foot-to-foot method. For sequential recordings, the time from the R wave to the foot of the carotid (t_1) and the femoral site (t_2) is measured, and t is calculated as $t_2 - t_1$.

regarding which technique is appropriate, but the method employed needs to be specified.

Doppler ultrasonography is another technique for measuring PWV. Here, the pulse transit time is calculated with the "foot-to-foot" method by recording the flow waveforms in the proximal aorta and the distal aorta or the common femoral artery.[138,139]

Magnetic resonance imaging (*MRI*) can be used to measure PWV in small or larger segments of the aorta. In this method, flow velocity signals are acquired along the ascending and the descending aorta, the pulse transit time, and the distance between the sites is measured to calculate PWV.[140,141]

Determinants of arterial elasticity

There are multiple predictors of arterial stiffness. Large epidemiological studies including the Anglo-Cardiff Collaborative Trial and the Framingham Heart Study have demonstrated that age is the single most important predictor of arterial elasticity.[142,143] Race and gender are also predictors of arterial stiffness. Data from the Atherosclerosis Risk In Communities (ARIC) Study and Bogalusa Heart Study showed that large artery stiffening may be more accelerated in African Americans than their white counterparts.[144,145] Arterial stiffness is also influenced by hemodynamic parameters (i.e., distending arterial pressure), anthropomorphics, and presence of traditional CV risk factors.[137,146,147] Table 16.2 lists the main demographic, lifestyle, and clinical characteristics that influence arterial stiffness.

Arterial stiffness and target organ damage

Arterial stiffness leads to increased arterial pressure and flow pulsatility that can cause damage to multiple organs, Fig. 16.5. Known mechanisms underlying organ

TABLE 16.2 Determinants and clinical correlations of arterial elastic properties.

1. Age
2. Gender
3. Cardiovascular risk factors
Hypertension
Dyslipidemia (hypercholesterolemia, hypertriglyceridemia)
Diabetes mellitus/impaired glucose tolerance
Smoking
Obesity
4. Cardiovascular diseases
Coronary artery disease
Peripheral artery disease
Heart failure
Cardiac syndrome X
5. Endocrinology and metabolic diseases
Metabolic syndrome
Hyperhomocysteinemia
Hypothyroidism
6. Nutrition and lifestyle
Coffee, caffeine
Chronic alcohol consumption
Sedentary lifestyle
Resistance training
7. Genetic factors
Family history of atherosclerotic disease
Polymorphisms of the renin-angiotensin system genes
Polymorphisms of the extracellular matrix proteins genes
8. Gynecological disorders
Menopause
Preeclampsia
Polycystic ovaries syndrome
9. Other disorders
Inflammation
Acute inflammation (i.e., acute infections)
Chronic inflammatory diseases (i.e., rheumatoid arthritis)
Subclinical, low-grade inflammation
End-stage renal disease
Sleep apnea

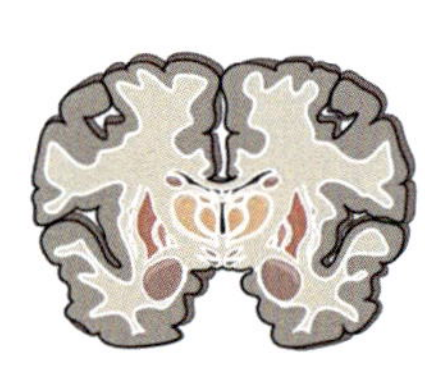

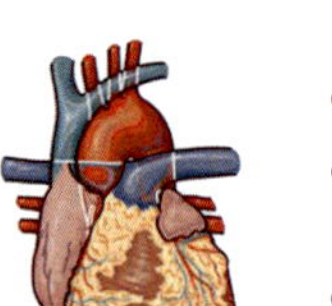

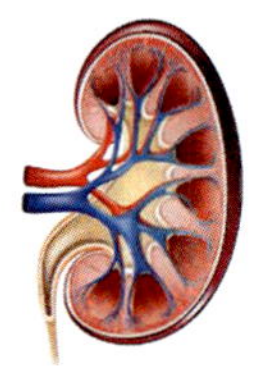

FIGURE 16.5 Potential Consequences of arterial stiffness on Target Organs. It can affect the heart through effects on afterload (particularly, modulation of the timing of wave reflections) and may damage low-resistance organs via excess arterial pulsatility. *Reproduced from Chirinos JA, Segers P, Hughes T, Townsend R. Large artery stiffness in health and disease: JACC state-of-the-art review.* J Am Coll Cardiol. *2019;74(9): 1237–1263.*

damage accompanying arterial stiffness include the impact of increased pulsatile load from the aorta and downstream wave reflections on the heart and other organs including the kidney and brain as the latter operate at low levels of microvascular resistance and require high levels of blood flow. Increased arterial stiffness can also lead to microvascular damage through barotrauma and excessive shear forces.[134]

Arterial stiffness and cardiac manifestations

When PWV increases, reflected waves arrive to the heart during systole rather than diastole, which impairs coronary perfusion due to reduced diastolic pressure and increased workload, causing a myocardial supply/demand mismatch. In addition, the increased pulsatile load during systole leads to left ventricular hypertrophy and fibrosis and impairs LV relaxation and diastolic dysfunction. In multiple clinical and community-based studies, arterial stiffness has been associated with heart failure with preserved ejection fraction.[134,146,148]

Arterial stiffness and renal manifestations

There is a link between the kidney and the arterial wall, where they interact closely with each other. A vicious cycle of arterial stiffening leading to renal dysfunction and of renal disease contributing to increased arterial stiffness ensues. Since the kidneys exhibit the highest flow rate and lowest vascular resistance of any large organ, they are extremely susceptible to barotrauma that results from arterial stiffness.[149] In turn, impaired renal function leads to dysregulation of bone and mineral metabolism[150] and the renin–angiotensin–aldosterone system, elevation of inflammatory biomarkers (e.g., high-sensitivity C-reactive protein, Interleukin-6, tumor necrosis factor-α),[151] reduced ability to handle sodium load,[152] and sympathetic hyperactivity,[153] all of which can worsen arterial stiffness. Thus, in the chronic renal insufficiency cohort (CRIC) study, a large longitudinal study of patients with chronic kidney disease in the United States, participants with higher arterial stiffness had more rapid progression to end-stage kidney disease, independent of measured blood pressure.[154]

Arterial stiffness and the brain

The brain is densely vascularized and requires high blood flow and low resistance arterial circuitry that is highly susceptible to barotrauma arising from arterial stiffness. Arterial stiffness has been associated with cognitive impairment and decline, incident dementia and imaging findings suggestive of cerebral small-vessel disease (lacunar infarcts, white matter hyperintensities, cerebral microbleeds, and enlarged perivascular spaces) and regional brain atrophy.[155–159] In addition, there have been emerging data suggesting that arterial stiffness is associated with β-amyloid deposition which is the pathologic hallmark of Alzheimer's disease.[160,161]

Arterial stiffness and prognosis

Arterial stiffness is a lifelong marker of vascular aging that is both a precursor and ultimately a predictor of incident hypertension[162] and as such represents a noninvasive, inexpensive, and easy to measure marker that can be monitored clinically for improved cardiac risk assessment and targeted therapeutically. From a risk assessment standpoint, arterial stiffness may provide important information to evaluate younger individuals with a family history of isolated systolic hypertension as a substantial variability in arterial stiffness is inheritable.[163] In addition, measurement of arterial stiffness may be beneficial for special populations in whom the pooled cohort equations are unreliable since arterial stiffness measurements directly characterize subclinical vascular damage over years.[134]

Arterial stiffness is an independent predictor of longer term CV outcomes.[134,146–148,164] In the Framingham Heart Study, a crude index of arterial stiffness, the brachial artery pulse pressure, was a predictor of the risk of coronary artery disease.[165] In a recent metaanalysis of prospective studies including 17,635 participants, each standard deviation change in carotid-femoral PWV was associated with a 35%, 54%, and 45% increase in the risk of coronary heart disease, stroke, and overall CV disease, respectively.[164] Carotid-femoral PWV predicts CV events in both men and women, in smokers, diabetics, those with renal disease, and individuals without clinical hypertension.[134,164] Table 16.3 lists potential applications for arterial stiffness in clinical decision-making.[5]

Arterial stiffness as a target for treatment

From a therapeutic standpoint, studies with antihypertensive medications support using arterial stiffness as a treatment goal. For instance, in patients with end-stage renal disease, the administration of antihypertensive treatment was clinically beneficial only in subjects who experienced a decrease of aortic PWV.[154] Similarly, in the Conduit Artery Function Evaluation (CAFÉ) Study, a substudy of the Anglo-Scandinavian Cardiac Outcomes Trial (ASCOT), a higher treatment-related decrease of central pulse pressure, that was not evident in brachial pressure measurements, was independently associated with clinical benefit and reduced CV events in hypertensive patients.[166] These studies suggest that the evaluation of central hemodynamics is clinically useful and may aid clinical decision making.

Interventions that reduce arterial stiffness include drugs prescribed for the treatment of hypertension and heart failure.[2] Drug classes include vasodilators, aldosterone blockers, β-blockers, and diuretics each of which have different effects on arterial stiffness properties. Diuretics and β-blockers have little direct effect on arterial properties.[2] Selective and nonselective aldosterone blockers lower PWV in certain patient groups by increasing nitric oxide bioactivity and improving endothelial vasodilator dysfunction.[167,168] Vasodilating drugs such as hydralazine have little effect on aortic wave reflection.[169] Nitrates relax smooth muscle cells in large conduit muscular arteries, and as such, decrease arterial stiffness.[170–172] Angiotensin-converting enzyme inhibitors, angiotensin receptor blockers, and calcium channel blockers appear to have little effect on arterial stiffness,[173] but there are studies questioning these findings.[174–176] For example, a metanalysis showed that angiotensin-converting enzyme inhibitor therapy lowered arterial stiffness.[177] Following this, some studies have demonstrated that angiotensin-converting enzyme inhibition may

TABLE 16.3 Suggested clinical applications of measurements of arterial stiffness in primordial and primary prevention of cardiovascular disease.

Clinical scenario	Rationale	Impact on clinical management
Hypertension		
ACC/AHA stage 1 Hypertension (130–139/80–89 mm Hg) with PCE-calculated 10-year ASCVD risk ~10% without diabetes or CKD	Can be useful to refine risk stratification when PCE-calculated 10-year ASCVD risk is close to the threshold for treatment, after an informed clinician–patient discussion.	• Initiation of pharmacologic antihypertensive therapy
Stage 2 isolated systolic hypertension (>140 mm Hg) in very young adults with paucity of other cardiovascular risk factors	The combination of high pulse pressure amplification (with normal central systolic pressure) and low or normal arterial stiffness for age support a low CV risk.	• Withholding of pharmacologic antihypertensive therapy
Other CV risk assessment scenarios		
Refinement of cardiovascular risk assessment in nondiabetic adults 40–75 years of age at intermediate PCE-calculated 10-year ASCVD risk	In this group of patients, risk-based decisions for preventive interventions may be uncertain and arterial stiffness measurements can be used to refine risk assessment (particularly if various "risk-enhancing" clinical parameters do not clearly favor a specific course of action).	• Guide clinician–patient risk discussion • Guide decision making regarding initiation of pharmacologic therapy (i.e., statins)
Refinement of cardiovascular risk assessment in middle-aged nondiabetic adults at borderline PCE-calculated 10-year risk of ASCVD (5% to <7.5%) who also have other factors that increase their ASCVD risk ("risk enhancers")	In this group of patients, arterial stiffness measurements may be useful to improve risk-based decisions as an alternative or as a "gate-keeper" for coronary calcium score testing, particularly when concerns about radiation exposure (younger age, overweight/obese) or about cost are present.	• Guide clinician–patient risk discussion • Guide decision making regarding further testing (coronary calcium score) or initiation of therapy (i.e., statins)
Assessment of CV risk in special populations'	PCE-calculated 10-year risk estimations can provide notoriously miscalibrated estimates in non-U.S. populations, particularly those at earlier stages of the epidemiologic transition. This may also apply to immigrants from those populations in the U.S.	• Guide clinician–patient risk discussions and various interventions • An abnormal PWV for age indicates subclinical arterial damage and suggests a higher risk. Expert clinical judgment must guide result interpretation and decision making depending on the specific clinical scenario.

ACC/AHA, American College of Cardiology/American Heart Association; *ASCVD*, atherosclerotic cardiovascular disease; *CV*, cardiovascular; *PCE*, pulled cohort equations.
Adapted from Chirinos JA, Segers P, Hughes T, Townsend R. Large-artery stiffness in health and disease: JACC state-of-the-art review. J Am Coll Cardiol. *2019;74(9):1237–1263.*

decrease arterial stiffness beyond lowering blood pressure.[178,179] However, future studies continue to be required to determine the optimal drugs to lower arterial stiffness.

Conclusions

In health, the endothelium releases several substances including NO that maintains vascular tone and homeostasis.[3,180] Its dysfunction results from exposure to ASCVD risk factors.[181,182] Endothelial function is impaired in subjects exposed to risk factors,[183,184] and abnormal endothelial function is associated with future development of hypertension and diabetes,[132,185,186] and with up to fourfold increased CVD risk.[110,187–190] Importantly, restoration of endothelial function is associated with improved outcomes and its worsening with worse outcomes.[89,131]

Aging and exposure to CVD risk factors leads to arterial wall thickening and compensatory remodeling, adversely affecting the internal elastic properties of the vessel wall that is called arterial stiffness.[4] Noninvasive and reproducible techniques allow estimation of central aortic pressures and stiffness.[5,6] Aortic PWV, an estimate of the speed of the pressure wave traveling along the aorta, is regarded as a direct measure of large artery stiffness.[4–6,191] Impaired arterial elastic properties are increasingly recognized as independent predictors of incident CVD events (myocardial infarction, stroke, revascularization, and mortality).[7–12]

Appendix A. Supplementary data

Supplementary data to this article can be found online at https://doi.org/10.1016/B978-0-12-822546-2.00019-8.

References

1. Quyyumi AA, Dakak N, Mulcahy D, et al. Nitric oxide activity in the atherosclerotic human coronary circulation. *J Am Coll Cardiol.* 1997;29(2):308–317.
2. Ashfaq S, Beinart SC, Abramson JL, et al. Plasma glutathione redox state: a novel marker of oxidative stress, correlates with early atherosclerosis in humans. *J Am Coll Cardiol.* 2003;41(Suppl A):293A–294A.
3. Quyyumi AA. Endothelial function in health and disease: new insights into the genesis of cardiovascular disease. *Am J Med.* 1998; 105(1A):32S–39S.
4. Zieman SJ, Melenovsky V, Kass DA. Mechanisms, pathophysiology, and therapy of arterial stiffness. *Arterioscler Thromb Vasc Biol.* 2005;25(5):932–943.
5. Wilkinson IB, Fuchs SA, Jansen IM, et al. Reproducibility of pulse wave velocity and augmentation index measured by pulse wave analysis. *J Hypertens.* 1998;16(12):2079–2084.
6. O'Rourke MF, Adji A. An updated clinical primer on large artery mechanics: implications of pulse waveform analysis and arterial tonometry. *Curr Opin Cardiol.* 2005;20(4):275–281.
7. Vlachopoulos C, Aznaouridis K, Stefanadis C. Prediction of cardiovascular events and all-cause mortality with arterial stiffness: a systematic review and meta-analysis. *J Am Coll Cardiol.* 2010; 55(13):1318–1327.
8. London GM, Blacher J, Pannier B, Guerin AP, Marchais SJ, Safar ME. Arterial wave reflections and survival in end-stage renal failure. *Hypertension.* 2001;38(3):434–438.
9. Nurnberger J, Keflioglu-Scheiber A, Opazo Saez AM, Wenzel RR, Philipp T, Schafers RF. Augmentation index is associated with cardiovascular risk. *J Hypertens.* 2002;20(12):2407–2414.
10. Laurent S, Boutouyrie P, Asmar R, et al. Aortic stiffness is an independent predictor of all-cause and cardiovascular mortality in hypertensive patients. *Hypertension.* 2001;37(5):1236–1241.
11. Laurent S, Katsahian S, Fassot C, et al. Aortic stiffness is an independent predictor of fatal stroke in essential hypertension. *Stroke.* 2003;34(5):1203–1206.
12. Mitchell GF, Hwang S-J, Vasan RS, et al. Arterial stiffness and cardiovascular events: the Framingham Heart Study. *Circulation.* 2010;121(4):505–511.
13. Flammer AJ, Luscher TF. Human endothelial dysfunction: EDRFs. *Pflugers Arch.* 2010;459(6):1005–1013.
14. Prasad A, Zhu J, Halcox JP, Waclawiw MA, Epstein SE, Quyyumi AA. Predisposition to atherosclerosis by infections: role of endothelial dysfunction. *Circulation.* 2002;106(2):184–190.
15. Quyyumi AA. Does acute improvement of endothelial dysfunction in coronary artery disease improve myocardial ischemia? A double-blind comparison of parenteral D- and L-arginine. *J Am Coll Cardiol.* 1998;32(4):904–911.
16. McDermott DH, Halcox JP, Schenke WH, et al. Association between polymorphism in the chemokine receptor CX3CR1 and coronary vascular endothelial dysfunction and atherosclerosis. *Circ Res.* 2001;89(5):401–407.
17. Zhu J, Quyyumi AA, Wu H, et al. Increased serum levels of heat shock protein 70 are associated with low risk of coronary artery disease. *Arterioscler Thromb Vasc Biol.* 2003;23(6):1055–1059.
18. Zhu J, Quyyumi AA, Rott D, et al. Antibodies to human heat-shock protein 60 are associated with the presence and severity of coronary artery disease: evidence for an autoimmune component of atherogenesis. *Circulation.* 2001;103(8):1071–1075.
19. Channon KM. Tetrahydrobiopterin: regulator of endothelial nitric oxide synthase in vascular disease. *Trends Cardiovasc Med.* 2004; 14(8):323–327.
20. Schmidt TS, Alp NJ. Mechanisms for the role of tetrahydrobiopterin in endothelial function and vascular disease. *Clin Sci (Lond).* 2007;113(2):47–63.
21. Dudzinski DM, Igarashi J, Greif D, Michel T. The regulation and pharmacology of endothelial nitric oxide synthase. *Annu Rev Pharmacol Toxicol.* 2006;46:235–276.
22. Aicher A, Heeschen C, Mildner-Rihm C, et al. Essential role of endothelial nitric oxide synthase for mobilization of stem and progenitor cells. *Nat Med.* 2003;9(11):1370–1376.
23. Vasa M, Fichtlscherer S, Adler K, et al. Increase in circulating endothelial progenitor cells by statin therapy in patients with stable coronary artery disease. *Circulation.* 2001;103(24): 2885–2890.
24. Walter DH, Rittig K, Bahlmann FH, et al. Statin therapy accelerates reendothelialization: a novel effect involving mobilization and incorporation of bone marrow-derived endothelial progenitor cells. *Circulation.* 2002;105(25):3017–3024.
25. Celermajer DS, Sorensen KE, Gooch VM, et al. Non-invasive detection of endothelial dysfunction in children and adults at risk of atherosclerosis. *Lancet.* 1992;340(8828):1111–1115.
26. Corretti MC, Anderson TJ, Benjamin EJ, et al. Guidelines for the ultrasound assessment of endothelial-dependent flow-mediated vasodilation of the brachial artery: a report of the International Brachial Artery Reactivity Task Force. *J Am Coll Cardiol.* 2002; 39(2):257–265.
27. Joannides R, Haefeli WE, Linder L, et al. Nitric oxide is responsible for flow-dependent dilatation of human peripheral conduit arteries in vivo. *Circulation.* 1995;91(5):1314–1319.
28. Deanfield J, Donald A, Ferri C, et al. Endothelial function and dysfunction. Part I: methodological issues for assessment in the different vascular beds: a statement by the Working Group on Endothelin and Endothelial Factors of the European Society of Hypertension. *J Hypertens.* 2005;23(1):7–17.
29. Anderson TJ, Uehata A, Gerhard MD, et al. Close relation of endothelial function in the human coronary and peripheral circulations. *J Am Coll Cardiol.* 1995;26(5):1235–1241.
30. Mollace V, Muscoli C, Masini E, Cuzzocrea S, Salvemini D. Modulation of prostaglandin biosynthesis by nitric oxide and nitric oxide donors. *Pharmacol Rev.* 2005;57(2):217–252.
31. Moncada S, Higgs EA, Vane JR. Human arterial and venous tissues generate prostacyclin (prostaglandin x), a potent inhibitor of platelet aggregation. *Lancet.* 1977;1(8001):18–20.
32. Bonetti PO, Lerman LO, Lerman A. Endothelial dysfunction: a marker of atherosclerotic risk. *Arterioscler Thromb Vasc Biol.* 2003; 23(2):168–175.
33. Halcox JP, Narayanan S, Cramer-Joyce L, Mincemoyer R, Quyyumi AA. Characterization of endothelium-derived hyperpolarizing factor in the human forearm microcirculation. *Am J Physiol Heart Circ Physiol.* 2001;280(6):H2470–H2477.
34. Feletou M, Vanhoutte PM. Endothelium-derived hyperpolarizing factor: where are we now? *Arterioscler Thromb Vasc Biol.* 2006;26(6): 1215–1225.

35. Fleming I, Busse R. Endothelium-derived epoxyeicosatrienoic acids and vascular function. *Hypertension*. 2006;47(4):629–633.
36. Busse R, Edwards G, Feletou M, Fleming I, Vanhoutte PM, Weston AH. EDHF: bringing the concepts together. *Trends Pharmacol Sci*. 2002;23(8):374–380.
37. Shimokawa H, Yasutake H, Fujii K, et al. The importance of the hyperpolarizing mechanism increases as the vessel size decreases in endothelium-dependent relaxations in rat mesenteric circulation. *J Cardiovasc Pharmacol*. 1996;28(5):703–711.
38. Nishikawa Y, Stepp DW, Chilian WM. Nitric oxide exerts feedback inhibition on EDHF-induced coronary arteriolar dilation in vivo. *Am J Physiol Heart Circ Physiol*. 2000;279(2):H459–H465.
39. Taddei S, Versari D, Cipriano A, et al. Identification of a cytochrome P450 2C9-derived endothelium-derived hyperpolarizing factor in essential hypertensive patients. *J Am Coll Cardiol*. 2006; 48(3):508–515.
40. Taddei S, Virdis A, Ghiadoni L, Versari D, Salvetti A. Endothelium, aging, and hypertension. *Curr Hypertens Rep*. 2006;8(1): 84–89.
41. Yanagisawa M, Kurihara H, Kimura S, et al. A novel potent vasoconstrictor peptide produced by vascular endothelial cells. *Nature*. 1988;332(6163):411–415.
42. Schiffrin EL. State-of-the-Art lecture. Role of endothelin-1 in hypertension. *Hypertension*. 1999;34(4 Pt 2):876–881.
43. Yachie A, Niida Y, Wada T, et al. Oxidative stress causes enhanced endothelial cell injury in human heme oxygenase-1 deficiency. *J Clin Invest*. 1999;103(1):129–135.
44. Kawashima A, Oda Y, Yachie A, Koizumi S, Nakanishi I. Heme oxygenase-1 deficiency: the first autopsy case. *Hum Pathol*. 2002; 33(1):125–130.
45. Chen YH, Lin SJ, Lin MW, et al. Microsatellite polymorphism in promotor of heme oxygenase-1 gene is associated with susceptibility to coronary artery disease in type 2 diabetic patients. *Hum Genet*. 2002;111(1):1–8.
46. Kaneda H, Ohno M, Taguchi J, et al. Heme oxygenase-1 gene promoter polymorphism is associated with coronary artery disease in Japanese patients with coronary risk factors. *Arterioscler Thromb Vasc Biol*. 2002;22(10):1680–1685.
47. Jones DP. Extracellular redox state: refining the definition of oxidative stress in aging. *Rejuvenation Res*. 2006;9(2):169–181.
48. Harrison D, Griendling KK, Landmesser U, Hornig B, Drexler H. Role of oxidative stress in atherosclerosis. *Am J Cardiol*. 2003; 91(3A):7A–11A.
49. Jones DP. Radical-free biology of oxidative stress. *Am J Physiol Cell Physiol*. 2008;295(4):C849–C868.
50. Go YM, Jones DP. Thiol/disulfide redox states in signaling and sensing. *Crit Rev Biochem Mol Biol*. 2013;48(2):173–181.
51. Jones DP, Liang Y. Measuring the poise of thiol/disulfide couples in vivo. *Free Radic Biol Med*. 2009;47(10):1329–1338.
52. Ashfaq S, Abramson JL, Jones DP, et al. Endothelial function and aminothiol biomarkers of oxidative stress in healthy adults. *Hypertension*. 2008;52(1):80–85.
53. Ashfaq S, Abramson JL, Jones DP, et al. The relationship between plasma levels of oxidized and reduced thiols and early atherosclerosis in healthy adults. *J Am Coll Cardiol*. 2006;47(5): 1005–1011.
54. Patel RS, Al Mheid I, Morris AA, et al. Oxidative stress is associated with impaired arterial elasticity. *Atherosclerosis*. 2011;218(1): 90–95.
55. Go YM, Jones DP. Intracellular proatherogenic events and cell adhesion modulated by extracellular thiol/disulfide redox state. *Circulation*. 2005;111(22):2973–2980.
56. Nkabyo YS, Ziegler TR, Gu LH, Watson WH, Jones DP. Glutathione and thioredoxin redox during differentiation in human colon epithelial (Caco-2) cells. *Am J Physiol Gastrointest Liver Physiol*. 2002;283(6):G1352–G1359.
57. Droge W. The plasma redox state and ageing. *Ageing Res Rev*. 2002; 1(2):257–278.
58. Samiec PS, Drews-Botsch C, Flagg EW, et al. Glutathione in human plasma: decline in association with aging, age-related macular degeneration, and diabetes. *Free Radic Biol Med*. 1998;24(5): 699–704.
59. Ghasemzadeh N, Patel RS, Eapen DJ, et al. Oxidative stress is associated with increased pulmonary artery systolic pressure in humans. *Hypertension*. 2014;63(6):1270–1275.
60. Bettermann EL, Hartman TJ, Easley KA, et al. Higher mediterranean diet quality scores and lower body mass index are associated with a less-oxidized plasma glutathione and cysteine redox status in adults. *J Nutr*. 2018;148(2):245–253.
61. Boger RH, Maas R, Schulze F, Schwedhelm E. Elevated levels of asymmetric dimethylarginine (ADMA) as a marker of cardiovascular disease and mortality. *Clin Chem Lab Med*. 2005;43(10): 1124–1129.
62. Vallance P, Leone A, Calver A, Collier J, Moncada S. Accumulation of an endogenous inhibitor of nitric oxide synthesis in chronic renal failure. *Lancet*. 1992;339(8793):572–575.
63. Vallance P, Leiper J. Cardiovascular biology of the asymmetric dimethylarginine:dimethylarginine dimethylaminohydrolase pathway. *Arterioscler Thromb Vasc Biol*. 2004;24(6):1023–1030.
64. Kielstein JT, Boger RH, Bode-Boger SM, et al. Asymmetric dimethylarginine plasma concentrations differ in patients with end-stage renal disease: relationship to treatment method and atherosclerotic disease. *J Am Soc Nephrol*. 1999;10(3): 594–600.
65. Feng Q, Lu X, Fortin AJ, et al. Elevation of an endogenous inhibitor of nitric oxide synthesis in experimental congestive heart failure. *Cardiovasc Res*. 1998;37(3):667–675.
66. Saitoh M, Osanai T, Kamada T, et al. High plasma level of asymmetric dimethylarginine in patients with acutely exacerbated congestive heart failure: role in reduction of plasma nitric oxide level. *Heart Vessels*. 2003;18(4):177–182.
67. Usui M, Matsuoka H, Miyazaki H, Ueda S, Okuda S, Imaizumi T. Increased endogenous nitric oxide synthase inhibitor in patients with congestive heart failure. *Life Sci*. 1998;62(26):2425–2430.
68. Gorenflo M, Zheng C, Werle E, Fiehn W, Ulmer HE. Plasma levels of asymmetrical dimethyl-L-arginine in patients with congenital heart disease and pulmonary hypertension. *J Cardiovasc Pharmacol*. 2001;37(4):489–492.
69. Savvidou MD, Hingorani AD, Tsikas D, Frolich JC, Vallance P, Nicolaides KH. Endothelial dysfunction and raised plasma concentrations of asymmetric dimethylarginine in pregnant women who subsequently develop pre-eclampsia. *Lancet*. 2003; 361(9368):1511–1517.
70. Fickling SA, Williams D, Vallance P, Nussey SS, Whitley GS. Plasma concentrations of endogenous inhibitor of nitric oxide synthesis in normal pregnancy and pre-eclampsia. *Lancet*. 1993; 342(8865):242–243.
71. Holden DP, Fickling SA, Whitley GS, Nussey SS. Plasma concentrations of asymmetric dimethylarginine, a natural inhibitor of nitric oxide synthase, in normal pregnancy and preeclampsia. *Am J Obstet Gynecol*. 1998;178(3):551–556.
72. Ridker PM, Brown NJ, Vaughan DE, Harrison DG, Mehta JL. Established and emerging plasma biomarkers in the prediction of first atherothrombotic events. *Circulation*. 2004;109(25 Suppl 1):IV6–IV19.
73. Hasdai D, Lerman A. The assessment of endothelial function in the cardiac catheterization laboratory in patients with risk factors for atherosclerotic coronary artery disease. *Herz*. 1999;24(7): 544–547.
74. Ludmer PL, Selwyn AP, Shook TL, et al. Paradoxical vasoconstriction induced by acetylcholine in atherosclerotic coronary arteries. *N Engl J Med*. 1986;315(17):1046–1051.

75. Hess OM, Buchi M, Kirkeeide R, et al. Potential role of coronary vasoconstriction in ischaemic heart disease: effect of exercise. *Eur Heart J*. 1990;11(Suppl B):58–64.
76. Puri R, Liew GY, Nicholls SJ, et al. Coronary β2-adrenoreceptors mediate endothelium-dependent vasoreactivity in humans: novel insights from an in vivo intravascular ultrasound study. *Eur Heart J*. 2012;33(4):495–504.
77. Gordon JB, Ganz P, Nabel EG, et al. Atherosclerosis influences the vasomotor response of epicardial coronary arteries to exercise. *J Clin Invest*. 1989;83(6):1946–1952.
78. Yeung AC, Vekshtein VI, Krantz DS, et al. The effect of atherosclerosis on the vasomotor response of coronary arteries to mental stress. *N Engl J Med*. 1991;325(22):1551–1556.
79. Zeiher AM, Drexler H, Wollschlaeger H, Saurbier B, Just H. Coronary vasomotion in response to sympathetic stimulation in humans: importance of the functional integrity of the endothelium. *J Am Coll Cardiol*. 1989;14(5):1181–1190.
80. Beltrame JF, Crea F, Camici P. Advances in coronary microvascular dysfunction. *Heart Lung Circ*. 2009;18(1):19–27.
81. Camici PG, Crea F. Coronary microvascular dysfunction. *N Engl J Med*. 2007;356(8):830–840.
82. Linder L, Kiowski W, Buhler FR, Luscher TF. Indirect evidence for release of endothelium-derived relaxing factor in human forearm circulation in vivo. Blunted response in essential hypertension. *Circulation*. 1990;81(6):1762–1767.
83. Brunner H, Cockcroft JR, Deanfield J, et al. Endothelial function and dysfunction. Part II: Association with cardiovascular risk factors and diseases. A statement by the Working Group on Endothelins and Endothelial Factors of the European Society of Hypertension. *J Hypertens*. 2005;23(2):233–246.
84. Endemann DH, Schiffrin EL. Endothelial dysfunction. *J Am Soc Nephrol*. 2004;15(8):1983–1992.
85. Hayward CS, Kraidly M, Webb CM, Collins P. Assessment of endothelial function using peripheral waveform analysis: a clinical application. *J Am Coll Cardiol*. 2002;40(3):521–528.
86. Oliver JJ, Webb DJ. Noninvasive assessment of arterial stiffness and risk of atherosclerotic events. *Arterioscler Thromb Vasc Biol*. 2003;23(4):554–566.
87. Wilkinson IB, Hall IR, MacCallum H, et al. Pulse-wave analysis: clinical evaluation of a noninvasive, widely applicable method for assessing endothelial function. *Arterioscler Thromb Vasc Biol*. 2002;22(1):147–152.
88. Dawes M, Chowienczyk PJ, Ritter JM. Effects of inhibition of the L-arginine/nitric oxide pathway on vasodilation caused by beta-adrenergic agonists in human forearm. *Circulation*. 1997;95(9): 2293–2297.
89. Thijssen DHJ, Bruno RM, van Mil A, et al. Expert consensus and evidence-based recommendations for the assessment of flow-mediated dilation in humans. *Eur Heart J*. 2019;40(30): 2534–2547.
90. Kuvin JT, Patel AR, Sliney KA, et al. Assessment of peripheral vascular endothelial function with finger arterial pulse wave amplitude. *Am Heart J*. 2003;146(1):168–174.
91. Bonetti PO, Pumper GM, Higano ST, Holmes Jr DR, Kuvin JT, Lerman A. Noninvasive identification of patients with early coronary atherosclerosis by assessment of digital reactive hyperemia. *J Am Coll Cardiol*. 2004;44(11):2137–2141.
92. Nohria A, Gerhard-Herman M, Creager MA, Hurley S, Mitra D, Ganz P. Role of nitric oxide in the regulation of digital pulse volume amplitude in humans. *J Appl Physiol (1985)*. 2006;101(2): 545–548.
93. Young A, Garcia M, Sullivan SM, et al. Impaired peripheral microvascular function and risk of major adverse cardiovascular events in patients with coronary artery disease. *Arterioscler Thromb Vasc Biol*. 2021;41(5):1801–1809.
94. Schnabel RB, Schulz A, Wild PS, et al. Noninvasive vascular function measurement in the community: cross-sectional relations and comparison of methods. *Circ Cardiovasc Imaging*. 2011;4(4): 371–380.
95. Hamburg NM, Keyes MJ, Larson MG, et al. Cross-sectional relations of digital vascular function to cardiovascular risk factors in the Framingham Heart Study. *Circulation*. 2008;117(19): 2467–2474.
96. Vita JA, Treasure CB, Nabel EG, et al. Coronary vasomotor response to acetylcholine relates to risk factors for coronary artery disease. *Circulation*. 1990;81(2):491–497.
97. Cai H, Harrison DG. Endothelial dysfunction in cardiovascular diseases. The role of oxidant stress. *Circ Res*. 2000;87:840–844.
98. Dzau VJ. Theodore Cooper Lecture: tissue angiotensin and pathobiology of vascular disease. A unifying hypothesis. *Hypertension*. 2001;37(4):1047–1052.
99. Lerman A, Edwards BS, Hallett JW, Heublein DM, Sandberg SM, Burnett Jr JC. Circulating and tissue endothelin immunoreactivity in advanced atherosclerosis. *N Engl J Med*. 1991;325(14):997–1001.
100. Lavi S, Bae JH, Rihal CS, et al. Segmental coronary endothelial dysfunction in patients with minimal atherosclerosis is associated with necrotic core plaques. *Heart*. 2009;95(18):1525–1530.
101. Stone GW, Maehara A, Lansky AJ, et al. A prospective natural-history study of coronary atherosclerosis. *N Engl J Med*. 2011; 364(3):226–235.
102. Hasdai D, Gibbons RJ, Holmes Jr DR, Higano ST, Lerman A. Coronary endothelial dysfunction in humans is associated with myocardial perfusion defects. *Circulation*. 1997;96(10):3390–3395.
103. Celermajer DS, Sorensen KE, Bull C, Robinson J, Deanfield JE. Endothelium-dependent dilation in the systemic arteries of asymptomatic subjects relates to coronary risk factors and their interaction. *J Am Coll Cardiol*. 1994;24:1468–1474.
104. Halcox JP, Schenke WH, Zalos G, et al. Prognostic value of coronary vascular endothelial dysfunction. *Circulation*. 2002;106(6): 653–658.
105. Chan SY, Mancini GB, Kuramoto L, Schulzer M, Frohlich J, Ignaszewski A. The prognostic importance of endothelial dysfunction and carotid atheroma burden in patients with coronary artery disease. *J Am Coll Cardiol*. 2003;42(6):1037–1043.
106. Fichtlscherer S, Breuer S, Zeiher AM. Prognostic value of systemic endothelial dysfunction in patients with acute coronary syndromes: further evidence for the existence of the "vulnerable" patient. *Circulation*. 2004;110(14):1926–1932.
107. Gokce N, Keaney Jr JF, Hunter LM, et al. Predictive value of non-invasively determined endothelial dysfunction for long-term cardiovascular events in patients with peripheral vascular disease. *J Am Coll Cardiol*. 2003;41(10):1769–1775.
108. Heitzer T, Baldus S, von Kodolitsch Y, Rudolph V, Meinertz T. Systemic endothelial dysfunction as an early predictor of adverse outcome in heart failure. *Arterioscler Thromb Vasc Biol*. 2005; 25(6):1174–1179.
109. Perticone F, Ceravolo R, Pujia A, et al. Prognostic significance of endothelial dysfunction in hypertensive patients. *Circulation*. 2001;104(2):191–196.
110. Schachinger V, Britten MB, Zeiher AM. Prognostic impact of coronary vasodilator dysfunction on adverse long-term outcome of coronary heart disease. *Circulation*. 2000;101(16):1899–1906.
111. Suwaidi JA, Hamasaki S, Higano ST, Nishimura RA, Holmes Jr DR, Lerman A. Long-term follow-up of patients with mild coronary artery disease and endothelial dysfunction. *Circulation*. 2000;101(9):948–954.
112. Bonetti PO, Barsness GW, Keelan PC, et al. Enhanced external counterpulsation improves endothelial function in patients with symptomatic coronary artery disease. *J Am Coll Cardiol*. 2003; 41(10):1761–1768.

113. Szmitko PE, Fedak PW, Weisel RD, Stewart DJ, Kutryk MJ, Verma S. Endothelial progenitor cells: new hope for a broken heart. *Circulation*. 2003;107(24):3093–3100.
114. Murakami T, Mizuno S, Ohsato K, et al. Effects of troglitazone on frequency of coronary vasospastic-induced angina pectoris in patients with diabetes mellitus. *Am J Cardiol*. 1999;84(1):92–94. A98.
115. Brevetti G, Silvestro A, Schiano V, Chiariello M. Endothelial dysfunction and cardiovascular risk prediction in peripheral arterial disease: additive value of flow-mediated dilation to ankle-brachial pressure index. *Circulation*. 2003;108(17): 2093–2098.
116. Dimmeler S, Aicher A, Vasa M, et al. HMG-CoA reductase inhibitors (statins) increase endothelial progenitor cells via the PI 3-kinase/Akt pathway. *J Clin Invest*. 2001;108(3):391–397.
117. Vasa M, Fichtlscherer S, Aicher A, et al. Number and migratory activity of circulating endothelial progenitor cells inversely correlate with risk factors for coronary artery disease. *Circ Res*. 2001; 89(1):E1–E7.
118. Erbs S, Beck EB, Linke A, et al. High-dose rosuvastatin in chronic heart failure promotes vasculogenesis, corrects endothelial function, and improves cardiac remodeling–results from a randomized, double-blind, and placebo-controlled study. *Int J Cardiol*. 2011;146(1):56–63.
119. Treasure CB, Klein JL, Weintraub WS, et al. Beneficial effects of cholesterol-lowering therapy on the coronary endothelium in patients with coronary artery disease. *N Engl J Med*. 1995;332(8): 481–487.
120. Luscher TF, Vanhoutte PM, Raij L. Antihypertensive treatment normalizes decreased endothelium-dependent relaxations in rats with salt-induced hypertension. *Hypertension*. 1987;9(6 Pt 2): III193–III197.
121. Investigators E. Effect of nifedipine and cerivastatin on coronary endothelial function in patients with coronary artery disease: the ENCORE I Study (Evaluation of nifedipine and cerivastatin on recovery of coronary endothelial function). *Circulation*. 2003;107(3): 422–428.
122. Flammer AJ, Hermann F, Wiesli P, et al. Effect of losartan, compared with atenolol, on endothelial function and oxidative stress in patients with type 2 diabetes and hypertension. *J Hypertens*. 2007;25(4):785–791.
123. Taddei S, Virdis A, Ghiadoni L, Sudano I, Salvetti A. Effects of antihypertensive drugs on endothelial dysfunction: clinical implications. *Drugs*. 2002;62(2):265–284.
124. Taddei S, Virdis A, Ghiadoni L, Uleri S, Magagna A, Salvetti A. Lacidipine restores endothelium-dependent vasodilation in essential hypertensive patients. *Hypertension*. 1997;30(6):1606–1612.
125. Cangiano E, Marchesini J, Campo G, et al. ACE inhibition modulates endothelial apoptosis and renewal via endothelial progenitor cells in patients with acute coronary syndromes. *Am J Cardiovasc Drugs*. 2011;11(3):189–198.
126. Xiaozhen H, Yun Z, Mei Z, Yu S. Effect of carvedilol on coronary flow reserve in patients with hypertensive left-ventricular hypertrophy. *Blood Press*. 2010;19(1):40–47.
127. Meaney E, Vela A, Samaniego V, et al. Metformin, arterial function, intima-media thickness and nitroxidation in metabolic syndrome: the mefisto study. *Clin Exp Pharmacol Physiol*. 2008;35(8): 895–903.
128. Syngle A, Vohra K, Sharma A, Kaur L. Endothelial dysfunction in ankylosing spondylitis improves after tumor necrosis factor-alpha blockade. *Clin Rheumatol*. 2010;29(7):763–770.
129. Mehta A, Mavromatis K, Ko YA, et al. Rationale and design of the granulocyte-macrophage colony stimulating factor in peripheral arterial disease (GPAD-3) study. *Contemp Clin Trials*. 2020;91: 105975.
130. Kitta Y, Obata JE, Nakamura T, et al. Persistent impairment of endothelial vasomotor function has a negative impact on outcome in patients with coronary artery disease. *J Am Coll Cardiol*. 2009; 53(4):323–330.
131. Modena MG, Bonetti L, Coppi F, Bursi F, Rossi R. Prognostic role of reversible endothelial dysfunction in hypertensive postmenopausal women. *J Am Coll Cardiol*. 2002;40(3):505–510.
132. Rossi R, Chiurlia E, Nuzzo A, Cioni E, Origliani G, Modena MG. Flow-mediated vasodilation and the risk of developing hypertension in healthy postmenopausal women. *J Am Coll Cardiol*. 2004; 44(8):1636–1640.
133. Wilkinson IB, MacCallum H, Cockcroft JR, Webb DJ. Inhibition of basal nitric oxide synthesis increases aortic augmentation index and pulse wave velocity in vivo. *Br J Clin Pharmacol*. 2002;53(2): 189–192.
134. Chirinos JA, Segers P, Hughes T, Townsend R. Large-artery stiffness in health and disease: JACC state-of-the-art review. *J Am Coll Cardiol*. 2019;74(9):1237–1263.
135. Vlachopoulos C, Aznaouridis K, Stefanadis C. Clinical appraisal of arterial stiffness: the Argonauts in front of the Golden Fleece. *Heart*. 2006;92(11):1544–1550.
136. Cohn JN, Quyyumi AA, Hollenberg NK, Jamerson KA. Surrogate markers for cardiovascular disease: functional markers. *Circulation*. 2004;109(25 Suppl 1):IV31–IV46.
137. Vlachopoulos C, O'Rourke M, Nichols WW. *McDonald's Blood Flow in Arteries: Theoretical, Experimental and Clinical Principles*. CRC press; 2011.
138. Lehmann ED, Hopkins KD, Rawesh A, et al. Relation between number of cardiovascular risk factors/events and noninvasive Doppler ultrasound assessments of aortic compliance. *Hypertension*. 1998;32(3):565–569.
139. Jiang B, Liu B, McNeill KL, Chowienczyk PJ. Measurement of pulse wave velocity using pulse wave Doppler ultrasound: comparison with arterial tonometry. *Ultrasound Med Biol*. 2008;34(3): 509–512.
140. van der Meer RW, Diamant M, Westenberg JJ, et al. Magnetic resonance assessment of aortic pulse wave velocity, aortic distensibility, and cardiac function in uncomplicated type 2 diabetes mellitus. *J Cardiovasc Magn Reson*. 2007;9(4):645–651.
141. Wiesmann F, Petersen SE, Leeson PM, et al. Global impairment of brachial, carotid, and aortic vascular function in young smokers: direct quantification by high-resolution magnetic resonance imaging. *J Am Coll Cardiol*. 2004;44(10):2056–2064.
142. Mitchell GF, Guo CY, Benjamin EJ, et al. Cross-sectional correlates of increased aortic stiffness in the community: the Framingham Heart Study. *Circulation*. 2007;115(20):2628–2636.
143. McEniery CM, Yasmin, Hall IR, Qasem A, Wilkinson IB, Cockcroft JR. Normal vascular aging: differential effects on wave reflection and aortic pulse wave velocity. The Anglo-Cardiff Collaborative Trial (ACCT). *J Am Coll Cardiol*. 2005;46(9):1753–1760.
144. Din-Dzietham R, Couper D, Evans G, Arnett DK, Jones DW. Arterial stiffness is greater in African Americans than in whites: evidence from the Forsyth County, North Carolina, ARIC cohort. *Am J Hypertens*. 2004;17(4):304–313.
145. Chen W, Srinivasan SR, Bond MG, et al. Nitric oxide synthase gene polymorphism (G894T) influences arterial stiffness in adults: the Bogalusa Heart Study. *Am J Hypertens*. 2004;17(7):553–559.
146. Chirinos JA, Kips JG, Jacobs DR, et al. Arterial wave reflections and incident cardiovascular events and heart failure: MESA (Multiethnic Study of Atherosclerosis). *J Am Coll Cardiol*. 2012; 60(21):2170–2177.
147. Townsend RR, Wilkinson IB, Schiffrin EL, et al. Recommendations for improving and standardizing vascular research on arterial stiffness: a Scientific statement from the American Heart Association. *Hypertension*. 2015;66(3):698–722.

148. Chirinos JA, Segers P, Duprez DA, et al. Late systolic central hypertension as a predictor of incident heart failure: the Multi-ethnic Study of Atherosclerosis. *J Am Heart Assoc*. 2015;4(3):e001335.
149. Hashimoto J, Ito S. Central pulse pressure and aortic stiffness determine renal hemodynamics: pathophysiological implication for microalbuminuria in hypertension. *Hypertension*. 2011;58(5):839–846.
150. Scialla JJ, Leonard MB, Townsend RR, et al. Correlates of osteoprotegerin and association with aortic pulse wave velocity in patients with chronic kidney disease. *Clin J Am Soc Nephrol*. 2011;6(11):2612–2619.
151. Peyster E, Chen J, Feldman HI, et al. Inflammation and arterial stiffness in chronic kidney disease: findings from the CRIC study. *Am J Hypertens*. 2017;30(4):400–408.
152. Sanders PW. Effect of salt intake on progression of chronic kidney disease. *Curr Opin Nephrol Hypertens*. 2006;15(1):54–60.
153. Mehdi UF, Adams-Huet B, Raskin P, Vega GL, Toto RD. Addition of angiotensin receptor blockade or mineralocorticoid antagonism to maximal angiotensin-converting enzyme inhibition in diabetic nephropathy. *J Am Soc Nephrol*. 2009;20(12):2641–2650.
154. Guerin AP, Blacher J, Pannier B, Marchais SJ, Safar ME, London GM. Impact of aortic stiffness attenuation on survival of patients in end-stage renal failure. *Circulation*. 2001;103(7):987–992.
155. Rabkin SW. Arterial stiffness: detection and consequences in cognitive impairment and dementia of the elderly. *J Alzheimers Dis*. 2012;32(3):541–549.
156. Poels MM, Zaccai K, Verwoert GC, et al. Arterial stiffness and cerebral small vessel disease: the Rotterdam Scan Study. *Stroke*. 2012;43(10):2637–2642.
157. Pase MP, Beiser A, Himali JJ, et al. Aortic stiffness and the risk of incident mild cognitive impairment and dementia. *Stroke*. 2016;47(9):2256–2261.
158. Meyer ML, Palta P, Tanaka H, et al. Association of central arterial stiffness and pressure pulsatility with mild cognitive impairment and dementia: the Atherosclerosis Risk in Communities Study-Neurocognitive Study (ARIC-NCS). *J Alzheimers Dis*. 2017;57(1):195–204.
159. Henskens LH, Kroon AA, van Oostenbrugge RJ, et al. Increased aortic pulse wave velocity is associated with silent cerebral small-vessel disease in hypertensive patients. *Hypertension*. 2008;52(6):1120–1126.
160. Hughes TM, Kuller LH, Barinas-Mitchell EJ, et al. Arterial stiffness and β-amyloid progression in nondemented elderly adults. *JAMA Neurol*. 2014;71(5):562–568.
161. Hughes TM, Wagenknecht LE, Craft S, et al. Arterial stiffness and dementia pathology: atherosclerosis risk in Communities (ARIC)-PET study. *Neurology*. 2018;90(14):e1248–e1256.
162. Kaess BM, Rong J, Larson MG, et al. Aortic stiffness, blood pressure progression, and incident hypertension. *J Am Med Assoc*. 2012;308(9):875–881.
163. Pierce GL. Mechanisms and subclinical consequences of aortic stiffness. *Hypertension*. 2017;70(5):848–853.
164. Ben-Shlomo Y, Spears M, Boustred C, et al. Aortic pulse wave velocity improves cardiovascular event prediction: an individual participant meta-analysis of prospective observational data from 17,635 subjects. *J Am Coll Cardiol*. 2014;63(7):636–646.
165. Franklin SS, Khan SA, Wong ND, Larson MG, Levy D. Is pulse pressure useful in predicting risk for coronary heart Disease? The Framingham heart study. *Circulation*. 1999;100(4):354–360.
166. Williams B, Lacy PS, Thom SM, et al. Differential impact of blood pressure-lowering drugs on central aortic pressure and clinical outcomes: principal results of the Conduit Artery Function Evaluation (CAFE) study. *Circulation*. 2006;113(9):1213–1225.
167. Boesby L, Elung-Jensen T, Strandgaard S, Kamper A-L. Eplerenone attenuates pulse wave reflection in chronic kidney disease stage 3–4-A randomized controlled study. *PLoS One*. 2013;8(5):e64549.
168. Kithas PA, Supiano MA. Spironolactone and hydrochlorothiazide decrease vascular stiffness and blood pressure in geriatric hypertension. *J Am Geriatr Soc*. 2010;58(7):1327–1332.
169. Tsoucaris D, Benetos A, Legrand M, London GM, Safar ME. Proximal and distal pulse pressure after acute antihypertensive vasodilating drugs in Wistar-Kyoto and spontaneously hypertensive rats. *J Hypertens*. 1995;13(2):243–249.
170. Cecelja M, Jiang B, Spector TD, Chowienczyk P. Progression of central pulse pressure over 1 decade of aging and its reversal by nitroglycerin: a twin study. *J Am Coll Cardiol*. 2012;59(5):475–483.
171. Pauca AL, Kon ND, O'Rourke MF. Benefit of glyceryl trinitrate on arterial stiffness is directly due to effects on peripheral arteries. *Heart*. 2005;91(11):1428–1432.
172. Stokes GS, Barin ES, Gilfillan KL. Effects of isosorbide mononitrate and AII inhibition on pulse wave reflection in hypertension. *Hypertension*. 2003;41(2):297–301.
173. Mackenzie IS, McEniery CM, Dhakam Z, Brown MJ, Cockcroft JR, Wilkinson IB. Comparison of the effects of antihypertensive agents on central blood pressure and arterial stiffness in isolated systolic hypertension. *Hypertension*. 2009;54(2):409–413.
174. Ait-Oufella H, Collin C, Bozec E, et al. Long-term reduction in aortic stiffness: a 5.3-year follow-up in routine clinical practice. *J Hypertens*. 2010;28(11):2336–2341.
175. Benetos A, Adamopoulos C, Bureau JM, et al. Determinants of accelerated progression of arterial stiffness in normotensive subjects and in treated hypertensive subjects over a 6-year period. *Circulation*. 2002;105(10):1202–1207.
176. Van Bortel LM, Struijker-Boudier HA, Safar ME. Pulse pressure, arterial stiffness, and drug treatment of hypertension. *Hypertension*. 2001;38(4):914–921.
177. Shahin Y, Khan JA, Chetter I. Angiotensin converting enzyme inhibitors effect on arterial stiffness and wave reflections: a meta-analysis and meta-regression of randomised controlled trials. *Atherosclerosis*. 2012;221(1):18–33.
178. Mallareddy M, Parikh CR, Peixoto AJ. Effect of angiotensin-converting enzyme inhibitors on arterial stiffness in hypertension: systematic review and meta-analysis. *J Clin Hypertens*. 2006;8(6):398–403.
179. Ong K-T, Delerme S, Pannier B, et al. Aortic stiffness is reduced beyond blood pressure lowering by short-term and long-term antihypertensive treatment: a meta-analysis of individual data in 294 patients. *J Hypertens*. 2011;29(6):1034–1042.
180. Hamburg NM, Benjamin EJ. Assessment of endothelial function using digital pulse amplitude tonometry. *Trends Cardiovasc Med*. 2009;19(1):6–11.
181. Busse R, Fleming I. Regulation and functional consequences of endothelial nitric oxide formation. *Ann Med*. 1995;27(3):331–340.
182. Rubanyi GM. The role of endothelium in cardiovascular homeostasis and diseases. *J Cardiovasc Pharmacol*. 1993;22(Suppl 4):S1–S14.
183. Roman MJ, Naqvi TZ, Gardin JM, Gerhard-Herman M, Jaff M, Mohler E. Clinical application of noninvasive vascular ultrasound in cardiovascular risk stratification: a report from the American Society of Echocardiography and the Society of Vascular Medicine and Biology. *J Am Soc Echocardiogr*. 2006;19(8):943–954.
184. Colacurci N, Manzella D, Fornaro F, Carbonella M, Paolisso G. Endothelial function and menopause: effects of raloxifene administration. *J Clin Endocrinol Metab*. 2003;88(5):2135–2140.

185. Bairey Merz CN, Shaw LJ, Reis SE, et al. Insights from the NHLBI-Sponsored Women's Ischemia Syndrome Evaluation (WISE) Study: Part II: gender differences in presentation, diagnosis, and outcome with regard to gender-based pathophysiology of atherosclerosis and macrovascular and microvascular coronary disease. *J Am Coll Cardiol*. 2006;47(3 Suppl):S21–S29.
186. Elesber AA, Redfield MM, Rihal CS, et al. Coronary endothelial dysfunction and hyperlipidemia are independently associated with diastolic dysfunction in humans. *Am Heart J*. 2007;153(6): 1081–1087.
187. Rossi R, Nuzzo A, Origliani G, Modena MG. Prognostic role of flow-mediated dilation and cardiac risk factors in post-menopausal women. *J Am Coll Cardiol*. 2008;51(10):997–1002.
188. Lerman A, Zeiher AM. Endothelial function: cardiac events. *Circulation*. 2005;111(3):363–368.
189. Jaffe R, Charron T, Puley G, Dick A, Strauss BH. Microvascular obstruction and the no-reflow phenomenon after percutaneous coronary intervention. *Circulation*. 2008;117(24):3152–3156.
190. Ong P, Athanasiadis A, Hill S, Vogelsberg H, Voehringer M, Sechtem U. Coronary artery spasm as a frequent cause of acute coronary syndrome: the CASPAR (coronary artery spasm in patients with acute coronary syndrome) study. *J Am Coll Cardiol*. 2008;52(7):523–527.
191. Nichols WW, O'Rourke MF. *McDonald's Blood Flow in Arteries: Theoretical, Experimental, and Clinical Principles*. 5th ed. London: Hodder Arnold; 2005.

SECTION 2

Realizing the promise of new high-content technologies

CHAPTER

17

Angiodiversity—A tale retold by comparative transcriptomics

Xiaowu Gu and Ondine Cleaver

Department of Molecular Biology, University of Texas Southwestern Medical Center, Dallas, TX, United States

Introduction

Endothelial cells (ECs) are cells that line both blood and lymphatic vessels and together constitute the vascular system which pervades nearly all the tissues to sustain life. Both the blood and lymphatic vessels form hierarchical trees of large and small caliber vessels. Across these varied vessels and throughout different organs, ECs exhibit dramatically different structural, functional, and molecular properties.[1–7] This observed endothelial heterogeneity gives rise to a host of questions. How do ECs coordinate their own growth with developing organs and achieve regional specializations adapted to different tissue landscapes and functional requirements? If ECs are so different from organ to organ, is their only commonality that they line channels that carry blood? What induces their dramatic regional differences?

An emerging concept in the past two decades that has provided key insight into the development and function of this vascular heterogeneity is the idea of EC-tissue signaling reciprocity.[2,5,6,8–11] In this context, ECs receive molecular cues (i.e., angiogenic factors) secreted by nonvascular cells (e.g., mesenchymal cells, epithelial cells, immune cells, neuronal cells, niche stem cells, etc.) within respective tissues, which in turn respond to EC-derived factors (i.e., angiocrine factors). How these cells converse and integrate trophic signals to ultimately orchestrate organ-vascular development remains an area of active research.

Given the intimate association of vascular and nonvascular cell types, it is clear that ECs are poised to transmit regionally specific signals to the organs and tissues they pervade. Conversely, and not surprisingly, dysfunctional ECs can drive and aggravate aging, cancers, and other cardiovascular/noncardiovascular diseases.

In the last few years, innovations in model systems, imaging platforms, gene-editing techniques, and rapidly evolving next-generation sequencing methods have expanded our understanding of just how amazingly diverse the cells of the vascular system can be. We see the depth and breadth of diversity across organ systems, at the resolution of a single cell, and in the genetic and epigenetic levels. In this focused review, we will discuss coordinated EC-tissue interaction as an emerging paradigm in organogenesis, both in terms of generating endothelial heterogeneity and also in terms of positioning ECs to provide locally relevant signals. First, we will provide a brief overview of blood and lymphatic vascular formation, and we will summarize our current knowledge regarding endothelial heterogeneity. We will then review recent vasculomic studies centered on the brain and the kidney as excellent contexts to study EC heterogeneity. Lastly, we will conclude by highlighting current efforts to engineer vascularized tissues and propose new directions in EC-tissue interaction research.

Diverse origins of ECs and various modes of vessel formation

Vasculogenesis and angiogenesis are the two primary mechanisms for generating new blood vessels, both during development and in the context of disease. Understanding and differentiating between these modes of blood vessel formation is critical, as the molecular and cellular mechanisms that underlie these processes are very different.

The Vasculome
https://doi.org/10.1016/B978-0-12-822546-2.00029-0

Vasculogenesis

Vasculogenesis defines the de novo blood vessel formation from EC precursor cells or angioblasts.[12–14] During embryogenesis, angioblasts emerge from lateral plate mesoderm (LPM), migrate, then aggregate de novo into linear groups of cells, and coalesce into vascular cords. How cells know where to go is driven by both positive and negative sets of cues in the microenvironment. These cues govern the patterning of blood vessels and help coordinate formation of the primitive vascular plexus. These primitive linear aggregates of pre-ECs then undergo morphological changes, such as junction reorganization, cellular thinning, and lumen opening, to create the first primitive blood vessels.[15,16] The process of opening central lumens to carry blood is a multistep process called tubulogenesis.[17,18] The first vessels in the embryo primarily constitute the dorsal aortae (DA) and cardinal veins (CVs), and most subsequently formed vessels will emerge via coordinated vasculogenesis and angiogenic sprouting.

EC fate

Interestingly, most ECs of the early vascular plexus emerge prior to blood flow, and display markers of both arterial and venous fate. Some ECs in the primitive DA are fated to switch from arterial to venous fate, and form CVs.[19] Furthermore, the venous cells located in the anterior CV transdifferentiate to become lymphatic (LECs).[20] These findings underscore the concept of EC heterogeneity even in the earliest blood vessels, despite a common origin from the LPM-derived angioblasts. Erythro-myeloid progenitors may be an alternative source to ECs,[21] but this contribution has recently been questioned.[22]

Angiogenesis

Angiogenesis refers to a general process whereby new vessels grow from preexisting ones. In sprouting angiogenesis, ECs sprout ("tip cells") and extend ("stalk cells") from existing vessels into avascular tissues, following environmental cues.[23–25] A classic example of sprouting angiogenesis is postnatal retinal vascularization.[26–28] Another mode of angiogenesis is intussusceptive angiogenesis or nonsprouting angiogenesis. It was first discovered in postnatal rat lung microvasculature and defines a process in which existing vessels split to make new ones.[29,30] In tumors, three additional modes of vessel formation have been proposed, including vascular mimicry, vessel co-option, and direct EC differentiation from cancer stem cell-like tumor cells.[24]

Lymphangiogenesis

The formation of the lymphatic vascular network primarily occurs in a similar manner to blood vessel sprouting angiogenesis and is termed lymphangiogenesis.[23,31] A venous origin of LECs is by far the most prevalent and is evolutionarily conserved.[20,32,33] In mouse, initial LECs arise and bud from the CV (i.e., transdifferentiation) to form the primitive lymph sacs, from where Prox1-expressing LECs continue to sprout and generate the entire lymphatic vasculature. Interestingly, the zebrafish posterior CV (where their lymphatics are derived from) harbors specialized angioblasts capable of forming LECs, as well as ECs destined to become both arterial and venous.[34] Furthermore, several nonvenous LEC progenitor populations have been discovered across multiple species.[7,32] They contribute to lymphatic development by a vasculogenic process, or lymphvasculogenesis, although the precise identities of some progenitors remain to be determined. Venous and nonvenous sources of LECs may cooperate via distinct mechanisms (i.e., lymphangiogenesis vs. lymphvasculogenesis) to spatiotemporally sculpt local lymphatics. For example, in the mouse heart, dorsal and ventral surface lymphatic vessels arise from venous ECs and the second heart field, respectively.[35,36] Recently, Stone and Stainier observed incorporation of paraxial mesoderm (PXM)-derived cells into the CV before specification and emergence of LECs from the same site, suggesting a significant contribution from PXM to CV and subsequent lymphatic vessel development.[7] A substantial portion of PXM-derived LECs is found in many organs (e.g., heart, lung, skin, liver, and meninges). Interestingly, PXM contributes little to blood vessels, except in skin and liver.

Vascular expansion during growth and remodeling angiogenesis

Following vasculogenesis and initial angiogenesis, the initial primitive vascular plexus is then progressively elaborated and transformed into a complex and hierarchical vascular network. As the embryo grows dramatically in size over the course of weeks and months, and as organs are formed and expanded, blood vessels expand coordinately with each tissue they pervade. Intrinsic genetic influence as well as extrinsic stimuli and cues, such as molecular signals, blood flow, and cell–cell interactions, continuously shape the vasculature to fit the needs of each new cell that arises in each emerging organ. Diversity of ECs and blood vessels therefore arises in a stepwise and increasingly complex manner over time. We and others aim to understand this complex tree of cell fate and differentiation decisions within the vasculature. In addition, we aim to build a "vasculome," with spatial and temporal resolution, and to understand the nodes that define where EC diversity arises.

Discovery of endothelial heterogeneity

As early as the 1950–60s, ECs were recognized displaying not only diverse intracellular features but also incredible ultrastructural heterogeneity.[37] Morphologically, microvascular ECs across different tissues fall into the following three categories, which are determined by elegant electron microscopy studies (reviewed in Refs. 1,5,38). Largely, their categorization is based on continuity of the ECs and on configuration of the associated basement membrane (BM) (Fig. 17.1).

1. *Continuous endothelium*: Continuous ECs are connected to one another via tightly organized cell junctions. They are also surrounded by a continuous BM and lack fenestrae. This configuration renders ECs generally less permeable and continuous ECs form a "barrier" between the blood and tissue cells. Continuous ECs make up all the arteries, veins, and most capillaries in the brain, lung, heart, and skin.
2. *Fenestrated endothelium*: Similar to continuous ECs, fenestrated ECs retain an intact BM and is characterized by the presence of fenestrae or fenestrations. These are transcellular pores of 60–80 nm in diameter that traverse a single EC; they allow rapid transcapillary passage of small substances. Fenestrated ECs are found in tissues active in filtration and secretion (e.g., choroid plexus, intestine villi, kidney peritubular capillaries, glomeruli, and endocrine/exocrine organs) and in specific brain regions that release hormones and sense peripheral signals (e.g., circumventricular organs (CVOs)). Most fenestrae are spanned by diaphragms that enable refined regulation of permeability, with one exception being the adult kidney glomerular ECs.
3. *Discontinuous endothelium*: Distinct from the previous two types, the discontinuous EC is provided with a segmented layer of BM, intercellular gaps, and nondiaphragmed fenestrae with a typical size range of 100–200 nm. This configuration allows diffusion of both small solutes and macromolecules and is a feature of liver and bone marrow sinusoidal ECs.

While these categorizations of the endothelium provided an initial appreciation of its heterogeneity, many more structural cellular features have now been identified in ECs. Together, these characteristics provide further landmarks for delineating regional differences in vascular beds.

Interest in angiodiversity,[39,40] however, did not end with these early studies. In 1987, Mary Gerritsen linked endothelial heterogeneity to its many different functions and varied metabolism, and she noted how culture of ECs could drastically change both their structure and biochemistry.[41] In the late 90s, with the onset of transgenic reporters, Wang et al. reported the differential expression of the ligand ephrin-B2 as preferentially expressed in arterial ECs, while its receptor EphB4 is expressed in venous ECs,[42] thereby demonstrating a clear molecular difference between different types of blood vessels. In 1997 and again in 1999, Pasqualini et al. used phage display to not only reveal inherent

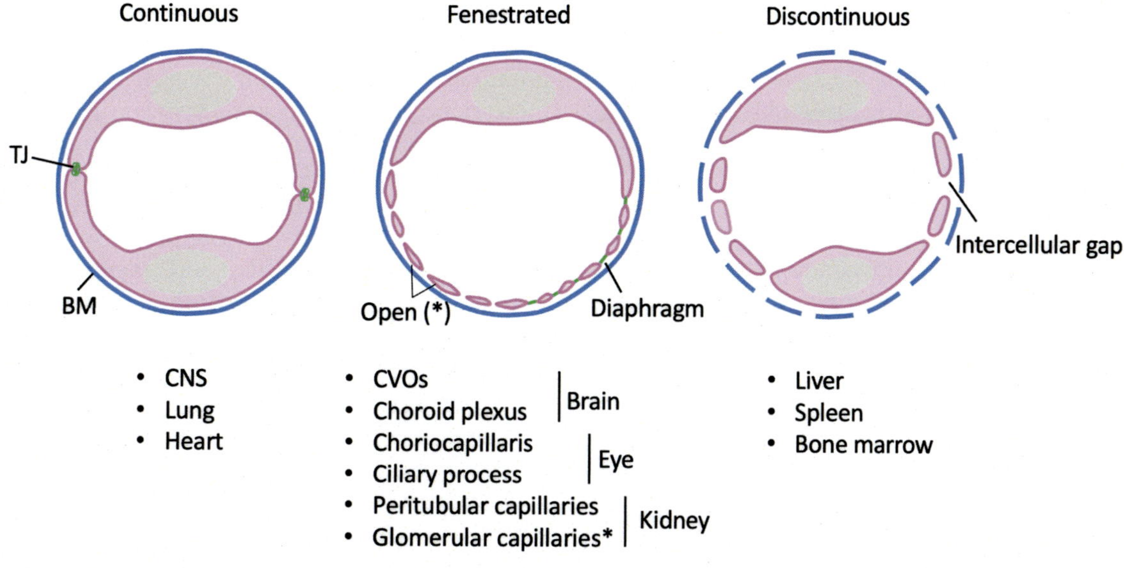

FIGURE 17.1 **Morphological configurations of microvascular ECs.** Illustration of endothelial heterogeneity. Ultrastructural continuity differences in endothelial cells in continuous, fenestrated, or discontinuous vessels, in the different tissues listed.

molecular differences between healthy and tumor blood vessels but they also used this system to deliver anti-tumor agents in a targeted manner, reducing toxicity normally associated with treatment.[43,44] Altogether, these studies laid the groundwork for decades of focused investigations into what made vascular cells different across different vascular beds.

Endothelial heterogeneity and organ-specific vasculature

One easily recognizable difference between blood vessels is that of vessel size and anatomy. ECs of blood and lymphatic vessels are functionally partitioned within the vascular hierarchy (arteries/veins/arterioles/venules/capillaries). In addition, they exhibit incredible structural, functional, and molecular heterogeneity throughout the body, displaying striking organ-specificity. Capillary beds within different organs and tissues are specialized in various ways such that the energy requirements and functions of the underlying tissues are satisfied (i.e., interorgan heterogeneity).[3,5,9] Within the same organ or vascular bed, ECs are geographically and molecularly distinct in order to fulfill regionalized tasks and modulate local environment (i.e., intraorgan heterogeneity).[1,45–49] At the transcriptional level, there are likely as many different types of ECs as there are tissues, or even different individual cell types.

Structural endothelial features—caveolae versus fenestrae

Understanding how ECs develop and how they regulate certain microdomains in space and time will be a crucial step toward understanding the origins and functional implications of endothelial heterogeneity. Here, we focus on subcellular structures observed in ECs that can be used as landmarks for delineating signatures of different vascular beds.

Beside fenestrae, ECs contain several other subcellular structures including caveolae, vesiculo-vacuolar organelles (VVOs), and transendothelial channels (TECs).[5,38] These structures in part determine permeability and are differentially distributed in tissue-specific blood vessels and/or in vessel segments. For example, caveolae are abundant in lung capillaries but are rare in BBB-containing ECs,[50] even though both lung and brain capillary ECs are continuous. Brain capillaries are devoid of fenestrae. VVOs are rarely observed in normal capillary ECs but are concentrated in postcapillary venules and in tumor microvasculature.[51] TECs are more frequently found in fenestrated ECs than in continuous vessels. However, none of the structures are mutually exclusive—a fenestrated EC can also have caveolae and TECs. Caveolae and VVOs can be present in the same cell.

Caveolae, also known as plasmalemma vesicles, are specialized lipid rafts involved in vesicular trafficking, cholesterol metabolism, mechanotransduction, and cell signaling.[52,53] These membrane invaginations with sizes of 50–100 nm have a characteristic flask-shaped appearance and lack a clear electron-dense coating observed with clathrin-coated pits. Like fenestrae, caveolae frequently have diaphragms flanking the neck region. Caveolins (i.e., caveolin-1/Cav1, Cav2, and Cav3) are the major scaffold proteins of caveolae: Cav1 is obligatory for caveolar biogenesis in nearly all cells except skeletal muscle cells whose caveolar formation exclusively requires Cav3. In the past decade, the cavin family of proteins (composed of Cavin1, Cavin2, Cavin3, and Cavin4) was discovered as another obligatory structural components required for caveolar assembly.[52] Therefore, Cav1 and Cavin1 expression levels in a particular cell type generally correlate with its caveolar numbers and function.

Unlike caveolae, the biogenesis of fenestrae remains poorly understood. A major roadblock is a lack of marker protein(s), which would otherwise enable systematic investigation of fenestral biology using genetic and biochemical toolkits.[54] A molecule named plasmalemma vesicle-associated protein (PLVAP, or PV1) is sometimes mistaken as a fenestral marker. In fact, PLVAP is only responsible for making diaphragms bridging fenestrae, caveolae, and TECs, and it is not required for fenestral biogenesis.[55,56] For example, lung capillary ECs are continuous and abundantly express PLVAP, which correlates with numerous diaphragmed caveolae in lung vasculature. In addition, these nanoscale EC structures including fenestrae are preserved without the presence of diaphragms in Plvap deficient mice.[57,58] EC-targeted PLVAP reexpression in constitutive Plvap null mice fully restores fenestral diaphragms and functions.[57] More precisely, PLVAP marks transcellular permeability, and mediates fenestral/caveolar function by controlling diaphragm assembly.

Throughout the body, fenestrated capillaries are typically present in tissues engaged in filtration and secretion (e.g., glomeruli, choroid plexus, and pancreatic islets) or in blood-borne signal sensing (e.g., CVOs). Futures studies focusing on fenestral formation and/or maintenance will need to consider context, including special physiological sites and vascular beds.

Paracrine VEGF signaling, fenestral biogenesis and maintenance

Morphological studies of the choriocapillaris in the eye offered some initial clues regarding formation of distinct EC features. Reports identified choriocapillaris fenestrations that were polarized toward the retinal pigment epithelium (RPE). Interestingly, when retinal ECs—both continuous and nonfenestrated—are encapsulated by RPE under certain degenerative conditions, they develop fenestrations.[59] These observations indicate that the RPE may secrete an inductive signal enabling fenestral formation. Following the discovery of vascular endothelial growth factor (VEGF), Breier et al. characterized and reported strong VEGF mRNA expression in choroid plexus epithelial cells and kidney podocytes, which both are positioned adjacent to fenestrated ECs.[60] This spatial arrangement resembles that of choriocapillaris and RPE—a predominant VEGF source in the eye.[61,62] Furthermore, ECs express specific receptors to VEGF such as Flk1 and Flt1.[63] Altogether, paracrine VEGF signaling is likely critical in the establishment and/or the maintenance of fenestrations in ECs next to epithelial cells.

Is VEGF a de facto inducer of fenestrations? Indeed, topical VEGF application to adult rat cremaster muscle for as little as 10 min was able to transform continuous ECs to the fenestrated type,[64] likely through rapid cytoskeletal rearrangements. Further supporting the idea that paracrine VEGF signaling influences fenestral induction, nonfenestrated ECs remarkably develop fenestrations when co-cultured with choroid plexus epithelial cells endogenously expressing VEGF at high levels, or with mammary gland epithelial cells exogenously overexpressing VEGF.[65] Unlike the quick and efficient response observed in vivo, ECs cultured in vitro took more than 24 h for fenestrations to develop and the magnitude of fenestral induction was very low, averaging less than 1 fenestration per cell.[64,65] The disparity may be attributed to experimental context, such as presence of additional factors in vivo which are not present in vitro, different VEGF concentrations (e.g., 200 ng/mL vs. 50–100 ng/mL), or tissue-specific differences of ECs with underlying competence difference (e.g., muscle vs. adrenal cortex ECs). Importantly, genetic ablations and/or pharmacological interventions to disrupt VEGF/VEGFR signaling reduce fenestrations and cause functional deficits in almost all fenestrated vascular beds (e.g., choriocapillaris, choroid plexus, glomeruli, pituitary, pancreatic islets, etc.),[66–71] lending strong support to idea of paracrine VEGF dependency in the maintenance of fenestrations. Paradoxically, recombinant VEGF-A delivery to monkey eyes failed to induce fenestrations; instead, it upregulated numerous caveolae and hyperpermeability in the retina.[72] Furthermore, fenestral formation is observed in retinal ECs defective in Wnt signaling,[73,74] which likely is secondary to heightened VEGF signaling. Altogether, the data may suggest a layered mechanism and a dependence on cellular competence in driving fenestral formation via paracrine VEGF signaling.

Modeling fenestral formation in vitro could be a powerful and high-throughput approach to identify key players and ultimately define sequential steps during assembly of capillary fenestrae in vivo. An efficient generation of fenestrae to a magnitude comparable to that in vivo would ensure consistency and quantitative measurements. Growth factors and other extracellular signals, ECM components and cytoskeleton have been identified to play a part in fenestral induction and/or modulation. Most notably, Ioannidou et al., established a platform in which latrunculin A (i.e., a potent inhibitor of actin polymerization) rapidly induces fenestrations in bEND5 cells (i.e., a mouse endothelioma cell line) within hours and up to a density reported in kidney capillaries in situ, about 30 fenestrae per square micron.[75] The fenestra-composed patches are called sieve plates. A major advantage of this system is that latrunculin A promptly redistributes PLVAP from caveolae to fenestrae, enabling fluorescence-based readout (i.e., PLVAP staining) of fenestral formation.[75] The same group recently identified a linkage of Na,K-ATPase, PLVAP, and actin/fodrin membrane cytoskeleton in fenestral biogenesis, which is differentially regulated by ERM protein moesin and actin binding protein annexin II.[54] Remarkably, topical application of ouabain, a specific Na,K-ATPase inhibitor to rat cremaster muscle in vivo induces fenestrations within 10 min in the continuous vessels and results in increased permeability,[54] an effect previously reported in VEGF topical treatment.[64] While the functions of moesin, annexin II, and Na,K-APTase in fenestrated vascular beds across the body remain to be determined in vivo, this platform may be of use for future screening candidate players in fenestral morphogenesis and may stimulate future efforts to elucidate molecular underpinning of fenestral assembly.

Endothelial specialization in the CNS

As anatomical constituents of the blood–brain barrier (BBB) and the analogous blood–retinal barrier (BRB), capillary ECs supplying the central nervous system (CNS) are endowed with unique properties. In particular, these ECs exhibit (a) specialized tight junctions (TJs), (b) inhibited transcytosis, (c) expression of specific influx and efflux transporters, (d) presence of leukocyte adhesion molecules, and (e) high mitochondria number—that mark them distinct from any other vascular bed in the body.[76,77] Within the proximity of

ECs exist diverse cell types including pericytes, glial and neuronal cells, and perivascular fibroblasts that promote, support, and regulate the barrier properties by direct interaction and paracrine signaling, forging the concept of neurovascular unit.[76,77] During BBB/BRB development, Wnt/β-catenin signaling is a major pathway to govern CNS angiogenesis and barriergenesis: neuroepithelium-derived Wnts (e.g., Wnt7a, 7b), and glia-produced norrin (in retina and cerebellum) bind to frizzled 4 receptor and co-receptors (e.g., LRP5/6, TSPN12, GPR124, and RECK) on BBB/BRB ECs to elicit canonical Wnt/β-catenin signaling, turning on barrier inductive genes.[77]

Why and how the brain vessels become so unique at the molecular level has intrigued the field for decades. Complementing classic morphogenetic studies, large-scale cross-organ sequencing studies have been invaluable to provide an unbiased, holistic view of interorgan heterogeneity and have facilitated our quest to understand the special brain vasculature. The sequencing techniques are constantly evolving, from early DNA microarray to bulk RNA-seq and to recent innovations enabling transcriptomic and epigenomic profiling at single cell or single nucleus resolution. Here we will highlight a few studies representative of the incremental progress utilizing various sequencing platforms and that bring novel insights to interorgan and intraorgan heterogeneity of CNS ECs.

In an early and elegant microarray study by Daneman et al., ECs were isolated to high purity via fluorescence-activated cell sorting (FACS) from cerebral cortex, liver, and lung from Tie2-GFP mice.[76] This combinatorial use of reporter mice and FACS-based cell enrichment is now a common method for next-generation multiomics studies. By generating comprehensive microarray datasets that cross-compare gene expression between adult brain parenchymal cells, vascular cells (ECs and pericytes) and ECs, between brain, lung, and liver ECs, as well as between adult and early postnatal brain ECs, the authors identified large sets of BBB-enriched transcripts. These included tight junctional components, transporters, receptors, cytoplasmic signaling proteins and enzymes, transcription factors, and secreted molecules.

In line with their function as a tight barrier, BBB ECs were shown to express high levels of TJ molecules such as Jam4 and Occludin and tricellular junctional components, namely Marveld2/tricellulin and angulin-1/LSR.76 At the protein level, tricellulin and angulin/LSR have previously been shown to be selectively enriched in BBB ECs.[78] In addition, angulin-1/LSR expression has been shown to coincide with BBB maturation during development and its loss results in defective BBB in mice,[79] supporting a pivotal role of tricellular junction in BBB development and function. As Wnt/β-catenin signaling is active and required in BBB vasculature, several Wnt/β-catenin targets and regulators are significantly enriched in BBB ECs over peripheral ECs including Apcdd1, Zic3, Foxq1, Foxf2, Lef1, and Slc2a1/Glut1. Apcdd1 is a newly identified Wnt inhibitor and has recently been shown to regulate BRB maturation.[80,81]

Like in any large datasets, the Daneman et al. microarray study uncovered many novel genes whose links with BBB formation and performance had yet been clearly established. One such gene is Mfsd2a, or major facilitator super family domain containing 2a (discussed below). Mfsd2a was selected from a microarray comparison of E14.5 brain and lung ECs, prior to BBB sealing around E15.5. Mfsd2a is exclusively expressed in BBB ECs and has been identified as the only known brain-specific docosahexaenoic acid (DHA) transporter.[82,83] Genetic ablation of Mfsd2a in mice leads to drastic reduction in brain DHA level, hippocampal neuronal loss, and microcephaly, as well as persisting vascular permeability lasting into adulthood.[82] The overall brain vascular pattern and endothelial TJ structure/function remain unchanged, suggesting an intact paracellular pathway. Intriguingly, the hyperpermeability correlates with increased caveolae numbers in brain ECs: HRP-laden caveolae are often observed in the abluminal side of ECs, indicating that Mfsd2a may suppress caveolae-mediated transcellular permeability to promote BBB maturation.

A subsequent study showed that, as expected, the vascular leakage observed upon loss of Mfsd2a was prevented in Mfsd2a/Cav1 double knockout mice, where no caveolae are present due to the loss of Cav-1.[50] Moreover, overexpression of Mfsd2a in ECs in vitro and in vivo cell-autonomously reduces caveolar vesicles. Interestingly, caveolar biogenesis and resulting permeability are precisely determined by the lipid transport function of Mfsd2a. Lipidomic analysis of acutely isolated brain and lung ECs identified an enrichment of multiple DHA-containing phospholipid species in brain capillaries. As caveolar integrity is in part determined by the cholesterol content,[84] the polyunsaturated fatty acids transported by Mfsd2a may increase membrane fluidity in brain ECs to displace cholesterol and caveolar components (e.g., Cav1 and cavin1) and create an unstable environment for caveolae formation.[77] Of important note, Mfsd2a is undetectable in brain ECs devoid of pericyte contacts,[82,85] suggesting a pericyte-directed role in Mfsd2a expression and vesicle formation. Pinocytic vesicles have previously been observed in pericyte loss models.[76,86] As CNS capillaries are estimated to have the highest pericyte coverage,[86] this may confer additional regulation in transcytosis suppression in healthy BBB-containing ECs.

Exploring the CNS vasculome

Newer sequencing and genetic tools accelerated and expanded the visualization of interorgan heterogeneity in multiple dimensions. Hupe et al. developed a novel mouse model in which a mCherry tag could be added to ribosomes in a Cre-dependent manner, enabling affinity or FACS enrichment of a particular cell type.[87] They then used this model to create whole-organ and endothelial transcriptomes of multiple developing organs (e.g., forebrain, heart, lung, kidney, etc) by bulk RNA-seq.[88] In contrast to all other queried organs, Brain ECs express unique transporters (e.g., Mfsd2a, Slc38a5, Slc2a1, Abcb1a, etc) and transcription factors (e.g., Foxf2, Foxq1, Lef1, Foxl2, Ppard, and Zfp551), which are involved in Wnt/β-catenin signaling and BBB formation/maturation.[88] The expression of these BBB-specific transcription factors and transporters are markedly reduced when *Ctnnb1* (coding for β-catenin) is deleted from ECs. In addition, the authors managed to purify about 60 forebrain ECs for scRNA-seq and observed a differential distribution of both pan- and BBB-enriched EC genes within forebrain ECs, indicating an existing intracompartmental EC heterogeneity. In spite of the technical limitations (e.g., poor EC capture, limited bioinformatic rendering), this study nonetheless demonstrates the strength of scRNA-seq to explore EC heterogeneity.

Building the blood–brain barrier

Sabbagh et al. purified ECs from early postnatal mouse brain, liver, lung, and kidney and integrated bulk RNA-seq, Methyl-seq, and ATAC-seq to build an excellent and searchable database for visualization of gene expression, methylation, and chromatin accessibility across the four different vascular beds.[45] In line with previous studies,[76,88] the most differentially expressed genes in brain ECs are the ones related to BBB and Wnt/β-catenin signaling such as Apcdd1, Slc2a1/Glut1, Slco1c1, Mfsd2a, Zic3, Lef1, and Foxq1 (Fig. 17.2). Methyl-seq shows a typical inverse correlation between tissue-specific DNA methylation and gene expression in ECs. For example, poorly methylated regions in Slc2a1 and Mfsd2a loci are observed in brain ECs, but not in non-CNS ECs, which retain high levels of methylation. Conversely, hypomethylation at the Fabp4 locus is specific to liver and kidney ECs that highly express Fabp4.[45] ATAC-seq peaks (i.e., regions of accessible chromatin) and hypomethylation regions tend to overlap and are useful to identify cis-regulatory elements (e.g., promoters and enhancers).

Could epigenetic heterogeneity be a major contributor, or a determinant in EC tissue diversity? Motif enrichment analysis identified both common and EC subtype-specific motifs.[45] For example, brain ECs are highly enriched for motifs of Forkhead box (FOX) (e.g., Foxc1, Foxf2, and Foxq1), TCF/LEF (e.g., Tcf7 and Lef1), and ZIC (e.g., Zic3) transcription factor families, which share known links with Wnt/β-catenin signaling and BBB development. Motifs matching Gata4 and Maf transcription factors are specific to liver ECs; Gata4 and Maf govern liver sinusoidal EC differentiation.[89,90] Altogether, CNS ECs and non-CNS ECs clearly have different transcription profiles and epigenomic landscapes.

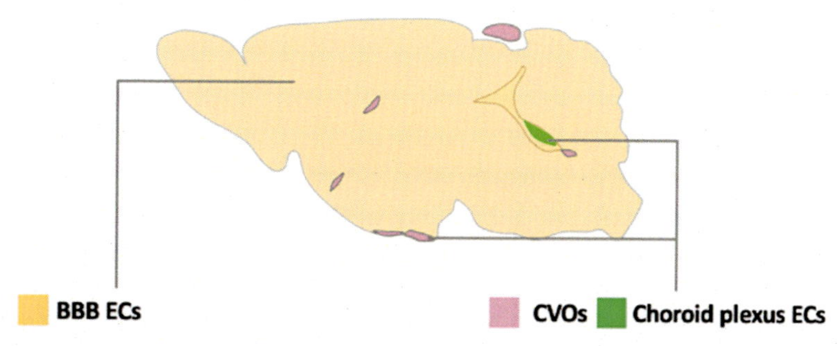

FIGURE 17.2 **EC regionalization in the brain.** Comparison of endothelial structural features and regionalized gene expression in different types of brain vessels. Blood–brain barrier (BBB) ECs exhibit specialized tight junctions (TJs), and genes involved in transcytosis suppression, specialized solute transport, and high Wnt/β-catenin signaling. Circumventricular organs (CVOs) and the choroid plexus lack BBB features, but display fenestrations, high VEGF signaling, and low Wnt/β-catenin signaling.

Wnt/β-catenin signaling delineates CNS EC regionalization

Within the CNS (and ocular) compartments, there are several specialized vascular beds that lack BBB, including CVOs, choroid plexus, choriocapillaris, and ciliary processes. In stark contrast to BBB-containing ECs, ECs in these sites are characterized by the presence of fenestrations, high permeability, and a very low level of Wnt/β-catenin signaling (Fig. 17.2).[91,92] They share a high degree of similarity to ECs from peripheral organs in gene expression. Healthy BBB ECs highly express TJ protein Cldn5 and completely lack Plvap expression ($Cldn5^{+}Plvap^{-}$). By contrast, these ECs are negative for Cldn5 expression, but are enriched for Plvap ($Cldn5^{-}Plvap^{+}$).[91,93] Remarkably, such leaky ECs retain great plasticity and can acquire BBB-like properties (e.g., fewer fenestrations, upregulation of BBB genes and downregulation of peripheral EC genes, and reduced permeability) in response to persisting β-catenin signaling.[91–93]

Intriguingly, current data suggest that only ECs positioned proximal to neuroepithelial derivatives are capable of molecular and phenotypical conversion upon β-catenin stabilization. For example, choroid plexus capillary ECs are embedded within choroid plexus epithelium whereas choriocapillaris in the eye is positioned next to RPE. The choroid plexus epithelium and RPE arise from the neuroepithelium during development. This general principle is best illustrated by one of the CVOs, the pituitary gland. Anatomically, it is divided into the anterior and posterior lobes, both of which are heavily vascularized but share different developmental history. The anterior pituitary arises from Rathke's pouch (i.e., nonneural epithelium), whereas the posterior pituitary has a neuroepithelial origin. As a result, ECs in posterior pituitary, but not ECs in anterior pituitary are rapidly and efficiently switched to a barrier-like state.[91]

What makes ECs in these particular sites so specialized and malleable to β-catenin activation? Are they developmentally primed possibly by a core set of factors encoded by neuroepithelium? Differences in non-EC cells such as mural cells, astrocytes, fibroblasts, or neurons may sculpt EC regionalization via paracrine signaling. Single cell genomic approach is well suited to begin to address these questions.

A recent study transcriptionally profiled the median eminence (ME), which is a structure at the base of the hypothalamus which serves an interface between neural and peripheral endocrine systems. Using scRNA-seq, authors aimed to determine how ECs achieve regionalization within the brain.[94] As a type of CVO involved in the secretion of hypothalamus-produced hormones, ME capillaries are $Plvap^{+}Cldn5^{-}$ and possess numerous fenestrations.[91] Transcriptionally, arterial, venous, and capillary EC clusters are observed in both ME and size-matched brain cortex samples. Authors found a $Plvap^{+}$ EC subcluster that is unique to ME and significantly differs from BBB.[94] Interestingly, Endomucin/Emcn, a typical capillary and venous EC marker,[95,96] was strongly expressed in ME ECs, but not in cortical ECs. Barrier genes including Slc2a1/Glut1 and Mfsd2a were restricted to BBB.[91,92] ME ECs displayed low Wnt activity as well as heightened VEGF signaling, which may function synergistically to maintain fenestration and high permeability. Importantly, this study also molecularly dissected vessel-associated astrocytes, pericytes, and fibroblasts in ME and cortex, and identified unique interaction patterns of mural cells with blood vessels.[94] As mural cells exhibit phenotypic and functional heterogeneity across the body,[97] differences in perivascular cell types may contribute to EC diversity via cell–cell communication.

EC arteriovenous zonation in the CNS

As outlined earlier, the most easily recognizable intra-organ endothelial heterogeneity is identified between ECs of different vascular segments (i.e., arterial-capillary-venous (ACV) hierarchy), and/or in various cellular states (e.g., angiogenesis, proliferation, metabolism, etc) within a given tissue.[1] In seminal work from Vanlandewijck et al., authors sequenced 1100 FACS-captured brain ECs from adult Cldn5-GFP mice (10–19 weeks old) and constructed a molecular atlas of the arteriovenous zonation.[98] Known arterial (e.g., Bmx, Sema3g, Vegfc, and Efnb2) and venous (e.g., Nr2f2) genes are assigned to opposite ends corresponding to the ACV axis. Capillary genes join these two larger categories.

The differentially expressed genes across the ACV hierarchy fell into six major patterns: (1) arterial, (2) capillary, (3) venous, (4) arterial and capillary, (5) arterial and venous, and (6) capillary and venous. Of note, Notch pathway components such as Dll4, Hey1, and Hes7 were found to be enriched in arterial and capillary ECs. Genes that followed arterial and venous expression pattern include Trpv4, Car7, P2ry1, Ace, Vwf, and Vcam1. Genes enriched in capillary and venous ECs include Slc38a5, Slc16a1, Nos2, Tfrc, Ramp3, Tbx4, and Aplnr. Furthermore, it was observed that transcription factors are generally biased toward arterial ECs, whereas solute transporters are distributed more in the capillary and venous ECs, indicating a functional partition in different vascular segments (i.e., functional heterogeneity). Importantly, the authors clearly demonstrated that at the transcriptional level, EC arteriovenous zonation is a single seamless continuum. By contrast, the mural cell (pericytes and smooth muscle cells) transcriptome is more distinct. Only two subclasses could be identified along the ACV axis: (1)

arterial/arteriolar smooth muscle cells and (2) pericytes/venous smooth muscle cells. Importantly, EC arteriovenous zonation is, at least in part, dependent on mural cells. Mäe et al. isolated brain ECs from mice engineered to lack pericytes (and smooth muscle cells) and identified a skewing in arterial and venous EC genes along the ACV axis.[85] Interestingly, two novel EC clusters emerge when pericytes are lost, one enriched for angiogenic genes (e.g., Angpt2, Mcam, and Trp53i11), and the other for growth factor signaling-related genes (e.g., Fgfbp1, Ctgf, Bmp5, and Bsg). During development, Fgfbp1 has recently been shown to promote BBB maturation through mediating proper ColIV deposition and maintaining Wnt/β-catenin signaling.[99] It is plausible that in response to pericyte loss, adult brain ECs upregulate developmental pathways to maintain BBB integrity.

The transcriptional continuity of ECs in arteriovenous zonation is similarly observed in several other scRNA-seq studies.[45,100,101] Notably, one group developed an innovative approach involving in vivo fluorescence-conjugated VCAM1 antibody labeling approach to improve brain EC capture.[100] Vcam1 marks both arterial and venous ECs.[98] In good agreement with functional heterogeneity, the Vcam1^{+} arterial EC cluster expresses Notch signaling components whereas Vcam1^{+} venous ECs are enriched for proinflammatory genes. The Vcam1^{-} EC cluster is enriched for capillary genes involved in metabolism and transport.[100,101] In addition, Chen et al. identified and validated a number of novel genes specific to arterial (e.g., Mgp, Clu, Stmn2, and Cdh13) and venous ECs (e.g., Il1r1, Cfh, Ctsc, and Tmsb10) (Table 17.1). Of interest, the identification of cell surface markers such as Cdh13 and Il1r may expand toolkits for enriching arterial and venous ECs, respectively.

Profound alterations of EC programs during aging

Remarkably, vascular segments differentially react to physiological aging. Compared to arterial and venous ECs, capillary ECs bear the most significant transcriptional changes during aging.[100,101] In aged capillaries, genes associated with innate immunity, antigen processing, VEGF and TGFβ signaling, matrix remodeling and hypoxia as well as oxidative stress pathways are prominently upregulated. Notably, young mice receiving plasma from aged mice exhibit the age-related transcriptional changes in arteriovenous zonation. Even more significantly, EC transcriptome in aged animals is reversed when given plasma from young mice, suggesting that the brain ECs are capable of sensing circulatory cues.[101] The identities of the potential EC sensors remain to be determined.

Non-EC cell types (e.g., pericytes, glia, and neurons) in the neurovascular unit that modulate endothelial phenotypes and functions may equally endure age-related changes. It was recently reported that physiological aging induces sporadic blood flow occlusion, blood vessel regression, pericyte dropout, and neuronal dysfunction.[110] Apoptosis-induced segmental loss of vascular smooth muscle cells from retinal arteries was also observed in aged mice.[111] Therefore, it would be of future interest to parse out whether and how each of the neurovascular unit components drive the age-related EC responses.

Transcriptional analysis of brain ECs reveals new endothelial cell types

Sabbagh et al. examined the intracompartmental EC heterogeneity in P7 developing brain by scRNA-seq and observed similar arteriovenous zonation with clusters corresponding to arterial, venous, capillary-arterial, capillary-venous ECs.[45] Most transcripts in a specific arteriovenous cluster are also present in the same EC cluster in adult brain. Furthermore, two new EC subtypes emerged with tip cell and mitotic cell genes, respectively, suggesting an active angiogenic program. Besides common tip cell markers such as Cxcr4, Angpt2, Esm1, Plaur, Lcp2, and Apln,[95,96] this emerging tip cell cluster was also enriched for Piezo2, Scn1b, Kcne3, Pcdh17, Clec1a, Smoc2, Mcam, and Trp53i11[45] (Table 17.1). Of note, the tip cell localization of Mcam and Trp53i11 was further validated on developing retina by in situ hybridization. These two genes also mark the angiogenic EC cluster in adult brain devoid of pericyte coverage.[85]

Recently, Zarkada and colleagues characterized temporal retinal EC heterogeneity using scRNA-seq of P6 and P10 retinas.[109] They found that this heterogeneity corresponded to angiogenic waves along the superficial layer and into the deep layer, respectively. In good agreement with prior work on BRB maturation,[112] ECs from P10 retinas have upregulated genes expressions associated with proliferation, AV specification, and BRB maturation compared to that in P6.[109] Most notably, they observed two distinct tip cell clusters: superficial or S-tip cell, and diving, or D-tip cell. As expected, P6 retinas have mostly the S-tip cells whereas the D-tip cell cluster mainly consists of P10 tip cells. Transcriptionally, S-tip cells are enriched in Esm1, Angpt2, Cav1, and Plvap, and D-tip cells have elevated Wnt/β-catenin signaling and highly express Apod, Cldn5, Spock2, and Mfsd2a. Pathway analysis also revealed reduced metabolism and increased ECM deposition in D-tip cells. Importantly, D-tip cells exhibit high TGFβ signaling which is required for their differentiation and for vascularization of inner retina.[109]

TABLE 17.1 Summary list of EC heterogeneity and brain/kidney-focused vasculomic studies (2010–present).
Detailed list of studies discussed in this review that aimed to profile ECs, including biological sources used (mouse/human), strains, age and type of tissues, data accession numbers, as well as associated useful websites when available.

Technique	Species	Strain	Age	Sex	Organ	Data accession no.	Website	Reference
Microarray	Mouse	Tie2GFP (JAX #003658)	P2-8, P60-70	n.s.	Brain, liver, lung	n.a.	n.a.	76
Microarray	Mouse	CD1, Dll4 ±	P6	n.s.	Retina	E-MTAB-347	n.a.	95
Microarray	Mouse	CD1	P1-1.5	n.s.	Retina	GSE19284	n.a.	96
Microarray	Mouse	Tie2GFP (JAX #003658)	P7	n.s.	Brain	GSE52564	Brain RNA-seq	102
Microarray	Mouse	Tie2GFP (JAX #003658)	E13.5	n.s.	Brain cortex, lung	GSE56777	n.a.	82
Microarray	Mouse	Pdgfbret/ret	7–12 mo	n.s.	Brain	GSE15892	Brain mural cell transcriptome	86
RNA-seq	Mouse	Pdgfrb-eGFP; NG2-dsRed	P6	Mixed	Brain	GSE75668		103
RNA-seq	Mouse	mCherryTRAP	E14.5	n.s.	Brain, kidney	GSE79306	n.a.	87
			E11.5–E17.5		Brain-less head, forebrain, heart, lung, limb, kidney and liver			88
RNA-seq	Mouse	Tie2GFP (JAX #003658)	P7	Male	Brain, liver, lung and kidney	GSE111839	Vascular endothelial cell transomics resource database	45
ATAC-seq								
Methyl-seq								
scRNA-seq					Brain			
RNA-seq			8–16 wk		Primary brain ECs			104
ATAC-seq								
RNA-seq	Human	n.a.	16–20 wk	n.s.	Kidney, liver, heart and lung	GSE114607	n.a.	107
scRNA-seq	Mouse	Cldn5-GFP and Pdgfrb-eGFP; NG2-dsRed	10–19 wk	Mixed	Brain and lung	GSE99235	Betsholtz lab vascular single cell database	98
						GSE98816		103
RNA-seq	Mouse	Flk1-GFP	E12.5, E15.5, E18.5	n.s.	Kidney, lung and pancreas	n.a.	n.a.	108
RNA-seq	Mouse	C57BL/6J (JAX 00064)	E13 - E17, P2, P4, 6–8 wk	Male for adult stages	Kidney, heart, lung and liver	GSE129005	n.a.	46
scRNA-seq			E17, P2, P7, 6–8 wk		Kidney			

RNA-seq	Mouse	Tie2GFP (JAX #003658), Rpl22ribotag (JAX 029977)	P30–P90	Male	Cerebellum, pituitary gland, cortex and choroid plexus	GSE122117	Vascular endothelial cell transomics resource database	91
ATAC-seq		Pdgfb-CreERT2; Ctnnb1flex3/+						
scRNA-seq	Mouse	C57BL/6N (Charles River Laboratory #027)	9 wk	Male	Median eminence and size-matched cortex	n.a.	n.a.	94
scRNA-seq	Mouse	C57BL/6N (Charles River Laboratory #027)	12 wk	Male	Kidney	n.a.	Carmeliet lab EC atlas	47
scRNA-seq			8 wk		Brain, kidney, heart, lung, liver, and many other organs			48
RNA-seq	Mouse	C57BL/6J (JAX 00064)	3 mo, 19 mo	Male	Hippocampus and cortex	GSE127758	n.a.	100
scRNA-seq						GSE146395		101
RNA-seq	Mouse	Rpl22ribotag (JAX 011029), Tie2Cre (JAX 004128)	10 wk	Male	Brain, kidney, heart, and lung	GSE138630	n.a.	105
RNA-seq	Mouse	Rpl22ribotag, Cdh5CreERT2	8–10 wk	n.s.	Brain, heart and lung	GSE136648	Ribotag EC heterogeneity at baseline and during inflammation	106
scRNA-seq	Mouse	Alk5f/f, Cdh5CreERT2	P6, P10	n.s.	Retina	GSE175895	n.a.	109

The renal endothelium

At the organismal level, the kidney is a rich source of peripheral ECs frequently included in vascular heterogeneity studies to contrast BBB-containing ECs. At the tissue level, the kidney vasculature is extremely heterogeneous and remarkably regionalized; but surprisingly, it has thus far received relatively less attention. In the following section, we will briefly discuss kidney vascular patterning and then focus on recent advances in kidney endothelial interorgan and intraorgan heterogeneity.

Regionalization and specialization of kidney vasculature

The main job of the kidney is to act as a filter to efficiently remove metabolic wastes, retain water and vital solutes, and ultimately make urine to help regulate body fluid homeostasis.[113] The human kidneys, for instance, filter approximately 180 L of blood per day and only excrete about 2 L of urine. The basic operational unit of the kidney is the nephron. A healthy adult human kidney is estimated to contain over 1 million nephrons. Each nephron is comprised of a glomerulus, and a set of highly structured and functionally divided epithelial tubules that travel across the cortex and the medulla. They eventually merge to the collecting ducts and drain to the ureter.[113] Of note, the oxygenation level between the cortex and the medulla varies significantly; the medulla is a very hypoxic environment and is especially prone to ischemic injury.[114]

The functional realization of the nephrons is tightly coupled to the kidney vasculature, which is developed with an unusual ACV hierarchy and is organized following the path of blood flow.[47,108,115] Blood enters the kidney through the renal arteries that first branch into the segmental arteries, then interlobar arteries, arcuate arteries, and interlobular arteries. Interlobular arteries divide into numerous afferent arterioles that feed into the glomerular capillaries within the renal corpuscles, where the filtration takes place. Next, blood leaves glomeruli via efferent arterioles, which supply the cortical peritubular capillaries and medullary capillaries (i.e., vasa recta). This second set of capillary bed surrounds different nephron segments over its entire tubular length and engages in reabsorption and secretion between blood and epithelial compartments. Finally, blood is returned to venous circulation by way of interlobular veins, arcuate veins, interlobar veins, and renal veins, which drain into the inferior vena cava. The arrangement of two capillary beds—first glomerular capillaries and then peritubular capillaries and vasa recta is unique. As they vascularize different parts of the nephrons to fulfill specific tasks, each set of the capillary ECs experiences different cues and stimuli, and interacts with different non-EC cell types, which in return likely shapes EC phenotypes via cell–cell signaling. Kidney EC heterogeneity primarily arises from differences in renal capillary ECs.[47]

Glomerular ECs form a compact capillary loop, or a vascular tuft, to maximize the contact surface area of blood flow and subsequent blood filtration. Mature glomerular capillary ECs are provided with nondiaphragmed fenestrations.[116] In addition, two distinct cell types—mesangial cells and podocytes are situated adjacent to glomerular ECs.[108,113] Mesangial cells are a type of specialized stromal cells which are located within the glomerular capillary tuft (i.e., mesangium) and function like pericytes to support glomerular ECs. Podocytes are highly specialized epithelial cells and are fundamentally involved in glomerular filtration barrier (GFB) formation. First, they make and share the glomerular basement membrane (GBM) with glomerular ECs. Second, they extend foot processes to enwrap glomerular ECs at GBM side and form slit diaphragms.[117,118]

Unlike mature glomerular ECs, peritubular capillary ECs vascularize the proximal and distal convoluted tubules in the kidney cortex and possess PLVAP$^+$ diaphragmed fenestrations.[47] Vasa recta running in parallel to the loop of Henle in the medulla are divided to the descending vasa recta (DVR) and ascending vasa recta (AVR) based on directionality.[119] The arterial-like DVRs are lined with continuous ECs and enwrapped with mural cells, whereas the AVRs are similarly provided with diaphragmed fenestrae observed in cortical peritubular capillaries.[120] Interestingly, the AVRs express both venous blood and lymphatic EC markers and operate as blood-lymphatic hybrid vessels, whose mixed identity and patterning require angiopoietin/Tie2-signaling.[121] The primary function of peritubular capillary and vasa recta is to uptake nutrients and water from the nephron tubules back to the circulation, and release ions and metabolic wastes to the tubules for urine formation.[47]

Exploring the kidney vasculome

In the last decade, there have been a number of large-scale kidney transcriptomic studies. However, these have primarily focused on nephron epithelial cells and other non-EC cell types[122–130] (also see Kidney Interactive Transcriptomics: http://humphreyslab.com/SingleCell/). As ECs account for only a small fraction of total kidney cells, they often escape attention in such studies. In the following sections, we will discuss recent studies that focused specifically on kidney EC molecular and functional heterogeneity.

Our lab recently compared mouse embryonic lung, pancreas, and kidney ECs isolated from Flk1-GFP mice at three developmental stages (E12.5, E15.5, and E18.5).[108] We isolated organ-specific ECs from Flk1-GFP mice and analyzed gene expression by bulk RNA-seq. Kidney ECs showed over 400 differentially expressed genes across these three time points. Analysis of each gene using online in situ hybridization databases such as Genepaint, allowed us (1) to screen for potentially novel kidney EC markers and (2) to derive spatial information for each gene to establish an expression atlas of kidney EC heterogeneity. Notably, Rsad2 and Gimap4 emerge as novel kidney EC-dominant transcripts by our pipeline, and we note their localization to cortical regions and areas proximal to the collecting ducts. In alignment with kidney anatomy, we observed zonal expressions of select EC genes localized to cortical, corticomedullary, and medullary regions. For example, we noted that Gpihbp1 and Cyp26b1 show restricted expression in the outer cortex; Slfn5 is localized to the corticomedullary region. We also assessed many EC genes known to have a general and/or specific expression pattern in the arteriovenous zonation by immunostaining of E15.5, P1, and P5 kidneys. Pan-endothelial genes such as Pecam/CD31, Vegfr2, Cdh5, and Icam2 are consistently expressed at all time points with little or no change across different zones. Of note, P5 arteries have reduced VEGFR2 expression. Another interesting observation is an absent or low Tie2 expression in glomerular ECs. ECs genes specific to an EC subtype, on the other hand, show more variability across the kidney. For instance, EMCN is expressed in venous and capillary ECs whereas CX40 marks arterial ECs. Interestingly, low EMCN expression is transiently detected in nascent $CX40^+$ arterial ECs of E13.5 kidney, which may reflect an immature arterial EC state during AV specification. Of note, we observed no or low Tie2 expression in glomerular ECs in three PLVAP show downregulated and restricted expression in postnatal glomerular ECs, which correlates with the fact that mature glomerular ECs lack diaphragms.[116]

Mining the vascular epigenetic landscape

A different group examined not only transcripts in kidney ECs but also chromatin accessibility and epigenetic marks to predict renal EC gene expression. Using bulk RNA-seq, ATAC-seq, and Methyl-seq, Sabbagh et al. characterized the differences between kidney ECs versus brain ECs purified from Tie2-GFP mice.[45] Transcriptionally, kidney ECs were found to be enriched for Rsad2, Fabp4, Irx3, Hoxa7, Cldn15, Pbx1, and Meis2. Likewise, low methylation and increased chromatin accessibility are observed in some of the gene loci. Interestingly, Pbx1 and Meis2 are strongly expressed in kidney stromal cells during kidney development.[131,132] Given that interstitial cells and ECs are tightly associated with each other, it remains to be determined whether Pbx1 and Meis2 are true EC transcripts or if they are introduced as potential mesenchymal mRNA contamination during tissue dissociation and/or EC isolation steps. In line with our observation on Tie2 protein expression, Sabbagh et al.[45] also noted missing GFP signals—surrogate for Tie2 expression—in glomeruli, which may indicate an incomplete recovery of glomerular ECs in this dataset, and may undermine the compatibility of Tie2-GFP, or Tie2-Cre mice as genetic tools for unbiased enrichment of kidney ECs in the future.

Transcriptomic analysis of embryonic and adult kidney

The Rafii laboratory recently harnessed scRNA-seq to analyze purified embryonic and adult kidney ECs and established a molecular taxonomy of kidney vasculature.[46] Using this approach, distinct molecular signatures were observed in all vascular segments traveling along the nephron tubules, from large arteries/afferent arterioles (Aqp1, Sox17, Cxcl12, Gja5, Mest, Fbln5), glomerular capillaries (Ehd3, Mapt, Igfbp5, Sox17, Mest), efferent arterioles (Sox17, Cxcl12, Igfbp7, Plvap, Mest, Fbln5), peritubular capillaries (Plvap, Igfbp5, low Ehd3 and Aplnr), DVRs (Aqp1, Slc14a1, Sox17, Igfbp7, Fbln5, Gja5) to AVRs and veins (Igfbp7, Plvap, Nr2f2, Igfbp5, Ehd3) (Fig. 17.3).[46] Interestingly, efferent arterioles leaving the glomeruli do not express the flow-responsive arterial marker Gja5/Cx40, which marks afferent arterioles. As we previously showed a lack of VEGFR2 expression in P5 kidney arteries, Barry et al.[46] noted an absence of Vegfr2/Kdr transcript from the arterial-like DVRs. Remarkably, they identified an embryonic EC cluster, or vascular progenitors (VPs), which are enriched for Plvap, Nr2f2, Aplnr, and a low level of Sox17 and Igfbp5. Notably, the VPs seem to give rise to diverse EC subtypes (i.e., heterogeneity) during kidney development as predicted by pseudotime trajectory analysis and by protein expression validation. The initiation of AV specification is noted to occur around E13.5 in the developing kidney, a process likely mediated by remodeling of the preexisting embryonic capillaries.[108] This event is recapitulated bioinformatically that the putative VPs first branch to renal arteries and veins before differentiating to other defined kidney EC subtypes such as glomerular ECs and vasa recta.[46]

This invaluable dataset offered a direct validation of renal vascular markers, including EC-derived transporters, secreted proteins, and transcription factors. For example, aligned with their glucose uptake function, peritubular capillaries specifically express the glucose transporter Slc21a/Glut1 at high levels. DVRs play critical roles in water reabsorption as well as urea secretion, and they are shown to be enriched for the water-specific

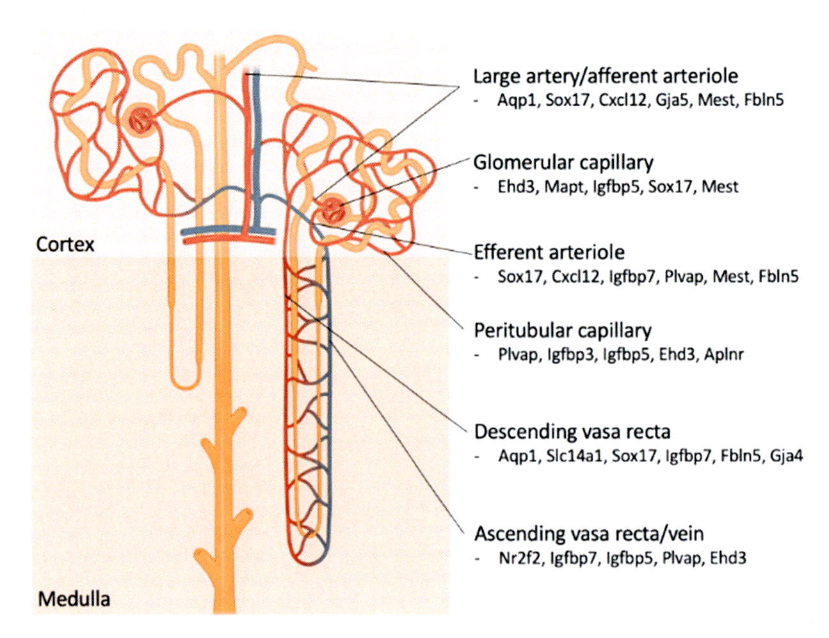

FIGURE 17.3 **EC regionalization in the kidney.** Numerous genes have been identified as being expressed in a regionalized manner along the kidney nephron. Genes listed were identified in studies discussed in this review based on numerous reports including.[46,47,108]

channel Aqp1 and the urea transporter Slc14a1. In addition, many potential angiocrine factors are found to be differentially enriched across EC subtypes. Glomerular ECs express Vegfa, Fgf1, Pdgfb, Sema5a, Dll4, Dkk2, Igfbp4, and Igfbp5; peritubular capillary ECs are enriched for Ltbp4, Igfbp3, and Igfbp5 and DVRs highly express Igfbp7 and Igfbp3. It will be of future interest to determine whether and how the EC-derived factors regulate kidney development in a paracrine manner. Nonetheless, this highlights an array of signaling networks between ECs and non-EC types within the developing kidney to synergistically promote vascular specification and kidney development. Moreover, distinct regulons are observed across different vessel types that potentially contributing to vascular specialization. Endothelial-specific inactivation of Tbx3, a transcription factor specifically enriched in glomerular ECs, results in varying degrees of glomerular pathologies such as fibrosis, hypoplasia, and microaneurysms.[46]

Inventory of the adult kidney vasculature

The Carmeliet laboratory took a different approach that preselects kidney EC subtypes by mechanical separation of kidneys to cortex, glomeruli, and medulla.[47] This pipeline results in three general types of renal endothelial cells (RECs), namely cortex RECs (cRECs), glomerular RECs (gRECs), and medullary RECs (mRECs). As expected, the three types of RECs have very different transcriptomic profiles. In particular, gRECs show as a separate cluster, away from cRECs and mRECs, which may share some transcriptional continuity. A typical gREC marker is Ehd3, which was previously validated in the glomerulus by mRNA and protein localization.[46,133] Both cRECs and mRECs highly express Plvap, likely reflecting the enrichment of fenestrated peritubular capillaries and AVRs in these cell populations. Of interest, several novel REC subtype-specific marker genes are identified: Plat and Lpl for gRECs, Igfbp3 and Npr3 for cRECs, and Igfbp7 and Cryab for mRECs. These transcripts are not uniformly distributed based on TSNE plot visualization, likely demonstrating differential expression in the different REC subtypes. Consistent with the Barry et al. study, Dumas et al. observed distinct regulatory networks enriched in REC subtypes that may drive vascular specification.

Next, they reclustered each REC subtype to reveal intracompartment heterogeneity and to reconstruct molecular hierarchy of the kidney vasculature. Within the gREC cluster, they identified a novel EC subcluster which expresses CD9, Rdx, Podxl, and Sgk1 and marks the transitional cell populations between capillary and efferent arteriolar ECs. Within mREC cluster, they uncovered papillary AVR and DVR subclusters enriched in hyperosmotic stress-related genes such as S100a6, S100a4, Fxyd2, and Cryab. In contrast to the medullary DVRs, papillary DVRs do not express the water channel Aqp1. Interestingly, several new capillary subclusters are identified in mRECs and cRECs to contain signature genes indicating angiogenic or interferon response pathways, thus referred to as activated capillaries. The angiogenic capillaries are marked by expression of Aplnr, Apln, Col4a1, and Trp53i11, whereas interferon signaling-related capillaries highly express Irf7, Ifit1, and Ifit3. It warrants future research to determine the functional importance of activated capillaries especially in healthy versus stressed kidneys.

Metabolic adaption of renal ECs revealed by transcriptomic analysis

As kidney blood vessels are central for blood filtration and urine formation, how do they cope with environmental adversity such as dehydration? The study from the Carmeliet laboratory tackled this question by modeling this type of stress by limiting water intake for up to 48 h and assessing global transcriptional changes in REC subtypes at a 12-hour interval by scRNA-seq.[47] Of the three EC compartments, mRECs exhibit most changes from 36 to 48 h of water deprivation in comparison to control mRECs, and to other REC subtypes of either experimental regime. The capillary subclusters in mRECs are especially affected by dehydration. As mRECs are primarily comprised of AVRs and DVRs important for water reabsorption and urine concentration, the significant transcriptional changes observed in mREC capillary subclusters backs up the anatomical and functional characterization of mRECs. Notably, mRECs from 12- and 24-hour water deprivation groups, and from 36- and 48-hour groups cluster together, respectively, suggesting mRECs undergo two molecularly distinct phases to adapt to hyperosmolarity.

Of interest, scRNA-seq analysis of ECs over the course of 48-hour water deprivation shows no change in overall composition of mREC subclusters. However, mRECs from the 48-hour dehydrated group exhibit upregulation of many genes associated with hypoxia, OXPHOS, and glycolysis pathway, suggesting a drastic alteration in mREC metabolic profiles. Specifically, mRECs in the papillary region (likely the papillary AVRs and DVRs), but not mRECs in the medullary zone, are enriched for OXPHOS genes. Thus, OXPHOS and glycolysis may represent major metabolic adaptation mechanisms utilized by mRECs to endure acute dehydration. In summary, this study and other transcriptomic analyses of the kidney vasculature highlight scRNA-seq as a powerful tool for determination of intraorgan EC heterogeneity, both in developing and adult kidneys, and for delineation of stress-coping mechanisms of kidney ECs experiencing adversity.

Harnessing in vivo understanding of the vasculature to instruct in vitro tissue generation

An important motivation for understanding endothelial heterogeneity is to be able to distinguish a kidney or a brain blood vessel at the molecular level. Characterizing vessels in vivo, and being able to recognize their signature transcriptomes, will help us assess whether a vessel has formed properly in engineered tissues. Case in point are the efforts to generate "organs in a dish" or "organs on a chip." These are sought either as organ-specific platforms to model development and disease and to test pharmacological drugs,[134–137] or as bona fide replacement tissues for patients with diseased organs (such as polycystic kidney disease or chronic kidney disease).[138–140] Remarkably, the Zheng laboratory recently discovered that freshly isolated ECs from human fetal kidney, heart, lung, and liver can retain their respective in vivo structural, molecular, and functional features even after five passages, which would be tremendously useful for characterizing intrinsic and extrinsic factors delineating EC heterogeneity and even more so for innovating in vitro tissue engineering including vascular bed-specific microfluidic devices.[107,141]

There have been valiant efforts to use organoid-based methods based on self-organization and directed differentiation of kidney cell types to produce kidney-like clusters of cells.[142–144] However, while nephron epithelial cells display a relatively faithful array of kidney cell types, the associated vasculature does not pattern correctly and the signatures needed to assess whether the correct vessels are forming at the right place and time are not yet in hand.[145] Efforts to improve vascularization in engineered organs such as the kidney and in organ-on-chip models nonetheless continue unabated,[146–149] and a deeper understanding of vascular transcriptional profiles will undoubtedly help push these efforts forward.

Where do we go from here?

Our appreciation and understanding of EC heterogeneity has evolved significantly from early ultrastructural observations to deep comparative molecular characterization. How do ECs end up so strikingly different in different tissues? In 1987, Mary Gerritsen surmised that the "plasticity of ECs' responsiveness enables them to maintain their integrity in a world of constant challenges." This is not an outlandish possibility—that ECs are inherently adaptable, because they have to be. When they emerge in embryogenesis, perhaps this is how they acquire such striking diversity, they adapt to their changing microenvironments and they diverge over developmental time, each under the influence of their own regionalized niche.

However, not only have we begun to understand how ECs "get their groove," but there are increasing numbers of studies that identify how ECs in turn alter their surroundings via angiocrine signals. In many instances, it is clear that ECs actively signal to the tissues they pervade.[150] Numerous studies indicate that ECs organize tissue development and function through

EC-tissue interaction. The idea of a constant and constructive reciprocal endothelial-tissue molecular crosstalk has therefore gained more fuel in recent decades. Viewing this molecular conversation through the prism of angiodiversity makes the vasculature an increasingly attractive subject for deeper analysis.

One of the exciting approaches to investigate EC heterogeneity is the recent development of large-scale sequencing techniques, as reviewed here and recently by others.[151,152] Over the past decade, many invaluable "transcript-centric" vasculomes have been generated from various species, organs, and developmental stages (Table 17.2). The next decade will likely see a deluge of such data, which will require bioinformatic corralling to make sense of. In that regard, Paik et al. explored the large data generated by Tabula Muris consortium and identified unique tissue-specific EC genes as well as ligand-receptor expression patterns that highlight potential EC-tissue interaction to induce or maintain organotypic ECs.[153] In the meantime, we face a challenge regarding the validation of each individual novel transcript identified by deep sequencing approaches, as it is a labor-intensive process. The development of spatial transcriptomics is a high throughput and breakthrough development in that regard.[154] In addition, "protein-centric" and "lipid-centric" vasculomic studies have recently been reported, offering new perspectives in posttranslational modification, vascular niche signaling, and EC specialization.[155,156]

It is our sincere hope that such large and informative datasets can be harnessed to characterize the depth and expanse of angiodiversity. In addition, we hope that such knowledge can then in turn be tapped to surgically target therapies to specific tissues or even specific cells, thereby reducing overall toxicity and increasing treatment efficacy. Judah Folkman vividly analogized ECs to zip codes, a proposition especially attractive for vascular-directed drug delivery and precision medicine.[40] Resonating with that analogy, the vasculature was recently conceptualized as an address system for mapping the entire human body.[157] Indeed, the NIH has spearheaded a Common Coordinate Framework (CCF), within which the Vascular Common Coordinate Framework (VCCF) will identify and characterize every

TABLE 17.2 EC arteriovenous zonation in the adult and developing brain.
List of genes identified in ECs along the arteriovenous axis, as reported by studies listed.

Arterial	Arterial venous	Arterial capillary	Capillary	Venous capillary	Venous
Bmx	Vwf	DII4	Mfsd2a	Slcl6a1	Nr2f2
Vegfc	Vcam1	Hey1	Slcla1	Nos2	Slc38a5
Sema3g	Trpv4	Hes7		Tfrc	Il1r1
Gkn3	Car7	Efnb2		Ramp3	Cfh
Igfbp3	P2ry1	Unc5b		Tbx4	
Stmn2	Ace			Aplnr	
Cdh13				Car4	
Mgp					

Arterial	Capillary-A	Capillary-V	Venous	Mitotic cell	Tip cell
Bmx	Slcla1	Lrrc55	Gpihbp1	Aurkb	Apod
Vegfc	Slcola4	Nkd1	Nrp2	Cenpe	Angpt2
Sema3g	Cxcr4	Atp1b1	Txnip	Kif4	Plaur
Gkn3	Spock2	Chn2	Car14	Bub1	Pcdhl7
Igfbp3	Ddc	Myc	Mafb		Piezo2
Tm4sf1	Sema3c		Ccl19		Trp53ill
Msx1	Kcnj2				Mcam
Cxcl12					

Source: Sabbagh MF, Heng JS, Luo C, et al 2018. Transcriptional and epigenomic landscapes of CNS and non-CNS vascular endothelial cells. Elife 2018;7.
Source: Based on Vanlandewijck M, He L, Mae MA, et al. A molecular atlas of cell types and zonation in the brain vasculature. Nature 2018;554:475–480; Yousef H, Czupalla CJ, Lee D, et al. Aged blood impairs hippocampal neural precursor activity and activates microglia via brain endothelial cell VCAM1. Nat Med. 2019;25:988–1000; Chen MB, Yang AC, Yousef H, et al. Brain endothelial cells are exquisite sensors of age-related circulatory cues. Cell Rep. 2020;30:4418–4432.

single blood vessel in the body, anatomically and molecularly.[157] Recognizing such regional vascular codes will lay the foundation for a deeper understanding of not only the vast array of different blood vessels but also the tissues with which they are intimately associated. We look forward the explosion of vascular information that will arise from these many efforts and to the many discoveries that lay ahead.

Acknowledgments

We thank the ReBuilding the Kidney consortium for constant stimulating input over the past five years. We also thank the Vascular Common Coordinate Framework (VCCF) for sharing ideas, as well as Zorina S. Galis for launching the idea of the "Vasculome," the Vascular Common Coordinate Framework concept, and constantly breathing energy into the field. We are grateful to the entire Cleaver lab for discussions and ideas along the way. Thank you to Steve Spurgin for critical reading of the manuscript. We also sincerely apologize to all those whose work are not included here due to space limits.

Funding

This work was supported by the American Heart Association postdoctoral fellowship 18POST34030187 to X.G., and National Institute of Diabetes and Digestive and Kidney Diseases awards DK106743, DK106743, and DK079862 to O.C.

References

1. Ribatti D, Nico B, Vacca A, Roncali L, Dammacco F. Endothelial cell heterogeneity and organ specificity. *J Hematother Stem Cell Res*. 2002;11:81–90.
2. Cleaver O, Melton DA. Endothelial signaling during development. *Nat Med*. 2003;9:661–668.
3. Augustin HG, Koh GY. Organotypic vasculature: from descriptive heterogeneity to functional pathophysiology. *Science*. 2017;357.
4. Aird WC. Phenotypic heterogeneity of the endothelium: II. Representative vascular beds. *Circ Res*. 2007;100:174–190.
5. Aird WC. Phenotypic heterogeneity of the endothelium: I. Structure, function, and mechanisms. *Circ Res*. 2007;100:158–173.
6. Aird WC. Endothelial cell heterogeneity. *Cold Spring Harb Perspect Med*. 2012;2:a006429.
7. Stone OA, Zhou B, RED-Horse K, Stainier DYR. Endothelial ontogeny and the establishment of vascular heterogeneity. *Bioessays*. 2021;43:e2100036.
8. Rafii S, Butler JM, Ding BS. Angiocrine functions of organ-specific endothelial cells. *Nature*. 2016;529:316–325.
9. Potente M, Makinen T. Vascular heterogeneity and specialization in development and disease. *Nat Rev Mol Cell Biol*. 2017;18: 477–494.
10. Daniel E, Cleaver O. Vascularizing organogenesis: lessons from developmental biology and implications for regenerative medicine. *Curr Top Dev Biol*. 2019;132:177–220.
11. Azizoglu DB, Cleaver O. Blood vessel crosstalk during organogenesis-focus on pancreas and endothelial cells. *Wiley Interdiscip Rev Dev Biol*. 2016;5:598–617.
12. Risau W, Flamme I. Vasculogenesis. *Annu Rev Cell Dev Biol*. 1995; 11:73–91.
13. Fish JE, Wythe JD. The molecular regulation of arteriovenous specification and maintenance. *Dev Dynam*. 2015;244:391–409.
14. Qiu J, Hirschi KK. Endothelial cell development and its application to regenerative medicine. *Circ Res*. 2019;125:489–501.
15. Chong DC, Koo Y, Xu K, Fu S, Cleaver O. Stepwise arteriovenous fate acquisition during mammalian vasculogenesis. *Dev Dynam*. 2011;240:2153–2165.
16. Azizoglu DB, Chong DC, Villasenor A, et al. Vascular development in the vertebrate pancreas. *Dev Biol*. 2016;420:67–78.
17. Xu K, Cleaver O. Tubulogenesis during blood vessel formation. *Semin Cell Dev Biol*. 2011;22:993–1004.
18. Xu K, Sacharidou A, Fu S, et al. Blood vessel tubulogenesis requires Rasip1 regulation of GTPase signaling. *Dev Cell*. 2011;20: 526–539.
19. Lindskog H, Kim YH, Jelin EB, et al. Molecular identification of venous progenitors in the dorsal aorta reveals an aortic origin for the cardinal vein in mammals. *Development*. 2014;141: 1120–1128.
20. Srinivasan RS, Dillard ME, Lagutin OV, et al. Lineage tracing demonstrates the venous origin of the mammalian lymphatic vasculature. *Genes Dev*. 2007;21:2422–2432.
21. Plein A, Fantin A, Denti L, Pollard JW, Ruhrberg C. Erythro-myeloid progenitors contribute endothelial cells to blood vessels. *Nature*. 2018;562:223–228.
22. Feng T, Gao Z, Kou S, et al. No evidence for erythro-myeloid progenitor-derived vascular endothelial cells in multiple organs. *Circ Res*. 2020;127:1221–1232.
23. Adams RH, Alitalo K. Molecular regulation of angiogenesis and lymphangiogenesis. *Nat Rev Mol Cell Biol*. 2007;8:464–478.
24. Carmeliet P, Jain RK. Molecular mechanisms and clinical applications of angiogenesis. *Nature*. 2011;473:298–307.
25. Herbert SP, Stainier DY. Molecular control of endothelial cell behaviour during blood vessel morphogenesis. *Nat Rev Mol Cell Biol*. 2011;12:551–564.
26. Ruhrberg C, Gerhardt H, Golding M, et al. Spatially restricted patterning cues provided by heparin-binding VEGF-A control blood vessel branching morphogenesis. *Genes Dev*. 2002;16: 2684–2698.
27. Dorrell MI, Aguilar E, Friedlander M. Retinal vascular development is mediated by endothelial filopodia, a preexisting astrocytic template and specific R-cadherin adhesion. *Invest Ophthalmol Vis Sci*. 2002;43:3500–3510.
28. Gerhardt H, Golding M, Fruttiger M, et al. VEGF guides angiogenic sprouting utilizing endothelial tip cell filopodia. *J Cell Biol*. 2003;161:1163–1177.
29. Patan S, Alvarez MJ, Schittny JC, Burri PH. Intussusceptive microvascular growth: a common alternative to capillary sprouting. *Arch Histol Cytol*. 1992;55(Suppl):65–75.
30. Burri PH, Hlushchuk R, Djonov V. Intussusceptive angiogenesis: its emergence, its characteristics, and its significance. *Dev Dynam*. 2004;231:474–488.
31. Oliver G, Kipnis J, Randolph GJ, Harvey NL. The lymphatic vasculature in the 21(st) century: novel functional roles in homeostasis and disease. *Cell*. 2020;182:270–296.
32. Ulvmar MH, Makinen T. Heterogeneity in the lymphatic vascular system and its origin. *Cardiovasc Res*. 2016;111:310–321.
33. Yaniv K, Isogai S, Castranova D, Dye L, Hitomi J, Weinstein BM. Live imaging of lymphatic development in the zebrafish. *Nat Med*. 2006;12:711–716.
34. Nicenboim J, Malkinson G, Lupo T, et al. Lymphatic vessels arise from specialized angioblasts within a venous niche. *Nature*. 2015; 522:56–61.
35. Klotz L, Norman S, Vieira JM, et al. Cardiac lymphatics are heterogeneous in origin and respond to injury. *Nature*. 2015;522:62–67.
36. Lioux G, Liu X, Temino S, et al. A second heart field-derived vasculogenic niche contributes to cardiac lymphatics. *Dev Cell*. 2020; 52:350–363 e6.
37. Florey L. The endothelial cell. Br Med J. 1966;2:487–490.

38. Tse D, Stan RV. Morphological heterogeneity of endothelium. *Semin Thromb Hemost*. 2010;36:236–245.
39. Koch PS, Lee KH, Goerdt S, Augustin HG. Angiodiversity and organotypic functions of sinusoidal endothelial cells. *Angiogenesis*. 2021;24:289–310.
40. Folkman J. Angiogenic zip code. *Nat Biotechnol*. 1999;17:749.
41. Gerritsen ME. Functional heterogeneity of vascular endothelial cells. *Biochem Pharmacol*. 1987;36:2701–2711.
42. Wang HU, Chen ZF, Anderson DJ. Molecular distinction and angiogenic interaction between embryonic arteries and veins revealed by ephrin-B2 and its receptor Eph-B4. *Cell*. 1998;93: 741–753.
43. Pasqualini R, Koivunen E, Ruoslahti E. Alpha v integrins as receptors for tumor targeting by circulating ligands. *Nat Biotechnol*. 1997;15:542–546.
44. Koivunen E, Arap W, Valtanen H, et al. Tumor targeting with a selective gelatinase inhibitor. *Nat Biotechnol*. 1999;17:768–774.
45. Sabbagh MF, Heng JS, Luo C, et al. Transcriptional and epigenomic landscapes of CNS and non-CNS vascular endothelial cells. *Elife*. 2018;7.
46. Barry DM, Mcmillan EA, Kunar B, et al. Molecular determinants of nephron vascular specialization in the kidney. *Nat Commun*. 2019;10:5705.
47. Dumas SJ, Meta E, Borri M, et al. Single-cell RNA sequencing reveals renal endothelium heterogeneity and metabolic adaptation to water deprivation. *J Am Soc Nephrol*. 2020;31:118–138.
48. Kalucka J, DE Rooij L, Goveia J, et al. Single-cell transcriptome atlas of murine endothelial cells. *Cell*. 2020;180:764–779. e20.
49. Dumas SJ, Meta E, Borri M, et al. Phenotypic diversity and metabolic specialization of renal endothelial cells. *Nat Rev Nephrol*. 2021;17:441–464.
50. Andreone BJ, Chow BW, Tata A, et al. Blood-brain barrier permeability is regulated by lipid transport-dependent suppression of caveolae-mediated transcytosis. *Neuron*. 2017;94:581–594 e5.
51. Dvorak AM, Kohn S, Morgan ES, Fox P, Nagy JA, Dvorak HF. The vesiculo-vacuolar organelle (VVO): a distinct endothelial cell structure that provides a transcellular pathway for macromolecular extravasation. *J Leukoc Biol*. 1996;59:100–115.
52. Parton RG, DEL Pozo MA. Caveolae as plasma membrane sensors, protectors and organizers. *Nat Rev Mol Cell Biol*. 2013;14: 98–112.
53. Gu X, Reagan AM, Mcclellan ME, Elliott MH. Caveolins and caveolae in ocular physiology and pathophysiology. *Prog Retin Eye Res*. 2017;56:84–106.
54. Ju M, Ioannidou S, Munro P, et al. A Na,K-ATPase-Fodrin-Actin membrane cytoskeleton complex is required for endothelial fenestra biogenesis. *Cells*. 2020;9.
55. Stan RV, Kubitza M, Palade GE. PV-1 is a component of the fenestral and stomatal diaphragms in fenestrated endothelia. *Proc Natl Acad Sci U S A*. 1999;96:13203–13207.
56. Stan RV, Ghitescu L, Jacobson BS, Palade GE. Isolation, cloning, and localization of rat PV-1, a novel endothelial caveolar protein. *J Cell Biol*. 1999;145:1189–1198.
57. Stan RV, Tse D, Deharvengt SJ, et al. The diaphragms of fenestrated endothelia: gatekeepers of vascular permeability and blood composition. *Dev Cell*. 2012;23:1203–1218.
58. Herrnberger L, Seitz R, Kuespert S, Bosl MR, Fuchshofer R, Tamm ER. Lack of endothelial diaphragms in fenestrae and caveolae of mutant Plvap-deficient mice. *Histochem Cell Biol*. 2012; 138:709–724.
59. Mancini MA, Frank RN, Keirn RJ, Kennedy A, Khoury JK. Does the retinal pigment epithelium polarize the choriocapillaris? *Invest Ophthalmol Vis Sci*. 1986;27:336–345.
60. Breier G, Albrecht U, Sterrer S, Risau W. Expression of vascular endothelial growth factor during embryonic angiogenesis and endothelial cell differentiation. *Development*. 1992;114:521–532.
61. Adamis AP, Shima DT, Yeo KT, et al. Synthesis and secretion of vascular permeability factor/vascular endothelial growth factor by human retinal pigment epithelial cells. *Biochem Biophys Res Commun*. 1993;193:631–638.
62. Blaauwgeers HG, Holtkamp GM, Rutten H, et al. Polarized vascular endothelial growth factor secretion by human retinal pigment epithelium and localization of vascular endothelial growth factor receptors on the inner choriocapillaris. Evidence for a trophic paracrine relation. *Am J Pathol*. 1999;155:421–428.
63. Breier G, Clauss M, Risau W. Coordinate expression of vascular endothelial growth factor receptor-1 (flt-1) and its ligand suggests a paracrine regulation of murine vascular development. *Dev Dynam*. 1995;204:228–239.
64. Roberts WG, Palade GE. Increased microvascular permeability and endothelial fenestration induced by vascular endothelial growth factor. *J Cell Sci*. 1995;108:2369–2379. Pt 6.
65. Esser S, Wolburg K, Wolburg H, Breier G, Kurzchalia T, Risau W. Vascular endothelial growth factor induces endothelial fenestrations in vitro. *J Cell Biol*. 1998;140:947–959.
66. Marneros AG, Fan J, Yokoyama Y, et al. Vascular endothelial growth factor expression in the retinal pigment epithelium is essential for choriocapillaris development and visual function. *Am J Pathol*. 2005;167:1451–1459.
67. Kamba T, Tam BY, Hashizume H, et al. VEGF-dependent plasticity of fenestrated capillaries in the normal adult microvasculature. *Am J Physiol Heart Circ Physiol*. 2006;290: H560–H576.
68. Maharaj AS, Walshe TE, Saint-Geniez M, et al. VEGF and TGF-beta are required for the maintenance of the choroid plexus and ependyma. *J Exp Med*. 2008;205:491–501.
69. Ford KM, Saint-Geniez M, Walshe TE, D'amore PA. Expression and role of VEGF–a in the ciliary body. *Invest Ophthalmol Vis Sci*. 2012;53:7520–7527.
70. Ford KM, SAINT-Geniez M, Walshe T, Zahr A, D'amore PA. Expression and role of VEGF in the adult retinal pigment epithelium. *Invest Ophthalmol Vis Sci*. 2011;52:9478–9487.
71. Parab S, Quick RE, Matsuoka RL. Endothelial cell-type-specific molecular requirements for angiogenesis drive fenestrated vessel development in the brain. *Elife*. 2021;10.
72. Hofman P, Blaauwgeers HG, Tolentino MJ, et al. VEGF-A induced hyperpermeability of blood-retinal barrier endothelium in vivo is predominantly associated with pinocytotic vesicular transport and not with formation of fenestrations. Vascular endothelial growth factor-A. *Curr Eye Res*. 2000;21:637–645.
73. Xu Q, Wang Y, Dabdoub A, et al. Vascular development in the retina and inner ear: control by Norrin and Frizzled-4, a high-affinity ligand-receptor pair. *Cell*. 2004;116:883–895.
74. Wang Z, Liu CH, Huang S, et al. Wnt signaling activates MFSD2A to suppress vascular endothelial transcytosis and maintain blood-retinal barrier. *Sci Adv*. 2020;6:eaba7457.
75. Ioannidou S, Deinhardt K, Miotla J, et al. An in vitro assay reveals a role for the diaphragm protein PV-1 in endothelial fenestra morphogenesis. *Proc Natl Acad Sci U S A*. 2006;103: 16770–16775.
76. Daneman R, Zhou L, Agalliu D, Cahoy JD, Kaushal A, Barres BA. The mouse blood-brain barrier transcriptome: a new resource for understanding the development and function of brain endothelial cells. *PLoS One*. 2010;5:e13741.
77. Langen UH, Ayloo S, Gu C. Development and cell biology of the blood-brain barrier. *Annu Rev Cell Dev Biol*. 2019;35:591–613.
78. Iwamoto N, Higashi T, Furuse M. Localization of angulin-1/LSR and tricellulin at tricellular contacts of brain and retinal endothelial cells in vivo. *Cell Struct Funct*. 2014;39:1–8.
79. Sohet F, Lin C, Munji RN, et al. LSR/angulin-1 is a tricellular tight junction protein involved in blood-brain barrier formation. *J Cell Biol*. 2015;208:703–711.

80. Shimomura Y, Agalliu D, Vonica A, et al. APCDD1 is a novel Wnt inhibitor mutated in hereditary hypotrichosis simplex. *Nature*. 2010;464:1043–1047.
81. Mazzoni J, Smith JR, Shahriar S, Cutforth T, Ceja B, Agalliu D. The wnt inhibitor Apcdd1 coordinates vascular remodeling and barrier maturation of retinal blood vessels. *Neuron*. 2017;96: 1055–1069 e6.
82. Ben-Zvi A, Lacoste B, Kur E, et al. Mfsd2a is critical for the formation and function of the blood-brain barrier. *Nature*. 2014;509: 507–511.
83. Nguyen LN, Ma D, Shui G, et al. Mfsd2a is a transporter for the essential omega-3 fatty acid docosahexaenoic acid. *Nature*. 2014; 509:503–506.
84. Breen MR, Camps M, Carvalho-Simoes F, Zorzano A, Pilch PF. Cholesterol depletion in adipocytes causes caveolae collapse concomitant with proteosomal degradation of cavin-2 in a switch-like fashion. *PLoS One*. 2012;7:e34516.
85. Mae MA, He L, Nordling S, et al. Single-cell analysis of blood-brain barrier response to pericyte loss. *Circ Res*. 2021;128:e46–e62.
86. Armulik A, Genove G, Mae M, et al. Pericytes regulate the blood-brain barrier. *Nature*. 2010;468:557–561.
87. Hupe M, Li MX, Gertow Gillner K, Adams RH, Stenman JM. Evaluation of TRAP-sequencing technology with a versatile conditional mouse model. *Nucleic Acids Res*. 2014;42:e14.
88. Hupe M, Li MX, Kneitz S, et al. Gene expression profiles of brain endothelial cells during embryonic development at bulk and single-cell levels. *Sci Signal*. 2017;10.
89. Geraud C, Schledzewski K, Demory A, et al. Liver sinusoidal endothelium: a microenvironment-dependent differentiation program in rat including the novel junctional protein liver endothelial differentiation-associated protein-1. *Hepatology*. 2010;52: 313–326.
90. de Haan W, Oie C, Benkheil M, et al. Unraveling the transcriptional determinants of liver sinusoidal endothelial cell specialization. *Am J Physiol Gastrointest Liver Physiol*. 2020;318: G803–G815.
91. Wang Y, Sabbagh MF, Gu X, Rattner A, Williams J, Nathans J. Beta-catenin signaling regulates barrier-specific gene expression in circumventricular organ and ocular vasculatures. *Elife*. 2019;8.
92. Benz F, Wichitnaowarat V, Lehmann M, et al. Low wnt/beta-catenin signaling determines leaky vessels in the subfornical organ and affects water homeostasis in mice. *Elife*. 2019;8.
93. Zhou Y, Wang Y, Tischfield M, et al. Canonical WNT signaling components in vascular development and barrier formation. *J Clin Invest*. 2014;124:3825–3846.
94. Pfau SJ, Langen UH, Fisher TM, et al. *Vascular and Perivascular Cell Profiling Reveals the Molecular and Cellular Bases of Blood-Brain Barrier Heterogeneity*. 2021.
95. DEL Toro R, Prahst C, Mathivet T, et al. Identification and functional analysis of endothelial tip cell-enriched genes. *Blood*. 2010; 116:4025–4033.
96. Strasser GA, Kaminker JS, Tessier-Lavigne M. Microarray analysis of retinal endothelial tip cells identifies CXCR4 as a mediator of tip cell morphology and branching. *Blood*. 2010;115:5102–5110.
97. Holm A, Heumann T, Augustin HG. Microvascular mural cell organotypic heterogeneity and functional plasticity. *Trends Cell Biol*. 2018;28:302–316.
98. Vanlandewijck M, He L, Mae MA, et al. A molecular atlas of cell types and zonation in the brain vasculature. *Nature*. 2018;554: 475–480.
99. Cottarelli A, Corada M, Beznoussenko GV, et al. Fgfbp1 promotes blood-brain barrier development by regulating collagen IV deposition and maintaining Wnt/beta-catenin signaling. *Development*. 2020;147.
100. Yousef H, Czupalla CJ, Lee D, et al. Aged blood impairs hippocampal neural precursor activity and activates microglia via brain endothelial cell VCAM1. *Nat Med*. 2019;25:988–1000.
101. Chen MB, Yang AC, Yousef H, et al. Brain endothelial cells are exquisite sensors of age-related circulatory cues. *Cell Rep*. 2020; 30:4418–4432 e4.
102. Zhang Y, Chen K, Sloan SA, et al. An RNA-sequencing transcriptome and splicing database of glia, neurons, and vascular cells of the cerebral cortex. *J Neurosci*. 2014;34:11929–11947.
103. He L, Vanlandewijck M, Raschperger E, et al. Analysis of the brain mural cell transcriptome. *Sci Rep*. 2016;11:35108.
104. Sabbagh MF, Nathans J. A genome-wide view of the de-differentiation of central nervous system endothelial cells in culture. *Elife*. 2020;9. e51276.
105. Cleuren ACA, van der Ent MA, Jiang H, et al. The in vivo endothelial cell translatome is highly heterogeneous across vascular beds. *Proc Natl Acad Sci USA*. 2019;116:23618–23624.
106. Jambusaria A, Hong Z, Zhang L, et al. Endothelial heterogeneity across distinct vascular beds during homeostasis and inflammation. *Elife*. 2020;9. e51413.
107. Marcu R, Choi YJ, Xue J, et al. Human organ-specific endothelial cell heterogeneity. *iScience*. 2018;4:20–35.
108. Daniel E, Azizoglu DB, Ryan AR, et al. Spatiotemporal heterogeneity and patterning of developing renal blood vessels. *Angiogenesis*. 2018;21:617–634.
109. Zarkada G, Howard JP, Xiao X. Specialized endothelial tip cells guide neuroretina vascularization and blood-retina-barrier formation. *Dev Cell*. 2021;56:2237–2251.
110. Gao X, Li J-L, Chen X, et al. *Reduction of Neuronal Activity Mediated by Blood-Vessel Regression in the Brain*. 2020.
111. Reagan AM, Gu X, Paudel S, et al. Age-related focal loss of contractile vascular smooth muscle cells in retinal arterioles is accelerated by caveolin-1 deficiency. *Neurobiol Aging*. 2018;71: 1–12.
112. Chow BW, Gu C. Gradual suppression of transcytosis governs functional blood-retinal barrier formation. *Neuron*. 2017;93: 1325–1333. e3.
113. Mcmahon AP. Development of the mammalian kidney. *Curr Top Dev Biol*. 2016;117:31–64.
114. Epstein FH. Oxygen and renal metabolism. *Kidney Int*. 1997;51: 381–385.
115. Mohamed T, Sequeira-Lopez MLS. Development of the renal vasculature. *Semin Cell Dev Biol*. 2018.
116. Satchell SC, Braet F. Glomerular endothelial cell fenestrations: an integral component of the glomerular filtration barrier. *Am J Physiol Ren Physiol*. 2009;296:F947–F956.
117. Miner JH. The glomerular basement membrane. *Exp Cell Res*. 2012;318:973–978.
118. Naylor RW, Morais M, Lennon R. Complexities of the glomerular basement membrane. *Nat Rev Nephrol*. 2021;17:112–127.
119. Kriz W. Structural organization of renal medullary circulation. *Nephron*. 1982;31:290–295.
120. Shaw I, Rider S, Mullins J, Hughes J, Peault B. Pericytes in the renal vasculature: roles in health and disease. *Nat Rev Nephrol*. 2018;14:521–534.
121. Kenig-Kozlovsky Y, Scott RP, Onay T, et al. Ascending vasa recta are angiopoietin/tie2-dependent lymphatic-like vessels. *J Am Soc Nephrol*. 2018;29:1097–1107.
122. Park J, Shrestha R, Qiu C, et al. Single-cell transcriptomics of the mouse kidney reveals potential cellular targets of kidney disease. *Science*. 2018;360:758–763.
123. Young MD, Mitchell TJ, Vieira Braga FA, et al. Single-cell transcriptomes from human kidneys reveal the cellular identity of renal tumors. *Science*. 2018;361:594–599.

124. Wu H, Malone AF, Donnelly EL, et al. Single-cell transcriptomics of a human kidney allograft biopsy specimen defines a diverse inflammatory response. *J Am Soc Nephrol*. 2018;29:2069–2080.
125. Lindstrom NO, Mcmahon JA, Guo J, et al. Conserved and divergent features of human and mouse kidney organogenesis. *J Am Soc Nephrol*. 2018.
126. Muto Y, Wilson PC, Ledru N, et al. Single cell transcriptional and chromatin accessibility profiling redefine cellular heterogeneity in the adult human kidney. *Nat Commun*. 2021;12:2190.
127. Miao Z, Balzer MS, Ma Z, et al. Single cell regulatory landscape of the mouse kidney highlights cellular differentiation programs and disease targets. *Nat Commun*. 2021;12:2277.
128. He B, Chen P, Zambrano S, et al. Single-cell RNA sequencing reveals the mesangial identity and species diversity of glomerular cell transcriptomes. *Nat Commun*. 2021;12:2141.
129. Wu H, Kirita Y, Donnelly EL, Humphreys BD. Advantages of single-nucleus over single-cell RNA sequencing of adult kidney: rare cell types and novel cell states revealed in fibrosis. *J Am Soc Nephrol*. 2019;30:23–32.
130. Karaiskos N, Rahmatollahi M, Boltengagen A, et al. A single-cell transcriptome atlas of the mouse glomerulus. *J Am Soc Nephrol*. 2018;29:2060–2068.
131. Schnabel CA, Godin RE, Cleary ML. Pbx1 regulates nephrogenesis and ureteric branching in the developing kidney. *Dev Biol*. 2003;254:262–276.
132. England AR, Chaney CP, Das A, et al. Identification and characterization of cellular heterogeneity within the developing renal interstitium. *Development*. 2020;147.
133. Patrakka J, Xiao Z, Nukui M, et al. Expression and subcellular distribution of novel glomerulus-associated proteins dendrin, ehd3, sh2d4a, plekhh2, and 2310066E14Rik. *J Am Soc Nephrol*. 2007;18: 689–697.
134. Benam KH, Dauth S, Hassell B, et al. Engineered in vitro disease models. *Annu Rev Pathol*. 2015;10:195–262.
135. Clevers H. Modeling development and disease with organoids. *Cell*. 2016;165:1586–1597.
136. Osaki T, Sivathanu V, Kamm RD. Vascularized microfluidic organ-chips for drug screening, disease models and tissue engineering. *Curr Opin Biotechnol*. 2018;52:116–123.
137. Wu Q, Liu J, Wang X, et al. Organ-on-a-chip: recent breakthroughs and future prospects. *Biomed Eng Online*. 2020;19:9.
138. Berthiaume F, Maguire TJ, Yarmush ML. Tissue engineering and regenerative medicine: history, progress, and challenges. *Annu Rev Chem Biomol Eng*. 2011;2:403–430.
139. Gehart H, Clevers H. Repairing organs: lessons from intestine and liver. *Trends Genet*. 2015;31:344–351.
140. Oxburgh L, Carroll TJ, Cleaver O, et al. (Re)Building a kidney. *J Am Soc Nephrol*. 2017;28:1370–1378.
141. Zhang T, Lih D, Nagao RJ, et al. Open microfluidic coculture reveals paracrine signaling from human kidney epithelial cells promotes kidney specificity of endothelial cells. *Am J Physiol Ren Physiol*. 2020;319:F41–F51.
142. Taguchi A, Kaku Y, Ohmori T, et al. Redefining the in vivo origin of metanephric nephron progenitors enables generation of complex kidney structures from pluripotent stem cells. *Cell Stem Cell*. 2014;14:53–67.
143. Morizane R, Lam AQ, Freedman BS, Kishi S, Valerius MT, Bonventre JV. Nephron organoids derived from human pluripotent stem cells model kidney development and injury. *Nat Biotechnol*. 2015;33:1193–1200.
144. Takasato M, Er PX, Chiu HS, et al. Kidney organoids from human iPS cells contain multiple lineages and model human nephrogenesis. *Nature*. 2015;526:564–568.
145. Ryan AR, England AR, Chaney CP, et al. Vascular deficiencies in renal organoids and ex vivo kidney organogenesis. *bioRxiv*. 2021.
146. Homan KA, Gupta N, Kroll KT, et al. Flow-enhanced vascularization and maturation of kidney organoids in vitro. *Nat Methods*. 2019;16:255–262.
147. Haase K, Kamm RD. Advances in on-chip vascularization. *Regen Med*. 2017;12:285–302.
148. Zhang S, Wan Z, Kamm RD. Vascularized organoids on a chip: strategies for engineering organoids with functional vasculature. *Lab Chip*. 2021;21:473–488.
149. Shirure VS, Hughes CCW, George SC. Engineering vascularized organoid-on-a-chip models. *Annu Rev Biomed Eng*. 2021;23: 141–167.
150. Pasquier J, Ghiabi P, Chouchane L, Razzouk K, Rafii S, Rafii A. Angiocrine endothelium: from physiology to cancer. *J Transl Med*. 2020;18:52.
151. Jakab M, Augustin HG. Understanding angiodiversity: insights from single cell biology. *Development*. 2020;147.
152. Chavkin NW, Hirschi KK. Single cell analysis in vascular biology. *Front Cardiovasc Med*. 2020;7:42.
153. Paik DT, Tian L, Williams IM, et al. Single-cell RNA sequencing unveils unique transcriptomic signatures of organ-specific endothelial cells. *Circulation*. 2020;142:1848–1862.
154. Longo SK, Guo MG, Ji AL, Khavari PA. Integrating single-cell and spatial transcriptomics to elucidate intercellular tissue dynamics. *Nat Rev Genet*. 2021.
155. Inverso D, Shi J, Lee KH, et al. A spatial vascular transcriptomic, proteomic, and phosphoproteomic atlas unveils an angiocrine Tie-Wnt signaling axis in the liver. *Dev Cell*. 2021;56:1677–1693. e10.
156. Clark AR, Randall EC, Lopez BGC, et al. Spatial distribution of transcytosis relevant phospholipids in response to omega-3 dietary deprivation. *ACS Chem Biol*. 2021;16:106–115.
157. Weber GM, Ju Y, Borner K. Considerations for using the vasculature as a coordinate system to map all the cells in the human body. *Front Cardiovasc Med*. 2020;7:29.

CHAPTER

18

Using pattern recognition and discriminant analysis of functional perfusion data to create "angioprints" for normal and perturbed angiogenic microvascular networks

Eva K. Lee[1,2] *and Zorina S. Galis*[3]

[1]Center for Operations Research in Medicine and Healthcare, Georgia Institute of Technology, Atlanta, GA, United States
[2]Center for Bioinformatics and Computational Genomics, Georgia Institute of Technology, Atlanta, GA, United States
[3]Vascular Researcher, Bethesda, MD, United States

Introduction

A major function of the vasculature is to provide oxygen and nutrients to the entire body. The arteries conduct the oxygenated blood pumped out by the heart, and the veins return it for recycling; however, the actual oxygen and nutrient delivery occurs at the level of the microvasculature by the smallest vessels. While these tiny structures evade our normal perception, they provide many square miles for blood-tissue exchanges that keep alive any and all the tissues of the body. Not surprisingly, pathological conditions that interfere with normal microvascular perfusion result in tissue ischemia, causing irreversible tissue damage with serious consequences, such as limb loss in diabetes, or debilitating and deadly heart attacks and strokes. Beneficial to tissue ischemia is the angiogenic response, which involves formation of new capillaries. However, this physiological process is impaired in older patients and with preexisting disease, also due to genetic predisposition; thus, there is an interest in strategies that enhance angiogenesis.[1,2] On the other hand, excessive angiogenesis fuels progression of serious pathological conditions, such as cancer, macular degeneration, or atherosclerosis, where therapeutic inhibition of angiogenesis is a key strategy for clinical management.

A major limitation for progress in these areas of research comes from the difficulty of detecting and monitoring natural or induced changes in the functional perfusion of microvasculature. Most current methods are unable to provide both functional and accurate structural information. Commonly, the native and angiogenic microvasculature are investigated experimentally using two-dimensional, nonfunctional histological examination of tissue sections. On the other hand, most perfusion measurement methods, providing functional information, cannot reveal the microstructural basis of measured effects.[3] Techniques for three-dimensional analysis of microvasculature are time and labor intensive[4,5] or require the use of sophisticated, expensive equipment[6–8] making them prohibitive for widespread use.

Fluorescence microangiography was developed[9] specifically to provide high-resolution functional information of microvasculature perfusion within thick tissue slices or even whole mouse organs. The technique was used to reveal the specific normal microvascular patterns in various organs.[10] Two major advantages of this approach are: (1) images obtained have exceptional resolution, thus allowing for identification of microstructural and functional details; (2) this is a quantifiable technique. However, the quantification is time consuming and operator-dependent when a human operator must manually trace the contours to be measured using image analysis software, as done in a previous report.[9]

We hypothesized that the specific microvascular functional details revealed by fluorescence microangiography

The Vasculome
https://doi.org/10.1016/B978-0-12-822546-2.00036-8

could be systematically analyzed via an automated pattern recognition algorithm, and that the resulting patterns could be used to develop classification rules for fingerprinting native and angiogenic microvascular networks. The results will lay foundations for new insights into diagnosis, monitoring, and drug discovery for diseases associated with changes in microvascular perfusion.

Overall experimental design

To develop and validate prediction rules for fingerprinting of native and angiogenic microvascular networks, we used the experimental data previously collected in a mouse model of angiogenesis induced by hind limb ischemia. Briefly, the hind limb ischemia-induced angiogenesis model was employed in two commonly used mouse strains, C57BL6/J inbred strain (Jackson Labs, Bar Harbor, ME) and the 129SvEv inbred strain (Taconic, Germantown, NY), and their matrix metalloproteinase (MMP)-9-deficient counterparts. Ischemia-induced angiogenesis was produced as described in Couffinhal et al.[3] The animals were sacrificed 1-, 3-, 7-, or 14-days postsurgery, and fluorescence angiography images of the ischemic muscle were taken and analyzed during these intervals over a period of 14 days.[9] The Emory University Institutional Animal Care and Use Committee approved all animal protocols. Fig. 18.1 shows the information regarding the use of the mice specimen set.

The computational platform consists of a pattern recognition algorithm and a classification model and solution engine. The pattern recognition algorithm was designed to analyze the microvascular networks in the images collected with the confocal microscope. Specifically, the algorithm analyzed objectively the shapes, branches, and curvatures of the microvascular networks in the entire image field. This overcomes the tedious manual approach in which only 4 to 6 sample regions were collected per image. The images were first filtered to eliminate the noise. This was followed by binarization and skeletonization.[11] Next, vascular branches were quantified via a crossing number equation, designed specifically for handling these microvascular images. Fig. 18.2 shows the computational schema of the image processor and the automatic pattern recognition algorithm. Fig. 18.3 illustrates a sample of the branching information obtained via our automatic algorithm from specimens in each of the groups shown in Fig. 18.1.

The patterns in the microangiographic images were then used as input into the classification model for developing the predictive rule. The two groups are "wild-type" and "MMP-9 deficient." The training sample consisted of 10 MMP-9-deficient individual animals (specimens), and 11 wild-type mice. The accuracy of the resulting rule was estimated by both ten-fold and take-one-out cross validation. In addition to developing a predictive rule to discriminate between the normal and anomalous angiogenic responses to obtain an unbiased estimate on the robustness of the rule, we also performed "blind-prediction" on specimens from 6 MMP-9-deficient mice treated by wild-type bone marrow transplant to restore the angiogenic response.

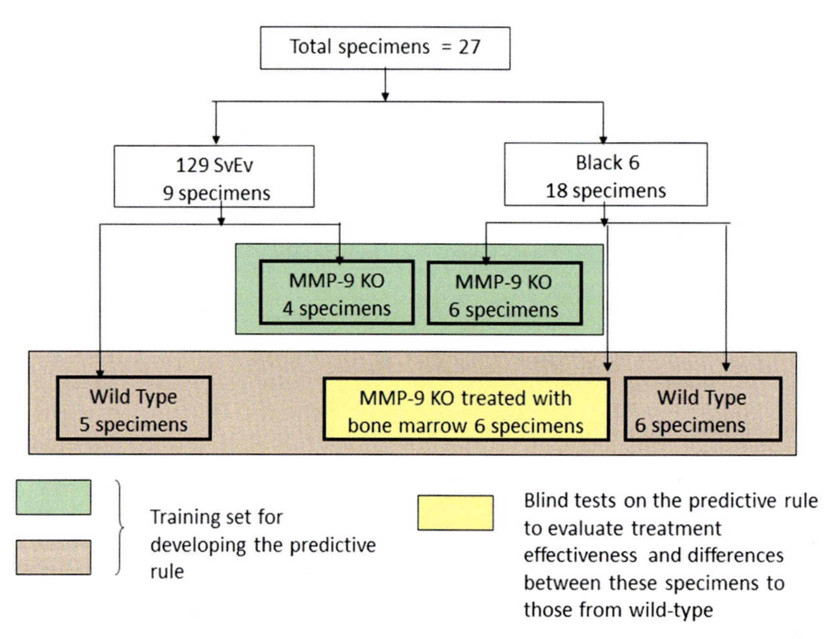

FIGURE 18.1 Information regarding the use of the mice specimen set. The angiogenic response was induced and analyzed in wild-type mice (normal response) and MMP-9-deficient mice (anomalous response) from two commonly used strains, 129 SvEv and C57BL6/J (Black 6). The color-coding indicates the various sets used for developing the described image-based pattern recognition and for classification.

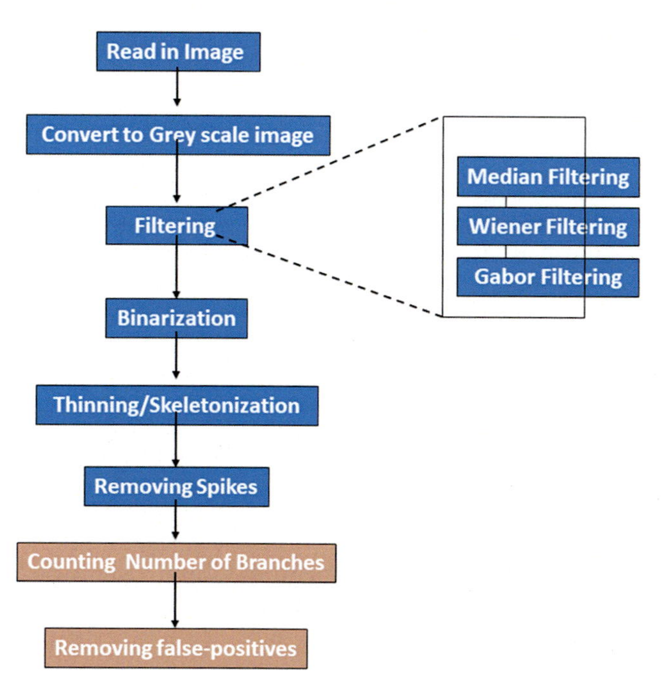

FIGURE 18.2 The computational schema of the automatic image processor and the pattern recognition algorithm.

The classification tool was an in-house discrete optimization-based discriminant analysis model.[12–15] Specifically, a nonlinear binary optimization model, DAMIP, was designed for constrained discriminant analysis in which a decision variable of 1 or 0 is used to denote whether an entity is correctly classified or not. The DAMIP predictive model has the following distinct features[12–15]: (1) the ability to classify any number of distinct groups; (2) allows incorporation of continuous-time and heterogeneous types of attributes as input; (3) a high-dimensional data transformation that reduces noise and errors in biological and clinical data; (4) a reserved judgment region that safeguards against overtraining (which tends to lead to high misclassification rates from the resulting predictive rule); and (5) successive multistage classification capability to handle entities placed in the reserved judgment region.

Results

The automated pattern recognition algorithm consistently identifies useful image patterns and capillary branching. The crossing number (CN) method designed for our pattern recognition algorithm consistently and accurately determined the vascular branching automatically and objectively, as opposed to tedious manual expert counting. The aim of the pattern recognition algorithm was to (1) duplicate the number of branches counted in an angiogenesis image by an experienced eye, and (2) thoroughly analyze entire images instead of performing random analysis as in the manual procedure. This requires an algorithm to simulate the neural functions of the human brain in dealing with complex noisy images. Methods existed in the literature and

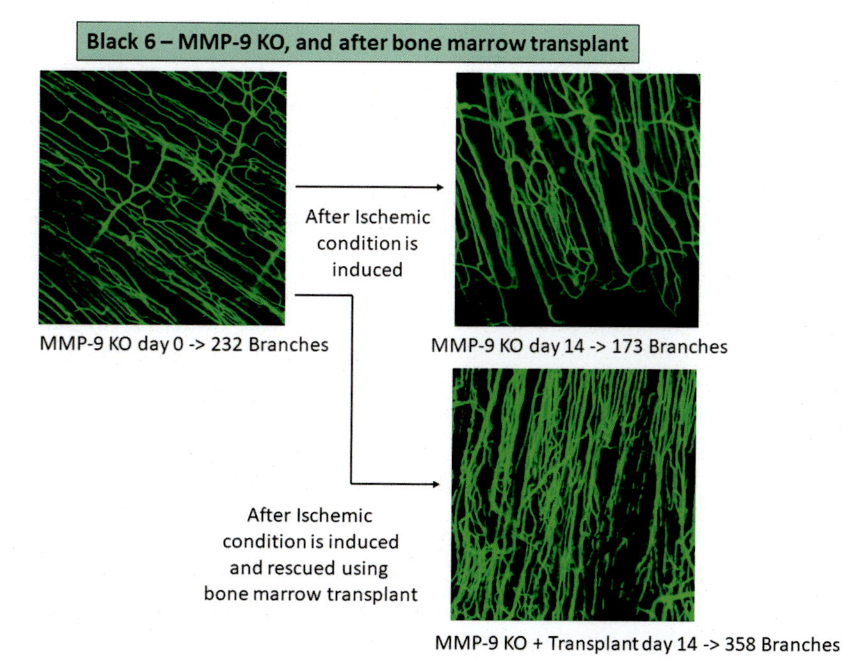

FIGURE 18.3 Fluorescence microangiography images of three MMP-9 KO mice specimens: Day 0, 14 days after ischemic is induced, and rescued by bone marrow transplant after ischemic is induced. The branching information is obtained via the automated pattern recognition algorithm.

standard libraries can convert a gray scale image into binary and reduce the thickness of the different branches to a single pixel for a consistent basis of comparing branches (skeletonizing). The novelty of our approach consists of the branch-counting, the pixel-removal phases, and the rapid computational speed. The algorithm calculates the number of lines emanating from any branching point via a CN equation, which was designed specifically for this angiogenic study. Furthermore, accuracy of the number of branches counted requires the removal of branches of length equal to a single pixel (produced while skeletonizing an image). Our algorithm involves a mechanism to detect and remove these spurious branches that appeared at different points along a skeletonized capillary.

The predictive model shows superior classification ability on angiogenesis status. Based on the attributes obtained via our pattern recognition algorithm, the DAMIP classification model generates a predictive rule with an unbiased estimate of correct classification of 100% on the MMP-9-deficient anomalous angiogenic activities and 91% on the wild-type normal angiogenic activities. Equally promising, blind prediction resulted in a correct prediction rate of around 95%. This illustrates the ability of classifying and quantifying the functional angiogenesis and microvasculature activities via pattern analysis of fluorescence microangiography images. Compared to other discriminant analysis techniques (e.g., neural networks, regression, support vector machine, and decision tree), the approach described herein offers a more robust classification mechanism that separates the two groups with higher correct classification rates.

The rate of angiogenic activities provides strong discriminatory power for "fingerprinting" native, healthy microvascular networks. Among the quantitative information obtained via fluorescence angiographic images, we have found that the rate of angiogenic activities over a time horizon is important for developing a superior classification rule. Over the 15-day time horizon, the removal of the capillary branching information for Day 0 and Day 14 severely reduces the discriminatory rate to roughly 60%. There is no significance in the shapes of the capillaries; rather the speed in which capillaries are formed appear to correlate well with the category to which the specimen belongs (wild-type, or MMP-9 deficient). Specifically, for the genetically perturbed group, the ratio of the number of branches on day 0 to that on day 14 ranges from 0.3 to 0.99; whereas in the native group, this ratio ranges from 0.7 to 2.0. Such mixed clinical attributes normally present challenges to existing classification methods. Although there is some overlap in the two intervals, together with the capillary characteristics (e.g., thickness, curvature shape), branching rate is a powerful discriminatory factor for distinguishing the two groups. Thus, analysis of branching characteristics—including rate of formation—via our pattern recognition algorithm offers a potential mechanism for developing robust predictive rules for diagnosis, monitoring, and prognosis of microvascular perfusion-dependent diseases.

The predictive rule can be used to monitor the effect of treatment. Based on the blind prediction, our predictive rule classified the specimens obtained from MMP-9-deficient mice treated with bone marrow transplant as "wild-type" category. This indicated that the treatment was successful in restoring the normal tissue perfusion in the MMP-9 genetically deficient animals. Reviewing the microvascular activities manually or by using biochemical assays of tissue homogenates (not shown) confirmed our predictive accuracy to be 100%. The predictive rule thus can serve as a reliable, fast, and convenient means for monitoring and evaluating progress of angiogenic activities and of their treatment.

Discussion

Our study offers an elegant solution for the automated fingerprinting of different microvascular networks that could be used to investigate the potential perturbing effects of conditions such as aging, genetic deficiency, diabetes, and cancer on the microvascular perfusion and structure in relevant tissues. The study reported herein is a proof-of-principle investigation designed and performed on a well-defined animal model system to allow for objective and unbiased quantification of microvascular activities based on functional perfusion images obtained via fluorescence angiography. The pattern recognition algorithm and the DAMIP classification model developed allowed us to obtain functional "angioprints" of the native, healthy microvascular networks and discriminate them against diseased or abnormal microvascular networks. Such information would be valuable for early detection and monitoring of functional abnormalities before they produce macro and clinical-level and lasting effects, which may include improper perfusion of tissue, or support of tumor development.

Fluorescence microangiography allows for visualization analysis of tissue and cell level characteristics and collection of quantitative information from microvascular beds that facilitate classification and prediction of various native healthy microvascular networks versus their perturbed or diseased states. When we compared fluorescence microangiography to the commonly used method of endothelial specific staining to investigate the angiogenic response, we were able to reveal the most relevant aspect for tissue perfusion—the changes in the functionality of the microvascular network, and we were also able to identify and quantify the microstructural basis for these functional changes.[9] Specifically, we were able to identify quantitative changes in microvascular branching via our newly developed pattern recognition algorithm. This type of detailed information may be very useful for monitoring the effect of experimental and clinical interventions aimed at manipulating the microvasculature.

Further fine-tuning of our pattern recognition algorithms and classification models for prediction of functional perfusion status and its changes may be obtained by developing additional experimental models to specifically examine the effect of other conditions or treatments upon the angiogenic response. For example, it is conceivable that the algorithm may discriminate between alterations in the normal perfusion or the angiogenic response in a native healthy specimen compared to groups with impairment due to age, chemical, or other genetic deficiency. Similarly, it can be applied to analyze angiogenic responses as a result of various treatments. This will serve two important goals. First, the identification of discriminatory patterns/attributes that distinguish functional tissue perfusion status will lead to a better understanding of the basic mechanisms underlying this process. Because therapeutic control of angiogenesis could influence physiological and pathological processes such as wound and tissue repair, aging, heart disease, cancer progression and metastasis, or macular degeneration, the ability to understand it under different conditions will offer new insights into diagnosis, as well as developing and monitoring novel therapeutic interventions. Second, our study and the results presented herein form the foundation of a valuable diagnostic tool for changes in the functionality of the microvasculature and for discovery of drugs that alter the angiogenic response. The methods can be applied to tumor diagnosis, monitoring, and prognosis. In particular, we suggest that it will be possible to derive microangiographic fingerprints to acquire specific microvascular patterns associated with early stages of tumor development. Such "angioprinting" could become an extremely helpful early diagnostic modality, especially for easily accessible tumors such as skin cancer.

Normal microvascular patterns are directly related to the tissue's metabolic needs, and thus are a characteristic of each tissue. Changes in vascular structures are central to numerous physiological and pathophysiological processes. Thus, better understanding of microvascular networks, their associated patterns, and the underlying processes has great potential to advance medical practice.[5] Since the first human study using fluorescence induced by near infrared light for intraoperative imaging of the anastomotic site was reported to improve the success of the surgical procedure,[16] numerous studies have been conducted to correlate microvascular networks with underlying disease and health conditions.[2,5,17,18] However, a comprehensive understanding of the mechanisms controlling microvascular structures and patterns during physiological and pathophysiological processes of health, disease, and recovery remain elusive. Our method presented herein is generalizable and has broad potential to help advances this important area of investigation. The automatic pattern recognition offers a rapid computational engine to analyze microvascular patterns objectively and comprehensively. Machine learning facilitates uncovering of correlation of the patterns toward various physiological and pathophysiological phenomena. Machine learning and big data analytics are increasingly playing critical roles in advancing and transforming medicine and healthcare delivery systems. The voluminous and evolving heterogeneous multisource data from electronic medical records, images, and biological and "omics" experimentations present complex challenges yet unique opportunities for modeling and computational advances.

This work demonstrates the application of machine learning in understanding the foundations of microvascular patterns. The machine learning model, DAMIP, maximizes the number of correctly classified cases; thus, it is robust and not skewed by outliers and errors committed by observation values. The modeling framework allows users to classify any number of groups with the ability to handle heterogeneous types of mixed and evolving data simultaneously. DAMIP is *NP-hard*. Furthermore, the model has many appealing characteristics including: (1) the resulting DAMIP classification rule is *strongly universally consistent*, given that the Bayes optimal rule for classification is known[19,20]; (2) the misclassification rates using the DAMIP method are consistently lower than other classification approaches in both simulated data and real-world data; (3) the classification rules from DAMIP appear to be insensitive to mixed sample sizes, thus affording better predictive results than other ML models; (4) the DAMIP model generates stable classification rules on imbalanced data, regardless of the proportions of training entities from each group[14,15,21–25]; and (5) the multistage ability of DAMIP can classify patients into the same outcome groups under different conditions and features. This is powerful as it can uncover multiple pathways that achieve the same (good) outcome. This affords stakeholders flexibility to explore preferences and options and select the ones most conducive for implementation success.

In Brooks and Lee (2010) and Brooks and Lee (2014), we have shown that DAMIP is difficult to solve.[19,20] We applied the hypergraphic structures that Lee, Wei, and

Maheshwary derived to efficiently solve these instances.[26]

Our work also demonstrates that effective automatic pattern recognition remains vital to mining through raw data to extract important characteristics for input into the machine learning model. The holistic and rapid image processing offers a good global characterization of the tissues. This facilitates broader applications of our work. Since the "angioprints" from fluorescence angiography are highly tissue-specific,[10] in the future, microvascular pattern recognition and classification should be developed as a characteristic for each healthy tissue. Such knowledge can serve as a baseline of health and be used as a rapid diagnostic tool. This can serve as a powerful approach for detecting potential health issues or perturbations for early intervention, and as a means for monitoring treatment prognosis and restoring health and tissue functions.

Methods

To develop and validate predictive rules for fingerprinting of native and angiogenic microvascular networks, we used experimental data collected in a mouse model of angiogenesis induced by hind limb ischemia. Perfusion of the hind limb muscle was analyzed in wild-type and MMP-9-deficient mice as previously described.[9] Fluorescence angiography images of the ischemic muscle were taken over a period of 14 days at several time points. A pattern recognition algorithm was designed to analyze the microvascular networks in the images collected with the confocal microscope. The patterns were then used as input into the classification models for developing the predictive rule.

Experimental data

Mouse strains

Two commonly used mouse strains, C57BL6/J inbred strain (Jackson Labs, Bar Harbor, ME) and the 129SvEv inbred strain (Taconic, Germantown, NY), and their MMP-9-deficient counterparts[9] for the hind limb ischemia-induced angiogenesis model were employed.

Mouse models of experimental angiogenesis

Ischemia-induced angiogenesis was induced as described in Couffinhal T et al.[3] Experimental animals were sacrificed 1-, 3-, 7-, or 14-days postsurgery and compared to nonischemic animals by functional fluorescent microangiographic analysis.[9] The Emory University Institutional Animal Care and Use Committee approved all animal protocols.

Functional fluorescence microangiography

Mice were sacrificed with a lethal dose of ketamine (174 mg/kg) and xylazine (26 mg/kg), followed by cutting of the inferior vena cava, and perfusion with 10 mL heparinized (10 U/mL) saline injected through the left ventricle using a 23 G infusion set to remove all blood from functional vessels. Yellow-green fluorospheres (λ_{ex} 505 nm, λ_{em} 515 nm, Molecular Probes) were infused (0.1 μm fluorospheres at a 1:20 dilution in heparinized saline). For examination of angiogenesis, the adductor muscle group was dissected and examined by confocal microscopy or processed by tissue homogenization for measurement of vascular perfusion capacity.

The fluorescent microangiographies were obtained using a Zeiss LSM/NLO 510 confocal/multiphoton microscope (Zeiss, Thornwood, NY). Tissues were placed on cover slips and micrographs were taken using the argon laser ($\lambda_{ex} = 488$ nm) with collection filter ($\lambda_{em} = 510$). Images were captured and analyzed, including the three-dimensional reconstruction, using LSM software (Zeiss).

Automated pattern recognition and discrete optimization-based classification model for discriminant analysis

The discriminant analysis approach for our study involves two major components: (1) a pattern recognition algorithm and quantitative analysis of fluorescence microangiography images to identify capillary attributes, and (2) development of a classification rule for future predictions of healthy and diseased microvasculature based on the attributes (patterns) determined by (1).

Image-based pattern recognition algorithm

We studied the potential utility of fluorescence microangiography for vascular hierarchical analysis. The method we used is related to image segmentation where microvascular characteristics (shapes, branches, and curvatures) are identified against the background to obtain numerical measurements. These image patterns, together with information such as microvascular density, functional vessel capacity, and branching form the core attributes to be used for classification. In the process, the most complex part of the image analysis involves the determination of the amount of branching in the capillaries. The existing approach involves manually counting the microvascular branches per 100× field for 4 to 6 randomly collected samples per image. Herein, an automated procedure was developed which examined the entire image thoroughly to determine the branches objectively and rapidly.

The automated algorithm takes as input the color image and creates a gray scale image for analysis. *Filtering* is first performed to eliminate noise. Three filters were employed in this process: the "median filter," the "Wiener filter," and "Gabor filter." Median filtering is similar to using an averaging filter, where each output pixel is set to an average of the pixel values in the neighborhood of the corresponding input pixel. However, with median filtering, the value of an output pixel is determined by the median of the neighborhood pixels rather than the mean, as the former (median) is much less sensitive than the latter (mean) to extreme values (outliers). Hence, median filtering performs better by removing these outliers without reducing the sharpness of the image. The Wiener filter (a type of linear filter) is then applied to the image adaptively, tailoring itself to the local image variance. The Wiener filter performs little smoothing where the variance is large, and more smoothing where the variance is small. This strategy often produces better results than standard linear filtering. The adaptive filter is more selective than a comparable linear filter, preserving edges and other high-frequency parts of an image. Next, a modified Gabor filter was implemented to remove all cloud-like structures or blobs. In the spatial domain, a 2D Gabor filter is a Gaussian kernel function modulated by a sinusoidal plane wave. Gabor filters are often used for enhancement of fingerprint images.[27] Our implementation works well empirically on the fluorescence microangiographic images.

Skeletonization: After filtering, the image is then *skeletonized*. A thinning operation is performed on the image so that the branches are accessible easily for further quantization. Skeletonization removes pixels so that an object without holes shrinks to a minimally connected stroke, and an object with holes shrinks to a connected ring halfway between each hole and the outer boundary.

Removal of single pixels: Single pixels that are not extended as proper branches are next removed, as their presence may give rise to false positives in the counting of branches (the next and last step).

Counting branches via CN concept: The method employed to count branches is known as the *CN concept*.[11,17,18] This method involves the use of the skeleton image where the ridge flow pattern is eight-connected. The minutiae are extracted by scanning the local neighborhood of each ridge pixel in the image using a *3 x 3* window. For our angiogenesis study, specifically the CN value of a pixel is computed by summing up the pixel values in its eight neighboring pixels, as shown in Fig. 18.4. The ridge pixel can then be classified as a ridge ending, bifurcation or nonminutiae point. For example, a ridge pixel with a CN value of two corresponds to a ridge ending, and a CN value of four corresponds to a bifurcation.

Eliminating false minutiae: This algorithm tests the validity of each minutia point by scanning the skeleton image and examining the local neighborhood around the point.

The information on minutiae, capillary density, functional vessel capacity, quantified endothelial cell surface marker level, and microvascular perfusion are then used as attributes for developing the predictive rules. Hence, corresponding to each tissue microvasculature, there is a "vector" which characterizes its states for each of these

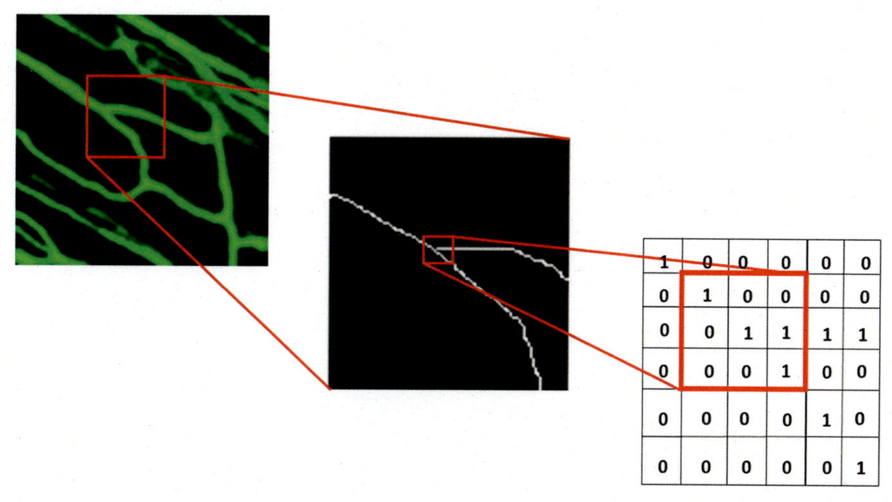

FIGURE 18.4 Calculation of crossing number (CN) for pixel (i,j) from its eight neighboring pixels. The values determine the minutiae resulting from pixel (i,j).

discriminatory attributes at a specific time. The input to the predictive model includes all the attribute information for each tissue, and the associated group (Wild-type vs. MMP-9 deficient) to which it belongs. Fig. 18.3 illustrates a sample of the branching information obtained from our automatic pattern recognition for three MMP-9 KO mice specimens: Day 0, 14 days after ischemic is induced, and rescued by bone marrow transplant after ischemic is induced.

Development of classification rules for prediction of normal and anomalous angiogenic microvasculature

We developed a method for classifying each specimen into response categories (Wild-type mouse = Normal, vs. MMP-9-deficient = anomalous) based on the quantitative attributes obtained from the fluorescence microangiographic image analysis via our pattern recognition algorithms. The "attribute vectors" were used as input for discriminant analysis on a training group to generate a classification rule. In our study, we tested the discriminatory ability of the attributes obtained via fluorescence microangiography on determining the status of the angiogenic response in the individual specimen. Here, the two groups are "Wild-type" and "MMP-9 deficient." The training sample consisted of 10 MMP-9-deficient individual animals (specimens), and 11 Wild-type specimens. The accuracy of the resulting rule was estimated by both ten-fold and take-one-out cross validation. In addition to developing a predictive rule to discriminate between the normal and anomalous angiogenic responses to obtain an unbiased estimate on the robustness of the rule, we also performed "blind-prediction" on specimens from six MMP-9-deficient mice that have been treated with bone marrow transplant to rescue the angiogenic response.

The classification tool was an in-house large-scale discrete optimization discriminant analysis model based on mixed-integer programming (DAMIP).[12–15] Specifically, DAMIP is a nonlinear binary optimization model designed for constrained discriminant analysis in which binary variables (i.e., variables that can only assume the values 0 or 1) are used to denote whether training entities are correctly classified or not.[12,13,15] The objective function maximizes the number of entities classified into the correct groups. DAMIP has the following distinct features simultaneously: (1) It can classify any number of groups. (2) It can accommodate any types or combinations of heterogeneous data sources including continuous time data as input. (3) It includes a high-dimensional data transformation that reduces noise and errors in biological data. (4) An entity can be placed into each of the groups or the reserved judgment region. (5) It allows users to preset desirable misclassification levels, which can be specified as overall errors for each group, pairwise errors, or overall errors for all groups together. (6) With the reserved judgment in place, DAMIP offers a multistage classification framework where next stage classification can be carried out on entities that are placed in the reserved judgment region. Applications of this approach to broad bio/medical problems including tumor shape prediction, ultrasound assisted drug delivery, epigenetics for cancer prediction, and vaccine immunogenicity prediction, have yielded between 80% and 95% accuracy based on cross validation tests and actual blind predictions.[14,15,21–25,28] Compared to well-known approaches such as linear discriminant analysis, decision tree, naïve Bayesian, support vector machine, logistic regression, random forest, neural networks, and nearest shrunken centroid, uniformly, DAMIP produces robust predictive rules that perform well on blind-data, even when the training set is small or the data is imbalanced.[21–24,28–33]

To obtain an unbiased estimate of the reliability and quality of the derived classification rule, *ten-fold* and *take-one-out cross validation* were performed. In ten-fold cross validation, the set of entities is randomly partitioned into 10 subsets of approximately equal size. Ten trials are then run, each of which involves a distinct training set made up of nine of the 10 subsets and a test set made up of the remaining subset. The classification rule obtained via a given training set is applied to each entity in the associated test set to determine to which group the rule allocates it. Take-one-out works by using all but one entity for developing the predictive rule. The resulting rule is tested on the take-out entity. This is repeated until each entity has been taken-out and tested exactly once. In our study, there are 21 training specimens, each of which corresponds to a microvasculature status. In the ten-fold cross validation, in each trial, 18–19 entities are used as the training set to derive a classification rule, and the remaining 2–3 entities are used as the test set to gauge the accuracy of the derived rule. As one rotates through the 10 trials, each of the 21 entities is held back exactly once for use in a test set. The cumulative classification rate over the 10 trials is reported in the "classification matrix."

Blind prediction performed on the rescued mice specimens were designed to verify if there is a significant difference between these specimens versus the normal response of the wild-type mice, thus evaluating the effectiveness of the treatment (bone marrow transplant), and the use of the predictive rule as a monitoring device.

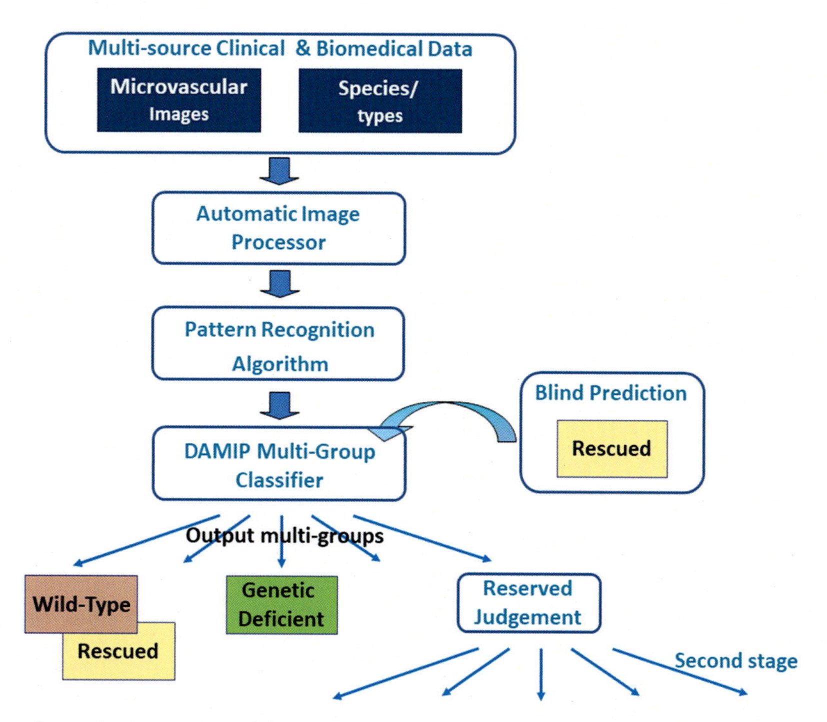

FIGURE 18.5 This figure shows the integration of the automatic image processor, the pattern recognition algorithm, and the classification model and solution engine into a general-purpose machine learning framework that allows classification of multiple groups. The reserved judgment facilitates multistage analysis. The solid boxes correspond to the pertinent input data and classification performed for this study.

Reviewing the microvascular activities manually or by using biochemical assays of tissue homogenates confirmed our predictive accuracy to be 100%.

Fig. 18.5 shows the integration of the image processor, pattern recognition algorithm, and the classification model and solution engine into a general-purpose machine learning framework that allows classification of multiple groups. The reserved judgment facilitates multistage analysis.

Acknowledgments

The experimental work was performed at Emory University and was supported through National Institutes of Health (NIH) RO1 HL64689, the American Heart Association Established Investigator Award No. 0040087N, and the National Science Foundation (NSF) Award EEC-9731643 to Dr. Zorina S. Galis. Dr. Lee's pattern recognition and machine learning work was performed at Georgia Institute of Technology and was partially supported by NSF Award CMMI-0800057 and IIP-0832390. We wish to thank Dr. Chad Johnson, supported through the NSF Award EEC-9731643, Dr. Susan Lessner, supported through NIH National Research Service Award No. 1 F32 HL68449-01, for their contributions to the experimental work. Findings and conclusions in this chapter are those of the authors and do not necessarily reflect the views of the National Institutes of Health, the National Science Foundation, and the American Heart Association.

References

1. Carmeliet P, Jain RK. Angiogenesis in cancer and other diseases. *Nature*. 2000;407:249–257.
2. Popovic N, Vujosevic S, Popovic T. Regional patterns in retinal microvascular network geometry in health and disease. *Sci Rep*. 2019;9:16340. https://doi.org/10.1038/s41598-019-52659-8.
3. Couffinhal T, Silver M, Zheng LP, Kearney M, Witzenbichler B, Isner JM. Mouse model of angiogenesis. *Am J Pathol*. 1998;152: 1667–1679.
4. Brey EM, King TW, Johnston C, McIntire LV, Reece GR, Patrick Jr CW. A technique for quantitative three–dimensional analysis of microvascular structure. *Microvasc Res*. 2002;63: 279–294.
5. Tata DA, Anderson BJ. A new method for the investigation of capillary structure. *J Neurosci Methods*. 2002;113:199–206.
6. Manegold PC, Hutter J, Pahernik SA, Messmer K, Dellian M. Platelet-endothelial interaction in tumor angiogenesis and microcirculation. *Blood*. 2003;101:1970–1976.
7. Minnich B, Bartel H, Lametschwandtner A. How a highly complex three-dimensional network of blood vessels regresses: the gill blood vascular system of tadpoles of Xenopus during metamorphosis. A SEM study on microvascular corrosion casts. *Microvasc Res*. 2002;64:425–437.
8. Toyota E, Fujimoto K, Ogasawara Y, et al. Dynamic changes in three-dimensional architecture and vascular volume of transmural coronary microvasculature between diastolic- and systolic-arrested rat hearts. *Circulation*. 2002;105:621–626.
9. Johnson C, Sung H, Lessner S, Fini E, Galis ZS. Matrix metalloproteinase-9 is required for capillary branching during ischemia-induced angiogenesis. *Circ Res*. 2004;94:262–268.
10. Bosetti F, Galis ZS. "Small blood vessels: big health problems?": scientific recommendations of the National Institutes of Health Workshop. *J Am Heart Assoc*. 2016;5:e004389. https://doi.org/10.1161/JAHA.116.004389.
11. Thai R. *Fingerprint Image Enhancement and Minutiae Extraction* (Ph.D. thesis). School of Computer Science and Software Engineering, University of Western Australia; 2003. www.cs.uwa.edu.au.

12. Gallagher RJ, Lee EK, Patterson DA. An optimization model for constrained discriminant analysis and numerical experiments with iris, thyroid, and heart disease datasets. *Proc AMIA Annu Fall Symp*. 1996:209–213.
13. Gallagher RJ, Lee EK, Patterson DA. Constrained discriminant analysis via 0/1 mixed integer programming. *Ann Oper Res*. 1997;74:65–88.
14. Lee EK. Large-scale optimization-based classification models in medicine and biology. *Ann Biomed Eng*. 2007;35(6):1095–1109.
15. Lee EK, Gallagher RJ, Patterson DA. A linear programming approach to discriminant analysis with a reserved judgment region. *Informs J Comput*. 2001;15:23–41.
16. Holm C, Mayr M, Höfter E, Dornseifer U, Ninkovic M. Assessment of the patency of microvascular anastomoses using microscope-integrated near-infrared angiography: a preliminary study. *Microsurgery*. 2009;29(7):509–514.
17. Hong L, Wan Y, Jain A. Fingerprint image enhancement: algorithm and performance evaluation. *IEEE Trans Pattern Anal Mach Intell*. 1998;20:777–789.
18. Pratt WK,W. *Digital Image Processing*. 2nd ed. New York: John Wiley and Sons; 1991.
19. Brooks JP, Lee EK. Analysis of the consistency of a mixed integer programming-based multi-category constrained discriminant model. *Ann Oper Res Data Min*. 2010;174(1):147–168.
20. Brooks JP, Lee EK. Solving a multigroup mixed-integer programming-based constrained discrimination model. *Informs J Comput*. 2014;26(3):567–585.
21. Lee EK, Fung AY, Brooks JP, Zaider M. Automated planning volume definition in soft- tissue sarcoma adjuvant brachytherapy. *Phys Med Biol*. 2002;47:1891–1910.
22. T:/PGN/ELSEVIER_BOOK/BARHOUM-9780128239155/WEB/FM-CTR-02/FM-CTR-02-9780128239155.3dLee EK, Nakaya H, Yuan F, et al. Machine learning framework for predicting vaccine immunogenicity. *Interfaces*. 2016;46(5):368–390. The Daniel H. Wagner Prize for Excellence in Operations Research Practice.
23. Lee EK, Gallagher RJ, Campbell A, Prausnitz M. Prediction of ultrasound-mediated disruption of cell membranes using machine learning techniques and statistical analysis of acoustic spectra. *IEEE Trans Biomed Eng*. 2004;51:1–9.
24. Lee EK, Wu TL, Goldstein F, Levey A. Predictive model for early detection of mild cognitive impairment and Alzheimer's disease. *Optim Data Anal Biomed Inf*. 2012;63:83–97. Fields Institute Communications.
25. Lee EK, Wang Y, Hagen MS, Wei X, Davis RA, Egan BM. Machine learning: multi-site evidence-based best practice discovery. *Int Workshop Mach Learn, Optim Big Data*. 2016:1–15.
26. Lee EK, Wei X, Maheshwary S. Facets of conflict hyper graphs. *SIAM J Optim*. 2020 (Submitted).
27. Yang JW, Lifeng Liu LF, Tianzi Jiang TZ, Fan Y. A modified Gabor filter design method for fingerprint image enhancement. *Pattern Recognit Lett*. 2003;24:1805–1817.
28. Kazmin D, Nakaya HI, Lee EK, et al. Systems analysis of protective immune responses to RTS,S malaria vaccination in humans. *Proc Natl Acad Sci U S A*. 2017;114(9):2425–2430.
29. Feltus FA, Lee EK, Costello JF, Plass C, Vertino PM. Predicting aberrant CpG island methylation. *Proc Natl Acad Sci U S A*. 2003; 100(21):12253–12258.
30. McCabe MT, Lee EK, Vertino PM. A multi-factorial signature of DNA sequence and polycomb binding predicts aberrant CpG island methylation. *Cancer Res*. 2009;69(1):282–291.
31. Nakaya HI, Wrammert J, Lee EK, et al. Systems biology of seasonal influenza vaccination in humans. *Nat Immunol*. 2011;12:786–795.
32. Nakaya HI, Hagan T, Kwissad M, et al. Systems analysis of immunity to influenza vaccination across multiple years and in diverse populations reveals shared molecular signatures. *Immunity*. 2015; 43(6):1186–1198.
33. Querec TD, Akondy R, Lee EK, et al. Systems biology approach predict immunogenicity of the yellow fever vaccine in humans. *Nat Immunol*. 2009;10:116–125.

Further reading

1. Helthuis JHG, van Doormaal TPC, Hillen B, et al. Branching pattern of the cerebral arterial tree. *Anat Rec*. 2019;302(8):1434–1446. https://doi.org/10.1002/ar.23994. Epub 2018 Dec 5. PMID: 30332725; PMCID: PMC6767475.
2. Secomb TW, Pries AR. Microvascular plasticity: angiogenesis in health and disease. Preface. *Microcirculation*. 2016;23(2):93–94.

CHAPTER

19

Artificial intelligence for the vasculome

Luis Eduardo Juarez-Orozco[1,2], *Ming Wai Yeung*[2], *Jan Walter Benjamins*[2], *Fatemeh Kazemzadeh*[2], *Gonçalo Hora de Carvalho*[2] *and Pim van der Harst*[1,2]

[1]**Department of Cardiology, Heart and Lung Division, University Medical Centre Utrecht, Utrecht, the Netherlands**
[2]**Department of Cardiology, University of Groningen, University Medical Center Groningen, Groningen, the Netherlands**

Introduction

The interrelatedness of the vasculature in biological systems conveys an academic blind spot that has undeniably affected our ability to cope in ever more integrative ways with its pathological states.

In medical sciences, knowledge parcellation has caused divergence between the evaluation and treatment of different vascular beds based on mere location (e.g., in the central nervous system, the coronary artery tree, and the renal system), sheer caliber (i.e., large, medium, and small caliber arteries), and their functional role (i.e., arterial, capillary, and venous structures). This picture neglects a simple, yet crucial element of interdependency between all vessels and their particular (dynamic) profile across individuals. This notion has been explored in genetics and neurosciences delivering the well-known concepts of the *genome* and *connectome*. Adequate integration of core anatomic and (patho) physiologic correlations between blood vessels and their common mediating elements and mechanisms across the entire organism can serve as the base for a comprehensive approach to the deterioration or dysfunction of this general complex system, i.e., the *vasculome* (Fig. 19.1 shows the conventional schematic of the vascular system with functional characteristics that vary across different organs and locations).

Recently, a new era of artificial intelligence (AI) development has captured the attention of both medical researchers and clinicians. Due to the convergence of increasing computational power with parallel computing capabilities (such as graphic processing units and tensor processing units) and the streamlining of cloud-computing services, as well as the maturation of big datasets, the study and characterization of complex systems through their intrinsic linear and nonlinear interdependencies has exploded. This has allowed for the automation and optimization of a vast number of tasks ranging from image recognition in medical diagnosis to the discovery of novel patterns that could better inform on disease prognosis.

The present chapter explores the new basic principle-based concept of the *vasculome* from the perspective of the possibilities posed by modern AI implementations for its study and integration. First, we provide a theoretical overview regarding machine learning–based AI, then we summarize research and clinical implementations in the field of vascular medicine at different levels of specialization, and finally, we discuss how knowledge emerging from these applications could support the adoption and integral study of the *vasculome* and its pathological states.

Machine learning–based artificial intelligence

The concept of AI is not a new one as it emerged roughly half a century ago when mathematical modeling started fueling computational principles opening the theoretical door for self-sustained learning. AI refers then to the theory and application of systems (models) able to perform tasks traditionally understood to require human-level intellect.[1] Although open-ended and somehow recursive, such definition underlines the applicational nature of the concept and the expected performance of its implementation (usually beyond that of human capabilities).

Enthusiasm for AI has been intermittent, the prior era of AI development during the 80s and 90s provided part

The Vasculome
https://doi.org/10.1016/B978-0-12-822546-2.00033-2

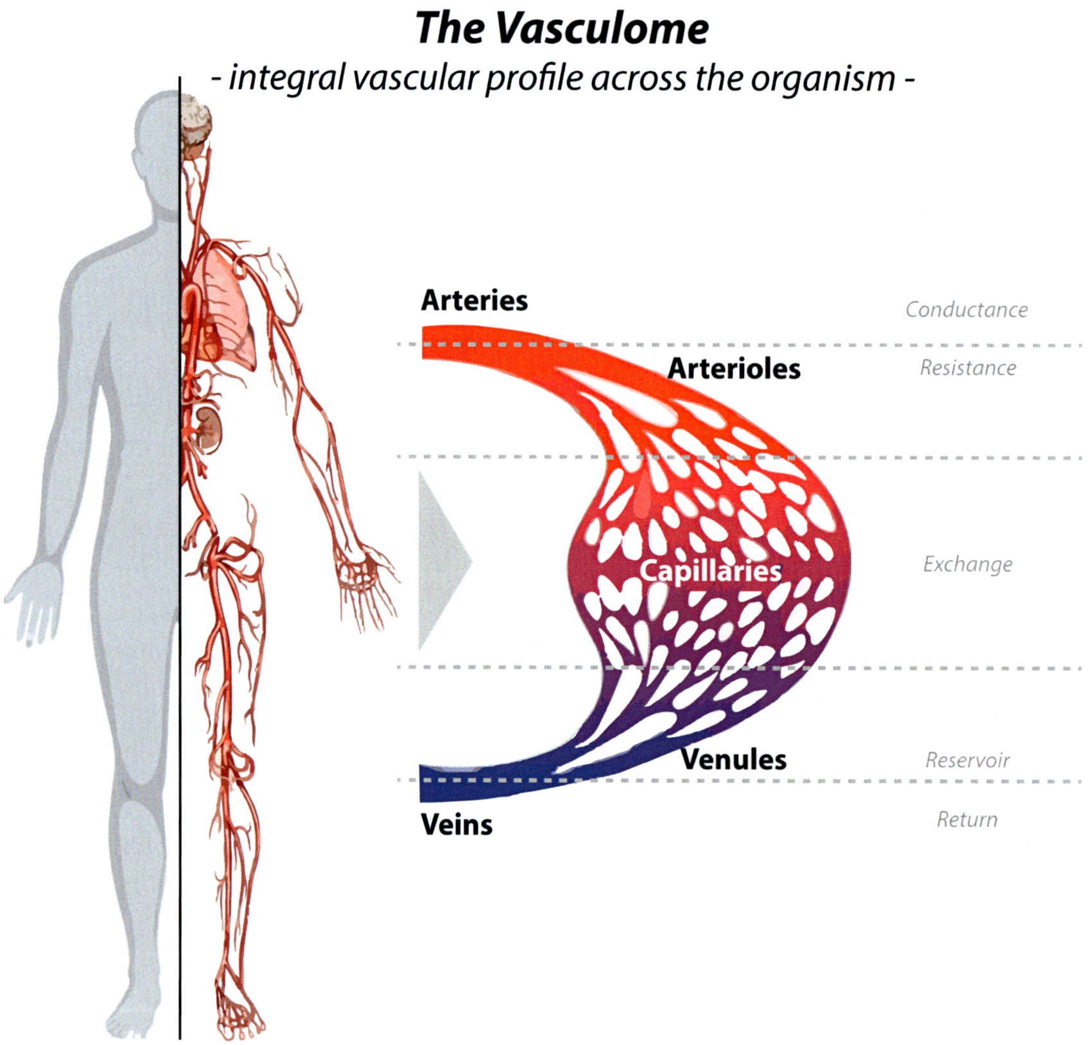

FIGURE 19.1 The vascular system and the *vasculome* concept.

of the development of machine learning algorithms (vide infra). This meant a leap forward from rule-based programming but was unfortunately limited by the computational hardware capacities of the time. The latest era of AI has arisen from the convergence of two crucial factors, namely: increasingly powerful computational capacity that allow for massive parallel processing and the expanding availability of large datasets that provide a suitable target for complex analytics.[2] In fact, AI has permeated into numerous areas setting the current standard for all data-driven interfaces ranging from search engines to driverless cars.

So modern-day AI is fueled by (machine) learning algorithms. But how do they work and how are they viewed from data science perspective? Machine learning modeling, as any statistical approach, engages an input (independent variables or *features*) through one or several mathematical operators to generate an output (dependent variable or *prediction/estimation*). Such output will have an associated magnitude of error defined by the distance between the output prediction and the selected reference. The characterization of this error translates the performance of the algorithm or model as a tool to produce accurate and adequate outputs. But what differentiates machine learning from more "traditional" statistics is the iterative nature of the modeling process. This means that once the error is quantified, the model aims to reduce it though adjustment of its operators. In this way, when exposed to a new run of the data, the model will be able to produce better results (output) every time, a process called *training*. The limit to this optimization is marked by the *generalization* potential of the model. This means how far will the model perform well when exposed to new or unseen data. *Testing* in unseen data reflects the true performance of a machine learning model.

Therefore, effective machine learning seeks to *learn* enough about the complex interrelations (i.e., extract relevant features) within an informative dataset (that is large enough to be representative and varied) to be able to perform the same task in another dataset with comparable performance.

Learning is thus an iterative process that can be approached through mathematical modeling and several types of it are recognized in this area. These are discussed ahead.

Types of learning

Three types of learning are considered in AI, namely supervised, unsupervised, and reinforcement (Fig. 19.2). In supervised learning (the most common utilized in observational datasets), the model is given input-output pairs (thus a reference) to optimize a function that maps these accordingly. For example, the inputs may be structured or imaging computed tomography (CT) or magnetic resonance (MR) data of the heart/coronary arteries or brain and its circulation that have already been labeled (based on expert interpretation, or ideally on the effective gold standard) as having a particular condition such as obstructive atherosclerosis, aneurysmal changes, or a left ventricular thrombus.[3] The output of the model for each individual (training) case is compared with its corresponding (reference) label and the error calculated. With this estimation, parameters within the model are adjusted such that the error can be minimized through the optimization of a loss function in an iterative manner.

In unsupervised learning, the algorithm explores patterns from unlabeled input (i.e., with no reference label and thus with no known output). Given that each unlabeled input is characterized by a very specific set of attributes (with expands with the number of features characterized per case), the algorithm naturally identifies cardinal patterns that differentiate between sets of similar datapoints. This is typically exemplified by clustering, which in the pathological setting may be interpreted to represent discrete known or novel risk and disease subgroups. A beautiful example of this type of learning has been the suggestion, based on

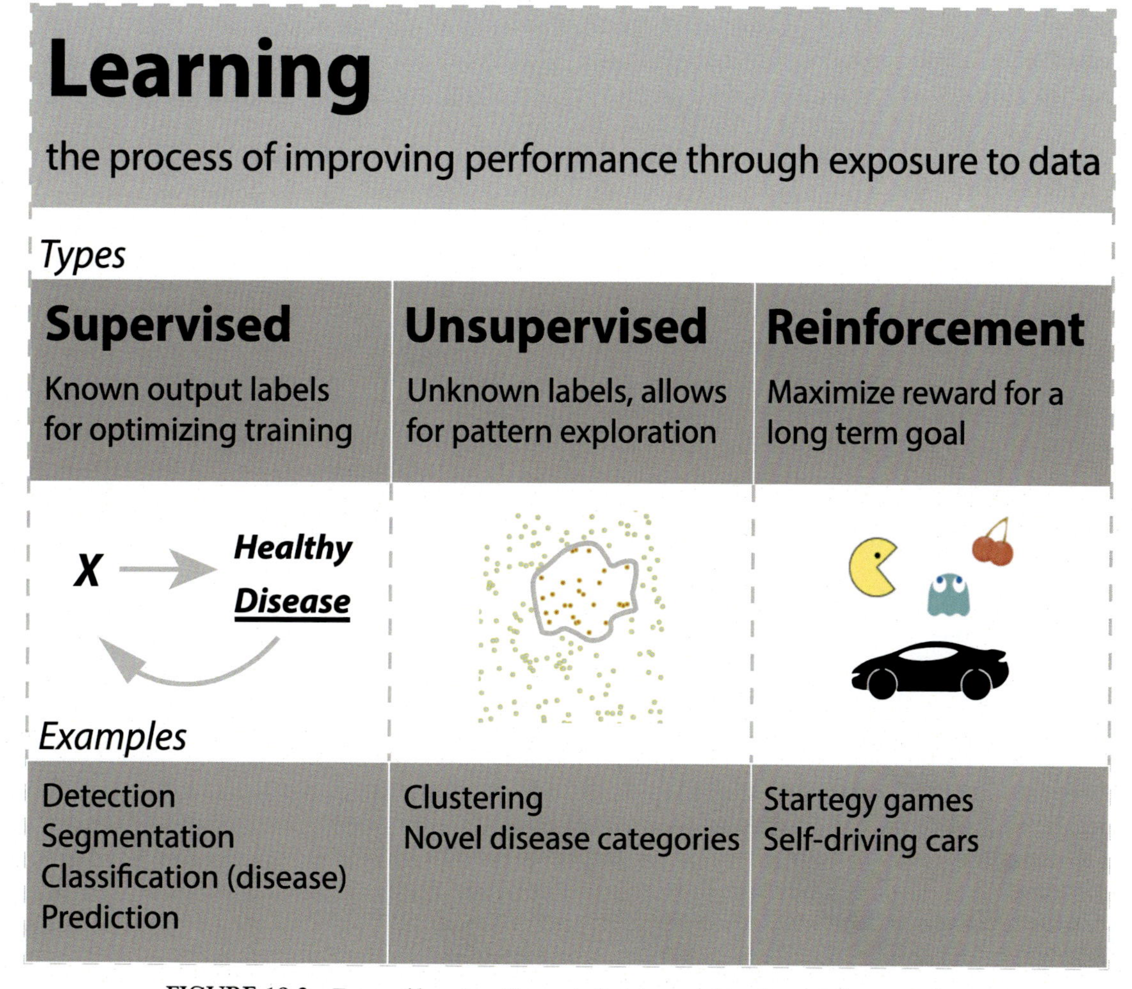

FIGURE 19.2 Types of learning: Concept, divisions, and implementation examples.

echocardiographic and clinical data, of phenogroups of patients likely to respond to cardiac resynchronization therapy.[4]

Reinforcement learning gauges rewards and punishments for a specific goal in the form of maximizing the cumulative reward. In this type of learning, the algorithm takes actions while interacting with the environment, it improves on the target task (or multiple tasks simultaneously) through positive of negative feedback on the cumulative gain achieved. The model decides which of the actions prior to the reinforcement were most responsible for the result and should alter its actions with the goal of maximizing cumulative rewards. Reinforcement learning is suitable for problems that involve continuous interactions between the agent and the environment surrounding it. Iconic examples include strategy games such as Chess and Go (*DeepMind*), and the implementation of self-driving cars.

Statistical machine learning algorithms

Traditional statistical methods like linear and logistic regression are also machine learning algorithms and are conventionally utilized owing to their interpretability and ease of implementation. Regularized forms of the basic regression model, such as Ridge, Lasso, and Elastic net regression, can be employed in analysis of data with a large number of features. Both Ridge and Lasso regression shrink the coefficients of features by applying either an L1 or L2 norm (the cost function) to produce a sparse model; alternatively, a combination of both error optimization terms with additional finetuning may be adopted to generate the final model, a technique known as elastic net regularization.

Support vector machine (SVM) is a popular method for classification tasks, typically trained in a supervised manner but may also be extended to its unsupervised version (support vector clustering). In an SVM model, individual data points are first mapped to a feature space, sometimes to one of higher dimensions via a kernel separation to facilitate data point separation through a hyperplane based on support vectors.

Decision trees and their ensemble version (a technique that combines output of multiple models to derive the final output with the aim of improved robustness and performance than individual models) random forest are supervised learning methods in which each new incoming data point is classified in a manner that resembles a flowchart with decisions made sequentially to arrive at the class label. Selection of features as well as decision rules are optimized on a labeled training set often using a heuristic algorithm. To increase the generalizability, a consensus prediction can be drawn from an ensemble of decision trees, such as in the case of random forests method.

Deep learning

Deep learning refers to the algorithms based on artificial neural networks with multiple intermediate layers (denoting their depth) with different types of activating functions being the most common convolution. Deep learning involves the use of multilayered artificial neural network consisting of connections with tunable weights or strengths. These connections exist in layers which transform the input stepwise. More specifically, individual artificial neurons at each layer perform arithmetic operations with the given incoming signals (input) which is typically the sum of all products between input (from artificial neurons at previous layer) and the weights of corresponding connections; the integrated signal then passes through a typically nonlinear thresholding function before being sent out to the artificial neurons at next layer; such transformation is repeated sequentially to generate the final output of the model (Fig. 19.3). This results in a complex mapping of input to output thus capable of modeling of more complicated relationships than any linear model.

These convolutional neural networks have seen numerous applications positioning deep learning at the top of the performance scale in tasks involving image recognition and natural language processing. With sufficient training data, deep learning is able to harness useful representations and features automatically and directly from raw data, bypassing the need for feature engineering and extraction. Deep learning offers advantage in high-dimensional data such as images and therefore its experimental application in medical imaging has rapidly grown with applications ranging from detection and segmentation, to interimage modality translation and diagnostic detection.

In practice, numerous factors should be considered in the decision to approach specific tasks through machine learning. It is essential to have clear understanding of the problem at hand and its context. Emerging evidence has suggested that simpler statistical modeling can offer adequate performance and provide more interpretability than machine learning methods for some endeavors, while in other cases results in image recognition approached through deep learning have proved unparalleled. Furthermore, it seems that machine learning algorithms can be useful in integrating data from different sources beyond the capabilities of linear modeling.[5] This is especially interesting in the medical research where various aspects of a pathological process (disease) are usually characterized in a range of clinical, biochemical, and imaging attributes.

Thereon, it is crucial to examine the comparative utility of several algorithms in achieving specific objectives given the preprocessing, data, and computational needs of each particular approach. Additionally, assessment of

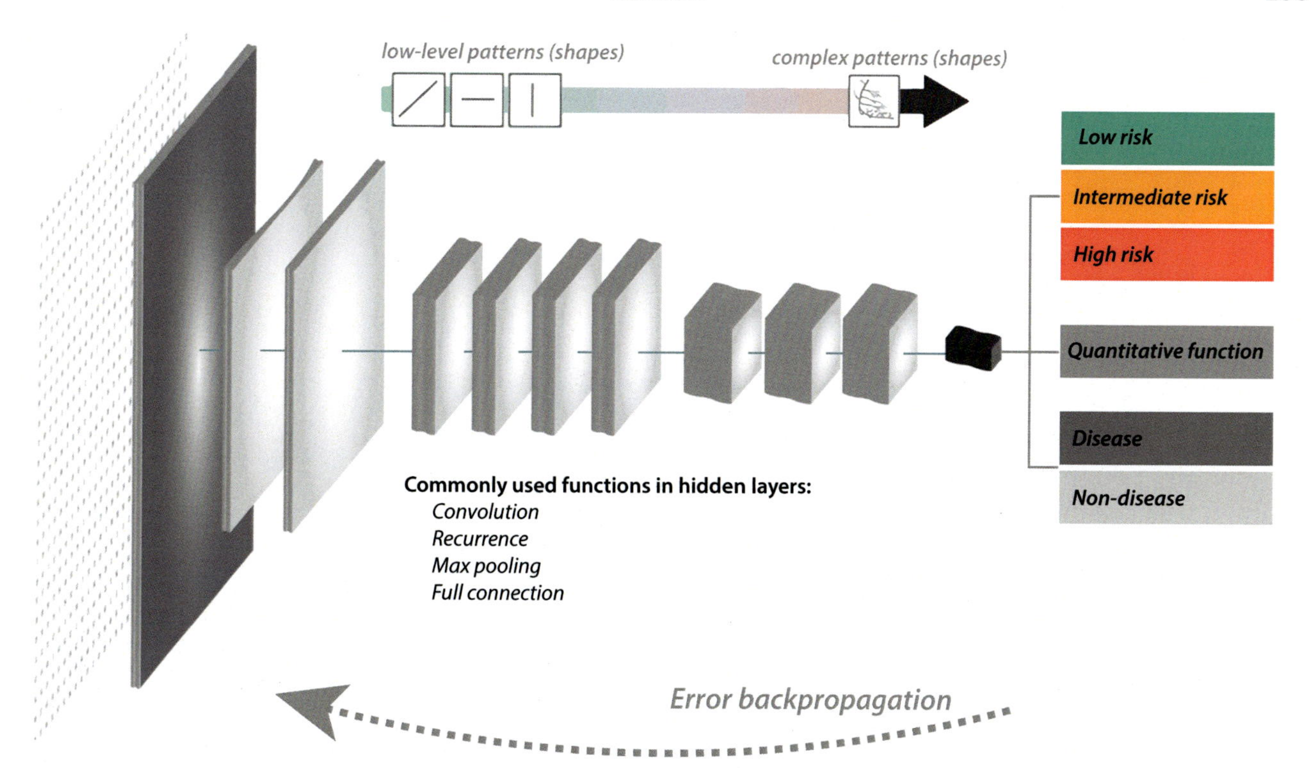

FIGURE 19.3 Deep learning: Schematic architecture of a convolutional neural network. From the input layer on the left side, the values of every pixel in a given image can be processed forward through the hidden layers (represented as narrower and thicker panels) and finally arrive to the fully connected output layer which is tailored for the specific task at hand (e.g., image classification, estimation of risk of specific outcomes, or prediction of associated risk factors). Standard architectures are formed by blocks that are composed in turn by different layers. For example, a block can be composed by a convolutional, a batch-normalization, and a nonlinear activation layer. Certain blocks also include a max pooling layer, which reduces the spatial dimension of the input. *Modified from Juarez-Orozco LE. Machine learning in the evaluation of myocardial ischemia through nuclear cardiology. Curr Cardiovasc Imaging Rep. 2019;12:5.*

the intrinsic characteristics of the target population such as the expected magnitude of relevant features and expected relationships which are to be extracted along with the dataset size can also yield good clues for the most optimal methods. Lastly, the decision will also depend on data/labels availability which relates to feasibility and maximum achievable performance of methods as well as potential consecutive improvements in the long term. In solving complex problems, one may also consider a workflow that makes use of a combination of different learning approaches that suits best for respective tasks.[6] We have coined the concept of method hybridization in the expectation that future implementations can make additive use of the particular advantages of both statistical and machine learning methods.

The vasculome and the role of machine learning–based AI in its study

So, the *vasculome* can refer to the integrative and patient-specific profile of all sensitive attributes regarding the entire cardiovascular system. In this sense, a vascular *fingerprint* could help us characterize the causes, status, and consequences of pathological processes affecting it, while also providing a general overview of its health determinants (see Fig. 19.4).

The cardiovascular system is the product of more than 500 million years of evolution and fine tuning from a single layered tube in an open system to the intricated pulsatile closed system with a multiple chambered pump of today. This *convergence* (in evolutionary biology, this refers to the independent tendency to evolve similarly under equal environmental conditions[7] attests to its physical advantages, while also suggesting that, in its current version, the *vasculome* may not be optimized for the longevity we have procured ourselves in the last centuries. Parallelly, the fractal-like nature of vascular beds strongly contrasts with the anatomical, mechanical, and physiological differences reflected at every level of their organization from robust and elastic conductance vessels to fine continuous-flow capillaries. Therefore, in order to study the *vasculome*, one must frame it from the complementary standpoint of the purpose, anatomy and function of the cardiovascular system. In this sense, AI can provide a powerful tool to approach the general modeling of this system from big-data and to identify its individual profile as a marker or predictor of health and disease. Below, we will comment on

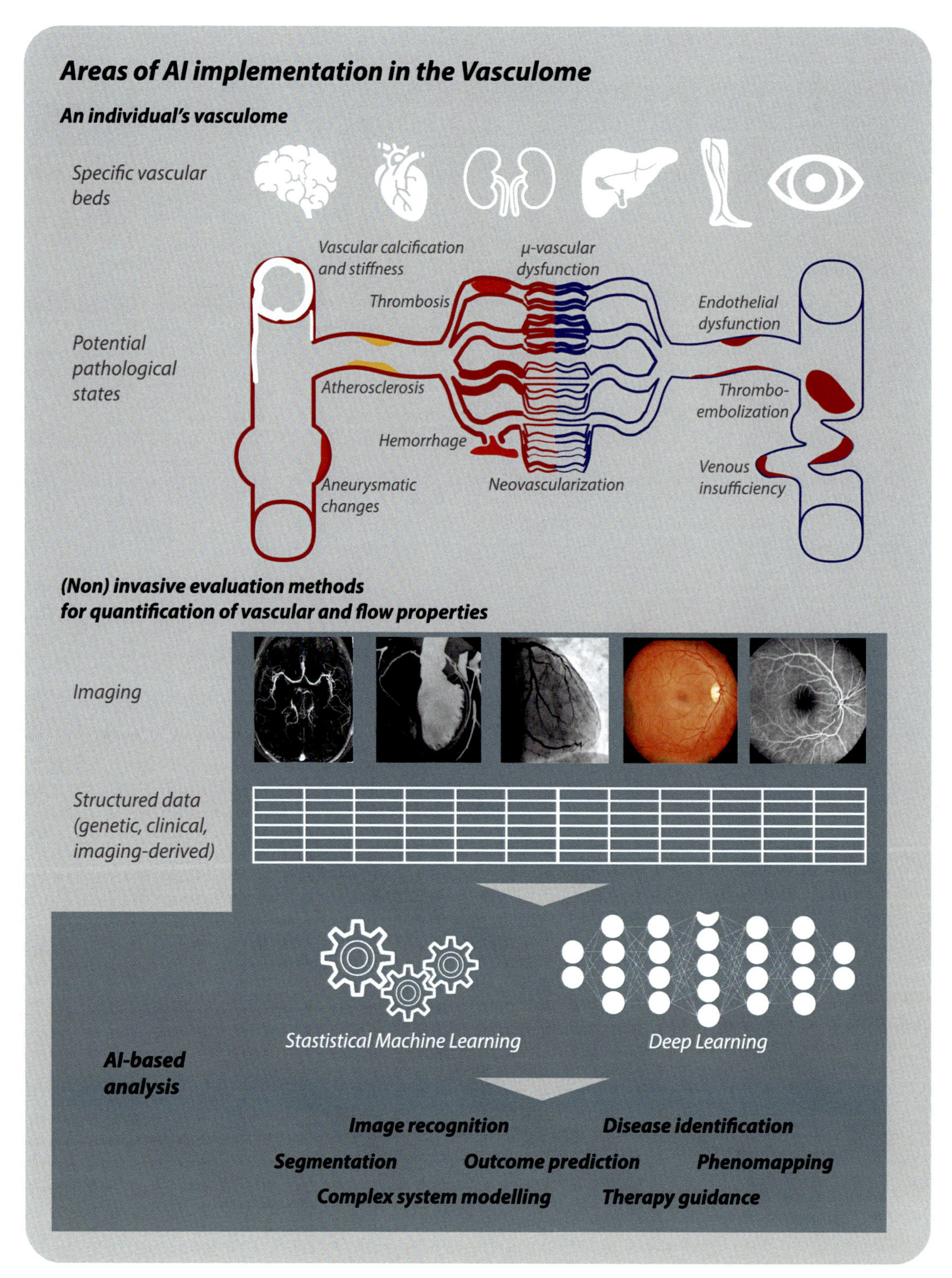

FIGURE 19.4 The conceptual structure for AI implementation on the study of the *vasculome* through the analysis of clinical and imaging data.

developments in the intersection of machine learning–based AI and (cardio) vascular research starting from the heart and proceeding distally into the vasculature.

Cardiac and vascular imaging constitute a clear target in AI analytics due to the increasing size of big datasets and registries. Deep learning has been successfully implemented in the evaluation of cardiac MR for segmentation and functional evaluation of biventricular function and strain.[8] Automated deep learning imaging analysis achieves a pixelwise image segmentation by applying a number of convolutional filters onto an input image. The model learns image features from fine to coarse levels using convolutions and combines multi-scale features for predicting the label class at each pixel. Deep learning utilizes kernels (specific patterns of weights that convey relevant key features found in the image, resembling a filter for image features crucial for a specific classification). This process of applying the kernel to the pixels of the image (or to the mapping spatially organized units in a subsequent layer) is called convolution, a term borrowed from signal processing. The term is quite synonymous to cross-correlation, which is a measure of similarity of two series as a function of the displacement of one relative to the other. Advances in AI analytics for coronary CT have been the most notable given the size of these databases. Notably, beyond segmentation and extraction, even functional properties of coronary circulation have been derived from anatomical imaging (e.g., fractional flow reserve) through deep learning.[9] It is therefore expected that complementary estimations such as the emerging wall shear stress, which has been related to atherosclerotic plaque progression,[10] will add to the body of determinations that AI can streamline in our search to better understand the dynamics of coronary artery disease.

Currently, quality acquisitions may hold the key to fully model physical properties of the heart and vascular system. Interesting advances have reported on methods to model fluid structure, dynamics, and artery wall stiffness from MRI images.[11] Ultimately, with enough dynamic information on the vascular beds, a general model to interact and experiment with the intrinsic properties of the *vasculome* could be generated. Moreover, it could allow us to better understand the status of the *vasculome* as both a proxy of health and a predictor of risk of response to therapy.

In vascular imaging, machine learning analytics have also been successful in the estimation of aortic flow in a fully automated fashion[12] as well as its corresponding 3D segmentations.[13] And regarding pathological states, deep learning has been a successful component of novel analytical pipelines in order to, for example, identify and segment type B aortic dissection in CT imaging.[14] Of note, novel deep learning–based CT reconstruction has attracted attention as an alternative to iterative reconstruction with promising results for imaging vessels.[15]

At the level of the brain, machine learning strategies have showed value in stroke risk stratification based on plaque morphology using carotid ultrasound data, while deep learning has been used to identify consequences of vascular dysfunction[16] as in small vessel disease. Furthermore, deep learning analysis of MR angiography has demonstrated value in automating cerebral aneurysms achieving a generalizable sensitivity of 93%.[17]

A landmark study in 2016 reported the utilization of deep learning in the evaluation of retinal fundus images for diabetic retinopathy and other related eye diseases.[18] Notably, these images containing vascular structures have also been analyzed in order to predict the presence or absence of systemic cardiovascular risk factors individually.[19] These reports elegantly hint the underlying mechanistic relationship that the *vasculome* concept aims to capture: similar caliber vessels in distinct part of the organism will tend to respond to the chronic effect of adverse risk factors in a similar way such that either vasculature will constitute a proxy of vascular health. The ultimate goal is to build models of biological systems that faithfully reflect the underlying physiology in order to generate mechanistic predictions.

Until 2019 there was no systematic description of all vessels of all sizes in an entire organ in three dimensions. This was in part due to lack of good automated analysis methods for large 3D imaging datasets; technology showed a large variance in signal intensity and a bad signal-to-noise ratio; most common techniques cannot reliably differentiate between vessel walls and background in whole-organ scans; relevant datasets are at the terabyte level, where standard imaging processing techniques break down, requiring intensive manual labor for fine tuning. Todorov and colleagues developed a Vessel Segmentation & Analysis Pipeline (VesSAP) to face the problems mentioned above as they apply specifically to the brain. This consisted of a deep learning approach to automated analysis of an entire mouse brain vasculature.[20] Their approach is described in three steps. Firstly, histological treatment of the brain tissue, so as to enhance vasculature imaging down to the capillary level using two dyes: wheat germ agglutinin and Evans blue. Secondly, automatic segmentation and tracing of the whole-brain vasculature data is accomplished using deep learning in simulated vasculature data and then by tapering the system with some expert manually annotated examples through transfer learning. Lastly, extraction of vascular features for hundreds of brain regions and their annotation and clustering to a gold-standard brain map. The convolutional steps in their approach integrate the 3D voxel neighborhood, while at the last convolution, each voxel has 50 features and these are combined with a kernel size of one. A sigmoidal activation is used to estimate the likelihood of a voxel representing a vessel. The technique was proven robust against variations in signal intensities and structures, outperforming previous filter-based methods and reaching the quality of segmentation achieved by human annotators. It achieved similar performance as the deep learning U-net and V-net architectures, but required substantially less trainable parameters while outperforming all other architectures in terms of speed due to less trainable parameters necessary. While tissue preparation represent a clear limiting step for implementation, this approach opens the door to conceptualize novel solutions for the study of the *vasculome* as a concept. For example, transfer learning could be applied to analyze cardiovascular-specific data given that vascular structures are in fact the common axis of study across different organs that will possibly demonstrate its own *vasculome* print. Therefore, it would be possible, at least conceptually, to apply this approach to a number of organ-specific vascular prints that later may not need tissular input anymore.

In summary, the current approach to the study of the (cardio) vascular system remains parcellated. However, independent efforts are aggregating and our understanding of the possibilities that machine learning–based AI offers in the study of blood and vascular tissue properties (such as *endothelial dysfunction, elasticity/rigidity, neovascularization, atherosclerosis, aneurysmatic changes, thromboembolic events, dissection, and rupture*) and its underlying dynamics is rapidly increasing. In essence, the characterization of the heart and the multiple vascular beds throughout the organism sets a favorable environment for the warranted integration of a novel conceptual approach to (cardio) vascular health and disease though the *vasculome*. We believe that supervised learning has much to deliver as we harness the potential of the currently available big datasets. And ultimately, unsupervised learning of such data may identify distinct profiles to link an individual's *vasculome* to specific system diseases or risk thereof as well as to particular prognostic profiles. Ideally, it could also provide orientation on adequate risk mitigation strategies based on identified pathological mechanisms that may differ between individuals even if the pathological expression in the vascular beds may be similar. AI has arrived at the right to boost these data and framework integration, with unparalleled performance to analyze and gain insight from the massive stream of information at our disposal.

Conclusion

A novel conceptual framework for the study of the (cardio) vascular system is warranted. Although traditionally parcellated, the vascular fingerprint or *vasculome* of an organ or an individual may allow us to reformat our understanding of its health and disease. Machine learning–based AI will continue to permeate into different domains in the research of this integrative concept offering unparalleled performance to model and gain insight from the large amounts of information contained in clinical and imaging data.

References

1. Juarez-Orozco LE. *Curr Cardiovasc Imaging Rep*. 2019;12(5).
2. Benjamins JW. *Neth Heart J*. 2019 Sep;27(9):392–402.
3. Salerno M, Kramer CM. Advances in cardiovascular MRI for diagnostics: applications in coronary artery disease and cardiomyopathies. *Expert Opin Med Diagn*. 2009;3(6):673–687. 10.1517/17530050903140514.
4. Cikes M. *Eur J Heart Fail*. 2019 Jan;21(1):74–85.
5. Benjamins JW. *Int J Cardiol*. 2021 Apr 6. S0167-5273(21)00645-8.
6. Slart RHJA. *Eur J Nucl Med Mol Imaging*. 2021 May;48(5): 1399–1413.
7. Pontarotti P, ed. *Evolutionary Biology 2016: Convergent Evolution, Evolution of Complex Traits, Concepts and Methods*. Cham: Springer International; 2016.
8. Ruijsink B. *JACC Cardiovasc Imaging*. 2020 Mar;13(3):684–695.
9. Coenen A. *Circ Cardiovasc Imaging*. 2018 Jun;11(6):e007217.
10. Costopoulos C. *Eur Heart J*. 2019 May 7;40(18):1411–1422.
11. Bertoglio C. *Int J Numer Method Biomed Eng*. 2012 Apr;28(4): 434–455.
12. Bratt A. *J Cardiovasc Magn Reson*. 2019 Jan 7;21(1):1.
13. Berhane H. *Magn Reson Med*. 2020 Oct;84(4):2204–2218.
14. Cao. *Eur J Radiol*. 2019 Dec;121:108713.
15. Park C. *Eur Radiol*. 2021 May;31(5):3156–3164.
16. Cuadrado-Godia E. *J Stroke*. 2018 Sep;20(3):302–320.
17. Ueda D. *Radiology*. 2019 Jan;290(1):187–194.
18. Ting DSW. *JAMA*. 2017 Dec 12;318(22):2211–2223.
19. Poplin R. *Nat Biomed Eng*. 2018 Mar;2(3):158–164.
20. Todorov MI. *Nat Methods*. 2020 Apr;17(4):442–449.

PART III

Vasculome dynamics: in health and in sickness

SECTION 1

Cooperating during development and organogenesis to create vasculome diversity

CHAPTER

20

Vascular development and organogenesis: depots of diversity among conduits of connectivity define the vasculome

Victoria L. Bautch[1,2,3,4]

[1]Department of Biology, University of North Carolina at Chapel Hill, Chapel Hill, NC, United States [2]Lineberger Comprehensive Cancer Center, University of North Carolina at Chapel Hill, Chapel Hill, NC, United States [3]Curriculum in Cell Biology and Physiology, University of North Carolina at Chapel Hill, Chapel Hill, NC, United States [4]McAllister Heart Institute, University of North Carolina at Chapel Hill, Chapel Hill, NC, United States

Introduction

The science fiction movie "*Fantastic Voyage*" follows the saga of a miniaturized submarine and crew that enters the bloodstream of a famous scientist. The goal is to travel through blood vessels and dissolve a blood clot in the brain, and the film provides visually striking images of the journey through the heart and blood vessels of the lungs, inner ear, and brain.[1] This story graphically illustrates the connectivity of the blood vessels in our body, as do measurements showing that each person has about 60,000 miles of blood vessels that circulate six quarts of blood through the body three times per minute. Blood vessels are found virtually everywhere in the body (some exceptions are the cornea of the eye and cartilage), and no cell is very far from a blood vessel that provides oxygen and nutrients. This access requires a connected set of conduits that share properties required to transport blood and provide a barrier between the circulation and tissues. However, diversity in both vessel structure and function is also necessary to accommodate the needs of different tissues and organs. This diversity is evident as soon as the first vessels form and expand during development, since genetic cues set up artery and vein identity prior to the onset of significant blood flow.[2] Expanding vessel networks initiate cross-talk with organ primordia that leads to structural and functional vascular specialization while also critically contributing to organogenesis. For example, during lung development, an intricate dance between vascular endothelial cells and lung epithelial cells results in the coordinated branching of both tissues required for gas exchange,[3] while in the developing kidney an equally intricate communication leads to development of the glomeruli and associated networks to remove waste from the blood[4] (Fig. 20.1). The last several decades have provided partial answers to questions such as the following: How does the intricate and ubiquitous circulatory system form? How do vessels specialize locally while remaining linked to the larger circulation? And how do blood vessels influence organ development and vice versa? This chapter will address some common attributes of these processes and highlight some of the specialized cross-talk that leads to regional and tissue specialization of the circulatory system.

Blood vessels are comprised of endothelial cells that initiate blood vessel formation and expansion, along with support cells called pericytes, and in larger vessels, smooth muscle cells, and sometimes fibroblasts provide structural support and tone. Together, the endothelial cells and support cells secrete proteins that form a matrix and basement membrane; these extracellular structures are also specialized within tissues and contribute to organogenesis. However, since blood vessels form, expand, and remodel by virtue of endothelial cell behaviors and responses, here I focus on the role of endothelial cells in the combined specialization and integration of the vasculature that has been captured in the term "vasculome." I will first examine some

The Vasculome
https://doi.org/10.1016/B978-0-12-822546-2.00005-8

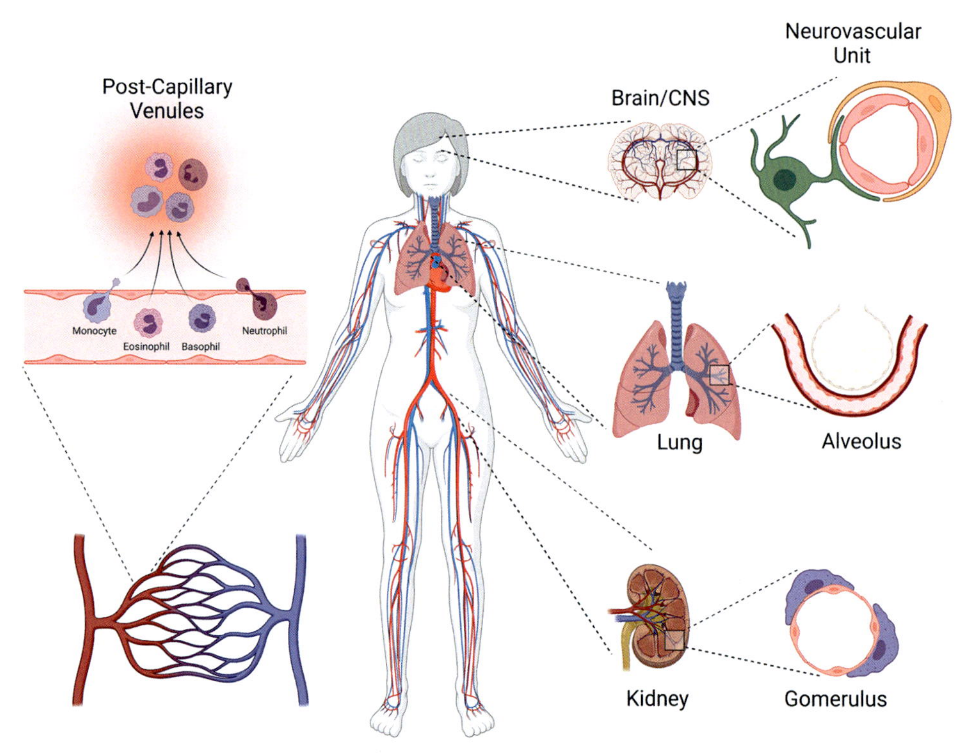

FIGURE 20.1 **The vasculome.** The integrated circulatory system consists of organ depots (right side) where endothelial cells interact with tissue-specific cells; it also consists of locales such as postcapillary venules (left side) in multiple tissues that provide the microenvironment for vessel–tissue interactions such as leukocyte extravasation.

seminal historical discoveries that inform our current understanding of the vasculome. I discuss some common aspects of blood vessel–organ cross-talk, and then highlight a few examples of specialization in vessel–organ crosstalk during development. Finally, I examine several vascular developmental processes that illustrate how vessels specialize using cues that are provided by the intact circulatory system. Rather than a comprehensive analysis of specific developmental programs, the goal of this chapter is to convey an appreciation for both the commonalities and the differences that characterize the developing and functioning vasculome.

Historical perspective of the vasculome concept

Since ancient times, it has been appreciated that different blood vessels carry different cargoes. Galen (129 CE-c.200) found that some conduits were visually bright red, while others were darker; he attributed this to different "humors" without realizing that the different systems were interconnected. Although we now know that blood oxygenation status leads to the color differences (oxygenated blood is red), it was not until 1000 years later that Ibn al-Nafis (1213–88) and later Leonardo da Vinci (1452–1519) established that the circulatory system was continuous, that it was connected within tissues by a meshwork of small conduits called capillaries, and that blood was pumped from the heart through the circulation and returned via larger vessels.[5,6] These observations were codified by William Harvey (1578–1657) in his treatise *De Motu Cordis* (1628), where he also described some of the first experiments investigating the circulation, such as using a ligature (primitive blood pressure cuff) to show that blood flowed outward from the heart via the arteries.[7] The presence of red blood cells within vessels provided visual cues for early descriptions of blood vessel patterns and temporal changes documented by Marcello Malpighi (1628–1694),[8] and India ink injections were used to visualize patent vessels and confirm that the heart and circulation was a single closed unit.[9,10] As modern science moved from a focus on description-based data to include molecular and genetic manipulation of genes and pathways, elucidation of cellular processes and mechanisms underlying vascular development initially lagged because blood vessels were difficult to isolate for analysis of RNA and protein. Vascular endothelial cells are embedded within vessels of every tissue and contribute only minimally to bulk analyses, yet they are found in all tissues, making comparative bulk analysis uninformative. This issue was partially mitigated when conditions for endothelial cell culture in vitro were established by Gimbrone and colleagues,[11] and these advances over time allowed for identification of a set of common endothelial cell attributes and markers. We now know that normal cell culture conditions lead to the loss of many location-specific endothelial cell properties and markers.[12,13] However, cell culture remains a valuable tool for understanding behaviors and responses shared by endothelial cells found in disparate locations in the body, and recent advances have identified culture conditions that better mimic tissue environments in vivo, including serum-free defined medium, coculture with accessory cells, and reconstruction of the 3D environment and mechanical forces experienced by endothelial cells in vivo.[14–16]

Starting about 30 years ago, endothelial cell–selective markers such as PECAM1 (Platelet-Endothelial Cell Adhesion Molecule 1) and vWF (von Willebrand Factor) were identified, and antibodies and RNA probes were developed that revealed patent and nonpatent vessels in animal models and human tissue sections. Alongside these markers, genetic tools were designed to selectively manipulate genes in endothelial cells in vivo using dual expression systems based on activation-response (i.e., Gal4-UAS) and/or drug responsiveness (i.e., the "Tet-on" or "Tet-off" system).[17] The advent of tissue-targeted deletion via Cre-Lox recombination allowed for functional interrogation of genes and pathways during vascular development via endothelial cell–selective loss of function.[18] Further refinement of inducible recombination drivers that target gene deletion to subsets of endothelial cells continues to expand our knowledge and appreciation of the similarities and differences among endothelial cells in blood vessels, while fluorescent probes that allow for live imaging have opened the door to dynamic analysis of vascular development.[19,20] These experimental refinements have recently culminated in the development of single-cell RNA sequencing and other single-cell -omics technologies, and it is now possible to document individual transcriptomes of different endothelial cells within an organ or tissue and over time. These analyses have confirmed that (1) a subset of genes contribute to an expression profile shared by most or all endothelial cells, regardless of location or local function in the body and (2) this common core is overlaid by expression heterogeneity, and some of this heterogeneity is functionally relevant for normal function and in disease,[21,22] leading to the concept of the vasculome, defined as all the vasculature of an organism. Thus, we have come full circle, from a reductionist approach to endothelial cell biology based on cell culture outputs to an explosion of information and detail, with the challenge now being to assign functional relevance to both the shared and distinct information, and to better understand how this information is coopted in disease.

Normal vasculome function and diseases that lead to or are accompanied by vasculome dysfunction, are often best understood in terms of the developmental processes involved in forming the vasculome. For example, humans born with a predisposition to develop Hereditary Hemorrhagic Telangiectasia (HHT) have mutations in genes that regulate BMP signaling in endothelial cells.[23,24] This disease leads to hemorrhage and arteriovenous malformations, but the focal yet dispersed nature of the lesions makes HHT difficult to dissect in human patients. However, animal models show that the lesions form as a result of an angiogenic stimulus accompanied by blood flow and likely a somatic mutation that leads to mosaic loss of function, highlighting that both genetic changes, changes in external signal generation and mechanical inputs, are important for HHT lesion formation.[25] Thus, we will review vascular development to better understand the function and properties of the vasculome.

Vascular development and organogenesis: where it all starts

Blood vessel formation, one of the most strikingly elegant developmental processes, gives rise to the intricate vascular network that circulates blood and nutrients in the body. Some attributes of vascular development are as follows: (1) blood vessels arise early in development, (2) vessels function as soon as they are formed, and (3) functioning embryonic vessels are constantly sculpted and expanded as development proceeds, in response to the changing needs of developing organs and tissues. These attributes result in a network that forms and functions while changing dramatically, and it is absolutely necessary for viability and function of the organism throughout life. Vessels also direct organogenesis, by providing oxygen and nutrients to support expansion, and by providing temporal and spatial cues for organ patterning and differentiation. Once overall development slows down during the early postnatal period, emphasis in the circulation shifts toward setting up and maintaining vessel homeostasis and preserving the organ-specific attributes of endothelial cells. During this transition, blood vessel proliferation and sprouting initiated by endothelial cells is suppressed, and the barrier that regulates oxygen/nutrient exchange and prevents leak is stabilized. However, most vessels are able to respond quickly and induce new blood vessel formation in response to angiogenic cues throughout life; this property allows for physiological vascular responses during wound healing and pregnancy, and it contributes to pathology in cancer as tumor-induced angiogenesis or in response to chronic inflammation. Other chapters in this book focus on early endothelial cell differentiation and patterning events, so here I briefly summarize this information before discussing common attributes of vascular–organ cross-talk.

Most endothelial cells differentiate from progenitor cells, called angioblasts, that arise primarily from the lateral plate mesoderm in mammals shortly after gastrulation, although some angioblasts derive from somatic mesoderm.[26–28] Angioblasts coalesce around the embryonic midline and differentiate into endothelial cells as they form a cord linked to the developing heart; the cord subsequently lumenizes to form the dorsal aorta. The dorsal aorta arches as it leaves the heart, then descends into the trunk, where it bifurcates to arteries that lead to the liver, kidney, and other internal organs, along with vessels that service the developing limbs. Aortic arches (also called branchial arches) form from the aortic sac ventral to the dorsal aorta, and they remodel extensively to give rise to "great arteries" that provide circulation to the lungs and organs of the head and neck, including the pulmonary and carotid arteries. Among numerous genes involved in the initial differentiation of endothelial cells, transcription factors of the FOX family and the ETS family play crucial roles,[29] and the ETS family transcription factor ETV2 stands out as being required for most endothelial cell differentiation in vivo and in culture.[30,31] Additional angioblasts differentiate from mesoderm as organ primordia begin to develop, and these tissue-resident angioblasts form capillaries that link to the larger arteries and veins. A few tissues, such as the neural ectoderm that comprises the central nervous system (CNS), do not generate additional endothelial cells, and all vascular tissue colonizes the CNS from outside. Finally, recent work identifies a population of endothelial cells that derive from erythro-myeloid hematopoietic cells and contribute to the vasculature of some but not all organs,[32] while a subset of endothelial cells become "hemogenic" in vivo and delaminate from embryonic blood vessels to produce hematopoietic cells.[33–35] Thus, common attributes of endothelial cells in the majority of blood vessels contributing to the vasculome are derivation from embryonic mesoderm, and activation of transcriptional programs that include FOX and ETS-type factors for differentiation.

Most blood vessels share responses to signals that initiate in organs and impact nascent vessels to orchestrate organ-directed vascular development. Among several signaling hubs, the VEGF-A/VEGFR signaling pathway stands out as necessary, and in many instances sufficient, to initiate expansion of the vasculome to developing organs.[36–38] VEGF-A ligand is upregulated downstream of cellular hypoxia, which increases as embryonic cells proliferate and begin to form organ primordia. The relative lack of oxygen stabilizes the

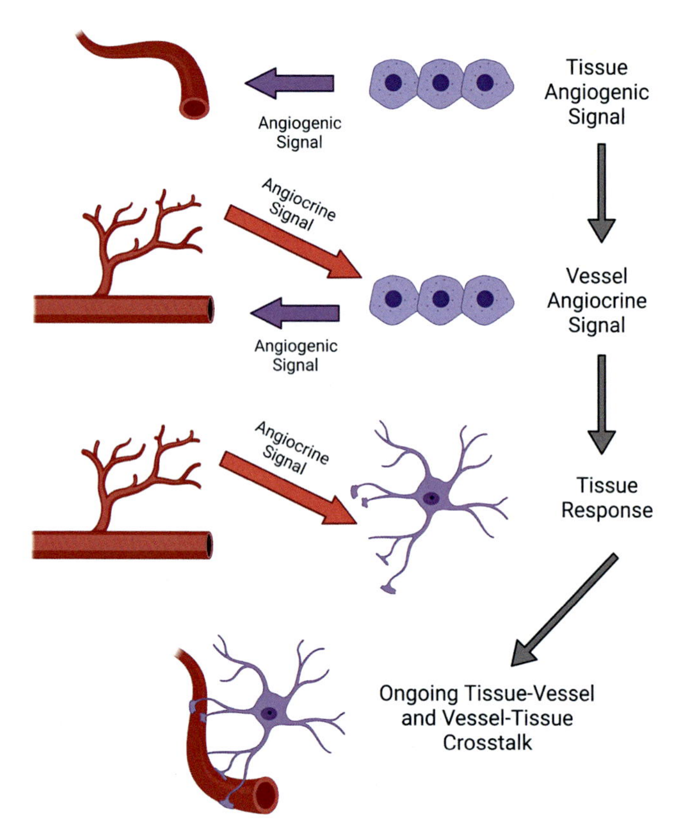

FIGURE 20.2 Angiogenic–angiocrine cross-talk. Schematic showing that tissue-derived angiogenic signals recruit new blood vessels that then produce angiocrine signals that affect tissue differentiation, patterning and function. The cross-talk can be iterated and continues throughout life.

HIF1α transcription factors that form heterodimers with ARNT/HIF1β subunits to bind gene promoters and induce transcription, including that of the *VEGFA* gene.[39] VEGF-A is spliced and secreted so as to form a version of a gradient from the source tissue to the nearest target, which are endothelial cells that express a VEGF-A receptor called VEGFR2/Flk1/KDR (Fig. 20.2). This signaling intersects with other pathways such as Wnt, Notch, and chemokines to facilitate vascularization of organ primordia. Thus, VEGF signaling promotes angiogenic sprouting and provides survival cues to endothelial cells, leading to vessel network expansion.

Once nascent sprouts form in organ primordia during development, two aspects of endothelial cell biology start the vessel–organ communication that will continue throughout development and life. First, endothelial cells express proteins that modulate the effects of incoming organ-derived signals on their own cellular processes. An excellent example is the expression of a second VEGF-A receptor called VEGFR1/Flt1 that is alternatively spliced to produce both membrane-localized and secreted isoforms.[40] VEGFR1 binds VEGF-A with 10-fold higher affinity than does VEGFR2, the membrane-localized form does not efficiently signal to endothelial cell proliferation or migration pathways upon VEGF-A binding, and the signaling domain is not required for developmental angiogenesis.[41] These attributes allow both isoforms to act as a decoy or sink to bind VEGF-A and prevent its binding to VEGFR2 on endothelial cells, and thus modulate signaling through VEGFR2. Genetic loss of *VEGFR1/Flt1* in mouse leads to vessel overgrowth and embryonic lethality during early organogenesis and blood vessel dysmorphogenesis during embryonic stem cell differentiation to endothelial cells in primitive organoids.[42,43] Soluble VEGFR1/Flt1 also contributes to blood vessel patterning, likely by providing spatial information locally as vessel networks expand.[44,45] Thus, this endothelial cell-intrinsic cue critically interacts with organ-derived signals to modulate incoming signaling as the vasculature expands into new organs and tissues.

Endothelial cells also produce signals that directly impact organogenesis; these are called angiocrine factors and have been extensively reviewed.[46,47] Angiocrine factors exhibit some overlap but remarkable heterogeneity in different organ beds, and their functions and regulation are an area of active ongoing research. The general concept that vessel-derived signals shape organs is held in common during vascular development.[48,49] Angiocrine signals provide survival and patterning cues for organ development, thus completing a circle of inductive patterning from tissues to vessels to tissues during development, and this cycle is often iterated several times during organ development, with the same or different signals participating in the cross-talk (Fig. 20.2). One especially important aspect of angiocrine signaling, shared by some but not all vessel networks, is the construction of organ-specific stem cell niches near some vessels. Vascular cells in these niches provide signals to stem and progenitor cells that maintain multipotency.[50–53] The tissue-resident stem/progenitor cells divide to generate daughter cells that differentiate and populate organs. The vascular changes that accompany aging and disease likely perturb these organ depots of stem/progenitor cells via changes to the vascular niche and further exacerbate the dysfunction.

Blood vessel–organ cross-talk also influences the cellular properties of vessels and their overall function. Support cells called pericytes often hitchhike on nascent vessels as they invade organs, and pericytes are also recruited from local tissue sources of mesoderm, as are the smooth muscle cells that are recruited by larger vessels.[54] Pericytes share a basement membrane with endothelial cells that is contributed by both cell types. This basement membrane serves as a template for endothelial cells that reflects organ-induced remodeling of endothelial cell–cell junctions and instructs different types of barrier between blood and tissues.[46] These interactions in turn provide structural cues to organ-derived cells to set up distinct relationships with the vasculature. The commonality is the presence of this cross-talk in many different organs, while the signals often differ, and the outcomes are tailored to the needs of the tissue.

Tissues also have the property of stiffness, defined as the deformability of the extracellular matrix. Tissue stiffness is influenced by stromal cells, such as fibroblasts, that often have tissue-specific properties.[55] Tissue stiffness combined with a force vector for tension provides cues for the orientation and patterning of vessels, especially the microvasculature. These effects can be reproduced in vitro, where constraining the long axis but not the short axis of a collagen matrix leads to neovessel patterning parallel to the long axis.[56]

The circulatory system responds to the expanding needs of developing organs for oxygen and nutrients by blood flow-dependent remodeling of the vessels within organs. This remodeling results in a hierarchical patterned vasculature that is tailored to the size and function of the particular organ. In general, once a primary plexus forms in an organ, location-specific needs within the organ stimulate localized changes in blood flow that lead to formation of larger diameter conduits (arteries and veins), while other vessels remain as capillaries or regress. In fact, blood flow is necessary to maintain the initial fate of large vessels that is genetically determined, especially that of larger arteries.[57] This important vessel–organ signaling requires a patent circulation and maintenance of blood flow throughout the entire circulatory system.

Vascular development and organogenesis: location-dependent crosstalk

Along with generic responses of endothelial cells to tissue signals, blood flow, and extracellular matrix, each organ provides unique signals and microenvironmental cues that set up local specialized vascular domains within the larger circulation. One of the most elegant examples of tissue-influenced vascular patterning is the neurovascular cross-talk that leads to the patterning of the vessels relative to domains of the spinal cord, which is part of the CNS. As mentioned above, CNS is ectoderm derived and does not produce angioblasts, so blood vessels are recruited from outside the CNS via sprouting angiogenesis. Vascular sprouts initially enter in a stereotypical pattern that is influenced by neural tissue–derived VEGF-A and other signals.[58–60] Once vessel sprouts have entered the CNS, they expand into certain areas while avoiding other areas, likely in response to repulsive cues produced by the neural cells. This early patterning allows for oxygenation of the neuroepithelium, while preserving cohorts of neural cells in specialized domains important for neural function. Evidence for neurovascular cross-talk was revealed by the finding that endothelial cell expression of Dab1/Reelin is important for proper neuronal patterning in the spinal cord.[61] A specific example of bidirectional neurovascular cross-talk in the spinal cord starts with neural progenitor cells expressing the ligand angiopoietin-1 (Ang1) that binds and activates the Tie2 receptor on endothelial cells, which in turn leads to endothelial cell production of TGFβ1 that signals neural precursor cells to an oligodendrocyte fate.[62]

At the cellular level, a second unique aspect of neurovascular cross-talk is the development of the blood–brain barrier, which protects the CNS from overt inflammation and swelling. The blood–brain barrier stringently limits the size of small molecules leaving the circulation and prevents leukocyte extravasation. The blood–brain barrier results from complex cross-signaling among four different cell types that comprise the neurovascular unit[58,63–65] (Fig. 20.1). The neurovascular unit consists of two vascular-derived and two neural-derived cell types; vascular endothelial cells and pericytes come from the vessels, while neurons and astroglia provide the neural components, and together these cell types signal to each other so that the barrier is formed and maintained, and also so that the changing metabolic status of the neurons translates into local changes in blood flow; this relationship is the biological basis for fMRI (functional Magnetic Resonance Imaging) that reveals localized areas of brain activity via blood flow changes. The molecular cross-talk involved in these relationships is extensive and includes many known signaling pathways that are also used outside the CNS.

Another example of organ–tissue cross-talk that leads to specialization is the intimate relationship of the vascular endothelial cells and different cell types in the kidney.[4,66] The kidney consists of a medullary region and an extensive cortex that contains 1,000,000 nephrons consisting of a glomerulus and associated structures. The glomerulus is composed of capillary endothelial cells that are fenestrated and share a basement membrane with podocytes; this relationship allows for the transfer of solutes and fluid to the kidney tubules (Fig. 20.1). The blood then enters peritubular capillaries that surround and align with the cortical renal tubules carrying the glomerular filtrate, and here the vessels reabsorb some solutes and fluids. Finally, the blood enters vessels called the vasa recta which first descends, then ascends the medulla alongside the medullary renal tubules; the vasa recta vessels deliver oxygen and nutrients to the medulla and also contribute to urine concentration by reabsorbing water. Thus, every aspect of kidney development occurs in conjunction with blood vessels, and kidney development does not proceed properly absent blood vessels. The vessels in the different locales express strong location-specific markers while sharing a common set of markers.[67]

Circulation-dependent input for local vascular specialization

Some or most of the above examples of tissue–organ cross-talk can be partially recapitulated upon removal of the system from the rest of the circulation, either by coculture of relevant cell types or by ex vivo development of tissues. However, some aspects of vascular specialization require the entire circulatory system to be intact to faithfully recapitulate developmental processes, thus truly highlighting the importance of the integrated closed circulatory system. One example is the dramatic changes to the circulation that occur at birth. Prior to birth, very little blood circulates through the fetal lungs, and the embryo receives oxygenated blood from the placenta that mixes with deoxygenated blood. This mixture flows into the right atrium of the heart and then into the left atrium via the foramen ovale (opening between the atrial chambers with a flap), and from there to the left ventricle and out the dorsal aorta. At birth, placenta removal cuts off placental blood flow, and the decreased pressure in the vena cava, along with lung capillary dilation due to breath-derived oxygen, sets up a pressure gradient in the heart atria that leads to closure of the foramen ovale flap, forcing blood from the right atrium into the right ventricle and from there into the pulmonary artery leading to the lungs. This pressure change, along with reduced prostaglandin secretion from placenta removal, allows blood to flow through the lungs and become oxygenated. At the same time, the ductus arteriosus, which connects the pulmonary artery and the dorsal aorta in the fetus, is closed off because of increased oxygenated blood in the dorsal aorta along with reduced prostaglandins, further preventing the mixing of oxygenated and deoxygenated blood; the ductus arteriosus subsequently degenerates. Thus, much of the circulation changes within a few minutes of birth to separate pulmonary and systemic flow through the heart, utilizing pressure changes initiated in the placenta and lungs. These changes are critical to life and would not happen in this way absent a closed circulatory system.

Another example of vascular specialization that requires an intact circulation is the maturation of postcapillary venules as sites of leukocyte extravasation into tissues (called transendothelial migration (TEM))[68,69] (Fig. 20.1). Postcapillary venules are found downstream of capillaries, and their properties and gene expression patterns support the elegant leukocyte–endothelial cell interactions that allow neutrophils and other immune cells to move from the bloodstream into the interstitial space. Briefly, P-selectin on endothelial cells binds to a neutrophil ligand to initiate leukocyte rolling in the postcapillary venule. This interaction leads to interactions of β2 integrin complexes on leukocytes (i.e., LFA1, VLA4) with adhesion molecules such as ICAM1, ICAM2, and VCAM on endothelial cells that results in leukocyte capture and arrests the rolling behavior. At this point, leukocytes sample the endothelial monolayer via several mechanisms[70] and eventually move through the endothelial cell monolayer using a paracellular (between) or transcellular (through) route. These areas of leukocyte rolling, capture, and extravasation are found in the same spatial location in different organs, indicating that spatially restricted inputs such as the shear flow vector and local stiffness imparted by the extracellular matrix/basement membrane are important for this specialized niche.[71] Within the postcapillary venule niche, additional endothelial cell and/or matrix heterogeneity sets up local "hotspots" for leukocyte extravasation.[70] How these hotspots are set up and maintained is an area of current investigation.

The endothelial cells of postcapillary venules have a distinct expression pattern.[22] Among the genes upregulated in this niche is a chemokine receptor called ACKR1 (atypical chemokine receptor-1) that presents the chemokine CXCL2 that is secreted by neutrophils and required for proper leukocyte TEM.[72,73] A version of this niche also exists in lymph nodes and secondary lymphoid organs, called high endothelial venules (HEV). At these sites, endothelial cells express enzymes that generate leukocyte antigens via addition of sulfated carbohydrates such as 6-sulfo silyl Lewis^{X} that binds L-selectin on lymphocytes; these interactions allow naïve leukocytes to enter HEV and access sites of antigen presentation.[74] The link between postcapillary venules and HEV is suggested by the common expression of ACKR1 at both sites.[75] In aggregate, these examples reveal how the continuous and connected nature of the vasculome allows for regional specialization based on changing mechanical forces imparted by the blood flow itself combined with the local microenvironment.

The vasculome: conclusions and future directions

The development and patterning of the vasculature is largely driven by endothelial cells. This process is paradoxical in requiring a closed circulatory system for many aspects of development, while also responding to different signals and groupings of signals to articulate tissue- and organ-specific vascular patterns and functions. These contrasting processes set up the vasculome—a group of cells that share enough common characteristics to provide a closed circulatory system, but are diverse enough to service distinct metabolic and functional needs of organs and tissues. Our ability to examine the development and function of the vasculome has expanded dramatically both in terms of depth and breadth, and this capacity has opened new doors for ongoing and future research.

Several important questions remain. First, the heterogeneity of the vasculome makes it difficult to generalize mechanisms and disease outcomes in some cases. Pathways such as YAP/TAZ have different roles in the CNS versus bone vasculature.[76] Likewise, diseases with a genetic predisposition imparted by the germline, such as HHT, have tissue-selective lesion patterns, so that small HHT lesions cluster in mucosal membranes. Another vascular genetic hemorrhagic disease, cerebral cavernous malformations, has lesions largely restricted to the CNS.[77] Conversely, the connected nature of the circulatory system stymies targeted therapies for disease; the chemicals that attack tumor cells and the antiangiogenesis drugs that prevent tumor neovascularization have systemic effects that lead to severe and often life-threatening side effects. Finally, investigations and new knowledge regarding how the circulation forms, functions, and is affected by disease must be put into the context of the continuity of the system overlaid with the context of the particular process under investigation, leading to more interesting challenges for the future.

Acknowledgments

I thank Bautch Lab members and colleagues for fruitful discussions, and I apologize to colleagues whose excellent work could not be directly cited due to space constraints. Illustrations were made using Biorender.com.

Funding

NIH R35 HL139950 and NIH R01 GM129074.

References

1. Fleischer R. *Fantastic Voyage*. 1966.
2. Chong DC, Koo Y, Xu K, Fu S, Cleaver O. Stepwise arteriovenous fate acquisition during mammalian vasculogenesis. *Dev Dynam*. 2011;240:2153–2165.
3. Lazarus A, Del-Moral PM, Ilovich O, Mishani E, Warburton D, Keshet E. A perfusion-independent role of blood vessels in determining branching stereotypy of lung airways. *Development*. 2011; 138:2359–2368.
4. Herzlinger D, Hurtado R. Patterning the renal vascular bed. *Semin Cell Dev Biol*. 2014;36:50–56.
5. Sterpetti AV. The revolutionary studies by Leonardo on blood circulation were too advanced for his times to be published. *J Vasc Surg*. 2015;62:259–263.
6. West JB. Ibn al-Nafis, the pulmonary circulation, and the islamic golden age. *J Appl Physiol*. 2008;105:1877–1880.

7. Pasipoularides A. Historical perspective: Harvey's epoch-making discovery of the circulation, its historical antecedents, and some initial consequences on medical practice. *J Appl Physiol*. 2013;114: 1493–1503.
8. West JB. Marcello Malpighi and the discovery of the pulmonary capillaries and alveoli. *Am J Physiol Lung Cell Mol Physiol*. 2013; 304:L383–L390.
9. Ashton N, Blach R. Studies on developing retinal vessels VIII. Effect of oxygen on the retinal vessels of the ratling. *Br J Ophthalmol*. 1961;45:321–340.
10. Bergeron L, Tang M, Morris SF. A review of vascular injection techniques for the study of perforator flaps. *Plast Reconstr Surg*. 2006; 117:2050–2057.
11. Gimbrone Jr MA, Cotran RS, Folkman J. Human vascular endothelial cells in culture : growth and DNA synthesis. *J Cell Biol*. 1974;60: 673–684.
12. Calabria AR, Shusta EV. A genomic comparison of in vivo and in vitro brain microvascular endothelial cells. *J Cerebr Blood Flow Metabol*. 2007;28:135–148.
13. Cleuren ACA, van der Ent MA, Jiang H, et al. The in vivo endothelial cell translatome is highly heterogeneous across vascular beds. *Proc Natl Acad Sci USA*. 2019;116:23618.
14. Dewey Jr CF, Bussolari SR, Gimbrone Jr MA, Davies PF. The dynamic response of vascular endothelial cells to fluid shear stress. *J Biomech Eng*. 1981;103:177–185.
15. James D, Nam H-S, Seandel M, et al. Expansion and maintenance of human embryonic stem cell–derived endothelial cells by TGFβ inhibition is Id1 dependent. *Nat Biotechnol*. 2010;28:161–166.
16. Orlidge A, D'Amore PA. Inhibition of capillary endothelial cell growth by pericytes and smooth muscle cells. *J Cell Biol*. 1987; 105:1455–1462.
17. Mallo M. Controlled gene activation and inactivation in the mouse. *Front Biosci*. 2006;11:313–327.
18. Payne S, De Val S, Neal A. Endothelial-specific Cre mouse models. *Arterioscler Thromb Vasc Biol*. 2018;38:2550–2561.
19. Asnani A, Peterson RT. The zebrafish as a tool to identify novel therapies for human cardiovascular disease. *Dis Model Mech*. 2014;7:763–767.
20. Laviña B, Gaengel K. New imaging methods and tools to study vascular biology. *Curr Opin Hematol*. 2015;22.
21. Aird WC. Endothelial cell heterogeneity. *Cold Spring Harb Perspect Med*. 2012;2.
22. Kalucka J, de Rooij LPMH, Goveia J, et al. Single-cell transcriptome atlas of murine endothelial cells. *Cell*. 2020;180:764–779.e20.
23. Johnson DW, Berg JN, Baldwin MA, et al. Mutations in the activin receptor–like kinase 1 gene in hereditary haemorrhagic telangiectasia type 2. *Nat Genet*. 1996;13:189–195.
24. McAllister KA, Baldwin MA, Thukkani AK, et al. Six novel mutations in the endoglin gene in hereditary hemorrhagic telangiectasia type 1 suggest a dominant-negative effect of receptor function. *Hum Mol Genet*. 1995;4:1983–1985.
25. Snodgrass RO, Chico TJA, Arthur HM. Hereditary haemorrhagic telangiectasia, an inherited vascular disorder in need of improved evidence-based pharmaceutical interventions. *Genes*. 2021:12.
26. Ambler CA, Nowicki JL, Burke AC, Bautch VL. Assembly of trunk and limb blood vessels involves extensive migration and vasculogenesis of somite-derived angioblasts. *Dev Biol*. 2001;234:352–364.
27. Cleaver O, Krieg PA. Molecular mechanisms of vascular development. In: Harvey RP, Rosenthal N, eds. *Heart Development*. New York: Academic Press; 1999:221–252.
28. Pardanaud L, Luton D, Prigent M, Bourcheix LM, Catala M, Dieterlen-Lievre F. Two distinct endothelial lineages in ontogeny, one of them related to hemopoiesis. *Development*. 1996;122: 1363–1371.
29. De Val S, Black BL. Transcriptional control of endothelial cell development. *Dev Cell*. 2009;16:180–195.
30. Ginsberg M, James D, Ding B-S, et al. Efficient direct reprogramming of mature amniotic cells into endothelial cells by ETS factors and TGFβ suppression. *Cell*. 2012;151:559–575.
31. Lammerts van Bueren K, Black BL. Regulation of endothelial and hematopoietic development by the ETS transcription factor Etv2. *Curr Opin Hematol*. 2012;19.
32. Plein A, Fantin A, Denti L, Pollard JW, Ruhrberg C. Erythro-myeloid progenitors contribute endothelial cells to blood vessels. *Nature*. 2018;562:223–228.
33. Canu G, Ruhrberg C. First blood: the endothelial origins of hematopoietic progenitors. *Angiogenesis*. 2021. https://doi.org/10.1007/s10456-021-09783-9.
34. Gao L, Tober J, Gao P, Chen C, Tan K, Speck NA. RUNX1 and the endothelial origin of blood. *Exp Hematol*. 2018;68:2–9.
35. Gritz E, Hirschi KK. Specification and function of hemogenic endothelium during embryogenesis. *Cell Mol Life Sci*. 2016;73: 1547–1567.
36. Bautch VL. VEGF-directed blood vessel patterning: from cells to organism. *Cold Spring Harb Perspect Med*. 2012;2. https://doi.org/10.1101/cshperspect.a006452.
37. Ferrara N, Carver-Moore K, Chen H, et al. Heterozygous embryonic lethality induced by targeted inactivation of the VEGF gene. *Nature*. 1996;380:439–442.
38. Shalaby F, Rossant J, Yamaguchi TP, et al. Failure of blood-island formation and vasculogenesis in Flk-1-deficient mice. *Nature*. 1995;376:62–66.
39. Covello KL, Simon MC. HIFs, hypoxia, and vascular development. Curr Topic Dev Biol. In: Schatten GP, ed. *Developmental Vascular Biology*. Vol. 62. Academic Press; 2004:37–54.
40. Kendall RL, Thomas KA. Inhibition of vascular endothelial cell growth factor activity by an endogenously encoded soluble receptor. *Proc Natl Acad Sci USA*. 1993;90:10705–10709.
41. Hiratsuka S, Minowa O, Kuno J, Noda T, Shibuya M. Flt-1 lacking the tyrosine kinase domain is sufficient for normal development and angiogenesis in mice. *Proc Natl Acad Sci USA*. 1998;95: 9349–9354.
42. Fong GH, Rossant J, Gertsenstein M, Breitman ML. Role of the Flt-1 receptor tyrosine kinase in regulating the assembly of vascular endothelium. *Nature*. 1995;376:66–70.
43. Kearney JB, Ambler CA, Monaco KA, Johnson N, Rapoport RG, Bautch VL. Vascular endothelial growth factor receptor Flt-1 negatively regulates developmental blood vessel formation by modulating endothelial cell division. *Blood*. 2002;99:2397–2407.
44. Chappell JC, Darden J, Payne LB, Fink K, Bautch VL. Blood vessel patterning on retinal astrocytes requires endothelial Flt-1 (VEGFR-1). *J Dev Biol*. 2019;7:18.
45. Chappell JC, Taylor SM, Ferrara N, Bautch VL. Local guidance of emerging vessel sprouts requires soluble Flt-1. *Dev Cell*. 2009;17: 377–386.
46. Augustin HG, Koh GY. Organotypic vasculature: from descriptive heterogeneity to functional pathophysiology. *Science*. 2017;357. eaal2379.
47. Rafii S, Butler JM, Ding B-S. Angiocrine functions of organ-specific endothelial cells. *Nature*. 2016;529:316–325.
48. Lammert E, Cleaver O, Melton D. Induction of pancreatic differentiation by signals from blood vessels. *Science*. 2001;294:564.
49. Matsumoto K, Yoshitomi H, Rossant J, Zaret KS. Liver organogenesis promoted by endothelial cells prior to vascular function. *Science*. 2001;294:559.
50. Bautch VL. Stem cells and the vasculature. *Nat Med*. 2011;17: 1437–1443.
51. Coskun S, Hirschi KK. Establishment and regulation of the HSC niche: roles of osteoblastic and vascular compartments. *Birth Defects Res C Embryo Today Rev*. 2010;90:229–242.
52. Licht T, Keshet E. The vascular niche in adult neurogenesis. *Mech Dev*. 2015;138:56–62.

53. Shen Q, Wang Y, Kokovay E, et al. Adult SVZ stem cells lie in a vascular niche: a quantitative analysis of niche cell-cell interactions. *Cell Stem Cell*. 2008;3:289–300.
54. Majesky MW, Dong XR, Regan JN, Hoglund VJ, Schneider M. Vascular smooth muscle progenitor cells. *Circ Res*. 2011;108: 365–377.
55. Chang CC, Krishnan L, Nunes SS, et al. Determinants of microvascular network topologies in implanted neovasculatures. *Arterioscler Thromb Vasc Biol*. 2012;32:5–14.
56. Hoying JB, Utzinger U, Weiss JA. Formation of microvascular networks: role of stromal interactions directing angiogenic growth. *Microcirculation*. 2014;21:278–289.
57. le Noble F, Moyon D, Pardanaud L, et al. Flow regulates arterial-venous differentiation in the chick embryo yolk sac. *Development*. 2004;131:361–375.
58. Bautch VL, James JM. Neurovascular development: the beginning of a beautiful friendship. *Cell Adhes Migrat*. 2009;3:199–204.
59. James J, Gewolb C, Bautch VL. Neurovascular development uses VEGF-A signaling to regulate blood vessel ingression into the neural tube. *Development*. 2009;136:833–841.
60. Vieira JR, Shah B, Ruiz de Almodovar C. Cellular and molecular mechanisms of spinal cord vascularization. *Front Physiol*. 2020;11. https://doi.org/10.3389/fphys.2020.599897.
61. Segarra M, Aburto MR, Cop F, et al. Endothelial Dab1 signaling orchestrates neuro-glia-vessel communication in the central nervous system. *Science*. 2018;361:eaao2861.
62. Paredes I, Vieira JR, Shah B, et al. Oligodendrocyte precursor cell specification is regulated by bidirectional neural progenitor–endothelial cell crosstalk. *Nat Neurosci*. 2021;24:478–488.
63. Armulik A, Genove G, Mae M, et al. Pericytes regulate the blood-brain barrier. *Nature*. 2010 [advance online publication].
64. Engelhardt B, Liebner S. Novel insights into the development and maintenance of the blood–brain barrier. *Cell Tissue Res*. 2014;355: 687–699.
65. Segarra M, Aburto MR, Acker-Palmer A. Blood–brain barrier dynamics to maintain brain homeostasis. *Trends Neurosci*. 2021;44: 393–405.
66. Munro DAD, Hohenstein P, Davies JA. Cycles of vascular plexus formation within the nephgrogenic zone of the developing mouse kidney. *Sci Rep*. 2017;7:3273.
67. Barry DM, McMillan EA, Kunar B, et al. Molecular determinants of nephron vascular specialization in the kidney. *Nat Commun*. 2019; 10:5705.
68. Carman CV, Springer TA. Trans-cellular migration: cell–cell contacts get intimate. *Curr Opin Cell Biol*. 2008;20:533–540.
69. Vestweber D. How leukocytes cross the vascular endothelium. *Nat Rev Immunol*. 2015;15:692–704.
70. Grönloh MLB, Arts JJG, van Buul JD. Neutrophil transendothelial migration hotspots – mechanisms and implications. *J Cell Sci*. 2021: 134. https://doi.org/10.1242/jcs.255653.
71. Schaefer A, Hordijk PL. Cell-stiffness-induced mechanosignaling – a key driver of leukocyte transendothelial migration. *J Cell Sci*. 2015;128:2221–2230.
72. Girbl T, Lenn T, Perez L, et al. Distinct compartmentalization of the chemokines CXCL1 and CXCL2 and the atypical receptor ACKR1 determine Discrete stages of neutrophil diapedesis. *Immunity*. 2018; 49, 1062.e6–1076.e6.
73. Thiriot A, Perdomo C, Cheng G, et al. Differential DARC/ACKR1 expression distinguishes venular from non-venular endothelial cells in murine tissues. *BMC Biol*. 2017;15:45.
74. Blanchard L, Girard J-P. High endothelial venules (HEVs) in immunity, inflammation and cancer. *Angiogenesis*. 2021. https://doi.org/10.1007/s10456-021-09792-8.
75. Middleton J, Americh L, Gayon R, et al. A comparative study of endothelial cell markers expressed in chronically inflamed human tissues: MECA-79, Duffy antigen receptor for chemokines, von Willebrand factor, CD31, CD34, CD105 and CD146. *J Pathol*. 2005; 206:260–268.
76. Sivaraj KK, Dharmalingam B, Mohanakrishnan V, et al. YAP1 and TAZ negatively control bone angiogenesis by limiting hypoxia-inducible factor signaling in endothelial cells. *eLife*. 2020;9:e50770.
77. Lampugnani MG, Malinverno M, Dejana E, Rudini N. Endothelial cell disease: emerging knowledge from cerebral cavernous malformations. *Curr Opin Hematol*. 2017;24.

C H A P T E R

21

Normal vascular identity (arteries, veins, and lymphatics) and malformations

Luis Gonzalez[1,2] *and Alan Dardik*[1,2,3,4]

[1]Department of Cellular and Molecular Biology, Yale School of Medicine, New Haven, CT, United States
[2]Vascular Biology and Therapeutics Program, Yale School of Medicine, New Haven, CT, United States
[3]Department of Surgery, Yale School of Medicine, New Haven, CT, United States [4]VA Connecticut Healthcare System, West Haven, CT, United States

Introduction

The circulatory system is one of the first functional organ systems formed during vertebrate embryonic development and can be divided into three main vessel systems. The arterial system carries blood away from the heart, with larger arteries feeding into progressively smaller diameter arteries, arterioles, and capillaries. The venous system transports blood back to the heart originating from the capillaries and returning blood through progressively larger venules and veins. The lymphatic system collects lymph from tissues and returns it to the circulation via drainage into lymphatic ducts which empty into the subclavian veins. Structural differences used to define vessels as arteries, veins, or lymphatics are well established.[1–4] Several molecular determinants differentiate nascent vessels into arteries, veins, or lymphatics during embryological development of the circulatory system and are retained on vessels throughout adult life as markers of vascular identity.[5–8] However, expression of these molecular markers has been shown to be plastic in adults, with altered expression in response to pathology and surgical interventions.[6,9–13] Additionally, vascular pathologies such as vascular malformations or Kaposi's sarcoma are characterized by mutations or alterations in expression of these molecular markers.[14–16] This chapter will focus on molecular markers of vascular identity and how alterations and/or mutations may lead to a variety of vascular anomalies.

Vasculogenesis: arteriovenous differentiation

Vasculogenesis, de novo formation of blood vessels, generates the primitive capillary plexus[17] and is restricted to embryogenesis. Vasculogenesis, which is achieved through differentiation of individual angioblasts and endothelial progenitor cells, has been studied extensively.[18–22] During vasculogenesis, angioblasts differentiate into arterial or venous endothelial cells (ECs) prior to coalescencing into capillary-like vessels as well as the dorsal aorta.[23–25] Vascular endothelial growth factor (VEGF) and Notch are required for angioblast differentiation into arterial EC,[26] while the transcription factor COUP-TFII is necessary to block Notch expression in venous EC.[27] In the absence of blood flow during these early stages of development, hemodynamic forces are unable to contribute to the de novo formation of blood vessels, and therefore, these data challenge the original assumption that blood flow drives differentiation. The primitive capillary plexus subsequently undergoes remodeling throughout embryogenesis[28] and facilitates the development of new vessels via budding from this rudimentary vascular network, termed sprouting angiogenesis.[29] Unlike vasculogenesis, angiogenesis can continue throughout adulthood through similar processes utilized in response to hypoxia, tumor growth, and wound healing.[30–33]

The Vasculome
https://doi.org/10.1016/B978-0-12-822546-2.00016-2

Angiogenesis: sprouting angiogenesis

The rapidly developing embryo outgrows its blood supply leading to remodeling of the primitive capillary plexus to meet the increasing oxygen demand via the development of new vessels, using sprouting angiogenesis from the preexisting vasculature. Several proangiogenic factors common to repair, homeostasis, and tumor angiogenesis initiate formation of these new vessels.[34] Arteriovenous sprouting occurs in response to different factors.[35] The *Vegf* family are the most well-studied factors, which together with Notch signaling regulate new vessel outgrowth in a variety of different sites during embryogenesis.[34] VEGF directly induces sprouting via formation of tip cells which sprout from the dorsal aorta and ultimately anastomose in the dorsal region of the embryo,[36–38] resulting in the interconnection of arterial blood vessels but without venous drainage.[3]

Bone morphogenic protein (BMP) signaling similarly stimulates sprouting from the cardinal vein[39,40] and some of these sprouts anastomose to the arterial vessels, creating a venous drainage system.[3] In contrast to early stages of embryogenesis, sprouting frequently arises from veins and not from arteries during later stages of organogenesis.[41–45] These studies suggest migration of venous EC toward arteries and suggest the ability of venous EC to be reprogrammed into arteries. These findings also show distinct mechanisms of vessel formation during early and late embryogenesis.

Establishing vascular identity

Three distinct types of vessels are established during the development of the circulatory system: the arterial, venous, and lymphatic networks. At the onset of vessel formation and in some instances before the establishment of blood flow, EC heterogeneity is initiated by the specification of arterial–venous fates. Several studies have assessed arteriovenous fates via expression analysis of marker genes (Table 21.1).[27,46–48] In the avian yolk sac capillary plexus, NRP1 and NRP2 are segregated in blood islands and later in the arterial and venous regions prior to the onset of blood flow.[49,50] Similar observations are found in the mouse yolk sac with ephrin-B2 and eph-B4 expression segregated in arteries and veins, respectively.[46] Interestingly, mouse intraembryonic vessels express arterial markers, such as EPHB4, NRP1, Notch1/4, and DLL4, which precede expression of venous markers and are found in both arterial EC as well as EC that ultimately give rise to veins.[51] Shortly after establishment of the dorsal aorta, the ECs that give rise to veins start to coalesce to form the cardinal vein and begin to express venous markers, such as NRP2, VEGFR-3, EPHB4, and COUP-TFII.[51] Therefore, within the mouse embryo, there exists a specific spatiotemporal sequence that dictates arteriovenous differentiation.

TABLE 21.1 Molecular markers of vessel identity.

Artery	Vein	Lymphatic
Alk1	APJ	LYVE-1
CD44	COUP-TFII	Podoplanin
Connexin-37	Eph-B4	Prox1
Connexin-40	Nrp2	Sox18
Dll4		VEGFR-3
Depp		
Ephrin-B2		
Hey2		
IGFB-5P		
Jagged1		
Nrp1		
Notch-1		
Notch-4		
Unc5b		
VEGFR-2		

Arteries

Shh, VEGF-A, and Notch signaling are integral to specify arterial and venous fates of the dorsal aorta and cardinal vein, respectively.[52–55] Shh is a morphogen that regulates epithelial–mesenchymal interactions and patterning during embryonic development.[56,57] Shh also plays an important role during blood vessel development, as the absence of Shh activity leads to defects in arterial–venous differentiation, including loss of arterial markers.[52] In zebrafish, Shh is released from the notochord[55] which induces VEGF-A expression from the mesoderm.[53]

Vegf family members are critical regulators of vascular development during embryogenesis and act as ligands activating their receptors on EC to promote the growth of blood vessels and vascular networks. Exposure to high levels of VEGF-A in immediately adjacent angioblasts induces arterial differentiation and ephrin-B2 expression via VEGFR-2 signaling,[58] ultimately giving rise to the dorsal aorta.[53,55] VEGF-A suppresses venous identity in adult EC via inhibition of Eph-B4 and COUP-TFII expression and induction of Dll4 expression (Fig. 21.1A).[59,60] In addition, studies in

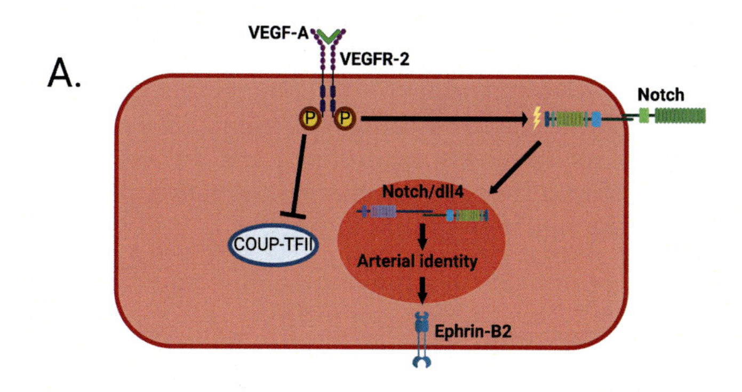

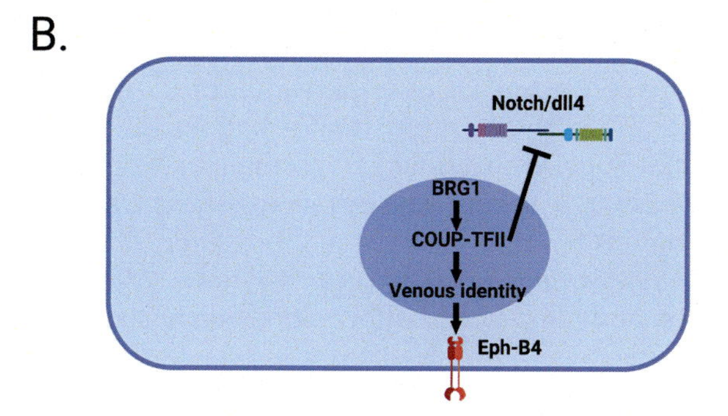

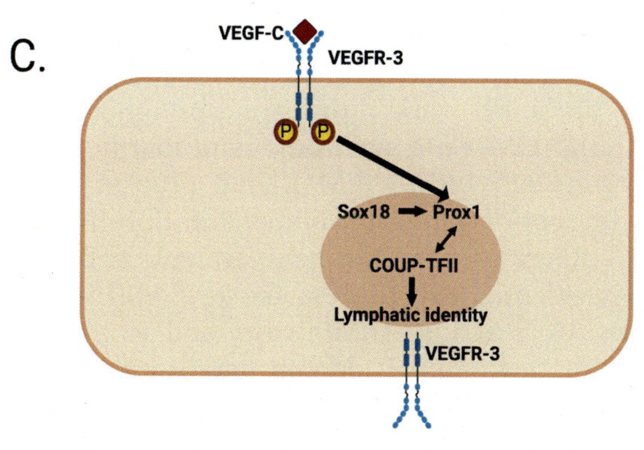

FIGURE 21.1 Molecular determinants of endothelial cell identity. (A) Arterial identity results from binding of VEGF-A to its receptor, VEGFR-2 results in Notch activation and cleavage; the cytoplasmic domain of Notch is cleaved and binds to dll4; the Notch-dll4 complex translocates to the nucleus, resulting in ephrin-B2 expression. (B) Venous identity results from BRG1 induction of COUP-TFII expression; COUP-TFII inhibits the Notch-dll4 complex and results in the inhibition of ephrin-B2 and the expression of Eph-B4. (C) Lymphatic identity results from binding of VEGF-C to its receptor, VEGFR-3, and in combination of Sox18 resulting in Prox1 expression; Prox1 forms a heterodimer with COUP-TFII to induce lymphatic identity.

murine embryonic stem cells exposed to high or low concentrations of VEGF-A result in expression of arterial markers (Ephrin-B2 and NRP1) or venous markers (COUP-TFII and NRP2), respectively.[54]

Notch is a highly conserved family of proteins in vertebrates, which express five Notch ligands (Dll1, 3 and 4, and Jagged-1 and -2) and four Notch receptors (Notch1–4).[61] Notch receptors are transmembrane proteins that are cleaved upon binding a ligand and translocate to the nucleus.[61] Dll4, Notch1, and Notch4 are expressed exclusively in arteries[48]; Dll4 activation of Notch signaling induces expression of ephrin-B2 while simultaneously suppressing COUP-TFII–mediated expression of Eph-B4.[59,60,62]

Ephrin-B2 is initially expressed in mice at day E8.5,[46] however, its sole expression is not sufficient for proper establishment of the circulatory system.[63] Rather, bidirectional signaling via the ephrin-B2–Eph-B4 axis is necessary for proper vascular development.[64] These findings suggest that the Shh–VEGF–Notch signaling pathway is critical to induce arterial differentiation of EC precursors during embryogenesis.

Veins

Venous cell fate was classically believed to be the default pathway of vascular development due to the absence of Notch signaling activation but was later suggested to be driven by COUP-TFII–mediated inhibition of Notch signaling.[27] COUP-TFII is a transcription factor critical to establish venous identity during embryogenesis and is highly expressed in the endothelium of embryonic and adult veins.[27,65,66] Expression of COUP-TFII is induced during mouse embryonic development via BRG1, a chromatin remodeling ATPase.[67] COUP-TFII expression in the developing venous endothelium downregulates VEGF and its coreceptor Nrp1 resulting in inhibition of Notch signaling that results in increased Eph-B4 expression and lack of ephrin-B2 expression (Fig. 21.1B).[27,68]

Eph-B4 serves as a principal marker of venous identity, as EC destined to become veins can be identified in mice as early as day E9.0 via their expression of Eph-B4. Ephrin-B2 and Eph-B4 signaling are important for the establishment of the dorsal aorta and cardinal vein, respectively, from a single precursor vessel.[64] In zebrafish embryos, angioblasts form a precursor vessel that expresses both ephrin-B2 and Eph-B4; the dorsal domain is specified by Notch signaling to become arterial and the ventral domain becomes venous.[64] Via Ephrin-B2–Eph-B4 bidirectional signaling, ventral migration of Eph-B4-positive EC leads to the separation of the dorsal aorta from the cardinal vein by repulsion from the Ephrin-B2-positive EC.[64] Eph-B4 continues to be expressed on the venous endothelium throughout adulthood.[10]

Lymphatics

The third EC subtype that arises after arteriovenous differentiation is the lymphatic EC. Lymphatic EC fate is specified in mice at day E9.75 in a subset of venous ECs that form the walls of the anterior cardinal vein.[69] Lymphatic specification is driven by Prox1, a transcription factor that binds to COUP-TFII to form a complex necessary to regulate several lineage-specific genes including VEGFR-3, NRP1, and FGFR-3.[70] Prox1 is both necessary and sufficient to induce lymphatic differentiation of EC.[71–73] Prox1 expression is regulated by the transcription factors Sox18 and COUP-TFII (Fig. 21.1C).[74] Although transient expression of Sox18 is detectable in embryonic arteries, these vessels do not express Prox1[75]; rather, the convergent expression of Sox18 and COUP-TFII is necessary for Prox1 expression.[74,76,77] Once Prox1 expression is induced in the venous EC of the cardinal vein, these cells downregulate genes specific to blood EC[72] and upregulate genes specific to lymphatic EC,[78–80] including VEGFR-3 that allows lymphatic EC to respond to VEGF-C signaling.[81,82]

VEGFR-3 is initially expressed in murine veins and mesenchyme beginning at day E8.5[83] but by day E14.5 VEGFR-3 expression is restricted to lymphatic vessels.[79] VEGF-C preferentially binds VEGFR-3 and is coexpressed with VEGFR-3 during lymphangiogenesis. In response to VEGF-C signaling from the nearby mesenchyme, lymphatic EC bud off from the cardinal vein at day E10.5.[84] During this process, platelets form clots between the cardinal vein and the nascent lymphatic vasculature, leading to separation of the blood and lymphatic circulations.[85–87] Upon completion of this process, lymphatic EC coalesce to form the lymphatic sacs that extend via sprouting to form the lymphatic system.[88,89]

LYVE-1 has been used as a marker for lymphatic vessels,[90] but its expression is not restricted to lymphatic vessels.[91] However, LYVE-1 plays a vital role during lymphangiogenesis, as MT1-MMP–mediated cleavage of LYVE-1 on lymphatic EC results in the inhibition of lymphangiogenesis in mice.[92]

Vascular malformations

Vascular anomalies are a spectrum of disorders that range from a simple birthmark to life-threatening disorders; vascular anomalies primarily affect people from infancy to young adulthood.[93] Terminology used to describe and classify vascular anomalies is critical for proper diagnosis and treatment; the International Society for the Study of Vascular Anomalies has developed a classification system to categorize vascular anomalies into two groups: (1) vascular malformations and (2) vascular tumors.[94,95] The distinction between the two is based on histopathological assessment; vascular tumors are true neoplasms, whereas vascular malformations are comprised of an abnormally formed circulatory and/or lymphatic system.

Vascular malformations are divided into four classifications: *simple malformations*, *combined malformations*, *malformations of major named vessels*, and *malformations associated with other anomalies*. Vascular malformations are believed to arise due to developmental errors during embryogenesis, such as abnormal signaling processes that control apoptosis, maturation, and differentiation of vascular cells; some of the malformations involve genes that regulate vascular identity.

Simple malformations

Most simple malformations are composed primarily of one type of vessel (e.g., capillaries, lymphatics, or veins) except for arteriovenous malformations (AVM), which are comprised of arteries, capillaries, and veins (Table 21.2). Within simple malformations, there are four major subclassifications of vascular malformations based on their flow characteristics: slow-flow (capillary malformation [CM], lymphatic malformation [LM] and venous malformation [VM]) and fast-flow (AVM). By definition, only fast-flow malformations have an arterial component.

Capillary malformations

CM mainly affect the papillary dermis of the skin and the mucosa, appearing as pink to red macules that are present at birth and persist throughout life (Table 21.2).[96] CM-AVM syndrome is primarily due to mutations in RASA1 (CM-AVM1),[97] but a subset is associated with loss-of-function mutations in Eph-B4 (CM-AVM2).[98] Reverse signaling between ephrin-B2 and Eph-B4 is essential to suppress the RAS–MEK–ERK and RAS–AKT–mTORC1 pathways, and to establish arteriovenous endothelial identity and corresponding vessel formation.[99,100] Thus, in some cases, the lack of Eph-B4 or RASA1 activity in venous EC results in constitutive activation of the RAS–MEK–ERK and RAS–AKT–mTORC1 pathways,[101] leading to abnormal EC differentiation and disorganized vascular development.

Lymphatic malformations

LM are often evident at birth and appear as small, crimson dome-shaped nodules (Table 21.2). Primary lymphedemas are a subtype of LM due to primary dysgenesis of the lymphatic system. Nonne–Milroy syndrome is characterized by lymphedema due to congenital abnormalities in the lymphatic system; some cases of Nonne–Milroy syndrome are attributed to mutations in the FLT4 gene.[102,103] The FLT4 gene encodes for the VEGFR-3 protein, which is exclusively expressed in lymphatic vessels by day E14.5 and is critical for lymphangiogenesis.[79] Mutations in the FLT4 gene disrupt the growth, migration, and survival of lymphatic EC and result in small or absent lymphatic vessels. Hypotrichosis–lymphedema–telangiectasia

TABLE 21.2 Simple vascular malformations and affected genes.

Vascular malformation	Affected genes
Capillary malformations (CM)	
Nevus simplex/salmon patch, "angel kiss" "stork bite"	
Cutaneous and/or mucosal CM ("port-wine" stain)	
Nonsyndromic CM	**GNAQ**
CM with CNS and/or ocular anomalies (Sturge–Weber syndrome)	**GNAQ**
CM with bone and/or soft tissue overgrowth	**GNA11**
Diffuse CM with overgrowth (DCMO)	**GNA11**
Reticulate CM	
CM of MIC-CAP (microcephaly–capillary malformation)	**STAMBP**
CM of MCAP (megalencephaly–capillary malformation-polymicrogyria)	**PIK3CA**
CM of CM-AVM	**RASA1/EPHB4**
Cutis marmorata telangiectatica congenita (CMTC)	
Hereditary hemorrhagic telangiectasia (HHT)	
HHT1	**ENG**
HHT2	**ACVRL1 (ALK1)**
HHT3	
Juvenile polyposis/HHT	**SMAD4**
Lymphatic malformations (LM)	
Common (cystic) LM	
Macrocystic LM	**PIK3CA**
Microcystic LM	
Mixed cystic LM	
Generalized lymphatic anomaly (GLA)	
Kaposiform lymphangiomatosis (KLA)	
LM in Gorham–Stout disease	
Channel type LM	
"Acquired" progressive lymphatic anomaly	
Primary lymphedema (various forms)	
Nonne–Milroy syndrome	**FLT4/VEGFR-3**
Primary hereditary lymphedema	**VEGF-C**
Primary hereditary lymphedema	**GJC2/Connexin 47**
Lymphedema–distichiasis	**FOXC2**
Hypotrichosis–lymphedema–telangiectasia	**SOX18**
Primary lymphedema with myelodysplasia	**GATA2**

Continued

TABLE 21.2 Simple vascular malformations and affected genes.—cont'd

Vascular malformation	Affected genes
Primary generalized lymphatic anomaly (Hennekam lymphangiectasia–lymphedema syndrome)	**CCBE1**
Microcephaly with or without chorioretinopathy, lymphedema, or mental retardation syndrome	**KIF11**
Lymphedema–choanal atresia	**PTPN14**
Venous malformation (VM)	
Common VM	**TEK (TIE2)/ PIK3CA**
Familial VM cutaneo-mucosal (VMCM)	**TEK (TIE2)**
Blue rubber bleb nevus (Bean) syndrome VM	**TEK (TIE2)**
Glomuvenous malformation (GVM)	**Glomulin**
Cerebral cavernous malformation (CCM)	
CCM1	**KRIT1**
CCM2	**Malcavernin**
CCM3	**ELMO2**
Familial intraosseous vascular malformation (VMOS)	**MAP3K3**
Verrucous venous malformation ***(formerly verrucous hemangioma)***	
Arteriovenous malformations (AVM) and congenital AVF	
Sporadic	**MAP2K1**
In HHT	See above
In CM-AVM	See above

syndrome is characterized by Sox18 mutation; Sox18 regulates Prox1 expression which in turns regulates VEGFR-3 expression. These LM syndromes are due to disruption of the VEGFR-3 signaling pathway, which is critical for proper lymphangiogenesis.[82]

Venous malformations

VM are present at birth and present clinically as bluish, soft, compressible lesions that commonly present on the face, limbs, or trunk (Table 21.2).[104] Cerebral cavernous malformations (CCM) are characterized by a collection of capillaries in the brain that are enlarged and irregular in structure; a subset of CCM is linked to loss-of-function mutations in CCM1, which codes for KRIT1.[105,106] The role of KRIT1 is not well understood, but data from mice embryos lacking the CMM1 gene suggest that KRIT1 plays an essential role in arterial morphogenesis and identity.[107] These $CCM1^{-/-}$ mice developed similar vascular defects as seen in patients with CCM and were associated with a loss of arterial

EC markers, specifically ephrin-B2, Dll4 and Notch4, and a similar loss of Notch4 expression was observed in patients with CCM.[107] Additional studies assessing molecular markers in VM showed that the expression of Eph-B4 is not altered, but ephrin-B2 is ectopically expressed in venous EC.[108]

Arteriovenous malformations

AVM and congenital arteriovenous fistula (AVF) tend to present sporadically or in patients presenting with hereditary hemorrhagic telangiectasia (HHT) syndrome or CM-AVM (Table 21.2).[109] HHT is characterized by vessel patterning defects resulting in telangiectasias with bleeding in the skin and mucosal membranes, as well as AVM in the brain, lung, and liver.[110,111] Loss-of-function mutations in either ENG or ACVRL1 (ALK1) result in HHT-1 and -2, respectively.[112,113] ENG and ALK1 are coreceptors active in the TGFβ/BMP signaling pathway.[114] Inactivation of ENG or ALK1 in mice and zebrafish shows that ENG silencing affects VEGFR-2 trafficking to increase its downstream signaling,[115] while VEGFR-2 phosphorylation and activity were increased upon ALK1 silencing.[116] VEGFR-2 plays a critical role during vasculogenesis[117] and upon binding to its ligand, VEGF-A, results in the upregulation of ephrin-B2 expression.[58] Interestingly, ephrin-B2 expression is not altered in AVM found in HHT-1 but is downregulated in HHT-2.[118]

Combined vascular malformations

Combined vascular malformations have two or more types of vascular malformations in a single lesion. Similar to the simple vascular malformations, combined vascular malformations can be divided into slow-flow and fast-flow malformations and are frequently associated with complex clinical syndromes.[95,119] Presentation is variable and depends on the type of malformation, as well as location, size, and relation to other structures.

Malformations of major named vessels

These malformations affect the large caliber axial or conducting veins, arteries, or lymphatics. Anomalies include abnormalities (such as aplasia, hypoplasia, stenosis, ectasia, or aneurysm) in vessel origin, course, number, length, diameter, and/or valves.[95] Abnormal communications, such as congenital AVF or persistent embryonic vessels, are also included in this group of malformations.

Vein of Galen aneurysmal malformations (VGAM) are among the most common and severe neonatal brain AVM that develops before birth.[120,121] This anomaly results in a cerebral AVF of the median porencephalic vein (MPV), a precursor of the Vein of Galen. VGAM are largely sporadic, but a subset of cases are hereditary and due to mutations in Eph-B4.[122,123] The role of Eph-B4 in arteriovenous specification is well established in model systems,[124,125] and Eph-B4 knockdown studies in zebrafish embryos disrupt the angioarchitecture of the dorsal longitudinal vein, the homolog of the human MPV,[126] further corroborating that Eph-B4 germline mutations contribute to a spectrum of vascular malformations and that EPHB4 is a genuine VGAM risk gene.

Vascular malformations associated with other anomalies

Vascular malformations of all types (simple, combined, involving any type of vessel) may be associated with nonvascular anomalies, most commonly bone, soft tissue, or viscera overgrowth. Many of these associations have been characterized as eponymous syndromes (Table 21.3).

Vascular tumors

Vascular tumors are soft tissue growths of vascular origin, formed from blood vessels or lymphatic vessels. They can be divided into benign, locally aggressive or borderline, and overtly malignant (Table 21.4).

Infantile hemangiomas (IH) are the most common tumors in infancy and occur in approximately 5%–10% of the population.[127,128] In contrast to vascular malformations, IH are usually inconspicuous at birth and are characterized by predictable biological behavior of rapid postnatal proliferation followed by slow spontaneous involution.[129] IH are distinguished from other vascular tumors and malformations by their specific expression of erythrocyte-type GLUT-1.[130] Additionally, IH EC express LYVE-1 during the proliferative phase that is then downregulated during involution.[131] Also of note, IH EC are also Prox-1$^-$, SLC$^-$, VEGFR-3$^+$, CD31$^+$, and CD34$^+$,[131] an immunophenotype that resembles that of the cardinal vein during early normal embryogenesis.[132] This is the stage leading to the development of mature blood vessels or to acquiring lymphatic bias.[132] As IH regress, the EC attain a more mature immunophenotype, LYVE-1$^-$, Prox-1$^-$, SLC$^-$, VEGFR-3$^-$, CD31$^+$, and CD34$^+$, and develop into mature blood vessels. These findings suggest that the ECs of IH are arrested in an early stage of vascular development, which may explain their rapid proliferation.

Kaposi sarcoma (KS) is a soft tissue tumor that occurs in several distinct populations with varying presentations. Its most well-known form occurs in patients in

TABLE 21.3 Vascular malformations associated with other anomalies with affected genes.

Syndrome	Vascular malformation/anomaly	Affected genes
Klippel–Trenaunay syndrome	CM + VM ± LM + limb overgrowth	**PIK3CA**
Parkes Weber syndrome	CM + AVF + limb overgrowth	**RASA1**
Servelle–Martorell syndrome	Limb VM + bone undergrowth	
Sturge-Weber syndrome	Facial + leptomeningeal CM + eye anomalies ± bone and/or soft tissue overgrowth	**GNAQ**
	Limb CM + congenital nonprogressive limb overgrowth	**GNA11**
Maffucci syndrome	VM ± spindle-cell hemangioma + enchondroma	**IDH1/IDH2**
M-CM/MCAP	Macrocephaly + CM	**PIK3CA**
MIC-CAP	Microcephaly + CM	**STAMBP**
CLOVES syndrome	LM + VM + CM ± AVM + lipomatous overgrowth	**PIK3CA**
Proteus syndrome	CM, VM, and/or LM + asymmetrical somatic overgrowth	**AKT1**
Bannayan–Riley–Ruvalcaba syndrome	AVM + VM + macrocephaly, lipomatous overgrowth	**PTEN**
CLAPO syndrome	Lower lip CM + face and neck LM + asymmetry and partial/generalized overgrowth	**PIK3CA**

TABLE 21.4 Classification of vascular tumors with affected genes.

Vascular tumors	Affected genes
Benign vascular tumors	
Infantile hemangioma/hemangioma of infancy	
Congenital hemangioma	**GNAQ/GNA11**
Rapidly involuting CH (RICH)	
Noninvoluting CH (NICH)	
Partially involuting (PICH)	
Tufted angioma	**GNA14**
Spindle-cell hemangioma	**IDH1/IDH2**
Epithelioid hemangioma	**FOS**
Pyogenic granuloma	**BRAF/RAS/GNA14**
Locally aggressive or borderline vascular tumors	
Kaposiform hemangioendothelioma	**GNA14**
Retiform hemangioendothelioma	
Papillary intralymphatic angioendothelioma (PILA), Dabska tumor	
Composite hemangioendothelioma	
Pseudomyogenic hemangioendothelioma	**FOSB**
Polymorphous hemangioendothelioma	
Kaposi sarcoma	
Malignant vascular tumors	
Angiosarcoma	**MYC** (postradiation)
Epithelioid hemangioendothelioma	**CAMTA1/TFE3**

an immunosuppressed state.[133] Interestingly, HHV-8 is present in all forms of KS.[134] KS is a disease of EC origin,[135] and cells of KS are of arterial phenotype with abundant expression of ephrin-B2 and little to no expression of Eph-B4.[15] Venous ECs transfected with HHV-8 undergo a phenotypic switch from Eph-B4 to ephrin-B2 expression.[15] Eph-B4 suppresses tumor growth in various cancers via the antioncogenic Abl–Crk pathway,[136,137] and inhibition of ephrin-B2, using the extracellular domain of Eph-B4 fused with human serum albumin, inhibits migration, and invasion of KS cells in vitro.[138]

Angiosarcomas are rare and highly malignant vascular tumors characterized by rapid proliferation with extensively infiltrating anaplastic cells of EC origin.[139,140] Lymphedema, radiation, and carcinogens are possible risk factors.[141] Angiosarcomas express the typical vascular markers CD34, CD31, Fli1, ERF, and occasionally podoplanin, a lymphatic marker.[142–144] A subset of patients with skin angiosarcoma expressed Eph-B4, suggesting a role for Eph-B4 in the pathogenesis and behavior of angiosarcoma.[6] Eph-B4 upstream signaling is implicated as well, as analysis of Notch-1 knockout mice showed increased Eph-B4 expression in liver sinusoidal endothelium resulting in increased angiogenesis, proliferation, and development of angiosarcomas.[145] Additionally, high level of MYC amplification and coamplification of the FLT4 gene has been observed in radiation-induced angiosarcomas,[146] which may play vital roles in tumor progression and can serve as potential therapeutic targets.

Switching of arterial–venous identity

Vascular identity may be plastic, depending on the environment. Despite the complexity that regulates arteriovenous specification, several studies have shown the existence of cell fate plasticity during embryogenesis.[51,147–149] Several vascular markers are also differentially expressed during pathological conditions in adults, such as cancer.[6,150] And although previous work has shown that the initiation of arterial specification to be flow independent, flow may be critical to maintain arteriovenous cell fates; ligated arteries can differentiate into and anastomose with veins[148] and flow is required to maintain expression of arterial markers in the mouse dorsal aorta.[51,149] Interestingly, vascular identity markers change after surgical procedures in adults; for example, venous identity is altered after placement of a vein into an arterial or fistula environment.[11,12,151,152] Therefore, hemodynamics and interactions with neighboring EC may influence, to some extent, the plasticity of EC identity even during adulthood.

Vein grafts

Vein grafts are the preferred conduit used by vascular surgeons to treat severe arterial occlusive disease. Placement of the vein graft into arterial circulation exposes the vein to arterial pressure, flow, and increased oxygen tension resulting in venous adaptive remodeling.[153] Human studies have shown that Eph-B4 is reduced in vein grafts compared with native veins.[152] Additionally, mouse and rat vein graft models show similar reduction in Eph-B4 expression.[152–154] In a mouse vein graft model, transiently increased expression of VEGF-A may be a mechanism by which vein grafts downregulate Eph-B4 expression during long-term adaptation to the arterial environment.[152] Although venous identity is lost during vein graft adaptation, arterial identity is not gained as neither ephrin-B2, Dll4, nor Notch4 were expressed within the vein graft.[152] However, a recent study showed acquisition of Dll4 and ephrin-B2 in a young human vein graft (Table 21.5), suggesting plasticity of identity at least in young adults.[155]

AVF

AVF are the preferred mode of vascular access for hemodialysis with superior patency compared to grafts or catheters.[156] Successful AVF maturation requires venous adaptation to the fistula environment which is characterized by altered hemodynamic forces, disturbed flow, and increased venous wall oxygen tension; these environmental changes are different than the hemodynamics of the arterial environment.[157] Studies in an AVF mouse model showed that the vein acquires arterial markers, such as ephrin-B2, Nrp1, Notch1, and Dll4 (Table 21.5).[11] These data show that after AVF creation the vein gains dual arterial–venous identity within the fistula environment which is distinct from the loss of venous identity observed during vein graft remodeling. Vessel remodeling may be fundamentally different in response to alterations in hemodynamics, such as disturbed flow observed in AVF that is typically absent during vein graft remodeling; hemodynamics may responsible for the altered expression patterns of Eph-B4 and ephrin-B2.[6]

Vascular patches

Pericardial vascular patches are commonly used as a surgical patch material for vascular reconstruction of both arteries and veins. Bovine pericardial patches are decellularized prior to use and thus comprise homogenous extracellular matrix scaffolds that have low rates of induction of the host immune system.[158,159] Therefore, pericardial patches lack vascular identity prior to implantation, but interestingly, patches acquire the

TABLE 21.5 Expression of molecular markers depend on vascular environment.

Vessel	Preoperative	Postoperative
Vein graft	Eph-B4	None
AVF	Eph-B4	Eph-B4, Dll4, Ephrin-B2, Nrp1
Vascular patches		
Pericardial patch		
Arterial environment	None	Ephrin-B2, Notch-4
Venous environment	None	Eph-B4, COUP-TFII
AVF environment	None	Eph-B4, COUP-TFII, Ephrin-B2, Notch-4
Autologous patch		
Arterial environment		
Artery	Ephrin-B2, Notch-4, Dll4	Ephrin-B2, Notch-4, Dll4
Vein	Eph-B4, COUP-TFII	Ephrin-B2, Notch-4, Dll4
Venous environment		
Artery	Ephrin-B2, Notch-4, Dll4	Eph-B4, COUP-TFII
Vein	Eph-B4, COUP-TFII	Eph-B4, COUP-TFII

vascular identity of the vascular environment into which they are implanted.[13] Patches express ephrin-B2 and Notch-4 in an arterial environment,[12,13] whereas they express Eph-B4 and COUP-TFII in a venous environment,[13,151] a context-dependent gain of identity (Table 21.5). Interestingly, a patch placed into a vein and then additionally exposed to a fistula environment gained dual arterial–venous identity, similar to that of an AVF.[11,13] Acquisition of vascular identity in pericardial patches occurs via host cell infiltration since the patches are acellular prior to implantation.

Although prosthetic patches are commonly used in vascular surgery, autologous tissue remains the gold standard for vessel closure. In a rat patch angioplasty model, autologous jugular vein or carotid artery patches placed in an arterial environment expressed ephrin-B2, Notch-4, and Dll4 but not Eph-B4 or COUP-TFII.[160] Conversely, autologous tissue patches placed in a venous environment expressed Eph-B4 and COUP-TFII but not ephrin-B2, Notch-4, or Dll4.[160] These findings show that acquisition of vascular identity is dependent upon the environment and not dependent of the particular biomaterial (Table 21.5).

Summary

Over the last century, there has been significant identification of the events required for normal vessel development including the mechanisms that underlie determination of embryonic vascular identity. During embryogenesis, progenitor cells express molecular determinants that specify differential development into arterial, venous, or lymphatic vessels. Arterial determinants include VEGF-A, Notch-4, Dll4, and ephrin-B2; venous determinants include COUP-TFII and eph-B4; and lymphatics determinants include Prox1 and VEGFR-3. Mutations in these determinants and signaling pathways during embryogenesis may be a mechanism for several vascular anomalies. These molecular determinants persist in adults as markers of vascular identity; altered expression of these markers may reflect environmental changes and may induce pathology such as vascular tumors. Changes in vascular identity after surgical procedures show that vascular identity remains plastic even in adults.

Acknowledgments

This work was supported by US National Institute of Health (NIH) grant R01-HL144476 as well as with the resources and the use of facilities at the VA Connecticut Healthcare System, West Haven, CT.

References

1. dela Paz NG, D'Amore PA. Arterial versus venous endothelial cells. *Cell Tissue Res*. 2009;335(1):5–16.
2. Niklason L, Dai G. Arterial venous differentiation for vascular bioengineering. *Annu Rev Biomed Eng*. 2018;20:431–447.
3. Red-Horse K, Siekmann AF. Veins and arteries build hierarchical branching patterns differently: bottom-up versus top-down. *Bioessays*. 2019;41(3):e1800198.
4. Breslin JW, Yang Y, Scallan JP, Sweat RS, Adderley SP, Murfee WL. Lymphatic vessel network structure and physiology. *Compr Physiol*. 2018;9(1):207–299.

5. Wolf K, Hu H, Isaji T, Dardik A. Molecular identity of arteries, veins, and lymphatics. *J Vasc Surg*. 2019;69(1):253–262.
6. Hashimoto T, Tsuneki M, Foster TR, et al. Membrane-mediated regulation of vascular identity. *Birth Defects Res C Embryo Today*. 2016;108(1):65–84.
7. Drake CJ, Fleming PA. Vasculogenesis in the day 6.5 to 9.5 mouse embryo. *Blood*. 2000;95(5):1671–1679.
8. Jain RK. Molecular regulation of vessel maturation. *Nat Med*. 2003; 9(6):685–693.
9. Wang M, Collins MJ, Foster TR, et al. Eph-B4 mediates vein graft adaptation by regulation of endothelial nitric oxide synthase. *J Vasc Surg*. 2017;65(1):179–189.
10. Model LS, Hall MR, Wong DJ, et al. Arterial shear stress reduces Eph-B4 expression in adult human veins. *Yale J Biol Med*. 2014; 87(3):359–371.
11. Protack CD, Foster TR, Hashimoto T, et al. Eph-B4 regulates adaptive venous remodeling to improve arteriovenous fistula patency. *Sci Rep*. 2017;7(1):15386.
12. Li X, Jadlowiec C, Guo Y, et al. Pericardial patch angioplasty heals via an Ephrin-B2 and CD34 positive cell mediated mechanism. *PLoS One*. 2012;7(6):e38844.
13. Bai H, Hu H, Guo J, et al. Polyester vascular patches acquire arterial or venous identity depending on their environment. *J Biomed Mater Res A*. 2017;105(12):3422–3431.
14. Wetzel-Strong SE, Detter MR, Marchuk DA. The pathobiology of vascular malformations: insights from human and model organism genetics. *J Pathol*. 2017;241(2):281–293.
15. Masood R, Xia G, Smith DL, et al. Ephrin B2 expression in Kaposi sarcoma is induced by human herpesvirus type 8: phenotype switch from venous to arterial endothelium. *Blood*. 2005;105(3): 1310–1318.
16. Urness LD, Sorensen LK, Li DY. Arteriovenous malformations in mice lacking activin receptor-like kinase-1. *Nat Genet*. 2000;26(3): 328–331.
17. Goldie LC, Nix MK, Hirschi KK. Embryonic vasculogenesis and hematopoietic specification. *Organogenesis*. 2008;4(4):257–263.
18. Childs S, Chen JN, Garrity DM, Fishman MC. Patterning of angiogenesis in the zebrafish embryo. *Development*. 2002;129(4): 973–982.
19. Zhong TP, Childs S, Leu JP, Fishman MC. Gridlock signalling pathway fashions the first embryonic artery. *Nature*. 2001; 414(6860):216–220.
20. Fouquet B, Weinstein BM, Serluca FC, Fishman MC. Vessel patterning in the embryo of the zebrafish: guidance by notochord. *Dev Biol*. 1997;183(1):37–48.
21. Jin SW, Beis D, Mitchell T, Chen JN, Stainier DY. Cellular and molecular analyses of vascular tube and lumen formation in zebrafish. *Development*. 2005;132(23):5199–5209.
22. Kohli V, Schumacher JA, Desai SP, Rehn K, Sumanas S. Arterial and venous progenitors of the major axial vessels originate at distinct locations. *Dev Cell*. 2013;25(2):196–206.
23. Risau W, Flamme I. Vasculogenesis. *Annu Rev Cell Dev Biol*. 1995; 11:73–91.
24. Lindskog H, Kim YH, Jelin EB, et al. Molecular identification of venous progenitors in the dorsal aorta reveals an aortic origin for the cardinal vein in mammals. *Development*. 2014;141(5): 1120–1128.
25. Pardanaud L, Dieterlen-Lievre F. Emergence of endothelial and hemopoietic cells in the avian embryo. *Anat Embryol (Berl)*. 1993; 187(2):107–114.
26. Liu ZJ, Shirakawa T, Li Y, et al. Regulation of Notch1 and Dll4 by vascular endothelial growth factor in arterial endothelial cells: implications for modulating arteriogenesis and angiogenesis. *Mol Cell Biol*. 2003;23(1):14–25.
27. You LR, Lin FJ, Lee CT, DeMayo FJ, Tsai MJ, Tsai SY. Suppression of Notch signalling by the COUP-TFII transcription factor regulates vein identity. *Nature*. 2005;435(7038):98–104.
28. Patel-Hett S, D'Amore PA. Signal transduction in vasculogenesis and developmental angiogenesis. *Int J Dev Biol*. 2011;55(4-5): 353–363.
29. Garcia MD, Larina IV. Vascular development and hemodynamic force in the mouse yolk sac. *Front Physiol*. 2014;5:308.
30. DiPietro LA. Angiogenesis and wound repair: when enough is enough. *J Leukoc Biol*. 2016;100(5):979–984.
31. Li J, Zhang YP, Kirsner RS. Angiogenesis in wound repair: angiogenic growth factors and the extracellular matrix. *Microsc Res Tech*. 2003;60(1):107–114.
32. Papetti M, Herman IM. Mechanisms of normal and tumor-derived angiogenesis. *Am J Physiol Cell Physiol*. 2002;282(5): C947–C970.
33. Gerhardt H. VEGF and endothelial guidance in angiogenic sprouting. *Organogenesis*. 2008;4(4):241–246.
34. Adams RH, Eichmann A. Axon guidance molecules in vascular patterning. *Cold Spring Harb Perspect Biol*. 2010;2(5):a001875.
35. Hogan BM, Schulte-Merker S. How to plumb a pisces: understanding vascular development and disease using zebrafish embryos. *Dev Cell*. 2017;42(6):567–583.
36. Zygmunt T, Trzaska S, Edelstein L, et al. 'In parallel' interconnectivity of the dorsal longitudinal anastomotic vessels requires both VEGF signaling and circulatory flow. *J Cell Sci*. 2012;125(Pt 21): 5159–5167.
37. Covassin LD, Villefranc JA, Kacergis MC, Weinstein BM, Lawson ND. Distinct genetic interactions between multiple Vegf receptors are required for development of different blood vessel types in zebrafish. *Proc Natl Acad Sci U S A*. 2006;103(17):6554–6559.
38. Isogai S, Lawson ND, Torrealday S, Horiguchi M, Weinstein BM. Angiogenic network formation in the developing vertebrate trunk. *Development*. 2003;130(21):5281–5290.
39. Wiley DM, Kim JD, Hao J, Hong CC, Bautch VL, Jin SW. Distinct signalling pathways regulate sprouting angiogenesis from the dorsal aorta and the axial vein. *Nat Cell Biol*. 2011;13(6):686–692.
40. Kim JD, Lee HW, Jin SW. Diversity is in my veins: role of bone morphogenetic protein signaling during venous morphogenesis in zebrafish illustrates the heterogeneity within endothelial cells. *Arterioscler Thromb Vasc Biol*. 2014;34(9):1838–1845.
41. Bussmann J, Wolfe SA, Siekmann AF. Arterial-venous network formation during brain vascularization involves hemodynamic regulation of chemokine signaling. *Development*. 2011;138(9): 1717–1726.
42. Fujita M, Cha YR, Pham VN, et al. Assembly and patterning of the vascular network of the vertebrate hindbrain. *Development*. 2011; 138(9):1705–1715.
43. Xu C, Hasan SS, Schmidt I, et al. Arteries are formed by vein-derived endothelial tip cells. *Nat Commun*. 2014;5:5758.
44. Red-Horse K, Ueno H, Weissman IL, Krasnow MA. Coronary arteries form by developmental reprogramming of venous cells. *Nature*. 2010;464(7288):549–553.
45. Hen G, Nicenboim J, Mayseless O, et al. Venous-derived angioblasts generate organ-specific vessels during zebrafish embryonic development. *Development*. 2015;142(24):4266–4278.
46. Wang HU, Chen ZF, Anderson DJ. Molecular distinction and angiogenic interaction between embryonic arteries and veins revealed by ephrin-B2 and its receptor Eph-B4. *Cell*. 1998;93(5): 741–753.
47. Adams RH, Wilkinson GA, Weiss C, et al. Roles of ephrinB ligands and EphB receptors in cardiovascular development: demarcation of arterial/venous domains, vascular morphogenesis, and sprouting angiogenesis. *Genes Dev*. 1999;13(3):295–306.

48. Villa N, Walker L, Lindsell CE, Gasson J, Iruela-Arispe ML, Weinmaster G. Vascular expression of Notch pathway receptors and ligands is restricted to arterial vessels. *Mech Dev.* 2001; 108(1-2):161–164.
49. Herzog Y, Kalcheim C, Kahane N, Reshef R, Neufeld G. Differential expression of neuropilin-1 and neuropilin-2 in arteries and veins. *Mech Dev.* 2001;109(1):115–119.
50. Herzog Y, Guttmann-Raviv N, Neufeld G. Segregation of arterial and venous markers in subpopulations of blood islands before vessel formation. *Dev Dyn.* 2005;232(4):1047–1055.
51. Chong DC, Koo Y, Xu K, Fu S, Cleaver O. Stepwise arteriovenous fate acquisition during mammalian vasculogenesis. *Dev Dyn.* 2011;240(9):2153–2165.
52. Lawson ND, Scheer N, Pham VN, et al. Notch signaling is required for arterial-venous differentiation during embryonic vascular development. *Development.* 2001;128(19):3675–3683.
53. Lawson ND, Vogel AM, Weinstein BM. Sonic hedgehog and vascular endothelial growth factor act upstream of the Notch pathway during arterial endothelial differentiation. *Dev Cell.* 2002;3(1):127–136.
54. Lanner F, Sohl M, Farnebo F. Functional arterial and venous fate is determined by graded VEGF signaling and notch status during embryonic stem cell differentiation. *Arterioscler Thromb Vasc Biol.* 2007;27(3):487–493.
55. Odenthal J, Haffter P, Vogelsang E, et al. Mutations affecting the formation of the notochord in the zebrafish, *Danio rerio*. *Development.* 1996;123:103–115.
56. Echelard Y, Epstein DJ, St-Jacques B, et al. Sonic hedgehog, a member of a family of putative signaling molecules, is implicated in the regulation of CNS polarity. *Cell.* 1993;75(7):1417–1430.
57. Chiang C, Litingtung Y, Lee E, et al. Cyclopia and defective axial patterning in mice lacking Sonic hedgehog gene function. *Nature.* 1996;383(6599):407–413.
58. Masumura T, Yamamoto K, Shimizu N, Obi S, Ando J. Shear stress increases expression of the arterial endothelial marker ephrinB2 in murine ES cells via the VEGF-Notch signaling pathways. *Arterioscler Thromb Vasc Biol.* 2009;29(12):2125–2131.
59. Yang C, Guo Y, Jadlowiec CC, et al. Vascular endothelial growth factor-A inhibits EphB4 and stimulates delta-like ligand 4 expression in adult endothelial cells. *J Surg Res.* 2013;183(1):478–486.
60. Kang J, Yoo J, Lee S, et al. An exquisite cross-control mechanism among endothelial cell fate regulators directs the plasticity and heterogeneity of lymphatic endothelial cells. *Blood.* 2010;116(1): 140–150.
61. Siebel C, Lendahl U. Notch signaling in development, tissue homeostasis, and disease. *Physiol Rev.* 2017;97(4):1235–1294.
62. Swift MR, Pham VN, Castranova D, Bell K, Poole RJ, Weinstein BM. SoxF factors and Notch regulate nr2f2 gene expression during venous differentiation in zebrafish. *Dev Biol.* 2014; 390(2):116–125.
63. Gerety SS, Wang HU, Chen ZF, Anderson DJ. Symmetrical mutant phenotypes of the receptor EphB4 and its specific transmembrane ligand ephrin-B2 in cardiovascular development. *Mol Cell.* 1999; 4(3):403–414.
64. Herbert SP, Huisken J, Kim TN, et al. Arterial-venous segregation by selective cell sprouting: an alternative mode of blood vessel formation. *Science.* 2009;326(5950):294–298.
65. Pereira FA, Qiu Y, Zhou G, Tsai MJ, Tsai SY. The orphan nuclear receptor COUP-TFII is required for angiogenesis and heart development. *Genes Dev.* 1999;13(8):1037–1049.
66. Cui X, Lu YW, Lee V, et al. Venous endothelial marker COUP-TFII regulates the distinct pathologic potentials of adult arteries and veins. *Sci Rep.* 2015;5:16193.
67. Davis RB, Curtis CD, Griffin CT. BRG1 promotes COUP-TFII expression and venous specification during embryonic vascular development. *Development.* 2013;140(6):1272–1281.
68. Fancher TT, Muto A, Fitzgerald TN, et al. Control of blood vessel identity: from embryo to adult. *Ann Vasc Dis.* 2008;1(1):28–34.
69. Srinivasan RS, Dillard ME, Lagutin OV, et al. Lineage tracing demonstrates the venous origin of the mammalian lymphatic vasculature. *Genes Dev.* 2007;21(19):2422–2432.
70. Lee S, Kang J, Yoo J, et al. Prox1 physically and functionally interacts with COUP-TFII to specify lymphatic endothelial cell fate. *Blood.* 2009;113(8):1856–1859.
71. Hong YK, Harvey N, Noh YH, et al. Prox1 is a master control gene in the program specifying lymphatic endothelial cell fate. *Dev Dyn.* 2002;225(3):351–357.
72. Petrova TV, Makinen T, Makela TP, et al. Lymphatic endothelial reprogramming of vascular endothelial cells by the Prox-1 homeobox transcription factor. *EMBO J.* 2002;21(17):4593–4599.
73. Wigle JT, Harvey N, Detmar M, et al. An essential role for Prox1 in the induction of the lymphatic endothelial cell phenotype. *EMBO J.* 2002;21(7):1505–1513.
74. Srinivasan RS, Geng X, Yang Y, et al. The nuclear hormone receptor COUP-TFII is required for the initiation and early maintenance of Prox1 expression in lymphatic endothelial cells. *Genes Dev.* 2010;24(7):696–707.
75. Pennisi D, Gardner J, Chambers D, et al. Mutations in Sox18 underlie cardiovascular and hair follicle defects in ragged mice. *Nat Genet.* 2000;24(4):434–437.
76. Yamazaki T, Yoshimatsu Y, Morishita Y, Miyazono K, Watabe T. COUP-TFII regulates the functions of Prox1 in lymphatic endothelial cells through direct interaction. *Genes Cells.* 2009;14(3): 425–434.
77. Francois M, Caprini A, Hosking B, et al. Sox18 induces development of the lymphatic vasculature in mice. *Nature.* 2008; 456(7222):643–647.
78. Schacht V, Ramirez MI, Hong YK, et al. T1alpha/podoplanin deficiency disrupts normal lymphatic vasculature formation and causes lymphedema. *EMBO J.* 2003;22(14):3546–3556.
79. Kaipainen A, Korhonen J, Mustonen T, et al. Expression of the fms-like tyrosine kinase 4 gene becomes restricted to lymphatic endothelium during development. *Proc Natl Acad Sci U S A.* 1995;92(8):3566–3570.
80. Petrova TV, Karpanen T, Norrmen C, et al. Defective valves and abnormal mural cell recruitment underlie lymphatic vascular failure in lymphedema distichiasis. *Nat Med.* 2004;10(9): 974–981.
81. Kukk E, Lymboussaki A, Taira S, et al. VEGF-C receptor binding and pattern of expression with VEGFR-3 suggests a role in lymphatic vascular development. *Development.* 1996;122(12): 3829–3837.
82. Tammela T, Alitalo K. Lymphangiogenesis: molecular mechanisms and future promise. *Cell.* 2010;140(4):460–476.
83. Tammela T, Zarkada G, Wallgard E, et al. Blocking VEGFR-3 suppresses angiogenic sprouting and vascular network formation. *Nature.* 2008;454(7204):656–660.
84. Kreuger J, Nilsson I, Kerjaschki D, Petrova T, Alitalo K, Claesson-Welsh L. Early lymph vessel development from embryonic stem cells. *Arterioscler Thromb Vasc Biol.* 2006;26(5):1073–1078.
85. Bertozzi CC, Schmaier AA, Mericko P, et al. Platelets regulate lymphatic vascular development through CLEC-2-SLP-76 signaling. *Blood.* 2010;116(4):661–670.
86. Carramolino L, Fuentes J, Garcia-Andres C, Azcoitia V, Riethmacher D, Torres M. Platelets play an essential role in separating the blood and lymphatic vasculatures during embryonic angiogenesis. *Circ Res.* 2010;106(7):1197–1201.
87. Welsh JD, Kahn ML, Sweet DT. Lymphovenous hemostasis and the role of platelets in regulating lymphatic flow and lymphatic vessel maturation. *Blood.* 2016;128(9):1169–1173.
88. Oliver G, Alitalo K. The lymphatic vasculature: recent progress and paradigms. *Annu Rev Cell Dev Biol.* 2005;21:457–483.

89. Wigle JT, Oliver G. Prox1 function is required for the development of the murine lymphatic system. *Cell*. 1999;98(6):769–778.
90. Banerji S, Ni J, Wang SX, et al. LYVE-1, a new homologue of the CD44 glycoprotein, is a lymph-specific receptor for hyaluronan. *J Cell Biol*. 1999;144(4):789–801.
91. Gordon EJ, Gale NW, Harvey NL. Expression of the hyaluronan receptor LYVE-1 is not restricted to the lymphatic vasculature; LYVE-1 is also expressed on embryonic blood vessels. *Dev Dyn*. 2008;237(7):1901–1909.
92. Wong HL, Jin G, Cao R, Zhang S, Cao Y, Zhou Z. MT1-MMP sheds LYVE-1 on lymphatic endothelial cells and suppresses VEGF-C production to inhibit lymphangiogenesis. *Nat Commun*. 2016;7: 10824.
93. Fereydooni A, Dardik A, Nassiri N. Molecular changes associated with vascular malformations. *J Vasc Surg*. 2019;70(1), 314.e311–326.e311.
94. Mulliken JB, Glowacki J. Hemangiomas and vascular malformations in infants and children: a classification based on endothelial characteristics. *Plast Reconstr Surg*. 1982;69(3):412–422.
95. Wassef M, Blei F, Adams D, et al. Vascular anomalies classification: recommendations from the International Society for the study of vascular anomalies. *Pediatrics*. 2015;136(1):e203–e214.
96. Colletti G, Valassina D, Bertossi D, Melchiorre F, Vercellio G, Brusati R. Contemporary management of vascular malformations. *J Oral Maxillofac Surg*. 2014;72(3):510–528.
97. Eerola I, Boon LM, Mulliken JB, et al. Capillary malformation-arteriovenous malformation, a new clinical and genetic disorder caused by RASA1 mutations. *Am J Hum Genet*. 2003;73(6): 1240–1249.
98. Amyere M, Revencu N, Helaers R, et al. Germline loss-of-function mutations in EPHB4 cause a second form of capillary malformation-arteriovenous malformation (CM-AVM2) Deregulating RAS-MAPK signaling. *Circulation*. 2017;136(11): 1037–1048.
99. Lin FJ, Tsai MJ, Tsai SY. Artery and vein formation: a tug of war between different forces. *EMBO Rep*. 2007;8(10):920–924.
100. Kaenel P, Hahnewald S, Wotzkow C, Strange R, Andres AC. Overexpression of EphB4 in the mammary epithelium shifts the differentiation pathway of progenitor cells and promotes branching activity and vascularization. *Dev Growth Differ*. 2014;56(4): 255–275.
101. Xiao Z, Carrasco R, Kinneer K, et al. EphB4 promotes or suppresses Ras/MEK/ERK pathway in a context-dependent manner: implications for EphB4 as a cancer target. *Cancer Biol Ther*. 2012; 13(8):630–637.
102. Ferrell RE, Levinson KL, Esman JH, et al. Hereditary lymphedema: evidence for linkage and genetic heterogeneity. *Hum Mol Genet*. 1998;7(13):2073–2078.
103. Irrthum A, Karkkainen MJ, Devriendt K, Alitalo K, Vikkula M. Congenital hereditary lymphedema caused by a mutation that inactivates VEGFR3 tyrosine kinase. *Am J Hum Genet*. 2000; 67(2):295–301.
104. Vogel SA, Hess CP, Dowd CF, et al. Early versus later presentations of venous malformations: where and why? *Pediatr Dermatol*. 2013;30(5):534–540.
105. Craig HD, Gunel M, Cepeda O, et al. Multilocus linkage identifies two new loci for a mendelian form of stroke, cerebral cavernous malformation, at 7p15-13 and 3q25.2-27. *Hum Mol Genet*. 1998; 7(12):1851–1858.
106. Laberge-le Couteulx S, Jung HH, Labauge P, et al. Truncating mutations in CCM1, encoding KRIT1, cause hereditary cavernous angiomas. *Nat Genet*. 1999;23(2):189–193.
107. Whitehead KJ, Plummer NW, Adams JA, Marchuk DA, Li DY. Ccm1 is required for arterial morphogenesis: implications for the etiology of human cavernous malformations. *Development*. 2004;131(6):1437–1448.
108. Diehl S, Bruno R, Wilkinson GA, et al. Altered expression patterns of EphrinB2 and EphB2 in human umbilical vessels and congenital venous malformations. *Pediatr Res*. 2005;57(4):537–544.
109. McDonald J, Bayrak-Toydemir P, Pyeritz RE. Hereditary hemorrhagic telangiectasia: an overview of diagnosis, management, and pathogenesis. *Genet Med*. 2011;13(7):607–616.
110. Azuma H. Genetic and molecular pathogenesis of hereditary hemorrhagic telangiectasia. *J Med Invest*. 2000;47(3-4):81–90.
111. Guttmacher AE, Marchuk DA, White Jr RI. Hereditary hemorrhagic telangiectasia. *N Engl J Med*. 1995;333(14):918–924.
112. Berg JN, Guttmacher AE, Marchuk DA, Porteous ME. Clinical heterogeneity in hereditary haemorrhagic telangiectasia: are pulmonary arteriovenous malformations more common in families linked to endoglin? *J Med Genet*. 1996;33(3):256–257.
113. Johnson DW, Berg JN, Baldwin MA, et al. Mutations in the activin receptor-like kinase 1 gene in hereditary haemorrhagic telangiectasia type 2. *Nat Genet*. 1996;13(2):189–195.
114. Lebrin F, Goumans MJ, Jonker L, et al. Endoglin promotes endothelial cell proliferation and TGF-beta/ALK1 signal transduction. *EMBO J*. 2004;23(20):4018–4028.
115. Jin Y, Muhl L, Burmakin M, et al. Endoglin prevents vascular malformation by regulating flow-induced cell migration and specification through VEGFR2 signalling. *Nat Cell Biol*. 2017;19(6): 639–652.
116. Ola R, Dubrac A, Han J, et al. PI3 kinase inhibition improves vascular malformations in mouse models of hereditary haemorrhagic telangiectasia. *Nat Commun*. 2016;7:13650.
117. Shalaby F, Rossant J, Yamaguchi TP, et al. Failure of blood-island formation and vasculogenesis in Flk-1-deficient mice. *Nature*. 1995;376(6535):62–66.
118. Sorensen LK, Brooke BS, Li DY, Urness LD. Loss of distinct arterial and venous boundaries in mice lacking endoglin, a vascular-specific TGFbeta coreceptor. *Dev Biol*. 2003;261(1):235–250.
119. Carqueja IM, Sousa J, Mansilha A. Vascular malformations: classification, diagnosis and treatment. *Int Angiol*. 2018;37(2):127–142.
120. Long DM, Seljeskog EL, Chou SN, French LA. Giant arteriovenous malformations of infancy and childhood. *J Neurosurg*. 1974;40(3):304–312.
121. Deloison B, Chalouhi GE, Sonigo P, et al. Hidden mortality of prenatally diagnosed vein of Galen aneurysmal malformation: retrospective study and review of the literature. *Ultrasound Obstet Gynecol*. 2012;40(6):652–658.
122. Vivanti A, Ozanne A, Grondin C, et al. Loss of function mutations in EPHB4 are responsible for vein of Galen aneurysmal malformation. *Brain*. 2018;141(4):979–988.
123. Duran D, Zeng X, Jin SC, et al. Mutations in chromatin modifier and ephrin signaling genes in vein of Galen malformation. *Neuron*. 2019;101(3):429–443 e424.
124. Zhang J, Hughes S. Role of the ephrin and Eph receptor tyrosine kinase families in angiogenesis and development of the cardiovascular system. *J Pathol*. 2006;208(4):453–461.
125. Mosch B, Reissenweber B, Neuber C, Pietzsch J. Eph receptors and ephrin ligands: important players in angiogenesis and tumor angiogenesis. *J Oncol*. 2010;2010:135285.
126. Aurboonyawat T, Suthipongchai S, Pereira V, Ozanne A, Lasjaunias P. Patterns of cranial venous system from the comparative anatomy in vertebrates. Part I, introduction and the dorsal venous system. *Interv Neuroradiol*. 2007;13(4):335–344.
127. Burns AJ, Navarro JA, Cooner RD. Classification of vascular anomalies and the comprehensive treatment of hemangiomas. *Plast Reconstr Surg*. 2009;124(1 Suppl):69e–81e.
128. Lowe LH, Marchant TC, Rivard DC, Scherbel AJ. Vascular malformations: classification and terminology the radiologist needs to know. *Semin Roentgenol*. 2012;47(2):106–117.
129. Bruckner AL, Frieden IJ. Hemangiomas of infancy. *J Am Acad Dermatol*. 2003;48(4):477–493. quiz 494-476.

130. North PE, Waner M, Mizeracki A, Mihm Jr MC. GLUT1: a newly discovered immunohistochemical marker for juvenile hemangiomas. *Hum Pathol*. 2000;31(1):11–22.
131. Dadras SS, North PE, Bertoncini J, Mihm MC, Detmar M. Infantile hemangiomas are arrested in an early developmental vascular differentiation state. *Mod Pathol*. 2004;17(9):1068–1079.
132. Oliver G, Detmar M. The rediscovery of the lymphatic system: old and new insights into the development and biological function of the lymphatic vasculature. *Genes Dev*. 2002;16(7):773–783.
133. Stanescu L, Foarfa C, Georgescu AC, Georgescu I. Kaposi's sarcoma associated with AIDS. *Rom J Morphol Embryol*. 2007; 48(2):181–187.
134. Kemeny L, Gyulai R, Kiss M, Nagy F, Dobozy A. Kaposi's sarcoma-associated herpesvirus/human herpesvirus-8: a new virus in human pathology. *J Am Acad Dermatol*. 1997;37(1): 107–113.
135. Zhang YM, Bachmann S, Hemmer C, et al. Vascular origin of Kaposi's sarcoma. Expression of leukocyte adhesion molecule-1, thrombomodulin, and tissue factor. *Am J Pathol*. 1994;144(1): 51–59.
136. Noren NK, Foos G, Hauser CA, Pasquale EB. The EphB4 receptor suppresses breast cancer cell tumorigenicity through an Abl-Crk pathway. *Nat Cell Biol*. 2006;8(8):815–825.
137. Berclaz G, Flutsch B, Altermatt HJ, et al. Loss of EphB4 receptor tyrosine kinase protein expression during carcinogenesis of the human breast. *Oncol Rep*. 2002;9(5):985–989.
138. Scehnet JS, Ley EJ, Krasnoperov V, et al. The role of Ephs, Ephrins, and growth factors in Kaposi sarcoma and implications of EphrinB2 blockade. *Blood*. 2009;113(1):254–263.
139. Mark RJ, Poen JC, Tran LM, Fu YS, Juillard GF. Angiosarcoma. A report of 67 patients and a review of the literature. *Cancer*. 1996; 77(11):2400–2406.
140. Lahat G, Dhuka AR, Hallevi H, et al. Angiosarcoma: clinical and molecular insights. *Ann Surg*. 2010;251(6):1098–1106.
141. Coldwell DM, Baron RL, Charnsangavej C. Angiosarcoma. Diagnosis and clinical course. *Acta Radiol*. 1989;30(6):627–631.
142. Folpe AL, Chand EM, Goldblum JR, Weiss SW. Expression of Fli-1, a nuclear transcription factor, distinguishes vascular neoplasms from potential mimics. *Am J Surg Pathol*. 2001;25(8): 1061–1066.
143. DeYoung BR, Swanson PE, Argenyi ZB, et al. CD31 immunoreactivity in mesenchymal neoplasms of the skin and subcutis: report of 145 cases and review of putative immunohistologic markers of endothelial differentiation. *J Cutan Pathol*. 1995;22(3):215–222.
144. Miettinen M, Lindenmayer AE, Chaubal A. Endothelial cell markers CD31, CD34, and BNH9 antibody to H- and Y-antigens – evaluation of their specificity and sensitivity in the diagnosis of vascular tumors and comparison with von Willebrand factor. *Mod Pathol*. 1994;7(1):82–90.
145. Dill MT, Rothweiler S, Djonov V, et al. Disruption of Notch1 induces vascular remodeling, intussusceptive angiogenesis, and angiosarcomas in livers of mice. *Gastroenterology*. 2012;142(4): 967–977. e962.
146. Guo T, Zhang L, Chang NE, Singer S, Maki RG, Antonescu CR. Consistent MYC and FLT4 gene amplification in radiation-induced angiosarcoma but not in other radiation-associated atypical vascular lesions. *Genes Chromosomes Cancer*. 2011;50(1):25–33.
147. Moyon D, Pardanaud L, Yuan L, Breant C, Eichmann A. Plasticity of endothelial cells during arterial-venous differentiation in the avian embryo. *Development*. 2001;128(17):3359–3370.
148. le Noble F, Moyon D, Pardanaud L, et al. Flow regulates arterial-venous differentiation in the chick embryo yolk sac. *Development*. 2004;131(2):361–375.
149. Nguyen HL, Lee YJ, Shin J, et al. TGF-beta signaling in endothelial cells, but not neuroepithelial cells, is essential for cerebral vascular development. *Lab Invest*. 2011;91(11):1554–1563.
150. Holash J, Maisonpierre PC, Compton D, et al. Vessel cooption, regression, and growth in tumors mediated by angiopoietins and VEGF. *Science*. 1999;284(5422):1994–1998.
151. Bai H, Wang M, Foster TR, et al. Pericardial patch venoplasty heals via attraction of venous progenitor cells. *Physiol Rep*. 2016; 4(12).
152. Kudo FA, Muto A, Maloney SP, et al. Venous identity is lost but arterial identity is not gained during vein graft adaptation. *Arterioscler Thromb Vasc Biol*. 2007;27(7):1562–1571.
153. Mitra AK, Gangahar DM, Agrawal DK. Cellular, molecular and immunological mechanisms in the pathophysiology of vein graft intimal hyperplasia. *Immunol Cell Biol*. 2006;84(2):115–124.
154. Muto A, Yi T, Harrison KD, et al. Eph-B4 prevents venous adaptive remodeling in the adult arterial environment. *J Exp Med*. 2011; 208(3):561–575.
155. Bai H, Wang Z, Li M, et al. Adult human vein grafts retain plasticity of vessel identity. *Ann Vasc Surg*. 2020;68:468–475.
156. Santoro D, Benedetto F, Mondello P, et al. Vascular access for hemodialysis: current perspectives. *Int J Nephrol Renovasc Dis*. 2014; 7:281–294.
157. Lu DY, Chen EY, Wong DJ, et al. Vein graft adaptation and fistula maturation in the arterial environment. *J Surg Res*. 2014;188(1): 162–173.
158. Aguiari P, Iop L, Favaretto F, et al. In vitro comparative assessment of decellularized bovine pericardial patches and commercial bioprosthetic heart valves. *Biomed Mater*. 2017;12(1):015021.
159. Umashankar PR, Arun T, Kumari TV. Short duration gluteraldehyde cross linking of decellularized bovine pericardium improves biological response. *J Biomed Mater Res A*. 2011;97(3):311–320.
160. Bai H, Guo J, Liu S, et al. Autologous tissue patches acquire vascular identity depending on the environment. *Vasc Investig Ther*. 2018;1(1):14–23.

CHAPTER

22

Sprouting angiogenesis in vascular and lymphatic development

Anne Eichmann[1,2,3] *and Jinyu Li*[1,2]

[1]Cardiovascular Research Center, Department of Internal Medicine, Yale University School of Medicine, New Haven, CT, United States [2]Department of Cellular and Molecular Physiology, Yale University School of Medicine, New Haven, CT, United States [3]Inserm U970, Paris Cardiovascular Research Center, Paris, France

Vascular development

Vascular development is initiated upon gastrulation, starting with de novo specification and differentiation of endothelial cells from mesodermal precursors, which subsequently coalesce into a primitive vascular plexus in a process called vasculogenesis. Following the onset of embryonic heartbeat, a combination of genetic and hemodynamic factors promotes expression of arterial and venous-specific genes in primitive endothelial cells.[1] The subsequent growth and expansion of the vascular plexus from preexisting endothelial cells is named angiogenesis, and denotes a complex process of proliferation, splitting, and sprouting of blood vessels.[2] Angiogenesis accompanies tissue and organ growth and, once complete, leads to formation of a vascular network of considerable length—about 60,000 miles of vessels in every human being, and 400 miles of vessels in every human brain. Various growth factors regulate the angiogenic development of vasculature, including VEGF (vascular endothelial growth factor) signaling, axon guidance receptors, Notch signaling, and BMP (bone morphogenetic protein) signaling, that will be covered in this chapter.

Vasculature in different organs develops morphological and functional specializations, such as the tight blood–brain barrier between brain capillaries and neural cells, the highly permeable vasculature in lung and liver allowing gas and nutrient exchanges, respectively, or the renal glomerular filtration apparatus. These functional endothelial specializations are induced by signals from organ-specific cells, such as podocytes in the kidney, which produce soluble VEGFR-1 to modulate the permeability of glomerular vasculature.[3] Reciprocally, endothelial cells also produce paracrine (also called angiocrine) factors critical for organ development, maintenance, repair, and regeneration.[4–6] For example, lung endothelial cells secret Tsp1 (thrombospondin-1) to promote alveolar differentiation from stem cells.[7] Understanding mechanisms driving organ-specific angiogenic processes is thus another critical need for future research. Single cell RNA sequencing of organ endothelial cells revealed heterogeneous gene expression between different organs,[8] providing a molecular road map that will ultimately allow understanding how organ-specific vasculature development is orchestrated (see chapter by O Cleaver).

Sprouting angiogenesis

Sprout formation is a highly dynamic process that entails collective migration toward a source of growth factors, with endothelial cells polarizing and migrating in an ordered fashion. In response to angiogenic cues, single endothelial cells located at the leading extremity of the angiogenic network, called tip cells, are selected to lead sprout outgrowth, while follower stalk cells form endothelial tubes, proliferate to elongate sprouts, and initiate blood flow to provide oxygen and nutrient supply. Tip cells use aerobic glycolysis to drive actin-dependent migratory machinery.[9,10] Tip cell formation entails a change in endothelial cell polarity from an apico-basal configuration within the lumenized vessel to an antero-posterior one where the cell extends filopodia toward the extracellular matrix (ECM) in view of breaking away from its neighbors.[11] To liberate the tip cell, the ECM has to be degraded, which is achieved

The Vasculome
https://doi.org/10.1016/B978-0-12-822546-2.00006-X

by production of matrix-degrading proteases that are enriched in tip cells. The rear end of the tip cell remains attached to follower stalk cells, which proliferate and intercalate to elongate the forming sprout. Anastomosis of tip cells with neighboring cells leads to formation of a perfused vascular loop and reestablishment of endothelial quiescence, and repeated rounds of sprouting and anastomosis occur during tissue vascularization.[12,13]

The division of labor between formation of tip and stalk cells is essential for formation of properly patterned vascular networks and is controlled by numerous integrated signaling systems that will be described in this chapter. Understanding mechanisms driving angiogenic sprouting and subsequent formation of mature arteries and veins is critical to revascularize ischemic tissue areas or prevent pathological blood vessel growth. Furthermore, signaling pathways driving vascular branching are conserved in other branched tubular systems. One example is the nervous system, where outgrowth and orientation of axons is regulated by a set of guidance cues that also operate in endothelial cells, as discussed below.[14]

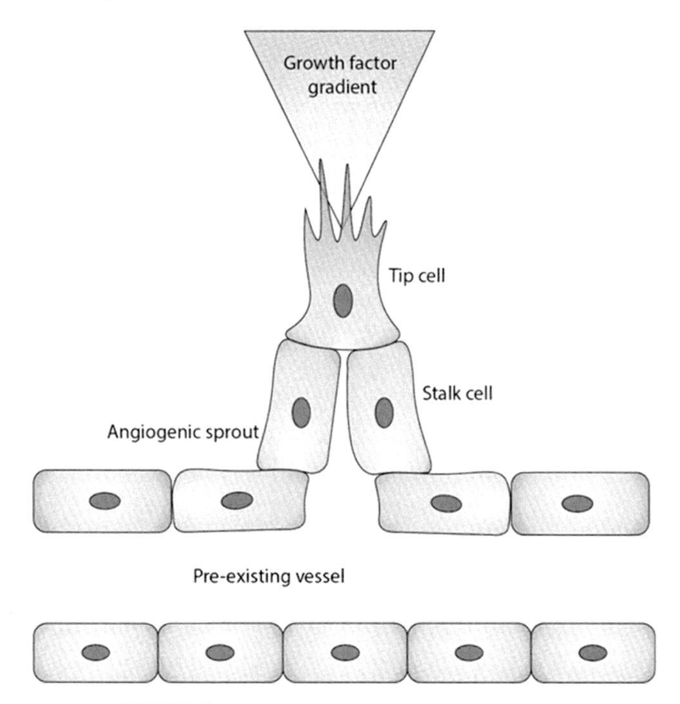

FIGURE 22.1 Structure of a vascular sprout.

Vascular endothelial growth factor and regulation of tip cell formation

Tissues in need of oxygen supply secrete angiogenic growth factors, the best studied of which is a secreted polypeptide, VEGF-A. VEGF-A expression is upregulated by hypoxia, and downregulated in normoxia, providing a direct molecular link between oxygen demand and tissue vascular density.[15] Once secreted by hypoxic tissues, VEGF-A binds to a tyrosine kinase receptor, VEGF receptor-2 (VEGFR-2) on the endothelial cell surface. Ligand binding triggers receptor dimerization and activation of endothelial cell responses, including migration, proliferation, and survival.

The importance of VEGF–VEGFR-2 signaling for vascular development was recognized by genetic inactivation of either the ligand or the receptor in mice, which led to early embryonic lethality due to a paucity of endothelial cells and failure to generate a circulatory system.[16,17] More recent studies have specifically addressed the role of the VEGF signaling system in tip cell formation, using zebrafish embryos and pathway blocking reagents as well as cell-type specific, temporally inducible gene inactivation in mice. Pathway blockade leads to aborted tip cell formation in all tissue contexts examined, demonstrating that VEGFR-2 activation is critical for tip cell formation (Fig. 22.1).[18,19] In more sophisticated assays, labeled endothelial cells with different genetic backgrounds were mixed and tested for their ability to reach the sprout tip position. When competing with wildtype cells, cells lacking one of two VEGFR-2 alleles are unable to migrate to the sprout tip, which is instead occupied by wildtype cells. Of note, VEGFR-2 heterozygous mice are healthy and fertile, demonstrating that lack of one receptor allele is compatible with normal vasculature development, but that tip cell formation critically depends on full VEGFR-2 signaling activation.[20] Because VEGFR-2 signaling activation is also required for endothelial cell proliferation and survival, a critical question is how the different downstream effects are regulated at the molecular level. The VEGFR-2 kinase domain contains several conserved intracellular tyrosine residues activating distinct signaling pathways, including the phospholipase Cγ (PLC-γ)–ERK1/2 pathway, PI3K–AKT–mTOR signaling, the SRC pathway, and small GTPases of the Rho, RAC, and CDC42 family.[21]

Among those pathways, signaling through tyrosine (Y) 1173 in mouse VEGFR-2 has a critical role in vascular development, as embryos bearing a point mutation in this residue exhibit early embryonic lethality and vascular defects comparable to those seen in VEGFR-2 knockout mice.[22] The human VEGFR-2 1175 homologue of mouse Y1173 is essential for phosphorylation of PLC-γ and downstream ERK activation. Mutations of VEGFR-2 and PLC-γ as well as pharmacological ERK inhibition lead to aborted tip cell formation during intersegmental vessel sprouting in zebrafish embryos, suggesting a critical role of this pathway in tip cell formation.[22,23] However, monitoring VEGF and PLC-γ-induced Ca^{2+} oscillations in zebrafish showed pathway activation in both tip and stalk cells.[24] In mice, lack of

ERK signaling also inhibits endothelial cell proliferation and arteriogenesis.[25] Thus, this pathway is required, but not sufficient for sprouting tip cells.

PI3K and AKT are activated indirectly by VEGFR-2, either via SRC and VE-Cadherin or by the tyrosine kinase AXL.[26,27] Mice carrying an inducible deletion of the p110α catalytic subunit of PI3K develop sprouting defects and suppressed endothelial migration, correlated with reduced activation of the small GTPase RhoA.[28] Mice carrying inducible, endothelial cell–specific deletions of *Akt1* also develop reduced sprouting, but they exhibit additional vascular defects including reduced proliferation and enhanced vessel regression.[29] Among the known AKT targets likely to be involved in endothelial proliferation is the transcription factor FOXO1. AKT phosphorylation of FOXO1 promotes its nuclear export, hence restricting FOXO1 activity.[30] Increased FOXO1 activity in temporally inducible, endothelial-specific FOXO1 overexpressing mice leads to strongly reduced angiogenesis and endothelial proliferation, coupled with Myc-dependent altered metabolic state of the endothelium.[31] Conversely, *Foxo1* loss of function leads to significantly enhanced endothelial proliferation and overgrowth that can be rescued by Myc deletion.[32] These studies highlight the importance of core metabolic pathways as regulators that drive the angiogenic process, as reviewed elsewhere in detail.[33]

SRC activation by VEGFR-2 is dependent on pY951, which binds to the T cell-specific adapter (TSAd), which in turn binds to SRC. Activated SRC phosphorylates VE-cadherin, leading to its endocytosis and disruption of adherens junctions, thereby increasing vascular permeability.[34] Mice carrying an inactivating Y–F substitution at the corresponding VEGFR-2 Y949 phosphosite are unable to develop VEGF-induced permeability and display angiogenesis defects in adult tissues, but not in the developing retina.[35,36] Phosphorylation of VEGFR-2 Y949 is inhibited by Robo4-UNC5B signaling, an endothelial guidance receptor system.[37,38] Tsad-deficient mice displayed impaired angiogenesis during tracheal vessel development, but not during retinal vasculogenesis.[39] While TSAd deficiency did not affect the ability of cells to reach the capillary tip position, a block in VEGF-induced junctional Src activity may impair rearrangement of endothelial cells in the sprout, thereby restricting sprout elongation.[34] The tissue specificity of this effect highlights the possible role of vascular bed-specific cues in the regulation of sprouting angiogenesis that remain to be better understood.

Guidance receptor signaling

Pathways that regulate axon guidance also modulate angiogenesis, including Slit and Robo (Roundabout) receptors, Neuropilin (NRP) receptors, and ephrin and Eph receptors.[40] Hence, axons and endothelial cells share expression of attractive and repulsive guidance cues for navigation. Guidance receptors control angiogenesis in concert with angiogenic growth factors by modulating receptor endocytosis and cytoskeletal behavior. NRP1 forms a complex with VEGFRs upon stimulation by VEGF, and guides the endosomal translocation of VEGFR-2, which is crucial for VEGF-induced activation of ERK signaling.[21] Deletion of NRP1 in mice leads to embryonic lethality due to defects of the vasculature and the CNS.[41] Surprisingly, absence of either VEGF binding site or cytoplasmic domain of NRP1 does not cause defects of vasculogenesis or angiogenesis.[21]

Small GTPases of the Rho, Rac, and Cdc42 family are involved in cell shape, cell migration, and polarization. Among those, VEGF-dependent activation of Rac and Cdc42 was shown to depend on the presence of the guidance molecule Slit2.[42] Slit2 signals through Robo1 and 2 receptors on endothelial cells, leading to recruitment of the NCK adaptor to the Robo1 cytoplasmic domain. Genetic deletion of Slit2, Robo1 and 2, or Nck adaptors leads to defects in endothelial sprouting and antero-posterior polarity without affecting endothelial lumen formation. Mechanistically, lack of Slit-Robo-Nck signaling also impairs VEGF-induced activation of Rac1 and CDC42, without affecting VEGF-induced ERK activation.[43] Furthermore, Robo1 and 2 receptors are required for the EndophilinA2-dependent VEGFR-2 internalization in response to ligand activation, but not for clathrin-dependent endocytosis. These two distinct endocytic processes lead to sorting and trafficking of VEGFR-2 receptor and activation of different downstream signaling. EndophilinA2-mediated VEGFR-2 endocytosis leads to activation of PAK/p38, which promotes endothelial cell polarity and migration, thus, explaining the effect of Slit-Robo on VEGFR-2 signaling.[44] These studies reveal a guidance pathway that can be targeted to prevent VEGF-induced migration without affecting other VEGF-induced cell behaviors such as proliferation.

Other guidance pathways that are relevant for sprouting angiogenesis include ephrinB2. Lack of ephrinB2 in endothelial cell prevents sprout formation by inhibiting VEGFR-2 endocytosis.[45,46] EphrinB2 regulates the movement of VEGFR-2 from the plasma membrane into endothelial cells by interacting with disabled homologue 2 (DAB2) and the cell polarity regulator partitioning defective 3 homologue (PAR3) and can be inhibited by DAB2 atypical PKC phosphorylation.[47] Overall, these studies highlight endocytosis of VEGFR-2 as an important signaling control mechanism that is at present incompletely understood.

The Notch directive: tip or stalk?

Notch signaling is an evolutionarily conserved pathway that controls cell fate specification in many tissues. Notch signaling is initiated when Notch receptors expressed on one cell bind to their membrane-bound ligands on an adjacent cell. In mammals, there are four single-pass transmembrane Notch receptors (Notch1–4), and five Notch ligands including Delta-like 1, 3, and 4 (Dll1, Dll3, Dll4) and Jagged1 and 2 (Jag1, Jag2). The most prominent family members operating in endothelium are Notch 1, Dll4, and Jag1. Ligand binding induces conformational change in Notch receptors, leading to the cleavage of the Notch intracellular domain (NICD), which then translocates to the nucleus, where it binds a transcriptional complex including RBP-J, CSL, and MAML to induce effector genes, including Hes and Hey transcription factors.[48–50]

Genetic or pharmacological blockade of Dll4-Notch signaling during mouse development leads to excessive formation of tip cells and development of an abnormally dense vascular network composed of immature and nonfunctional vessels that exhibit increased proliferation and express tip cell–specific genes more uniformly.[51,52] Likewise, Dll4-Notch signaling inhibition in zebrafish embryos leads to ectopic branching and increased motility of endothelial cells in intersegmental sprouts.[53] These studies have given rise to a model where VEGFR-2-induced tip cell Dll4 signals through Notch1 in neighboring cells to alter their gene expression and reduce sensitivity to VEGF, thereby preventing excessive tip cell formation and ultimately inducing stalk cell behavior (Fig. 22.2). Antagonism between tip cell VEGF–VEGFR-2 signaling and Notch signaling in neighboring cells is a core auto-regulated pathway that induces differential endothelial cell behavior and regulates vascular branching morphogenesis.

Notch signaling intensity and duration is tightly regulated at multiple levels, including by ligand expression, with Jag1 and Dll4 exhibiting different expression patterns and acting antagonistically during tip cell sprouting. In the mouse retinal vasculature, Jagged1 is almost exclusively found on stalk cells, while Dll4 is highest in tips.[54] Additional modes of Notch regulation are receptor glycosylation, cleavage, regulation of NICD stability, and transcriptional activation of downstream targets. This is accomplished by a number of different factors, including guidance receptor signaling through Sema3E/PlexinD1, interaction with the basement membrane, cell autonomous regulation of protein stability, and integration with the metabolic state of the cell.[55]

Deletion of Notch1 receptor in mice leads to lethality of mice around E9.5, caused by vascular underdevelopment, hypersprouting and poor remodeling of the primitive vascular plexus.[56] Arteries and veins in the Notch1 deleted mice form arterial–venous malformation (AVMs).[56,57] Although Notch4 null mice are viable, Notch1 and Notch4 double mutants display a more severe phenotype than mice with Notch1 gene deletion alone, suggesting that both of the receptors play a role in angiogenesis.[56] Similarly, deletion of Notch ligands Jag1[58] and Dll4[59] leads to severe developmental defects of embryonic and yolk sac vasculature. Combined loss of Hey1 and Hey2 also leads to lethality after E9.5 due to the lack of vascular remodeling and massive hemorrhage.[50] In pathological conditions, disruption of Dll4-Notch interactions in the tumor vasculature also induces excessive sprouting and branching, thereby compromising blood transport through this abnormal vessel network, which decreases experimental tumor growth and might be a target for antiangiogenic cancer therapy. However, not all vascular beds respond to Notch inhibition by hypersprouting. In the postnatal long bones, Notch signaling inhibition impaired bone vessel growth, in turn leading to stunted bone growth and decreased bone mass.[60] Thus, Notch signaling promotes vessel proliferation in bone. Mechanisms leading to opposing Notch actions in different tissue vascular beds are not well understood.

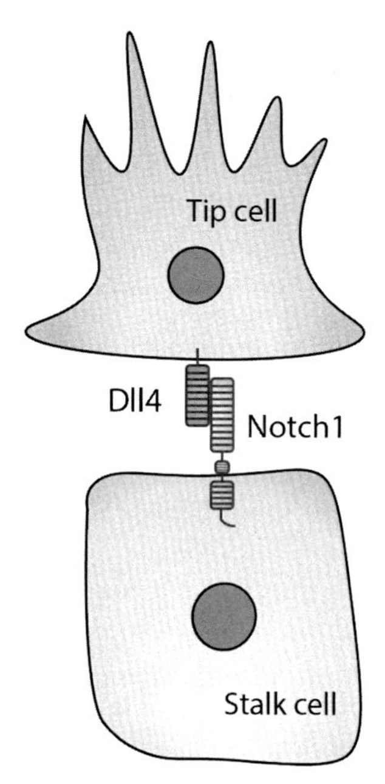

FIGURE 22.2 Regulation of tip-stalk cell by Notch signaling.

An important question is how Notch activation induces changes in cell behavior mechanistically. Notch alters expression of VEGF receptors, most prominently by increasing expression of VEGFR-1,[61] which acts as a decoy to prevent VEGF signaling through VEGFR-2. In competition assays with wildtype cells, cells expressing one copy of VEGFR-1 are preferentially found in the tip position.[20] Thus, upregulation of VEGFR-1 in stalk cells by activated Notch renders these cells less sensitive to VEGF, thereby limiting sprouting.

VEGF signaling also induces tip cell VEGFR-3 expression, which promotes both angiogenesis and lymphangiogenesis. Blocking binding of the VEGFR-3 ligand VEGF-C limits tip cell formation during sprouting angiogenesis as well as lymphangiogenesis,[62] which will be discussed later in this chapter. Paradoxically, however, in competition assays with wildtype cells, VEGFR-3 heterozygous cells are preferentially found in the tip position. Moreover, temporally inducible, endothelial-specific VEGFR-3 deletion in mice leads to hypersprouting, suggesting ligand-independent VEGFR-3 signaling activation.[63] Indeed, VEGFR-3 kinase phosphorylation can be induced by stimulation with collagen 1 in cultured endothelial cells in an Src-dependent manner. VEGFR-3 deleted endothelial cells have a reduction of Notch pathway activation, and their hypersprouting phenotype can be rescued by activation of residual Notch signaling using a Jag1 peptide.[63] Together, these studies have suggested a bimodal model of VEGFR-3 signaling, whereby ligand-independent ("passive") receptor activation in tip cells induces Dll4 expression and subsequent Notch activation in neighboring stalk cells, while ligand-dependent ("active") signaling promotes angiogenesis.

Importantly, while all three VEGF receptors contribute to tip-stalk competition, Notch signaling inhibition in cells deficient in each VEGF receptor normalizes their contribution to the tip position, firmly placing VEGF signaling upstream of Notch (Fig. 22.3). This raises the question of downstream Notch effectors required to induce stalk cell behavior. Recent studies suggest that BMPs signaling is an important effector of this process.

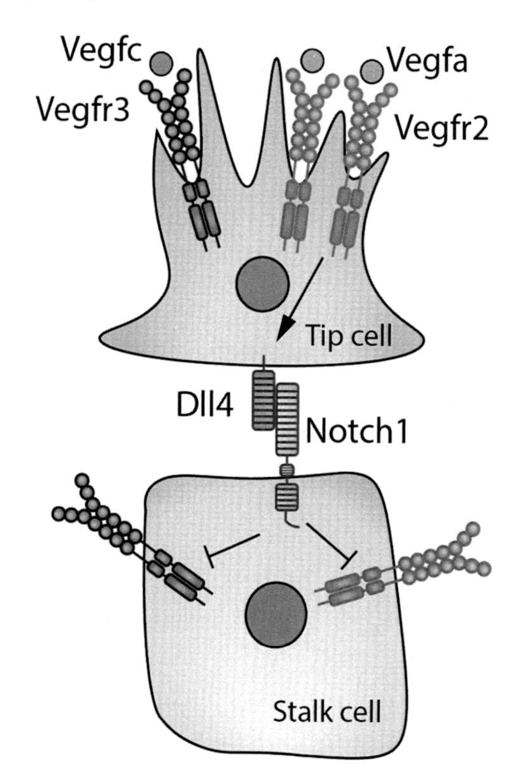

FIGURE 22.3 Regulation of tip-stalk cell by Notch and VEGF signaling.

Refinement of sprouting angiogenesis by BMP signals

BMPs are secreted ligands that belong to the transforming growth factor β (TGF-β) superfamily. BMP ligands bind and activate serine/threonine kinase receptor heterotetramer of two type I and two type II receptors. The receptors are single transmembrane-spanning proteins consisting of an extracellular cysteine-rich domain and an intracellular serine–threonine kinase domain. In mammals, there are seven type I receptors ALK1–7 (activin receptor-like kinase) and five type II receptors.[64,65] Moreover, endoglin and TGFBR3 are accessory receptors that have no intrinsic enzymatic activity but regulate the interaction of TGF-β ligands to their receptors. Upon ligand binding, the type II receptor phosphorylates and activates the type I receptor, which then phosphorylates downstream SMAD transcription factors.[64,66]

BMP9 is a potent regulator for angiogenesis that inhibits VEGF-induced endothelial cell migration and proliferation.[67] BMP9 circulates within bloodstream and binds with high affinity to Alk1 and endoglin, as well as to BMPR-II, which are abundant in vascular endothelium.[67,68] Alk1 activation leads to phosphorylation of cytoplasmic SMAD1, 5, and 8 proteins, which form complexes with SMAD4 and enter nucleus to regulate gene expression. Heterozygous mutations of Alk1 and endoglin in human cause hereditary hemorrhagic telangiectasia (HHT),[69,70] which is an inherited autosomal dominant vascular disorder. HHT patients develop AVMs, which are prone to ruptures and cause frequent epistaxis and GI bleeding. Deletion of *Alk1* in mice also leads to AVMs and primary capillary plexus remodeling defects, as well as embryonic lethality at E11.5.[71] Similarly, blocking Alk1 signaling in postnatal mice resulted in AVMs and retinal hypervascularization.[72] Mechanistically, Alk1 cross-talks to Notch signaling to inhibit angiogenesis. Inhibition of Notch signaling exaggerates the phenotype in Alk1 deleted mice, while activation Alk1 with BMP9 rescues Notch-deficient mice. In stalk cells, Alk1 signaling activates SMADs, which works synergistically with Notch signaling to induce expression of Notch targets, such as Hey1 and Hey2.[72] In addition, Alk1 signaling has also been shown to activate ERK, which mediates endothelial proliferation.[72,73] Therefore, Alk1 signaling coordinates Notch and VEGF signaling and confines the formation of tip cells and angiogenic sprouting (Fig. 22.4). Furthermore, deletion of the SMAD4 in endothelial cells abolished BMP9/Alk1-induced AKT activation and resulted in AVMs.[74] Similarly, endoglin depletion causes AVMs, endothelial hyperproliferation, and remodeling defects of the capillary plexus.[75] Deletion of endoglin also causes vascular abnormality in vitro and AVMs in adult mice brain.[76–78] In contrast, BMPR-II deletion alone in cultured endothelial cells does not abolish BMP9 signaling, but impairs BMP9 promoted proliferation.[79,80]

Lymphatic development

The balance of tissue fluid is regulated by the lymphatic system, which collects and returns interstitial fluid to blood circulation. Although traditionally considered a passive route for immune cells and fluids, recent studies reveal the heterogeneity and the diverse function of lymphatic vessels. Anatomically, the structure of large lymphatic vessels is similar to that of veins, and the lymphatic capillaries are made up of single, nonfenestrated endothelial cell layers.[81,82] In mice, lymphatic endothelial cell specification occurs at around E9.75, after the establishment of blood circulation.[83] Similar to blood endothelium, lymphatics develop via two different processes, termed lymphangiogenesis and lymphovasculogenesis.

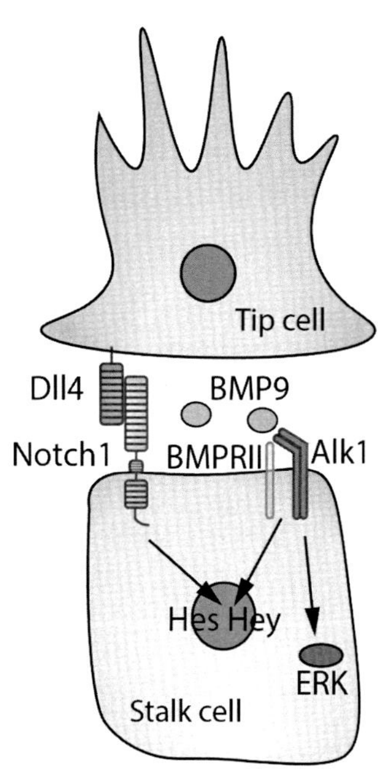

FIGURE 22.4 Regulation of angiogenesis by BMP signaling.

In lymphangiogenesis, a subpopulation of blood endothelial cells in the cardinal vein undergoes transdifferentiation to lymphatic endothelium. These cells migrate out of the vein and form the primitive lymph sacs that extend via lymphangiogenesis. Although the cardinal vein was believed to be the origin of lymphatic endothelial cells, recent studies revealed diverse origins of lymphatic endothelial cells. Two studies found nonvenous origins of lymphatic endothelial cells in skin and heart, respectively.[84,85] Later, progenitors within the capillary bed are found to give rise to dermal lymphatics.[86] Moreover, a progenitor population located within somites can give rise to both muscle cells and lymphatic endothelial cells.[87,88] cKit-expressing hemogenic endothelial cells are found to give rise to part of mesenteric lymphatics.[89] These different progenitor populations form lymphatic vessels via a process defined as lymphovasculogenesis. Notably, lymphatic endothelial cells of different organs show unique characteristics and gene expression profile, highlighting the organo-specificity of lymphatic vessels in addition to heterogenous blood vessels.[90]

The homeobox gene *Prox1* is believed to be the master regulator for lymphangiogenesis. Prox1 is expressed by a subset of venous endothelial cells, where budding and sprouting of lymphatics initiates. Between E9.75 and E14.5, these *Prox1* positive cells upregulate the expression of lymphatic-specific markers, such as VEGFR3 receptor, Lyve1 (lymphatic vessel endothelial hyaluronan receptor 1) and Pdpn (podoplanin). By expressing VEGFR3, the newly specified lymphatic endothelial cells respond to VEGF-C signals and bud off from the cardinal vein.[91] Deletion of Prox1 in mice leads to embryonic lethality at E15 due to lack of lymphatic vascular development, while vasculogenesis and angiogenesis remain unaffected.[92] Moreover, when artificially expressed in blood endothelial cells, Prox1 inhibits gene expression by blood endothelial cells and promotes lymphatic marker expression, such as VEGFR-3 and Pdpn.[93,94] VEGF-C knockout mice also die embryonically due to lack of lymphatic vessels and fluid accumulation within tissues. Altered VEGF-C level in the heterozygous also resulted in lymphatic hypoplasia and lymphoedema within the skin.[95] VEGF-C is synthesized as a precursor, which requires proteolytic processing for full activation. Secreted CCBE1 (collagen- and calcium-binding EGF domain-containing protein 1) and the protease ADAMTS3 (a disintegrin and metalloprotease with thrombospondin motifs 3) mediate the activation of VEGFC, which enables a spatial and temporal control in the context of lymphangiogenesis.[96–98]

Further research showed that a complex signaling network regulates the initiation of lymphatic development via Prox1. Several molecules involved in blood endothelial angiogenesis also play a role in lymphangiogenesis, such as Sox18 and COUP-TFII (Nr2f2). Sox18 binds directly to the promoter region of *Prox1* gene and activates its transcription. Sox18 null mice lack the ability to differentiate into lymphatic endothelial cells. Whereas overexpression of Sox18 in blood endothelial cells promotes Prox1 expression and lymphatic fate.[99] COUP-TFII binds to the regulatory region of Prox1 and is indispensable for Prox1 expression.[100] In order to induce the expression of VEGFR-3, Prox1 binds to cyclin E1, which promotes lymphatic cell proliferation. Loss-of-function experiments show that COUP-TFII is required for Prox1 to upregulate VEGFR-3 expression. Surprisingly, overexpression of COUP-TFII suppresses Prox1 function by affecting cyclin E1 and VEGFR-3 expression,[101] thus, forming a complex regulatory network of Prox1 and lymphangiogenesis.

Other signaling pathways involved in blood vessel development are found to be also important for sprouting, branching, and remodeling of lymphatics.[2,5] Nrp2 receptor can interact with VEGFR3 and regulate lymphangiogenesis.[102] Deletion of Nrp2 reduces lymphatic endothelial cell proliferation and leads to absence of small lymphatic vessels and capillaries.[103] Tbx1 binds to the enhancer of *VEGFR-3* gene and regulates its expression. Deletion of Tbx1 causes lymphangiogenesis defects and perinatal lethality of mice. But Tbx1 is not required for the initiation of lymphatic endothelial cell fate, instead, is important for lymphatic growth and maintenance.[104] Angiopoietin2-deficient mice show defects in the pattern of lymphatic vessels and smooth muscle cell coverage. Angiopoietin1 can rescue the lymphatic defects in angiopoietin2-deficient mice by promoting lymphatic endothelial cell proliferation.[105] Similarly, ephrin-B2 is required for lymphatic endothelial cell proliferation and formation of a hierarchical lymphatic network and lymphatic valves.[106] Notch signaling is found to restrict VEGF-C-induced lymphangiogenesis by regulating tip-stalk cell selection of the lymphatic vasculature.[107]

References

1. Qiu J, Hirschi Karen K. Endothelial cell development and its application to regenerative medicine. *Circ Res*. 2019;125:489–501.
2. Adams RH, Alitalo K. Molecular regulation of angiogenesis and lymphangiogenesis. *Nat Rev Mol Cell Biol*. 2007;8:464–478.
3. Jin J, Sison K, Li C, et al. Soluble FLT1 binds lipid microdomains in podocytes to control cell morphology and glomerular barrier function. *Cell*. 2012;151:384–399.
4. Augustin HG, Koh GY. Organotypic vasculature: from descriptive heterogeneity to functional pathophysiology. *Science*. 2017;357.
5. Potente M, Mäkinen T. Vascular heterogeneity and specialization in development and disease. *Nat Rev Mol Cell Biol*. 2017;18: 477–494.
6. Rafii S, Butler JM, Ding B-S. Angiocrine functions of organ-specific endothelial cells. *Nature*. 2016;529:316–325.
7. Lee J-H, Bhang DH, Beede A, et al. Lung stem cell differentiation in mice directed by endothelial cells via a BMP4-NFATc1-thrombospondin-1 axis. *Cell*. 2014;156:440–455.
8. Kalucka J, de Rooij L, Goveia J, et al. Single-cell transcriptome atlas of murine endothelial cells. *Cell*. 2020;180, 764.e720-779.e720.
9. De Bock K, Georgiadou M, Schoors S, et al. Role of PFKFB3-driven glycolysis in vessel sprouting. *Cell*. 2013;154:651–663.
10. Schoors S, De Bock K, Cantelmo AR, et al. Partial and transient reduction of glycolysis by PFKFB3 blockade reduces pathological angiogenesis. *Cell Metab*. 2014;19:37–48.
11. Lizama CO, Zovein AC. Polarizing pathways: balancing endothelial polarity, permeability, and lumen formation. *Exp Cell Res*. 2013;319:1247–1254.
12. Geudens I, Gerhardt H. Coordinating cell behaviour during blood vessel formation. *Development*. 2011;138:4569–4583.
13. Ruehle MA, Eastburn EA, LaBelle SA, et al. Extracellular matrix compression temporally regulates microvascular angiogenesis. *Sci Adv*. 2020;6:eabb6351.
14. Eichmann A, Thomas JL. Molecular parallels between neural and vascular development. *Cold Spring Harb Perspect Med*. 2013;3: a006551.
15. Rattner A, Williams J, Nathans J. Roles of HIFs and VEGF in angiogenesis in the retina and brain. *J Clin Invest*. 2019;129:3807–3820.
16. Carmeliet P, Ferreira V, Breier G, et al. Abnormal blood vessel development and lethality in embryos lacking a single VEGF allele. *Nature*. 1996;380:435–439.
17. Ferrara N, Carver-Moore K, Chen H, et al. Heterozygous embryonic lethality induced by targeted inactivation of the VEGF gene. *Nature*. 1996;380:439–442.
18. Gerhardt H, Golding M, Fruttiger M, et al. VEGF guides angiogenic sprouting utilizing endothelial tip cell filopodia. *J Cell Biol*. 2003;161:1163–1177.
19. Isogai S, Horiguchi M, Weinstein BM. The vascular anatomy of the developing zebrafish: an atlas of embryonic and early larval development. *Dev Biol*. 2001;230:278–301.
20. Jakobsson L, Franco CA, Bentley K, et al. Endothelial cells dynamically compete for the tip cell position during angiogenic sprouting. *Nat Cell Biol*. 2010;12:943–953.
21. Simons M, Gordon E, Claesson-Welsh L. Mechanisms and regulation of endothelial VEGF receptor signalling. *Nat Rev Mol Cell Biol*. 2016;17:611–625.
22. Sakurai Y, Ohgimoto K, Kataoka Y, Yoshida N, Shibuya M. Essential role of Flk-1 (VEGF receptor 2) tyrosine residue 1173 in vasculogenesis in mice. *Proc Natl Acad Sci U S A*. 2005;102:1076–1081.
23. Takahashi T, Yamaguchi S, Chida K, Shibuya M. A single autophosphorylation site on KDR/Flk-1 is essential for VEGF-A-dependent activation of PLC-γ and DNA synthesis in vascular endothelial cells. *EMBO J*. 2001;20:2768–2778.
24. Yokota Y, Nakajima H, Wakayama Y, et al. Endothelial Ca^{2+} oscillations reflect VEGFR signaling-regulated angiogenic capacity in vivo. *ELife*. 2015;4:e08817.
25. Ren B, Deng Y, Mukhopadhyay A, et al. ERK1/2-Akt1 crosstalk regulates arteriogenesis in mice and zebrafish. *J Clin Invest*. 2010;120:1217–1228.
26. Carmeliet P, Lampugnani M-G, Moons L, et al. Targeted deficiency or cytosolic truncation of the VE-cadherin gene in mice

impairs VEGF-mediated endothelial survival and angiogenesis. *Cell*. 1999;98:147–157.
27. Ruan GX, Kazlauskas A. Axl is essential for VEGF-A-dependent activation of PI3K/Akt. *EMBO J*. 2012;31:1692–1703.
28. Graupera M, Guillermet-Guibert J, Foukas LC, et al. Angiogenesis selectively requires the p110α isoform of PI3K to control endothelial cell migration. *Nature*. 2008;453:662–666.
29. Chen J, Somanath PR, Razorenova O, et al. Akt1 regulates pathological angiogenesis, vascular maturation and permeability in vivo. *Nat Med*. 2005;11:1188–1196.
30. Zhuang G, Yu K, Jiang Z, et al. Phosphoproteomic analysis implicates the mTORC2-FoxO1 axis in VEGF signaling and feedback activation of receptor tyrosine kinases. *Sci Signal*. 2013;6:ra25.
31. Dharaneeswaran H, Abid MR, Yuan L, et al. FOXO1-mediated activation of Akt plays a critical role in vascular homeostasis. *Circ Res*. 2014;115:238–251.
32. Wilhelm K, Happel K, Eelen G, et al. FOXO1 couples metabolic activity and growth state in the vascular endothelium. *Nature*. 2016;529:216–220.
33. Potente M, Carmeliet P. The link between angiogenesis and endothelial metabolism. *Annu Rev Physiol*. 2017;79:43–66.
34. Sun Z, Li X, Massena S, et al. VEGFR2 induces c-Src signaling and vascular permeability in vivo via the adaptor protein TSAd. *J Exp Med*. 2012;209:1363–1377.
35. Li X, Padhan N, Sjöström EO, et al. VEGFR2 pY949 signalling regulates adherens junction integrity and metastatic spread. *Nat Commun*. 2016;7:1–16.
36. Smith RO, Ninchoji T, Gordon E, et al. Vascular permeability in retinopathy is regulated by VEGFR2 Y949 signaling to VE-cadherin. *ELife*. 2020;9:e54056.
37. Koch AW, Mathivet T, Larrivée B, et al. Robo4 maintains vessel integrity and inhibits angiogenesis by interacting with UNC5B. *Dev Cell*. 2011;20:33–46.
38. Zhang F, Prahst C, Mathivet T, et al. The Robo4 cytoplasmic domain is dispensable for vascular permeability and neovascularization. *Nat Commun*. 2016;7:13517.
39. Matsumoto T, Bohman S, Dixelius J, et al. VEGF receptor-2 Y951 signaling and a role for the adapter molecule TSAd in tumor angiogenesis. *EMBO J*. 2005;24:2342–2353.
40. Adams RH, Eichmann A. Axon guidance molecules in vascular patterning. *Cold Spring Harb Perspect Biol*. 2010;2:a001875.
41. Kawasaki T, Kitsukawa T, Bekku Y, et al. A requirement for neuropilin-1 in embryonic vessel formation. *Development*. 1999; 126:4895–4902.
42. Rama N, Dubrac A, Mathivet T, et al. Slit2 signaling through Robo1 and Robo2 is required for retinal neovascularization. *Nat Med*. 2015;21:483–491.
43. Dubrac A, Genet G, Ola R, et al. Targeting NCK-mediated endothelial cell front-rear polarity inhibits neovascularization. *Circulation*. 2016;133:409–421.
44. Genet G, Boyé K, Mathivet T, et al. Endophilin-A2 dependent VEGFR2 endocytosis promotes sprouting angiogenesis. *Nat Commun*. 2019;10:2350.
45. Sawamiphak S, Seidel S, Essmann CL, et al. Ephrin-B2 regulates VEGFR2 function in developmental and tumour angiogenesis. *Nature*. 2010;465:487–491.
46. Wang Y, Nakayama M, Pitulescu ME, et al. Ephrin-B2 controls VEGF-induced angiogenesis and lymphangiogenesis. *Nature*. 2010;465:483–486.
47. Nakayama M, Nakayama A, van Lessen M, et al. Spatial regulation of VEGF receptor endocytosis in angiogenesis. *Nat Cell Biol*. 2013;15:249–260.
48. Phng L-K, Gerhardt H. Angiogenesis: a team effort coordinated by notch. *Dev Cell*. 2009;16:196–208.
49. Ohtsuka T, Ishibashi M, Gradwohl G, Nakanishi S, Guillemot F, Kageyama R. Hes1 and Hes5 as notch effectors in mammalian neuronal differentiation. *EMBO J*. 1999;18:2196–2207.
50. Fischer A, Schumacher N, Maier M, Sendtner M, Gessler M. The Notch target genes Hey1 and Hey2 are required for embryonic vascular development. *Genes Dev*. 2004;18:901–911.
51. Suchting S, Freitas C, le Noble F, et al. The Notch ligand Delta-like 4 negatively regulates endothelial tip cell formation and vessel branching. *Proc Natl Acad Sci U S A*. 2007;104:3225–3230.
52. Jakobsson L, Franco CA, Bentley K, et al. Endothelial cells dynamically compete for the tip cell position during angiogenic sprouting. *Nat Cell Biol*. 2010;12:943–953.
53. Lawson ND, Scheer N, Pham VN, et al. Notch signaling is required for arterial-venous differentiation during embryonic vascular development. *Development*. 2001;128:3675–3683.
54. Benedito R, Roca C, Sörensen I, et al. The notch ligands Dll4 and Jagged1 have opposing effects on angiogenesis. *Cell*. 2009;137: 1124–1135.
55. Gu C, Yoshida Y, Livet J, et al. Semaphorin 3E and plexin-D1 control vascular pattern independently of neuropilins. *Science*. 2005; 307:265–268.
56. Krebs LT, Xue Y, Norton CR, et al. Notch signaling is essential for vascular morphogenesis in mice. *Genes Dev*. 2000;14:1343–1352.
57. Krebs LT, Starling C, Chervonsky AV, Gridley T. Notch1 activation in mice causes arteriovenous malformations phenocopied by ephrinB2 and EphB4 mutants. *Genesis*. 2010;48:146–150.
58. Xue Y, Gao X, Lindsell CE, et al. Embryonic lethality and vascular defects in mice lacking the Notch ligand Jagged1. *Hum Mol Genet*. 1999;8:723–730.
59. Krebs LT, Shutter JR, Tanigaki K, Honjo T, Stark KL, Gridley T. Haploinsufficient lethality and formation of arteriovenous malformations in Notch pathway mutants. *Genes Dev*. 2004;18: 2469–2473.
60. Ramasamy SK, Kusumbe AP, Wang L, Adams RH. Endothelial Notch activity promotes angiogenesis and osteogenesis in bone. *Nature*. 2014;507:376–380.
61. Harrington LS, Sainson RC, Williams CK, et al. Regulation of multiple angiogenic pathways by Dll4 and Notch in human umbilical vein endothelial cells. *Microvasc Res*. 2008;75:144–154.
62. Tammela T, Zarkada G, Wallgard E, et al. Blocking VEGFR-3 suppresses angiogenic sprouting and vascular network formation. *Nature*. 2008;454:656–660.
63. Tammela T, Zarkada G, Nurmi H, et al. VEGFR-3 controls tip to stalk conversion at vessel fusion sites by reinforcing Notch signalling. *Nat Cell Biol*. 2011;13:1202–1213.
64. Atri D, Larrivée B, Eichmann A, Simons M. Endothelial signaling and the molecular basis of arteriovenous malformation. *Cell Mol Life Sci*. 2014;71:867–883.
65. Goumans MJ, Ten Dijke P. TGF-β signaling in control of cardiovascular function. *Cold Spring Harb Perspect Biol*. 2018;10.
66. Attisano L, Cárcamo J, Ventura F, Weis FM, Massagué J, Wrana JL. Identification of human activin and TGF beta type I receptors that form heteromeric kinase complexes with type II receptors. *Cell*. 1993;75:671–680.
67. Scharpfenecker M, van Dinther M, Liu Z, et al. BMP-9 signals via ALK1 and inhibits bFGF-induced endothelial cell proliferation and VEGF-stimulated angiogenesis. *J Cell Sci*. 2007;120: 964–972.
68. David L, Mallet C, Mazerbourg S, Feige JJ, Bailly S. Identification of BMP9 and BMP10 as functional activators of the orphan activin

receptor-like kinase 1 (ALK1) in endothelial cells. *Blood*. 2007;109: 1953–1961.
69. Johnson DW, Berg JN, Baldwin MA, et al. Mutations in the activin receptor-like kinase 1 gene in hereditary haemorrhagic telangiectasia type 2. *Nat Genet*. 1996;13:189–195.
70. Gallione CJ, Klaus DJ, Yeh EY, et al. Mutation and expression analysis of the endoglin gene in hereditary hemorrhagic telangiectasia reveals null alleles. *Hum Mutat*. 1998;11:286–294.
71. Urness LD, Sorensen LK, Li DY. Arteriovenous malformations in mice lacking activin receptor-like kinase-1. *Nat Genet*. 2000;26: 328–331.
72. Larrivée B, Prahst C, Gordon E, et al. ALK1 signaling inhibits angiogenesis by cooperating with the Notch pathway. *Dev Cell*. 2012;22:489–500.
73. David L, Mallet C, Vailhé B, Lamouille S, Feige JJ, Bailly S. Activin receptor-like kinase 1 inhibits human microvascular endothelial cell migration: potential roles for JNK and ERK. *J Cell Physiol*. 2007;213:484–489.
74. Ola R, Künzel SH, Zhang F, et al. SMAD4 prevents flow induced arteriovenous malformations by inhibiting casein kinase 2. *Circulation*. 2018;138:2379–2394.
75. Mahmoud M, Allinson KR, Zhai Z, et al. Pathogenesis of arteriovenous malformations in the absence of endoglin. *Circ Res*. 2010; 106:1425–1433.
76. Choi EJ, Walker EJ, Shen F, et al. Minimal homozygous endothelial deletion of Eng with VEGF stimulation is sufficient to cause cerebrovascular dysplasia in the adult mouse. *Cerebrovasc Dis*. 2012;33: 540–547.
77. Nolan-Stevaux O, Zhong W, Culp S, et al. Endoglin requirement for BMP9 signaling in endothelial cells reveals new mechanism of action for selective anti-endoglin antibodies. *PLoS One*. 2012; 7:e50920.
78. Satomi J, Mount RJ, Toporsian M, et al. Cerebral vascular abnormalities in a murine model of hereditary hemorrhagic telangiectasia. *Stroke*. 2003;34:783–789.
79. Upton PD, Davies RJ, Trembath RC, Morrell NW. Bone morphogenetic protein (BMP) and activin type II receptors balance BMP9 signals mediated by activin receptor-like kinase-1 in human pulmonary artery endothelial cells. *J Biol Chem*. 2009;284: 15794–15804.
80. Morrell NW, Bloch DB, Ten Dijke P, et al. Targeting BMP signalling in cardiovascular disease and anaemia. *Nat Rev Cardiol*. 2016;13: 106.
81. Pepper MS, Skobe M. Lymphatic endothelium: morphological, molecular and functional properties. *J Cell Biol*. 2003;163:209.
82. Oliver G, Kipnis J, Randolph GJ, Harvey NL. The lymphatic vasculature in the 21st century: novel functional roles in homeostasis and disease. *Cell*. 2020;182:270–296.
83. Udan RS, Culver JC, Dickinson ME. Understanding vascular development. *Wiley Interdiscip Rev Dev Biol*. 2013;2:327–346.
84. Martinez-Corral I, Ulvmar MH, Stanczuk L, et al. Nonvenous origin of dermal lymphatic vasculature. *Circ Res*. 2015;116: 1649–1654.
85. Klotz L, Norman S, Vieira JM, et al. Cardiac lymphatics are heterogeneous in origin and respond to injury. *Nature*. 2015;522: 62–67.
86. Pichol-Thievend C, Betterman KL, Liu X, et al. A blood capillary plexus-derived population of progenitor cells contributes to genesis of the dermal lymphatic vasculature during embryonic development. *Development*. 2018;145:dev160184.
87. Wilting J, Aref Y, Huang R, et al. Dual origin of avian lymphatics. *Dev Biol*. 2006;292:165–173.
88. Stone OA, Stainier DYR. Paraxial mesoderm is the major source of lymphatic endothelium. *Dev Cell*. 2019;50:247–255. e243.
89. Stanczuk L, Martinez-Corral I, Ulvmar MH, et al. cKit lineage hemogenic endothelium-derived cells contribute to mesenteric lymphatic vessels. *Cell Rep*. 2015;10:1708–1721.
90. Ulvmar MH, Mäkinen T. Heterogeneity in the lymphatic vascular system and its origin. *Cardiovasc Res*. 2016;111:310–321.
91. Gutierrez-Miranda L, Yaniv K. Cellular origins of the lymphatic endothelium: implications for cancer lymphangiogenesis. *Front Physiol*. 2020;11:577584.
92. Wigle JT, Oliver G. Prox1 function is required for the development of the murine lymphatic system. *Cell*. 1999;98:769–778.
93. Hong YK, Harvey N, Noh YH, et al. Prox1 is a master control gene in the program specifying lymphatic endothelial cell fate. *Dev Dyn*. 2002;225:351–357.
94. Petrova TV, Mäkinen T, Mäkelä TP, et al. Lymphatic endothelial reprogramming of vascular endothelial cells by the Prox-1 homeobox transcription factor. *EMBO J*. 2002;21:4593–4599.
95. Karkkainen MJ, Haiko P, Sainio K, et al. Vascular endothelial growth factor C is required for sprouting of the first lymphatic vessels from embryonic veins. *Nat Immunol*. 2004;5:74–80.
96. Bui HM, Enis D, Robciuc MR, et al. Proteolytic activation defines distinct lymphangiogenic mechanisms for VEGFC and VEGFD. *J Clin Invest*. 2016;126:2167–2180.
97. Jeltsch M, Jha SK, Tvorogov D, et al. CCBE1 enhances lymphangiogenesis via A disintegrin and metalloprotease with thrombospondin motifs-3-mediated vascular endothelial growth factor-C activation. *Circulation*. 2014;129:1962–1971.
98. Le Guen L, Karpanen T, Schulte D, et al. Ccbe1 regulates Vegfc-mediated induction of Vegfr3 signaling during embryonic lymphangiogenesis. *Development*. 2014;141:1239–1249.
99. François M, Caprini A, Hosking B, et al. Sox18 induces development of the lymphatic vasculature in mice. *Nature*. 2008;456: 643–647.
100. Srinivasan RS, Geng X, Yang Y, et al. The nuclear hormone receptor COUP-TFII is required for the initiation and early maintenance of Prox1 expression in lymphatic endothelial cells. *Genes Dev*. 2010;24:696–707.
101. Yamazaki T, Yoshimatsu Y, Morishita Y, Miyazono K, Watabe T. COUP-TFII regulates the functions of Prox1 in lymphatic endothelial cells through direct interaction. *Gene Cell*. 2009;14:425–434.
102. Xu Y, Yuan L, Mak J, et al. Neuropilin-2 mediates VEGF-C–induced lymphatic sprouting together with VEGFR3. *J Cell Biol*. 2010;188:115–130.
103. Yuan L, Moyon D, Pardanaud L, et al. Abnormal lymphatic vessel development in neuropilin 2 mutant mice. *Development*. 2002;129: 4797.
104. Chen L, Mupo A, Huynh T, et al. Tbx1 regulates Vegfr3 and is required for lymphatic vessel development. *J Cell Biol*. 2010;189: 417–424.
105. Dellinger M, Hunter R, Bernas M, et al. Defective remodeling and maturation of the lymphatic vasculature in Angiopoietin-2 deficient mice. *Dev Biol*. 2008;319:309–320.
106. Mäkinen T, Adams RH, Bailey J, et al. PDZ interaction site in ephrinB2 is required for the remodeling of lymphatic vasculature. *Genes Dev*. 2005;19:397–410.
107. Zheng W, Tammela T, Yamamoto M, et al. Notch restricts lymphatic vessel sprouting induced by vascular endothelial growth factor. *Blood*. 2011;118:1154–1162.

SECTION 2

Physiological and pathological remodeling of the vasculome

CHAPTER

23

Enablers and drivers of vascular remodeling

Jay D. Humphrey

Department of Biomedical Engineering and Vascular Biology and Therapeutics Program, Yale University, New Haven, CT, United States

Introduction

The primary function of the cardiovascular system is biotransport, including transport of oxygen and nutrients to tissues of the body and transport of carbon dioxide and waste products from tissues of the body. Most immune cells are similarly transported within the circulatory system and so too hormones and other key biomolecules. The vascular tree evolved to facilitate biotransport via complementary structural and functional diversity along its length, typically ordered from large (elastic), medium-sized (muscular), and small (arterioles) arteries that conduct blood to the capillary bed and small (venules), medium-sized, and large (capacitance) veins that return blood to the heart. It is in the capillaries where most mass transport occurs; it is in the postcapillary venules where most immune cells are delivered to tissues.

In this chapter, however, we focus on arteries and, in particular, their remarkable ability to grow (change mass) and remodel (change microstructure) throughout life, including both advantageous adaptations and disabling disease progression. It is within arteries that significant diseases such as atherosclerosis and aneurysms manifest. Importantly, all three primary cell types of the arterial wall—endothelial cells (ECs) that line each blood vessel as a monolayer and contribute to the inner (intimal) layer of the wall, smooth muscle cells (SMCs) of the middle (medial) layer, and fibroblasts (FBs) of the outer (adventitial) layer—are exquisitely sensitive to changes in their local mechanical environment.[13] Hence, most of our attention will be directed to mechanobiological enablers and drivers of arterial remodeling.

Brief on vascular mechanics

Because vascular cells are highly mechanosensitive, it is critical to quantify the mechanical environment in which they function and how it changes. Notwithstanding the complexity of vascular mechanics (Appendix 1), including flow of blood within complex geometries and large deformations of multilayered vessels having complex compositions,[6,13] simple relationships capture mean values of the three components of stress of primary importance:

$$\tau_w = \frac{4\mu Q}{\pi a^3}, \tag{23.1}$$

$$\sigma_\theta = \frac{Pa}{h}, \tag{23.2}$$

$$\sigma_z = \frac{f}{\pi h(2a+h)}, \tag{23.3}$$

where τ_w is the flow-induced wall shear stress (with μ fluid viscosity, Q volumetric flowrate, and a luminal radius), σ_θ is the pressure-induced circumferential wall stress (with P the pressure that tends to distend the vessel and h the thickness of the vascular wall), and σ_z is the axial wall stress (with f the total axial force that

The Vasculome
https://doi.org/10.1016/B978-0-12-822546-2.00004-6

acts on the vascular wall, tending to extend it). These relations are derived from Newton's second law of motion ($f = ma$), that is, balance of linear momentum. Importantly, they describe stress at points in space, which is thus a local quantity. It is common to think of τ_w as that component of stress most affecting the mechanobiology of ECs and σ_θ, σ_z as components of intramural stress most affecting that of SMCs and FBs.

To these three basic relations, we add two simple relations that describe global hemodynamics, first,

$$\Delta P = QR, \tag{23.4}$$

where

$$R = 8\mu l/\pi a^4, \tag{23.5}$$

which reveals that the volumetric flow rate Q is driven by a pressure drop $\Delta P = P_{upstream} - P_{downstream} > 0$ but affected by the resistance to flow R (where l is the length over which pressure drops). Because R increases nonlinearly with decreases in luminal radius a, most of the resistance to flow in the arterial tree tends to arise in arterioles (often referred to as resistance vessels). Second, consider another (highly simplified) relation, namely, for the speed at which the pressure wave travels along the arterial tree,

$$PWV = \pm\sqrt{\frac{Eh}{2\rho_f a}}, \tag{23.6}$$

where E is a simplified metric for the material stiffness of the vascular wall and ρ_f is the mass density of the contained fluid. This Moens–Korteweg equation reveals that it is primarily the structural stiffness (Eh) that dictates this so-called pulse wave velocity (PWV): given the same luminal radius a, the pulse wave would travel at the same speed in a thin, stiff vessel or a thick, compliant vessel as long as the product Eh was the same in both. The square root reveals that the wave can travel in forward or reverse directions, the former driven by the heart, the latter due to distal reflection sites such as bifurcations and branches.

Biological and physiological consequences of mechanics

Observations in patients and animal models reveal that the vasculature tends to remodel so as to maintain the aforementioned load-induced mechanical stresses near target (i.e., homeostatic) values.[7,9,16] Although increases in axial loading elicit the most rapid remodeling responses, consider the two most common and easily measured changes. Let sustained fold changes in blood flow be denoted by the parameter ε (e.g., $\varepsilon = 1.1$ for a sustained 10% increase in flowrate Q) and sustained fold changes in blood pressure be denoted by the parameter γ. If flow-induced wall shear stress and pressure-induced circumferential wall stress are initially perturbed due to sustained changes in flow and pressure ($Q \rightarrow \varepsilon Q_o$ and $P \rightarrow \gamma P_o$, where the subscript o denotes an original (homeostatic) value or set point), then from Eqs. 23.1 and 23.2

$$\tau_w = \frac{4\mu Q_o}{\pi a_o^3} \rightarrow \frac{4\mu(\varepsilon Q_o)}{\pi a^3}, \tag{23.7}$$

$$\sigma_\theta = \frac{P a_o}{h_o} \rightarrow \frac{(\gamma P_o)a}{h} \tag{23.8}$$

Hence, to restore wall shear stress and circumferential stress toward normal values under the action of these altered hemodynamics requires, respectively, that[14]

$$a \rightarrow \varepsilon^{1/3} a_o, \tag{23.9}$$

$$h \rightarrow \gamma \varepsilon^{1/3} h_o \tag{23.10}$$

As an example, if a 10% increase in volumetric flowrate is sustained at an unchanging local pressure (remembering that it is a pressure drop, not local pressure, that drives flow; Eq. 23.4), then luminal radius must increase by 3% (i.e., $1.1^{1/3} \sim 1.03$) to restore wall shear stress to its original value. Interestingly, the wall thickness must similarly increase by 3% because a local dilation causes an initially isochoric decrease in wall thickness that must be resolved if circumferential stress is to be restored. In contrast, an increase in blood pressure alone should cause a proportional increase in wall thickness, with no change in luminal radius. The latter implies that the passive pressure-induced distension of the arterial wall would need to be compensated by an appropriate active vasoconstriction. Hence, we recognize that distending pressure P and SMC contractility C together dictate arterial geometry and thus wall stress, namely, Eq. 23.2 should be thought of as $\sigma_\theta = Pa(P,C)/h(P,C)$ where the dependence of radius on pressure reflects the classical nonlinear "pressure-diameter" behavior. If an artery obeys the geometric relations defined by Eqs. 23.9 and 23.10 it is said to be mechano-adaptive; if not, it is mechanically maladaptive. We consider below some molecular and cellular mechanisms underlying adaptive versus maladaptive changes.

First, however, recall that our simple relations for load-induced mechanical stresses (Eqs. 23.1–23.3), and associated geometric remodeling (Eqs. 23.9 and 23.10), are local. Importantly, such local changes yet affect global relations. For example, consider the important case of a sustained

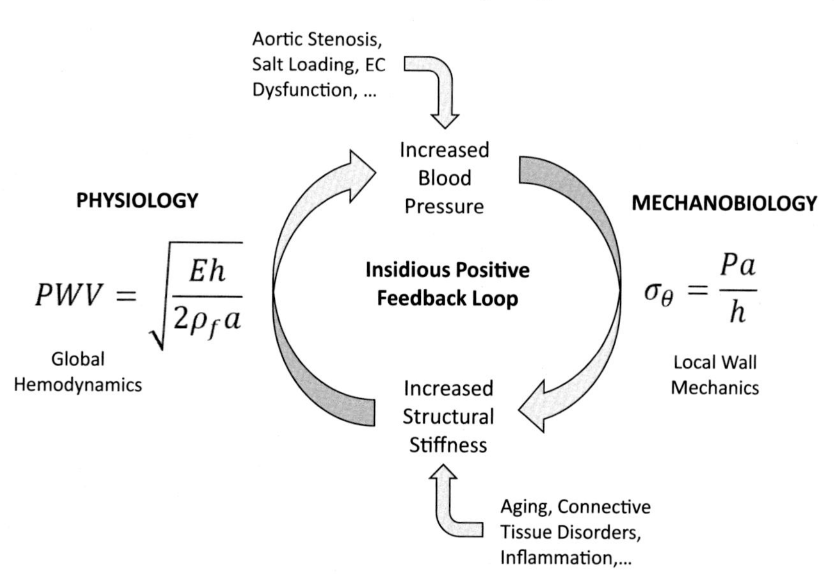

FIGURE 23.1 Schema showing how local mechanobiological responses can lead to global pathophysiological consequences via positive feedback. *Adapted from Humphrey JD, Harrison DG, Figueroa CA, Lacolley P, Laurent S. Central artery stiffness in hypertension and aging: a problem with cause and consequence.* Circ Res. *2016;118:379–381.*

increase in blood pressure (i.e., hypertension) with preserved volumetric flow ($\gamma > 1$, $\varepsilon = 1$). Whereas local mechao-adaptations require that the wall thicken while preserving luminal radius (Eqs. 23.9 and 23.10), Moens–Korteweg (Eq. 23.6) reveals that *PWV* will increase unless the material stiffness reduces proportional to the increase in wall thickness. It is increasingly evident, however, that circumferential material stiffness (e.g., *E*) is a highly mechano-regulated quantity, across vessels and species.[30,32] Indeed, failed regulation of material stiffness is an indicator or initiator of different vascular diseases, including thoracic aortic aneurysm.[4] With preserved or increased material stiffness, a local mechano-adaptation to increased pressure thus increases *PWV*, which results in an earlier reflection of the pulse wave from distal reflection sites, which in turn results in the reverse pulse wave returning to the proximal aorta earlier in the cardiac cycle and thereby augmenting central pulse pressure (noting that it is pulse pressure or at least systolic pressure, not mean arterial pressure, that drives hypertensive remodeling of central arteries).[11] This consequent increase in pulse pressure can, in turn, promote mechanoadaptive attempts to thicken the wall further, which would increase *PWV*, and so forth. Hence, this simple example illustrates how a local mechano-adaptive response has global consequences, in this case resulting in a possible insidious positive feedback loop (Fig. 23.1) that drives disease progression.[19]

It is important to note, however, that mechanoadaptations to sustained increases in pressure (Eqs. 23.8 and 23.10) appear to operate best in muscular arteries, as evidenced in human hypertension and aging[23] as well as diverse animal studies. That is, elastic arteries tend to dilate while adapting wall thickness in hypertension, muscular arteries tend to maintain their lumen while adapting wall thickness, and arterioles tend to experience luminal narrowing while adapting wall thickness. Importantly, any dilatation of elastic arteries will help slow the pulse wave, since luminal radius is in the denominator of Moens–Korteweg, whereas luminal encroachment of the arterioles appears to reduce penetration of the pulse pressure into the microcirculation of end organs. Hence, both of these "nonideal" local adaptations may have some global benefit—there are clearly compensatory local-global interactions. Yet, luminal encroachment in arterioles also increases the total peripheral resistance via Eq. 23.5, which increases mean blood pressure and contributes to the aforementioned positive feedback (Fig. 23.1). There exist many other cases of differential local and global effects of arterial remodeling. Whether advantageous or not, most remodeling is enabled by the mechano-sensitive cells and driven by their gene products, thus focusing attention on the (local) mechanobiology.

Mechanobiology and SMC phenotype

There is extensive information on mechano-sensitive changes in gene expression in ECs, SMCs, and FBs. Perhaps best known (cf.[7]) are changes in wall shear stress that induce altered production of endothelial-derived nitric oxide (NO) synthase, which catalyzes the production of NO as a byproduct of the conversion of L-arginine to L-citrulline; similar shear stress–induced

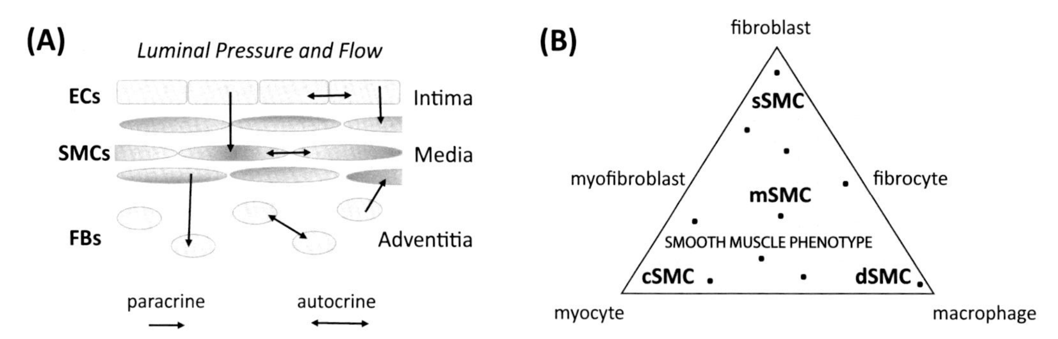

FIGURE 23.2 Schema of different cell types of the arterial wall (EC-endothelial, SMC-smooth muscle, FB-fibroblasts), and their interactions, emphasizing phenotypic heterogeneity: (A) different gray gradients suggest different SMC phenotypes; (B) points within the SMC-space as would be revealed by scRNAseq. SMC phenotypes: sSMC-synthetic, mSMC-maintenance, cSMC-contractile, dSMC-degradative.

changes drive the production of endothelin-1 (ET-1). NO production increases in response to increases in wall shear stress above basal values; it is a potent vasodilator, which helps to restore wall shear stress toward its set point in cases of increased flow. NO is also antiinflammatory and it slows extracellular matrix (ECM) synthesis by SMCs and FBs, hence delaying the entrenchment of remodeling in response to increased flow. ET-1 is produced in response to decreases in wall shear stresses below basal values; it is a potent vasoconstrictor, which helps to restore wall shear stress toward normal in cases of reduced flow. ET-1 is also proinflammatory and it hastens ECM synthesis by intramural cells, hence speeding the entrenchment of remodeling in response to decreased flow. NO and ET-1 are transported via local mechanisms from the ECs to the underlying SMCs and possibly FBs (Fig. 23.2A), as dictated by balance of mass relations.

Notwithstanding the importance of ECs, which play key roles in regulating local vasoactivity as well as endothelial permeability, hemostasis, and inflammation, SMCs serve as central nodes both in mechano-regulating the structure, properties, and biomechanical function of the medial layer and endowing the wall with contractile capacity.[5,21] Perhaps not surprisingly, it was long thought that SMCs exhibit one of two primary phenotypes: synthetic in development and many diseases or contractile in normalcy in maturity. It is now clear, however, that the phenotypic spectrum is much broader, depending on both regional location along the arterial tree and health/disease status. In particular, SMCs within elastic arteries are primarily responsible for mechano-sensing and mechano-regulating the ECM of the media, which consists primarily of elastic fibers organized as concentric laminae (fashioned during the perinatal period), glycosaminoglycans, and fibrillar collagens. For this reason these SMCs exhibit primarily a "maintenance phenotype" (Fig. 23.2B), which includes low-level synthetic activity, degradative activity, and acto-myosin activity to mechano-regulate the matrix that is deposited within the extant ECM. The elastic laminae endow large arteries with resilience, thus enabling their primary function: to store elastic energy during systole and to use this energy during diastole to work on the blood and augment flow within a pulsatile system. In contrast, glycosaminoglycans contribute a (radial) compressive stiffness that appears to facilitate SMC mechano-sensing of the stress in, or stiffness of, the tensile load-bearing elastic laminae.[29] The fibrillar type III collagen contributes to the stiffness and strength of the media, though most of the strength of the arterial wall comes from the fibrillar type I collagen of the adventitia,[3] which appears to be mechano-regulated primarily by the FBs.

In contrast, SMCs within muscular arteries and arterioles exhibit primarily a "contractile phenotype" (Fig. 23.2B), though one that differs in degree between muscular arteries and arterioles since the latter exhibit an endothelial-independent pressure-sensitive myogenic contractile response.[26] Recall from above that this myogenic response may protect the microcirculatory beds of end organs from penetrating pulse pressures. Muscular arteries have but an internal and external elastic laminae and a thicker adventitia that consists largely of type I collagen fibers. Arterioles consist primarily of a one-to-two SMC-thick medial layer enclosed within a thin protective adventitia. That SMCs of elastic arteries (which experience highly pulsatile blood pressures on the order of 120/80 mm Hg at rest) and arterioles (with less pulsatile blood pressures on the order of 45 mm Hg) exhibit different phenotypes should not be surprising given their different functions—efficient elastic conduits and vasoactive flow-regulating distributors, respectively. It should similarly not be surprising that SMCs can exhibit strong synthetic or degradative phenotypes given the importance of ECM turnover in the maintenance, remodeling, or repair of any vascular wall. What is perhaps unexpected, however, is that increasing data from single cell RNA sequencing reveal multiple natural clusters (phenotypes) of SMCs within a given region of the arterial tree, suggesting considerable SMC heterogeneity in health and especially disease. In this spirit, we can imagine SMCs of a maintenance phenotype (sensing) as enablers of remodeling for they sense and signal a need, and SMCs of a synthetic and degradative phenotype as drivers of remodeling for they effect the requisite changes in ECM composition and organization, as an example.

SMCs interact with the surrounding ECM via various cell surface receptors, especially integrins.[28] These heterodimeric transmembrane protein complexes can be denoted $\alpha_x\beta_y$, where x and y take different values and signify ligand-specific interactions. For example, $\alpha_1\beta_1$ and $\alpha_2\beta_1$ facilitate interactions with fibrillar collagen, whereas $\alpha_5\beta_1$ facilitates interactions with fibronectin. The latter tends to promote a proinflammatory response, often via transcription factor NF-κB. Importantly, SMCs in elastic arteries connect to the elastic laminae via microfibrillar connections consisting largely of the elastin-associated glycoprotein fibrillin-1; it appears that $\alpha_5\beta_1$ and $\alpha_v\beta_3$ facilitate these interactions, thus allowing SMCs of elastic arteries to mechano-sense loads borne by the energy storing elastic laminae. Mutations to the gene that encodes fibrillin-1 lead to Marfan syndrome, which affects the aortic root and ascending aorta, among other tissues, leading to aneurysmal dilatation, dissection, and rupture. An emerging hypothesis is that either loss of functional fibrillin-1 connections between SMCs and the elastic laminae or loss of the actomyosin activity needed to probe the mechanical state of the laminae via these microfibrillar connections result in dysfunctional mechanosensing and/or mechanoregulation of ECM within the media, leading to a progressive medial degeneration responsible for the aortopathy.[18] Yet, in other cases, loss of integrin-mediated mechanosensing appears to be somewhat beneficial. Mice having α_1 integrin subunit deletion in elastic arteries reduce an adverse fibrotic response to angiotensin II-induced hypertension,[25] whereas mice having β_1 integrin subunit deletion show slight early protection against supravalvular aortic stenosis in Williams syndrome.[27] Integrins similarly enable mechano-sensitive responses by arterioles, with the myogenic response to increased pressure also depending on $\alpha_5\beta_1$ and $\alpha_v\beta_3$. Hence, SMCs exhibiting different phenotypes within different regions of the arterial tree (e.g., maintenance phenotype in elastic arteries and contractile phenotype in arterioles) use the same integrins to assess their local mechanical environment but to drive different mechanobiological responses, with dysfunction leading to disease in some cases and attenuating disease in others, the latter of which likely reflects a compensatory response.

ECM turnover in evolving mechanical states

Whether in an elastic artery, muscular artery, or arteriole, SMC contractility plays key roles in vascular remodeling.[26,31] Recall from above that an ideal mechano-adaptation to a sustained increase in pressure requires increased SMC contractility to return the pressure-distended wall to its original radius, thereby restoring wall shear stress as well. Such complementary contractility and ECM turnover appears to function best in muscular arteries, at least in hypertension and aging. Yet, dramatic cases of coexisting SMC contractility and ECM turnover also accompany disease processes. For example, extravasation of blood from a ruptured intracranial saccular aneurysm promotes thrombus on the adventitial surface of nearby cerebral (muscular) arteries. Due to the host of factors released by an early thrombus, including thrombin, thromboxane, and serotonin, the vessel undergoes what appears to be repeated vasoconstriction and ECM turnover, sequentially entrenching the vessel at a much reduced diameter within 3–5 days. Such "cerebral vasospasm" is life threatening and cannot be reversed by vasodilator therapy alone due to the ECM turnover in vaso-constricted states. Rather, vasospasm tends to reverse only as the thrombus resolves and with it the exogenous source of vasoconstrictors, which also hasten matrix production by intramural cells.[1] Microvascular remodeling can be even faster. A mere 4 h of sustained vasoconstriction of an arteriole can entrench the vessel at the reduced diameter, due in part to reorganization of cytoskeleton within the SMCs to optimize their contractile force-length behavior. While this short duration is not sufficient for marked turnover of ECM in the vaso-altered state, it appears that tissue transglutaminase can cross-link the ECM in this new state, thus enabling rapid remodeling.[2] Cross-linking of ECM is central to all remodeling, with transglutaminases fundamental to short-term remodeling of extant ECM and lysyl oxidase important in longer term remodeling that includes deposition of new ECM. Of course, degradation is similarly central to ECM remodeling and thus overall vessel adaptation or disease progression, with matrix metalloproteinases (MMPs) playing key roles.[12] Changes in mechanical stress can stimulate cellular production of new (pro)MMPs and facilitate the activation of pro-MMPs sequestered in the ECM, with altered mechanical stress on ECM also affecting its vulnerability to proteolytic activity. Thus, as in many cases in vascular mechanobiology, there are many checks and balances, with ECM turnover in changing mechanical states fundamental.

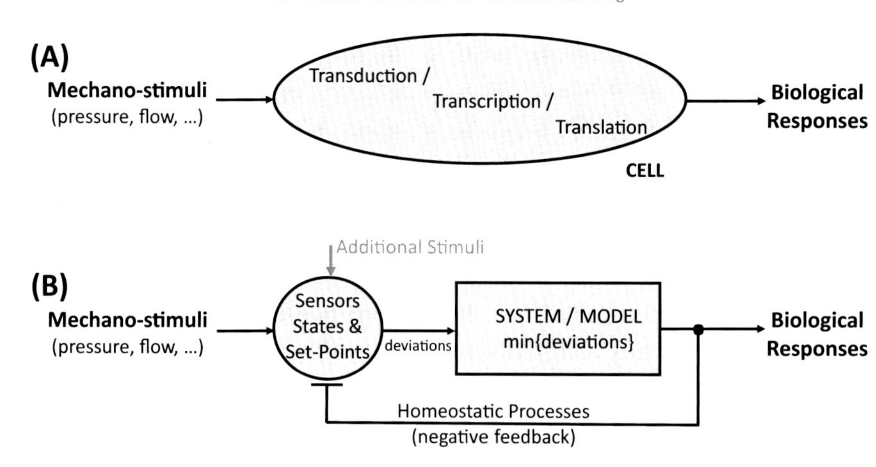

FIGURE 23.3 Schema of mechanobiological responses emphasizing: (A) mechano-transduction, transcription, and translation as mechanisms, and (B) mechanical homeostasis interpreted in terms of negative feedback control, which seeks to minimize deviations between sensed states and their set-points. Additional stimuli can include connective tissue disorders, inflammation, and drug treatments.

Mechanical homeostasis—an organizing principle

The concept of homeostasis was introduced by the physiologist Walter Cannon in the 1920s and is now a fundamental principle of biology. In brief, homeostasis is a process by which a particular biological or physiological quantity is regulated to remain near a target (homeostatic) value, often referred to as a set-point. Whereas tight control of body temperature (global) or interstitial fluid pH (local) are well-known homeostatic processes, the aforementioned control of mechanical quantities appears to be central to vascular biology.[15] Importantly, homeostasis is achieved via negative feedback, thought to result in stable, near optimal states. Hence, whereas we often think biologically of mechano-sensitive responses in terms of (mechano) transduction, (gene) transcription, and (protein) translation (Fig. 23.3A), it is similarly useful to think quantitatively in terms of mechanical homeostasis, namely, in terms of states, sensors, and set-points that are used by a biological system to evoke negative feedback control (Fig. 23.3B). In the context of most vascular remodeling, the biomechanical state is defined by the geometry, material properties, and applied loads, which are captured by the equations of vascular mechanics (Appendix 1); the sensors, which inform the cell of the local mechanical environment, include integrins, cadherins, and other specialized cell surface receptors; finally, the set-points are homeostatic targets that the cells seek to maintain particular quantities near (not necessarily at). Hence, the cell must compare appropriate metrics that define the state with the desired set-points. If deviations between the two are within an acceptable tolerance, the cell need not change its activity; if, instead, the deviations are outside of the accepted tolerance, this sets into motion various responses, including differential gene expression in an attempt to drive the state toward homeostatic.

Inasmuch as intramural cells attempt to effect appropriate ECM turnover via negative feedback to promote (asymptotically) stable homeostatic states that remain within a certain range,[17] such control is likely achieved via a so-called proportional feedback. Importantly, however, myriad observations in cell studies, animal models, and clinical cases suggest that such attempts can also include a resetting of set-points to promote (neutrally stable) adaptations rather than (asymptotically stable) restorations. Such a process has been termed "adaptive homeostasis" (Fig. 23.4A). Although adaptive homeostasis enables a broader range of homeostatic control,[10] with increased adaptivity also comes increased vulnerability. That is, it has been suggested that "homeostatic circuits with adjustable set-points are vulnerable to dysregulation precisely because they are designed to be adjustable."[20] It is suggested here that contextualizing enablers and drivers of vascular remodeling, repair, and disease within the rubric of homeostatic systems biology would increase our understanding considerably, especially in cases of coupled effectors (e.g., multiple hits) given the intrinsic complexity of understanding potentially synergistic or antagonistic processes—what, for example, are causes versus consequences is a ubiquitous unanswered question in many cases of vascular disease.

Finally, notwithstanding the importance of mechanical homeostasis in vascular biology and mechanics, inflammation plays a central role and must be considered as an additional "stimulus" (Fig. 23.3). Inflammation can drive many disease processes, including atherosclerotic plaque progression and many abdominal aortic aneurysms,[13,24] to name a few, yet it is also increasingly recognized that inflammatory cells also play fundamental roles in homeostasis. Resident

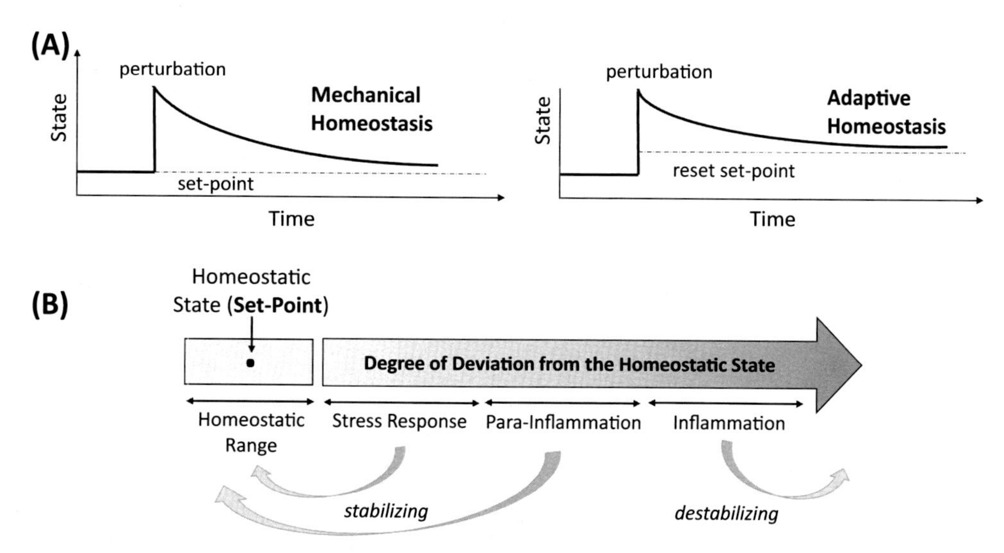

FIGURE 23.4 (A) Schema of mechanically homeostatic versus adaptively homeostatic responses, both stable but the former returns the state toward the original set-point and the later toward a new set-point. Being a preferred process (developed in response to life-threatening insults), inflammation can change homeostatic set-points and rates at which they are recovered. (B) Spectrum from mechanical homeostasis to supportive para-inflammatory responses to disruptive inflammation, the later driving unstable processes often leading to disease progression, as in atherosclerosis.

macrophages, for example, promote homeostasis by removing apoptotic cells or cellular debris and similarly by removing degraded or damaged ECM. Viewed in this way, inflammation can complement mechanical homeostasis (Fig. 23.4B), particularly in cases wherein the mechano-response is either delayed or insufficient to resolve the perturbation.[8] There is a pressing need to understand better the roles of mechanics and inflammation as enablers and drivers of both vascular remodeling and disease.[22]

Conclusion

The arterial tree is a complex, coordinated system consisting of many different parts having diverse functions that collectively promote near optimal function under normal conditions. Vascular cells, also diverse in type and even phenotype for a given type, similarly work together to establish, maintain, remodel, or repair the vessel wall, which consists of myriad ECM proteins, glycoproteins, and glycosaminoglycans that differentially endow the vessel with site-specific properties and function, again to ensure the overall global function of the system. We must continue to understand more deeply each component part—from molecule to cell to tissue to system, that is, out of many, one—but with a goal of understanding the integrated function that ultimately delineates human health and disease.

Appendix 1

Vascular mechanics is complicated by the complex geometry, nonlinear material properties, and time-varying hemodynamic loads. Nevertheless, continuing advances in computational methods and power allow the mechanical environment experienced by vascular cells to be understood with increasing clarity. Regardless of the many different situations of interest—from development to disease—there are multiple fundamental equations that dictate the one state (out of many, one). These equations include balance of mass, linear momentum, and energy augmented by appropriate descriptions of material behavior (constitutive relations) and initial boundary conditions subject to the constraints of balance of angular momentum and the second law of thermodynamics.[13] Fig. A1 shows, for example, how balance of linear momentum (Newton's second law of motion in cylindrical coordinates (r, θ, z), with t_{kl} representing the point-wise stress) is foundational for solving many problems in vascular mechanics, both standard problems of hemodynamics and wall stress and more contemporary simulations of evolving composition, geometry, and properties (cf).[22]

Acknowledgments

I wish to thank Dr. Zorina S. Galis for her scientific leadership in her lifelong engagement in and promotion of vascular research, specifically the important need to integrate knowledge across scales and

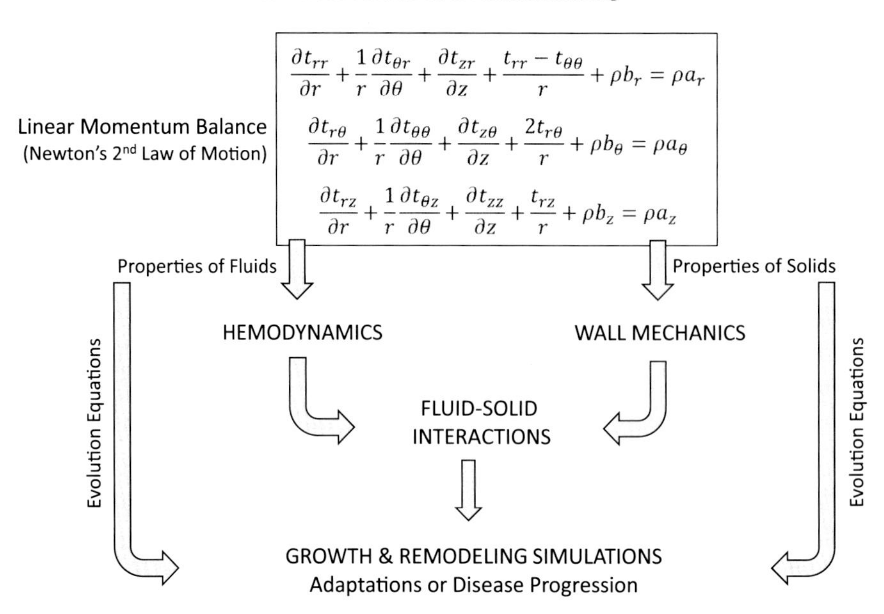

FIGURE A1 Schema of the workflow to perform coupled simulations of the hemodynamics and wall mechanics, including evolution equations for cell-mediated vascular adaptation, leading ultimately to fluid-solid-growth (FSG) analyses.

from local mechanisms to global manifestations. I also thank the NIH for grant support (R01 HL105297, U01 HL142518, P01 HL134605) and similarly the National Marfan Foundation.

References

1. Baek S, Valentin A, Humphrey JD. Biochemomechanics of cerebral vasospasm and its resolution: II. Constitutive relations and model simulations. *Ann Biomed Eng*. 2007;35:1498–1509.
2. Bakker ENTP, Buus CL, Spaan JAE, et al. Small artery remodeling depends on tissue-type transglutaminase. *Circ Res*. 2005;96: 119–126.
3. Bellini C, Ferruzzi J, Roccabianca S, DiMartino E, Humphrey JD. A microstructurally-motivated model of arterial wall mechanics with mechanobiological implications. *Ann Biomed Eng*. 2014;42: 488–502.
4. Bellini C, Bersi MR, Caulk A, et al. Comparison of 10 murine models reveals a distinct biomechanical phenotype in thoracic aortic aneurysms. *J R Soc Interface*. 2017;14:20161036.
5. Brown IAM, Diederich L, Good ME, et al. Vascular smooth muscle remodeling in conductive and resistance arteries in hypertension. *Arterioscler Thromb Vasc Biol*. 2018;38:1969–1985.
6. Chandran KB, Rittgers SE, Yoganathan AP. *Biofluid Mechanics: The Human Circulation*. CRC Press; 2012.
7. Chien S. Mechanotransduction and endothelial cell homeostasis: the wisdom of the cell. *Am J Physiol*. 2007;292:H1209–H1224.
8. Chovatiya R, Medzhitov R. Stress, inflammation, and defense of homeostasis. *Mol Cell*. 2014;54:281–288.
9. Dajnowiec D, Langille BL. Arterial adaptations to chronic changes in haemodynamic function: coupling vasomotor tone to structural remodelling. *Clin Sci (Lond)*. 2007;113:15–23.
10. Davies KJ. Adaptive homeostasis. *Mol Aspects Med*. 2016;49: 1–7.
11. Eberth JF, Gresham VC, Reddy AK, Popovic N, Wilson E, Humphrey JD. Importance of pulsatility in hypertensive carotid artery growth and remodeling. *J Hyperten*. 2009;27:2010–2021.
12. Galis Z, Khatri JJ. Matrix metalloproteinases in vascular remodeling and atherogenesis: the good, bad, and ugly. *Circ Res*. 2002;90: 251–262.
13. Humphrey JD. *Cardiovascular Solid Mechanics: Cells, Tissues, and Organs*. Springer; 2002.
14. Humphrey JD. Vascular adaptation and mechanical homeostasis at tissue, cellular, and sub-cellular levels. *Cell Biochem Biophys*. 2008; 50:53–78.
15. Humphrey JD. Mechanisms of arterial remodelling in hypertension: coupled roles of wall shear and intramural stress. *Hypertension*. 2008;52:195–200.
16. Humphrey JD, Eberth JF, Dye WW, Gleason RL. Fundamental role of axial stress in compensatory adaptations by arteries. *J Biomech*. 2009;42:1–8.
17. Humphrey JD, Dufrense E, Schwartz MA. Mechanotransduction and extracellular matrix homeostasis. *Nat Rev Mol Cell Biol*. 2014; 15:802–812.
18. Humphrey JD, Tellides G, Schwartz MA, Milewicz DM. Role of mechanotransduction in vascular biology: focus on thoracic aortic aneurysms and dissections. *Circ Res*. 2015;116:1448–1461.
19. Humphrey JD, Harrison DG, Figueroa CA, Lacolley P, Laurent S. Central artery stiffness in hypertension and aging: a problem with cause and consequence. *Circ Res*. 2016;118:379–381.
20. Kotas ME, Medzhitov R. Homeostasis, inflammation, and disease susceptibility. *Cell*. 2015;160:816–827.
21. Lacolley P, Regnault V, Segers P, Laurent S. Vascular smooth muscle cells and arterial stiffening: relevance in development, aging, and disease. *Physiol Rev*. 2017;97:1555–1617.
22. Latorre M, Humphrey JD. Modeling mechano-driven and inflammation-mediated aortic remodeling in hypertension. *Biomech Model Mechanobiol*. 2018;17:1497–1511.
23. Laurent S, Boutouyrie P. The structural factor of hypertension: large and small artery alterations. *Circ Res*. 2015;116:1007–1021.
24. Libby P. Inflammation in atherosclerosis. *Arterioscler Thromb Vasc Biol*. 2012;32:2045–2051.
25. Louis H, Kakou A, Regnault V, et al. Role of alpha1beta1-integrin in arterial stiffness and angiotensin-induced arterial wall hypertrophy in mice. *Am J Physiol Heart Circ Physiol*. 2007;293:H2597–H2604.
26. Martinez-Lemus LA, Hill MA, Meininger GA. The plastic nature of the vascular wall: a continuum of remodeling events contributing to control of arteriolar diameter and structure. *Physiology (Bethesda)*. 2009;24:45–57.

27. Misra A, Sheikh AQ, Kumar A, et al. Integrin beta3 inhibition is a therapeutic strategy for supravalvular aortic stenosis. *J Exp Med.* 2016;213:451–463.
28. Moiseeva EP. Adhesion receptors of vascular smooth muscle cells and their functions. *Cardiovasc Res.* 2001;52:372–386.
29. Roccabianca S, Bellini C, Humphrey JD. Computational modeling suggests good, bad, and ugly roles of glycosaminoglycans in arterial mechanics and mechanobiology. *J Roy Soc Interface.* 2014; 11:20140397.
30. Shadwick RE. Mechanical design in arteries. *J Exp Biol.* 1999;202: 3305–3313.
31. Valentin A, Cardamone L, Baek S, Humphrey JD. Complementary vasoactivity and matrix remodeling in arterial adaptations to altered flow and pressure. *J Roy Soc Interface.* 2009;6:293–306.
32. Wagenseil JE, Mecham RP. Vascular extracellular matrix and arterial mechanics. *Physiol Rev.* 2009;89:957–989.

CHAPTER

24

Extracellular matrix dynamics and contribution to vascular pathologies

Amanda L. Mohabeer[1,2] *and Michelle P. Bendeck*[1,2]

[1]Department of Laboratory Medicine and Pathobiology, University of Toronto, Toronto, ON, Canada [2]Translational Biology and Engineering Program, Ted Rogers Centre for Heart Research, University of Toronto, Toronto, ON, Canada

Atherosclerosis

Atherosclerosis is a leading cause of morbidity and mortality in North America. This chronic vascular disease affects elastic and large muscular arteries, and can cause ischemic heart disease, myocardial infarction, peripheral artery disease, and stroke. The pathology of atherosclerosis involves the accumulation of lipoproteins in the intimal layer and interactions between vessel wall cells and invading inflammatory cells, culminating in the development of fibrofatty plaques in the intima. Long thought to be a stable long-lasting component of arteries, we now recognize that in disease, the extracellular matrix (ECM) microenvironment is dynamic and rapidly changing. The accumulation and modification of the ECM contributes to most of the pathogenic processes in atherosclerosis.

Development of the atherosclerotic plaque

Fatty streaks are found in vessels even in early childhood, and are thought to be the precursors of atherosclerotic plaques. They are composed of extracellular lipids and lipid-laden foam cells which accumulate in the subendothelial layer, and with time these lesions progress to diffuse intimal thickenings containing vascular smooth muscle cells (VSMCs) which have migrated from the medial layer. The classic atherosclerotic lesion is the fibro-fatty lipid plaque. Plaques are covered by a layer of endothelial cells with a basement membrane. The plaque is protected by a thick fibrous cap composed of collagen, elastin, and VSMCs, and contains a central necrotic core with foam cells, proteoglycans, and lipids. As atherosclerosis progresses, matrix metalloproteinases (MMPs) produced by inflammatory cells erode the endothelial surface, and/or disrupt the plaque fibrous cap leading to plaque rupture. Rupture is responsible for acute myocardial infraction and sudden cardiac death. The robust action of MMPs in plaques found in the abdominal aorta is responsible for degradation of the underlying media leading to abdominal aortic aneurysms (AAAs), which will be discussed in detail in Section Aneurysms of this chapter. Another late term complication of atherosclerosis is plaque calcification, where transdifferentiated cells deposit minerals in the plaque, stiffen the vessels, and complicate the use of interventional treatments such as angioplasty and stenting. Vascular calcification will be discussed in more detail in Section Vascular calcification of this chapter.

How the dynamic ECM influences plaque pathology

Proteoglycans bind lipids, signal cell migration, and facilitate collagen fibrillogenesis

Proteoglycans are produced in abundance in atherosclerosis, and since they are negatively charged, they bind avidly to positively charged lipoproteins. Proteoglycans present in the subendothelial intimal layer include biglycan, versican, and perlecan which sequester lipids in this layer during development of the fatty streak.[1] Decorin is associated with collagen in the fibrous cap where it mediates collagen fibrillogenesis and thus confers protection and stability.[2] Hyaluronan is present in both early and advanced lesions and it signals to enhance the proliferation and migration of VSMCs.[3] Antibodies that bind to and block the glycosaminoglycan chains of hyaluronan reduce macrophage

The Vasculome
https://doi.org/10.1016/B978-0-12-822546-2.00012-5

infiltration and plaque growth in mouse models of atherosclerosis.[1]

Fibronectin is increased with shear stress and promotes EC adhesion molecule expression

Fibronectin (FN) in the atherosclerotic plaque is derived from cellular and plasma sources.[4] FN accumulates in the subendothelial matrix of vessels with disturbed blood flow, and triggers NFκB signaling leading to increased expression of adhesion molecules which allow the binding and infiltration of monocytes.[5,6] Studies in mouse models of atherosclerosis show that deletion of cellular FN splice variants results in decreased inflammatory cell accumulation and attenuated plaque growth.[7]

Fibrillar collagens strengthen the fibrous cap, while nonfibrillar collagens and degradation products activate VSMCs and macrophages

VSMCs synthesize collagen types I, III, VIII and XV after the transition from a quiescent to synthetic phenotype during atherogenesis.[9,10] Collagens are bioactive in signaling to VSMCs and inflammatory cells in the plaque, acting via integrins (α1β1, α2β1, α10β1) and discoidin domain receptor tyrosine kinases (DDR1, DDR2).[9] Collagens play both protective and harmful roles in atherosclerosis. Fibrillar collagens I and III strengthen the fibrous cap and protect against plaque rupture. Collagen VIII, expressed in atherosclerosis, promotes VSMC migration and proliferation to form a strong fibrous cap.[10,11] Collagen VIII is also produced by macrophages in response to inflammatory mediators in the plaque.[12] The structure of the matrix also affects cell behavior during disease, for example, binding to intact fibrillar collagen I maintains VSMC quiescence, while degraded or denatured collagen stimulates VSMC migration and proliferation, and can also promote macrophage activation and invasion.[13–15] Collagen XV promotes VSMC migration and induces immune cell activation, adhesion, migration, and cytokine expression in the advanced atherosclerotic plaque.[16] Previous work from our lab has shown that DDR1 promotes VSMC migration after vessel injury, and enhances the accumulation of macrophages in the plaque.[17,18]

Elastin and degradation products influence VSMC phenotype

Elastin is also abundant in the arterial wall, and vascular integrity and function are critically dependent upon its highly ordered structure of circumferential aligned lamellae which provide strength and elasticity. Elastin synthesis is upregulated during the development of atherosclerotic plaques, but assembly is erratic, and the matrix is subject to fragmentation by proteases in the dynamic plaque environment. Moreover, binding of lipids and calcium phosphate crystals exacerbate elastin degradation. Elastin degradation products (EDPs) act via the elastin receptor complex on VSMCs and can stimulate transition to the synthetic phenotype, and also to an osteoblast-like phenotype contributing to vascular calcification.[19] EDPs also stimulate angiogenesis which could play a role in plaque neovascularization and hemorrhage from the fragile vasa vasorum.[20]

Matrix metalloproteinases and plaque rupture

Stable atherosclerotic plaques are characterized by the presence of a thick fibrous cap which protects the lipid-rich necrotic core. The cap is composed of VSMCs and a few macrophages embedded in a dense matrix of collagen and elastin. Plaque rupture or fissure occurs when proteases degrade the ECM, destabilizing the matrix scaffold. MMPs are a family of zinc-dependent enzymes that degrade a full spectrum of ECM proteins. Studies in humans have detected the expression of MMPs by macrophages located in shoulder regions of plaques and proposed that these destabilize the cap and cause plaque rupture.[21,22] Numerous studies have been conducted in mice using genetic deletion or overexpression of various MMPs and their endogenous tissue inhibitors (TIMPs). Early studies showed that MMPs-2, 9, and MT1-MMP were proatherogenic: genetic deletion of these MMPs limited plaque growth and the accumulation of SMCs and macrophages.[23,24] By contrast, MMPs-1 and 3 were antiatherogenic, preventing plaque expansion by limiting the excessive accumulation of collagen.[25,26] However, there has been some contradictory evidence published. Overexpression of exogenous MMP-9 resulted in fibrous cap thinning and rupture.[27,28] However, other studies suggested that if endogenous MMP-9 was deleted, rupture was actually increased, and only deletion or inhibition of MMP-12 resulted in protection from rupture.[29,30] Recently, the expression of MMP-12 by macrophages has been correlated with arterial stiffening, atherosclerosis, and inflammation in mice,[31] and MMP-12 is a candidate locus for atherosclerosis in humans.[32]

The action of MMPs is balanced by the action of the endogenous TIMPs. Overexpression of TIMP-2 but not TIMP-1 protected against plaque rupture,[33] while deletion of TIMP-4 resulted in increased atherosclerosis burden.[34] TIMP-3 is also atheroprotective[35] and ex vivo gene therapy to express TIMP-3 reduced neointima formation in saphenous vein grafts.[36] Mutations in TIMP-3 were associated with high serum levels of TIMP-3 and rupture susceptible carotid plaques in humans.[37] It is important to remember that MMPs play diverse and complex roles in atherosclerosis, not only causing degradation of ECM leading to plaque rupture, but also contributing to vascular pathology by releasing growth factors and cytokines from the ECM, with especially prominent roles in TGFβ, TNFα, and ADAM signaling. Most of this research has been conducted in genetic mouse models and there are different roles for MMPs/TIMPs at different stages of disease. In humans mutations in the genes encoding MMP-3, MMP-12 and TIMP-3 are correlated with atherosclerosis, but very little is known about causation.

Vascular calcification

Vascular calcification occurs as a late term complication in aging, atherosclerosis, diabetes, and chronic kidney disease. It is remarkably prevalent, occurs in the aorta, coronary and peripheral arteries, and there is no medical treatment for the disorder.[38–41] Moreover, the presence of calcification reduces the options for interventional treatments, as it is difficult to perform angioplasty or stenting of calcified vessel segments. Many factors influence calcification including inflammation, oxidative stress, alterations in systemic metabolism (hyperglycemia, hyperlipidemia, hyperphosphatemia), alterations in the ECM, and mechanical stress.[42] It is now recognized that vascular calcification is an active process involving inflammation and transdifferentiation of vessel wall and certain progenitor cells into osteochondrocytic-like cells. Cells release extracellular vesicles containing enzymes which mediate mineralization, or cell apoptotic debris provides a nidus for calcium–phosphate crystallization.[43,44] A central molecular mechanism involves the activation of the master osteochondrocytic transcription factor (TF) Runx2 which causes the transdifferentiation of VSMCs.[45–47] Silencing Runx2 prevents, while over-expression promotes calcification by VSMCs in vitro.[48–51] ApoE-deficient mice with VSMC-specific deletion of Runx2 fail to calcify atherosclerotic plaques.[52] Runx2 regulates the transcription of several osteochondrocytic genes including alkaline phosphatase, osteocalcin, type I collagen, osteopontin, MMP-9, and RANKL.[50] Runx2 activity is controlled at several levels. Expression is induced by BMP-2 and oxidative and mechanical stress.[50,53,54] The signaling kinases Akt and MAPK phosphorylate and activate Runx2, while GS3Kβ inhibits phosphorylation, and changes in phosphorylation impact the association of Runx2 with transcriptional cofactors.[47,50,53]

ECM signals for calcification

Direct signaling from molecules in the ECM can influence cell phenotype and thus affect osteogenesis and calcification. ECM signaling is important because it has the potential to act along two axes, transducing biophysical signals from a progressively stiffening matrix, and initiating biochemical signals as the composition of the matrix changes.

Collagen and fibronectin

Calcification is actively mediated by VSMCs which deposit mineral within the collagen and fibronectin-rich ECM. Fibronectin modulates calcification and blocking antibodies against α5β1 integrin reduced VSMC-mediated calcification in vitro.[55] We have shown that the collagen-binding receptor DDR1 regulates atherosclerotic and diabetic vascular calcification. Deficiency of DDR1 attenuates calcification in the atherosclerotic plaques and media of *Ldlr*$^{-/-}$ mice.[56,57] Growing VSMCs in culture we found that DDR1 acts via a PI3K/AKT pathway to activate Runx2 causing transdifferentiation to an osteochondrocytic phenotype.[57]

Elastin degradation products

With aging, diabetes, and atherosclerosis, elastic fiber integrity is compromised by proteolytic degradation by MMPs, neutrophil elastases, and cathepsins. Elastin breaks have been correlated with regions of calcification, and EDPs cause calcification.[58–60] Doxycycline inhibits MMP-mediated elastin degradation leading to the attenuation of arterial calcification.[61,62]

AGE/RAGE, MGP, and osteopontin

In diabetes, high plasma and interstitial glucose levels cause the formation of advanced glycation end-products (AGEs). Glycation commonly occurs on collagens because of their long half-life in the tissue,[63] and makes the collagen fibrils stiffer, altering matrix mechanics and cell–matrix interactions. Glycated molecules also bind as ligands to receptors for advanced glycation end-products on VSMCs where they stimulate osteochondrocytic transdifferentiation and vascular calcification.[64,65] Matrix Gla protein (MGP) is a secreted ECM protein which was implicated as an inhibitor of vascular calcification when studies revealed that gene knockout in mice leads to spontaneous aortic calcification and death.[66] MGP inhibits BMP signaling and can inhibit hydroxyapatite mineral formation.[67,68] Osteopontin is a matrix protein synthesized by VSMCs in phenotypic transition; it is upregulated in calcified vessels and serves as an inhibitor of calcification by interacting with hydroxyapatite to inhibit mineralization.[69] Knockout of the osteopontin gene leads to increased calcification in spontaneously calcifying MGP-deficient mice, and mice fed a high phosphate diet.[70,71]

Matrix stiffness sensing by the cytoskeleton and vascular calcification

During atherosclerosis, there are pronounced changes in the mechanics of the cellular microenvironment as fibrillar collagen and fibronectin accumulates and hydroxyapatite crystals are deposited in the ECM. Atomic force microscopy and micropipette aspiration were used to measure elastic moduli in the vessel walls of mice, rabbits, and humans, and show values ranging from 5 kPa (healthy media) to >300 kPa (mineralized plaque).[72–75] Mechanical signals from a stiffening ECM promote the osteogenic differentiation of mesenchymal precursor cells,[76] and cardiac valvular interstitial cells also respond to increases in stiffness by transdifferentiating to cause fibrosis and calcify the matrix.[77–79] Integrins serve as mechanoreceptors which signal by activating Rho GTPases and enabling actomyosin-based cell contraction to resist the stiffening ECM. DDR1 is another mechanosensor for the collagen matrix.[80–83] Using a mechanotunable cell culture system, we have recently shown that a DDR1–RhoA axis senses matrix stiffness to regulate Runx2 activation and calcification in VSMCs.[84]

Several transcription factors (TFs) which are responsive to changes in matrix stiffness are involved in VSMC phenotypic transition and calcification. These include Yes-associated protein (YAP) and its homologue Taffazin (TAZ), and myocardin-related transcription factor-A (MRTFA). All three TFs are released from binding complexes with G-actin and translocate to the nucleus upon assembly of actin stress fibers. YAP blocks serum response factor (SRF) transcriptional activity to mediate the phenotypic transition of VSMCs from the contractile to the synthetic phenotype which precedes vascular calcification.[85,86] MRTFA is a transcriptional coactivator for SRF which maintains the VSMC contractile phenotype; however, under calcifying conditions, Runx2 interferes with MRTFA to downregulate VSMC-specific genes and upregulate osteochondrocytic genes.[87] Thus, there is a feedforward loop activated with tissue fibrosis and stiffening which potentiates the development of vascular calcification.

Arterial stiffening

Defining arterial stiffening: causes and consequences

Arterial stiffening is a pathological condition which severely hinders the ability of central and conduit arteries to maintain their compliance and cushioning effects on the pressure waves generated during the cardiac cycle. In 2010, the Framingham Heart Study reported two fundamental conclusions: (1) arterial stiffening is a cholesterol-independent risk factor for cardiovascular disease and (2) increased aortic pulse wave velocity (PWV), which is indicative of stiffening, was associated with a nearly 50% increase in cardiovascular disease risk.[88] Deleterious clinical outcomes of arterial stiffness include recurrent stroke and dementia,[8,89] and failure of the kidneys[90] and heart.[91,92]

Aging is a major cause of arterial stiffening. Over the lifespan of a human, elastic arteries are subjected to over three billion cycles of expansion and recoil.[93] This repetitive loading and unloading generates fracture fatigue of the elastic fibers resulting in their gradual loss, fragmentation, and damage. With improper elastic fiber function, the mechanical load (pressurized blood) is now transferred onto collagen proteins, which normally only engage at the tail end of physiological pressures.[94] Collagen fibers are 100–1000× stiffer than elastic ones,[93] and confer tensile strength to the vessel wall and resistance to overexpansion.

Extracellular matrix remodeling in arterial stiffening, the old favorite

Historically, this field has emphasized ECM remodeling as the principle determinant of stiffening. Elastic fibers are degraded by elastases (serine proteases, cysteine proteases, and MMPs) leading to a reduction in the elastin to collagen ratio of the vessel.[93,95] Elastases generate EDPs which are biologically active signaling molecules that can negatively impact the pathogenesis of stiffening by worsening the severity of comorbidities including atherosclerosis, aneurysmal disease, and diabetes, triggering a feedforward loop.[93] Moreover, frayed elastic lamellae have enhanced susceptibility to vascular calcification in the media of large arteries which contributes to vascular stiffening.[96–98] Elastic and collagen fibers in the vessel wall can also undergo glycation which results in a stiffer protein.[99] Furthermore, AGEs form intra- and extracellular cross-links with both collagen and elastin proteins, reducing their functionality and increasing the susceptibility to hydrolytic degradation.[100–103]

Collagen VIII in arterial stiffening: an area for future investigation

Collagen VIII is a nonfibrillar matrix protein which is upregulated during vessel injury and atherosclerosis, functioning as a provisional matrix protein to promote VSMC migration, outward remodeling, and formation of a thick fibrous cap.[10] Collagen VIII is also upregulated in arterial stiffening; it is prevalent in aged vessels of nonhuman primates and positively correlates with high PWV in human patients.[104,105] Furthermore, the expression of collagen VIII is increased in atherosclerosis and in the vessels of elastin haploinsufficient ($Eln^{+/-}$) mice, two conditions characterized by arterial stiffening.[106,107] Recently, *Col8a1* was found to be the most upregulated gene in $Eln^{-/-}$ mice,[108] which die at birth due to obstructive arterial disease and demonstrate severely stiffened vessels.[109] Collectively, these studies demonstrate an association between collagen VIII and stiffness, but have yet to determine a causative relationship.

Cellular dysfunction in arterial stiffening, the new favorite

The biophysical and biochemical alterations in the vessel wall caused by matrix remodeling can trigger cellular dysfunction. Endothelial cells (ECs) and VSMCs are mechanosensitive and respond to changes in ECM stiffness by upregulating RhoA/ROCK activity which triggers stress fiber formation, cytoskeletal tension, and intracellular stiffness.[110–113] In the EC, uncoupling of nitric oxide synthesis due to quenching by AGEs and extrinsic factors preferentially leads to the production of reactive oxygen species (ROS).[110,114] ROS are detrimental since they can initiate inflammatory and stress-signaling cascades that potentiate matrix degradation, abnormal vascular tone, and VSMC dedifferentiation, augmenting a feedforward loop with arterial stiffness.

In the VSMC, an emerging literature outlines the fundamental role of cellular mechanisms regulating arterial stiffening including mechanotransduction, mitochondrial oxidative stress, metabolic, genetic, and epigenetic factors.[115] For instance, the Hippo pathway transcriptional regulators YAP and TAZ are mechanosensitive proteins which translocate to the nucleus after activation of RhoA/ROCK signaling. In response to ECM stiffness, nuclear YAP/TAZ turns on focal adhesion genes leading to the stabilization of the cytoskeleton at the cell membrane, promoting intracellular tension and stiffness.[116] With respect to genome mutations, mutant E3-ubiqutin ligase cullin-3 has reduced inhibition of RhoA, leading to hypertension and increased PWV.[117] Persistent RhoA activity can also enhance arterial stiffening by augmenting downstream signaling cascades including stress fiber formation and cellular stiffening. Lastly, the field of epigenetics in arterial stiffening is rapidly growing, especially due to its predominant role in aging.[118] Of importance, Elia et al.[119] report that ubiquitin-like with PHD and RING finger domains 1 (UHRF1) is a master epigenetic regulator of VSMC phenotype and plasticity. UHRF1 represses the expression of VSMC differentiation genes via DNA and histone methylation, leading to cells adapting a matrix synthesizing phenotype, which ultimately increases ECM stiffness. This protein may therefore be an interesting new candidate for therapeutic intervention for vascular pathologies.

Comorbidities associated with arterial stiffening: atherosclerosis and diabetes

Atherosclerosis and arterial stiffening have overlapping pathogenesis, but whether this is a correlative or causative relationship remains unclear.[120] Selwaness et al.[121] reported that increased aortic stiffness was associated with greater likelihood of developing carotid artery atherosclerosis, suggesting that cause–effect outcomes may be regionally localized. Several studies have identified the importance of EDPs, a by-product of vessel stiffening, in exacerbating atherosclerosis. EDPs were found to increase plaque size in atherosclerotic mice[122] and promote VSMC dedifferentiation into synthetic[123,124] and osteoblast-like phenotypes[59,125] in vitro. Moreover, when fibrillin-1 heterozygous mutant (*C1039G+/−*) mice which are characterized by elastic fiber fragmentation were crossed with *ApoE−/−* mice, it resulted in arterial stiffening, plaque rupture, and lesions which recapitulated human atherogenesis,[126,127] suggesting a causative role for EDPs in disease progression.

Arterial stiffness is a major complication of diabetes mellitus (DM), and an independent predictor of cardiovascular mortality in diabetic patients.[90,128] Both chronic hyperglycemia and hyperinsulinemia promote renin–angiotensin–aldosterone system activation to increase vascular tone and blood pressure, potentiating stiffness via cell signaling cascades.[129] Furthermore, the generation of glycated proteins and AGEs is increased in diabetes due to improper glucose handling, which contribute to stiffening of the ECM.[130] Diabetes-associated arterial stiffening also leads to increased protease-mediated ECM degradation. Serum levels of the elastase MMP-12 were elevated in diabetic patients and positively correlated with stiffness, suggesting it as a potential biomarker for diabetic arterial stiffening.[131]

Sex differences in arterial stiffening: more studies needed

Several human studies have revealed sex-specific differences in vascular stiffness between males and females relating to premenopausal protection, rate of deterioration, and compounding cardiovascular pathologies.[132] However, despite the premenopausal protective nature conferred to females by estrogen, research has indicated that arterial stiffening is greater in postmenopausal women compared to age-matched men, and is associated with the development of metabolic syndrome and cardiovascular comorbidities.[133,134] Hawes et al.[135] determined that adult female mice have lower amounts of collagen and elastin in their arteries when normalized to total vessel protein compared to males, indicating predisposed differences in wall composition. This unequivocally influences cardiovascular remodeling and disease progression. In nonhuman primates, aortic stiffening was greater in male than female monkeys, but the same study reported no sex-specific differences in arterial stiffening in rats, which suggest differences between species.[136] Moreover, estrogen was determined to be protective against arterial stiffening by inhibiting oxLDL-induced MMP-12 activity in both mouse and human macrophages.[31] Therapeutics targeting MMP-12 could therefore be beneficial in limiting disease in both men and postmenopausal women; however, studies researching arterial stiffening in both male and female specimens are urgently needed in cardiovascular research.

Aneurysms

Defining aneurysms

Aneurysms represent abnormal bulging or dilatation of the arterial wall. A true aneurysm is expansion across all three layers of the vessel (intima, media, and adventitia), whereas a false or pseudoaneurysm results from vessel wall injury wherein a local hematoma forms in the adventitial wall.[137] Hallmark features of vessel wall weakening and subsequent aneurysmal growth include apoptosis of medial VSMCs, ECM degradation, and inflammation.[138] Aneurysm dissections result from an acute tear in the intima which enables blood to penetrate the vessel wall, causing layer separation and rupture.[139] Detection and diagnosis of aneurysms is challenging because growth is largely asymptomatic, and there is limited availability of screening tools to determine true prevalence.[140]

Aneurysms within the vascular tree

Aneurysms occur in the aortic, peripheral, and cerebral arteries. Cerebral or intracranial aneurysms predominantly occur in the anterior circulation of the circle of Willis,[141] and have low risk of rupture.[142] However, when they do rupture, subarachnoid hemorrhage significantly increases mortality. Peripheral aneurysms occurring in the popliteal, femoral, or pulmonary arteries, and rupture less frequently.[143,144] Finally, aortic aneurysms (AAs), which include both the thoracic and abdominal regions, are frequent, with the highest prevalence at the infrarenal aorta.[137] The only available treatment for AAs is surgery: open or endovascular repair.[145] As such, there is a critical need for the identification of clinical biomarkers for improved detection, diagnosis, and therapeutic development.[137,140]

Thoracic aortic disease

Thoracic aortic disease (TAD) encompasses both thoracic AAs and dissections. Dissections can be classified as Type A or B; Type A dissections appear in the ascending aorta with or without dissection in the descending region and have a 50% sudden death occurrence, whereas Type B exclusively present in the descending thoracic aorta with much lower odds of rupture.[146] Risk factors associated with TAD can include the presence of a bicuspid valve and pregnancy, but the main drivers are hypertension and genetics.[147] At least 20% of TAD cases are familial, with a mutation event occurring in a single gene.[148,149] Collectively, these are known as heritable thoracic aortic disease (HTAD). 11 genes have a definitive relationship with HTAD, though over 20 other genes are suggested to be related.[150] These hallmark genes code for (a) VSMC contractile proteins: alpha (α) smooth muscle actin (*ACTA2*), smooth muscle myosin heavy chain 11 (*MYH11*), myosin light chain kinase (*MYLK*), and protein kinase cGMP-dependent type 1 (*PRKG1*); (b) ECM proteins: fibrillin-1 (*FBN1*), collagen III (*COL3A1*), lysyl oxidase (*LOX*); and (c) regulators of transforming growth factor beta (TGFβ) signaling: TGFβ-receptor type 1 (*TGFβR1*), TGFβ-receptor type 2 (*TGFβR2*), TGFβ2 (*TGFβ2*), and mothers against decapentaplegic drosophila homolog 3 (*SMAD3*). It is noteworthy that individuals with HTAD do not have increased susceptibility of developing abdominal aortic anuerysms (AAAs),[147] emphasizing that distinct mechanisms account for the regional prevalence of aneurysms.

Elastin-contractile unit uncoupling in TAD

The thoracic aorta (both ascending and descending) comprises the major elastic vessels which receive highly pressurized blood from the heart during each cardiac cycle. The expansion and recoil of these elastic vessels regulate blood pressure and blood flow. These unique properties of the aorta are due to their elastin-contractile units: structural connections established between elastic lamellae and VSMCs via focal adhesions.[151] These units allow for the propagation of biomechanical forces across the vessel wall, and produce a coordinated response between VSMC contractility and elastic tension. Consequently, mutations in genes which impair VSMC contraction (*ACTA2*, *MYH11*, *MYLK*,

PRKG1) result in improper mechanotransduction and ultimately TAD.[152,153] Such dysfunctional signaling and secondary responses trigger activation of VSMC repair mechanisms, but lead to apoptosis of VSMCs and matrix degradation—hallmarks of aneurysmal disease.[147] Furthermore, hypertension alone can also trigger dysfunctional signaling and elastin-contractile unit uncoupling due to the increased biomechanical force, even in the absence of any genetic mutations.[147,153]

Marfan syndrome and Loeys–Dietz syndrome

Mutations in fibrillin-1 cause Marfan syndrome (MFS), a connective tissue disorder that results in skeletal, ocular, and cardiovascular defects.[154,155] At least 2000 unique mutations in fibrillin-1 are linked to MFS.[156] Fibrillin-1 is an essential component of the microfibril scaffold of elastic fibers and establishes a signaling connection between VSMCs and elastic lamellae through integrin engagement at focal adhesions.[153] Furthermore, fibrillin-1 sequesters latent TGFβ complexes to regulate its release and signaling. Mutated fibrillin-1 severely weakens the vessel wall by reducing the structural connections between VSMCs and elastic fibers,[157] and by dysregulating TGFβ signaling,[158] thereby increasing the susceptibility to aneurysms. Patients with MFS typically present with aortic root Type A dissections.[154] Similar to MFS, but more aggressive in phenotype, is Loeys–Dietz syndrome (LDS). Individuals with LDS present with skeletal, neurocognitive, and cardiovascular defects,[159] which most commonly result from mutations in TGFβR1, TGFβR2, TGFβ2, and SMAD3.[150] Collectively, these proteins function in canonical TGFβ signaling which regulates VSMC differentiation and maturation during vessel development and matrix remodeling.[160,161] Consequently, mutations in this family of proteins can result in disruption of VSMC quiescence state, weakened contractility, and reduced biomechanical signal transduction.[147,158,162] Individuals with LDS not only present with aneurysms in their aorta, but also in their peripheral and cerebral vessels, making this disease much more severe than MFS.[159] Moreover, there is heightened prevalence of aneurysmal dissection and rupture in LDS patients.

Abdominal aortic aneurysms

AAA is a polygenic, multifactorial disease which differs from TAD. Tobacco use is the leading risk factor of AAAs, but others include hypertension, hyperlipidemia, family history, and male sex.[137,163] A subset of inherited factor gene variants associated with AAA include interleukin 6 receptor, LDL receptor, and MMP-9.[164] The main factors causing aneurysmal disease are ECM degradation and inflammation: two mechanisms which work synergistically in the development and progression of AAAs.[165,166] An acute injury to the vessel wall leads to the formation of an intraluminal thrombus (ILT), or blood clot.[167] This ILT creates an inflammatory environment wherein oxidation and activation of matrix degrading enzymes drive proteolysis of the vessel wall.[137,138] Elastases cause vessel dilatation and aneurysm, whereas collagenases promote dissection and rupture since they directly reduce the tensile strength of the vessel wall.[168] Chronic inflammation produces cytokines and chemokines which trigger phenotypic switching and apoptosis of the VSMCs in the media layer.[166,169,170] Collectively, these processes progressively weaken the vessel wall and ultimately lead to rupture.

Diabetes and aneurysms

Aneurysms occur less frequently in individuals with diabetes mellitus (DM),[171] a surprising finding since most cardiovascular diseases are positively linked to DM. The potential mechanisms underlying the protection are twofold. First, the increased blood glucose present in patients with DM increases glycation of ECM proteins which stiffen the matrix[99] and produce AGEs, which can form cross-links with the ECM,[100] leaving the matrix less susceptible to proteolysis.[101] Second, metformin (a common therapeutic for DM) reduces the progression of AAAs in animal studies[172,173] and in human patients.[173–175] Metformin reduces monocyte infiltration and differentiation to macrophages,[172] thereby limiting inflammation, and also protects against medial elastin degradation.[173] From these promising findings, there are currently randomized controlled clinical trials (*clinicaltrials.gov*: NCT03507413, NCT04224051) ongoing to investigate the use of metformin as a therapeutic for aneurysmal disease.

Limitations and areas needing further study

Most of our knowledge to date comes from studies of mouse models of vascular disease, particularly atherosclerosis. There are important experimental limitations to these studies; the systemic embryonic deletion of genes is confounded by expression over different time courses during the development of disease and potential roles in multiple tissues. This has been a significant limitation in defining the roles of various MMPs and TIMPs, where a particular MMP may play a different role depending upon the time of expression (in early or advanced disease) or in the specific cell type expressing the enzyme. Future experiments using inducible, tissue-specific, or CRISPR knockouts are needed to sort out these problems.

This chapter has focused on the transdifferentiation of VSMCs to osteochondrocytic-like cells which mediate vascular calcification. The identification of active mechanisms of calcification opens the door to the generation of new targeted therapeutics to prevent or reverse the pathology. Recent research has uncovered strong evidence for the role of adventitial progenitor cells and the clonal expansion of VSMCs contributing to atherosclerotic and aneurysmal vascular disease. However, little is known of the interplay of these cells in complex processes like vascular stiffening and calcification, and this will be an important area for future research.

Since the Framingham Heart Study in 2010, studies related to understanding the mechanisms driving arterial stiffening have grown exponentially. However, despite what is known about the association of stiffening with other cardiovascular pathologies, measuring arterial stiffness in the clinic is still not standard practice. With novel findings related to epigenetic regulation of cellular phenotypes, it is exciting to consider their potential for inspiring therapeutics to limit the onset of disease.

There has been significant progress in the study of human aneurysmal disease, defining mutations in several ECM genes that are associated with disease in MFS and LDS, expanding the study of mechanisms behind these diseases and uncovering new roles for inflammatory mechanisms. This has led to important clinical trials investigating AGE and TGFβ inhibitors, and even metformin. Progress in identifying human gene mutations associated with atherosclerosis has been slower, probably because of the multigenic nature of this complex disease which progresses over the human life span. As new studies emerge across the disciplines discussed in this chapter, it is essential that future studies focus on using both male and female subjects. This will allow us to progressively work toward reducing the sex and gender gap which currently exists in cardiovascular research.

References

1. Allahverdian S, Ortega C, Francis GA. Smooth muscle cell-proteoglycan-lipoprotein interactions as drivers of atherosclerosis. *Handb Exp Pharmacol.* 2020. https://doi.org/10.1007/164_2020_364 [Published online first: Epub Date].
2. Fischer JW, Kinsella MG, Clowes MM, Lara S, Clowes AW, Wight TN. Local expression of bovine decorin by cell-mediated gene transfer reduces neointimal formation after balloon injury in rats. *Circ Res.* 2000;86(6):676–683. https://doi.org/10.1161/01.res.86.6.676 [Published online first: Epub Date].
3. Krolikoski M, Monslow J, Pure E. The CD44-HA axis and inflammation in atherosclerosis: a temporal perspective. *Matrix Biol.* 2019;78-79:201–218. https://doi.org/10.1016/j.matbio.2018.05.007 [Published online first: Epub Date].
4. Holm Nielsen S, Jonasson L, Kalogeropoulos K, et al. Exploring the role of extracellular matrix proteins to develop biomarkers of plaque vulnerability and outcome. *J Intern Med.* 2020;287(5):493–513. https://doi.org/10.1111/joim.13034 [Published online first: Epub Date].
5. Orr AW, Sanders JM, Bevard M, Coleman E, Sarembock IJ, Schwartz MA. The subendothelial extracellular matrix modulates NF-kappaB activation by flow: a potential role in atherosclerosis. *J Cell Biol.* 2005;169(1):191–202. https://doi.org/10.1083/jcb.200410073 [Published online first: Epub Date].
6. Al-Yafeai Z, Yurdagul Jr A, Peretik JM, Alfaidi M, Murphy PA, Orr AW. Endothelial FN (fibronectin) deposition by alpha5beta1 integrins drives atherogenic inflammation. *Arterioscler Thromb Vasc Biol.* 2018;38(11):2601–2614. https://doi.org/10.1161/ATVBAHA.118.311705 [Published online first: Epub Date].
7. Doddapattar P, Jain M, Dhanesha N, Lentz SR, Chauhan AK. Fibronectin containing extra domain A induces plaque destabilization in the innominate artery of aged apolipoprotein E-deficient mice. *Arterioscler Thromb Vasc Biol.* 2018;38(3):500–508. https://doi.org/10.1161/ATVBAHA.117.310345 [Published online first: Epub Date].
8. Rabkin SW. Arterial stiffness: detection and consequences in cognitive impairment and dementia of the elderly. *J Alzheimers Dis.* 2012;32(3):541–549. https://doi.org/10.3233/JAD-2012-120757 [Published online first: Epub Date].
9. Adiguzel E, Ahmad PJ, Franco C, Bendeck MP. Collagens in the progression and complications of atherosclerosis. *Vasc Med.* 2009;14(1):73–89.
10. Lopes J, Adiguzel E, Gu S, et al. Type VIII collagen mediates vessel wall remodeling after arterial injury and fibrous cap formation in atherosclerosis. *Am J Pathol.* 2013;182:2241–2253.
11. MacBeath JRE, Kielty CM, Shuttleworth CA. Type VIII collagen is a product of vascular smooth-muscle cells in development and disease. *Biochem J.* 1996;319:993–998.
12. Weitkamp B, Cullen P, Plenz G, Robenek H, Rauterberg J. Human macrophages synthesize type VIII collagen in vitro and in the atherosclerotic plaque. *Faseb J.* 1999;13:1445–1457.
13. Koyama H, Raines EW, Bornfeldt KE, Roberts JM, Ross R. Fibrillar collagen inhibits arterial smooth muscle cell proliferation through regulation of Cdk2 inhibitors. *Cell.* 1996;87:1069–1078.
14. Lepidi S, Kenagy RD, Raines EW, et al. MMP9 production by human monocyte-derived macrophages is decreased on polymerized type I collagen. *J Vasc Surg.* 2001;34(6):1111–1118.
15. Kamohara H, Yamashiro S, Galligan C, Yoshimura T. Discoidin domain receptor 1 isoform-a (DDR1a) promotes migration of leukocytes in three-dimensional collagen lattices. *Faseb J.* 2001;15:2724–2726.
16. Durgin BG, Cherepanova OA, Gomez D, et al. Smooth muscle cell-specific deletion of Col15a1 unexpectedly leads to impaired development of advanced atherosclerotic lesions. *Am J Physiol Heart Circ Physiol.* 2017;312(5):H943–H958. https://doi.org/10.1152/ajpheart.00029.2017 [Published online first: Epub Date].
17. Hou G, Vogel W, Bendeck MP. The discoidin domain-receptor tyrosine kinase DDR1 in arterial wound repair. *J Clin Invest.* 2001;107(6):727–735.
18. Franco C, Hou G, Ahmad PJ, et al. Discoidin domain receptor 1 (ddr1) deletion decreases atherosclerosis by accelerating matrix accumulation and reducing inflammation in low-density lipoprotein receptor-deficient mice. *Circ Res.* 2008;102(10):1202–1211. https://doi.org/10.1161/CIRCRESAHA.107.170662 [Published online first: Epub Date].
19. Wahart A, Hocine T, Albrecht C, et al. Role of elastin peptides and elastin receptor complex in metabolic and cardiovascular diseases. *FEBS J.* 2019;286(15):2980–2993. https://doi.org/10.1111/febs.14836 [Published online first: Epub Date].

20. Robinet A, Fahem A, Cauchard JH, et al. Elastin-derived peptides enhance angiogenesis by promoting endothelial cell migration and tubulogenesis through upregulation of MT1-MMP. *J Cell Sci*. 2005;118(Pt 2):343–356. https://doi.org/10.1242/jcs.01613 [Published online first: Epub Date].
21. Lee RT, Schoen FJ, Loree HM, Lark MW, Libby P. Circumferential stress and matrix metalloproteinase 1 in human coronary atherosclerosis - implications for plaque rupture. *Arterioscler Thromb Vasc Biol*. 1996;16(8):1070–1073.
22. Galis ZS, Sukhova GK, Lark MW, Libby P. Increased expression of matrix metalloproteinases and matrix degrading activity in vulnerable regions of human atherosclerotic plaque. *J Clin Invest*. 1994;94:2493–2503.
23. Kuzuya M, Nakamura K, Sasaki T, Cheng XW, Itohara S, Iguchi A. Effect of MMP-2 deficiency on atherosclerotic lesion formation in ApoE-deficient mice. *Arterioscler Thromb Vasc Biol*. 2006;26(5): 1120–1125.
24. Luttun A, Lutgens E, Manderveld A, et al. Loss of matrix metalloproteinase-9 or matrix metalloproteinase-12 protects apolipoprotein E-deficient mice against atherosclerotic media destruction but differentially affects plaque growth. *Circulation*. 2004;109(11):1408–1414.
25. Lemaitre V, O'Byrne TK, Borczuk AC, Okada Y, Tall AR, D'Armiento J. ApoE knockout mice expressing human matrix metalloproteinase-1 in macrophages have less advanced atherosclerosis. *J Clin Invest*. 2001;107(10):1227–1234.
26. Silence J, Lupu F, Collen D, Lijnen HR. Persistence of atherosclerotic plaque but reduced aneurysm formation in mice with stromelysin-1 (MMP-3) gene inactivation. *Arterioscler Thromb Vasc Biol*. 2001;21(9):1440–1445.
27. Gough PJ, Gomez IG, Wille PT, Raines EW. Macrophage expression of active MMP-9 induces acute plaque disruption in ApoE-deficient mice. *J Clin Invest*. 2006;116(1):59–69.
28. de NR, Verkleij CJ, von der Thusen JH, et al. Lesional overexpression of matrix metalloproteinase-9 promotes intraplaque hemorrhage in advanced lesions but not at earlier stages of atherogenesis. *Arterioscler Thromb Vasc Biol*. 2006;26(2):340–346.
29. Johnson JL, George SJ, Newby AC, Jackson CL. Divergent effects of matrix metalloproteinases 3, 7, 9, and 12 on atherosclerotic plaque stability in mouse brachiocephalic arteries. *Proc Natl Acad Sci U S A*. 2005;102(43):15575–15580.
30. Johnson JL, Devel L, Czarny B, et al. A selective matrix metalloproteinase-12 inhibitor retards atherosclerotic plaque development in apolipoprotein EGÇôKnockout mice. *Arterioscler Thromb Vasc Biol*. 2011;31(3):528–535.
31. Liu SL, Bajpai A, Hawthorne EA, et al. Cardiovascular protection in females linked to estrogen-dependent inhibition of arterial stiffening and macrophage MMP12. *JCI Insight*. 2019;4(1). https://doi.org/10.1172/jci.insight.122742 [Published online first: Epub Date].
32. Mahdessian H, Perisic Matic L, Lengquist M, et al. Integrative studies implicate matrix metalloproteinase-12 as a culprit gene for large-artery atherosclerotic stroke. *J Intern Med*. 2017;282(5): 429–444. https://doi.org/10.1111/joim.12655 [Published online first: Epub Date].
33. Johnson JL, Baker AH, Oka K, et al. Suppression of atherosclerotic plaque progression and instability by tissue inhibitor of metalloproteinase-2: involvement of macrophage migration and apoptosis. *Circulation*. 2006;113(20):2435–2444.
34. Hu M, Jana S, Kilic T, et al. Loss of TIMP4 (tissue inhibitor of metalloproteinase 4) promotes atherosclerotic plaque deposition in the abdominal aorta despite suppressed plasma cholesterol levels. *Arterioscler Thromb Vasc Biol*. 2021. https://doi.org/10.1161/ATVBAHA.120.315522. ATVBAHA120315522, [Published Online First: Epub Date].
35. Stohr R, Cavalera M, Menini S, et al. Loss of TIMP3 exacerbates atherosclerosis in ApoE null mice. *Atherosclerosis*. 2014;235(2): 438–443. https://doi.org/10.1016/j.atherosclerosis.2014.05.946 [Published online first: Epub Date].
36. George SJ, Wan S, Hu J, MacDonald R, Johnson JL, Baker AH. Sustained reduction of vein graft neointima formation by ex vivo TIMP-3 gene therapy. *Circulation*. 2011;124(11 Suppl):S135–S142. https://doi.org/10.1161/CIRCULATIONAHA.110.012732 [Published online first: Epub Date].
37. Zheng Z, He X, Zhu M, et al. Tissue inhibitor of the metalloproteinases-3 gene polymorphisms and carotid plaque susceptibility in the Han Chinese population. *Int J Neurosci*. 2018;128(10):920–927. https://doi.org/10.1080/00207454.2018.1436544 [Published online first: Epub Date].
38. Budoff MJ, Nasir K, Katz R, et al. Thoracic aortic calcification and coronary heart disease events: the multi-ethnic study of atherosclerosis (MESA). *Atherosclerosis*. 2011;215(1):196–202. https://doi.org/10.1016/j.atherosclerosis.2010.11.017 [Published online first: Epub Date].
39. Criqui MH, Denenberg JO, Ix JH, et al. Calcium density of coronary artery plaque and risk of incident cardiovascular events. *J Am Med Assoc*. 2014;311(3):271–278. https://doi.org/10.1001/jama.2013.282535 [Published online first: Epub Date].
40. Santos RD, Rumberger JA, Budoff MJ, et al. Thoracic aorta calcification detected by electron beam tomography predicts all-cause mortality. *Atherosclerosis*. 2010;209(1):131–135. https://doi.org/10.1016/j.atherosclerosis.2009.08.025 [Published online first: Epub Date].
41. Huang CL, Wu IH, Wu YW, et al. Association of lower extremity arterial calcification with amputation and mortality in patients with symptomatic peripheral artery disease. *PLoS One*. 2014; 9(2):e90201. https://doi.org/10.1371/journal.pone.0090201 [Published online first: Epub Date].
42. Ngai D, Lino M, Bendeck MP. Cell-matrix interactions and matricrine signaling in the pathogenesis of vascular calcification. *Front Cardiovasc Med*. 2018;5:174. https://doi.org/10.3389/fcvm.2018.00174 [Published online first: Epub Date].
43. Hutcheson JD, Goettsch C, Bertazzo S, et al. Genesis and growth of extracellular-vesicle-derived microcalcification in atherosclerotic plaques. *Nat Mater*. 2016;15(3):335–343. https://doi.org/10.1038/nmat4519 [Published online first: Epub Date].
44. Kapustin AN, Shanahan CM. Emerging roles for vascular smooth muscle cell exosomes in calcification and coagulation. *J Physiol*. 2016;594(11):2905–2914. https://doi.org/10.1113/JP271340 [Published online first: Epub Date].
45. Sallam T, Cheng H, Demer LL, Tintut Y. Regulatory circuits controlling vascular cell calcification. *Cell Mol Life Sci*. 2013;70(17): 3187–3197. https://doi.org/10.1007/s00018-012-1231-y [Published online first: Epub Date].
46. Speer MY, Li X, Hiremath PG, Giachelli CM. Runx2/Cbfa1, but not loss of myocardin, is required for smooth muscle cell lineage reprogramming toward osteochondrogenesis. *J Cell Biochem*. 2010; 110(4):935–947. https://doi.org/10.1002/jcb.22607 [Published online first: Epub Date].
47. Speer MY, Yang HY, Brabb T, et al. Smooth muscle cells give rise to osteochondrogenic precursors and chondrocytes in calcifying arteries. *Circ Res*. 2009;104:733–741.
48. Naik V, Leaf EM, Hu JH, et al. Sources of cells that contribute to atherosclerotic intimal calcification: an in vivo genetic fate mapping study. *Cardiovasc Res*. 2012;94:545–554. https://doi.org/10.1093/cvr/cvs126 [Published online first: Epub Date].
49. Byon CH, Sun Y, Chen J, et al. Runx2-upregulated receptor activator of nuclear factor {kappa}B ligand in calcifying smooth muscle cells promotes migration and osteoclastic differentiation of macrophages. *Arterioscler Thromb Vasc Biol*. 2011;31(6):1387–1396.

50. Byon CH, Javed A, Dai Q, et al. Oxidative stress induces vascular calcification through modulation of the osteogenic transcription factor Runx2 by AKT signaling. *J Biol Chem*. 2008;283(22): 15319–15327. https://doi.org/10.1074/jbc.M800021200 [Published online first: Epub Date].
51. Chen NX, Duan D, O'Neill KD, et al. The mechanisms of uremic serum-induced expression of bone matrix proteins in bovine vascular smooth muscle cells. *Kidney Int*. 2006;70(6):1046–1053. https://doi.org/10.1038/sj.ki.5001663 [Published online first: Epub Date].
52. Sun Y, Byon CH, Yuan K, et al. Smooth muscle cell-specific runx2 deficiency inhibits vascular calcification. *Circ Res*. 2012;111(5): 543–552. https://doi.org/10.1161/CIRCRESAHA.112.267237 [Published online first: Epub Date].
53. Simmons CA, Nikolovski J, Thornton AJ, Matlis S, Mooney DJ. Mechanical stimulation and mitogen-activated protein kinase signaling independently regulate osteogenic differentiation and mineralization by calcifying vascular cells. *J Biomech*. 2004; 37(10):1531–1541.
54. Zebboudj AF, Imura M, Bostrom K. Matrix GLA protein, a regulatory protein for bone morphogenetic protein-2. *J Biol Chem*. 2002; 277(6):4388–4394.
55. Watson KE, Parhami F, Shin V, Demer LL. Fibronectin and collagen I matrixes promote calcification of vascular cells in vitro, whereas collagen IV matrix is inhibitory. *Arterioscler Thromb Vasc Biol*. 1998;18:1964–1971.
56. Ahmad PJ, Trcka D, Xue S, et al. Discoidin domain receptor-1 deficiency attenuates atherosclerotic calcification and smooth muscle cell-mediated mineralization. *Am J Pathol*. 2009;175(6): 2686–2696.
57. Lino M, Wan MH, Rocca AS, et al. Diabetic vascular calcification mediated by the collagen receptor discoidin domain receptor 1 via the phosphoinositide 3-kinase/Akt/runt-related transcription factor 2 signaling Axis. *Arterioscler Thromb Vasc Biol*. 2018;38: 1878–1889. https://doi.org/10.1161/ATVBAHA.118.311238 [Published online first: Epub Date].
58. Wanga S, Hibender S, Ridwan Y, et al. Aortic microcalcification is associated with elastin fragmentation in Marfan syndrome. *J Pathol*. 2017;243(3):294–306. https://doi.org/10.1002/path.4949 [Published online first: Epub Date].
59. Simionescu A, Philips K, Vyavahare N. Elastin-derived peptides and TGF-beta1 induce osteogenic responses in smooth muscle cells. *Biochem Biophys Res Commun*. 2005;334(2):524–532.
60. Pai A, Leaf EM, El-Abbadi M, Giachelli CM. Elastin degradation and vascular smooth muscle cell phenotype change precede cell loss and arterial medial calcification in a uremic mouse model of chronic kidney disease. *Am J Pathol*. 2011;178(2):764–773. https://doi.org/10.1016/j.ajpath.2010.10.006 [Published online first: Epub Date].
61. Qin X, Corriere MA, Matrisian LM, Guzman RJ. Matrix metalloproteinase inhibition attenuates aortic calcification. *Arterioscler Thromb Vasc Biol*. 2006;26(7):1510–1516.
62. Bouvet C, Moreau S, Blanchette J, de BD, Moreau P. Sequential activation of matrix metalloproteinase 9 and transforming growth factor beta in arterial elastocalcinosis. *Arterioscler Thromb Vasc Biol*. 2008;28(5):856–862.
63. Kasper M, Funk RH. Age-related changes in cells and tissues due to advanced glycation end products (AGEs). *Arch Gerontol Geriatr*. 2001;32(3):233–243. https://doi.org/10.1016/s0167-4943(01) 00103-0 [Published online first: Epub Date].
64. Hofmann Bowman MA, Gawdzik J, Bukhari U, et al. S100A12 in vascular smooth muscle accelerates vascular calcification in apolipoprotein E-null mice by activating an osteogenic gene regulatory program. *Arterioscler Thromb Vasc Biol*. 2011;31(2):337–344. https://doi.org/10.1161/ATVBAHA.110.217745 [Published online first: Epub Date].
65. Ren X, Shao H, Wei Q, Sun Z, Liu N. Advanced glycation end-products enhance calcification in vascular smooth muscle cells. *J Int Med Res*. 2009;37(3):847–854. https://doi.org/10.1177/147323000903700329 [Published online first: Epub Date].
66. Luo G, Ducy P, McKee MD, et al. Spontaneous calcification of arteries and cartilage in mice lacking matrix GLA protein. *Nature*. 1997;386(6620):78–81.
67. Malhotra R, Burke MF, Martyn T, et al. Inhibition of bone morphogenetic protein signal transduction prevents the medial vascular calcification associated with matrix Gla protein deficiency. *PLoS One*. 2015;10(1):e0117098. https://doi.org/10.1371/journal.-pone.0117098 [Published online first: Epub Date].
68. Xue W, Wallin R, Olmsted-Davis EA, Borras T. Matrix GLA protein function in human trabecular meshwork cells: inhibition of BMP2-induced calcification process. *Invest Ophthalmol Vis Sci*. 2006;47(3):997–1007. https://doi.org/10.1167/iovs.05-1106 [Published online first: Epub Date].
69. Ahmed S, O'Neill KD, Hood AF, Evan AP, Moe SM. Calciphylaxis is associated with hyperphosphatemia and increased osteopontin expression by vascular smooth muscle cells. *Am J Kidney Dis*. 2001; 37(6):1267–1276.
70. Speer MY, McKee MD, Guldberg RE, et al. Inactivation of the osteopontin gene enhances vascular calcification of matrix Gla protein-deficient mice: evidence for osteopontin as an inducible inhibitor of vascular calcification in vivo. *J Exp Med*. 2002;196(8): 1047–1055.
71. Paloian NJ, Leaf EM, Giachelli CM. Osteopontin protects against high phosphate-induced nephrocalcinosis and vascular calcification. *Kidney Int*. 2016;89(5):1027–1036. https://doi.org/10.1016/j.kint.2015.12.046 [Published online first: Epub Date].
72. Matsumoto T, Abe H, Ohashi T, Kato Y, Sato M. Local elastic modulus of atherosclerotic lesions of rabbit thoracic aortas measured by pipette aspiration method. *Physiol Meas*. 2002; 23(4):635–648.
73. Tracqui P, Broisat A, Toczek J, Mesnier N, Ohayon J, Riou L. Mapping elasticity moduli of atherosclerotic plaque in situ via atomic force microscopy. *J Struct Biol*. 2011;174(1):115–123.
74. Chai CK, Akyildiz AC, Speelman L, et al. Local axial compressive mechanical properties of human carotid atherosclerotic plaques-characterisation by indentation test and inverse finite element analysis. *J Biomech*. 2013;46(10):1759–1766. https://doi.org/10.1016/j.jbiomech.2013.03.017 [Published online first: Epub Date].
75. Akyildiz AC, Speelman L, van Velzen B, et al. Intima heterogeneity in stress assessment of atherosclerotic plaques. *Interface Focus*. 2018;8(1):20170008. https://doi.org/10.1098/rsfs.2017.0008 [Published online first: Epub Date].
76. Engler AJ, Sen S, Sweeney HL, Discher DE. Matrix elasticity directs stem cell lineage specification. *Cell*. 2006;126(4):677–689.
77. Yip CY, Chen JH, Zhao R, Simmons CA. Calcification by valve interstitial cells is regulated by the stiffness of the extracellular matrix. *Arterioscler Thromb Vasc Biol*. 2009;29(6):936–942. https://doi.org/10.1161/ATVBAHA.108.182394 [Published online first: Epub Date].
78. Wyss K, Yip CYY, Mirzaei Z, Jin X, Chen JH, Simmons CA. The elastic properties of valve interstitial cells undergoing pathological differentiation. *J Biomech*. 2012;45(5):882–887.
79. Chen JH, Chen WL, Sider KL, Yip CY, Simmons CA. beta-catenin mediates mechanically regulated, transforming growth factor-beta1-induced myofibroblast differentiation of aortic valve interstitial cells. *Arterioscler Thromb Vasc Biol*. 2011;31(3):590–597. https://doi.org/10.1161/ATVBAHA.110.220061 [Published online first: Epub Date].
80. Ghosh S, Ashcraft K, Jahid MJ, et al. Regulation of adipose oestrogen output by mechanical stress. *Nat Commun*. 2013;4:1821.
81. Sun X, Gupta K, Wu B, et al. Tumor-extrinsic discoidin domain receptor 1 promotes mammary tumor growth by regulating adipose

stromal interleukin 6 production in mice. *J Biol Chem*. 2018;293(8): 2841–2849. https://doi.org/10.1074/jbc.RA117.000672 [Published online first: Epub Date].
82. Coelho NM, Arora PD, van Putten S, et al. Discoidin domain receptor 1 mediates myosin-dependent collagen contraction. *Cell Rep*. 2017;18(7):1774–1790. https://doi.org/10.1016/j.celrep.2017.01.061 [Published online first: Epub Date].
83. Staudinger LA, Spano SJ, Lee W, et al. Interactions between the discoidin domain receptor 1 and b1 integrin regulate attachment to collagen. *Biology Open*. 2013;2:1148–1159.
84. Ngai D, Lino M, Rothenberg KE, Simmons CA, Fernandez-Gonzalez R, Bendeck MP. DDR1 (discoidin domain receptor-1)-RhoA Axis senses matrix stiffness to promote vascular calcification. *Arterioscler Thromb Vasc Biol*. 2020. https://doi.org/10.1161/ATVBAHA.120.314697. ATVBAHA120314697, [Published online first: Epub Date].
85. Xie C, Guo Y, Zhu T, Zhang J, Ma PX, Chen YE. Yap1 protein regulates vascular smooth muscle cell phenotypic switch by interaction with myocardin. *J Biol Chem*. 2012;287(18):14598–14605. https://doi.org/10.1074/jbc.M111.329268 [Published online first: Epub Date].
86. Feng X, Liu P, Zhou X, et al. Thromboxane A2 activates YAP/TAZ protein to induce vascular smooth muscle cell proliferation and migration. *J Biol Chem*. 2016;291(36):18947–18958. https://doi.org/10.1074/jbc.M116.739722 [Published online first: Epub Date].
87. Tanaka T, Sato H, Doi H, et al. Runx2 represses myocardin-mediated differentiation and facilitates osteogenic conversion of vascular smooth muscle cells. *Mol Cell Biol*. 2008;28(3): 1147–1160. https://doi.org/10.1128/MCB.01771-07 [Published online first: Epub Date].
88. Mitchell GF, Hwang SJ, Vasan RS, et al. Arterial stiffness and cardiovascular events: the Framingham heart study. *Circulation*. 2010; 121(4):505–511. https://doi.org/10.1161/CIRCULATIONAHA.109.886655 [Published online first: Epub Date].
89. Ding J, Mitchell GF, Bots ML, et al. Carotid arterial stiffness and risk of incident cerebral microbleeds in older people: the Age, Gene/Environment Susceptibility (AGES)-Reykjavik study. *Arterioscler Thromb Vasc Biol*. 2015;35(8):1889–1895. https://doi.org/10.1161/ATVBAHA.115.305451 [Published online first: Epub Date].
90. McEniery CM, Wilkinson IB, Avolio AP. Age, hypertension and arterial function. *Clin Exp Pharmacol Physiol*. 2007;34(7):665–671. https://doi.org/10.1111/j.1440-1681.2007.04657.x [Published online first: Epub Date].
91. Safar ME, Levy BI, Struijker-Boudier H. Current perspectives on arterial stiffness and pulse pressure in hypertension and cardiovascular diseases. *Circulation*. 2003;107(22):2864–2869. https://doi.org/10.1161/01.CIR.0000069826.36125.B4 [Published online first: Epub Date].
92. Sethi S, Rivera O, Oliveros R, Chilton R. Aortic stiffness: pathophysiology, clinical implications, and approach to treatment. *Integr Blood Press Control*. 2014;7:29–34. https://doi.org/10.2147/IBPC.S59535 [Published online first: Epub Date].
93. Duca L, Blaise S, Romier B, et al. Matrix ageing and vascular impacts: focus on elastin fragmentation. *Cardiovasc Res*. 2016;110(3): 298–308. https://doi.org/10.1093/cvr/cvw061 [Published online first: Epub Date].
94. Wagenseil JE, Mecham RP. Elastin in large artery stiffness and hypertension. *J Cardiovasc Transl Res*. 2012;5(3):264–273. https://doi.org/10.1007/s12265-012-9349-8 [Published online first: Epub Date].
95. Cocciolone AJ, Hawes JZ, Staiculescu MC, Johnson EO, Murshed M, Wagenseil JE. Elastin, arterial mechanics, and cardiovascular disease. *Am J Physiol Heart Circ Physiol*. 2018;315(2): H189–H205. https://doi.org/10.1152/ajpheart.00087.2018 [Published online first: Epub Date].
96. Cattell MA, Anderson JC, Hasleton PS. Age-related changes in amounts and concentrations of collagen and elastin in normotensive human thoracic aorta. *Clin Chim Acta*. 1996;245(1):73–84. https://doi.org/10.1016/0009-8981(95)06174-6 [Published online first: Epub Date].
97. Spina M, Garbin G. Age-related chemical changes in human elastins from non-atherosclerotic areas of thoracic aorta. *Atherosclerosis*. 1976;24(1-2):267–279. https://doi.org/10.1016/0021-9150(76)90082-4 [Published online first: Epub Date].
98. Watanabe M, Sawai T, Nagura H, Suyama K. Age-related alteration of cross-linking amino acids of elastin in human aorta. *Tohoku J Exp Med*. 1996;180(2):115–130. https://doi.org/10.1620/tjem.180.115 [Published online first: Epub Date].
99. Lee AT, Cerami A. Role of glycation in aging. *Ann N Y Acad Sci*. 1992;663:63–70. https://doi.org/10.1111/j.1749-6632.1992.tb38649.x [Published online first: Epub Date].
100. Singh VP, Bali A, Singh N, Jaggi AS. Advanced glycation end products and diabetic complications. *Korean J Physiol Pharmacol*. 2014;18(1):1–14. https://doi.org/10.4196/kjpp.2014.18.1.1 [Published online first: Epub Date].
101. Verzijl N, DeGroot J, Thorpe SR, et al. Effect of collagen turnover on the accumulation of advanced glycation end products. *J Biol Chem*. 2000;275(50):39027–39031. https://doi.org/10.1074/jbc.M006700200 [Published online first: Epub Date].
102. Winlove CP, Parker KH, Avery NC, Bailey AJ. Interactions of elastin and aorta with sugars in vitro and their effects on biochemical and physical properties. *Diabetologia*. 1996;39(10):1131–1139. https://doi.org/10.1007/BF02658498 [Published online first: Epub Date].
103. Konova E, Baydanoff S, Atanasova M, Velkova A. Age-related changes in the glycation of human aortic elastin. *Exp Gerontol*. 2004;39(2):249–254. https://doi.org/10.1016/j.exger.2003.10.003 [Published online first: Epub Date].
104. Qiu H, Depre C, Ghosh K, et al. Mechanism of gender-specific differences in aortic stiffness with aging in nonhuman primates. *Circulation*. 2007;116(6):669–676. https://doi.org/10.1161/CIRCULATIONAHA.107.689208 [Published online first: Epub Date].
105. Lyck Hansen M, Beck HC, Irmukhamedov A, Jensen PS, Olsen MH, Rasmussen LM. Proteome analysis of human arterial tissue discloses associations between the vascular content of small leucine-rich repeat proteoglycans and pulse wave velocity. *Arterioscler Thromb Vasc Biol*. 2015;35(8):1896–1903. https://doi.org/10.1161/ATVBAHA.114.304706 [Published online first: Epub Date].
106. Lopes J, Adiguzel E, Gu S, et al. Type VIII collagen mediates vessel wall remodeling after arterial injury and fibrous cap formation in atherosclerosis. *Am J Pathol*. 2013;182(6):2241–2253. https://doi.org/10.1016/j.ajpath.2013.02.011 [Published online first: Epub Date].
107. Hinton RB, Adelman-Brown J, Witt S, et al. Elastin haploinsufficiency results in progressive aortic valve malformation and latent valve disease in a mouse model. *Circ Res*. 2010;107(4):549–557. https://doi.org/10.1161/CIRCRESAHA.110.221358 [Published online first: Epub Date].
108. Staiculescu MC, Cocciolone AJ, Procknow JD, Kim J, Wagenseil JE. Comparative gene array analyses of severe elastic fiber defects in late embryonic and newborn mouse aorta. *Physiol Genom*. 2018;50(11):988–1001. https://doi.org/10.1152/physiolgenomics.00080.2018 [Published online first: Epub Date].
109. Wagenseil JE, Ciliberto CH, Knutsen RH, Levy MA, Kovacs A, Mecham RP. Reduced vessel elasticity alters cardiovascular structure and function in newborn mice. *Circ Res*. 2009;104(10): 1217–1224. https://doi.org/10.1161/CIRCRESAHA.108.192054 [Published online first: Epub Date].
110. Huveneers S, Daemen MJ, Hordijk PL. Between Rho(k) and a hard place: the relation between vessel wall stiffness, endothelial

contractility, and cardiovascular disease. *Circ Res*. 2015;116(5): 895–908. https://doi.org/10.1161/CIRCRESAHA.116.305720 [Published online first: Epub Date].
111. Klein EA, Yin L, Kothapalli D, et al. Cell-cycle control by physiological matrix elasticity and in vivo tissue stiffening. *Curr Biol*. 2009;19(18):1511–1518. https://doi.org/10.1016/j.cub.2009.07.069 [Published online first: Epub Date].
112. Hsu BY, Bae YH, Mui KL, Liu SL, Assoian RK. Apolipoprotein E3 inhibits Rho to regulate the mechanosensitive expression of Cox2. *PLoS One*. 2015;10(6):e0128974. https://doi.org/10.1371/journal.pone.0128974 [Published online first: Epub Date].
113. Lim SM, Trzeciakowski JP, Sreenivasappa H, Dangott LJ, Trache A. RhoA-induced cytoskeletal tension controls adaptive cellular remodeling to mechanical signaling. *Integr Biol (Camb)*. 2012;4(6):615–627. https://doi.org/10.1039/c2ib20008b [Published online first: Epub Date].
114. Taddei S, Virdis A, Ghiadoni L, et al. Age-related reduction of NO availability and oxidative stress in humans. *Hypertension*. 2001; 38(2):274–279. https://doi.org/10.1161/01.hyp.38.2.274 [Published online first: Epub Date].
115. Lacolley P, Regnault V, Laurent S. Mechanisms of arterial stiffening: from mechanotransduction to epigenetics. *Arterioscler Thromb Vasc Biol*. 2020;40(5):1055–1062. https://doi.org/10.1161/ATVBAHA.119.313129 [Published online first: Epub Date].
116. Nardone G, Oliver-De La Cruz J, Vrbsky J, et al. YAP regulates cell mechanics by controlling focal adhesion assembly. *Nat Commun*. 2017;8:15321. https://doi.org/10.1038/ncomms15321 [Published online first: Epub Date].
117. Agbor LN, Ibeawuchi SC, Hu C, et al. Cullin-3 mutation causes arterial stiffness and hypertension through a vascular smooth muscle mechanism. *JCI Insight*. 2016;1(19):e91015. https://doi.org/10.1172/jci.insight.91015 [Published online first: Epub Date]
118. Laurent S, Boutouyrie P, Cunha PG, Lacolley P, Nilsson PM. Concept of extremes in vascular aging. *Hypertension*. 2019;74(2): 218–228. https://doi.org/10.1161/HYPERTENSIONAHA.119.12655 [Published online first: Epub Date].
119. Elia L, Kunderfranco P, Carullo P, et al. UHRF1 epigenetically orchestrates smooth muscle cell plasticity in arterial disease. *J Clin Invest*. 2018;128(6):2473–2486. https://doi.org/10.1172/JCI96121 [Published online first: Epub Date].
120. Wang YX, Fitch RM. Vascular stiffness: measurements, mechanisms and implications. *Curr Vasc Pharmacol*. 2004;2(4):379–384. https://doi.org/10.2174/1570161043385448 [Published online first: Epub Date].
121. Selwaness M, van den Bouwhuijsen Q, Mattace-Raso FU, et al. Arterial stiffness is associated with carotid intraplaque hemorrhage in the general population: the Rotterdam study. *Arterioscler Thromb Vasc Biol*. 2014;34(4):927–932. https://doi.org/10.1161/ATVBAHA.113.302603 [Published online first: Epub Date].
122. Gayral S, Garnotel R, Castaing-Berthou A, et al. Elastin-derived peptides potentiate atherosclerosis through the immune Neu1-PI3Kγ pathway. *Cardiovasc Res*. 2014;102(1):118–127. https://doi.org/10.1093/cvr/cvt336 [Published online first: Epub Date].
123. Ooyama T, Fukuda K, Oda H, Nakamura H, Hikita Y. Substratum-bound elastin peptide inhibits aortic smooth muscle cell migration in vitro. *Arteriosclerosis*. 1987;7(6):593–598. https://doi.org/10.1161/01.atv.7.6.593 [Published online first: Epub Date].
124. Mochizuki S, Brassart B, Hinek A. Signaling pathways transduced through the elastin receptor facilitate proliferation of arterial smooth muscle cells. *J Biol Chem*. 2002;277(47):44854–44863. https://doi.org/10.1074/jbc.M205630200 [Published online first: Epub Date].
125. Simionescu A, Simionescu DT, Vyavahare NR. Osteogenic responses in fibroblasts activated by elastin degradation products and transforming growth factor-beta1: role of myofibroblasts in vascular calcification. *Am J Pathol*. 2007;171(1):116–123. https://doi.org/10.2353/ajpath.2007.060930 [Published online first: Epub Date].
126. Van Herck JL, De Meyer GR, Martinet W, et al. Impaired fibrillin-1 function promotes features of plaque instability in apolipoprotein E-deficient mice. *Circulation*. 2009;120(24):2478–2487. https://doi.org/10.1161/CIRCULATIONAHA.109.872663 [Published online first: Epub Date].
127. Van der Donckt C, Van Herck JL, Schrijvers DM, et al. Elastin fragmentation in atherosclerotic mice leads to intraplaque neovascularization, plaque rupture, myocardial infarction, stroke, and sudden death. *Eur Heart J*. 2015;36(17):1049–1058. https://doi.org/10.1093/eurheartj/ehu041 [Published online first: Epub Date].
128. Stehouwer CD, Henry RM, Ferreira I. Arterial stiffness in diabetes and the metabolic syndrome: a pathway to cardiovascular disease. *Diabetologia*. 2008;51(4):527–539. https://doi.org/10.1007/s00125-007-0918-3 [Published online first: Epub Date].
129. Nickenig G, Roling J, Strehlow K, Schnabel P, Bohm M. Insulin induces upregulation of vascular AT1 receptor gene expression by posttranscriptional mechanisms. *Circulation*. 1998;98(22): 2453–2460. https://doi.org/10.1161/01.cir.98.22.2453 [Published online first: Epub Date].
130. Aronson D. Cross-linking of glycated collagen in the pathogenesis of arterial and myocardial stiffening of aging and diabetes. *J Hypertens*. 2003;21(1):3–12. https://doi.org/10.1097/00004872-200301000-00002 [Published online first: Epub Date].
131. Goncalves I, Bengtsson E, Colhoun HM, et al. Elevated plasma levels of MMP-12 are associated with atherosclerotic burden and symptomatic cardiovascular disease in subjects with type 2 diabetes. *Arterioscler Thromb Vasc Biol*. 2015;35(7):1723–1731. https://doi.org/10.1161/ATVBAHA.115.305631 [Published online first: Epub Date].
132. Ogola BO, Zimmerman MA, Clark GL, et al. New insights into arterial stiffening: does sex matter? *Am J Physiol Heart Circ Physiol*. 2018;315(5):H1073–H1087. https://doi.org/10.1152/ajpheart.00132.2018 [Published online first: Epub Date].
133. AlGhatrif M, Strait JB, Morrell CH, et al. Longitudinal trajectories of arterial stiffness and the role of blood pressure: the Baltimore Longitudinal Study of Aging. *Hypertension*. 2013;62(5):934–941. https://doi.org/10.1161/HYPERTENSIONAHA.113.01445 [Published online first: Epub Date].
134. Safar ME, Regnault V, Lacolley P. Sex differences in arterial stiffening and central pulse pressure: mechanistic insights? *J Am Coll Cardiol*. 2020;75(8):881–883. https://doi.org/10.1016/j.jacc.2019.12.041 [Published online first: Epub Date].
135. Hawes JZ, Cocciolone AJ, Cui AH, et al. Elastin haploinsufficiency in mice has divergent effects on arterial remodeling with aging depending on sex. *Am J Physiol Heart Circ Physiol*. 2020; 319(6):H1398–H1408. https://doi.org/10.1152/ajpheart.00517.2020 [Published online first: Epub Date].
136. Qiu H, Tian B, Resuello RG, et al. Sex-specific regulation of gene expression in the aging monkey aorta. *Physiol Genom*. 2007;29(2): 169–180. https://doi.org/10.1152/physiolgenomics.00229.2006 [Published online first: Epub Date].
137. Sakalihasan N, Michel JB, Katsargyris A, et al. Abdominal aortic aneurysms. *Nat Rev Dis Primers*. 2018;4(1):34. https://doi.org/10.1038/s41572-018-0030-7 [Published online first: Epub Date].
138. Michel JB, Martin-Ventura JL, Egido J, et al. Novel aspects of the pathogenesis of aneurysms of the abdominal aorta in humans.

Cardiovasc Res. 2011;90(1):18–27. https://doi.org/10.1093/cvr/cvq337 [Published online first: Epub Date].

139. Nienaber CA, Clough RE, Sakalihasan N, et al. Aortic dissection. *Nat Rev Dis Primers*. 2016;2:16071. https://doi.org/10.1038/nrdp.2016.71 [Published online first: Epub Date].
140. Golledge J. Abdominal aortic aneurysm: update on pathogenesis and medical treatments. *Nat Rev Cardiol*. 2019;16(4):225–242. https://doi.org/10.1038/s41569-018-0114-9 [Published online first: Epub Date].
141. Schievink WI. Intracranial aneurysms. *N Engl J Med*. 1997;336(1): 28–40. https://doi.org/10.1056/NEJM199701023360106 [Published online first: Epub Date].
142. Toth G, Cerejo R. Intracranial aneurysms: review of current science and management. *Vasc Med*. 2018;23(3):276–288. https://doi.org/10.1177/1358863X18754693 [Published online first: Epub Date].
143. Lawrence PF, Lorenzo-Rivero S, Lyon JL. The incidence of iliac, femoral, and popliteal artery aneurysms in hospitalized patients. *J Vasc Surg*. 1995;22(4):409–415. https://doi.org/10.1016/s0741-5214(95)70008-0. Discussion 15-6, [Published online first: Epub Date].
144. Gupta M, Agrawal A, Iakovou A, Cohen S, Shah R, Talwar A. Pulmonary artery aneurysm: a review. *Pulm Circ*. 2020;10(1). https://doi.org/10.1177/2045894020908780, 2045894020908780. [Published online first: Epub Date].
145. Investigators IT. Comparative clinical effectiveness and cost effectiveness of endovascular strategy v open repair for ruptured abdominal aortic aneurysm: three year results of the IMPROVE randomised trial. *Br Med J*. 2017;359:j4859. https://doi.org/10.1136/bmj.j4859 [Published online first: Epub Date].
146. Prakash SK, Haden-Pinneri K, Milewicz DM. Susceptibility to acute thoracic aortic dissections in patients dying outside the hospital: an autopsy study. *Am Heart J*. 2011;162(3):474–479. https://doi.org/10.1016/j.ahj.2011.06.020 [Published online first: Epub Date].
147. Pinard A, Jones GT, Milewicz DM. Genetics of thoracic and abdominal aortic diseases. *Circ Res*. 2019;124(4):588–606. https://doi.org/10.1161/CIRCRESAHA.118.312436 [Published online first: Epub Date].
148. Albornoz G, Coady MA, Roberts M, et al. Familial thoracic aortic aneurysms and dissections–incidence, modes of inheritance, and phenotypic patterns. *Ann Thorac Surg*. 2006;82(4):1400–1405. https://doi.org/10.1016/j.athoracsur.2006.04.098 [Published online first: Epub Date].
149. Biddinger A, Rocklin M, Coselli J, Milewicz DM. Familial thoracic aortic dilatations and dissections: a case control study. *J Vasc Surg*. 1997;25(3):506–511. https://doi.org/10.1016/s0741-5214(97)70261-1 [Published online first: Epub Date].
150. Renard M, Francis C, Ghosh R, et al. Clinical validity of genes for heritable thoracic aortic aneurysm and dissection. *J Am Coll Cardiol*. 2018;72(6):605–615. https://doi.org/10.1016/j.jacc.2018.04.089 [Published online first: Epub Date].
151. Davis EC. Smooth muscle cell to elastic lamina connections in developing mouse aorta. Role in aortic medial organization. *Lab Invest*. 1993;68(1):89–99.
152. Humphrey JD, Milewicz DM, Tellides G, Schwartz MA. Cell biology. Dysfunctional mechanosensing in aneurysms. *Science*. 2014;344(6183):477–479. https://doi.org/10.1126/science.1253026 [Published online first: Epub Date].
153. Humphrey JD, Schwartz MA, Tellides G, Milewicz DM. Role of mechanotransduction in vascular biology: focus on thoracic aortic aneurysms and dissections. *Circ Res*. 2015;116(8):1448–1461. https://doi.org/10.1161/CIRCRESAHA.114.304936 [Published online first: Epub Date].
154. Pyeritz RE, McKusick VA. The Marfan syndrome: diagnosis and management. *N Engl J Med*. 1979;300(14):772–777. https://doi.org/10.1056/NEJM197904053001406 [Published online first: Epub Date].
155. Dietz HC, Cutting GR, Pyeritz RE, et al. Marfan syndrome caused by a recurrent de novo missense mutation in the fibrillin gene. *Nature*. 1991;352(6333):337–339. https://doi.org/10.1038/352337a0 [Published online first: Epub Date].
156. Pinard A, Salgado D, Desvignes JP, et al. WES/WGS reporting of mutations from cardiovascular "actionable" genes in clinical practice: a key role for UMD knowledgebases in the era of big databases. *Hum Mutat*. 2016;37(12):1308–1317. https://doi.org/10.1002/humu.23119 [Published online first: Epub Date].
157. Bunton TE, Biery NJ, Myers L, Gayraud B, Ramirez F, Dietz HC. Phenotypic alteration of vascular smooth muscle cells precedes elastolysis in a mouse model of Marfan syndrome. *Circ Res*. 2001;88(1):37–43. https://doi.org/10.1161/01.res.88.1.37 [Published online first: Epub Date].
158. Neptune ER, Frischmeyer PA, Arking DE, et al. Dysregulation of TGF-beta activation contributes to pathogenesis in Marfan syndrome. *Nat Genet*. 2003;33(3):407–411. https://doi.org/10.1038/ng1116 [Published online first: Epub Date].
159. Loeys BL, Chen J, Neptune ER, et al. A syndrome of altered cardiovascular, craniofacial, neurocognitive and skeletal development caused by mutations in TGFBR1 or TGFBR2. *Nat Genet*. 2005;37(3):275–281. https://doi.org/10.1038/ng1511 [Published online first: Epub Date].
160. Sinha S, Hoofnagle MH, Kingston PA, McCanna ME, Owens GK. Transforming growth factor-beta1 signaling contributes to development of smooth muscle cells from embryonic stem cells. *Am J Physiol Cell Physiol*. 2004;287(6):C1560–C1568. https://doi.org/10.1152/ajpcell.00221.2004 [Published online first: Epub Date].
161. Owens GK, Kumar MS, Wamhoff BR. Molecular regulation of vascular smooth muscle cell differentiation in development and disease. *Physiol Rev*. 2004;84(3):767–801. https://doi.org/10.1152/physrev.00041.2003 [Published online first: Epub Date].
162. Inamoto S, Kwartler CS, Lafont AL, et al. TGFBR2 mutations alter smooth muscle cell phenotype and predispose to thoracic aortic aneurysms and dissections. *Cardiovasc Res*. 2010;88(3):520–529. https://doi.org/10.1093/cvr/cvq230 [Published online first: Epub Date].
163. Quintana RA, Taylor WR. Cellular mechanisms of aortic aneurysm formation. *Circ Res*. 2019;124(4):607–618. https://doi.org/10.1161/CIRCRESAHA.118.313187 [Published online first: Epub Date].
164. Jones GT, Tromp G, Kuivaniemi H, et al. Meta-analysis of genome-wide association studies for abdominal aortic aneurysm identifies four new disease-specific risk loci. *Circ Res*. 2017;120(2): 341–353. https://doi.org/10.1161/CIRCRESAHA.116.308765 [Published online first: Epub Date].
165. Hellenthal FA, Buurman WA, Wodzig WK, Schurink GW. Biomarkers of AAA progression. Part 1: extracellular matrix degeneration. *Nat Rev Cardiol*. 2009;6(7):464–474. https://doi.org/10.1038/nrcardio.2009.80 [Published online first: Epub Date].
166. Hellenthal FA, Buurman WA, Wodzig WK, Schurink GW. Biomarkers of abdominal aortic aneurysm progression. Part 2: inflammation. *Nat Rev Cardiol*. 2009;6(8):543–552. https://doi.org/10.1038/nrcardio.2009.102 [Published online first: Epub Date].
167. Haller SJ, Crawford JD, Courchaine KM, et al. Intraluminal thrombus is associated with early rupture of abdominal aortic aneurysm. *J Vasc Surg*. 2018;67(4). https://doi.org/10.1016/j.jvs.2017.08.069, 1051.e1–1058.e1, [Published online first: Epub Date].
168. Dobrin PB, Baker WH, Gley WC. Elastolytic and collagenolytic studies of arteries. Implications for the mechanical properties of aneurysms. *Arch Surg*. 1984;119(4):405–409. https://doi.org/

10.1001/archsurg.1984.01390160041009 [Published online first: Epub Date].
169. Loppnow H, Libby P. Proliferating or interleukin 1-activated human vascular smooth muscle cells secrete copious interleukin 6. *J Clin Invest*. 1990;85(3):731–738. https://doi.org/10.1172/JCI114498 [Published online first: Epub Date].
170. Alexander MR, Murgai M, Moehle CW, Owens GK. Interleukin-1β modulates smooth muscle cell phenotype to a distinct inflammatory state relative to PDGF-DD via NF-κB-dependent mechanisms. *Physiol Genomics*. 2012;44(7):417–429. https://doi.org/10.1152/physiolgenomics.00160.2011 [Published online first: Epub Date].
171. Golledge J, Muller J, Daugherty A, Norman P. Abdominal aortic aneurysm: pathogenesis and implications for management. *Arterioscler Thromb Vasc Biol*. 2006;26(12):2605–2613. https://doi.org/10.1161/01.ATV.0000245819.32762 [Published online first: Epub Date].
172. Vasamsetti SB, Karnewar S, Kanugula AK, Thatipalli AR, Kumar JM, Kotamraju S. Metformin inhibits monocyte-to-macrophage differentiation via AMPK-mediated inhibition of STAT3 activation: potential role in atherosclerosis. *Diabetes*. 2015;64(6):2028–2041. https://doi.org/10.2337/db14-1225 [Published online first: Epub Date].
173. Fujimura N, Xiong J, Kettler EB, et al. Metformin treatment status and abdominal aortic aneurysm disease progression. *J Vasc Surg*. 2016;64(1):46–54. https://doi.org/10.1016/j.jvs.2016.02.020. e8, [Published online first: Epub Date].
174. Golledge J, Moxon J, Pinchbeck J, et al. Association between metformin prescription and growth rates of abdominal aortic aneurysms. *Br J Surg*. 2017;104(11):1486–1493. https://doi.org/10.1002/bjs.10587 [Published online first: Epub Date].
175. Itoga NK, Rothenberg KA, Suarez P, et al. Metformin prescription status and abdominal aortic aneurysm disease progression in the U.S. veteran population. *J Vasc Surg*. 2019;69(3):710–716. https://doi.org/10.1016/j.jvs.2018.06.194. e3, [Published online first: Epub Date].

CHAPTER

25

Lymphatic vascular anomalies and dysfunction

Christian El Amm[1], *Federico Silva-Palacios*[2], *Xin Geng*[3] *and R. Sathish Srinivasan*[3,4]

[1]Department of Surgery, University of Oklahoma Health Sciences Center, Oklahoma City, OK, United States [2]Department of Internal Medicine, University of Oklahoma Health Sciences Center, Oklahoma City, OK, United States [3]Cardiovascular Biology Research Program, Oklahoma Medical Research Foundation, Oklahoma City, OK, United States [4]Department of Cell Biology, University of Oklahoma Health Sciences Center, Oklahoma City, OK, United States

Introduction

Lymphatic vasculature consists of lymphatic capillaries, collecting lymphatic vessels, lymphatic valves (LVs), lymph nodes (LNs), and lymphovenous valves (LVVs)[1,2] (Fig. 25.1). Lymphatic endothelial cells (LECs) are the building blocks of all these structures. Lymphatic capillaries absorb interstitial fluid, digested lipids, and extravasated immune cells (collectively called as lymph) from nearly all tissues. Lymphatic capillaries are made of LECs with discontinuous "button-like" junctions that allow that entry of lymph into the vessels. Lymph collected by capillaries is transported upstream to larger diameter vessels known as collecting lymphatic vessels. Collecting lymphatic vessels have continuous "zipper-like" junctions that prevent lymph leakage. LVs within collecting lymphatic vessels prevent backflow. In addition, collecting lymphatic vessels are surrounded by a thin layer of mural cells known as lymphatic muscle cells with contractile property to propel the fluid forward against gravitational and hydrostatic pressures.

Lymphatic vessels are not observed within normal bones, the brain or spinal cord. Lymphatic vessels play diverse roles in other organs in normal and pathological contexts. LNs are sites where blood and lymphatic vasculatures interact to initiate immune response against pathogens. LECs are not necessary for the formation of rudimentary LNs anlage.[3] However, complex cross-talk between LECs and LN-forming cells regulates the growth, maturation, and functioning of LNs.[4,5] Furthermore, six different types of LECs exist within mature LNs to precisely regulate the migration of various immune cells in and out of the LNs.[6,7] Lacteals are specialized lymphatic capillaries of the intestinal villi.[8] Lacteals absorb digested lipids and transport it to blood circulation through thoracic duct. In addition, intestinal lymphatic vessels are in close proximity to intestinal microbiota and are therefore critical for initiating adaptive immune response against potential pathogens. Lungs are highly enriched with lymphatic vessels. Clearance of pulmonary fluids by lymphatic vessels is necessary for lung inflation immediately after birth.[9] Pulmonary lymphatic vessels are also necessary to prevent or resolve inflammation.[10] Furthermore, promotion of pulmonary lymphatic vessel growth reduces fibrosis[11] and organ rejection after transplantation.[12] Meningeal lymphatic vessels are located under the skull and they are necessary for the drainage of cerebrospinal fluid and immune cells from brain parenchyma.[1] Lymphatic vessels have also been identified around the spinal cord and here too the lymphatic vessels are regulating the drainage of cerebrospinal fluid.[13,14] Finally, extensive networks of lymphatic vessels are present in all three layers of the heart (pericardium, myocardium, and endocardium). Nearly 6 decades ago, interruption of cardiac lymphatics in dogs was found to cause subendocardial hemorrhage, ventricular wall thickening, and signs of myocardial infarction.[15] More recently, promoting lymphatic vessel growth was found to improve cardiac function.[16] Furthermore, cardiac lymphatic vessels secrete reelin to promote the growth of cardiomyocytes during embryogenesis.[17] This lymphangiocrine signal also regulates neonatal cardiac regeneration and is cardioprotective after myocardial infarction.

Thus, lymphatic vessels play numerous roles in various tissues to maintain overall health. Lymphatic

The Vasculome
https://doi.org/10.1016/B978-0-12-822546-2.00025-3

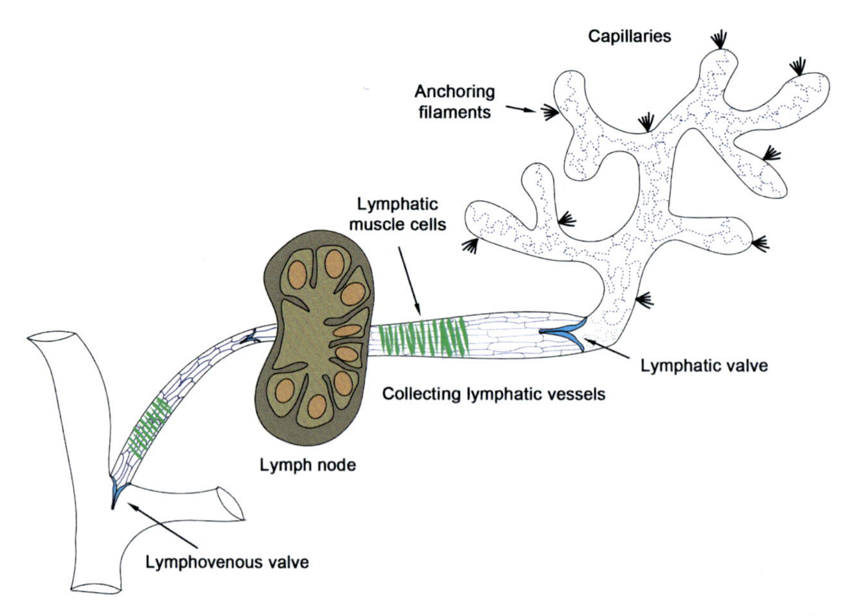

FIGURE 25.1 **Basic architecture of the lymphatic vasculature.** Lymph enters the lymphatic vasculature through the button-like junctions of lymphatic capillaries. Anchoring filaments assist in the opening and closing of the capillary endothelial cells. Lymph enters the collecting lymphatic vessels with zipper-like junctions that prevent leakage. Lymphatic valves regulate the unidirectional flow of fluid and contractile property of lymphatic muscle cells provides the propulsive force. Lymph nodes filter out antigens and activate immune response if necessary. Finally, lymph is returned to blood circulation through lymphovenous valves at the junction of jugular and subclavian veins.

vascular anomalies arise due to defects such as hypoplasia, hyperplasia, or abnormal presence of lymphatic vessels in tissues such as bones. These defects could have significant impact on health and even result in death. The goal of this chapter is to describe the various lymphatic disorders and the associated molecular mechanisms. Different terms have been used to define the same lymphatic disorder. Hence, the International Society for the Study of Vascular Anomalies (ISSVA) has provided guidelines for classifying vascular anomalies in a rational manner.[18] Gordon and colleagues have further refined these guidelines for lymphatic disorders.[19] In this chapter, we have used the classification by Gordon et al. with a few modifications to define the lymphatic anomalies and describe the molecular mechanisms behind these pathologies (Fig. 25.2).

Lymphatic vascular tumors

Vascular tumors (hemangiomas) consist of rapidly proliferating endothelial cells. We describe two diseases that are associated with LEC proliferation.

Kaposiform hemangioendothelioma

Kaposiform hemangioendothelioma (KHE) is a congenital hemangioma that is fully formed at birth. KHE is a locally invasive tumor that forms in deep tissues and it is associated with high morbidity and mortality due to compressive effect on visceral organs and consumptive coagulopathy known as Kasabach–Merritt phenomenon.[20,21] KHE consists of numerous glomeruloid-shaped nodules of capillary endothelial cells that are surrounded by spindle-shaped (kaposiform) LECs that are positive for the homeobox transcription factor PROX1, glycoprotein podoplanin, and lymphatic vessel endothelial hyaluronan receptor 1. The proliferating capillary endothelial cells are negative for GLUT1, which is a marker for hemangiomas. The etiology of KHE is not fully understood although in a small number of patients somatic activating mutations have been found in the G protein Gα14.[20]

Lymphangioleiomyomatosis

Lymphangioleiomyomatosis (LAM) is currently not included in the ISSVA classification. We are nevertheless describing it here due to its tumor-like characteristic and strong relationship with the lymphatic vasculature.[22] LAM occurs mostly in women, and it is caused by mutations in tumor suppressors tuberous sclerosis (TSC)1 or TSC2, which are GTPase activating proteins that inhibit mTOR1 signaling pathway. Loss of TSC1 or TSC2 activity results in the upregulation of mTORC1 activity in smooth muscle actin$^+$ LAM cells of unknown origin. Due to the strong female predisposition of LAM and due to the expression of progesterone receptor LAM cells thought to originate from the uterus. However, other sources such as mural cell lineage have not been fully excluded. LAM cell clusters express VEGF-D to promote the growth of lymphatic vessels. Subsequently, LAM cell clusters intravasate into the lymphatic vessels,

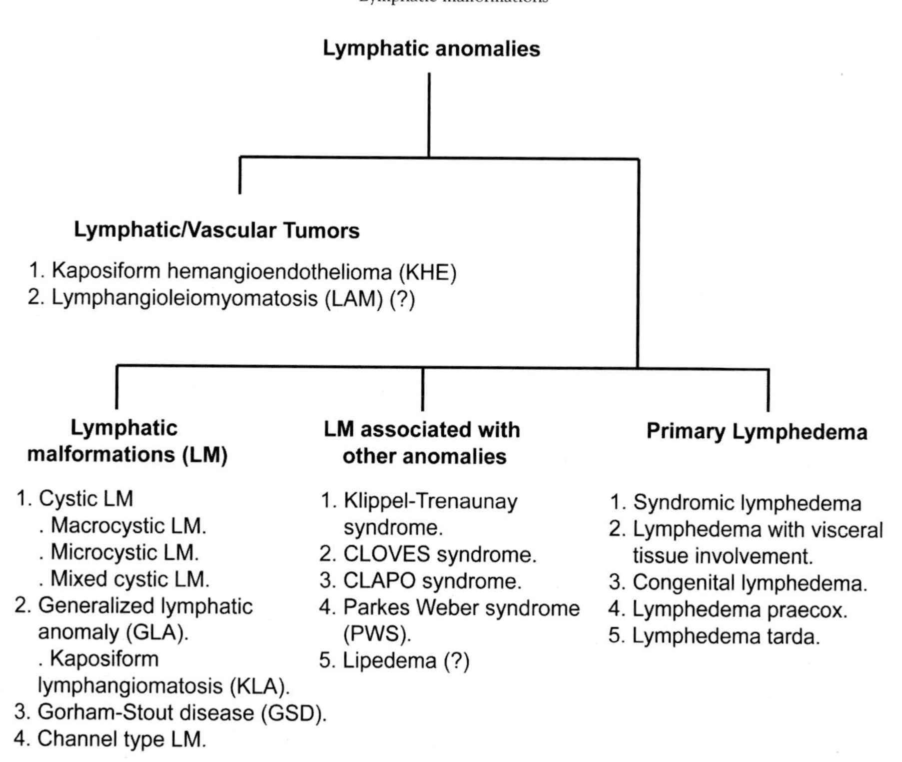

FIGURE 25.2 **Classification of lymphatic anomalies.** The disorders with the question marker are not included in the ISSVA classification.

proliferate rapidly, and block lymphatic drainage resulting in chylous fluid accumulation in peritoneal, pleural, and pericardial cavities. LAM cells are also disseminated to the lung where they form thin-walled cysts surrounded by LECs that produce matrix metalloproteinases that results in the destruction of lung parenchyma. Inhibition of mTORC1 activity by the rapamycin analog sirolimus is effective in controlling the complications of LAM.

Lymphatic malformations

Congenital lymphatic malformations (LMs) are relatively uncommon conditions, occurring in approximately 1 in 4000 live births. In contrast to vascular tumors, in LMs the LECs do not proliferate excessively. LMs are further classified based on the lymphatic vascular defects and the tissues that are additionally impacted (Fig. 25.2).

Common or cystic LM

These LMs are made of dilated or cyst-like lymphatic vessels.[18] These typically present as a soft subcutaneous mass with poor demarcation. They can occur anywhere in the body but have a predilection for the cervical, axillary, and inguinal basins. They are further subdivided into macrocystic, microcystic, or mixed type LMs. Due to their location and size, macrocystic LMs are sometimes diagnosed prenatally as they can complicate vaginal delivery (Fig. 25.3). However, they are easier to treat by aspiration or sclerotherapy. Microcystic LMs with cutaneous involvement present a typical blister-like appearance called "lymphangioma circumscriptum" (Fig. 25.4).

Somatic mutations in *PIK3CA* are associated with cystic LMs.[23] Several gain-of-function mutations have been identified in exon 9 or 20 resulting in the hyperactivation of AKT and mTORC1 signaling pathways. Using mouse models, it was recently demonstrated that ectopic hyperactivation of PIK3CA in embryonic LEC progenitors results in macrocystic LMs, while hyperactivation in adult LECs results in microcystic LMs.[24] Vascular endothelial growth factor receptor 3 (VEGFR3) and its coreceptor neuropilin 2 were upregulated in the LECs of lesions. Furthermore, LEC junctions were defective as indicated by the downregulation of adherens junction molecule VE-Cadherin. Importantly, inhibition of VEGFR3 signaling using "VEGF-C trap" together with rapamycin-mediated inhibition of mTORC1 in adult mice reduced lymphatic vascular density in microcystic lesions without affecting normal lymphatic vessels.

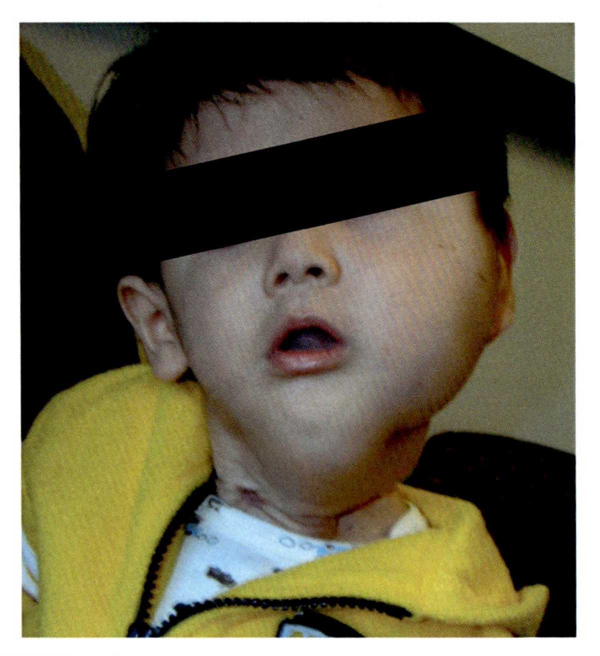

FIGURE 25.3 Macrocystic lymphatic malformation. Cystic hygroma in the neck of a young patient.

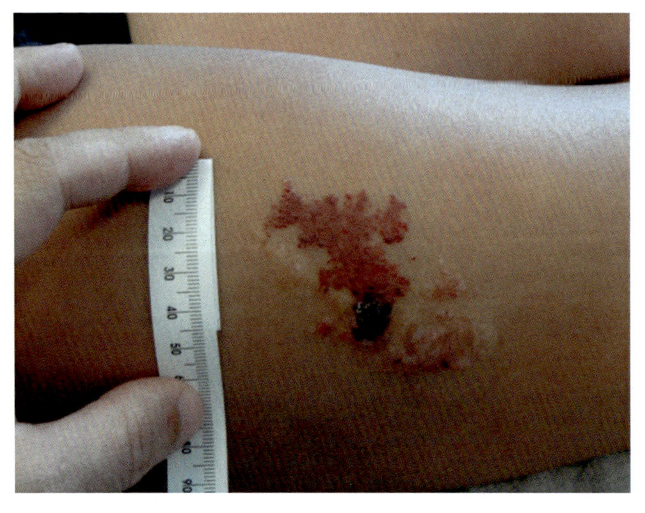

FIGURE 25.4 Lymphangioma circumscriptum. Blister-like microcystic lymphatic malformations in the leg of a patient. Some of the cysts are filled with blood, others are filled with a clear fluid.

Generalized lymphatic anomaly

Generalized lymphatic anomaly (GLA) is a lymphatic disorder that results in macrocystic LM in multiple visceral organs.[18,25] GLA can cause lesions in the spleen and liver and chylous fluid build-up (chylous ascites, pleural effusion). Additionally, lymphatic vessels could infiltrate appendicular and axial bones. Dilated lymphatic vessels could be observed in the medulla of bones without damage to cortical bones.[26]

Hyperactivating somatic mutations in *PIK3CA* are observed in the lesions of patients with GLA.[26] 1%−33% of cells in the lesions were found to carry mutated *PIK3CA*. Ectopic expression of mutated PIK3CA in mouse LECs recapitulates the phenotypes of GLA patients including the infiltration of lymphatic vessels into the bones.[26] Lymphangiography revealed that the excessive lymphatic vessels in mutant animals were defective. Rapamycin inhibited the growth of lymphatic vessels and ameliorated lymphatic function.

Kaposiform lymphangiomatosis (KLA) is an aggressive form of GLA in which patients experience effusions, which could be chylous or hemorrhagic, splenic cysts, bone lesions, respiratory symptoms, bleeding, and subcutaneous mass.[25,27] Thrombocytopenia and tortuous blood vessels in the visceral pleura are also observed. D-dimers (indicator of coagulopathy), angiopoietin-2, and soluble VEGFR3 are increased in the plasma of KLA patients. A defining characteristic of KLA is the presence of spindle-shaped LECs around lymphatic vessels in the lesions. Unlike KHE, there is no capillary proliferation in KLA. A combination of sirolimus (PI3K inhibitor), prednisone (antiinflammatory steroid), vincristine (antitumor), and zolendronate (osteoclast inhibitor) was used to successfully treat a KLA patient with unknown molecular etiology.[28] Activating mutations in NRAS have been identified in patients with KLA.[29,30] It remains to be seen whether MEK inhibitors could be used to treat these KLA patients.

Gorham-Stout disease

Gorham−Stout disease (GSD) is also known as vanishing bone disease.[25,31,32] GSD patients experience plural effusions, but normally do not have macrocytic lesions. Importantly, the cortical bones of GSD patients undergo progressive osteolysis resulting in fractures and even complete disappearance of bones. Unlike GLA, axial bones (cranium, sternum, and ribs) are more commonly affected in GSD patients. The cause of this painful disease is currently unknown. However, at least one GLA patient carrying activating mutation in PIK3CA was reported to have cortical bone loss indicating that hyperactivation of PI3K pathway could be involved.[26] In addition, ectopic overexpression of VEGF-C in bones resulted in a phenotype that closely resembled GSD.[33] Mechanistically, VEGF-C induced the growth of lymphatic vessels into the bones, which produced M-CSF (macrophage colony stimulating factor). In turn, M-CSF activated osteoclasts that resulted in bone reabsorption. Consistent with these findings, bone loss in VEGF-C overexpressing mice could be attenuated by inhibiting osteoclasts with zoledronic acid.

Channel type LM

This disease is also known as central conducting lymphatic anomaly (CCLA) or lymphangiectasia, which causes the dysfunction of thoracic duct and cisterna chyli resulting in retrograde flow and abnormal lymph drainage. Symptoms include pericardial effusions, respiratory distress, lower extremity, and lower abdominal lymphedema.

Gain-of-function mutation in ARAF and loss-of-function mutation in EPHB4 have been identified in patients with CCLA.[34,35] Intriguingly, ARAF mutation results in the upregulation of mitogen-activated protein kinase (MAPK) signaling, while the EPHB4 mutation results in the downregulation of MAPK signaling. However, loss of EPHB4 upregulates mTORC1 pathway through unknown mechanisms.[34] An FDA-approved MAPK inhibitor was able to dramatically ameliorate the lymphatic vascular defects in patients with ARAF mutation.[35] Whether rapamycin or MAPK inhibitors could ameliorate CCLA caused by mutations in EPHB4 is yet to be determined.

Lymphatic malformations associated with other vascular malformations or overgrowth syndromes

Occasionally, LMs are associated with other vascular malformations or organ overgrowth. Whole or parts of limbs, digits, or adipose tissue deposits could be involved.

Klippel-Trenaunay Syndrome (KTS)

Klippel–Trenaunay syndrome (KTS) associates capillary, venous, and LMs with limb overgrowth.[36] Overgrowth can affect digits, parts, or the whole limb. Occasionally, abnormal vascular channels ("embryonal veins") are observed, and these can divert enough blood flow leading to hypoplasia of the typical channels (humeral and femoral veins). Treatment involves surgical debulking, pulse dye laser, and select epiphysiodesis to arrest growth of involved limbs or digits once adult size is reached (Fig. 25.5).

CLOVES syndrome (congenital lipomatous overgrowth, vascular malformations, epidermal nevi, and spinal/skeletal anomalies and/or scoliosis)

CLOVES is a similar condition affecting limbs. CLOVES affects lymphatic, venous, capillary, and arteriovenous malformations with adipose tissue hypertrophy ("lipomatous overgrowth").[37] Treatment involves embolization (for arteriovenous malformations), sclerotherapy (for the venous component), surgical debulking, liposuction and pulse-dye laser as needed (Fig. 25.6).

CLAPO syndrome (capillary malformation of the lower lip, lymphatic malformation, asymmetry, and partial or generalized overgrowth)

CLAPO syndrome associates lower lip capillary malformation and cervical LM, with asymmetrical somatic overgrowth (Fig. 25.7).[38]

KTS, CLOVES, and CLAPO syndromes involve mutation of the *PIK3CA* gene, and are now classified under PIK3CA-related Overgrowth Syndromes.[39]

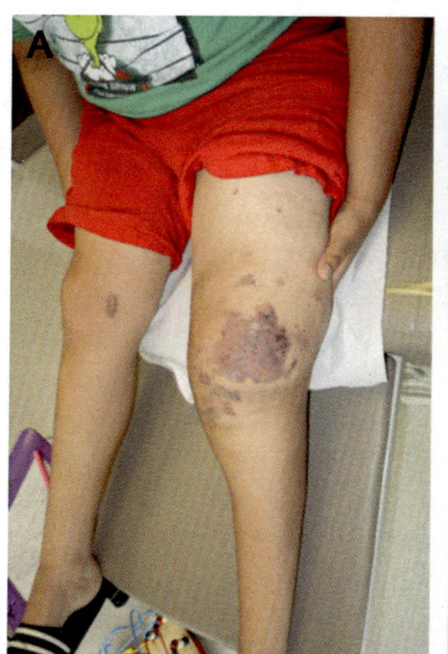

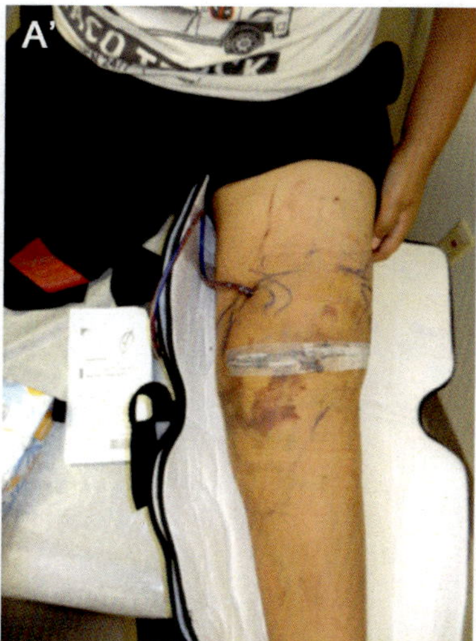

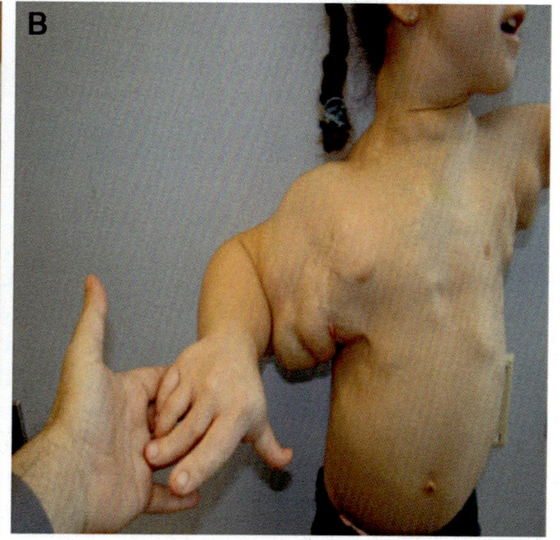

FIGURE 25.5 **Klippel–Trenaunay syndrome.** (A) Before and (A′) after combined surgical debulking to improve contour and tibial epiphysiodesis to control overgrowth. Note that debulking improves contour and allows a better fit into surgical splints. (B) Macrodactyly with large bilateral axillary lymphatic malformations.

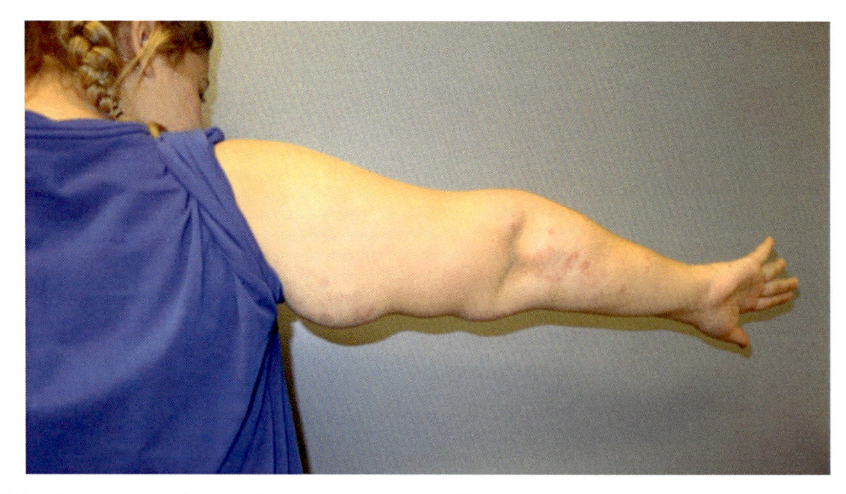

FIGURE 25.6 **CLOVES syndrome.** Notice the vascular malformations in the skin and the lipomatous overgrowth.

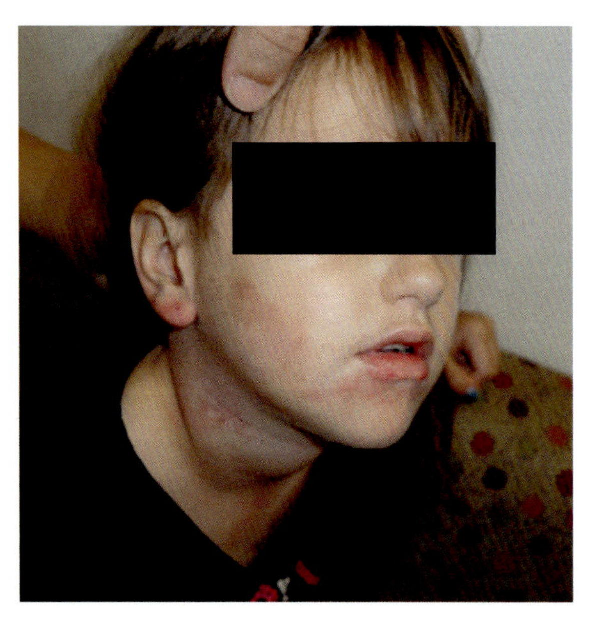

FIGURE 25.7 **CLAPO syndrome.** Notice the capillary malformation in the cheek and lower lip with cervical lymphatic malformation.

Parkes Weber syndrome

This disease is caused by mutations in the Ras GTPase activating protein RAS P21 Protein Activator 1 (RASA1).[40,41] The symptoms include lymphedema, capillary malformation, and arteriovenous malformation. Mechanistically, RASA1 was found to inhibit MAPK signaling and lymphatic vascular patterning and valve morphogenesis.[42–45] In fact, the diseases that we have described so far are primarily caused by hyperactivation of either PI3K or MAPK signaling pathways (Fig. 25.8).

Lipedema

Lipedema is not included in the ISSVA classification or the classification by Gordon and colleagues. While clinically similar to lymphedema, lipedema involves adipose tissue hypertrophy, with symmetrical nonpitting enlargement in the lower extremities (Fig. 25.9).[46] Arms, scalp, and other locations may be involved. Lipedema stops at the ankles, respecting the feet. Consequently, "Stemmer's sign" is negative: The skin fold over the second or third toe can be pinched without difficulty. While chronic lymphedema may lead to adipose hypertrophy ("lipo-lymphedema"), lipedema is not associated with preliminary lymphedema. The patients are almost exclusively female. The etiology of lipedema is unclear, although a role for estrogen is suspected. Local lymphatic dysfunction and capillary fragility are implicated. Due to clinical (and possibly molecular) overlap, lipedema is often included in the study of LMs.

Primary lymphedema

In the absence of a causative factor such as infection, filariasis, or surgery, patients with primary lymphedema often have a strong family history and frequently an autosomal dominant transmission pattern. Commonly, they present with pitting edema in one or both lower extremities, although the trunk, face, genitalia, or upper extremities may be involved. The skin eventually thickens and presents dyskeratosis and papillomatosis, leading to a "woody" texture. The skin is fragile and slow to heal from minor scratches or insect bites. An early sign is "Stemmer's sign": The skin fold over the second or third toe cannot be pinched due to thickening of the skin. Primary lymphedema is classified as follows.[19]

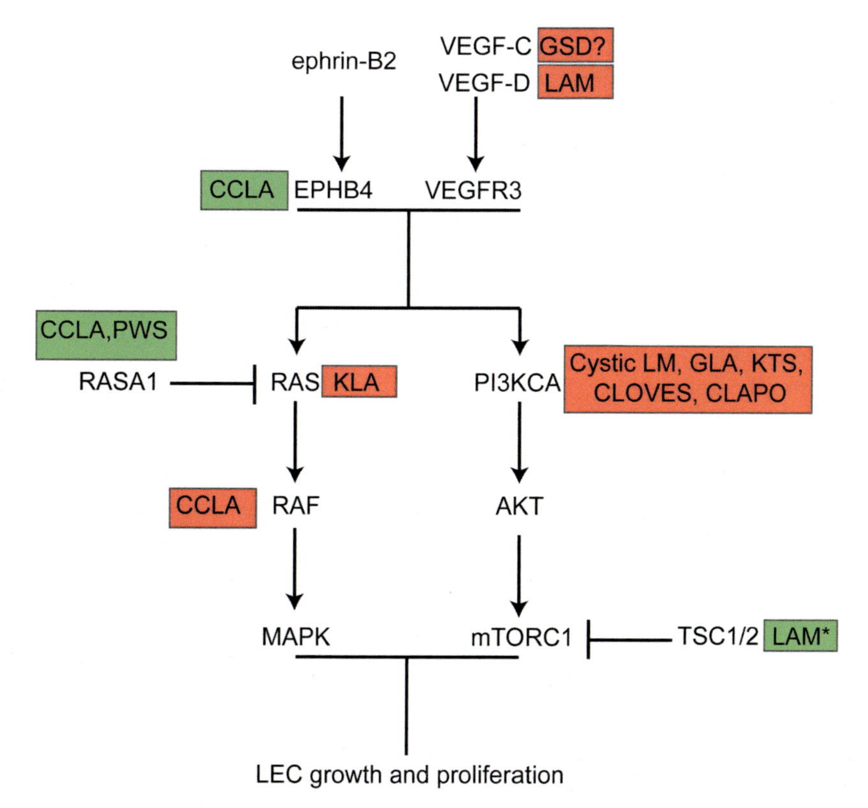

FIGURE 25.8 **Involvement of MAPK and PI3K pathways in LMs.** Diseases caused by the loss-of-function mutations in genes are within green boxes and those caused by gain-of-function mutations are within red boxes. A mouse model for VEGF-C overexpression in bones causes GSD. Whether VEGF-C is overexpressed in human GSD patients is unknown. Mutations in TSC1/2 are observed in the tumor cells of LAM.

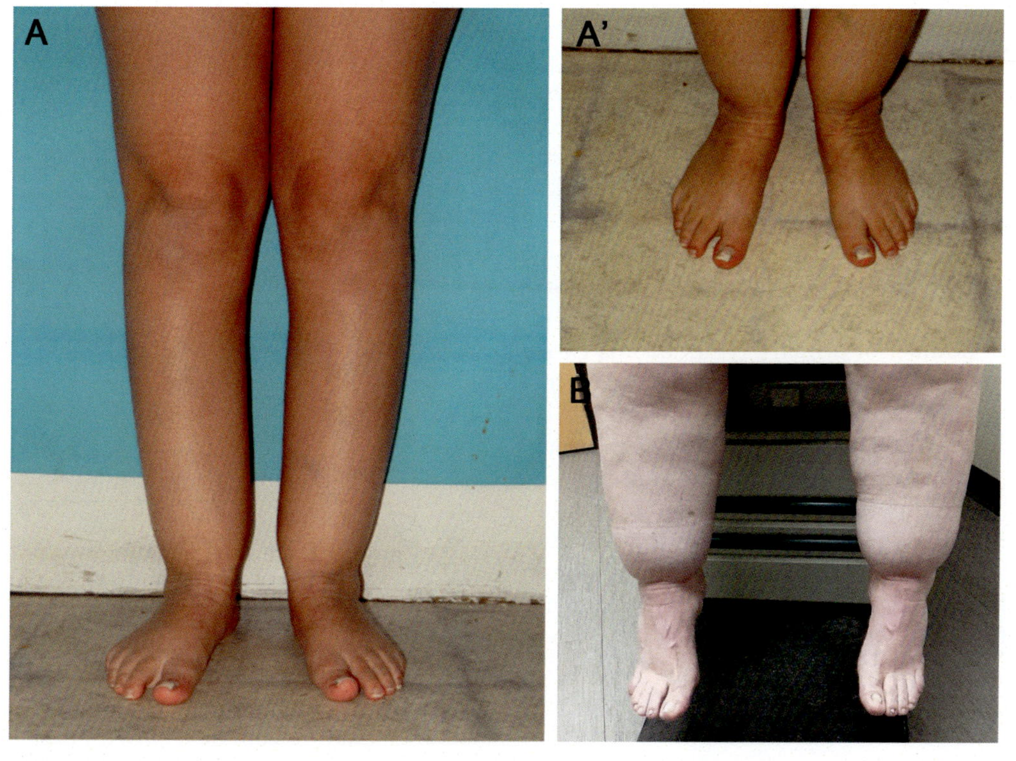

FIGURE 25.9 **Lipedema.** (A) and (B) are pictures from two patients. Note disproportionate deposit below the knees, and sparing of the feet.

Syndromic lymphedema

Primary lymphedema could be associated with syndromes such as Noonan syndrome or Turner syndrome.[47,48] Noonan syndrome is a "RASopathy" caused by hyperactivation of RAS/MAPK pathway. Mutations in PTPN11 are most frequently (>50%) associated with Noonan syndrome. Other common genes include *SOS1*, *RAF*, *KRAS*, *HRAS*, and *BRAF*. Learning disability and webbed shoulders are some of the associated disorders. Turner syndrome is caused by the absence of one sex chromosome (45, X). The lymphedema causing gene is currently unknown.

Lymphedema with visceral involvement

This group of lymphatic anomalies includes lymphatic-related nonimmune hydrops fetalis, lymphedema, intestinal lymphangiectasia, and pulmonary lymphangiectasia. Bone defects have not been reported in these patients. Causative loss-of-function mutations have been identified in SOX18, CCBE1, FAT4, ADAMTS3, and PIEZO1.[19] Gordon and colleagues have also included the previously described CCLA in this group.

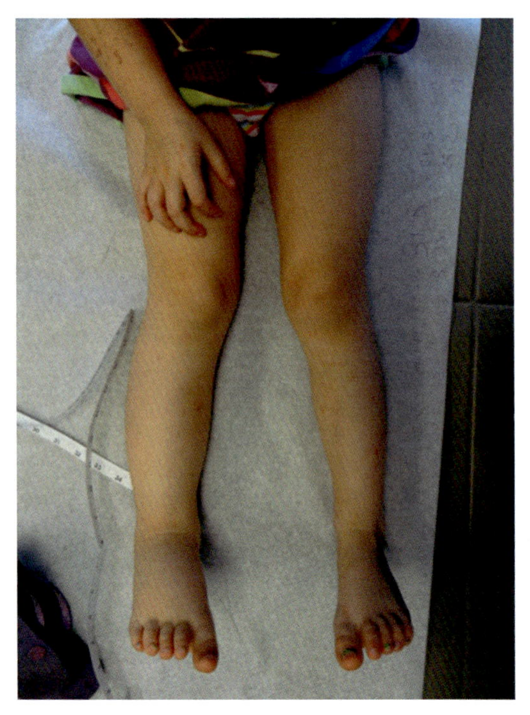

FIGURE 25.10 **Congenital lymphedema.** A 3-year-old girl with Milroy disease. Notice the swelling in the right leg.

Congenital lymphedema

In the absence of systemic involvement, primary lymphedema of the limbs is classified based on the time of onset. Congenital lymphedema is observed before birth or presents shortly after birth. Skin changes are pronounced, and the nails may be upturned. Milroy disease or Nonne–Milroy disease is caused by loss-of-function mutation in FLT4/VEGFR3 (Fig. 25.10). Loss-of-function mutations in VEGF-C, which is the ligand for VEGFR3, are also associated with congenital lymphedema.[19]

Lymphedema praecox

Predominant number of primary lymphedema cases are identified around adolescence.[19] Yellowing of the nails is sometimes observed. Meige's disease of unknown etiology, lymphedema–distichiasis syndrome caused by heterozygous mutations in the transcription factor forkhead box protein C2 (FOXC2), Emberger syndrome caused by heterozygous mutations in the transcription factor GATA2, and heterozygous missense mutations in the gap junction molecule GJC2/Connexin-47 are associated with lymphedema praecox.

Lymphedema tarda

This form of lymphedema occurs after the age of 35 years. Onset of this disease could be triggered by infections or cellulitis.[49] Lymphedema tarda is inherited although causal genes are unknown.

Most of the listed lymphedema-associated genes could be grouped into two categories: genes that regulate VEGF-C signaling and those that regulate shear stress response in LECs (Fig. 25.11). The shear stress pathway could also enhance VEGF-C signaling thereby suggesting that VEGF-C signaling could be the central player that prevents the onset of lymphedema.

While we have listed the most commonly observed mutations in lymphedema, mutations in *ANGPT2*, *ITGA9*, or approximately 30 other genes could also cause lymphedema.[50] Even so, these genes account for only 30% of congenital cases of lymphedema. Hence, further work is needed to fully understand the molecular mechanisms of lymphedema.

Conclusions

We have attempted to describe the various lymphatic anomalies that are observed in human patients. Despite careful classification, we acknowledge that there is significant overlap in diagnostic criteria. For example, some GLA patients have cortical bone lysis that resembles GSD, some GLA patients have intralesional bleeding that resembles KLA, and LM patients frequently have lymphedema. Furthermore, mutations in some genes (e.g., PIK3CA) result in a variety of

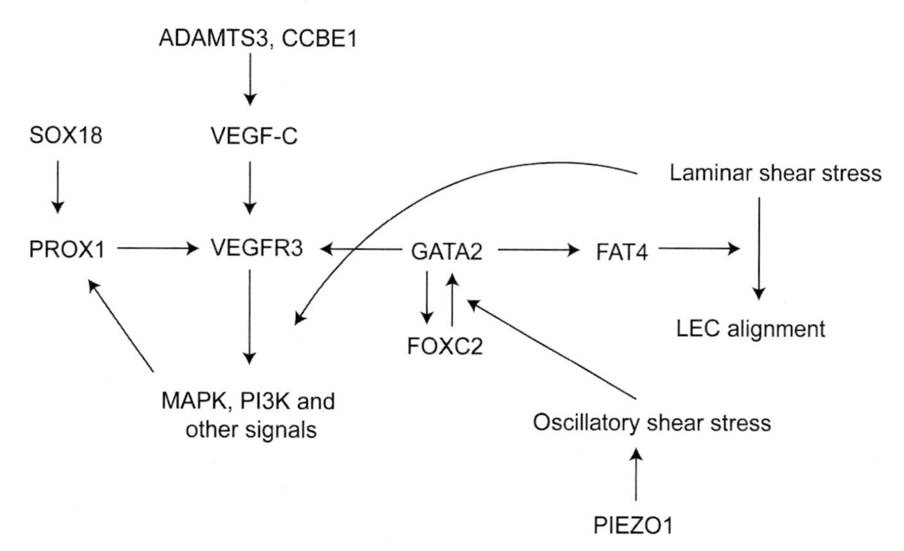

FIGURE 25.11 Functional relationship between genes that are associated with primary lymphedema. All except PROX1 are molecules that are found mutated in patients with lymphedema. SOX18, PROX1, GATA2, and FOXC2 are transcription factors; ADAMTS3 and CCBE1 are proteases that process and activate VEGF-C; VEGFR3 is a receptor tyrosine kinase that is the cognate receptor for VEGF-C; PIEZO1 is a mechanosensitive ion channel and FAT4 is a proto-cadherin.

disorders. Likewise, mutations in genes that presumably function in the same pathway (e.g., ARAF, NRAS, and RASA1) could also result in distinct anomalies. It is argued that the time of the onset of mutation or the cells (LEC progenitors vs. mature LECs) that acquire the mutation could dictate the disease outcome. However, we cannot exclude the possibility that these differences are due to the myriad genetic and environmental variations between the individuals. Hence, as stated recently, we need to be open to the possibility that the current classifications could be artificial.[35] Identification of the disease-causing mutations could be more helpful than that of disease classification to determine the treatment option. The Next Generation Sequencing technology is developing rapidly and it has revealed that most LMs are caused by abnormal activation of PI3K or MAPK pathways. It is satisfying to report that drugs targeting these pathways are being successfully used to treat patients. Further understanding of the pathological mechanisms could assist in tailoring the treatment to the individual patient. However, technical challenges do exist. For example, in case of diseases with somatic variations, only a small number of cells within the lesions appear to carry the causative mutations, and in some cases, mutations could not be identified.[26] Lymphatic vasculature is a network of interconnected vessels. Therefore, damage to an upstream structure (such as the LVV) could affect lymphatic drainage and cause lesions at downstream vessels. We need approaches to identify and biopsy such upstream structures. Approaches to repair these upstream structures could also be highly effective in treating the lymphatic disorders.

Acknowledgment

This work is supported by NIH/NHLBI (R01HL131652 and R01HL133216 to RSS), NIH/NIGMS COBRE (P20 GM103441 to XG; PI: Dr. Rodger McEver), and Oklahoma Center for Adult Stem Cell Research (4340 to RSS). We thank Dr. Michael Dellinger (UT Southwestern Medical Center and The Lymphatic Malformation Institute) for several helpful discussions. We are grateful to the patients who agreed to share their pictures for this work. We apologize to the authors of numerous articles that we are unable to cite due to space restrictions.

References

1. Oliver G, Kipnis J, Randolph GJ, Harvey NL. The lymphatic vasculature in the 21(st) century: Novel functional roles in homeostasis and disease. *Cell*. 2020;182(2):270–296.
2. Tammela T, Alitalo K. Lymphangiogenesis: molecular mechanisms and future promise. *Cell*. 2010;140(4):460–476.
3. Vondenhoff MF, van de Pavert SA, Dillard ME, et al. Lymph sacs are not required for the initiation of lymph node formation. *Development*. 2009;136(1):29–34.
4. Onder L, Morbe U, Pikor N, et al. Lymphatic endothelial cells control initiation of lymph node organogenesis. *Immunity*. 2017;47(1), 80.e84–92.e84.
5. Bovay E, Sabine A, Prat-Luri B, et al. Multiple roles of lymphatic vessels in peripheral lymph node development. *J Exp Med*. 2018; 215(11):2760–2777.
6. Jalkanen S, Salmi M. Lymphatic endothelial cells of the lymph node. *Nat Rev Immunol*. 2020;20(9):566–578.

7. Xiang M, Grosso RA, Takeda A, et al. A single-cell transcriptional roadmap of the mouse and human lymph node lymphatic vasculature. *Front Cardiovasc Med*. 2020;7:52.
8. Bernier-Latmani J, Petrova TV. Intestinal lymphatic vasculature: structure, mechanisms and functions. *Nat Rev Gastroenterol Hepatol*. 2017;14(9):510–526.
9. Jakus Z, Gleghorn JP, Enis DR, et al. Lymphatic function is required prenatally for lung inflation at birth. *J Exp Med*. 2014;211(5): 815–826.
10. Baluk P, Tammela T, Ator E, et al. Pathogenesis of persistent lymphatic vessel hyperplasia in chronic airway inflammation. *J Clin Invest*. 2005;115(2):247–257.
11. Baluk P, Naikawadi RP, Kim S, et al. Lymphatic proliferation ameliorates pulmonary fibrosis after lung injury. *Am J Pathol*. 2020; 190(12):2355–2375.
12. Cui Y, Liu K, Monzon-Medina ME, et al. Therapeutic lymphangiogenesis ameliorates established acute lung allograft rejection. *J Clin Invest*. 2015;125(11):4255–4268.
13. Jacob L, Boisserand LSB, Geraldo LHM, et al. Anatomy and function of the vertebral column lymphatic network in mice. *Nat Commun*. 2019;10(1):4594.
14. Ma Q, Decker Y, Muller A, Ineichen BV, Proulx ST. Clearance of cerebrospinal fluid from the sacral spine through lymphatic vessels. *J Exp Med*. 2019;216(11):2492–2502.
15. Miller AJ, Pick R, Katz LN, Jones C, Rodgers J. Ventricular endomyocardial changes after impairment of cardiac lymph flow in dogs. *Br Heart J*. 1963;25(2):182–190.
16. Klotz L, Norman S, Vieira JM, et al. Cardiac lymphatics are heterogeneous in origin and respond to injury. *Nature*. 2015;522(7554): 62–67.
17. Liu X, De la Cruz E, Gu X, et al. Lymphoangiocrine signals promote cardiac growth and repair. *Nature*. 2020;588(7839):1–7.
18. Wassef M, Blei F, Adams D, et al. Vascular anomalies classification: recommendations from the International Society for the study of vascular anomalies. *Pediatrics*. 2015;136(1):e203–e214.
19. Gordon K, Varney R, Keeley V, et al. Update and audit of the St George's classification algorithm of primary lymphatic anomalies: a clinical and molecular approach to diagnosis. *J Med Genet*. 2020; 57(10):653–659.
20. Ji Y, Chen S, Yang K, Xia C, Li L. Kaposiform hemangioendothelioma: current knowledge and future perspectives. *Orphanet J Rare Dis*. 2020;15(1):39.
21. Le Huu AR, Jokinen CH, Rubin BP, et al. Expression of prox1, lymphatic endothelial nuclear transcription factor, in Kaposiform hemangioendothelioma and tufted angioma. *Am J Surg Pathol*. 2010;34(11):1563–1573.
22. Henske EP, McCormack FX. Lymphangioleiomyomatosis – a wolf in sheep's clothing. *J Clin Invest*. 2012;122(11):3807–3816.
23. Castillo SD, Baselga E, Graupera M. PIK3CA mutations in vascular malformations. *Curr Opin Hematol*. 2019;26(3):170–178.
24. Martinez-Corral I, Zhang Y, Petkova M, et al. Blockade of VEGF-C signaling inhibits lymphatic malformations driven by oncogenic PIK3CA mutation. *Nat Commun*. 2020;11(1):2869.
25. Ozeki M, Fukao T. Generalized lymphatic anomaly and Gorham-Stout disease: overview and recent insights. *Adv Wound Care*. 2019;8(6):230–245.
26. Rodriguez-Laguna L, Agra N, Ibanez K, et al. Somatic activating mutations in PIK3CA cause generalized lymphatic anomaly. *J Exp Med*. 2019;216(2):407–418.
27. Croteau SE, Kozakewich HP, Perez-Atayde AR, et al. Kaposiform lymphangiomatosis: a distinct aggressive lymphatic anomaly. *J Pediatr*. 2014;164(2):383–388.
28. Crane J, Manfredo J, Boscolo E, et al. Kaposiform lymphangiomatosis treated with multimodal therapy improves coagulopathy and reduces blood angiopoietin-2 levels. *Pediatr Blood Cancer*. 2020; 67(9):e28529.
29. Barclay SF, Inman KW, Luks VL, et al. A somatic activating NRAS variant associated with kaposiform lymphangiomatosis. *Genet Med*. 2019;21(7):1517–1524.
30. Ozeki M, Aoki Y, Nozawa A, et al. Detection of NRAS mutation in cell-free DNA biological fluids from patients with kaposiform lymphangiomatosis. *Orphanet J Rare Dis*. 2019;14(1):215.
31. Dellinger MT, Garg N, Olsen BR. Viewpoints on vessels and vanishing bones in Gorham-Stout disease. *Bone*. 2014;63:47–52.
32. Lala S, Mulliken JB, Alomari AI, Fishman SJ, Kozakewich HP, Chaudry G. Gorham-Stout disease and generalized lymphatic anomaly – clinical, radiologic, and histologic differentiation. *Skeletal Radiol*. 2013;42(7):917–924.
33. Hominick D, Silva A, Khurana N, et al. VEGF-C promotes the development of lymphatics in bone and bone loss. *ELife*. 2018;7:e34323.
34. Li D, Wenger TL, Seiler C, et al. Pathogenic variant in EPHB4 results in central conducting lymphatic anomaly. *Hum Mol Genet*. 2018;27(18):3233–3245.
35. Li D, March ME, Gutierrez-Uzquiza A, et al. ARAF recurrent mutation causes central conducting lymphatic anomaly treatable with a MEK inhibitor. *Nat Med*. 2019;25(7):1116–1122.
36. Cohen Jr MM. Klippel-Trenaunay syndrome. *Am J Med Genet*. 2000; 93(3):171–175.
37. Osborn AJ, Dickie P, Neilson DE, et al. Activating PIK3CA alleles and lymphangiogenic phenotype of lymphatic endothelial cells isolated from lymphatic malformations. *Hum Mol Genet*. 2015; 24(4):926–938.
38. Lopez-Gutierrez JC, Lapunzina P. Capillary malformation of the lower lip, lymphatic malformation of the face and neck, asymmetry and partial/generalized overgrowth (CLAPO): report of six cases of a new syndrome/association. *Am J Med Genet A*. 2008; 146A(20):2583–2588.
39. Le Cras TD, Goines J, Lakes N, et al. Constitutively active PIK3CA mutations are expressed by lymphatic and vascular endothelial cells in capillary lymphatic venous malformation. *Angiogenesis*. 2020;23(3):425–442.
40. Lapinski PE, Doosti A, Salato V, North P, Burrows PE, King PD. Somatic second hit mutation of RASA1 in vascular endothelial cells in capillary malformation-arteriovenous malformation. *Eur J Med Genet*. 2018;61(1):11–16.
41. Burrows PE, Gonzalez-Garay ML, Rasmussen JC, et al. Lymphatic abnormalities are associated with RASA1 gene mutations in mouse and man. *Proc Natl Acad Sci U S A*. 2013;110(21):8621–8626.
42. Chen D, Geng X, Lapinski PE, Davis MJ, Srinivasan RS, King PD. RASA1-driven cellular export of collagen IV is required for the development of lymphovenous and venous valves in mice. *Development*. 2020;147(23):dev192351.
43. Chen D, Teng J, North P, Lapinski PE, King PD. RASA1-dependent cellular export of collagen IV controls blood and lymphatic vascular development. *J Clin Invest*. 2019;129(9):3545–3561.
44. Lapinski PE, Lubeck BA, Chen D, et al. RASA1 regulates the function of lymphatic vessel valves in mice. *J Clin Invest*. 2017;127(7):2569–2585.
45. Lapinski PE, Kwon S, Lubeck BA, et al. RASA1 maintains the lymphatic vasculature in a quiescent functional state in mice. *J Clin Invest*. 2012;122(2):733–747.
46. Ho YC, Srinivasan RS. Lymphatic vasculature in energy homeostasis and obesity. *Front Physiol*. 2020;11:3.
47. Atton G, Gordon K, Brice G, et al. The lymphatic phenotype in Turner syndrome: an evaluation of nineteen patients and literature review. *Eur J Hum Genet*. 2015;23(12):1634–1639.
48. Joyce S, Gordon K, Brice G, et al. The lymphatic phenotype in Noonan and Cardiofaciocutaneous syndrome. *Eur J Hum Genet*. 2016;24(5):690–696.
49. Burgos JA, Luginbuhl A. Images in clinical medicine. Lymphedema tarda. *N Engl J Med*. 2009;360(10):1015.
50. Brouillard P, Boon L, Vikkula M. Genetics of lymphatic anomalies. *J Clin Invest*. 2014;124(3):898–904.

PART IV

The vasculome as perpetrator and victim in local and systemic diseases

CHAPTER

26

Endothelial dysfunction: basis for many local and systemic conditions

Thomas Münzel[1,2], Omar Hahad[1,2] and Andreas Daiber[1,2]

[1]Department of Cardiology, Medical Center of the Johannes Gutenberg University, Mainz, Germany [2]German Center for Cardiovascular Research (DZHK), Mainz, Germany

Introduction

The contribution of different risk factors to the global disease burden and their impact on life expectancy has shifted over the last 20 years from risks for communicable childhood diseases toward those for noncommunicable adulthood diseases that are more frequently observed in the elderly,[1,2] with a major contribution of cardiovascular risk factors such as arterial hypertension and cardiovascular disease such as ischemic heart disease (Fig. 26.1A).[3–5] This shift is largely due to demographic changes, improved clinical prevention of childhood mortality, reductions in several preventable causes of death, and lower exposures to some risk factors (e.g., improved water quality and sanitation). However, there are large regional differences in the extent of these epidemiological and socio-economic changes and their impact on the importance of different risk factors, disease burden, and mortality (e.g., poverty and childhood diseases continue as leading risk factors in sub-Saharan Africa).

Based on the aforementioned data on a major role of cardiovascular disease and risk factors for global deaths and disability-adjusted life years, physicians and pharmacologists have for several decades searched for a reliable and in particular an early predictor of cardiovascular mortality. One of the most promising candidates still is the measurement of endothelial function (encompassing production of the different endothelium-derived messengers that help to control vascular tone, blood flow, immune cell, and platelet activity/adhesion, thereby regulating perfusion and/or blood pressure), which can be considered as an important biomarker correlating positively with classical markers of inflammation, obesity and cardiovascular risk such as C-reactive protein (CRP), adiponectin, and brain natriuretic peptide (BNP).[6–8] Classic cardiovascular risk factors such as arterial hypertension,[9] hypercholesterolemia,[10] diabetes mellitus,[11] and chronic smoking[12] but also novel cardiovascular risk factors such as noise,[13] air pollution,[14] and e-cigarette smoking[15] are all associated with endothelial dysfunction. The presence of several risk factors produces synergistic effects on endothelial function as well as the associated cardiovascular prognosis.[16] Previous studies confirmed that hypercholesterolemia or chronic smoking leads to moderate impairment of endothelial function (reduction of the maximal acetylcholine-dependent vasodilation by ~ 30%), whereas the presence of both risk factors caused severe endothelial dysfunction (reduction of the maximal acetylcholine-dependent vasodilation by ~ 60%).[17]

Physiological role of endothelial (vascular) function

The endothelium, a single-layered continuous cell sheet lining the luminal vessel wall, separates the intravascular (blood) from the interstitial compartment and the vascular smooth muscle. Based on cell count (6×10^{13}), mass (1.5 kg), and surface area, (1000 m^2) the endothelium is an autonomous organ. Though for a long time regarded as a passive barrier for blood cells and macrosolutes, this view completely changed with the advent of endothelial autacoids like prostacyclin (PGI_2)[18] and nitric oxide (•NO),[19] as well as with the

The Vasculome
https://doi.org/10.1016/B978-0-12-822546-2.00011-3

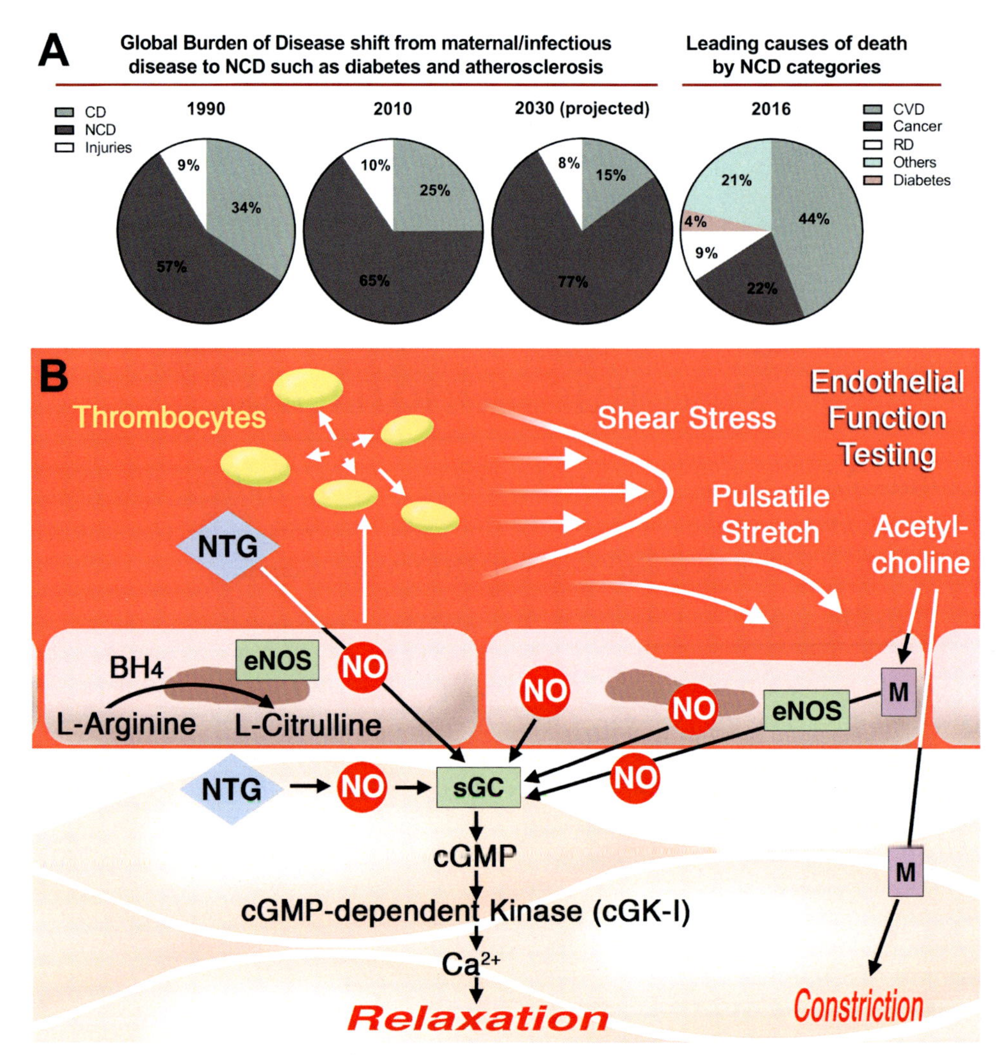

FIGURE 26.1 **(A) Global burden of (noncommunicable) disease and endothelial function as a prognostic parameter for cardiovascular events and mortality.**[168] (A) The proportional distribution of global deaths attributable to communicable diseases, noncommunicable diseases (NCDs), and injuries for 1990, 2010, and projected for 2030. 1990 and 2010 based on data from Lozano et al.[3]; 2030 projected by WHO data.[4] Global WHO status report on noncommunicable diseases 2016 estimating the leading causes of death by different categories of NCDs. Global Health Observatory (GHO) data.[5] (B) Regulation of vascular tone by the endothelium. The endothelial nitric oxide synthase (eNOS) synthesizes NO by a two-step oxidation of the amino acid L-arginine thereby leading to the formation of L-citrulline. NO is released into the bloodstream inhibiting platelet aggregation and the release of vasoconstricting factors such as serotonin and thromboxane. NO diffuses also into the media and activates the soluble guanylate cyclase (sGC). The resulting second messenger cGMP in turn activates the cGMP-dependent kinase, which mediates decreases in intracellular Ca^{2+} concentrations causing vasorelaxation. The physiological stimuli to release NO are shear stress and pulsatile stretch. Intraarterial infusion is used in the clinics to assess endothelial function. Infused into the forearm (brachial artery), acetylcholine (ACh) causes a dose-dependent vasodilation. In the coronary artery, the response (vasoconstriction vs. vasodilation) strictly depends on the functional integrity of the endothelium. In the presence of cardiovascular risk factors and endothelial dysfunction, ACh will cause vasoconstriction due to stimulation of muscarinergic receptors in the media. muscarinic acetylcholine receptor (M), nitroglycerin (NTG). *(A) Pie charts were adapted from Daiber A., Kuntic M., Hahad O., et al. Effects of air pollution particles (ultrafine and fine particulate matter) on mitochondrial function and oxidative stress - Implications for cardiovascular and neurodegenerative diseases. Arch Biochem Biophys. 2020;696:108662. With permission of Elsevier Inc. Copyright © 2020, Elsevier Inc. All rights reserved. (B) Adapted from Munzel T. Endothelial dysfunction: pathophysiology, diagnosis and prognosis. Dtsch Med Wochenschr. 2008; 133(47):2465e2470. With permission of Georg Thieme Verlag KG Stuttgart. Copyright © 2008, Rights Managed by Georg Thieme Verlag KG Stuttgart • New York. All rights reserved.*

discovery of integrins and other surface signals.[20] It is now evident that the endothelium is not only at the cross-bridges of communication between blood and tissue by interaction with surrounding cells by a plethora of signaling routes. One of the prominent communication lines is established by the so-called L-arginine-•NO-cyclic guanosine monophosphate (GMP) pathway.[21] This signaling cascade starts with endothelial

NO synthase (eNOS, NOSIII), which generates •NO and L-citrulline from L-arginine and O_2 in response to receptor-dependent agonists (bradykinin, acetylcholine, adenosine triphosphate (ATP)) and physicochemical stimuli (shear, stretch) (Fig. 26.1B).[22]

•NO diffuses to the adjacent smooth muscle where it interacts with different receptor molecules, of which the soluble guanylate cyclase (sGC) is the best characterized and presumably most important one with regard to control of vessel tone and smooth muscle proliferation. Activation by •NO requires sGC heme-iron to be in the ferrous (II) state. Upon •NO binding, cGMP formation will increase substantially. cGMP in turn activates the cGMP-dependent kinase I, which in turn will increase the open probability of Ca^{2+}-activated K^+ (BK)-channels, thereby inducing a hyperpolarization of the smooth muscle cells and inhibition of agonist-induced Ca^{2+} influx. In addition, activated cyclic GMP-dependent kinase I (cGK-Ib) phosphorylates the inositol trisphosphate (IP3) receptor—associated G-kinase substrate, thereby inhibiting agonist-induced Ca^{2+} release and smooth muscle contraction. Another cGK-I substrate found in many cell types is the 46/50 kD vasodilator-stimulated phosphoprotein (VASP). cGK-I phosphorylates VASP specifically at serine 239, and this reaction can be exploited as a biochemical monitor for the integrity and activity of the •NO-cGMP pathway.[23]

Mechanisms of endothelial dysfunction

Oxidative stress

Although the mechanisms underlying endothelial dysfunction may be multifactorial, there is a growing body of evidence that increased production of reactive oxygen species (ROS) may contribute considerably to this phenomenon.[24,25] ROS production has been demonstrated to occur in the endothelial cell layer and also within the media and adventitia, all of which may impair NO signaling within vascular tissue to endothelium-dependent but also endothelium-independent vasodilators. More recent experimental but also clinical studies point to the pathophysiological importance of the xanthine oxidase (XO), the vascular (NOX1) and phagocytic (NOX2) NADPH oxidase, mitochondria, and an uncoupled endothelial nitric oxide as significant enzymatic superoxide source (Fig. 26.2). Therefore, it is of great importance to reveal the mechanisms underlying endothelial dysfunction with focus on oxidative stress, also by routine detection of suitable redox biomarkers in clinical studies.[26] The clinical methods for assessment of endothelial function are of prognostic and diagnostic relevance,[16] also in order to test the efficacy of therapeutic strategies for improvement of endothelial dysfunction by reducing oxidative stress.[27–29]

Primary sources of ROS such as NADPH oxidases and the mitochondrial respiratory chain can generate secondary sources of ROS by oxidative conversion of xanthine dehydrogenase to XO or NO-producing endothelial nitric oxide synthase (eNOS) to the uncoupled superoxide-generating enzyme[30] (summarized in Fig. 26.2). NADPH oxidases are highly specialized ROS producing protein complexes with almost no other biological function.[31] Interestingly, NOX2, the phagocytic isoform of the NADPH oxidase, has been demonstrated to be markedly upregulated in the presence of new cardiovascular risk factors including noise,[32] air pollution,[33] and e-cigarette smoking.[15] Interestingly, NOX2 knockout experiments almost completely abolished endothelial dysfunction, oxidative stress, and inflammation in the vasculature and the brain pointing to a crucial role of NOX2 in mediating cardiovascular and cerebral side effects of these environmental stressors.[15,33,34]

Mitochondria, under normal conditions, produce ROS as a byproduct of the excessive use of oxygen, transport of electrons from reducing factors along the respiratory complexes, and escape of these electrons to molecular oxygen (up to 1% of the total flow of electrons by estimation).[35] In pathophysiological situations, e.g., upon ischemia/reperfusion, the (redox) activation of $p66^{Shc}$ or monoamine oxidases, activation of mitochondrial ATP-sensitive potassium channels, and opening of the mitochondrial permeability transition pore (mPTP) can dramatically increase the formation and release of ROS from mitochondria.[36] Also an ROS-induced ROS formation via redox-based amplification mechanisms was described.[37,38] ROS can exert beneficial and detrimental effects, which makes their therapeutic targeting so difficult and so far mostly unsuccessful (reviewed extensively in[39]). They can act protective via preconditioning-like mechanisms, kill pathogens during host defense, or even contribute to essential cellular redox signaling (e.g., in cell differentiation, proliferation, and migration). Vice versa, after exceeding a certain threshold, ROS confer adverse effects including oxidative damage of various biomolecules, activation of inflammatory, and cell death (e.g., apoptotic) pathways.

The endothelium-derived relaxing factor, previously identified as NO[19] or a closely related compound,[40] has potent antiatherosclerotic properties. NO released from endothelial cells works in concert with prostacyclin to inhibit platelet aggregation:[40] it inhibits the adhesion of neutrophils to endothelial cells and the expression of adhesion molecules. NO in high concentrations inhibits the proliferation of smooth muscle cells.[41] Therefore, under all conditions where an absolute or relative NO deficit is encountered, the process of atherosclerosis is being initiated or accelerated. The half-life of NO and therefore its biological activity is decisively determined

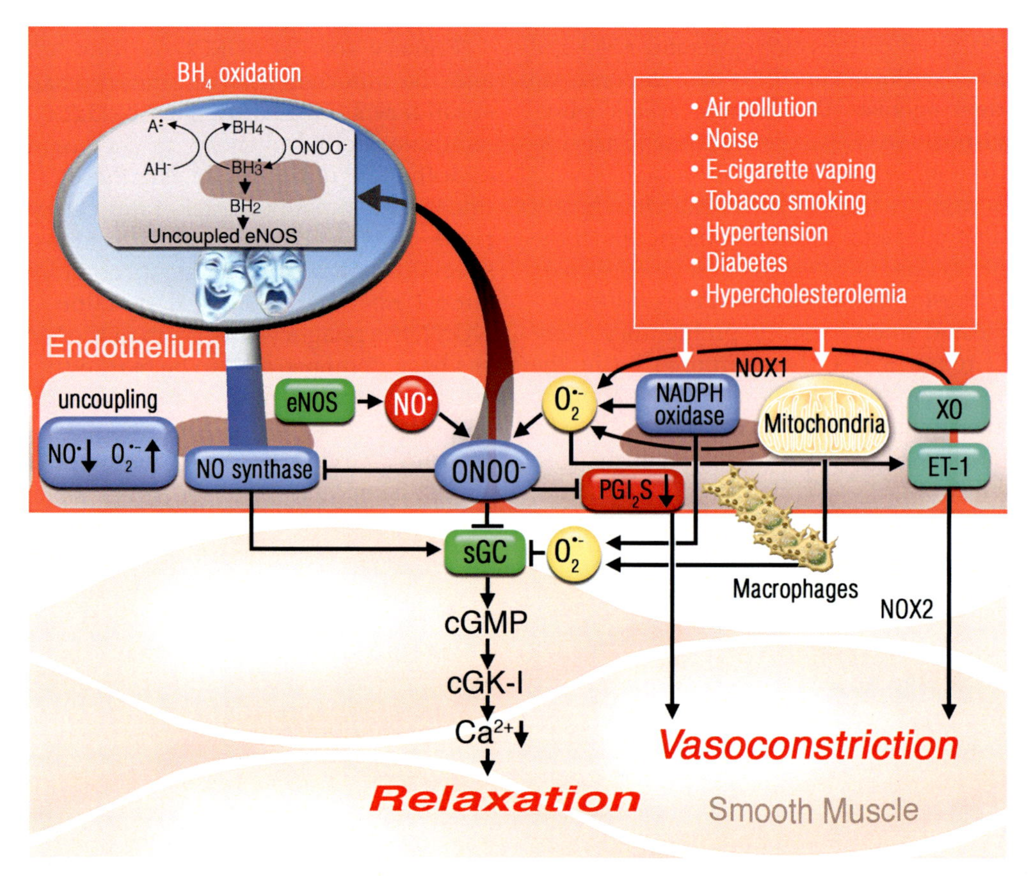

FIGURE 26.2 Mechanisms underlying endothelial dysfunction and the functional consequences of decreased vascular bioavailability of nitric oxide (•NO).[169] An unbalanced production of nitric oxide and superoxide ($^{\cdot}O_2^-$) secondary to cardiovascular risk factor exposure leads to inappropriate formation of superoxide by the nicotinamide dinucleotide phosphate oxidase, xanthine oxidase, and the mitochondria. Superoxide immediately reacts with nitric oxide to form the highly reactive intermediate peroxynitrite, which causes vascular dysfunction through different mechanisms including nitric oxide synthase uncoupling secondary to BH_4 oxidation (inset), tyrosine nitration of the prostacyclin synthase (PGI_2S), and subsequent inhibition of PGI2 formation and by inhibiting the activity of the soluble guanylate cyclase (sGC) and therefore nitric oxide signaling. Superoxide per se is also known to potently stimulate the expression of endothelin-1 (ET-1) in endothelial and smooth muscle cells, shifting the balance from vasodilation more toward vasoconstriction.[170] Additional consequences are adhesion and infiltration of the vascular wall with inflammatory cells such as macrophages (containing NADPH oxidase isoform NOX2) and neutrophils, which are further increasing oxidative stress and are further worsening endothelial dysfunction. Importantly, there is a close relation of endothelial function in the human coronary and peripheral circulations, thus endothelial dysfunction of the brachial artery can be considered as a readout for coronary endothelial dysfunction.[58] *Updated and reused from Munzel T, Gori T, Bruno RM, Taddei S. Is oxidative stress a therapeutic target in cardiovascular disease?* Eur Heart J *2010;31(22):2741–2748. With permission of Oxford University Press. Copyright © 2010, Oxford University Press.*

by oxygen-derived free radicals such as superoxide.[42] Superoxide rapidly reacts with NO to form the highly reactive intermediate peroxynitrite ($ONOO^-$).[43] The rapid bimolecular reaction between NO and superoxide yielding peroxynitrite (rate constant: $5–10 \times 10^9$ $M^{-1}s^{-1}$) is about 3–4 times faster than the dismutation of superoxide by the superoxide dismutase.[44] Therefore, peroxynitrite formation represents a major potential pathway of NO reactivity depending on the rates of tissue superoxide production. Peroxynitrite in high concentrations is cytotoxic and may cause oxidative damage to proteins, lipids, and DNA.[45] Studies also indicate that peroxynitrite may have deleterious effects on activity and function of the prostacyclin synthase[46] and the eNOS.[47,48] Other ROS, such as the dismutation product of superoxide, hydrogen peroxide, and the hypochlorous acid released by activated neutrophils, are not free radicals, but have a powerful oxidizing capacity, which will further contribute to oxidative stress and eNOS uncoupling within vascular tissue.

Inflammation

According to recent clinical studies, cardiovascular risk is significantly increased in patients with chronic autoimmune diseases such as systemic lupus erythematosus, rheumatoid arthritis, and severe psoriasis.[49–52] It is not surprising that psoriasis has been proposed as a risk factor for future cardiovascular events independently of the classical cardiovascular risk factors such as smoking, obesity, and diabetes.[53] The guidelines of the European League against Rheumatism strongly emphasize the increased cardiovascular risk associated, e.g., with inflammatory arthritis.[54] Chronic inflammation causes a significant increase in the intima-media thickness and is associated with impaired vascular function in patients with rheumatoid arthritis.[55]

Prognostic value of endothelial function and its measurement

As mentioned, endothelial dysfunction is the first clinical correlate of atherosclerosis that can be diagnosed.[9,10] Since most cardiovascular diseases are either related to or are a direct consequence of atherosclerosis, endothelial dysfunction is an early predictor of subsequent cardiovascular events or mortality[56,57] (Fig. 26.3A and B). Patients with peripheral arterial occlusive disease,[58,59] coronary artery disease,[60–62] or heart failure[63] demonstrate impaired endothelial function. Endothelial dysfunction is also clearly associated with oxidative stress,[60] which is also another feature in the development of atherosclerosis.[64–66] Levels of oxidized thiols were also negatively correlated with cumulative survival in a large-scale clinical trial of patients undergoing coronary angiography, thus providing strong evidence for a role of oxidative stress in cardiovascular disease-associated mortality.[67]

The most reliable parameters with prognostic value currently are circulating markers such as BNP or CRP as well as risk scores that are based on calculations considering different risk factors/markers such as Framingham Risk Score, Synergy Between PCI With Taxus, and Cardiac Surgery Score.[68,69] During the last 15 years, a number of clinical studies supported the prognostic importance of coronary and peripheral measurement of endothelial dysfunction in patients with coronary artery disease,[61] peripheral arterial occlusive disease,[58] arterial hypertension,[70] postmenopausal women,[71] and heart failure[63] but also in healthy subjects.[72,73] Of interest is a study assessing endothelial function (measured by flow-mediated dilation, FMD) for the prediction of future cardiovascular events in patients undergoing coronary bypass surgery: the results were expressed in tertiles representing the lowest event rate (normal FMD >8%), an intermediate event rate (FMD = 4%–8%), and the highest event rate (FMD <4%).[56] Another large cohort study identified the hyperemic velocity, a stimulus for FMD and also a marker of microvascular function, but not FMD itself, as a prognostic marker for future cardiovascular events.[74]

There are several established techniques to determine endothelial function in vivo (Fig. 26.3C–F). We mainly focus on those techniques that are routinely applied in clinical studies or emerging techniques that may be of relevance for human diagnostics in the future. Some more sophisticated methods, including biochemical markers and bioassays, measurement of endothelial-derived microparticles, and progenitor cells, are discussed elsewhere.[75] A helpful correlation matrix of multiple parameters of endothelial function, arterial stiffness, and end-points of atherosclerosis can be found elsewhere.[76] The early stages of atherosclerosis are characterized by endothelial dysfunction, while the later stages result in arterial stiffness, both of which can be quantified by different techniques. Pulse wave velocity (PWV) has emerged as an established method to examine arterial stiffness and has a strong correlation with cardiovascular outcomes.[77] Applanation tonometry and Doppler ultrasound are also available clinical methods for evaluating global PWV.[77–79] However, the gold standard of measuring PWV remains flow meter or catheter-based pressure probes. Nevertheless, their invasiveness limits their use in clinical studies.[80] In recent years, magnetic resonance imaging has become more accessible and is another minimally invasive method to determine PWV.[81–83]

Determination of endothelial function by intracoronary infusion of acetylcholine and forearm plethysmography

Acetylcholine binds to muscarinic receptors on endothelial cells leading to eNOS activation, elevation of vascular •NO production, and vasodilation of blood vessels.[84] The relaxation of coronary vessels upon infusion of acetylcholine can be detected by quantitative coronary angiography or measurement of coronary blood flow[61] (Fig. 26.3C). Similarly, acetylcholine infusion can be applied to any peripheral resistance vessel (e.g., in the forearm) and coupled with ultrasound measurement of blood flow, which is termed forearm plethysmography[85] (Fig. 26.3D).

Flow-mediated dilation of the brachial artery and finger-pulse plethysmography (ENDO-PAT)

A noninvasive technique for measurement of endothelial function in vivo is quantification of FMD[86] (Fig. 26.3E). In this process, a baseline diameter of the brachial artery is first recorded by ultrasound; blood flow is then interrupted by inflation of a blood pressure cuff for 5 min, and restoration of blood flow leads to a reactive hyperemia that is associated with increased release of •NO. The subsequent vasodilation is again detected by ultrasound. Maximal dilation of the brachial artery is induced by sublingual application of nitroglycerin (NMD) and then used for normalization of FMD values. A newer noninvasive technique to detect endothelial function in humans is peripheral arterial tonometry, where a finger probe assesses digital volume changes and pulse waves that can be detected after induction of reactive hyperemia[87] (Fig. 26.3F). Other advanced techniques for the diagnosis of endothelial dysfunction are low flow-mediated constriction[88,89] and passive leg movement as a potentially practical assessment of systemic vascular function via •NO-dependent mechanisms.[90,91] The most traditional technique for the ex vivo assessment of endothelial dysfunction is the isometric tension method as first described in animal tissues by the Nobel laureate Robert Furchgott,[92] or in human vessels from bypass surgery.[93] An additional novel ex vivo technique to assess endothelial dysfunction is based on the identification of atherosclerotic lesions by Raman spectroscopy.[94] The advantages or disadvantages of these methods were discussed in full detail elsewhere.[16] Clearly, the noninvasive techniques are easier to apply to a large study population and the risks of side effects (e.g., puncture-induced bleeding complications) are lower.

Therapeutic prevention of endothelial (vascular) dysfunction

Besides therapy with established cardiovascular drugs or antioxidant approaches (summarized in Fig. 26.4),[95] there is emerging evidence that epigenetic pathways are redox regulated and play a causal role in cardiovascular diseases by directly affecting endothelial function (for review see[96,97]). Due to their reversibility, epigenetic processes represent ideal targets for therapeutic approaches. Treatments based on modulation of microRNAs, histone methylation/acetylation, or DNA methylation were reviewed previously.[27] Also lifestyle changes can improve endothelial function substantially, as shown for healthy diets (e.g., Mediterranean diet), exercise, and intermittent fasting.[27]

Established drugs improving endothelial dysfunction

Angiotensin-converting enzyme (ACE) inhibitors, AT1 receptor (angiotensin II type-1 receptor) antagonists, statins, and many other cardiovascular drugs display pleiotropic indirect antioxidant properties (e.g., inhibition of Nox enzymes and secondary to this prevention of eNOS uncoupling) and antiinflammatory effects leading to improved endothelial function (for review see[98–101]). Similar beneficial effects were described for endothelin-1 receptor antagonists.[102,103] Gemfibrozil is a lipid-lowering drug that also activates sGC in a nitric oxide- and hem-independent fashion, which may explain the more pronounced cardiovascular benefit observed with this drug as compared to other members of the fibrate group of drugs.[104]

Antioxidant therapy

As shown in Fig. 26.4 for the concept of oxidative stress in CVD in general, classical cardiovascular drugs decrease the risk factor-mediated activation of RAAS and inflammatory pathways, which largely defines their antioxidant profile.[101,105] In addition, cholesterol lowering or direct antioxidant and antiinflammatory effects by other pleiotropic properties may come into play. Additional redox therapeutic strategies including targeting specific ROS sources (e.g., inhibitors of NADPH oxidases and XO,[106] mitochondrial dysfunction/ROS formation[107,108]) are under investigation in clinical studies. Repair drugs or mimetics of vasoactive products that target oxidatively modified enzymatic regulators of the vascular tone such as eNOS, prostacyclin synthase, and soluble guanylyl cyclase have partially been translated into clinical therapy. Examples are the successful activators/stimulators of cGMP-producing soluble guanylyl cyclase,[109] phosphodiesterase (PDE) inhibitors that prevent degradation of cGMP,[110] and nitric oxide replacement drugs such as organic nitrates[111] as well as prostacyclin replacement drugs such as iloprost.[112] Endothelin-1 antagonization by $ET_{A/B}$-receptor blockade confers improved prognosis in patients with pulmonary hypertension,[113] which may be explained at least in part by decreased oxidative stress and inflammation, as reported by animal studies.[114,115]

On the other hand, despite encouraging preclinical data for pharmacological tetrahydrobiopterin supplementation/genetic inductions and eNOS enhancers, the strategy of preventing oxidative inhibition or uncoupling of eNOS remains largely unexplored in patients with CVD.[29] Similarly, modulators of mitochondrial function and ROS formation remain largely unexplored in patients with CVD, apart from some drugs that are used for the treatment of metabolic disease or diabetes such as glibenclamide.[116] Another example is the experimental clinical study on the inhibitor of mPTP cyclosporine A and beneficial effects in patients with myocardial infarction,[117] although for this drug, antiinflammatory actions may also come into play via immunosuppressive effects. Activators of Nrf2, which regulates antioxidant enzymes such as heme oxygenase-1 and peroxiredoxins, were evaluated in over 90 clinical trials for multiple indications and represent a leading therapeutic principle-based induction of endogenous antioxidant pathways.[118] The reasons for the failure of large clinical trials using oral antioxidant therapy are numerous and include, among others, the limited uptake of classic oral antioxidants in tissues facing oxidative stress, access to intracellular sites of ROS production, the limited reactivity toward specific ROS (e.g., hydrogen peroxide or superoxide), and most importantly, interference with essential physiological ROS signaling (for review, see[100,105,119]).

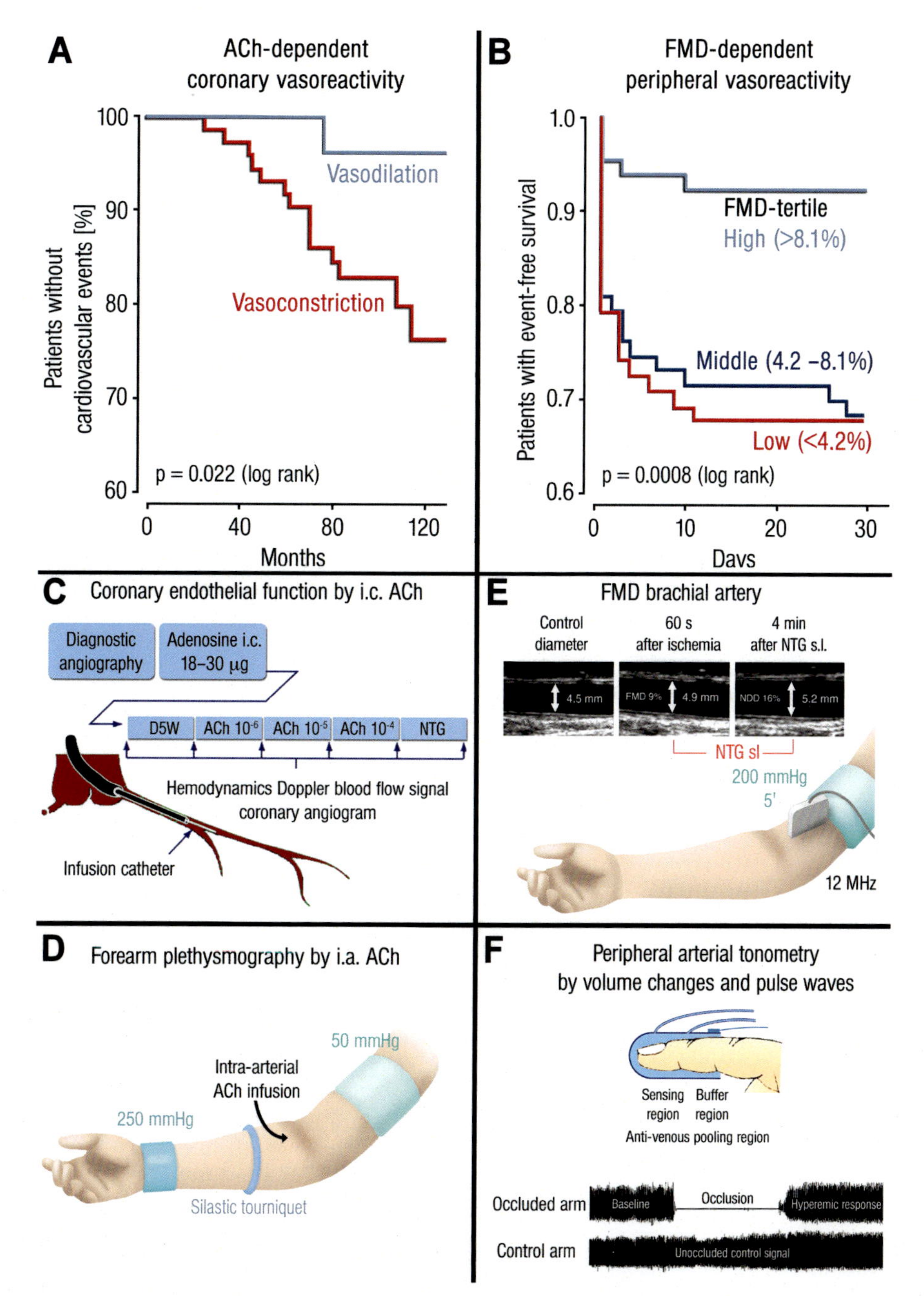

FIGURE 26.3 Endothelial function as a prognostic parameter for cardiovascular events and mortality.[27] (A): Kaplan–Meier curves for the correlation between acetylcholine (ACh)-induced coronary vasoreactivity and prognosis of patients with coronary heart disease (n = 147). Patients displaying ACh-induced vasodilation had a better prognosis than those with ACh-dependent vasoconstriction. Cardiovascular events included lethal and nonlethal myocardial infarction, stroke, coronary and peripheral revascularization, and symptoms of unstable *Angina pectoris.* (B) Kaplan–Meier-curves illustrating the correlation between flow-mediated dilation (FMD) of capacitance vessels in the forearm and event-free survival of patients after coronary or peripheral bypass surgery (n = 187). Patients displaying higher FMD values (indicator of endothelial function) had a better prognosis than those with lower FMD values. Cardiovascular events included lethal and non-lethal myocardial infarction, symptoms of unstable Angina pectoris, atrial fibrillation and increased troponin-I. (C) Acetylcholine (ACh)-dependent vasoreactivity of coronary vessels caused by intracoronary ACh infusion. Vasodilation and stenotic areas are monitored by angiographic imaging and Doppler ultrasound (blood flow). (D) ACh-dependent vasoreactivity of capacity vessels of the forearm upon intraarterial ACh infusion. Vasodilation is recorded by Doppler ultrasound (diameter and blood flow). (E) Flow-mediated dilation (FMD) of capacity vessels of the forearm (brachial artery) upon occlusion/ischemia and reperfusion/hyperemia. Vasodilation is recorded by Doppler ultrasound (diameter and blood flow). Maximal vasoreactivity/dilation is determined by sublingual administration of nitroglycerin (NTG). (F) Peripheral arterial tonometry measures volume changes and pulse waves by a finger probe. Additional details for the determination of pulse wave velocity (PWV) are explained in the text. *(A, B) Approximated from .Gokce N., Keaney J.F. Jr., Hunter L.M., Watkins M.T., Menzoian J.O., Vita J.A. Risk stratification for postoperative cardiovascular events via noninvasive assessment of endothelial function: a prospective study. Circulation. 2002;105(13):1567–1572 and Schachinger V, Britten MB, Zeiher AM. Prognostic impact of coronary vasodilator dysfunction on adverse long-term outcome of coronary heart disease.* Circulation. *2000;101(16):1899–1906. (C-F) Adapted from Daiber A., Steven S., Weber A., et al. Targeting vascular (endothelial) dysfunction. Br J Pharmacol. 2017;174(12):1591–1619. With permission of The British Pharmacological Society. © 2016 The British Pharmacological Society.*

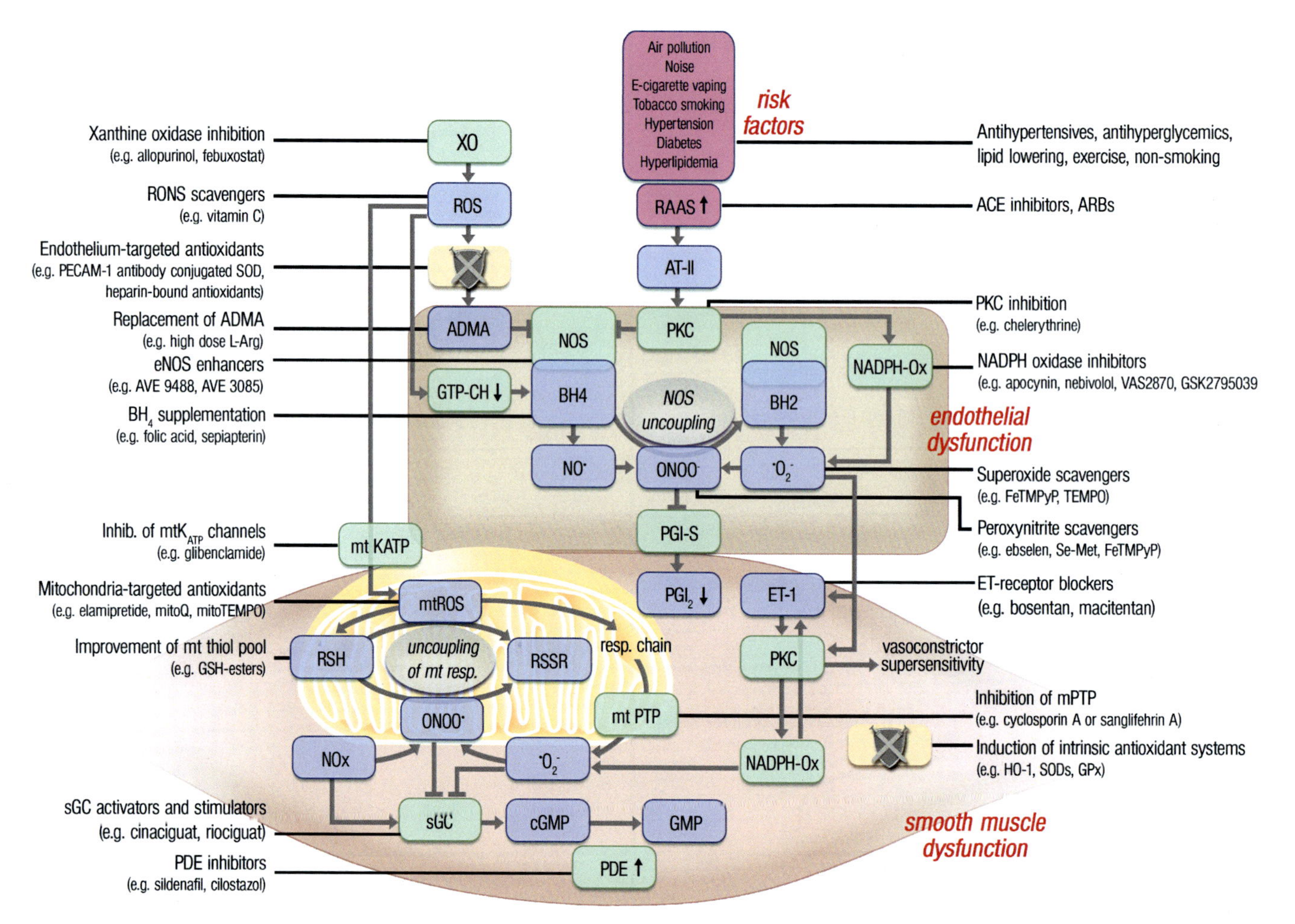

FIGURE 26.4 **Scheme illustrating the mechanisms underlying vascular (endothelial) dysfunction by oxidative stress[171] and their modulation by clinical and experimental interventions.** Known cardiovascular risk factors (e.g., tobacco smoking, hypertension, hyperlipidemia, diabetes, air pollution, noise, and e-cigarette smoking) activate the renin–angiotensin–aldosterone–system (RAAS) leading to elevated angiotensin-II levels as well as increased endothelial and smooth muscle superoxide ($O_2\bullet^-$) formation from NADPH oxidase activation by protein kinase C (PKC) and from the mitochondria. Superoxide reacts with •NO, thereby decreasing •NO bioavailability in favor of peroxynitrite ($ONOO^-$) formation. Peroxynitrite causes uncoupling of endothelial NOS due to oxidation of tetrahydrobiopterin (BH_4) to BH_2 and nitration/inactivation of prostacyclin synthase (PGI_2S). Direct proteasome-dependent degradation of the BH_4 synthase GTP-cyclohydrolase (GTP-CH) further contributes to eNOS uncoupling. Uncoupled NOS produces superoxide instead of •NO and nitrated PGI_2S produces no prostacyclin (PGI_2) but activated cyclooxygenase-2 (due to increased peroxide tone) generates vasoconstrictive prostaglandin H_2. Inhibition of smooth muscle soluble guanylyl cyclase (sGC) by superoxide and peroxynitrite contributes to vascular dysfunction as well as increased inactivation of cyclic GMP (cGMP) by phosphodiesterases (PDE) and oxidative stress increases the sensitivity to vasoconstrictors such as endothelin-1 (ET-1), which vice versa increases NADPH oxidase activity. Mitochondrial ROS formation is modulated by oxidative activation of ATP-dependent potassium channels (K_{ATP}) leading to altered mitochondrial membrane potential and permeability. Upon uncoupling of mitochondrial respiratory complexes, the mitochondrial permeability transition pore (mPTP) may be oxidatively opened allowing mtROS to escape to the cytosol activating the PKC-NADPH-Ox system. Reported (non)pharmacological interventions are shown for the different mechanisms along with examples of experimental or clinical drugs. *Updated from Chen AF, Chen DD, Daiber A, et al. Free radical biology of the cardiovascular system.* Clin Sci. *2012;123(2): 73–91 and reused from Daiber A, Andreadou I, Oelze M, Davidson SM, Hausenloy DJ. Discovery of new therapeutic redox targets for cardioprotection against ischemia/reperfusion injury and heart failure.* Free Radic Biol Med. *2021;163:325–343. With permission of Elsevier Inc.*

Diverse approaches have been designed to achieve specific delivery of antioxidants. For example, conjugation with polyethylene glycol (PEG), PEG-containing carriers, or liposomes improves the bioavailability of antioxidants and their ability to detoxify ROS. However, encouraging data from in vitro and whole animal studies have not yet translated to clinical practice[120,121] (see also "Antioxidant therapy" section above). Current efforts to mitigate oxidative stress have yielded rather mixed outcomes in clinical studies.[122] This may be explained by the fact that endothelial ROS are marginally accessible to untargeted antioxidants.[122] Most reactive and hence damaging ROS are short-lived species that act at angstrom to nanometer distances,[123] increasing the need for precise antioxidant delivery to quench endothelial ROS.[124]

Antiinflammatory therapy

Inflammation is an independent risk factor for the development of cardiovascular disease.[125] Levels of the acute phase protein CRP are elevated in various inflammatory diseases. The PROVE IT-TIMI 22 study demonstrated that high CRP levels in patients with acute coronary syndrome predicted death from myocardial infarction.[126] Furthermore, it is now generally agreed that in addition to atherosclerosis, arterial hypertension is also a low-grade inflammatory disease.[127,128] Several clinical trials recently investigated whether antiinflammatory treatment can improve cardiovascular outcomes, e.g., methotrexate therapy (TETHYS trial and CIRT trial)[129,130] and blockade of the cytokine IL-1β with canakinumab for the management of cardiovascular disease (CANTOS trial),[131] with mixed results, yet supporting the benefit of targeted antiinflammatory interventions, and begging questions about most effective targets Another new antiinflammatory therapy could be based on the restoration of a disturbed glycocalyx, which is associated with higher susceptibility to triggers of atherosclerosis and leukocyte/platelet adhesion.[132–134]

Pharmacological targeting of oxidatively impaired eNOS and sGC

There is overwhelming evidence supporting an important pathophysiological role of eNOS dysregulation/uncoupling in several diseases, making targeting of this enzyme, in particular in its uncoupled state, an attractive therapeutic target. eNOS-directed pharmaceutical attempts have so far included the development of so-called eNOS enhancers, compounds that upregulate the expression of eNOS at the mRNA and protein level.[135–137] The results indicate that sole overexpression of eNOS without upregulation of its cofactor sapropterin (BH_4) will ultimately lead to its uncoupling and so worsen disease conditions rather than improving it. In line with this, Channon and coworkers demonstrated that eNOS overexpression leads to an increase in •NO formation only when the BH_4 synthase GTP-cyclohydrolase 1 (GCH-1) is also upregulated.[138] BH_4 treatment or treatment with analogues such as folic acid and sepiapterin improved endothelial dysfunction in, e.g., chronic smokers and diabetes via recoupling eNOS[139,140] and in numerous ex vivo studies[141,142] (for review, see[143]). The PHACeT Trial reported benefits of endothelial •NO synthase gene-enhanced progenitor cell therapy for pulmonary arterial hypertension.[144]

Recent studies have identified sGC as an important therapeutic target since oxidation of the enzyme at numerous sites impairs activation by •NO. The discovery of compounds such as hem-dependent sGC stimulators and hem-independent sGC activators that can stimulate the enzyme in a •NO-independent manner has led to the development of a clinical program for the treatment of patients with pulmonary (arterial) hypertension,[145] chronic thromboembolic pulmonary hypertension,[146] and heart failure[147] (for review, see[109]). The sGC activator cinaciguat normalized vascular oxidative stress and coronary endothelial function in a rat model of myocardial infarction,[148] a canine model of cardiopulmonary bypass[149] and prevented endothelial dysfunction of isolated rat aorta when challenged with peroxynitrite.[150] Inhibitors of PDEs act downstream of sGC to increase cGMP levels and are used for multiple indications (for review, see[151]). A recent metaanalysis of subjects with type 2 diabetes mellitus (T2DM) suggests a beneficial effect of a PDE5 inhibitor (sildenafil) therapy on endothelial function.[152] Vardenafil also improved endothelial function of isolated rat aorta upon ex vivo challenges with peroxynitrite[153] and hypochlorite.[154]

Nitric oxide and other substitution therapies

Nitrovasodilators such as organic nitrates act in •NO substitution therapy and have been used for more than a century for the relief of anginal pain that is caused by stenosis or spasm of the coronary arteries associated with endothelial dysfunction (for review, see[155]). However, as is the case for eNOS enhancer therapy, chronic substitution of •NO alone does not produce beneficial effects since •NO under oxidative stress conditions rapidly yields the potent oxidant peroxynitrite[156] (for review, see[157]). The target enzyme receptor for •NO, sGC, can also be inhibited under conditions of oxidative stress.[109] Therefore, a combined antioxidant and •NO substitution therapy seems to be a more promising strategy, as shown for pentaerithrityl tetranitrate, the only organic nitrate that improves oxidative stress, nitrate tolerance, and endothelial dysfunction.[158,159] Inorganic nitrite and nitrate are emerging options for •NO substitution therapy as they have beneficial effects in hypertension, myocardial infarction, and acute heart failure (for review, see[160–162]). Accumulation of the endogenously produced eNOS inhibitor asymmetric dimethyl arginine (ADMA) under oxidative stress conditions is another trigger of endothelial dysfunction that could be remedied by treatment with high doses of L-arginine which directly replaces ADMA at the level of eNOS enzyme (for review, see[163]) or by prevention of endothelial ADMA accumulation through activation of the cationic amino acid extrusion transporter.[164] Another new therapeutic option is supplementation with an H_2S donor that not only confers direct antioxidant effects but which also acts synergistically with •NO via different signaling pathways.[165,166] Prostacyclin substitution, e.g., by iloprost, is not generally used for the treatment of cardiovascular disease but more specifically for the therapy of severe pulmonary hypertension.[167]

Conclusions and clinical implications

The prognostic value of endothelial function measurements for risk stratification of patients for future cardiovascular events is still under debate, although it is pretty convincing that endothelial dysfunction is an important biomarker in the presence of cardiovascular risk factors and cardiovascular diseases. It is likely that parameters reflecting vascular function more globally (including smooth muscle and perivascular adipose tissue function) such as vascular stiffness index and intima/media thickness ratio may be better predictors of future cardiovascular events. New diagnostic tools may improve the "imaging" of vascular (endothelial) dysfunction and so herald a new method to even predict cardiovascular events. Targeting vascular (endothelial) dysfunction still represents an attractive pharmacological approach due to its central involvement in the progression of most cardio- and cerebrovascular diseases as well as inflammatory processes. In addition to the classical cardiovascular drugs that have beneficial effects on vascular (endothelial) function such as ACE inhibitors and statins, new therapeutic strategies based on site-specific drug delivery, antiinflammatory, as well as site- and source-specific antioxidant interventions, are being developed. Moreover, physical exercise and healthy nutrition represent a main stay in nonpharmacological cardiovascular therapy. Direct targeting of vascular homeostasis (e.g., •NO/cGMP signaling cascade by •NO donors, PDE inhibitors, and other vasodilators) is still not fully explored, but emerging developments such as sGC activators and stimulators have therapeutic potential.

Acknowledgments

We are indebted to Margot Neuser for expert graphical assistance. Text excerpts were reused from *Daiber A, Hahad O, Andreadou I, Steven S, Daub S, Munzel T. Redox-related biomarkers in human cardiovascular disease - classical footprints and beyond. Redox Biol. 2021;42:101875.* (with permission by the authors published by Elsevier),[26] *Münzel T, Daiber A, Ullrich V, Mülsch A.Vascular consequences of endothelial nitric oxide synthase uncoupling for the activity and expression of the soluble guanylyl cyclase and the cGMP-dependent protein kinase. Arterioscler Thromb Vasc Biol. 2005;25(8):1551–1557.* (with permission by Wolters Kluwer Health),[172] *Daiber A, Steven S, Weber A, et al. Targeting vascular (endothelial) dysfunction. Br J Pharmacol. 2017;174(12):1591–1619.* (with permission by the British Pharmacological Society)[27] and *Daiber A, Andreadou I, Oelze M, Davidson SM, Hausenloy DJ. Discovery of new therapeutic redox targets for cardioprotection against ischemia/reperfusion injury and heart failure. Free Radic Biol Med. 2021;163:325–343.* (with permission by Elsevier).[95]

References

1. Lim SS, Vos T, Flaxman AD, et al. A comparative risk assessment of burden of disease and injury attributable to 67 risk factors and risk factor clusters in 21 regions, 1990-2010: a systematic analysis for the Global Burden of Disease Study 2010. *Lancet*. 2012; 380(9859):2224–2260.
2. Murray CJ, Ezzati M, Flaxman AD, et al. Gbd 2010: design, definitions, and metrics. *Lancet*. 2012;380(9859):2063–2066.
3. Lozano R, Naghavi M, Foreman K, et al. Global and regional mortality from 235 causes of death for 20 age groups in 1990 and 2010: a systematic analysis for the Global Burden of Disease Study 2010. *Lancet*. 2012;380(9859):2095–2128.
4. WHO. *Projections of Mortality and Causes of Death, 2016 to 2060*; 2021. https://www.who.int/healthinfo/global_burden_disease/projections/en/.
5. WHO. Global Health Observatory Data on Causes of Non-communicable Disease in 2016; n.d. https://www.who.int/gho/ncd/mortality_morbidity/en/.
6. Gonzalez MA, Selwyn AP. Endothelial function, inflammation, and prognosis in cardiovascular disease. *Am J Med*. 2003; 115(Suppl 8A):99S–106S.
7. Pauriah M, Khan F, Lim TK, et al. B-type natriuretic peptide is an independent predictor of endothelial function in man. *Clin Sci (Lond)*. 2012;123(5):307–312.
8. Okui H, Hamasaki S, Ishida S, et al. Adiponectin is a better predictor of endothelial function of the coronary artery than HOMA-R, body mass index, immunoreactive insulin, or triglycerides. *Int J Cardiol*. 2008;126(1):53–61.
9. Panza JA, Quyyumi AA, Brush Jr JE, Epstein SE. Abnormal endothelium-dependent vascular relaxation in patients with essential hypertension. *N Engl J Med*. 1990;323(1):22–27.
10. Vita JA, Treasure CB, Nabel EG, et al. Coronary vasomotor response to acetylcholine relates to risk factors for coronary artery disease. *Circulation*. 1990;81(2):491–497.
11. Calver A, Collier J, Vallance P. Inhibition and stimulation of nitric oxide synthesis in the human forearm arterial bed of patients with insulin-dependent diabetes. *J Clin Invest*. 1992;90(6):2548–2554.
12. Celermajer DS, Sorensen KE, Georgakopoulos D, et al. Cigarette smoking is associated with dose-related and potentially reversible impairment of endothelium-dependent dilation in healthy young adults. *Circulation*. 1993;88(5 Pt 1):2149–2155.
13. Munzel T, Sorensen M, Daiber A. Transportation noise pollution and cardiovascular disease. *Nat Rev Cardiol*. 2021;18(9):619–636.
14. Munzel T, Miller MR, Sorensen M, Lelieveld J, Daiber A, Rajagopalan S. Reduction of environmental pollutants for prevention of cardiovascular disease: it's time to act. *Eur Heart J*. 2020; 41(41):3989–3997.
15. Kuntic M, Oelze M, Steven S, et al. Short-term e-cigarette vapour exposure causes vascular oxidative stress and dysfunction: evidence for a close connection to brain damage and a key role of the phagocytic NADPH oxidase (NOX-2). *Eur Heart J*. 2020; 41(26):2472–2483.
16. Munzel T, Sinning C, Post F, Warnholtz A, Schulz E. Pathophysiology, diagnosis and prognostic implications of endothelial dysfunction. *Ann Med*. 2008;40(3):180–196.
17. Heitzer T, Ylä-Herttuala S, Luoma J, et al. Cigarette smoking potentiates endothelial dysfunction of forearm resistance vessels in

patients with hypercholesterolemia: role of oxidized LDL. *Circulation*. 1996;93:1346–1353.
18. Moncada S, Korbut R, Bunting S, Vane JR. Prostacyclin is a circulating hormone. *Nature*. 1978;273(5665):767–768.
19. Palmer RM, Ferrige AG, Moncada S. Nitric oxide release accounts for the biological activity of endothelium-derived relaxing factor. *Nature*. 1987;327(6122):524–526.
20. Stupack DG, Cheresh DA. Integrins and angiogenesis. *Curr Top Dev Biol*. 2004;64:207–238.
21. Busse R, Fleming I. Regulation of NO synthesis in endothelial cells. *Kidney Blood Press Res*. 1998;21(2–4):264–266.
22. Fleming I, Busse R. Molecular mechanisms involved in the regulation of the endothelial nitric oxide synthase. *Am J Physiol Regul Integr Comp Physiol*. 2003;284(1):R1–R12.
23. Oelze M, Mollnau H, Hoffmann N, et al. Vasodilator-stimulated phosphoprotein serine 239 phosphorylation as a sensitive monitor of defective nitric oxide/cGMP signaling and endothelial dysfunction. *Circ Res*. 2000;87(11):999–1005.
24. Karbach S, Wenzel P, Waisman A, Munzel T, Daiber A. eNOS uncoupling in cardiovascular diseases–the role of oxidative stress and inflammation. *Curr Pharmaceut Des*. 2014;20(22):3579–3594.
25. Schulz E, Jansen T, Wenzel P, Daiber A, Munzel T. Nitric oxide, tetrahydrobiopterin, oxidative stress, and endothelial dysfunction in hypertension. *Antioxid Redox Sign*. 2008;10(6):1115–1126.
26. Daiber A, Hahad O, Andreadou I, Steven S, Daub S, Munzel T. Redox-related biomarkers in human cardiovascular disease - classical footprints and beyond. *Redox Biol*. 2021;42:101875.
27. Daiber A, Steven S, Weber A, et al. Targeting vascular (endothelial) dysfunction. *Br J Pharmacol*. 2017;174(12):1591–1619.
28. Daiber A, Chlopicki S. Revisiting pharmacology of oxidative stress and endothelial dysfunction in cardiovascular disease: evidence for redox-based therapies. *Free Radic Biol Med*. 2020;157: 15–37.
29. Daiber A, Xia N, Steven S, et al. New therapeutic implications of endothelial nitric oxide synthase (eNOS) function/dysfunction in cardiovascular disease. *Int J Mol Sci*. 2019;20(1).
30. Schulz E, Wenzel P, Munzel T, Daiber A. Mitochondrial redox signaling: interaction of mitochondrial reactive oxygen species with other sources of oxidative stress. *Antioxid Redox Sign*. 2014; 20(2):308–324.
31. Schmidt HH, Stocker R, Vollbracht C, et al. Antioxidants in translational medicine. *Antioxid Redox Sign*. 2015;23(14):1130–1143.
32. Munzel T, Daiber A, Steven S, et al. Effects of noise on vascular function, oxidative stress, and inflammation: mechanistic insight from studies in mice. *Eur Heart J*. 2017;38(37):2838–2849.
33. Xu X, Yavar Z, Verdin M, et al. Effect of early particulate air pollution exposure on obesity in mice: role of p47phox. *Arterioscler Thromb Vasc Biol*. 2010;30(12):2518–2527.
34. Kroller-Schon S, Daiber A, Steven S, et al. Crucial role for Nox2 and sleep deprivation in aircraft noise-induced vascular and cerebral oxidative stress, inflammation, and gene regulation. *Eur Heart J*. 2018;39(38):3528–3539.
35. Robinson BH. The role of manganese superoxide dismutase in health and disease. *J Inherit Metab Dis*. 1998;21(5):598–603.
36. Daiber A, Di Lisa F, Oelze M, et al. Crosstalk of mitochondria with NADPH oxidase via reactive oxygen and nitrogen species signalling and its role for vascular function. *Br J Pharmacol*. 2017;174(12): 1670–1689.
37. Zorov DB, Filburn CR, Klotz LO, Zweier JL, Sollott SJ. Reactive oxygen species (ROS)-induced ROS release: a new phenomenon accompanying induction of the mitochondrial permeability transition in cardiac myocytes. *J Exp Med*. 2000;192(7):1001–1014.
38. Zorov DB, Juhaszova M, Sollott SJ. Mitochondrial reactive oxygen species (ROS) and ROS-induced ROS release. *Physiol Rev*. 2014; 94(3):909–950.
39. Egea J, Fabregat I, Frapart YM, et al. European contribution to the study of ROS: a summary of the findings and prospects for the future from the COST action BM1203 (EU-ROS). *Redox Biol*. 2017;13:94–162.
40. Myers PR, Minor Jr RL, Guerra Jr R, Bates JN, Harrison DG. Vasorelaxant properties of the endothelium-derived relaxing factor more closely resemble S-nitrosocysteine than nitric oxide. *Nature*. 1990;345(6271):161–163.
41. Garg UC, Hassid A. Nitric oxide-generating vasodilators and 8-bromo-cyclic guanosine monophosphate inhibit mitogenesis and proliferation of cultured rat vascular smooth muscle cells. *J Clin Invest*. 1989;83(5):1774–1777.
42. Gryglewski RJ, Moncada S, Palmer RM. Bioassay of prostacyclin and endothelium-derived relaxing factor (EDRF) from porcine aortic endothelial cells. *Br J Pharmacol*. 1986;87(4):685–694.
43. Beckman JS. Oxidative damage and tyrosine nitration from peroxynitrite. *Chem Res Toxicol*. 1996;9(5):836–844.
44. Kissner R, Nauser T, Bugnon P, Lye PG, Koppenol WH. Formation and properties of peroxynitrite as studied by laser flash photolysis, high-pressure stopped-flow technique, and pulse radiolysis. *Chem Res Toxicol*. 1997;10(11):1285–1292.
45. Beckman JS, Koppenol WH. Nitric oxide, superoxide, and peroxynitrite: the good, the bad, and ugly. *Am J Physiol*. 1996;271(5 Pt 1): C1424–C1437.
46. Zou MH, Ullrich V. Peroxynitrite formed by simultaneous generation of nitric oxide and superoxide selectively inhibits bovine aortic prostacyclin synthase. *FEBS Lett*. 1996;382(1–2): 101–104.
47. Kuzkaya N, Weissmann N, Harrison DG, Dikalov S. Interactions of peroxynitrite, tetrahydrobiopterin, ascorbic acid, and thiols: implications for uncoupling endothelial nitric-oxide synthase. *J Biol Chem*. 2003;278(25):22546–22554.
48. Zou MH, Shi C, Cohen RA. Oxidation of the zinc-thiolate complex and uncoupling of endothelial nitric oxide synthase by peroxynitrite. *J Clin Invest*. 2002;109(6):817–826.
49. Soltesz P, Kerekes G, Der H, et al. Comparative assessment of vascular function in autoimmune rheumatic diseases: considerations of prevention and treatment. *Autoimmun Rev*. 2011;10(7):416–425.
50. Murdaca G, Colombo BM, Cagnati P, Gulli R, Spano F, Puppo F. *Endothelial Dysfunction in Rheumatic Autoimmune Diseases*. Atherosclerosis; 2012.
51. Vena GA, Vestita M, Cassano N. Psoriasis and cardiovascular disease. *Dermatol Ther*. 2010;23(2):144–151.
52. Hak AE, Karlson EW, Feskanich D, Stampfer MJ, Costenbader KH. Systemic lupus erythematosus and the risk of cardiovascular disease: results from the nurses' health study. *Arthritis Rheum*. 2009;61(10):1396–1402.
53. Mehta NN, Azfar RS, Shin DB, Neimann AL, Troxel AB, Gelfand JM. Patients with severe psoriasis are at increased risk of cardiovascular mortality: cohort study using the General Practice Research Database. *Eur Heart J*. 2010;31(8):1000–1006.
54. Peters MJ, Symmons DP, McCarey D, et al. EULAR evidence-based recommendations for cardiovascular risk management in patients with rheumatoid arthritis and other forms of inflammatory arthritis. *Ann Rheumat Dis*. 2010;69(2):325–331.
55. Sodergren A, Karp K, Boman K, et al. Atherosclerosis in early rheumatoid arthritis: very early endothelial activation and rapid progression of intima media thickness. *Arthr Res & Ther*. 2010;12(4):R158.
56. Gokce N, Keaney Jr JF, Hunter LM, Watkins MT, Menzoian JO, Vita JA. Risk stratification for postoperative cardiovascular events via noninvasive assessment of endothelial function: a prospective study. *Circulation*. 2002;105(13):1567–1572.
57. Gokce N, Keaney Jr JF, Hunter LM, et al. Predictive value of noninvasively determined endothelial dysfunction for long-term cardiovascular events in patients with peripheral vascular disease. *J Am Coll Cardiol*. 2003;41(10):1769–1775.

58. Anderson TJ, Uehata A, Gerhard MD, et al. Close relation of endothelial function in the human coronary and peripheral circulations. *J Am Coll Cardiol*. 1995;26(5):1235–1241.
59. Boger RH, Bode-Boger SM, Thiele W, Creutzig A, Alexander K, Frolich JC. Restoring vascular nitric oxide formation by L-arginine improves the symptoms of intermittent claudication in patients with peripheral arterial occlusive disease. *J Am Coll Cardiol*. 1998;32(5):1336–1344.
60. Heitzer T, Schlinzig T, Krohn K, Meinertz T, Munzel T. Endothelial dysfunction, oxidative stress, and risk of cardiovascular events in patients with coronary artery disease. *Circulation*. 2001;104(22): 2673–2678.
61. Schachinger V, Britten MB, Zeiher AM. Prognostic impact of coronary vasodilator dysfunction on adverse long-term outcome of coronary heart disease. *Circulation*. 2000;101(16):1899–1906.
62. Suwaidi JA, Hamasaki S, Higano ST, Nishimura RA, Holmes Jr DR, Lerman A. Long-term follow-up of patients with mild coronary artery disease and endothelial dysfunction. *Circulation*. 2000;101(9):948–954.
63. Heitzer T, Baldus S, von Kodolitsch Y, Rudolph V, Meinertz T. Systemic endothelial dysfunction as an early predictor of adverse outcome in heart failure. *Arterioscler Thromb Vasc Biol*. 2005; 25(6):1174–1179.
64. Laufs U, Wassmann S, Czech T, et al. Physical inactivity increases oxidative stress, endothelial dysfunction, and atherosclerosis. *Arterioscler Thromb Vasc Biol*. 2005;25(4):809–814.
65. Harrison D, Griendling KK, Landmesser U, Hornig B, Drexler H. Role of oxidative stress in atherosclerosis. *Am J Cardiol*. 2003; 91(3A):7A–11A.
66. Miller Jr FJ, Gutterman DD, Rios CD, Heistad DD, Davidson BL. Superoxide production in vascular smooth muscle contributes to oxidative stress and impaired relaxation in atherosclerosis. *Circ Res*. 1998;82(12):1298–1305.
67. Patel RS, Ghasemzadeh N, Eapen DJ, et al. Novel biomarker of oxidative stress is associated with risk of death in patients with coronary artery disease. *Circulation*. 2016;133(4):361–369.
68. Matsuzawa Y, Sugiyama S, Sumida H, et al. Peripheral endothelial function and cardiovascular events in high-risk patients. *J Am Heart Assoc*. 2013;2(6):e000426.
69. Verma S, Wang CH, Lonn E, et al. Cross-sectional evaluation of brachial artery flow-mediated vasodilation and C-reactive protein in healthy individuals. *Eur Heart J*. 2004;25(19):1754–1760.
70. Perticone F, Ceravolo R, Pujia A, et al. Prognostic significance of endothelial dysfunction in hypertensive patients. *Circulation*. 2001;104(2):191–196.
71. Sharma S, Mells JE, Fu PP, Saxena NK, Anania FA. GLP-1 analogs reduce hepatocyte steatosis and improve survival by enhancing the unfolded protein response and promoting macroautophagy. *PLoS One*. 2011;6(9):e25269.
72. Shechter M, Shechter A, Koren-Morag N, Feinberg MS, Hiersch L. Usefulness of brachial artery flow-mediated dilation to predict long-term cardiovascular events in subjects without heart disease. *Am J Cardiol*. 2014;113(1):162–167.
73. Maruhashi T, Soga J, Fujimura N, et al. Relationship between flow-mediated vasodilation and cardiovascular risk factors in a large community-based study. *Heart*. 2013;99(24):1837–1842.
74. Anderson TJ, Charbonneau F, Title LM, et al. Microvascular function predicts cardiovascular events in primary prevention: long-term results from the Firefighters and Their Endothelium (FATE) study. *Circulation*. 2011;123(2):163–169.
75. Lekakis J, Abraham P, Balbarini A, et al. Methods for evaluating endothelial function: a position statement from the European society of cardiology working group on peripheral circulation. *Eur J Cardiovasc Prev Rehabil*. 2011;18(6):775–789.
76. Frolow M, Drozdz A, Kowalewska A, Nizankowski R, Chlopicki S. Comprehensive assessment of vascular health in patients; towards endothelium-guided therapy. *Pharmacol Rep*. 2015;67(4):786–792.
77. Reference Values for Arterial Stiffness' Collaboration. Determinants of pulse wave velocity in healthy people and in the presence of cardiovascular risk factors: 'establishing normal and reference values'. *Eur Heart J*. 2010;31(19):2338–2350.
78. Nelson MR, Stepanek J, Cevette M, Covalciuc M, Hurst RT, Tajik AJ. Noninvasive measurement of central vascular pressures with arterial tonometry: clinical revival of the pulse pressure waveform? *Mayo Clin Proc*. 2010;85(5):460–472.
79. Lehmann ED, Parker JR, Hopkins KD, Taylor MG, Gosling RG. Validation and reproducibility of pressure-corrected aortic distensibility measurements using pulse-wave-velocity Doppler ultrasound. *J Biomed Eng*. 1993;15(3):221–228.
80. Wentland AL, Grist TM, Wieben O. Review of MRI-based measurements of pulse wave velocity: a biomarker of arterial stiffness. *Cardiovasc Diagn Ther*. 2014;4(2):193–206.
81. Boese JM, Bock M, Schoenberg SO, Schad LR. Estimation of aortic compliance using magnetic resonance pulse wave velocity measurement. *Phys Med Biol*. 2000;45(6):1703–1713.
82. Dogui A, Redheuil A, Lefort M, et al. Measurement of aortic arch pulse wave velocity in cardiovascular MR: comparison of transit time estimators and description of a new approach. *J Magn Reson Imag*. 2011;33(6):1321–1329.
83. Bar A, Skorka T, Jasinski K, Chlopicki S. MRI-based assessment of endothelial function in mice in vivo. *Pharmacol Rep*. 2015;67(4): 765–770.
84. Ludmer PL, Selwyn AP, Shook TL, et al. Paradoxical vasoconstriction induced by acetylcholine in atherosclerotic coronary arteries. *N Engl J Med*. 1986;315(17):1046–1051.
85. Heitzer T, Just H, Munzel T. Antioxidant vitamin C improves endothelial dysfunction in chronic smokers. *Circulation*. 1996; 94(1):6–9.
86. Ramsey M, Goodfellow J, Bellamy M, Lewis M, Henderson A. Non-invasive detection of endothelial dysfunction. *Lancet*. 1996; 348(9020):128–129.
87. Bonetti PO, Barsness GW, Keelan PC, et al. Enhanced external counterpulsation improves endothelial function in patients with symptomatic coronary artery disease. *J Am Coll Cardiol*. 2003; 41(10):1761–1768.
88. Gori T, Dragoni S, Lisi M, et al. Conduit artery constriction mediated by low flow a novel noninvasive method for the assessment of vascular function. *J Am Coll Cardiol*. 2008;51(20):1953–1958.
89. Gori T, Muxel S, Damaske A, et al. Endothelial function assessment: flow-mediated dilation and constriction provide different and complementary information on the presence of coronary artery disease. *Eur Heart J*. 2012;33(3):363–371.
90. Hughes WE, Kruse NT. A 'passive' movement into the future of assessing endothelial dysfunction? *J Physiol*. 2016;594(6):1525–1526.
91. Groot HJ, Trinity JD, Layec G, et al. The role of nitric oxide in passive leg movement-induced vasodilatation with age: insight from alterations in femoral perfusion pressure. *J Physiol*. 2015;593(17): 3917–3928.
92. Furchgott RF, Zawadzki JV. The obligatory role of endothelial cells in the relaxation of arterial smooth muscle by acetylcholine. *Nature*. 1980;288(5789):373–376.
93. Schulz E, Tsilimingas N, Rinze R, et al. Functional and biochemical analysis of endothelial (dys)function and NO/cGMP signaling in human blood vessels with and without nitroglycerin pretreatment. *Circulation*. 2002;105(10):1170–1175.
94. Baranska M, Kaczor A, Malek K, et al. Raman microscopy as a novel tool to detect endothelial dysfunction. *Pharmacol Rep*. 2015;67(4):736–743.
95. Daiber A, Andreadou I, Oelze M, Davidson SM, Hausenloy DJ. Discovery of new therapeutic redox targets for cardioprotection against ischemia/reperfusion injury and heart failure. *Free Radic Biol Med*. 2021;163:325–343.

96. Mikhed Y, Gorlach A, Knaus UG, Daiber A. Redox regulation of genome stability by effects on gene expression, epigenetic pathways and DNA damage/repair. *Redox Biol.* 2015;5:275–289.
97. Kim GH, Ryan JJ, Archer SL. The role of redox signaling in epigenetics and cardiovascular disease. *Antioxidants Redox Signal.* 2013; 18(15):1920–1936.
98. Drexler H, Hornig B. Endothelial dysfunction in human disease. *J Mol Cell Cardiol.* 1999;31(1):51–60.
99. Warnholtz A, Munzel T. Why do antioxidants fail to provide clinical benefit? *Curr Control Trials Cardiovasc med.* 2000;1(1):38–40.
100. Gori T, Munzel T. Oxidative stress and endothelial dysfunction: therapeutic implications. *Ann Med.* 2011;43(4):259–272.
101. Steven S, Munzel T, Daiber A. Exploiting the pleiotropic antioxidant effects of established drugs in cardiovascular disease. *Int J Mol Sci.* 2015;16(8):18185–18223.
102. Thorin E, Clozel M. The cardiovascular physiology and pharmacology of endothelin-1. *Adv Pharmacol.* 2010;60:1–26.
103. Clozel M. Effects of bosentan on cellular processes involved in pulmonary arterial hypertension: do they explain the long-term benefit? *Ann Med.* 2003;35(8):605–613.
104. Sharina IG, Sobolevsky M, Papakyriakou A, et al. The fibrate gemfibrozil is a NO- and haem-independent activator of soluble guanylyl cyclase: in vitro studies. *Br J Pharmacol.* 2015;172(9): 2316–2329.
105. Chen AF, Chen DD, Daiber A, et al. Free radical biology of the cardiovascular system. *Clin Sci (Lond).* 2012;123(2):73–91.
106. Casas AI, Nogales C, Mucke HAM, et al. On the clinical pharmacology of reactive oxygen species. *Pharmacol Rev.* 2020;72(4): 801–828.
107. Murphy MP, Smith RA. Targeting antioxidants to mitochondria by conjugation to lipophilic cations. *Annu Rev Pharmacol Toxicol.* 2007;47:629–656.
108. Smith RA, Hartley RC, Murphy MP. Mitochondria-targeted small molecule therapeutics and probes. *Antioxid Redox Sign.* 2011; 15(12):3021–3038.
109. Evgenov OV, Pacher P, Schmidt PM, Hasko G, Schmidt HH, Stasch JP. NO-independent stimulators and activators of soluble guanylate cyclase: discovery and therapeutic potential. *Nat Rev Drug Discov.* 2006;5(9):755–768.
110. Blanco-Rivero J, Xavier FE. Therapeutic potential of phosphodiesterase inhibitors for endothelial dysfunction- related diseases. *Curr Pharmaceut Des.* 2020;26(30):3633–3651.
111. Munzel T, Daiber A, Gori T. More answers to the still unresolved question of nitrate tolerance. *Eur Heart J.* 2013;34(34):2666–2673.
112. Gryglewski RJ. Prostacyclin among prostanoids. *Pharmacol Rep.* 2008;60(1):3–11.
113. Pulido T, Adzerikho I, Channick RN, et al. Macitentan and morbidity and mortality in pulmonary arterial hypertension. *N Engl J Med.* 2013;369(9):809–818.
114. Oelze M, Knorr M, Kroller-Schon S, et al. Chronic therapy with isosorbide-5-mononitrate causes endothelial dysfunction, oxidative stress, and a marked increase in vascular endothelin-1 expression. *Eur Heart J.* 2013;34(41):3206–3216.
115. Steven S, Oelze M, Brandt M, et al. Pentaerythritol tetranitrate in vivo treatment improves oxidative stress and vascular dysfunction by suppression of endothelin-1 signaling in monocrotaline-induced pulmonary hypertension. *Oxid Med Cell Longev.* 2017; 2017:4353462.
116. Gangji AS, Cukierman T, Gerstein HC, Goldsmith CH, Clase CM. A systematic review and meta-analysis of hypoglycemia and cardiovascular events: a comparison of glyburide with other secretagogues and with insulin. *Diabetes Care.* 2007;30(2):389–394.
117. Piot C, Croisille P, Staat P, et al. Effect of cyclosporine on reperfusion injury in acute myocardial infarction. *N Engl J Med.* 2008; 359(5):473–481.
118. Cuadrado A, Manda G, Hassan A, et al. Transcription factor NRF2 as a therapeutic target for chronic diseases: a systems medicine approach. *Pharmacol Rev.* 2018;70(2):348–383.
119. Schmidt HH, Stocker R, Vollbracht C, et al. Antioxidants in translational medicine. *Antioxid Redox Sign.* 2015;23(14):1130–1143.
120. Barnard ML, Baker RR, Matalon S. Mitigation of oxidant injury to lung microvasculature by intratracheal instillation of antioxidant enzymes. *Am J Physiol.* 1993;265(4 Pt 1):L340–L345.
121. Freeman BA, Turrens JF, Mirza Z, Crapo JD, Young SL. Modulation of oxidant lung injury by using liposome-entrapped superoxide dismutase and catalase. *Fed Proc.* 1985;44(10): 2591–2595.
122. Bowler RP, Arcaroli J, Crapo JD, Ross A, Slot JW, Abraham E. Extracellular superoxide dismutase attenuates lung injury after hemorrhage. *Am J Respir Crit Care Med.* 2001;164(2):290–294.
123. Parthasarathy S, Khan-Merchant N, Penumetcha M, Khan BV, Santanam N. Did the antioxidant trials fail to validate the oxidation hypothesis? *Curr Atheroscler Rep.* 2001;3(5):392–398.
124. Shuvaev VV, Han J, Yu KJ, et al. PECAM-targeted delivery of SOD inhibits endothelial inflammatory response. *FASEB J.* 2011;25(1): 348–357.
125. Kaptoge S, Seshasai SR, Gao P, et al. Inflammatory cytokines and risk of coronary heart disease: new prospective study and updated meta-analysis. *Eur Heart J.* 2014;35(9):578–589.
126. Ridker PM, Cannon CP, Morrow D, et al. Pravastatin or atorvastatin E, infection therapy-thrombolysis in myocardial infarction I. Creactive protein levels and outcomes after statin therapy. *N Engl J Med.* 2005;352(1):20–28.
127. Libby P. Inflammation in atherosclerosis. *Nature.* 2002;420(6917): 868–874.
128. Wenzel P, Knorr M, Kossmann S, et al. Lysozyme M-positive monocytes mediate angiotensin II-induced arterial hypertension and vascular dysfunction. *Circulation.* 2011;124:1370–1381.
129. Everett BM, Pradhan AD, Solomon DH, et al. Rationale and design of the Cardiovascular Inflammation Reduction Trial: a test of the inflammatory hypothesis of atherothrombosis. *Am Heart J.* 2013;166(2):199–207 e15.
130. Moreira DM, Lueneberg ME, da Silva RL, Fattah T, Mascia Gottschall CA. Rationale and design of the TETHYS trial: the effects of methotrexate therapy on myocardial infarction with ST-segment elevation. *Cardiology.* 2013;126(3):167–170.
131. Ridker PM, Thuren T, Zalewski A, Libby P. Interleukin-1beta inhibition and the prevention of recurrent cardiovascular events: rationale and design of the Canakinumab Anti-inflammatory Thrombosis Outcomes Study (CANTOS). *Am Heart J.* 2011; 162(4):597–605.
132. Drake-Holland AJ, Noble MI. Update on the important new drug target in cardiovascular medicine - the vascular glycocalyx. *Cardiovasc Hematol Disord Drug Targets.* 2012;12(1):76–81.
133. Tarbell JM, Cancel LM. The glycocalyx and its significance in human medicine. *J Intern Med.* 2016;280(1):97–113.
134. Meuwese MC, Mooij HL, Nieuwdorp M, et al. Partial recovery of the endothelial glycocalyx upon rosuvastatin therapy in patients with heterozygous familial hypercholesterolemia. *J Lipid Res.* 2009;50(1):148–153.
135. Fraccarollo D, Widder JD, Galuppo P, et al. Improvement in left ventricular remodeling by the endothelial nitric oxide synthase enhancer AVE9488 after experimental myocardial infarction. *Circulation.* 2008;118(8):818–827.
136. Frantz S, Adamek A, Fraccarollo D, et al. The eNOS enhancer AVE 9488: a novel cardioprotectant against ischemia reperfusion injury. *Basic Res Cardiol.* 2009;104(6):773–779.
137. Wohlfart P, Xu H, Endlich A, et al. Antiatherosclerotic effects of small-molecular-weight compounds enhancing endothelial nitric-oxide synthase (eNOS) expression and preventing eNOS uncoupling. *J Pharmacol Exp Therapeut.* 2008;325(2):370–379.

138. Crabtree MJ, Tatham AL, Al-Wakeel Y, et al. Quantitative regulation of intracellular endothelial nitric-oxide synthase (eNOS) coupling by both tetrahydrobiopterin-eNOS stoichiometry and biopterin redox status: insights from cells with tet-regulated GTP cyclohydrolase I expression. *J Biol Chem*. 2009;284(2):1136–1144.
139. Heitzer T, Brockhoff C, Mayer B, et al. Tetrahydrobiopterin improves endothelium-dependent vasodilation in chronic smokers: evidence for a dysfunctional nitric oxide synthase. *Circ Res*. 2000; 86(2):E36–E41.
140. Heitzer T, Krohn K, Albers S, Meinertz T. Tetrahydrobiopterin improves endothelium-dependent vasodilation by increasing nitric oxide activity in patients with Type II diabetes mellitus. *Diabetologia*. 2000;43(11):1435–1438.
141. Gori T, Burstein JM, Ahmed S, et al. Folic acid prevents nitroglycerin-induced nitric oxide synthase dysfunction and nitrate tolerance: a human in vivo study. *Circulation*. 2001;104(10): 1119–1123.
142. Tiefenbacher CP, Bleeke T, Vahl C, Amann K, Vogt A, Kubler W. Endothelial dysfunction of coronary resistance arteries is improved by tetrahydrobiopterin in atherosclerosis. *Circulation*. 2000;102(18):2172–2179.
143. Antoniades C, Antonopoulos AS, Bendall JK, Channon KM. Targeting redox signaling in the vascular wall: from basic science to clinical practice. *Curr Pharmaceut Des*. 2009;15(3):329–342.
144. Granton J, Langleben D, Kutryk MB, et al. Endothelial NO-synthase gene-enhanced progenitor cell therapy for pulmonary arterial hypertension: the PHACeT trial. *Circ Res*. 2015;117(7): 645–654.
145. Ghofrani HA, Galie N, Grimminger F, et al. Riociguat for the treatment of pulmonary arterial hypertension. *N Engl J Med*. 2013; 369(4):330–340.
146. Ghofrani HA, D'Armini AM, Grimminger F, et al. Riociguat for the treatment of chronic thromboembolic pulmonary hypertension. *N Engl J Med*. 2013;369(4):319–329.
147. Lapp H, Mitrovic V, Franz N, et al. Cinaciguat (BAY 58-2667) improves cardiopulmonary hemodynamics in patients with acute decompensated heart failure. *Circulation*. 2009;119(21):2781–2788.
148. Korkmaz S, Radovits T, Barnucz E, et al. Pharmacological activation of soluble guanylate cyclase protects the heart against ischemic injury. *Circulation*. 2009;120(8):677–686.
149. Radovits T, Korkmaz S, Miesel-Groschel C, et al. Pre-conditioning with the soluble guanylate cyclase activator Cinaciguat reduces ischaemia-reperfusion injury after cardiopulmonary bypass. *Eur J Cardio Thorac Surg*. 2011;39(2):248–255.
150. Korkmaz S, Loganathan S, Mikles B, et al. Nitric oxide- and heme-independent activation of soluble guanylate cyclase attenuates peroxynitrite-induced endothelial dysfunction in rat aorta. *J Cardiovasc Pharmacol Therapeut*. 2013;18(1):70–77.
151. Boswell-Smith V, Spina D, Page CP. Phosphodiesterase inhibitors. *Br J Pharmacol*. 2006;147(Suppl 1):S252–S257.
152. Santi D, Giannetta E, Isidori AM, Vitale C, Aversa A, Simoni M. Therapy of endocrine disease. Effects of chronic use of phosphodiesterase inhibitors on endothelial markers in type 2 diabetes mellitus: a meta-analysis. *Eur J Endocrinol*. 2015;172(3): R103–R114.
153. Korkmaz S, Radovits T, Barnucz E, et al. Dose-dependent effects of a selective phosphodiesterase-5-inhibitor on endothelial dysfunction induced by peroxynitrite in rat aorta. *Eur J Pharmacol*. 2009;615(1–3):155–162.
154. Radovits T, Arif R, Bomicke T, et al. Vascular dysfunction induced by hypochlorite is improved by the selective phosphodiesterase-5-inhibitor vardenafil. *Eur J Pharmacol*. 2013;710(1–3):110–119.
155. Daiber A, Munzel T. Organic nitrate therapy, nitrate tolerance and nitrate induced endothelial dysfunction - emphasis on redox biology and oxidative stress. *Antioxid Redox Sign*. 2015;23(11): 899–942.
156. Warnholtz A, Mollnau H, Heitzer T, et al. Adverse effects of nitroglycerin treatment on endothelial function, vascular nitrotyrosine levels and cGMP-dependent protein kinase activity in hyperlipidemic Watanabe rabbits. *J Am Coll Cardiol*. 2002;40(7): 1356–1363.
157. Gori T, Parker JD. Long-term therapy with organic nitrates: the pros and cons of nitric oxide replacement therapy. *J Am Coll Cardiol*. 2004;44(3):632–634.
158. Munzel T, Meinertz T, Tebbe U, et al. Efficacy of the long-acting nitro vasodilator pentaerithrityl tetranitrate in patients with chronic stable angina pectoris receiving anti-anginal background therapy with beta-blockers: a 12-week, randomized, double-blind, placebo-controlled trial. *Eur Heart J*. 2014;35(14):895–903.
159. Munzel T, Daiber A, Gori T. Nitrate therapy: new aspects concerning molecular action and tolerance. *Circulation*. 2011;123(19): 2132–2144.
160. Bueno M, Wang J, Mora AL, Gladwin MT. Nitrite signaling in pulmonary hypertension: mechanisms of bioactivation, signaling, and therapeutics. *Antioxid Redox Sign*. 2013;18(14):1797–1809.
161. Bailey JC, Feelisch M, Horowitz JD, Frenneaux MP, Madhani M. Pharmacology and therapeutic role of inorganic nitrite and nitrate in vasodilatation. *Pharmacol Ther*. 2014;144(3):303–320.
162. Rassaf T, Ferdinandy P, Schulz R. Nitrite in organ protection. *Br J Pharmacol*. 2014;171(1):1–11.
163. Sydow K, Munzel T. ADMA and oxidative stress. *Atherosclerosis Suppl*. 2003;4(4):41–51.
164. Closs EI, Ostad MA, Simon A, et al. Impairment of the extrusion transporter for asymmetric dimethyl-l-arginine: a novel mechanism underlying vasospastic angina. *Biochem Biophys Res Commun*. 2012;423(2):218–223.
165. Cortese-Krott MM, Kuhnle GG, Dyson A, et al. Key bioactive reaction products of the NO/H_2S interaction are S/N-hybrid species, polysulfides, and nitroxyl. *Proc Natl Acad Sci USA*. 2015; 112(34):E4651–E4660.
166. Yuan S, Patel RP, Kevil CG. Working with nitric oxide and hydrogen sulfide in biological systems. *Am J Physiol Lung Cell Mol Physiol*. 2015;308(5):L403–L415.
167. Galie N, Manes A, Branzi A. Medical therapy of pulmonary hypertension. The prostacyclins. *Clin Chest Med*. 2001;22(3): 529–537 (x).
168. Daiber A, Kuntic M, Hahad O, et al. Effects of air pollution particles (ultrafine and fine particulate matter) on mitochondrial function and oxidative stress - Implications for cardiovascular and neurodegenerative diseases. *Arch Biochem Biophys*. 2020;696: 108662.
169. Munzel T. Endothelial dysfunction: pathophysiology, diagnosis and prognosis. *Dtsch Med Wochenschr*. 2008;133(47):2465–2470.
170. Munzel T, Gori T, Bruno RM, Taddei S. Is oxidative stress a therapeutic target in cardiovascular disease? *Eur Heart J*. 2010;31(22): 2741–2748.
171. Munzel T, Daiber A, Mulsch A. Explaining the phenomenon of nitrate tolerance. *Circ Res*. 2005;97(7):618–628.
172. Münzel T, Daiber A, Ullrich V, Mülsch A. Vascular consequences of endothelial nitric oxide synthase uncoupling for the activity and expression of the soluble guanylyl cyclase and the cGMP-dependent protein kinase. *Arterioscler Thromb Vasc Biol*. 2005; 25(8):1551–1557.

CHAPTER

27

The vascular phenotype in hypertension

Rhian M. Touyz[1], *Francisco J. Rios*[1], *Augusto C. Montezano*[1], *Karla B. Neves*[1], *Omotayo Eluwole*[2,3], *Muzi J. Maseko*[2], *Rheure Alves-Lopes*[1] and *Livia L. Camargo*[1]

[1]**Institute of Cardiovascular and Medical Sciences, BHF Glasgow Cardiovascular Research Centre, University of Glasgow, Glasgow, United Kingdom** [2]**Department of Physiology, University of Witwatersrand, Johannesburg, Gauteng, South Africa** [3]**Obafemi Awolowo University, Ile-Ife, Osun State, Nigeria**

Introduction

The vascular system plays a critical role in blood pressure regulation, primarily through changes in peripheral arterial resistance in attempts to maintain blood flow to organs in response to physiological demands.[1] This is associated with functional, mechanical, and structural changes of vessels that are short-lived and normalized once the triggering event has ceased. However, in pathological conditions, these modifications do not return to normal, leading to maladaptive vascular alterations including endothelial dysfunction, vascular hyperreactivity, and structural remodeling, processes that are tightly integrated.[2]

Arterial hypertension (high blood pressure), a major cause of morbidity and mortality globally, is a systemic and chronic disorder typically associated with impaired endothelial function, increased vasoconstriction, and arterial remodeling, vascular hallmarks of hypertension.[3,4] These processes, defined as the vascular phenotype of hypertension, are dynamic, functionally interlinked and occur at different times during the development of hypertension. Endothelial dysfunction is characterized by reduced endothelium-dependent vasodilation, loss of integrity of the endothelium, increased permeability, and promotion of a proinflammatory state, while hypercontractility is linked to increased contraction and reduced relaxation of vascular smooth muscle cells (VSMCs), which constitute the bulk of the vascular media.[5–9] Arterial remodeling, which occurs in large and small arteries, involves structural and architectural changes that influence the thickness of the vessel wall and lumen diameter and has been characterized as hypotrophic, eutrophic, and hypertrophic depending on all the wall thickness.[6,7] In eutrophic remodeling, there is no net change in vessel wall components while hypotrophic and hypertrophic remodeling are characteristically associated with reduced or increased vascular components (e.g., VSMCs, extracellular matrix, inflammatory cells, perivascular adipose tissue [PVAT]), respectively. An important additional factor that defines the vascular system in hypertension relates to mechanical changes. These phenomena, which are especially important in larger arteries, play an important role in arterial changes during aging and in established hypertension. A key mechanism contributing to vascular stiffness is fibrosis, which itself is a dynamic process that is initially adaptive to physiological demands.[8]

Vascular remodeling and increased arterial resistance, if not ameliorated or reversed, cause irreversible vascular damage, which worsens hypertension leading to target organ damage and consequent stroke, vascular dementia, heart failure, and kidney dysfunction.[9] While myriad molecular and cellular factors contribute to the vascular phenotype in hypertension, VSMCs play a major role due to their plasticity and multifunctional nature. This chapter provides a comprehensive overview of the (patho)physiological processes that regulate vascular function and structure in health and hypertension and delineates molecular mechanisms that define the vasculome in hypertension, with a particular focus on some new concepts related to VSMC signaling, inflammation, PVAT, and aging.

The Vasculome
https://doi.org/10.1016/B978-0-12-822546-2.00022-8

Vascular function: integration of vascular smooth muscle contraction/relaxation

The primary function of VSMCs includes contraction and relaxation, which are largely dependent on the activation and inactivation (phosphorylation/dephosphorylation) of contractile proteins. Calcium (Ca^{2+}) signaling plays a crucial role in these processes and triggers specific responses, although there are other molecular players that also play a role by fine-tuning VSMC contraction (Fig. 27.1). In pathological conditions, such as hypertension, these mechanisms are substantially altered, triggering a hypercontractile state and arterial wall remodeling.[10,11]

Functional changes, which manifest primarily as altered vascular reactivity, occur at the earliest stages of hypertension and influence the development and continuation of the condition. Precise mechanisms triggering altered vascular reactivity are not well understood but involve changes in the contraction/relaxation state of VSMCs.[11,12] Processes underlying this comprise alterations in Ca^{2+} signaling and impaired regulation of the vascular contractile machinery. In addition, a new paradigm has recently described, where crosstalk between endothelial cells and VSMCs regulates vascular tone. In particular, endothelial cell Ca^{2+} signaling has been shown to normally dampen VSMC contraction, an effect that is lost in hypertension.[13]

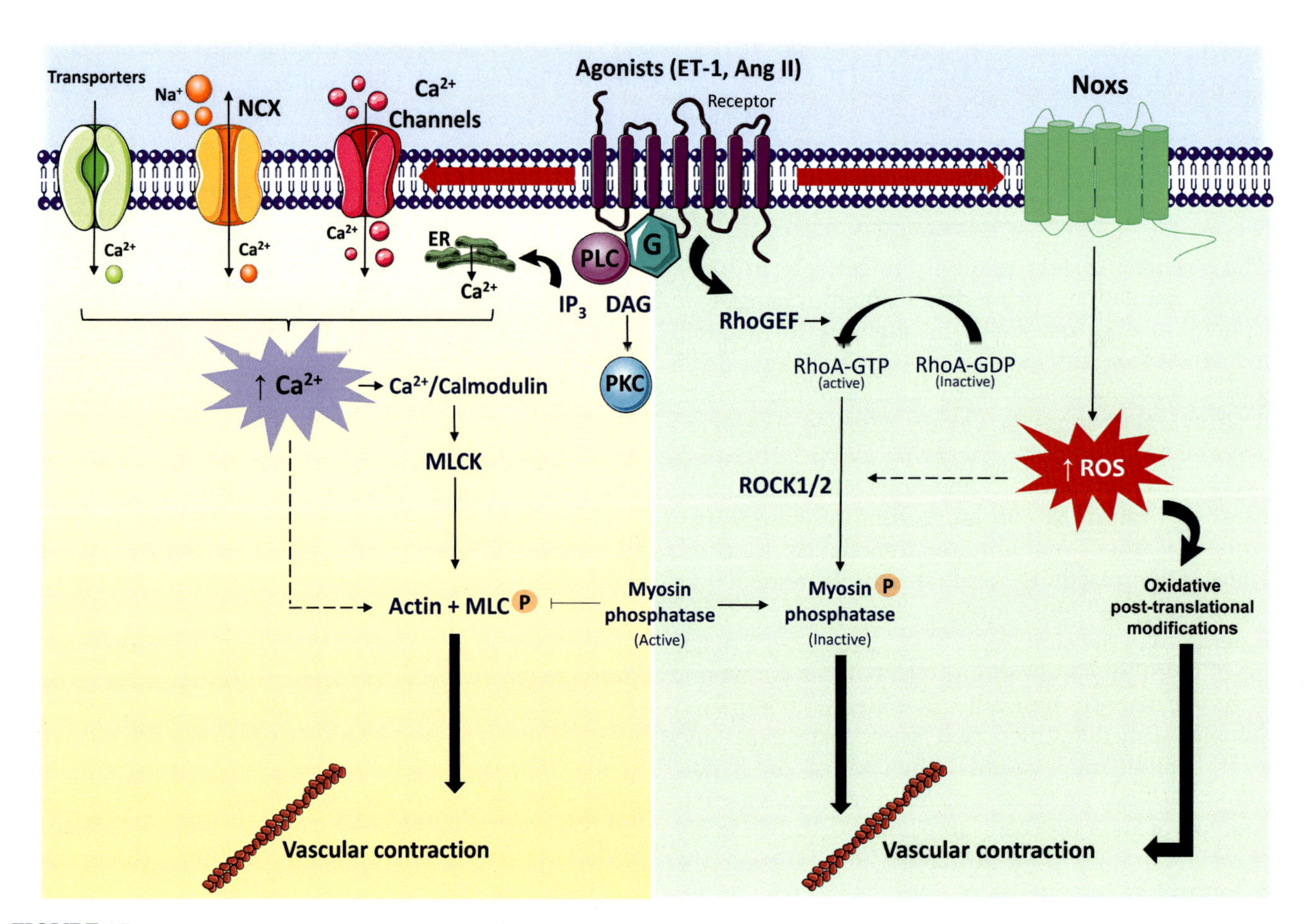

FIGURE 27.1 **Regulation of vascular smooth muscle cell contraction.** Ligand binding to cell membrane receptors induces activation of PLC via coupling through G proteins. PLC produces two second messengers: DAG and IP3. IP3 activates sarcoplasmic/endoplasmic reticulum receptors and promotes release of Ca^{2+}. DAG activates PKC, which phosphorylates target proteins that promote procontractile signaling. Ca^{2+} influx is also mediated by channels, pumps, and transporters on the plasma membrane. Intracellular Ca^{2+} binds to calmodulin and activates MLCK, which phosphorylates the light chain of myosin, and in conjunction with actin, initiates smooth muscle cell contraction. The increase in cytosolic Ca^{2+} is transient, and a Ca^{2+}-sensitizing mechanism is triggered to maintain the contractile state by inhibiting MLCP activity by Rho kinase. This involves activation of the small GTP-binding protein RhoA and subsequent increase in Rho kinase activity, culminating in inhibition of the myosin phosphatase and a contractile state. Noxs, which are important sources of vascular ROS, have also been associated with activation of Ca^{2+} channels and RhoA/ROCK. In addition, ROS is involved in posttranslational oxidative modifications of proteins, which can modulate vascular contraction. In hypertension these mechanisms are altered, leading to a hypercontractile state. *Ca^{2+}*, calcium; *DAG*, diacylglycerol; *ER*, endoplasmic reticulum; *IP3*, inositol 1,4,5-trisphosphate; *MLCK*, myosin light chain kinase; *MLCP*, myosin light chain phosphatase; *Noxs*, NADPH oxidases; *PKC*, protein kinase C; *PLC*, phospholipase C; *ROCK*, Rho kinase; *ROS*, reactive oxygen species.

Calcium-dependent mechanisms of vascular smooth muscle contraction

An increase in intracellular free concentration of Ca^{2+} ($[Ca^{2+}]_i$) is the key event in VSMC contraction.[14] Calcium signals and cytosolic Ca^{2+} concentrations are finely modified by intracellular sources, such as the sarcoplasmic reticulum, calcium-binding proteins, and plasma membrane calcium permeable channels, exchangers, and transporters. Mechanical, humoral, and neural stimuli lead to an increase in $[Ca^{2+}]_i$. The initial rapid increase in cytosolic Ca^{2+} is associated with Ca^{2+} release from the sarcoplasmic reticulum, whereas the later phase derives from the extracellular space through Ca^{2+} channels. Ligand–receptor interaction on the plasma membrane stimulates phospholipase C (PLC), which catalyzes the formation of the second messengers, inositol trisphosphate (IP_3) and diacylglycerol (DAG).[15,16] IP_3 binds to receptors on the sarcoplasmic reticulum (IP_3 receptor) and induces release of Ca^{2+} into the cytosol. DAG activates protein kinase C (PKC), which phosphorylates specific target proteins, including contractile proteins, regulatory proteins, channels, and pumps, furthering Ca^{2+} influx and increasing $[Ca^{2+}]_i$.[16] Calcium binds to calmodulin, and the calcium-calmodulin complex promotes a conformational change in myosin light chain kinase (MLCK) switching its inactive state to active state. Activated MLCK induces phosphorylation of MLC20 and stimulates myosin–actin interaction, leading to vascular contraction, processes induced by Ang II and other vasoactive agents.[17,18] In addition to numerous Ca^{2+} channels, such as voltage-operated channels, store-operated channels, receptor-operated channels, transient receptor potential cation (TRP) channels, and Ca^{2+}-permeable nonselective channels, activation of nuclear factor of activated T cells (NFAT), a transcription factor and major target of calcineurin, alters expression of plasma membrane potassium channels[19] promoting vascular contractility in hypertension. Cytokines, miRNAs, reactive oxygen species (ROS), and some cellular-derived factors (e.g., extracellular vesicles and endothelial progenitor cells) also regulate pathways that control $[Ca^{2+}]i$. Likewise, endothelium-derived vasoconstrictors (e.g., endothelin-1 [ET-1], prostanoids) regulate vascular contraction via activation of VSMC receptors leading to procontractile signaling.[19,20]

Calcium-independent mechanisms of vascular smooth muscle cell contraction

Calcium-independent processes also regulate VSMC contraction by priming the sensitivity of MLC to Ca^{2+}. Two key signaling pathways implicated in Ca^{2+} sensitization include the DAG-PLC-PKC and the RhoA-Rho kinase (ROCK) pathways.[21] DAG activates PKC, which in turn phosphorylates the C-kinase potentiated protein phosphatase 1 inhibitor, molecular mass 17 kDa (CPI-17), a smooth muscle-specific inhibitor of MLC phosphatase (MLCP), inhibiting MLCP phosphatase activity, and enabling persistent contraction. PKC also induces calponin and calmodulin phosphorylation, which are actin-binding proteins, promoting actin–myosin interaction and VSMC contraction.[19] Deletion of calponin genes decreases blood pressure and reduces vascular contraction in mice.[22] The other Ca^{2+}-sensitizing system is associated with activation of the small G protein, RhoA, and its downstream effector Rho kinase (ROCK), of which there are 2 isoforms, ROCK1 and ROCK2. ROCK influences Ca^{2+} sensitization through stimulation of myosin phosphatase target subunit 1 (MYPT1) phosphorylation or it phosphorylates ZIPK (also known as DAPK3), which also stimulates phosphorylation of MYPT1. MYPT1 phosphorylation in turn reduces its phosphatase activity leading to sustained contraction. ROCK also phosphorylates CPI-17. Hyperactivation of RhoA-ROCK signaling pathway culminates in decreased MLCP activation and consequently sustained vasoconstriction.[23,24]

Vascular smooth muscle cell relaxation

Counterbalancing vasoconstriction is vasorelaxation, which is a consequence of decreased $[Ca^{2+}]_i$ and an increased MLCP activity. Reduced $[Ca^{2+}]_i$ occurs due to inactivation of L-type Ca^{2+} channels and decreased Ca^{2+} influx and increased activity of Ca^{2+}-ATPase and the sodium-calcium exchanger (NCX), which remove Ca^{2+} from the cytosol. An activation of the sarcoplasmic/endoplasmic reticulum Ca^{2+}-ATPase (SERCA), which stimulates reuptake of Ca^{2+} into the sarcoplasmic reticulum, also occurs. MLCK becomes inactivated as Ca^{2+} dissociates from calmodulin and MLC20 is dephosphorylated by MLCP.[25]

Endothelial-derived nitric oxide (NO), via activation of endothelial nitric oxide synthase (eNOS), is another signaling pathway responsible for vascular relaxation. NO regulates cyclic guanosine monophosphate (cGMP) and adenosine monophosphate (cAMP) levels. cAMP, via protein kinase G and protein kinase A, increase MLCP activity and decrease Ca^{2+} sensitivity of the contractile machinery.[26] In VSMCs, NO also increases cGMP levels, another important cardiac and vascular second messenger, by stimulating the activity of soluble guanylate cyclase (sGC) leading to cytosolic Ca^{2+} depletion and VSMC relaxation.[26,27]

Cytoskeleton components and vascular function

Cytoskeletal proteins are critically involved in the regulation of VSMC contraction/relaxation, as well as in vascular structural changes that occur in hypertension. In particular, reorganization of actin and intermediate filaments and microtubules play an important role in vasoconstriction. Additionally, the maintenance of plasticity, a major factor controlling myogenic tone in hypertension, is crucially associated with a dynamic rearrangement of the actin cytoskeleton. Persistent vasoconstriction and associated actin polymerization contribute to VSMC architectural changes and vascular remodeling in hypertension.[28]

Oxidative stress and vascular function in hypertension

ROS and reactive nitrogen species are biologically important O_2 products and key players in vascular (patho)biology through their redox potential. In endothelial cells and VSMCs, the major source of ROS is the Nox family, comprising seven isoforms, with Nox 1, 2, 4, and 5 being expressed and functionally activated in human VSMCs.[29,30] We defined Nox5 as a procontractile ROS-generating oxidase, since it is a point of crosstalk between redox-dependent and Ca^{2+}-sensitive pathways that regulate VSMC contraction.[31] Various prohypertensive agents including Ang II, ET-1, and aldosterone activate Noxs, responses that are amplified in hypertension.[31,32] Increased Nox-derived ROS production augments Ca^{2+} signaling, modulates the actin cytoskeleton, and enhances ROCK signaling, thereby increasing vascular contraction and boosting vascular tone. ROS can also modulate procontractile signaling through posttranslational modification of proteins. In conditions associated with oxidative stress, such as in hypertension, cysteine and methionine residues, which are highly redox-sensitive, undergo oxidative modifications. Some of these modifications are irreversible resulting in cell death, vascular injury, and target organ damage. Actin and actin-binding proteins, myosin, and cofilin can be directly oxidized by ROS as well as many proteins that regulate VSMC Ca^{2+} homeostasis such as SERCA, Ca^{2+} channels, and ROCK.[33]

Vascular functional changes in hypertension

Hypertension, both human and experimental, is characterized by vascular hyperreactivity, increased vasoconstriction, and decreased endothelium-dependent vasorelaxation.[3,4,34–36] These findings are based on noninvasive vascular studies in humans, ex vivo examination of isolated resistance arteries, and cultured VSMCs from patients with hypertension and extensive preclinical data from animal models. Perturbed VSMC Ca^{2+} homeostasis is a major trigger for these phenomena with multiple factors contributing to abnormal Ca^{2+}-dependent procontractile signaling. High VSMC $[Ca^{2+}]_i$ is consistently observed in experimental (deoxycorticosterone acetate [DOCA]-salt, Ang II–infused, L-NAME-induced), genetic (spontaneously hypertensive rats [SHRs], SHR stroke-prone SHRsp), and human hypertension.[34–37] Calcium channels play a crucial role in abnormal Ca^{2+} handling during hypertension, which is further evidenced by the successful antihypertensive effects of L-type Ca^{2+} channel blockers observed clinically. Mechanisms involving enhanced expression and phosphorylation of subunits of L-type Ca^{2+} channels and other regulatory factors have been identified.[37] TRP channels (TRPC3, TRPC6, and TRPC7) and the NCX have also been shown to be associated with altered vascular function and Ca^{2+} handling in hypertension.[38] Novel findings identified TRPM2 (transient receptor potential cation channel, subfamily M, member 2) channel, a highly redox-sensitive channel permeable to Na^+ and Ca^{2+}, as an important regulator of VSMC Ca^{2+} homeostasis and vascular function during hypertension.[39] TRPM2 overexpression dysregulates $[Ca^{2+}]_i$ and contributes to vascular damage. Additionally, increased Ca^{2+} influx in VSMC from hypertensive patients is not observed when TRPM2 is downregulated, an effect associated with normalization of vascular hypercontractility.[39] Many of the ion channels in VSMCs are redox-sensitive. In hypertension, vascular oxidative stress promotes cytoskeletal reorganization and increased $[Ca^{2+}]i$, processes that are associated with activation of Nox5 in human VSMCs.[29,30]

Increased activation of RhoA/ROCK via various mechanisms, such as alterations in the Rho guanine nucleotide exchange factors (Rho-GEFs), has also been implicated in vascular dysfunction in hypertension.[25,26] Clinical studies show increased phosphorylation of ROCK-regulated MYPT1 in patients with essential hypertension. ROCK2 also plays a major role in Ang II–induced hypertension and cardiac hypertrophy, and pharmacological inhibition of ROCK is shown to improve vascular dysfunction in hypertension. There is also increasing evidence indicating a pathophysiological role for ROCK in pulmonary arterial hypertension. Although ROCK inhibitors are already used clinically to prevent cerebral vasospasm, often associated with subarachnoid hemorrhage and acute ischemic stroke, they are not yet approved to treat human essential hypertension but may be a promising approach.

Nonclassical systems have also been implicated in vascular hypercontractility in hypertension. In particular, activation of inflammatory pathways, tyrosine kinases, MAPKs, and transcription factors in VSMCs have been linked to vasoconstriction. Calcium-sensitive transcription factors, such as NFAT, modulate genes encoding contractile proteins, further affecting vascular hypercontractility in hypertension. Activated immune cells travel to vessels and target organs and cause damage by releasing bioactive factors including ROS, cytokines, and chemokines, which induce vascular dysfunction, remodeling, and rarefaction.[40] Damage-associated molecular patterns (DAMPs) and toll-like receptors (TLRs) have also been associated with altered vascular function in hypertension. Increased DAMP-TLR activation has been suggested to influence vascular inflammatory responses and reactivity in hypertension. In resistance arteries from SHR rats, TLR4 is upregulated and has been linked to increased contraction through regulation of DAMPs-dependent and cyclooxygenase-dependent signaling pathways.[40]

Maintaining the vascular contractile phenotype is highly dependent on the state of differentiation of VSMCs. Physiologically, VSMCs have a contractile phenotype, which becomes exaggerated in hypertension. Associated with the contractile changes, VSMCs undergo phenotypic dedifferentiation to a more synthetic, proliferative phenotype that plays an important role in structural changes associated with hypertension.[41]

Vascular remodeling in hypertension

Vascular remodeling is an adaptive process that vessels undergo to adjust to hemodynamic changes. This is a tightly regulated physiological process. However, when physiological remodeling becomes maladaptive and unable to compensate for hemodynamic changes, pathological architectural, structural, and mechanical changes occur, features that characterize hypertension. Changes in the balance between VSMC growth and apoptosis, rearrangement of VSMCs in the media, calcification, and production or degradation of components of the extracellular matrix are major contributors to remodeling.[42,43]

Clinical studies demonstrate that almost all hypertensive patients have evidence of arterial remodeling, whereas only some patients display endothelial dysfunction. Importantly remodeling of small resistance arteries is reversible with some, but not all, antihypertensive drugs.[44] In particular, inhibitors of the RAAS and calcium channel blockers correct structural alterations and vascular dysfunction, whereas drugs like thiazide diuretics and beta blockers are not as effective, despite similar blood pressure lowering.[44]

Arterial remodeling is a dynamic process and is influenced by the stimuli that trigger and sustain the process, such as humoral factors (vasoactive peptides, hormones, growth factors), mechanical processes (shear stress, stretch, pressure), chemical elements (oxidative stress), or blood pressure and blood flow.[42] Chronic elevation of blood pressure leads to changes in the media-to-lumen ratio due to enhanced VSMC mass and reorganization of cellular and extracellular matrix components.[43]

Vessels undergo different types of remodeling depending on the underlying pathophysiological process and vascular bed.[44–46] Mulvany et al. defined arterial remodeling as inward (reduced lumen diameter) or outward (increased lumen diameter) and hypertrophic (thick vascular wall), eutrophic (no changes in vessel wall thickness), or hypotrophic (thin vascular wall).[47,48] In hypertension, the two types of remodeling commonly observed are inward eutrophic remodeling and hypertrophic remodeling.[49] Inward eutrophic remodeling is characterized by reduction in outer and lumen diameter where the media cross-sectional area is unchanged, but the media-to-lumen ration is increased. This type of remodeling is observed in some experimental models of hypertension, such as the SHR[50] and 2-kidney, 1-clip Goldblatt rats,[51] and in patients with mild essential hypertension.[52,53] In patients with secondary hypertension[54] and salt-sensitive models of hypertension (DOCA-salt; salt-sensitive Dahl rats), arterial remodeling is characterized by reduction in lumen diameter and increased media-to-lumen ratio, typical of hypertrophic remodeling.[55,56]

Mechanisms of vascular remodeling in hypertension—role of Ang II

Remodeling involves multiple cell types and mechanisms. VSMCs are of major importance, as they regulate vascular tone and diameter to accommodate constant changes in hemodynamics, while in a contractile phenotype, characterized by the expression of specific proteins such as SM22α, αSMA and smoothelin.[40] Injurious signals, such as oxidative stress, shear stress, Ang II, ET-1, aldosterone and growth factors, and activation of proinflammatory and mitogenic signaling pathways, initiate a process of phenotypic switching, where VSMCs undergo dedifferentiation into other phenotypes, such as migratory, proliferative, secretory, osteogenic, or inflammatory.[57]

In hypertension, Ang II stimulates mitogenic signaling pathways in VSMCs promoting growth via activation of MAP kinases (ERK1/2, p38MAPK, ERK5) and tyrosine kinases (c-Src, JAK, FAK). These processes are amplified in hypertension rendering VSMCs to the proliferative/migratory phenotype.[58] Ang II–induced proliferative actions are also mediated through transactivation of growth receptors, such as the EGFR, via MMP-dependent shedding of heparin-binding EGF-like growth factor in the cell membrane.[59] Once in the synthetic and migratory phenotype, VSMCs produce MMPs, which influence VSMC adhesion to the ECM and surrounding cells facilitating migration.[57]

Most signaling events elicited by Ang II are redox-sensitive, and growing evidence shows that Ang II regulation of proliferation and migration is mediated through an increase in ROS and posttranslational oxidative modification of proteins.[58] Ang II also stimulates production of cytokines (Il-1β, Il-6, TNF-α), chemokines (monocyte chemoattractant protein 1 [MCP-1]), and adhesion molecules (ICAM-1, VCAM-1) through redox activation of transcription factors and proinflammatory nuclear factors, such as NF-κB and AP-1, inducing an inflammatory phenotype. These mediators stimulate immune responses that further contribute to vascular fibrosis, inflammation, and arterial remodeling.[49]

In some forms of hypertension, inward eutrophic remodeling is observed, suggesting that mechanisms other than proliferation contribute to vascular structural changes, such as apoptosis. In this situation, VSMC death compensates for increased proliferation with a net imbalance between apoptosis and cell growth. Markers of apoptosis, such as Bax/Bcl-2 and DNA fragmentation, are increased in models of hypertension including DOCA-salt rats, SHR, and Ang II–infused animals.[59–61] On the other hand, reduced apoptosis with exaggeration proliferation contributes to hypertrophic remodeling. These processes are dynamic with young SHR rats exhibiting decreased apoptosis in resistance arteries.[62]

The immune system and inflammation—a new paradigm in the vascular phenotype in hypertension

One of the major recent advances in hypertension research is the recognition that activation of the immune system and inflammation contributes to the pathophysiology of hypertension. Many prohypertensive stimuli are proinflammatory. Hypertension is associated with low-grade subclinical vascular inflammation characterized by increased adhesiveness of endothelial cells, migration of immune cells to the subendothelial space, and activation of resident macrophages and other immune cells in tissues[63] (Fig. 27.2). Extensive experimental evidence indicates the importance of the innate and adaptive immune systems in hypertension. Infiltration of immune cells is observed in the vascular media, PVAT, kidneys, and heart.[64–66] Genetic deletion of components of the immune system has demonstrated its important participation in hypertension and associated target organ damage.[66,67]

The immune system comprises innate immunity and adaptive immunity and physiologically they interact in response to injurious stimuli. Innate immunity is the first line of activation of the immune response and comprises epithelial cells, endothelial cells, macrophages, neutrophils, mast cells, natural killer, dendritic cells, and soluble proteins including the complement system, C-reactive proteins (CRPs), and pentraxins.[63] It exhibits low specificity, and activation is dependent on recognition of molecular patterns expressed in pathogens or endogenous modified molecules and dead cells. The acquired immunity has high specificity for antigens and is mediated by activated B and T lymphocytes. Hypertensive mediators such as Ang II, aldosterone, ET-1, and catecholamines exhibit proinflammatory properties through receptor-mediated signaling in immune cells and production of proinflammatory cytokines, chemokines, and eicosanoids, which lead to target organ damage.[64]

Inflammation and the vasculome in hypertension

Central to the process of vascular inflammation is the endothelium. In a proinflammatory environment, endothelial cells express adhesion molecules such as ICAM-1 (intercellular adhesion molecule 1), VCAM-1 (vascular cell adhesion molecule 1), P-selectin, and E-selectin. Activated endothelial cells also produce chemokines, including CCL2 (MCP-1), CCL5 (RANTES), CXCL10, and CXCL1 that interact with specific receptors expressed in immune cells (neutrophils, monocytes, and lymphocytes)[68] (Fig. 27.3). This interaction creates a gradient environment that acts synergically with adhesion molecules to attract leukocytes to the vascular wall. Physiologically circulating leukocytes do not traverse the endothelium. However, in an inflammatory environment, they are attracted to adhesion molecules expressed by endothelial cells resulting in adhesion of leukocytes to the endothelium, a process that is followed by transmigration to the subendothelial space. The transmigration is mediated by interactions of CD31/PECAM (platelet/endothelial cell adhesion molecule 1), expressed in endothelial cells, leukocytes, and platelets.

Vasoactive agents (Ang II, ET-1, aldosterone, growth factors) induce proinflammatory signaling through redox-sensitive MAPK/ERK1/2/P38MAPK pathways.[63] In the subendothelial space, monocytes differentiate to macrophages and together with transmigrated neutrophils, they amplify the local response, activating inflammatory transcription factors AP-1 (activator protein 1), NF-κB (nuclear factor kappa B), and STAT3 (signal transducer and activator of transcription 3).[69]

While leukocyte migration into the vascular media has been observed in hypertension, the exact role of chemokines is unclear because experimental models deficient in chemokines (CCL2, CCL5, CXCL1) or chemokine receptors (CCR2, CXCR3, CXCR2) failed to consistently show a blood pressure–associated response.[70] On the other hand, mice deficient in adhesion molecules such as ICAM-1 and P-selectin showed reduced blood pressure and cardiovascular damage induced by Ang II.[71,72] Soluble forms of adhesion

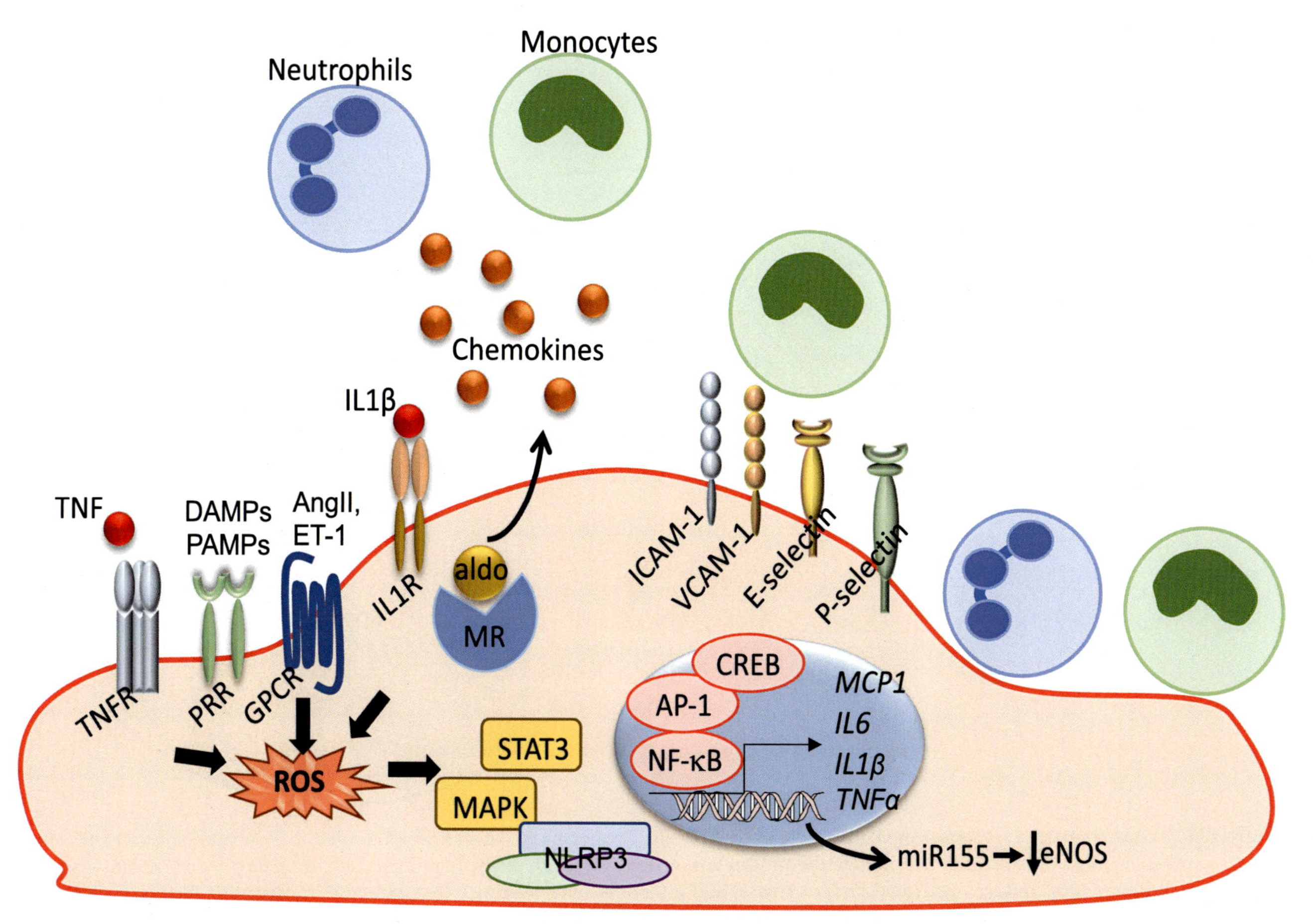

FIGURE 27.2 **Perivascular adipose tissue (PVAT) regulation of vascular tone in hypertension.** In health, PVAT has an anticontractile effect by releasing adipocyte-derived relaxing factors promoting endothelium-dependent and independent vasorelaxation. Among the substances secreted by PVAT are adipokines (adiponectin, leptin), nitric oxide (NO), hydrogen peroxide (H_2O_2), hydrogen sulfide (H_2S), methyl palmitate (PAME), and angiotensin 1−7 (Ang 1−7). Molecular mechanisms of adipokine-induced effects involve increase in NO production in endothelial cells and opening of K^+ channels in VSMC promoting vasorelaxation. In hypertension and obesity-related hypertension, there is an increase in the release of procontractile adipokines (chemerin, resistin, and vistatin), Ang II, and reactive oxygen species (ROS) that contributes to the loss of PVAT anticontractile effect. In addition, low-grade inflammation is observed, accompanied by increase in proinflammatory cytokines (TNF-α, INF-γ, IL-6, IL-17) involved in endothelial dysfunction and increased VSMC contraction, proliferation, and migration. PVAT dysfunction contributes to increased peripheral resistance and blood pressure elevation in obesity-related hypertension.

molecules (sVCAM-1, sICAM-1, sP-selectin, and sE-selectin) are associated with endothelial cell dysfunction and their increased plasma concentration is observed in patients with hypertension and atherosclerosis.[73] A small clinical study in hypertensive patients showed that Ca^{2+} channel blockers reduce not only blood pressure but also circulating levels of P-selectin and E-selectin.[74] Soluble adhesion factors have thus been considered as biomarkers of endothelial injury and hypertension.

Plasma concentrations of proinflammatory mediators such as CRP, TNF-α, and IL-6 are associated with vascular stiffness in hypertension.[75] TNF-α induces ROS production, activation of transcription factors (NF-κB, CREB, AP1), and inhibits NO production by increased expression of miR155, which interferes with mRNA expression of eNOS.[76] This pathway was evidenced by observing reduced blood pressure, cardiac hypertrophy, and fibrosis in TNF-α−deficient animals in DOCA/salt-hypertensive animals, a model of salt-sensitive hypertension with severe vascular and renal dysfunction and target organ damage.[76] Disruption of TNF-α signaling using a biologic agent that binds the free cytokine (etanercept) prevented Ang II−induced hypertension and ROS production in mice.[77] Moreover endothelial dysfunction in human placental vessels induced by TNF-α was reduced by aspirin, indicating the upstream effects of cyclooxygenases in this pathway.[78] IL-6 is a pleiotropic cytokine that activates endothelial cells and VSMCs through the JAK/STAT3 pathway. Phosphorylated STAT3 dimers translocate to the nucleus where they regulate proinflammatory gene expression, including production of MCP-1, which activates monocytes and promotes their adhesion to the inflamed endothelium. Pharmacologic inhibition of STAT3 reduced blood pressure and inflammation and

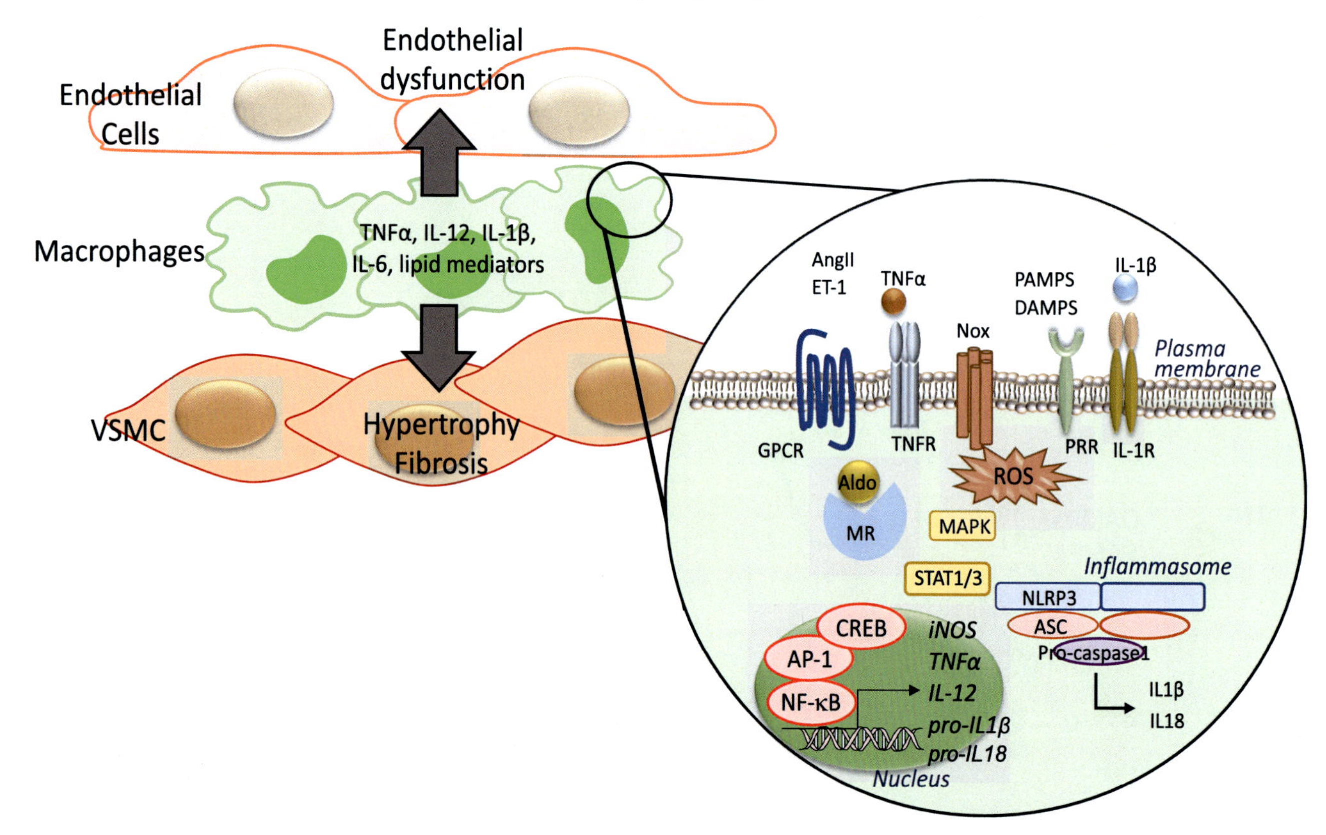

FIGURE 27.3 **Inflammatory responses mediated by endothelial cells.** Prohypertensive and proinflammatory mediators interact with specific receptors expressed in endothelial cells and induce the production of ROS (reactive oxygen species). ROS influence redox-sensitive signaling molecules, including MAPK (mitogen-activated protein kinase) and inflammasome, leading to activation of transcription factors, such as STAT3 (signal transducer and activator of transcription 3), NF-κB (nuclear factor-κB), AP-1 (activator protein 1), and CREB (cAMP-response element binding protein). These processes increase the expression of adhesion molecules ICAM-1 (intercellular adhesion molecule 1), VCAM-1 (vascular cell adhesion molecule 1), E-selectin and P-selectin, and the release of chemokines (MCP-1 [monocyte chemoattractant protein 1]) and cytokines (IL-6, IL-1β, TNF-α), which led to migration of leukocytes to the subendothelial space and increase the inflammatory response. Inflammatory mediators induce the expression of microRNA miR155, which reduces the expression of eNOS, leading to endothelial dysfunction. *Ang II*, angiotensin II; *aldo*, aldosterone; *DAMPs*, danger-associated molecular patterns; *ET-1*, endothelin-1; *GPCR*, G protein–coupled receptors; *MR*, mineralocorticoid receptor; *NLRP3*, NOD-, LRR-, and pyrin domain-containing protein 3; *PAMPs*, pathogen-associated molecular patterns; *PRR*, pattern recognition receptor.

ameliorated endothelial dysfunction. Patients with primary aldosteronism have higher IL-6.[79] The deleterious effects of IL-6 on target organ damage have been demonstrated in several models of hypertension.[79,80]

In addition to the endothelium, VSMC and fibroblasts are involved in the vascular inflammatory response in hypertension. VSMCs express receptors for several cytokines, including IL-6, TNF-α, IL-1β, and IL-10 that are produced by macrophages and endothelial cells during the inflammatory response. TNF-α stimulates VSMC to produce ROS through Nox1 activation.[81] This inflammatory pathway is shared by several signaling molecules that are important triggers of vascular inflammation, including platelet-derived growth factor, Ang II, and IL-1β. However, the final response to proinflammatory mediators differs between endothelial cells and VSMCs. For example, endothelial cells activated by TNF-α trigger a process that may result in cell death, while VSMCs respond by increased proliferation and migration.[82]

Eicosanoids, such as prostaglandins and thromboxane A2, are also involved in vascular inflammation and dysfunction in hypertension.[83] With advances in OMICS techniques and mass spectrometry, it has been possible to study the entire population of eicosanoids in human plasma. A recent large study with 10,958 participants identified 545 distinct eicosanoids, in which 187 were associated with systolic blood pressure.[84] These observations identify putative novel inflammatory lipid mediators in hypertension.

Innate immunity, vascular dysfunction, and hypertension

The innate immune system reacts rapidly and with low specificity by the interaction of pattern recognition receptors (PRRs) with general structures referred as pathogen-associated molecular patterns expressed in pathogens or DAMPS for endogenous structures.[85] The innate immune system comprises numerous cell types including vascular macrophages (M1 macrophages [classical activated proinflammatory] and M2 macrophages [alternative antiinflammatory]) that are activated by vasoactive agonists in hypertension.[85]

Almost all cell types of the vasculature express PRRs, and accordingly vascular cells participate in innate immune responses. PRRs comprise many families of receptors such as TLRs, C-type lectin receptors, nucleotide-binding domain, leucine-rich repeat-containing protein receptors (NLRs), and the scavenger receptor CD36.[86] Activation of these receptors triggers a complex signaling cascade that ultimately activates transcription factors including NF-κB, AP-1, ETS domain-containing protein Elk-1 (ELK-1), activating transcription factor 2 (ATF2), the phosphoprotein p53, and members of the interferon-regulatory factor family, leading to a massive production of cytokines, chemokines, and eicosanoids that further induce vascular inflammation.[85,86]

Another component of innate immunity is the inflammasome, a complex of proteins that generate IL-1β and IL-18, which promote vascular inflammation. Inflammasome activation is dependent on the intracellular molecule NLRP3 (nucleotide-binding and oligomerization domain-like receptor family pyrin domain-containing 3), which drives formation of the inflammasome complex that includes ASC (apoptosis-associated speck-like protein) and procaspase-1. Activation of the inflammasome pathway has been observed in experimental models of hypertension, such as Ang II and aldosterone-salt infusion, DOCA-salt, and pulmonary hypertension, and has been associated with endothelial dysfunction, target organ damage, vascular remodeling, and fibrosis.[63–65,87] The inflammasome is redox-sensitive and oxidative stress in hypertension is a potent inducer of the system.

Adaptive immunity, vascular inflammation, and hypertension

Adaptive immunity involves specialized lymphocytes with high specificity. These lymphocytes have a single type of receptor but an almost unlimited repertoire of variants that can recognize different antigens. Dendritic cells are at the interface of innate and adaptive immunity. The two main populations of T lymphocytes, CD^{4+} or T helper cells and CD^{8+} or cytotoxic T cells, have been associated with vascular dysfunction in hypertension.[87,88] Experimental models of hypertension that are deficient in both B and T lymphocytes exhibit reduced pressor responses to Ang II.[63–65,87,88] In these models, increased blood pressure was restored after adoptive transfer of T lymphocytes, but not B lymphocytes, thus confirming the pathogenic role of T lymphocytes in hypertension.[63–66,87–89] Isolevuglandins (IsoLGs) are products of peroxidation of arachidonic acid and have been suggested to be important triggers of the immune response in hypertension.[64] In experimental hypertension, T lymphocytes were activated by dendritic cells presenting IsoLG adducted peptides. Activated T cells produce inflammatory cytokines including IL-17A, TNF-α, and IFN-γ which lead to hypertension and target organ damage.[87] These processes are amplified by high salt diet and aldosterone.[65,90] In Ang II–induced hypertension, mice deficient in T CD^{8+} lymphocytes were protected against endothelial dysfunction and blood pressure elevation.[89] On the other hand, regulatory T cells (Treg) seem to be vasoprotective in hypertension because adoptive transfer of Tregs in Ang II–induced hypertensive mice resulted in blood pressure lowering and prevention of vascular inflammation and target organ damage.[88]

Despite growing experimental evidence demonstrating that innate and adaptive immunity are involved in vascular inflammation and target organ damage in hypertension, there is still a paucity of information in human hypertension. However, a few studies have reported increased circulating cytokines and chemokines in patients with essential hypertension.[70,91]

Perivascular adipose tissue influences vascular function in hypertension

PVAT is the adipose tissue surrounding most blood vessels in the body, except the cerebral and pulmonary vasculature. Initially considered to function as connective tissue to support vessels, PVAT is now recognized as an endocrine tissue that secretes diverse vasoactive molecules (adipokines).[92] Normally, PVAT produces vasoprotective adipokines important in maintenance of vascular tone. However, under pathological conditions, such as in hypertension, diabetes, and obesity, there is a change in the PVAT secretome leading to release of procontractile and proinflammatory substances contributing to vascular dysfunction[93] (Fig. 27.4).

Adipocytes are classified into white or brown according to the morphological and functional characteristics.[94] White adipocytes are spherical cells, rich in lipids, and have few mitochondria. The main function of white adipose tissue (WAT) is energy storage in the form of fatty acids and is found in subcutaneous and visceral adipose tissue.[95] Brown adipocytes are smaller elliptical cells with many mitochondria that express high levels of uncoupling protein 1 (UCP-1), important in thermogenesis.[96] A third type of UCP-1–positive adipocytes has been identified termed beige adipocytes. These are multilocular adipocytes, with large lipid

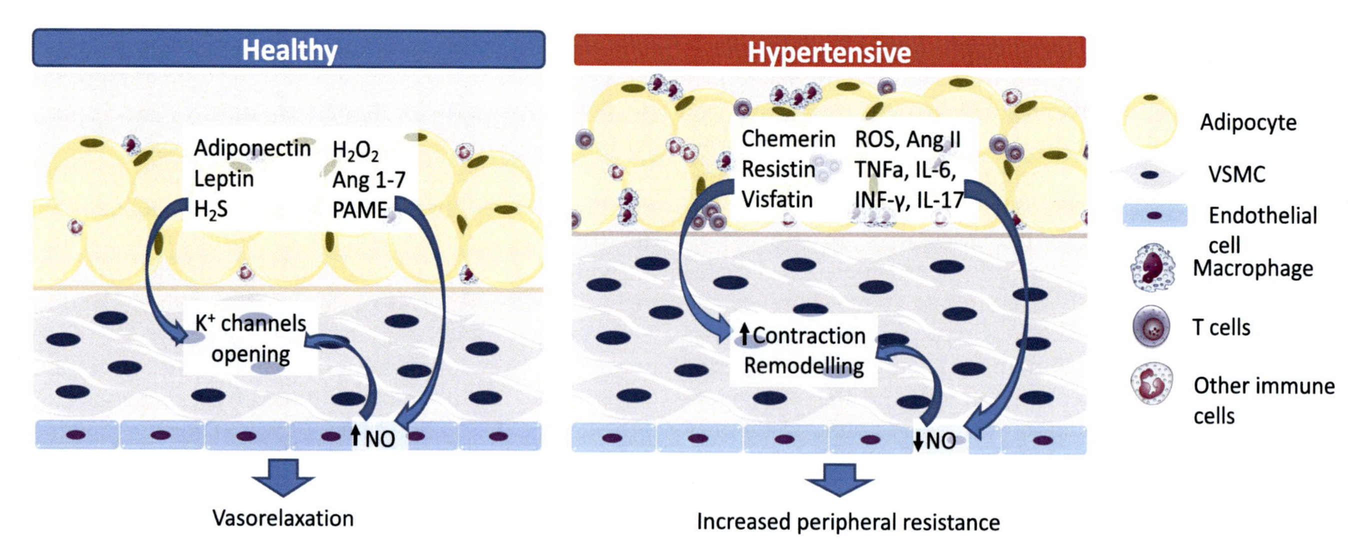

FIGURE 27.4 **Inflammatory responses mediated by macrophages in the vascular wall.** In the subendothelial space, monocytes differentiate into macrophages that interact with prohypertensive and proinflammatory mediators and activate proinflammatory signaling pathways. These processes increase the production of cytokines (IL-6, TNF-α, IL-12, pro–IL-1β and pro–IL-18) and induce expression of iNOS. Intracellular inflammatory processes lead to activation of the inflammasome complex (NLRP3 [NOD-, LRR-, and pyrin domain-containing protein 3], ASC [apoptosis-associated speck-like protein containing a CARD], procaspase 1)], which converts pro–IL-1β and pro–IL-18 to functional IL-1β and IL-18. Inflammatory mediators have significant impact in the vasculature by inducing endothelial dysfunction and VSMC hypertrophy. These processes contribute to vascular injury in hypertension. *Ang II*, angiotensin II; *aldo*, aldosterone; *DAMPs*, danger-associated molecular patterns; *ET-1*, endothelin-1; *GPCR*, G protein–coupled receptors; *MR*, mineralocorticoid receptor; *NLRP3, NOD-, LRR-*, and pyrin domain-containing protein 3; *PAMPs*, pathogen-associated molecular patterns; *PRR*, pattern recognition receptor; *VSMCs*, vascular smooth muscle cells.

droplets and low expression of thermogenic genes; however, upon stimulation by cold exposure, β3-adrenergic receptor agonists, or exercise, they assume more of a brown-like phenotype with upregulation of UCP-1, thus allowing thermogenesis.[97] Beige adipose tissue (BeAT) can also assume a white phenotype in warm conditions.[97] Adipocyte composition may have an impact in cardiovascular diseases, as accumulation of visceral WAT is associated with hypertension and obesity is characterized by an increase in WAT and decrease in BAT.[94] BAT is cardiovascular protective.[98,99]

PVAT regulation of vascular function in hypertension

PVAT releases several vasoactive substances that promote vasoconstriction or vasodilation. In health, PVAT has an anticontractile effect by releasing adipocyte-derived relaxing factors promoting endothelium-dependent and independent vasorelaxation, while in hypertension, PVAT releases proinflammatory and procontractile adipokines.[100,101]

In hypertension there is a loss of the protective anticontractile effect of PVAT contributing to vascular dysfunction.[102,103] The PVAT anticontractile effect seems to be proportional to its mass, as reduction of adipose tissue in lipoatrophic mice increases vasoconstriction and blood pressure.[104] Additionally, experimental models of hypertension such as Ang II–induced, SHR, and DOCA-salt hypertensive rats demonstrated a reduction in PVAT mass and adipocyte size.[105–107] However, the cross-sectional area of PVAT from thoracic aorta was similar in normotensive WKY and SHR rats and transfer of bath solution incubated with PVAT-intact vessel promoted less relaxation in SHR[108] suggesting that alterations in PVAT in hypertension are more complex and may involve impairment in function rather than just alterations in fat mass.

In contrast, increased PVAT content as observed in obesity is also accompanied by loss of the protective anticontractile effect and PVAT inflammation affecting vascular homeostasis and blood pressure. Obesity is strongly associated with hypertension and obese patients are more likely to develop high blood pressure. In obesity-related hypertension, there is a decrease in levels of vasodilator molecules, an increase in the release of procontractile substances and development of low-grade inflammation in PVAT.[92,109,110] These phenomena likely relate to changes in the adipokine secretome.

Vascular aging in hypertension

Age-related decline in function is a physiological process occurring in all systems, including the vascular system. Many of the vascular changes that occur during physiological aging are also observed in hypertension at

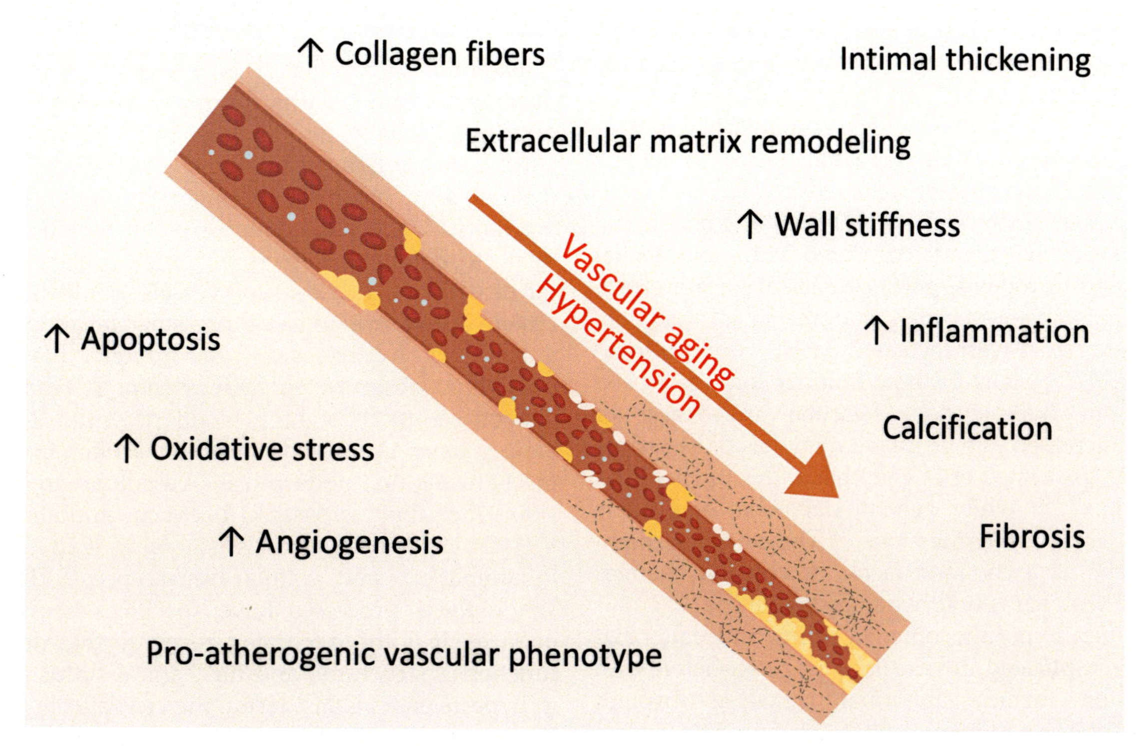

FIGURE 27.5 **Schematic demonstrating the vascular phenotype of aging and hypertension.** During biological aging and development of hypertension, arteries undergo functional, structural, and mechanical changes. Hypertension amplifies aging-associated vascular dysfunction and remodeling.

relatively younger ages (Fig. 27.5). Accordingly hypertension has been considered a condition of "premature vascular aging."[111] On the other hand, aging itself is an important risk factor for hypertension and associated vascular changes.[111–113]

Extracellular matrix components and vascular structural changes with aging

Extracellular matrix remodeling, fibrosis, calcification, and inflammation play a major role in the development of vascular stiffness associated with aging and hypertension.[111–115] This is usually associated with reduced elastin content and decreased elasticity.[113] The process of ECM mineralization and calcification are hallmarks of vascular aging and are amplified by hypertension.[116] Calcification involves dedifferentiation of VSMCs to an osteogenic phenotype and is induced by prohypertensive stimuli.[117] We demonstrated that these processes are ameliorated by Mg^{2+}, which opposes Ca^{2+}-induced actions.[118] Levels of calpain-1, a ubiquitous, cytosolic Ca^{2+}-activated neutral protease, are significantly increased in aging, which leads to MMP2 activation, increased collagen I and III production, and vascular calcification, processes that are also stimulated by oxidative stress.[119]

Vascular functional changes with aging and hypertension

In addition to structural changes, aging promotes altered endothelium-dependent relaxation and hyperreactivity, phenomena also observed in hypertension. Mechanisms underlying aging-associated endothelial dysfunction include decreased bioavailability of NO, oxidative stress, and mitochondrial dysfunction.[120] NF-κB also seems to be important, since NF-κB inhibition improves endothelial-dependent dilation in aged humans to that observed in young healthy adults.[121] Gene targets of NF-κB that regulate endothelial function in aging include IL-6, RAGE, TNF-α, and MCP-1.[122] In vivo, vessels from the senescence-accelerated mouse (SAM-P8) demonstrate hypercontractility in response to phenylephrine compared to controls, and in vitro, as with aging and hypertension, SAM-P8 VSMCs undergo phenotypic changes to become stiff and promigratory.[123]

p66Shc, vascular aging, and blood pressure

Of the many signaling molecules implicated in vascular aging is p66Shc. Genetic ablation of p66Shc in mice prolongs life span by 30% and reduces blood pressure in hypertensive mice with associated reduction in vascular production of ROS.[124] Oxidative stress-induced phosphorylation of p66Shc leads to accumulation of p66Shc in mitochondria, followed by increased production of mitochondrial H_2O_2, which causes cellular swelling and apoptosis.[125] In endothelial cells, p66Shc facilitates the transcription of a key regulator of endothelial function repression Kruppel-like factor 2 (KLF2), promotes eNOS uncoupling, and reduces NO availability.[126] In humans, increased levels of p66Shc in circulating monocytes are negatively correlated with endothelial function, and in rodents, p66Shc-deficient mice are protected against age-associated endothelial dysfunction, effected associated with increased Ser1177 phosphorylation of eNOS.[127] Many of these findings have also been observed in hypertension. Vascular expression of p66Shc is increased in experimental models of hypertension and is associated with endothelial dysfunction and remodeling,[128,129] while genetic depletion of p66Shc results in decreased vascular tone.[130] These findings highlight p66shc as a common factor in vascular changes associated with aging and hypertension.

In addition to p66Shc, other signaling molecules that have been implicated in vascular aging associated with hypertension include abnormal signaling through mTOR, adenosine monophosphate protein kinase, and sirtuins.[131] Moreover, with aging, endothelial cells and VSMCs undergo cellular senescence contributing to impaired regenerative and angiogenic capacity further contributing to the vasculome in hypertension.[132,133]

Conclusions

Essential hypertension is a chronic disorder that develops over time. It is common, affecting up to 35% of adults globally, and is a major cause of morbidity (stroke, heart failure, ischemic heart disease, kidney dysfunction, vascular dementia) and premature mortality.[3,4] The consequences of high blood pressure relate to damaging effects on target organs including the heart, brain, kidneys, and vessels. While the vascular system contributes to blood pressure elevation, it is also a target of hypertension. Accordingly, the vascular phenotype observed in hypertension is both a cause and consequence of high blood pressure. Recent advances have unraveled novel molecular and cellular mechanisms that underpin the vasculome in hypertension, including crosstalk between endothelial and VSMCs, Ca^{2+} and Rho kinase signaling, oxidative stress, inflammation, and cellular senescence[134] (Fig. 27.6). While these processes have specific vascular consequences, it is the integrated response that defines the functional, structural, and mechanical status of vessels in hypertension. Considering the central role of vessels in the pathophysiology and complications of hypertension, it is not surprising that they constitute important disease- and therapeutic-specific targets. Current clinical studies are focusing on (1) identifying

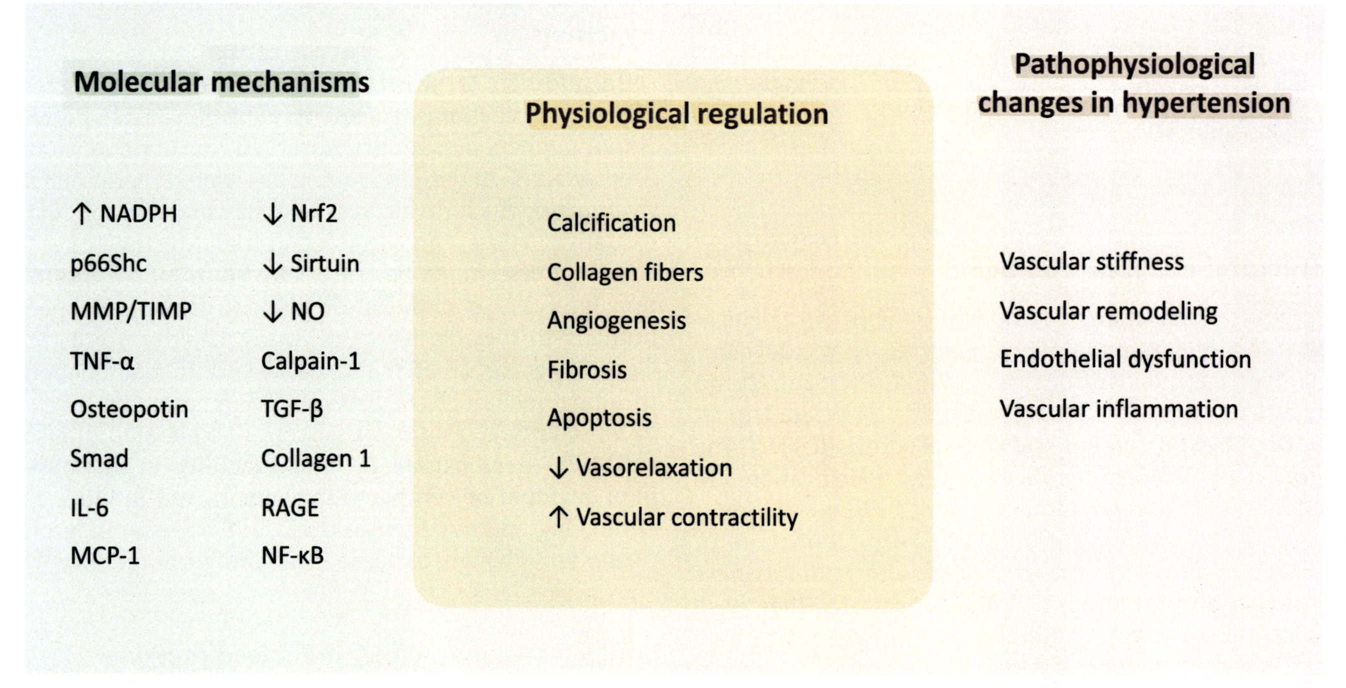

FIGURE 27.6 **Molecular mechanisms, physiological regulation, and pathophysiological changes associated with vascular changes in aging and hypertension.** Aging-associated activation of proinflammatory, redox-sensitive, and growth/apoptotic signaling pathways lead to physiological regulation in the vascular resulting in vascular stiffness, remodeling, inflammation, and endothelial dysfunction. *IL-6*, interleukin 6; *MCP-1*, monocyte chemoattractant protein 1; *MMP*, matrix metalloproteinases; *NADPH*, nicotinamide adenine dinucleotide phosphate; *NF-κB*, nuclear factor kappa beta; *NO*, nitric oxide; *Nrf2*, nuclear factor erythroid 2–related factor 2; *RAGE*, receptor for advanced glycation end products; *TGF-β*, transforming growth factor beta; *TIMPs*, tissue inhibitors of metalloproteinases; *TNF-α*, tumor necrosis factor alpha.

biomarkers of the vascular phenotype before and during development of hypertension, (2) establishment of tools and methodologies to predict effects of vascular dysfunction (impaired endothelial function, vascular hypercontractility) remodeling (arterial stiffness), and (3) assessment of vasoprotective agents with potential blood pressure–lowering effects. Ultimately, targeting strategies to promote and maintain vascular health and to ameliorate the vascular phenotype in hypertension will ensure better blood pressure control, prevention of target organ damage, and a healthier population at large.

Acknowledgments

The authors are funded by grants from the British Heart Foundation (BHF) (RE/13/5/30177; 18/6/34217). RMT is supported through a BHF Chair award (CH/12/29762).

References

1. Joyner MJ, Limberg JK. Blood pressure regulation: every adaptation is an integration? *J Appl Physiol*. 2014;114(3):445–450.
2. Smith BE, Madigan VM. Understanding the haemodynamics of hypertension. *Curr Hypertens Rep*. 2018;20(4):29.
3. Mills KT, Stefanescu A, He J. The global epidemiology of hypertension. *Nat Rev Nephrol*. April 2020;16(4):223–237.
4. Schiffrin EL. How structure, mechanics, and function of the vasculature contribute to blood pressure elevation in hypertension. *Can J Cardiol*. 2020;36(5):648–658.
5. Saxena T, Ali AO, Saxena M. Pathophysiology of essential hypertension: an update. *Expert Rev Cardiovasc Ther*. December 2018; 16(12):879–887.
6. Zanoli L, Briet M, Empana JP, et al. Association for research into arterial structure, physiology (ARTERY) society, the European Society of Hypertension (ESH) working group on vascular structure and function, and the European Network for Noninvasive Investigation of large arteries. Vascular consequences of inflammation: a position statement from the ESH working group on vascular structure and function and the ARTERY society. *J Hypertens*. September 2020;38(9):1682–1698.
7. Dumor K, Shoemaker-Moyle M, Nistala R, Whaley-Connell A. Arterial stiffness in hypertension: an update. *Curr Hypertens Rep*. July 4, 2018;20(8):72.
8. Pi X, Xie L, Patterson C. Emerging roles of vascular endothelium in metabolic homeostasis. *Circ Res*. 2018;123(4):477–494.
9. Bohr DF, Webb RC. Vascular smooth muscle function and its changes in hypertension. *Am J Med*. October 5, 1984;77(4A):3–16.
10. Goulopoulou S, Webb RC. Symphony of vascular contraction: how smooth muscle cells lose harmony to signal increased vascular resistance in hypertension. *Hypertension*. 2014;63:e33–39.
11. Touyz RM, Alves-Lopes R, Rios FJ, et al. Vascular smooth muscle contraction in hypertension. *Cardiovasc Res*. 2018;114:529–539.
12. Fisher SA. Vascular smooth muscle phenotypic diversity and function. *Physiol Genom*. 2010;42A(3):169–187.
13. Allen BG, Walsh MP. The biochemical basis of the regulation of smooth-muscle contraction. *Trends Biochem Sci*. 1994;19:362–368.
14. Walsh MP. Vascular smooth muscle myosin light chain diphosphorylation: mechanism, function, and pathological implications. *IUBMB Life*. 2011;63:987–1000.
15. Brozovich FV, Nicholson CJ, Degen CV, Gao YZ, Aggarwal M, Morgan KG. Mechanisms of vascular smooth muscle contraction and the basis for pharmacologic treatment of smooth muscle disorders. *Pharmacol Rev*. 2016;68:476–532.
16. Lin Q, Zhao G, Fang X, et al. P(3) receptors regulate vascular smooth muscle contractility and hypertension. *JCI Insight*. 2016; 1(17):e89402.
17. Nieves-Cintron M, Amberg GC, Nichols CB, Molkentin JD, Santana LF. Activation of nfatc3 down-regulates the beta1 subunit of large conductance, calcium-activated k^+ channels in arterial smooth muscle and contributes to hypertension. *J Biol Chem*. 2007;282:3231–3240.
18. Touyz RM, Schiffrin EL. Angiotensin II regulates vascular smooth muscle cell pH, contraction, and growth via tyrosine kinase-dependent signaling pathways. *Hypertension*. 1997;30(2 Pt 1): 222–229.
19. Hill MA, Meininger GA. Small artery mechanobiology: roles of cellular and non-cellular elements. *Microcirculation*. 2016;23: 611–613.
20. Feletou M, Verbeuren TJ, Vanhoutte PM. Endothelium-dependent contractions in shr: a tale of prostanoid tp and ip receptors. *Br J Pharmacol*. 2009;156:563–574.
21. Loirand G, Pacaud P. The role of rho protein signaling in hypertension. *Nat Rev Cardiol*. 2010;7:637–647.
22. Feng HZ, Wang H, Takahashi K, Jin JP. Double deletion of calponin 1 and calponin 2 in mice decreases systemic blood pressure with blunted length-tension response of aortic smooth muscle. *J Mol Cell Cardiol*. 2019;129:49–57.
23. Ringvold HC, Khalil RA. Protein kinase c as regulator of vascular smooth muscle function and potential target in vascular disorders. *Adv Pharmacol*. 2017;78:203–301.
24. Behuliak M, Bencze M, Vaneckova I, Kunes J, Zicha J. Basal and activated calcium sensitization mediated by rhoa/rho kinase pathway in rats with genetic and salt hypertension. *Biomed Res Int*. 2017;2017:8029728.
25. Webb RC. Smooth muscle contraction and relaxation. *Adv Physiol Educ*. 2003;27:201–206.
26. Uehata M, Ishizaki T, Satoh H, et al. Calcium sensitization of smooth muscle mediated by a rho-associated protein kinase in hypertension. *Nature*. 1997;389:990–994.
27. Vanhoutte PM, Shimokawa H, Feletou M, Tang EH. Endothelial dysfunction and vascular disease - a 30th anniversary update. *Acta Physiol*. 2017;219:22–96.
28. Martinez-Lemus LA, Hill MA, Meininger GA. The plastic nature of the vascular wall: a continuum of remodeling events contributing to control of arteriolar diameter and structure. *Physiology*. 2009;24:45–57.
29. Camargo LL, Harvey AP, Rios FJ, et al. Vascular Nox (NADPH oxidase) compartmentalization, protein hyperoxidation, and endoplasmic reticulum stress response in hypertension. *Hypertension*. 2018l;72(1):235–246.
30. Montezano AC, Touyz RM. Reactive oxygen species, vascular noxs, and hypertension: focus on translational and clinical research. *Antioxidants Redox Signal*. 2014;20:164–182.
31. Montezano AC, De Lucca Camargo L, Persson P, et al. NADPH oxidase 5 is a pro-contractile Nox isoform and a point of cross-talk for calcium and redox signaling-implications in vascular function. *J Am Heart Assoc*. 2018;7(12):e009388.
32. Griendling KK, Camargo LL, Rios FJ, Alves-Lopes R, Montezano AC, Touyz RM. Oxidative stress and hypertension. *Circ Res*. 2021;128(7):993–1020.
33. Xu Q, Huff LP, Fujii M, Griendling KK. Redox regulation of the actin cytoskeleton and its role in the vascular system. *Free Radic Biol Med*. 2017;109:84–107.

34. Safar ME. Arterial stiffness as a risk factor for clinical hypertension. *Nat Rev Cardiol*. 2018;15(2):97–105.
35. Carbone ML, Bregeon J, Devos N, et al. Angiotensin ii activates the rhoa exchange factor arhgef1 in humans. *Hypertension*. 2015; 65:1273–1278.
36. Tsounapi P, Saito M, Kitatani K, et al. Fasudil improves the endothelial dysfunction in the aorta of spontaneously hypertensive rats. *Eur J Pharmacol*. 2012;691:182–189.
37. Takeya K, Wang X, Sutherland C, et al. Involvement of myosin regulatory light chain diphosphorylation in sustained vasoconstriction under pathophysiological conditions. *J Smooth Muscle Res*. 2014;50:18–28.
38. Martín-Bórnez M, Galeano-Otero I, Del Toro R, Smani T. TRPC and TRPV channels' role in vascular remodeling and disease. *Int J Mol Sci*. 2020;21(17):6125.
39. Alves-Lopes R, Neves KB, Anagnostopoulou A, et al. Crosstalk between vascular redox and calcium signaling in hypertension involves trpm2 (transient receptor potential melastatin 2) cation channel. *Hypertension*. 2020;75:139–149.
40. Wu J, Saleh MA, Kirabo A, et al. Immune activation caused by vascular oxidation promotes fibrosis and hypertension. *J Clin Invest*. 2016;126(1):50–67.
41. Badran A, Nasser SA, Mesmar J, et al. Reactive oxygen species: modulators of phenotypic switch of vascular smooth muscle cells. *Int J Mol Sci*. 2020 20;21(22):8764.
42. Gibbons GH, Dzau VJ. The emerging concept of vascular remodeling. *N Engl J Med*. May 19, 1994;330(20):1431–1438.
43. Feihl F, Liaudet L, Waeber B, Levy BI. Hypertension: a disease of the microcirculation? *Hypertension*. December 2006;48(6): 1012–1017.
44. Schiffrin EL. Remodeling of resistance arteries in essential hypertension and effects of antihypertensive treatment. *Am J Hypertens*. 2004,17(12 Pt 1).1192–1200.
45. Mulvany MJ, Hansen OK, Aalkjaer C. Direct evidence that the greater contractility of resistance vessels in spontaneously hypertensive rats is associated with a narrowed lumen, a thickened media, and an increased number of smooth muscle cell layers. *Circ Res*. December 1978;43(6):854–864.
46. Renna NF, de Las Heras N, Miatello RM. Pathophysiology of vascular remodeling in hypertension. *Int J Hypertens*. 2013;2013: 808353.
47. Mulvany MJ, Baumbach GL, Aalkjaer C, et al. Vascular remodeling. *Hypertension*. September 1996;28(3):505–506.
48. Intengan HD, Schiffrin EL. Vascular remodeling in hypertension: roles of apoptosis, inflammation, and fibrosis. *Hypertension*. September 2001;38(3 Pt 2):581–587.
49. Deng LY, Schiffrin EL. Effects of endothelin-1 and vasopressin on resistance arteries of spontaneously hypertensive rats. *Am J Hypertens*. 1992;5:817–822.
50. Li JS, Knafo L, Turgeon A, Garcia R, Schiffrin EL. Effect of endothelin antagonism on blood pressure and vascular structure in renovascular hypertensive rats. *Am J Physiol*. 1996;40:H88–H93.
51. Korsgaard N, Aalkjær C, Heagerty AM, Izzard AS, Mulvany MJ. Histology of subcutaneous small arteries from patients with essential hypertension. *Hypertension*. 1993;22:523–526.
52. Schiffrin EL, Deng LY, Larochelle P. Morphology of resistance arteries and comparison of effects of vasoconstrictors in mild essential hypertensive patients. *Clin Invest Med*. 1993;16:177–186.
53. Rizzoni D, Porteri E, Castellano M, et al. Vascular hypertrophy and remodeling in secondary hypertension. *Hypertension*. 1996; 28:785–790.
54. Deng LY, Schiffrin EL. Effects of endothelin on resistance arteries of DOCA-salt hypertensive rats. *Am J Physiol*. 1992;262: H1782–H1787.
55. D'Uscio LV, Barton M, Shaw S, Moreau P, Luscher TF. Structure and function of small arteries in salt-induced hypertension: effects of chronic endothelin-subtype-A-receptor blockade. *Hypertension*. 1997;30:905–911.
56. van Varik BJ, Rennenberg RJMW, Reutelingsperger CP, Kroon AA, de Leeuw PW, Schurgers LJ. Mechanisms of arterial remodeling: lessons from genetic diseases. *Front Genet*. December 13, 2012;3:290.
57. Savoia C, Burger D, Nishigaki N, Montezano A, Touyz RM. Angiotensin II and the vascular phenotype in hypertension. *Expet Rev Mol Med*. March 30, 2011;13:e11.
58. Montezano AC, Nguyen Dinh Cat A, Rios FJ, Touyz RM. Angiotensin II and vascular injury. *Curr Hypertens Rep*. June 2014; 16(6):431.
59. Touyz RM, Cruzado M, Tabet F, Yao G, Salomon S, Schiffrin EL. Redox-dependent MAP kinase signaling by Ang II in vascular smooth muscle cells: role of receptor tyrosine kinase transactivation. *Can J Physiol Pharmacol*. 2003;81(2):159–167.
60. Montezano AC, Tsiropoulou S, Dulak-Lis M, Harvey A, Camargo Lde L, Touyz RM. Redox signaling, Nox5 and vascular remodeling in hypertension. *Curr Opin Nephrol Hypertens*. September 2015;24(5):425–433.
61. Diep QN, Li JS, Schiffrin EL. In vivo study of AT1 and AT2 angiotensin receptors in apoptosis of rat blood vessels. *Hypertension*. 1999;34:617–624.
62. Dickhout JG, Lee RMKW. Apoptosis in the muscular arteries from young spontaneously hypertensive rats. *J Hypertens*. 1999;17: 1413–1419.
63. Mian MO, Paradis P, Schiffrin EL. Innate immunity in hypertension. *Curr Hypertens Rep*. 2014;16(2):413.
64. Madhur MS, Elijovich F, Alexander MR, et al. Hypertension: do inflammation and immunity hold the key to solving this epidemic? *Circ Res*. 2021;128(7):908–933.
65. Guzik TJ, Hoch NE, Brown KA, et al. Role of the T cell in the genesis of angiotensin II induced hypertension and vascular dysfunction. *J Exp Med*. 2007;204(10):2449–2460.
66. Wenzel P, Knorr M, Kossmann S, et al. Lysozyme M-positive monocytes mediate angiotensin II-induced arterial hypertension and vascular dysfunction. *Circulation*. 2011;124(12):1370–1381.
67. Wynn TA, Vannella KM. Macrophages in tissue repair, regeneration, and fibrosis. *Immunity*. 2016;44(3):450–462.
68. Dikalova AE, Pandey A, Xiao L, et al. Mitochondrial deacetylase Sirt3 reduces vascular dysfunction and hypertension while Sirt3 depletion in essential hypertension is linked to vascular inflammation and oxidative stress. *Circ Res*. 2020;126(4):439–452.
69. Muller DN, Mervaala EM, Schmidt F, et al. Effect of bosentan on NF-kappaB, inflammation, and tissue factor in angiotensin II-induced end-organ damage. *Hypertension*. 2000;36(2):282–290.
70. Martynowicz H, Janus A, Nowacki D, Mazur G. The role of chemokines in hypertension. *Adv Clin Exp Med*. 2014;23(3):319–325.
71. Lin QY, Lang PP, Zhang YL, et al. Pharmacological blockage of ICAM-1 improves angiotensin II-induced cardiac remodeling by inhibiting adhesion of LFA-1(+) monocytes. *Am J Physiol Heart Circ Physiol*. 2019;317(6):H1301–H1311.
72. Wang Q, Wang H, Wang J, et al. Angiotensin II-induced hypertension is reduced by deficiency of P-selectin glycoprotein ligand-1. *Sci Rep*. 2018;8(1):3223.
73. Tchalla AE, Wellenius GA, Travison TG, et al. Circulating vascular cell adhesion molecule-1 is associated with cerebral blood flow dysregulation, mobility impairment, and falls in older adults. *Hypertension*. 2015;66(2):340–346.
74. Sanada H, Midorikawa S, Yatabe J, et al. Elevation of serum soluble E- and P-selectin in patients with hypertension is reversed by benidipine, a long-acting calcium channel blocker. *Hypertens Res*. 2005;28(11):871–878.
75. Mahmud A, Feely J. Arterial stiffness is related to systemic inflammation in essential hypertension. *Hypertension*. 2005;46(5): 1118–1122.

76. Cai R, Hao Y, Liu YY, Huang L, Yao Y, Zhou MS. Tumor necrosis factor alpha deficiency improves endothelial function and cardiovascular injury in deoxycorticosterone acetate/salt-hypertensive mice. *Biomed Res Int*. 2020;2020:3921074.
77. Sriramula S, Francis J. Tumor Necrosis factor - alpha is essential for angiotensin II-induced ventricular remodeling: role for oxidative stress. *PLoS One*. 2015;10(9):e0138372.
78. Kim J, Lee KS, Kim JH, et al. Aspirin prevents TNF-alpha-induced endothelial cell dysfunction by regulating the NF-kappaB-dependent miR-155/eNOS pathway: role of a miR-155/eNOS axis in preeclampsia. *Free Radic Biol Med*. 2017;104:185–198.
79. Chou CH, Hung CS, Liao CW, et al. IL-6 trans-signalling contributes to aldosterone-induced cardiac fibrosis. *Cardiovasc Res*. 2018; 114(5):690–702.
80. Johnson AW, Kinzenbaw DA, Modrick ML, Faraci FM. Small-molecule inhibitors of signal transducer and activator of transcription 3 protect against angiotensin II-induced vascular dysfunction and hypertension. *Hypertension*. 2013;61(2):437–442.
81. Choi H, Ettinger N, Rohrbough J, Dikalova A, Nguyen HN, Lamb FS. LRRC8A channels support TNFalpha-induced superoxide production by Nox1 which is required for receptor endocytosis. *Free Radic Biol Med*. 2016;101:413–423.
82. Choi S, Park M, Kim J, et al. TNF-alpha elicits phenotypic and functional alterations of vascular smooth muscle cells by miR-155-5p-dependent down-regulation of cGMP-dependent kinase 1. *J Biol Chem*. 2018;293(38):14812–14822.
83. Martinez-Revelles S, Avendano MS, Garcia-Redondo AB, et al. Reciprocal relationship between reactive oxygen species and cyclooxygenase-2 and vascular dysfunction in hypertension. *Antioxidants Redox Signal*. 2013;18(1):51–65.
84. Palmu J, Watrous JD, Mercader K, et al. Eicosanoid inflammatory mediators are robustly associated with blood pressure in the general population. *J Am Heart Assoc*. 2020;9(19):e017598.
85. Gong T, Liu L, Jiang W, Zhou R. DAMP-sensing receptors in sterile inflammation and inflammatory diseases. *Nat Rev Immunol*. 2020;20(2):95–112.
86. Sturtzel C. Endothelial cells. *Adv Exp Med Biol*. 2017;1003:71–91.
87. Rothman AM, MacFadyen J, Thuren T, et al. Effects of interleukin-1beta inhibition on blood pressure, incident hypertension, and residual inflammatory risk: a secondary analysis of CANTOS. *Hypertension*. 2020;75(2):477–482.
88. Caillon A, Paradis P, Schiffrin EL. Role of immune cells in hypertension. *Br J Pharmacol*. 2019;176(12):1818–1828.
89. Barhoumi T, Kasal DA, Li MW, et al. T regulatory lymphocytes prevent angiotensin II-induced hypertension and vascular injury. *Hypertension*. 2011;57(3):469–476.
90. Kasal DA, Barhoumi T, Li MW, et al. T regulatory lymphocytes prevent aldosterone-induced vascular injury. *Hypertension*. 2012; 59(2):324–330.
91. Youn JC, Yu HT, Lim BJ, et al. Immunosenescent $CD8^+$ T cells and C-X-C chemokine receptor type 3 chemokines are increased in human hypertension. *Hypertension*. 2013;62(1):126–133.
92. Saxton SN, Clark BJ, Withers SB, Eringa EC, Heagerty AM. Mechanistic links between obesity, diabetes, and blood pressure: role of perivascular adipose tissue. *Physiol Rev*. 2019;99:1701–1763.
93. Bielecka-Dabrowa A, Bartlomiejczyk MA, Sakowicz A, Maciejewski M, Banach M. The role of adipokines in the development of arterial stiffness and hypertension. *Angiology*. 2020;71(8): 754–761.
94. Giordano A, Smorlesi A, Frontini A, Barbatelli G, Cinti S. White, brown and pink adipocytes: the extraordinary plasticity of the adipose organ. *Eur J Endocrinol*. 2014;170:R159–R171.
95. Heinonen S, Jokinen R, Rissanen A, Pietiläinen KH. White adipose tissue mitochondrial metabolism in health and in obesity. *Obes Rev*. 2020;21(2):e12958.
96. Virtanen KA, Lidell ME, Orava J, et al. Functional brown adipose tissue in healthy adults. *N Engl J Med*. 2009;360:1518–1525.
97. Wu J, Bostrom P, Sparks LM, et al. Beige adipocytes are a distinct type of thermogenic fat cell in mouse and human. *Cell*. 2012;150: 366–376.
98. Becher T, Palanisamy S, Kramer DJ, et al. Brown adipose tissue is associated with cardiometabolic health. *Nat Med*. 2021;27: 58–65.
99. Friederich-Persson M, Nguyen Dinh Cat A, Persson P, Montezano AC, Touyz RM. Brown adipose tissue regulates small artery function through NADPH oxidase 4-derived hydrogen peroxide and redox-sensitive protein kinase G-1alpha. *Arterioscler Thromb Vasc Biol*. 2017;37:455–465.
100. Nosalski R, Guzik TJ. Perivascular adipose tissue inflammation in vascular disease. *Br J Pharmacol*. 2017;174:3496–3513.
101. Neves KB, Nguyen Dinh Cat A, Lopes RA, et al. Chemerin regulates crosstalk between adipocytes and vascular cells through Nox. *Hypertension*. September 2015;66(3):657–666.
102. Hu H, Garcia-Barrio M, Jiang ZS, Chen YE, Chang L. Roles of perivascular adipose tissue in hypertension and atherosclerosis. *Antioxidants Redox Signal*. 2021;34:736–749.
103. Lee YC, Chang HH, Chiang CL, et al. Role of perivascular adipose tissue-derived methyl palmitate in vascular tone regulation and pathogenesis of hypertension. *Circulation*. 2011;124:1160–1171.
104. Takemori K, Gao YJ, Ding L, et al. Elevated blood pressure in transgenic lipoatrophic mice and altered vascular function. *Hypertension*. 2007;49:365–372.
105. Lee RM, Ding L, Lu C, Su LY, Gao YJ. Alteration of perivascular adipose tissue function in angiotensin II-induced hypertension. *Can J Physiol Pharmacol*. 2009;87:944–953.
106. Galvez B, de Castro J, Herold D, et al. Perivascular adipose tissue and mesenteric vascular function in spontaneously hypertensive rats. *Arterioscler Thromb Vasc Biol*. 2006;26:1297–1302.
107. Ruan CC, Zhu DL, Chen QZ, et al. Perivascular adipose tissue-derived complement 3 is required for adventitial fibroblast functions and adventitial remodeling in deoxycorticosterone acetate-salt hypertensive rats. *Arterioscler Thromb Vasc Biol*. 2010; 30:2568–2574.
108. Lu C, Su LY, Lee RM, Gao YJ. Alterations in perivascular adipose tissue structure and function in hypertension. *Eur J Pharmacol*. 2011;656:68–73.
109. Guzik TJ, Skiba DS, Touyz RM, Harrison DG. The role of infiltrating immune cells in dysfunctional adipose tissue. *Cardiovasc Res*. 2017;113:1009–1023.
110. Mikolajczyk TP, Nosalski R, Szczepaniak P, et al. Role of chemokine RANTES in the regulation of perivascular inflammation, T-cell accumulation, and vascular dysfunction in hypertension. *FASEB J*. 2016;30:1987–1999.
111. Harvey A, Montezano AC, Touyz RM. Vascular biology of ageing-implications in hypertension. *J Mol Cell Cardiol*. 2015;83:112–121.
112. Ungvari Z, Tarantini S, Donato AJ, Galvan V, Csiszar A. Mechanisms of vascular aging. *Circ Res*. 2018;123:849–867.
113. Lakatta EG. The reality of aging viewed from the arterial wall. *Artery Res*. 2013;7:73–80.
114. Jacob MP. Extracellular matrix remodeling and matrix metalloproteinases in the vascular wall during aging and in pathological conditions. *Biomed Pharmacother*. 2003;57:195–202.
115. Nosalski R, Mikolajczyk T, Siedlinski M, et al. Nox1/4 inhibition exacerbates age dependent perivascular inflammation and fibrosis in a model of spontaneous hypertension. *Pharmacol Res*. 2020;161:105235.
116. Durham AL, Speer MY, Scatena M, Giachelli CM, Shanahan CM. Role of smooth muscle cells in vascular calcification: implications in atherosclerosis and arterial stiffness. *Cardiovasc Res*. 2018;114(4): 590–600.

117. Tyson J, Bundy K, Roach C, et al. Mechanisms of the osteogenic switch of smooth muscle cells in Vascular calcification: WNT signaling, BMPs, mechanotransduction, and EndMT. *Bioengineering (Basel)*. 2020;7(3):88.
118. Montezano AC, Zimmerman D, Yusuf H, et al. Vascular smooth muscle cell differentiation to an osteogenic phenotype involves TRPM7 modulation by magnesium. *Hypertension*. 2010;56(3): 453–462.
119. Jiang L, Zhang J, Monticone RE, et al. Calpain-1 regulation of matrix metalloproteinase 2 activity in vascular smooth muscle cells facilitates age-associated aortic wall calcification and fibrosis. *Hypertension*. 2012;60:1192–1199.
120. Ungvari Z, Sonntag WE, Csiszar A. Mitochondria and aging in the vascular system. *J Mol Med (Berl)*. 2010;88:1021–1027.
121. Lopes RA, Neves KB, Tostes RC, Montezano AC, Touyz RM. Downregulation of nuclear factor erythroid 2-related factor and associated antioxidant genes contributes to redox-sensitive vascular dysfunction in hypertension. *Hypertension*. 2015;66: 1240–1250.
122. Ungvari Z, Bailey-Downs L, Gautam T, et al. Age-associated vascular oxidative stress, nrf2 dysfunction, and nf-{kappa}b activation in the nonhuman primate macaca mulatta. *J Gerontol A Biol Sci Med Sci*. 2011;66:866–875.
123. Llorens S, de Mera RM, Pascual A, et al. The senescence-accelerated mouse (sam-p8) as a model for the study of vascular functional alterations during aging. *Biogerontology*. 2007;8:663–672.
124. Migliaccio E, Giorgio M, Mele S, et al. The p66shc adaptor protein controls oxidative stress response and life span in mammals. *Nature*. 1999;402:309–313.
125. Yamamori T, White AR, Mattagajasingh I, et al. P66shc regulates endothelial no production and endothelium-dependent vasorelaxation: implications for age-associated vascular dysfunction. *J Mol Cell Cardiol*. 2005;39:992–995.
126. Kumar A, Hoffman TA, Dericco J, Naqvi A, Jain MK, Irani K. Transcriptional repression of kruppel like factor-2 by the adaptor protein p66shc. *FASEB J*. 2009;23:4344–4352.
127. Miao Q, Wang Q, Dong L, Wang Y, Tan Y, Zhang X. The expression of p66shc in peripheral blood monocytes is increased in patients with coronary heart disease and correlated with endothelium-dependent vasodilatation. *Heart Ves*. 2015;30:451–457.
128. Kumar S. P66shc and vascular endothelial function. *Biosci Rep*. 2019;39.
129. Miller B, Palygin O, Rufanova VA, et al. p66Shc regulates renal vascular tone in hypertension-induced nephropathy. *J Clin Invest*. 2016;126(7):2533–2546.
130. Gierhardt M, Pak O, Sydykov A, et al. Genetic deletion of p66shc and/or cyclophilin D results in decreased pulmonary vascular tone. *Cardiovasc Res*. 2020. https://doi.org/10.1093/cvr/cvaa310. cvaa310. [Online ahead of print].
131. Xu S, Yin M, Koroleva M, et al. Sirt6 protects against endothelial dysfunction and atherosclerosis in mice. *Aging (Albany NY)*. 2016;8:1064–1082.
132. Baker DJ, Childs BG, Durik M, et al. Naturally occurring p16(ink4a)-positive cells shorten healthy lifespan. *Nature*. 2016; 530:184–189.
133. Alghatrif M, Lakatta EG. The conundrum of arterial stiffness, elevated blood pressure, and aging. *Curr Hypertens Rep*. 2015;17: 12–18.
134. Touyz RM, Schiffrin EL. A compendium on hypertension: new advances and future impact. *Circ Res*. 2021;128(7):803–807.

CHAPTER

28

Functional roles of lymphatics in health and disease

Xiaolei Liu and Guillermo Oliver

Center for Vascular and Developmental Biology, Feinberg Cardiovascular and Renal Research Institute, Feinberg School of Medicine, Northwestern University, Chicago, IL, United States

Introduction

The lymphatic vasculature is a one-way route facilitating the return of tissue fluid and macromolecules to the blood circulation. The main functional roles of the lymphatic system are to maintain fluid homeostasis, drain interstitial fluid, reabsorb fat and small macromolecules from the tissue extracellular spaces, and return them to the venous circulation.[1,2] Malfunction of the lymphatic vasculature is primarily associated to primary or secondary lymphedema, a disfiguring and disabling clinical condition characterized by localized interstitial fluid accumulation, tissue fibrosis, subcutaneous fat accumulation, and swelling of the extremities. While primary lymphedema is often a consequence of genetic defects that affect the development and normal function of the lymphatic vasculature, secondary lymphedema is the more common medical presentation and is caused by lymphatic damage during surgery, infection, or trauma.[3,4] Lymphatic vessels have also been recognized as immune-friendly paths for tumor metastasis and immune trafficking, therefore lymphatics have been involved in cancer progression and immune response.[1,2] Recent studies have broadened these traditional views about the functional role of lymphatics in health and disease, and a number of newly identified physiological and pathological activities are reviewed below and are summarized in Fig. 28.1.

Lymphatics in adipose metabolism and obesity

Obesity is currently a major global health problem and is a key risk factor for a number of metabolic and cardiovascular diseases.[5] Diseases associated with obesity are the second-leading cause of death in the United States.[6] The main leading cause of obesity is excessive dietary intake. Dietary fats are absorbed in the intestinal villi, the enterocytes-lined extensions of the gut wall. Gut villi contain one or two central lymphatic vessels named lacteals. In mammals, dietary lipids are packaged in enterocytes into triglyceride-rich particles named chylomicrons. Lacteals take up chylomicrons and other interstitial fluid components from the villi that line the small intestine and transport them to the submucosal and mesenteric collecting lymphatics. Intestinal lymph is transported via the mesenteric lymph nodes and thoracic duct to the blood circulation at the level of the left subclavian vein.[7–9]

The lymphatic vasculature in the gut plays unique functions in fat absorption and lipid homeostasis, and defects in lipid absorption by the lacteals[10–12] or in lipid transport by the lymphatic vasculature[13,14] have been linked to metabolic disease and obesity. In humans, impairment of lymphatic function leads to enhanced adipose tissue accumulation in patients with congenital and secondary lymphedema.[15,16] The *Chy* mouse is considered a model of lymphedema. The phenotype is consequence of

The Vasculome
https://doi.org/10.1016/B978-0-12-822546-2.24001-X

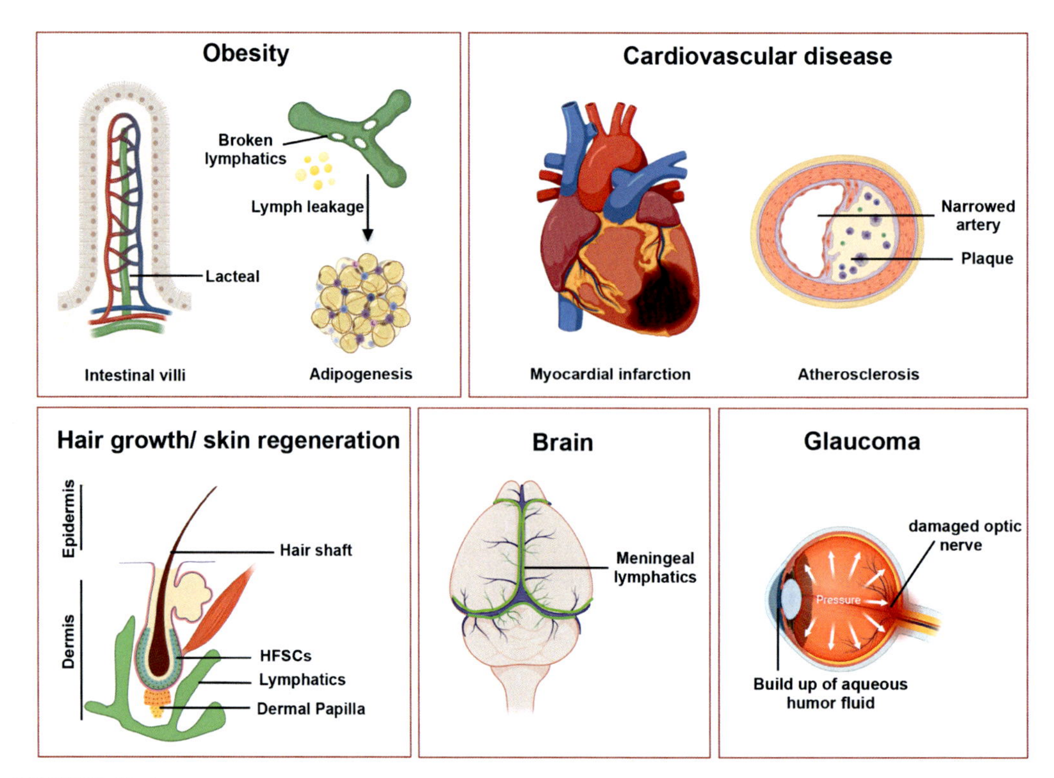

FIGURE 28.1 Major novel functions of lymphatics in health and disease. *Figure is created with BioRender.com.*

heterozygous inactivating mutations in vascular endothelial growth factor receptor (VEGFR)3. The *Chy* mice show excessive subcutaneous adipose tissue accumulation adjacent to dysfunctional hypoplastic lymphatic vessels.[17,18] In addition, a connection between lymphatic defects and obesity was reported for *Prox1* heterozygous mice. Prox1 is a master regulator of lymphatic development and lymphatic endothelial cell (LEC) fate identity,[19] and adult Prox1+/− mice develop excessive fat accumulation and obesity consequence of defective and leaky lymphatic vessels, particularly mesenteric lymphatics.[13] In vitro studies showed that lymph is adipogenic,[13,14] results suggesting that a defective lymphatic vasculature can promote adipose accumulation and obesity.

Recent work in mice showed that different to zipper junctions that tightly seal endothelial cells in collecting lymphatics,[20] entry of chylomicrons into lacteals is facilitated by the presence of specialized discontinuous button-like junctions between lacteal endothelial cells.[10] The authors showed that experimental transformation of button into zipper junctions arrested chylomicron uptake and protected mice from becoming obese when maintained on a high fat diet.[10,21] Mice with deletion of vascular endothelial growth factor receptor 1 and neuropilin 1 in endothelial cells are resistant to high-fat diet−promoted obesity, as a consequence of increased vascular endothelial growth factor (VEGF)A availability and signaling through VEGFR2 that promotes button-to-zipper transformation and reduce chylomicron uptake by lacteals.[10]

Lacteal maintenance requires constant VEGFC-dependent signaling, as genetic deletion of VEGFC in mice leads to lacteal regression, defective lipid absorption, and protection against diet-induced obesity.[12] Similarly, genetic deletion of VEGFR3 or blockage of VEGFR3 signaling in mice also affects lacteal maintenance and fat absorption.[11,22] Those responses are likely mediated by the Notch ligand delta-like protein 4 (Dll4), as lacteal LEC Dll4 expression is dependent on active VEGFR2 or VEGFR3 signaling, and lymphatic-specific deletion of *Dll4* impacts lacteal maintenance causing impaired chylomicron uptake and transport.[11] Adrenomedullin (Adm) is a small peptide that together with its G-protein-coupled receptor calcitonin receptor-like receptor participates in a variety of physiological processes including lymphangiogenesis and lacteal maintenance.[23] Together, all these results provided experimental evidence linking lymphatic malfunction to fat accumulation and obesity, and suggesting that lacteals could be targeted to prevent obesity.[10−12]

Lymphatics in skin regeneration

Dermal lymphatics are essential for skin fluid homeostasis, regeneration, and immune cell trafficking. Dermal lymphatics in the skin are organized in two plexuses, a deep one in the lower dermis made up of larger vessels, and a superficial one of lymphatic capillaries in the upper dermis.[24] It was previously shown that lymphatic capillaries branches extend to the hair follicles (HFs) and drain into the subcutaneous collecting lymphatic vessels, likely through connections of blind capillaries with the dermal papillae.[25]

In the adult skin, HF stimulatory and inhibitory signals modulate the continuous pattern of HF growth and regeneration.[26] The expansion and involution of the HF-associated blood vascular network is involved in this process,[27] as for example, VEGFA-induced angiogenesis is known to promote HF growth.[28,29]

By using a K14-VEGFC transgenic mouse model with skin-specific overexpression of the lymphangiogenic growth factor VEGFC, it was shown that HFs remain longer in the anagen growth phase.[30] Consistently, intradermal injections of recombinant VEGFC protein also promote anagen induction. Conversely, HF entry into the catagen progression was accelerated in K14-sVEGFR3 transgenic mice expressing a soluble form of the extracellular domain of VEGFR3 inhibiting VEGFR3 signaling and lacking cutaneous lymphatic vessels.[30] Interestingly, this study also found that LEC conditioned media promoted cell proliferation of human dermal papilla cells and secreted a number of factors that potentially induce HF stem cell activation.[30]

Recent work extended those results by showing that superficial dermal lymphatics in mouse and human exhibit a nearby and dynamic spatial association with the hair follicle stem cells (HFSCs) niche and more importantly that they play additional functional roles during skin regeneration.[30–32]

Using high resolution imaging, the authors were able to determine that the proximity of the lymphatics to the bulge region of the HF where HFSCs are normally located, changes during cyclical phases of hair growth and quiescence such that lymphatics become closely associated with HFSCs during the resting state, but become dissociated and have reduced drainage capacity during the proliferation and growth phases.[31,32]

They determined that reciprocal signaling between lymphatics and HFSCs is critical for the coordination of HFSCs quiescence, HF cycling, and skin regeneration, as demonstrated by results revealing that disrupting lymphatic vessel integrity in mice promoted hair HF proliferation and regeneration. Furthermore, Gur-Cohen S et al. showed that once HFs regenerate, resting HFSCs express angiopoietin-like protein 7 promoting lymphatic drainage; instead, activated HFSCs switch to Angiopoietin like (Angptl)4, triggering transient lymphatic dissociation and reduced drainage.[32] They concluded that dynamic lymphatic remodeling, driven by HFSCs, regulates the regenerative process such that HFSC-driven signals promote lymphatic fitness that keeps SC quiescence.[32] Instead, activated HFSC-driven signals reshape and disrupt lymphatic capillaries promoting regeneration.[32] During stem cell activation, or upon pharmacological or genetic induction of HF growth, lymphatics transiently expand their caliber, suggesting increased tissue drainage capacity.[32]

These new results determined that dermal lymphatic vessels dynamically reorganize by increasing their caliber upon activation of HFSCs, and demonstrate that lymphatics are important for HF development and skin regeneration. In the future, the analysis of other possible functional interactions between lymphatics and stem cells during repair and regenerative processes could lead to new valuable approaches in regenerative medicine.

Lymphatics in the central nervous system

The human brain is enveloped in a three-layered membranous structure called the meninges. Its outmost layer, the dura mater, lines the skull and contains sinuses—venous structures that drain the blood from the brain before it leaves the cranium. Meningeal lymphatic vessels are positioned along the dural sinuses and drain brain-derived soluble waste to deep cervical lymph nodes.[33–36] The presence of meningeal lymphatics was already reported in the 18th century by Paolo Mascagni[37]; however, their functional importance was not appreciated until recently.[33–36,38] Using advanced imaging tools and molecular markers specific to lymphatic vessels, detailed structural, and functional characterization of meningeal lymphatics has been reported.

Development of meningeal lymphatics starts after birth.[33] Similar to lacteals,[12] meningeal lymphatics also requires constant VEGFC signaling for development and maintenance, and blockage of VEGFC–VEGFR3 activity impairs meningeal lymphatic development and leads to their regression in adult mice.[33] Using in vivo imaging and molecular tracer injections into adult mice, studies have shown that the cerebrospinal fluid (CSF), which is produced by the epithelial cells of the choroid plexus within the brain ventricles, is largely drained by the lymphatic vessels of the meninges and further drained into the deep cervical lymph nodes and later into superficial cervical lymph nodes.[33–35,39] Impaired clearance of CSF waste and disruption of meningeal lymphatics has been associated with several neurological disorders including Alzheimer's disease,[34] Parkinson's disease,[40] aging,[34,39] and multiple sclerosis.[36,41]

Alzheimer's disease

Alzheimer's disease is characterized by the aggregation of β-amyloid plaques resulting from increased production or lack of clearance from brain tissues.[42,43] In transgenic mouse models of Alzheimer's disease, ablations of meningeal lymphatics promote β-amyloid deposition in the meninges, and this phenotype resembles the human pathology, indicating that meningeal lymphatic dysfunction may be an aggravating factor in Alzheimer's disease pathology.[34] In addition, in the brain parenchyma, high-density lipoproteins (HDLs) are secreted by astrocytes, and excess lipoproteins and β-amyloid are transported by the glymphatic system, a pathway by which water and solutes are dynamically exchanged between CSF and interstitial fluid.[44] A recent study has shown a likely connection between the glymphatic system and meningeal lymphatics by revealing glymphatic/meningeal lymphatics function as a new route for removal of lipoproteins and β-amyloid in the brain.[34] In mice, impaired meningeal lymphatic function affects glymphatic macromolecule perfusion, and stimulation of meningeal lymphatic vessel growth by VEGFC treatment improves glymphatic function.[34] Glymphatic/meningeal lymphatic pathway should be further investigated as a whole for its potential use in reversing β-amyloid accumulation and promoting clearance.

Aging

Aging is characterized to be the principal risk factor for a number of neurological disorders, including Alzheimer's disease.[45,46] Studies have also reported that aging is associated with reduced diameter and branching of meningeal lymphatics and impaired drainage.[47,48] It has been reported that in aging mice, a decreased drainage of CSF into deep cervical lymph nodes correlates with a decrease in meningeal lymphatic vessel diameter, integrity, and coverage.[34,39] It was also shown that glymphatic function decreases dramatically during aging.[49] Interestingly, treatment of aging mice with VEGFC increases meningeal lymphatic diameter and leads to a significant increase in CSF drainage.[34] Moreover, VEGFC treatment also improved glymphatic flow and enhanced brain perfusion,[34] and these improvements are abolished by surgical ligation of meningeal lymphatics.[34] These findings highlight the importance of meningeal lymphatic drainage in aging and age-related diseases, and future studies in are needed to determine whether improving their function would have a beneficial role in aging.

Lymphatic function in aqueous drainage and glaucoma

In the eye, aqueous humor (AH) is found in the anterior and posterior chambers of the eye and provides nutrients to these avascular systems, and removes metabolic by-products, significantly contributing to the regulation of ocular tissue homeostasis. AH is constantly produced by the ciliary epithelium in the posterior chamber, flows into the anterior chamber, and exits the eye mainly through the trabecular meshwork into the Schlemm's canal (SC), and is drained into episcleral veins via aqueous veins.[50] The SC is a unique vascular structure that is responsible for controlling the balance between production and drainage of the aqueous back to the systemic circulation.[51] Severe obstruction of the outflow leads to increased intraocular pressure (IOP), the most important risk factor for the onset and progression of glaucoma. Glaucoma is a degenerative, age-related eye disease that is the second leading cause of blindness worldwide. In glaucoma, aqueous drainage into the SC is reduced, such that aqueous outflow resistance rises and IOP increases, resulting in optic neuropathy.[52] The absence or hypoplasia of the SC or alterations to its mechanobiology have been implicated in primary congenital glaucoma.[53–55] Work in the last few years has revealed that the SC is lymphatic-like structure that exhibits mixed blood EC and LEC molecular and morphological characteristics.[56–60]

Lineage tracing studies showed that in mice the SC originates from the choroidal vein and its ECs acquire a lymphatic fate via the upregulation of Prox1.[56,57,60,61] However, the SC can be considered a hybrid vessel sharing lymphatic and blood endothelial molecular and morphological features.[56,60] Similar to the primary lymphatic plexus, the SC forms a blind-ended tube with a continuous, monolayer of ECs, and is surrounded by a discontinuous basement membrane with no pericyte or smooth muscle cell coverage.[58,62,63] Many lymphatic-related molecular pathways have been identified to be important for SC development and maintenance. Upregulation of Prox1 expression is required for SC development, and Prox1 upregulation is stimulated by the initiation of the flow within the primitive SC.[60] Delivery of VEGFC into the adult eye leads to sprouting, proliferation, and growth of SC endothelial cells, and a single injection of recombinant VEGFC also induces SC growth and is associated with a sustained decrease in IOP in adult mice.[56] These studies link the importance of VEGFC/VEGFR3 signaling in SC development. In addition, loss-of-function mutations in the receptor tyrosine kinase Tie2/TEK or its ligand ANGPT1 cause primary congenital glaucoma in humans and mice due to failure of SC development.[61,64,65] Intriguingly, deletion of a single *Ptprb* allele (gene encoding phosphatase PTPRB, also known as the Vascular Endothelial Protein Tyrosine Phosphatase, VE-PTP) in mice leads to increased TEK activation and is sufficient for normal SC development and prevents ocular hypertension and retinal ganglion cell loss.[66] Future studies are needed to determine whether pharmacological modulation of these lymphatic-related pathways could provide novel therapeutic approaches for glaucoma treatment in the future.

Lymphatics in cardiovascular disease

The heart contains an extensive lymphatic network of capillaries and collecting lymphatics that plays important roles in maintaining tissue fluid balance and immune surveillance, key features in myocardial infarction (MI). MI is the most common type of heart injury.

In the past several years, the development of cardiac lymphatics in mouse and zebrafish has been well characterized. In zebrafish, cardiac lymphatics develop toward the ventricle in the juvenile-adult stages, using coronary arteries to guide their expansion down the ventricle.[67] In mice, the first cardiac lymphatic sprouts appear particularly over the dorsal side of the heart at around E14.5.[68] These cardiac lymphatics develop into an extensive lymphatic plexus of lymphatic covering most of the heart by E18.5, and continue remodeling and establishing a mature network 2–3 weeks after birth in mice.[68,69] In adults, lymphatic vessels are mostly quiescent, but altered cardiac lymphatic function or lymphatic malformations have been shown to aggravate various cardiovascular conditions.[67,68,70–75] In the adult heart, cardiac lymphatics form a network that helps to regulate and maintain fluid balance and immune surveillance, two key features observed during and after MI, whereas abnormal cardiac lymph flow appears to promote cardiac edema. Recent data suggest that natural or therapeutic formation of new lymphatics correlates with improved systolic function after MI; it can delay atherosclerotic plaque formation, facilitate the healing process after MI, and be a natural response to fluid accumulation into the myocardium during cardiac edema.[68,71,73–77] Impairment of cardiac lymphatic flow has also been shown to be associated with atrial fibrillation, fibrosis, and pulmonary arterial hypertension.[78,79] Despite these findings, the functional roles of cardiac lymphatics in health and disease have only recently started to be investigated. Here, we only discuss their role in atherosclerosis and MI.

Atherosclerosis

Atherosclerosis is characterized by chronic inflammation and accumulation of fat, cholesterol, and immune cells inside the arterial wall.[80] This can result in the narrowing and hardening of arterial walls, limiting blood flow from the heart.[81] Although the presence of lymphatics in the arterial wall was described over 100 years ago,[82] the detailed anatomy of lymphatics during atherosclerotic progression has only been characterized recently.[72,83,84] Lymphatic vessels can be visualized at atherosclerotic sites in the adventitial layer of coronary arteries,[83] and a positive correlation between the number of adventitial lymphatics and the severity of atherosclerotic disease has been proposed.[72] Work in mice showed that lymphatics are the main route for the transport of HDL particles into the bloodstream (reverse cholesterol transport, RCT),[85,86] indicating an essential functional role of lymphatics in atherosclerosis.[85–87] Disruption of lymphatic function by surgical ablation of collecting lymphatic vessels greatly impairs RCT.[85] Similarly, inhibition of VEGFR3-dependent lymphangiogenesis aggravates atherosclerotic plaque formation and increases intimal and adventitial T cell density.[83] These results suggest a beneficial role for periadventitial lymphatics in limiting cholesterol accumulation and chronic plaque inflammation during atherosclerosis.[83] Lymphatics could provide a protective pathway for lipid and inflammatory cell efflux from the arterial wall, counteracting the development of atherosclerotic plaques.[88] Consistent with these data, *Chy* mice and sVEGFR3-overexpressing mice with impaired lymphatic transport also show increased plasma cholesterol levels, although accelerated atherosclerotic lesion formation only occurred in sVEGFR3-overexpressing mice compared to control mice.[89] Therefore, defective lymphangiogenesis might contribute to the build-up of atherosclerotic lesions in large arteries as a consequence of the accumulation of lipids and activated immune cells. Restoration of lymphatic function and lymphangiogenesis by injection of VEGFC or VEGFC152S, a mutant form that only activates VEGFR3 signaling, improves cholesterol clearance and RCT transport.[85,90] A more recent study has also revealed that arterial delivery of VEGFC stabilizes atherosclerotic lesions.[91] It will be important to determine whether therapeutic lymphangiogenesis targeting the arterial wall can slow down fat deposition and tissue inflammation conferring protection against atherosclerosis.[92]

Myocardial infarction

MI is a life-threatening condition and is the most common type of heart injury that occurs as a result of decreased oxygenated blood supply to parts of the heart. After MI, a significant influx of circulating monocytes and activated macrophages occur in the infarcted regions, and the immune cell efflux depends on lymphatic vessels to circulate back to the lymph nodes.[93]

Stimulation of lymphangiogenesis by VEGFC treatment in a mouse MI model improved heart function and promoted clearance of immune cells in the injured heart.[68] Similarly, using a rat ischemia–reperfusion MI model, selective stimulation of lymphangiogenesis by intramyocardial-targeted delivery of VEGFC152S, a rat VEGFR3 selective ligand, accelerated lymphangiogenesis in the infarcted area and improved myocardial fluid balance, macrophage efflux, reducing fibrosis, and cardiac dysfunction.[73] Therefore, increased lymphangiogenesis in the infarcted area provides a pathway for immune cells efflux in the injured heart, and this process is dependent on lymphatic endothelial hyaluronan receptor 1 (LYVE1).[93] Conversely, aggravated cardiac function and cardiac remodeling have been reported in lymphatic insufficient mouse models after MI.[94] In this study, *Chy* mice and sVEGFR3-overexpressing mice, two mouse models with defective VEGFR3 signaling, show altered cardiac lymphatic structure but normal cardiac function in healthy hearts.[94] However, sVEGFR3-overexpressing mice show higher mortality with reduced capability to respond to lymphangiogenic signals and a modified structure of the infarcted area after MI. Interestingly, inflammatory cell infiltration is not altered in these mice, despite the dilated lymphatics and altered structures in the infarcted region.[94] Other signaling pathways that regulate lymphangiogenesis have also shown to be important for cardiac repair after MI. A recent study showed that overexpression of Adm, a known cardioprotective peptide previously reported to be essential for proper cardiovascular and lymphatic development in mice,[95] improves cardiac lymphangiogenesis post-MI in mice, with less cardiac edema and improved heart function.[96] Notably, this study also showed for the first time sex-dependent differences in cardiac lymphatic numbers in uninjured mice.[96]

Functional inactivation of apelin, which is a bioactive peptide known to play roles in angiogenesis and cardiac contractility,[97,98] and in pathological lymphangiogenesis,[99,100] results in abnormal dilated and leaky lymphatics associated with a proinflammatory status after MI in mice.[101] Conversely, overexpression of apelin is sufficient to restore lymphatic function and reduce matrix remodeling and inflammation after MI.[101] In addition, a recent study showed that deletion of integrin-linked kinase (ILK) in mouse embryos lead to hyperactive VEGFR3 signaling and LEC-specific deletion of *Ilk* in adult mice showed increased cardiac lymphangiogenesis post-MI.[102] Finally, studies in zebrafish have also demonstrated that cardiac lymphatics undergo increased lymphangiogenesis in response to injury, and that zebrafish with defects in VEGFC/VEGFR3 signaling display severely impaired regeneration capabilities.[67,74] Taken together, these studies show a promising functional role of cardiac lymphangiogenesis in improving heart function after injury, and provide an alternative therapeutic strategy for treatment of cardiovascular disease.

Conclusions

In summary, studies in the recent decades allowed great progress in understanding the functional roles of lymphatics in health and disease, and hold promise for new findings in the future. In addition, the diagnosis and treatment of lymphatic-related medical conditions have improved. Importantly, our recent work identified platelet factor 4 (PF4/CXCL4) as a reliable biomarker to diagnose lymphatic vasculature dysfunction.[103] This could facilitate a better diagnosis and treatment of patients with lymphatic diseases, particularly those with asymptomatic lymphatic defects and associated comorbidities.

Acknowledgments

This work was supported by NIH grant (RO1HL073402-19) to G.O, and AHA grant (18CDA34110356) to X.L.

References

1. Oliver G. Lymphatic vasculature development. *Nat Rev Immunol*. 2004;4(1):35–45.
2. Oliver G, Kipnis J, Randolph GJ, Harvey NL. The lymphatic vasculature in the 21st century: novel functional roles in homeostasis and disease. *Cell*. 2020;182(2):270–296.
3. Rockson SG. Lymphedema. *Vasc Med*. 2016;21(1):77–81.
4. Witte MH, Bernas MJ, Martin CP, Witte CL. Lymphangiogenesis and lymphangiodysplasia: from molecular to clinical lymphology. *Microsc Res Tech*. 2001;55(2):122–145.
5. Escobedo N, Oliver G. The lymphatic vasculature: its role in adipose metabolism and obesity. *Cell Metab*. 2017;26(4):598–609.
6. Mokdad AH, Marks JS, Stroup DF, Gerberding JL. Actual causes of death in the United States, 2000. *J Am Med Assoc*. 2004; 291(10):1238.
7. Bernier-Latmani J, Petrova TV. Intestinal lymphatic vasculature: structure, mechanisms and functions. *Nat Rev Gastroenterol Hepatol*. 2017;14(9):510–526.
8. Randolph GJ, Miller NE. Lymphatic transport of high-density lipoproteins and chylomicrons. *J Clin Invest*. 2014;124(3):929–935.
9. Dixon JB. Mechanisms of chylomicron uptake into lacteals. *Ann N Y Acad Sci*. 2010;1207(Supp l1):E52–E57.
10. Zhang F, Zarkada G, Han J, et al. Lacteal junction zippering protects against diet-induced obesity. *Science*. 2018;361(6402): 599–603.
11. Bernier-Latmani J, Cisarovsky C, Demir CS, et al. DLL4 promotes continuous adult intestinal lacteal regeneration and dietary fat transport. *J Clin Invest*. 2015;125(12):4572–4586.
12. Nurmi H, Saharinen P, Zarkada G, Zheng W, Robciuc MR, Alitalo K. VEGF-C is required for intestinal lymphatic vessel maintenance and lipid absorption. *EMBO Mol Med*. 2015;7(11): 1418–1425.
13. Harvey NL, Srinivasan RS, Dillard ME, et al. Lymphatic vascular defects promoted by Prox1 haploinsufficiency cause adult-onset obesity. *Nat Genet*. 2005;37(10):1072–1081.

14. Escobedo N, Proulx ST, Karaman S, et al. Restoration of lymphatic function rescues obesity in Prox1-haploinsufficient mice. *JCI Insight*. 2016;1(2):e85096.
15. Tavakkolizadeh A, Wolfe KQ, Kangesu L. Cutaneous lymphatic malformation with secondary fat hypertrophy. *Br J Plast Surg*. 2001;54(4):367–369.
16. Wang Y, Oliver G. Current views on the function of the lymphatic vasculature in health and disease. *Genes Dev*. 2010;24(19): 2115–2126.
17. Karkkainen MJ, Saaristo A, Jussila L, et al. A model for gene therapy of human hereditary lymphedema. *Proc Natl Acad Sci U S A*. 2001;98(22):12677–12682.
18. Rutkowski JM, Markhus CE, Gyenge CC, Alitalo K, Wiig H, Swartz MA. Dermal collagen and lipid deposition correlate with tissue swelling and hydraulic conductivity in murine primary lymphedema. *Am J Pathol*. 2010;176(3):1122–1129.
19. Wigle JT, Oliver G. Prox1 function is required for the development of the murine lymphatic system. *Cell*. 1999;98(6):769–778.
20. Yao LC, Baluk P, Srinivasan RS, Oliver G, McDonald DM. Plasticity of button-like junctions in the endothelium of airway lymphatics in development and inflammation. *Am J Pathol*. 2012; 180(6):2561–2575.
21. McDonald DM. Tighter lymphatic junctions prevent obesity. *Science*. 2018;361(6402):551–552.
22. Shew T, Wolins NE, Cifarelli V. VEGFR-3 signaling regulates triglyceride retention and absorption in the intestine. *Front Physiol*. 2018;9:1783.
23. Davis RB, Kechele DO, Blakeney ES, Pawlak JB, Caron KM. Lymphatic deletion of calcitonin receptor-like receptor exacerbates intestinal inflammation. *JCI Insight*. 2017;2(6):e92465.
24. Skobe M, Detmar M. Structure, function, and molecular control of the skin lymphatic system. *J Investing Dermatol Symp Proc*. 2000; 5(1):14–19.
25. Forbes G. Lymphatics of the skin, with a note on lymphatic watershed areas. *J Anat*. 1938;72(Pt 3):399–410.
26. Plikus MV, Chuong CM. Macroenvironmental regulation of hair cycling and collective regenerative behavior. *Cold Spring Harb Perspect Med*. 2014;4(1):a015198.
27. Mecklenburg L, Tobin DJ, Müller-Röver S, et al. Active hair growth (anagen) is associated with angiogenesis. *J Invest Dermatol*. 2000;114(5):909–916.
28. Yano K, Brown LF, Detmar M. Control of hair growth and follicle size by VEGF-mediated angiogenesis. *J Clin Invest*. 2001;107(4): 409–417.
29. Detmar M. Molecular regulation of angiogenesis in the skin. *J Invest Dermatol*. 1996;106(2):207–208.
30. Yoon SY, Dieterich LC, Karaman S, et al. An important role of cutaneous lymphatic vessels in coordinating and promoting anagen hair follicle growth. *PLoS One*. 2019;14(7):e0220341.
31. Peña-Jimenez D, Fontenete S, Megias D, et al. Lymphatic vessels interact dynamically with the hair follicle stem cell niche during skin regeneration in vivo. *EMBO J*. 2019;38(19):e101688.
32. Gur-Cohen S, Yang H, Baksh SC, et al. Stem cell-driven lymphatic remodeling coordinates tissue regeneration. *Science*. 2019; 366(6470):1218–1225.
33. Aspelund A, Antila S, Proulx ST, et al. A dural lymphatic vascular system that drains brain interstitial fluid and macromolecules. *J Exp Med*. 2015;212(7):991–999.
34. Da Mesquita S, Louveau A, Vaccari A, et al. Functional aspects of meningeal lymphatics in ageing and Alzheimer's disease. *Nature*. 2018;560(7717):185–191.
35. Louveau A, Smirnov I, Keyes TJ, et al. Structural and functional features of central nervous system lymphatic vessels. *Nature*. 2015;523(7560):337–341.
36. Louveau A, Herz J, Alme MN, et al. CNS lymphatic drainage and neuroinflammation are regulated by meningeal lymphatic vasculature. *Nat Neurosci*. 2018;21(10):1380–1391.
37. Bucchieri F, Farina F, Zummo G, Cappello F. Lymphatic vessels of the dura mater: a new discovery? *J Anat*. 2015;227(5):702–703.
38. Antila S, Karaman S, Nurmi H, et al. Development and plasticity of meningeal lymphatic vessels. *J Exp Med*. 2017;214(12): 3645–3667.
39. Ahn JH, Cho H, Kim J-H, et al. Meningeal lymphatic vessels at the skull base drain cerebrospinal fluid. *Nature*. 2019;572(7767): 62–66.
40. Zou W, Pu T, Feng W, et al. Blocking meningeal lymphatic drainage aggravates Parkinson's disease-like pathology in mice overexpressing mutated α-synuclein. *Transl Neurodegener*. 2019;8:7.
41. Van Zwam M, Huizinga R, Heijmans N, et al. Surgical excision of CNS-draining lymph nodes reduces relapse severity in chronic-relapsing experimental autoimmune encephalomyelitis. *J Pathol*. 2009;217(4):543–551.
42. Takahashi RH, Nagao T, Gouras GK. Plaque formation and the intraneuronal accumulation of β-amyloid in Alzheimer's disease. *Pathol Int*. 2017;67(4):185–193.
43. Bateman RJ, Xiong C, Benzinger TLS, et al. Clinical and biomarker changes in dominantly inherited Alzheimer's disease. *N Engl J Med*. 2012;367(9):795–804.
44. Iliff JJ, Wang M, Liao Y, et al. A paravascular pathway facilitates CSF flow through the brain parenchyma and the clearance of interstitial solutes, including amyloid β. *Sci Transl Med*. 2012; 4(147):147ra111.
45. Brookmeyer R, Abdalla N, Kawas CH, Corrada MM. Forecasting the prevalence of preclinical and clinical Alzheimer's disease in the United States. *Alzheimer's Dement*. 2018;14(2):121–129.
46. Erkkinen MG, Kim MO, Geschwind MD. Clinical neurology and epidemiology of the major neurodegenerative diseases. *Cold Spring Harb Perspect Biol*. 2018;10(4):a033118.
47. Chevalier S, Ferland G, Tuchweber B. Lymphatic absorption of retinol in young, mature, and old rats: influence of dietary restriction. *FASEB J*. 1996;10(9):1085–1090.
48. Hos D, Bachmann B, Bock F, Onderka J, Cursiefen C. Age-related changes in murine limbal lymphatic vessels and corneal lymphangiogenesis. *Exp Eye Res*. 2008;87(5):427–432.
49. Kress BT, Iliff JJ, Xia M, et al. Impairment of paravascular clearance pathways in the aging brain. *Ann Neurol*. 2014;76(6):845–861.
50. Goel M, Picciani RG, Lee RK, Bhattacharya SK. Aqueous humor dynamics: a review. *Open Ophthalmol J*. 2010;4:52–59.
51. Andrés-Guerrero V, García-Feijoo J, Konstas AG. Targeting schlemm's canal in the medical therapy of glaucoma: current and future considerations. *Adv Ther*. 2017;34(5):1049–1069.
52. Almasieh M, Wilson AM, Morquette B, Cueva Vargas JL, Di Polo A. The molecular basis of retinal ganglion cell death in glaucoma. *Prog Retin Eye Res*. 2012;31(2):152–181.
53. Overby DR, Zhou EH, Vargas-Pinto R, et al. Altered mechanobiology of Schlemm's canal endothelial cells in glaucoma. *Proc Natl Acad Sci U S A*. 2014;111(38):13876–13881.
54. Perry LP, Jakobiec FA, Zakka FR, Walton DS. Newborn primary congenital glaucoma: histopathologic features of the anterior chamber filtration angle. *J AAPOS*. 2012;16(6):565–568.
55. Smith RS, Zabaleta A, Kume T, et al. Haploinsufficiency of the transcription factors FOXC1 and FOXC2 results in aberrant ocular development. *Hum Mol Genet*. 2000;9(7):1021–1032.
56. Aspelund A, Tammela T, Antila S, et al. The Schlemm's canal is a VEGF-C/VEGFR-3-responsive lymphatic-like vessel. *J Clin Invest*. 2014;124(9):3975–3986.
57. Kizhatil K, Ryan M, Marchant JK, Henrich S, John SWM. Schlemm's canal is a unique vessel with a combination of blood vascular and lymphatic phenotypes that forms by a novel developmental process. *PLoS Biol*. 2014;12(7):e1001912.
58. Ramos RF, Hoying JB, Witte MH, Daniel Stamer W. Schlemm's canal endothelia, lymphatic, or blood vasculature? *J Glaucoma*. 2007; 16(4):391–405.

59. Truong TN, Li H, Hong YK, Chen L. Novel characterization and live imaging of Schlemm's canal expressing Prox-1. *PLoS One*. 2014;9(5):e98245.
60. Park DY, Lee J, Park I, et al. Lymphatic regulator PROX1 determines Schlemm's canal integrity and identity. *J Clin Invest*. 2014; 124(9):3960–3974.
61. Thomson BR, Heinen S, Jeansson M, et al. A lymphatic defect causes ocular hypertension and glaucoma in mice. *J Clin Invest*. 2014;124(10):4320–4324.
62. Gong H, Tripathi RC, Tripathi BJ. Morphology of the aqueous outflow pathway. *Microsc Res Tech*. 1996;33(4):336–367.
63. Hamanaka T, Bill A, Ichinohasama R, Ishida T. Aspects of the development of Schlemm's canal. *Exp Eye Res*. 1992;55(3):479–488.
64. Souma T, Tompson SW, Thomson BR, et al. Angiopoietin receptor TEK mutations underlie primary congenital glaucoma with variable expressivity. *J Clin Invest*. 2016;126(7):2575–2587.
65. Thomson BR, Souma T, Tompson SW, et al. Angiopoietin-1 is required for Schlemm's canal development in mice and humans. *J Clin Invest*. 2017;127(12):4421–4436.
66. Thomson BR, Carota IA, Souma T, Soman S, Vestweber D, Quaggin SE. Targeting the vascular-specific phosphatase PTPRB protects against retinal ganglion cell loss in a pre-clinical model of glaucoma. *Elife*. 2019;8:e48474.
67. Harrison MR, Feng X, Mo G, et al. Late developing cardiac lymphatic vasculature supports adult zebrafish heart function and regeneration. *Elife*. 2019;8:e42762.
68. Klotz L, Norman S, Vieira JM, et al. Cardiac lymphatics are heterogeneous in origin and respond to injury. *Nature*. 2015;522(7554): 62–67.
69. Juszyński M, Ciszek B, Stachurska E, Jabłońska A, Ratajska A. Development of lymphatic vessels in mouse embryonic and early postnatal hearts. *Dev Dyn*. 2008;237(10):2973–2986.
70. Bradham RR, Parker EF, Barrington BA, Webb CM, Stallworth JM. The cardiac lymphatics. *Ann Surg*. 1970;171(6):899–902.
71. Ishikawa Y, Akishima-Fukasawa Y, Ito K, et al. Lymphangiogenesis in myocardial remodelling after infarction. *Histopathology*. 2007;51(3):345–353.
72. Kholová I, Dragneva G, Čermáková P, et al. Lymphatic vasculature is increased in heart valves, ischaemic and inflamed hearts and in cholesterol-rich and calcified atherosclerotic lesions. *Eur J Clin Invest*. 2011;41(5):487–497.
73. Henri O, Pouehe C, Houssari M, et al. Selective stimulation of cardiac lymphangiogenesis reduces myocardial edema and fibrosis leading to improved cardiac function following myocardial infarction. *Circulation*. 2016;133(15):1484–1497.
74. Gancz D, Raftrey BC, Perlmoter G, et al. Distinct origins and molecular mechanisms contribute to lymphatic formation during cardiac growth and regeneration. *Elife*. 2019;8:e44153.
75. Kong XQ, Wang LX, Kong DG. Cardiac lymphatic interruption is a major cause for allograft failure after cardiac transplantation. *Lymphat Res Biol*. 2007;5(1):45–47.
76. Cui Y, Urschel JD, Petrelli NJ. The effect of cardiopulmonary lymphatic obstruction on heart and lung function. *Thorac Cardiovasc Surg*. 2001;49(1):35–40.
77. Dashkevich A, Bloch W, Antonyan A, Fries JUW, Geissler HJ. Morphological and quantitative changes of the initial myocardial lymphatics in terminal heart failure. *Lymphat Res Biol*. 2009;7(1): 21–27.
78. Lupinski RW. Aortic fat pad and atrial fibrillation: cardiac lymphatics revisited. *ANZ J Surg*. 2009;79(1-2):70–74.
79. Laine GA, Allen SJ. Left ventricular myocardial edema. Lymph flow, interstitial fibrosis, and cardiac function. *Circ Res*. 1991; 68(6):1713–1721.
80. Moore KJ, Sheedy FJ, Fisher EA. Macrophages in atherosclerosis: a dynamic balance. *Nat Rev Immunol*. 2013;13(10):709–721.
81. Shi G-P, Bot I, Kovanen PT. Mast cells in human and experimental cardiometabolic diseases. *Nat Rev Cardiol*. 2015;12(11):643–658.
82. Hoggan G, Hoggan FE. The lymphatics of the walls of the larger blood-vessels and lymphatics. *J Anat Physiol*. 1882;17(Pt 1):1–23.
83. Rademakers T, Van Der Vorst EPC, Daissormont ITMN, et al. Adventitial lymphatic capillary expansion impacts on plaque T cell accumulation in atherosclerosis. *Sci Rep*. 2017;7:45263.
84. Hatakeyama K, Kaneko MK, Kato Y, et al. Podoplanin expression in advanced atherosclerotic lesions of human aortas. *Thromb Res*. 2012;129(4):e70–e76.
85. Lim HY, Thiam CH, Yeo KP, et al. Lymphatic vessels are essential for the removal of cholesterol from peripheral tissues by SR-BI-mediated transport of HDL. *Cell Metab*. 2013;17(5):671–684.
86. Martel C, Li W, Fulp B, et al. Lymphatic vasculature mediates macrophage reverse cholesterol transport in mice. *J Clin Invest*. 2013;123(4):1571–1579.
87. Hjelms E, Nordestgaard BG, Stender S, Kjeldsen K. A surgical model to study in vivo efflux of cholesterol from porcine aorta evidence for cholesteryl ester transfer through the aortic wall. *Atherosclerosis*. 1989;77(2-3):239–249.
88. Alitalo K. The lymphatic vasculature in disease. *Nat Med*. 2011; 17(11):1371–1380.
89. Vuorio T, Kurkipuro J, Heinonen SE, et al. Lymphatic vessel insufficiency in hypercholesterolemic mice alters lipoprotein levels and promotes atherogenesis. *Arterioscler Thromb Vasc Biol*. 2014;34(6): 1162–1170.
90. Milasan A, Smaani A, Martel C. Early rescue of lymphatic function limits atherosclerosis progression in Ldlr mice. *Atherosclerosis*. 2019;283:106–119.
91. Silvestre-Roig C, Lemnitzer P, Gall J, et al. Arterial delivery of VEGF-C stabilizes atherosclerotic lesions. *Circ Res*. 2020. CIrcRESAHA.120.317186.
92. Brakenhielm E, Alitalo K. Cardiac lymphatics in health and disease. *Nat Rev Cardiol*. 2019;16(1):56–68.
93. Vieira JM, Norman S, Villa del Campo C, et al. The cardiac lymphatic system stimulates resolution of inflammation following myocardial infarction. *J Clin Invest*. 2018;128(8): 3402–3412.
94. Vuorio T, Ylä-Herttuala E, Laakkonen JP, Laidinen S, Liimatainen T, Ylä-Herttuala S. Downregulation of VEGFR3 signaling alters cardiac lymphatic vessel organization and leads to a higher mortality after acute myocardial infarction. *Sci Rep*. 2018;8(1):16709.
95. Caron KM, Smithies O. Extreme hydrops fetalis and cardiovascular abnormalities in mice lacking a functional Adrenomedullin gene. *Proc Natl Acad Sci U S A*. 2012;98(2):615–619.
96. Trincot CE, Xu W, Zhang H, et al. Adrenomedullin induces cardiac lymphangiogenesis after myocardiali and regulates cardiac edema via connexin 43. *Circ Res*. 2019;124(1):101–113.
97. Dai T, Ramirez-Correa G, Gao WD. Apelin increases contractility in failing cardiac muscle. *Eur J Pharmacol*. 2006;553(1-3): 222–228.
98. Ashley EA, Powers J, Chen M, et al. The endogenous peptide apelin potently improves cardiac contractility and reduces cardiac loading in vivo. *Cardiovasc Res*. 2005;65(1):73–82.
99. Kim JD, Kang Y, Kim J, et al. Essential role of apelin signaling during lymphatic development in Zebrafish. *Arterioscler Thromb Vasc Biol*. 2014;34(2):338–345.
100. Berta J, Hoda MA, Laszlo V, et al. Apelin promotes lymphangiogenesis and lymph node metastasis. *Oncotarget*. 2014;5(12): 4426–4437.
101. Tatin F, Renaud-Gabardos E, Godet AC, et al. Apelin modulates pathological remodeling of lymphatic endothelium after myocardial infarction. *JCI Insight*. 2017;2(12):e93887.
102. Urner S, Planas-Paz L, Hilger LS, et al. Identification of ILK as a critical regulator of VEGFR 3 signalling and lymphatic vascular growth. *EMBO J*. 2019;38(2):e99322.
103. Ma W, Gil HJ, Escobedo N, et al. Platelet factor 4 is a biomarker for lymphatic-promoted disorders. *JCI Insight*. 2020;5(13):e135109.

CHAPTER

29

Extracellular matrix genetics of thoracic and abdominal aortic diseases

Kaveeta Kaw[a], *Anita Kaw*[a] *and Dianna M. Milewicz*

Division of Medical Genetics, Department of Internal Medicine, McGovern Medical School, University of Texas Health Science Center at Houston, Houston, TX, United States

Aortic aneurysm is a dilation of the aorta due to weakening of the wall, often occurring without any presenting symptoms. Undiagnosed aortic aneurysms can result in fatal complications, specifically aortic dissection, which can otherwise be prevented by early recognition and repair of the aortic aneurysm.[1] Despite the availability of an established treatment for aortic aneurysms, the complications associated with aortic aneurysms, dissections, and ruptures were identified as the 19th leading cause of death in the United States by the Centers for Disease Control.[2] Thus, using genetic variants to identify individuals who are predisposed to aortic aneurysm and dissection is paramount in preventing associated complications and premature death.

Aneurysms can occur in any segment of the aorta; however, the modes of inheritance differ between aneurysms in the thoracic and abdominal aortic segments. Abdominal aortic aneurysms (**AAAs**) have multifactorial etiology, including both genetic and environmental risk factors. Importantly, no single high-risk gene has been recognized to explicitly cause this disease.

In contrast to AAAs, syndromes such as Marfan syndrome (**MFS**) and Loeys–Dietz syndrome (**LDS**) are caused by pathogenic variants in a single gene that confer a highly prenetrant risk for an thoracic aortic aneurysm and dissection (**TAAD**). Furthermore, approximately 20% of patients with TAAD have a family history of the disease without syndromic features, implicating a significant role for genetics in nonsyndromic thoracic aortic disease.[3] Collectively, these single gene variants predispose to TAAD in an autosomal dominant inheritance, referred to as heritable thoracic aortic disease (**HTAD**). The genes that are mutated to predispose to HTAD provide critical information about the structures and signaling pathways in the thoracic aorta that prevent aneurysms and dissections.

Thoracic aortic aneurysms predispose to acute aortic dissections

Thoracic aortic aneurysms commonly occur in the ascending aorta and aortic root with asymptomatic enlargement, therefore going undetected before progressing to dissection or rupture. In the event of an aortic dissection, blood passes through a tear in the innermost layer of the aortic wall and further separates the aortic layers. The tear can propagate anterograde into the ascending and descending aorta or retrograde in which a rupture into the pericardial sac can lead to pericardial tamponade, a major cause of death in patients who die prior to the hospital admission.[4] Aortic rupture and dissections have a high risk of morbidity and mortality. Management via medications such as β-adrenergic receptor blockers and angiotensin receptor type 1 antagonists can slow the progression of the aneurysm. However, prophylactic aneurysm repair remains the mainstay for prevention of premature death. For example, repair of an aortic root aneurysm has been shown to increase life expectancy in Marfan patients by at least 25 years.[5,6] Hypertension and genetic risk are the major risk factors for thoracic aortic disease, highlighting the importance of biomechanical forces on

[a] These authors contributed equally to this work.

The Vasculome
https://doi.org/10.1016/B978-0-12-822546-2.00008-3

the ascending thoracic aorta in the pathogenesis of the disease.

Heritable thoracic aortic disease genetics highlights the importance of maintaining an aortic structural component, the elastin-contractile unit

The aorta is the major conduit of blood flow from the heart to the rest of the body. Given that the aorta bears the highest biomechanical forces of pulsatile blood from the heart, the extracellular matrix (**ECM**) plays a critical role in maintaining the integrity of the aorta. Specifically, the ECM is the noncellular component of the aorta that provides structural support, aids in the movement of blood, conducts mechanical signals to cells, and stores growth factors and cytokines that influence aortic remodeling. The proteins that comprise the ECM of the aorta include elastin, fibronectin, fibrillin, fibulin, collagen types I–VI, and proteoglycans.[7] In addition, the aorta is composed of three main cell types—endothelial cells, smooth muscle cells (**SMCs**), and fibroblasts—which inhabit three different layers of the aorta. The innermost layer of the aorta is the tunica intima, which is comprised of a single layer of endothelial cells that are supported by a thin sheet of specialized ECM called the basement membrane and the internal elastic laminae. The tunica media is the thick, middle layer which contains concentric layers of SMCs, collagen, and elastic fibers. The tunica adventitia is the outermost layer which contains mainly collagen and fibroblast cells and is supported by the external elastic laminae with collagen fibers.

Of importance, vascular SMCs, the major cell type in the aortic wall, are quiescent cells that express SMCs-specific contractile genes, including *ACTA2* and *MYH11*, in order to contract to maintain blood pressure and regulate blood flow. Specifically, the medial layer, which consists of up to 50 layers of SMCs and elastic lamellae in the human aorta, confers the elastic recoil.[8] The elastic recoil is specifically due to the elastin fibers and aids in maintaining blood flow during diastole (Windkessel effect). The elastin fibers have oblique extensions, and at the end of the extensions are microfibrils. Microfibril extensions from the extracellular elastin lamellae link to the intracellular contractile filaments in SMCs via focal adhesions (or dense plaques) to translate mechanical forces between elastin and SMCs through integrin receptors.[8] Importantly, this "elastin-contractile unit" is the functional unit in the aorta through which elastin fibers conduct mechanical signals to SMCs, maintaining SMC response to mechanical strain. Thus, disruption of mechanosensing by SMCs has the potential to influence the integrity of the aortic wall.

Thoracic aortic aneurysms and dissections (**TAAD**) are marked by medial degeneration, a pathology that is characterized by fragmentation and loss of elastin fibers, loss of SMCs, and increased proteoglycan deposition. Genes harboring pathogenic variants that predispose to TAAD encode components of the elastin-contractile unit, and these mutations are predicted or have been shown to disrupt the elastin-contractile unit, leading to disruption of SMC mechanosensing homeostasis, which subsequently leads to harmful remodeling of the aorta, and aneurysm and dissection (Fig. 29.1). The following genes are confirmed to cause HTAD when mutated.[10] Extracellular components encoded by some causative genes include fibrillin-1 (*FBN1*), which is the major protein in the microfibrils at the tips of the oblique elastin extensions that anchor to the SMC cell membrane at dense plaques, lysyl oxidase (*LOX*), which crosslinks collagen and elastin, and procollagen type III α1 (*COL3A1*), which provides structural support to the aorta. The smooth muscle–specific isoform of α-actin (*ACTA2*) and smooth muscle–specific isoform of myosin heavy chain (*MYH11*) are the major proteins in the thin and thick filaments, respectively, of the contractile unit in SMCs. *MYLK* encodes myosin light chain kinase (MLCK), which phosphorylates the regulatory light chain (**RLC**) to move the myosin motor domain, and *PRKG1* encodes protein kinase G (PKG-1), which regulates SMC relaxation through dephosphorylation of the RLC. Genes for proteins involved in the canonical TGF-β signaling pathway are also mutated to predispose to TAAD, including *SMAD3* (mothers against decapentaplegic homolog 3), *TGFB2* (TGF-β2), *TGFBR1* (TGF-β receptor type 1), and *TGFBR2* (TGF-β receptor type 2).[9] Table 29.1 lists the validated genes predisposing to HTAD with the corresponding encoded protein and domain affected by the mutation.

Pathogenic genetic variants that disrupt the extracellular matrix

It has been known for many years that alterations in genes related to the ECM can strongly predispose to TAAD. The first HTAD gene identified was *FBN1*, which encodes fibrillin-1 and, when altered, causes MFS, an autosomal dominant disorder characterized by skeletal, ocular, cardiovascular, and pulmonary manifestations.[11] Despite pleiotropic manifestations, the leading cause of premature death in MFS is acute aortic dissections. MFS is caused by heterozygous mutations in *FBN1*.[9] Fibrillin-1 is a large secreted cysteine-rich protein that assembles into microfibrils and connects SMCs to the

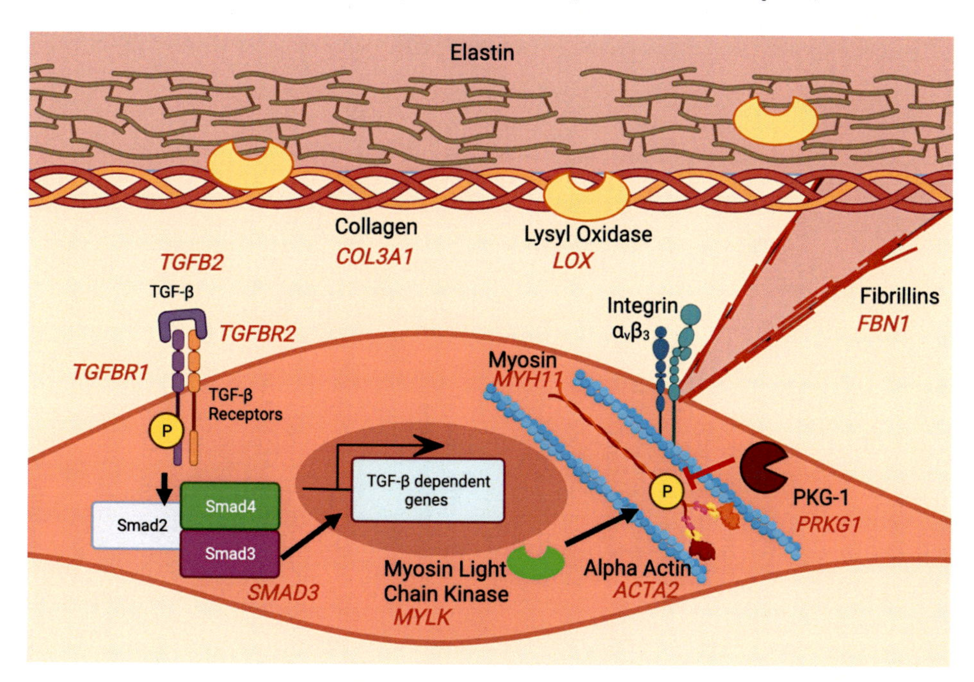

FIGURE 29.1 Genes implicated in heritable thoracic aortic disease demonstrate the importance of maintenance of the smooth muscle cell elastin-contractile unit in aortic wall integrity. Microfibril extensions from the extracellular elastin lamellae link to the intracellular contractile filaments (actin and myosin) in SMCs to transmit mechanical forces between elastin and SMCs through integrin receptors. This "elastin-contractile unit" is the functional unit in the aorta through which SMCs potentially sense mechanical forces on the aorta. Human genetic studies have demonstrated that the smooth muscle contractile apparatus, ECM, and TGF-β signaling pathway are critical in maintenance of aortic wall integrity. Validated genes predisposing to thoracic aortic disease are displayed in red with their corresponding protein. *PKG-1*, type I cGMP-dependent protein kinase; *Smad*, mothers against decapentaplegic drosophila homolog. *Adapted from Pinard A, Jones GT, Milewicz DM. Genetics of thoracic and abdominal aortic diseases.* Circ Res. *February 15, 2019;124(4):588–606. https://dx.doi.org/10.1161/circresaha.118.312436. Illustration Credit: BioRender.*

elastin lamellae. Fibrillin-1 contains 47 calcium-binding epidermal growth factor (EGF) motifs and 7 cysteine-containing TB (TGF-β1 binding) motifs. EGF domains facilitate fibrillin-1 binding to integrin $\alpha_v\beta_3$ on the surface of SMCs, thus regulating the interaction between fibrillin-1 and SMCs in the elastin-contractile unit.[12] *FBN1* pathogenic variants causing MFS are primarily missense variants, but frameshift mutations, exon splicing errors, and nonsense mutations can also cause the disease.[13] The downstream effects of these mutations are reduced levels of assembled fibrillin-1 protein or reduced incorporation of fibrillin-1 into the ECM.[14–17] For example, pathogenic variants that disrupt the cysteine codon in EGF motifs result in misfolding or destabilization of the fibrillin-1 protein.[18] As a key component of the elastin-contractile unit, alterations in fibrillin-1 are predicted to alter this apparatus through decreasing the amount of fibrillin-1–containing microfibrils that can link to the SMCs, leading to impaired homeostatic mechanosensing of SMCs in the aortic wall.[19]

Pathogenic variants in *LOX*, which encodes a copper-dependent lysyl oxidase that crosslinks collagens and elastin, have also been shown to predispose to TAAD.[20,21] These heterozygous mutations result in decreased lysyl oxidase activity and are proposed to destabilize collagen and elastin fibrils and disrupt ECM integrity and elasticity. Patients develop fusiform thoracic aortic aneurysms and ascending aortic dissections and have a few Marfan-like systemic features.

Pathogenic variants that disrupt the SMC contractile unit

Maintenance of the contractile function of aortic SMCs is crucial to preventing the formation of TAAD. Heterozygous mutations in the major components of the SMC contractile unit and the kinases that regulate contraction lead to ascending aortic aneurysms and subsequent dissections.[22] SMCs produce contractile force via ATP-dependent interactions between thick myosin filaments, consisting of myosin heavy chain (encoded by *MYH11*), and thin actin filaments, consisting of polymers of smooth muscle alpha actin (encoded by *ACTA2*). MLCK (encoded by *MYLK*) regulates phosphorylation of the smooth muscle myosin RLC kinase, which triggers SMC contraction, while PRKG-1

TABLE 29.1 Validated genes which predispose to heritable thoracic aortic disease.

Gene name	Protein	Inheritance	Associated syndrome	Domain location for missense, inframe deletion, or duplication	Pathways impacted
FBN1	Fibrillin-1	AD	Marfan syndrome	Loss-of-function mutations, missense mutations in EGF-like domains (disruption of cysteines and amino acids involved in calcium binding to the domain)	Extracellular matrix (ECM)
COL3A1	Procollagen type III α1	AD	Vascular Ehlers–Danlos syndrome	Triple helical domain (disrupt glycines)	ECM
LOX	Lysyl oxidase	AD	Musculoskeletal manifestations of Marfan syndrome	Loss of function or missense variants in the catalytic domain	ECM
MYH11	Smooth muscle myosin heavy chain 11	AD		Inframe deletions in coiled coil domain and rare missense mutations	SMC contraction
MYLK	Myosin light chain kinase	AD		Loss of function and missense mutations in the kinase domain	SMC contraction
ACTA2	Smooth muscle actin α2	AD	Smooth muscle dysfunction syndrome	[a]	SMC contraction
PRKG1	Protein kinase cGMP-dependent type 1			One gain-of-function mutation, p.Arg177Gln	SMC contraction
TGFB2	Transforming growth factor β2	AD	Loeys–Dietz syndrome 4	Loss of function, missense mutation in the furin cleavage site, and cytokine domain	TGF-β
TGFBR1	Transforming growth factor β receptor type I	AD	Loeys–Dietz syndrome 1	Intracellular kinase domain	TGF-β
TGFBR2	Transforming growth factor β receptor type II	AD	Loeys–Dietz syndrome 2	Intracellular kinase domain	TGF-β
SMAD3	Mothers against decapentaplegic drosophila homolog 3	AD	Loeys–Dietz syndrome 3	Loss-of-function mutations and primarily missense mutations in the MH2 domain	TGF-β

[a]*Pathogenic variants are found throughout the gene.*
AD, autosomal dominant.
Adapted from Pinard A, Jones GT, Milewicz DM. Genetics of thoracic and abdominal aortic diseases. Circ Res. *February 15, 2019;124(4):588–606. https://dx.doi.org/10.1161/circresaha.118.312436.*

(encoded by *PRKG1*) regulates SMC relaxation through dephosphorylation of the RLC.

Heterozygous missense mutations in *ACTA2* predispose to thoracic aortic aneurysms and dissections, and subsets of *ACTA2* mutations also lead to early-onset coronary artery disease or moyamoya-like cerebrovascular disease.[23] Furthermore, *ACTA2* variants which alter arginine 179 residue (p.R179H, p.R179C, p.R179L, and p.R179S) also predispose to multisystemic disease called smooth muscle dysfunction syndrome.[24] These systemic features include patent ductus arteriosus, TAAD, early-onset stroke, pulmonary hypertension, and additional complications in smooth muscle–dependent organs due to global SMC disease. Importantly, *ACTA2* mutations are responsible for 14% of nonsyndromic HTAD.[25] *ACTA2* is a highly conserved gene, which encodes smooth muscle (**SM**)-specific isoform of α-actin, the most highly expressed protein in SMCs. SM α-actin polymerizes to form the thin filaments of the SMC contractile apparatus. Mutations that disrupt SM α-actin are located in all four subdomains of the protein and are predicted to alter actin monomers, leading to disruption of SMC contraction and force generation and consequently development of aneurysms and dissections. *In vitro* studies of *ACTA2* missense variants, including p.Arg258Cys, p.Arg179His, and p.Arg149Cys, indicate that these mutations alter actin by at least one of the following: disrupting actin filament stability, increasing susceptibility of filaments to severing by cofilin, enhancing interaction with profilin, which is predicted to increase monomeric actin, and reducing the interaction between smooth muscle myosin and mutant α-actins.[26,27] Together, these alterations disrupt SMC mechanosensing by impairing the ability of the SMC to contract in response to pulse pressures.

Mutations in *MYH11*, which encodes SM-specific myosin heavy chain, predispose to thoracic aortic disease associated with patent ductus arteriosus.[28] Like SM α-actin, SM myosin heavy chain is a key component of the SMC contractile apparatus. The majority of mutations in *MYH11* occur in the C-terminal region of the coiled coil or rod domain and are predicted to disrupt myosin polymerization into thick filaments. Interestingly, deletions of a region of 16p13.1, which contains *MYH11*, do not predispose to thoracic aortic disease while duplications of this same region increase the risk of thoracic aortic disease.[29,30]

Missense mutations in *MYLK* or haploinsufficiency of this gene, which encodes MLCK, predispose to aortic dissection with minimal enlargement of the aorta prior to dissection.[31,32] MLCK is a ubiquitously expressed kinase that phosphorylates the RLC of smooth muscle and nonmuscle myosin to initiate contraction. Only the short form of MLCK is expressed in the human aorta, and missense mutations or mutations leading to nonsense mediated decay are predicted to disrupt the kinase activity and reduce RLC phosphorylation, thus disrupting aortic SMC contraction.[33]

Only one gain-of-function rare variant (p.Arg177Gln) has been identified in *PRKG1*, which encodes PKG-1, a type 1 cGMP-dependent protein kinase that mediates SMC relaxation. This rare variant leads to early-onset thoracic aortic disease, with presentation of the disease at a mean age of 31 years in 63% of carriers. Patients may also present with additional aneurysms in the descending thoracic and abdominal aorta, as well as coronary arteries.[24] While the mutation disrupts binding of cGMP to PKG-1, the altered kinase becomes constitutively active, leading to decreased force generation in SMCs through dephosphorylation of the RLC.

Mutations in the contractile apparatus of SMCs disrupt the ability of the SMCs to contract in response to pulsatile blood flow from the heart, thus disrupting homeostasis of the aorta. Mutations in ECM components and the SMC contractile apparatus alter the SMC force generation by disrupting the contractile unit or altering the ECM that anchor the SMCs so they can generate force. Consequently, the majority of the rare variants that lead to HTAD disrupt the elastin-contractile unit and are thus predicted to alter homeostatic force generation and mechanosensing by SMCs, leading to harmful

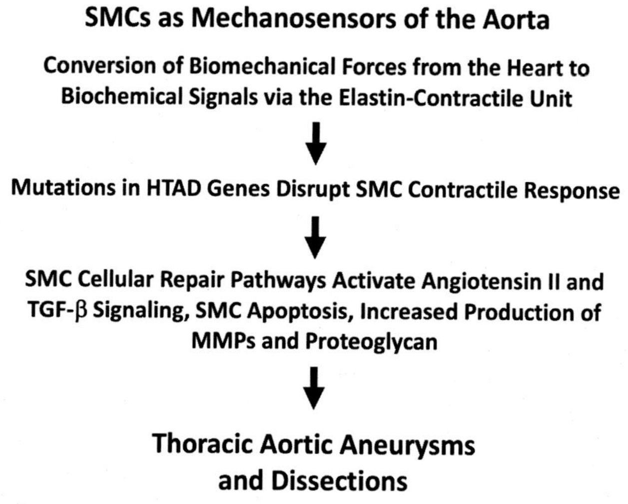

SMCs as Mechanosensors of the Aorta

Conversion of Biomechanical Forces from the Heart to Biochemical Signals via the Elastin-Contractile Unit

Mutations in HTAD Genes Disrupt SMC Contractile Response

SMC Cellular Repair Pathways Activate Angiotensin II and TGF-β Signaling, SMC Apoptosis, Increased Production of MMPs and Proteoglycan

Thoracic Aortic Aneurysms and Dissections

FIGURE 29.2 **Proposed mechanism of thoracic aortic aneurysms and dissections.** Hypertension and genetic risk are the major drivers of thoracic aortic disease. The mutations which comprise HTAD are predicted to disrupt the elastin-contractile unit through which SMCs respond to pulsatile blood flow.[19,34] At the cellular level, hypertension may alter the forces across the elastin-contractile unit, similarly to how genetic variants alter the SMC contractile forces, leading to downstream remodeling pathways. *Adapted from Pinard A, Jones GT, Milewicz DM. Genetics of thoracic and abdominal aortic diseases.* Circ Res. *February 15, 2019;124(4):588–606. https://dx.doi.org/10.1161/circresaha.118.312436. Illustration Credit: BioRender.*

remodeling of the aorta and aneurysm and dissection (Fig. 29.2). The hypothesis that altered mechanosensing of SMCs leads to aortic aneurysm and dissection also explains why hypertension is a major risk factor for thoracic aortic disease. At the cellular level, hypertension alters the forces across the elastin-contractile unit and leads to activation of the same downstream remodeling pathways.

HTAD genes disrupting TGF-β signaling

Pathogenic variants in *TGFBR2*, which encodes the TGF-β receptor type II, were initially recognized in patients with features of Marfan syndrome.[35] Proteins of the canonical TGF-β signaling pathway such as TGF-β receptor type I (*TGFBR1*), SMAD3 (*SMAD3*), SMAD4 (*SMAD4*), and one of the three TGF-β ligands, TGF-β2 (*TGFB2*), also predispose to thoracic aortic disease when altered.[35–39] Interestingly, aneurysm formation associated with these genes extends beyond the aorta and includes the aortic arterial branches, as well as intracranial arteries. Systemic features like translucent skin with delayed wound healing characteristic of Ehlers–Danos syndrome have also been identified in these patients. Mutations in *TGFBR1* and *TGFBR2* have been associated with Loeys-Dietz syndrome (LDS), a connective tissue disorder characterized by Marfan skeletal manifestations, craniofacial features, and aggressive and early onset of both thoracic aortic disease and aneurysms and dissections in other arteries.[40] Importantly, HTAD families with an absence of syndromic features have also been identified with genetic alterations in these genes encoding components of the canonical TGF-β signaling pathway. Thus, genes that predispose to HTAD with syndromic features, such as *TGFBR1* and *TGFBR2*, can also predispose to familial TAAD without syndromic features.[41,42]

Mutations in the components of the canonical TGF-β signaling pathway have been shown to or are predicted to decrease TGF-β signaling.[37,43,44] TGF-β is a cytokine which is sequestered in the ECM. The TGF-β signaling cascade is initiated when this cytokine binds to the TGF-β type II receptor, which recruits and phosphorylates the TGF-β type 1 receptor and subsequently leads to phosphorylation of SMAD2 and SMAD3 (mothers against decapentaplegic homolog 2 and 3, respectively). SMAD2 and SMAD3 complex with SMAD4, and this complex translocates to the nucleus to act as a transcription factor for TGF-β dependent genes, such as SMC contractile genes. Genetic mutations that predispose to HTAD in this pathway have been identified, including genes encoding one of the three TGF-β ligand family members, *TGFB2* (TGF-β2), the TGF-β type II and I receptors, *TGFBR2* and *TGFBR1*, and *SMAD3*. These mutations are predicted to decrease TGF-β signaling through haploinsufficiency or missense mutations that disrupt protein function.[37,43,45] Missense mutations in *TGFBR1* and *TGFBR2* are located in the intracellular kinase domain and lead to reduced kinase activity. Patients with mutations in *TGFB2*, a syndrome termed LDS4, share some phenotypic features observed in MFS and LDS and consist of 1% of HTAD patients.[37] *SMAD3* mutations, primarily in the MH2 domain, which mediates oligomerization of SMAD3 with SMAD4, result in reduced SMAD complex formation.[39] These mutations lead to aneurysm-osteoarthritis syndrome, an autosomal dominant condition with early-onset joint abnormalities and aortic aneurysms and dissections including the arterial tree.[38,39] With some syndromic features similar to LDS and MFS, aneurysm-osteoarthritis syndrome has been termed LDS type 3. While mutations in other genes such as *SMAD2*, *SMAD4*, and *TGFB3* have also been identified, additional data are needed to validate these genes for HTAD.

Although initial studies of a Marfan mouse model were found overactivation of TGF-β signaling, the mutations causing HTAD in the TGF-β signaling genes are predicted or have been shown to decrease TGF-β signaling.[37,43,44,46] SMCs treated with TGF-β increase expression of the SMC differentiation genes, including the contractile genes. Thus, loss-of-function mutations in the TGF-β signaling pathway may disrupt the formation of contractile units in SMCs, decreasing the ability of SMCs to contract to maintain tissue homeostasis and resulting in structural changes, pathogenic aortic remodeling, and TAAD.

Sporadic thoracic aortic disease

Patients without family history of aortic disease and who lack syndromic features such as Marfanoid features comprise approximately 80% of all thoracic aortic disease cases and are classified as patients with sporadic thoracic aortic disease (**STAD**). Genome-wide association studies (**GWAS**) identified *FBN1* as a common variant which confers 1.6–1.8-fold increased risk for STAD.[47] Duplications of 16p13.1, a region containing nine genes including *MYH11*, were also identified as a genetic alteration that increased the risk for TAAD.[29] However, compared to the 1.6-fold increased risk for TAAD associated with *FBN1* common variants, duplications of *MYH11* have a 12-fold risk for TAAD, warranting closer clinical management of these patients and early screening for aortic disease. These studies indicate that both rare and common variants in *FBN1* and *MYH11* are associated with thoracic aortic disease.

The genetics of abdominal aortic aneurysms

AAA is a common condition that shares risk factors with coronary heart disease and other occlusive atherosclerotic diseases. Twin studies indicate a strong genetic component in both predisposition and progression of AAA, estimating heritability at ~70%.[48] Population studies indicate that approximately 20% of AAA patients report having a first-degree relative with the disease compared to the 2%–10% of patients without AAA.[49–53] Positive family history doubles the risk of an individual having AAA.[54] The pattern of inheritance for AAA appears autosomal in most patients; however, there is evidence of recessive and dominant inheritance.[55] There are currently very few shared genetic risk factors between thoracic and abdominal aneurysmal phenotypes.[56,57]

A total of 10 genetic loci associated with AAA were identified in a meta-analysis of six GWAS from five countries (United States, United Kingdom, Netherlands, Iceland, and New Zealand) with validation analysis in a further eight independent cohorts.[58] The loci and their nearest gene(s) are 9p21 (*CDKN2B-AS1/ANRIL*),[57] 9q33 (*DAB2IP*),[59] 12q13 (*LRP1*),[60] 1q21.3 (*IL6R*),[61] 1p13.3 (*PSRC1/CELSR2/SORT1*),[62] and 19p13.2 (*LDLR*)[63] followed by the four loci identified in the meta-GWAS, 1q32.3 (*SMYD2*), 13q12.11 (*LINC00540*), 20q13.12 (*PLTP/PCIF1/MMP9/ZNF335*), and 21q22.2 (*ERG*).[58] Unlike single gene rare variants that confer a high risk for thoracic aortic disease, many of the common genetic variants that predispose to AAA are in intragenic regions and are not predicted to directly alter the functional element in a gene. However, analyses of gene associations using tools such as chromatin interaction databases have highlighted genes that regulate major pathways, such as metalloproteinase activity, tissue inflammation, and lipid accumulation, that are known biological processes involved in AAA. More recently, a smaller but still significant GWAS identified 14 new loci associated with AAA. Many of the loci identified are associated with genes involved in lipid metabolism (*LDLR, PCSK9, LPA, APOE, APOA5, TRIB1*).[64]

The extracellular matrix in genetics of aortic disease

ECM remodeling is a shared pathology in thoracic and abdominal aortic disease. Human genetic studies have demonstrated that the smooth muscle contractile apparatus, ECM, and TGF-β signaling pathway are critical in maintenance of aortic wall integrity. A subset of HTAD genes encode for proteins in the SMC elastin-contractile unit, the functional unit through which SMCs act as vascular mechanosensors of the aorta. Thus, these genes imply that impairment of the SMC elastin-contractile unit may be the major driver of aortic aneurysms and dissections. Disruption of mechanosensing may upregulate repair pathways in SMCs such as angiotensin II and TGF-β signaling, SMC apoptosis, and increased production of matrix metalloproteinase and proteoglycans, altering the ECM and leading to pathologic aortic remodeling (Fig. 29.2).[9] Thus, ECM homeostasis remains an important factor in aortic wall integrity and prevention of aortic aneurysms.

Acknowledgments

Figures were created with BioRender.com.

Sources of funding

This work was supported by the National Heart, Lung and Blood Institute (RO1 HL146583) to D.M. Milewicz. K. Kaw is supported by Training Interdisciplinary Pharmacology Scientists (TIPS) grant funded by the National Institutes of Health (grant T32GM120011) to the University of Texas Health Science Center at Houston, and A. Kaw by the National Center for Advancing Translational Sciences of the National Institutes of Health under Award Numbers TL1TR003169 and UL1TR003167.

Disclosures

None.

References

1. Huynh TT, Starr JE. Diseases of the thoracic aorta in women. *J Vasc Surg*. April 2013;57(4 suppl l). https://doi.org/10.1016/j.jvs.2012.08.126, 11s-7s.
2. Hoyert DL, Arias E, Smith BL, Murphy SL, Kochanek KD. Deaths: final data for 1999. *Natl Vital Stat Rep*. September 21, 2001;49(8):1–113.
3. Biddinger A, Rocklin M, Coselli J, Milewicz DM. Familial thoracic aortic dilatations and dissections: a case control study. *J Vasc Surg*. March 1997;25(3):506–511. https://doi.org/10.1016/s0741-5214(97)70261-1.
4. Prakash SK, Haden-Pinneri K, Milewicz DM. Susceptibility to acute thoracic aortic dissections in patients dying outside the hospital: an autopsy study. *Am Heart J*. September 2011;162(3):474–479. https://doi.org/10.1016/j.ahj.2011.06.020.
5. Finkbohner R, Johnston D, Crawford ES, Coselli J, Milewicz DM. Marfan syndrome. Long-term survival and complications after aortic aneurysm repair. *Circulation*. February 1, 1995;91(3):728–733. https://doi.org/10.1161/01.cir.91.3.728.
6. Silverman DI, Burton KJ, Gray J, et al. Life expectancy in the Marfan syndrome. *Am J Cardiol*. January 15, 1995;75(2):157–160. https://doi.org/10.1016/s0002-9149(00)80066-1.
7. Bou-Gharios G, Ponticos M, Rajkumar V, Abraham D. Extracellular matrix in vascular networks. *Cell Prolif*. 2004;37(3):207–220. https://doi.org/10.1111/j.1365-2184.2004.00306.x.
8. Davis EC. Smooth muscle cell to elastic lamina connections in developing mouse aorta. Role in aortic medial organization. *Lab Invest*. January 1993;68(1):89–99.
9. Pinard A, Jones GT, Milewicz DM. Genetics of thoracic and abdominal aortic diseases. *Circ Res*. February 15, 2019;124(4):588–606. https://doi.org/10.1161/circresaha.118.312436.

10. Renard M, Francis C, Ghosh R, et al. Clinical validity of genes for heritable thoracic aortic aneurysm and dissection. *J Am Coll Cardiol*. August 7, 2018;72(6):605–615. https://doi.org/10.1016/j.jacc.2018.04.089.
11. Pyeritz RE, McKusick VA. The Marfan syndrome: diagnosis and management. *N Engl J Med*. April 5, 1979;300(14):772–777. https://doi.org/10.1056/nejm197904053001406.
12. Corson GM, Chalberg SC, Dietz HC, Charbonneau NL, Sakai LY. Fibrillin binds calcium and is coded by cDNAs that reveal a multi-domain structure and alternatively spliced exons at the 5′ end. *Genomics*. August 1993;17(2):476–484. https://doi.org/10.1006/geno.1993.1350.
13. Pinard A, Salgado D, Desvignes JP, et al. WES/WGS reporting of mutations from cardiovascular "actionable" genes in clinical practice: a key role for UMD knowledgebases in the era of big databases. *Hum Mutat*. December 2016;37(12):1308–1317. https://doi.org/10.1002/humu.23119.
14. Hollister DW, Godfrey M, Sakai LY, Pyeritz RE. Immunohistologic abnormalities of the microfibrillar-fiber system in the Marfan syndrome. *N Engl J Med*. July 19, 1990;323(3):152–159. https://doi.org/10.1056/nejm199007193230303.
15. Milewicz DM, Pyeritz RE, Crawford ES, Byers PH. Marfan syndrome: defective synthesis, secretion, and extracellular matrix formation of fibrillin by cultured dermal fibroblasts. *J Clin Invest*. January 1992;89(1):79–86. https://doi.org/10.1172/jci115589.
16. Whiteman P, Hutchinson S, Handford PA. Fibrillin-1 misfolding and disease. *Antioxidants Redox Signal*. 2006;8(3–4):338–346. https://doi.org/10.1089/ars.2006.8.338.
17. Suk JY, Jensen S, McGettrick A, et al. Structural consequences of cysteine substitutions C1977Y and C1977R in calcium-binding epidermal growth factor-like domain 30 of human fibrillin-1. *J Biol Chem*. December 3, 2004;279(49):51258–51265. https://doi.org/10.1074/jbc.M408156200.
18. Dietz HC, Saraiva JM, Pyeritz RE, Cutting GR, Francomano CA. Clustering of fibrillin (FBN1) missense mutations in Marfan syndrome patients at cysteine residues in EGF-like domains. *Hum Mutat*. 1992;1(5):366–374. https://doi.org/10.1002/humu.1380010504.
19. Bunton TE, Biery NJ, Myers L, Gayraud B, Ramirez F, Dietz HC. Phenotypic alteration of vascular smooth muscle cells precedes elastolysis in a mouse model of Marfan syndrome. *Circ Res*. January 19, 2001;88(1):37–43. https://doi.org/10.1161/01.res.88.1.37.
20. Lee VS, Halabi CM, Hoffman EP, et al. Loss of function mutation in LOX causes thoracic aortic aneurysm and dissection in humans. *Proc Natl Acad Sci USA*. August 2, 2016;113(31):8759–8764. https://doi.org/10.1073/pnas.1601442113.
21. Guo DC, Regalado ES, Gong L, et al. LOX mutations predispose to thoracic aortic aneurysms and dissections. *Circ Res*. March 18, 2016; 118(6):928–934. https://doi.org/10.1161/circresaha.115.307130.
22. Milewicz DM, Østergaard JR, Ala-Kokko LM, et al. De novo ACTA2 mutation causes a novel syndrome of multisystemic smooth muscle dysfunction. *Am J Med Genet*. October 2010; 152a(10):2437–2443. https://doi.org/10.1002/ajmg.a.33657.
23. Guo DC, Papke CL, Tran-Fadulu V, et al. Mutations in smooth muscle alpha-actin (ACTA2) cause coronary artery disease, stroke, and Moyamoya disease, along with thoracic aortic disease. *Am J Hum Genet*. May 2009;84(5):617–627. https://doi.org/10.1016/j.ajhg.2009.04.007.
24. Guo DC, Regalado E, Casteel DE, et al. Recurrent gain-of-function mutation in PRKG1 causes thoracic aortic aneurysms and acute aortic dissections. *Am J Hum Genet*. August 8, 2013;93(2):398–404. https://doi.org/10.1016/j.ajhg.2013.06.019.
25. Guo DC, Pannu H, Tran-Fadulu V, et al. Mutations in smooth muscle alpha-actin (ACTA2) lead to thoracic aortic aneurysms and dissections. *Nat Genet*. December 2007;39(12):1488–1493. https://doi.org/10.1038/ng.2007.6.
26. Chen J, Kaw K, Lu H, et al. Resistance of Acta2(R149C/+) mice to aortic disease is associated with defective Release of mutant smooth muscle α-actin from the Chaperonin-containing TCP1 folding complex. *J Biol Chem*. September 30, 2021:101228. https://doi.org/10.1016/j.jbc.2021.101228.
27. Lu H, Fagnant PM, Bookwalter CS, Joel P, Trybus KM. Vascular disease-causing mutation R258C in ACTA2 disrupts actin dynamics and interaction with myosin. *Proc Natl Acad Sci U S A*. August 4, 2015;112(31):E4168–E4177. https://doi.org/10.1073/pnas.1507587112.
28. Zhu L, Vranckx R, Khau Van Kien P, et al. Mutations in myosin heavy chain 11 cause a syndrome associating thoracic aortic aneurysm/aortic dissection and patent ductus arteriosus. *Nat Genet*. March 2006;38(3):343–349. https://doi.org/10.1038/ng1721.
29. Kuang SQ, Guo DC, Prakash SK, et al. Recurrent chromosome 16p13.1 duplications are a risk factor for aortic dissections. *PLoS Genet*. June 2011;7(6):e1002118. https://doi.org/10.1371/journal.pgen.1002118.
30. Kwartler CS, Chen J, Thakur D, et al. Overexpression of smooth muscle myosin heavy chain leads to activation of the unfolded protein response and autophagic turnover of thick filament-associated proteins in vascular smooth muscle cells. *J Biol Chem*. May 16, 2014; 289(20):14075–14088. https://doi.org/10.1074/jbc.M113.499277.
31. Shalata A, Mahroom M, Milewicz DM, et al. Fatal thoracic aortic aneurysm and dissection in a large family with a novel MYLK gene mutation: delineation of the clinical phenotype. *Orphanet J Rare Dis*. March 15, 2018;13(1):41. https://doi.org/10.1186/s13023-018-0769-7.
32. Wallace SE, Regalado ES, Gong L, et al. MYLK pathogenic variants aortic disease presentation, pregnancy risk, and characterization of pathogenic missense variants. *Genet Med*. January 2019;21(1): 144–151. https://doi.org/10.1038/s41436-018-0038-0.
33. Herring BP, Dixon S, Gallagher PJ. Smooth muscle myosin light chain kinase expression in cardiac and skeletal muscle. *Am J Physiol Cell Physiol*. November 2000;279(5):C1656–C1664. https://doi.org/10.1152/ajpcell.2000.279.5.C1656.
34. Chen J, Peters A, Papke CL, et al. Loss of smooth muscle α-actin leads to NF-κB-Dependent increased sensitivity to angiotensin II in smooth muscle cells and aortic enlargement. *Circ Res*. June 9, 2017;120(12):1903–1915. https://doi.org/10.1161/circresaha.117.310563.
35. Mizuguchi T, Collod-Beroud G, Akiyama T, et al. Heterozygous TGFBR2 mutations in Marfan syndrome. *Nat Genet*. August 2004; 36(8):855–860. https://doi.org/10.1038/ng1392.
36. Loeys BL, Schwarze U, Holm T, et al. Aneurysm syndromes caused by mutations in the TGF-beta receptor. *N Engl J Med*. August 24, 2006;355(8):788–798. https://doi.org/10.1056/NEJMoa055695.
37. Boileau C, Guo DC, Hanna N, et al. TGFB2 mutations cause familial thoracic aortic aneurysms and dissections associated with mild systemic features of Marfan syndrome. *Nat Genet*. July 8, 2012; 44(8):916–921. https://doi.org/10.1038/ng.2348.
38. Regalado ES, Guo DC, Villamizar C, et al. Exome sequencing identifies SMAD3 mutations as a cause of familial thoracic aortic aneurysm and dissection with intracranial and other arterial aneurysms. *Circ Res*. September 2, 2011;109(6):680–686. https://doi.org/10.1161/circresaha.111.248161.
39. van de Laar IM, Oldenburg RA, Pals G, et al. Mutations in SMAD3 cause a syndromic form of aortic aneurysms and dissections with early-onset osteoarthritis. *Nat Genet*. February 2011;43(2):121–126. https://doi.org/10.1038/ng.744.
40. Loeys BL, Chen J, Neptune ER, et al. A syndrome of altered cardiovascular, craniofacial, neurocognitive and skeletal development caused by mutations in TGFBR1 or TGFBR2. *Nat Genet*. March 2005;37(3):275–281. https://doi.org/10.1038/ng1511.

41. Tran-Fadulu V, Pannu H, Kim DH, et al. Analysis of multigenerational families with thoracic aortic aneurysms and dissections due to TGFBR1 or TGFBR2 mutations. *J Med Genet*. September 2009; 46(9):607–613. https://doi.org/10.1136/jmg.2008.062844.
42. Pannu H, Fadulu VT, Chang J, et al. Mutations in transforming growth factor-beta receptor type II cause familial thoracic aortic aneurysms and dissections. *Circulation*. July 26, 2005;112(4):513–520. https://doi.org/10.1161/circulationaha.105.537340.
43. Inamoto S, Kwartler CS, Lafont AL, et al. TGFBR2 mutations alter smooth muscle cell phenotype and predispose to thoracic aortic aneurysms and dissections. *Cardiovasc Res*. December 1, 2010; 88(3):520–529. https://doi.org/10.1093/cvr/cvq230.
44. Lindsay ME, Schepers D, Bolar NA, et al. Loss-of-function mutations in TGFB2 cause a syndromic presentation of thoracic aortic aneurysm. *Nat Genet*. July 8, 2012;44(8):922–927. https://doi.org/10.1038/ng.2349.
45. Horbelt D, Guo G, Robinson PN, Knaus P. Quantitative analysis of TGFBR2 mutations in Marfan-syndrome-related disorders suggests a correlation between phenotypic severity and Smad signaling activity. *J Cell Sci*. December 15, 2010;123(Pt 24): 4340–4350. https://doi.org/10.1242/jcs.074773.
46. Habashi JP, Judge DP, Holm TM, et al. Losartan, an AT1 antagonist, prevents aortic aneurysm in a mouse model of Marfan syndrome. *Science*. April 7, 2006;312(5770):117–121. https://doi.org/10.1126/science.1124287.
47. LeMaire SA, McDonald ML, Guo DC, et al. Genome-wide association study identifies a susceptibility locus for thoracic aortic aneurysms and aortic dissections spanning FBN1 at 15q21.1. *Nat Genet*. September 11, 2011;43(10):996–1000. https://doi.org/10.1038/ng.934.
48. Wahlgren CM, Larsson E, Magnusson PK, Hultgren R, Swedenborg J. Genetic and environmental contributions to abdominal aortic aneurysm development in a twin population. *J Vasc Surg*. January 2010;51(1):3–7. https://doi.org/10.1016/j.jvs.2009.08.036. discussion 7.
49. Johansen K, Koepsell T. Familial tendency for abdominal aortic aneurysms. *JAMA*. October 10, 1986;256(14):1934–1936.
50. Jones GT, Hill BG, Curtis N, et al. Comparison of three targeted approaches to screening for abdominal aortic aneurysm based on cardiovascular risk. *Br J Surg*. August 2016;103(9):1139–1146. https://doi.org/10.1002/bjs.10224.
51. Kuivaniemi H, Shibamura H, Arthur C, et al. Familial abdominal aortic aneurysms: collection of 233 multiplex families. *J Vasc Surg*. February 2003;37(2):340–345. https://doi.org/10.1067/mva.2003.71.
52. Ogata T, MacKean GL, Cole CW, et al. The lifetime prevalence of abdominal aortic aneurysms among siblings of aneurysm patients is eightfold higher than among siblings of spouses: an analysis of 187 aneurysm families in Nova Scotia, Canada. *J Vasc Surg*. November 2005;42(5):891–897. https://doi.org/10.1016/j.jvs.2005.08.002.
53. Rossaak JI, Hill TM, Jones GT, Phillips LV, Harris EL, van Rij AM. Familial abdominal aortic aneurysms in the Otago region of New Zealand. *Cardiovasc Surg*. June 2001;9(3):241–248. https://doi.org/10.1016/s0967-2109(00)00140-x.
54. Larsson E, Granath F, Swedenborg J, Hultgren R. A population-based case-control study of the familial risk of abdominal aortic aneurysm. *J Vasc Surg*. January 2009;49(1):47–50. https://doi.org/10.1016/j.jvs.2008.08.012. discussion 51.
55. Sandford RM, Bown MJ, London NJ, Sayers RD. The genetic basis of abdominal aortic aneurysms: a review. *Eur J Vasc Endovasc Surg*. April 2007;33(4):381–390. https://doi.org/10.1016/j.ejvs.2006.10.025.
56. van 't Hof FN, Ruigrok YM, Lee CH, et al. Shared genetic risk factors of intracranial, abdominal, and thoracic aneurysms. *J Am Heart Assoc*. July 14, 2016;(7):5. https://doi.org/10.1161/jaha.115.002603.
57. Helgadottir A, Thorleifsson G, Magnusson KP, et al. The same sequence variant on 9p21 associates with myocardial infarction, abdominal aortic aneurysm and intracranial aneurysm. *Nat Genet*. February 2008;40(2):217–224. https://doi.org/10.1038/ng.72.
58. Jones GT, Tromp G, Kuivaniemi H, et al. Meta-analysis of genome-wide association studies for abdominal aortic aneurysm identifies four new disease-specific risk loci. *Circ Res*. January 20, 2017;120(2): 341–353. https://doi.org/10.1161/circresaha.116.308765.
59. Gretarsdottir S, Baas AF, Thorleifsson G, et al. Genome-wide association study identifies a sequence variant within the DAB2IP gene conferring susceptibility to abdominal aortic aneurysm. *Nat Genet*. August 2010;42(8):692–697. https://doi.org/10.1038/ng.622.
60. Bown MJ, Jones GT, Harrison SC, et al. Abdominal aortic aneurysm is associated with a variant in low-density lipoprotein receptor-related protein 1. *Am J Hum Genet*. November 11, 2011;89(5): 619–627. https://doi.org/10.1016/j.ajhg.2011.10.002.
61. Harrison SC, Smith AJ, Jones GT, et al. Interleukin-6 receptor pathways in abdominal aortic aneurysm. *Eur Heart J*. December 2013; 34(48):3707–3716. https://doi.org/10.1093/eurheartj/ehs354.
62. Jones GT, Bown MJ, Gretarsdottir S, et al. A sequence variant associated with sortilin-1 (SORT1) on 1p13.3 is independently associated with abdominal aortic aneurysm. *Hum Mol Genet*. July 15, 2013;22(14):2941–2947. https://doi.org/10.1093/hmg/ddt141.
63. Bradley DT, Hughes AE, Badger SA, et al. A variant in LDLR is associated with abdominal aortic aneurysm. *Circ Cardiovasc Genet*. October 2013;6(5):498–504. https://doi.org/10.1161/circgenetics.113.000165.
64. Klarin D, Verma SS, Judy R, et al. Genetic architecture of abdominal aortic aneurysm in the million veteran program. *Circulation*. October 27, 2020;142(17):1633–1646. https://doi.org/10.1161/circulationaha.120.047544.

CHAPTER

30

Peripheral arterial disease (pathophysiology, presentation, prevention/management)

Michael S. Conte and Alexander S. Kim

Department of Surgery, Division of Vascular and Endovascular Surgery, University of California San Francisco, San Francisco, CA, United States

Introduction

Peripheral arterial disease (PAD) is a manifestation of systemic atherosclerotic occlusive disease leading to impaired blood flow to the lower extremities. PAD can be symptomatic, with clinical presentations ranging from intermittent pain with impaired physical activity to ischemic pain at rest and tissue loss that threaten the viability of the affected limb. The majority of individuals with PAD are asymptomatic, with disease detected through physical exam and noninvasive studies such as the ankle–brachial index (ABI). PAD is estimated to affect upwards of 200 million people worldwide, and more than 8.5 million adult Americans.[1] The vast and ever-increasing prevalence of PAD can be attributed to the aging of the global population, compounded by the significant uptrend of key vascular risk factors, including diabetes mellitus (DM), smoking, hypertension, dyslipidemia, and obesity. Recent data from the American Heart Association's 2021 statistical update suggest worrisome trends of health factors and behaviors in the United States that portend a growing PAD burden for the decades ahead.[2]

PAD: epidemiology and risk factors

Demographic, socioeconomic, and environmental factors

Age is the strongest predictor for PAD. PAD is scarce in individuals under 50 years of age, and rises in prevalence with each decade, affecting greater than 20% of individuals over 80 years old[3–7] (Fig. 30.1). Globally, PAD affects men and women at approximately equal rates in high-income countries. In low- and middle-income countries, rates broken down by decile appear to be higher in women.[8] In the United States, African American and Native American populations demonstrate higher prevalence of PAD across age categories (Fig. 30.2).[2] Lower levels of income and education are associated with PAD prevalence and severity across race and ethnicity.[9–11] Multivariable analysis controlling for other risk factors, including age, sex, and other predictors discussed below such as smoking, reduces the strength but does not eliminate this association. This correlation may be attributable to social determinants such as chronic exposure to mental and psychosocial stress, reduced health literacy, and access to healthy food and housing security. Recently data suggest that exposures to pollutants such as motor vehicle exhaust, power plant fumes, and industrial processing waste may be associated with increased cardiovascular related morbidity and mortality.[12–14]

Smoking

Cigarette smoking is a strong independent factor which is associated with double, and in some studies quadruple, the risk of PAD.[3,6,15] Cessation of smoking is a critical strategy in reducing the risk of PAD and other atherosclerotic diseases such as coronary heart disease and stroke; even 1 year of cessation is associated with 4% reduced risk of PAD. However, the impacts

The Vasculome
https://doi.org/10.1016/B978-0-12-822546-2.00031-9

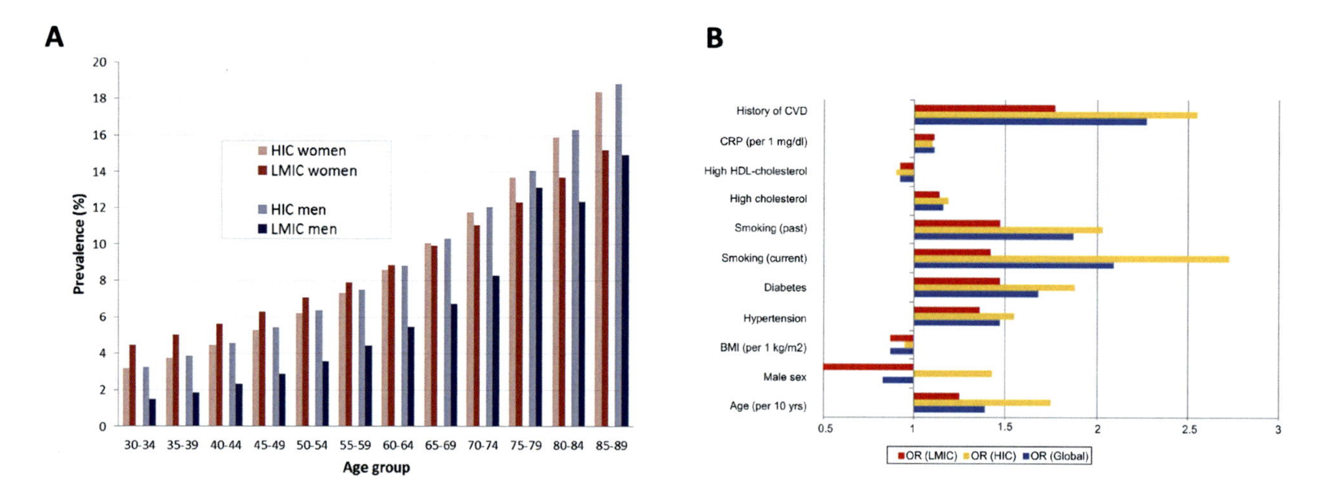

FIGURE 30.1 (A) Global prevalence of PAD stratified by age and separated by high income versus low-middle income countries, and gender. *Reprinted with permission from Conte MS, Bradbury AW, Kolh P, et al. Global vascular guidelines on the management of chronic limb-threatening ischemia. J Vasc Surg. 2019;69(6). https://doi.org/10.1016/j.jvs.2019.02.016, 3S. e40–125S.e40. Adapted from data in Fowkes FGR, Rudan D, Rudan I, et al. Comparison of global estimates of prevalence and risk factors for peripheral artery disease in 2000 and 2010: a systematic review and analysis. Lancet. 2013; 382(9901): 1329–1340. https://doi.org/10.1016/S0140-6736(13) 61249-0.* (B) Odds ratio of risk of PAD by comorbidities, separated by low-middle income versus high-income versus global population. *Adapted from data in Criqui MH, Denenberg JO, Langer RD, Fronek A. The epidemiology of peripheral arterial disease: importance of identifying the population at risk. Vasc Med. 1997;2(3):221–226. https://doi.org/10.1177/1358863X9700200310.*

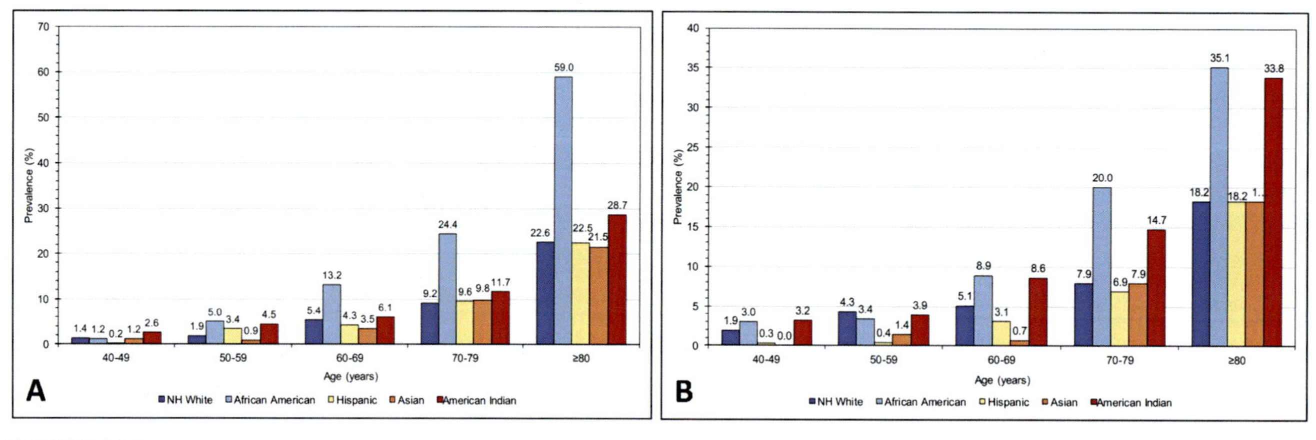

FIGURE 30.2 (A) Estimates of prevalence of peripheral artery disease in males by age and ethnicity, United States, 2000. (B) Estimates of prevalence of peripheral artery disease in females by age and ethnicity, United States, 2000. *Reprinted with permission from Virani SS, Alonso A, Aparicio HJ, et al. Heart disease and stroke statisticsd2021 update: a report from the American Heart Association. Circulation. 2021;143(8). https://doi.org/10.1161/CIR.0000000000000950. Adapted from data in Allison MA, Ho E, Denenberg JO, et al. Ethnic-specific prevalence of peripheral arterial disease in the United States. Am J Prev Med. 2007;32(4):328–333. https://doi.org/10.1016/j.amepre.2006.12.010.*

of smoking are long lasting, as even those who quit smoking for 20–30 years have 1.7x risk of PAD compared to never-smokers.[16]

Diabetes

DM raises the risk for PAD with a similar magnitude as smoking, and this risk increases with age and duration of DM.[3] In those with an established PAD diagnosis, the presence of DM portends a fivefold increased rate of amputation and reduced long-term survival.[5] The presence of DM also negatively impacts the outcomes of lower extremity revascularization, as it is associated with increased hospitalization, major adverse limb events (MALEs), and major amputation.[17] Recent trends in the prevalence of diabetes and prediabetes in the United States demonstrated continued growth with approximately 13% of the adult population currently affected.[18] Moreover, the rate of diabetes related nontraumatic lower extremity amputations in the United States increased by 50% between 2009 and 2015 after a prior decade of decline.[19]

When comparing type 1 (insulin deficient) and type 2 (insulin resistant) DM, patients with type 1 are more likely to present at a younger age with advanced PAD and complicated lower extremity wounds, and more likely to require major amputation.[20] The combination of smoking and DM is strongly associated with significant PAD-related morbidity and mortality. In patients with DM, glucose-lowering therapy is an important component of managing PAD and reducing the risk of major adverse cardiovascular events (MACE) and MALEs. However, careful consideration should be taken in selecting the optimal glucose lowering agent, as a recent study of canagliflozin, a sodium–glucose transporter 2 inhibitor was associated with a twofold risk of amputation in patients with PAD.[21] Maintaining normoglycemia and hemoglobin A1c (HgbA1c) levels less than 8% is recommended management strategies for PAD.

Hypertension, dyslipidemia, metabolic syndrome

Hypertension is highly prevalent and associated with PAD (less so than smoking and diabetes), but due to its association with age, this association appears weaker on more vigorous statistical analyses. Derangements in lipid profile have also been associated with PAD. An increased total cholesterol to high-density lipoprotein cholesterol (HDL-C) ratio is a predictor for PAD.[22,23] Elevated triglyceride levels and low-density lipoprotein cholesterol (LDL-C) have also been linked to PAD, but their role in PAD requires further investigation.[24,25] Obesity in relation to a constellation of other physiologic derangements (hypertension, elevated triglycerides, low HDL, elevated fasting blood glucose), also referred to as "metabolic syndrome," appears to be of more significant in relation to PAD compared to obesity as an independent factor.[26,27]

Inflammation

A common characteristic of many of the above risk factors for PAD is a state of elevated systemic inflammation. Markers of inflammation, such as circulating levels of C-reactive protein (CRP) and fibrinogen, have been strongly associated with PAD.[3,23] They have also been shown to correlate with PAD severity and increased morbidity in patients with symptomatic disease, i.e., those with chronic limb-threatening ischemia (CLTI) have both the worst burden of disease and the most elevated levels of circulating proinflammatory biomarkers including CRP (CRP).[23,28(p6),29(p6)] Previous studies have found that the expressions of inflammatory cytokines such as interleukin-6 and adhesion molecules (e.g., soluble VCAM-1[28]) are also elevated in this population, but their use as clinical indicators of disease severity and/or prognosis (as well as other candidate biomarkers such as D-dimer, B2-macroglobulin, asymmetric dimethylarginine, and cystatin C) remains to be elucidated.[29–32]

Diagnosis and initial evaluation

Evaluation of a patient for suspected PAD begins with a thorough history and physical examination. As mentioned above, lower extremity symptoms range from intermittent claudication (IC) with exertion to severe ischemic ulceration or gangrene. The physical examination should be utilized to determine the location of disease. This includes careful examination of wounds, auscultation for bruits, and palpation of pulses.

Physiologic assessment and imaging

If PAD is suspected, the next step in diagnosis includes noninvasive physiologic assessments which include the ABI. This index is measured separately for each limb as the ratio of the highest systolic ankle pressure (dorsalis pedis or posterior tibial arteries) to the higher of the two brachial artery pressures.[33] By convention, the lower of the two leg ABI values is used to categorize the individual. In healthy young adults, a normal ABI ranges from 1.1 to 1.3. If either leg has an index of <0.9, this is considered PAD. If a score is on the lower end of normal, and PAD is suspected, a postexercise ABI decrease of >20% indicates the presence of PAD. The toe–brachial index is a variation of the ABI which utilizes highest systolic pressure in the hallux rather than the ankle arteries, and clinical studies have indicated that this study may be more accurate and predictive of outcomes in patients with CLTI.[34] These tests can be unreliable in the presence of dense calcifications which make the vessels noncompressible or in the setting of prior amputation or significant wounds/tissue loss. In this case, alternate studies which measure tissue perfusion, such as transcutaneous oximetry or segmental perfusion pressure, may be more suitable.[35]

Following diagnosis with the above assessments, it may be appropriate to define the anatomic pattern of disease in patients who are deemed as potential candidates for revascularization. The most common noninvasive imaging modalities for PAD include duplex ultrasonography (DUS), computed tomography angiography, and magnetic resonance angiography.[36,37] Catheter-based digital subtraction angiography remains the gold standard for resolution, and it also offers the opportunity for simultaneous therapeutic interventions. A brief summary of the advantages and limitations is summarized in Table 30.1.

Spectrum of disease

PAD comprises a broad spectrum of clinical presentations. Several classification systems have been developed and utilized to stratify disease severity to help direct evaluation and management (Table 30.2). For example, the Fontaine system is a clinical staging system which categorizes disease severity into four stages ranging from asymptomatic to tissue loss. The Rutherford system further divides IC into three stages of severity, and CLTI into three stages (ischemic rest pain, minor tissue loss, and major tissue loss). However, the Rutherford system is limited in discriminating limb risk across the CLTI spectrum, particularly in the setting of diabetes and neuropathy. The more recent Society for Vascular Surgery Threatened Limb Classification System [wound, ischemia, foot infection; WIfI] system homes in on the key components that define CLTI to help stratify amputation risk and direct management strategies in this medically complex population.[38] For the purposes of this review, we will stratify PAD presentation into its rudimentary classifications: asymptomatic disease, IC, CLTI, and acute limb ischemia (ALI).

Clinical presentations, staging, and natural history

Asymptomatic disease

Most patients with PAD diagnosed by ABI (ankle–brachial index) are asymptomatic at the time of presentation.[39] This is likely due to milder disease, adequacy of collateral circulation, sedentary lifestyle, or limited mobility from other factors. Despite the lack of symptoms, PAD still carries an association with increased cardiovascular morbidity and mortality[40–42] (Fig. 30.3). Although the natural course of progression from asymptomatic PAD to IC is not well described, a metaanalysis estimates that 7% of patients with asymptomatic disease advance to IC within five years.[41] Importantly, individuals with asymptomatic PAD have significant functional limitations as measured by objective testing (e.g.,

TABLE 30.1 Summary of imaging and physiologic assessment tools.

Clinical need	Assessment tool	Advantages	Limitations
Presence/severity of limb ischemia	Ankle pressure and ankle–brachial index (ABI)	Noninvasive, inexpensive	Evaluates perfusion only to ankle level, readings falsely elevated with arterial calcification
	Toe pressure and toe–brachial index (TBI)	Noninvasive measure of perfusion at the toe level; improved TBI following intervention associated with healing	Limited use in setting of toe wound, prior forefoot amputation, toe vessel calcification
	Transcutaneous oximetry (TcPO2)	Measures local skin oxygen levels; higher TcPO2 associated with improved healing	Quality depends on operator experience, skin perfusion may vary with temperature
	Segmental perfusion pressure (SPP)	Evaluates capillary opening pressure, low scores predict failure of wound healing	Quality depends on operator experience, skin perfusion may vary with temperature
Anatomic imaging for revascularization planning	Duplex ultrasonography (DUS)	Noninvasive, inexpensive	Quality depends on operator experience, limited in evaluation of proximal vasculature, calcifications
	Computed tomography angiography (CTA)	High resolution cross-sectional imaging of arterial anatomy	Limited by calcifications, reduced resolution in distal pedal vessels, requires injection of contrast
	Magnetic resonance angiography (MRA)	High resolution cross-sectional imaging of arterial anatomy	Unable to use if patient has incompatible medical devices, may require sedation, overestimates stenosis, liable to venous contamination
	Catheter-based digital subtraction angiography (DSA)	Highest resolution, may combine with pressure gradient measurement, offers opportunity for simultaneous treatment	Invasive, requires contrast, access site complications

TABLE 30.2 Comparison and descriptions of Fontaine and Rutherford clinical staging systems used to describe the various presentations of PAD.

Fontaine grade	Rutherford category	Clinical description
I	0	Asymptomatic
IIa	1	Mild claudication—able to ambulate without stopping to recover
IIb	2	Moderate claudication—between 1 and 3
	3	Severe claudication—must stop ambulation due to severe pain
III	4	Pain at rest
IV	5	Minor tissue loss—nonhealing ulcer, focal areas of gangrene limited to digits or distal aspect of foot
	6	Major tissue loss—extensive ulcers and gangrene which preclude possibility of functional foot salvage

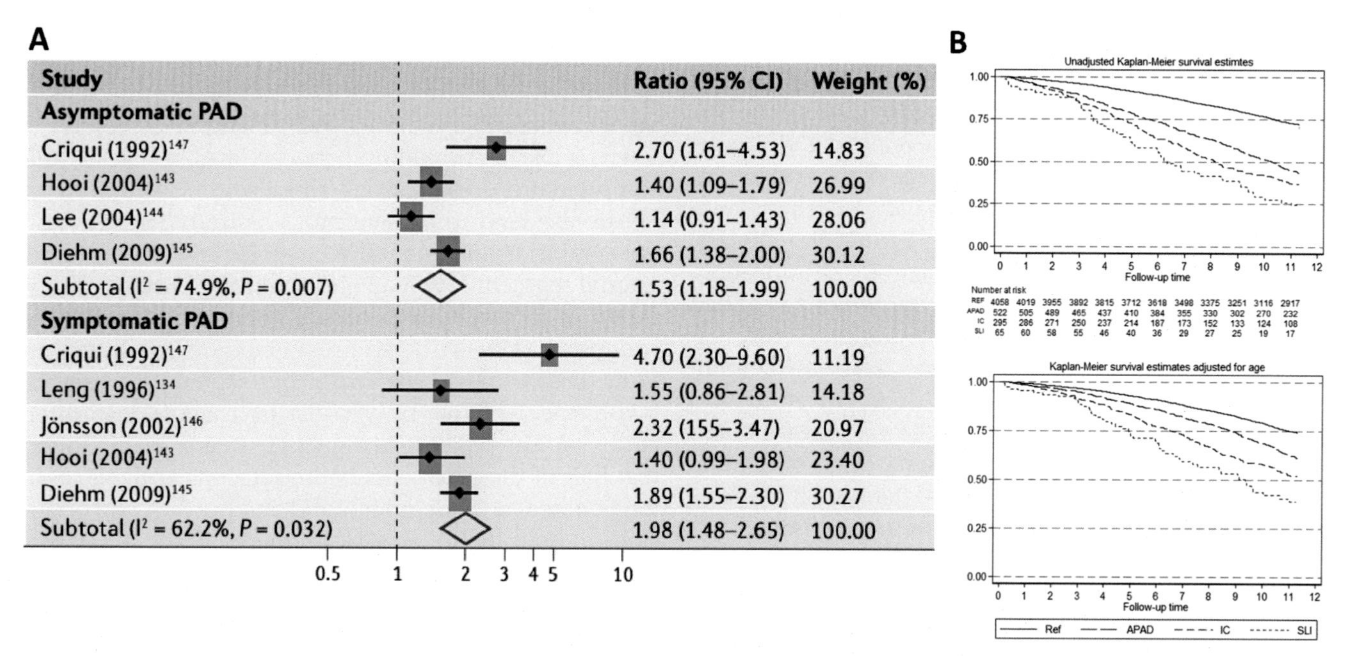

FIGURE 30.3 (A) Summary of studies analyzing mortality odds ratios separated by asymptomatic and symptomatic PAD. Asymptomatic PAD was defined as ankle–brachial index <0.9 without clinical symptoms. Symptomatic PAD was defined as intermittent claudication or critical limb ischemia. *Reprinted with permission from Sigvant B, Lundin F, Wahlberg E. The risk of disease progression in peripheral arterial disease is higher than expected: a metaanalysis of mortality and disease progression in peripheral arterial disease. Eur J Vasc Endovasc Surg. 2016;51(3):395–403. https://doi.org/10.1016/j.ejvs.2015.10.022. Updated from Fowkes FGR, Aboyans V, Fowkes FJI, McDermott MM, Sampson UKA, Criqui MH. Peripheral artery disease: epidemiology and global perspectives. Nat Rev Cardiol. 2017;14(3): 156-170. https://doi.org/10.1038/nrcardio.2016.179.* (B) Unadjusted and age-adjusted Kaplan–Meier survival curves separately for varying severity of PAD. *APAD*, asymptomatic; *IC*, intermittent claudication; *SLI*, severe limb ischemia. *Reprinted with permission from Sartipy F, Sigvant B, Lundin F, Wahlberg E. Ten year mortality in different peripheral arterial disease stages: a population based observational study on outcome. Eur J Vasc Endovasc Surg. 2018; 55(4):529–536. https://doi.org/10.1016/j.ejvs.2018.01.019.*

6-minute walk), and demonstrate progressive functional decline in their walking ability over time.[43]

Intermittent claudication

IC is the most common symptom of PAD in which leg pain is reproduced by exertion and relieved with rest. Although IC can affect all muscles throughout the leg, calf pain accounts for the majority of presentations. Approximately 20% of patients presenting with IC will suffer a major cardiovascular event within 5 years, resulting in 10%–15% mortality.[1] Recently, a systematic review demonstrated that approximately 21% of patients with IC experienced limb-related complications within 5 years, with most deterioration occurring within the first year of diagnosis.[41] The risk of IC advancing to CLTI is most closely associated with diabetes and smoking status. Risk of major amputation, however, is low overall among individuals with IC (<1%/year); thus education, optimization of risk factors, evidence-based medical therapies, and exercise should be the focus of treatment in these patients.[37,44]

Chronic Limb-Threatening Ischemia

CLTI is an advanced stage of PAD characterized by the presence of either rest pain or tissue loss (gangrene or nonhealing wound) for at least 2 weeks. This state implies inadequate perfusion to maintain minimal tissue survival requirements. In the Fontaine and Rutherford classification schemes, rest pain and tissue loss are stratified into different stages, and minor (Stage 5) and major (Stage 6) tissue loss are further separated in the Rutherford classification. However, in practice, CLTI includes a broad spectrum of presentations ranging from a non-salvageable foot to tissue loss of varying severity that may or may not include infection. In order to address this wide variety of presentations and better define prognosis and indications for revascularization, the Society for Vascular Surgery (SVS) developed the Lower Extremity Threatened Limb Classification System to stratify limb risk by grading Wound, Ischemia, and foot Infection severity ("WIfI").[38] Multiple studies have now validated this staging system for stratification of risk of amputation or likelihood of wound healing. WIfI stage 1 is associated with minimal risk of amputation, while stages 3–4 are associated with increased risk of limb loss and frequent need for revascularization.[45–48] Like other clinical disease staging systems, WIfI staging should be repeated over time to gage responses to treatment and determine if alternative strategies or repeat interventions are warranted.[49]

Acute limb ischemia and major adverse limb events

ALI is an abrupt reduction in blood flow to the limb with a duration of less than 2 weeks.[3,9] ALI can occur due to embolization from a proximal source (e.g., cardiac), trauma, dissection, and aneurysmal disease. However, it can also occur due to thrombosis of preexisting lesions or occlusion of stents or bypass grafts in the setting of chronic PAD. ALI is associated with 10%–15% rate of major amputation and 15%–40% 1-year mortality.[3,9] Recent studies have highlighted important risk factors for ALI among PAD patients including lower ABI, diabetes, and prior limb revascularization.[50,51]

MALE is a composite of adverse outcomes including ALI, major amputations, or urgent revascularization commonly utilized as an outcome measure in research related to PAD. Those with a history of PAD and prior revascularization or amputation are at significantly increased risk of MALE compared to those with PAD who have no history of revascularization or amputation.[52] The risk of MALE is also significantly greater among patients with polyvascular disease (disease in multiple vascular beds, e.g., coronary artery, cerebrovascular disease). Although treatment with standard guideline-recommended therapies for PAD (e.g., antiplatelet therapy) reduces the risk of MALE in these patients, risk remains elevated compared to those without polyvascular disease, which highlights the importance of intensification of therapeutic regimens selectively in these patients.[53,54]

Disease patterns and risk factor correlations

PAD encompasses a broad range and severity of anatomic lesions spanning a topography from the infrarenal aorta to the pedal vessels. The anatomic distribution of disease is classically described by the level of disease, including aortoiliac (AI), femoropopliteal (FP), and tibiopedal (TP). Common patterns of disease are associated with different risk factors. AI and FP disease is strongly associated with smoking, while diabetes is more commonly associated with FP and TP disease.[55–58] There is also a correlation between level of disease and symptomatic presentation. Those with isolated AI or FP disease tend to present more typically with IC, and the symptoms are generally one joint level distal to the arterial segment involved (i.e., thigh and buttock claudication in AI disease, calf claudication in FP disease). In contrast, CLTI usually requires hemodynamically significant disease in at least two major segments, or severe involvement of the TP vessels. In a single institution study of 450 patients with CLTI, Rueda et al. found that half of these patients had occlusions at the level of the popliteal or TP segments and another 30% had occlusions of both the FP and TP segments.[59] These patterns of disease were especially common in those with diabetes and/or end-stage renal disease.

Pathophysiology and pathobiology

Endothelial dysfunction

The arterial wall is a highly complex and organized structure designed to efficiently carry oxygenated blood to end organs. In healthy arteries, the endothelial lining acts as a protective barrier between quiescent vascular architecture and circulating blood elements, helping to maintain laminar flow. The endothelium is metabolically active, elaborating vasculo-protective factors such as nitric oxide (NO) which enable physiologic responses such as reactive hyperemia. When this layer is disrupted or activated, a thrombo-inflammatory and vasoconstrictive cascade occurs as the physiologic response to defend against hemorrhage or invading pathogens. Platelets and other circulating factors interact with the

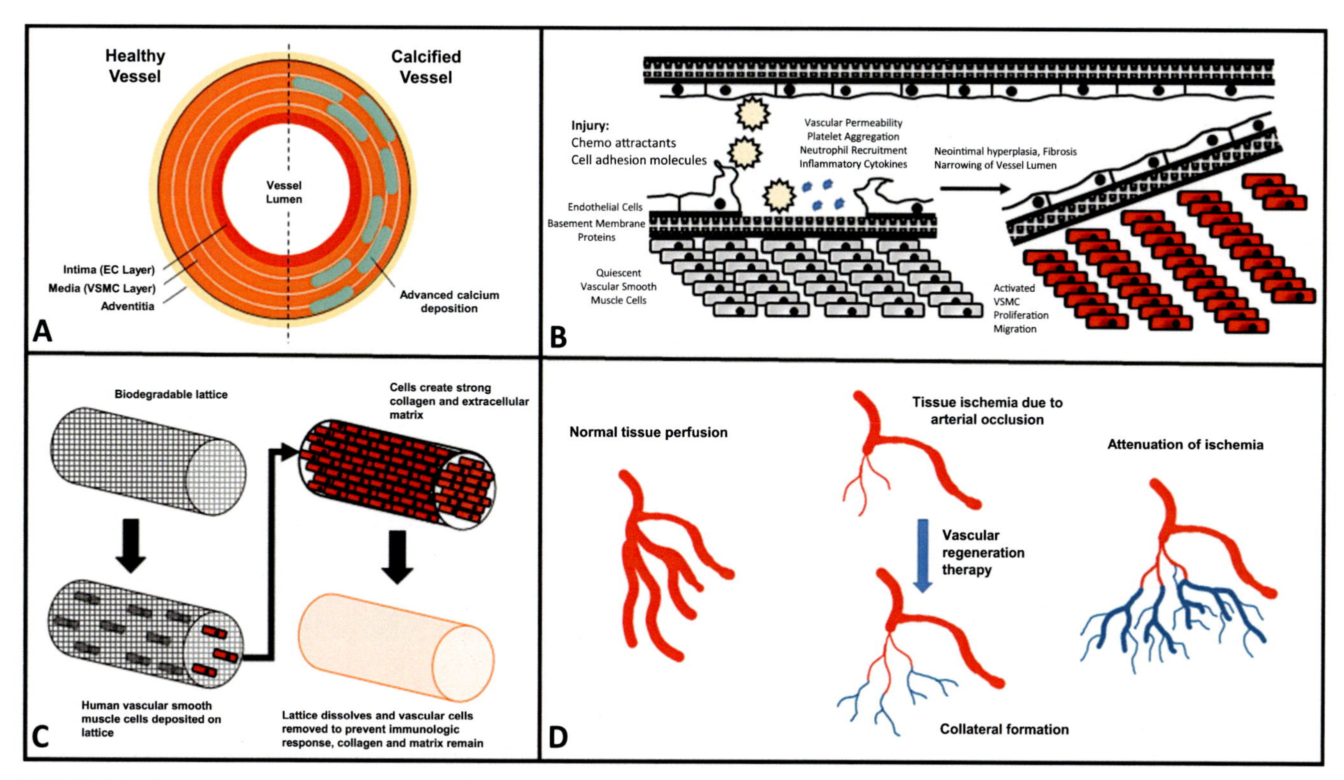

FIGURE 30.4 Major biological hurdles and developing therapies. (A) Vascular calcifications. (B) Vascular injury response and remodeling. (C) Bioengineered vascular conduits. (D) Vascular regeneration.

endothelium, which then expresses proinflammatory cytokines and adhesion molecules, leading to increased EC–leukocyte interactions. In patients with PAD, this response may be prolonged leading to chronic vascular inflammation, vascular smooth muscle activation, lesion progression, and thromboembolism (Fig. 30.4).

Vascular smooth muscle activation

Vascular Smooth Muscle Cells (VSMCs) make up the bulk of the vessel wall and play a critical role in vascular health and disease. Normally, they are quiescent, contributing to structural integrity and maintaining vascular tone. When activated due to inflammation or injury, they contribute to lesion progression and vessel wall remodeling. Inflammatory cytokines and growth factors activate VSMCs, inducing a dedifferentiated state characterized by increased migration, proliferation, and further secretion of proinflammatory signaling molecules. In PAD, this process leads to intimal hyperplasia of the vessel wall following vascular interventions such as angioplasty, stenting, or bypass surgery.

Vascular injury response, restenosis

Following acute mechanical injury to the artery, loss of the protective endothelial barrier, interaction with platelets and other circulating thrombo-inflammatory elements, and resultant VSMC activation lead to neointimal hyperplasia and treatment failure following vascular intervention. VSMCs migrate toward the lumen, proliferate, and assume a synthetic phenotype. Therefore, reendothelialization, platelet activation, inflammation, VSMC migration, and proliferation are considered potential therapeutic targets for the prevention of restenosis.

Restenosis of treated arteries or bypass grafts leads to recurrent symptoms, adverse clinical events, and the need for repeat procedures. This remains a major obstacle in the arena of PAD intervention. Repeat interventions tend to be more complex and result in inferior durability and outcomes compared to primary procedures.[60] Therefore, improving the primary patency of vascular interventions is paramount to improve long-term outcomes. Factors associated with higher risk for reinterventions include female gender, continued smoking, dyslipidemia, diabetes, and elevated inflammatory markers. Risk factors for restenosis also differ by the type of intervention. For bypass surgery, the quality of the vascular conduit is the most critical factor. Diabetes, anatomy, and lesion complexity (e.g., length, occlusive nature, and calcification) appear to impact endovascular therapies more adversely.

The current paradigm of pharmacologic treatment for preventing restenosis primarily targets VSMC proliferation and inflammation. In the setting of PAD, paclitaxel is the primary compound included in drug eluting devices. Paclitaxel inhibits microtubule polymerization thereby blocking mitosis and proliferation (G2/M inhibitor). Sirolimus, and other -limus agents (everolimus, zotarolimus), are immunosuppressant and antiproliferative (G1/S inhibition), but are currently only used in coronary drug eluting devices. These compounds successfully reduce mid-term rates of restenosis, but they are nonspecific and cytotoxic. They inhibit healing of the endothelium and may thus perpetuate a thrombogenic environment which necessitates the use of antiplatelet therapy. Another significant challenge for these compounds is optimizing their pharmacokinetics and bioavailability for PAD. Oftentimes they are incorporated into synthetic vehicles coated onto stents or balloons, so they may be lost into the systemic circulation during treatment and lead to adverse off-target effects. The search for the next generation of antirestenosis treatment is a tall order; investigators should consider a therapeutic profile which not only inhibits VSMC activation, but is also antithrombotic, noncytotoxic, and more conducive to vessel healing and resolving inflammation. Multiple candidate compound classes have been investigated. Recent studies have characterized specialized proresolving lipid mediators which attenuate inflammation and reduce neointimal hyperplasia without deleterious effects (immune suppression, toxicity) in cellular and animal models of vascular injury and inflammation.[61] Further studies of these and other novel compounds are required to develop new alternatives for human clinical applications.

Thrombosis and thromboembolism

Contrary to traditional paradigms which assumed the pathogenesis of atherosclerosis was similar in all vascular arterial beds, recent literature suggest that thrombosis and embolization contribute more significantly to PAD compared to coronary and cerebrovascular disease.[62] Studies of below-knee arteries obtained from amputation specimens showed a greater burden of thrombus compared to atherosclerotic lesions examined in coronary arteries.[63] At the genomic level, a factor V variant of a novel loci associated with atherosclerosis was found to be specific for PAD.[64] This variant was not associated with other atherosclerotic diseases such as coronary artery or cerebrovascular disease. Further analysis of this variant in patients with PAD demonstrated that the variant was also associated with greater burden of disease.

Antithrombotic therapeutics have also been associated with improved limb outcomes that are not seen with antiplatelet medications alone. Vorapaxar, a PAR-1 inhibitor, was associated with reduced healthcare utilization for ALI/urgent vascular interventions/major amputations.[65] Low-dose rivaroxaban, a direct factor Xa inhibitor, was shown to reduce MALEs (severe limb ischemia requiring intervention, including amputation) in patients with symptomatic PAD in the COMPASS trial and was also beneficial in patients with symptomatic PAD undergoing both endovascular and surgical interventions in the VOYAGER trial.[52,66] On the other hand, warfarin treatment did not reduce MALE and was associated with increased rates of intracranial hemorrhage, and other attempts at factor Xa inhibition at higher doses have not proven beneficial.[67–70]

Diabetes and hyperglycemia

The major pathophysiologic components of diabetic vasculopathy include hyperglycemia, dyslipidemia, and insulin resistance. These disturbances exacerbate PAD by contributing to vascular inflammation and interfering with endothelial cell (EC), VSMC, and circulating blood cell function.[71] Progression and severity of PAD correlate with worse glycemic control and duration of DM.[55]

DM is associated with elevated systemic proinflammatory cytokines, including TNF-alpha and IL-6.[72] These signaling molecules activate NF-kB in the endothelium, which results in a proinflammatory phenotype through the upregulation of adhesion molecules and subsequent EC–leukocyte and EC–platelet interactions.

Hyperglycemia, insulin resistance, and dyslipidemia impair NO production through various pathways, including the inhibition of endothelial nitric oxide synthase, protein kinase C activation, and inhibition of phosphatidylinositol (PI)-3 kinase.[73,74] Dysfunctional NO regulation impairs EC function and VSMC relaxation.

Chronic uncontrolled DM and resulting hyperglycemia is associated with the production of advance glycation end products (AGEs), mediated through the Maillard reaction which binds reducing sugars to free amino groups. AGEs impair cellular function and interact with receptors which enhance the expression of proinflammatory factors.[75,76] These interactions also lead to increased leukocyte recruitment and transformation into foam cells, a major contributor to the formation of atheromatous plaques.[77] In VSMC, activation of NFkB, inhibition of PI-3 kinase, and activation of AGE receptors result in an atherogenic phenotype and impair collagen synthesis.[77,78] As a result, DM both promotes atherogenesis, but also destabilizes plaques, increasing the risk for rupture and thromboembolism.

Calcification

Calcification represents a major hurdle of treating advanced stages of PAD. Calcification can occur in the medial or intimal layers of the vessel. Although the pathophysiology of calcification remains incompletely defined, medial and intimal calcification appear to have different patterns of etiology and follow a different natural history. Diabetes is a significant risk factor for medial artery calcification, and this risk increases with longer duration of disease. Another significant risk factor for medial artery calcification is chronic kidney disease. The combination of the two nearly doubles the risk.[79] Several studies have demonstrated an association between the presence of medial calcification and adverse outcomes in patients with PAD. In patients with CLTI, the presence of calcifications regardless of anatomy was associated with increased all-cause mortality. However, below-knee calcification (typically medial calcification), but not calcification of the femoral–popliteal segment (typically intimal calcification), was associated with increased risk of amputation.[80] Medial calcification of the tibial artery is associated with a higher risk of amputation and presence of a foot wound.[81,82] Recent studies have also highlighted the impact of medial calcification involving the pedal arteries. These findings are likely due to limited revascularization options in those with medial calcifications, as well as the association between medial calcification and increased severity of vascular disease.[82,83] At the level of intervention, differences in outcomes between the different levels of disease (medial vs. intimal calcification) arise from the fact that intimal calcification in larger arteries can be removed by endarterectomy or atherectomy, whereas medial calcification cannot be effectively treated.

Microvascular disease

PAD affecting vessels ranging from capillaries to arterioles, approximately 100 uM in diameter, is considered microvascular disease (MVD). These vessels are critical for nutrient and oxygen exchange, cell signaling, and wound healing.[84] Underlying chronic diseases such as hypertension, diabetes, and chronic viral infections lead to the development of MVD.[85,86] MVD is associated with increased rates of amputation, even while controlling for diabetes, renal disease, and smoking, suggesting that MVD is an important independent risk factor for amputation, regardless of the way the patient developed MVD.[87] An important downstream consequence of MVD is neuropathy, especially in patients with diabetes. Neuropathy results in the blunting of sensation in the feet, which increases the risk of repetitive and severe injuries in tissue with impaired ability to heal. Thus far, no treatments for MVD exist, and further elucidation required of the pathogenesis and treatment of this disease.

Disease management

Optimal medical therapy, risk reduction, and exercise

The key components of medical therapy for PAD include cessation of aggravating factors such as smoking, optimal control of systemic illnesses which contribute to PAD such as hypertension and diabetes, exercise therapy, and the use of adjunctive medical therapies. Exercise therapy can improve functional status and quality of life in patients with PAD while also reducing all-cause and cardiovascular-related mortality.[88–90] Medical therapies include the use of agents such as antiplatelet medications, statins, and antihypertensive drugs (particularly ACE inhibitors and ARBs), which are of demonstrated benefit for the secondary prevention of adverse cardiovascular outcomes.[37,91–94] However, for patients with more severe PAD, revascularization, in combination with the above primary management strategies, is likely a more effective treatment option for limb symptoms.

Revascularization

The advent of endovascular technology and procedures has increased the scope of options available to patients with PAD and has the benefit of reducing perioperative risks in those with significant comorbidities.[95] Technologies used in endovascular revascularization include plain and drug-coated angioplasty balloons, bare-metal, covered, and drug-eluting stents, and atherectomy devices. Open surgical revascularization options include endarterectomy and bypass.

Modern vascular practice incorporates a selective approach to revascularization depending on individual patient risk, symptom severity, anatomic complexity, and treatment goals. "Hybrid" approaches incorporate both open and endovascular techniques concurrently, made possible by the increasing availability of improved intraoperative imaging, devices, and training. This flexibility allows for the treatment of complex patterns of disease in a single setting. The expected technical success, procedural risk, and anatomic durability of the various interventional options for PAD vary considerably based on the aforementioned factors and is beyond the scope of this review. Ongoing improvements in medical therapies, imaging, device and surgical technologies make this a highly dynamic field. These approaches are broadly summarized in Table 30.3.

Surveillance

Continued anatomic patency of the primary vascular intervention is a critical factor in determining the long-

TABLE 30.3 Summary of revascularization strategies in PAD.

Techniques		Common applications			Advantages	Limitations
		AI	FP	IP		
Endovascular revascularization	Balloon angioplasty	x	x	x	- Minimally invasive - Low morbidity - Repeatable - Favorable outcomes in larger arteries, short to intermediate length lesions, and stenosis (vs. occlusion)	- Longer lesion length, small vessel diameter, severe calcification - Common femoral artery and popliteal artery disease is less favorable - Reduced anatomic durability for femoropopliteal and infrapopliteal interventions - In-stent restenosis is difficult to treat
	Drug-coated balloon		x			
	Bare-metal and covered stents	x	x			
	Drug-eluting stents		x			
	Atherectomy		x	x		
Open surgical revascularization	Endarterectomy	Common femoral artery			- Flexibility to address diverse anatomic patterns and lesions - Can be combined with endovascular revascularization in hybrid approaches - Improved anatomical durability	- Invasive and increased risk for patient - Wound morbidity and systemic complications - Adequate quality autogenous vein is absent in 20%–40% of patients who require a distal bypass - Poor outcomes for nonautogenous conduits in below-knee bypass
	Open bypass with prosthetic or autogenous vein conduits	All anatomic sites				

AI, aortoiliac; *FP*, femoropopliteal; *IP*, infrapopliteal.

term clinical success of revascularization. The outcomes for patients undergoing infrainguinal bypass were significantly worse at 1 year for those who previously had an intervention that failed (e.g., previously endovascular intervention or bypass).[96] Therefore, surveillance is critical to identify interventions at increased risk of failure to prolong the long-term health of the limb and patient. The SVS recommends clinical exam, ABI, and DUS within 1 month of intervention as a baseline, and repeat studies at 3 months for surveillance.[97] For stents, surveillance should then be performed every 6 months, as an occluded stent is much more difficult to treat compared to a stenotic stent. For bypass grafts, surveillance should be performed at 6 and 12 months, then yearly to avoid graft failure. Surveillance should be expedited in the setting of new or recurrent symptoms.

Postprocedural considerations and complications

Complications related to vascular procedures range from local to systemic. For large open procedures that carry a risk of significant hemorrhage or fluid shifts, cardiac complications such as myocardial infarction are among the most significant. Wound-related complications, such as infection, are the most common overall. Complications arising from endovascular access include bleeding, pseudoaneurysm, arteriovenous fistula formation, and infection. Thrombosis of bypass grafts in the first 30 days following surgery is uncommon (5%) but can occur due to technical error or use of a poor-quality conduit. Embolization to distal sites can occur, often without consequence, but can compromise deep tissue perfusion on rare occasions.[98] For endovascular interventions incorrect use or use of inappropriately sized endografts or balloons may result in rupture and hemorrhage.

PAD therapies in development

Vascular regeneration

Those with severe PAD often have complex and advanced disease for which there are no surgical options. Vascular regenerative strategies have focused on the enhancement of collateral arteries (arteriogenesis) and nutrient vessel development (angiogenesis). These can be facilitated by the use of proangiogenic factors such as fibroblast growth factor, vascular endothelial growth factor, and hepatocyte growth factor.[99] To date these strategies have not yet demonstrated improved outcomes in large scale trials.[100,101] Gene therapy using either viral or nonviral vectors aims to modulate cellular function and promote vascular regeneration. Cell therapies utilize mesenchymal stem cells and other blood and bone marrow–derived progenitor cells for their regenerative and antiinflammatory properties, whereby autologous cells are harvested and implanted into an ischemic limb to promote angiogenesis. These methods appear safe, but neither approach has yet been proven in adequate clinical trials and they currently remain in the investigational realm.[102–104]

Bioengineered vascular conduits and scaffolds

A significant portion (20%–40%) of PAD patients who require surgical bypass do not have adequate quality and caliber autogenous vein available for bypass grafting. In the setting of distal bypass to tibial or pedal vessels, widely used prosthetic grafts and cryopreserved allografts result in notably inferior outcomes. To address these shortcomings, vascular tissue engineers have made significant progress over recent decades to overcome issues with structural integrity, thrombogenicity, and immune response. Some strategies include use of human-derived cells and extracellular matrices for scaffolds, ex vivo flow conditioning, and decellularization with repopulation using autologous cells. These biological conduits may provide higher quality options for those without suitable veins or require bypass in infected or otherwise challenging operative fields.[105–107]

Bioabsorbable vascular scaffolds that also deliver therapeutics are also in advanced stages of preclinical and clinical development. These stent-like devices provide mechanical buttressing in the setting of acute vascular injury to reduce recoil and prevent dissection, delivery drugs such as sirolimus to reduce neointimal hyperplasia, and may then be resorbed over a period of up to 18 months.[108] These devices may provide the advantages of drug eluting stent, while also eliminating the harmful inflammatory effects of a chronic implant within the vascular wall.

Conclusions

As the global population continues to age and the prevalence of PAD increases, the need for improved multimodal therapies to address the spectrum of clinical disease, particularly CLTI, becomes ever apparent. Although there are a wide range of tools currently available, more high-quality clinical research is needed to determine the most effective approach for specific clinical scenarios. Modulation of vascular healing following intervention offers opportunity for further innovation and discovery to improve long-term outcomes. MVD, calcification, and vascular regeneration present significant scientific challenges in PAD that mandate continued basic and translational research.

References

1. Fowkes FGR, Aboyans V, Fowkes FJI, McDermott MM, Sampson UKA, Criqui MH. Peripheral artery disease: epidemiology and global perspectives. *Nat Rev Cardiol*. 2017;14(3): 156–170. https://doi.org/10.1038/nrcardio.2016.179.
2. Virani SS, Alonso A, Aparicio HJ, et al. Heart disease and stroke statistics—2021 update: a report from the American Heart Association. *Circulation*. 2021;143(8). https://doi.org/10.1161/CIR.0000000000000950.
3. Selvin E, Erlinger TP. Prevalence of and risk factors for peripheral arterial disease in the United States: results from the National Health and Nutrition Examination survey, 1999–2000. *Circulation*. 2004;110(6):738–743. https://doi.org/10.1161/01.CIR.0000137913.26087.F0.
4. Diehm K, Lawall. Epidemiology of peripheral arterial disease. *Vasa*. 2004;33(4):183–189. https://doi.org/10.1024/0301-1526.33.4.183.
5. Hirsch AT. Peripheral arterial disease detection, awareness, and treatment in primary care. *J Am Med Assoc*. 2001;286(11):1317. https://doi.org/10.1001/jama.286.11.1317.
6. Meijer WT, Hoes AW, Rutgers D, Bots ML, Hofman A, Grobbee DE. Peripheral arterial disease in the elderly: the Rotterdam study. *Arterioscler Thromb Vasc Biol*. 1998;18(2):185–192. https://doi.org/10.1161/01.ATV.18.2.185.
7. Criqui MH, Fronek A, Barrett-Connor E, Klauber MR, Gabriel S, Goodman D. The prevalence of peripheral arterial disease in a defined population. *Circulation*. 1985;71(3):510–515. https://doi.org/10.1161/01.CIR.71.3.510.
8. Fowkes FGR, Rudan D, Rudan I, et al. Comparison of global estimates of prevalence and risk factors for peripheral artery disease in 2000 and 2010: a systematic review and analysis. *Lancet*. 2013; 382(9901):1329–1340. https://doi.org/10.1016/S0140-6736(13)61249-0.
9. Allison MA, Ho E, Denenberg JO, et al. Ethnic-specific prevalence of peripheral arterial disease in the United States. *Am J Prev Med*. 2007;32(4):328–333. https://doi.org/10.1016/j.amepre.2006.12.010.
10. Pande RL, Creager MA. Socioeconomic inequality and peripheral artery disease prevalence in US adults. *Circ Cardiovasc Qual Outcomes*. 2014;7(4):532–539. https://doi.org/10.1161/CIRCOUTCOMES.113.000618.
11. Vart P, Coresh J, Kwak L, Ballew SH, Heiss G, Matsushita K. Socioeconomic status and incidence of hospitalization with lower-extremity peripheral artery disease: atherosclerosis risk in communities study. *J Am Heart Assoc*. 2017;6(8). https://doi.org/10.1161/JAHA.116.004995.
12. Brook RD. Is air pollution a cause of cardiovascular disease? Updated review and controversies. *Rev Environ Health*. 2007; 22(2). https://doi.org/10.1515/REVEH.2007.22.2.115.
13. Hoffmann B, Moebus S, Kröger K, et al. Residential exposure to urban air pollution, ankle–brachial index, and peripheral arterial disease. *Epidemiology*. 2009;20(2):280–288. https://doi.org/10.1097/EDE.0b013e3181961ac2.
14. Navas-Acien A, Selvin E, Sharrett AR, Calderon-Aranda E, Silbergeld E, Guallar E. Lead, cadmium, smoking, and increased risk of peripheral arterial disease. *Circulation*. 2004;109(25): 3196–3201. https://doi.org/10.1161/01.CIR.0000130848.18636.B2.
15. Conen D, Everett BM, Kurth T, et al. Smoking, smoking cessation, and risk for symptomatic peripheral artery disease in women: a cohort study. *Ann Intern Med*. 2011;154(11):719. https://doi.org/10.7326/0003-4819-154-11-201106070-00003.
16. Ding N, Sang Y, Chen J, et al. Cigarette smoking, smoking cessation, and long-term risk of 3 major atherosclerotic diseases. *J Am Coll Cardiol*. 2019;74(4):498–507. https://doi.org/10.1016/j.jacc.2019.05.049.
17. Hess CN, Fu JW, Gundrum J, et al. Diabetes mellitus and risk stratification after peripheral artery revascularization. *J Am Coll Cardiol*. 2021;77(22):2867–2869. https://doi.org/10.1016/j.jacc.2021.04.011.
18. Centers for Disease Control and Prevention. *National Diabetes Statistics Report*. 2020.
19. Geiss LS, Li Y, Hora I, Albright A, Rolka D, Gregg EW. Resurgence of diabetes-related nontraumatic lower-extremity amputation in the young and middle-aged adult U.S. Population. *Diabetes Care*. 2019;42(1):50–54. https://doi.org/10.2337/dc18-1380.
20. Jain N, Agarwal MA, Jalal D, Dokun AO. Individuals with peripheral artery disease (PAD) and type 1 diabetes are more likely to undergo limb amputation than those with PAD and type 2 diabetes. *J Clin Med*. 2020;9(9). https://doi.org/10.3390/jcm9092809.
21. Bonaca MP, Hamburg NM, Creager MA. Contemporary medical management of peripheral artery disease. *Circ Res*. 2021;128(12): 1868–1884. https://doi.org/10.1161/CIRCRESAHA.121.318258.
22. Criqui MH, Vargas V, Denenberg JO, et al. Ethnicity and peripheral arterial disease: the San Diego population study. *Circulation*. 2005;112(17):2703–2707. https://doi.org/10.1161/CIRCULATIONAHA.105.546507.
23. Ridker PM, Stampfer MJ, Rifai N. Novel risk factors for systemic atherosclerosis: a comparison of C-reactive protein, fibrinogen, homocysteine, lipoprotein(a), and standard cholesterol screening as predictors of peripheral arterial disease. *J Am Med Assoc*. 2001; 285(19):2481. https://doi.org/10.1001/jama.285.19.2481.
24. Fowkes FGR. Epidemiology of atherosclerotic arterial disease in the lower limbs. *Eur J Vasc Surg*. 1988;2(5):283–291. https://doi.org/10.1016/S0950-821X(88)80002-1.
25. Criqui MH, Denenberg JO, Langer RD, Fronek A. The epidemiology of peripheral arterial disease: importance of identifying the population at risk. *Vasc Med*. 1997;2(3):221–226. https://doi.org/10.1177/1358863X9700200310.
26. Conen D, Rexrode KM, Creager MA, Ridker PM, Pradhan AD. Metabolic syndrome, inflammation, and risk of symptomatic peripheral artery disease in women: a prospective study. *Circulation*. 2009;120(12):1041–1047. https://doi.org/10.1161/CIRCULATIONAHA.109.863092.
27. Vidula H, Liu K, Criqui MH, et al. Metabolic syndrome and incident peripheral artery disease – the Multi-Ethnic Study of Atherosclerosis. *Atherosclerosis*. 2015;243(1):198–203. https://doi.org/10.1016/j.atherosclerosis.2015.08.044.
28. Edlinger C, Lichtenauer M, Wernly B, et al. Disease-specific characteristics of vascular cell adhesion molecule-1 levels in patients with peripheral artery disease. *Heart Ves*. 2019;34(6):976–983. https://doi.org/10.1007/s00380-018-1315-1.
29. Cooke JP, Wilson AM. Biomarkers of peripheral arterial disease. *J Am Coll Cardiol*. 2010;55(19):2017–2023. https://doi.org/10.1016/j.jacc.2009.08.090.
30. Hiatt WR, Zakharyan A, Fung ET, et al. A validated biomarker panel to identify peripheral artery disease. *Vasc Med*. 2012;17(6): 386–393. https://doi.org/10.1177/1358863X12463491.
31. Pradhan AD, Rifai N, Ridker PM. Soluble intercellular adhesion molecule-1, soluble vascular adhesion molecule-1, and the development of symptomatic peripheral arterial disease in men. *Circulation*. 2002;106(7):820–825. https://doi.org/10.1161/01.CIR.0000025636.03561.EE.
32. Joosten MM, Pai JK, Bertoia ML, et al. β2-Microglobulin, cystatin C, and creatinine and risk of symptomatic peripheral artery disease. *J Am Heart Assoc*. 2014;3(4). https://doi.org/10.1161/JAHA.114.000803.

33. Aboyans V, Criqui MH, Abraham P, et al. Measurement and interpretation of the ankle-brachial index: a scientific statement from the American Heart Association. *Circulation*. 2012;126(24):2890–2909. https://doi.org/10.1161/CIR.0b013e318276fbcb.
34. Salaun P, Desormais I, Lapébie F-X, et al. Comparison of ankle pressure, systolic toe pressure, and transcutaneous oxygen pressure to predict major amputation after 1 Year in the COPART cohort. *Angiology*. 2019;70(3):229–236. https://doi.org/10.1177/0003319718793566.
35. Fife CE, Smart DR, Sheffield PJ, Hopf HW, Hawkins G, Clarke D. Transcutaneous oximetry in clinical practice: consensus statements from an expert panel based on evidence. *Undersea Hyperb Med J Undersea Hyperb Med Soc Inc*. 2009;36(1):43–53.
36. Conte MS, Bradbury AW, Kolh P, et al. Global vascular guidelines on the management of chronic limb-threatening ischemia. *J Vasc Surg*. 2019;69(6). https://doi.org/10.1016/j.jvs.2019.02.016, 3S.e40–125S.e40.
37. Gerhard-Herman MD, Gornik HL, Barrett C, et al. 2016 AHA/ACC guideline on the management of patients with lower extremity peripheral artery disease: a report of the American college of cardiology/American Heart Association Task Force on clinical practice guidelines. *Circulation*. 2017;135(12):e726–e779. https://doi.org/10.1161/CIR.0000000000000471.
38. Mills JL, Conte MS, Armstrong DG, et al. The society for vascular surgery lower extremity threatened limb classification system: risk stratification based on wound, ischemia, and foot infection (WIfI). *J Vasc Surg*. 2014;59(1). https://doi.org/10.1016/j.jvs.2013.08.003, 220.e2–234.e2.
39. Fowkes FGR, Housley E, Cawood EHH, Macintyre CCA, Ruckley CV, Prescott RJ. Edinburgh artery study: prevalence of asymptomatic and symptomatic peripheral arterial disease in the general population. *Int J Epidemiol*. 1991;20(2):384–392. https://doi.org/10.1093/ije/20.2.384.
40. Criqui MH, Langer RD, Fronek A, et al. Mortality over a period of 10 years in patients with peripheral arterial disease. *N Engl J Med*. 1992;326(6):381–386. https://doi.org/10.1056/NEJM199202063260605.
41. Sigvant B, Lundin F, Wahlberg E. The risk of disease progression in peripheral arterial disease is higher than expected: a meta-analysis of mortality and disease progression in peripheral arterial disease. *Eur J Vasc Endovasc Surg*. 2016;51(3):395–403. https://doi.org/10.1016/j.ejvs.2015.10.022.
42. Sartipy F, Sigvant B, Lundin F, Wahlberg E. Ten year mortality in different peripheral arterial disease stages: a population based observational study on outcome. *Eur J Vasc Endovasc Surg*. 2018;55(4):529–536. https://doi.org/10.1016/j.ejvs.2018.01.019.
43. Polonsky TS, McDermott MM. Lower extremity peripheral artery disease without chronic limb-threatening ischemia: a review. *J Am Med Assoc*. 2021;325(21):2188. https://doi.org/10.1001/jama.2021.2126.
44. Conte MS, Pomposelli FB, Clair DG, et al. Society for Vascular Surgery practice guidelines for atherosclerotic occlusive disease of the lower extremities: management of asymptomatic disease and claudication. *J Vasc Surg*. 2015;61(3). https://doi.org/10.1016/j.jvs.2014.12.009, 2S.e1–41S.e1.
45. Cull DL, Manos G, Hartley MC, et al. An early validation of the society for vascular surgery lower extremity threatened limb classification system. *J Vasc Surg*. 2014;60(6):1535–1542. https://doi.org/10.1016/j.jvs.2014.08.107.
46. Zhan LX, Branco BC, Armstrong DG, Mills JL. The Society for Vascular Surgery lower extremity threatened limb classification system based on Wound, Ischemia, and foot Infection (WIfI) correlates with risk of major amputation and time to wound healing. *J Vasc Surg*. 2015;61(4):939–944. https://doi.org/10.1016/j.jvs.2014.11.045.
47. Causey MW, Ahmed A, Wu B, et al. Society for Vascular Surgery limb stage and patient risk correlate with outcomes in an amputation prevention program. *J Vasc Surg*. 2016;63(6). https://doi.org/10.1016/j.jvs.2016.01.011, 1563.e2–1573.e2.
48. Darling JD, McCallum JC, Soden PA, et al. Predictive ability of the Society for Vascular Surgery Wound, Ischemia, and foot Infection (WIfI) classification system following infrapopliteal endovascular interventions for critical limb ischemia. *J Vasc Surg*. 2016;64(3):616–622. https://doi.org/10.1016/j.jvs.2016.03.417.
49. Conte MS, Mills JL, Bradbury AW, White JV. Implementing global chronic limb-threatening ischemia guidelines in clinical practice: utility of the society for vascular surgery threatened limb classification system (WIfI). *J Vasc Surg*. 2020;72(4):1451–1452. https://doi.org/10.1016/j.jvs.2020.06.049.
50. Hess CN, Wang TY, Weleski Fu J, et al. Long-term outcomes and associations with major adverse limb events after peripheral artery revascularization. *J Am Coll Cardiol*. 2020;75(5):498–508. https://doi.org/10.1016/j.jacc.2019.11.050.
51. Hess CN, Debus ES, Nehler MR, et al. Reduction in acute limb ischemia with rivaroxaban versus placebo in peripheral artery disease after lower extremity revascularization: insights from VOYAGER PAD. *Circulation*. 2021;121:055146. https://doi.org/10.1161/CIRCULATIONAHA.121.055146.
52. Kaplovitch E, Eikelboom JW, Dyal L, et al. Rivaroxaban and aspirin in patients with symptomatic lower extremity peripheral artery disease: a subanalysis of the COMPASS randomized clinical trial. *JAMA Cardiol*. 2020. https://doi.org/10.1001/jamacardio.2020.4390.
53. Armstrong EJ, Chen DC, Westin GG, et al. Adherence to guideline-recommended therapy is associated with decreased major adverse cardiovascular events and major adverse limb events among patients with peripheral arterial disease. *J Am Heart Assoc*. 2014;3(2). https://doi.org/10.1161/JAHA.113.000697.
54. Gutierrez JA, Mulder H, Jones WS, et al. Polyvascular disease and risk of major adverse cardiovascular events in peripheral artery disease: a secondary analysis of the EUCLID trial. *JAMA Netw Open*. 2018;1(7):e185239. https://doi.org/10.1001/jamanetworkopen.2018.5239.
55. Jude EB, Oyibo SO, Chalmers N, Boulton AJM. Peripheral arterial disease in diabetic and nondiabetic patients: a comparison of severity and outcome. *Diabetes Care*. 2001;24(8):1433–1437. https://doi.org/10.2337/diacare.24.8.1433.
56. Haltmayer M, Mueller T, Horvath W, Luft C, Poelz W, Haidinger D. Impact of atherosclerotic risk factors on the anatomical distribution of peripheral arterial disease. *Int Angiol J Int Union Angiol*. 2001;20(3):200–207.
57. Van Der Feen C, Neijens FS, Kanters SDJM, WPThM M, Stolk RP, Banga JD. Angiographic distribution of lower extremity atherosclerosis in patients with and without diabetes: original article. *Diabet Med*. 2002;19(5):366–370. https://doi.org/10.1046/j.1464-5491.2002.00642.x.
58. Diehm N, Shang A, Silvestro A, et al. Association of cardiovascular risk factors with pattern of lower limb atherosclerosis in 2659 patients undergoing angioplasty. *Eur J Vasc Endovasc Surg*. 2006;31(1):59–63. https://doi.org/10.1016/j.ejvs.2005.09.006.
59. Rueda CA, Nehler MR, Perry DJ, et al. Patterns of artery disease in 450 patients undergoing revascularization for critical limb ischemia: implications for clinical trial design. *J Vasc Surg*. 2008;47(5):995–1000. https://doi.org/10.1016/j.jvs.2007.11.055.
60. Jones WS, Patel MR, Dai D, et al. High mortality risks after major lower extremity amputation in Medicare patients with peripheral

artery disease. *Am Heart J*. 2013;165(5). https://doi.org/10.1016/j.ahj.2012.12.002, 809.e1–815.e1.
61. Kim AS, Conte MS. Specialized pro-resolving lipid mediators in cardiovascular disease, diagnosis, and therapy. *Adv Drug Deliv Rev*. July 2020. https://doi.org/10.1016/j.addr.2020.07.011. S0169409X20300958.
62. Aronow HD, Beckman JA. Parsing atherosclerosis: the unnatural history of peripheral artery disease. *Circulation*. 2016;134(6):438–440. https://doi.org/10.1161/CIRCULATIONAHA.116.022971.
63. Narula N, Olin JW, Narula N. Pathologic disparities between peripheral artery disease and coronary artery disease. *Arterioscler Thromb Vasc Biol*. 2020;40(9):1982–1989. https://doi.org/10.1161/ATVBAHA.119.312864.
64. VA Million Veteran Program, Klarin D, Lynch J, et al. Genome-wide association study of peripheral artery disease in the Million Veteran Program. *Nat Med*. 2019;25(8):1274–1279. https://doi.org/10.1038/s41591-019-0492-5.
65. Qamar A, Morrow DA, Creager MA, et al. Effect of vorapaxar on cardiovascular and limb outcomes in patients with peripheral artery disease with and without coronary artery disease: analysis from the TRA 2°P-TIMI 50 trial. *Vasc Med*. 2020;25(2):124–132. https://doi.org/10.1177/1358863X19892690.
66. Bonaca MP, Bauersachs RM, Anand SS, et al. Rivaroxaban in peripheral artery disease after revascularization. *N Engl J Med*. 2020; 382(21):1994–2004. https://doi.org/10.1056/NEJMoa2000052.
67. Oral anticoagulant and antiplatelet therapy and peripheral arterial disease. *N Engl J Med*. 2007;357(3):217–227. https://doi.org/10.1056/NEJMoa065959.
68. Mega JL, Braunwald E, Wiviott SD, et al. Rivaroxaban in patients with a recent acute coronary syndrome. *N Engl J Med*. 2012;366(1): 9–19. https://doi.org/10.1056/NEJMoa1112277.
69. Eikelboom JW, Connolly SJ, Bosch J, et al. Rivaroxaban with or without aspirin in stable cardiovascular disease. *N Engl J Med*. 2017;377(14):1319–1330. https://doi.org/10.1056/NEJMoa1709118.
70. Alexander JH, Lopes RD, James S, et al. Apixaban with antiplatelet therapy after acute coronary syndrome. *N Engl J Med*. 2011; 365(8):699–708. https://doi.org/10.1056/NEJMoa1105819.
71. Thiruvoipati T. Peripheral artery disease in patients with diabetes: epidemiology, mechanisms, and outcomes. *World J Diabetes*. 2015; 6(7):961. https://doi.org/10.4239/wjd.v6.i7.961.
72. Pradhan AD. C-reactive protein, interleukin 6, and risk of developing type 2 diabetes mellitus. *J Am Med Assoc*. 2001;286(3):327. https://doi.org/10.1001/jama.286.3.327.
73. Steinberg H, Baron A. Vascular function, insulin resistance and fatty acids. *Diabetologia*. 2002;45(5):623–634. https://doi.org/10.1007/s00125-002-0800-2.
74. Prepared with the assistance of Cosentino F, Beckman JA Creager MA, Lüscher TF. Diabetes and vascular disease: pathophysiology, clinical consequences, and medical therapy: Part I. *Circulation*. 2003;108(12):1527–1532. https://doi.org/10.1161/01.CIR.0000091257.27563.32.
75. Miyazawa T, Nakagawa K, Shimasaki S, Nagai R. Lipid glycation and protein glycation in diabetes and atherosclerosis. *Amino Acids*. 2012;42(4):1163–1170. https://doi.org/10.1007/s00726-010-0772-3.
76. Bierhaus A, Nawroth PP. Multiple levels of regulation determine the role of the receptor for AGE (RAGE) as common soil in inflammation, immune responses and diabetes mellitus and its complications. *Diabetologia*. 2009;52(11):2251–2263. https://doi.org/10.1007/s00125-009-1458-9.
77. Peripheral arterial disease in people with diabetes. *Diabetes Care*. 2003;26(12):3333–3341. https://doi.org/10.2337/diacare.26.12.3333.
78. Geng Y-J, Libby P. Progression of atheroma: a struggle between death and procreation. *Arterioscler Thromb Vasc Biol*. 2002;22(9): 1370–1380. https://doi.org/10.1161/01.ATV.0000031341.84618.A4.
79. Krishnan P, Moreno PR, Turnbull IC, et al. Incremental effects of diabetes mellitus and chronic kidney disease in medial arterial calcification: synergistic pathways for peripheral artery disease progression. *Vasc Med*. 2019;24(5):383–394. https://doi.org/10.1177/1358863X19842276.
80. Konijn LCD, Takx RAP, de Jong PA, et al. Arterial calcification and long-term outcome in chronic limb-threatening ischemia patients. *Eur J Radiol*. 2020;132:109305. https://doi.org/10.1016/j.ejrad.2020.109305.
81. Guzman RJ, Brinkley DM, Schumacher PM, Donahue RMJ, Beavers H, Qin X. Tibial artery calcification as a marker of amputation risk in patients with peripheral arterial disease. *J Am Coll Cardiol*. 2008;51(20):1967–1974. https://doi.org/10.1016/j.jacc.2007.12.058.
82. Liu IH, Wu B, Krepkiy V, et al. Pedal arterial calcification score correlates with risk of major amputation in chronic limb-threatening ischemia. *J Vasc Surg*. 2020;72(3):e337. https://doi.org/10.1016/j.jvs.2020.07.035.
83. Ferraresi R, Ucci A, Pizzuto A, et al. A novel scoring system for small artery disease and medial arterial calcification is strongly associated with major adverse limb events in patients with chronic limb-threatening ischemia. *J Endovasc Ther Off J Int Soc Endovasc Spec*. 2021;28(2):194–207. https://doi.org/10.1177/1526602820966309.
84. Behroozian A, Beckman JA. Microvascular disease increases amputation in patients with peripheral artery disease. *Arterioscler Thromb Vasc Biol*. 2020;40(3):534–540. https://doi.org/10.1161/ATVBAHA.119.312859.
85. Venkatramani J, Mitchell P. Ocular and systemic causes of retinopathy in patients without diabetes mellitus. *BMJ*. 2004;328(7440): 625–629. https://doi.org/10.1136/bmj.328.7440.625.
86. Raile K, Galler A, Hofer S, et al. Diabetic nephropathy in 27,805 children, adolescents, and adults with type 1 diabetes: effect of diabetes duration, A1C, hypertension, dyslipidemia, diabetes onset, and sex. *Diabetes Care*. 2007;30(10):2523–2528. https://doi.org/10.2337/dc07-0282.
87. Beckman JA, Duncan MS, Damrauer SM, et al. Microvascular disease, peripheral artery disease, and amputation. *Circulation*. 2019; 140(6):449–458. https://doi.org/10.1161/CIRCULATIONAHA.119.040672.
88. Hiatt WR, Wolfel EE, Meier RH, Regensteiner JG. Superiority of treadmill walking exercise versus strength training for patients with peripheral arterial disease. Implications for the mechanism of the training response. *Circulation*. 1994;90(4):1866–1874. https://doi.org/10.1161/01.CIR.90.4.1866.
89. Sakamoto S, Yokoyama N, Tamori Y, Akutsu K, Hashimoto H, Takeshita S. Patients with peripheral artery disease who complete 12-week supervised exercise training program show reduced cardiovascular mortality and morbidity. *Circ J*. 2009;73(1):167–173. https://doi.org/10.1253/circj.CJ-08-0141.
90. Chang P, Nead KT, Olin JW, Myers J, Cooke JP, Leeper NJ. Effect of physical activity assessment on prognostication for peripheral artery disease and mortality. *Mayo Clin Proc*. 2015;90(3):339–345. https://doi.org/10.1016/j.mayocp.2014.12.016.
91. Norgren L, Hiatt WR, Dormandy JA, Nehler MR, Harris KA, Fowkes FGR. Inter-society consensus for the management of peripheral arterial disease (TASC II). *Eur J Vasc Endovasc Surg*. 2007;33(1):S1–S75. https://doi.org/10.1016/j.ejvs.2006.09.024.
92. Hirsch AT, Haskal ZJ, Hertzer NR, et al. ACC/AHA 2005 practice guidelines for the management of patients with peripheral arterial disease (lower extremity, renal, mesenteric, and abdominal aortic): a collaborative report from the American Association for Vascular Surgery/Society for Vascular Surgery,* Society for Cardiovascular Angiography and Interventions, Society for Vascular Medicine and Biology, Society of Interventional Radiology, and the Acc/Aha Task Force on Practice Guidelines (Writing

Committee to Develop Guidelines for the Management of Patients with Peripheral Arterial Disease): Endorsed by The American Association of Cardiovascular and Pulmonary Rehabilitation; National Heart, Lung, and Blood Institute; Society for Vascular Nursing; Transatlantic Inter-Society Consensus; and Vascular Disease Foundation. *Circulation*. 2006;113(11). https://doi.org/10.1161/CIRCULATIONAHA.106.174526.

93. Diehm N, Schmidli J, Setacci C, et al. Chapter III: management of cardiovascular risk factors and medical therapy. *Eur J Vasc Endovasc Surg*. 2011;42:S33–S42. https://doi.org/10.1016/S1078-5884(11)60011-7.
94. Aboyans V, Ricco J-B, Bartelink M-LEL, et al. 2017 ESC guidelines on the diagnosis and treatment of peripheral arterial diseases, in collaboration with the European Society for Vascular Surgery (ESVS). *Eur Heart J*. 2018;39(9):763–816. https://doi.org/10.1093/eurheartj/ehx095.
95. Hiramoto JS, Teraa M, de Borst GJ, Conte MS. Interventions for lower extremity peripheral artery disease. *Nat Rev Cardiol*. 2018; 15(6):332–350. https://doi.org/10.1038/s41569-018-0005-0.
96. Jones DW, Schanzer A, Zhao Y, et al. Growing impact of restenosis on the surgical treatment of peripheral arterial disease. *J Am Heart Assoc*. 2013;2(6). https://doi.org/10.1161/JAHA.113.000345.
97. Zierler RE, Jordan WD, Lal BK, et al. The Society for Vascular Surgery practice guidelines on follow-up after vascular surgery arterial procedures. *J Vasc Surg*. 2018;68(1):256–284. https://doi.org/10.1016/j.jvs.2018.04.018.
98. Boc A, Blinc A, Boc V. Distal embolization during percutaneous revascularization of the lower extremity arteries. *Vasa*. 2020; 49(5):389–394. https://doi.org/10.1024/0301-1526/a000877.
99. Gu Y, Cui S, Wang Q, et al. A randomized, double-blind, placebo-controlled phase II study of hepatocyte growth factor in the treatment of critical limb ischemia. *Mol Ther*. 2019;27(12):2158–2165. https://doi.org/10.1016/j.ymthe.2019.10.017.
100. Belch J, Hiatt WR, Baumgartner I, et al. Effect of fibroblast growth factor NV1FGF on amputation and death: a randomised placebo-controlled trial of gene therapy in critical limb ischaemia. *Lancet*. 2011;377(9781):1929–1937. https://doi.org/10.1016/S0140-6736(11)60394-2.
101. Miao Y-L, Wu W, Li B-W, et al. Clinical effectiveness of gene therapy on critical limb ischemia: a meta-analysis of 5 randomized controlled clinical trials. *Vasc Endovascular Surg*. 2014;48(5–6): 372–377. https://doi.org/10.1177/1538574414539397.
102. Forster R, Liew A, Bhattacharya V, Shaw J, Stansby G, Cochrane Vascular Group. Gene therapy for peripheral arterial disease. *Cochrane Database Syst Rev*. 2018. https://doi.org/10.1002/14651858.CD012058.pub2.
103. Frangogiannis NG. Cell therapy for peripheral artery disease. *Curr Opin Pharmacol*. 2018;39:27–34. https://doi.org/10.1016/j.coph.2018.01.005.
104. Iafrati MD, Hallett JW, Geils G, et al. Early results and lessons learned from a multicenter, randomized, double-blind trial of bone marrow aspirate concentrate in critical limb ischemia. *J Vasc Surg*. 2011;54(6):1650–1658. https://doi.org/10.1016/j.jvs.2011.06.118.
105. Niklason LE, Lawson JH. Bioengineered human blood vessels. *Science*. 2020;370(6513):eaaw8682. https://doi.org/10.1126/science.aaw8682.
106. Gutowski P, Gage SM, Guziewicz M, et al. Arterial reconstruction with human bioengineered acellular blood vessels in patients with peripheral arterial disease. *J Vasc Surg*. 2020;72(4): 1247–1258. https://doi.org/10.1016/j.jvs.2019.11.056.
107. Kirkton RD, Santiago-Maysonet M, Lawson JH, et al. Bioengineered human acellular vessels recellularize and evolve into living blood vessels after human implantation. *Sci Transl Med*. 2019;11(485):eaau6934. https://doi.org/10.1126/scitranslmed.aau6934.
108. Varcoe RL, Menting TP, Thomas SD, Lennox AF. Long-term results of a prospective, single-arm evaluation of everolimus-eluting bioresorbable vascular scaffolds in infrapopliteal arteries. *Catheter Cardiovasc Interv*. 2021;97(1):142–149. https://doi.org/10.1002/ccd.29327.

CHAPTER

31

Venous diseases including thromboembolic phenomena

Andrea T. Obi[1], Daniel D. Myers[2], Peter K. Henke[1], Suman Sood[3] and Thomas W. Wakefield[1]

[1]Section of Vascular Surgery, Department of Surgery, University of Michigan, Ann Arbor, MI, United States [2]Unit for Laboratory Animal Medicine, University of Michigan, Ann Arbor, MI, United States [3]Division of Hematology/Oncology, Department of Internal Medicine, University of Michigan, Ann Arbor, MI, United States

Venous thromboembolism is a common and deadly problem.[1] In this chapter, we will examine issues of venous thrombogenesis, thrombus resolution, animal models available to study these issues, and the future of venous thrombosis research.

Venous thrombogenesis with special emphasis on selectins

For over 100 years, venous thrombogenesis has been driven by the concept of Virchow's triad of endothelial injury, blood hypercoagulability, and circulatory stasis. In 1992, inflammation was added as it was shown that if the inflammatory glycoprotein P-selectin was inhibited, thrombosis decreased in a primate AV fistula model.[2] Early work suggested that P-selectin was upregulated at the level of the vein wall after deep vein thrombosis (DVT) as early as a few hours post-thrombosis, while E-selectin was expressed later.[3] In a mouse with elevated levels of soluble(s) P-selectin in a model of complete IVC stasis, animals with elevated sP-selectin had a 50% increase in thrombus mass (thrombus weight/IVC length), while those animals with genes deleted for selectins had a significant decrease in thrombus mass.[4] This was shown to be caused by the release of procoagulant microparticles and neutrophil extracellular traps (NETs).[5]

The ability of multiple forms of P-selectin inhibitors to affect thrombosis in two different primate models has been studied, first a model of IVC thrombosis and second a model of iliac vein thrombosis. In five studies, vein reopening was significantly higher with inhibitors to P-selectin and its receptor P-selectin glycoprotein ligand-1 (PSGL-1) compared to saline, and similar to enoxaparin. Inflammation, reflected as gadolinium enhancement in the vein wall by magnetic resonance venography (MRV), was significantly decreased in the P-selectin-treated group when compared to saline, with no significant differences to enoxaparin treated animals. In addition, there was less prolongation in TCT with P-selectin/PSGL-1 inhibitors compared to enoxaparin ($P < .0001$).[6]

Aptamers have been compared in our primate iliac vein thrombosis model, one an anti-P-selectin aptamer which blocks soluble and bound P-selectin/PSGL-1 interactions, and a second aptamer which blocks *Von Willebrand factor* (vWF) A1 domain/platelet glycoprotein 1b (GP1b) interactions. In treatment, we found the anti-P-selectin aptamer resulted in the most open iliac vein at day 21 with 72% recanalization compared to 50% for enoxaparin, 13% for control, and 0% for the anti-VWF while in prophylaxis, both aptamers produced similar good results, although neither totally prevented initial thrombus formation.[7] Additionally, inhibition of P-selectin resulted in less vein wall fibrosis as indicated by a decrease in vein wall collagen immunostaining. These beneficial results of P-selectin inhibition occurred without any increase in coagulation times, suggesting less bleeding potential.

E-selectin is also a key regulator of thrombus formation and fibrin content.[3,8–11] In studies using an E-selectin inhibitor (GMI-1271), in a mouse model of thrombosis that used stasis with some preserved flow,

The Vasculome
https://doi.org/10.1016/B978-0-12-822546-2.00024-1

the inhibitor was equivalent to enoxaparin for limiting thrombosis, but with a significant decrease in tail vein bleeding time.[12] This and other data led to phase I and II clinical trials in which GMI-1271 was given to normal volunteers in a dose-dependent fashion as a one-time dose, then as a daily dose for five consecutive days (all compared to enoxaparin or saline), and in a protocol to treat calf vein DVT, sponsored by the NIH VITA program. Biomarkers of inflammation and coagulation were measured, along with markers of cell adhesion, and leukocyte and platelet activation.[13] In this study, there were: (1) No serious adverse events; (2) GMI-1271 did not affect TEG parameters; (3) Lower levels of E-selectin in GMI-1271 treated volunteers as expected for on-target effects; and (4) Lower leukocyte and platelet activation in GMI-1271 treated volunteers, evidenced by MPO and MAC-1 levels. Two patients with calf vein DVT were treated with GMI-1271 with a 5-day IV course. Both patients had immediate relief of pain and significant increase in vein recanalization by day 19.

In a recent study, the effects of GMI-1271 in a nonhuman primate model of balloon stasis–induced proximal iliac vein thrombosis were studied in a second trial sponsored by the NIH VITA program.[14,15] In Phase 1 starting on postocclusion Day 2, animals were treated with GMI-1271, 25 mg/kg, now SC, once daily until Day 21 (n = 4). Nontreated control (CTR) animals received no treatment (n = 5). Phase 1 data demonstrated the effectiveness of GMI-1271 to allow for improved thrombus resolution with less vein wall changes and no bleeding potential. In phase 2, animals were treated with GMI-1271 with or without low molecular weight heparin (LMWH). E-selectin inhibition alone was found superior to E-selectin inhibition plus LMWH and LMWH alone regarding vein recanalization, intimal fibrosis and intimal thickness, without any significant changes in markers of coagulation.[16] All of these data suggest that E-selectin has a significant role in both thrombogenesis and pain/swelling after DVT (post-thrombotic syndrome), and may be an excellent pharmacologic target.[17]

Venous thrombosis and resolution

In general, DVT is thought to start at the inner area of venous valve cusps[18] that physically thicken with aging.[19] At these high risk locations, thrombus propagation can occur, likely promoted by impaired oscillatory shear stress, and loss of antithrombotic endothelial surface phenotype. The antithrombotic phenotype is maintained by high levels of thrombomodulin, endothelial protein C receptor, and tissue factor pathway inhibitor, and low levels of prothrombotic proteins vWF, P-selectin, and intracellular adhesion molecule-1.[20] Furthermore, the hypoxemic microenvironment present in the valve cusp may also drive changes in the endothelial phenotype, thereby promoting a procoagulant state (Fig. 31.1A).[21]

Imaging of DVT in humans suggests that there are regions of total stasis and nonstasis within the same affected venous segment, such as the femoral or iliofemoral vein.[22–24] Moreover, human specimen evaluation of chronically occluded femoral vein has shown a collagen-rich and cellular histology, similar to the experimental histological appearance noted in chronic mouse models of late VT.[25,26]

Venous thrombosis (VT) and its resolution is an inflammatory process.[27–31] Specifically, experimental DVT models suggest the process is driven by the mechanism of thrombogenesis (stasis or nonstasis), leukocytes such as neutrophils (PMN) and monocyte/macrophage (Mo/MΦ), the duration of thrombus-vein wall contact, and other factors such as proinflammatory cytokines, chemokines, and matrix metalloproteinases (MMPs).[27,32,33]

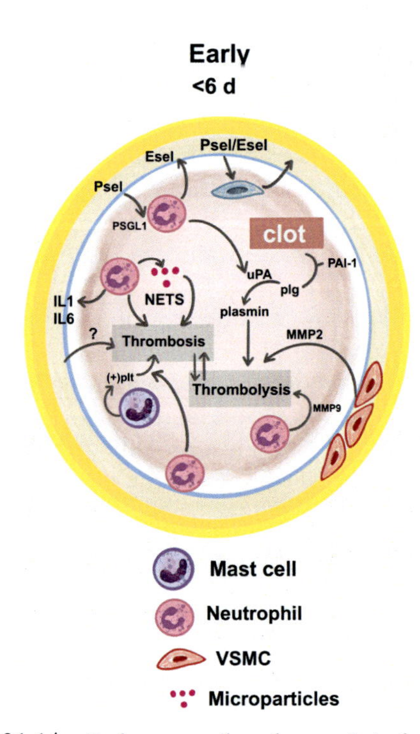

FIGURE 31.1A Early venous thrombogenesis is characterized by the interaction of multiple factors beyond the clotting cascade, platelets, and tissue factor. Neutrophils (PMN) trapped and chemoattracted into the thrombus may drive thrombosis as well as contribute to its dissolution. Cellular factors from PMNs include MMP-9 and NETs, as well as proinflammatory cytokines. Microparticles promote thrombogenesis via selectin-platelet and leukocyte interactions. Mast cells may augment thrombosis via histamine–platelet–vWF interactions. The vein wall may recruit these cells via cell adhesion molecules such as P-selectin, E-selectin, as well as secrete MMP2 that contributes to thrombus breakdown.

Leukocyte roles

Venous thrombus resolution resembles sterile wound healing,[34,35] with well-regulated phases of PMN and Mo/MΦ influx, followed by later fibrosis. Concurrently, the postthrombotic vein walls stiffen and thicken with interstitial matrix and myofibroblast accumulation, likely mediated by these leukocyte and other vein wall cellular interactions.[36–40]

Polymorphonuclear cells are present in early, recently formed venous thrombus. The role of the PMN is multifactorial and dictated by the thrombotic model, and environment, as PMN depletion may not affect thrombogenesis, may increase VT size, or may be associated with smaller VT, depending on the experimental model and setting.[41–43] Neutrophils release several factors, including plasminogen activators that may promote thrombus resolution, but, in a model-dependent fashion, may promote thrombosis via NETs, elastase, and chemokines.[44]

Monocytes/macrophages (Mo/MΦ) are the critical second leukocyte type to infiltrate the thrombus, beginning at day 4 experimentally, and these cells peak at day 8 in a mouse model of complete stasis thrombosis (Fig. 31.1B).[45,46] Mo/MΦ drive VT resolution by clearance of apoptotic and necrotic cells and matrix debris,[35,44,47–49] as well as having profibrinolytic activity and promoting thrombus neovascularization.[37,50–53] However, Mo/MΦ are not required for early stasis or nonstasis thrombogenesis, as suggested from a conditional Mo/MΦ (CD11b-DTr) depletion model.[54]

Mo/MΦ are classified by their inflammatory or antiinflammatory functions.[55–57] For example, cell surface Ly6C^{hi}, CCR2^{++}, and CX$_3$CR1^{+} antigen expression characterize classically activated, proinflammatory Mo/MΦ. Conversely, cell surface Ly6C^{lo}, CCR2^{-}, and CX$_3$CR1^{++} antigen expression characterize alternatively activated antiinflammatory Mo/MΦ, with prohealing and inflammation-resolving activities. The microenvironment of the VT seems to determine whether the Mo/MΦ is proinflammatory that drives tissue damage or proresolution/healing that promotes tissue homeostasis.[49,58–63] It is important to note that this process is highly dynamic in tissues, and a pure M_1 (proinflammatory) or M_2 (antiinflammatory) phenotype derived in cell culture is not replicated in vivo.[60,62,64,65] Nonetheless, early proinflammatory Mo/MΦ activities such as uPA release[50] and factors such as p53-mediated downstream effects[66] are likely important early for resolution and set the stage for a later healing response.

A recent study in a nonstasis VT model suggests that Mo/MΦ subtype are important in resolution.[67] Utilizing a lysM^{+} leukocyte conditional depletion strategy,

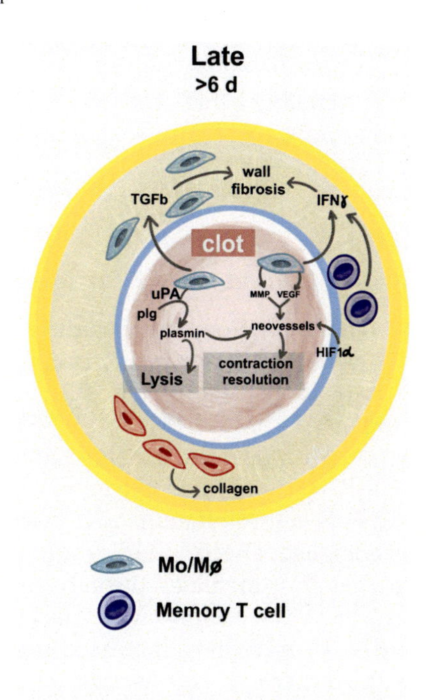

FIGURE 31.1B Later thrombus resolution is driven by influx of monocyte/macrophages, as well as the vein wall response. The plasmin axis via urokinase plasminogen activator plasminogen is critical for fibrinolysis, as well as MMPs. Contraction, neovascularization, and incorporation into the vein wall occur over time. Factors such as VEGF, HIF1α, and TGFβ contribute to thrombus maturation as well as vein wall fibrotic responses, with increased matrix and collagen. T-cells may also drive monocyte/macrophage responses, in part via IFNγ.

thrombosis resolution was enhanced when this cell line was depleted, with decreased PMNs and increased IL-4 and IL-10, suggesting a shift to a prohealing response. To further investigate this, Tbx21$^{-/-}$ mice, lacking ability to produce interferon gamma (IFNγ), are skewed toward a TH$_2$ immune response. Postthrombosis, these mice had decreased thrombus collagen, increased neovascular channels, and decreased IL-12 and CCL-2 at day 14 in the same nonstasis model. Consistently, others have shown that VT resolution is accelerated with IFNγ deletion, and consequently increased MMP2 and MMP9.[68]

The role of lymphocytes in VT resolution is now somewhat clearer using a nonstasis model of VT.[69] Memory T-cells play a role in postthrombotic vein wall inflammation with T-cells peaking late, at 21 days. Reducing primarily effector T-cells after thrombosis demonstrated reduced levels of IL-1, IL-6, and IL-18, cytokines all inducible by IFNγ. The T-cell-depleted mice showed decreased thrombus size, decreased IFNγ, and increased neovascular channels at 21 days. The direct role of the cells on vein wall thickening or injury was not explicitly examined.

Lastly, mast cells may direct venous thrombogenesis, in part via histamine effects, as well as vWF and platelet recruitment.[70] Whether this cell can be targeted clinically is unclear.

Cellular factors in fibrinolysis, neovascularization, and matrix turnover

The local venous environment postthrombosis contains growth factors such as TGFβ, proinflammatory cytokines such as IL-1,[71] interleukin (IL)-6, and activation of the uPA/PAI-1/MMP axis.[39,72–74] For example, using a nonstasis VT model, endothelin-1 may modulate $TGF\beta_1$–$TGF\beta_1$-R signaling, and impair VT resolution at 21 days.[75] IL-6 is a proinflammatory cytokine, with diverse physiological effects. Our group and others have shown that IL-6 mediates thrombogenesis, with decreased early VT size in mice lacking IL-6.[76] Conversely, IL-$6^{-/-}$ mice have increased mid and later VT size, with decreased neovascular channels and decreased MMP2/9 activity, suggesting bimodal effects.[77]

Matrix remodeling enzymes have been shown through multiple studies to modulate thrombus resolution, primarily in the stasis VT model.[37] Using a stasis VT model, MMP-$9^{-/-}$ mice had less Mo/MΦ cell infiltration, less collagen, yet an increased VT size (corrected for body weight),[78] with the effects observed at the later time points of 8–14 day. Moreover, MMP-2 deletion was also shown in a stasis VT model to have impaired thrombus resolution at 8 days, decreased TNF, and IL-1β but increased mononuclear cell influx.[40] The VT was larger, suggesting the role of MMP-2/9 in VT resolution, in addition to the vein wall effects. Others have shown the VT resolution is accelerated with IFNγ deletion, and consequently increased MMP2 and MMP9.[68]

Not surprisingly, plasmin over expression, using a global PAI-$1^{-/-}$ mouse, was associated with smaller VT size.[39] In PAI-$1^{-/-}$ mice, administration of low molecular weight heparin postthrombosis showed VT resolution was increased, with no impact on thrombogenesis. There was no change in PAI-1 levels, but PAI-$1^{-/-}$ mice had increased MMP-2, TIMP-1, IL-13, and increased vein wall Mo/MΦ influx and smaller VT. This study was further expanded using PAI-1 over expressing mice that had the opposite effect, with decreased Mo/MΦ influx, and increased VT size.

Toll-like receptor (TLR)-9, an important receptor for sterile inflammatory processes, and $TLR9^{-/-}$ mice have increased thrombogenesis, with larger thrombi at 2 days.[79] $TLR9^{-/-}$ mice also had decreased mid-term (8d) VT resolution, with increased thrombus matrix cellular breakdown products such as uric acid, cathepsin-G, high mobility group box (HMGB)-1 as compared with controls.[80] Others have shown that HMGB-1 can directly modulate thrombogenesis, likely through platelet actions and NET production.[81]

The local venous environment is hypoxic and may also be a driver of neovascularization. A major angiogenic growth factor is hypoxia-inducible factor-1α (HIF-Iα). Murine stasis models of VT have shown that thrombosis stimulates HIF-Iα, which accelerates thrombus resolution.[82] VEGF has also been associated with increased thrombus recanalization, and increased Mo/MΦ and smaller venous thrombi from a rat model of DVT.[83] Thus, thrombus resolution is in part dependent on neovascularization.

Recurrent DVT is a strong risk factor for PTS,[84] and larger VT after the second thrombotic injury. A recently published model was developed that showed increased levels of TGFβ, IL-6, and MMPs in the secondary (e.g., recurrent) VT postthrombotic thrombus/vein wall.[85] This study also delineated the incorporation of the primary thrombus into the vein wall, contributing to the thickness and architecture of the postthrombotic vein. The exact mechanism of how a postthrombotic wall may contribute to recurrent thrombogenesis is not clear.

Several of the aforementioned mediators of thrombosis and resolution may be attractive targets to accelerate resolution without anticoagulation.

Animal models for venous research: applicability to clinical venous thrombosis

Animal models of venous thrombosis (VT) must consistently manipulate at least two arms of Virchow's Triad: (1) blood stasis, (2) endothelial injury, and (3) blood hypercoagulability.[86]

Pre-clinical/basic science models

Murine models of venous thrombosis

There are several recent review articles which highlight the details of commonly utilized mouse models of venous thrombosis.[87–94] These animal models are valuable to elucidate hemostatic and thrombotic mechanisms that evaluate safety profiles of therapeutic interventions, dosing regimens, and imaging studies. Advantages include murine availability, short generation time, and ease of employing gene targeting technologies (Table 31.1).

Mouse complete stasis: Blood stasis in a vessel is achieved by complete ligation of the inferior vena cava (IVC) in the region caudal to the renal vein. This model mimics complete occlusion of the vessel in the clinical setting. Molecular and imaging methodologies allow study of the interactions between the vein wall and the

TABLE 31.1 Pre-clinical/basic science models of venous thrombosis.

Model	Vascular site	Advantages	Disadvantages	References
Pre-clinical/basic science				
Mouse complete stasis	IVC	Cell–cell interactions at occlusive vein wall thrombus interface, histopathology	Therapeutic agent testing variable due to lack of blood flow	3,46
Mouse stenosis	IVC	Produces a laminar thrombus with flow, imaging, cell–cell interactions, histopathology	Therapeutic agent testing variable, produces an inconsistent thrombus	42,95–99
Mouse electrolysis	IVC, FV	Produces a consistent laminar thrombus with flow, imaging, therapeutic agent testing, histopathology	Requires specialized equipment, vein wall injury at site of needle	100–103
Mouse chemical – ferric chloride	MV, SV, EJV	Imaging, platelet, acute thrombosis, cell–cell interactions, histopathology	Transmural injury, chemical toxicity	104–106
Mouse chemical – Rose Bengal	EJV	Imaging, endothelial injury, acute thrombosis, histopathology	Requires specialized equipment, produces an inconsistent thrombus, chemical toxicity	93,107
Rat complete stasis	IVC	Imaging, endothelial injury, acute, and chronic occlusive thrombosis, histopathology	Therapeutic agent testing variable due to lack of blood flow	108–112
Rat stenosis	IVC	Produces a laminar thrombus with flow, imaging, cell–cell interactions, histopathology	Produces an inconsistent thrombus	113–115
Preclinical/translational				
Rabbit complete stasis	EJV	Imaging, acute occlusive thrombosis, histopathology, therapeutic agent testing, used in PK/PD studies	Therapeutic agent testing variable unless blood flow is restored in this model	116,117
Rabbit stenosis	IVC	Imaging, acute nonocclusive thrombosis, histopathology, therapeutic agent testing, used in PK/PD studies	Produces an inconsistent thrombus	118,119
Rabbit ferric chloride	EJV	Imaging, platelet, acute thrombosis, cell–cell interactions, therapeutic agent testing, used in PK/PD studies, histopathology	Transmural injury, chemical toxicity	120

Continued

TABLE 31.1 Pre-clinical/basic science models of venous thrombosis.—cont'd

Model	Vascular site	Advantages	Disadvantages	References
Canine complete stasis	IVC, FV	Imaging, acute, and chronic occlusive thrombosis, histopathology, therapeutic agent testing, used in PK/PD studies	Requires specialized equipment, vein wall injury at site of needle; Ethical considerations require strong scientific justification for use	121–123
Canine electrolysis	EJV	Imaging, acute nonocclusive thrombosis, histopathology, platelet and neutrophil studies, therapeutic agent testing	Requires specialized equipment, vein wall injury at site of needle; Ethical considerations require strong scientific justification for use	124,125
Swine complete stasis	CVC, IV	Imaging, acute occlusive thrombosis, histopathology, therapeutic agent testing, used in PK/PD studies	Requires highly trained personnel	126,127
Nonhuman primate complete stasis	IVC, IV, FV	Imaging, acute, and chronic occlusive thrombosis, histopathology, therapeutic agent testing, used in PK/PD studies	Requires highly trained personnel; Ethical considerations require strong scientific justification for use	7,14,128–132
Review articles				
Rodent basic science				87–94
Nonrodent translational				87,94,133,134
Research Reproducibility/ comparative Pathology/statistical consultation				133–139

CVC, caudal vena cava; *EJV*, external jugular vein; *FV*, femoral vein; *IVC*, inferior vena cava; *MV*, mesenteric venules; *PK/PD*, pharmacokinetics/pharmacodynamics; *SV*, saphenous vein.

occlusive thrombus, evaluating progression from acute to chronic stages.[3,46,87,88,91,93]

Mouse stenosis: A ligature is placed around the IVC to reduce blood flow combined with a temporary endothelial compression producing laminar thrombus. In this model, the ligature is tied down on a suture or needle on top of the vessel in the transverse plane and the needle or suture removed, decreasing blood flow by approximately 80%–90%. The degree of stenosis is inconsistent leading to thrombus variability.[42,91,93,95–99]

Mouse Electrolysis: The EIM is an alternative to the stenosis model.[140] In our laboratory, we modified the model described in dogs to form the EIM model in the mouse IVC.[91,93,95,96,100] A 25-gauge stainless steel needle, attached to a 30-gauge silver-coated copper wire, is inserted into the subcutaneous tissue (cathode) and caudal IVC (anode). A direct current from the copper wire through the stainless steel needle generates free radicals that induce endothelial cell activation, producing a consistent nonocclusive laminar thrombus.[91,101,140] The electrolysis model has also been well characterized using the mouse femoral vein.[102,103]

Mouse Chemical - Ferric Chloride: A small piece of filter paper saturated with ferric chloride solution is directly applied to the exposed vein and removed after 2–3 min. The oxidative damage induced causes a transmural injury to vessel wall, resulting in the formation of an occlusive thrombus within a matter of minutes.[104–106]

Mouse Chemical - Rose Bengal: This model involves the intravenous administration of a photo reactive substance, Rose Bengal, followed by illumination of an exposed venous segment with green laser light (540 nm).[141,142] The photochemical reaction that produces singlet oxygen damages the vascular endothelium and stimulates thrombus formation.[107] The chemical models do not closely represent the common clinical presentation of VT.[93]

Rat models of venous thrombosis

Rat Complete Stasis: The IVC is isolated as described in the mouse models. After occlusion, the IVC causes stasis-induced venous thrombosis. The stretching of the IVC during thrombosis imitates vein wall injury significant enough to promote tissue factor expression in endothelial cells and leukocytes, producing thrombosis.[108–112]

Rat Stenosis: This model is achieved by tying a suture around the IVC, just caudal to the renal veins, onto a 3–0 silk. The 3–0 silk was then removed to create a >90% stenosis in the IVC that creates a laminar venous thrombosis.[113–115]

Preclinical/translational models

Models are imperative in the evaluation of novel drug therapies, imaging or interventional therapeutics, the development of new endovascular or surgical therapeutics, and the immunologic and histopathological analysis of blood and venous tissue.[133,135,143] The Food and Drug Administration (FDA) requires preclinical product testing in rodent (mice, rats) and a nonrodent species (rabbit, canine, swine, nonhuman primate) for proof of concept and safety data before clinical trials can be initiated (Table 31.1).[136,144,145]

Rabbit models of venous thrombosis

Rabbit Compete Stasis: These models commonly study a surgically dissected external jugular vein. Using vascular clamps, a 2–4 cm segment of the external jugular vein is isolated between the clamps to induce stasis. The rabbit can easily accommodate venous and arterial access for therapeutic drug administration and serial blood collections for PK/PD studies.[116,117]

Rabbit Stenosis: This model is created by the introduction of cotton threads attached to copper wire introduced into the abdominal vena cava via the femoral vein that induces a nonocclusive fibrin rich thrombus.[118,119]

Rabbit Ferric Chloride: This model is created by surgically isolating a segment of the external jugular vein. Unlike in the mouse, the rabbit is large enough to obtain Doppler flow readings to document venous blood flow and time to occlusion postthrombus induction. Thrombosis is induced by placing a piece of filter paper saturated with 70% $FeCl_3$ on the external jugular vein.[120]

Canine model of venous thrombosis

Canine Complete Stasis and Canine Electrolysis: Early evaluations defining the role of neutrophils and platelets during venous thrombosis were performed in canines.[146–150] The external jugular vein (by electrolysis),[124,125] femoral vein,[121] and IVC (by balloon catheter occlusion) have been utilized to study venous thrombosis.[122,123]

Swine deep vein thrombosis model

Swine Complete Stasis: Swine models of venous thrombosis are useful and are still under development.[134] Swine have similar vein size and anatomy when compared to humans, thus making it an ideal model for medical device and endovenous techniques. The most common models isolate a segment of the iliac vein or caudal IVC by opposing balloon catheters to create a stasis-induced thrombus. Swine are considered hypercoagulable in comparison with humans and other species.[126,127]

Nonhuman primate models of venous thrombosis

Extensive reviews have been done on primate models of thrombosis.[87,133] Well-designed large animal studies justify investing resources on study design, as opposed to a failed clinical trials due to wrong dose, wrong animal model, or wrong indication.[135,143] The phylogenetic similarities of nonhuman primates to humans enable immunologic assays, drug pharmacokinetics, and antithrombotic compounds to be evaluated in these models.[7,14,128–131,151,152] Their hematologic and vascular homology to humans allow the determination of adverse effects of investigational agents on coagulation,[153,154] platelet function,[153,155–157] and fibrinolysis prior to clinical trials.[7,14,130,131,151,158–160]

Nonhuman primate complete stasis

Baboon IVC balloon occlusion model: This model is used as a model of venous stasis thrombosis induced by balloon catheter occlusion of the IVC.[128,129] Opposing balloon catheters are placed to induce a 6-hour period of complete blood stasis in the IVC.[128,129]

Baboon Iliac vein balloon occlusion model: This model allows for the evaluation of lower extremity venous thrombosis in the iliac vein with the contralateral vein serving as a naive control. First, inflow veins such as the hypogastric must be ligated to isolate the iliac vein segment. Balloon catheters are placed and positioned at the distal and proximal ends of the iliac vein. The balloons are inflated for 6 h and ultrasound imaging is used to determine the cessation of blood flow. After occlusive thrombosis is confirmed by ultrasound imaging, the balloons are deflated, removed, and the primate recovered and studied temporally using imaging modalities, coagulation, and molecular assays.[7,14,128–131]

Baboon Common femoral vein occlusion model: This model is used to evaluate venous thrombosis, tissue inflammation, and diagnostic imaging of platelets.[132] A segment of the right common femoral vein is completely occluded with a nonreactive ligature immediately proximal to the profunda femoris vein.

Research reproducibility/comparative pathology/statistical consultation

Careful experimental design and planning are necessary to employ standards that ensure reproducible science and transparent reporting.[136] In 2010, the Arrive Guidelines (Animal Research: Reporting of *In Vivo* Experiments) were implemented to improve reporting of results in animal research,[137,138] providing recommendations on reporting in vivo experiments to promote standardization.[134] Defining endpoints and using comparative pathology to validate animal end points related to human venous pathology is imperative. Thus, it is essential to work with a comparative pathologist and a statistician to determine the translational potential of animal models of venous disease to evaluate their applicability to the human clinical condition.[139] In general, preclinical research requires product testing in rodent and or nonrodent species (commonly rabbits) for proof of concept and safety data before large animal models are employed like canines, swine, ruminants, and nonhuman primate.[133,135]

New areas of venous research in thrombogenesis, resolution, and novel approaches for the future

Despite a high burden of disease, VTE remains poorly studied. A call to action in 2008 by the United States Surgeon General aimed to raise awareness, develop evidence-based practices, and address knowledge gaps. 12 years later, many gaps still remained, as recognized by a recent summary of research priorities issued jointly by the American Heart Association and the International Society on Thrombosis and Haemostasis.[161] Highlighted below are some of the major identified gaps in VTE care ranging from the preclinical to postimplementation setting, recognizing emerging trends and possible novel approaches for advancement of the field. The need for applicable animal models was addressed in the previous section.

Biomarkers

The sole biomarker in clinical use for the diagnosis and treatment of VTE is D-dimer. A low D-dimer in combination with a clinical score such as a low Wells' score can effectively exclude VTE. While sensitive, D-dimer is not specific—it is elevated in a wide range of medical conditions and increases with age. Identification of a single or combination of biomarkers would make VTE testing accessible to healthcare providers without access to a diagnostic ultrasound vascular laboratory. Recently, in a prospective cohort of 294 patients, soluble(s) P selectin was found to have a specificity of 97.5% and positive predictive value (PPV) of 91%, compared to D-dimer specificity of 65% and PPV 69%.[162] In another cohort of 178 patients, sP selectin >90 in combination with a Wells' score of ≥2 established a diagnosis of DVT with a specificity of 96% and a PPV of 100% and could exclude DVT (<60 ng/mL and Wells <2) with a sensitivity of 99% and negative predictive value of 96%.[163] However, use of sP selectin remains investigational at this time. Another promising biomarker is fibrin monomer complex (FMC). In comparison to D-dimer which results from thrombus degradation, FMC is reflective of early thrombus formation and thus has the benefit of potentially being a better predictor of VTE formation. As an example, FMC concentration measured postoperatively at 24 h was a superior predictor of VTE when compared to D-dimer.[164] Other biomarkers under investigation for the diagnosis of VTE include Factor VIII and circulating DNA.[165,166]

Beyond the narrow scope of determining VTE diagnosis, unbiased assessment of genomics, proteomics, and metabolomics involved in VTE is necessary in developing methodology for personalized assessment of VTE risk and response to treatment. Recently, a large consortium GWAS analysis of the Million Veterans Project identified a polygenic risk score for VTE risk and identified overlap with arterial risk factors.[167] Such findings in combination with identification of at-risk epigenetic modifiers and functional circulating proteins may be central to identifying (1) patients most likely to benefit from prophylaxis and treatment, (2) to develop postthrombotic syndrome and to require additional intervention, (3) risk—benefit balance of longterm anticoagulation, and (4) likely to benefit from antiinflammatory, statin, or other adjunct therapy. The HERDOO2 clinical scoring system has been validated to accurately identify women at low risk of recurrent VTE after a single episode and may safely discontinue anticoagulation after short-term (standard) treatment.[168] Clinical prediction models may be used to identify patients at low risk of recurrent VTE after a single episode who may safely discontinue anticoagulation after short-term (standard) treatment, but have limitations. One can imagine the cost and morbidity associated with anticoagulant therapy that could be avoided if a similar clinical prediction model could be extended to men, who are at higher risk of recurrent VTE.

Specific clinical scenarios

While the "gold standard" for clinical trial design remains the randomized control trial, such trials can be difficult to undertake for a variety of reasons: cost, competitive peer-reviewed funding, regulatory requirements, and recruitment of patients. The recently completed CACTUS trial, in which low-risk patients were randomized to undergo serial imaging compared to standard anticoagulation for treatment of symptomatic calf vein DVT, underscores this problem: the trial was underpowered (total enrollment 259, required 572 for adequate statistical power).[169] A multitude of nuances remain unanswered regarding the optimal treatment strategy for distal DVT, for example, how to best treat the incidental diagnosis in an asymptomatic patient with risk factors for extension, as well as optimal treatment duration and intensity. Other similar practical questions remain: what is optimal perioperative bridging strategy, balancing bleeding risk, and thrombosis risk, for patients undergoing major surgery within 3 months of VTE?[170] For superficial thrombophlebitis of >5 cm, which patient population should be treated with anticoagulant therapy to achieve the best cost–benefit ratio?[171] For many of these questions, the answer may lie with leverage of organizations such as the invent-VTE global collaboration.[172] This not-for-profit collaborative trial group utilizes a multinational platform to engage like-minded investigators to engage in answering patient oriented key clinical research questions via "Wiki" funding or underwriting from pharmaceutical companies. A second option is the use of large-scale patient registries which allow prospective data collection on a specific condition. Patient registries allow for observational study of less common disease processes and can be beneficial in understanding the real world results of therapy rather than the outcomes of a narrowly selected population of an RCT.[173] For example, the VQI venous stent registry launched in October 2019 will allow the long-term follow-up of patients undergoing venous stenting for obstruction, which will provide unbiased insight beyond the short-term follow-up of the clinical trials required for FDA approval.

Medical and surgical evolution

In his book, *"The Structure of Scientific Revolutions"* Thomas Kuhn makes the argument that science is advanced not in a steady, linear fashion, but rather via large paradigm shifts once a critical mass of data has accumulated. Despite the rapid acceleration of new thrombectomy devices, venous stents, and anticoagulant drugs brought to market in the last decade, the paradigm of VTE treatment has remained the same: treat with anticoagulant/fibrinolytic therapy, apply compression, and remove obstruction. In parallel fields such as cancer, inflammatory bowel disease, and autoimmune disorder, biologics have revolutionized treatment: 6 of the 10 best-selling pharmaceuticals today are biologics and half of all clinical trials involve the use of injectable proteins related to antibodies.[174] With an unprecedented level of interest from industry, rapid endovenous technology evolution, sophisticated drug delivery platforms, and increased understanding of the interplay between immunity and thrombosis, the treatment paradigm for VTE is primed for a massive revolution in the upcoming years. Focus on defining the preclinical and clinic problems as described and a broader understanding of the epidemiologic underpinnings of VTE are crucial to ensuring this progress.

References

1. Deitelzweig SB, Johnson BH, Lin J, Schulman KL. Prevalence of clinical venous thromboembolism in the USA: current trends and future projections. *Am J Hematol*. 2011;86:217–220.
2. Palabrica T, Lobb R, Furie BC, et al. Leukocyte accumulation promoting fibrin deposition is mediated in vivo by P-selectin on adherent platelets. *Nature*. 1992;359:848–851.
3. Myers Jr D, Farris D, Hawley A, et al. Selectins influence thrombosis in a mouse model of experimental deep venous thrombosis. *J Surg Res*. 2002;108:212–221.
4. Myers DD, Hawley AE, Farris DM, et al. P-selectin and leukocyte microparticles are associated with venous thrombogenesis. *J Vasc Surg*. 2003;38:1075–1089.
5. Etulain J, Martinod K, Wong SL, Cifuni SM, Schattner M, Wagner DD. P-selectin promotes neutrophil extracellular trap formation in mice. *Blood*. 2015;126:242–246.
6. Ramacciotti E, Myers Jr DD, Wrobleski SK, et al. P-selectin/PSGL-1 inhibitors versus enoxaparin in the resolution of venous thrombosis: a meta-analysis. *Thromb Res*. 2010;125:e138–e142.
7. Diaz JA, Wrobleski SK, Alvarado CM, et al. P-selectin inhibition therapeutically promotes thrombus resolution and prevents vein

wall fibrosis better than enoxaparin and an inhibitor to von Willebrand factor. *Arterioscler Thromb Vasc Biol*. 2015;35:829–837.
8. Sullivan VV, Hawley AE, Farris DM, et al. Decrease in fibrin content of venous thrombi in selectin-deficient mice. *J Surg Res*. 2003; 109:1–7.
9. Jilma B, Marsik C, Kovar F, Wagner OF, Jilma-Stohlawetz P, Endler G. The single nucleotide polymorphism Ser128Arg in the E-selectin gene is associated with enhanced coagulation during human endotoxemia. *Blood*. 2005;105:2380–2383.
10. Jilma B, Kovar FM, Hron G, et al. Homozygosity in the single nucleotide polymorphism Ser128Arg in the E-selectin gene associated with recurrent venous thromboembolism. *Arch Intern Med*. 2006;166:1655–1659.
11. Chase SD, Magnani JL, Simon SI. E-selectin ligands as mechanosensitive receptors on neutrophils in health and disease. *Ann Biomed Eng*. 2012;40:849–859.
12. Culmer DL, Dunbar ML, Hawley AE, et al. E-selectin inhibition with GMI-1271 decreases venous thrombosis without profoundly affecting tail vein bleeding in a mouse model. *Thromb Haemostasis*. 2017;117:1171–1181.
13. Devata S, Angelini D, Blackburn S, et al. Use of GMI-1271, an E-selectin antagonist in healthy subjects and in 2 patients with calf vein thrombosis. *Pract Thromb Haemost*. 2020:1–12.
14. Myers Jr D, Lester P, Adili R, et al. A new way to treat proximal deep venous thrombosis using E-selectin inhibition. *J Vasc Surg Venous Lymphat Disord*. 2020;8:268–278.
15. Raffetto JD. Understanding venous thrombosis pathways to effect pharmacologic treatment and resolution of venous thrombosis without increasing bleeding risk. *J Vasc Surg Venous Lymphat Disord*. 2020;8:279.
16. Myers DD Jr, Ning J, Lester P, Adili R, Hawley A, Durham L, Dunivant V, Reynolds G, Crego K, Sood S, Sigler R, Fogler WE, Magnani JL, Holinstat M, Wakefield T. E-selectin inhibitor is superior to low-molecular-weight heparin for the treatment of experimental venous thrombosis. J Vasc Surg Venous Lymphat Disord. 2021 Apr 17:S2213-333X(21)00195-5. doi: 10.1016/j.jvsv.2020.12.086. Epub ahead of print. PMID: 33872819.
17. Bittar LF, Silva LQD, Orsi FLA, et al. Increased inflammation and endothelial markers in patients with late severe post-thrombotic syndrome. *PLoS One*. 2020;15:e0227150.
18. Sevitt S. The structure and growth of valve-pocket thrombi in femoral veins. *J Clin Pathol*. 1974;27:517–528.
19. van Langevelde K, Sramek A and Rosendaal FR. The effect of aging on venous valves. Arterioscler Thromb Vasc Biol. 30:2075-2080.
20. Welsh JD, Hoofnagle MH, Bamezai S, et al. Hemodynamic regulation of perivalvular endothelial gene expression prevents deep venous thrombosis. *J Clin Invest*. 2019;129:5489–5500.
21. Brooks EG, Trotman W, Wadsworth MP, et al. Valves of the deep venous system: an overlooked risk factor. *Blood*. 2009;114: 1276–1279.
22. Labropoulos N, Waggoner T, Sammis W, Samali S, Pappas PJ. The effect of venous thrombus location and extent on the development of post-thrombotic signs and symptoms. *J Vasc Surg*. 2008;48: 407–412.
23. Deatrick KB, Elfline M, Baker N, et al. Postthrombotic vein wall remodeling: preliminary observations. *J Vasc Surg*. 2011;53: 139–146.
24. Meissner MH, Caps MT, Zierler BK, et al. Determinants of chronic venous disease after acute deep venous thrombosis. *J Vasc Surg*. 1998;28:826–833.
25. Laser A, Elfline M, Luke C, et al. Deletion of cysteine-cysteine receptor 7 promotes fibrotic injury in experimental post-thrombotic vein wall remodeling. *Arterioscler Thromb Vasc Biol*. 2014;34:377–385.
26. Comerota AJ, Oostra C, Fayad Z, et al. A histological and functional description of the tissue causing chronic postthrombotic venous obstruction. *Thromb Res*. 2015;135:882–887.
27. Wakefield TW, Myers DD, Henke PK. Mechanisms of venous thrombosis and resolution. *Arterioscler Thromb Vasc Biol*. 2008;28: 387–391.
28. Swystun LL, Liaw PC. The role of leukocytes in thrombosis. *Blood*. 2016;128:753–762.
29. Henke PK, Wakefield T. Thrombus resolution and vein wall injury: dependence on chemokines and leukocytes. *Thromb Res*. 2009;123(suppl 4):S72–S78.
30. Saha P, Humphries J, Modarai B, et al. Leukocytes and the natural history of deep vein thrombosis: current concepts and future directions. *Arterioscler Thromb Vasc Biol*. 2011;31:506–512.
31. Engelmann B, Massberg S. Thrombosis as an intravascular effector of innate immunity. *Nat Rev Immunol*. 2013;13:34–45.
32. Wakefield TW, Myers DD, Henke PK. Role of selectins and fibrinolysis in VTE. *Thromb Res*. 2009;123(suppl 4):S35–S40.
33. Budnik I, Brill A. Immune factors in deep vein thrombosis initiation. *Trends Immunol*. 2018;39:610–623.
34. Soehnlein O, Lindbom L. Phagocyte partnership during the onset and resolution of inflammation. *Nat Rev Immunol*. 2010;10: 427–439.
35. Nathan C, Ding A. Nonresolving inflammation. *Cell*. 2010;140: 871–882.
36. Baldwin JF, Sood V, Elfline MA, et al. The role of urokinase plasminogen activator and plasmin activator inhibitor-1 on vein wall remodeling in experimental deep vein thrombosis. *J Vasc Surg*. 2012;56:1089–1097.
37. Henke PK, Varma MR, Moaveni DK, et al. Fibrotic injury after experimental deep vein thrombosis is determined by the mechanism of thrombogenesis. *Thromb Haemostasis*. 2007;98:1045–1055.
38. Oostra C, Henke P, Luke C, Acino R, Lurie F, Comerota AJ. Human chronic post-thrombotic intraluminal venous obstruction involves neovascularization. *J Vasc Surg Venous Lymphat Disord*. 2014;2:110.
39. Obi AT, Diaz JA, Ballard-Lipka NL, et al. Plasminogen activator-1 overexpression decreases experimental postthrombotic vein wall fibrosis by a non-vitronectin-dependent mechanism. *J Thromb Haemostasis*. 2014;12:1353–1363.
40. Deatrick KB, Luke CE, Elfline MA, et al. The effect of matrix metalloproteinase 2 and matrix metalloproteinase 2/9 deletion in experimental post-thrombotic vein wall remodeling. *J Vasc Surg*. 2013;58:1375–1384 e2.
41. Henke PK, Varma MR, Deatrick KB, et al. Neutrophils modulate post-thrombotic vein wall remodeling but not thrombus neovascularization. *Thromb Haemostasis*. 2006;95:272–281.
42. von Bruhl ML, Stark K, Steinhart A, et al. Monocytes, neutrophils, and platelets cooperate to initiate and propagate venous thrombosis in mice in vivo. *J Exp Med*. 2012;209:819–835.
43. Obi AT, Andraska E, Kanthi Y, et al. Endotoxaemia-augmented murine venous thrombosis is dependent on TLR-4 and ICAM-1, and potentiated by neutropenia. *Thromb Haemostasis*. 2017;117: 339–348.
44. Fuchs TA, Brill A, Duerschmied D, et al. Extracellular DNA traps promote thrombosis. *Proc Natl Acad Sci USA*. 2010;107:15880–15885.
45. Henke PK, Pearce CG, Moaveni DM, et al. Targeted deletion of CCR2 impairs deep vein thombosis resolution in a mouse model. *J Immunol*. 2006;177:3388–3397.
46. Henke PK, Varga A, De S, et al. Deep vein thrombosis resolution is modulated by monocyte CXCR2-mediated activity in a mouse model. *Arterioscler Thromb Vasc Biol*. 2004;24: 1130–1137.
47. Massberg S, Grahl L, von Bruehl ML, et al. Reciprocal coupling of coagulation and innate immunity via neutrophil serine proteases. *Nat Med*. 2010;16:887–896.

48. Stark K, Philippi V, Stockhausen S, et al. Disulfide HMGB1 derived from platelets coordinates venous thrombosis in mice. *Blood*. 2016;128:2435–2449.
49. Chang CF, Goods BA, Askenase MH, et al. Erythrocyte efferocytosis modulates macrophages towards recovery after intracerebral hemorrhage. *J Clin Invest*. 2018;128:607–624.
50. Singh I, Burnand KG, Collins M, et al. Failure of thrombus to resolve in urokinase-type plasminogen activator gene-knockout mice: rescue by normal bone marrow-derived cells. *Circulation*. 2003;107:869–875.
51. Sood V, Luke CE, Deatrick KB, et al. Urokinase plasminogen activator independent early experimental thrombus resolution: MMP2 as an alternative mechanism. *Thromb Haemostasis*. 2010; 104:1174–1183.
52. Henke PK, Wakefield TW, Kadell AM, et al. Interleukin-8 administration enhances venous thrombosis resolution in a rat model. *J Surg Res*. 2001;99:84–91.
53. Humphries J, Gossage JA, Modarai B, et al. Monocyte urokinase-type plasminogen activator up-regulation reduces thrombus size in a model of venous thrombosis. *J Vasc Surg*. 2009;50: 1127–1134.
54. Kimball AS, Obi AT, Luke CE, et al. Ly6CLo monocyte/macrophages are essential for thrombus resolution in a murine model of venous thrombosis. *Thromb Haemostasis*. 2020;120:289–299.
55. Murray PJ, Wynn TA. Protective and pathogenic functions of macrophage subsets. *Nat Rev Immunol*. 2011;11:723–737.
56. Murray PJ, Allen JE, Biswas SK, et al. Macrophage activation and polarization: nomenclature and experimental guidelines. *Immunity*. 2014;41:14–20.
57. Wynn TA, Chawla A, Pollard JW. Macrophage biology in development, homeostasis and disease. *Nature*. 2013;496:445–455.
58. Gundra UM, Girgis NM, Ruckerl D, et al. Alternatively activated macrophages derived from monocytes and tissue macrophages are phenotypically and functionally distinct. *Blood*. 2014;123: e110–e122.
59. Leuschner F, Rauch PJ, Ueno T, et al. Rapid monocyte kinetics in acute myocardial infarction are sustained by extramedullary monocytopoiesis. *J Exp Med*. 2012;209:123–137.
60. Hilgendorf I, Gerhardt LM, Tan TC, et al. Ly-6Chigh monocytes depend on Nr4a1 to balance both inflammatory and reparative phases in the infarcted myocardium. *Circ Res*. 2014;114: 1611–1622.
61. Dal-Secco D, Wang J, Zeng Z, et al. A dynamic spectrum of monocytes arising from the in situ reprogramming of CCR^{2+} monocytes at a site of sterile injury. *J Exp Med*. 2015;212:447–456.
62. Arnold L, Henry A, Poron F, et al. Inflammatory monocytes recruited after skeletal muscle injury switch into antiinflammatory macrophages to support myogenesis. *J Exp Med*. 2007;204: 1057–1069.
63. Anzai A, Choi JL, He S, et al. The infarcted myocardium solicits GM-CSF for the detrimental oversupply of inflammatory leukocytes. *J Exp Med*. 2017;214:3293–3310.
64. Varga T, Mounier R, Horvath A, et al. Highly dynamic transcriptional signature of distinct macrophage subsets during sterile inflammation, resolution, and tissue repair. *J Immunol*. 2016;196: 4771–4782.
65. Hamers AAJ, Dinh HQ, Thomas GD, et al. Human monocyte heterogeneity as revealed by high-dimensional mass cytometry. *Arterioscler Thromb Vasc Biol*. 2019;39:25–36.
66. Mukhopadhyay S, Antalis TM, Nguyen KP, Hoofnagle MH, Sarkar R. Myeloid p53 regulates macrophage polarization and venous thrombus resolution by inflammatory vascular remodeling in mice. *Blood*. 2017;129:3245–3255.
67. Schonfelder T, Brandt M, Kossmann S, et al. Lack of T-bet reduces monocytic interleukin-12 formation and accelerates thrombus resolution in deep vein thrombosis. *Sci Rep*. 2018;8:3013.
68. Nosaka M, Ishida Y, Kimura A, et al. Absence of IFN-gamma accelerates thrombus resolution through enhanced MMP-9 and VEGF expression in mice. *J Clin Invest*. 2011;121:2911–2920.
69. Luther N, Shahneh F, Brahler M, et al. Innate effector-memory T-cell activation regulates post-thrombotic vein wall inflammation and thrombus resolution. *Circ Res*. 2016;119:1286–1295.
70. Ponomaryov T, Payne H, Fabritz L, Wagner DD, Brill A. Mast cells granular contents are crucial for deep vein thrombosis in mice. *Circ Res*. 2017;121:941–950.
71. Yadav V, Chi L, Zhao R, et al. Ectonucleotidase tri(di) phosphohydrolase-1 (ENTPD-1) disrupts inflammasome/interleukin 1beta-driven venous thrombosis. *J Clin Invest*. 2019;129: 2872–2877.
72. Deatrick KB, Eliason JL, Lynch EM, et al. Vein wall remodeling after deep vein thrombosis involves matrix metalloproteinases and late fibrosis in a mouse model. *J Vasc Surg*. 2005;42:140–148.
73. Deatrick KB, Obi A, Luke CE, et al. Matrix metalloproteinase-9 deletion is associated with decreased mid-term vein wall fibrosis in experimental stasis DVT. *Thromb Res*. 2013;132:360–366.
74. Obi AT, Diaz JA, Ballard-Lipka NL, et al. Low-molecular-weight heparin modulates vein wall fibrotic response in a plasminogen activator inhibitor 1-dependent manner. *J Vasc Surg Venous Lymphat Disord*. 2014;2:441–450 e1.
75. Bochenek ML, Leidinger C, Rosinus NS, et al. Activated endothelial TGFbeta1 signaling promotes venous thrombus nonresolution in mice via endothelin-1: potential role for chronic thromboembolic pulmonary hypertension. *Circ Res*. 2020;126:162–181.
76. Metz AK, Luke CE, Dowling A, Henke PK. Acute experimental venous thrombosis impairs venous relaxation but not contraction. *J Vasc Surg*. 2020;71, 1006.e1–1012 e1.
77. Nosaka M, Ishida Y, Kimura A, et al. Crucial involvement of IL-6 in thrombus resolution in mice via macrophage recruitment and the induction of proteolytic enzymes. *Front Immunol*. 2019;10: 3150.
78. Nguyen KP, McGilvray KC, Puttlitz CM, Mukhopadhyay S, Chabasse C, Sarkar R. Matrix metalloproteinase 9 (MMP-9) regulates vein wall biomechanics in murine thrombus resolution. *PLoS One*. 2015;10:e0139145.
79. Henke PK, Mitsuya M, Luke CE, et al. Toll-like receptor 9 signaling is critical for early experimental deep vein thrombosis resolution. *Arterioscler Thromb Vasc Biol*. 2011;31:43–49.
80. Dewyer NA, El-Sayed OM, Luke CE, et al. Divergent effects of Tlr9 deletion in experimental late venous thrombosis resolution and vein wall injury. *Thromb Haemostasis*. 2015;114:1028–1037.
81. Dyer MR, Chen Q, Haldeman S, et al. Deep vein thrombosis in mice is regulated by platelet HMGB1 through release of neutrophil-extracellular traps and DNA. *Sci Rep*. 2018;8:2068.
82. Evans CE, Humphries J, Waltham M, et al. Upregulation of hypoxia-inducible factor 1 alpha in local vein wall is associated with enhanced venous thrombus resolution. *Thromb Res*. 2011; 128:346–351.
83. Modarai B, Humphries J, Burnand KG, et al. Adenovirus-mediated VEGF gene therapy enhances venous thrombus recanalization and resolution. *Arterioscler Thromb Vasc Biol*. 2008;28: 1753–1759.
84. Kahn SR, Comerota AJ, Cushman M, et al, American Heart association Council on Peripheral vascular disease CoCC, Council on C and Stroke N. The postthrombotic syndrome: evidence-based prevention, diagnosis, and treatment strategies: a scientific statement from the American Heart Association. *Circulation*. 2014; 130:1636–1661.
85. Andraska EA, Luke CE, Elfline MA, et al. Pre-clinical model to study recurrent venous thrombosis in the inferior vena cava. *Thromb Haemostasis*. 2018;118:1048–1057.

86. Virchow R. *Gesammelte Abhandlungen zur wissenschaftlichen Medicin*. Frankfurt: A M: Verlag Von Meidinger Sohn & Comp.; 1856.
87. Jagadeeswaran P, Cooley BC, Gross PL, Mackman N. Animal models of thrombosis from Zebrafish to nonhuman primates: use in the elucidation of new pathologic pathways and the development of antithrombotic drugs. *Circ Res*. 2016;118:1363–1379.
88. Diaz JA, Saha P, Cooley B, et al. Choosing a mouse model of venous thrombosis. *Arterioscler Thromb Vasc Biol*. 2019;39:311–318.
89. Grover SP, Evans CE, Patel AS, Modarai B, Saha P, Smith A. Assessment of venous thrombosis in animal models. *Arterioscler Thromb Vasc Biol*. 2016;36:245–252.
90. Day SM, Reeve JL, Myers DD, Fay WP. Murine thrombosis models. *Thromb Haemostasis*. 2004;92:486–494.
91. Diaz JA, Obi AT, Myers Jr DD, et al. Critical review of mouse models of venous thrombosis. *Arterioscler Thromb Vasc Biol*. 2012;32:556–562.
92. Mackman N. Mouse models of venous thrombosis are not equal. *Blood*. 2016;127:2510–2511.
93. Schonfelder T, Jackel S, Wenzel P. Mouse models of deep vein thrombosis. *Gefässchirurgie*. 2017;22:28–33.
94. Albadawi H, Witting AA, Pershad Y, et al. Animal models of venous thrombosis. *Cardiovasc Diagn Ther*. 2017;7:S197–S206.
95. Burnand KG, Gaffney PJ, McGuinness CL, Humphries J, Quarmby JW, Smith A. The role of the monocyte in the generation and dissolution of arterial and venous thrombi. *Cardiovasc Surg*. 1998;6:119–125.
96. Singh I, Smith A, Vanzieleghem B, et al. Antithrombotic effects of controlled inhibition of factor VIII with a partially inhibitory human monoclonal antibody in a murine vena cava thrombosis model. *Blood*. 2002;99:3235–3240.
97. Brandt M, Schonfelder T, Schwenk M, et al. Deep vein thrombus formation induced by flow reduction in mice is determined by venous side branches. *Clin Hemorheol Microcirc*. 2014;56:145–152.
98. Brill A, Fuchs TA, Chauhan AK, et al. von Willebrand factor-mediated platelet adhesion is critical for deep vein thrombosis in mouse models. *Blood*. 2011;117:1400–1407.
99. Brill A, Fuchs TA, Savchenko AS, et al. Neutrophil extracellular traps promote deep vein thrombosis in mice. *J Thromb Haemostasis*. 2012;10:136–144.
100. Romson JL, Haack DW, Lucchesi BR. Electrical induction of coronary artery thrombosis in the ambulatory canine: a model for in vivo evaluation of anti-thrombotic agents. *Thromb Res*. 1980;17:841–853.
101. Diaz JA, Wrobleski SK, Hawley AE, Lucchesi BR, Wakefield TW, Myers Jr DD. Electrolytic inferior vena cava model (EIM) of venous thrombosis. *J Vis Exp*. 2011:e2737.
102. Cooley BC. In vivo fluorescence imaging of large-vessel thrombosis in mice. *Arterioscler Thromb Vasc Biol*. 2011;31:1351–1356.
103. Cooley BC, Szema L, Chen CY, Schwab JP, Schmeling G. A murine model of deep vein thrombosis: characterization and validation in transgenic mice. *Thromb Haemostasis*. 2005;94:498–503.
104. Chauhan AK, Kisucka J, Lamb CB, Bergmeier W, Wagner DD. von Willebrand factor and factor VIII are independently required to form stable occlusive thrombi in injured veins. *Blood*. 2007;109:2424–2429.
105. Buyue Y, Whinna HC, Sheehan JP. The heparin-binding exosite of factor IXa is a critical regulator of plasma thrombin generation and venous thrombosis. *Blood*. 2008;112:3234–3241.
106. Yang M, Kirley TL. Engineered human soluble calcium-activated nucleotidase inhibits coagulation in vitro and thrombosis in vivo. *Thromb Res*. 2008;122:541–548.
107. Eitzman DT, Westrick RJ, Nabel EG, Ginsburg D. Plasminogen activator inhibitor-1 and vitronectin promote vascular thrombosis in mice. *Blood*. 2000;95:577–580.
108. Zhou J, May L, Liao P, Gross PL, Weitz JI. Inferior vena cava ligation rapidly induces tissue factor expression and venous thrombosis in rats. *Arterioscler Thromb Vasc Biol*. 2009;29:863–869.
109. Henke PK, DeBrunye LA, Strieter RM, et al. Viral IL-10 gene transfer decreases inflammation and cell adhesion molecule expression in a rat model of venous thrombosis. *J Immunol*. 2000;164:2131–2141.
110. Herbert JM, Bernat A, Maffrand JP. Importance of platelets in experimental venous thrombosis in the rat. *Blood*. 1992;80:2281–2286.
111. Berry CN, Girard D, Lochot S, Lecoffre C. Antithrombotic actions of argatroban in rat models of venous, 'mixed' and arterial thrombosis, and its effects on the tail transection bleeding time. *Br J Pharmacol*. 1994;113:1209–1214.
112. Seth P, Kumari R, Dikshit M, Srimal RC. Effect of platelet activating factor antagonists in different models of thrombosis. *Thromb Res*. 1994;76:503–512.
113. Myers Jr DD, Henke PK, Bedard PW, et al. Treatment with an oral small molecule inhibitor of P selectin (PSI-697) decreases vein wall injury in a rat stenosis model of venous thrombosis. *J Vasc Surg*. 2006;44:625–632.
114. Baxi S, Crandall DL, Meier TR, et al. Dose-dependent thrombus resolution due to oral plaminogen activator inhibitor (PAI)-1 inhibition with tiplaxtinin in a rat stenosis model of venous thrombosis. *Thromb Haemostasis*. 2008;99:749–758.
115. Thanaporn P, Myers DD, Wrobleski SK, et al. P-selectin inhibition decreases post-thrombotic vein wall fibrosis in a rat model. *Surgery*. 2003;134:365–371.
116. Wienen W, Stassen JM, Priepke H, Ries UJ, Hauel N. Antithrombotic and anticoagulant effects of the direct thrombin inhibitor dabigatran, and its oral prodrug, dabigatran etexilate, in a rabbit model of venous thrombosis. *J Thromb Haemostasis*. 2007;5:1237–1242.
117. Matthiasson SE, Lindblad B, Bergqvist D. Prevention of experimental venous thrombosis in rabbits with different low molecular weight heparins, dermatan sulphate and hirudin. *Haemostasis*. 1995;25:124–132.
118. Hollenbach S, Sinha U, Lin PH, et al. A comparative study of prothrombinase and thrombin inhibitors in a novel rabbit model of non-occlusive deep vein thrombosis. *Thromb Haemostasis*. 1994;71:357–362.
119. Sinha U, Ku P, Malinowski J, et al. Antithrombotic and hemostatic capacity of factor Xa versus thrombin inhibitors in models of venous and arteriovenous thrombosis. *Eur J Pharmacol*. 2000;395:51–59.
120. Marsh Lyle E, Lewis SD, Lehman ED, Gardell SJ, Motzel SL, Lynch Jr JJ. Assessment of thrombin inhibitor efficacy in a novel rabbit model of simultaneous arterial and venous thrombosis. *Thromb Haemostasis*. 1998;79:656–662.
121. Dupe RJ, English PD, Smith RA, Green J. Acyl-enzymes as thrombolytic agents in dog models of venous thrombosis and pulmonary embolism. *Thromb Haemostasis*. 1984;51:248–253.
122. Trerotola SO, McLennan G, Eclavea AC, et al. Mechanical thrombolysis of venous thrombosis in an animal model with use of temporary caval filtration. *J Vasc Intervent Radiol*. 2001;12:1075–1085.
123. Trerotola SO, McLennan G, Davidson D, et al. Preclinical in vivo testing of the Arrow-Trerotola percutaneous thrombolytic device for venous thrombosis. *J Vasc Intervent Radiol*. 2001;12:95–103.
124. Hennan JK, Hong TT, Shergill AK, Driscoll EM, Cardin AD, Lucchesi BR. Intimatan prevents arterial and venous thrombosis

in a canine model of deep vessel wall injury. *J Pharmacol Exp Therapeut*. 2002;301:1151–1156.
125. Rebello SS, Blank HS, Rote WE, Vlasuk GP, Lucchesi BR. Antithrombotic efficacy of a recombinant nematode anticoagulant peptide (rNAP5) in canine models of thrombosis after single subcutaneous administration. *J Pharmacol Exp Therapeut*. 1997; 283:91–99.
126. Palsgaard-Van Lue A, Strom H, Lee MH, et al. Cellular, hemostatic, and inflammatory parameters of the surgical stress response in pigs undergoing partial pericardectomy via open thoracotomy or thoracoscopy. *Surg Endosc*. 2007;21:785–792.
127. Maxwell AD, Owens G, Gurm HS, Ives K, Myers Jr DD, Xu Z. Noninvasive treatment of deep venous thrombosis using pulsed ultrasound cavitation therapy (histotripsy) in a porcine model. *J Vasc Intervent Radiol*. 2011;22:369–377.
128. Wakefield TW, Strieter RM, Schaub R, et al. Venous thrombosis prophylaxis by inflammatory inhibition without anticoagulation therapy. *J Vasc Surg*. 2000;31:309–324.
129. Myers Jr DD, Schaub R, Wrobleski SK, et al. P-selectin antagonism causes dose-dependent venous thrombosis inhibition. *Thromb Haemostasis*. 2001;85:423–429.
130. Myers Jr DD, Wrobleski SK, Longo C, et al. Resolution of venous thrombosis using a novel oral small-molecule inhibitor of P-selectin (PSI-697) without anticoagulation. *Thromb Haemostasis*. 2007;97:400–407.
131. Meier TR, Myers Jr DD, Wrobleski SK, et al. Prophylactic P-selectin inhibition with PSI-421 promotes resolution of venous thrombosis without anticoagulation. *Thromb Haemostasis*. 2008; 99:343–351.
132. Dormehl IC, Jacobs DJ, Pretorius JP, Maree M, Franz RC. Baboon (*Papio ursinus*) model to study deep vein thrombosis using 111-indium-labeled autologous platelets. *J Med Primatol*. 1987;16: 27–38.
133. Myers Jr DD. Nonhuman primate models of thrombosis. *Thromb Res*. 2012;129(suppl 2):S65–S69.
134. Schwein A, Magnus L, Chakfe N, Bismuth J. Critical review of large animal models for central deep venous thrombosis. *Eur J Vasc Endovasc Surg*. 2020.
135. Leadley Jr RJ, Chi L, Rebello SS, Gagnon A. Contribution of in vivo models of thrombosis to the discovery and development of novel antithrombotic agents. *J Pharmacol Toxicol Methods*. 2000;43:101–116.
136. National Research Council (U.S.). *Committee for the Update of the Guide for the Care and Use of Laboratory Animals*. In: *Guide for the care and use of laboratory animals*. vol. xxv. Institute for Laboratory Animal Research (U.S.) and National Academies Press (U.S.); 2011:220.
137. Danos O, Davies K, Lehn P, Mulligan R. The ARRIVE guidelines, a welcome improvement to standards for reporting animal research. *J Gene Med*. 2010;12:559–560.
138. Drummond GB, Paterson DJ, McGrath JC. ARRIVE: new guidelines for reporting animal research. *J Physiol*. 2010;588:2517.
139. Everitt JI, Treuting PM, Scudamore C, et al. Pathology study design, conduct, and reporting to achieve rigor and reproducibility in translational research using animal models. *ILAR J*. 2018;59:4–12.
140. Diaz JA, Hawley AE, Alvarado CM, et al. Thrombogenesis with continuous blood flow in the inferior vena cava. A novel mouse model. *Thromb Haemostasis*. 2010;104:366–375.
141. Kikuchi S, Umemura K, Kondo K, Saniabadi AR, Nakashima M. Photochemically induced endothelial injury in the mouse as a screening model for inhibitors of vascular intimal thickening. *Arterioscler Thromb Vasc Biol*. 1998;18:1069–1078.
142. Matsuno H, Uematsu T, Nagashima S, Nakashima M. Photochemically induced thrombosis model in rat femoral artery and evaluation of effects of heparin and tissue-type plasminogen activator with use of this model. *J Pharmacol Methods*. 1991;25: 303–317.
143. Olejniczak KGP, Bass R. Preclinical testing strategies. *Drug Inf J*. 2001;35:321–336.
144. Welfare NIoHUSOoLA. *Public Health Service Policy on Humane Care and Use of Laboratory Animals*. 2002.
145. Prior H, Baldrick P, Beken S, et al. Opportunities for use of one species for longer-term toxicology testing during drug development: a cross-industry evaluation. *Regul Toxicol Pharmacol*. 2020; 113:104624.
146. Ritchie WG, Lynch PR, Stewart GJ. The effect of contrast media on normal and inflamed canine veins. A scanning and transmission electron microscopic study. *Invest Radiol*. 1974;9: 444–455.
147. Schaub RG, Lynch PR, Stewart GJ. The response of canine veins to three types of abdominal surgery: a scanning and transmission electron microscopic study. *Surgery*. 1978;83:411–424.
148. Stewart GJ, Schaub RG, Niewiarowski S. Products of tissue injury. Their induction of venous endothelial damage and blood cell adhesion in the dog. *Arch Pathol Lab Med*. 1980;104:409–413.
149. Stewart GJ. Neutrophils and deep venous thrombosis. *Haemostasis*. 1993;23(suppl 1):127–140.
150. Stewart GJ, Ritchie WG, Lynch PR. Venous endothelial damage produced by massive sticking and emigration of leukocytes. *Am J Pathol*. 1974;74:507–532.
151. Feingold HM, Pivacek LE, Melaragno AJ, Valeri CR. Coagulation assays and platelet aggregation patterns in human, baboon, and canine blood. *Am J Vet Res*. 1986;47:2197–2199.
152. Valeri CR, Ragno G. The survival and function of baboon red blood cells, platelets, and plasma proteins: a review of the experience from 1972 to 2002 at the Naval Blood Research Laboratory, Boston, Massachusetts. *Transfusion*. 2006;46:1S–42S.
153. Abildgaard CF, Harrison J, Johnson CA. Comparative study of blood coagulation in nonhuman primates. *J Appl Physiol*. 1971; 30:400–405.
154. Kelly CA, Gleiser CA. Selected coagulation reference values for adult and juvenile baboons. *Lab Anim Sci*. 1986;36:173–175.
155. Melaragno AJ, Abdu W, Katchis R, Doty A, Valeri CR. Liquid and freeze preservation of baboon platelets. *Cryobiology*. 1981;18: 445–452.
156. Vecchione JJ, Melaragno AJ, Weiblen BJ, Halkett JA, Callow AD, Valeri CR. Repeated intravenous administrations of human albumin and human fibrinogen in the baboon: survival measurements. *J Med Primatol*. 1982;11:91–105.
157. Callow AD, Ledig CB, O'Donnell TF, et al. A primate model for the study of the interaction of 111In-labeled baboon platelets with Dacron arterial prostheses. *Ann Surg*. 1980;191:362–366.
158. Hampton JW, Matthews C. Similarities between baboon and human blood clotting. *J Appl Physiol*. 1966;21:1713–1716.
159. Savage B, McFadden PR, Hanson SR, Harker LA. The relation of platelet density to platelet age: survival of low- and high-density 111indium-labeled platelets in baboons. *Blood*. 1986;68:386–393.
160. Todd ME, McDevitt E, Goldsmith EI. Blood-clotting mechanisms of nonhuman primates. Choice of the baboon model to simulate man. *J Med Primatol*. 1972;1:132–141.
161. Cushman M, Barnes GD, Creager MA, et al, American Heart Association Council on Peripheral Vascular D, Council on arteriosclerosis T, Vascular B, Council on C, Stroke N, Council on clinical C, Council on E, Prevention, the International Society on T and Haemostasis. Venous Thromboembolism Research Priorities: a scientific statement from the American Heart Association

and the International Society on Thrombosis and Haemostasis. *Circulation*. 2020;142:e85–e94.
162. Vandy FC, Stabler C, Eliassen AM, et al. Soluble P-selectin for the diagnosis of lower extremity deep venous thrombosis. *J Vasc Surg Venous Lymphat Disord*. 2013;1:117–1125.
163. Ramacciotti E, Blackburn S, Hawley AE, et al. Evaluation of soluble P-selectin as a marker for the diagnosis of deep venous thrombosis. *Clin Appl Thromb Hemost*. 2011;17:425–431.
164. Mitani G, Takagaki T, Hamahashi K, et al. Associations between venous thromboembolism onset, D-dimer, and soluble fibrin monomer complex after total knee arthroplasty. *J Orthop Surg Res*. 2015;10:172.
165. Jenkins PV, Rawley O, Smith OP, O'Donnell JS. Elevated factor VIII levels and risk of venous thrombosis. *Br J Haematol*. 2012; 157:653–663.
166. Audu CO, Gordon AE, Obi AT, Wakefield TW, Henke PK. Inflammatory biomarkers in deep venous thrombosis organization, resolution, and post-thrombotic syndrome. *J Vasc Surg Venous Lymphat Disord*. 2020;8:299–305.
167. Klarin D, Busenkell E, Judy R, et al. Genome-wide association analysis of venous thromboembolism identifies new risk loci and genetic overlap with arterial vascular disease. *Nat Genet*. 2019;51:1574–1579.
168. Rodger MA, Le Gal G, Anderson DR, et al. Validating the HERDOO2 rule to guide treatment duration for women with unprovoked venous thrombosis: multinational prospective cohort management study. *BMJ*. 2017;356:j1065.
169. Righini M, Galanaud JP, Guenneguez H, et al. Anticoagulant therapy for symptomatic calf deep vein thrombosis (CACTUS): a randomised, double-blind, placebo-controlled trial. *Lancet Haematol*. 2016;3:e556–e562.
170. Smith M, Wakam G, Wakefield T, Obi A. New trends in anticoagulation therapy. *Surg Clin North Am*. 2018;98:219–238.
171. Gianesini S, Obi A, Onida S, et al. Global guidelines trends and controversies in lower limb venous and lymphatic disease: narrative literature revision and experts' opinions following the vWINter international meeting in Phlebology, Lymphology & Aesthetics, 23-25 January 2019. *Phlebology*. 2019;34:4–66.
172. Rodger M, Langlois N, Middeldorp S, et al. Initial strides for invent-VTE: towards global collaboration to accelerate clinical research in venous thromboembolism. *Thromb Res*. 2018;163: 128–131.
173. Lurie F, Obi A, Schul M, Hofmann LV, Kasper G, Wakefield T. Venous disease patient registries available in the United States. *J Vasc Surg Venous Lymphat Disord*. 2018;6:118–125.
174. Monaco C, Nanchahal J, Taylor P, Feldmann M. Anti-TNF therapy: past, present and future. *Int Immunol*. 2015;27:55–62.

PART V

The vasculome in diagnosis, prevention, and treatment of other diseases

C H A P T E R

32

Targeting vascular zip codes: from combinatorial selection to drug prototypes

Tracey L. Smith[1,2], *Richard L. Sidman*[3], *Wadih Arap*[1,4] *and Renata Pasqualini*[1,2]

[1]Rutgers Cancer Institute of New Jersey, Newark, NJ, United States [2]Division of Cancer Biology, Department of Radiation Oncology, Rutgers New Jersey Medical School, Newark, NJ, United States [3]Department of Neurology, Harvard Medical School, Boston, MA, United States [4]Division of Hematology/Oncology, Department of Medicine, Rutgers New Jersey Medical School, Newark, NJ, United States

Introduction

Despite major advances from the Human Genome Project, the molecular diversity of receptors in the human vasculature remains largely unknown. Data from the Mouse and Human Genome Projects indicate that the phenotypic complexity of humans relative to other mammalian species derives from expression patterns of proteins at different tissue sites, levels, or times rather than from a greater number of genes.[1–3] Therefore, the concept of the human vascular map was formulated, wherein we proposed to utilize phage display technology to identify proteins that are specifically or selectively expressed in the vasculature of individual tissues or organ systems under normal or pathological conditions.[4,5] Endothelial cells, which are primarily targeted by the in vivo phage display methodology, are known to differ in their morphological properties in different tissues, but relatively few corresponding molecular differences (termed human vascular zip codes) had been exploited for clinical applications. Our methodology, direct selection of combinatorial libraries in cancer patients, sought to enable accelerated identification, validation, and prioritization of such differentially expressed receptors. Combinatorial peptide library selection from multiple patients has been reported.[6,7] This work is ongoing, as technological advances have enabled greater depth and breadth of data acquisition and analyses from subsequent patients. A robust drug development pipeline is also in place to evaluate the translational applications for the vascular-targeted agents identified. Five examples are presented below, but, first, we introduce phage and phage display technology.

A brief history of phage display

During his 2018 Nobel Lecture, Dr. George P. Smith presented phage display libraries as versatile tools that meet any goal defined by each researcher utilizing them. He also detailed the impetus for the transition of his lab to phage display: a sabbatical in a phage biology lab at Duke under the guidance Dr. Bob Webster, which was a fortuitous event for those of us who came after.[8]

Briefly, phage are viruses that naturally infect bacteria but have no native affinity for mammalian cells. Smith's original idea for exploiting phage as a unique selection tool was to genetically manipulate gene III of filamentous phage genomes derived from fd-tet. This would enable a foreign peptide to be displayed on the phage capsid. While engineering peptides displayed on a natural protein or "fusion protein" was routine in the early 1980s,[9] Smith went further and developed a phage display vector suitable for efficient engineering for generating libraries (the fUSE vector series),[10] which shaped the field for the decades to come. In this "filamentous fusion phage," about five copies of pIII

The Vasculome
https://doi.org/10.1016/B978-0-12-822546-2.25001-6

typically cap the tip of one end of the virions, which are less than 10 nm wide and about 1 μm long. Upon engineering the phage to display a foreign peptide, the infective properties of pIII are maintained, while the displayed peptides can interact with antibodies and other proteins. This interaction is the critical step, and once Smith confirmed functionality of the displayed peptide, he turned to peptide libraries and affinity selection.[8]

In the early years of phage display, the panning or affinity selection experiments were performed on antibodies immobilized on plastic to "map" antigen epitopes.[11,12] Soon thereafter, researchers began using phage display technology to affinity select the proteins most relevant to their research interests. Because the phage remains infective, bound phage can be recovered and propagated by bacterial infection. Multiple rounds of selection can be performed until a population of highly selective binders is obtained; the amino acid sequence of the peptides is then determined by sequencing the DNA corresponding to the insert in the phage genome. The potential range of applications for this technique is quite broad, and considerable progress has been made toward developing novel screening methods to serve specific purposes, including panning on cells, in mice or other animal models, and even in humans. Our goal was to identify molecular zip codes of human diseases and exploit peptide ligands for targeted drug delivery.

In vivo phage display and vascular zip codes

To identify vascular addresses, or zip codes, an in vivo selection method was developed in mice in which peptides that home to specific vascular beds are selected after intravenous administration of phage displaying random peptide libraries.[4] This in vivo selection of peptide phage libraries enabled the identification of ligands that are capable of homing to specific receptors of target tissues throughout the circulation (Fig. 32.1). Peptides homing to selective vascular beds in mice were first reported in 1996 for normal blood vessels[4] and in 1998 for tumor blood vessels.[5] Soon after, various peptide libraries had been screened for ligands selectively homing to just about every murine tissue, including lymph nodes, brain, kidney, lung, skin, pancreas, intestine, uterus, adrenal gland, retina, muscle, prostate, breast, placenta, and fat tissue.[4,13–16] Even organ microcompartments can be specifically targeted to reveal differentially expressed zip codes within specialized subcellular regions, such as markers of pancreatic islets not accessible on exocrine pancreatic tissue.[17] These and other studies have revealed a surprising degree of specialization in various normal tissues, which was a somewhat contentious finding for the time.

In vivo phage display is a remarkable tool also for the discovery of molecular markers in many diseases with a vascular component, which can then be targeted systemically to deliver a therapeutic or diagnostic payload. For example, specific molecular markers of the endothelium in cancer, obesity, and retinal diseases have been identified for selective delivery of imaging or therapeutic agents. An emerging trend is the identification of markers related to chronic or acute damage to blood vessels or the tissues they sustain. Specifically, in vivo phage display has been used to identify potential markers of atherosclerosis,[18,19] neurodegenerative diseases,[20] and traumatic brain injury.[21]

Our preference for in vivo selection is driven by the desire to identify receptors in their native—oftentimes disease-specific—conditions or orientations, including

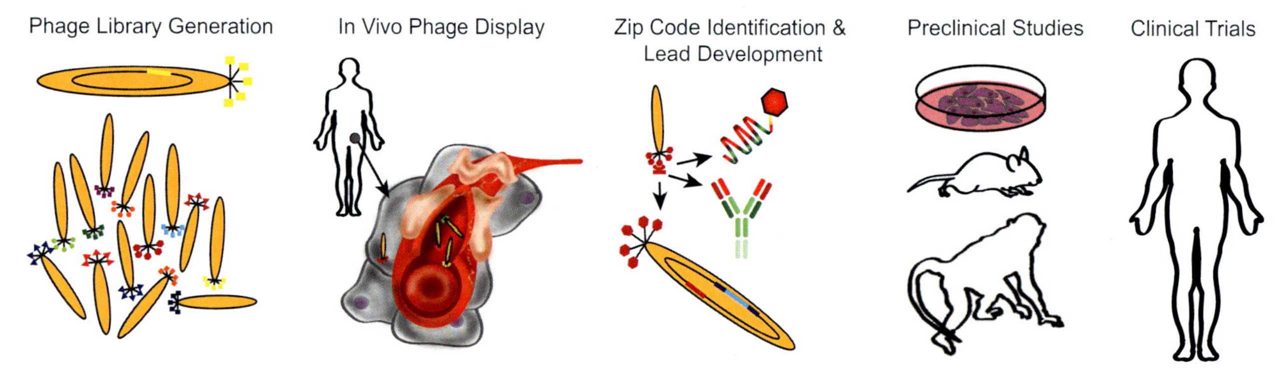

FIGURE 32.1 From phage library generation to clinical trials. First, peptide phage libraries are engineered by introducing sequences in the gene encoding pIII, which enables the display of random peptides on the capsid in ~5 copies per phage particle. The phage library can be administered to live organisms, including rodents, nonhuman primates, or even humans. Bound phage are recovered after circulating and the receptors bound identified as vascular zip codes. Strategies to deliver a therapeutic payload are employed and tested in preclinical studies and, ultimately, in clinical trials. The interleukin-11R/BMTP-11 story followed this discovery pipeline, from zip code discovery in a patient with prostate cancer to a Phase 1 trial of a targeted apoptotic fusion peptide.

microanatomical contexts and environment-dependent receptor features.[22] This is important because correct delivery of imaging agents or therapeutic payloads typically requires both accessibility and cell internalizing properties. One example of this is the transferrin (Tf) receptor targeting peptide CRTIGPSVC, which was found to reliably target and internalize into tumor cells specifically in a model of human-derived malignant glioblastoma. Further work showed that the CRTIGPSVC peptide could cross the blood–brain barrier (BBB) by binding apo-transferrin (apo-Tf), but not the iron-containing closed form of transferrin or the transferrin receptor, thereby partially recapitulating the conformational changes to transferrin typically observed after iron binding to enable transport across the BBB.[23] These unique findings indicate that phage-targeting properties are largely dictated by receptor accessibility to the circulating ligand peptide. Additional known factors such as increased vascular permeability leading to greater retention within tumors are likely to contribute to the enhanced targeting by CRTIGPSVC-directed phage particles. Notably, in a commentary associated with that publication, Nathanson and Mischel thoughtfully summarized the unique utility of in vivo phage display when they wrote, "This observation, in and of itself, demonstrates the benefit of phage display approaches for uncovering molecular mechanisms, since the peptides seem to retain key biological function."[24]

Despite efforts to systematically map protein interactions in model organisms, including human-derived biological systems, transitioning findings from rodents or even nonhuman primates to human patients remained challenging, particularly in the setting of complex biological processes such as blood vessel formation and cancer metastasis. Therefore, a series of phage peptide library screenings in human subjects was undertaken. This revealed novel peptides that localized nonrandomly to different organs and diseased tissues, and revealed associated vascular zip codes, including clinically relevant molecular markers of disease.

Human vascular mapping project

The molecular mapping of human blood vessels is ongoing. After designing and implementing ethics guidelines for research with brain dead or terminal wean patients ineligible for organ transplantation but wishing to participate in research studies,[25,26] the first patient screening was published in 2002.[6] Following a short 15-minute infusion of the phage library, tissue samples were collected by biopsy or surgical excision and any phage present in the samples were recovered by bacterial infection. After sequencing the peptide inserts, frequency distribution and other data analyses were completed to reveal the candidate ligand, and the RRAGGS peptide motif was validated for homing to the prostate. This peptide is a mimic of interleukin-11 and binds its receptor IL-11Rα.

After confirming it was feasible to complete synchronous rounds of selection in mice,[27] this approach was utilized in three human patients. The synchronous in vivo phage display approach allows for serial selection of homing peptides to multiple organs in parallel, so phage recovery from various organs in a single screen in patients became feasible. The systematic approach adopted for the isolation of tissue-specific and pan-vascular peptides in living human patients required three rounds of serial selection of a phage display random 7mer cyclic peptide library. The rapid evolution of next-generation sequencing and data analytics fortuitously coincided with this programmatic timeline, so phage inserts were sequenced by Sanger and novel pyrosequencing methodologies.[28] Monte Carlo simulations and high-throughput tripeptide analyses were used to evaluate positive selection of peptides compared to the random peptide library, and tissue-specific and pan-vascular peptide candidates were selected for further biostatistical and biochemical analyses.

Ultimately, four ligand–receptor pairs were validated from the hundreds of thousands of sequences compiled from selections in these patients. Two ligand–receptor pairs (integrin α4/annexin A4 and cathepsin B/apolipoprotein E3) were previously unknown and were pan-vascular targets shared among multiple human tissues. The other two interacting pairs had a restricted and specific distribution in white fat (prohibitin/annexin A2 [ANX2]) or cancer (RAGE/proteinase 3 [PR3] in bone metastases).[7] Importantly, RAGE expressed on prostate tumor cells interacts with PR3-expressing cells of the bone marrow microenvironment to mediate bone metastases during prostate cancer progression.[29] Implications for clinical interventions are being pursued for this target, but of particular relevance here is the other tissue-specific vascular marker that was validated in white fat.

ANX2 and prohibitin were validated as a ligand–receptor pair in human white adipose tissue (WAT) vasculature, confirming earlier findings in mice.[16] Evolutionary conservation of prohibitin indicated this likelihood, so peptides enriched in WAT from the human studies were compared against the ANXA2 and prohibitin sequences, revealing dozens of ligands that matched exposed regions in these two proteins, including CKGGRAKDC.[7] Further analyses demonstrated that ANX2 and prohibitin are coexpressed on blood vessels in fatty tissue. These proteins form a complex with the fatty acid transporter CD36 to direct fatty acid transport in WAT, implicating this pathway for intervention in obesity and metabolic diseases.[30]

A comprehensive human vascular map is still far from completion. As efforts continue toward a global map of functional ligand−receptors in human blood vessels for research and medical applications, advances in computing power and next-generation sequencing as well as access to autopsy material from phage screenings in patients will lead to the rapid expansion of the number of available peptide ligands for validation. The work presented here provides an initial blueprint on this ultimate quest, while the zip codes identified along the way are being evaluated for clinical translation.

Translational/clinical applications

Translational applications, such as first-in-human clinical trials, will ultimately determine the value of in vivo phage display technology and the strategies introduced here to identify vascular zip codes. Indeed, the Food and Drug Administration (FDA) granted "safe-to-proceed" status for our first vascular-targeted IND in 2009. A first-in-human study has already been completed for a fusion peptide targeting the IL-11Rα zip code in prostate cancer patients. A second IND was approved for a prohibitin-targeting peptide. These programs are detailed below. Three other agents are also in the pre-IND stage, all of which will be introduced briefly.

Of critical importance is the variety of targeting agents and payloads that can be utilized clinically for imaging and/or treatment. Peptide ligands conjugated to an apoptotic moiety are easily anticipated. Other options include antibodies or antibody−drug conjugates (ADCs) and gene therapy vectors, whose feasibility will also be explored.

Bone Metastasis Targeting Peptidomimetic-11 (BMTP-11)

In the screening performed on the first patient mentioned above, a homing peptide from prostate biopsies was selected after intravenous administration of a phage display library. This sequence (CGRRAGGSC) turned out to mimic a motif of interleukin-11 (IL-11) and to bind to interleukin-11 receptor alpha (IL-11Rα).[6] Subsequent studies demonstrated that IL-11Rα is a potential therapeutic target in human prostate cancer, including a comprehensive immunohistochemical (IHC) analysis of primary and metastatic prostate cancer samples, which showed IL-11Rα expression increases during disease progression and was highest in bone metastases.[31]

To therapeutically exploit this zip code, a fusion peptide formulation targeting IL-11Rα was developed by coupling the CGRRAGGSC ligand to an apoptosis-inducing agent, D(KLAKLAKKLAKLAK). $_D$(KLAKLAK)$_2$ is a potent antimicrobial peptide that disrupts mitochondrial membranes upon receptor-mediated cell internalization, which would result in the selective killing of the associated vasculature.[32] Importantly, the all-D enantiomer $_D$(KLAKLAK)$_2$ is able to avoid degradation by proteases. This moiety was termed bone metastasis targeting peptidomimetic-11 (BMTP-11) and efforts were soon underway to complete the required preclinical studies to get FDA approval to evaluate drug activity and localization in a first-in-human study of BMTP-11 in patients with metastatic prostate cancer.

A comprehensive package of IND-enabling studies was completed in rodents and nonhuman primates. Studies revealed the biologic properties of the CGRRAGGSC peptide, including the binding interactions and receptor complex formation,[33] antitumor activity in mouse models, and in vivo biodistribution and stability in rodents and monkeys.[34] To identify an appropriate clinical dosing strategy, dose range finding studies were performed in mice, rats, and cynomolgus monkeys; detectable clinical signs frequently included markers of kidney dysfunction or damage, in a dose-dependent manner, which were supported by the histopathological evaluation revealing kidney lesions. Elevated BUN and creatinine found in the rodent studies were observed early in monkeys but returned to baseline over the course of the study, suggesting an effective adaptive and reparative response by the kidneys, so a GLP-compliant safety study mimicking the proposed clinical protocol dose and schedule regimen was carried out in nonhuman primates. Kidney damage was noted, but the mild necrotic damage was reversed over the course of the study.[34] This finding, along with the dose dependency of the observations, encouraged further investigation in human patients and garnered support from the FDA for a Phase 1 trial (NCT00872157). In addition to the traditional safety objectives, the first-in-human study was designed to confirm ligand-directed targeting of the first-in-class candidate drug BMTP-11 and to provide a sense of BMTP-11 activity that would support the design of a larger dose selection trial.

The objectives of the clinical study were (1) to determine the acute toxicity of BMTP-11 in patients with prostate cancer, (2) to determine the pharmacokinetics and localization of BMTP-11, and (3) to assess the effects of BMTP-11 on tumor tissue (via serum PSA determinations). So, starting just about a decade after a phage library was administered to the first end-of-life patient enrolled in the human vascular mapping study, six prostate cancer patients with high-volume, castrate-resistant bone metastases were enrolled in a trial to test the first-in-class vascular-targeting drug derived from that work. Two patients presented with dose-limiting renal toxicity, but biopsies from all six patients showed BMTP-11 localization to bone metastases and evidence of cell death associated with BMTP-11 activity.[34]

The kidney damage seems to be the result of BMTP-11 by-products containing the D(KLAKLAK)$_2$ moiety, as renal toxicity has similarly been observed in studies of all peptides conjugated to D(KLAKLAK)$_2$, including adipotide, described below.[34–36] Finally, extensive analyses of IL-11Rα expression reveal its potential utility in other cancer types, including lung,[37] leukemia and lymphoma,[38] and osteosarcoma[39,40] in addition to prostate cancer. Revised dosing levels or schedules could reveal a threshold with a better safety profile for clinical use in an oncology setting, all the more likely since efforts to systemically target and destroy fat cells utilizing a very low dose daily rather than higher doses weekly generated promising results.

Adipotide

While the prohibitin zip code was first identified in WAT of mice,[16] it was validated in the synchronous selection from three human subjects as a marker of human WAT.[7] This prompted our proposal to develop an entirely new conceptual approach against obesity: targeted induction of apoptosis in blood vessels supplying fat cells. The impact of a novel treatment for obesity and associated comorbidities, including cancer, would potentially be enormous, so a prohibitin-targeting peptide (named adipotide) was designed as a prototype for a new class of drugs that specifically target the vascular endothelium of white fat.

A fusion peptide linked the CKGGRAKDC fat-localizing peptide with D(KLAKLAK)$_2$, the helix-forming sequence that disrupts mitochondrial membranes and induces apoptosis.[16] This novel anti-obesity agent—but not the targeting sequence alone, nor other controls tested including unconjugated D(KLAKLAK)$_2$—induced apoptosis in differentiated adipocytes in vitro and significant weight loss in obese mice, rats, and nonhuman primates. This was possibly due to a central (presumably hypothalamic) feedback control loop of satiation triggered by release of free lipids into the blood stream, supporting real-time observations of reduced food intake.[16,35,41]

A comprehensive series of studies was completed in monkeys in anticipation of human trials. Efficacy studies showed daily adipotide treatment induced moderate weight loss, decreased body fat, decreased appetite, and improved insulin sensitivity in naturally obese baboons and rhesus monkeys. The striking improvement in insulin response was particularly good news. Glucose tolerance tests performed before and after adipotide treatment resulted in a reduced insulin response and the average insulinogenic index (a method for quantifying insulin resistance) almost halved in the treated monkeys. Insulin resistance is a common adverse side effect of obesity in humans, and could lead to dangerous comorbidities and complications.[35] While efficacy studies in obese animals were informative, human trials require a complete toxicological assessment, including in normal, or lean, animals. As with BMTP-11, the comprehensive preclinical studies included dose range finding studies in lean mice, rats, and cynomolgus monkeys, pharmacokinetic studies, and a safety study in rhesus monkeys that mimicked the daily dosing schedule proposed for the human trial. Notably, adipotide treatment was generally well tolerated in all three species of nonhuman primates tested. The primary side effect, renal toxicity, was predictable, dose dependent, and reversible.[35]

There is a complex relationship between obesity and prostate cancer.[42] Studies have demonstrated an association of obesity with progression and a more aggressive clinical course after a prostate cancer diagnosis.[43–45] Thus, we proposed to test the hypothesis that targeting WAT in obese cancer patients with adipotide would cause fat loss and improved metabolic syndrome and, potentially, lessen morbidity or mortality from obesity and comorbidities including cancer. For this first-in-man study of this new drug class, we enrolled obese patients with biopsy-proven castrate-resistant prostate cancer and established the following clinical objectives: (1) define a safe and effective dose for subsequent studies, (2) characterize the metabolomic consequences of adipotide treatment, and (3) assess changes to PSA and other cancer-associated indicators. The complex mechanistic interplay among obesity, metabolic syndrome, and cancer provided a useful avenue for carefully introducing this vascular-targeted IND to the FDA. This plan was ultimately successful, and FDA approval was granted for a first-in-human, phase 1 evaluation of a single cycle of adipotide in patients with castrate-resistant prostate cancer and no standard treatment options (NCT01262664). Multiple patients successfully completed the full study dosing cycle, and the promising results will be published with a metabolomics profile supporting the clinical findings.

Other zip codes

Other vascular zip codes relevant for oncology and vascular diseases have been identified. 78 kDa glucose-regulated protein (GPR78) and ephrin type-A receptor 5 (EphA5) are critical oncology zip codes, and, although initially identified in tumor cells, VEGF has been successfully targeted in retinopathies, another group of diseases associated with angiogenic vasculature.

GRP78, a promising marker of breast and prostate cancers as well as leukemia and lymphoma, has been targeted with a fusion peptide containing D(KLAKLAK)$_2$, with novel gene delivery vectors, and with antibodies conjugated to toxic moieties, each with pros and cons, revealing the breadth of recent attempts to exploit this zip code. Briefly, GPR78 is part of an evolutionarily conserved endoplasmic reticulum-linked stress response mechanism that provides cellular survival signals during environmental and physiologic stress.[46] GRP78 was traditionally thought to be exclusively involved in processing unfolded proteins, however, we and others found that altered glucose metabolism of cancer cells and glucose starvation in poorly vascularized tumors are associated with GRP78 overexpression at the tumor cell surface,[47–49] and GRP78 expression is correlated with disease progression and worse outcomes in breast cancer and prostate cancer.[50,51]

As stated above, phage libraries were first studied as tools to map antibody epitopes. Epitope-mapping (or "fingerprinting") the circulating pool of human antibodies elicited against tumors in prostate cancer patients led to the isolation of GRP78 as a molecular marker of prostate cancer,[47] which was found to be expressed at the surface of human prostate and breast tumor cells and enable internalization of circulating ligands.[52,53] Two GRP78-targeting peptides, WIFPWIQL and CSNTRVAPC, have been evaluated for translational potential in two formats, and a series of human antibodies against GRP78 were selected for use as ADCs.

First, WIFPWIQL was fused to the proapoptotic D(KLAKLAK)$_2$ peptide to form what was termed bone metastasis targeting peptidomimetic-78 (BMTP-78).[52,54] BMTP-78 successfully localized to tumors in mouse models of breast and prostate cancer, suppressing tumor growth, inhibiting metastasis formation, and extending survival.[52,53] Encouraging preclinical data prompted further toxicological analyses toward clinical development. Like BMTP-11 and adipotide, rodent studies showed renal damage, which was reversible at lower doses. However, unlike BMTP-11 and adipotide, several monkeys developed life-threatening arrhythmias after dosing, leading to termination of all BMTP-78 testing because of its unacceptable cardiac toxicity profile in nonhuman primates.[36] GRP78 is still a remarkable tumor zip code, so other delivery strategies and payloads were considered.

Next, adeno-associated virus/phage (AAVP) is a hybrid viral vector that combines the favorable biological attributes of a eukaryotic virus (AAV) with those of prokaryotic phage to yield chimeric particles that function as targeted gene delivery tools for molecular imaging and/or therapy.[55,56] AAVP displaying either WIFPWIQL or CSNTRVAPC targeting peptides were engineered to deliver *Herpes simplex* virus thymidine kinase (HSVtk), which functions as both an imaging tool in the presence of a radiolabeled substrate[57] and a suicide gene in the presence of the prodrug ganciclovir (GCV), which is converted to a toxic compound. Both functions of GRP78-targeting AAVP-HSVtk were evaluated in mouse models of unfavorable breast cancer phenotypes and aggressive variant prostate cancer, where mice bearing tumors were imaged, treated, and monitored to evaluate tumor response following GCV therapy.[58,59] The remarkable success of these studies and others reinforced ongoing efforts to get a version of AAVP into clinical trials.[23,60–64]

The third translational priority for targeting the GRP78 zip code required exploiting a robust selection pipeline implementing a naive single chain fragment (scFv) phage antibody library, yeast display, and deep sequencing analyses to select hundreds of different antibodies that target a desired antigen. The selection of phage-displayed accessible recombinant targeted antibodies, or SPARTA, discovery methodology was used to identify and validate monoclonal antibodies against GRP78 that could home to tumors in vivo and induce cell death when administered as ADC formulations in vitro.[65]

The SPARTA methodology was also used to identify clinically relevant antibodies against EphA5, a marker of lung cancer. EphA5 was originally discovered by screening a combinatorial peptide phage library against cultured lung tumor cells from the NCI-60 panel[66] using an in vitro phage display method developed for cultured cells.[67] A complete IHC analysis of EphA5 expression in a large cohort of lung cancer patients showed that EphA5 is widely expressed in human lung cancer.[68] Since EphA5 was first identified in vitro, accessibility in vivo needed to be confirmed. An antibody against EphA5, termed 11C12, successfully homed to tumors in xenograft models of lung cancer in mice and rats, and demonstrated remarkable activity as a radiosensitizer in combination with radiotherapy.[68]

SPARTA methodology then identified recombinant human antibodies against EphA5 that home to lung tumor models in vivo, and pseudo-ADC formulations tested in vitro elicited promising results suitable for rapid translation.[65,69] Thus, ADC design and preclinical validation with various linker styles and toxic payloads will be pursued toward clinical development.

Finally, a vascular zip code known as a tumor marker was validated in retinopathies, which are eye diseases associated with pathologic angiogenesis, including retinopathy of prematurity, diabetic retinopathy, and age-related macular degeneration (AMD). Hypoxic conditions induce the overexpression of growth factors, including vascular endothelial growth factor (VEGF), and a mouse model of oxygen-induced retinopathy revealed mechanisms of disease in retinitis pigmentosa and in retinopathy-of-prematurity in mice and patients.[70] This knowledge supported the hypothesis that ligands identified from combinatorial phage libraries that target tumor zip codes may serve to target angiogenic vasculature in retinal diseases. The CPQPRPLC peptide ligand was found to bind endothelial cells stimulated with VEGF; it maps to a region of VEGF-B and binds to VEGF receptor 1 (VEGFR-1) and to neuropilin-1 (NRP-1), but not to other VEGF receptor family members.[67] Anti-VEGF antibody therapy with ranibizumab or off-label bevacizumab requires an injection directly into the eye. Anecdotal evidence indicates agents suitable for systemic administration would be appreciated by patients. Toward that end, the minimal functional tripeptide motif RPL was found to be necessary and sufficient for receptor binding.[71] Retro-inverted D amino acids were used to synthesize the cyclic $_{D}$(CLPRC) peptidomimetic ultimately named vasotide, which maintains binding properties while avoiding proteolytic degradation.[72] Vasotide treatment administered by eye drops or intraperitoneal injection decreased retinal angiogenesis in rodent and nonhuman primate models, including a mouse retinopathy-of-prematurity model, a mouse knockout model of the AMD subtype called retinal angiomatous proliferation, and a laser-induced monkey model of human wet AMD.[72,73] Thus, vasotide could ultimately serve as a less invasive clinical option to mitigate pathological angiogenesis and the consequential vision impairment in this common disease.

Conclusions

In summary, it is clear that a large-scale analysis of proteins expressed in blood vessels of healthy and diseased organs can uncover many novel and unique molecular markers in the human vasculature. As such, generation and annotation of a comprehensive zip code map based on accessible ligand–receptor pairs in blood vessels may be used as a starting point for functional discovery of clinically relevant markers in the human vasculature. Here we report five vascular zip codes that have been validated for translational applications. Continuing efforts to validate new peptide ligands and associated vascular zip codes from patient screens will be an invaluable resource for targeted drug delivery.

Conflict of interest statement

Drs. Arap and Pasqualini are founders and equity stockholders of PhageNova Bio and MBrace Therapeutics. Dr. Pasqualini is a consultant for PhageNova Bio and serves as the Chief Scientific Officer of PhageNova Bio and MBrace Therapeutics. Drs. Arap and Pasqualini are inventors on issued and pending patent applications related to technology described here and will be entitled to royalties if licensing or commercialization occurs. These arrangements are managed in accordance with the established institutional conflict of interest policies of Rutgers, The State University of New Jersey.

References

1. Cheng Y, Ma Z, Kim BH, et al. Principles of regulatory information conservation between mouse and human. *Nature*. 2014;515(7527): 371–375.
2. Lin S, Lin Y, Nery JR, et al. Comparison of the transcriptional landscapes between human and mouse tissues. *Proc Natl Acad Sci U S A*. 2014;111(48):17224–17229.
3. Yue F, Cheng Y, Breschi A, et al. A comparative encyclopedia of DNA elements in the mouse genome. *Nature*. 2014;515(7527): 355–364.
4. Pasqualini R, Ruoslahti E. Organ targeting in vivo using phage display peptide libraries. *Nature*. 1996;380(6572):364–366.
5. Arap W, Pasqualini R, Ruoslahti E. Cancer treatment by targeted drug delivery to tumor vasculature in a mouse model. *Science*. 1998;279(5349):377–380.
6. Arap W, Kolonin MG, Trepel M, et al. Steps toward mapping the human vasculature by phage display. *Nat Med*. 2002;8(2):121–127.
7. Staquicini FI, Cardo-Vila M, Kolonin MG, et al. Vascular ligand-receptor mapping by direct combinatorial selection in cancer patients. *Proc Natl Acad Sci U S A*. 2011;108(46):18637–18642.
8. Smith GP. Phage display: simple evolution in a petri dish (Nobel lecture). *Angew Chem Int Ed Engl*. 2019;58(41):14428–14437.
9. Smith GP. Filamentous fusion phage: novel expression vectors that display cloned antigens on the virion surface. *Science*. 1985; 228(4705):1315–1317.
10. Parmley SF, Smith GP. Antibody-selectable filamentous fd phage vectors: affinity purification of target genes. *Gene*. 1988;73(2):305–318.
11. Cwirla SE, Peters EA, Barrett RW, Dower WJ. Peptides on phage: a vast library of peptides for identifying ligands. *Proc Natl Acad Sci U S A*. 1990;87(16):6378–6382.
12. Scott JK, Smith GP. Searching for peptide ligands with an epitope library. *Science*. 1990;249(4967):386–390.
13. Rajotte D, Arap W, Hagedorn M, Koivunen E, Pasqualini R, Ruoslahti E. Molecular heterogeneity of the vascular endothelium revealed by in vivo phage display. *J Clin Invest*. 1998;102(2): 430–437.

14. Trepel M, Arap W, Pasqualini R. Modulation of the immune response by systemic targeting of antigens to lymph nodes. *Cancer Res*. 2001;61(22):8110–8112.
15. Arap W, Haedicke W, Bernasconi M, et al. Targeting the prostate for destruction through a vascular address. *Proc Natl Acad Sci U S A*. 2002;99(3):1527–1531.
16. Kolonin MG, Saha PK, Chan L, Pasqualini R, Arap W. Reversal of obesity by targeted ablation of adipose tissue. *Nat Med*. 2004;10(6): 625–632.
17. Yao VJ, Ozawa MG, Trepel M, Arap W, McDonald DM, Pasqualini R. Targeting pancreatic islets with phage display assisted by laser pressure catapult microdissection. *Am J Pathol*. 2005; 166(2):625–636.
18. Kelly KA, Nahrendorf M, Yu AM, Reynolds F, Weissleder R. In vivo phage display selection yields atherosclerotic plaque targeted peptides for imaging. *Mol Imag Biol*. 2006;8(4):201–207.
19. Chung J, Shim H, Kim K, et al. Discovery of novel peptides targeting pro-atherogenic endothelium in disturbed flow regions -Targeted siRNA delivery to pro-atherogenic endothelium in vivo. *Sci Rep*. 2016;6:25636.
20. Mann AP, Scodeller P, Hussain S, et al. Identification of a peptide recognizing cerebrovascular changes in mouse models of Alzheimer's disease. *Nat Commun*. 2017;8(1):1403.
21. Mann AP, Scodeller P, Hussain S, et al. A peptide for targeted, systemic delivery of imaging and therapeutic compounds into acute brain injuries. *Nat Commun*. 2016;7:11980.
22. Ozawa MG, Zurita AJ, Dias-Neto E, et al. Beyond receptor expression levels: the relevance of target accessibility in ligand-directed pharmacodelivery systems. *Trends Cardiovasc Med*. 2008;18(4): 126–132.
23. Staquicini FI, Ozawa MG, Moya CA, et al. Systemic combinatorial peptide selection yields a non-canonical iron-mimicry mechanism for targeting tumors in a mouse model of human glioblastoma. *J Clin Invest*. 2011;121(1):161–173.
24. Nathanson D, Mischel PS. Charting the course across the blood-brain barrier. *J Clin Invest*. 2011;121(1):31–33.
25. Pentz RD, Flamm AL, Pasqualini R, Logothetis CJ, Arap W. Revisiting ethical guidelines for research with terminal wean and brain-dead participants. *Hastings Cent Rep*. 2003;33(1):20–26.
26. Pentz RD, Cohen CB, Wicclair M, et al. Ethics guidelines for research with the recently dead. *Nat Med*. 2005;11(11): 1145–1149.
27. Kolonin MG, Sun J, Do KA, et al. Synchronous selection of homing peptides for multiple tissues by in vivo phage display. *Faseb J*. 2006; 20(7):979–981.
28. Dias-Neto E, Nunes DN, Giordano RJ, et al. Next-generation phage display: integrating and comparing available molecular tools to enable cost-effective high-throughput analysis. *PLoS One*. 2009; 4(12):e8338.
29. Kolonin MG, Sergeeva A, Staquicini DI, et al. Interaction between tumor cell surface receptor RAGE and proteinase 3 mediates prostate cancer metastasis to bone. *Cancer Res*. 2017;77(12):3144–3150.
30. Salameh A, Daquinag AC, Staquicini DI, et al. Prohibitin/annexin 2 interaction regulates fatty acid transport in adipose tissue. *JCI Insight*. 2016;1(10):e86351.
31. Zurita AJ, Troncoso P, Cardo-Vila M, Logothetis CJ, Pasqualini R, Arap W. Combinatorial screenings in patients: the interleukin-11 receptor alpha as a candidate target in the progression of human prostate cancer. *Cancer Res*. 2004;64(2):435–439.
32. Ellerby HM, Arap W, Ellerby LM, et al. Anti-cancer activity of targeted pro-apoptotic peptides. *Nat Med*. 1999;5(9):1032–1038.
33. Cardo-Vila M, Zurita AJ, Giordano RJ, et al. A ligand peptide motif selected from a cancer patient is a receptor-interacting site within human interleukin-11. *PLoS One*. 2008;3(10):e3452.
34. Pasqualini R, Millikan RE, Christianson DR, et al. Targeting the interleukin-11 receptor alpha in metastatic prostate cancer: a first-in-man study. *Cancer*. 2015;121(14):2411–2421.
35. Barnhart KF, Christianson DR, Hanley PW, et al. A peptidomimetic targeting white fat causes weight loss and improved insulin resistance in obese monkeys. *Sci Transl Med*. 2011;3(108), 108ra112.
36. Staquicini DI, D'Angelo S, Ferrara F, et al. Therapeutic targeting of membrane-associated GRP78 in leukemia and lymphoma: preclinical efficacy in vitro and formal toxicity study of BMTP-78 in rodents and primates. *Pharmacogenomics J*. 2018;18(3):436–443.
37. Cardo-Vila M, Marchio S, Sato M, et al. Interleukin-11 receptor is a candidate target for ligand-directed therapy in lung cancer: analysis of clinical samples and BMTP-11 preclinical activity. *Am J Pathol*. 2016;186(8):2162–2170.
38. Karjalainen K, Jaalouk DE, Bueso-Ramos C, et al. Targeting IL11 receptor in leukemia and lymphoma: a functional ligand-directed study and hematopathology analysis of patient-derived specimens. *Clin Cancer Res*. 2015;21(13):3041–3051.
39. Lewis VO, Ozawa MG, Deavers MT, et al. The interleukin-11 receptor alpha as a candidate ligand-directed target in osteosarcoma: consistent data from cell lines, orthotopic models, and human tumor samples. *Cancer Res*. 2009;69(5):1995–1999.
40. Lewis VO, Devarajan E, Cardo-Vila M, et al. BMTP-11 is active in preclinical models of human osteosarcoma and a candidate targeted drug for clinical translation. *Proc Natl Acad Sci U S A*. 2017; 114(30):8065–8070.
41. Seeley RJ. Treating obesity like a tumor. *Cell Metab*. 2012;15(1):1–2.
42. Allott EH, Masko EM, Freedland SJ. Obesity and prostate cancer: weighing the evidence. *Eur Urol*. 2013;63(5):800–809.
43. Snowdon DA, Phillips RL, Choi W. Diet, obesity, and risk of fatal prostate cancer. *Am J Epidemiol*. 1984;120(2):244–250.
44. Avgerinos KI, Spyrou N, Mantzoros CS, Dalamaga M. Obesity and cancer risk: emerging biological mechanisms and perspectives. *Metabolism*. 2019;92:121–135.
45. Kelly SP, Lennon H, Sperrin M, et al. Body mass index trajectories across adulthood and smoking in relation to prostate cancer risks: the NIH-AARP Diet and Health Study. *Int J Epidemiol*. 2019;48(2): 464–473.
46. Luo B, Lee AS. The critical roles of endoplasmic reticulum chaperones and unfolded protein response in tumorigenesis and anticancer therapies. *Oncogene*. 2013;32(7):805–818.
47. Mintz PJ, Kim J, Do KA, et al. Fingerprinting the circulating repertoire of antibodies from cancer patients. *Nat Biotechnol*. 2003;21(1): 57–63.
48. Li J, Lee AS. Stress induction of GRP78/BiP and its role in cancer. *Curr Mol Med*. 2006;6(1):45–54.
49. Farshbaf M, Khosroushahi AY, Mojarad-Jabali S, Zarebkohan A, Valizadeh H, Walker PR. Cell surface GRP78: an emerging imaging marker and therapeutic target for cancer. *J Control Release*. 2020;328: 932–941.
50. Daneshmand S, Quek ML, Lin E, et al. Glucose-regulated protein GRP78 is up-regulated in prostate cancer and correlates with recurrence and survival. *Hum Pathol*. 2007;38(10):1547–1552.
51. Conner C, Lager TW, Guldner IH, et al. Cell surface GRP78 promotes stemness in normal and neoplastic cells. *Sci Rep*. 2020;10(1):3474.
52. Arap MA, Lahdenranta J, Mintz PJ, et al. Cell surface expression of the stress response chaperone GRP78 enables tumor targeting by circulating ligands. *Cancer Cell*. 2004;6(3):275–284.
53. Miao YR, Eckhardt BL, Cao Y, et al. Inhibition of established micrometastases by targeted drug delivery via cell surface-associated GRP78. *Clin Cancer Res*. 2013;19(8):2107–2116.
54. Blond-Elguindi S, Cwirla SE, Dower WJ, et al. Affinity panning of a library of peptides displayed on bacteriophages reveals the binding specificity of BiP. *Cell*. 1993;75(4):717–728.

55. Hajitou A, Trepel M, Lilley CE, et al. A hybrid vector for ligand-directed tumor targeting and molecular imaging. *Cell*. 2006; 125(2):385–398.
56. Hajitou A, Rangel R, Trepel M, et al. Design and construction of targeted AAVP vectors for mammalian cell transduction. *Nat Protoc*. 2007;2(3):523–531.
57. Soghomonyan S, Hajitou A, Rangel R, et al. Molecular PET imaging of HSV1-tk reporter gene expression using [18F]FEAU. *Nat Protoc*. 2007;2(2):416–423.
58. Dobroff AS, D'Angelo S, Eckhardt BL, et al. Towards a transcriptome-based theranostic platform for unfavorable breast cancer phenotypes. *Proc Natl Acad Sci U S A*. 2016;113(45): 12780–12785.
59. Ferrara F, Staquicini DI, Driessen WHP, et al. Targeted molecular-genetic imaging and ligand-directed therapy in aggressive variant prostate cancer. *Proc Natl Acad Sci U S A*. 2016;113(45):12786–12791.
60. Hajitou A, Lev DC, Hannay JA, et al. A preclinical model for predicting drug response in soft-tissue sarcoma with targeted AAVP molecular imaging. *Proc Natl Acad Sci U S A*. 2008;105(11):4471–4476.
61. Paoloni MC, Tandle A, Mazcko C, et al. Launching a novel preclinical infrastructure: comparative oncology trials consortium directed therapeutic targeting of TNFalpha to cancer vasculature. *PLoS One*. 2009;4(3):e4972.
62. Tandle A, Hanna E, Lorang D, et al. Tumor vasculature-targeted delivery of tumor necrosis factor-alpha. *Cancer*. 2009;115(1):128–139.
63. Smith TL, Yuan Z, Cardo-Vila M, et al. AAVP displaying octreotide for ligand-directed therapeutic transgene delivery in neuroendocrine tumors of the pancreas. *Proc Natl Acad Sci U S A*. 2016; 113(9):2466–2471.
64. Staquicini FI, Smith TL, Tang FHF, et al. Targeted AAVP-based therapy in a mouse model of human glioblastoma: a comparison of cytotoxic versus suicide gene delivery strategies. *Cancer Gene Ther*. 2020;27(5):301–310.
65. D'Angelo S, Staquicini FI, Ferrara F, et al. Selection of phage-displayed accessible recombinant targeted antibodies (SPARTA): methodology and applications. *JCI Insight*. 2018;3(9):e98305.
66. Kolonin MG, Bover L, Sun J, et al. Ligand-directed surface profiling of human cancer cells with combinatorial peptide libraries. *Cancer Res*. 2006;66(1):34–40.
67. Giordano RJ, Cardo-Vila M, Lahdenranta J, Pasqualini R, Arap W. Biopanning and rapid analysis of selective interactive ligands. *Nat Med*. 2001;7(11):1249–1253.
68. Staquicini FI, Qian MD, Salameh A, et al. Receptor tyrosine kinase EphA5 is a functional molecular target in human lung cancer. *J Biol Chem*. 2015;290(12):7345–7359.
69. Tang FHF, Davis D, Arap W, Pasqualini R, Staquicini FI. Eph receptors as cancer targets for antibody-based therapy. *Adv Cancer Res*. 2020;147:303–317.
70. Lahdenranta J, Pasqualini R, Schlingemann RO, et al. An anti-angiogenic state in mice and humans with retinal photoreceptor cell degeneration. *Proc Natl Acad Sci U S A*. 2001;98(18): 10368–10373.
71. Giordano RJ, Anobom CD, Cardo-Vila M, et al. Structural basis for the interaction of a vascular endothelial growth factor mimic peptide motif and its corresponding receptors. *Chem Biol*. 2005;12(10): 1075–1083.
72. Giordano RJ, Cardo-Vila M, Salameh A, et al. From combinatorial peptide selection to drug prototype (I): targeting the vascular endothelial growth factor receptor pathway. *Proc Natl Acad Sci U S A*. 2010;107(11):5112–5117.
73. Sidman RL, Li J, Lawrence M, et al. The peptidomimetic Vasotide targets two retinal VEGF receptors and reduces pathological angiogenesis in murine and nonhuman primate models of retinal disease. *Sci Transl Med*. 2015;7(309), 309ra165.

CHAPTER

33

Angiosome concept for vascular interventions

V.A. Alexandrescu, A. Kerzmann, E. Boesmans, C. Holemans and J.O. Defraigne

Cardiovascular and Thoracic Surgery Department, CHU Sart-Tilman University Hospital, Liège, Belgium

Introduction

General considerations of the angiosome concept

After the initial work of Taylor and Palmer published in 1987, the angiosome concept (AC) in the field of plastic and reconstructive surgery was increasingly developed.[1] In these previous studies, the anatomy of the structures responsible for blood supply to the different regions of the human body, from the skin to the deeper layers, was assessed.[1,2] Results showed the reproducible patterns of arterial and venous allotments with distinct topographic orientation in the human body.[1,2]

Studies published in the last decade in vascular pathology revealed the potential benefit of the AC in the management of chronic limb-threatening ischemia (CLTI) and in topographic inferior limb revascularization.[3–5]

The AC appears increasingly cited in the current treatment of CLTI and limb salvage. However, its current utilization by bypass or transcatheter techniques only began in recent years.[6–8] Angiosome-guided direct revascularization has been increasingly utilized with particular soar in the field of endovascular interventional techniques.[3,5–8]

However, there are still some unanswered questions about the current technical feasibility of this strategy in the actual management of CLTI, definitions for direct (DR) versus indirect revascularization (IR), and indications for the use of bypass versus endovascular techniques (EVTs) in specific high-risk groups of patients.[3–5]

In this chapter, we performed a succinct review of the main benefits of angiosome-guided direct revascularization and the unanswered questions focusing on this continually evolving concept in the current vascular practice guidelines.

Angiosome concept: anatomical and pathophysiological data

In the initial work of Taylor et al., 44 angiosomes and appended source arteries (SA) in the human body were described. Of them, 6 maintain the normal vascularization in the lower leg and foot.[1] Adjacent angiosomes are linked by a vast collateral web containing numerous small-to-large collaterals,[1,2,9] arterial–arterial communicants, and thousands of millimetric choke vessels (CV)[1,2,9] with important compensatory roles.[1,6,7] In cases in which the main angiosomal arteries are interrupted or occluded, this rescue system redirects the blood flow via available collaterals toward the neighboring angiosomes.[1,2,6,7,9] The diameter and topography of the compensatory collaterals vary based on anatomic location, patient's age, and type of CLTI aggression.[6,10–14]

Anatomy of the distal leg angiosomes

Based on the initial description of Taylor, SAs have related collaterals and CVs present with their own specific and reproducible regional distribution to tissues.[1,9,10]

Angiosomal SA and their collateral system

The angiosomal topographic partition was initially pictured as a continuous three-dimensional (3D) network that is harmoniously dispensed to tissue,[1,2,9] and that holds several levels of dichotomy[13] toward specific parts of the skin and deep tissue region.[1,10] Each

The Vasculome
https://doi.org/10.1016/B978-0-12-822546-2.00020-4

angiosome has correspondent « arteriosomes » and « venosomes »,[1,2] which share harmonious patterns of vascular architecture.[1,2] This flow arrangement indicates a fractal distribution of flow to specific limb regions.[6,13]

Angiosomes were initially described as distinct *anatomical* entities, named from the Greek term *angeion* (meaning vessel) and *somite*, or *soma* (indicating the section of the body).[10] However, their *clinical* significance conveys concomitant functional features that are dependent on each angiosome's *perimeter of anastomotic vessels*.[10,15]

Primary *SA* of the distal leg and foot

Based on the characteristics of the six-foot angiosomes,[1] the SA and underlying tissue territories are depicted as follows (Fig. 33.1):

The posterior tibial artery provides flow to its *medial calcaneal* branch and appended angiosome, adjoining the *medial* and *lateral plantar* arteries and subsequent angiosomes, the plantar heel territory, and the entire medial and lateral plantar regions of the foot and toes.

The anterior tibial artery transitions into the *dorsalis pedis* artery below the ankle level and supplies its appended dorsalis pedis angiosome that covers the dorsum of the foot down to the dorsal toe territories.

The peroneal artery provides flow to its *lateral calcaneal* artery and angiosome to a more restricted zone in the posterolateral heel and to its *anterior perforating branch*. Moreover, it irrigates the lateral and anterior upper ankle and appended angiosome.[1,2,9]

From a practical perspective, the anterior tibial artery nourishes the anterior ankle and the dorsal aspect of the foot and toes. Meanwhile, the posterior tibial artery provides flow to the medial, posteromedial ankle, and heel territories and equally to the entire sole and the plantar side of the toes. The peroneal artery irrigates the anterolateral upper ankle zone and the lateral and plantar heel regions.[1,9,15]

C*ollateral network* surrounding the foot angiosomes

Before reaching the capillary system, the interangiosomal collaterals can be differentiated in *large-* (approximately 1 mm in diameter), *medium-* (<1 mm), and *small*-sized caliber (<0.5 mm).[13,15,16] Taylor et al. additionally described the *cutaneous perforator* branches that provide flow to each 3D tissue block, specifically supplying the skin.[15] These *cutaneous arteries* (CAs) emerge directly from the main SA and provide direct flow to the skin.[15] Other indirect or spent terminal ramifications were referred to as *cutaneous perforators* (CPs),[10,15] and they are derived from the deep tissue layers in continuity with the artery that is the initial source of perfusion.[10,15] CVs, CAs, and CPs include tiny vessels (approximately 0.5 mm in diameter) that often can be detected on routine

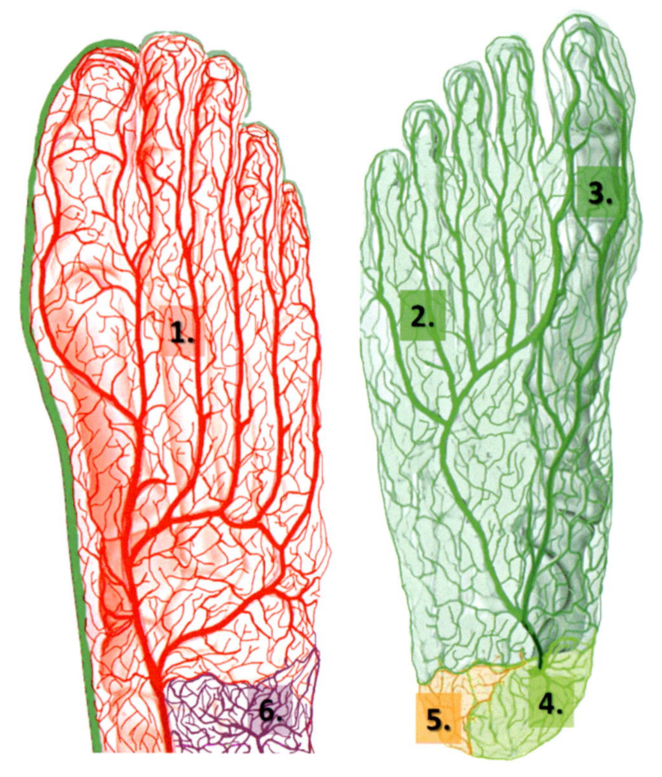

FIGURE 33.1 Schematic dorsal (left side), and plantar (right side) distribution of the foot angiosomes: 1. Dorsalis Pedis angiosome (Anterior Tibial territory), 2. Lateral Plantar angiosome (Posterior Tibial territory), 3. Medial Plantar angiosome (Posterior Tibial territory), 4. Medial calcaneal angiosome (Posterior Tibial territory), 5. Lateral calcaneal angiosome (Peroneal territory), 6. Anterior perforator branch angiosome (Peroneal territory).

angiography.[12,13,16] However, the visual accuracy is limited for vascular structures <500 μm.[17]

Specific CLTI pathologies such as diabetes mellitus or renal insufficiency inflict foot collateral destruction (from CAs and CPs, down to the small CVs and capillaries) and enhance a notable risk for tissue loss and major amputation.[6,18–21]

Main connections between the foot angiosomes

Apart from the accepted anatomical variations (9%),[22–24] specific groups of collaterals, which provide prompt flow compensation, were identified at the foot level.[1,9,13] Numerous *large* collaterals retain a specific weight in supplying neighboring foot angiosomes in CLTI.[6,9,12,13] Moreover, they play a pivotal role in topographic *wound-targeted* revascularization.[5–7] These vital branches comprise the *foot arches*,[6,16,25] forefoot *metatarsal perforators*,[6,9,16] and anterior or posterior arcuate artery *interconnections*.[6,9,16] Other sizable *arterial–arterial* branches, such as the dorsal foot-to-plantar, or the peroneal perforators to the posterior, or the anterior tibial arteries were evoked, such as "*rescue*" midfoot, or heel collaterals.[6,9,13]

From a topographic perspective, the communicants between the posterior tibial and peroneal artery via their lateral and medial calcaneal branches, along with the posterior peroneal branch, play a major compensatory role in providing blood supply in *ischemic heel ulcers*. If available, these collaterals yield equal valuable flow shifts when intentional hindfoot or heel DR is performed.[3–7]

The interconnections relying on the dorsalis pedis (the anterior tibial artery flow) to the plantar arteries (the posterior tibial artery circulation) comprise the medial or lateral tarsal branches, metatarsal perforators, or paired metatarsal anterior and posterior interdigital collaterals.[9,15,16] They also significantly contribute in maintaining viable forefoot and toe perfusion during ischemic threat.[9,15,16]

Lastly, the lateral and medial communicants in both plantar arteries (the posterior tibial circulation) link the lateral and medial tarsal arteries (the anterior tibial flow). Moreover, they provide support between the dorsal and plantar foot perfusion in patients exempted from wide CLTI collateral extinction.[9,10,15,16]

Foot angiosomes as fractal levels of perfusion in the inferior limb

All individuals possess a specific inherited collateral reserve that compensates blood flow between bordering angiosomes. This remarkable self-regulating vascular web continuously undergoes dynamic adaptations to various endogenous and exogenous stimuli.

In a larger picture, the entire inferior limb vasculature can be described as balanced and reproducible patterns of peripheral tissue irrigation.[11,16] In this harmonious scaffolding, each arterial trunk gradually divides into inferior levels of segmentation to generate a wider cross-sectional area of flow toward the peripheral tissues.[16] Each staged dichotomy constantly creates branches smaller than its parent trunk.[10,16] For every arterial bifurcation, the assembled sectional area of the derived branches is greater than that of the primary vessel.[1,10,16]

From the iliac inflow down to the myriad of distal foot ramifications, a remarkable sequential distribution of blood supply was observed.[13,16] These characteristic levels of dichotomic flow dispensation (Table 33.1) can be stratified as per rank of tissue perfusion and can be summarized as follows:[13,16]

Level I gathers the primary inferior limb arterial bundles of irrigation (i.e., iliac and common femoral vessels). *Level II* assembles the following branches in the thigh and calf (i.e., the superficial and profunda femoris and the tibial arteries), and *Level III* joins specific ramifications for precise cutaneous and deep tissue territories in the leg and foot.[13] This third level holds a peculiar interest in CLTI revascularization. It contains the angiosomal SA, large (1 mm) collaterals, foot arches, and correspondent metatarsal communicants.[13] It also provides specific clinical applications for angiosome-targeted or wound-directed revascularization (WDR).[6,13,16] The next level, *level IV*, assembles the *medium* and *small* collaterals ($\leq$0.5 mm in diameter), including the CAs, CPs, and CVs.[13,16] The subsequent division ranks further assemble the microcirculatory network containing *level V* (the arterioles) and *level VI* (the capillary system) that hold countless micrometer–diameter vessels.[13,16]

Main arterial anatomical variants of the lower leg

Specific lower limb arterial variants, most of which concerning *Level III* of flow segmentation, were

TABLE 33.1 The inferior limb specific levels for dichotomic blood irrigation.

Inferior limb levels of perfusion	Type of inferior limb arteries	Arterial segmentation
Level I.	The original arterial and venous bundles of the inferior limb	The iliac and common femoral vessels
Level II.	The first rank of arterial division: Main branches in the thigh and calf	The superficial and profunda femoris, The three tibial trunks
Level III.	The second rank of arterial division: specific tissue sectors large branches	The pedal and angiosomal branches, The foot arches, The large collaterals (around 1 mm).
Level IV.	The third rank of arterial division: Smaller interconnections between different inferior limb regions	The medium-sized collaterals (0.5–1 mm), The small collaterals (<0.5 mm), The "choke-vessels", The skin and muscular perforators.
Level V.	The microcirculatory arteriolar ramifications	The arterioles
Level VI.	The capillary tier	The capillaries

identified.[13,16] Native variations of the leg arteries were observed in approximatively 7.9%–10% in individuals in the general population.[22,23] Among these atypical presentations, hypoplastic or aplastic posterior tibial arteries were described in 3.3% cases, whereas the anterior tibial artery anomalies were reported in 1.5% of subjects.[23] The high (popliteal) emergence of the anterior tibial artery (5.6%)[23] associated or not with abnormal dorsalis pedis paths was observed in about 6% of individuals.[23,24] The presence of one popliteal artery variation on the targeted leg for revascularization may predict about 21% of other possible ipsilateral vascular abnormalities and up to 48% of eventual contralateral arterial variants.[22–24] The precise identification of these variations may help interventionists establish a diligent flow reconstruction in planning WDR.[6,12,16,22]

Pathophysiological data of angiosomal flow

The basic pathophysiological mechanisms of limb ischemia may be associated to *acute* (brisk presentations) or *chronic* (slow unfolding) tissue ischemia.[11,26] These two clinical entities are dependent on three major factors: *time* of ischemic threat, number and size of available *compensatory collaterals*, and individual *cardiac output*.[25,26]

The amount of collateral network is not uniformly allocated in the whole angiosomes of the human body.[9,15] For example, compared with forefoot, thigh, myocardial, or pulmonary angiosomes, the hindfoot and heel angiosomes have fewer compensatory native collaterals, CVs, and CPs.[9,13,16] In hemodynamic terms, about 16 collaterals with a diameter of 0.25 cm may match the flow of 625 collaterals with a diameter of 0.1 cm to provide peripheral resistance as low as that of an unobstructed artery with a diameter of 0.5 cm.[25] A few large collaterals played a more efficient role in flow compensation than hundreds of small collaterals, arterioles, and capillaries.[25]

Two major processes trigger collateral development during ischemic threat. These processes include *angiogenesis* (sprouting capillary development enhanced by hypoxia and macrophages) and *arteriogenesis* (remodeling and enlargement of preexisting collaterals enhanced by the vessel's shear stress and by reactionary inflammatory cells).[11,26] Arteriogenesis is essentially stimulated by pulsatile pressure flow in the collateral bed, and it can determine an increase in the diameter and length of the appended arterioles.[11,25,27,28] Moreover, it is influenced by the release of specific endothelial factors and by the local migration of macrophages.[26,29–31]

In the treatment of CLTI, compared to angioplasty, bypass facilitates a higher volume of blood flow in the peripheral collateral system and pulsatile pressure flow.[32] This phenomenon may be extremely beneficial for surgical treatment, regardless of whether revascularization has an angiosome-oriented topography.[13,32]

Angiogenesis and arteriogenesis processes can be significantly inhibited by the CLTI condition itself[29] and by associated pathologies, such as metabolic syndrome[30,31] and renal insufficiency.[31–34] Normal inferior limb perfusion does not express enlarged collaterals, unless a reactional response to ischemic conditions is requested.[26]

In addition to the well-known devastating features of acute ischemia–reperfusion syndrome after acute hypoxic tissue damage,[26,35] countless intermediate functional patterns of chronic tissue reperfusion were observed.[26,34]

In accordance with previous studies on plastic reconstructive surgery,[9,15] interventional cardiology,[36] vascular surgery,[25] and neurosurgery,[37] several phases of flow redistribution were noted before and after the retrieval of chronic ischemic conditions.[11,25,26] These functional stages are based on specific time intervals, and they expand according to the intensity and duration of CLTI aggression.[11,25–28]

Flow compensation during preischemic conditions

An impressive flow compensation system was recognized in ischemic conditions according to adjacent angiosomes and appended collaterals and CVs.[9,10,15]

Since flow pressure in specific SA significantly decreases, the CVs between adjacent angiosomes progressively open and convey maximal compensatory capacity. In CLTI circumstances, advancing alteration in SAs and parallel collateral decay lead to the gradual activation of the remaining branches and CVs.[13,19–21]

Postischemic reperfusion stages

CLTI injury inhibits large BTK arterial trunks and various amounts of collaterals.[12,13,25] When hypoxic burden is relieved after revascularization, a cascade of pathophysiological changes is enhanced, and it can be schematized as the reperfusion stages (subject to changes upon each individual lasting collateral network), which are as follows:[11,27–29]

The initiatory flow redistribution stage involves large and medium collaterals around the ischemic angiosome. This starting phase operates via the lasting permeable branches (of all sizes), scattered around the ischemic zone, and allows rapid rescue flow toward low-resistances territories.[11,30] This stage lasts for hours. Flow mainly follows the surviving channels with low resistances and native angiosome partition.[34]

The *average* flow dispensation phase, which is the next stage, can be further observed over the medium-to-small collaterals (including the CVs, CPs, and arterioles).[34,38–40] Some of these connections are open and visible on perioperative angiographic examinations. Meanwhile, others express higher flow resistances and only progressively become functional during these two periods (the « dormant collaterals »).[34,39,40]

Both initial phases can persist for several days and may be juxtaposed to the delay phenomenon, as described in previous publications about skin flap surgery.[9,15,41] Both starting flow redistribution phases are conditioned by changing flow resistances; thus, they may have a minimal topographic correlation with standardized anatomical angiosome orientation.[34,39,40]

The *retarded* postischemic phase, which is the third step, can last for several weeks.

It is essentially characterized by consolidation of flow, via available collaterals and arterial–arterial interconnections, which are reorganized throughout the whole process of arteriogenesis and angiogenesis.[11,26,38,40] Flow still dwells partially characteristic to the basic marks of anatomic angiosomes (which vary based on the amount of preserved collaterals). The true connections between neighboring angiosomes now have a new physiological entity, which is referred to as *the functional angiosome*.[42]

Functional angiosome

In 2017, 30 years after the first anatomical angiosomal description, Taylor et al.[1,42] completed their pioneering work by defining the *physiological*, or the functional angiosome (FA).[42] This novel hemodynamic entity completed previous anatomical studies[15] and identified the volume of tissue that can be clinically isolated on a specific source vessel.[42] Moreover, it represents the area of perfusion that one particular SA can afford throughout the true collaterals beyond its anatomical territory.[15,42] This functional area stretches upon a wider clinical region of the foot by capturing adjacent angiosome territories, owing to *efficient* connections via CVs or, more likely, via true anastomoses.[42]

True connections assist the broader FA to operate and remodel CVs throughout the angiogenesis and arteriogenesis processes,[27] starting with *the second phase* of flow reperfusion.

Collateral flow reserve and microvascular resistances between angiosomes

CLTI encompasses a multilevel arterial disease associated with variable collateral loss that invariably increases distal limb flow resistances.[30,38,39] Remote foot collateral resistances steadily and proportionally increase to the degree of lumen narrowing, systemic changes in heart rate, and collateral destruction rate.[25,27–30]

These significant hemodynamic interactions between flow and hindered collateral network are not new. A previous research by Macci et al. has reported these discrepancies in 1996 using the Windsor perfusion index,[43] which was recently updated by parallel studies conducted by Mangi et al.[44] or Ikeoda et al.[45] These studies focus on peripheral vascular flow reserve (VFR) and peripheral fractional flow resistances (PFFRs).[44,45] These latest indicators have been assessed recently also in CLTI. However, in the current literature, data on these indicators are still limited. Clinical correlations between PFFR and routine noninvasive examinations including the ankle–brachial index (ABI) and duplex imaging were already established.[44] PFFRs and VFR are also correlated with microvascular $TcPO_2$ or Laser Doppler skin perfusion pressure (SPP).[45] Both methods can equally merge to describe a derived collateral index for specific tissue perfusion (initially evoked in interventional cardiology).[46]

These studies provide a better understanding of regional collateral reserve in ischemic foot syndrome and adding parallel useful associations with evoked phases of reperfusion.

Clinical implementation of the angiosome model in the current treatment of CLTI

Defining angiosome-directed revascularization

Each CLTI presentation is unique due to its specific anatomical patterns, countless collateral pathophysiological changes, and distinct individual risk factors for tissue healing.[6,29–31] The implementation of the true AC in current vascular practice undoubtedly implies detailed macro- and microvascular individual evaluation (Fig. 33.2).

Although initial CLTI studies defined DR mainly as intentional reperfusion in SA,[3–6] surgical and endovascular studies in recent years broadened the purpose of DR by including other Level III and IV arterial ramifications. The foot arches, large- and medium-sized collaterals, and CVs completed DR purposes in topographically oriented foot reperfusion.[6–9,12,47] However, in contemporary literature, there is no clear consensus regarding the definition of DR versus IR or WDR.[48–55]

Some authors have observed improvement in clinical results via direct angiosome SA revascularization.[3–9] Meanwhile, others have described similar results using available collateral-enhanced reperfusion (with or without wound orientation), which is referred to as *indirect* revascularization.[9,47] More refined clinical data have shown improvement in healing and limb preservation in

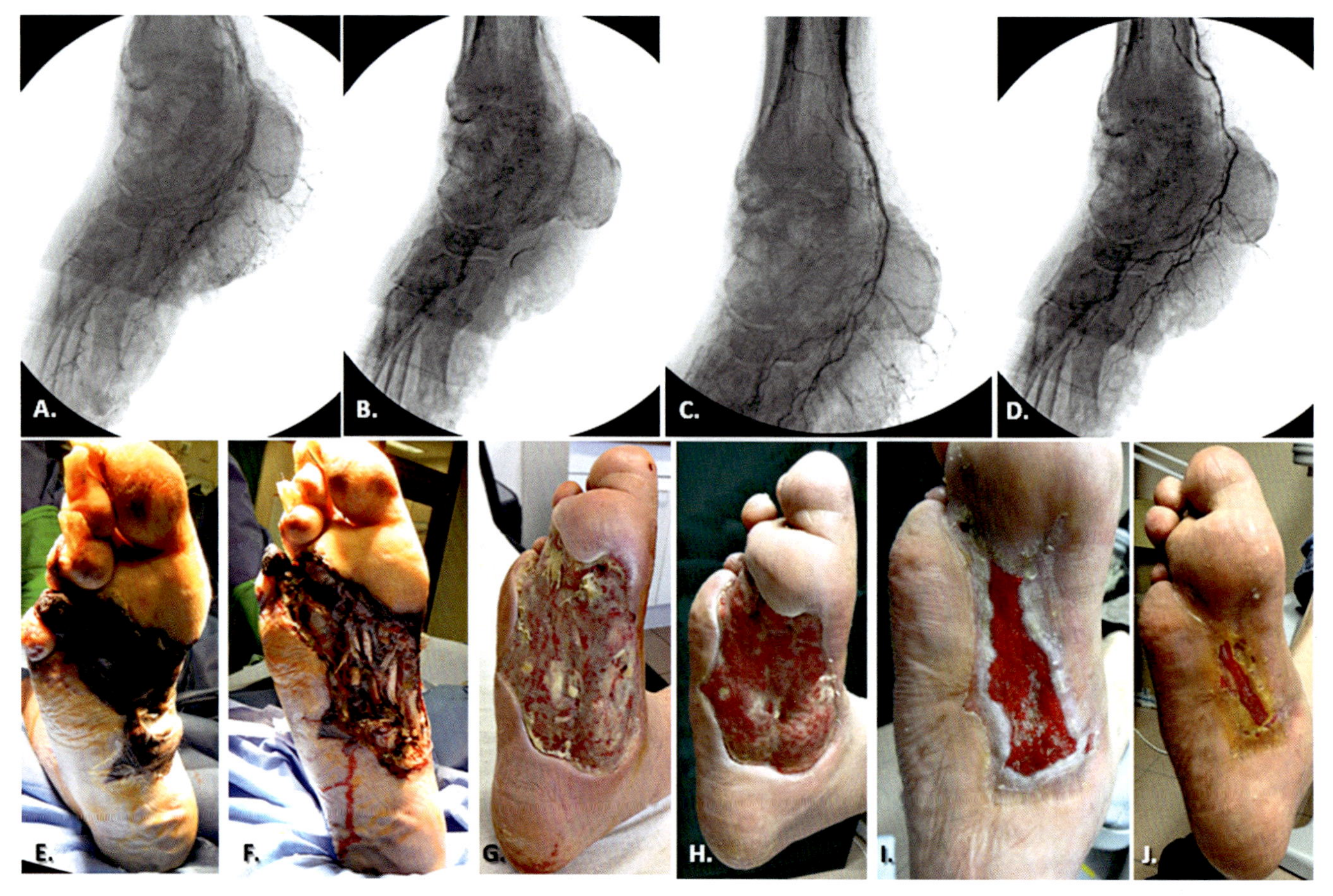

FIGURE 33.2 Clinical application of the angiosme strategy in CLTI and diabetic neuro-ischemic foot treatment: (A and E) initial angiographic and clinical presentation of a Rutherford 5, CLTI - GLASS stage 3, Wagner 4, diabetic neuro-ischemic foot wound, (B) Intraoperative aspect of related endovascular approach. According to the ulcer's location, a targeted recanalization of the posterior tibial artery adding the angiosomal lateral and medial plantar branches was planned. A 0.014 in. guidewire (Terumo Japan) was placed at the posterior tibial artery bifurcation and the initial part of the plantar arteries, as to allow subsequent angioplasties in these specific plantar regions. (C) Angiographic result after intentional angioplasty at the posterior tibial and plantar arteries junction, (D) The procedure completion angiography, demonstrating a correct flow throughout the plantar arteries and their appended angiosomes (including the wound's sole territory). (F) Intraprocedural aspect of tissue debridement for extended plantar ulcer and deep tissue necrosis, performed during the same initial interventional stage. (G–J) Sequential clinical aspects of wound healing and plantar tissue regeneration after targeted "direct plantar revascularization." Wound cicatrization features at 8, 14, 21, and 26 weeks postoperatively, owning regular multidisciplinary team follow-up.

individuals with CLTI on DR angioplasty,[3,5–8] DR bypasses,[4,9] or both.[47,56] However, parallel studies did not show statistical differences between DR and IR in terms of clinical success and limb preservation rates.[57–60] Analogous revascularization series highlight the key role played by permeable pedal arteries,[12,32,50] foot arches,[6,9,32,50–52] or available large foot collaterals[6,9,12,13,32,47] in tissue healing regardless if the angiosome location is targeted.[61] The accuracy in using limb salvage as an indicator for DR/IR clinical success remains controversial.[19,31,48,53] The limb salvage refers to a heterogeneous group of limb-threatening factors and issues, in addition to vascular DR/IR anatomic and hemodynamic effects.[19,31,32,53]

Špillerová et al.[62] showed that the clinical results and the prognosis of DR versus IR primarily depend on the protocol for the type of revascularization in each study.[62] Due to the limited prospective data in the literature,[6,56] several recent metaanalyses have provided a better understanding of these current concepts.[8,63,64]

Biancari et al.[63] conducted a systematic review of 1290 limbs. Results showed that since DR is feasible, it was found to enhance better wound recovery and had a higher limb salvage rates than IR for both EVT and bypass techniques.[63] In two analogous analyses of 1868 and 779 cases, Bosanquet et al.[8] and Huang et al.[65] revealed that DR can improve tissue healing and limb salvage compared with IR.[8,65] However, in all studies, compared to IR, DR was not considered superior in terms of survival and reintervention rates in these patients. Based on the same type of analysis, Jongsma et al.[66] found that the angiosome model may be less applicable for bypass surgery due to distal leg anastomosis that is generally performed on the less affected pedal artery.[32,66] Thus, limb preservation in CLTI may be less affected by DR or IR for bypass compared with EVTs.[66,67] These results were in accordance with those of Dilaver et al.[64] and with other analogous studies in this field.[32,67]

Stimson et al. conducted another remarkable updated metaanalysis.[68] Results showed that the AC may be useful for both EVT and bypass in CLTI revascularization. In particular, endovascular DR (Fig. 33.1) is considered superior to IR.[68] Alternatively, open surgery appears less dependent on anatomical angiosome partition since adequate foot arches and large collaterals are still patent. Similar publications showed that lasting foot collaterals may not follow the classical angiosome sectorization in the diabetic foot syndrome.[6,8,19,68]

The importance of uniform standardization of DR/IR in surgical and endovascular practice is again emphasized.[53,64,66–70]

Technical feasibility of intentional direct revascularization

Most angiosome-targeted arteries (SA) for revascularization are described to harbor severe atherosclerotic and calcific occlusive lesions,[6,16,40] equivalent to GLASS stage III degree of anatomical severity.[32] Endovascular DR is often associated with long chronic total occlusions (CTO) and dense calcifications recanalizations.[6,40,49] Alternatively, comparable challenges in bypass DR in terms of selecting permeable runoff branches in extensively diseased pedal SA were described.[4,16,47,49]

Based on modern interventional standards, the feasibility of endovascular DR can vary from 61% to 88% in different studies.[3,5–8,12,49] The technical success of EVT may be also correlated with the number of treated angiosomes.[49] In a recent study, the technical feasibility rate in one specific angiosome revascularization was 69%.[49] However, DR could be performed in 86% of cases with two targeted foot angiosomes, in 85% for three, and only in 25% of limbs requiring four angiosome EVT reperfusion.[49]

The feasibility of bypass or endovascular DR indicates that the diligent use of all available foot collaterals, arterial–arterial communicants, and, eventually, permeable foot arches remains essential.[6,12,51,53,62–66]

Chronic limb-threatening ischemia and chronic angiosome-threatening ischemia: two competitive or rather complementary notions

In 1982, the term of critical limb ischemia (CLI) was first introduced,[54] and it referred to a heterogeneous population with ischemic inferior limb presentations, including diabetic and renal patients.[32,55,71]

Recent publications show that at the collateral and arteriolar levels[13] (levels III–V), not all foot territories may have shear equivalent ischemic burden.[19,30,31,40] The predominant infragenicular and inframalleolar forms of CLTI were particularly observed in diabetic and renal patients.[30,39] These distal limb anatomical patterns of CLTI are not new,[72] and difficulties in achieving an accurate clinical[73] and hemodynamic diagnosis[40,71] were already described.[40,71–73] These specific CLTI patterns are commonly associated with extensive tibio-pedal CTOs (Fig. 33.2) and calcifications and often indicate severe concomitant neuropathic autonomic denervation in diabetic patients.[31,74,75]

In these particular cases, sole macrocirculatory assessment based on ABI,[76] TBI,[77] Computed Tomography angiography, or Magnetic Resonance angiography[19,40] can be only partially helpful, combined to digital subtraction angiography.[17,19,71,76] Meticulous microcirculatory assessment[40,71] can add complementary information about the precise dispensation of the ischemic load throughout each region of the CLTI foot.[40,53,71] Particularly in diabetic patients, the concomitant location of dominant neuropathic and true neuroischemic ulcers may not always be easy to ascertain in each region of the threatened limb.[18,40,53,73]

In these patients, a more precise microvascular CLTI diagnostic as topographic angiosome-threatening ischemia assignment may be useful in current practice.[40,53] As pointed in recent publications, this simultaneous microcirculatory evaluation could associate methods like Indocyanine green dye-based fluorescent angiography imaging,[34,40] to parallel microcirculatory exams[18,32] that more specifically explore Levels III-V angiosomal and intraangiosomal ramifications. Levels III–V although represent the anatomical ground of derived "chronic angiosomal threatening ischemia" (CATI) notion.[34,40,53] Among focused diagnostic methods, $TcPO_2$, transcutaneous laser Doppler (SPP), hyperspectral imaging, or PET/SPECT scan nuclear imaging can all afford useful information in microcirculatory assessment of specific ischemic foot regions.[40]

The CATI concept showing that a more detailed quantification of the ischemic burden in each foot region influences CLTI perception is not new.[34,40,68–71]

A more refined CATI assessment that includes the features of each diabetic foot's *end-artery occlusive disease*,[31] every disrupted *arterio-* and *angiogenesis* processes,[19,30,31,39] in addition to standard CLI[54] or CLTI[32] characteristics (Fig. 33.2), proved to be particularly useful in diabetic patients.[19,34,53,69–71]

Current investigation and perspectives for angiosome-targeted revascularization in the treatment of CLTI

Similar to new strategies of multidisciplinary interest, the AC should be validated in larger prospective and multicentric clinical studies for current vascular applications.

However, unlike other new medical theories, the accuracy of angiosome-guided revascularization is dependent on the rigorous control of *several risk factors* among participants associated with clinical failure. These factors include individual anatomic and hemodynamic SA condition,[32,53] local wound characteristics,[32,40] specific anatomical[15,16] and pathophysiological features of the local collaterals,[34,42] and individual patient's specificities, that embodies all CLTI presentations.

Since there is no current consensus about foot collateral evaluation and utilization, the best diagnostic method for assessing CATI and CLTI features[34,53] was not yet standardized. Although balanced comparison between DR/IR owing multidisciplinary team approaches is observed in only 20% of contemporary studies,[53,64] a centralized perception of the angiosome strategy based on unitary recommendations in daily practice is still expected nowadays.

Finally, equivalent research criteria for tissue recovery, time for healing, and complete rehabilitation[34,53] indicators, all with undeniable clinical utility, should be more clearly categorized by future publications.[34,53]

The recent Global Vascular Guidelines document and recommendations[32] states that angiosome-guided revascularization may be of real importance "in the setting of endovascular intervention for midfoot and hindfoot lesions but is likely to be irrelevant for ischemic rest pain and of marginal value for most forefoot lesions and minor ulcers."[32] However, the precise role of topographic foot bypass,[64,67,68,78] or multivessel (tibial) revascularization, remains equally unknown in current CLTI practice.[32,68–70]

It appears reasonable that for selected high-risk patients with specific high-risk tissue regions (i.e., WIfI stages 3 and 4)[32] to benefit from first endovascular DR,[32] if this technique can be safely performed (concerning GLASS stages I–II), without compromising runoff to the foot for eventual bypass target.[32,79]

Summary

Despite lack of pertinent data in current CLTI treatment, angiosome-guided revascularization appears preferable for inducing better wound cicatrization and limb salvage, whenever technically achievable. Higher levels of evidence for the use of direct revascularization in the daily management of CLTI based on standardized definitions, anatomic and functional diagnostics, and uniform indications for each type of treatment are required. Larger multicentric and prospective studies are therefore expected before routine application of this strategy in current clinical practice. Since direct revascularization proves difficult to achieve, alternative indirect reperfusion via available foot collaterals is recommended.

Acknowledgments

I would like to acknowledge the members of the Cardiovascular Surgery department of CHU Sart-Tilman Hospital in Liège, for their manifold support during many years of data collection, tireless study, subject analysis, and previous thesis redaction on the same topic that grounded the present updated chapter work.

References

1. Taylor GI, Palmer JH. The vascular territories (angiosomes) of the body: experimental study and clinical applications. *Br J Plast Surg*. 1987;40(2):113–141.
2. Taylor GI, Caddy CM, Watterston PA, Crock JG. The venous territories (venosomes) of the human body: experimental study and clinical implications. *Plast Reconstr Surg*. 1990;86(2):185–213.
3. Iida O, Nanto S, Uematsu M, et al. Importance of the angiosome concept for endovascular therapy in patients with critical limb ischemia. *Catheter Cardiovasc Interv.* 2010;75:830–836.
4. Neville RF, Attinger CE, Bulan EJ, Ducic I, Thomassen M, Sidawy AN. Revascularization of a specific angiosome for limb salvage: does the target artery matter? *Ann Vasc Surg*. 2009;23(3): 367–373.
5. Alexandrescu VA, Hubermont G, Philips Y, et al. Selective primary angioplasty following an angiosome model of reperfusion in the treatment of Wagner 1–4 diabetic foot lesions: practice in a multidisciplinary diabetic limb service. *J Endovasc Ther.* 2008;15:580–593.
6. Alexandrescu VA, Brochier S, Limgba A, et al. Healing of diabetic neuroischemic foot wounds with vs without wound-targeted revascularization: preliminary observations from an 8-year prospective dual-center registry. *J Endovasc Ther.* 2020;27(1):20–30.
7. Zheng XT, Zeng RC, Huang JY, et al. The use of the angiosome concept for treating infrapopliteal critical limb ischemia through interventional therapy and determining the clinical significance of collateral vessels. *Ann Vasc Surg*. 2016;32:41–49.
8. Bosanquet DC, Glasbey JC, Williams IM, Twine CP. Systematic review and meta-analysis of direct versus indirect angiosomal revascularization of infrapopliteal arteries. *Eur J Vasc Endovasc Surg*. 2014;48(1):88–97.
9. Attinger CE, Evans KK, Bulan E, Blume P, Cooper P. Angiosomes of the foot and ankle and clinical implications for limb salvage: reconstruction, incisions, and revascularization. *Plast Reconstr Surg*. 2006; 117(7 Suppl):261S–293S.
10. Taylor GI, Pan WR. Angiosomes of the leg: anatomic study and clinical implications. *Plast Reconstr Surg*. 1998;102(3):599–616.
11. Schaper W. Collateral circulation, past and present. *Basic Res Cardiol*. 2009;104(1):5–21.
12. Osawa S, Terashi H, Tsuji Y, Kitano I, Sugimoto K. Importance of the six angiosomes concept through arterial-arterial connections in CLI. *Int Angiol*. 2013;32(4):375–385.
13. Alexandrescu VA, Pottier M, Balthazar S, Azdad K. The foot angiosomes as integrated level of lower limb arterial perfusion:

amendments for chronic limb threatening ischemia presentations. *J Vasc Endovasc Ther*. 2019;4(1):7–14.
14. Suh HS, Oh TS, Lee HS, et al. A new approach for reconstruction of diabetic foot wounds using the angiosome ans supermicrosurgery concept. *Plast Reconstr Surg*. 2016;38(4):702–709.
15. Taylor GI, Corlett RJ, Dhar SC, Ashton MW. The anatomical (angiosome) and clinical territories of cutaneous perforating arteries: development of the concept and designing safe flaps. *Plast Reconstr Surg*. 2011;127(4):1447–1459.
16. Alexandrescu VA, Kerzmann A, Melvin R, Limgba A. Understanding the tibial-pedal anatomy: practical points for current presentations. *Vasc Disease Manag*. 2019;16(12):E1 79–E1 82.
17. Diehm N. Commentary. Intra-arterial digital subtraction angiography: what you see is not always what you get. *J Endovasc Ther*. 2015;63:252–253.
18. Hingorani A, LaMuraglia GM, Henke P, et al. The management of the diabetic foot: a clinical practice guideline by the Society for Vascular Surgery in collaboration with the American Podiatric Medical Association and the Society for Vascular Medicine. *J Vasc Surg*. 2016;63:3S–21S.
19. Jörneskog G. Why critical limb ischemia criteria are not applicable to diabetic foot and what the consequences are. *J Scan Surg*. 2012; 101(2):114–118.
20. Alexandrescu VA, Hubermont G. Does peripheral neuropathy have a clinical impact on the endovascular approach as a primary treatment for limb-threatening ischemic foot wounds in diabetic patients? *J Diabetes Metab*. 2012;S5:01–07.
21. Williams PL, Warwick R, Dyson M, Bannister LH. Angiology, blood vessels. In: Williams PL, Warwick R, Dyson M, Bannister LH, eds. *Churchill Livingstone."Gray's Anatomy."*. 37th ed. 1989:682–694. New York, London.
22. Kropman RH, Kiela G, Moll FL, de Vries JP. Variations in the anatomy of the popliteal artery and its side branches. *Vasc Endovasc Surg*. 2011;45(6):536–540.
23. Abou-Foul AK, Borumandi F. Anatomical variations of lower limb vasculature and implications for free fibula flap: systematic review and critical analysis. *Microsurgery*. 2016;36(2):165–172.
24. Yamada T, Gloviczki P, Bower TC, Naessens JM, Carmichael SW. Variations of the arterial anatomy of the foot. *Am J Surg*. 1993; 166(2):130–135.
25. Summer DS. Hemodynamics and rheology of vascular disease: applications to diagnosis and treatment. In: Ascer E, Hollier LH, Strandness Jr D, Towne JB, eds. *Haimovici's Vascular Surgery Principles and Techniques*. 4th ed. Blackwell Science; 1996:104–124.
26. Simon F, Oberhuber A, Floros N, et al. Acute limb ischemia-much more than just a lack of oxygen. *Int J Mol Sci*. 2018:374. https://doi.org/10.3390/ijms19020374.
27. Wang L, Zhou ZW, Yang LH, et al. Vasculature characterization of a multiterritory perforator flap: an experimental study. *J Reconstr Microsurg*. 2017;33(4):292–297.
28. Ziegler MA, Distasi MR, Bills RG, et al. Marvels, mysteries and misconceptions of vascular compensation to peripheral artery occlusion. *Microcirculation*. 2010;17(1):3–20.
29. Hillier C, Sayers RD, Watt PA, Naylor R, Bell PR, Thurston H. Altered small artery morphology and reactivity in critical limb ischaemia. *Clin Sci (Lond)*. 1999;96(2):155–163.
30. Waltenberg J. Impaired collateral vessel development in diabetes: potential cellular mechanisms and therapeutic implications. *Cardiovasc Res*. 2001;49:554–560.
31. O'Neal LW. Surgical pathology of the diabetic foot and clinicopathologic correlations. In: Bowker JH, Pfeifer MA, eds. *Levin and O'Neal's the Diabetic Foot*. 7th ed. Mosby Elsevier; 2007:367–401.
32. Conte MS, Bradbury AW, Kolh P, et al. Global vascular guidelines on the management of chronic limb-threatening ischemia (vol 69, 2019). *J Vasc Surg*. 2019;70:662.
33. Ken Chang K, Yoon S, Sheth N, et al. Rapid vs. delayed infrared responses after ischemia reveal recruitment of different vascular beds. *Quant Infrared Thermogr J*. 2015;12(2):173–183.
34. Alexandrescu VA, Defraigne JO. Angiosome system and principle, Chapter 77. In: Lanzer P, ed. *Textbook of Catheter-Based Cardiovascular Interventions*. Springer; 2018:1343–1360.
35. Gillani S, Cao J, Suzuki T, Hak DJ. The effect of ischemia reperfusion injury on skeletal muscle. *Injury*. 2012;43(6):670–675.
36. Guarini G, Capozza PG, Huqi A, Morrone D, Chilian WM, Marzilli M. Microvascular function/dysfunction downstream a coronary stenosis. *Curr Pharm Des*. 2013;19(13):2366–2374.
37. Vagal A, Aviv R, Sucharew H, et al. Collateral clock is more important than time clock for tissue fate: a natural history study of acute ischemic strokes. *Stroke*. 2018;49(9):2102–2107.
38. Nichols WW, O'Rurke MF, Vlachopoulos C. Generalized and metabolic arterial disease. In: Nichols WW, O'Rurke MF, Vlachopoulos C, eds. *McDonald's Blood Flow in Arteries. Theoretical, Experimental and Clinical Principles*. 6th ed. Hodder Arnold; 2011:523–534.
39. Kamenskaia OV, Klinkova AS, Karpenko AA, Karas'kov AM, Zeĭdlits GA. Peripheral microcirculation in patients with lower-limb atherosclerosis on the background of metabolic syndrome. *Angiol Sosud Khir*. 2014;20(4):22–26.
40. Alexandrescu VA, London V. Angiosomes: the cutaneous and arterial evaluation in CLI patients. (Chapter 5). In: Mustapha JA, ed. *Critical Limb Ischemia: CLI Diagnosis and Interventions*. Chicago: HMP; 2015:71–88.
41. Rozen WM, Grinsell D, Koshima I, Ashton MW. Dominance between angiosome and perforator territories: a new anatomical model for the design of perforator flaps. *J Reconstr Microsurg*. 2010;26(8):539–545.
42. Taylor GI, Corlett RJ, Ashton MW. The functional angiosome: clinical implications of the anatomical concept. *Plast Reconstr Surg*. 2017;140(4):721–733.
43. Macchi C, Giannelli F, Cecchi F, et al. Collateral circulation in occlusion of lower limbs arteries: an anatomical study and statistical research in 35 old subjects. *Ital J Anat Embryol*. 1996;101(2):89–95.
44. Mangi MA, Kahloon R, Elzanaty A, Zafrullah F, Eltahawy E. The use of fractional flow reserve for physiological assessment of indeterminate lesions in peripheral artery disease. *Cureus*. 2019;11(4): e4445. https://doi.org/10.7759/cureus.4445.
45. Ikeoka K, Hoshida S, Watanabe T, et al. Pathophysiological significance of velocity-based microvascular resistance at maximal hyperemia in peripheral artery disease. *J Atheroscler Thromb*. 2018; 25:1128–1136.
46. Berry C, Corcoran D, Hennigan B, Watkins S, Layland J, Oldroyd G. Fractional flow reserve-guided management in stable coronary disease and acute myocardial infarction: recent developments. *Eur Heart J*. 2015;36:3155–3164.
47. Varela C, Acín F, de Haro J, Bleda S, Esparza L, March JR. The role of foot collateral vessels on ulcer healing and limb salvage after successful endovascular and surgical distal procedures according to an angiosome model. *Vasc Endovascular Surg*. 2010;44:654–660.
48. Alexandrescu VA. Myths and proofs of angiosome applications in CLI: where do we stand? *J Endovasc Ther*. 2014;21:616–624.
49. Špillerová K, Sörderström M, Albäck A, Venermo M. The feasibility of angiosome-targeted endovascular treatment in patients with critical limb ischaemia and foot ulcer. *Ann Vasc Surg*. 2015;30: 270–276.
50. Bargellini I, Piaggesi A, Cicorelli A, et al. Predictive value of angiographic scores for the integrated management of the ischemic diabetic foot. *J Vasc Surg*. 2013;57:1204–1212.
51. Settembre N, Biancari F, Spillerova K, Albäck A, Söderström M, Venermo M. Competing risk analysis of the impact of pedal arch status and angiosome-targeted revascularization in chronic limb threatening ischemia. *Ann Vasc Surg*. 2020;S0890–5096(20). https://doi.org/10.1016/j.avsg.2020.03.042, 30315-0.

52. Rashid H, Slim H, Zayed H, et al. The impact of arterial pedal arch quality and angiosome revascularization on foot tissue loss healing and infrapopliteal bypass outcome. *J Vasc Surg*. 2013;57:1219–1226.
53. Alexandrescu VA, Sinatra T, Maufroy C. Current issues and interrogations in angiosome wound targeted revascularization for chronic limb threatening ischemia: a review. *World J Cardiovasc Dis*. 2019;9:168–192.
54. Bell PRF, Charlesworth D, DePalma RG, et al. The definition of critical ischemia of a limb. *Br J Surg*. 1982;69(Suppl):S2–S3.
55. Gulati A, Botnaru I, Garcia LA. Critical limb ischemia and its treatments: a review. *J Cardiovasc Surg*. 2015;56(5):775–785.
56. Kabra A, Suresh KR, Vivekanand V, Vishnu M, Sumanth R, Nekkanti M. Outcome of angiosome and non-angiosome targeted revascularization in critical lower limb ischemia. *J Vasc Surg*. 2013; 57(1):44–49.
57. Azuma N, Uchida H, Kokubo T, Koya A, Akasaka N, Sasajima T. Factors influencing wound Healing of critical ischemic foot after bypass surgery: is the angiosome important in selecting bypass target artery? *Eur J Vasc Endovasc Surg*. 2012;43(3):322–328.
58. Ortí PB, Vázquez RR, Minguell PP, Garcia SV, Munõz EMR, Vilardell PL. Percutaneous revascularization of specific angiosome in critical limb ischemia. *Angeologia*. 2011;63:11–17.
59. Pavé M, Benadiba L, Berger L, Gouicem D, Hendricks M, Plissonnier D. Below-the-knee angioplasty for critical limb ischemia: results of a series of 157 procedures and impact of the angiosome concept. *Ann Vasc Surg*. 2016;36:199–207.
60. Soares Rde A, Brochado Neto FC, Matielo MF, et al. Concept of angiosome does not affect limb salvage in infrapopliteal angioplasty. *Ann Vasc Surg*. 2016;32:34–40.
61. Park SW, Kim JS, Yun IJ, et al. Clinical outcomes of endovascular treatments for critical limb ischemia with chronic total occlusive lesions limited to below-the-knee arteries. *Acta Radiol*. 2013;54(7): 785–789.
62. Špillerová K, Biancari F, Settembre N, Albäck A, Venermo M. The prognostic significance of different definitions for angiosome-targeted lower limb revascularization. *Ann Vasc Surg*. 2017;40: 183–189.
63. Biancari F, Juvonen T. Angiosome-targeted lower limb revascularization for ischemic foot wounds: systematic review and meta-analysis. *Eur J Vasc Endovasc Surg*. 2014;47(5):517–522.
64. Dilaver N, Twine CP, Bosanquet DC. Direct vs indirect angiosomal revascularization of infrapopliteal arteries, an updated systematic review and meta-nalysis. *Eur J Vasc Endovasc Surg*. 2018;56(6): 834–848.
65. Huang TY, Huang TS, Wang YC, Huang PF, Yu HC, Yeh CH. Direct revascularization with the angiosome concept for lower limb ischemia: a systematic review and meta-analysis. *Medicine*. 2015; 94(34):e1427.
66. Jongsma H, Bekken JA, Akkersdijk GP, Hoeks SE, Verhagen HJ, Fioole B. Angiosome-directed revascularization in patients with critical limb ischemia. *J Vasc Surg*. 2017;65:1208–1219.
67. Spillerova K, Biancari F, Leppäniemi A, Albäck A, Söderström M, Venermo M. Differential impact of bypass surgery and angioplasty on angiosome-targeted infrapopliteal revascularization. *Eur J Vasc Endovasc Surg*. 2015;49(4):412–419.
68. Stimpson AL, Dilaver N, Bosanquet DC, Ambler GK, Twine CP. Angiosome specific revascularisation: does the evidence support it? *Eur J Vasc Endovasc Surg*. 2019;57(2):311–317.
69. Hata Y, Iida O, Mano T. Is angiosome-guided endovascular therapy worthwhile? *Ann Vasc Dis*. 2019;12(2):315–318.
70. Fujii M, Terashi H. Angiosome and tissue healing. *Ann Vasc Dis*. 2019;12(2):147–150.
71. Vallabhaneni R, Kalbaugh CA, Kouri A, Farber A, Marston WA. Current accepted hemodynamic criteria for critical limb ischemia do not accurately stratify patients at high risk for limb loss. *J Vasc Surg*. 2016;63:105–113.
72. Walden R, Adar R, Mozes M. Gangrene of toes with normal peripheral pulses. *Ann Surg*. 1977;185(3):269–272.
73. Aerden D, Denecker N, Gallala S, Debing E, Van den Brande P. Wound morphology and topography in the diabetic foot: hurdles in implementing angiosome-guided revascularization. *Int J Vasc Med*. 2014;2:1–5.
74. Gilbey SG, Walters H, Edmonds ME, et al. Vascular calcification, autonomic neuropathy, and peripheral blood flow in patients with diabetic nephropathy. *Diabet Med*. 1989;1:37–42.
75. Forst T, Pfützner A, Kann P, Lobmann R, Schäfer H, Beyer J. Association between diabetic-autonomic-C-fibre-neuropathy and medial wall calcification and the significance in the outcome of trophic foot lesions. *Exp Clin Endocrinol Diabetes*. 1995;103(2): 94–98.
76. Aerden D, Massaad D, von Kemp K, van Tussenbroek F, Debing E. The ankle-brachial index and the diabetic foot: a troublesome marriage. *Ann Vasc Surg*. 2011;25(6):770–777.
77. Tehan PE, Santos D, Chuter VH. A systematic review of the sensitivity and specificity of the toe-brachial index for detecting peripheral artery disease. *Vasc Med*. 2016;21(4):382–389.
78. Harth C, Randon C, Vermassen F. Impact of angiosome targeted femorodistal bypass surgery on healing rate and outcome in chronic limb threatening ischaemia. *Eur J Vasc Endovasc Surg*. 2020;60(1):68–75. https://doi.org/10.1016/j.ejvs.2020.03.011.
79. Alexandrescu VA, Houbiers A. Limb-based patency for chronic limb-threatening ischemia treatment: do we face a threshold for redefining current revascularization practice? *J Endovasc Ther*. 2020:1–4. https://doi.org/10.1177/1526602820926953.

CHAPTER

34

RNA therapies for cardiovascular disease

Ageliki Laina[1,2], *Nikolaos I. Vlachogiannis*[2,3], *Kimon Stamatelopoulos*[1,2] *and Konstantinos Stellos*[2,3]

[1]Department of Clinical Therapeutics, National and Kapodistrian University of Athens, School of Medicine, Athens, Greece [2]Biosciences Institute, Vascular Biology and Medicine Theme, Faculty of Medical Sciences, Newcastle University, Newcastle upon Tyne, United Kingdom [3]Department of Cardiology, Newcastle Hospitals NHS Foundation Trust, Newcastle upon Tyne, United Kingdom

Introduction to RNA therapeutics

Until recently, therapeutic approaches focused solely on the targeting of proteins with the most efficient approach comprising the development of monoclonal antibodies (mAbs).[1] mAbs have indeed revolutionized the therapeutic approach to multiple chronic inflammatory disorders. Moreover, by targeting specific cell populations (e.g., anti-CD20 [B cells]) or proteins (e.g., anticytokine therapies such as anti-TNF, anti-IL-6R, anti-IL-1), mAbs have expanded our understanding on their role in disease progression. In total, it is estimated that out of the approximately 20,000 human proteins, less than 700 have actually been targeted to date.[2] However, mAbs are only able to target proteins expressed on cell surface or released in the circulation, as the large size of mAbs prevents them from entering the cells, thus having an even more limited target repertoire.[2] These limitations along with the increasing understanding of the regulatory role of small RNAs (e.g., small interfering RNAs; siRNAs) in the stability and expression of mRNAs have led the scientific community to explore the RNA world for new therapeutics.[3,4]

According to the central dogma of molecular biology, RNA was historically considered a passive intermediate carrier of information between the genetic code (DNA) and the functional end-product (proteins).[5] However, recent research suggests that RNA is not merely a carrier of genetic information, but an essential regulator of gene expression and cellular function.[6] RNA-binding proteins, which bind to consensus RNA motif nucleotide sequences and usually recognize specific RNA secondary structures (single-stranded or double-stranded RNA), account for more than 10% of the protein-coding genes, underlining the importance of this class of proteins through evolution.[7] Their function can affect multiple aspects of RNA metabolism including splicing, RNA stability, and localization ultimately defining the proteome and thus providing an additional regulatory level of gene expression.[6,7] It is noteworthy that what was until recently considered as junk DNA/RNA plays a key regulatory role in gene expression. Various RNA species of noncoding RNA (ncRNA), such as micro-RNAs, long noncoding RNAs, and circular RNAs, have been implicated either in the maintenance of homeostatic mechanisms or, upon their dysregulation, in the pathogenesis or progression of human disease.[8] The emerging therapeutic potential of targeting RNA molecules has led to the development of a new class of therapeutic agents, collectively termed as RNA therapeutics. These agents are divided in two main categories: (1) agents that target messenger RNAs (mRNAs), which have already reached the clinic; and (2) agents targeting ncRNAs, which in turn control the expression of several mRNAs and thus bear the potential to target several genes by using one therapeutic molecule.[3]

Cardiovascular diseases (CVDs) comprise the leading cause of death worldwide, despite the development of multiple therapeutic agents and aggressive, early interventions such as primary percutaneous coronary intervention. Thus, it is not surprising that the cardiovascular field is at the forefront of the development of RNA-based therapies.[3,9] The emerging agents targeting ncRNAs are still at experimental level and have been summarized before in these excellent reviews.[10–12] To date, three main classes of RNA-based therapies have

The Vasculome
https://doi.org/10.1016/B978-0-12-822546-2.00003-4

reached clinical stage development, which we will briefly describe below (excellently reviewed in Ref. [4]).

Antisense oligonucleotides

Antisense oligonucleotides (ASOs) comprise a class of synthetic oligonucleotides designed to modulate gene expression usually by downregulating target genes. ASOs are short, single-stranded DNA molecules consisting of approximately 20 bases in length that bind to the target RNA in a sequence-specific manner. ASOs can target various classes of RNA inside the cell including pre-mRNA or mature mRNA, but also ncRNAs. ASOs can lead to degradation of targeted mRNAs either by recruiting RNase H ("Gapmers"; Fig. 34.1A) or by blocking polyadenylation sites, thereby accelerating mRNA decay. Moreover, they may prevent binding of splicing factors, thereby inducing alternative splicing or may block translation by targeting sites of translation initiation.[4] On the other hand, although not used till now, ASOs could also target inefficient upstream open reading frames to increase translation efficiency.[13] Finally, ASOs that target mature miRNAs ("antagomirs") are currently being developed.[14]

First-generation ASOs were designed with phosphorothioate linkages between the nucleotides in order to induce firmer bonds that are more resistant to degradation by intra/extracellular endonucleases.[15] Moreover, phosphorothioate linkages were later shown to facilitate cellular uptake and bioavailability in vivo.[16] In newer generation ASOs, the introduction of other modifications such as 2′-O-methyl (2′-OMe), 2′-O-methoxyethyl (2′-MOE), and locked nucleic acids (LNAs; additional methylene group placed between 2′ oxygen and 4′ carbon of the ribose ring) led to better targeting of mRNAs while reducing undesired immune activation.[17]

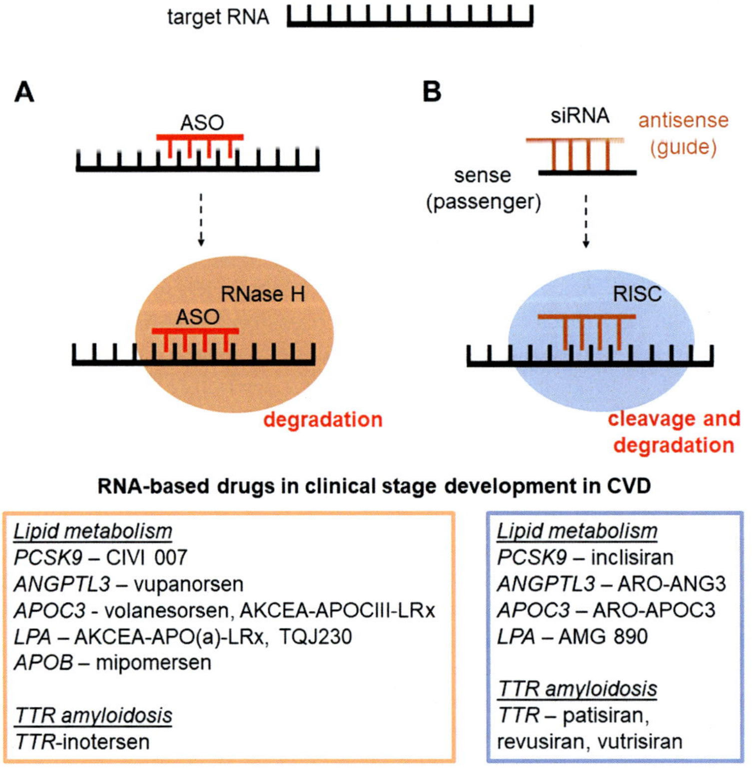

FIGURE 34.1 **Simplified graphic representation of the two most common classes of RNA-based therapies in cardiovascular disease (CVD).** (A) Antisense oligonucleotides (ASOs) are short, single-stranded DNA molecules consisting of approximately 20 bases in length that bind to the target RNA in a sequence-specific manner. ASOs can target various classes of RNA inside the cell including pre-mRNA or mature mRNA, but also noncoding RNAs. One of the most commonly used mechanisms of action of ASOs is degradation of targeted mRNAs by recruitment of RNase H. ASOs at clinical stage development in CVD target various components of lipid metabolism (PCSK9, ANGPTL3, APOC3, LPA, APOB) and TTR in amyloidosis. (B) Small interfering RNAs (siRNAs) are also approximately 19–25 nucleotides long; however, unlike ASOs, they contain both a specific mRNA sequence and its complement (antisense) strand and therefore are double-stranded. siRNAs are loaded into the RNA-induced silencing complex (RISC). AGO2, a component of RISC, cleaves the sense strand of siRNA, while the remaining antisense strand binds with perfect complementarity to the target mRNA. Once the single-stranded siRNA (part of the RISC complex) binds to its target mRNA, it induces mRNA cleavage. The mRNA is now cut and recognized as abnormal by the cell. This causes degradation of the mRNA and in turn no translation of the mRNA into amino acids and then proteins.

Small interfering RNAs

The second large class of RNA-based therapeutics are siRNAs. siRNAs are also approximately 19–25 nucleotides long; however, unlike ASOs, they contain both a specific RNA sequence and its complement (antisense) strand and therefore are double-stranded. siRNAs are loaded into the RNA-induced silencing complex (RISC). Argonaute2 (AGO2), a component of RISC, cleaves the sense strand of siRNA, while the remaining antisense strand binds with perfect complementarity to the target mRNA resulting in its cleavage and subsequent degradation (Fig. 34.1B).[18–20] Endogenous RNA interference is widely conserved among species[21] and siRNAs have a prolonged half-life thanks to their incorporation in RISC. Similar to ASOs, chemical modifications of the 2′ position of the ribose prevent digestion of siRNAs by endonucleases, while they also reduce recognition by innate immune receptors preventing the unwanted immune activation.[22,23] Moreover, siRNAs can be encapsulated in lipid-based nanoparticles which can further prevent their quick degradation or renal excretion.[17]

Given their highly efficient and prolonged action, the main challenge in the clinical application of siRNAs is tissue specificity to avoid systematic adverse effects.[17] One widespread approach to tackle this issue is conjugation of siRNAs with triantennary N-acetylgalactosamine (GalNAc).[24,25] These groups are specifically recognized by hepatocytes, which express the respective asialoglycoprotein receptor (ASGPR).[24,25] Given the central role of the liver in multiple metabolic pathways, this targeted approach can be used for various targets, while preventing the systematic adverse effects of classical therapeutic regimens.[17,25]

Aptamer-based therapeutics

Aptamers comprise a new promising class of RNA therapeutics. They are synthetic single-stranded DNA or RNA molecules that fold into defined architectures and bind to target molecules.[26] Aptamers that bind their target with high affinity and specificity are selected from large libraries after exponential enrichment through a process called SELEX.[26–28] Aptamers can be used both for diagnostic and therapeutic purposes as they can target proteins, viruses, or bacteria. Their ability to form multidimensional interactions with a very specific target has led to their characterization as an RNA analog to mAbs. However, their small size allows them to easily enter cells, therefore expanding their targetome to thousands of intracellular molecules. Aptamers are at the same time easier and cheaper to produce than mAbs.[29] Finally, aptamers can be combined to target multiple proteins at the same time or can be used as scavengers to carry other RNA therapeutics such as siRNAs.[26] Their high specificity, easy production, and capacity to target intracellular components render aptamers a promising alternative to the currently widely used mAbs.[30] However, no aptamer-based therapies have successfully reached the clinic in CVD till now.

RNA therapeutics in lipid disorders

Proprotein convertase subtilisin/kexin type 9

Proprotein convertase subtilisin/kexin type 9 (PCSK9) reduces low-density lipoprotein (LDL) receptor (LDLR) expression by promoting endosomal and lysosomal degradation of the PCSK9-LDLR complex, thus preventing LDLR recycling to the hepatocyte membrane. This leads to a reduction in LDL uptake and transport into hepatocytes and an increase in circulating LDL levels.[31,32] The benefit of PCSK9 inhibition on reducing LDL-C and CVD has been established in two large cardiovascular (CV) outcome trials using fully human mAbs.[33,34] The addition of evolocumab to standard statin therapy in a clinical trial of 27,564 high-risk patients with stable atherosclerotic CVD (FOURIER study) was associated with a 59% reduction in LDL-C at 48 weeks and resulted in 15% risk reduction in the primary outcome, driven largely by reductions in nonfatal myocardial infarction (MI), stroke, and revascularization.[33] Similarly, alirocumab in the ODYSSEY Outcomes trial was evaluated in patients with recent acute coronary syndrome (ACS) and with elevated LDL-C levels despite high-intensity statin therapy. Alirocumab resulted in 61% reduction of baseline LDL-C and 15% in major adverse CV events, with the greatest benefit noted in individuals with a baseline LDL-C >100 mg/dL, after a median follow-up of 2.8 years.[34] An alternative approach targeting PCSK9 is inclisiran, a long-acting, GalNAc-conjugated synthetic siRNA directed against PCSK9 mRNA.[35] GalNAc modification facilitates efficient and specific liver drug uptake, resulting in fewer off-target effects and achieving a pharmacodynamic effect at markedly lower doses.[36] Unlike inclisiran that inhibits hepatic PCSK9 production inside the cell, mAbs bind extracellular PCSK9 and thus do not affect any biological action of PCSK9 inside the cell. Another disadvantage of anti-PCSK9 mAbs is the high manufacturing cost and inconvenient dosage scheme accounting for 12 or 24 yearly subcutaneous (sc) injections versus twice a year in the case of inclisiran.

Inclisiran has been shown to significantly decrease LDL-C in two phase III studies.[37,38] In both studies, approximately 90% of the participants were treated with a background statin. Inclisiran 300 mg or matching placebo was administered at baseline, at 3 months, and every 6 months thereafter. In 482 adults with heterozygous familial hypercholesterolemia (HeFH) (ORION-9, NCT03397121) already receiving high-intensity statins along with ezetimibe, LDL-C at day 510 was 39.7% lower than baseline in the inclisiran group compared to 8.2% increase in the placebo group.[37] The ORION-10 (NCT03399370) and the ORION-11 (NCT03400800) were similar in design studies. Patients with atherosclerotic CVD or risk equivalent (type 2 diabetes, FH or a 10-year risk >20% as assessed by Framingham Risk Score) and elevated LDL-C despite maximally tolerated doses of lipid-lowering therapies were included. Inclisiran reduced LDL-C by 52.3% and 49.9% at 18 months, respectively.[38] In all three studies, adverse events were similar in both treatment arms, and injection site reactions were more commonly reported in those receiving inclisiran, with none being severe or persistent[37,38] (Table 34.1).

Ongoing clinical trials accessing the therapeutic value of targeting PCSK9 mRNA

The effects of inclisiran on clinical outcomes among people with CVD are currently under investigation in ORION-4 (NCT03705234) study conducted at approximately 150 sites across the United Kingdom and United States, where 15,000 participants over 55 years old with preexisting CVD will be randomized in inclisiran or placebo (Table 34.1).[47] Inclisiran (300 mg) will be administered s.c. at day 1, at 3 months, and every 6 months thereafter. Patients will be followed for a median of 5 years for CV death, MI, fatal or nonfatal ischemic stroke, and urgent coronary revascularization procedure. Estimated completion study date is in 2024.[47] Safety, tolerability, and efficacy of inclisiran are currently examined in ORION-5 (NCT03851705), a phase III double-blind, placebo-controlled, randomized trial in subjects with homozygous FH (HoFH) (Table 34.1).[48] Patients with a genetic or clinical diagnosis of HoFH and on maximally tolerated statins or other lipid-lowering therapy with fasting LDL-C $\geq$ 130 mg/dL and triglycerides (TG) $<$ 400 mg/dL will participate in the study. The primary outcome will be percentage change in LDL-C from baseline to day 150.[48] An active comparator open-label phase II study (ORION-3, NCT03060577, an extension trial of ORION-1) in patients with clinical CVD, risk equivalent, or HeFH despite maximum tolerated LDL-C—lowering drug assigned to inclisiran or evolocumab is currently ongoing (Table 34.1).[49] Preliminary results from ORION-3 were presented at the 2019 National Lipid Association Scientific Sessions. Participants that received inclisiran twice a year demonstrated a 51% reduction in LDL-C from baseline at ~22 months, with minimal adverse effects.[49] ORION-8 (NCT03814187) is a phase III open-label long-term extension of ORION-5, -9, -10, and -11 designed to assess the long-term effect of inclisiran in subjects with high CV risk and elevated LDL-C (Table 34.1).[50] Patients with established CVD, risk equivalent, HeFH, or HoFH and elevated LDL-C despite maximally tolerated lipid-lowering therapy will be enrolled. Subjects will receive inclisiran 300 mg s.c. at day 1, at 3 months, and then every 6 months. The primary outcome will assess the proportion of subjects reaching an LDL-C<70 mg/dL or <100 mg/dL by day 1,080.[50] The study is estimated to be completed in 2023.

A long-acting PCSK9 third-generation LNA ASO directed against PCSK9, CIVI007, is currently being assessed in a phase IIa trial (NCT04164888) (Table 34.1). A total of 48 individuals with fasting LDL-C$\geq$100 mg/dL or 70 mg/dL (depending on CVD history) and fasting TG $<$ 400 mg/dL on stable statin therapy will be enrolled.[51]

Angiopoietin-like 3

Angiopoietin-like 3 (ANGPTL3) is secreted by the liver and regulates plasma lipoprotein metabolism by acting as an inhibitor of lipoprotein lipase (LPL) and endothelial lipase, which are key enzymes in the metabolism of very-low-density-lipoprotein (VLDL) and high-density lipoprotein (HDL)-cholesterol, respectively.[52] Mouse studies have shown that ANGPTL3 also has direct effects on VLDL production.[53] Loss-of-function mutations are associated with beneficial effect on lipid and glucose metabolism and reduced risk of coronary heart disease.[52,54,55] A monoclonal antibody against ANGPTL3, evinacumab, has been shown to decrease TG, LDL-C, and Lp(a) in patients with homozygous familial hypercholesterolemia.[56] Vupanorsen (AKCEA-ANGPTL3-LRx) is a GalNAc-conjugated ASO targeting hepatic ANGPTL3 mRNA. Overweight patients with hypertriglyceridemia, type 2 diabetes, and hepatic steatosis participated in a phase II double-blind dose-ranging placebo-controlled study and were followed-up for 6 months (NCT03371355) (Table 34.1).[39] Vupanorsen significantly reduced TG levels and had a favorable safety profile. In specific, triglycerides decreased by 36% at 40 mg and 53% at 80 mg of vupanorsen administered every 4 weeks and 47% at 20 mg of vupanorsen administered every week, compared to 16% pooled reduction observed in the placebo group. The use of statins was similar but low in both study drug and placebo group (45% vs. 37%, respectively). No changes in liver fat or HbA1c were observed. Of note, the study included a small number of participants (n = 105).[39]

TABLE 34.1 Clinical development of RNA targeted therapies for lipid disorders.

Drug name	Class/delivery system	Target	Phase	n	Trial No.	Indication	Dose	Endpoint/outcome	Adverse reactions
Inclisiran	GalNAc siRNA	PCSK9	II	490	NCT03060577 ORION-3	ASCVD or ASCVD RE	300 mg sc Q6 months versus evolocumab 140 mg Q2W	% LDL-C change at day 210	*Expected February 2022*
			III	15000	NCT03705234 ORION-4	ASCVD or ASCVD RE	300 mg sc Q3-6 months	MACE	*Expected December 2024*
			III	56	NCT03851705 ORION-5	HoFH	300 mg sc	% LDL-C change baseline—day 150	*Expected September 2021*
			III	3300	NCT03814187 ORION-8	ASCVD or ASCVD RE or HeFH/HoFH	300 mg sc Q3-6 months	Proportion of subjects reaching LDL-C <70 mg/dL baseline—day 1080	*Expected August 2023*
			III	482 1561 1617	NCT03397121 ORION-9[37] NCT03399370 ORION-10[38] NCT03400800 ORION-11[38]	HeFH ASCVD ASCVD or RE	300 mg sc Q3-6 months	~50% LDL-C reduction from baseline—day 510	Injection site reaction
CIVI 007	GalNAc ASO	PCSK9	II	48	NCT04164888	LDL-C ≥70 mg/dL TG < 400 mg/dL ± CVD	Different doses and dosing regimens	AEs related to treatment up to 2 months	*Expected September 2020*
Vupanorsen	LICA ASO	ANGPTL3	II	105	NCT03371355[39]	TG > 150 mg/dL T2DM HFF>8% 40>BMI>27	40/80 mg sc Q4W or 20 mg QW	%TG change baseline—6 months 40 mg (36%)/80 mg (53%)/20 mg (47%)	Injection site reaction
ARO-ANG3	GalNAc siRNA	ANGPTL3	I/IIa	93	NCT03747224	Healthy Dyslipidemia	Single and multiple ascending dose	AEs related to treatment up to 113 (±3 days) post-dose	*Expected May 2021*
Volanesorsen	ASO	ApoC-III	III	67 114	NCT02211209 APPROACH[40] NCT02300233 COMPASS[62]	FCS TG ≥ 500 mg/dL	300 mg sc QW	>70% TG decrease baseline—13 weeks	Injection site reaction Thrombocytopenia
			III	40	NCT02527343 BROADEN	Familial partial lipodystrophy	300 mg sc QW	%TG change baseline—13 weeks	*Expected September 2021*

Continued

TABLE 34.1 Clinical development of RNA targeted therapies for lipid disorders.—cont'd

Drug name	Class/delivery system	Target	Phase	n	Trial No.	Indication	Dose	Endpoint/outcome	Adverse reactions
AKCEA-APOCIII-LRx	GalNAc ASO	ApoC-III	I/IIa	56	NCT02900027[41]	Healthy; mild TG increase	Single and multiple ascending dose	Up to 77% TG decrease	Well tolerated
			II	114	NCT03385239	Hypertriglyceridemia and CVD	Different doses and dosing regimens	TG change baseline—6 months	*Expected February 2020*
ARO-APOC3	GalNAc siRNA	ApoC-III	I/IIa	112	NCT03783377	Healthy hypertriglyceridemia, FCS	Single and multiple ascending dose	AEs related to treatment up to 113 (±3 days) post-dose	*Expected April 2021*
Mipomersen[42–45]	ASO	ApoB	III	51 58 124	NCT00607373 NCT00794664 NCT00706849	HoFH HeFH ± CVD	200 mg sc QW	FDA approval 2013 withdrawn 2019	Injection site reaction Liver toxicity Steatohepatitis
AKCEA-APO(a)-LRx	LICA ASO	Lp(a)	II	286	NCT03070782[46]	CVD Lp(a) > 60 mg/dL	20, 40, or 60 mg Q4W; 20 mg Q2W; 20 mg QW	>70% Lp(a) decrease baseline—6 months	Injection site reaction UTI Myalgia Headache
TQJ230	LICA ASO	Lp(a)	III	7680	NCT04023552	High CVD-risk; Lp(a) > 70 mg/dL	80 mg sc monthly	Time to MACE (4 years)	*Expected June 2024*
AMG 890	GalNAc siRNA	Lp(a)	II	290	NCT04270760	Elevated Lp(a)	Dose ranging	%Lp(a) change baseline—week 48	*Expected April 2023*

2′-MOE, 2′-O-methoxyethyl; *AEs*, adverse events; *ANGPTL3*, angiopoietin-like 3; *Lp(a)*, lipoprotein a; *ApoC-III*, apolipoprotein C-III; *ASCVD*, atherosclerotic cardiovascular disease; *BMI*, body mass index; *CVD*, cardiovascular disease; *FCS*, familial chylomicronemia syndrome; *GalNAc*, N-acetylgalactosamine; *HeFH*, heterozygous familial hypercholesterolemia; *HFF*, hepatic fat fraction; *HoFH*, homozygous familial hypercholesterolemia; *LDL*, low-density lipoprotein; *LICA*, ligand conjugated antisense; *LNA*, locked nucleic acid; *MACE*, major adverse cardiac events; *PCSK9*, proprotein convertase subtilisin/kexin type 9; *RE*, risk equivalent; *sc*, subcutaneous; *T2DM*, type 2 diabetes mellitus; *TG*, triglycerides; *UTI*, urinary tract infection.

Ongoing clinical trials accessing the therapeutic value of targeting angiopoietin-like 3 mRNA

The siRNA targeting ANGPTL3 mRNA, ARO-ANG3, is currently being evaluated in a phase I/IIa trial in healthy individuals with TG > 100 mg/dL and LDL-C>70 mg/dL (NCT03747224) (Table 34.1).[57] Interim analysis showed reductions in LDL-C of 39%–42% in 22 patients with hypercholesterolemia and decreased triglycerides by 79% in 5 patients with hypertriglyceridemia. Favorable safety profile and tolerability were observed.[57]

Apolipoprotein C-III

Another key regulator of triglyceride metabolism is apolipoprotein C-III (ApoC-III) which exerts its atherogenic action by attenuating lipolysis of triglyceride-rich lipoproteins through LPL inhibition, resulting in increased circulating levels of VLDL and chylomicrons.[58] Increased levels of ApoC-III are found in patients with hypertriglyceridemia and have been causally associated with metabolic syndrome and insulin resistance.[59] On the contrary, carriers of mutations disrupting ApoC-III function presented 40% lower risk for coronary heart disease compared to noncarriers.[60] Volanesorsen is a second-generation ASO designed to target ApoC-III mRNA.[61] Patients with familial chylomicronemia syndrome (FCS) (n = 67), a rare condition characterized by reduced or absent LPL activity, marked fasting, and postprandial chylomicronemia with TG levels 10–100 times above normal level of 150 mg/dL, were randomized to volanesorsen at a dose of 300 mg s.c. once a week or placebo in the phase III APPROACH study (NCT02211209) (Table 34.1).[40] Volanesorsen resulted in 84% reduction of plasma ApoC-III levels at 3 months and 77% decrease in mean TG levels. Among common adverse events were injection site reactions and dose-dependent thrombocytopenia. European Medicines Agency has recently approved Waylivra (volanesorsen) as an adjunct to diet in adult patients with genetically confirmed FCS at high risk for pancreatitis, in whom response to diet and TG lowering therapy is not sufficient. Waylivra is administered once weekly at first and after 3 months, patients with adequate TG reduction, receive the drug once every 2 weeks. Blood platelet count should be monitored regularly during treatment with Waylivra.[40]

In the COMPASS study (NCT02300233), a phase III randomized, double-blind, placebo-controlled trial, patients with hypertriglyceridemia (fasting TG > 500 mg/dL) were randomized 2:1 to receive 300 mg volanesorsen s.c. once a week or placebo for 26 weeks (Table 34.1).[62] Treatment with volanesorsen yielded a more than 70% reduction in TG levels in patients with or without FCS and reduced the occurrence of pancreatitis. There were no serious platelet events in the study. The most commonly reported adverse event was injection site reactions, occurring in 23.5% of volanesorsen injections. The ASO sequence was later conjugated with GalNAc in order to improve liver-specific uptake facilitating dose reductions. The AKCEA-APOCIII-LRx (NCT02900027) was evaluated in a phase I/IIa dose escalating study in 56 healthy individuals with mildly elevated TG levels, resulting in reduction of ApoC-III by up to 92% and TG levels by up to 77% (Table 34.1).[41] Additionally, it promoted a favorable lipid profile by significantly lowering total cholesterol, VLDL-C, and non–HDL-C levels and also increasing HDL-C. The modified ASO was well tolerated with no significant adverse events at the injection site or significant effects on platelet count.

The ASO toxicity profile led to ARO-APOC3, a hepatocyte-targeted siRNA designed to induce deep and durable gene-specific silencing while avoiding off-target effects. AROAPOC31001 (NCT03783377) is a phase I/IIa trial assessing safety and tolerability of RNAi therapeutic in healthy individuals with fasting TG levels above 80 mg/dL, resulting in reduced circulating ApoC-III levels, even at lowest dose schemes (Table 34.1).[63] Besides mild and transient local injection site reactions and moderate transient transaminase elevation in one subject, no adverse events were reported. Studies examining the effect of ARO-APOC3 chronic therapy in patients with severe hypertriglyceridemia and FCS are warranted.

Ongoing clinical trials accessing the therapeutic value of targeting apolipoprotein C-III mRNA

Results are awaited in 2021 from the BROADEN phase III study examining volanesorsen in patients with familial partial lipodystrophy (NCT02527343) (Table 34.1).[64] AKCEA-APOCIII-LRx is currently being assessed in patients with hypertriglyceridemia and established CVD in a phase II randomized, double-blind, placebo-controlled, dose-ranging study (NCT03385239) (Table 34.1).[65]

Apolipoprotein B

Mipomersen is a second-generation ASO directed against ApoB mRNA, administered s.c. once weekly.[42] Mipomersen received approval for HoFH in 2013 as an adjunct to lipid-lowering drugs and diet. Mipomersen lowered LDL-C and ApoB levels by 28% and 31%, respectively.[43] However, its use has been associated with increased risk of injection site reactions, flu-like symptoms, and, inherent to its pharmacological mechanism, liver toxicity and steatohepatitis (Table 34.1).[44]

Two different regimens of mipomersen SQ (200 mg once weekly and 70 mg thrice weekly) versus placebo were tested for 60 weeks in 309 patients with severe HeFH or with HeFH and CVD history in the FOCUS-FH study. Mipomersen was effective in lowering ApoB-containing lipoproteins; however, this benefit was offset by limited tolerability and increased hepatic transaminase levels in about 21% of patients.[45]

Apolipoprotein(a)

Lipoprotein(a) (Lp(a)) consists of an LDL particle with apolipoprotein a (APO(a)), covalently bound to its apoB100 backbone.[66] Lp(a) levels are genetically determined and, when elevated, are a risk factor for CVD and aortic valve stenosis.[67,68] Currently, there are no approved pharmacologic therapies to lower Lp(a) levels. Results from phase I and II studies of an ASO in patients with normal Lp(a) values as well as in patients with elevated Lp(a) concentrations have shown a reduction of >90%.[69,70] A phase II placebo-controlled study (NCT03070782) was conducted in patients with established CVD and Lp(a) levels above 60 mg/dL who received the ASO AKCEA-APO(a)-LRx, administered in ascending doses at intervals of 1–4 weeks[46] (Table 34.1). A dose-dependent reduction in Lp(a) was observed, with mean percentage decrease of 72% at 60 mg every 4 weeks, and 80% at 20 mg every week, as compared with 6% with placebo. Among most frequently encountered adverse events were injection site reactions, headache, myalgia, and urinary tract infection.[46]

Ongoing clinical trials accessing the therapeutic value of targeting APO(a) mRNA

Based on these encouraging data, HORIZON trial will assess the effect of once monthly s.c. injection of 80 mg TQJ230 (also known as AKCEA-APO(a)-LRx ASO) on CV endpoints (cardiovascular death, nonfatal MI, nonfatal stroke, and urgent coronary revascularization requiring hospitalization) in patients at increased CVD risk and Lp(a) levels above 70 mg/dL (NCT04023552) during a 4-year follow-up period. Study completion date is estimated to be in 2024 (Table 34.1).[71]

The GalNAc-conjugated siRNA AMG 890 that targets apolipoprotein(a) mRNA is currently under evaluation in double-blind, randomized, phase I and II studies to assess efficacy, safety, and tolerability in patients with elevated Lp(a) (NCT03626662, NCT04270760) (Table 34.1).[72,73]

Emerging RNA therapeutics for cardiac transthyretin amyloidosis

Transthyretin-mediated cardiomyopathy (ATTR-CM) is an increasingly recognized cause of heart failure with preserved ejection fraction (HFpEF) and is caused either by a mutation in the TTR gene (hATTR-CM) or by age-related TTR misfolding (wild-type ATTR-CM).[74–78] Patients with cardiac involvement experience progressive symptoms of heart failure and cardiac arrhythmias, high hospitalization rates, with death typically occurring 2.5–5 years after diagnosis.[77,78] Several therapeutics, including the use of siRNA and ASO drugs, showed suppression of TTR synthesis in the liver.

Patisiran

Patisiran (ALN-TTR02) is a siRNA formulated in lipid nanoparticles that specifically target TTR mRNA resulting in reduced TTR protein levels. Phase I and phase II studies of patisiran in healthy volunteers and in patients with hATTR polyneuropathy showed a dose-dependent and robust mean reduction in serum TTR levels up to 90%.[79,80] Patisiran (brand name Onpattro) was also approved by the European Commission for the treatment of hATTR in adults with stage 1 or 2 polyneuropathy based on the results of APOLLO study, a phase III randomized, double-blind, placebo-controlled clinical trial (NCT01960348).[81] In a prespecified subpopulation of patients from the APOLLO study with evidence of cardiac amyloid involvement (n = 126; 56% of total population), patisiran improved cardiac structure and function parameters, providing evidence that patisiran may halt or possibly reverse the progression of hATTR amyloid heart disease. In specific, patisiran decreased mean left ventricular wall thickness, improved global longitudinal strain, increased cardiac output, and reduced N terminal pro–B-type natriuretic peptide (NT-proBNP) levels, as compared to placebo[82] (Table 34.2). Moreover, a post hoc analysis of safety data showed that rates of all-cause death and cardiac hospitalization were decreased in the patisiran group compared with the placebo group (10.1 vs. 18.7 per 100 patient-years, hazard ratio (HR): 0.54; 95% CI: 0.28–1.01) (Table 34.2). However, the small sample size and the fact that solely left ventricular wall thickness was used to define cardiac involvement should be acknowledged as study limitations. Furthermore, only patients with hereditary ATTR polyneuropathy and patients with NYHA I-II participated in the APOLLO study; thus, there is no evidence whether patisiran is also effective in patients with wild-type ATTR-CM or patients with advanced heart failure.

Revusiran

Revusiran is a siRNA directed against TTR mRNA sequence present in both wild-type and mutant TTR variants, which are conjugated with GalNAc ligand that facilitates hepatocyte uptake.[83,84] Revusiran was previously studied in a phase I study in healthy volunteers[83] and a phase II study in patients with ATTR amyloidosis with cardiomyopathy.[84] The phase III ENDEAVOR study was a multicenter, randomized, placebo-controlled, double-blind study designed to evaluate the efficacy and safety of revusiran in patients with hATTR amyloidosis with cardiomyopathy (NCT02319005).[85] However, the trial was discontinued after a median follow-up of 6.71 months due to an imbalance of deaths in the revusiran group (18 patients, 12.9%) compared with the placebo group (2 patients, 3.0%) during the on-treatment period (Table 34.2). The majority of those who died were ≥75 years old with advanced heart failure and most deaths were deemed as cardiovascular due to heart failure. However, after thorough investigation, no clear causative mechanism for the observed mortality imbalance between treatment arms could be identified.

Vutrisiran

Vutrisiran (ALN-TTRsc02) is a second-generation GalNAc−siRNA conjugate targeting TTR mRNA, characterized by greater metabolic stability, potency, and durability compared to first-generation revusiran. The efficacy and safety of vutrisiran is evaluated in HELIOS-B, an ongoing phase III study (NCT04153149). Overall, 655 patients with hereditary and wild-type cardiomyopathy will be enrolled and will be assigned to either 25 mg vutrisiran administered s.c. or placebo. Primary endpoint of the study is the composite of all-cause mortality and recurrent cardiovascular events (CV hospitalizations and urgent heart failure visits). Results are expected to be reported in 2025 (Table 34.2).[86]

Inotersen

Recently, it has been shown that progression of TTR amyloid peripheral neuropathy is inhibited by treatment with inotersen, an ASO inhibitor,[87] which was recently approved to stabilize neuropathy in hereditary patients with or without cardiomyopathy. A subsequent open-label study investigated the efficacy of inotersen in patients with ATTR cardiomyopathy NYHA class I-III.[88] Patients received weekly 300 mg inotersen s.c., which was well tolerated and resulted in improved cardiac function, as assessed by 6-minute walk test and global systolic strain. Furthermore, left ventricular mass on MRI and intraventricular septum on echo did not progress overtime in most subjects. Of interest, left ventricular mass on MRI actually decreased in 8 of 10 patients (Table 34.2).

Future perspectives

While RNA therapeutics have been under development for almost 30 years, advances in medicinal chemistry and technology have only recently enabled their successful transition into clinical practice. Major issues such as tissue- or even cell-specific delivery, off-target

TABLE 34.2 Clinical development of RNA-based therapies for transthyretin cardiac amyloidosis.

Drug name	Delivery system	Phase	Trial no	Dose	Outcome	Current state
Patisiran (ALN-TTR02)	LNP siRNA	III APOLLO	NCT01960348 n = 126	0.3 mg/kg patisiran or placebo IV Q3W for 18 months	Decreased mean LV wall thickness, GLS NT-proBNP, cardiac hospitalization, all-cause death	Completed
Revusiran (ALN-TTRSC)	GalNAc-siRNA	III ENDEAVOR	NCT02319005 n = 206	500 mg sc daily for 5 days and then QW or placebo	6-min walk test distance and serum TTR reduction	Terminated due to mortality imbalance between treatment arms
Vutrisiran (ALN-TTRsc02)	GalNAc−siRNA	III HELIOS-B	NCT04153149 n = 655	25 mg vutrisiran sc Q3M or placebo	All-cause mortality, CV hospitalizations, and urgent heart failure visits	Ongoing Expected June 2025
Inotersen (IONIS-TTRRx)	ASO	III	NCT03702829 n = 50	300 mg sc QW	LV strain, 6MWT, BNP, LVM	Ongoing Expected March 2022

6MWT, 6 minute walk test; *ASO*, antisense oligonucleotide; *CV*, cardiovascular; *GLS*, global longitudinal strain; *LNPs*, lipid-based nanoparticles; *LV*, left ventricular; *LVM*, left venticular mass; *NT-proBNP*, N terminal pro−B-type natriuretic peptide; *sc*, subcutaneous; *TTR*, transthyretin.

effects, and immunogenicity are being addressed with the introduction of chemical modifications or novel (nano)carriers.[17] Nanoparticles conjugated with peptides to ensure cell-specific delivery, or encompassing novel polymeric formulations that favor uptake by specific tissues, have enabled the delivery of RNA drugs in additional tissues beyond the liver, such as the endothelial-enriched lung tissue.[89] In another promising approach, researchers have used exosomes to deliver RNA drugs to pancreatic tissue,[90] which are better protected from phagocytosis compared to the classical liposomal nanocarriers. Beyond the technical advancements of current RNA therapeutic classes regarding delivery, safety, and efficacy, a novel RNA therapeutic strategy has emerged. The intensive study of gene editing[91–93] and RNA-based modifications in the last years[94–97] has enabled the development of CRISPR-Cas13b RNA editing.[98] CRISPR-Cas13b, which carries the enzymatic region of adenosine deaminase acting on RNA-2 (ADAR2), promises to "correct" pathogenic single nucleotide mutations at the RNA level,[98] without the adverse effects classically linked to genomic CRISPR manipulations.

In conclusion, the emerging field of RNA therapeutics holds tremendous potential to specifically target genes that are causatively involved in the pathogenesis of CVD. Future studies are warranted to investigate the therapeutic potential of RNA-based therapies of previously undruggable targets, specific alleles or isoforms, intracellular disease-related molecules, or even base mutations in a manner that ensures tissue-specific drug delivery and prevents unwanted systematic effects. With the development of RNA therapeutics, the precision medicine era has already arisen highlighting the need of training the new generation of physicians to understand RNA biology and medicine.

Funding source

K. Stellos is supported by grants from the European Research Council (ERC) under the European Union's Horizon 2020 research and innovation program (MODVASC, grant agreement No. 759248) and the German Research Foundation DFG (SFB834 B12, project number 75732319).

References

1. Rajewsky K. The advent and rise of monoclonal antibodies. *Nature*. 2019;575(7781):47–49. https://doi.org/10.1038/d41586-019-02840-w.
2. Santos R, Ursu O, Gaulton A, et al. A comprehensive map of molecular drug targets. *Nat Rev Drug Discov.* 2017;16(1):19–34. https://doi.org/10.1038/nrd.2016.230.
3. Landmesser U, Poller W, Tsimikas S, Most P, Paneni F, Lüscher TF. From traditional pharmacological towards nucleic acid-based therapies for cardiovascular diseases. *Eur Heart J*. 2020;41(40): 3884–3899. https://doi.org/10.1093/eurheartj/ehaa229.
4. Lieberman J. Tapping the RNA world for therapeutics. *Nat Struct Mol Biol*. 2018;25(5):357–364. https://doi.org/10.1038/s41594-018-0054-4.
5. Crick F. Central dogma of molecular biology. *Nature*. 1970; 227(5258):561–563. https://doi.org/10.1038/227561a0.
6. Van Nostrand EL, Freese P, Pratt GA, et al. A large-scale binding and functional map of human RNA-binding proteins. *Nature*. 2020; 583(7818):711–719. https://doi.org/10.1038/s41586-020-2077-3.
7. Gerstberger S, Hafner M, Tuschl T. A census of human RNA-binding proteins. *Nat Rev Genet*. 2014;15(12):829–845. https://doi.org/10.1038/nrg3813.
8. Esteller M. Non-coding RNAs in human disease. *Nat Rev Genet*. 2011;12(12):861–874. https://doi.org/10.1038/nrg3074.
9. Laina A, Gatsiou A, Georgiopoulos G, Stamatelopoulos K, Stellos K. RNA therapeutics in cardiovascular precision medicine. *Front Physiol*. 2018;9:953. https://doi.org/10.3389/fphys.2018.00953.
10. Huang C-K, Kafert-Kasting S, Thum T. Preclinical and clinical development of noncoding RNA therapeutics for cardiovascular disease. *Circ Res*. 2020;126(5):663–678. https://doi.org/10.1161/CIRCRESAHA.119.315856.
11. Lucas T, Dimmeler S. RNA therapeutics for treatment of cardiovascular diseases: promises and challenges. *Circ Res*. 2016;119(7): 794–797. https://doi.org/10.1161/CIRCRESAHA.116.308730.
12. De Majo F, De Windt LJ. RNA therapeutics for heart disease. *Biochem Pharmacol*. 2018;155:468–478. https://doi.org/10.1016/j.bcp.2018.07.037.
13. Liang X-H, Shen W, Sun H, Migawa MT, Vickers TA, Crooke ST. Translation efficiency of mRNAs is increased by antisense oligonucleotides targeting upstream open reading frames. *Nat Biotechnol*. 2016;34(8):875–880. https://doi.org/10.1038/nbt.3589.
14. Krützfeldt J, Rajewsky N, Braich R, et al. Silencing of microRNAs in vivo with "antagomirs.". *Nature*. 2005;438(7068):685–689. https://doi.org/10.1038/nature04303.
15. Putney SD, Benkovic SJ, Schimmel PR. A DNA fragment with an alpha-phosphorothioate nucleotide at one end is asymmetrically blocked from digestion by exonuclease III and can be replicated in vivo. *Proc Natl Acad Sci USA*. 1981;78(12):7350–7354. https://doi.org/10.1073/pnas.78.12.7350.
16. Eckstein F. Phosphorothioates, essential components of therapeutic oligonucleotides. *Nucleic Acid Therapeut*. 2014;24(6):374–387. https://doi.org/10.1089/nat.2014.0506.
17. Roberts TC, Langer R, Wood MJA. Advances in oligonucleotide drug delivery. *Nat Rev Drug Discov.* 2020;19(10):673–694. https://doi.org/10.1038/s41573-020-0075-7.
18. Matranga C, Tomari Y, Shin C, Bartel DP, Zamore PD. Passenger-strand cleavage facilitates assembly of siRNA into ago2-containing RNAi enzyme complexes. *Cell*. 2005;123(4):607–620. https://doi.org/10.1016/j.cell.2005.08.044.
19. Liu J, Carmell MA, Rivas FV, et al. Argonaute2 is the catalytic engine of mammalian RNAi. *Science*. 2004;305(5689):1437–1441. https://doi.org/10.1126/science.1102513.
20. Martinez J, Patkaniowska A, Urlaub H, Lührmann R, Tuschl T. Single-stranded antisense siRNAs guide target RNA cleavage in RNAi. *Cell*. 2002;110(5):563–574. https://doi.org/10.1016/S0092-8674(02)00908-X.
21. Elbashir SM, Harborth J, Lendeckel W, Yalcin A, Weber K, Tuschl T. Duplexes of 21-nucleotide RNAs mediate RNA interference in cultured mammalian cells. *Nature*. 2001;411(6836):494–498. https://doi.org/10.1038/35078107.
22. Judge AD, Sood V, Shaw JR, Fang D, McClintock K, MacLachlan I. Sequence-dependent stimulation of the mammalian innate immune response by synthetic siRNA. *Nat Biotechnol*. 2005;23(4): 457–462. https://doi.org/10.1038/nbt1081.

23. Morrissey DV, Lockridge JA, Shaw L, et al. Potent and persistent in vivo anti-HBV activity of chemically modified siRNAs. *Nat Biotechnol.* 2005;23(8):1002–1007. https://doi.org/10.1038/nbt1122.
24. Nair JK, Willoughby JLS, Chan A, et al. Multivalent N-acetylgalactosamine-conjugated siRNA localizes in hepatocytes and elicits robust RNAi-mediated gene silencing. *J Am Chem Soc.* 2014;136(49):16958–16961. https://doi.org/10.1021/ja505986a.
25. Springer AD, Dowdy SF. GalNAc-siRNA conjugates: leading the way for delivery of RNAi therapeutics. *Nucleic Acid Therapeut.* 2018;28(3):109–118. https://doi.org/10.1089/nat.2018.0736.
26. Keefe AD, Pai S, Ellington A. Aptamers as therapeutics. *Nat Rev Drug Discov.* 2010;9(7):537–550. https://doi.org/10.1038/nrd3141.
27. Tuerk C, Gold L. Systematic evolution of ligands by exponential enrichment: RNA ligands to bacteriophage T4 DNA polymerase. *Science.* 1990;249(4968):505–510. https://doi.org/10.1126/science.2200121.
28. Ellington AD, Szostak JW. In vitro selection of RNA molecules that bind specific ligands. *Nature.* 1990;346(6287):818–822. https://doi.org/10.1038/346818a0.
29. Zhou J, Rossi J. Aptamers as targeted therapeutics: current potential and challenges. *Nat Rev Drug Discov.* 2017;16(3):181–202. https://doi.org/10.1038/nrd.2016.199.
30. Zhang Y, Lai BS, Juhas M. Recent advances in aptamer discovery and applications. *Molecules.* 2019;24(5). https://doi.org/10.3390/molecules24050941.
31. Abifadel M, Varret M, Rabès J-P, et al. Mutations in PCSK9 cause autosomal dominant hypercholesterolemia. *Nat Genet.* 2003;34(2):154–156. https://doi.org/10.1038/ng1161.
32. Nozue T. Lipid lowering therapy and circulating PCSK9 concentration. *J Atherosclerosis Thromb.* 2017;24(9):895–907. https://doi.org/10.5551/jat.RV17012.
33. Sabatine MS, Giugliano RP, Keech AC, et al. Evolocumab and clinical outcomes in patients with cardiovascular disease. *N Engl J Med.* 2017;376(18):1713–1722. https://doi.org/10.1056/NEJMoa1615664.
34. Schwartz GG, Steg PG, Szarek M, et al. Alirocumab and cardiovascular outcomes after acute coronary syndrome. *N Engl J Med.* 2018;379(22):2097–2107. https://doi.org/10.1056/NEJMoa1801174.
35. Fitzgerald K, White S, Borodovsky A, et al. A highly durable RNAi therapeutic inhibitor of PCSK9. *N Engl J Med.* 2017;376(1):41–51. https://doi.org/10.1056/NEJMoa1609243.
36. Kosmas CE, DeJesus E, Morcelo R, Garcia F, Montan PD, Guzman E. Lipid-lowering interventions targeting proprotein convertase subtilisin/kexin type 9 (PCSK9): an emerging chapter in lipid-lowering therapy. *Drugs Context.* 2017;6:212511. https://doi.org/10.7573/dic.212511.
37. Raal FJ, Kallend D, Ray KK, et al. Inclisiran for the treatment of heterozygous familial hypercholesterolemia. *N Engl J Med.* 2020;382(16):1520–1530. https://doi.org/10.1056/NEJMoa1913805.
38. Ray KK, Wright RS, Kallend D, et al. Two phase 3 trials of inclisiran in patients with elevated LDL cholesterol. *N Engl J Med.* 2020;382(16):1507–1519. https://doi.org/10.1056/NEJMoa1912387.
39. Gaudet D, Karwatowska-Prokopczuk E, Baum SJ, et al. Vupanorsen, an N-acetyl galactosamine-conjugated antisense drug to ANGPTL3 mRNA, lowers triglycerides and atherogenic lipoproteins in patients with diabetes, hepatic steatosis, and hypertriglyceridaemia. *Eur Heart J.* 2020;41(40):3936–3945. https://doi.org/10.1093/eurheartj/ehaa689.
40. Witztum JL, Gaudet D, Freedman SD, et al. Volanesorsen and triglyceride levels in familial chylomicronemia syndrome. *N Engl J Med.* 2019;381(6):531–542. https://doi.org/10.1056/NEJMoa1715944.
41. Alexander VJ, Xia S, Hurh E, et al. N-acetyl galactosamine-conjugated antisense drug to $APOC_3$ mRNA, triglycerides and atherogenic lipoprotein levels. *Eur Heart J.* 2019;40(33):2785–2796. https://doi.org/10.1093/eurheartj/ehz209.
42. Hair P, Cameron F, McKeage K. Mipomersen sodium: first global approval. *Drugs.* 2013;73(5):487–493. https://doi.org/10.1007/s40265-013-0042-2.
43. Santos RD, Duell PB, East C, et al. Long-term efficacy and safety of mipomersen in patients with familial hypercholesterolaemia: 2-year interim results of an open-label extension. *Eur Heart J.* 2015;36(9):566–575. https://doi.org/10.1093/eurheartj/eht549.
44. Fogacci F, Ferri N, Toth PP, Ruscica M, Corsini A, Cicero AFG. Efficacy and safety of mipomersen: a systematic review and meta-analysis of randomized clinical trials. *Drugs.* 2019;79(7):751–766. https://doi.org/10.1007/s40265-019-01114-z.
45. Reeskamp LF, Kastelein JJP, Moriarty PM, et al. Safety and efficacy of mipomersen in patients with heterozygous familial hypercholesterolemia. *Atherosclerosis.* 2019;280:109–117. https://doi.org/10.1016/j.atherosclerosis.2018.11.017.
46. Tsimikas S, Karwatowska-Prokopczuk E, Gouni-Berthold I, et al. Lipoprotein(a) reduction in persons with cardiovascular disease. *N Engl J Med.* 2020;382(3):244–255. https://doi.org/10.1056/NEJMoa1905239.
47. A Randomized Trial Assessing the Effects of Inclisiran on Clinical Outcomes Among People with Cardiovascular Disease (ORION-4); https://clinicaltrials.gov/ct2/show/NCT03705234. Accessed 25 October 2020.
48. A Study of Inclisiran in Participants with Homozygous Familial Hypercholesterolemia (HoFH) (ORION-5); https://clinicaltrials.gov/ct2/show/NCT03851705. Accessed 25 October 2020.
49. An Extension Trial of Inclisiran Compared to Evolocumab in Participants With Cardiovascular Disease and High Cholesterol (ORION-3); https://clinicaltrials.gov/ct2/show/NCT03060577. Accessed 25 October 2020.
50. Trial to Assess the Effect of Long Term Dosing of Inclisiran in Subjects With High CV Risk and Elevated LDL-C (ORION-8); https://clinicaltrials.gov/ct2/show/NCT03814187. Accessed 25 October 2020.
51. Phase 2a Study to Assess CIVI 007 in Patients on a Background of Statin Therapy; https://clinicaltrials.gov/ct2/show/NCT04164888. Accessed 25 October 2020.
52. Stitziel NO, Khera AV, Wang X, et al. ANGPTL3 deficiency and protection against coronary artery disease. *J Am Coll Cardiol.* 2017;69(16):2054–2063. https://doi.org/10.1016/j.jacc.2017.02.030.
53. Graham MJ, Lee RG, Brandt TA, et al. Cardiovascular and metabolic effects of ANGPTL3 antisense oligonucleotides. *N Engl J Med.* 2017;377(3):222–232. https://doi.org/10.1056/NEJMoa1701329.
54. Romeo S, Yin W, Kozlitina J, et al. Rare loss-of-function mutations in ANGPTL family members contribute to plasma triglyceride levels in humans. *J Clin Invest.* 2009;119(1):70–79. https://doi.org/10.1172/JCI37118.
55. Musunuru K, Pirruccello JP, Do R, et al. Exome sequencing, ANGPTL3 mutations, and familial combined hypolipidemia. *N Engl J Med.* 2010;363(23):2220–2227. https://doi.org/10.1056/NEJMoa1002926.
56. Raal FJ, Rosenson RS, Reeskamp LF, et al. Evinacumab for homozygous familial hypercholesterolemia. *N Engl J Med.* 2020;383(8):711–720. https://doi.org/10.1056/NEJMoa2004215.
57. Study of ARO-ANG 3 in Healthy Volunteers and In Dyslipidemic Patients; https://clinicaltrials.gov/ct2/show/NCT03747224. Accessed 25 October 2020.
58. Huff MW, Hegele RA. Apolipoprotein C-III: going back to the future for a lipid drug target. *Circ Res.* 2013;112(11):1405–1408. https://doi.org/10.1161/CIRCRESAHA.113.301464.
59. Baldi S, Bonnet F, Laville M, et al. Influence of apolipoproteins on the association between lipids and insulin sensitivity: a cross-sectional analysis of the RISC Study. *Diabetes Care.* 2013;36(12):4125–4131. https://doi.org/10.2337/dc13-0682.
60. TG and HDL Working Group of the Exome Sequencing Project, National Heart, Lung, and Blood Institute, Crosby J, Peloso GM,

Auer PL, et al. Loss-of-function mutations in APOC3, triglycerides, and coronary disease. *N Engl J Med*. 2014;371(1):22–31. https://doi.org/10.1056/NEJMoa1307095.
61. Graham MJ, Lee RG, Bell TA, et al. Antisense oligonucleotide inhibition of apolipoprotein C-III reduces plasma triglycerides in rodents, nonhuman primates, and humans. *Circ Res*. 2013;112(11):1479–1490. https://doi.org/10.1161/CIRCRESAHA.111.300367.
62. The COMPASS Study: A Study of Volanesorsen (Formally ISIS-APOCIIIRx) in Patients with Hypertriglyceridemia; https://www.clinicaltrials.gov/ct2/show/NCT02300233. Accessed 25 October 2020.
63. Schwabe C, Scott R, Sullivan D, et al. RNA interference targeting apolipoprotein C-III with ARO-$APOC_3$ in healthy volunteers mimics lipid and lipoprotein findings seen in subjects with inherited apolipoprotein C-III deficiency. *Eur Heart J*. 2020;41(ehaa946.3330). https://doi.org/10.1093/ehjci/ehaa946.3330.
64. The BROADEN Study: A Study of Volanesorsen (Formerly ISIS-APOCIIIRx) in Patients with Familial Partial Lipodystrophy. https://clinicaltrials.gov/ct2/show/NCT02527343. Accessed 25 October 2020.
65. Study of ISIS 678354 (AKCEA-APOCIII-LRx) in Patients with Hypertriglyceridemia and Established Cardiovascular Disease (CVD). https://clinicaltrials.gov/ct2/show/NCT03385239. Accessed 25 October 2020.
66. Fless GM, Rolih CA, Scanu AM. Heterogeneity of human plasma lipoprotein (a). Isolation and characterization of the lipoprotein subspecies and their apoproteins. *J Biol Chem*. 1984;259(18): 11470–11478.
67. Waldeyer C, Makarova N, Zeller T, et al. Lipoprotein(a) and the risk of cardiovascular disease in the European population: results from the BiomarCaRE consortium. *Eur Heart J*. 2017;38(32): 2490–2498. https://doi.org/10.1093/eurheartj/ehx166.
68. Kamstrup PR, Tybjærg-Hansen A, Nordestgaard BG. Elevated lipoprotein(a) and risk of aortic valve stenosis in the general population. *J Am Coll Cardiol*. 2014;63(5):470–477. https://doi.org/10.1016/j.jacc.2013.09.038.
69. Viney NJ, van Capelleveen JC, Geary RS, et al. Antisense oligonucleotides targeting apolipoprotein(a) in people with raised lipoprotein(a): two randomised, double-blind, placebo-controlled, dose-ranging trials. *Lancet*. 2016;388(10057):2239–2253. https://doi.org/10.1016/S0140-6736(16)31009-1.
70. Tsimikas S, Viney NJ, Hughes SG, et al. Antisense therapy targeting apolipoprotein(a): a randomised, double-blind, placebo-controlled phase 1 study. *Lancet*. 2015;386(10002):1472–1483. https://doi.org/10.1016/S0140-6736(15)61252-1.
71. Assessing the Impact of Lipoprotein (a) Lowering with TQJ230 on Major Cardio- Vascular Events in Patients with CVD (Lp(a)-HORIZON); https://clinicaltrials.gov/ct2/show/NCT04023552. Accessed 25 October 2020.
72. Randomized Study to Evaluate Efficacy, Safety, and Tolerability of AMG 890 in Subjects with Elevated Lipoprotein(a); https://clinicaltrials.gov/ct2/show/NCT0427 0760. Accessed 25 October 2020.
73. Safety, Tolerability, Pharmacokinetics and Pharmacodynamics Study of AMG 890 in Subjects with Elevated Plasma Lipoprotein(a); https://clinicaltrials.gov/ct2/show/NCT03626662. Accessed 25 October 2020.
74. Ruberg FL, Berk JL. Transthyretin (TTR) cardiac amyloidosis. *Circulation*. 2012;126(10):1286–1300. https://doi.org/10.1161/CIRCULATIONAHA.111.078915.
75. Ruberg FL, Grogan M, Hanna M, Kelly JW, Maurer MS. Transthyretin amyloid cardiomyopathy: JACC state-of-the-art review. *J Am Coll Cardiol*. 2019;73(22):2872–2891. https://doi.org/10.1016/j.jacc.2019.04.003.
76. Kapoor M, Rossor AM, Laura M, Reilly MM. Clinical presentation, diagnosis and treatment of TTR amyloidosis. *J Neuromuscul Dis*. 2019;6(2):189–199. https://doi.org/10.3233/JND-180371.
77. Yamamoto H, Yokochi T. Transthyretin cardiac amyloidosis: an update on diagnosis and treatment. *ESC Heart Fail*. 2019;6(6): 1128–1139. https://doi.org/10.1002/ehf2.12518.
78. González-López E, Gallego-Delgado M, Guzzo-Merello G, et al. Wild-type transthyretin amyloidosis as a cause of heart failure with preserved ejection fraction. *Eur Heart J*. 2015;36(38): 2585–2594. https://doi.org/10.1093/eurheartj/ehv338.
79. Coelho T, Adams D, Silva A, et al. Safety and efficacy of RNAi therapy for transthyretin amyloidosis. *N Engl J Med*. 2013;369(9): 819–829. https://doi.org/10.1056/NEJMoa1208760.
80. Suhr OB, Coelho T, Buades J, et al. Efficacy and safety of patisiran for familial amyloidotic polyneuropathy: a phase II multi-dose study. *Orphanet J Rare Dis*. 2015;10:109. https://doi.org/10.1186/s13023-015-0326-6.
81. Adams D, Gonzalez-Duarte A, O'Riordan WD, et al. Patisiran, an RNAi therapeutic, for hereditary transthyretin amyloidosis. *N Engl J Med*. 2018;379(1):11–21. https://doi.org/10.1056/NEJMoa1716153.
82. Solomon SD, Adams D, Kristen A, et al. Effects of patisiran, an RNA interference therapeutic, on cardiac parameters in patients with hereditary transthyretin-mediated amyloidosis. *Circulation*. 2019;139(4):431–443. https://doi.org/10.1161/CIRCULATIONAHA.118.035831.
83. Zimmermann TS, Karsten V, Chan A, et al. Clinical proof of concept for a novel hepatocyte-targeting GalNAc-siRNA conjugate. *Mol Ther*. 2017;25(1):71–78. https://doi.org/10.1016/j.ymthe.2016.10.019.
84. Gillmore JD, Falk RH, Maurer MS, et al. Phase 2, open-label extension (OLE) study of revusiran, an investigational RNAi therapeutic for the treatment of patients with transthyretin cardiac amyloidosis. *Orphanet J Rare Dis*. 2015;10(Suppl 1):O21. https://doi.org/10.1186/1750-1172-10-S1-O21.
85. Judge DP, Kristen AV, Grogan M, et al. Phase 3 multicenter study of revusiran in patients with hereditary transthyretin-mediated (hATTR) amyloidosis with cardiomyopathy (ENDEAVOUR). *Cardiovasc Drugs Ther*. 2020;34(3):357–370. https://doi.org/10.1007/s10557-019-06919-4.
86. HELIOS-B: A Study to Evaluate Vutrisiran in Patients with Transthyretin Amyloidosis With Cardiomyopathy; https://clinicaltrials.gov/ct2/show/NCT04153149. Accessed 31 October 2020.
87. Benson MD, Waddington-Cruz M, Berk JL, et al. Inotersen treatment for patients with hereditary transthyretin amyloidosis. *N Engl J Med*. 2018;379(1):22–31. https://doi.org/10.1056/NEJMoa1716793.
88. Rakhi DN, Benson Merrill D. Potential reversal of transthyretin amyloid cardiomyopathy with TTR specific antisense oligonucleotide therapy. *J Am Coll Cardiol*. 2018;71(11_suppl.). https://doi.org/10.1016/S0735-1097(18)31201-4. A660-A660.
89. Dahlman JE, Barnes C, Khan O, et al. In vivo endothelial siRNA delivery using polymeric nanoparticles with low molecular weight. *Nat Nanotechnol*. 2014;9(8):648–655. https://doi.org/10.1038/nnano.2014.84.
90. Kamerkar S, LeBleu VS, Sugimoto H, et al. Exosomes facilitate therapeutic targeting of oncogenic KRAS in pancreatic cancer. *Nature*. 2017;546(7659):498–503. https://doi.org/10.1038/nature22341.
91. Stellos K, Musunuru K. Challenges and advances of CRISPR-Cas9 genome editing in therapeutics. *Cardiovasc Res*. 2019;115(2): e12–e14. https://doi.org/10.1093/cvr/cvy300.
92. Dorn LE, Tual-Chalot S, Stellos K, Accornero F. RNA epigenetics and cardiovascular diseases. *J Mol Cell Cardiol*. 2019;129:272–280. https://doi.org/10.1016/j.yjmcc.2019.03.010.

93. Stellos K. RNA in the spotlight: the dawn of RNA therapeutics in the treatment of human disease. *Cardiovasc Res*. 2017;113(12): e43–e44. https://doi.org/10.1093/cvr/cvx170.
94. Stellos K, Gatsiou A, Stamatelopoulos K, et al. Adenosine-to-inosine RNA editing controls cathepsin S expression in atherosclerosis by enabling HuR-mediated post-transcriptional regulation. *Nat Med*. 2016;22(10):1140–1150. https://doi.org/10.1038/nm.4172.
95. Higuchi M, Maas S, Single FN, et al. Point mutation in an AMPA receptor gene rescues lethality in mice deficient in the RNA-editing enzyme ADAR2. *Nature*. 2000;406(6791):78–81. https://doi.org/10.1038/35017558.
96. Gatsiou A, Vlachogiannis N, Lunella FF, Sachse M, Stellos K. Adenosine-to-inosine RNA editing in health and disease. *Antioxid Redox Signal*. 2018;29(9):846–863. https://doi.org/10.1089/ars.2017.7295.
97. Song Y, An O, Ren X, et al. RNA editing mediates the functional switch of COPA in a novel mechanism of hepatocarcinogenesis. *J Hepatol*. 2021;74(1):135–147. https://doi.org/10.1016/j.jhep.2020.07.021.
98. Cox DBT, Gootenberg JS, Abudayyeh OO, et al. RNA editing with CRISPR-cas13. *Science*. 2017;358(6366):1019–1027. https://doi.org/10.1126/science.aaq0180.

CHAPTER

35

The brain vasculome: an integrative model for CNS function and disease

Changhong Xing[1,2], Shuzhen Guo[1], Wenlu Li[1], Wenjun Deng[1,3], MingMing Ning[1,3], Josephine Lok[1,4], Ken Arai[1] and Eng H. Lo[1,3]

[1]Neuroprotection Research Laboratory, Department of Radiology and Neurology, Massachusetts General Hospital, Harvard Medical School, Boston, MA, United States [2]Department of Pathology, University of Texas Southwestern Medical Center, Dallas, TX, United States [3]Department of Neurology, Clinical Proteomics Research Center, Massachusetts General Hospital, Harvard Medical School, Boston, MA, United States [4]Pediatric Critical Care Medicine, Massachusetts General Hospital, Harvard Medical School, Boston, MA, United States

Introduction

Endothelial cells play a central role in the physiology of all organ systems.[1] Once regarded as passive conduits for the delivery of oxygen and nutrients, the microvasculature is now known to have diverse activities, with endocrine and paracrine functions that are unique to each organ. The transcriptional profile of the microvascular network has been termed the vasculome,[2] which is unique within each organ, and contributes to homeostasis under normal conditions and dysfunction under pathology. Conversely, an unhealthy vasculome contributes to the pathophysiology of a wide range of diseases, including central nervous system (CNS) injuries and disorders.[3–6] Perturbations in cerebral endothelial function may profoundly affect CNS homeostasis. After brain trauma, the upregulated pathways of the brain vasculome included those involved in regulation of inflammation and extracellular matrix (ECM) processes,[7] while downregulated pathways included those involved in regulation of metabolism, mitochondrial function, and transport systems. These findings suggest that mapping and dissecting the vasculome of the brain in health and disease may provide a novel database for investigating disease mechanisms, assessing therapeutic targets, and exploring new biomarkers for the CNS.

The importance of this concept is underscored by the observations that (1) the vasculome is perturbed after stroke, trauma, and neurodegeneration[3,7–10]; (2) vascular comorbidities significantly contribute to development of almost all CNS disorders[4,11,12]; and (3) reduction of vascular risk factors and/or administration of vascular therapeutics appear to be broadly effective in stroke, trauma, and neurodegeneration.[13–16] In this chapter, we will focus on the vasculome of the brain. We begin by asking how vasculomes are influenced by genetics, aging, comorbidities, and the circadian rhythm. Second, we discuss how cell–cell interaction between endothelium and astrocytes, pericytes, or smooth muscle cells may contribute to vasculome homeostasis. Then we survey different levels of measurement, i.e., how the vasculome may be mapped at the levels of the transcriptome, proteome, secretome and metabolome. Finally, we introduce the concept of a functional brain vasculome, i.e., how these different layers of regulation and analyses may subserve a broad range of CNS physiology.

Vasculome modifying factors

The vasculome is profoundly influenced by aging and vascular comorbidities such as hypertension and diabetes, all of which are well known to be risk factors for a wide variety of neurodegenerative and cerebrovascular disorders. In addition, the high level of complexity of endothelial cell development and the heterogeneity of organotypically differentiated endothelial cells imply the roles of genetics in the vasculome. Finally, the fact that adverse cardiovascular events including sudden

The Vasculome
https://doi.org/10.1016/B978-0-12-822546-2.00028-9

cardiac death and myocardial infarction occur most commonly in the morning soon after awakening[17–19] suggests that endogenous circadian rhythm may affect endothelial functions.

Genetics

Molecular regulation of endothelial cell differentiation and specialization during embryogenesis is a highly coordinated process, resulting in the transformation of the primitive vasculature into a functional circulatory network.[20] During embryogenesis, vessel formation is initiated during vasculogenesis, which is particularly important during vertebrate development as it lays the foundation for development of other organ systems.[20] Vasculogenesis is tightly regulated on the transcriptional level, and the E26 transformation-specific (ETS) transcription factors and Forkhead box (FOX) transcription factors have been recognized as essential regulators of this process.[20] Following de novo differentiation of endothelial cells and formation of primitive vascular plexi, there is rapid expansion and remodeling of circulatory networks.[20] During this process, primordial endothelial cells are specified into arterial, venous, hemogenic, and lymphatic subtypes. A number of signaling pathways and transcriptional regulators have been implicated in this process, including SRY-Box (Sox) genes, and VEGF/VEGFR2, Notch, Bone Morphogenetic Protein (BMP), and Transforming Growth Factor Beta (TGF-β) signaling pathways.[20]

In addition to the specification of endothelial cell (EC) subtypes, sequencing and microarray studies have revealed the high level of heterogeneity of tissue-specific vasculature. The vasculome is highly organ specific, and gene expression patterns in the brain vasculome significantly differed from those in the heart, lung, or kidney,[2,21,22] indicating that ECs from each tissue demonstrate a distinct transcriptional identity. Differential expression analysis identified 1692 genes which were differentially expressed in brain ECs, 1052 genes which were differentially expressed in lung ECs, and 570 genes which were differentially expressed in heart ECs.[21] As expected, blood brain barrier (BBB) genes were easily detected in the brain vasculome.[2] The significantly upregulated genes in the brain included T-box transcription factor (Tbx1) and the glucose transporter 1 (Slc2a1), as well as both ionotrophic and metabotropic glutamate receptors (Gria2, Gria3, Grin2b, and Grm5 for AMPA2, AMPA3, NMDA2B, and mGluR5, respectively).[2,21] Many signaling and regulatory pathways were found to be significantly enriched in the brain vasculome, including transporter activity, cell adhesion, plasma membrane, leukocyte transmigration, the Wnt signaling pathway,[2] GLUT-1 (glucose transporter type 1) and members of the ABC (adenosine triphosphate–binding cassette) transporter family,[22] as well as canonical neuronal functions such as synapse organization, neurotransmitter transport, axon development, and regulation of ion transmembrane transport.[21] The top 10 most significantly upregulated genes in the brain ECs included prostaglandin d synthase (Ptgds), adenosine 5′-triphosphatase (ATPase), Na^+/K^+ transporting, alpha two polypeptide (Atp1a2), basigin (Bsg), apolipoprotein E (ApoE), glutamate–ammonia ligase (Glul), apolipoprotein D (ApoD), pleiotrophin (Ptn), insulin like growth factor 2 (Igf2), osteonectin (Spock2), and glucose transporter 1 (Slc2a1).[21]

Identifying the unique pathways for the development of tissue-specific vasculomes expands the understanding of endothelial heterogeneity, yields a more comprehensive database of the regulatory genes and pathways, and provides insights for tissue engineering, regenerative medicine, and drug development.

Aging

Aging is associated with endothelial cell senescence, dysregulation of vascular tone, an increase in endothelium permeability, an impairment of angiogenesis and mitochondrial biogenesis, and an increase in arterial stiffness and systemic hypertension.[23] Comparing young and aging ECs, Mun et al. reported that 2351 genes were differentially altered under static and shear stress conditions.[24] Among these genes, there was significant correlation between transcriptional expression and senescence in argininosuccinate synthase 1, NOX4, aquaporin 1, p15, and p16.[24] When comparing 5-month old– 15-month old C57BL6 male mice, the heart vasculome appeared to be affected to a greater extent by hypertension and diabetes than by aging, whereas the brain vasculome was more affected by aging than hypertension and diabetes.[25] In the brain, aging seemed to upregulate pathways that were related to compensatory regulation of neuronal processes and to suppress immune-related processes.[25]

Comorbidities

Hypertension and diabetes are the most common comorbidities for vascular diseases. To explore the effects of hypertension and diabetes on the brain vasculome, Guo et al. compared 4-month old normotensive BPN male mice with hypertensive BPH male mice (for hypertension), and 3-month old diabetic db/db mice with their matching C57BLKS controls (for diabetes).[25] In the brain, hypertension activated pathways related to mitochondrial responses and apoptosis, and diabetes activated pathways related to insulin response and downregulated pathways of carbohydrate metabolism.[25] In spite of the overall pattern of distinct brain vasculome responses, immune-related processes seemed to be suppressed in both conditions. Altered immune pathways mainly involved cytokine/chemokine signaling, toll-like receptors, Fc gamma receptors, interferon-related, and complement-related cascades.[25] Suppression of these immune-related pathways appeared to be mediated by specific genes, with very little intersection with the conditions of aging, hypertension, and diabetes.[25] A small number of genes were uniformly affected by aging, hypertension, and diabetes—these included Was (Wiskott–Aldrich syndrome), Tab2 (TFG-beta-activated kinase 1/MAP3K7 binding protein 2), Irf3 (interferon regulatory factor 3), and Clcf1 (cardiotrophin like cytokine factor).[25] Cytokine/chemokine response and toll-like receptor signaling were significantly downregulated in both aging and hypertension.[25] Interferon response was mainly affected by diabetes, whereas hypertension inhibited the expression of complement factors.[25]

In a related study, the vasculomes of the brain and heart were compared between normotensive Wistar–Kyoto rats (WKY) and spontaneously hypertensive rats (SHRs).[26] In the SHR brain, signal transduction and immune responses were downregulated, whereas metabolism pathways showed a mixed response with both upregulation and downregulation.[26] Compared to WKY, 17 common Reactome pathways were changed in both brain and heart vasculomes of SHR, and many of them showed the same direction of change in both organs, including pathways that involve interferon signaling, interleukin signaling, and cytokine signaling. These similarities suggest that hypertension mediates common effects on the endothelium in the heart and the brain.[26]

Interestingly, exercise significantly renormalized subsets of genes and pathways that were affected by hypertension, suggesting that some of the beneficial physiologic effects of exercise may involve the rescue of diseased vasculomes in the brain and the heart.[26] Many hypertensive pathways were affected by exercise, including immune function, signal transduction, and metabolism. After 3 months of treadmill running, over two-thirds of these pathways appeared to have moved in the opposite direction. Three common pathways were perturbed by hypertension in SHRs and then renormalized by running exercise: potassium channels, GPCR ligand binding and class A/1 rhodopsin-like receptors.[26] In the brain, several pathways for the immune system including interferon induction and signaling and for smooth muscle contraction were all downregulated by hypertension, then renormalized by exercise, whereas small molecule transport, metabolite biosynthesis, and several neurotransmitter-related pathways were all upregulated by hypertension, then renormalized by exercise.[26]

Circadian rhythm

The endogenous circadian system contributes to cyclical changes in the vasculome over the course of the day. During the morning hours, the blood level of plasminogen activator inhibitor, a procoagulant protein, is increased. Sympathetic reactivity to exercise is also higher during this time, potentially contributing to a vulnerable period in the morning hours, with an increased risk of adverse cardiovascular events.[27,28] Similar to the morning, the night hours are also associated with changes in markers of cardiovascular disease. Vascular endothelial function (VEF) is a comprehensive marker of cardiovascular disease that is strongly associated with increased cardiovascular risk.[29] VEF is decreased during the night and the early morning hours, but the decrease was not found to be the result of sleep or inactivity during this time.[30] There is some evidence that the endogenous circadian system is responsible for the observed decline in VEF throughout the night.[29] Other cardiovascular risk factor–related proteins, such as malondialdehyde adducts and endothelin-1 (ET-1), also exhibit circadian rhythms with increases during the morning, peaks around noon, and declines thereafter. In conclusion, the endogenous circadian system affects VEF, oxidative stress, and ET-1, and may contribute to increased cardiovascular risk during the vulnerable morning period.[29]

The influence of the endogenous circadian rhythm on the BBB has been studied in the fruit fly, *Drosophila*. The BBB of *Drosophila* is structurally and functionally similar to the mammalian BBB. Zhang et al.[31] reported that xenobiotic permeability through the BBB of *Drosophila* demonstrates daily rhythmic changes and is associated with the oscillating expression of gap junction proteins in BBB perineural glia cells. These proteins regulate the intracellular concentration of magnesium ions in the ABC-like transporter-containing subperineural glia cells, leading to the rhythm change of xenobiotic efflux. Such periodic changes in BBB permeability produce cycles of different levels of drug accumulation in the *Drosophila* brain, associated with a higher level when these therapeutic drugs delivered at night.[31]

Recently, this group reported similar oscillations of ABCB1-mediated Rhodamine B efflux from the brain over the course of the day in wild-type mice, but not in endothelial-specific BmalI knock-out mice.[32] Further profiling of the transcriptome of fluorescence-activated cell sorting (FACS)–purified mouse brain endothelial cells at six different times of the day demonstrated that 45 genes were rhythmically expressed in control cells, while only six genes were rhythmically expressed in Bmal1 knock-out endothelial cells. The expression of the magnesium transporter Trpm7 in this transcriptome dataset was also analyzed for the presence of rhythmic expression patterns. Results showed a trend toward rhythmic expression, which did not reach statistical significance. However, when isolated mouse brain endothelial cells were studied using real-time PCR, the rhythmic expression of Trpm7 was confirmed. Using a human brain endothelial cell line hCMEC/D3, the authors demonstrated that BmalI could directly regulate the rhythm expression of Trpm7, which regulates the levels of intracellular magnesium and ABCB1-mediated efflux.[32] These examples highlight the important role that circadian cycling of ABCB1 plays in drug delivery.

Cell–Cell interactions

In 2001, the concept of the neurovascular unit was proposed at a workshop organized by the National Institute of Neurological Disorders and Stroke to encourage a broader approach in stroke research. The notion that stroke is no longer a purely "neuron death" disease led to a paradigm shift in which cell–cell interactions between neuronal, glial, and vascular components are considered to be an integral part of stroke pathophysiology.[33–40] In the 2 decades since the concept of the neurovascular unit was first introduced, the idea that cell–cell signaling underlies both brain function and dysfunction has been applied to many other neurological diseases besides stroke. It has supplied a framework that has been useful in examining not only pathological mechanisms of several CNS disorders but also the regulatory mechanisms of brain function. One of the most investigated aspects of the neurovascular unit is the cell-to-cell interactions between the cerebral endothelium and other neurovascular unit components, such as astrocytes, pericytes, and vascular smooth muscle cells. To date, a detailed map of how neurovascular unit components affect the brain vasculome has not yet been constructed. However, it is known that the cells that surround the cerebral blood vessels—the astrocytes, pericytes, and vascular smooth muscle cells—participate in the regulation of the cerebrovascular system, and it is likely that these cells also regulate the brain vasculome. To advance the understanding of the brain vasculome, an in-depth examination of cerebral endothelium-related cell–cell interactions will be required. This section provides a brief overview of the communications between endothelial cells and astrocytes, pericytes, and vascular smooth muscle cells.[41]

Astrocytes

Astrocytes constitute nearly half of all brain cells and outnumber neurons in the human brain. A single astrocyte extends thousands of fine membranous processes that ensheath microvessels. Recent research has revealed that the astrocyte performs a number of important cerebrovascular functions, beyond the traditionally recognized role of regulating extracellular ionic balance. Astrocytes control vascular tone and cerebral blood flow through their close associations with blood vessels and synapses,[42,43] which are enabled by their numerous fine processes. In response to neuronal activity, astrocytes signal to blood vessels directly through gap junctions, or indirectly through the release of soluble factors. This interaction results in an increase in regional blood flow and enhanced delivery of oxygen and glucose to active brain areas. In addition, astrocytes also regulate BBB integrity. In the event of a BBB injury, scar-forming reactive astrocytes play a critical role in sealing the BBB.[44] Conversely, in a mouse model of multiple sclerosis, reactive astrocytes were shown to release vascular endothelial growth factor-A, which causes BBB breakdown accompanied by lymphocyte infiltration, tissue damage, and clinical deficit.[45] Although the interactions between astrocytes and cerebral endothelial cells have been extensively studied for some time, it is only a recent recognition that astrocytes constitute a far more heterogenous group of cells than previously thought. Astrocytes are broadly divided into two main classes based on their morphology, antigenic phonotype, and location. In gray matter, protoplasmic astrocytes express low levels of glial fibrillary acidic protein (GFAP), and their processes ensheath synapses and blood vessels.[46–48] In white matter, fibrous astrocytes express high levels of GFAP and contact nodes of Ranvier and blood vessels.[46–48] Additionally, a translational profiling approach has now revealed a surprising amount of regional astrocyte heterogeneity. Data from translated mRNA confirm a significant amount of heterogeneity among cortical astrocytes, cerebellar astrocytes, and cerebellar Bergman glia in different brain regions.[49] Brain injury also alters the gene profiles of reactive astrocytes at different time points—with differences occurring in different brain regions and even within the same brain region.[47] However, it is not clear whether each astrocyte can transform its phenotype, or whether phenotypically different astrocyte populations exist in specific brain regions. How these astrocyte heterogeneities affect cerebrovascular dysfunction under diseased conditions needs to be clarified in future studies.

Pericytes

Pericytes are cells that surround the endothelial cell layers of the capillary network in the brain, and function in the regulation of cerebral blood flow.[50,51] Pericyte density and the extent of pericyte coverage of endothelial cells vary among different organs and vascular beds. The pericyte coverage in the CNS is higher than in other organs, with 1:1–3:1 ratio between endothelial cells and pericytes, resulting in an approximately 30% coverage of the endothelial abluminal surface.[52–55] Pericytes project elongated processes that ensheath the capillary wall. In areas without a basement membrane, interdigitations of pericyte and endothelial cell membrane make direct contacts containing cell–cell junction proteins such as N-cadherin and connexin, forming adherence and gap junctions.[56] In addition to their joint maintenance of the physical integrity of the BBB, pericytes and endothelial cells interact through the actions of several transduction cascades, including platelet derived growth factor subunit B (PDGF-B), transforming growth factor-β, Notch, sphingosine-1 phosphate, and angiopoietin signaling.[57] Within the neurovascular unit, pericytes are known to be important coordinators of neurovascular functions, including regulation of BBB integrity.[50,58,59] BBB forms early in embryogenesis, coinciding in time with initial pericyte recruitment and preceding astrocyte generation.[59] Beyond the perinatal period, pericytes continue to maintain BBB integrity in the adult and aging brain.[50,58] Pericyte-deficient mice lack functional PDGFRβ signaling and suffer from BBB breakdown, leading to a reduction in brain microcirculation as well as neuroinflammation and neurodegeneration in the adult and aging brain.[50] BBB breakdown resulting from loss of pericytes is also associated with the entry of several vasculotoxic and/or neurotoxic blood-derived macromolecules, such as fibrin, thrombin, plasmin, and hemoglobin-derived hemosiderin into the brain, resulting in an accumulation of iron, reactive oxygen species, and a dysregulation of the delicate balance between coagulation and fibrinolysis.[50,56] These observations underscore the contributions of the pericyte to normal BBB function.

Vascular smooth muscle cells

Vascular smooth muscle cells (VSMCs) constitute the major cells in the wall of blood vessels. The main role of vascular smooth muscle is to redistribute blood within the body by contracting and dilating in response to stimuli, thus changing the volume of blood vessels and the local blood pressure. Excessive vasoconstriction may lead to high blood pressure, and conversely, excessive vasodilation may lead to low blood pressure. Prolonged dysfunction of VSMCs exacerbates many vascular pathologies, including vasculitis, atherosclerosis, and hypertension.[60–63] In general, arteries have more smooth muscle cells within their walls than veins, which makes the wall thickness of arteries larger. VSMCs are especially important in the arteries as they regulate the flow of oxygenated blood from the heart to the organs and tissues, including the brain. In the brain, VSMCs surround cerebral blood vessels and are considered to be a major determinant of cerebral blood flow. However, it is unclear how VSMCs and pericytes differ in the roles they play in cerebral blood flow regulation. While still controversial, a recent study elegantly reported that the dynamics of cerebral blood flow surrounding the larger arterioles is exclusively controlled by smooth muscle cells.[64] In terms of BBB function, VSMCs are thought to have a definite role, as VSMC dysfunction induced by ablation of astrocytic laminin leads to BBB breakdown and hemorrhagic stroke.[65]

Cell–cell interactions within the neurovascular unit are key to the maintenance of neurovascular homeostasis. Because astrocytes, pericytes, and VSMCs are closely positioned to blood vessels, these 3 cell types are essential in regulating cerebrovascular function, highlighting the importance of understanding the relationship between the brain vasculome and these cell–cell interactions. It is important to note that pathological conditions disturb cell–cell trophic coupling, leading to neurovascular dysfunction, but these interactions often recover to some extent during the chronic phase to promote compensative regenerative responses. To better understand how brain vasculome contributes to the mechanisms of CNS pathology and recovery, future studies are warranted to investigate the roles of cell–cell interactions in the function of cerebrovascular system.[41]

Levels of vasculome mapping

Transcriptome

The transcriptome of endothelial cells has been extensively studied. Initial investigations of cultured endothelial cells were followed by studies of in vivo samples[66]; from microvessel fragments[7,67–69] purified endothelial cells purified from vessels by immunobinding or FACS.[2,25,70] The most recent advance has been the use of RNA sequencing on single endothelial cell[71–73]; harvested from arteries, veins, capillaries, and lymph nodes[74]; and from vessels of different tissues.[2,75–77]

Several comprehensive studies using single-cell RNA sequencing have been performed to compile a systematic analysis of endothelial cells from different tissues. In one of the studies, Kalucka et al. used 11 tissue types harvested from 8-week-old male C57BL6/J mice, which were enriched for viable CD45$^-$ CD31$^+$ endothelial cells using microbeads enrichment and FACS purification. Using single-cell RNA sequencing,[72] an atlas of 32,567 single endothelial cell transcriptomes were obtained, and 78 subclusters were identified. The data confirmed that endothelial cells from different tissues express common endothelial as well as tissue-specific markers. Endothelial cells from arterial, venous, and lymph nodes shared more common markers than capillary endothelial cells from heterogeneous tissues, suggesting that the tissue type, rather than the vessel type, contributes more to endothelial heterogeneity. In another study based on the Tabula Muris consortium,[78] Paik et al. compiled a data set using single-cell RNA sequencing of tissue-specific mouse endothelial cells.[71] The data showed that gender was a considerable source of heterogeneity in the endothelial transcriptome, with Lars2 being a highly enriched gene in endothelial cells from male mice.

The study of endothelial cells transcriptome under different pathological conditions is important, as endothelial cells are active participants in the progression of diseases. In a comparative study of endothelial transcriptomes from the brain and heart, during conditions of aging, hypertension, and diabetes, the transcriptomes showed strong tissue specificity, but there was a limited number of commonly changed genes under the tested conditions.[25] Another group studied lung tumor endothelial cells and murine choroidal endothelial cells under resting and neovascularization sprouting conditions to investigate therapeutic targets and identified two new metabolic angiogenic targets (SQLE and ALDH18A1) for tumor therapy.[79] In a study to explore the molecular mechanisms of BBB regulation, BBB-enriched genes were first identified by comparing endothelial cells from the brain, heart, kidney, lung, and liver. Subsequently, normal brain endothelial cells were compared to brain endothelial cells from disease models of stroke, multiple sclerosis, traumatic brain injury, and seizure, each at acute, subacute, and chronic stages. Similar changes in endothelial gene expressions were found in these different disease states, suggesting the existence of a core BBB dysfunction module that may be amenable to a common therapy.[70] In an effort to understand the circulatory markers of aging, Chen et al. used single-cell RNA sequencing to characterize FACS-purified endothelial cells from the hippocampus of 3-month old or 19-month old mice.[80] 642 genes were changed with aging, with capillary endothelial cells having the most aging-related changes compared to arterial or venous endothelial cells. Furthermore, short-term infusions of plasma from aged mice into young mice invoked key aspects of the aged transcriptome, while plasma from young mice reversed some of the signs of aging in brain endothelial cells. These data suggest that endothelial cells act as an exquisite sensor that responds to age-related circulatory cues at the transcriptional level. By examining hypertension-induced changes in the brain and heart at the transcriptome level in hypertensive SHR rat, Lan et al. reported that these changes could be reversed by the beneficial effect of exercise,[26] suggesting that the endothelial cell could be a potential therapeutic target in age-related brain and cardiac pathologies.

Proteome

In contrast to the large amount of available information regarding the vasculome at the level of mRNA transcription, studies at the proteomic level are more limited, with a focus on transporter proteins and membrane receptors. The source of samples and the type of proteomic techniques used may affect the conclusion of such reports. In this section, a quick survey of various studies using different techniques will be reported. Uchida et al. isolated microvessels from the brain cortexes of seven male donors; for liquid chromatography–tandem mass spectrometric quantification (LC-MS/MS), 114 membrane proteins were identified.[81] Chun et al. isolated fresh mouse brain microvessel fragments and characterized them using Multidimensional Protein Identification Technology, focusing on membrane proteins.[82] A total of 1143 proteins were identified, with a large number of transporters, tight junction proteins, and ECM proteins. An extensive comparison was made with previous protein and mRNA studies of the BBB, showing both consistencies and inconsistencies when compared with previously reported compositions and expression levels of BBB proteins and genes. Al Feteisi et al. used isolated rat brain microvessels for proteomic study, testing techniques to optimize microvessel isolation, including protein sample preparation and quantification methods.[83] Using a label-free proteomic method, nearly 2000 proteins were identified, and 1276 proteins were quantified in the isolated rat brain microvessels. Herland et al. used primary cell cultures of human brain microvascular endothelial cells, astrocytes, pericytes, and neuronal cell cultures for deep coverage proteomic analysis as well as for secreted metabolome characterization. Both targeted and untargeted mass spectrometry were undertaken to analyze small molecules (<550 Da) in these cells culture supernatants.[84] Correlated proteomics and metabolic pathways were identified, showing the connection between protein and metabolic expression in each type of cell, as well as the connections between different types of cells in the neurovascular unit (NVU). The effects on these cells of methamphetamine, an NVU-impairing brain stimulant, were further analyzed on proteomic and metabolic levels.[85] These studies highlight the feasibility as well as technical difficulties in studying the proteome of brain microvessels.

Secretome

A secretome is generally thought to be comprised of chemokines, cytokines, and growth factors; additionally, it may include ECM components, microvesicles, exosomes, as well as genetic material. The secretome may differ in the amount and type of substances secreted, depending on the tissue and the stimulus applied to the cells. Secretomes of cells are involved in physiological homeostasis, disease progression, and may have potential therapeutic implications for neurodegenerative disease, tissue regeneration, and brain repair.[86,86] As expected, the secretomes of different cells could act in synergistic as well as antagonistic directions in a given process. In a recent study, the secretomes of primary brain endothelial cells and pericytes were examined. The cells were isolated from human biopsy tissue, grown in culture, and incubated with a number of different inflammatory stimuli. These two types of cells demonstrated similar patterns of transcription factor activation but responded differently at the secretion level.[87] Most often they had overlapping but distinct secretory profiles in response to a given stimulus—for example, both types of cells responded to interleukin 1 beta (IL-1b); endothelial cells specifically responded to granulocyte-colony stimulating factor (G-CSF) and granulocyte-macrophage colony-stimulating factor (GM-CSF); and pericytes specifically responded to IGF-binding protein 2 (IGFBP2) and IGF-binding protein 3 (IGFBP3)3.[87] As expected, the secretome is altered after cellular injury. After oxygen and glucose deprivation (OGD), the secretome of immortalized human brain endothelial cell line hCMEC/D3 was analyzed by a quantitative proteomic approach with SILAC (stable isotope labeling with amino acids). OGD led to a change in the secretome in which 19 proteins were found differentially secreted between normal and OGD conditions.[88] The gender of the host has an effect on the secretome as well. Within the secretomes of serum-deprived HUVEC isolated from male or female donors, 20 proteins were found to be significantly more abundant in HUVECs from males and 3 proteins were more abundant in HUVECs from females.[89]

Secretomes may also contain extracellular vesicles (EVs), formed during the process of plasma membrane shedding. EVs contain particles enclosed by a lipid bilayer. The EVs from endothelial cells contribute to the physiological functions of endothelium and support the plasticity and adaptation of endothelial cells in a paracrine manner. They also contribute to the pathogenesis of many diseases, serve as a prognostic or diagnostic biomarkers, and even as a tool for drug delivery.[90] Virgintino et al. first observed EVs in the developing vascular system of human brain.[91] The microvesicles from tip endothelial cells of sprouting microvessels are tightly associated with endothelial filopodia and follow them during vessel sprouting. The authors suggested that these microvesicles may contain signaling molecules that provide guidance to tip cells and regulate angiogenesis. In another study, exosomes from uninjured cells were found to have beneficial effects on other endothelial cells during injury. Curcumin-primed exosomes from mouse brain endothelial cells alleviated hyperhomocysteinemia-induced endothelial dysfunction, reduced oxidative stress, and strengthened endothelial barrier function.[92] It is likely that the role of EVs in the brain secretome will be further expanded with additional research in the near future.

Metabolome

Metabolic dysregulation is prominent in many vascular-related pathological processes associated with BBB disturbance; however, the study of the brain endothelial cell metabolome is still very limited. Koronowski et al.[93] analyzed the metabolome of mouse brain cortex by gas chromatography time of flight-mass spectrometry (GC-TOF-MS), and identified purine metabolism as a pathway regulated by Sirtuin 5, protein kinase C epsilon (PKCε), and ischemic preconditioning in the brain. Elsheikha et al.[94] studied the metabolic changes induced in human brain microvascular endothelial by a parasitic infection. The use of targeted biochemical kits and unbiased Raman microspectroscopy provided measurements of metabolites in the culture media, while proton nuclear magnetic resonance (^{1}H NMR) techniques provided metabolite profiles from cell extracts.[95]

Huang et al.[96] utilized untargeted liquid chromatography—mass spectrometry (LC-MS) to compare the effects of different injury conditions on the metabolite profiles of primary rat astrocyte and endothelial cells. 372 metabolites were annotated: yielding data on the biochemical fingerprints of individual cells. Notably, each cell had a distinct metabolic signature; while as a group, endothelial cells and astrocyte had quite different basal metabolism compositions. With hypoxic/ischemic injury, their responses were divergent as well—astrocytes dynamically modulated their metabolism, whereas brain endothelial cells adhered to their normoxic characteristics, which differed from those of nonbrain endothelial cells. Some metabolites and pathways, such as ones related to glutathione (GSH), were reduced in brain endothelial cells but elevated or stabilized in astrocytes during injury. Astrocytes were found to secret and shuttle GSH to endothelial cells under normal condition as well as during the injury, in the process helping to maintain the redox homeostasis of endothelial cells and its barrier stability, providing a new therapeutic mechanism/target.[97] These investigations suggest that the metabolome of brain endothelial cells could serve as a powerful tool in the understanding of BBB changes during injury.

The transcriptome, proteome, secretome, and metabolome represent the multifaceted molecular dimensions of the vasculome. To explore the spatial dimension, Kirst et al.[98] used immunolabeling and tissue clearing to image the vascular network of adult mouse brains. Using techniques in light sheet microscopy, vascular graphs were built, composed of over 100 million vessel segments, then visualized and analyzed. Using datasets generated from over 20 mouse brains, the vascular network was labeled as arteries, veins, and capillaries and characterized according to their anatomical locations, regional variations, and functional correlates. The authors also analyzed brain-wide rearrangements of the vasculature in animal models of congenital deafness and ischemic stroke, revealing that vascular remodeling follow different paths in these models.

The studies described here illustrate the many on-going explorations of the brain vasculome. These synergistic efforts will no doubt contribute to further advances in the field.

The functional vasculome

The vasculome may provide a potentially useful conceptual framework for investigating how vascular mechanisms contribute to the pathomechanisms of CNS injury and disease. As discussed in the previous sections, the brain vasculome is centered at the endothelial interface, but regulation of this system is critically sustained by cell–cell interactions with adjacent cell types including astrocytes, pericytes, and smooth muscle cells. Mapping of the brain vasculome can then be performed at the transcriptional, proteomic, and metabolomic levels in order to define how neurovascular homeostasis can be modified by a wide range of factors including genetic background, aging, comorbidities, and circadian rhythms. Importantly, however, these efforts cannot just be limited to anatomic or database analyses. Instead, these different ways of mapping the vasculome should be interpreted as functional layers of integration that subserve physiology (please see Fig. 35.1).

The most well-investigated function of the vasculome is based on the regulation of blood flow and metabolism. The brain is an extremely high energy-consuming organ so blood flow must be closely regulated and coupled to metabolic needs during various functional activation states.[99] Dysfunction in flow metabolism coupling is not only an upstream trigger in stroke and acute trauma but now also recognized to mediate chronic neurodegeneration.[100,101] How this aspect of the functional brain vasculome is affected by aging and systemic disease should be a rich area for future studies.

A second major function of the brain vasculome comprises the maintenance of BBB permeability. Endothelial cells in the brain are connected with specialized tight junctions with a low rate of transcytosis.[102,103] Previous efforts to map the "BBB transcriptome or proteome" revealed that alterations in this system may include both common and unique "signatures."[104] Because the BBB is intimately connected to perivascular space, cell–cell interactions at this interface may also play vital roles in regulating not only passage of factors between circulating blood and brain but also the movement and clearance of solutes and fluids throughout the CNS. This connected system has been termed the glymphatics,[105] which provides another important aspect of the functional vasculome. Together, BBB and glymphatic function are significantly modulated by age, comorbidities, and circadian rhythm. For example, BBB permeability may change in aged or hypertensive brains,[106,107] whereas alterations in the circadian cycles of glymphatic flow may underlie susceptibility to neurodegeneration and dementia.[108–110] Hence, mapping these aspects of the functional vasculome may allow us to define pathways and targets for potential therapies.

Central to the concept of a functional brain vasculome is the idea that multiple cell types interact and provide cross-talk signaling between different compartments in the neurovascular unit.[38,111] Brain endothelial cells not only provide structure for blood vessels, but also act as trophic sources that sustain neurogenesis, oligodendrogenesis, and angiogenesis.[40,112–115] Endothelium has been long known to provide a vascular niche for neurogenesis,[116] and in nonneurogenic areas, the vasculome may provide trophic support for neuroprotection.[117] Similarly, in white matter, the vasculome may also be a key source of molecular signals for oligodendrocyte precursor cells.[118,119] Hence, further mapping of different layers of the vasculome may reveal novel targets for boosting plasticity in all CNS progenitor populations.

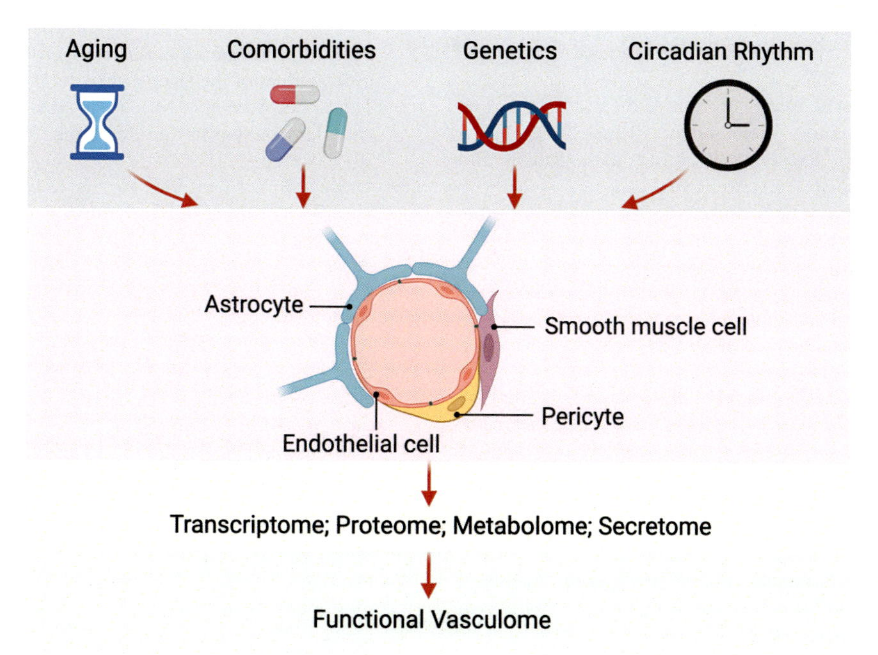

FIGURE 35.1 The brain vasculome provides an integrative model for defining the molecular architecture that connects the neurovascular system to CNS function and dysfunction. The vasculome is centered at the endothelial interface between brain and circulating blood. At the cellular level, vasculome homeostasis is modulated by cell–cell interactions between the endothelial compartment and adjacent cells including astrocytes, pericytes, and smooth muscle cells. The brain vasculome is profoundly influenced by systemic factors comprising genetics, aging, comorbidities, and circadian rhythms. Currently, efforts to map the brain vasculome are ongoing at the levels of the transcriptome, proteome, metabolome, and secretome. Ultimately, the goal is to connect the molecular and cellular vasculome to a functional vasculome that subserves the regulation of cerebral blood flow, blood–brain barrier, glymphatics, and cross-talk with the immune system.

Finally, it is now accepted that the neurovascular unit is not isolated but closely communicates with systemic biology.[120] In this regard, the vasculome serves as the interface between the brain and peripheral systems. The importance of central–peripheral cross-talk can already be seen in the effects of overall aging and vascular comorbidities on the transcriptomic expression of the brain vasculome.[25] In turn, injury and disease in the brain can also signal toward systemic biology. Induction of focal cerebral ischemia rapidly triggers responses in the lymphatic system,[121] spleen,[122] and systemic immunity.[123] Because the immune system is a key component in the pathophysiology of many CNS disorders,[124–126] mapping this connection between the brain vasculome and the immunome should be fertile ground for new science, especially in terms of understanding the complex influences of age, comorbidities, and biological rhythms.[25,127]

Conclusions and future directions

The brain vasculome is intricately connected with every aspect of brain physiology. Ongoing research efforts will enable a complete rendition of this integrative concept, connecting its molecular characteristics with anatomical features, cellular signaling, spatial topography, and temporal changes, and thus produce a multilayered database and model for investigating neurological function and pathologies.

Acknowledgments

Based in part on material and concepts proposed in Guo et al., PLoS One 2012, Guo et al., J Neurotrauma 2016, Guo et al., Neurobiol Dis 2019, Lan et al., Cond Med 2019. The authors apologize to colleagues whose vasculome-related work were not cited due to space limits.

Conflict of interest

The authors declare no potential conflicts of interest with respect to the research, authorship, and/or publication of this article.

References

1. Godo S, Shimokawa H. Endothelial functions. *Arterioscler Thromb Vasc Biol*. 2017;37:e108–e114.
2. Guo S, Zhou Y, Xing C, et al. The vasculome of the mouse brain. *PLoS One*. 2012;7:e52665.
3. Bosetti F, Galis ZS, Bynoe MS, et al. "Small blood vessels: big health problems?": scientific recommendations of the national institutes of health workshop. *J Am Heart Assoc*. 2016;5.
4. Gorelick PB, Scuteri A, Black SE, et al. Vascular contributions to cognitive impairment and dementia: a statement for healthcare professionals from the American Heart Association/American stroke association. *Stroke*. 2011;42:2672–2713.
5. Hu J, Dziumbla S, Lin J, et al. Inhibition of soluble epoxide hydrolase prevents diabetic retinopathy. *Nature*. 2017;552:248–252.

6. Koizumi K, Wang G, Park L. Endothelial dysfunction and amyloid-beta-induced neurovascular alterations. *Cell Mol Neurobiol*. 2016;36:155–165.
7. Guo S, Lok J, Zhao S, et al. Effects of controlled cortical impact on the mouse brain vasculome. *J Neurotrauma*. 2016;33:1303–1316.
8. Zhang ZG, Chopp M. Exosomes in stroke pathogenesis and therapy. *J Clin Invest*. 2016;126:1190–1197.
9. Acosta SA, Lee JY, Nguyen H, Kaneko Y, Borlongan CV. Endothelial progenitor cells modulate inflammation-associated stroke vasculome. *Stem Cell Rev Rep*. 2019;15:256–275.
10. Lau SF, Cao H, Fu AKY, Ip NY. Single-nucleus transcriptome analysis reveals dysregulation of angiogenic endothelial cells and neuroprotective glia in alzheimer's disease. *Proc Natl Acad Sci USA*. 2020;117:25800–25809.
11. Brown WR, Thore CR. Review: cerebral microvascular pathology in ageing and neurodegeneration. *Neuropathol Appl Neurobiol*. 2011;37:56–74.
12. Guo S, Tjarnlund-Wolf A, Deng W, et al. Comparative transcriptome of neurons after oxygen-glucose deprivation: potential differences in neuroprotection versus reperfusion. *J Cerebr Blood Flow Metabol*. 2018;38:2236–2250.
13. Hasnain M, Vieweg WV. Possible role of vascular risk factors in alzheimer's disease and vascular dementia. *Curr Pharmaceut Des*. 2014;20:6007–6013.
14. Pistoia F, Sacco S, Degan D, Tiseo C, Ornello R, Carolei A. Hypertension and stroke: epidemiological aspects and clinical evaluation. *High Blood Press Cardiovasc Prev*. 2016;23:9–18.
15. Pistoia F, Sacco S, Ornello R, Degan D, Tiseo C, Carolei A. Composite scores and other outcome measures in stroke trials. *Front Neurol Neurosci*. 2016;39:60–70.
16. Trigiani LJ, Hamel E. An endothelial link between the benefits of physical exercise in dementia. *J Cerebr Blood Flow Metabol*. 2017;37: 2649–2664.
17. Willich SN, Goldberg RJ, Maclure M, Perriello L, Muller JE. Increased onset of sudden cardiac death in the first three hours after awakening. *Am J Cardiol*. 1992;70:65–68.
18. Muller JE. Circadian variation in cardiovascular events. *Am J Hypertens*. 1999;12:35S–42S.
19. Muller JE, Ludmer PL, Willich SN, et al. Circadian variation in the frequency of sudden cardiac death. *Circulation*. 1987;75:131–138.
20. Qiu J, Hirschi KK. Endothelial cell development and its application to regenerative medicine. *Circ Res*. 2019;125:489–501.
21. Jambusaria A, Hong Z, Zhang L, et al. Endothelial heterogeneity across distinct vascular beds during homeostasis and inflammation. *Elife*. 2020;9.
22. Augustin HG, Koh GY. Organotypic vasculature: from descriptive heterogeneity to functional pathophysiology. *Science*. 2017;357.
23. Jia G, Aroor AR, Jia C, Sowers JR. Endothelial cell senescence in aging-related vascular dysfunction. *Biochim Biophys Acta Mol Basis Dis*. 2019;1865:1802–1809.
24. Mun GI, Lee SJ, An SM, Kim IK, Boo YC. Differential gene expression in young and senescent endothelial cells under static and laminar shear stress conditions. *Free Radic Biol Med*. 2009;47: 291–299.
25. Guo S, Deng W, Xing C, Zhou Y, Ning M, Lo EH. Effects of aging, hypertension and diabetes on the mouse brain and heart vasculomes. *Neurobiol Dis*. 2019;126:117–123.
26. Lan JGS, Shi J, Esposito E, et al. Effects of hypertension and exercise in the brain and heart vasculome of rats. *Cond Med*. 2019;2: 63–74.
27. Scheer FA, Hu K, Evoniuk H, et al. Impact of the human circadian system, exercise, and their interaction on cardiovascular function. *Proc Natl Acad Sci USA*. 2010;107:20541–20546.
28. Scheer FA, Shea SA. Human circadian system causes a morning peak in prothrombotic plasminogen activator inhibitor-1 (pai-1) independent of the sleep/wake cycle. *Blood*. 2014;123:590–593.
29. Thosar SS, Berman AM, Herzig MX, et al. Circadian rhythm of vascular function in midlife adults. *Arterioscler Thromb Vasc Biol*. 2019;39:1203–1211.
30. Thosar SS, Berman AM, Herzig MX, Roberts SA, Lasarev MR, Shea SA. Morning impairment in vascular function is unrelated to overnight sleep or the inactivity that accompanies sleep. *Am J Physiol Regul Integr Comp Physiol*. 2018;315:R986–R993.
31. Zhang SL, Yue Z, Arnold DM, Artiushin G, Sehgal A. A circadian clock in the blood-brain barrier regulates xenobiotic efflux. *Cell*. 2018;173:130–139 e110.
32. Zhang SL, Lahens NF, Yue Z, et al. A circadian clock regulates efflux by the blood-brain barrier in mice and human cells. *Nat Commun*. 2021;12:617.
33. Stanimirovic DB, Friedman A. Pathophysiology of the neurovascular unit: disease cause or consequence? *J Cerebr Blood Flow Metabol*. 2012;32:1207–1221.
34. del Zoppo GJ. Inflammation and the neurovascular unit in the setting of focal cerebral ischemia. *Neuroscience*. 2009;158:972–982.
35. Hawkins BT, Davis TP. The blood-brain barrier/neurovascular unit in health and disease. *Pharmacol Rev*. 2005;57:173–185.
36. Iadecola C. Neurovascular regulation in the normal brain and in alzheimer's disease. *Nat Rev Neurosci*. 2004;5:347–360.
37. Lo EH, Broderick JP, Moskowitz MA. Tpa and proteolysis in the neurovascular unit. *Stroke*. 2004;35:354–356.
38. Zacchigna S, Lambrechts D, Carmeliet P. Neurovascular signalling defects in neurodegeneration. *Nat Rev Neurosci*. 2008;9: 169–181.
39. Zlokovic BV. The blood-brain barrier in health and chronic neurodegenerative disorders. *Neuron*. 2008;57:178–201.
40. Arai K, Lo EH. Wiring and plumbing: oligodendrocyte precursors and angiogenesis in the oligovascular niche. *J Cerebr Blood Flow Metabol*. 2021. https://doi.org/10.1177/0271678X211014979.
41. Maki T, Hayakawa K, Pham LD, Xing C, Lo EH, Arai K. Biphasic mechanisms of neurovascular unit injury and protection in cns diseases. *CNS Neurol Disord Drug Targets*. 2013;12:302–315.
42. Iadecola C, Nedergaard M. Glial regulation of the cerebral microvasculature. *Nat Neurosci*. 2007;10:1369–1376.
43. Eroglu C, Barres BA. Regulation of synaptic connectivity by glia. *Nature*. 2010;468:223–231.
44. Bush TG, Puvanachandra N, Horner CH, et al. Leukocyte infiltration, neuronal degeneration, and neurite outgrowth after ablation of scar-forming, reactive astrocytes in adult transgenic mice. *Neuron*. 1999;23:297–308.
45. Argaw AT, Asp L, Zhang J, et al. Astrocyte-derived vegf-a drives blood-brain barrier disruption in cns inflammatory disease. *J Clin Invest*. 2012;122:2454–2468.
46. Barres BA. The mystery and magic of glia: a perspective on their roles in health and disease. *Neuron*. 2008;60:430–440.
47. Zhang Y, Barres BA. Astrocyte heterogeneity: an underappreciated topic in neurobiology. *Curr Opin Neurobiol*. 2010;20:588–594.
48. Cahoy JD, Emery B, Kaushal A, et al. A transcriptome database for astrocytes, neurons, and oligodendrocytes: a new resource for understanding brain development and function. *J Neurosci*. 2008;28:264–278.
49. Doyle JP, Dougherty JD, Heiman M, et al. Application of a translational profiling approach for the comparative analysis of cns cell types. *Cell*. 2008;135:749–762.
50. Bell RD, Winkler EA, Sagare AP, et al. Pericytes control key neurovascular functions and neuronal phenotype in the adult brain and during brain aging. *Neuron*. 2010;68:409–427.

51. Peppiatt CM, Howarth C, Mobbs P, Attwell D. Bidirectional control of cns capillary diameter by pericytes. *Nature*. 2006;443: 700–704.
52. Mathiisen TM, Lehre KP, Danbolt NC, Ottersen OP. The perivascular astroglial sheath provides a complete covering of the brain microvessels: an electron microscopic 3d reconstruction. *Glia*. 2010;58:1094–1103.
53. Sims DE. The pericyte–a review. *Tissue Cell*. 1986;18:153–174.
54. Armulik A, Genove G, Betsholtz C. Pericytes: developmental, physiological, and pathological perspectives, problems, and promises. *Dev Cell*. 2011;21:193–215.
55. Sa-Pereira I, Brites D, Brito MA. Neurovascular unit: a focus on pericytes. *Mol Neurobiol*. 2012;45:327–347.
56. Winkler EA, Bell RD, Zlokovic BV. Central nervous system pericytes in health and disease. *Nat Neurosci*. 2011;14:1398–1405.
57. Gaengel K, Genove G, Armulik A, Betsholtz C. Endothelial-mural cell signaling in vascular development and angiogenesis. *Arterioscler Thromb Vasc Biol*. 2009;29:630–638.
58. Armulik A, Genove G, Mae M, et al. Pericytes regulate the blood-brain barrier. *Nature*. 2010;468:557–561.
59. Daneman R, Zhou L, Kebede AA, Barres BA. Pericytes are required for blood-brain barrier integrity during embryogenesis. *Nature*. 2010;468:562–566.
60. Harman JL, Jorgensen HF. The role of smooth muscle cells in plaque stability: therapeutic targeting potential. *Br J Pharmacol*. 2019;176:3741–3753.
61. Wang D, Uhrin P, Mocan A, et al. Vascular smooth muscle cell proliferation as a therapeutic target. Part 1: molecular targets and pathways. *Biotechnol Adv*. 2018;36:1586–1607.
62. Bennett MR, Sinha S, Owens GK. Vascular smooth muscle cells in atherosclerosis. *Circ Res*. 2016;118:692–702.
63. Qureshi AI, Mendelow AD, Hanley DF. Intracerebral haemorrhage. *Lancet*. 2009;373:1632–1644.
64. Hill RA, Tong L, Yuan P, Murikinati S, Gupta S, Grutzendler J. Regional blood flow in the normal and ischemic brain is controlled by arteriolar smooth muscle cell contractility and not by capillary pericytes. *Neuron*. 2015;87:95–110.
65. Chen ZL, Yao Y, Norris EH, et al. Ablation of astrocytic laminin impairs vascular smooth muscle cell function and leads to hemorrhagic stroke. *J Cell Biol*. 2013;202:381–395.
66. Urich E, Lazic SE, Molnos J, Wells I, Freskgard PO. Transcriptional profiling of human brain endothelial cells reveals key properties crucial for predictive in vitro blood-brain barrier models. *PLoS One*. 2012;7:e38149.
67. Song HW, Foreman KL, Gastfriend BD, Kuo JS, Palecek SP, Shusta EV. Transcriptomic comparison of human and mouse brain microvessels. *Sci Rep*. 2020;10:12358.
68. Wallgard E, Larsson E, He L, et al. Identification of a core set of 58 gene transcripts with broad and specific expression in the microvasculature. *Arterioscler Thromb Vasc Biol*. 2008;28:1469–1476.
69. Enerson BE, Drewes LR. The rat blood-brain barrier transcriptome. *J Cerebr Blood Flow Metabol*. 2006;26:959–973.
70. Munji RN, Soung AL, Weiner GA, et al. Profiling the mouse brain endothelial transcriptome in health and disease models reveals a core blood-brain barrier dysfunction module. *Nat Neurosci*. 2019; 22:1892–1902.
71. Paik DT, Tian L, Williams IM, et al. Single-cell RNA sequencing unveils unique transcriptomic signatures of organ-specific endothelial cells. *Circulation*. 2020;142:1848–1862.
72. Kalucka J, de Rooij L, Goveia J, et al. Single-cell transcriptome atlas of murine endothelial cells. *Cell*. 2020;180:764–779 e720.
73. Vanlandewijck M, He L, Mae MA, et al. A molecular atlas of cell types and zonation in the brain vasculature. *Nature*. 2018;554: 475–480.
74. Brulois K, Rajaraman A, Szade A, et al. A molecular map of murine lymph node blood vascular endothelium at single cell resolution. *Nat Commun*. 2020;11:3798.
75. Sabbagh MF, Heng JS, Luo C, et al. Transcriptional and epigenomic landscapes of cns and non-cns vascular endothelial cells. *Elife*. 2018;7.
76. Hupe M, Li MX, Kneitz S, et al. Gene expression profiles of brain endothelial cells during embryonic development at bulk and single-cell levels. *Sci Signal*. 2017;10.
77. Daneman R, Zhou L, Agalliu D, Cahoy JD, Kaushal A, Barres BA. The mouse blood-brain barrier transcriptome: a new resource for understanding the development and function of brain endothelial cells. *PLoS One*. 2010;5:e13741.
78. Single-cell transcriptomics of 20 mouse organs creates a tabula muris. *Nature*. 2018;562:367–372.
79. Rohlenova K, Goveia J, Garcia-Caballero M, et al. Single-cell RNA sequencing maps endothelial metabolic plasticity in pathological angiogenesis. *Cell Metabol*. 2020;31:862–877 e814.
80. Chen MB, Yang AC, Yousef H, et al. Brain endothelial cells are exquisite sensors of age-related circulatory cues. *Cell Rep*. 2020; 30:4418–4432 e4414.
81. Uchida Y, Ohtsuki S, Katsukura Y, et al. Quantitative targeted absolute proteomics of human blood-brain barrier transporters and receptors. *J Neurochem*. 2011;117:333–345.
82. Chun HB, Scott M, Niessen S, et al. The proteome of mouse brain microvessel membranes and basal lamina. *J Cerebr Blood Flow Metabol*. 2011;31:2267–2281.
83. Al Feteisi H, Al-Majdoub ZM, Achour B, Couto N, Rostami-Hodjegan A, Barber J. Identification and quantification of blood-brain barrier transporters in isolated rat brain microvessels. *J Neurochem*. 2018;146:670–685.
84. Gaceb A, Paul G. Pericyte secretome. *Adv Exp Med Biol*. 2018;1109: 139–163.
85. Herland A, Maoz BM, FitzGerald EA, et al. Proteomic and metabolomic characterization of human neurovascular unit cells in response to methamphetamine. *Adv Biosyst*. 2020;4:e1900230.
86. Gaceb A, Barbariga M, Ozen I, Paul G. The pericyte secretome: potential impact on regeneration. *Biochimie*. 2018;155:16–25.
87. Smyth LCD, Rustenhoven J, Park TI, et al. Unique and shared inflammatory profiles of human brain endothelia and pericytes. *J Neuroinflammation*. 2018;15:138.
88. Llombart V, Garcia-Berrocoso T, Bech-Serra JJ, et al. Characterization of secretomes from a human blood brain barrier endothelial cells in-vitro model after ischemia by stable isotope labeling with aminoacids in cell culture (silac). *J Proteomics*. 2016;133:100–112.
89. Cattaneo MG, Banfi C, Brioschi M, Lattuada D, Vicentini LM. Sex-dependent differences in the secretome of human endothelial cells. *Biol Sex Differ*. 2021;12:7.
90. Mathiesen A, Hamilton T, Carter N, Brown M, McPheat W, Dobrian A. Endothelial extracellular vesicles: from keepers of health to messengers of disease. *Int J Mol Sci*. 2021;22.
91. Virgintino D, Rizzi M, Errede M, et al. Plasma membrane-derived microvesicles released from tip endothelial cells during vascular sprouting. *Angiogenesis*. 2012;15:761–769.
92. Kalani A, Kamat PK, Chaturvedi P, Tyagi SC, Tyagi N. Curcumin-primed exosomes mitigate endothelial cell dysfunction during hyperhomocysteinemia. *Life Sci*. 2014;107:1–7.
93. Koronowski KB, Khoury N, Morris-Blanco KC, Stradecki-Cohan HM, Garrett TJ, Perez-Pinzon MA. Metabolomics based identification of sirt5 and protein kinase c epsilon regulated pathways in brain. *Front Neurosci*. 2018;12:32.
94. Elsheikha HM, Alkurashi M, Kong K, Zhu XQ. Metabolic footprinting of extracellular metabolites of brain endothelium infected with neospora caninum in vitro. *BMC Res Notes*. 2014;7:406.

95. Al-Sandaqchi AT, Marsh V, Williams HEL, Stevenson CW, Elsheikha HM. Structural, functional, and metabolic alterations in human cerebrovascular endothelial cells during toxoplasma gondii infection and amelioration by verapamil in vitro. *Microorganisms*. 2020;8.
96. Huang SF, Fischer S, Koshkin A, Laczko E, Fischer D, Ogunshola OO. Cell-specific metabolomic responses to injury: novel insights into blood-brain barrier modulation. *Sci Rep*. 2020;10:7760.
97. Huang SF, Othman A, Koshkin A, et al. Astrocyte glutathione maintains endothelial barrier stability. *Redox Biol*. 2020;34:101576.
98. Kirst C, Skriabine S, Vieites-Prado A, et al. Mapping the fine-scale organization and plasticity of the brain vasculature. *Cell*. 2020;180: 780–795 e725.
99. Raichle ME, Gusnard DA. Appraising the brain's energy budget. *Proc Natl Acad Sci USA*. 2002;99:10237–10239.
100. Biju KC, Shen Q, Hernandez ET, Mader MJ, Clark RA. Reduced cerebral blood flow in an alpha-synuclein transgenic mouse model of Parkinson's disease. *J Cerebr Blood Flow Metabol*. 2020; 40:2441–2453.
101. Klinkmueller P, Kronenbuerger M, Miao X, et al. Impaired response of cerebral oxygen metabolism to visual stimulation in huntington's disease. *J Cerebr Blood Flow Metabol*. 2021;41: 1119–1130.
102. Ayloo S, Gu C. Transcytosis at the blood-brain barrier. *Curr Opin Neurobiol*. 2019;57:32–38.
103. Tjakra M, Wang Y, Vania V, et al. Overview of crosstalk between multiple factor of transcytosis in blood brain barrier. *Front Neurosci*. 2019;13:1436.
104. Lochhead JJ, Yang J, Ronaldson PT, Davis TP. Structure, function, and regulation of the blood-brain barrier tight junction in central nervous system disorders. *Front Physiol*. 2020;11:914.
105. Nedergaard M. Neuroscience. Garbage truck of the brain. *Science*. 2013;340:1529–1530.
106. Santisteban MM, Ahn SJ, Lane D, et al. Endothelium-macrophage crosstalk mediates blood-brain barrier dysfunction in hypertension. *Hypertension*. 2020;76:795–807.
107. Verheggen ICM, de Jong JJA, van Boxtel MPJ, et al. Increase in blood-brain barrier leakage in healthy, older adults. *Geroscience*. 2020;42:1183–1193.
108. Hablitz LM, Pla V, Giannetto M, et al. Circadian control of brain glymphatic and lymphatic fluid flow. *Nat Commun*. 2020;11:4411.
109. Mestre H, Mori Y, Nedergaard M. The brain's glymphatic system: current controversies. *Trends Neurosci*. 2020;43:458–466.
110. Rasmussen MK, Mestre H, Nedergaard M. The glymphatic pathway in neurological disorders. *Lancet Neurol*. 2018;17: 1016–1024.
111. Lo EH, Dalkara T, Moskowitz MA. Mechanisms, challenges and opportunities in stroke. *Nat Rev Neurosci*. 2003;4:399–415.
112. Chopp M, Zhang ZG, Jiang Q. Neurogenesis, angiogenesis, and mri indices of functional recovery from stroke. *Stroke*. 2007;38:827–831.
113. Miyamoto N, Pham LD, Seo JH, Kim KW, Lo EH, Arai K. Crosstalk between cerebral endothelium and oligodendrocyte. *Cell Mol Life Sci*. 2014;71:1055–1066.
114. Shen Q, Goderie SK, Jin L, et al. Endothelial cells stimulate self-renewal and expand neurogenesis of neural stem cells. *Science*. 2004;304:1338–1340.
115. Guo S, Lo EH. Dysfunctional cell-cell signaling in the neurovascular unit as a paradigm for central nervous system disease. *Stroke*. 2009;40:S4–S7.
116. Ohab JJ, Fleming S, Blesch A, Carmichael ST. A neurovascular niche for neurogenesis after stroke. *J Neurosci*. 2006;26: 13007–13016.
117. Guo S, Kim WJ, Lok J, et al. Neuroprotection via matrix-trophic coupling between cerebral endothelial cells and neurons. *Proc Natl Acad Sci USA*. 2008;105:7582–7587.
118. Hayakawa K, Seo JH, Pham LD, et al. Cerebral endothelial derived vascular endothelial growth factor promotes the migration but not the proliferation of oligodendrocyte precursor cells in vitro. *Neurosci Lett*. 2012;513:42–46.
119. Arai K, Lo EH. An oligovascular niche: cerebral endothelial cells promote the survival and proliferation of oligodendrocyte precursor cells. *J Neurosci*. 2009;29:4351–4355.
120. Marcet P, Santos N, Borlongan CV. When friend turns foe: central and peripheral neuroinflammation in central nervous system injury. *Neuroimmunol Neuroinflammation*. 2017;4:82–92.
121. Esposito E, Ahn BJ, Shi J, et al. Brain-to-cervical lymph node signaling after stroke. *Nat Commun*. 2019;10:5306.
122. Seifert HA, Hall AA, Chapman CB, Collier LA, Willing AE, Pennypacker KR. A transient decrease in spleen size following stroke corresponds to splenocyte release into systemic circulation. *J Neuroimmune Pharmacol*. 2012;7:1017–1024.
123. Zhang SR, Phan TG, Sobey CG. Targeting the immune system for ischemic stroke. *Trends Pharmacol Sci*. 2021;42:96–105.
124. Iadecola C, Anrather J. The immunology of stroke: from mechanisms to translation. *Nat Med*. 2011;17:796–808.
125. Greenhalgh AD, David S, Bennett FC. Immune cell regulation of glia during cns injury and disease. *Nat Rev Neurosci*. 2020;21: 139–152.
126. Heneka MT, McManus RM, Latz E. Inflammasome signalling in brain function and neurodegenerative disease. *Nat Rev Neurosci*. 2018;19:610–621.
127. Lo EH, Albers GW, Dichgans M, et al. Circadian biology and stroke. *Stroke*. 2021;52:2180–2190.

PART VI

Looking forward toward "Precision Health" for the vasculome

CHAPTER

36

Vasculome: defining and optimizing vascular health

Danny J. Eapen[1], Christian Faaborg-Andersen[2], Robert J. DeStefano[3], Angelos D. Karagiannis[3], Raymundo A. Quintana[1], Devinder Dhindsa[4], Munir Chaudhuri[3], Charles D. Searles, Jr.[1] and Laurence S. Sperling[4,5]

[1]Department of Medicine, Division of Cardiology, Emory University, Atlanta, GA, United States [2]Emory University School of Medicine, Atlanta, GA, United States [3]Department of Internal Medicine, Emory University School of Medicine, Atlanta, GA, United States [4]Emory Clinical Cardiovascular Research Institute, Division of Cardiology, Emory University School of Medicine, Atlanta, GA, United States [5]Katz Professor in Preventive Cardiology, Professor of Global Health, Hubert Department of Global Health, Rollins School of Public Health at Emory University, Atlanta, GA, United States

Introduction

Cardiovascular disease (CVD) remains a leading cause of morbidity and mortality in the United States and across the world. In 2017, approximately 17.8 million deaths were attributed to CVD globally which was a 21% increase compared to what was observed in 2007. The American Heart Association (AHA) estimates that by 2035, 45.1% of the U.S. population will have some form of CVD. The economic cost is significant, with total direct medical costs of CVD projected to increase from $318 billion in 2015 to $749 billion by 2035. Of this estimate, 55.5% will be attributable to hospital costs, 15.3% to medications, 15.0% to physicians, 7.2% to nursing home care, 5.5% to home health care, and 1.5% to other costs.[1]

Males have a greater overall prevalence of CVD at 51.2% compared to females at 44.7%. However, CVD becomes equivalent with age as female prevalence increases to 78.2% in the 60–79-year-old group compared to 77.2% of males.[2] This highlights two important points: first, cumulative exposure to cardiovascular risk factors over the life course is a major contributor to CVD development; and second, the premenopausal state carries protective CVD properties.[3] After menopause, CVD rates rise similar to those of males.

While CVD affects individuals of all races and ethnicities, a disproportionate rate is seen in non-Hispanic (NH) Black individuals who have the highest prevalence of disease with 60.1% of males and 57.1% of females affected, which is over 10 percentage points higher than all other races.[2]

CVD is usually caused by atherosclerosis, an inflammatory disease of large and medium-sized arteries. The four major vascular beds that comprise the bulk of atherosclerotic cardiovascular disease (ASCVD) include coronary, cerebrovascular, peripheral vascular, and the central aortic vascular bed. The pathophysiology of atherosclerosis typically involves a diffuse pan vascular process, and risk factors associated with disease in these vascular beds are often shared, as well as highly correlated. For example, in the GenePad study, in individuals >40 years of age with coronary artery disease (CAD), 32% of women and 20% of men had concomitant peripheral arterial disease (PAD).[4] Maintenance of vascular health is thus centered on the focused optimization of these shared risk factors. This chapter will discuss the current modalities available to assess cardiovascular health, as well as strategies to prevent vascular disease, in addition to maintaining and optimizing vascular health. The role of population health and public policy will be further explored as it will provide perhaps the most meaningful long-term role in preventing CVD both in the United States and globally.

The Vasculome
https://doi.org/10.1016/B978-0-12-822546-2.00026-5

TABLE 36.1 Modifiable risk factors and behaviors comprising the definitions of poor, intermediate, and ideal cardiovascular health.

Metric	Poor	Intermediate	Ideal
Blood pressure	SBP ≥140 or DBP ≥90 mm Hg	SBP 120–139 or DBP 80–89 mm Hg or treated to goal	SBP <120 or DBP <80 mm Hg
Total cholesterol	≥240 mg/dL	200–239 mg/dL or treated to goal	<200 mg/dL
Fasting glucose	≥126 mg/dL	100–125 mg/dL or treated to goal	<100 mg/dL
Smoking status	Current smoker	Former smoker or quit ≤12 months ago	Never smoker or quit >12 months ago
Physical activity	None	1–149 min/week moderate intensity or 1–74 min/week vigorous intensity or 1–149 min/week moderate + Vigorous intensity	≥150 min/week moderate intensity or ≥75 min/week vigorous intensity or ≥150 min/week moderate + Vigorous intensity
Body Mass index	≥30 kg/m^2	25–29.9 kg/m^2	<25 kg/m^2
Healthy diet score[a]	0–1 component	2–3 components	4–5 components

[a]*The Goals and Metrics Committee of the Strategic Planning Task Force selected five aspects of diet to define a healthy dietary score, which is detailed in their American Heart Association Special Report.*[5]
DBP, diastolic blood pressure; *kg/m²*, kilogram per meter squared; *mg/dL*, milligrams per deciliter; *min*, minutes; *mm HG*, millimeters of mercury; *SBP*, systolic blood pressure
Adapted from American Heart Association's Life's Simple 7.

Pillars of vascular health

The AHA has defined ideal cardiovascular health based on seven modifiable risk factors consisting of blood pressure, total cholesterol, fasting blood glucose, smoking, physical activity, body mass index, and diet (Table 36.1).[5] Ideal cardiovascular health is defined as the presence of optimal levels of all seven and poor cardiovascular health is defined as the presence of at least one poor health metric.

Ideal cardiovascular health is uncommon, particularly in the United States. The estimated prevalence of ideal cardiovascular health ranged from 0.5% to 12%.[6] In the National Health and Nutrition Examination Survey (NHANES), the proportion of American adults meeting all seven ideal cardiovascular health metrics declined over time from 2.0% [95% CI, 1.5%–2.5%] in 1988–1994 to 1.2% [95% CI, 0.8%–1.9%] in 2005–2010.[7] Women, NH whites, and those with higher education levels were more likely to meet a greater number of these cardiovascular health metrics than their male, ethnic minority, and less educated counterparts.

Epidemiological studies have demonstrated the direct association of ideal cardiovascular health with favorable long-term cardiovascular outcomes. In the Atherosclerosis Risk in Communities (ARIC) study consisting of 15,792 men and women 45–64 years of age without CVD at baseline, only 0.1% met all criteria for ideal cardiovascular health, with 82.5% of the study population determined to have poor cardiovascular health.[8] Incidence rates over a median follow-up of 18[7] years showed a graded relationship with the ideal, intermediate, and poor categories and with the number of ideal health metrics present: rates were one-tenth as high in those with six ideal health metrics (3.9 per 1000 person-years) compared with zero ideal health metrics (37.1 per 1000 person-years). Over the follow-up period, almost one-half of the cohort with no ideal health factors experienced a CVD event, in contrast to no events in patients who had all seven ideal health factors at baseline. NHANES also found similar associations between ideal cardiovascular health and outcomes in 44,959 adults in the United States.[9] Adults who met six or more cardiovascular health metrics had a hazard ratio for CVD mortality of 0.24 (95% CI, 0.13–0.47) compared to adults with one or less cardiovascular health metrics.

Assessing ASCVD risk

ASCVD risk factors can be classified into traditional and nontraditional subtypes. The initial set of traditional risk factors were outlined in the landmark Framingham Study and included age, male sex, hypertension, dyslipidemia, diabetes mellitus, and tobacco use.[10] A family history of ASCVD in a first-degree relative, defined as occurring before 55 years of age in males and 65 in females, was also recognized as a correlate of risk.[11] The 2013 ACC/AHA ASCVD Pooled Cohort Risk Calculator further designated race as an additional risk factor.[12] However, it was the 2018 AHA/ACC/Multi-Society Cholesterol guidelines that consolidated

TABLE 36.2 ASCVD risk enhancers according to the 2018 AHA/ACC/Multisociety guidelines on the management of blood cholesterol.

Family history of premature ASCVD	Males younger than 55; females younger than 65 years of age
Primary hypercholesterolemia	LDL-C 160–189 mg/dL; non-HDL-C 190–219 mg/dL
High-risk race/ethnicity	South Asian ancestry
Metabolic syndrome	$\geq$ 3/5, including: ↓ HDL-C, ↑ glucose, ↑ triglycerides, ↑ waist circumference, ↑ blood pressure,
Chronic kidney disease	eGFR 15–59 mL/min with or without albuminuria; Not treated with dialysis or kidney transplant
Chronic inflammatory conditions	HIV/AIDS, systemic lupus erythematosus, psoriasis, etc.
History of premature menopause and pregnancy-related CV conditions	Menopause before 40 years of age, preeclampsia, gestational hypertension, and diabetes
Socioeconomic status	Constellation of biologic, psychosocial, and behavioral patterns that ↑ CV risk ↓ income, racial discrimination
Lipid/Biomarkers	Persistent hypertriglyceridemia ($\geq$175 mg/dL) If measured: **1.** ↑ hsCRP ($\geq$ 2 mg/L) **2.** ↑ Lp(a) ($\geq$ 50 mg/dL) **3.** ↑ ApoB ($\geq$ 130 mg/dL) **4.** ABI <0.9

ABI, ankle-brachial index; *ApoB*, apolipoprotein B; *ASCVD*, atherosclerotic cardiovascular disease; *eGFR*, estimated glomerular filtration rate; *HIV/AIDS*, human immunodeficiency virus/acquired immunodeficiency syndrome; *hsCRP*, high-sensitivity C-reactive protein; *LDL-C/HDL-C*, high/low density lipoprotein cholesterol; *Lp(a)*, lipoprotein (a).

the most impactful nontraditional risk enhancing conditions into screening guidelines (Table 36.2).[13]

A number of these nontraditional ASCVD risk enhancers have been characterized including socioeconomic status (SES), autoimmune conditions, human immunodeficiency virus (HIV) status, and metabolic syndrome. SES has been associated with CVD and may confer a similar magnitude risk as a traditional risk enhancer.[14,15] This trend is underpinned by biologic, psychosocial, and behavioral patterns present in low-SES populations that increase the burden of CVD.[16] As an example, residing in a food desert (neighborhoods defined as low-income areas with low access to healthy food) is associated with an increased 10-year risk of CVD, and this trend is driven by a per capita income that falls below the national average in these same areas.[17] A growing body of literature has also described the relationship between racial discrimination and inferior delivery of cardiovascular care.[18–20] Autoimmune processes, such as rheumatoid arthritis (RA), systemic lupus erythematosus, and systemic sclerosis, cause endothelial dysfunction and chronic immune cell dysfunction that accelerate atherosclerosis. For instance, patients with RA are nearly twice as likely to develop CAD when compared to the general population.[21,22] In the HIV population, virus-mediated inflammation and immune activation contribute to an approximate 1.5 to 2-fold increase in the likelihood of myocardial infarction (MI) and ischemic heart disease after adjusting for traditional risk factors.[23–25] Metabolic syndrome, a complex pathophysiologic state defined by abnormal fasting glucose, hypertriglyceridemia, elevated blood pressure, low HDL, and abdominal obesity, is associated with a significant rise in CVD and mortality that tends to be both greater than the sum of its parts and clinically underrecognized.[26,27] Additionally, syndromes of pregnancy including gestational HTN and DM, psychosocial stressors, as well as the burden of extensive comorbidities, have all been recognized as important nontraditional ASCVD risk enhancers.

Biomarkers

Biomarkers such as high sensitivity C-reactive protein (hsCRP), Lipoprotein(a) (Lp(a)), Apolipoprotein B (ApoB), and high-density lipoprotein-C (HDL-C) have emerged as auxiliary prognostic tools, particular in those patients deemed to be at borderline or intermediate risk for CVD.[28]

hsCRP is an acute phase reactant that serves as a surrogate marker for systemic inflammation. It is broadly understood that inflammation is a risk factor in the development of ASCVD. In a study of over 50,000 patients, a serum hsCRP >3 mg/L predicted a significantly increased risk for ischemic heart disease when compared to patients with an hsCRP levels <1 mg/L.[29] Additionally, in the Justification for the Use of Statins in Primary Prevention: An Intervention Trial Evaluating Rosuvastatin trial, hsCRP levels were an independent predictor of cardiovascular events.[30] Several limitations to the use of hsCRP as a predictor of ASCVD include considerable inter- and intrapatient variability, as well as variability in those with comorbid conditions like RA or chronic kidney disease, or from certain ethnic backgrounds.

Lp(a), which consists of a low-density lipoprotein (LDL)-like particle covalently bound to apolipoprotein B100, is an independent risk factor for ASCVD. Indeed, a causal association has been demonstrated between genetically elevated Lp(a) and increased likelihood of MI.[31] Nonetheless, the role of Lp(a) in evaluating ASCVD risk remains incompletely characterized with guidelines varying from Europe to the United States. The 2018 AHA/ACC/Multisociety cholesterol guidelines state that an Lp(a) > 50 mg/dL is an ASCVD risk enhancing factor that may favor statin initiation in nondiabetic patients between 40 and 75 years old with an ASCVD risk score between 7.5% and 19.9%.[13] The National Lipid Association recommends Lp(a) screening in patients with a personal history of premature ASCVD, first degree relatives with premature ASCVD, and/or primary hypercholesterolemia (LDL>190).[32] In recent years, meaningful progress has been made in the development of antisense oligonucleotides that may significantly reduce the cardiovascular burden of genetically elevated Lp(a).[33]

ApoB is a structural protein embedded in LDL and very low-density lipoprotein (VLDL). A metaanalysis including more than 200,000 subjects found that apoB was the most potent marker of cardiovascular risk, surpassing LDL-C as a marker for ASCVD.[34,35] Drawbacks to using apoB as a prognosticator include its high cost as well as a considerably longer turnaround time compared to other biomarkers. For these reasons, apoB was not recommended for routine screening by the 2013 ACC/AHA guidelines on the assessment of cardiovascular risk.[12]

The cholesterol transported by HDL-C appears to be inversely related to ASCVD risk, though a causal relationship between HDL and decreased atherosclerotic burden has not been established.[36] Indeed, HDL-C forms part of the ASCVD pooled risk equation and there is evidence that increasing cholesterol efflux capacity may substantially reduce adverse CV events.[37]

In individuals with established CAD and at high risk based on established risk factors, the use of multimarker biomarker risk scores (BRSs) to further reclassify individuals into higher or lower CVD risk categories is a promising area of study. One such scoring algorithm utilizes a 4-biomarker risk score (BRS-4) that includes hsCRP, along with fibrin degradation product, heat shock protein-70 (HSP-70), and soluble urokinase plasminogen activator receptor (suPAR) to further reclassify individuals into higher and lower risk categories. A BRS of 1, 2, 3, or 4 was associated with a hazard ratio for all-cause death or MI of 1.81, 2.59, 6.17, and 8.80, respectively.[38]

Risk calculators

There are multiple CVD risk calculators used in clinical practice. In the United States, the most commonly used risk calculators are the 2008 Framingham General Cardiovascular Risk Score and the ACC/AHA Pooled Cohort ACSVD Risk Equations. Compared to its predecessors, the 2008 Framingham General Cardiovascular Risk Score added important cardiovascular outcomes like stroke, heart failure, and peripheral artery disease to its variables. This risk calculator assesses 10-year cardiovascular risk and classifies individuals into three categories: low, intermediate, or high risk.[39,40] The 2013 ACC/AHA Pooled Cohort ACSVD Risk Equations uses data on Caucasians and African American subjects derived primarily from the Framingham Study, the ARIC study, the Cardiovascular Health Study (CHS), and the Coronary Artery Risk Development in Young Adults (CARDIA) study. Family history of premature coronary disease is not assessed by this calculator and this could lead to underestimation of the true cardiovascular risk in certain populations. The Pooled Cohort Equations classifies patients into four different categories: low (5%), borderline (5 to < 7.5%), intermediate (7.5 to < 20%), and high (>20%) ASCVD risk.[12,41–45]

Cardiovascular imaging

Carotid intima–media thickness

The first studies looking into the utility and reproducibility of arterial wall intima–media thickness were done in the late 1980s using B-mode echocardiography.[46] Due to its accessibility and relative low-cost, carotid intima-media thickness (CIMT) was later studied as a noninvasive marker of subclinical atherosclerosis.[47,48] CIMT has been shown to predict cardiovascular events in multiple patient populations. In 1683 patients from the Rotterdam cohort, an increase in CIMT by 0.15 mm led to a 3.7% and a 10.5% increased risk of CAD and death within a 10–12-year time period, respectively.[49] In the Carotid Atherosclerosis Progression Study, which followed 37,197 patients for 5.5 years, an increase in CIMT by 0.1 mm was associated with an increased risk of MI and stroke of 10%–15% and 13%–18%, respectively.[50] The Malmo Diet and Cancer Study in Sweden followed >5000 patients for 10 years and found that a 0.15 mm increment in CIMT was associated with a 1.23 increased risk (95% confidence intervals: 1.07–1.41) of adverse coronary events after adjustment for cardiovascular risk factors.[51] The ARIC study monitored 15,792 patients in the United States for 4–7 years. The investigators found that an increase in CIMT by 0.1 mm was associated with a 1.13 (hazard ratio) increased risk of CAD.[52] The Individual Progression of Carotid Intima Media Thickness as a surrogate of vascular risk (PROG-IMT) project included 36, 984 patients which were followed for 7 years. Patients in this study had at least two CIMT evaluations, one at baseline and a subsequent measurement to assess CIMT progression. It was found that the mean baseline CIMT was associated with increased cardiovascular risk. However, no association was found between cardiovascular risk and CIMT progression.[53] Also, none or minimal additional predictive value was found when CIMT was used in conjunction with traditional cardiovascular risk prediction models.[54,55] There are significant differences in CIMT measurements across different studies. Some studies measure CIMT at the level of the internal carotid instead of the common carotid and atherosclerotic plaques are not always excluded from these measurements. The lack of standardization in CIMT measurements has led to varying study results and the 2013 AHA Guideline on the Assessment of Cardiovascular Risk recommending against its routine use in clinical practice for cardiovascular risk assessment.[56]

Coronary artery calcium

Coronary artery calcium (CAC) was initially assessed with chest X-rays and fluoroscopy.[57] Electron beam computed tomography (EBCT) with its ECG-gating, allowed more accurate quantification of CAC. EBCT was subsequently replaced in clinical practice by multidetector computed tomography (MDCT), the current modality of choice for CAC evaluation, which has a superior spatial resolution. A CAC score is derived from the calcified plaque area with a density >130 Hounsfield Units. Each area is multiplied by a weighting factor derived from the peak density of the lesion and summed for the whole coronary system.[58] CAC score has been studied in multiple populations. The Multi-Ethnic Study of Atherosclerosis (MESA) is a cohort of 6814 asymptomatic individuals from four ethnicities (38% Caucasian, 28% African American, 22% Hispanic, and 12% Chinese) without CVD who underwent CAC score evaluation. This study revealed CAC score was able to predict coronary events across all ethnicities. MESA also provided percentiles of CAC distribution across age, gender, and ethnicity.[59,60] The Heinz Nixdorf Recall (HNR) study screened 4487 individuals without clinical CAD in West Germany, and a cohort of 2063 subjects from the Rotterdam study underwent CAC assessment. Similar to MESA, CAC was found to predict coronary events in both cohorts.[61–63]

MESA study subjects with a 10-year ASCVD risk between 5% and 20% and a CAC score of 0 were found to have lower cardiovascular event rates than the guideline-supported threshold for statin benefit. A CAC score >0 was associated with higher cardiovascular event rates in support of the net clinical benefit of statins.[64] Therefore, CAC is currently recommended in clinical practice for cardiovascular risk reclassification in subjects ages 40–75-years-old with borderline (5 to <7.5%) and intermediate (7.5 to < 20%) ASCVD risk if the decision regarding statin therapy remains undetermined.[13]

In 4229 individuals followed in the MESA study, a CAC score of 0 predicted harm with aspirin therapy for primary prevention. Independent of the cardiovascular risk profile, a CAC score >100 predicted benefit in preventing coronary heart disease with regular low dose aspirin use. Thus, CAC assessment may have a clinical role in the decision to start aspirin for primary prevention.[65]

CAC progression has been linked to increased all-cause mortality and MI risk.[66,67] Thus, there is additional value in rescanning patients with a CAC of 0 to assess calcium deposition. In 3116 subjects from the MESA study with a baseline CAC of 0, the prevalence of CAC >0 at 10 years was 53%; CAC >10 was 36%; and CAC >100 was 8%. Using survival models, the warranty period of CAC = 0 in years was estimated. The investigators recommended rescanning individuals with low, borderline, intermediate risk, and high risk with diabetes every 6–7 years, 3–5 years, 3 years, and 3 years, respectively.[68]

Optimizing vascular health

Medications, supplements, vitamins

Optimizing a person's vasculome depends on their ability to avert proatherogenic processes like cholesterol deposition, inflammation, oxidation, and thrombosis. Multiple drugs targeting these proatherogenic pathways have been developed and are used for primary and secondary prevention of vascular disease.

Cholesterol deposition is a crucial step for developing atherosclerotic disease and results from a positive net of high cholesterol influx (increased LDL-C serum levels) and low cholesterol efflux (decreased HDL-C levels) inside the vascular wall.[69] Statins are HMG-CoA reductase inhibitors that inhibit cholesterol synthesis and subsequently reduce serum LDL-C.[70] Several large-scale trials including thousands of patients have proved statins to be efficacious in reducing rates of MI, unstable angina/NSTEMI, and stroke requiring hospitalization or revascularization.[71–74] Ezetimibe and PCSK9 inhibitors are newer lipid-lowering agents that have also been shown to significantly decrease LDL-C levels and improve cardiovascular outcomes.[75–77] On the contrary, randomized trials with Cholesteryl ester transfer protein inhibitors demonstrated no reduction in the risk of cardiovascular events despite significantly increased plasma HDL-C.[78,79]

Cytokines and vascular inflammation play a very important role in the initiation and progression of atherosclerosis.[80] Aside from their known LDL-C lowering effect, statins and PCSK9 inhibitors have been also found to have antiatherogenic effects via reducing vascular inflammation.[81,82] Antiinflammatory medications such as canakinumab (interleukin-1b inhibitor) and colchicine significantly reduced cardiovascular event recurrence compared to placebo in phase III randomized controlled trials.[83,84]

Excessive oxidative stress also contributes to the progression of vascular disease. However, large clinical trials with antioxidants (vitamin E and beta carotene) in high-risk individuals did not demonstrate any effect on improving cardiovascular outcomes.[85,86]

Aspirin is an antiinflammatory and antiplatelet agent that has been used for both primary and secondary CVD prevention. Aspirin can be beneficial for selected high-risk patients, between 40 and 70 years old for primary prevention,[87] as well as for patients following an MI, stroke, coronary stent placement, or coronary artery bypass graft surgery for secondary prevention.[88] Antiplatelet P2Y12 inhibitors (clopidogrel, prasugrel, ticagrelor) have been also effective in the secondary prevention setting in individuals with chronic CAD, previous MI, previous stroke, or PAD.[89,90]

Lifestyle approaches, population health, and public policy

Despite the uniqueness of each person's Vasculome, lifestyle optimization has universal benefits.[91] Approaches to lifestyle optimization are consistently evolving, but the risk factors identified by the landmark Framingham and Seven Countries studies continue to serve as targets to guide therapeutic lifestyle changes.[92,93]

Exercise directly benefits local vascular architecture through shear-stress induced endothelial nitric oxide (NO) synthase production, and outward vascular remodeling, evidenced by improvements in flow-mediated dilation and biomarker levels.[94–97] While exercise alone may only modestly improve individual risk factors (e.g., a decrease in glycosylated hemoglobin by 0.6%), the sum of the parts has a profound impact.[98] A metaanalysis accounting for over 100,000 patients measured cardiorespiratory fitness (CRF)—a surrogate for vascular health—and found higher CRF levels to be associated with a 13% lower all-cause mortality and 15% lower risk of CVD.[99,100] The current guidelines recommend at least 150 min of moderate-intensity exercise per week,[101] but data suggest that even just 15 min of moderate-intensity daily exercise can reduce all-cause mortality by 14% when compared to sedentary participants.[102] There are insufficient data to concretely recommend or dissuade a specific exercise modality, but some smaller studies question if prolonged vigorous exercise can lead to worse outcomes; marathon runners were noted to have higher CAC scores compared to their Framingham risk score-matched controls.[103]

Dietary patterns and factors have the potential to impact vascular health multiple times daily. Similar to exercise, nutrition has direct effects on endothelial function, and the favorable effects of a healthy dietary pattern on atherogenesis attenuate mortality from heart disease, stroke, and diabetes.[104,105] Early dietary recommendations and representations included the diet heart hypothesis, with its focus on low-fat consumption, and the food pyramid concept. These initial iterations were perhaps too specific and led to compensatory favoritism of processed carbohydrates.[106] The current guidance for optimal vascular health has shifted toward more comprehensive dietary recommendations. The Dietary Approaches to Stop Hypertension (DASH) trial prioritized fruits, vegetables, and limited saturated fat, which demonstrated a systolic blood pressure reduction by up to 11 mm Hg.[107] The Mediterranean dietary pattern—comparable to DASH but with an emphasis on unsaturated fat, modest red wine, and seafood—is another nutritional approach that has demonstrated a 30% relative risk reduction in primary cardiovascular events compared to a standard diet.[108] Through risk factor modification and antiinflammatory mediators, these healthy dietary patterns have ubiquitous vasculature benefits across genders and racial groups.[107] Additional data are needed to better understand the role of the ketogenic diet, intermittent fasting, and meal timing on cardiometabolic health.[109]

One of the single most detrimental substances to the microenvironment is tobacco. The resulting dysregulatory state leads to increased LDL oxidation and DNA damage, promoting atherogenesis.[110–112] Statins play a role in quelling this inflammatory process by reducing LDL, but the impact of statins may go above and beyond, with data suggesting a reduction in CD40 signaling and thus furthering interference of atherosclerosis.[113] Electronic cigarettes have been marketed as a less toxic alternative, and while further research is required, current studies suggest they too impair vascular function, though they seem to have less of an acute effect on NO bioavailability than traditional tobacco.[112] Nonetheless, it is well documented that smoking increases cardiovascular mortality, with a meta-analysis of nearly half a million participants suggesting smoking advances cardiovascular mortality by more than 5 years. Given its dose-dependent relationship, complete tobacco cessation is desirable, but stepwise reductions are equally paramount for vascular health.[114]

Extending beyond healthcare and health systems, public policy has the potential to play an integral role in risk factor modification and lifestyle optimization. Taxation of tobacco products has been shown to improve readiness and likelihood of quitting, though the price increase alone may not lead to sustainable changes.[115,116] Taxes on sugar-sweetened beverages have been associated with an initial reduction in consumption but with waning success at the 1 year mark.[117,118] Other studies indicate the need for more than deterrent approaches; for example, Massachusetts' health care reform required Medicaid to cover behavioral counseling and pharmacologic treatments for tobacco use, which resulted in a smoking rate decline of 26%.[119] While these approaches are promising, these studies suggest that multiple interventions are likely needed to improve population health.

Promoting and supporting lifestyle changes requires an integrative approach, both in the clinic and the community. While small actions can lead to long-term benefits, and the physician–patient conversation should not be underappreciated,[120] health promotion and focus on impactful policies are necessary complements. Improved visibility and intervention through the Million Hearts initiative—a public and private amalgam aimed at improving the management of blood pressure, cholesterol, and tobacco use—is an example of partnership in championing cardiovascular health and CVD prevention.[121,122] Cardiac rehabilitation (CR) can be viewed as a comprehensive environment for these interventions—a microcosm of integrative prevention. A metaanalysis examining the effect of CR on secondary prevention post-MI demonstrated a statistically significant reduction in MI, cardiac, and all-cause mortality.[123,124] CR has also demonstrated reductions in depression and its associated adverse cardiovascular effects, highlighting another important risk factor relevant to making successful life-related changes.[125] CR is currently underutilized due to financial barriers and limited awareness. Proven effective in secondary prevention, future studies are warranted to assess its impact on primary cardiac events given its antiischemic and antiatherogenic effects.[124,126]

Social determinants of health extend beyond traditional cardiac risk factors and play an instrumental role in vascular disease development. Income level is one such determinant, impacting a person's metabolic profile, access to interventions, and ultimately, affecting mortality and morbidity.[16] A large population-based study conducted in Taiwan's universal healthcare system demonstrated patients with lower SES were at increased risk of post-MI mortality.[127] Grouping SES by zip code in the United States, similar disparities were demonstrated among coronary angiogram timing; those with lower income presenting with MI experienced delayed imaging and fewer percutaneous coronary interventions.[128] Even initiation of preventive care is impacted. A large Danish study associated higher SES with an increased likelihood of receiving a statin and beta-blocker prescription post-MI.[129]

Education is a potential tool for improving health literacy and fostering wellness. A pooled analysis revealed an inverse relationship between the degree of a person's education and the prevalence of tobacco use, diabetes, and elevated total cholesterol.[130] CR participation echoes these discrepancies with lower enrollment rates for women, the uninsured, and those without a degree above high school.[131] Personal discrimination (both racial as well as other forms) is another source of overt disparities.[132] Studies suggest chronic exposure to discrimination is associated with a statistically significant presence of CAC in African American women.[133] Similar to income's effect, African Americans are less likely to be prescribed medically indicated statins compared to their white counterparts.[134]

While the Vasculome is complex and unique, optimization of vascular health is not intangible. Implementation of proven and effective strategies to foster and guide appropriate pharmacotherapy and lifestyle changes has the potential for meaningful impact on vascular risk.

Conclusion

A greater focus on primordial prevention of CVD is pivotal to the improvement in population-based vascular health. Vascular risk is influenced by genetics, environmental factors, and behavior. The AHA Simple 7 provides a foundational construct for optimizing vascular health: be active, maintain a healthy weight, and favorable cholesterol profile, do not smoke or use smokeless tobacco, adhere to a heart-healthy dietary pattern, and keep blood pressure and blood sugar controlled and in a normal range. Optimal vascular health must be a goal on an individual, family, community, and population level to address the current global pandemic of CVD.

References

1. International R. Projections of cardiovascular disease prevalence and costs: 2015–2035: technical report. *RTI International*. November 2016:3–25 [report prepared for the American heart association].
2. Virani SS, Alonso A, Benjamin EJ, et al. Heart disease and stroke statistics-2020 update: a report from the American heart association. *Circulation*. 2020;141:e139–e596.
3. Honigberg MC, Zekavat SM, Aragam K, et al. Association of premature natural and surgical menopause with incident cardiovascular disease. *J Am Med Assoc*. 2019;322(24):2411–2421.
4. Sadrzadeh Rafie AH, Stefanick ML, Sims ST, et al. Sex differences in the prevalence of peripheral artery disease in patients undergoing coronary catheterization. *Vasc Med*. 2010;15:443–450.
5. Lloyd-Jones DM, Hong Y, Labarthe D, et al. Defining and setting national goals for cardiovascular health promotion and disease reduction: the American Heart Association's strategic Impact Goal through 2020 and beyond. *Circulation*. 2010;121:586–613.
6. Younus A, Aneni EC, Spatz ES, et al. A systematic review of the prevalence and outcomes of ideal cardiovascular health in US and non-US populations. *Mayo Clin Proc*. 2016;91:649–670.
7. Yang Q, Cogswell ME, Flanders WD, et al. Trends in cardiovascular health metrics and associations with all-cause and CVD mortality among US adults. *J Am Med Assoc*. 2012;307:1273–1283.
8. Folsom AR, Yatsuya H, Nettleton JA, Lutsey PL, Cushman M, Rosamond WD. Community prevalence of ideal cardiovascular health, by the American Heart Association definition, and relationship with cardiovascular disease incidence. *J Am Coll Cardiol*. 2011;57:1690–1696.
9. Han L, You D, Ma W, et al. National trends in American heart association revised life's Simple 7 metrics associated with risk of mortality among US adults. *JAMA Network Open*. 2019;2: e1913131–e.
10. Kannel WB, Dawber TR, Kagan A, Revotskie N, Stokes 3rd J. Factors of risk in the development of coronary heart disease–six year follow-up experience. The Framingham Study. *Ann Intern Med*. 1961;55:33–50.
11. Lloyd-Jones DM, Nam BH, D'Agostino RB S, et al. Parental cardiovascular disease as a risk factor for cardiovascular disease in middle-aged adults: a prospective study of parents and offspring. *Jama*. 2004;291:2204–2211.
12. Goff Jr DC, Lloyd-Jones DM, Bennett G, et al. ACC/AHA guideline on the assessment of cardiovascular risk: a report of the American college of Cardiology/American heart association Task Force on practice guidelines. *Circulation*. 2013;129:S49–S73, 2014.
13. Grundy SM, Stone NJ, Bailey AL, et al. AHA/ACC/AACVPR/AAPA/ABC/ACPM/ADA/AGS/APhA/ASPC/NLA/PCNA guideline on the management of blood cholesterol: executive summary: a report of the American college of Cardiology/American heart association Task Force on clinical practice guidelines. *J Am Coll Cardiol*. 2018;73:3168–3209, 2019.
14. Franks P, Winters PC, Tancredi DJ, Fiscella KA. Do changes in traditional coronary heart disease risk factors over time explain the association between socio-economic status and coronary heart disease? *BMC Cardiovasc Disord*. 2011;11:28.
15. Stringhini S, Carmeli C, Jokela M, et al. Socioeconomic status and the 25×25 risk factors as determinants of premature mortality :a multicohort study and meta -analysis of 1·7 million men and women. *Lancet*. 2017;389:1229–1237.
16. Schultz WM, Kelli HM, Lisko JC, et al. Socioeconomic status and cardiovascular outcomes: challenges and interventions. *Circulation*. 2018;137:2166–2178.
17. Kelli HM, Hammadah M, Ahmed H, et al. Association between living in food deserts and cardiovascular risk. *Circ Cardiovasc Qual Outcomes*. 2017;10.
18. Capers Q, Sharalaya Z. Racial disparities in cardiovascular care: a review of culprits and potential solutions. *J Racial Ethn Health Disparities*. 2014;1:171–180.
19. Panza GA, Puhl RM, Taylor BA, Zaleski AL, Livingston J, Pescatello LS. Links between discrimination and cardiovascular health among socially stigmatized groups: a systematic review. *PLoS One*. 2019;14:e0217623.
20. Brewer LC, Cooper LA. Race, discrimination, and cardiovascular disease. *Virtual Mentor*. 2014;16:270–274.
21. Maradit-Kremers H, Crowson CS, Nicola PJ, et al. Increased unrecognized coronary heart disease and sudden deaths in rheumatoid arthritis: a population-based cohort study. *Arthritis Rheum*. 2005;52:402–411.
22. Solomon DH, Goodson NJ, Katz JN, et al. Patterns of cardiovascular risk in rheumatoid arthritis. *Ann Rheum Dis*. 2006;65: 1608–1612.
23. Durand M, Sheehy O, Baril JG, Lelorier J, Tremblay CL. Association between HIV infection, antiretroviral therapy, and risk of acute myocardial infarction: a cohort and nested case-control study using Québec's public health insurance database. *J Acquir Immune Defic Syndr*. 2011;57:245–253.
24. Freiberg MS, Chang CC, Kuller LH, et al. HIV infection and the risk of acute myocardial infarction. *JAMA Intern Med*. 2013;173: 614–622.
25. Obel N, Thomsen HF, Kronborg G, et al. Ischemic heart disease in HIV-infected and HIV-uninfected individuals: a population-based cohort study. *Clin Infect Dis*. 2007;44:1625–1631.
26. Mottillo S, Filion KB, Genest J, et al. The metabolic syndrome and cardiovascular risk a systematic review and meta-analysis. *J Am Coll Cardiol*. 2010;56:1113–1132.
27. Sperling LS, Mechanick JI, Neeland IJ, et al. The cardio metabolic health alliance: working toward a new care model for the metabolic syndrome. *J Am Coll Cardiol*. 2015;66:1050–1067.
28. Choi S. The potential role of biomarkers associated with ASCVD risk: risk-enhancing biomarkers. *J Lipid Atheroscler*. 2019;8: 173–182.
29. Zacho J, Tybjaerg-Hansen A, Jensen JS, Grande P, Sillesen H, Nordestgaard BG. Genetically elevated C-reactive protein and ischemic vascular disease. *N Engl J Med*. 2008;359:1897–1908.
30. Ridker PM, Genest J, Boekholdt SM, et al. HDL cholesterol and residual risk of first cardiovascular events after treatment with potent statin therapy: an analysis from the JUPITER trial. *Lancet*. 2010;376:333–339.
31. Kamstrup PR, Tybjaerg-Hansen A, Steffensen R, Nordestgaard BG. Genetically elevated lipoprotein(a) and increased risk of myocardial infarction. *Jama*. 2009;301:2331–2339.
32. Wilson DP, Jacobson TA, Jones PH, et al. Use of Lipoprotein(a) in clinical practice: a biomarker whose time has come. A scientific statement from the National Lipid Association. *J Clin Lipidol*. 2019;13:374–392.
33. Tsimikas S, Karwatowska-Prokopczuk E, Gouni-Berthold I, et al. Lipoprotein(a) reduction in persons with cardiovascular disease. *N Engl J Med*. 2020;382:244–255.
34. Di Angelantonio E, Sarwar N, Perry P, et al. Major lipids, apolipoproteins, and risk of vascular disease. *Jama*. 2009;302:1993–2000.
35. Sniderman AD, Williams K, Contois JH, et al. A meta-analysis of low-density lipoprotein cholesterol, non-high-density lipoprotein cholesterol, and apolipoprotein B as markers of cardiovascular risk. *Circ Cardiovasc Qual Outcomes*. 2011;4:337–345.
36. Rader DJ, Hovingh GK. HDL and cardiovascular disease. *Lancet*. 2014;384:618–625.

37. Rohatgi A, Khera A, Berry JD, et al. HDL cholesterol efflux capacity and incident cardiovascular events. *N Engl J Med*. 2014; 371:2383–2393.
38. Ghasemzedah N, Hayek SS, Ko YA, et al. Pathway-specific aggregate biomarker risk score is associated with burden of coronary artery disease and predicts near-term risk of myocardial infarction and death. *Circ Cardiovasc Qual Outcomes*. 2017;10.
39. D'Agostino RB S, Vasan RS, Pencina MJ, et al. General cardiovascular risk profile for use in primary care: the Framingham Heart Study. *Circulation*. 2008;117:743–753.
40. Khambhati J, Allard-Ratick M, Dhindsa D, et al. The art of cardiovascular risk assessment. *Clin Cardiol*. 2018;41:677–684.
41. The atherosclerosis risk in Communities (ARIC) study: design and objectives. The ARIC investigators. *Am J Epidemiol*. 1989; 129:687–702.
42. Third report of the national cholesterol education program (NCEP) expert panel on detection, evaluation, and treatment of high blood cholesterol in adults (adult treatment panel III) final report. *Circulation*. 2002;106:3143–3421.
43. Fried LP, Borhani NO, Enright P, et al. The cardiovascular health study: design and rationale. *Ann Epidemiol*. 1991;1:263–276.
44. Friedman GD, Cutter GR, Donahue RP, et al. CARDIA: study design, recruitment, and some characteristics of the examined subjects. *J Clin Epidemiol*. 1988;41:1105–1116.
45. Kannel WB, Feinleib M, McNamara PM, Garrison RJ, Castelli WP. An investigation of coronary heart disease in families. The Framingham offspring study. *Am J Epidemiol*. 1979;110:281–290.
46. Pignoli P, Tremoli E, Poli A, Oreste P, Paoletti R. Intimal plus medial thickness of the arterial wall: a direct measurement with ultrasound imaging. *Circulation*. 1986;74:1399–1406.
47. O'Leary DH, Bots ML. Imaging of atherosclerosis: carotid intima-media thickness. *Eur Heart J*. 2010;31:1682–1689.
48. Persson J, Stavenow L, Wikstrand J, Israelsson B, Formgren J, Berglund G. Noninvasive quantification of atherosclerotic lesions. Reproducibility of ultrasonographic measurement of arterial wall thickness and plaque size. *Arterioscler Thromb*. 1992;12:261–266.
49. Bots ML, Hoes AW, Hofman A, Witteman JC, Grobbee DE. Cross-sectionally assessed carotid intima-media thickness relates to long-term risk of stroke, coronary heart disease and death as estimated by available risk functions. *J Intern Med*. 1999;245: 269–276.
50. Lorenz MW, Markus HS, Bots ML, Rosvall M, Sitzer M. Prediction of clinical cardiovascular events with carotid intima-media thickness: a systematic review and meta-analysis. *Circulation*. 2007;115: 459–467.
51. Rosvall M, Janzon L, Berglund G, Engström G, Hedblad B. Incident coronary events and case fatality in relation to common carotid intima-media thickness. *J Intern Med*. 2005;257:430–437.
52. Chambless LE, Heiss G, Folsom AR, et al. Association of coronary heart disease incidence with carotid arterial wall thickness and major risk factors: the Atherosclerosis Risk in Communities (ARIC) Study, 1987-1993. *Am J Epidemiol*. 1997;146:483–494.
53. Lorenz MW, Polak JF, Kavousi M, et al. Carotid intima-media thickness progression to predict cardiovascular events in the general population (the PROG-IMT collaborative project): a meta-analysis of individual participant data. *Lancet*. 2012;379: 2053–2062.
54. Den Ruijter HM, Peters SA, Anderson TJ, et al. Common carotid intima-media thickness measurements in cardiovascular risk prediction: a meta-analysis. *Jama*. 2012;308:796–803.
55. Yeboah J, McClelland RL, Polonsky TS, et al. Comparison of novel risk markers for improvement in cardiovascular risk assessment in intermediate-risk individuals. *Jama*. 2012;308:788–795.
56. Goff Jr DC, Lloyd-Jones DM, Bennett G, et al. ACC/AHA guideline on the assessment of cardiovascular risk: a report of the American college of Cardiology/American heart association Task Force on practice guidelines. *J Am Coll Cardiol*. 2013;63: 2935–2959, 2014.
57. Detrano R, Markovic D, Simpfendorfer C, et al. Digital subtraction fluoroscopy: a new method of detecting coronary calcifications with improved sensitivity for the prediction of coronary disease. *Circulation*. 1985;71:725–732.
58. Agatston AS, Janowitz WR, Hildner FJ, Zusmer NR, Viamonte Jr M, Detrano R. Quantification of coronary artery calcium using ultrafast computed tomography. *J Am Coll Cardiol*. 1990;15:827–832.
59. Bild DE, Detrano R, Peterson D, et al. Ethnic differences in coronary calcification: the multi-ethnic study of atherosclerosis (MESA). *Circulation*. 2005;111:1313–1320.
60. McClelland RL, Chung H, Detrano R, Post W, Kronmal RA. Distribution of coronary artery calcium by race, gender, and age: results from the Multi-Ethnic Study of Atherosclerosis (MESA). *Circulation*. 2006;113:30–37.
61. Greenland P, Blaha MJ, Budoff MJ, Erbel R, Watson KE. Coronary calcium score and cardiovascular risk. *J Am Coll Cardiol*. 2018;72: 434–447.
62. Lehmann N, Erbel R, Mahabadi AA, et al. Value of progression of coronary artery calcification for risk prediction of coronary and cardiovascular events: result of the HNR study (Heinz Nixdorf Recall). *Circulation*. 2018;137:665–679.
63. Vliegenthart R, Oudkerk M, Hofman A, et al. Coronary calcification improves cardiovascular risk prediction in the elderly. *Circulation*. 2005;112:572–577.
64. Nasir K, Bittencourt MS, Blaha MJ, et al. Implications of coronary artery calcium testing among statin candidates according to American college of Cardiology/American heart association cholesterol management guidelines: MESA (Multi-Ethnic study of atherosclerosis). *J Am Coll Cardiol*. 2015;66:1657–1668.
65. Miedema MD, Duprez DA, Misialek JR, et al. Use of coronary artery calcium testing to guide aspirin utilization for primary prevention: estimates from the multi-ethnic study of atherosclerosis. *Circ Cardiovasc Qual Outcomes*. 2014;7:453–460.
66. Budoff MJ, Young R, Lopez VA, et al. Progression of coronary calcium and incident coronary heart disease events: MESA (Multi-Ethnic Study of Atherosclerosis). *J Am Coll Cardiol*. 2013;61: 1231–1239.
67. Raggi P, Callister TQ, Shaw LJ. Progression of coronary artery calcium and risk of first myocardial infarction in patients receiving cholesterol-lowering therapy. *Arterioscler Thromb Vasc Biol*. 2004;24:1272–1277.
68. Dzaye O, Dardari ZA, Cainzos-Achirica M, et al. warranty period of a calcium score of zero: comprehensive analysis from the multi-ethnic study of atherosclerosis. *JACC Cardiovasc Imag*. 2021;14(5): 990–1002.
69. Tabas I, Williams KJ, Borén J. Subendothelial lipoprotein retention as the initiating process in atherosclerosis: update and therapeutic implications. *Circulation*. 2007;116:1832–1844.
70. Adhyaru BB, Jacobson TA. Safety and efficacy of statin therapy. *Nat Rev Cardiol*. 2018;15:757–769.
71. Collins R, Reith C, Emberson J, et al. Interpretation of the evidence for the efficacy and safety of statin therapy. *Lancet*. 2016;388: 2532–2561.
72. Cannon CP, Braunwald E, McCabe CH, et al. Intensive versus moderate lipid lowering with statins after acute coronary syndromes. *N Engl J Med*. 2004;350:1495–1504.
73. Schwartz GG, Olsson AG, Ezekowitz MD, et al. Effects of atorvastatin on early recurrent ischemic events in acute coronary syndromes: the MIRACL study: a randomized controlled trial. *Jama*. 2001;285:1711–1718.
74. Ridker PM, Danielson E, Fonseca FAH, et al. Rosuvastatin to prevent vascular events in men and women with elevated C-reactive protein. *N Engl J Med*. 2008;359:2195–2207.

75. Cannon CP, Blazing MA, Giugliano RP, et al. Ezetimibe added to statin therapy after acute coronary syndromes. *N Engl J Med*. 2015; 372:2387–2397.
76. Sabatine MS, Giugliano RP, Keech AC, et al. Evolocumab and clinical outcomes in patients with cardiovascular disease. *N Engl J Med*. 2017;376:1713–1722.
77. Schwartz GG, Steg PG, Szarek M, et al. Alirocumab and cardiovascular outcomes after acute coronary syndrome. *N Engl J Med*. 2018;379:2097–2107.
78. Schwartz GG, Olsson AG, Abt M, et al. Effects of dalcetrapib in patients with a recent acute coronary syndrome. *N Engl J Med*. 2012;367:2089–2099.
79. Lincoff AM, Nicholls SJ, Riesmeyer JS, et al. Evacetrapib and cardiovascular outcomes in high-risk vascular disease. *N Engl J Med*. 2017;376:1933–1942.
80. Tousoulis D, Oikonomou E, Economou EK, Crea F, Kaski JC. Inflammatory cytokines in atherosclerosis: current therapeutic approaches. *Eur Heart J*. 2016;37:1723–1732.
81. Liao JK, Laufs U. Pleiotropic effects of statins. *Annu Rev Pharmacol Toxicol*. 2005;45:89–118.
82. Karagiannis AD, Liu M, Toth PP, et al. Pleiotropic anti-atherosclerotic effects of PCSK9 inhibitors from molecular biology to clinical translation. *Curr Atherosclerosis Rep*. 2018;20:20.
83. Ridker PM, Everett BM, Thuren T, et al. Antiinflammatory therapy with canakinumab for atherosclerotic disease. *N Engl J Med*. 2017;377:1119–1131.
84. Tardif J-C, Kouz S, Waters DD, et al. Efficacy and safety of low-dose colchicine after myocardial infarction. *N Engl J Med*. 2019; 381:2497–2505.
85. Yusuf S, Dagenais G, Pogue J, Bosch J, Sleight P. Vitamin E supplementation and cardiovascular events in high-risk patients. *N Engl J Med*. 2000;342:154–160.
86. Vivekananthan DP, Penn MS, Sapp SK, Hsu A, Topol EJ. Use of antioxidant vitamins for the prevention of cardiovascular disease: meta-analysis of randomised trials. *Lancet*. 2003;361:2017–2023.
87. Arnett DK, Blumenthal RS, Albert MA, et al. ACC/AHA guideline on the primary prevention of cardiovascular disease: a report of the American College of Cardiology/American Heart Association Task Force on clinical practice guidelines. *Circulation*. 2019; 140:e596–e646, 2019.
88. Smith SC, Benjamin EJ, Bonow RO, et al. AHA/ACCF secondary prevention and risk reduction therapy for patients with coronary and other atherosclerotic vascular disease: 2011 update. *Circulation*. 2011;124:2458–2473.
89. A randomised, blinded, trial of clopidogrel versus aspirin in patients at risk of ischaemic events (CAPRIE). CAPRIE Steering Committee. *Lancet*. 1996;348:1329–1339.
90. Knuuti J, Wijns W, Saraste A, et al. ESC Guidelines for the diagnosis and management of chronic coronary syndromes. *Eur Heart J*. 2019;41:407–477, 2020.
91. Guo S, Deng W, Xing C, Zhou Y, Ning M, Lo EH. Effects of aging, hypertension and diabetes on the mouse brain and heart vasculomes. *Neurobiol Dis*. 2019;126:117–123.
92. Keys A, Menotti A, Aravanis C, et al. The seven countries study: 2,289 deaths in 15 years. *Prev Med*. 1984;13:141–154.
93. Hubert HB, Feinleib M, McNamara PM, Castelli WP. Obesity as an independent risk factor for cardiovascular disease: a 26-year follow-up of participants in the Framingham Heart Study. *Circulation*. 1983;67:968–977.
94. Spence AL, Carter HH, Naylor LH, Green DJ. A prospective randomized longitudinal study involving 6 months of endurance or resistance exercise. Conduit artery adaptation in humans. *J Physiol*. 2013;591:1265–1275.
95. Hambrecht R, Adams V, Erbs S, et al. Regular physical activity improves endothelial function in patients with coronary artery disease by increasing phosphorylation of endothelial nitric oxide synthase. *Circulation*. 2003;107:3152–3158.
96. Feairheller DL, Diaz KM, Kashem MA, et al. Effects of moderate aerobic exercise training on vascular health and blood pressure in African Americans. *J Clin Hypertens*. 2014;16:504–510.
97. Green DJ, Smith KJ. Effects of exercise on vascular function, structure, and health in humans. *Cold Spring Harb Perspect Med*. 2018;8.
98. Boulé NG, Haddad E, Kenny GP, Wells GA, Sigal RJ. Effects of exercise on glycemic control and body mass in type 2 diabetes mellitus: a meta-analysis of controlled clinical trials. *J Am Med Assoc*. 2001;286:1218–1227.
99. Kodama S, Saito K, Tanaka S, et al. Cardiorespiratory fitness as a quantitative predictor of all-cause mortality and cardiovascular events in healthy men and women: a meta-analysis. *J Am Med Assoc*. 2009;301:2024–2035.
100. Lavie CJ, Arena R, Swift DL, et al. Exercise and the cardiovascular system: clinical science and cardiovascular outcomes. *Circ Res*. 2015;117:207–219.
101. Services USDoHaH. *Physical Activity Guidelines for Americans*. The Department; 2008, 2008.
102. Wen CP, Wai JP, Tsai MK, et al. Minimum amount of physical activity for reduced mortality and extended life expectancy: a prospective cohort study. *Lancet*. 2011;378:1244–1253.
103. Möhlenkamp S, Lehmann N, Breuckmann F, et al. Running: the risk of coronary events : prevalence and prognostic relevance of coronary atherosclerosis in marathon runners. *Eur Heart J*. 2008; 29:1903–1910.
104. Lara J, Ashor AW, Oggioni C, Ahluwalia A, Mathers JC, Siervo M. Effects of inorganic nitrate and beetroot supplementation on endothelial function: a systematic review and meta-analysis. *Eur J Nutr*. 2016;55:451–459.
105. Micha R, Peñalvo JL, Cudhea F, Imamura F, Rehm CD, Mozaffarian D. Association between dietary factors and mortality from heart disease, stroke, and type 2 diabetes in the United States. *J Am Med Assoc*. 2017;317:912–924.
106. Fischer NM, Pallazola VA, Xun H, Cainzos-Achirica M, Michos ED. The evolution of the heart-healthy diet for vascular health: a walk through time. *Vasc Med*. 2020;25:184–193.
107. Appel LJ, Moore TJ, Obarzanek E, et al. A clinical trial of the effects of dietary patterns on blood pressure. DASH Collaborative Research Group. *N Engl J Med*. 1997;336:1117–1124.
108. Martínez-González MA, Salas-Salvadó J, Estruch R, et al. Benefits of the mediterranean diet: insights from the PREDIMED study. *Prog Cardiovasc Dis*. 2015;58:50–60.
109. St-Onge MP, Ard J, Baskin ML, et al. Meal timing and frequency: implications for cardiovascular disease prevention: a scientific statement from the American heart association. *Circulation*. 2017;135:e96–e121.
110. Yanbaeva DG, Dentener MA, Creutzberg EC, Wesseling G, Wouters EF. Systemic effects of smoking. *Chest*. 2007;131:1557–1566.
111. Yamaguchi Y, Haginaka J, Morimoto S, Fujioka Y, Kunitomo M. Facilitated nitration and oxidation of LDL in cigarette smokers. *Eur J Clin Invest*. 2005;35:186–193.
112. Carnevale R, Sciarretta S, Violi F, et al. Acute impact of tobacco vs electronic cigarette smoking on oxidative stress and vascular function. *Chest*. 2016;150:606–612.
113. Schönbeck U, Gerdes N, Varo N, et al. Oxidized low-density lipoprotein augments and 3-hydroxy-3-methylglutaryl coenzyme A reductase inhibitors limit CD40 and CD40L expression in human vascular cells. *Circulation*. 2002;106:2888–2893.
114. Mons U, Müezzinler A, Gellert C, et al. Impact of smoking and smoking cessation on cardiovascular events and mortality among older adults: meta-analysis of individual participant data from prospective cohort studies of the CHANCES consortium. *BMJ*. 2015;350:h1551.

115. Dunlop SM, Cotter TF, Perez DA. Impact of the 2010 tobacco tax increase in Australia on short-term smoking cessation: a continuous tracking survey. *Med J Aust*. 2011;195:469–472.
116. Ross H, Blecher E, Yan L, Hyland A. Do cigarette prices motivate smokers to quit? New evidence from the ITC survey. *Addiction*. 2011;106:609–619.
117. Lee MM, Falbe J, Schillinger D, Basu S, McCulloch CE, Madsen KA. Sugar-sweetened beverage consumption 3 Years after the berkeley, California, sugar-sweetened beverage tax. *Am J Publ Health*. 2019;109:637–639.
118. Lawman HG, Bleich SN, Yan J, et al. One-year changes in sugar-sweetened beverage consumers' purchases following implementation of a beverage tax: a longitudinal quasi-experiment. *Am J Clin Nutr*. 2020;112:644–651.
119. Land T, Rigotti NA, Levy DE, et al. A longitudinal study of medicaid coverage for tobacco dependence treatments in Massachusetts and associated decreases in hospitalizations for cardiovascular disease. *PLoS Med*. 2010;7:e1000375.
120. Jepson RG, Harris FM, Platt S, Tannahill C. The effectiveness of interventions to change six health behaviours: a review of reviews. *BMC Publ Health*. 2010;10:538.
121. Frieden TR, Berwick DM. The "Million Hearts" initiative–preventing heart attacks and strokes. *N Engl J Med*. 2011;365:e27.
122. Knapper JT, Ghasemzadeh N, Khayata M, et al. Time to change our focus: defining, promoting, and impacting cardiovascular population health. *J Am Coll Cardiol*. 2015;66:960–971.
123. Lawler PR, Filion KB, Eisenberg MJ. Efficacy of exercise-based cardiac rehabilitation post-myocardial infarction: a systematic review and meta-analysis of randomized controlled trials. *Am Heart J*. 2011;162, 571-84.e2.
124. Sandesara PB, Lambert CT, Gordon NF, et al. Cardiac rehabilitation and risk reduction: time to "rebrand and reinvigorate". *J Am Coll Cardiol*. 2015;65:389–395.
125. Milani RV, Lavie CJ, Cassidy MM. Effects of cardiac rehabilitation and exercise training programs on depression in patients after major coronary events. *Am Heart J*. 1996;132:726–732.
126. Menezes AR, Lavie CJ, Milani RV, Forman DE, King M, Williams MA. Cardiac rehabilitation in the United States. *Prog Cardiovasc Dis*. 2014;56:522–529.
127. Wang JY, Wang CY, Juang SY, et al. Low socioeconomic status increases short-term mortality of acute myocardial infarction despite universal health coverage. *Int J Cardiol*. 2014;172: 82–87.
128. Yong CM, Abnousi F, Asch SM, Heidenreich PA. Socioeconomic inequalities in quality of care and outcomes among patients with acute coronary syndrome in the modern era of drug eluting stents. *J Am Heart Assoc*. 2014;3:e001029.
129. Rasmussen JN, Gislason GH, Rasmussen S, et al. Use of statins and beta-blockers after acute myocardial infarction according to income and education. *J Epidemiol Community Health*. 2007;61: 1091–1097.
130. Woodward M, Peters SA, Batty GD, et al. Socioeconomic status in relation to cardiovascular disease and cause-specific mortality: a comparison of Asian and Australasian populations in a pooled analysis. *BMJ Open*. 2015;5:e006408.
131. Sun EY, Jadotte YT, Halperin W. Disparities in cardiac rehabilitation participation in the United States: a systematic review and meta-analysis. *J Cardiopulm Rehabil Prev*. 2017;37:2–10.
132. Brewer LC, Cooper LA. Race, discrimination, and cardiovascular disease. *Virtual Mentor*. 2014;16:455–460.
133. Lewis TT, Everson-Rose SA, Powell LH, et al. Chronic exposure to everyday discrimination and coronary artery calcification in African-American women: the SWAN Heart Study. *Psychosom Med*. 2006;68:362–368.
134. Nanna MG, Navar AM, Zakroysky P, et al. Association of patient perceptions of cardiovascular risk and beliefs on statin drugs with racial differences in statin use: insights from the patient and provider assessment of lipid management registry. *JAMA Cardiol*. 2018;3:739–748.

CHAPTER

37

The Vasculome provides a body-wide cellular positioning system and functional barometer. The "*Vasculature as Common Coordinate Frame* (VCCF)" concept

Zorina S. Galis

Vascular Researcher, Bethesda, MD, United States

Key requirements for a common coordinate frame (CCF) for the human body

Researchers are producing an avalanche of biological and medical data as they apply new high-content investigative methods. Many efforts across the world are using multidimensional data obtained from single-cell methodologies to create maps of normal or diseased tissue at single-cell resolution that require the ability to correctly identify and precisely localize their cellular sources. Creating and using maps traditionally requires a coordinate system, and detailed reports have described the importance of identifying a human body common coordinate frame (CCF) and related challenges.[1,2]

An impetus to the field was given by a "Common Coordinate Framework Meeting" jointly organized by the Chan-Zuckerberg Initiative and the NIH in 2017. The discussions resulted in participants agreeing on a very useful list of key design requirements for an optimal CCF.[3]

A CCF for the human body should be able to meet most if not all of the following requirements:

1. ***Works across several scales (cell <> tissue <> body)***
2. ***Applicable to all (or most) body tissues***
3. ***Accounts for donor differences (e.g., size, shape, sex)***
4. ***Useful/Acceptable across specialty domains***

The overall group conclusion was that no one strategy of mapping can work for all body organs and across all scales, and that perhaps the best solution was linking multiple individual organ CCF systems within a hierarchical overarching description or physical map of the entire body.

At the end of the meeting, in direct response to the commonly agreed-upon design requirements, I put forward a potential solution: the "*Vasculature as CCF*" and briefly presented arguments to support the concept that we might be able to take advantage of the already built-in, nature-designed vasculature to define a human body CCF that is likely to meet all these key requirements. Further research and development needed to test this hypothesis were also suggested, and the "*Vasculature as CCF*" was captured as an alternative potential solution in the final group report.[3]

Evaluating the vasculature as CCF and road map for the human body

In subsequent years, in order to move forward the "Vasculature as CCF" concept, I presented it on several occasions in front of audiences with various expertise, including experimentalists, computationalists, and physicians, prompting additional challenging cross-disciplinary questions that needed to be addressed. I have worked to formulate responses that integrate many lessons from a life-long career investigating and seeking to better understand the vasculature. These were illustrated by gathering an extensive collection of examples from others and my own work on the vasculature in a variety of contexts, across expertise domains,

The Vasculome
https://doi.org/10.1016/B978-0-12-822546-2.00037-X

and across body scales. My proposal for the *Vasculature as CCF* is inextricably related to the concept of *the Vasculome*,[4] which encompasses not only the entire physical vasculature, but also the body of knowledge about its unified nature and its essential importance to the survival and function of the human body.

The arguments are briefly summarized here; however, this printed version cannot benefit from illustrations already published due to copyright restrictions. The reader is invited to watch some of these presentations available online,[5] as well as to check the original sources for some of the selected illustrations.

CCF *will work across scales*

Specifically, the CCF meeting discussions defined integration across scales as going from the "gross anatomy" scale of the human body and its organs (macroscale), transitioning through the "mesoscale," to the fine scale of the microscopic (histological and cellular) level of single cells.

The vasculature is nature built seamlessly across these body scales, tapering down from the largest vessel, the aorta, with an average normal maximum diameter of 3.5 cm and a foot long,[6] down to the medium and smaller organotypic (tissue-level) vasculature, that includes the single cell-size capillaries, with diameters under 10 microns. All vessels are physically connected through the contiguous endothelial layer made up of endothelial cells (ECs) that line their inside; hence, I have proposed the endothelium as the main "organizational principle" of the vasculature. This is an effort to reduce the complexity of the vasculature to a common denominator, the endothelium, that is key to both local and systemic vascular structure and function, yet still accounts for the important contributions of all other elements, including other cells types, associated with the various types of vasculature.

CCF *should be applicable to all (most) body tissues*

The vasculature is an integral and essential component of all body tissues, delivering oxygen and nutrients across the entire body, and thus comes in close proximity to all its cells. The normal distribution of the average distance between the cells within a tissue and their closest vascular structure is a characteristic for each tissue that depends on the local function and metabolic needs. Indeed, the vasculature organizes and frames all tissues—developmentally, structurally, and functionally. Intuitively, this can be illustrated by the ease of recognizing various organs solely based on seeing their vasculature, which can be highlighted in various anatomic preparations and imaging techniques.[7–9]

CCF *should account for donor differences*

As it is an integral and essential component of everyone's body, the vasculature is naturally adapted to different human body sizes and shapes, sexes, or ethnicities, across lifespan, from newborn to full size adult, changing with aging, and responding to exercise and variations in body mass index, etc. Many studies have investigated the connection between human body size and vasculature for various purposes, such as obtaining an accurate model of the normal organ-specific vasculature or predicting the clinical significance of abnormal enlargement of the aorta (aneurysm formation).[10,11]

CCF *should be useful and acceptable across specialty domains*

The human vasculature has been studied extensively across specialty domains, as illustrated within *The Vasculome* book,[4] and has been central to the practice of medicine for centuries.[12] While some of the more complex aspects of the vasculature's structure and function are less shared across all these domains (something that the Vasculome concept seeks to address), there is a lot of shared understanding about the vasculature, including a lot of shared terminology. Indeed, many of the main blood and lymphatic vessels, like "rock stars," are widely known by their unique names, and described in many existing sources of information, from the illustrations of anatomy textbooks to clinical surgical and radiological compendia, to computational models and software. New initiatives are making progress in collecting and unifying ontologies for the vasculature and making them widely available.[13,14]

Practical considerations for using the vasculature as a human body CCF

I want to acknowledge upfront that identifying and using a single CCF across the entire body is incredibly challenging. Many have previously reviewed in detail various challenges; some authors suggested the need to construct a hierarchical framework constituted from different coordinate systems at different scales.[2] In the same time, others have emphasized the challenge of working with a variety of specialized, domain-specific coordinate systems for various body organs, warning that prematurely dictating one over another would risk increasing complexity and losing their generality.[1]

While conceptually, the human vasculature seems to fit the key qualities desired in a human CCF, and perhaps can respond to the challenges just highlighted, generating practical approaches for correctly mapping every human cell within the body in relation to the vasculature is nontrivial. The details here are useful and perhaps somewhat daunting, as a lot of the most useful considerations require a deep understanding of the large body of knowledge associated with the vasculature, illustrated in the present book, *The Vasculome. From Many, One.*[4]

I have envisioned two main potential approaches for using the Vasculature as CCF that are not mutually exclusive, but could rather complete each other, as the CCF must work across the dimensional scale of the entire human body. These could be used iteratively to confirm each other and increase precision for localizing and placing individual cells on the 3D human body map. Conceptually speaking, these two approaches are not dissimilar to those used to complete the entire sequence of the human genome. The first approach used to complete the genome consistently worked to stitch together large chunks of independently but orderly sequenced DNA, while the other was a "shotgun approach" in which computers organized many small DNA pieces randomly sequenced, by analyzing their overlaps.[15] Both approaches contributed to the final correct order.

A. The vasculature can be considered as a "road map" for the body. As the vasculature is contiguous across scales, the entire vasculature could be coded within an orderly, hierarchical "road" system, as the one taken by blood to deliver oxygen to every cell. This could begin with the large and medium arteries visible to the naked eye, which connect and ensure oxygen delivery to all the body organs and their major parts (such as organ lobes), and seamlessly continue to the microscopic vascular structures, embedded within each tissue, coming close to all cells, where the major blood–tissue exchanges occur before the blood is returned for reoxygenation and the next delivery.

I. At the macroscale level, organs can be easily placed in their correct location on a vascular road map, as each organ is uniquely attached and accessed by vessels which have their own designated names, like cities being reachable by major highways. This vascular arrangement places all organs in their correct configuration at the whole-body level.

II. As the meso-level, through successive bifurcation within organs, the major blood vessels decrease in size and feed smaller areas of tissue. These can be defined as distinct suborgan vascular territories[16,17] investing parts of organs, much like vascular-defined tissue neighborhoods. The vascular territory concept has been used clinically, for instance, to assess the tissue area damaged through the blockage of a heart or brain artery, or during surgical interventions, to spare or recanalize specific areas of organs or limbs.[18] As the daughter vessels decrease in size, they no longer have unique names; however, they can be categorized based on their structure and size (e.g., artery, arteriole, capillary, etc.) and can be assigned a hierarchical level based on a Strahler number (or Horton–Strahler number), a measure of branching complexity, that has been used to describe the hierarchy of tributaries in a variety of systems, including rivers, blood vessels, trees, or social networks. The terminal branch level that has no "daughters" is numbered as "1"[19] or perhaps "0."[7] The microvascular patterns are tissue specific (Fig. 37.1).

III. As the vasculature seamlessly reaches the microscopic level, various blood and lymph vessel types can be recognized under the microscope based on their structure and size. Of note, ECs create a single-cell thin contiguous layer, the endothelium, that lines the inside of all the blood and lymph vessels (Fig. 37.2). This inner layer is the basis for the seamless connection of the entire vasculature, leading us to propose the endothelium as the "organizing principle" of the vasculature. The smallest, single-cell size vessels, called capillaries, form a rich network within tissues, and are understood to be the main site for exchanges between blood or lymph and tissues, and central to the various functional tissue units (FTUs) of various organs. Structural and functional differences can be distinguished among capillaries, based on the type of endothelium (continuous, fenestrated, discontinuous) that creates their lumen, but this requires the use of specialized methods, including techniques with single-cell resolution. Structural/functional heterogeneity of ECs in connection to the organ and type of vessel has been reported since the golden era of electron microscopy.[20] We highlighted functional differences through the use of intravascular tracers using electron microscopy[21] and later using two-photon microangiography[16] (Fig. 37.2). Many other technologies since used in various contexts have confirmed and expanded the understanding of endothelial heterogeneity.[22,23] For instance, the discovery of

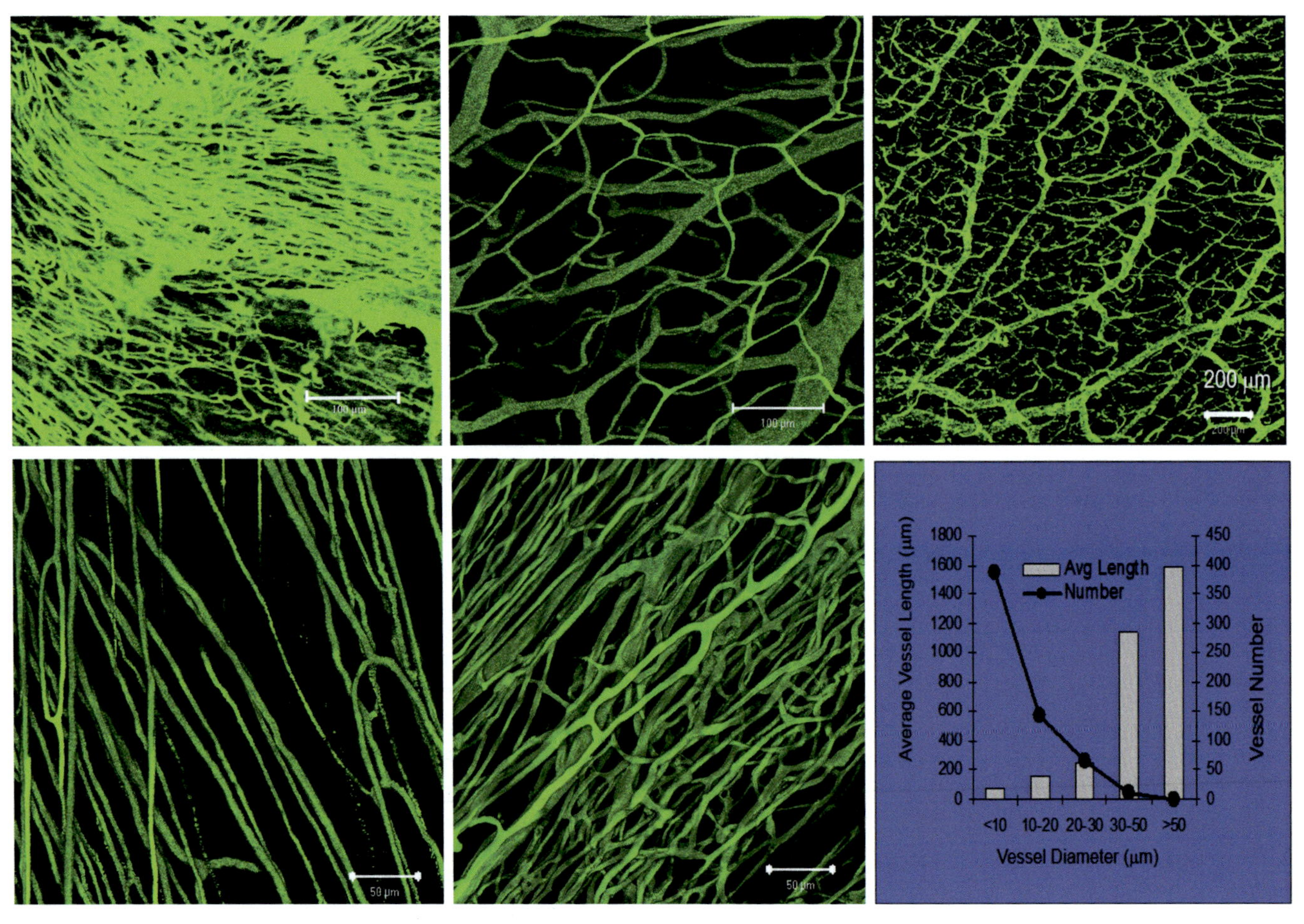

FIGURE 37.1 Vascular patterns are a characteristic of each tissues, related to its local metabolic needs and function, and are affected by various processes, such as hypoxia. Upper panels: Vascularization of heart (left), mesenteric muscle (middle), and intestinal wall (right) in a mouse genetically engineered to express green fluorescent protein (GFP) in the ECs. Lower panels: The normal pattern of skeletal muscle vascularization (left) is modified in response to ischemia (middle). Various characteristics of tissue-specific microvascular patterns can be quantified, e.g., number and distribution of vessels by length or diameter (right).

EC organ-specific "zip codes" through the use of phage display has already been exploited for delivery of therapeutic loads at specific vascular locations within various organs.[24] More recently, a variety of single-cell analyses have revealed an even richer and more subtle variety of ECs within different organs.[25] Such studies demonstrate that the molecular fingerprint of an EC not only contains a specific signature identifying its vessel type and tissue of origin, but also encodes its precise location along the continuum of the endothelial layer as it transitions from the arterial to the venous side,[26] which could be thought of as the EC zip code. One can envision that obtaining such positional information could help place individual ECs in their correct location on the integrated 3D road map of the entire vasculature.

Using this approach, the specific zip codes for the cellular addresses of other types of neighboring cells would consider their proximity to the closest hierarchical level/type of vasculature along the organ vascular road map. I illustrated a potential system to assign addresses for the kidney cells.[5] My original concept "Vasculature as CCF" has captured meanwhile the attention of investigators funded as part of the mapping component of the NIH Human Biomolecular Atlas Program (HuBMAP) and was further assessed and published.[27] Additional progress will be possible due to the ongoing work of a large group of international investigators representing several consortia who have assembled information about many cells types and structures of various organs,[13] including cataloging the hierarchical map and connections of the visible part of the human vasculature.[14] Importantly, the completion of the human vasculature road map will require the addition of the organotypic microvasculature and its integration within various organs.

B. Vasculature, more specifically ECs, as a body-wide cellular positioning system. A cellular GPS-like

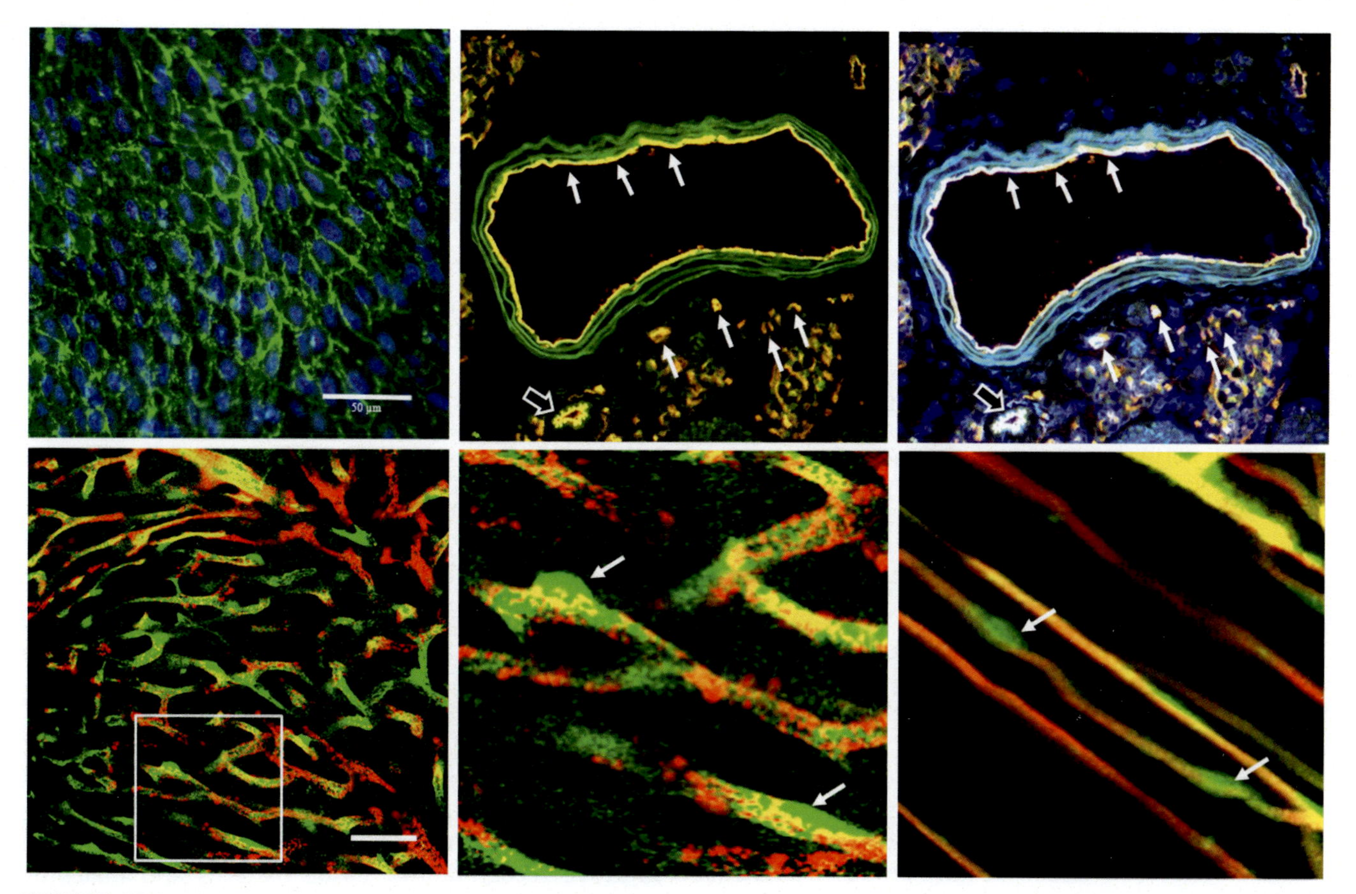

FIGURE 37.2 The endothelium. Upper panels: Immunofluorescent staining of "en-face" preparation of a mouse aorta (left). The inner contiguous mono-layer is created by endothelial cells (ECs), the green fluorescent signal detects presence of CD31 (an EC marker), and cell nuclei are counterstained blue (Hoechst). The thin endothelial layer (*thin arrows*) detected using anti-CD31 antibodies lines various vascular profiles in a histological section (middle and right panels): artery and arteriole (*thick arrows*, green autofluorescence is due to elastin), venule, and tiny capillaries (*thin arrows*). Same section is shown without and with blue counterstain of cell nuclei (Hoechst). Lower panels: The functional differences between two types of capillaries, as revealed through infusion of red microspheres in the vasculature of a mouse genetically engineered to express green fluorescent protein (GFP) in the ECs. Discontinuous capillaries of liver: the red microspheres escape through the EC gaps (left and middle panels, *arrows* point to some green ECs). Continuous capillaries of skeletal muscle: the red microspheres are well contained inside these capillaries (right panel, *arrows* point to some of the green ECs).

system would assign body-wide coordinates (or zip codes) for each cell based on the combination between cell's own molecular fingerprint and that of the closest EC, as revealed through single-cell analyses.

In the previous "road map" approach, the tissue sample location would be reported in connection to vascular hierarchy and requiring details about the site from which the tissue sample was extracted. When using the EC body-wide cellular positioning approach, this information should not be necessary to correctly "place" the analyzed cells in their original location within the body. Like the shotgun approach used to align the randomly generated sequences of the human genome, the experimental information obtained directly from analyzing a mixture of individual units, in this case single cells, would be assembled and reorganized in their original configuration within the tissue sample using computational methods, including some yet to be developed.

Our hypothesis stipulates that ECs within each sample being analyzed would be used to extract information about the sample's specific location. Some researchers may still be under the impression that ECs may be few and far in between, which would make this approach challenging. This is most likely due to their being familiar with histological sections in which ECs need to be highlighted using specific markers, and still appear either as a very thin layer inside larger vessels or perhaps as small "holes" (capillaries) Fig. 37.2. Actually, ECs are present in every tissue,[28] being the third most abundant human cell type.[29] By mass, the endothelium (the totality of ECs) could even be considered an independent organ.

Furthermore, recently published research has shown in several organs already that the single-cell signatures

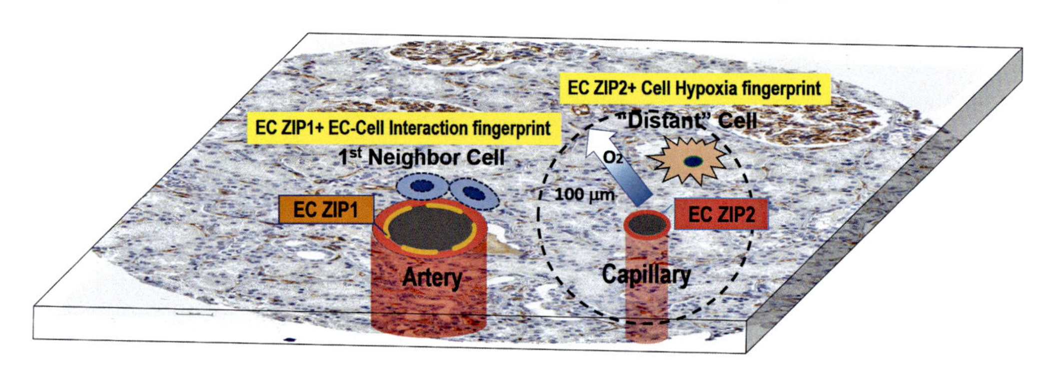

FIGURE 37.3 A proposed model for assigning addresses to individual cells using a body-wide system based on endothelial cell (EC) zip codes. Each EC has a molecular signature that indicates its position within the vasculature and the organ and can be considered EC's own zip code. Illustrated are two types of ECs, one lining a large artery (ZIP1) the other an organotypic capillary (ZIP2). The cellular address of EC's immediate neighbor cells will be a compilation between the EC zip and the cell–EC interaction fingerprint obtained by single-cell analyses. Most other cells are likely positioned within a radius of 100 µm (*dotted circle*) from their closest EC source of oxygen. Their cellular address combines the closest EC zip code and their own fingerprint, including oxygen-driven genes.

of various ECs include detailed information about their original context: within which type of vessel and which specific area of the organ.[30] Additional information extracted from ECs was shown to report on the functional status of their cellular neighborhoods and on systemic conditions.[31]

Assuming that information from the ECs can be used to correctly situate them in their original context (zip code), how could we use this information to report the exact location of *other* cells? We have proposed a schema (Fig. 37.3) that would identify their closest neighbors based on single-cell analyses that can detect direct interactions with ECs.[32] For the other cells, we propose that a good estimate of the maximum distance between most body cells and their closest EC (source of oxygen) would be 100 microns. This suggestion is based on previously published information about oxygen diffusion in various tissues.[16,33,34] The actual average distance is dependent on the specific cell type/tissue metabolic needs (as well as species). The 100 microns radius around an EC may only contain a handful of cells; the exact number would vary with tissue region and cell types that come in different sizes, each cell progressively further from the source of oxygen. Since the oxygen level is a major driver for many processes, it is possible to identify specific cell gene signatures that are associated with an oxygen gradient. These signatures could be used to help determine the distance between any cell and its source of oxygen. Previous studies of liver "zonation" have shown already the effects of distance from the central vein of the liver on gene expression in a panel of zonated landmark genes and used these to infer lobule coordinates of cells.[35]

Large amounts of single-cell data have been and continue to be generated using methods that require dissociation of tissues. Ongoing efforts are already developing illustrations based on the known overall organ anatomy and specific tissue patterns that can provide general spatial guidance for placing different cell types within these "anatomograms." For example, a pancreas illustration[36] can be used to highlight tissue areas where the various cell types identified in dissociated pancreatic tissue samples may have come from.[37] The accuracy of placing dissociated cells in their correct position would likely be increased by using Machine Learning (ML) and Artificial Intelligence (AI) to improve recognition of tissue patterns and matching these with the proposed EC body-wide positioning information generated through analytical methods. Spatial single-cell analytical methods that preserve tissue structure will still require the use of a CCF, and the use of EC zip codes should be useful also and likely easier to apply in these cases.

Vasculature as a barometer of functional tissue status and health

Normal vascular patterns at the meso level are recognizable as a fingerprint for each tissue; deviations from the normal patterns are indicative of processes ongoing within the tissue (Fig. 37.1).[16,38] We have shown that ML approaches can be used to recognize changes in the normal patterns of tissue-specific vasculature associated with genetic-triggered functional deficits in experimental models.[39] Analysis of EC gene expression shows not only their context (type of vessel, organ), but also indicates EC responses to local and systemic conditions.[40] For instance, analysis of the kidney revealed a great variety of ECs that reported on their precise positional and functional information.[40] This information is very important because the heterogeneity of ECs is key for the functional specificity of the FTUs of various tissues which are built around organotypic capillaries.[41] For instance, the central capillary of the kidney FTU is fenestrated to facilitate exchanges between blood and urine,[38] while the neurovascular unit of the brain is built around a single continuous capillary, which is essential for creating the blood–brain barrier.[42] These examples illustrate the inextricable connection between ECs and the state of their FTUs in various organs; thus, the ECs make for excellent local tissue function barometer. ECs are also sensitive reporters of local tissue transition to dysfunction,[43] but they also react to and could report on systemic conditions. Interestingly, Guo *et al.* compared endothelial signatures in two different organs: heart and brain and highlighted important differences in their responses to aging and systemic comorbidities.[31]

Parting thoughts

We propose that the vasculature offers the opportunity of a built-in reference system for the body and any of its constituent cells. This reference system is individualized, self-adapting, and self-standing, and enables the correct placement of each individual cell in its original location within the tissue sample and its donor.

There is a lot of angst about the perceived challenge of interindividual tissue variations. My professional experience has taught me that, under the microscope, we all are more alike than different. Knowledge of the "normal" patterns of our various tissues has been used for decades, perhaps centuries, as the basis for pathological diagnosis and clinical decisions. Differences observed in histological specimens are usually indicative of departure from normal tissue patterns, due to disease or perhaps age extremes (developmental or aged tissue), as for most tissues it is not possible to discern donors' individual characteristics, such as sex, race, or age within the normal "adult" range. This seems consistent with studies that showed that while individual cells, specifically ECs,[44] and various tissues display some sex differences, in normal tissues the expression of most of our genes is not significantly different among the two sexes, with notable differences associated with tissues of the reproductive systems.[45]

Anatomical variability occurs in the vasculature. Some specific thoughts (while trying to not oversimplify): the large and medium vasculature variations are most likely to present as inborn individual variations (malformations); however, these are likely to either have clinical consequences already revealed through the donor's medical record, or are functionally compensated.[46] In either case, the correct relation between the abnormal vasculature and neighboring cells would be built-in and revealed in the results of single-cell analyses. Variations in the normal meso- or microscale vascular patterns of tissues are recognizable through pathological examination and frequently very useful for medical diagnosis, as the type of vascular variation is a marker for various conditions, e.g., hypoxia, angiogenesis (aberrant vascular growth such as in tumor development), aging (when "vascular rarefaction" was shown to occur), that affect the health of surrounding tissue, hence are likely to induce stress signatures in ECs and neighboring cells.

We propose that when performed correctly, tissue preservation, preparation, and analysis will provide the visual and functional representations of the tissue and cells' "phenotype," which integrates all their activities and functional status at the time of harvesting. The microscopic inspections, including tissue organizational patterns, cell and nuclear sizes, and cellular and extracellular composition, are parameters that have been extensively used to make diagnoses used to dictate clinical management. Therefore, any single-cell analysis greatly benefits from—in fact should be accompanied by—pathological examination of the tissue sample, verifying its "normal" or diseased appearance across scales, from gross anatomy to microscopic examination.

Our knowledge about the human body is about to grow by leaps and bounds. While the development and validation of single-cell analyses still stand to benefit from the current large body of biomedical knowledge already accumulated about the human body, these high content multidimensional analyses are already increasing the precision of our information, providing deeper insights about body's inner workings. Achieving "precision" in vascular biology will be likely instrumental for single-cell analyses across the body because of the natural structural and functional integration of the vasculature within every tissue. Precision vascular biology will be necessary for understanding the normal range of functional variations at the cellular, FTU, and tissue levels, alerting about any subtle transitions toward dysfunction before it becomes clinical disease, and will be useful to build the basis for precision medicine.

Acknowledgments

I want to thank members of the previous Galis lab for all their contributions to our academic research, specifically to Yumiko Sakurai and Chad Johnson for the work that went into producing some of the figures used in this chapter. That work was supported through my National Heart Lung and Blood Institute (NHLBI) awards (HL71061 and HL64689).

Disclaimer

The author declares that all opinions are expressed in the absence of any commercial or financial relationships that could be construed as a potential conflict of interest. Zorina S. Galis is a Government employee; however, the chapter was written in a personal capacity and does not necessarily reflect the opinions or endorsement of the National Institutes of Health (NIH), or the Department of Human Health Services (HHS).

References

1. Halle M, Demeusy V, Kikinis R. The open anatomy browser: a collaborative web-based viewer for interoperable anatomy atlases. *Front Neuroinform*. 2017;11:22.
2. Rood JE, Stuart T, Ghazanfar S, et al. Toward a common coordinate framework for the human body. *Cell*. 2019;179(7):1455–1467.
3. NIH. *Common Coordinate Framework Meeting*; 2017. Available from: https://commonfund.nih.gov/sites/default/files/CCF%20summary%20final%20.pdf.
4. Galis ZS, ed. *The Vasculome. From Many, One*. Elsevier; 2022.
5. Galis ZS. *Using the vasculature as Common Frame Coordinate (CCF)*; 2019. Available from: https://www.youtube.com/watch?v=ZGYU_dsb0j4.
6. Mao SS, Ahmadi N, Shah B, et al. Normal thoracic aorta diameter on cardiac computed tomography in healthy asymptomatic adults: impact of age and gender. *Acad Radiol*. 2008;15(7):827–834.
7. Nordsletten DA, Blackett S, Bentley MD, Ritman EL, Smith NP. Structural morphology of renal vasculature. *Am J Physiol Heart Circ Physiol*. 2006;291(1):H296–H309.
8. Vasquez SX, Gao F, Su F, et al. Optimization of microCT imaging and blood vessel diameter quantitation of preclinical specimen vasculature with radiopaque polymer injection medium. *PLoS One*. 2011;6(4):e19099.
9. Zlokovic BV, Apuzzo ML. Strategies to circumvent vascular barriers of the central nervous system. *Neurosurgery*. 1998;43(4): 877–878.
10. Abadi E, Segars WP, Sturgeon GM, et al. Modeling lung architecture in the XCAT series of phantoms: physiologically based airways, arteries and veins. *IEEE Trans Med Imag*. 2018;37(3):693–702.
11. Paruchuri V, Salhab KF, Kuzmik G, et al. Aortic size distribution in the general population: explaining the size paradox in aortic dissection. *Cardiology*. 2015;131(4):265–272.
12. Bikfalvi A. History of the vascular system. In: Bikfalvi A, ed. *A Brief History of Blood and Lymphatic Vessels*. Cham: Springer International Publishing; 2017:5–33.
13. Borner K, Teichmann SA, Quardokus EM, et al. Anatomical structures, cell types and biomarkers of the Human Reference Atlas. *Nat Cell Biol*. 2021;23(11):1117–1128.
14. HuBMAP. ASCT+B Reporter. Available from: https://hubmap consortium.github.io/ccf-asct-reporter.
15. NHGRI. Shotgun Sequencing. Available from: https://www.genome.gov/genetics-glossary/Shotgun-Sequencing.
16. Bosetti F, Galis ZS, Bynoe MS, et al. Small blood vessels: big health problems?: scientific recommendations of the National Institutes of Health Workshop. *J Am Heart Assoc*. 2016;5(11).
17. Taylor GI, Palmer JH. The vascular territories (angiosomes) of the body: experimental study and clinical applications. *Br J Plast Surg*. 1987;40(2):113–141.
18. Majno P, Mentha G, Toso C, et al. Anatomy of the liver: an outline with three levels of complexity–a further step towards tailored territorial liver resections. *J Hepatol*. 2014;60(3):654–662.
19. Wikipedia, Strahler Number.
20. Florey. The endothelial cell. *Br Med J*. 1966;2(5512):487–490.
21. Galis Z, Ghitescu L, Simionescu M. Fatty acids binding to albumin increases its uptake and transcytosis by the lung capillary endothelium. *Eur J Cell Biol*. 1988;47(2):358–365.
22. Aird WC. Phenotypic heterogeneity of the endothelium: I. Structure, function, and mechanisms. *Circ Res*. 2007;100(2):158–173.
23. Aird WC. Phenotypic heterogeneity of the endothelium: II. Representative vascular beds. *Circ Res*. 2007;100(2):174–190.
24. Narasimhan K. Zip codes: deciphering vascular addresses. *Nat Med*. 2002;8(2):116.
25. Chavkin NW, Hirschi KK. Single cell analysis in vascular biology. *IEEE Trans Med Imag*. 2020;7(42).
26. Vanlandewijck M, He L, Mae MA, et al. A molecular atlas of cell types and zonation in the brain vasculature. *Nature*. 2018; 554(7693):475–480.
27. Weber GM, Ju Y, Borner K. Considerations for using the vasculature as a coordinate system to map all the cells in the human body. *Front Cardiovasc Med*. 2020;7:29.
28. Galis ZS. Editorial: where is waldo: contextualizing the endothelial cell in the era of precision biology. *Front Cardiovasc Med*. 2020;7(127).
29. Sender R, Fuchs S, Milo R. Revised estimates for the number of human and bacteria cells in the body. *PLoS Biol*. 2016;14(8): e1002533.
30. Xiang M, Grosso RA, Takeda A, et al. A single-cell transcriptional roadmap of the mouse and human lymph node lymphatic vasculature. *Front Cardiovasc Med*. 2020;7:52.
31. Guo S, Deng W, Xing C, et al. Effects of aging, hypertension and diabetes on the mouse brain and heart vasculomes. *Neurobiol Dis*. 2019;126:117–123.
32. Nitzan M, Karaiskos N, Friedman N, Rajewsky N. Gene expression cartography. *Nature*. 2019;576(7785):132–137.
33. De Santis V, Singer M. Tissue oxygen tension monitoring of organ perfusion: rationale, methodologies, and literature review. *Br J Anaesth*. 2015;115(3):357–365.
34. Tsai AG, Johnson PC, Intaglietta M. Oxygen gradients in the microcirculation. *Physiol Rev*. 2003;83(3):933–963.
35. Halpern KB, Shenhav R, Matcovitch-Natan O, et al. Single-cell spatial reconstruction reveals global division of labour in the mammalian liver. *Nature*. 2017;542(7641):352–356.
36. EMBL-EBI. Single Cell Expression Atlas. Wellcome; Available from: https://wwwdev.ebi.ac.uk/gxa/sc/experiments/E-GEOD-81547/results/anatomogram.
37. Enge M, Arda HE, Mignardi M, et al. Single-Cell analysis of human pancreas reveals transcriptional signatures of aging and somatic mutation patterns. *Cell*. 2017;171(2):321–330 e14.
38. Stolz DB, Sims-Lucas S. Unwrapping the origins and roles of the renal endothelium. *Pediatr Nephrol*. 2015;30(6):865–872.
39. Lee EK, Galis ZS. Using pattern recognition and discriminant analysis of functional perfusion data to create "angioprints" for normal and perturbed angiogenic microvascular networks. In: *The Vasculome. From Many, One*. Elsevier; 2022.
40. Dumas SJ, Meta E, Borri M, et al. Single-Cell RNA sequencing reveals renal endothelium heterogeneity and metabolic adaptation to water deprivation. *J Am Soc Nephrol*. 2020;31(1):118–138.
41. de Bono B, Grenon P, Baldock R, Hunter P. Functional tissue units and their primary tissue motifs in multi-scale physiology. *J Biomed Semant*. 2013;4(1):22.
42. McConnell HL, Kersch CN, Woltjer RL, Neuwelt EA. The translational significance of the neurovascular unit. *J Biol Chem*. 2017; 292(3):762–770.
43. Kovacic JC, Dimmeler S, Harvey RP, et al. Endothelial to mesenchymal transition in cardiovascular disease: JACC state-of-the-art review. *J Am Coll Cardiol*. 2019;73(2):190–209.
44. Boscaro C, Trenti A, Baggio C, et al. Sex differences in the proangiogenic response of human endothelial cells: focus on PFKFB3 and FAK activation. *Front Pharmacol*. 2020;11:587221.
45. Gershoni M, Pietrokovski S. The landscape of sex-differential transcriptome and its consequent selection in human adults. *BMC Biol*. 2017;15(1):7.
46. Schreiner W, Neumann F, Neumann M, End A, Roedler SM. Anatomical variability and functional ability of vascular trees modeled by constrained constructive optimization. *J Theor Biol*. 1997;187(2):147–158.

Index

Note: 'Page numbers followed by "*f*" indicate figures and "*t*" indicate tables.'

D

E

F

J

K

L

M

N

O

P

Q

R

W

X

Y

Z

Printed in the United States
by Baker & Taylor Publisher Services